W9-CUK-344

FUNDAMENTALS OF ADSORPTION

Proceedings of the Engineering Foundation Conference
held at Schloss Elmau, Bavaria, West Germany,
May 6-11, 1983

Editors
Alan L. Myers
and
Georges Belfort

Sponsored by:
Engineering Foundation

Cosponsored by:
American Institute of Chemical Engineers
Deutsche Vereinigung für Chemie- und Verfahrenstechnik

Financial Contributors:
National Science Foundation
(Grant No. CPE-8213953)
U.S. Environmental Protection Agency
Air Products and Chemicals, Inc.
Calgon Corporation
Engineering Foundation
EXXON Research & Engineering Company
Imperial Chemical Industries
Union Carbide Corporation
Universal Oil Products
Westvaco Corportion

Proceedings of this conference were published by the Engineering Foundation, New York, and distributed by the American Institute of Chemical Engineers, 345 East 47th Street, New York, NY 10017 U.S.A. Any findings, opinions, recommendations, or conclusions contained herein are those of the authors and do not necessarily represent the opinion of, nor imply endorsement by, the aforementioned co-sponsors or financial contributors to the conference.

Cover:

Schloss Elmau, site of the 1983 Engineering Foundation conference on Fundamentals of Adsorption, is located amidst the beauty of the Bavarian Alps. The next conference on this topic will be held in May, 1986 in Santa Barbara, California.

Library of Congress Catalog Card No.: 84-80753
ISBN 0-8169-0265-8

CONTENTS

ACKNOWLEDGMENTS

Many people contributed to the success of the first Fundamentals of Adsorption Conference. Without their considerable assistance, this meeting may not have been possible.

Financial support to help defray the costs of travel for participants from universities was received from Robert M. Wellek of the National Science Foundation; Courtney Riordan of the U.S. Environmental Protection Agency; Nance D. Kunz of Air Products and Chemicals, Inc.; Max N.Y. Lee of the Union Carbide Corporation; Paul V. Smith of the EXXON Research and Engineering Company; Robert V. Carrubba of the Calgon Corporation; Stanley A. Gembicki of UOP; Frank J. Ball of Westvaco Corporation; and B. W. Langley of Imperial Chemical Industries.

We also thank the members of the organizing board; G. T. Barnes of the University of Queensland; Douglas H. Everett of the University of Bristol; Max N. Y. Lee of the Union Carbide Corporation; David Nicholson of Imperial College, London; Alirio E. Rodrigues of the University of Porto; D. Schuhmann of CNRS, Montepellier; Shivaji Sircar of Air Products and Chemicals, Inc; William A. Steele of Pennsylvania State University; Motoyuki Suzuki of the University of Tokyo; Walter J. Weber, Jr. of the University of Michigan; and Robert M. Wellek of the National Science Foundation.

Special thanks are due to Sandford S. Cole and Harold A. Comerer of the Engineering Foundation for their encouragement and help in organizing the meeting and handling the arrangements at Schloss Elmau. Finally, we thank the participants for their contributions to the Conference and to the Proceedings contained in this book.

Alan L. Myers and Georges Belfort

INTRODUCTION

The Fundamentals of Adsorption Conference held at Schloss Elmau, Bavaria, West Germany, from May 6–11, 1983, provided an opportunity for chemists, chemical engineers and environmental engineers to exchange ideas in the field of adsorption. The program was designed to balance the interests and contributions of the academic and industrial communities, and to permit free, constructive exchange of information. Scientists from 18 countries attended the meeting. The format was relatively informal to encourage a congenial atmosphere for scientific exchange.

Increased interest in adsorption is a result of new theoretical advances and technology for gas and liquid separations by this method. The theoretical work reported at Elmau is focused on the effects of surface heterogeneity, and on the application of Monte Carlo "experiments" to unravel the complexities of intermolecular forces. The thermodynamic aspects of adsorption are developed to the point where reliable predictions of mixture equilibria can be made. More realistic models for calculating adsorber breakthrough curves are available. Several inventions described at this meeting allow bulk separations of both gas and liquid mixtures in fixed beds. These factors have contributed to the rapid increase in research and development devoted to adsorption.

The papers appearing in this book have been reviewed, and in many cases extensive revisions were made in response to the reviewers' comments and suggestions. We wish to thank the participants for their time and effort devoted to the reviewing process.

The first paper is the keynote address on the *Thermodynamics of Adsorption from Solution* delivered by Professor Douglas H. Everett. The rest of the papers appear alphabetically by first author, and an index of key words is provided.

A unique aspect of the Conference was the decision *not* to divide the papers into groups concerned with particular topics in adsorption, such as theory, experiment, kinetics, equilibrium, column design, etc. Instead, free time in the afternoons was provided to allow the formation of *ad hoc* groups on special subjects. Most participants attended all of the presentations, resulting in lively discussions.

The results of a survey taken at Schloss Elmau were overwhelmingly in favor of holding another international conference. Professors D. H. Everett (University of Bristol, Great Britain), A. I. Liapis (University of Missouri at Rolla, U.S.A.) and K. S. W. Sing (Brunel University, Great Britain) agreed to organize the next Fundamentals of Adsorption meeting. It is planned for May, 1986, in Santa Barbara, California.

The editors wish to thank all of the participants for their excellent presentations and papers, and especially for the attitude of open exchange of ideas which we call the *spirit of Elmau*.

Alan L. Myers and Georges Belfort

Editor's Biographies

ALAN L. MYERS is Professor of Chemical Engineering at the University of Pennsylvania. He is the author of 60 papers on thermodynamics and statistical mechanics, and his current research interests are adsorption and electrolyte solutions. Professor Myers has lectured extensively in the U.S. and in Europe, and serves as consultant with several companies. Appointments as Visiting Professor and Exchange Scientist include the Technical University of Graz, Austria, and the Institute of Physical Chemistry of the USSR Academy of Sciences in Moscow. His book *Introduction to Chemical Engineering and Computer Calculations*, coauthored with W. D. Seider, was published by Prentice-Hall in 1976, and has since been translated into several languages. After serving for three years in the U.S. Navy during the Korean War, he studied chemical engineering at the University of Cincinnati; that university recently conferred upon him their Distinguished Alumnus Award. After earning his Ph.D. degree at the University of California at Berkeley in 1964 with J. M. Prausnitz, he joined the faculty of the University of Pennsylvania. At Penn during the period from 1971 to 1980, he served as Graduate Chairman, and as Chairman of the Chemical Engineering Department. In 1983, he received the S. Reid Warren Award for distinguished teaching from the University of Pennsylvania.

GEORGES BELFORT is Professor of Chemical Engineering and Environmental Engineering at Rensselaer Polytechnic Institute. He received his Ph.D. and M.S. in Engineering at the University of California at Irvine, and his B.Sc. in Chemical Engineering from the University of Cape Town, South Africa. His current research interests include liquid-phase adsorption, fluid mechanics of dilute suspensions in porous ducts, synthetic membrane fouling, and biochemical separations using membranes and affinity techniques. He has taught in the School of Applied Science and Technology at the Hebrew University of Jerusalem and was a visiting lecturere at Cape Town and Northwestern universities. He is a member of the Board of Editors of the *Desalination Journal* and the *Journal of Biotechnology*, and has lectured in Japan, Europe and South Africa. He consults internationally and has coauthored more than 40 research papers and monograph chapters on adsorption and membrane separations.

KEYNOTE ADDRESS
THERMODYNAMICS OF ADSORPTION FROM SOLUTION

DOUGLAS H. EVERETT is Emeritus Professor of Physical Chemistry at the University of Bristol. He is the author of over 170 scientific papers and several books on the subject of thermodynamics and statistical mechanics. His research on adsorption is reflected in many standard monographs, such as the "IUPAC Manual of Symbols and Terminology in Colloid and Surface Chemistry" and the book *Surface Tension and Adsorption* coauthored with R. Defay, I. Prigogine and A. Bellemans. He is the senior reporter and author of several articles in the *Chemical Society Specialist Periodical Reports, Colloid Science*, the fourth volume of which appeared in 1983. Born in Hampton, not far from London, on December 26, 1916, he studied chemistry at the University of Reading, and his doctoral dissertation entitled "Some Aspects of Catalysis and Adsorption" was completed in 1942 at Oxford under the direction of R. P. Bell and C. N. Hinshelwood. After teaching at Oxford and University College, Dundee, he accepted in 1954 the appointment of Leverhulme Professor of Physical Chemistry at the University of Bristol, a post first held by J. W. McBain. At Bristol he built with a series of younger colleagues one of the leading centers in the world for research on surface and colloid phenomena.

Professor Everett served for several periods as Vice-President of the Faraday Society and the Faraday Division of the Chemical Society between 1958 and 1976, and was president from 1976 to 1978. In 1971 he was the first recipient of the Chemical Society Award in Surface and Colloid Chemistry. At the University of Bristol he has held the posts of Dean of Faculty of Science, and Pro-Vice-Chancellor. He was Chairman of the IUPAC Commission of Colloid and Surface Chemistry from 1969 to 1973. In 1980, in recognition of his numerous honors and achievements, he was elected Fellow of the Royal Society.

THERMODYNAMICS OF ADSORPTION FROM SOLUTION

Douglas H. Everett

Department of Physical Chemistry
School of Chemistry
University of Bristol U.K.

ABSTRACT

This paper surveys some of the major features in the development of the thermodynamics of adsorption from solution by solids over the past twenty years, and outlines some of the problems involved in the interpretation of experimental data in molecular terms.

The foundations of the thermodynamics of adsorption were laid by Willard Gibbs (1) who gave rigorous definitions of the relevant quantities needed to characterise the thermodynamic state of a system consisting of two or more phases separated by interfacial layers. His formulation paid special attention to fluid/fluid interfaces and although surface films on liquids were discussed he did not develop fully the formulae appropriate to either the liquid/vapour or the solid/liquid interface. The thermodynamics of the solution/vapour interface was not discussed in detail until the 1930's by Butler and others, (2) while, despite the interest in and importance of adsorption at the solid/liquid interface, a full thermodynamic treatment of this case was not developed until much later (3).

The reasons for the neglect of fundamental studies of adsorption from solution by solids are to be found in the difficulty of obtaining adequately precise and reproducible experimental data on well-characterised solids, and with preoccupation in earlier work with the practically important but somewhat narrow area of adsorption from dilute aqueous solutions. The theoretical interpretation of such work rarely went beyond the fitting of data to an appropriate adsorption isotherm, usually the Freundlich isotherm.

The progress which has been made in the last twenty years has resulted partly from the formulation of rigorous thermodynamic treatments, and partly from the development of more versatile and more precise methods of measuring adsorption from solution and the realisation that studies should be carried out on well-characterised solids and made over a range of temperatures so that information is obtained about the contributions of enthalpy and entropy factors to the equilibrium situation. Associated with such studies there has developed also an increasingly sophisticated study of the molecular interpretation of adsorption phenomena.

It is not the purpose of this paper to review in detail the different formulations of the thermodynamics of adsorption from solution most of which can be shown to be essentially equivalent. It is, however, worthwhile to comment briefly on the various ways in which definitions of adsorption can be realised in practice, and to quote without full proof some of the more important thermodynamic equations which are useful in analysing experimental data.

DEFINITIONS AND MEASUREMENT OF ADSORPTION FROM SOLUTION

In the neighbourhood of a solid surface the local concentrations of a solution differ from those in the bulk (Figure 1) so that the amount of a given component present in a system is different from that which would be present if the bulk concentration were maintained uniformly up the surface. The surface excess amount of a given component (2) defined in this way is denoted $n_2^{\sigma(n)}$, or when expressed in terms of the surface excess per unit area of solid, $\Gamma_2^{(n)}$, called the areal reduced surface excess (often simply the reduced adsorption (4),(5)).

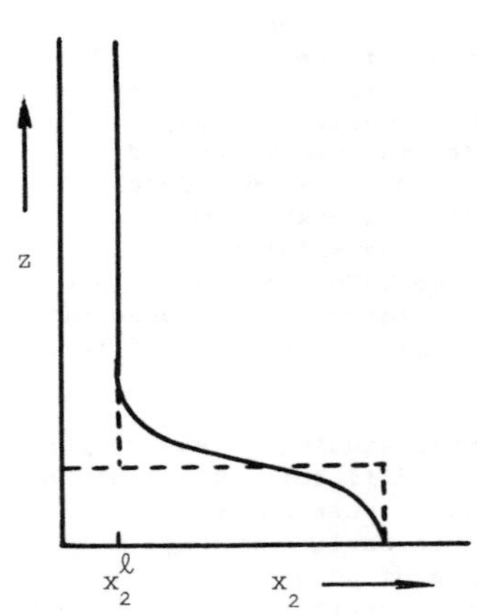

Figure 1:
Schematic representation of the variation of the local mole fraction of a solution (x_2) in the neighbourhood of a surface. z is the distance measured normal to the surface, and x_2^{ℓ} is the mole fraction in the bulk liquid. The dotted lines define the 'surface phase' used in the molecular interpretation of adsorption (see later).

This definition corresponds to choosing the Gibbs dividing surface to coincide with the surface of the solid, so that the adsorption of the solid is zero. For a binary solution it follows that since the local mole fractions x_1 and x_2 must always sum to unity, then

$$n_2^{\sigma(n)} = -n_1^{\sigma(n)} \text{ or } \Gamma_2^{(n)} = -\Gamma_1^{(n)}. \tag{1}$$

The Gibbs adsorption isotherm equation relates the adsorption of the various components to the changes in the interfacial tension (σ) of the solid/solution interface:

$$-d\sigma = \Gamma_1 d\mu_1 + \Gamma_2 d\mu_2, \tag{2}$$

where μ_1 and μ_2 are the chemical potentials of the two components which, at equilibrium, are constant throughout the system, including the surface layer. These chemical potentials are, however, related through the Gibbs-Duhem equation for the bulk solution:

$$x_1^{\ell} d\mu_1 + x_2^{\ell} d\mu_2 = 0. \tag{3}$$

Hence μ_1 can be eliminated from Equation (2) to give:

$$-d\sigma = \left[\Gamma_2 - \Gamma_1 \frac{x_2^{\ell}}{x_1^{\ell}}\right] d\mu_2. \tag{4}$$

The term in square brackets is, however, just $\Gamma_2^{(1)}$ the relative areal surface excess (relative adsorption), in the case in which components 1 and 2 are present in only one phase.

Thus,

$$-d\sigma = \Gamma_2^{(1)} \; d\mu_2. \tag{5}$$

However, since the solid surface has been chosen as the Gibbs dividing surface, $\Gamma_2 = -\Gamma_1$, and so

$$\Gamma_2^{(1)} = \Gamma_2^{(n)}/x_1^{\ell}, \tag{6}$$

and $$-d\sigma = \frac{\Gamma_2^{(n)}}{x_1^{\ell}} \; d\mu_2. \tag{7}$$

Equations (5) and (7) are the key equations used in the thermodynamic analysis of experimental data. It is seen that the experimental information needed consists of either $\Gamma_2^{(n)}$ or $\Gamma_2^{(1)}$ together with the chemical potentials in the equilibrium bulk liquid, as functions of x_2^{ℓ}.

The determination of $\Gamma_2^{(n)}$ is most commonly based (5) on the observation that if an amount n^o of solution of initial mole fraction x_2^o is equilibrated with a mass m of solid, then the concentration changes to x_2^{ℓ}, which if component 2 is positively adsorbed is less than x_2^o. If the whole of the solution at equilibrium were at the uniform mole fraction x_2^{ℓ}, it would contain an amount $n^o x_2^{\ell}$ of component 2, whereas the amount actually present is $n^o x_2^o$. The excess of component 2 which is present because of the solid interface is thus

$$n_2^{\sigma(n)} = n^o \; \Delta x_2^{\ell} \tag{8}$$

where $$\Delta x_2^{\ell} = x_2^o - x_2^{\ell}. \tag{9}$$

The specific surface excess is defined as $n_2^{\sigma(n)}/m$, while the areal surface excess is $\Gamma_2^{(n)} = n_2^{\sigma(n)}/m\, a_s$ where a_s is the specific surface area of the solid.

The batch-wise determination of successive points on an adsorption isotherm which is implied by this prescription has a long history (6). However, it is a tedious procedure which has many practical disadvantages.

Among these are the difficulty of outgassing the solid and transferring it to the solution without contamination by the atmosphere, the errors introduced in sampling successive batches of solid, the problem of extracting a sample of supernatant liquid especially when measurements are made at temperatures far from ambient, and finally that of analysing the liquid. Elaborate techniques have been devised to overcome these problems (7), but these only add to the complications of the method.

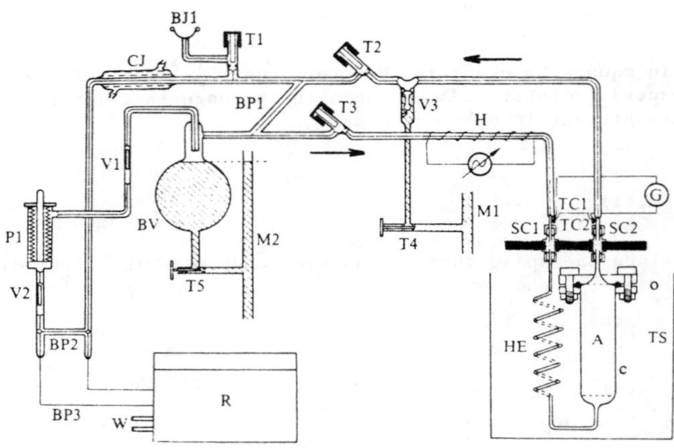

Figure 2: Circulation method of determining adsorption from solution: Solution is circulated by pump P1 over adsorbent in cell A in the thermostat TS, and through one arm of a differential refractometer R. Solution of the initial composition x_2^o is circulated through the reference cell W of the refracometer.

An alternative method (Figure 2) is to employ a technique in which the liquid is circulated over the solid sample and changes in concentration monitored by an appropriate method, for example by refractometry or spectrophotometry (8). In this way the experiment may be carried out in a closed system, with carefully purified and outgassed liquid, using a single sample of solid which can be evacuated and outgassed reproducibly between successive runs. Moreover, by changing the temperature of the solid sample measurements can be made over a wide range of temperature in a single run. One limitation on the precision of this method is that of calibrating the detector and ensuring that the calibration does not drift with time.

A related method is to use a chromatographic technique in which solution of given mole fraction x_2^o is passed through a column of solid and the concentration of the effluent (x_2) monitored. The experiment is continued until the effluent concentration has returned to x_2^o when

$$n_2^{\sigma(n)} = \int_0^{n_f} (x_2^o - x_2)\,dn, \tag{10}$$

where n_f is the amount of solution which has passed. In practice, one usually works in terms of molar concentrations and measures the volume of liquid passed when, in precise work, corrections have to be applied for the dead-space in the column and for variation in the molar volume of the solution with concentration (10).

It has recently been realised that one can modify the circulation technique to measure the relative adsorption directly (11). If provision is made in the system for the injection of component 2, then additions can be made until the solution composition is restored to its original value. If an amount Δn_2 is needed to achieve this, then the system contains a total amount of solution $n^o + \Delta n_2$. If this were uniform in concentration then the system would contain $(n^o + \Delta n_2)x_2^o$ of component 2. However, it actually contains $(n^o x_2^o + \Delta n_2)$ so that the surface excess is

$$n_2^{\sigma(n)} = (1 - x_2^o)\Delta n_2. \tag{11}$$

It follows from Equation (6) that Δn_2 is in fact equal to the relative surface excess:

$$\Delta n_2 = n_2^{\sigma(n)}/x_1^o = n_2^{\sigma(1)}. \tag{12}$$

In this method the need for calibration of the detector is eliminated. This technique has the additional advantage that it is no longer necessary to measure n^o. Moreover it enables measurements to be made up to the solubility limit of component 2 - using the earlier method one cannot get closer than Δx_2^ℓ to the saturation concentration. It may also be used to study the behaviour close to the phase separation point of systems showing a miscibility gap, where there are reasons to believe that the adsorption rises rapidly. By providing an injection point in the circulating system one also has the opportunity of studying the effect of pH on adsorption, or of correcting for changes in pH brought about by the adsorption process, in for example, ionic surfactant adsorption on clays.

In a further modification (12) (Figure 3) in which the column of adsorbent is replaced by a stirred suspension one can also titrate with powdered solid, adding successive amounts together with component 2 to maintain constant mole fraction. This emphasises the close relation of the two latter techniques to the alternative definition of the relative surface excesses (13):

$$n_2^{(1)} = \left(\frac{\partial n_2}{\partial m}\right)_{T,x_2^\ell}, \tag{13}$$

where dn_2 is the amount of component 2 which has to be added together with a mass dm of solid to maintain the bulk liquid concentration

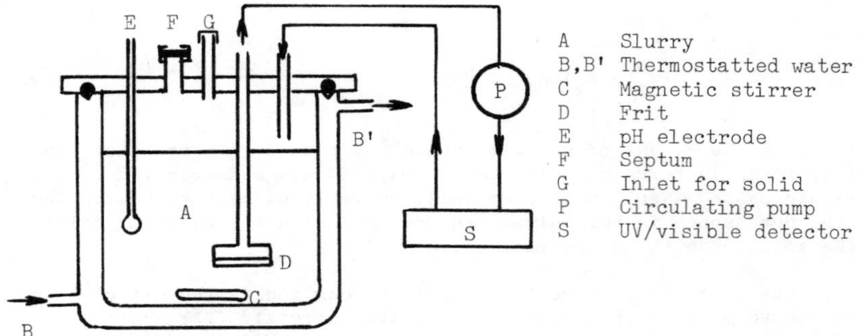

Figure 3: Circulation method in which the solid is used as a suspension rather than a packed column: solid can be added through the port G, and component 2 (or a concentrated solution) through the septum, F.

constant.

These developments in the techniques of measuring adsorption open up the feasibility of much more comprehensive studies than have been possible in the past.

THERMODYNAMIC ANALYSIS (14)

When adsorption has been measured over a range of concentrations then the change in the surface tension of the interface brought about by adsorption can be calculated:

$$\sigma - \sigma_2^* = -\int_{\mu_2^*}^{\mu} \Gamma_2^{(1)} \, d\mu_2, \tag{14}$$

where σ_2^* is the surface tension of the interface between pure component 2 and solid, and μ_2^* is the chemical potential of pure 2. Now

$$d\mu_2 = RT \, d \ln \gamma_2^{\ell} x_2^{\ell} , \tag{15}$$

where γ_2^{ℓ} is the activity coefficient of component 2 in the liquid, so that

$$\sigma - \sigma_2^* = -RT \int_{x_2=1}^{x_2} \Gamma_2^{(1)} \, d \ln \gamma_2^{\ell} x_2^{\ell}, \tag{16}$$

or (14a)

$$\sigma - \sigma_2^* = -RT \int_{x_2=1}^{x_2} \frac{\Gamma_2^{(n)}}{x_1^\ell x_2^\ell \gamma_2^\ell} \, d(\gamma_2^\ell x_2^\ell). \quad (17)$$

It must be emphasized that to make this calculation it is necessary to know the properties (in particular the activity coefficients) of the bulk phase, as well as the surface area of the solid. One of the major obstacles to a full thermodynamic analysis of adsorption data is the lack of accurate bulk activity coefficient data in the appropriate temperature ranges. Even with some of the most fully studied liquid mixtures, the relevant information is available only in the region of the normal boiling point and one has to rely on extrapolation formulae to obtain values near ambient temperatures.

If a system is studied over the whole concentration range then the difference between the surface tensions of the interface between the solid in contact with the two pure liquids can be calculated:

$$\sigma_1^* - \sigma_2^* = -RT \int_{x_2=1}^{x_2=0} \frac{\Gamma_2^{(n)}}{x_1^\ell x_2^\ell \gamma_2^\ell} \, d(\gamma_2^\ell x_2^\ell). \quad (18)$$

This equation provides the basis for a valuable criterion for assessing both the accuracy of experimental measurements and the consistency of the thermodynamic analysis, for if measurements are made on three mutually miscible liquid pairs 1 + 2, 2 + 3 and 3 + 1, the calculated surface tension difference should satisfy: (14a)

$$(\sigma_1^* - \sigma_2^*) + (\sigma_2^* - \sigma_3^*) + (\sigma_3^* - \sigma_1^*) = 0 . \quad (19)$$

A further useful application follows in the case in which 1 and 2, and 2 and 3 are completely miscible, while 3 and 1 are immiscible (15). It is now possible to calculate $(\sigma_3^* - \sigma_1^*)$. However, this surface tension difference is related to the contact angle at the liquid 3/liquid 1/ solid line of contact by Young's equation:

$$\sigma_3^* - \sigma_1^* = \sigma^{31} \cos\theta , \quad (20)$$

where σ^{31} is the interfacial tension between the pure liquids and θ is measured through component 1. In this way measurements of adsorption from solution can be used to investigate the contact angles for powdered or porous media. Of special interest is the case in which

$$\sigma_3^* - \sigma_1^* > \sigma^{31} \quad (21)$$

when pure liquid 1 will displace liquid 3 and wet the surface completely.

The relationship between adsorption from solution and adsorption of the pure vapours may also be studied by using the Gibbs adsorption isotherm which for adsorption from a pure vapour takes the form

$$\sigma - \sigma_s^* = -RT \int_{p_i=0}^{p_i} \Gamma_i \, d \ln p_i \,, \tag{22}$$

where σ_s^* is the surface tension of the pure solid/vacuum interface and Γ_i is the areal adsorption of vapour. If measurements are made up to the saturation vapour pressure then the integration yields $(\sigma_i^{*,sv} - \sigma_s^*)$, where $\sigma_i^{*,sv}$ is the surface tension of solid in equilibrium with saturated vapour. $\sigma_i^{*,sv}$ is related to the surface tension of the solid/liquid interface $(\sigma_i^{*,s\ell} \equiv \sigma_i^*)$ by

$$\sigma_i^{*,sv} = \sigma_i^{*,s\ell} + \sigma_i^{*,\ell v} \cos \theta \,, \tag{23}$$

where $\sigma_i^{*,\ell v}$ is the surface tension of the liquid/vapour interface and θ is the contact angle of liquid with solid measured through the liquid. It is often supposed that for many vapours adsorbed by solids $\theta = 0$, and the equilibrium situation at saturation is a duplex film with

$$\sigma_i^{*,sv} = \sigma_i^{*,s\ell} + \sigma_i^{*,\ell v} \quad \text{or} \quad \sigma_i^* = \sigma_i^{*,sv} - \sigma_i^{*,\ell v} . \tag{24}$$

If vapour adsorption is measured for both components 1 and 2, then the following relationship should be satisfied (16).

$$(\sigma_1^{*,s\ell} - \sigma_2^{*,s\ell}) = (\sigma_1^{*,sv} - \sigma_1^{*,\ell v}) - (\sigma_2^{*,sv} - \sigma_2^{*,\ell v}) . \tag{25}$$

Whether this equation also applies to adsorption by porous solids is not certain (14c).

The relationship between adsorption from solution and adsorption from mixed vapours involves a number of more subtle considerations which cannot be dealt with in detail here (14b),(17).

As indicated above, additional thermodynamic information about the adsorption equilibrium can be obtained if the measurements are made over a range of temperature. The key equation here is (14a), (14b),(18)

$$\frac{\partial(\sigma/T)}{\partial(1/T)} = \Delta_w \hat{h}, \tag{26}$$

where $\Delta_w \hat{h}$ is the enthalpy of immersion of unit area of solid in a volume of mixture large enough for the change in concentration arising from adsorption to be negligible. Since, however, only differences in surface tensions can be measured by adsorption methods, this equation has to be used in the form:

$$\frac{\partial}{\partial(1/T)} \left[\frac{\sigma - \sigma_2^*}{T} \right] = \Delta_w \hat{h} - \Delta_w \hat{h}_2^* \,, \tag{27}$$

or $$\frac{\partial}{\partial(1/T)}\left[\frac{\sigma_1^* - \sigma_2^*}{T}\right] = \Delta_w\hat{h}_1^* - \Delta_w\hat{h}_2^*. \tag{28}$$

The corresponding entropies of immersion are obtained either from $\partial\sigma/\partial T$ or from

$$(\sigma - \sigma_2^*) = (\Delta_w\hat{h} - \Delta_w\hat{h}_2) - T(\Delta_w\hat{s} - \Delta_w\hat{s}_2). \tag{29}$$

In attempting to relate these directly determinable experimental quantities to the thermodynamic properties of the interface, one must remember that the latter always involve arbitrary constants (15). Thus it is well known that the areal surface excess free energy $g^{\sigma(n)}$, is related to the surface tension by

$$\hat{g}^{\sigma(n)} = \sigma + \Gamma_2^{(n)}\,(\mu_2 - \mu_1), \tag{30}$$

or

$$\hat{g}^{\sigma(n)} = \sigma + \Gamma_2^{(n)}\left[(\mu_2^* - \mu_1^*) + RT\,\ln\,(\gamma_2^\ell x_2^\ell/\gamma_1^\ell x_1^\ell)\right], \tag{31}$$

where the arbitrary quantity $(\mu_2^* - \mu_1^*)$ is the difference in chemical potentials of pure 1 and 2.

Similarly it can be shown that $\hat{h}^{\sigma(n)}$ and $\hat{s}^{\sigma(n)}$, the surface excesses(relative to the enthalpy and entropy of the system if the molar enthalpy and entropies had remained constant up to the surface) are given by

$$\hat{h}^{\sigma(n)} = \Delta_w\hat{h} + \Gamma_2^{(n)}\,(h_2^\ell - h_1^\ell), \tag{32}$$

and $$\hat{s}^{\sigma(n)} = \Delta_w\hat{s} + \Gamma_2^{(n)}\,(s_2^\ell - s_1^\ell),$$

where h_2^ℓ, s_2^ℓ and h_1^ℓ, s_1^ℓ are respectively the partial molar enthalpies and entropies of the two components in solution. In principle these can be defined in terms of the enthalpies at absolute zero, but in practice this is not usually possible.

APPLICATIONS OF THE THERMODYNAMIC ANALYSIS

Some applications of Equation (17) are illustrated in Figure 4 (19). These examples show that the integrand can adopt widely different forms, and stress the importance of making precise measurements over the whole concentration range. Special attention has to be paid to the low concentration region since the abscissa is the product of concentration and activity coefficient; in those cases where the infinite dilution activity coefficient is high, measurements have to be extended to very low concentrations to make possible a reliable extrapolation of the curve.

Some results of calculations based on Equations (17), (27) and (29) are shown in Figure 5 (20). An important observation concerns the way in which the sign of $\sigma - \sigma_2^*$ is determined by the balance between

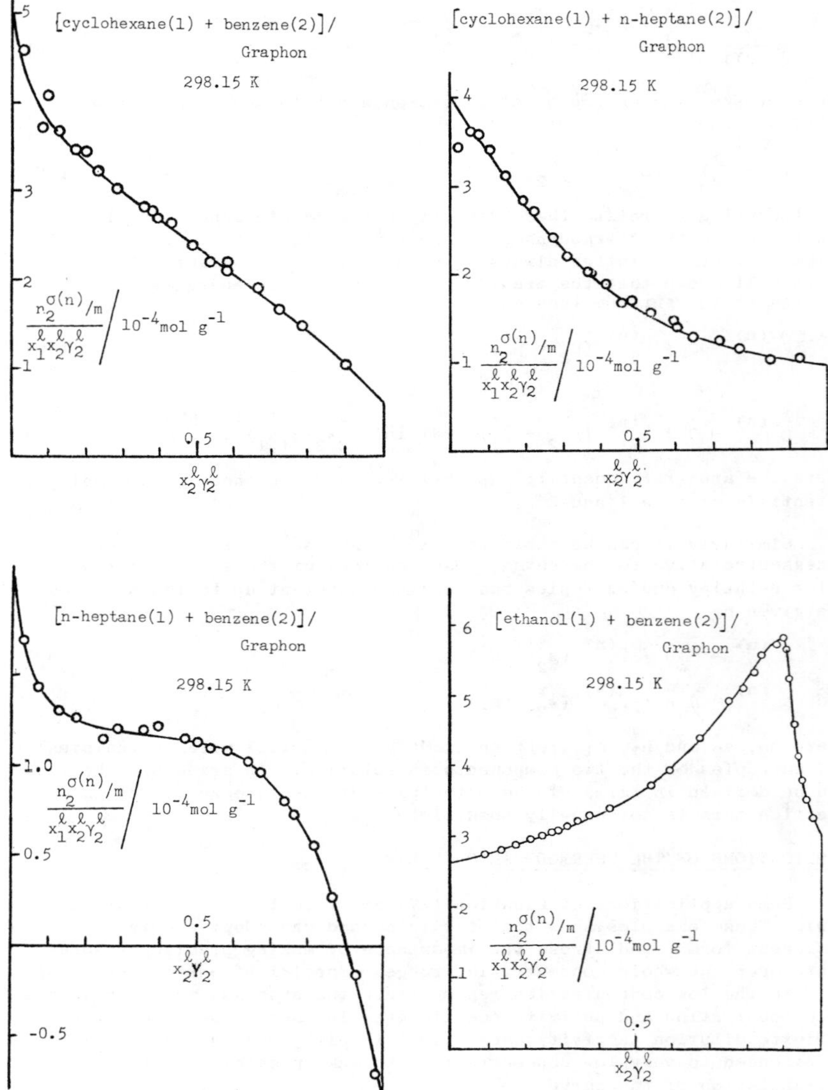

Figure 4: Typical curves of $(n_2^{\sigma(n)}/m)/(x_1^\ell x_2^\ell \gamma_2^\ell)$ as a function of $x_2^\ell \gamma_2^\ell$
(a)(cyclohexane + benzene)/Graphon
(b)(cyclohexane + n-heptane)/Graphon
(c)(n-heptane + benzene)/Graphon(d)(ethanol+benzene)/Graphon

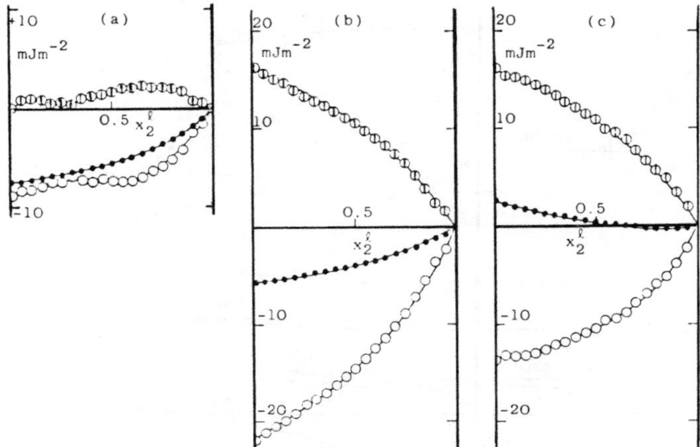

Figure 5: $(\sigma - \sigma_2^*)$, ●; $(\Delta_w \hat{h}_\ell - \Delta_w \hat{h}_2^*)$, ○; and $-T(\Delta_w \hat{s} - \Delta_w \hat{s}_2^*)$, ⦶ as functions of x_2^ℓ for the systems (a) [benzene (1) + cyclohexane (2)]/Graphon, (b) [n-heptane (1) + cyclohexane (2)]/Graphon, (c) [n-heptane (1) + benzene (2)]/Graphon.

enthalpy and entropy terms. Thus while in the cases of [benzene + cyclohexane]/Graphon and [n-heptane + cyclohexane]/Graphon the enthalpy determines the sign of σ (and, through its slope, the sign of the adsorption), the balance is more subtle in the case of [n-heptane + benzene]/Graphon and the dominance of the entropy effect at low benzene concentrations leads to a change in sign of the adsorption.

It may be commented that for several triads of liquids adsorbed by Graphon which we have studied Equation (19) is satisfied to within the experimental uncertainty, much of which we suspect arises from uncertainties in the bulk activity coefficients.

The use of adsorption from solution to investigate wettabilities is illustrated in Figure 6 in which the surface tensions of liquid/Graphon interfaces are shown taking the benzene/Graphon interface as reference zero (15). The meaning of this diagram is that if the difference between $\sigma_2^* - \sigma_1^*$ is greater than the interfacial tension σ^{12} then liquid 1 will displace liquid 2 from a Graphon surface; while if less then the contact angle can be calculated. This method has been tested (21) by measuring the contact angles at the hydrocarbon/water/freshly cleaved graphite surface using an interference microscope technique. When sufficient care is taken to purify the liquids and to outgas the graphite surface then contact angle hysteresis is minimised. The observed contact angle (22) for heptane/water/graphite of 7.2 $\pm$ 1.25° agrees well with that, 8°, calculated from adsorption measurements of the triad heptane, ethanol, water. In the case of benzene/water/graphite the adsorption measurements on both triads benzene, ethanol,

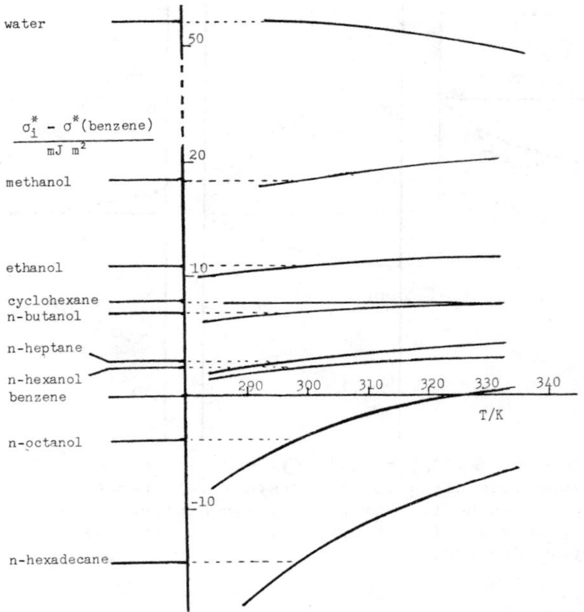

Figure 6: Interfacial tension of liquid i/graphite interface relative to that of the benzene/graphite interface. On the left are shown the values at 298 K while the curves on the right show the temperature dependence.

water and benzene, methanol, water lead to values of σ(water) - σ(benzene) greater than the interfacial tension between benzene and water, and hence predict that benzene will displace water from graphite. Direct observation indicated a contact angle of zero i.e. complete wetting by benzene. It is interesting to observe the way in which the surface tensions of hydrocarbon/graphite interfaces decrease with increases in chain length so that at a particular chain length they will displace water from graphite (Figure 7).

This particular application of the thermodynamics of adsorption from solution remains to be more fully exploited.

The correlation between adsorption from solution and pure vapour has been studied by several groups (23), and the indications are that Equation (25) is valid for non-porous adsorbents. However, in this work problems arise in the extrapolation of the vapour adsorption data to the saturation vapour pressure. One method, used by Larionov and his co-workers is to plot $(\sigma - \sigma_s^*)/\Gamma$ against $(\sigma - \sigma_s^*)$ and to extrapolate to $(\sigma - \sigma_s^*)/\Gamma \to 0$. It is not always clear from their papers how this was done, especially in those cases in which they introduced "corrections for capilliary condensation".

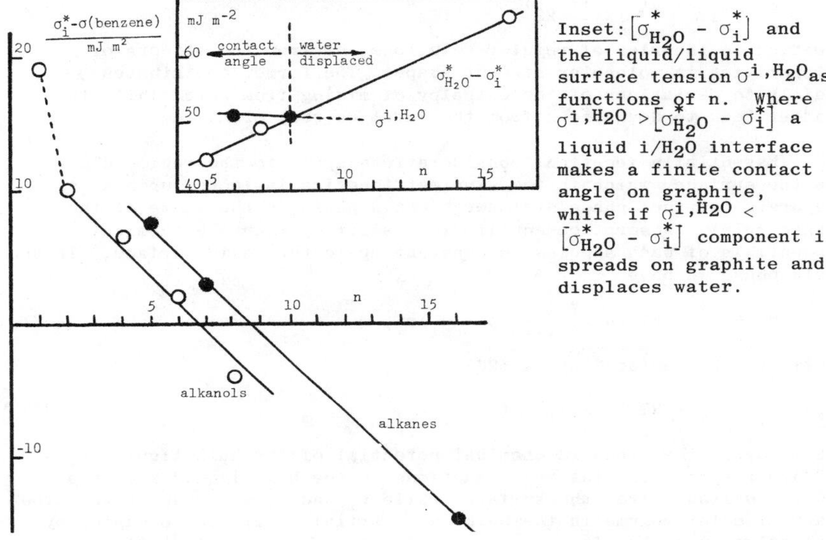

Inset: $[\sigma^*_{H_2O} - \sigma^*_i]$ and the liquid/liquid surface tension σ^{i,H_2O} as functions of n. Where $\sigma^{i,H_2O} > [\sigma^*_{H_2O} - \sigma^*_i]$ a liquid i/H_2O interface makes a finite contact angle on graphite, while if $\sigma^{i,H_2O} < [\sigma^*_{H_2O} - \sigma^*_i]$ component i spreads on graphite and displaces water.

Figure 7: $[\sigma_i^* - \sigma^*(\text{benzene})]$ as a function of number of carbon atoms (n) in n-alkanes (●) and in n-alkanols (○).

More recently a careful study by Davis (24) on the adsorption by Graphon from benzene + pentane and benzene + isopentane mixtures when combined with Polanco's (25) vapour adsorption data given further support to Equation (25), although uncertainties arise from the lack of activity coefficient data for the benzene + isopentane system.

MOLECULAR INTERPRETATION

Statistical mechanical theories of adsorption from solution have developed rapidly in the past twenty years. They have in the main evolved from the thermodynamic concepts of ideal behaviour and of deviations from such behaviour expressed in terms of activity coefficients. Ideal behaviour, in the thermodynamic sense, is linked with the properties of mixtures of molecules of types 1 and 2 of equal size whose energies of interaction (ε) at their mean separation in the liquid state satisfy the relation

$$\varepsilon_{12} = \frac{\varepsilon_{11} + \varepsilon_{22}}{2} \tag{34}$$

Deviations from ideality then arise either from deviations from Equation (34), expressed in terms of the interaction parameter

$$w = \varepsilon_{12} - \tfrac{1}{2}(\varepsilon_{12} + \varepsilon_{22}), \tag{35}$$

leading to theories of regular solutions, or from the effects of disparities in molecular size or shape. The former contributes mainly to deviations of the enthalpy of mixing from zero, while the latter lead to deviations from the ideal entropy of mixing.

Essentially identical considerations apply to the region close to the surface, which in the simplest theories is looked upon as a separate 'surface phase' (although not a phase in the sense of the phase rule). Adsorption equilibrium is set up when the chemical potentials of each species is constant up to the solid surface. In the bulk phase we have

$$\mu_i = \mu_i^* + RT \ln x_i^\ell \gamma_i^\ell + (p - p_i^*)v_i, \tag{36}$$

while in the surface layers (26)

$$\mu_i = \mu_i^* + RT \ln x_i^s \gamma_i^s - (\sigma - \sigma_i^*)a_i, \tag{37}$$

where μ_i^* is the standard chemical potential of the bulk liquid, x_i^ℓ and x_i^s are respectively the mole fractions in the bulk liquid and at a chosen distance from the surface, while v_i and a_i are respectively the partial molar volume in the bulk, and partial molar area occupied by molecules in the surface layer. In Equation (36) the last term can usually be neglected, whereas the last term in Equation (37) plays an essential role.

In general, x_i^s, γ_i^s and σ vary with the distance from the surface, but in the surface phase model they are regarded as constant within a defined thickness close to the surface, beyond which they change abruptly to, respectively, x_i^ℓ, γ_i^ℓ and zero. This picture is physically unrealistic and leads to equations which, to a lesser or greater extent, are thermodynamically inconsistent (27). Despite this, however, simple surface phase models have been extensively used to analyse experimental data. In the case of a binary mixture of molecules of equal size ($a_1=a_2=a$) they lead to the simple equilibrium condition (26)

$$K = \frac{x_2^s \gamma_2^s x_1^\ell \gamma_1^\ell}{x_1^s \gamma_1^s x_2^\ell \gamma_2^\ell}, \tag{38}$$

where $\ln K = -(\sigma_2^* - \sigma_1^*)a/RT.$ (39)

If the size ratio (a_2/a_1) is r then

$$K = \left(\frac{x_2^s \gamma_2^s}{x_2^\ell \gamma_2^\ell}\right) \left(\frac{x_1^\ell \gamma_1^\ell}{x_1^s \gamma_1^s}\right)^r, \tag{40}$$

and corresponds to the equilibrium constant for the exchange process between bulk and surface phases:

$$r(1)^s + (2)^\ell \rightleftharpoons r(1)^\ell + (2)^s \ . \tag{41}$$

To apply these equations it is necessary to calculate values of x_1^s, x_2^s from experimental measurements. To do this it is necessary to make some assumptions about the areas (a_1^o, a_2^o) occupied by the molecules, the surface area of the solid and the thickness (t) of the surface layer expressed in terms of molecular layers (28):

$$x_2^s = \frac{x_2 + \left(\frac{a_1^o}{t}\right) \Gamma_2^{(n)}}{1 - \frac{(a_2^o - a_1^o)}{t} \Gamma_2^{(n)}} \ . \tag{42}$$

In the simple case of $a_1^o = a_2^o = a$, $t = 1$ this reduces to

$$x_2^s = x_2^\ell + a\,\Gamma_2^{(n)} \ , \tag{43}$$

and this equation is often used as a rough approximation, although if its application leads to calculated values of $x_2^s > 1$, or regions in which $(\partial x_2^s/\partial x_2^\ell) < 0$, the assumption of $t = 1$ must be rejected (29).

Various assumptions may be employed concerning the activity coefficients. Thus it has variously been assumed that the activity coefficients $\gamma^s(x_2^s)$, $\gamma^\ell(x_2^\ell)$ are the same functions e.g. corresponding to regular solution behaviour (30), that they are in constant ratio (31), or that even when the bulk solution is non-ideal the surface is ideal (32). These contentions may be tested by calculating the surface activity coefficients from the equation (33)

$$\ln \gamma_2^s = \ln \frac{x_2^\ell \gamma_2^\ell}{x_2^s} + a \int_{x_2^\ell}^{x_2^\ell=1} \frac{\Gamma_2^{(n)}}{x_1^\ell x_2^\ell \gamma_2^\ell}\, d(x_2^\ell \gamma_2^\ell) . \tag{44}$$

The evidence derived from analyses of this kind does not support any general rules governing the relationship between bulk activity coefficients and surface activity coefficients calculated within the framework of the surface phase model. In a number of instances the surface phase appears to depart from ideality to a lesser extent than the bulk (34), but in other cases, adsorption from a near-ideal solution leads to strongly non-ideal behaviour in the surface (35).

The statistical mechanical interpretation of surface phase non-ideality follows generally the procedures developed for bulk systems in that the parameters determining such behaviour are identified with the interaction parameter w and the size ratio r. However, in treating the surface phase one has to take account of the fact that molecules in a given layer interact with molecules in adjacent layers in which the mole fractions are different from those in the layer under consideration (36). When applied to mixtures of molecules of the same size, the regular solution model can account for a range of types of behaviour, including the reversal of the sign of the adsorption when the bulk concentration is changed. In the parallel-layer model of adsorption of molecules of different size (in which all r-segments of the larger molecule lie in the surface layer) the configurational entropy term can be calculated by means analogous to the Flory-Huggins method (37). In general, segments will be distributed in adjacent layers out to a distance of r layers, and the calculation of the configurational term becomes much more complex (38).

It is clear that one cannot expect theories of this kind to deal satisfactorily with adsorption from bulk solutions whose behaviour is not adequately described by the analogous theory applied to the bulk mixture. Thus the interesting and complex adsorption isotherms observed in adsorption by graphite from hydrocarbon + alkanol solutions (39) cannot readily be understood until at least the bulk properties of these mixtures are accounted for satisfactorily. And even then one will need to provide a quantitative theory of the structuring phenomena (40) which undoubtedly occur in such systems in the close neighbourhood of the surface - whether this should be looked upon as a surface freezing phenomenon or whether it should be regarded as more nearly analogous to two-dimensional liquid-crystal formation has yet to be explored.

It is in this area,where one needs to seek guidance as to the most profitable theoretical line to follow, that recourse to thermodynamic data is likely to be useful. For example, in the case of alkane + benzene mixtures adsorbed by Graphon it was observed that the entropy of immersion in heptane is more negative than for immersion in benzene. This suggests immediately that this difference arises from the restrictions imposed by the surface on various rotations which are free in bulk liquid heptane. If this were so, then one might expect the effect to decrease successively for n-pentane and iso-pentane. This is confirmed by the results presented in Figure 8 (24). This figure also contains data for adsorption from n-butyl benzene + n-heptane: here restrictions in the motion of the n-butyl group in the adsorbed state reduces the difference between the entropies of adsorption of the two components.

The behaviour exhibited by the entropies of adsorption for these relatively simple systems is not taken into account in current theoretical treatments. One must conclude that although for a number of purposes discussion of adsorption from solution in terms of simple statistical mechanical theories is moderately satisfactory, more complete treatments are required. They will need, among other things, to break away from the surface phase model, (and in particular the monolayer model) and take a much more complete account of contributions to the entropy of adsorption from molecular configuration factors.

$\hat{s}^{\sigma(n)}$	areal (reduced) surface excess entropy
s_i^{ℓ}	partial molar entropy of component i in solution
$\Delta_w \hat{s}$	entropy of immersion of unit area of solid (areal entropy of immersion) in liquid mixture
$\Delta_w \hat{s}_i^*$	entropy of immersion of unit area of solid (areal entropy of immersion) in pure component i
t	thickness of adsorbed phase (in number of molecular layers)
T	Kelvin temperature
v_i	partial molar volume of component i in liquid
w	interation parameter (Equation 35)
x_i^{ℓ}	mole fraction of component i in liquid
x_i^s	mole fraction of component i in surface phase
x_i^o	initial mole fraction of component i in liquid
γ_i^{ℓ}	activity coefficient of component i in liquid
γ_i^s	activity coefficient of component i in surface phase
$\Gamma_i^{(n)}$	reduced adsorption of component i
$\Gamma_i^{(1)}$	relative adsorption of component i with respect to component 1.
Γ_i	areal adsorption of vapour of component i
ε_{ij}	energy of interaction of molecules i, j at their equilibrium separation
θ	contact angle at three-phase line of contact
μ_i	chemical potential of component i in solution
μ_i^*	chemical potential of pure liquid component i
σ	interfacial tension between solid and liquid
σ_i^*	interfacial tension between solid and pure liquid i
$\sigma^{i,j}$ (or $\sigma^{*,ij}$)	interfacial tension between pure liquids i,j
σ_s^*	surface tension of pure solid/vacuum interface
$\sigma_i^{*,sv}$	interfacial tension of saturated vapour of i/solid interface

$\sigma_i^{*,s\ell}$ interfacial tension of pure liquid i/solid interface

$\sigma_i^{*,\ell v}$ surface tension of pure liquid i/vapour interface

LITERATURE CITED

1. Gibbs, W., Collected Works (Longman 1928) Vol.1, p.219.

2. Butler, J.A.V., Proc.Roy.Soc., A135, 348 (1932); Schuchowitzky, A., Acta Physicochim., U.R.S.S., 19, 176 (1944); Belton, J.W. and Evans, M.G., Trans.Faraday Soc., 41, 1 (1945); Guggenheim, E.A., Trans.Faraday Soc., 41, 150 (1945).

3. Schuchowitzky, A., Acta Physicochim. U.R.S.S., 8, 531 (1938); Schay, G., Acta Chim.Acad.Sci.Hung., 10, 281 (1953); Siskova, M and Erdös, E., Coll.Czech.Chem.Comm., 25, 1729, 3086 (1960); Everett, D.H., Trans.Faraday Soc., 60, 1803 (1964); 61, 2478 (1965).

4. Defay, R., Prigogine, I., Bellemans, A., and Everett, D.H., Surface Tension and Adsorption, Longmans, London 1965.

5. IUPAC, Manual of Symbols and Terminology for Physicochemical Quantities and Units, Appendix II, Definitions, Terminology and Symbols in Colloid and Surface Science, Part I; Pure Appl.Chem., 31, 579 (1973).

6. Kipling, J.J., Adsorption from Solutions of Non-Electrolytes, Academic Press, London and New York, 1965.

7. Parfitt, G.D. and Willis, E., J.Phys.Chem., 63, 1780 (1964); Parfitt, G.D. and Thompson, P.C., Trans.Faraday Soc., 67, 3372 (1971).

8. Ash, S.G., Bown, R., and Everett, D.H., J.Chem.Thermodynamics, 5, 239 (1973); Kurbanbekov, E., Larionov, O.G., Chmutov, K.V., and Yudilevich, M.D., Zhur.fiz.Khim., 43, 1630 (1969) (Russ.J.Phys. Chem., 43, 916 (1969)).

9. Schay, G., Nagy, L.G. and Racz, G., Acta Chim.Acad.Sci.Hung., 71, 23 (1972); Sharma, S.C. and Fort, T., J.Coll.Interface Sci., 43, 36 (1973).

10. Wang, H., Duda, J.L. and Radke, C.J., J.Colloid Interface Sci., 66, 153 (1978).

11. Nunn, C. and Everett, D.H., J.C.S. Faraday I, in press.

12. Nunn, C., to be published.

13. Wagner, C., Nach Akad.Wiss.Gottingen, II, Math.Phys.Kl. 1973, 37.

14. see eg. Colloid Science, Specialist Periodical Reports, The Chemical Society, London, (D.H. Everett, Ed.) (a) Vol. 1 (1973) Chapt. 2; (b) Vol. 2 (1975) Chapt. 2; (c) Vol.3 (1979) Chapt. 2.

15. Everett, D.H., Pure Appl.Chem., 53, 2181 (1981).

16. Larionov, O.G., Chmutov, K.V. and Yudelevich, M.D., Zhur.fiz.Khim., 41, 2616 (1967).

17. Myers, A.L. and Prausnitz, J.M., A.I.Ch.E. Journal, 11, 121 (1965); Sircar, S. and Myers, A.L., Chem.Eng.Sci., 28, 489 (1973); Bering, B.P. and Serpinski, V.V., Izvest.Akad.Nauk.S.S.R., Ser. Khim., 1972, 166, 169, 171, (Bull.Acad.Sci., U.S.S.R., Chem.Ser., 1972, 152, 155, 158); Bering, B.P., Serpinski, V.V., and Surinova, S.I., Izvest.Akad.Nauk., Ser.Khim., 22, 3 (1973) (Bull.Acad.Sci., U.S.S.R., Chem.Ser., 22, 1 (1973)); Everett, D.H. in Adsorption at the Gas/Solid and Liquid/Solid Interface (J.Rouquerol and K.S.W. Sing Eds.) Elsevier, Amsterdam, p.1 (1982)

18. Sircar, S., Novosad, J. and Myers, A.L., I and EC. Fundamentals 11, 249 (1972); Sircar, S., and Myers, A.L., A.I.Ch.E. Symp. Ser. No. 117, 11 (1971); Schay, G., J.Coll.Interface Sci., 42, 478 (1973).

19. Data of Ash, S.G., Bown, R. and Everett, D.H., J.Chem.Soc., Faraday I, 71, 123 (1975); Brown, C.E., Everett, D.H. and Morgan, C.J., ibid., 71, 883 (1975).

20. Everett, D.H., J.Phys.Chem., 85, 3263 (1981).

21. Fletcher, A.J.P., Ph.D. Thesis, Bristol (1982).

22. Callaghan, I., Fletcher, A.J.P. and Everett, D.H., J.C.S. Faraday I, in press.

23. Sircar, S., and Myers, A.L., A.I.Ch.E. Journal, 19, 159 (1973); Larionov, O.G., Chmutov, K.V., and Yudilevich, M.D., Zhur.fiz. Khim., 41, 2616, (1967); Larionov, O.G., and coworkers, Zhur.fiz. Khim., 46, 545, 2166, 2966 (1972); 47, 1331, 1617, 1618, 1618, 1619, 2171 (1973) (Russ.J.Phys.Chem., 46, 318, 1242, 1694,(1972): 47, 756, 919, 920, 920, 921, 1231 (1973)): abstracts only cf. ref 14 (c) pp.82-93.

24. Davis, J., Ph.D. thesis, Bristol (1983).

25. Polanco, C., Ph.D. thesis, Bristol (1973).

26. Everett, D.H., Trans.Faraday Soc., 61, 2478 (1965); see also ref. 4.

27. Defay, R., and Prigogine, I., Trans.Faraday Soc., 46, 199 (1950).

28. cf. ref 14(a) p.67.

29. Rusanov, A.I., Phase Equilibrium and Surface Phenomena, Chimia, Leningrad 1967, Chap.VI.

30. e.g. Everett, D.H., Trans.Faraday Soc., 61, 2478 (1965).

31. Lane, J.E. in Adsorption from Solution (Ottewill, R.H., Rochester, C.H., and Smith, A.L. Eds.) Academic Press, London, New York etc. p.51 (1983).

32. e.g. Kiselev, A.V., and Khopina, V.V., Trans.Faraday Soc., 65, 1936 (1969) and earlier papers; Nagy, L.G. and Schay, G., Acta Chim. Acad.Sci.Hung., 39, 365 (1963); Myers, A.L. and Sircar, S., A.I.Ch.E. Journal, 17, 186 (1971).

33. Schay, G., Nagy, L.G. and Szekrenyesy, T., Periodica Polytechnica, 6, 91 (1962).

34. e.g. Ash, S. G., Bown, R., and Everett, D.H., J.Chem.Soc., Faraday I, 71, 123 (1975).

35. Everett, D.H. and Podoll, R.T., J.Coll.Interface Sci., 82, 14 (1981).

36. see Ref. 4, Chap.XII;

37. Ash, S.G., Findenegg, G.H. and Everett, D.H., Trans.Faraday Soc., 64, 2639 (1968).

38. Ash, S.G., Findenegg, G.H. and Everett, D.H., Trans.Faraday Soc., 64, 2645 (1968); 66, 708 (1970).

39. Everett, D.H., Prog.Coll.Polymer Sci., 65, 103 (1978).

40. Brown, C.E., Everett, D.H., Powell, A.V. and Thorne, P.E., Faraday Disc.Chem.Soc., 59, 97 (1975); Everett, D.H., Israel J.Chem. 14, 267 (1975).

41. House, W.A., in Colloid Science, Specialist Periodical Report, Royal Society of Chemistry (Everett, D.H. Ed.) Vol. 4 (1982), Chap. I.

42. Ref. 41, p.56; and e.g. Rudzinski, W., and Partyka, S., J.C.S. Faraday I, 77, 2577 (1981); Garbacz, J.K., Dabrowski, A., and Jaroniec, M., Thin Solid Films, 94, 79 (1982).

43. Dékány, I., Nagy, L.G. and Schay, G., J.Colloid Interface Sci., 66, 197 (1978); Dékány, I., and Nagy, L.G., Acta Phy.Chem.Acad. Sci.Hung., 23, 485 (1977).

SOME FACTS AND FANCIES ABOUT THE PHYSICAL ADSORPTION OF VAPORS

Arthur W. Adamson and Ruhullah Massoudi, Department of Chemistry, University of Southern California, Los Angeles, CA 90089-1062, U.S.A.

ABSTRACT

Several of the common assumptions made with respect to physical adsorption and contact angle phenomena are examined. These include chemical and topological surface uniformity, dominance of dispersion forces, lack of structural perturbation in the adsorbent surface region and in the adsorbed film, unimportance of vapor adsorption in non-wetting systems. Many of these assumptions have been coming under closer scrutiny, although the last is still widely held. Ellipsometrically determined adsorption data are given for adsorption of n-octane, 4-octyne, 1-octyne, water, and n-butylamine on smooth polytetrafluoroethylene, polycarbonate, polypropylene, and polyacetylene surfaces. In all cases there is significant adsorption even when the liquid adsorbate does not wet the smooth plastic surface; in some cases surface swelling occurs, indicating that the surface structure of the solid is perturbed on adsorption. Adsorbed films of 3,5-octadiyne polymerized by ultra-violet irradiation *in situ*, show a uniform film if the liquid wets the solid, but scattered droplets otherwise. The probable significance of this last observation is discussed in relation to contact angle and contact angle hysteresis.

There are a number of common assumptions that are made in the literature dealing with physical adsorption of vapors and with contact angle phenomena. In the past, some of them became so pervasive and ingrained that their presumptive nature seemed to be forgotten, hence the term "fancies". Certain of these assumptions are now receiving more careful attention, but others are yielding only slowly to reality. In the present paper, we summarize briefly several fancies, and then concentrate on a particular one, examined in the light of some current results.

Facts and Fancies

Solid surfaces are uniform. Most models of physical adsorption simplify matters by assuming a uniform adsorbent surface. This is true of the Langmuir and BET (Brunauer, Emmett, and Teller) adsorption isotherm equations, and of the Frenkel-Halsey-Hill variation of the Polanyi approach to adsorption, as well as of models assuming two-dimensional equations of state for the adsorbed film (see 1). However, it is by now fairly widely recognized that adsorbents typically are surface heterogeneous, especially those of technological interest, and deconvolution methods have been developed for obtaining site adsorption energy distributions (1 and papers in this symposium). More restrictive and therefor less satisfactory are the Freundlich and Dubinin-Radushkevich equations, which contain built-in assumptions as to the form of the site energy distribution (see 1).

The deconvolution methods for obtaining site energy distributions still contain an assumption, namely that heterogeneity occurs in patches sufficiently large that each type of patch has a local isotherm function which is independent of the occupancy of other patches. The more likely situation, however, is that "patches" due to chemical heterogeneity are small to the point of consisting of just one or two sites of a given kind. One needs to know the distribution of types of nearest neighbors to each type of site. The obtaining of such a distribution has so far seemed beyond reach. Until this can be done, however, calculated site energy distributions based on the patch model can only be approximate. One test, incidentally, of whether the distribution is patch-wise or individual site-wise is that in the latter case the deconvolution of adsorption isotherms will yield qualitatively different site energy distributions for small vs. large adsorbate molecules. The situation is illustrated in Fig. 1, which shows how a large adsorbed molecule will average nearest neighbor sites to give an apparently more uniform distribution than will a small molecule.

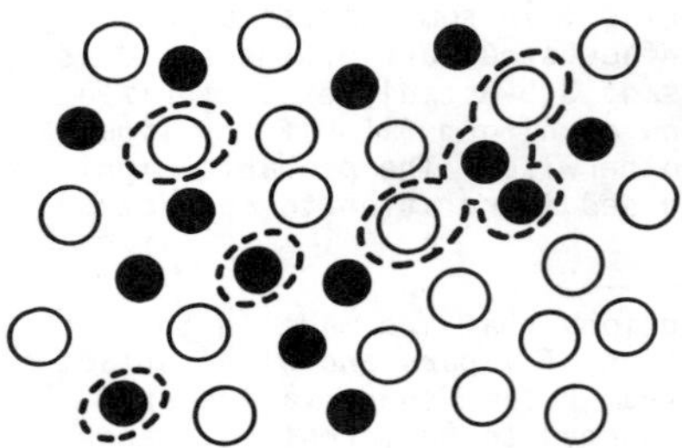

Figure 1. Schematic of a surface showing two kinds of adsorption sites. Dashed contours indicate adsorbed small and large molecules.

Topological heterogeneity or surface roughness is important in contact angle situations. This has been recognized in that explicit studies of the effect of roughening a surface has been made (1). Very often, however, surfaces that appear to be smooth are assumed to be smooth at the microscopic level. Figures 6 and 7, discussed later, illustrate that quite the contrary is likely to be true.

<u>Dispersion forces are dominant in physical adsorption</u>. The dispersion effect is a universal one, the attractive potential varying with the inverse sixth power of the distance between molecules (for distances large compared to molecular diameters). As a fair approximation, it is assumed to be pairwise additive, and integration yields an inverse cube dependence of the attractive potential between a molecule and a flat surface (1). This potential is the basis for the well known Frenkel-Halsey-Hill adsorption equation (1 to 4). Similarly, Good and Fowkes, in modeling interfacial tension, have assumed that only dispersion forces act across a liquid-liquid or a solid-liquid interface if one of the phases is "non-polar" (e.g. a hydrocarbon) (1,5).

The inverse cube dependence becomes a good approximation only for distances large compared to molecular diameters, however, and for intermediate separations, a smaller power or an exponential function is empirically better. In addition to questions about functional form, there are strong indications that non-dispersion type forces are important even with non-polar species. In a study of nitrogen adsorption on molecular solids, for example, we concluded that the ordering of adsorption energies was not consistent with a dispersion-only interaction and that polar forces originating in bond dipoles were important (6,7). The same conclusion was reached in the case of hexane adsorption on ice powder (6 to 8). In the thin multilayer region, induced dipole propagation (which gives an exponentially relaxing attractive potential) is probably more important than is commonly recognized (1).

Finally, the existence of characteristic isotherms denies the dominance of solid-adsorbate dispersion forces in the thin multilayer region. Here, the observation is that adsorption isotherms in this region are invariant to the nature of the adsorbent, to a first approximation (1). An important application of this observation has been in the de Boer *t*-plot (1).

Solid surfaces are not altered by adsorption. A pervasive assumption has been that the surface of a solid adsorbent is not structurally altered in adsorption. The advent of LEED (low energy electron diffraction) studies has shown the fallacy of this assumption in the case of chemisorption (9). Surface restructuring in chemisorption is now well accepted. The assumption persists, however, in the case of physical adsorption. Since even here adsorption energies may reach 10-15 kcal $mole^{-1}$, it is entirely possible that some surface restructuring takes place with "refractory" solids such as metals or oxides. Displacement of cations from normal positions in zeolites may occur on physical adsorption of hydrocarbons, as a related example.

In the case of molecular solids, there is strong evidence of surface restructuring in physical adsorption; here the adsorption bond and the cohesive binding of the solid are comparable. An example is the surface restructuring of ice on hexane adsorption (8). Polymers (polytetrafluoroethylene, PTFE, polyethylene, PE, etc.), while relatively rigid and high melting, are essentially molecular solids on a local scale since adjacent polymer strands interact mutually through ordinary van der Waals forces. Here again, surface disorder or structuring is to be expected on physical adsorption. It is this last situation that generally is ignored, particularly in the treatment of contact angle and wetting in which it is commonly assumed that a given polymer has a single characteristic surface tension, the "critical surface tension" (1). We have suggested instead that liquid-molecular solid interactions may range from type "A" with little surface restructuring to type "B" where it is extensive (10).

Multilayer adsorbed films are bulk liquid-like. A characteristic of Polanyi-type treatments of adsorption is the assumption that bulk liquid adsorbate condenses in the attractive potential field of the adsorbent (often taken to be inverse cube in distance $\underline{x}$). One writes

$$\phi\ (x) = kT \ln (P^\circ/P) \qquad (1)$$

where ϕ is the potential, P°, the vapor pressure of bulk liquid adsorbate, and P, the pressure of vapor in equilibrium with a given adsorbed film.

There can be no doubt, however, that in the thin multilayer region adsorbed films are perturbed structurally relative to bulk liquid (1, 11). We have recognized the situation empirically by taking the adsorbed films to have an effective vapor pressure P°'. That is, to say that a hypothetical bulk liquid of the same structure would have a vapor pressure P°', P°'>P°. A reasonable further assumption is that the excess free energy of structural perturbation, kT ln (P°'/P°), relaxes exponentially with film thickness.

<u>Adsorption is negligible if the liquid adsorbate is non-wetting.</u> This last of the series of assumptions or fancies is still widely held; it is the one to which the experimental part of this paper is addressed. The prevalence of the assumption is undoubtedly due to its convenience, even necessity, in the semi-empirical modeling of how an interfacial tension (such as γ_{SL}, S and L denoting solid and liquid phases) relates to the surface tensions of the separate phases (γ_S and γ_L). These quantities appear in the Young equation for contact angle,

$$\gamma_L \cos\theta = \gamma_{SV} - \gamma_{SL} = \gamma_S - \gamma_{SL} - \pi^\circ_{SV} \tag{2}$$

where SV refers to the solid vapor interface. Alternatively, $\gamma_{SV} = \gamma_S - \pi^\circ_{SV}$ where π°_{SV} is the reduction in surface tension of the solid due to the adsorbed films. That is, π°_{SV} is the equilibrium film pressure, and the specific assumption of convenience that is commonly made is that it is negligible if θ is non-zero. Film pressure relates to the adsorption isotherm through the Gibbs equation,

$$\pi^\circ_{SV} = kT \int_0^{P^\circ} \Gamma \, d \ln P \tag{3}$$

where Γ is the surface excess, $\Gamma = x/V$, where V is the molecular volume of the adsorbate. The assumption of negligible π°_{SV} thus implies negligible adsorption of vapor and that the solid-vapor surface tension can be approximated as γ_S.

The neglect of π°_{SV} is untenable at two levels. First, as illustrated in Fig. 2, consider a drop of liquid resting on a flat solid surface the system being enclosed and thermostatted. The liquid <u>must</u> be in equilibrium with its vapor pressure P°, and the uncovered portion of the solid must likewise be exposed to this vapor. Referring to

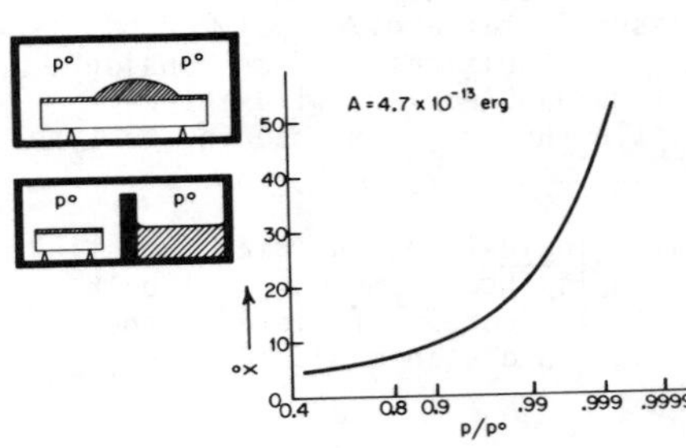

Figure 2. Left: the solid surface must be in equilibrium with the saturation pressure of the liquid, P° whether that is established by a separate dish of liquid or by a drop resting on the surface. Right: Film thickness vs. P/P° for a film obeying the equation ln (P°/P) = g/x^3 with g given by the indicated Hammaker constant A. The calculation is for a typical low-energy surface and a hydrocarbon vapor.

Equation (1), even the most minute attractive potential must give rise to some adsorption. More important, even if the function $\phi(x)$ is taken to be just that for dispersion, $\phi = g/x^3$, and a reasonable value is used for the constant $\underline{g}$, then π°_{SV} values result which are far from negligible compared to the $\gamma_L \cos\theta$ term in Equation (2).

At the experimental level, in this and preceding papers (12,13), direct ellipsometric measurements of vapor adsorption show non-negligible π°_{SV} values in non-wetting systems.

An Adsorption Model for Non-Wetting Systems.

It was noted above that Equation (1) implies adsorption even in non-wetting systems. In fact, Equation (1) implies infinite adsorption at $P = P^\circ$ and denies the possibility of a finite θ since a liquid drop would co-mingle with a macroscopically thick adsorbed film. Since non-wetting exists, adsorption at P° must be limited in such cases to some film thickness x_0. The theoretical problem is how to so limit adsorption. We have done this by invoking an exponentially relaxing term in $\ln (P^{\circ\prime}/P^\circ)$, obtaining the equation (14)

$$\ln (P^\circ/P) = g/x^3 + \varepsilon e^{-ax} - \beta e^{-\alpha x} \tag{4}$$

The $e^{-\alpha x}$ term arises from the structural perturbation. Of the two attractive potentials, the e^{-ax} term is taken to be the more important for thin films, and an empirically useful form of Equation (4) is then

$$\ln (P^\circ/P) = \varepsilon e^{-ax} - \beta e^{-\alpha x} \tag{5}$$

Depending on the choice of the constants g, ε, a, α, and β, adsorption isotherms generated by Equations (4) or (5) may cut the $P = P^\circ$ line at a finite film thickness, x_0, as illustrated in Fig. 3; this corresponds

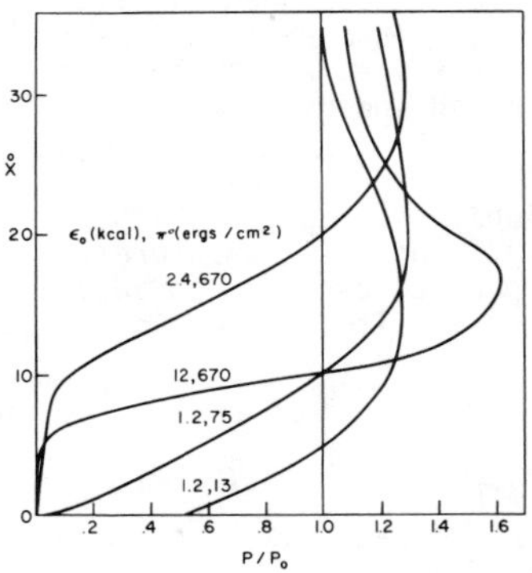

Figure 3. Adsorption isotherms of the type given by Equation (5) calculated with parameters such that $\theta = 90°$ in all cases. The parameter $\varepsilon_\circ$ is equal to $kT\,\varepsilon$, $T \simeq 20°C$, and $\pi^\circ = \pi^\circ_{SV}$.

to a contact angle situation. Integration over the experimentally unattainable region from $x = x_0$ to $x = \infty$ gives the spreading coefficient, S_{LSV},

$$S_{LSV} = \gamma_{SV} - (\gamma_{SL} + \gamma_L) = kT \int_{x=x_\circ}^{x=\infty} \Gamma \, d \ln P \qquad (6)$$

Combination with Equation (2) gives

$$S_{LSV} = \gamma_L (\cos \theta - 1) \qquad (7)$$

and if an adsorption isotherm is fitted to Equation (5), the analytical result is

$$S_{LSV} = (kT/V)e^{-ax_\circ} (1/a - 1/\alpha), \qquad (8)$$

where V is molecular volume.

Our experimental work has involved ellipsometric measurements of the adsorption of vapors on flat surfaces, along with contact angle measurements of the liquid adsorbate on the _same_ flat surface. The use of flat surfaces avoids the complication of capillary and interparticle condensation that becomes important with powdered adsorbents as P approaches P° (15). For non-wetting systems, the isotherms indeed cut the P - P° line at a finite film thickness. Cases previously studied in this Laboratory include water on various solids such as stearic acid coated copper, pyrolytic carbon, and a variety of polymers such as PE and PTFE. In addition, various organic vapors have been studied with a variety of polymer surfaces (12,13).

In all cases it was possible to fit the adsorption data to Equation (5) with values of the constants such that the observed θ was also given by Equation (9). Also, in all cases, the equilibrium film pressure π°_{SV} was not negligible compared to the $\gamma_L \cos \theta$ term in the Young equation. Agreement with a model does not prove the correctness of that model, however, and there has remained a concern that the actual situation might be more complex. There might, for example, be patchwise adsorption or "lakes" formed by Kelvin condensation on surface dimples or ripples. That the latter possibility is a real one was indicated by SEM photographs which showed that surfaces which were optically smooth and reflective were in fact rippled and dimpled on the tenth micron scale (13).

The present paper provides some additional systems for which adsorption and contact angle have been measured, including unsaturated hydrocarbons. The possibility of lakes is pursued by using an adsorbate, 3,5-octadiyne, which can be polymerized _in situ_.

EXPERIMENTAL

Materials

Trans-polyacetylene, PA or $(CH)_x$, was obtained in about 0.1 mm thick sheet form from A. G. MacDiarmid, University of Pennsylvania. Surface oxidation is a problem, and the sample was handled either under inert atmosphere or in vacuum. To give rigidity to the thin sample, a layer of gold was vacuum deposited on one side, which allowed cementing of the PA sheet to a glass slide with minimum likelihood that the epoxy cement used would perfuse the polymer and thus contaminate it. The PA sheet, so mounted, was cleaned ultrasonically in water, and dried under vacuum.

Polypropylene, PP, and polycarbonate, PC, were obtained from Port Plastics, Montebello, California, in 3 mm thick sheet form. Slips were cleaned ultrasonically under ethanol and then water, and air dried. The PP was then flow smoothed between optically smooth pyrex glass plates, under pressure and at 170 °C. PTFE was obtained from Bel-Art Products, Penquannock, New Jersey, and was cleaned as above, and flow smoothed at 190 °C. "Virgin" Teflon, VPTFE, with a military grade of P-22241 was obtained from Leed Plastics Corp., Los Angeles, California, supplied by TFE Welcome, North Carolina. After cleaning, the VPTFE was flow smoothed at 380 °C, between quartz plates, and further cleaned in a steam jet. Fluorinated ethylene propylene, FEP, was obtained from Port Plastics, and treated similarly to the PTFE. Cellulose acetate, CA, and cellulose acetate butyrate, CAB, were obtained from Lustro Plastics Co., Valencia, California, as optically smooth sheets, and were cleaned by _n_-heptane washing and vacuum dried before use.

Adsorbates 1-hexyne and 1-octyne were obtained from Alfa Products, and 1-heptene and _trans_-4-octene, from Sigma Chemical Co.; 2-octyne, 4-octyne, and 3,5-octadiyne were from Farchan Laboratories, Willoughby, Ohio. These hydrocarbons were kept protected from oxidation and were distilled if any indication of such was present (surface tension change, coloration). Nominally reagent grade butylamine and triethylamine (Matheson Cole) required distillation before use.

Measurements

Most of the contact angle measurements were carried out by means of Newmann's method in which meniscus height (in an enclosed, thermostatted chamber) was measured for the polymer slip by means of a telemicroscope, as previously described (12,13). These were advancing angles. For contact angle approaching or exceeding 90°, advancing angles were measured in an enclosed chamber for a small drop delivered onto the surface of the slip, using a goniometer telemicroscope. All values are for 20 °C.

Surface tensions were determined by means of the ring method, and were corrected by the use of standard tables (1). The measurements were made in an enclosed chamber, at room temperature, ca. 20 °C.

The adsorption measurements were made with the use of a Rudolph ellipsometer and sample chamber previously described (12,13). The pressure of adsorbate vapor was either measured directly by means of an MKS Baratron transducer or, for values close to P°, from the temperature of a thermostatted reservoir of liquid adsorbate connected to the sample chamber. SEM pictures were taken as previously described (13).

Surface Polymerization

We were interested in effecting an _in situ_ fixing of an adsorbed film, and noted that di-acetylenes may easily be polymerized by u.v. or gamma radiation either in the crystalline state or in monlayer form (16,17,18). Polymerization of vapor adsorbed films of 3,5-octadiyne was indeed possible. A slip of polymer was suspended above some of the neat liquid in a closed glass container having a quartz window on one side. Irradiation was by means of a Pen Ray lamp, which delivered a surface irradiation of about 3×10^{-9} einstein $cm^{-2}sec^{-1}$, mostly at 254 nm, as determined by ferrioxalate actinometry (19). The u.v. absorption spectrum of 3,5-octadiyne shows a broad maximum at 270 nm, and narrow maxima at 253 nm, 238 nm, and 228 nm, with respective molar extinction coefficients of 4.0×10^5, 5.5×10^5, 6.8×10^5, and 6.3×10^5 M^{-1} cm^{-1} as measured in octane solution. Irradiations were for periods of several minutes to several hours, and with the longer ones, a yellowing of the surface was apparent (which did not occur in the absence of the vapor).

The qualitative results on SEM examination were similar if the slip touched the liquid surface rather than being suspended above it. After the irradiation, the slip was vacuum dried to remove any monomer present. In the case of PTFE, steam cleaning largely removed the polymerized film. It was also possible to produce a yellowed surface on u.v. irradiation of solid 2,4-hexadiyne (but not of the neat liquid).

RESULTS

Surface Tension and Contact Angle

A number of surface tension and contact angle values were obtained in the course of this study. These are summarized in Table 1. Systems

Table 1. Surface Tension and Contact Angle Values[a] (20 °C).

Liquid	Surface Tension dyne/cm	Liquid	Substrate[a]	Contact Angle degrees
Water	72.75	3,5-octadiyne	PTFE	34
1-octyne	23.9	4-octyne		14
2-octyne	23.9	2-octyne		9
4-octyne	23.9	4-octyne	VPTFE	26
n-butylamine	23.9	water	VPTFE	105
n-butylamine	23.2	water	PA	49
trans-4-octyne	22.0	triethylamine		39
n-octane	20.8	n-butylamine		14
triethylamine	21.6	3,5-octadiyne	FEP	35
1-heptene	20.7	water	PC	82
		water	PP	101
		water	CA	58
		water	CAB	64

(a) See text

for which the contact angle was zero within our detection limit of about 1° were 1-octyne, 1-hexyne, 1-heptene, and n-octane on PTFE, 4-octyne on PA, 1-hexyne on PC. Also 3,5-octadiyne wet PE, CA, and CAB.

Adsorption Isotherms

The isotherms for PTFE and VPTFE are shown in Fig. 4. Desorption

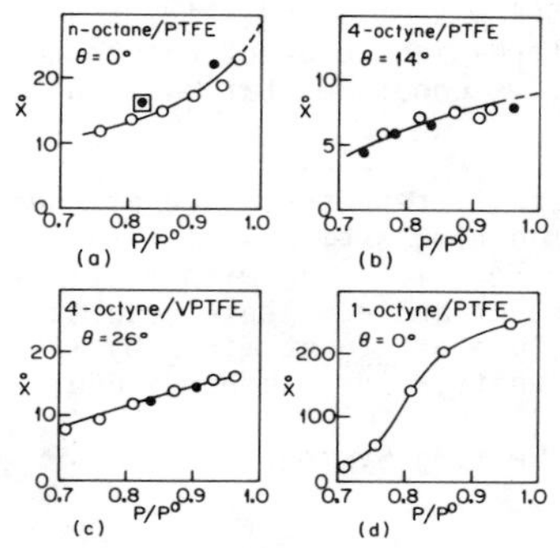

Figure 4. Adsorption of vapors on PTFE at 20 °C. Desorption points given by the full circles.

was reasonably reversible for the first three systems, although it tended to be slow, but probably for the trivial reason that the vapor was adsorbed on the walls and other surfaces of the sample chamber and the capacity of the connecting line was insufficient to remove it quickly. Surface swelling due to dissolving or absorption probably occurred in the case of 1-octyne on PTFE, Figure 4(d); the polarizer reading did not return to the original value on evacuation.

The two water isotherms shown in Fig. 5 were well behaved. In the case of 4-octyne on PA, a wetting system, shown in Fig. 5(c), desorption points fell below the adsorption ones and readsorption did not

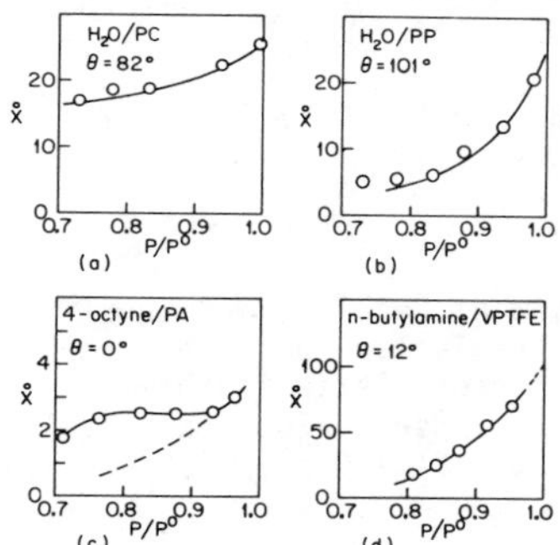

Figure 5. Adsorption of vapors at 20 °C

follow the original isotherms. Very likely there was some surface dissolving of the 4-octyne into the PA, which would change the optical parameters and falsify the interpretation of the ellipsometer readings. Irreversibility was also found for 1-hexyne on PC (not shown in the figures). This was again a wetting system, and here, film thicknesses on adsorption rose to some 500 Å. The desorption points gave apparent rising film thicknesses with decreasing P/P° and we again suspect surface swelling. After the adsorption-desorption run, and evacuation, the slip showed many fine surface cracks.

It was thought that alkyl amines might form an additional useful class of adsorbates since they were non-wetting on VPTFE. Although an adsorption isotherm for n-butylamine on VPTFE was obtained, Fig. 5(d), desorption points were off, and the optical readings did not return to the original ones after evacuation. Again, we conclude that surface swelling occurred.

The solid lines in the figures are visually representative of the data; no more sophisticated averaging or smoothing seemed justified. The reversible isotherms could be fitted to Equation (5), with fitting parameters which also gave the observed contact angle. These results are collected in Table 2, along with the values of π°_{SV} obtained by integrating the isotherms from P/P° = 0.8 to unity. The isotherms could

Table 2. Adsorption Isotherm Data for Non-Wetting Systems.

System	θ deg	ε^a $\mathring{a}_{1/2}$	β^a $\mathring{\alpha}_{1/2}$	$\mathring{x}^{\circ}$	$\mathring{x}'$ $P^{\circ\prime}/P^{\circ}$	π°_{SV}	$\gamma_L \cos\theta$ dyne cm^{-1}
water/PP	101	0.40 9.7	0.088 73	24	1.25 1.033	27	-14
water/PC	82	60 2.2	0.011 320	28	11.5 1.61	68	10
n-octane/PTFE	∿ 0	7.40 4.4	5.00 4.8	42	5.5 1.22	9.7	∿ 21
4-octyne/PTFE	14	3.02 3.6	1.83 4.6	12			23
4-octyne/VPTFE	26	2.40 4.5	0.436 11.7	18	3.5 1.13	3.1	21

(a) Energy relative to kT.

also be matched to the characteristic isotherm of Ref. 13, and the fitting parameters, P°' and x' are included in the Table. Note that the same π°_{SV} values could be obtained from the characteristic isotherm equation, $\ln(P^{\circ\prime}/P) = x'/x$ (13), from the relationship $\pi^{\circ}_{SV} = (kTx'/V)\ln(x^{\circ}/x_{0.8})$, where $x_{0.8}$ is the film thickness at P/P° = 0.8.

Surface Topologies

The ability to polymerize 3,5-octadiyne films *in situ* made it possible to gain some idea as to how adsorbed films are disposed on a surface in a wetting vs. a non-wetting situation. The nominal, ellipsometric thicknesses of the polymerized films were not determined because of uncertainty as to the proper optical constants to use. During long irradiation there was undoubtedly a continuing process of vapor deposition and polymerization; after one or two hours a noticeable yellowing of the surfaces occurred. There was no great qualitative change in the appearance of the SEM photographs, however, and it appeared that continued irradiation merely led to further development of a film topology rather than to a completely new one. A confirmation of this conclusion is discussed further below.

Figures 6ab compare the appearance of PE without and with a 30 min polymerized vapor adsorbed film. The flow-smoothed surface is relatively smooth although features of the order of 1000 Å are visible as a

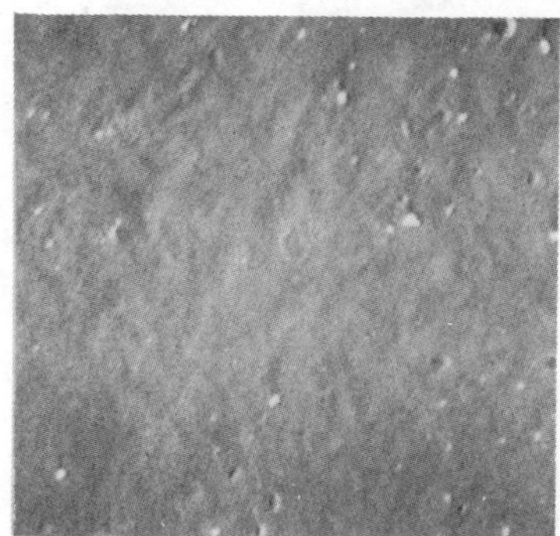

6a

6b

Figure 6. SEM photographs of a flow smoothed polyethylene surface at 10,000X. (a) Clean surface. (b) Surface irradiated for 30 min by 254 nm light while exposed to 3,5-octadiyne vapor.

graininess or dimpling. In Fig. 6b, this graininess is noticeably reduced, the appearance being one of an overlying blanket. This is a wetting system. Rather similar behavior was found for CA and CAB, also wet by 3,5-octadiyne.

Turning to the non-wetting systems, Fig. 7a shows the surface of VPTFE as received. The macroscopically smooth, mat surface is seen to be highly irregular on magnification. VPTFE sheet is formed under pressure from pellets, and the appearance of the surface suggests that partly adhering pellets massively populate the surface. The flow smoothed material is macroscopically mirror-like, but under magnification there is a fibrous structure, as shown in Fig. 7b. Figure 7c shows the film smoothed material after 1 hr. exposure to 3,5-octadiyne vapor while under irradiation. There is no indication of a blanket film, but rather the appearance is one of droplets with some interconnecting filaments. Quite similar photographs were obtained with longer exposure times, and also if the slip was first exposed to vapor in the dark for a period, and then irradiated while still in a vapor atmosphere . The appearance was also similar if the slip touched the liquid surface, the upper portion being irradiated. If samples of the above type were steam cleaned, virtually all traces of the film material disappeared.

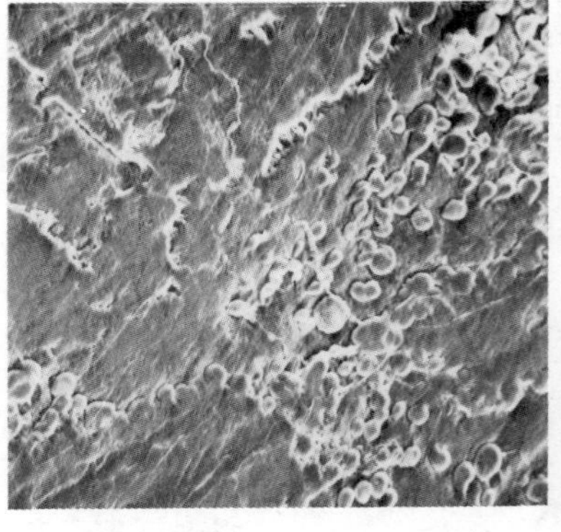

7a

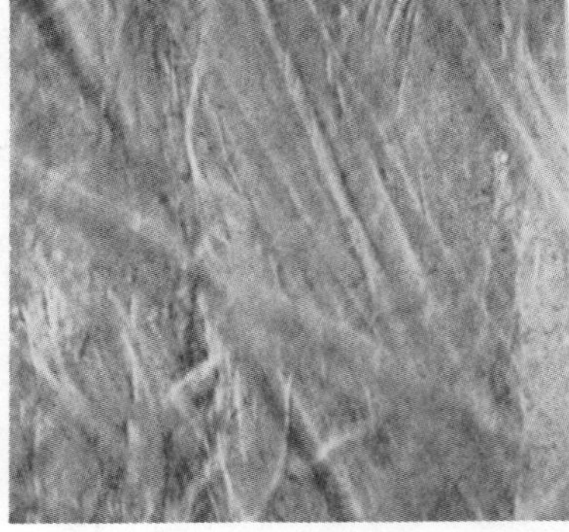

7b

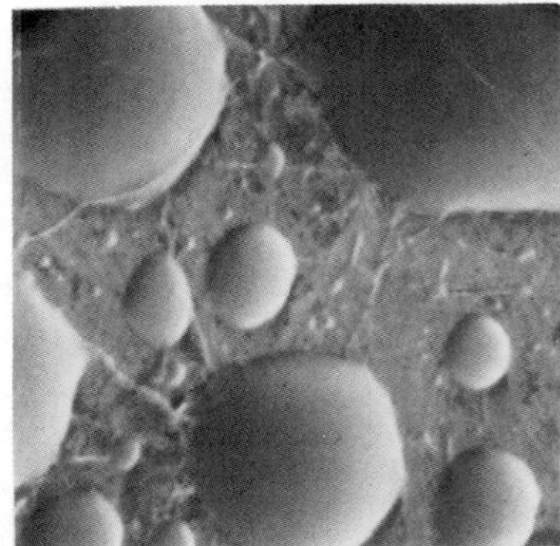

7c

Figure 7. SEM photographs of polytetrafluoroethylene surfaces. (a) Not flow smoothed, 1000x. (b) Flow smoothed, 10,000x. (c) Same as (b) but irradiated for 1 hr. by 254 nm light while exposed to 3,5-octadiyne vapor.

FEP polymer was also studied. As shown in Fig. 8a, the flow smoothed surface is more compact than that of VPTFE (the non flow smoothed surface showed a scaly roughness). The results of a 30 min irradiation were quite similar to those with VPTFE. A quite interesting phenomenon appeared with a 10 min irradiation, however, as shown in Figure 8b. One now sees that the larger droplets are collapsed and highly wrinkled. We infer that these were originally droplets of monomer and that the irradiation time was too short to polymerize the droplet completely. On evacuation, monomer from the interior of the droplets vaporized, leading to the collapsed appearance. An important conclusion is that the droplets formed as <u>monomer</u> droplets and that the polymerization merely fixes the topology of the adsorbed monomer.

8a

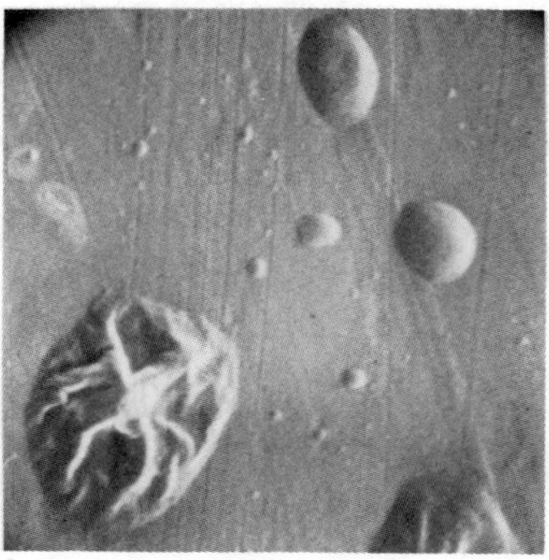
8b

Figure 8. SEM photographs of flow smoothed FEP surfaces at 10,000 X. (a) Clean surface. (b) Same as (a) but irradiated for 10 min by 254 nm light while exposed to 3,5-octadiyne vapor.

DISCUSSION

The results of the present study are of interest in several ways. First, additional systems are shown to exhibit appreciable vapor adsorption with a non-wetting bulk liquid adsorbate. Again, non-negligible π°_{SV} values are found. The relevent comparison is between the last two columns of Table 2. Clearly, any modeling of contact angle which neglects π°_{SV} in comparison to $\gamma_L \cos \theta$ is seriously flawed theoretically, although such treatments may have an empirical usefulness.

A second point is that in several of the present systems, there is strong evidence that the adsorbate penetrated into the surface - the ellipsometric readings did not return to the original, clean surface values (as they do in well behaved systems). Clearly, in these cases, the surface structure (and composition) of the polymer is affected. Such changes were not macroscopically or grossly evident; the polymer did not, for example, dissolve or obviously swell in the liquid adsorbates involved. A lessor degree of surface penetration or swelling would probably be undetectable ellipsometrically and could have been present in the apparently reversible systems. At this point, it seems reasonable to classify all polymer-organic liquid systems as class "B" in type. That is, it seems likely that interfacial structural changes are the rule rather than the exception in dealing with polymers in an adsorption or a contact angle situation. The fancy of the inert solid must indeed be just a fancy for such systems.

As noted earlier, ellipsometric data give an average film thickness and do not distinguish between a uniform film and one present in microscopic patches or lakes. Our SEM results appear to be important in indicating what the actual topological situation is. In the case of a wetting system, the vapor does appear to adsorb as a uniform blanket film covering the surface. The interesting results are those for the two non-wetting cases, 3,5-octadiyne on VPTFE and FEP. In both cases the "film" is seen actually to consist of myriad droplets each, incidentally, resting on the surface with about the macroscopic contact angle. A thin multilayer film could also be present since the SEM resolution was no more than about 500 Å. (Higher magnifications were not possible since surface melting of the polymer would begin to occur.)

To the extent that the above can be generalized some important conclusions suggest themselves. First, it must be accepted that surfaces considered to be smooth are generally grainy and dimpled on the 500 to 1000 Å (if not larger) scale. Second, in non-wetting systems, the advancing liquid front sees a heterogeneous terrain of lakes or droplets and the macroscopic contact angle must be a complex resultant of the intermingling of the bulk liquid front with such lakes. This heterogeneity may well be responsible for contact angle hysteresis since the terrain left by a receding liquid front will be different from that seen by the advancing front. Finally, the whole situation raises serious questions about the thermodynamic status of the Young equation when θ is a macroscopic angle and when γ_{SV} does not in fact apply to a uniform surface. The present observations, in fact, reinforce earlier concerns about the Young equation (20), as well as more recent ones (21).

ACKNOWLEDGEMENT

This investigation was supported in part by the U.S. National Science Foundation.

REFERENCES

1. Adamson, A.W., Physical Chemistry of Surfaces, 4th ed., Wiley, (1982).

2. Halsey, G.D. Jr., J. Chem. Phys., 16, 931 (1948).

3. Pierce, C., J. Phys. Chem., 63, 1076 (1959).

4. Vold, R.D. and Vold, M. Colloid and Interface Chemistry, Addison-Wesley, (1983).

5. Fowkes, F.M., Adv. Chem. 43, 99 (1964); J. Phys. Chem. 84, 510 (1980).

6. Adamson, A.W. and Orem, M.W., Progress in Surface and Membrane Science, 8, 285 (1974).

7. Dormant, L.M. and Adamson, A.W., J. Coll. Interface Sci., 28, 459 (1968).

8. Orem, M.W. and Adamson, A.W., J. Coll. Interface Sci., 31, 278 (1969).

9. Somorjai, G.A., Principles of Surface Chemistry, Prentice-Hall (1972).

10. Adamson, A.W., Shirley, F.P., and Kunichika, K. J. Coll. Interface Sci., 34, 461 (1970).

11. Etzler, F.M., J. Coll. Interface Sci., 92, 43 (1983).

12. Nayer, B.C. and Adamson, A.W., "Physicochemical Aspects of Polymer Surfaces," Plenum Press, (1983).

13. Tse, J. and Adamson, A.W., J. Coll. Interface Sci., 72, 515 (1979) and preceding papers.

14. Adamson, A.W., J. Coll. Interface Sci., 44, 273 (1973); 27, 180 (1968).

15. Wade, W.H. and Whalen, J. W., J. Phys. Chem., 82, 2898 (1968).

16. Day, D. and Ringsdorf, H., J. Polymer Sci., Polymer Letters, 16, 205 (1978).

17. Day, D. and Lando, J.B., J. Polymer Sci., Polymer Physics, 16, 1009 (1978).

18. Wegner, G., Adv. Chem. Ser., 129, 255 (1973).

19. Hatchard, G.C. and Parker, C.A., Proc. Roy. Soc., London, Ser. A, 235, 518 (1956).

20. Adamson, A.W. and Ling, I. Adv. in Chem. Ser., 43, 57 (1964).

21. Myers, A.L., AIChE Journal, 19, 415 (1973).

COMPARISON OF POROSITY RESULTS FROM NITROGEN ADSORPTION AND MERCURY PENETRATION

Bruce Adkins, Sayra Russell, Pasha Ganesan and Burtron Davis

Institute for Mining and Minerals Research
University of Kentucky, Lexington, Kentucky 40512

ABSTRACT

Pore volumes and pore size distributions were calculated from nitrogen adsorption and desorption curves and from mercury penetration curves obtained with a wetted nonporous alumina. The sample contained a large fraction of the pore volume in the narrow range of pore diameters amenable to measurement by both mercury penetration and nitrogen adsorption. The total pore volumes obtained by the two methods agree within 3%; the average pore sizes obtained also agree within 3%. These results support the view that, using accepted constants and equations, the two techniques yield similar porosity data.

The introduction of the BET equation (1) for the calculation of surface area and the use of chemisorption to obtain a measure of the surface coverage by an active component or promoter (2) provided a basis for comparing catalytic results obtained in different laboratories. These two techniques provided a reliable and reproducible means of characterizing surfaces; however, they do not define the nature of the porosity that accompanies the surface area (3).

A mathematical treatment presented in 1939 by Thiele (4) showed that the catalytic activity could be influenced by particle size and by porosity. With sufficiently small pores, the reaction rate was limited by the rate of diffusion of the reactant to the active site. Wheeler's two papers in the 1950's did much to define the influence of diffusion on the results of catalytic studies as well as the complexity of porosity models. The resurgent emphasis on coal liquefaction, and the need for catalysts with bimodal pore distributions, again demonstrates the need for pore volume analyses.

As early as 1914, Anderson (5) proposed that capillary adsorption could be used to obtain pore size distributions. However, it was Barrett, Joyner and Halenda (6) who developed the first widely used method for making these calculations. Many of the methods developed afterwards are reviewed by Broekhoff and Linsen (7). Dollimore and Heal (8) examined a large number of silica and alumina samples and obtained constants for an equation that corrects the Kelvin equation for an adsorbed layer of thickness $\underline{t}$.

Ritter and Drake (9) developed and applied a method of mercury penetration which could determine a macropore size distribution for a porous solid. Improvements on the initial design permitted operation to 60,000 psia in order to extend the calculations to pores as small as 20A. However, the wide distribution of smaller pore sizes in most materials causes the mercury penetration volume to increase slowly with pressure so that it becomes difficult to obtain an accurate micropore distribution. Thus, in practice, porosimetry is frequently limited to pore diameters larger than 100 A (10). Nitrogen adsorption isotherms permit one to obtain a size distribution in the range of 20 to 200 (at most up to 400) A. Even so, a few investigators have reported some success in obtaining correlations between the pore volumes obtained by mercury penetration and nitrogen adsorption (11-16). While good correlations could be obtained with low pore volume samples, large discrepancies were found in pore volumes obtained by the two methods for high pore volume silica (17).

Many materials used as adsorbents or catalyst supports have a bimodal pore size distribution. For nearly all of these materials, the porosity range is so extensive that it is not possible to measure the full range of porosity with a single experimental procedure. Thus, surface areas calculated from mercury penetration volumes are often lower than BET surface areas when large fractions of the pores are in the micro and small mesopore range (18). In many cases the experimentalist must use both gas adsorption and mercury penetration to cover the full pore size range; a scale factor must be used to combine the porosity results obtained by the two techniques. Thus, apart from a theoretical basis, practical considerations dictate the need for such a factor.

Degussa Aluminoxide C has been widely studied as an example of a nonporous material. As-received, it exhibits typical type II nitrogen adsorption characteristics with no evidence of limited adsorption in the higher relative pressure region. However, the adsorption properties of this material, when wetted and redried, do not resemble those of the as-received material; the isotherm changes to one having type IV character at a high relative pressure. The large pore volume reflected over this high pressure range suggests that this would be an excellent material to use in a study to obtain a scaling factor for these two procedures. This report describes the results of such a study.

EXPERIMENTAL

Aluminoxide C is manufactured by the Degussa Company by flame hydrolysis of anhydrous aluminum chloride. To make the wetted alumina, water was added until a paste was formed; the paste was then dried at 220°C. Selected samples were then calcined in air at 400°C for 4 to 6 hours; the isotherms for these differed little from those that were not calcined.

Mercury penetration was measured from 1 to 50,000 psia with a Micromeritics instrument. The combined volume of meso- and micropores was obtained from the plot of penetration volume vs. pressure by extending the flat portion of the curve on each side of the high pressure step and measuring the distance separating the lines at the midpoint of the step. The pore volume distribution was obtained from a plot of the slope of the penetration curve against pore radius, with contact angles of 130°, 140°, and 150° being used with the Washburn equation to equate pressure with pore size.

Rootare and Prenzlow (19) obtained an expression for calculating the surface area from mercury porosimetry data:

$$A = (0.02253/m) \int_{0}^{V_{max}} PdV \qquad (1)$$

where the integral is evaluated from the area enclosed by the experimental PV curve, m is the mass of material in grams, and the constant is calculated assuming 25°C, γ = 480 dynes/cm for mercury and θ = 130° in the Washburn equation.

Nitrogen adsorption and desorption isotherms were obtained using a Micromeritics Digisorb 2500 Automatic Surface Area and Pore Volume Analyzer. All samples were outgassed at 200°C for 4 to 6 hours under hard vacuum, and 21 equilibrium points were measured for both adsorption and desorption. Pore size distributions were calculated with a computer program which allows a variety of pore shapes; however, since the most widely accepted model for type IV isotherms seems to be open-ended cylinders, the results from this group of calculations will be presented.

The conditions governing the filling and emptying of these cylinders were taken from either the Cohan (20) or the Broekhoff-deBoer method (21). Methods A, B, and C are based on the Cohan formulation, where a cylindrical meniscus is assumed to govern filling; such a meniscus is unstable and leads to pore filling when the Kelvin equation is satisfied:

$$\ln(p/p_o)_a = (-\gamma V_m/RT)(1/r_p - t_a) \qquad (2)$$

where $(p/p_o)_a$ is the critical pressure for a pore of radius r_p, at which pressure a multilayer thickness of t_a is in equilbrium. On desorption, however, a hemispherical meniscus is present which becomes unstable when:

$$\ln(p/p_o)_d = (-\gamma V_m/RT)(2/r_p - t_d). \qquad (3)$$

Cohan noted that, if the approximation $t_a = t_d$ is made, the adsorption and desorption critical pressures are related by:

$$(p/p_o)_d = (p/p_o)_a^2. \qquad (4)$$

Method A uses Equations (2) and (3), with the multilayer expression being obtained by a fit of the Frenkel-Halsey-Hill (FHH) equation to adsorption data for as-received Degussa alumina between the relative pressures of 0.3 and 0.9:

$$t = 3.54 \times (-2.93/\ln(p/p_o))^{1/3}. \qquad (5)$$

The degree of interparticle contact (and resulting volumetric effects) should be much less for the as-received material; thus Equation (5) implicitly corrects for the effect of particle curvature on multilayer growth.

Methods B and C both use Equations (2) and (3). However, Method B uses the multilayer equation of Dollimore and Heal (8) obtained by analysis of 36 silicas and aluminas:

$$t = 3.54 \times (-5/\ln(p/p_o))^{1/3}. \qquad (6)$$

Method C corrects for multilayer growth with the t-curve of Broekhoff and deBoer (21); for $t < 10$ A, they advocated:

$$\ln(p/p_o) = 0.078 - 32.21/t^2 \qquad (7a)$$

and for $t > 10$ A:

$$\ln(p/p_o) = -37.09/t^2 + 0.3873 \exp(-0.1137\ t). \qquad (7b)$$

Equation 6 is based on the FHH equation; Equations 7a and 7b are of the same form as the Harkins-Jura equation.

The remaining calculations (Methods D and E) again use the open-ended cylinder model, except that a correction to the t-curve is made for the effect of cylindrical curvature on multilayer growth. The curvature correction adds a surface energy term based on dA/dn to the t-curve, where dA is the change in surface area upon adsorption of dn moles of condensate. In Method D, this correction is added to Equation (6) in a manner similar to that of Broekhoff and deBoer (21):

$$\ln(p/p_o) = 4.663/(r_p-t) - 221.8/t^3 \qquad (8)$$

and Equations 2 and 3 are again used for pore filling and emptying conditions. Method E uses the curvature-corrected forms of Equations (7a) and (7b), which gives (for $t < 10$ A):

$$\ln(p/p_o) = 0.078 - 32.21/t^2 - 4.663/(r_p-t) \qquad (9a)$$

and for $t > 10$ A: (9b)

$$\ln(p/p_o) = 0.3873 \exp(-0.1137\ t) - 37.09/t^2 - 4.663/(r_p-t).$$

Pore filling and emptying conditions for Methods D and E are taken from the Broekhoff-deBoer equations for adsorption (21) and desorption (22).

RESULTS

Electron Microscopy. Electron microscopic examination of the as-received Degussa Aluminoxide C shows the material to consist of aggregates of spherical primary particles. A typical TEM particle size distribution is shown in Figure 1. Taking the major particle diameter D_p to be that of the peak (130 A), the surface area can be estimated (for spheres) by:

$$S = 60000/\rho D_p \quad (10)$$

where D_p is in A, and ρ in g/cm^3. From this, a surface area of 118 m^2/g can be obtained, which agrees with the BET area of 110 m^2/g for the as-received alumina.

Both TEM and SEM revealed a more closely packed agglomeration of these primary particles in the wetted sample, which did not disperse in either acetone or methanol.

Nitrogen Adsorption. Typical isotherms are shown in Figure 2 for both as-received and wetted Aluminoxide C. The as-received alumina shows a type II adsorption isotherm. The wetted alumina, however, shows a type IV closed isotherm with a Gurvitsch plateau at about 480 cc/g (gas @ STP), which corresponds to a pore volume of about 0.74 cc/g. The BET surface areas for these samples show a small change on wetting, with values ranging from 115 to 130 m^2/g for the wetted alumina as compared to 105 to 110 m^2/g for the as-received. Typical BET c-values range from about 100 for the wetted to 240 for the as-received material. No reason can be given at present for the increase in surface area upon wetting, other than the presence of micropores or surface roughness. Calcination of the wetted alumina at 400°C results in little change in the isotherm; this fact discredits the possibility of major variations in surface hydroxyl density.

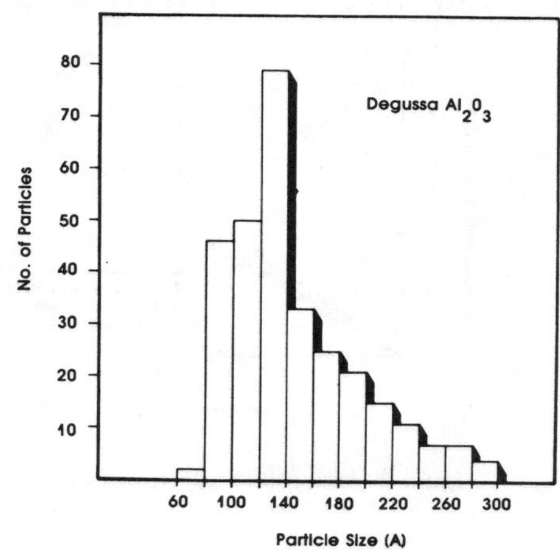

Figure 1.

TEM Particle Size Distribution for Aluminoxide C.

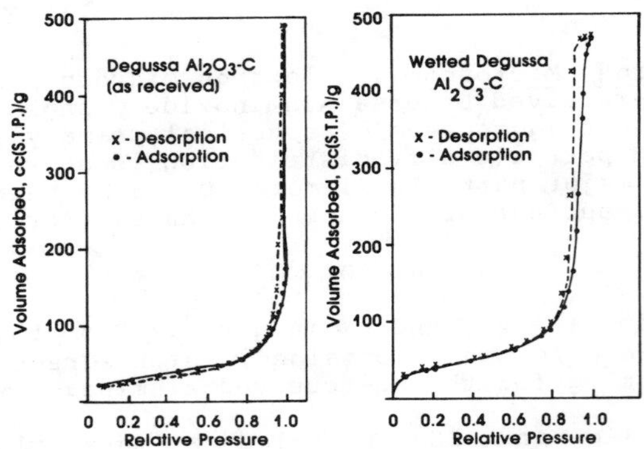

Figure 2. Nitrogen isotherms for Degussa Aluminoxide C.

Multilayer thicknesses were calculated assuming a flat surface, as is commonly done:

$$t = 15.5\ (V/S_{BET}) \tag{11}$$

These t-curves, along with that of Payne and Sing (23) for their 'non-porous alumina', are the basis of the t-plots shown in Figure 3, where the Broekhoff-deBoer curve is taken as the standard isotherm. Note that the Payne and Sing curve agrees well with that obtained here for our as-received alumina; also note that the straight line indicated by Eq. (9) for S_{BET} = 130 m^2/g falls directly over the t-plot for wetted alumina, except above a thickness of 10-12 A. This seems to indicate that the alumina used by Broekhoff and deBoer might have been exposed to water vapor. The deviation of the wetted sample from the straight line could reflect the fact that they extrapolated their data above a p/p_o of 0.9.

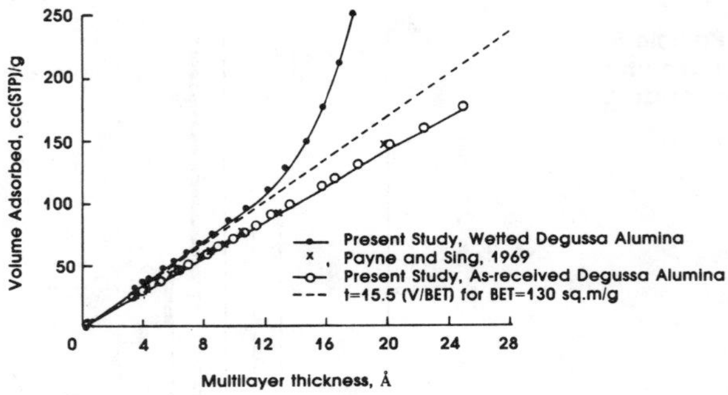

Figure 3. Adsorption t-plot with deBoer standard isotherm.

Pore volume distributions have been calculated from adsorption and desorption isotherms for the wetted alumina using the previously described methods. Typical distributions (dV/dr) calculated by these methods are shown in Figure 4. The major pore diameters (obtained from dV/dr maxima), cumulative pore volumes, and cumulative surface areas obtained by these methods are shown in Table 1. Also shown in Table 1 are numbers in parentheses for the cumulative pore volumes and surface areas. These sums differ from those above in that all negative values over the complete pore-size range are equated to zero; thus the difference in the two sums reflects the summation of 'negative volumes' or 'negative surface areas' calculated using each method.

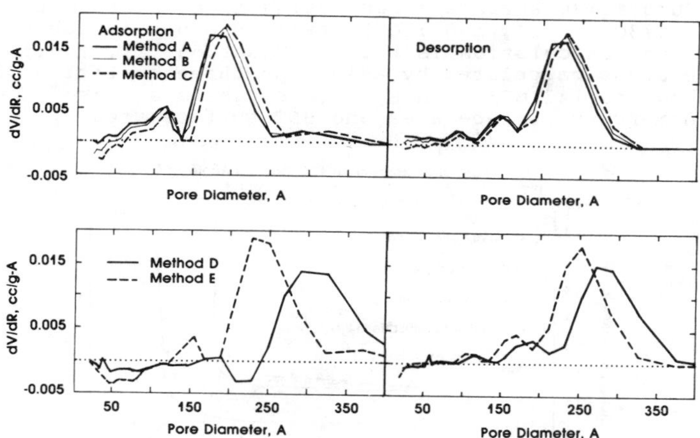

Figure 4. Adsorption-desorption pore size distributions for wetted Aluminoxide C.

Table 1

Nitrogen Adsorption Pore Calculations for Wetted Degussa Alumina

Method	Pore Diameter, A		Cum.Volume,cc/g (Cum.Vol.>0)		Cum.Surface,sq.m/g (Cum.Surf.>0)	
	Ads	Des	Ads	Des	Ads	Des
A	170	230	0.762 (0.762)	0.757 (0.757)	193 (193)	176 (176)
B	190	230	0.765 (0.780)	0.764 (0.764)	164 (183)	161 (161)
C	190	230	0.740 (0.772)	0.742 (0.746)	128 (136)	136 (141)
D	230	250	0.735 (0.835)	0.758 (0.764)	66.4 (136)	127 (135)
E	290	270	0.722 (0.840)	0.735 (0.762)	59.3 (103)	104 (117)

Mercury Penetration. Figure 5 represents the mercury penetration-depressurization curve for wetted alumina. Values calculated from the penetration curve are shown in Table 2.

The total pore volume, taken from the high pressure step, is 0.72 cc/g; this agrees well with those obtained from both the cumulative pore volumes and the Gurvitsch plateau. The average pore diameter, calculated with a contact angle of 130°, is 190 A; for a contact angle of 140°, this average diameter is 225 A. These values agree well with the range of 170-230 A obtained for both adsorption and desorption of nitrogen by the three Cohan methods (A, B, and C); the Broekhoff-deBoer methods yield higher values than mercury.

Cumulative surface areas, using three different contact angles (130°, 140°, and 150°) are also shown in Table 3; the method of calculation is that of Equation (1). The range of surface areas calculated by using the three contact angles is 109 m^2/g to 147 m^2/g; this range suggests a close agreement between mercury surface area and BET surface area.

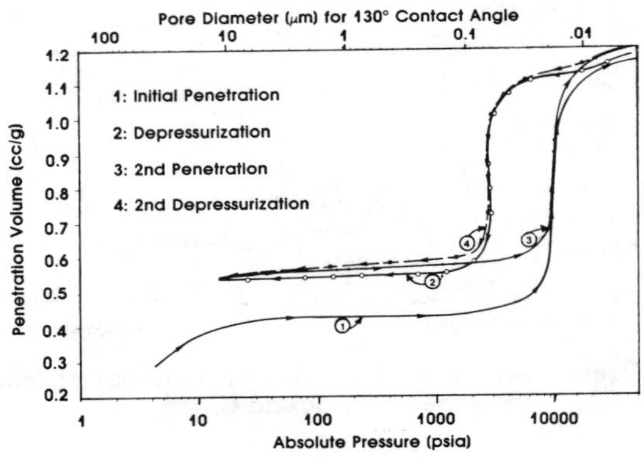

Figure 5. Mercury penetration for wetted Aluminoxide C.

Table 2

Hg Penetration Pore Calculations for Wetted Degussa Alumina

Pore Volume, cc/g	0.720
Pore Diameter, A	
(θ=130°)	190
(θ=140°)	225
Cum.Surface, sq.m/g	
(θ=130°)	147
(θ=140°)	123
(θ=150°)	109

DISCUSSION

Examination of the t-plot (Figure 3), based on the standard isotherm of Broekhoff and deBoer (obtained for a similar alumina), reveals that their t-curve agrees closely with the one obtained in this study for wetted alumina. The curve obtained by Payne and Sing correlates very well with the as-received Degussa alumina; this would suggest that Broekhoff and deBoer were possibly measuring condensation as well. The t-curve of Payne and Sing appears to be more valid, at least for Aluminoxide C; it would certainly be worthwhile to examine this for similar particle systems.

One use of a t-curve expression is a correction for multilayer growth in a pore volume calculation; thus, the effect of rather wide variations in t-curve expression on a calculated distribution is a useful characterization. For the Cohan open cylinder model, it has been shown that a significant change in the t-curve (within limits) results in little practical change in the pore size distribution. If, however, the Broekhoff-deBoer correction for the effect of cylindrical curvature on the t-curve is applied, results are obtained may reflect an appreciable 'negative pore volume'. If the reason for these 'negative volumes' were simply that the calculations were taken to low relative pressures, the desorption distribution should show this even moreso than adsorption. Since this is not the case, use of Equations (8) and (9) must be questioned for adsorption on spherical particle systems. In fact, the opposite sense of curvature (dA/dn for spheres) should be applied to spherical particles. In fitting the FHH curve to the as-received alumina, such a correction is implicitly made (assuming the effects of interparticle contact to be negligible). Since this option, when used with the Cohan model, reflected little change in the calculated pore size distributions, the extra effort is apparently of little value in porosity calculations. This also needs to be verified for similar systems.

The comparison of mercury penetration with nitrogen adsorption reveals that a scaling factor of about unity provides good agreement between the results from the Cohan pore model (nitrogen desorption data) and mercury penetration results ($\theta = 140^{\circ}$). The agreement indicates two things:

1. The approximation (at least from the volumetric standpoint) of the void spaces in a packing of primary spheres as open-ended cylinders is not unreasonable;
2. Mercury penetration and nitrogen data can be correlated well in this pore size range using simple models. This points to the utility of the simple models.

Since all pore size estimates were larger than the primary particle size, and the estimates were verified by two techniques, the packing which results from wetting the alumina is probably not a simple packing. This phenomenon needs to be investigated for other oxides as well.

Acknowledgment. This work was supported by the Kentucky Energy Cabinet and the University of Kentucky Institute for Mining and Minerals Research.

Literature Cited

1. S. Brunauer, P. H. Emmett and E. Teller, J. Am. Chem. Soc., 60, 309 (1938).

2. P. H. Emmett and S. Brunauer, J. Am. Chem. Soc., 59, 310 (1937).

3. P. H. Emmett and T. W. deWit, J. Am. Chem. Soc., 65, 1253 (1943).

4. E. W. Thiele, Ind. Eng. Chem., 31, 916 (1939).

5. J. S. Anderson, Z. physik. chem., 88, 191 (1914).

6. E. P. Barrett, L. G. Joyner and P. P. Halenda, J. Am. Chem. Soc., 73, 373 (1951).

7. J. C. P. Broekhoff and B. G. Linsen, "Physical and Chemical Aspects of Adsorbents and Catalysts," B. G. Linsen (ed.), Academic press, London, 1970, Chapter I.

8. D. Dollimore and G. R. Heal, J. Colloid and Interface Sci., 33, 508 (1970).

9. H. L. Ritter and L. C. Drake, Ind. Eng. Chem. Anal. Ed., 20, 665 (1948).

10. I. Halasz and K. Martin, Angewandte Chemie, Int. Ed., 17, 901 (1978).

11. L. A. deWitt and J. J. F. Scholten, J. Catal., 36, 36(1975).

12. M. M. Dubinin, M. M. Vishnyakova, E. G. Zhokovshaya, E. A. Leontev, V. M. Luk'yanovich and A. I. Sarakhov, Zhur. Fiz. Khim., 34, 2019 (1960).

13. L. G. Joyner, E. P. Barrett and R. Skold, J. Am. Chem. Soc., 73, 3155 (1951).

14. S. Brunauer, Chem. Eng. Progr. Symp. Ser., 65, 1, (1969).

15. C. N. Cochran and L. A. Cosgrove, J. Phys. Chem., 61, 1417 (1957).

16. S. M. Brown and E. W. Lard, Powder Technology, 9, 187 (1974).

17. I. D. Sills, L. A. G. Alymore and J. P. Quirk, Soil Sci. Soc. Am., Proc., 37, 353 (1973).

18a. M. P. Astier and K. S. W. Sing, Adsorption at the Gas-Solid and Liquid-Solid Interface. (J. Rouquerol, K. S. W. Sing, Eds.), Elsevier Scientific Publishing Co., Amsterdam (1982).

18b. C. H. Giles, D. C. Havard, W. McMillan, T. Smith, R. Wilson, Characterization of Porous Solids, (S. J. Gregg, K. S. W. Sing, H. F. Stoeckli, Eds.), Society of Chemical Industry, London (1979).

19. H. M. Rootare and C. F. Prenzlow, J. Phys. Chem., 71, 2733 (1967).

20. L. H. Cohan, J. Am. Chem. Soc., 60, 433 (1938).

21. J. C. P. Broekhoff and J. H. de Boer, J. Catal., 9, 8 (1967).

22. J. D. P. Broekhoff and J. H. de Boer, J. Catal., 10, 377 (1968).

23. D. A. Payne and K. S. W. Sing, Chem. and Ind., 918 (1969).

TRISIV ADSORBENT - THE OPTIMIZATION OF MOMENTUM AND MASS TRANSPORT VIA ADSORBENT PARTICLE SHAPE MODIFICATION

Joseph P. Ausikaitis
Union Carbide Corporation
Tarrytown Technical Center
Tarrytown, New York 10591

ABSTRACT

Historically, molecular sieve adsorbents have been available in three particle shapes - beads, pellets, and granules. Recently, an extended surface area adsorbent particle has been commercialized under the brand name TRISIV. This paper reviews the impact of particle size and shape upon the physical and performance characteristics of packed beds of adsorbent particles. Specifically, dimensional attributes, bed voidage, momentum and mass transport are discussed vis-a-vis particle size and shape.

INTRODUCTION

Individual molecular sieve crystals are typically 1 to 10 microns in size and their use in this form is limited to a very few applications. Most molecular sieve adsorbents are bonded into forms to give them the mechanical strength, size and suitable porosity to be used in commercial applications.

The intra and interparticle porosity of a packed bed of adsorbent is not only a function of the formulating and forming techniques used in manufacturing the adsorbent particles, but is a function of the size and shape of the particles as well. It is the latter two factors - size and shape - and their influence on mass and momentum transport in packed beds of molecular sieve adsorbents that is the subject of this paper.

PARTICLE SIZE AND SHAPE

Historically, there have been three basic molecular sieve adsorbent particle shapes available to the industry. These have been supplied in a range of sizes. These are beads (spheres), pellets (cylinders), and granules (mesh). In the area of catalysis, there has been renewed interest in particle shape modification to improve catalytic performance in fixed and fluidized beds (1,2,3,4). This work indicated that improved catalytic activity can be achieved with a concurrent reduction in pressure drop. In addition, the extended surfaces created by shape modification appeared to act as flow distributors enhancing both contacting efficiencies and diffusion in the active catalyst particles. Over the past five years, Union Carbide Corporation has developed and commercialized TRISIV, a trilobal or extended, cloverleaf shaped molecular sieve adsorbent for natural gas drying (5) and more recently for air prepurification. A review of commercial molecular sieve particle shapes and sizes along with their physical attributes will serve as background for the sections that follow.

Beads and Granules

These two particle shapes are lumped together because they are characterized by screen analyses. Granular materials have shapes that are irregular and can range from platelets to spheroids to cubes. In order to differentiate between beads and granules, shape factors and/or equivalent spherical diameters are used to account for the irregular shape. Table 1 is a list of commercial screen cuts and properties of beaded molecular sieve adsorbents. A screen cut is defined as a sized material which passes through, "-", a standard screen of a larger diameter opening and is retained, "+", on a screen with a smaller diameter opening. The average particle diameter, $\overline{D}_p$, in Table 1 is based on the arithmetic average of D_{max} and D_{min}, and these values are ±25 to ±40% of the average.

TABLE 1

PROPERTIES OF COMMERCIALLY AVAILABLE BEADED MOLECULAR SIEVE ADSORBENTS

SI UNITS (MM, MM^2, MM^3)

SCREEN SIZE	NO. SCREENS	D_{MAX}	D_{MIN}	$\overline{D}_p$	A_p	V_p	A_p/V_p	A_p/V_b*
(-)4.76X(+)2.38	4	4.76	2.38	3.57	40.0	23.82	1.68	1.06
(-)2.38X(+)1.68	2	2.38	1.68	2.03	12.95	4.38	2.96	[illegible].86
(-)2.00X(+).841	4	2.00	.841	1.42	6.34	1.50	4.22	2.63
(-)1.41X(+).595	4	1.41	0.595	1.00	3.16	0.528	5.99	3.77
(-)1.19X(+).420	5	1.19	0.420	0.805	2.04	0.273	7.45	4.70

*ϵ = 0.37 constant

$A_p = \pi a^2$

$V_p = \frac{\pi a^3}{6}$

Using the arithmetic average is a simplifying assumption since each commercial size shown is comprised of 2 to 5 intermediate screen cuts. In order to more properly determine the average particle diameter and resultant physical attributes, one must know the amount of beads of the intermediate sizes and weight them according to established methods (6). The potential variability in $\overline{D}_p$ is smaller than the difference between D_{max} and D_{min} and depends on the distribution of the intermediate sizes.

Pellets

A pellet or a cylinder is typically characterized by two factors, the diameter and the length-to-diameter ratio. The "equivalent" effective particle size is usually calculated by one of three methods. In the first method, sometimes referred to as Ergun's (or the hydraulic) diameter, the equivalent particle diameter is calculated as that of the equivalent sphere possessing the same specific surface area (surface area/unit volume) as the particle in question (D_p^{SV}). For a right circular cylindrical pellet, this becomes

$$D_p^{SV} = \frac{d}{2/3 + 1/3\,(d/\ell)}$$

where,

d = diameter of pellet

ℓ = length of pellet

In the second method, the equivalent particle diameter is calculated as the diameter of the sphere possessing the same surface area as the particle in question (D_p^{Area}). Thus the equivalent diameter becomes:

$$D_p^{Area} = d\sqrt{[(\ell/d) + 1/2]}$$

In the third method, the equivalent particle diameter is calculated as the diameter of the sphere having the same volume as the particle (D_p^{Vol}). For pellets this becomes

$$D_p^{Vol} = d\sqrt[3]{\frac{3}{2}\,(\ell/d)}$$

Table 2 presents a list of commercial pelletized molecular sieve adsorbents and dimensional qualities for a typical average ℓ/d ratio of 1.7. The effective diameter is presented for the three methods described above.

TABLE 2

DIMENSIONAL PROPERTIES OF PELLETED MOLECULAR SIEVE ADSORBENTS

SI UNITS (MM, MM^2, MM^3)

NOM. SIZE	d	ℓ/d	A_p	V_p	D_p^{SV}	D_p^{Area}	D_p^{Vol}	A_p/V_p	A_p/V_B **
0.8 mm	0.794	1.7	4.357	0.667	0.921	1.177	1.085	6.529	4.113
1.61 *	1.587	1.7	17.42	5.342	1.841	2.356	2.170	3.261	2.054
2.4	2.381	1.7	39.20	18.03	2.762	3.530	3.261	2.175	1.370
3.2 *	3.176	1.7	69.68	42.74	3.688	4.712	4.328	1.631	1.027
6.3	6.349	1.7	278.86	341.8	7.361	9.418	8.687	0.814	0.513

* Most commonly offered sizes

** $\epsilon = 0.37$

$A_p = \pi d^2\,[(\ell/d) + 1/2]$

$V_p = \frac{\pi d^3}{4}\,(\ell/d)$

As with beads, there is variability in the size of pellets. However, the variabiity is quite different. The prime dimensional variable for pellets is the length-to-diameter ratio (ℓ /d). The value ℓ /d is distributed around a mean of about 1.7 but can range from 1.0 to 4.0. Calculating D_p^{SV} for the extremes results in a maximum variation of ± 15% for each pellet size. This compares to ± 25% to ± 40% maximum variation at the extremes for the various bead cuts. The diameter of the pellet can also change but this variation is very small due to the nature of the manufacturing process.

TRISIV

The dimensional proportions of the TRISIV molecular sieve adsorbent is given in Figure 1.

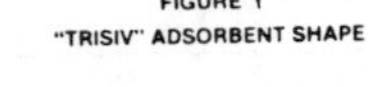

FIGURE 1
"TRISIV" ADSORBENT SHAPE

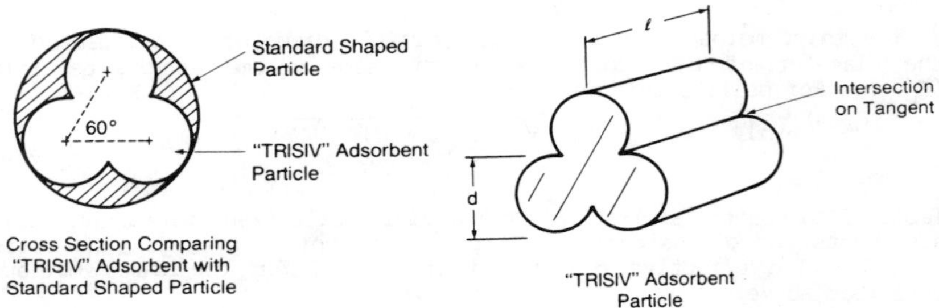

As with a pellet, the shape is characterized primarily by the diameter of each lobe and the length to lobe diameter ratio. The equivalent spherical particle diameter based on specific volume is given by:

$$D_p^{SV} = \frac{14.379\ d^3 \times (\ \ell/d)}{7.854\ d^2\ (\ \ell/d) + 4.793\ d^2}$$

where,

ℓ = length of equal length lobes

d = diameter of equal single lobes

The equivalent diameter based on surface area is given by:

$$D_p^{Area} = \sqrt{2.5\ d^2\ (\ \ell/d) + 1.526\ d^2}$$

The equivalent diameter based on volume is given by:

$$D_p^{Vol} = d \sqrt[3]{4.577\ (\ell/d)}$$

Table 3 presents a list of TRISIV pelletized molecular sieve adsorbents having this particular shape and the corresponding dimensional qualities for a commercial typical average ℓ/d ratio of 3.2. Worthy of note is the large difference between the effective particle size calculated by the three techniques.

TABLE 3

DIMENSIONAL PROPERTIES OF TRISIV MOLECULAR SIEVE ADSORBENTS

SI UNITS (MM, MM^2, MM^3)

NOM. SIZE	d	ℓ/d	A_p	V_p	D_p^{SV}	D_p^{Area}	D_p^{Vol}	A_p/V_p	A_p/V_B **
0.8 mm	0.794	3.2	18.86	3.83	1.219	2.438	1.939	4.919	3.099
1.6 *	1.588	3.2	75.44	30.68	2.438	4.907	3.871	2.459	1.549
3.2	3.176	3.2	301.7	245.5	4.877	9.815	7.772	1.229	0.774

* Most commonly available size

** ϵ = 0.37

Extrudates of TRISIV molecular sieve adsorbents vary in size much the same way as pellets. The typical ℓ/d ratio is larger than standard pellets because the diameter of each equal lobe is used as the reference value. The mean ℓ/d ratio of 3.2 will also vary but typically over a narrower range of about 3.0 to 4.0. Calculating D_p^{SV} for these extremes results in a maximum variation of about ± 2.5%. The total particle surface area of a volume of a packed bed can be calculated as:

$$A/V_B = A_p(1-\varepsilon)/V_p$$

where,

A = total external packed bed particle surface area

V_B = volume of bed

V_p = particle volume

A_p = external particle surface area

ε = interparticle void fraction

However, the internal surface area of a molecular sieve adsorbent particle is often more important than the external surface area. The vast majority of the adsorptive surface area is microporous and independent of the size and shape of the particle. This microporous surface area is contained within the cumulative internal volume of the particle and is accessible at some distance from the external surface. The cumulative internal pore volume does vary as a function of distance from external surface and as a function of particle shape. Figure 2 is a plot of the cumulative volume contained within a fractional distance (depth) from the external surface for beads, pellets, and TRISIV adsorbents where the depth is normalized as the radius of an equivalent volume sphere (D_p^{Vol}).

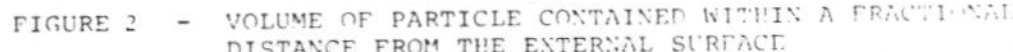
FIGURE 2 - VOLUME OF PARTICLE CONTAINED WITHIN A FRACTIONAL DISTANCE FROM THE EXTERNAL SURFACE

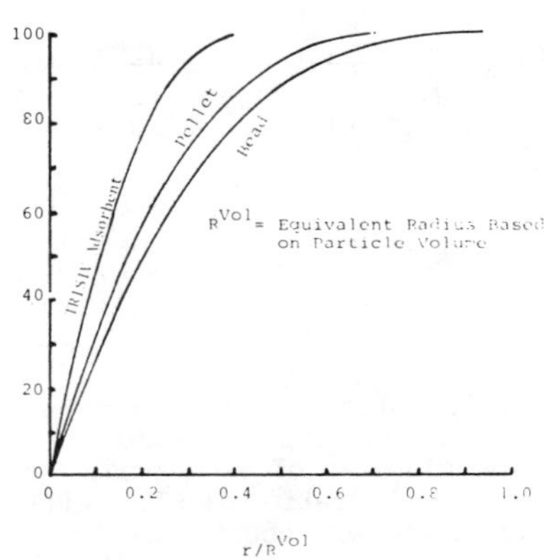

This effectively equates the magnitude of "r", with the depth of penetration, since the distance to the center of the particle (r=R) is a variable for each shape. Note that 80% of the adsorbent is within a 0.2 r/R value for the shape of TRISIV molecular sieves while in the case of the bead this does not occur until an r/R value of 0.4 has been reached.

INTERPARTICLE VOIDS

The distribution of voids within the individual molecular sieve particles and between the particles within packed beds of molecular sieve pellets has been previously documented (8). The intraparticle void fraction is generally independent of the size and shape of the particle. The interparticle void fraction within an aggregate is dependent on four factors of varying importance. For particles of uniform surface roughness, these factors are: (1) the size of the particle; (2) the size distribution of the particles; (3) the shape of the particles, and (4) the ratio of the column diameter to the particle diameter. The theory and mechanical properties of packed particles lie within the science of powder mechanics. Excellent general reviews on bed packing appear in the literature (9, 10). The impact of size and shape on interparticle bed voidage and the resultant effect on performance will be discussed next.

The basic objectives in packing an adsorbent column are to attain uniformity in bed packing, to attain an expected bulk density, and to attain plane and parallel top and bottom faces. The first of these objectives is normally the most difficult to achieve. There are several reasons for this. During bed filling, when adsorbent particles are transferred from containers to the adsorption column, particle segregation can occur. Segregation can occur by size and/or by density. In density segregation, heavier particles will remain under a fill point while lighter ones will roll or be pushed off to the side. Since molecular sieve adsorbents have uniform piece density, this is not a serious problem and segregation by size is more prevalent. In size segregation, the fine particles separate from coarse particles. The tendency is for coarse particles to collect near the outer wall, that is, to roll away from the fill point. Smaller particles may later "percolate" into these newly created voids. At the walls, the packing will never be uniform. High voidage occurs within 1 to 2 particle diameter from the wall. In large commercial beds which are not "dense loaded" (11), some slow settling will occur with the assistance of slight pressure fluctuation on the faces. This settling can either increase or decrease the overall uniformity in packing.

The segregation, percolation and settling of particles in packed beds is dependent on how readily the particles can move relative to each other. This is true for both small beds, which can be vibrated during filling, and for large beds, which are filled by a variety of techniques. The relative movement depends on the coefficient of friction between the particles within the aggregate. An expression which relates the coefficient of friction to particle properties has been suggested (12) as follows:

$$\mu = \frac{a}{f_s^2} + b\left(\frac{\gamma}{D_p}\right)^{0.5} + c\,\rho_p + d \qquad (1)$$

where,

μ	=	coefficient of friction
a,b,c,d	=	constants
f_s	=	shape factor (<1.0 for non-spheres)
γ	=	roughness index
D_p	=	particle diameter
ρ_p	=	particle density

Thus the ease of settling and size segregation decreases with denser, smaller, and irregularly shaped particles.

The ultimate degree of settling which could occur (but seldom does in an unvibrated bed), is determined by the maximum theoretical density attainable by the packing. This is based on theoretical geometrical relationship for aggregates (12,13). Molecular sieve adsorbent particles of all shapes are available as "sized" composites. This sizing limits the particle size range within an aggregate to much less than an order of magnitude. Therefore, the degree of the potential bed settling is limited to a maximum of about 0.05 in void fraction and typically to a value less than 0.02 from its initial unsettled state ($\varepsilon \cong 0.35$ to 0.40).

The impact of a nonuniform bed packing can vary. A random distribution of packing irregularities is less important when the length and diameter of the bed are large. In a right circular cylindrical bed configuration, uniform axial settling and densification will not result in poor flow distribution. This will only cause some variation in dynamics in the axial direction. As a packed bed settles, the interparticle void fraction decreases and the bed height decreases. The net result is a larger pressure drop across the entire bed length since the effect of a void fraction decrease on pressure drop is dominant. Thus, another objective for a packed adsorbent bed would be to stabilize the potential settling to maintain a stable predictable aggregate of particles as long as the desired bulk density has been achieved.

Uneven settling, especially in the radial direction, is worse. The most severe situation occurs when a region of high voidage extends the entire length of the bed and causes a flow "by-pass". The less radical "wall channeling", caused by normal high voidage at the walls, is proportional to the ratio of perimeter to cross sectional area. Small diameter columns should always be packed using the criteria that the column inside diameter should be 1.5 to 2 orders of magnitude greater than the particle size of the adsorbent. Some analogous and theoretical effects of gas maldistribution on the performance of packed beds of catalyst particles have appeared in the literature (14).

PRESSURE DROP

Pressure losses through packed beds of adsorbent particles can be predicted using established techniques of either Ergun (15) or Leva (16). For particles of irregular shape, Leva's correlation is perhaps more conveniently used. In this correlation, the pressure drop per unit length of packed bed is defined as

$$\frac{\Delta p}{L} = \frac{2f_m' G^2 (1-\varepsilon)^{3-n}}{D_p^{Vol}\, g_c\, \rho\, \phi_s^{3-n}\, \varepsilon^3} \qquad (2)$$

where,

f_m'	=	modified friction factor
G	=	mass flux
ε	=	void fraction
D_p^{Vol}	=	equivalent spherical diameter based on volume
g_c	=	gravitational constant
ϕ_s	=	shape factor
n	=	exponent dependent on N_{Re}'

The friction factor and exponential factor are functions of the modified Reynolds number.

$$N_{Re}' = \frac{D_p^{sv}\, G}{\phi_s\, \mu}$$

where,

D_p^{sv}	=	equivalent spherical diameter
μ	=	viscosity

The effect of particle size on pressure drop is straightforward. In the laminar region and the turbulent flow region, the pressure drop is roughly inversely proportional to the particle size.

The effect of particle shape on the pressure drop is more complex. The shape factor, ϕ_s, is defined as the ratio of the surface area of a sphere with a diameter equal to D_p^{Vol} divided by the actual surface area of the particle. Using the values in Tables 2 and 3, and published values for granules, the values below are appropriate.

Material	ϕ_s
Beads	1.0
Pellets	0.85
TRISIV	0.63
Granules	0.45 to 0.65

The shape factor is raised to the power 3-n in the calculation of pressure drop in equation (2). The exponential factor, n, increases from about 1.0 to 2.0 as the flow increases from laminar into the turbulent region (N_{Re}' =10 to 10^4). Exact values can be found elsewhere (17). For particles with ϕ_S less than 1.0, the difference in pressure drop relative to beads increases with decreasing flowrate down to N'_{Re}=10. There are several implications of this for the design of commercial adsorbent beds that warrant further discussion.

High Pressure Drop

Excess pressure drop can cause either excess force on the bottom of the bed or fluidization. Short of these catastrophic results, there is always an energy loss (pressure drop). The criticality of this pressure drop is dependent upon the specific process application and relative cost of the other energy inputs to the process. This same pressure drop per unit length in a packed bed can also make a positive contribution on performance. The resultant viscous and kinetic energy losses from local mixing and turbulence between particles acts as a radial flow distributor.

Low Pressure Drop

Very low pressure drop can result in poor low distribution and, in the limit, "channeling". Low flowrates in adsorption units are most frequently encountered during the regeneration steps. Usually, during regeneration, only a small fraction of the flowrate experienced on the turbulent adsorption step passes through the bed. Flows in the transition and laminar ($N'_{Re} \cong$ 10 to 100) region are common.

Particle shapes with lower ϕ_S values can compensate to prevent poor flow distribution at low regeneration gas flowrates by creating additional pressure drop relative to particles with high ϕ_S values. Alternatively, a low ϕ_S particle can provide the minimum required pressure drop on regeneration without a high pressure drop penalty on the subsequent adsorption step. The desirability of irregular particle shapes is therefore dependent on the process application and the relative importance of the design of the adsorption and regeneration steps.

MASS TRANSFER

Mass Transfer in a packed bed of molecular sieve adsorbents occurs sequentially. Adsorbate is transferred from the bulk fluid phase, through an external film, through an intraparticle (intercrystal) region of macro and mesopores, and finally into the microporous zeolite crystal. The influence of particle size and shape on mass transfer depends on the location of the controlling resistances. These can be categorized as fluid film, macropore and micropore.

Micropore Control

Micropore controlled mass transfer is extremely complex and is not of concern for the purposes of this paper. Once the adsorbate gets to the crystal surface, the sorption into the crystal is independent of the size and shape of the particle in which it is contained.

Intraparticle Control

Macropore or intraparticle controlled mass transfer occurs within the particle and is dependent on the length of the path the adsorbate must follow to reach the center of the particle. Many researchers (19,20) have shown that the intraparticle mass transfer coefficient is inversely proportional to the square of the particle diameter. Smaller diameter particles result in larger intraparticle mass transfer coefficients and transfer rates. For particles other than true spheres, the hydraulic diameter D_p^{sv} has been shown to be the most appropriate (21). Thus intraparticle mass transfer rates are larger for particle shapes with small D_p^{sv} values.

Film Control

For gas film controlled mass transfer, j_D factors (Colburn factor) have been used to correlate mass transfer coefficients. These correlations result in mass transfer coefficients expressed as functions of the Schmidt and Reynolds numbers. The j_D factor is generally defined as:

$$j_D = \frac{k_c \quad \rho}{G} N_{Sc}^{2/3}$$

where,

k_c = mass transfer coefficient based on adsorbate concentration

ρ = fluid density

G = mass flux

N_{Sc} = Schmidt number

An estimation of the influence of particle size and shape can be made by using the more rigorous correlation of j_D - factors (23) which is the following:

$$j_D = \frac{a \quad \phi_{ea}}{N_{Re_b}^{n}}$$

where,

a = constant

n = exponential factor

where,

ϕ_{ea} = shape factor based on effective surface area

The modified Reynolds' number is defined as

$$N_{Re_b} = \frac{D_p^{Area} G}{\mu} \cdot \frac{1}{6(1-\varepsilon)\ \phi_{ea}}$$

For laminar or turbulent flow for both gases and liquids, the values of a and n are different. The value of n is always less than 1.0 and will be somewhat larger for laminar flow. Actual values can be found elsewhere (22). For smaller particle sizes of the same shape, the j_D - factor and, therefore, k_c will be larger. The proportionate increase with decreasing particle size will be somewhat larger in the laminar region than in the turbulent flow regions. For a constant particle size (D_p^{Area}), a value of ϕ_{ea} less than 1.0 will decrease the j_D-factor significantly. Typical values of ϕ_{ea} are 1.0 for spheres, 0.91 for pellets, 0.86 for granules and an estimated 0.8 to 0.85 for TRISIV adsorbent. The shape factor, ϕ_{ea}, can be thought of as parameter which compensates for the "shading" of the surface area. Particle shapes other than spheres have more than point contact within an aggregate and are thereby "shaded" from some fluid contact. Thus, the effect of irregular particle shapes is to reduce the film coefficient.

The rate of fluid film mass transfer is dependent on the product of the mass transfer coefficient and the external surface area involved. Using the appropriate ϕ_{ea} values and surface areas per unit volume based on constant D_p^{Area}, particle shapes can be compared in terms of the mass transfer rate assuming other parameters are constant. In so doing, the fluid film mass transfer rate in packed beds is roughly constant for beads, pellets, and TRISIV adsorbent. The lower coefficient is offset by the larger surface area for irregular particle shapes. From this analysis, one might predict a slightly higher fluid film mass transfer rate for beads in the turbulent region and for TRISIV in the laminar region.

In general and for the majority of dynamic adsorption applications, the rate of mass transfer is dependent on particle size and shape. This is due to the fact that the primary resistances to mass transfer are not entirely micropore controlled. Fluid film resistances are the most variable resistance to mass transport and may dominate at very low flowrates or when liquid coats the surface of an adsorbent particle. At moderate to high flowrates, film resistances are smaller relative to surface and intraparticle resistances. In these instances particle size and shape determine the effective particle size and the length of the diffusional path to the center of the particle. For the same D_p^{SV}, the TRISIV molecular sieve adsorbent has a greater percentage of its internal sorptive volume close to the surface and should thereby enhance mass transport in this region.

CONCLUSIONS & SIGNIFICANCE

Both the size and shape of adsorbent particles affect the performance of packed beds of molecular sieves. By itself, the characterization of particle size is fundamental to the prediction of performance. Particle shape can affect the determination of and the variability in the average effective particle size and potentially the uniformity in bed packing. In this respect, regular shapes are not necessarily more uniform, nor more predictable, than irregular particle shapes.

Pressure drops for irregular particle shapes are larger for equivalent particle size. The difference in pressure drop between regular and irregular shapes increases as flowrate is reduced. This phenomenon can actually provide a balancing effect to improve flow distribution in adsorption systems which have large flowrate variations between adsorption and desorption cycles.

Mass transfer can be affected by irregular particle shapes depending on the location of the controlling resistances. In general, the effect is favorable with the greatest benefits occurring in the intraparticle region. This is simply due to the proximity of the internal adsorptive volume and surface area to the fluid/particle interface.

The shape of the TRISIV molecular sieve adsorbent provides a unique combination of performance properties in one particle. Pressure drop or mass transfer characteristics, taken separately, could be matched with conventional shapes of various sizes. The unique properties of TRISIV when compared against standard molecular sieve beads and pellets for adsorption unit design cases can sometimes result in savings in the,overall evaluated cost (i.e. investment plust operating) and that these savings can be significant relative to the cost of the adsorbent.

LITERATURE CITED

1. Gustafson, W. R., "Hydrotreating of Petroleum Distillates Using Shaped Catalyst Particles," U.S. Patent 3,990,964 Nov. 9, 1976.

2. Gustafson, W. R., "Hydroforming Petroleum Fractions in Gas Phase Using Shaped Catalyst particles," U.S. Patent 3,857,780, Dec. 31, 1974.

3. Hoekstra, G. B., and Jacobs, R. B., "Process and Catalyst for Hydroprocessing a Resid Hydrocarbon," U.S. Patent 3,674,680, July 4, 1972.

4. Richardson, R. L. Riddick, F. C., and Ishikawa, M., "New Resid HDS Process Has Long Run," Oil & Gas Journal, May 28, 1978.

5. "New Sieve Shaves Drying Costs at Dorchester," Oil & Gas Journal, Feb. 22, 1982.

6. Leva, M., "Fluidization," p. 61, McGraw-Hill, New York (1959).

7. Breck, D. W., "Zeolite Molecular Sieves," p. 751, John Wiley & Sons, New York (1974).

8. Ibid 7., pg. 751 & 752.

9. Brown, R. L., and J. C. Richards, "Principles of Powder Mechanics," International Ser. of Monographs in Chem. Eng., (10), Pergamon Press, New York (1970).

10. Hirschhorn, J. S., "Introduction to Powder Metallurgy," Colonial Press, New York (1969).

11. Snow, A. I., et. al., "Distribution Method Makes Catalyst More Effective," The Oil and Gas Journal, p. 76-79, August 14, 1972.

12. Furnas, C. C., "Grading Aggregates," Ind. & Eng. Chem., 23 (9) p. 1052-1058 (1931).

13. McGeary, R. K., "Mechanical Packing of Spherical Particles," J. Am. Ceram. Soc., 44 (10), pg. 513-22 (1961).

14. Craven, P., "The Effect of Gas Maldistribution on Catalyst Performance, "British Chemical Engineering," 15 (7), p. 918-919 (1970).

15. Ergun, S., "Fluid Flow Through Packed Columns," Chem. Eng. Prog., 48 (2), p. 89-94 (1952).

16. Perry, J. H., C. H. Chilton, and S. D. Kirkpatrick, "Chemical Engineers' Handbook," 4th ed., p. 5-50, McGraw-Hill, New York (1963).

17. Ibid 9., p. -5-51.

18. Mehta, D. and M. C. Hawley, "Wall Effect in Packed Columns," Ind. & Eng. Chem. Design and Development, 8 (2), p. 280-282 (1969).

19. Glueckauf, E., "Theory of Chromotography," Trans. Faraday Ser., 51, p. 1540 (1955).

20. Antonson, C. R. and J. S. Dranoff, "Adsorption of Ethane on Type 4A and 5A Molecular Sieve Particles," Developments in Physical Adsorption - AICHE Sym. Ser., 65 (96), p. 27-33 (1969).

21. Antonson, C. R. and J. S. Dranoff, "Nonlinear Equilibria and Particle Shape Effects in Intraparticle Diffusion Controlled Adsorption," AICHE Sym. Series, 65 (96), p. 23 (1969).

22. Gamson, B. W., G. Thodos, and O. A. Hougen, "Heat Mass, and Momentum Transfer in the Flow of Gases Through Granular Solids," Trans. AICHE, 39 (1), p. 1-35 (1943).

23. Lightfoot, E. N., R. J. Sanchez-Palma and D. O. Edwards, "Chromotography and Allied Fixed Bed Separation Processes," New Chem. Eng. Separation Techniques, Ed. H. M. Schoem, Interscience, New York (1962).

HIGH PRESSURE ADSORPTION OF METHANE ON POROUS CARBONS

S.S. BARTON, J.R. DACEY and D.F. QUINN

Dept. of Chemistry and Chemical Engineering
Royal Military College of Canada
Kingston, Ontario.

ABSTRACT

High pressure adsorption of natural gas may increase its storage density. Methane was absorbed on six microporous carbons over a temperature range -30 to 25°C and a pressure range to 200 atmospheres (20 MPa). Increased storage density is greatest at <80 atmospheres and is trivial >150 atmospheres.

INTRODUCTION

High density storage of gaseous hydrocarbon mixtures such as natural gas is an important consideration in the efficient use of these gases as fuels. One possible method of achieving a higher storage density is to adsorb the gas onto a microporous solid.

Natural gas is about ninety-five percent methane and methane (CP Grade) was used as the model gas in this study. Methane cannot be liquified at ambient temperatures, and so it is stored as a compressed gas. As such it has a storage density of 0.16 Kg/dm^3 at 20 MPa.

Since the critical temperature of methane is -82.1^{o}C, the density of methane stored in porous media at ambient temperatures cannot be expected to approach that of liquid methane, at its boiling point, (≃0.41 Kg/dm^3 at -160°C.).

A high pressure volumetric adsorption apparatus was constructed and the amount of methane adsorbed by six microporous carbons measured at -30, 0 and 25°C, and by two molecular sieves at 25°C, at pressures up to 20 MPa. Surface areas, densities and pore volumes of the carbons were measured in an attempt to relate the amount of methane adsorbed to some physical parameter of the carbon.

EXPERIMENTAL

Characterisation of the Carbon samples

Nitrogen and methane adsorption isotherms were measured at liquid nitrogen temperature by the traditional volumetric method using a glass apparatus with twin five bulb gas burettes (1). Surface areas were determined from the adsorption data using the BET equation (2). Using the same apparatus at room temperature, helium densities were obtained for the carbons on the assumption that no helium was adsorbed.

A mercury penetration porosimeter (American Instrument Company) was used to determine pore size distribution in the range 90μm to 3nm. Samples were evacuated at room temperature before being subjected to increasing intrusion pressures to 60,000 p.s.i. (410 MPa). Low and high pressure densities and a macropore volume were obtained from these measurements.

Apparent bulk densities were measured following a method similar to ASTM D2854. This gives the lowest value for density because the voids between the particles are considered as available volume.

High Pressure Apparatus

The apparatus used is shown in Figure 1 and is similar to that described by Masters and Gesser (3). Standard "Swagelok" and "Cajon"

FIGURE 1 HIGH PRESSURE ADSORPTION APPARATUS

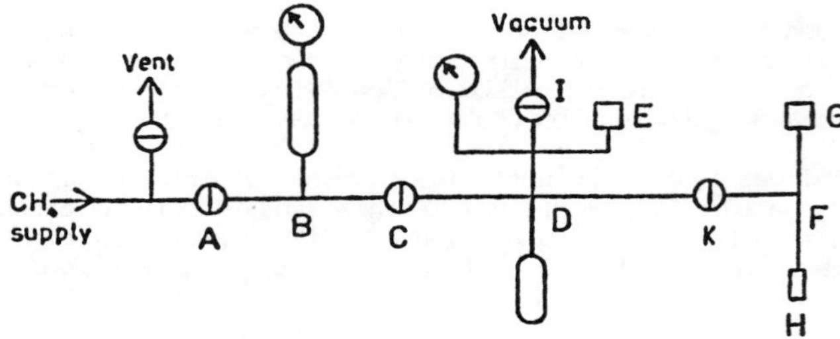

fittings, 1/4 inch stainless steel tubing and "Hoke" valves (Type 2325 F4Y) all capable of withstanding pressure in excess of 20 MPa (200 atm) were used. Valve K separating adsorption section F from the manifold D was an Ohno UH200 bellows seal valve. A rupture disc (3600 p.s.i. or 25.7 MPa) was incorporated for safety. Rough pressure measurements were made with Bourdon dial gauges and accurate pressures were obtained using National Semiconductor pressure transducers E (0-3000 p.s.i.a., Type LX1460AS). The transducers were calibrated against a 0-1000 p.s.i.a. Bell and Howell transducer of known sensitivity. Degassing was carried out by an Edwards High Vacuum oil diffusion pump (Model 63/150M Diffstak). Vacuum readings were made on an Edwards Penning gauge (Model CP25). The volumes of manifold D and the dead space volumes were obtained by low pressure helium expansion from a calibrated glass bulb. A calibrated 0.15 p.s.i.a. transducer was connected to D for this purpose. The second Monel storage bulb was used as a cold finger to condense methane in order to obtain pressures above the storage tank pressure.

High Pressure Isotherms

Approximately 5 grams of adsorbent, weighed accurately in air, was sealed in the sample holder and de-gassed at 200°C for a minimum of four hours. After calibration of the dead space in the sample section with helium, the sample was again evacuated thoroughly.

Methane adsorption was measured by admitting small doses from manifold D to the sample section F. The amount of methane adsorbed being taken as the decrease in methane from manifold D minus the increase in methane in the gas phase in the sample section F.

Since methane deviates considerably from ideality at pressures above a few atmospheres correction was made using the virial equation from the data of Douslin et al (4).

Equilibrium was achieved in less than one hour on all the samples, and was reached faster at higher pressures. Some desorption isotherms were measured. No detectable hysteresis was observed.

RESULTS AND DISCUSSION

The choice of the six carbons used in this study was made on the basis of their already known characteristics.

Two of the carbons, Saran (D) and the activated Carbone-Lorraine cloth, are largely microporous, and contain few macropores (5). These materials should give results which are predictable using the method proposed by Ozawa et al (6). Another carbon, Saran (B) (7) prepared in a different way from the same starting material (PVDC) as Saran (D) but having a macropore structure as well as a micropore structure, was examined.

Three commercially produced carbons were studied. BPL and Norit R1 carbons are produced from coal and wood respectively and the very high surface area Amoco GX-32 carbon is made by treating bituminous coal at high temperatures with potassium hydroxide.

Isotherms for both methane and nitrogen were measured at 77 K on all six carbons. Since both these molecules have about the same cross sectional area, 0.157 and 0.162 nm^2 respectively, similar areas are to be expected. Areas remained unchanged after the carbons had been subjected to pressures up to 200 atmospheres.

By applying the equations of Pierce (8) and Dubinin (9) to the isotherms, the micropore volume of each carbon was calculated. The results show remarkably good agreement between both methods. The simple Gurvitsch point volumes (10) for these isotherms agree well to the Dubinin and Pierce micropore volumes.

Helium density measurements at room temperature gave values higher than that of graphite (2.24 Kg/dm^3) suggesting that some adsorption of helium had taken place.

Two densities for each carbon were obtained from mercury penetration measurements. One, the bulk density was measured at low mercury pressures (20 to 40 p.s.i.) where the mercury would fill any interstices between the particles but not penetrate into pores. The other was obtained from high mercury pressures (60,000 p.s.i.) under which the mercury penetrates into macropores but does not enter micropores

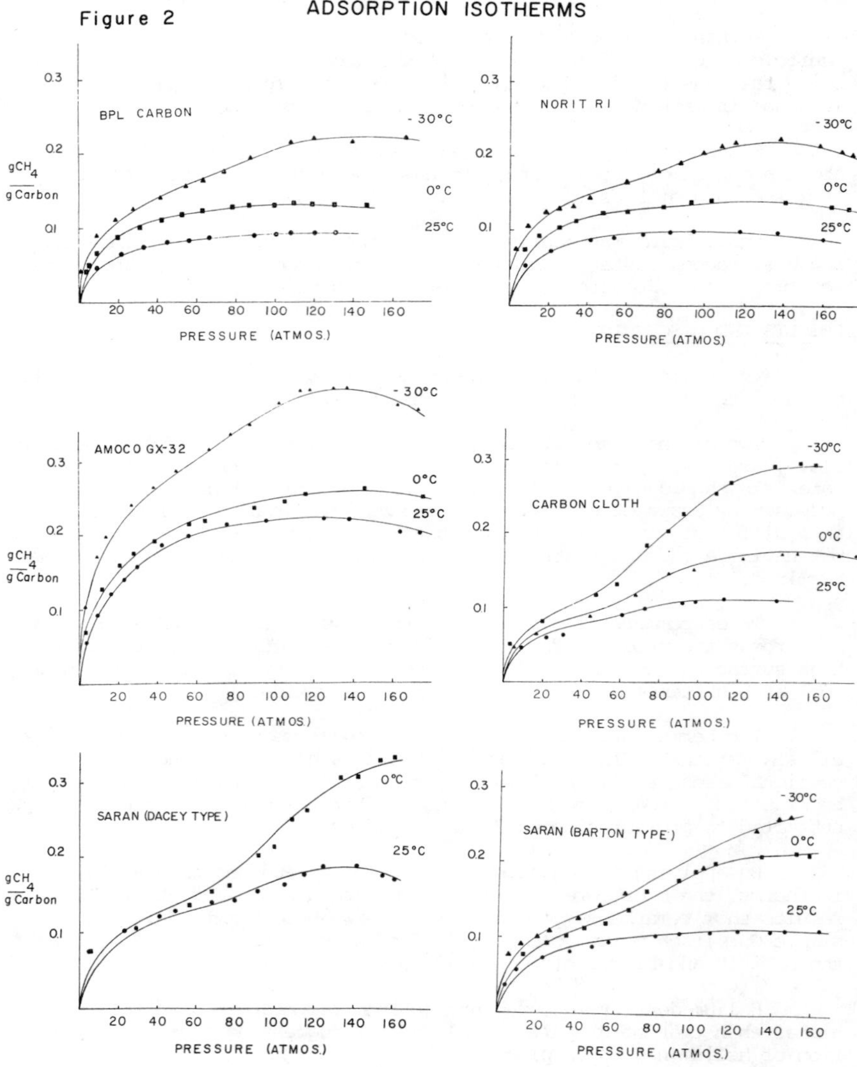
Figure 2
ADSORPTION ISOTHERMS
BPL CARBON
- 30°C
0° C
25°C
NORIT RI
-30°C
0°C
25°C
AMOCO GX-32
- 30°C
0°C
25°C
CARBON CLOTH
-30°C
0°C
25°C
SARAN (DACEY TYPE)
0°C
25°C
SARAN (BARTON TYPE)
-30°C
0°C
25°C
gCH_4 / g Carbon
PRESSURE (ATMOS.)
0.1
0.2
0.3
20
40
60
80
100
120
140
160

(diameter <3 nm).

Using these densities both macropore and micropore volumes can be calculated as follows:

$$\text{Macropore volume} = (d_{HP}-d_{LP})/d_{HP}d_{LP} \quad (1)$$

$$\text{Micropore volume} = (d_{He}-d_{HP})/d_{He}d_{HP} \quad (2)$$

The calculated values from all these measurements are collected in Table 1.

TABLE 1

CARBON	BPL	NORIT R1	SARAN (B)	SARAN (D)	CARBONE LORRAINE	AMOCO GX-32
SOURCE	BITUMINOUS COAL	WOOD, STEAM ACTIVATED	ALCOHOLIC KOH REDUCTION OF PVDC POWDER	COMPRESSED PVDC HEATED TO 700°C	CARBONISED RAYON, AIR ACTIVATED	HIGH TEMP. KOH TREATED COAL
BET AREA METERS^2/G						
NITROGEN	1030	1240	900	1050	640	2500
METHANE	950	1210	870	890	620	2300
DENSITY KG/DM^3						
APPARENT	0.44	0.39	0.32	1.12		0.30
BULK (HG 30 PSI)	0.80	0.71	0.51	1.12	0.86	0.45
HG (60000 PSI)	1.26	1.15	1.05	1.22	0.98	1.00
MACROPORE VOLUME DM^3/KG	0.45	0.55	1.00	0.08	0.14	1.22
MICROPORE VOLUME DM^3/KG						
HELIUM	0.35	0.42	0.51	0.37	0.49	0.55
PIERCE (N_2)	0.42	0.54	0.39	0.46	0.26	1.10
DUBININ (N_2)	0.47	0.63	0.39	0.47	0.29	1.06
DUBININ (CH_4)	0.51	0.61	0.41	0.44	0.29	1.09
GURVITSCH (N_2)	0.43	0.57	0.37	0.45	0.28	1.10
GURVITSCH (CH_4)	0.44	0.56	0.35	0.44	0.30	1.13

In an earlier paper (11) we have discussed the difficulties encountered in correlating gas adsorption data with mercury porosimetry data, particularly with highly porous carbons such as the Amoco product.

At temperatures above critical, methane condensation is unlikely, so the density of the adsorbed gas will be less than the liquid density at its normal boiling point. Methane is an extremely stable molecule with low bond polarity and London dispersion forces are most likely to be involved in adsorption. Solids with extensive microporous structures are considered to have potential energy wells which would be more favorable to adsorption of methane than a plane surface. Thus from the micropore volumes in Table 1, Amoco GX-32 would be expected to show the greatest adsorption and the Carbone-Lorraine cloth the least, the other carbons being between these two.

The high pressure isotherms are shown in Figure 2. The scale used is the same for all the plots. The isotherms for BPL and Norit R1 carbons are very similar and both carbons have similar macro and micropore volumes as well as BET surface areas. The Amoco GX-32, as expected, clearly shows the greatest adsorption, commensurate with its large micropore volume and large surface area. The isotherms obtained from the Carbone-Lorraine cloth were unexpected, this carbon has a very low surface area and small micropore volume but adsorbed large quantities of methane under high pressure. This finding may be explained by considering the micropores to be "ink-bottle" in nature. Helium is able to enter the pores and so giving the large helium micropore volume reported. The pores are not accessible to nitrogen or methane at low pressures so that apparently low micropore volumes are observed. At high pressures however methane can be forced into the pores with resulting high adsorption. The Saran (D) carbon, which like the cloth, has only a micropore structure, also showed substantial adsorption, greater than the adsorption on Saran (B) which has considerable macropore volume.

All the isotherms show the expected maximum adsorption peak. It is important to point out that it is "apparent" adsorption that is being measured and that adsorption is still taking place at pressures above P_{max} except that now the rate of increase in the density of the adsorbed phase is not as great as the rate of increase in the density of the gas phase above it (13).

Initially the isotherms appear to be Type I in nature but on closer inspection all show the "knee" typical of a Type IV isotherm. Some of the methane isotherms of Ozawa (6) seem to also show this behaviour.

Menon (14) has suggested that the observed maximum for high pressure isotherms is independent of the adsorbent and that P_{max} can be calculated empirically for any adsorbate from

$$P_{max} = (T/T_c)\ P_c \qquad (3)$$

where T_c and P_c are the critical temperature and the critical pressure. This equation predicts that P_{max} values for the methane are 112, 94 and 74 atmospheres at 25°, 0° and -30°C respectively. The present results give P_{max} values of 100 to 140 atmospheres at 25°C in reasonable agreement with the predicted values. However, as the temperature decreases, contrary to prediction, the experimental P_{max} values increase. Thus when compared with an existing theory the results do not show a consistent pattern and are not easily analysed or interpreted.

Attempts were made to fit the Dubinin-Astakhov equation (15)

$$\ln(\ln W_o/W) = n(\ln\Sigma - \ln E) \qquad (4)$$

where W_o = pore volume

W = volume adsorbed (STP)

Σ = RT ln (f_s/f)

f_s = fugacity of CH_4 at P_{max} = $(T/T_c)\ P_c$

f = fugacity of CH_4 at equilibrium pressure P and temperature T

which is based on the Polanyi potential theory to the results. Pressures were replaced by fugacities to correct for deviations from ideality, but no linear correlation was found. A plot of $\ln(\ln W_o/W)$ versus $\ln\Sigma$ should be linear of slope n and intercept n ln E.

Ozawa et al (6) have pointed out that this equation is only valid over a limited pressure range and have suggested a generalised Dubinin-Astakhov equation. By plotting ln W against Σ/β, where β, an "affinity coefficient" = $\alpha_{CH_4}/\alpha_{N_2}$ = 1.22 for CH_4. The resulting curve is then fitted to a second or third degree polynomial giving,

$$\ln W = \ln W_o + k_1(\Sigma/\beta) + k_2(\Sigma/\beta)^2 + k_3(\Sigma/\beta)^3 + \ldots \quad (5)$$

whose intercept will give the pore volume W_o the coefficients k_1, k_2 and k_3 are independent of the adsorbate for a given adsorbent. For any given carbon adsorbent, the pore volume will remain virtually constant over a limited temperature range. On applying the Ozawa equation to our isotherms, it was found that the calculated W_o varies with temperature, (see Table 2). This could be a result of the heat of liquifaction not being equal to the heat of adsorption at P_{max}.

Isosteric heats of adsorption, also given in Table 2 were obtained, in the usual way, from the isotherms for the maximum adsorption at 25°C. The heats thus do not refer to quite the same amount of adsorption for all samples. It is at once apparent that with the exception of BPL carbon the heats of adsorption all lie very close to the value for the heat of liquifaction of methane (-8.5 kJ-mole^{-1}). It appears that the observed variation of the Ozawa W_o with temperature does not result from ΔH_a being larger than ΔH_L.

From a practical point of view, on the basis of these heat values, it may be stated that if during use, a reasonably slow desorption of methane takes place and the container is in good thermal contact with the surroundings, no undue cooling of the carbon-methane system will occur.

For comparison, and to check the validity of the experimental procedure, the adsorption isotherm for methane at 25°C was measured on a Linde 5A molecular sieve to 200 atmospheres. The result from this compares favorably with that given by Ozawa (12) up to their P_{max} of about 60 atmospheres. Our isotherm showed the same adsorption but P_{max} was observed over a pressure range from 60 to 180 atmospheres before showing the expected apparent desorption. A Linde 13X molecular sieve adsorbent gave a very similar isotherm with a very large P_{max} range. Maximum apparent adsorption was 5.1 mgs CH_4 per g, a little lower than the 5A adsorbent.

TABLE 2 OZAWA PARAMETERS AND HEATS OF ADSORPTION

CARBON	BPL	NORIT R1	SARAN (B)	SARAN (D)	CARBONE LORRAINE	AMOCO GX-32
25 °C						
W_o	0.46	0.59	0.56	0.62	0.59	1.10
$10^3 K_1$	-0.71	-0.63	-1.10	-1.53	-1.82	-0.54
$10^7 K_2$	-0.55	-1.64	4.09	12.49	10.72	-6.32
$10^{10} K_3$	-0.42	0.50	-1.39	-4.52	-3.48	1.83
0 °C						
W_o	0.70	0.71	1.03	1.37	0.82	1.27
$10^3 K_1$	-0.90	-0.66	-2.38	-3.67	-1.81	-0.66
$10^7 K_2$	0.46	-3.96	14.22	26.61	-1.97	-3.94
$10^{10} K_3$	-0.46	2.58	-3.43	-6.32	5.56	1.17
-30 °C						
W_o	1.15	1.15	1.37		1.60	2.08
$10^3 K_1$	-2.12	-1.67	-3.54		-3.87	-1.47
$10^7 K_2$	13.43	7.20	34.57		24.65	2.92
$10^{10} K_3$	4.27	-1.67	13.36		-5.45	-0.67
$\overline{\Delta H_A}$ KJ/MOL	-23	-13	-10	-8	-6	-10

Storage Density

Since the objective of this study was to find a method of increasing the storage density of methane a more direct and practical use of the above results must be made. With a high pressure density of about 1Kg/dm^3 and a typical micropore volume of 0.5 dm^3/Kg, approximately half the volume occupied by a carbon is available for methane adsorption, thus for any benefit at all in storage density over compressed gas, the density of the adsorbed methane must be more than double that of the gas phase above it.

The amount of methane which was stored per unit volume in the vessel filled with carbon adsorbent was calculated from the adsorption isotherms,

$$n(\text{total}) = n(\text{ads}) + n(\text{gas}) \qquad (6)$$

and this was compared to the amount of methane per unit volume with no adsorbent present. The results are presented in Figure 3.

It is realised that this approach is arbitrary. The values could vary considerably. At low pressures (<80 atmospheres) the storage densities could be higher if the carbon were more efficiently packed into the vessel to reduce void space.

What these results do show is that, taking as an example Amoco GX-32 at 25°C and 40 atmospheres pressure, to store the same amount of methane per unit volume as a compressed gas, would require a pressure of more than 100 atmospheres. Alternatively, a vessel at 40 atmospheres

STORAGE DENSITY VS. PRESSURE

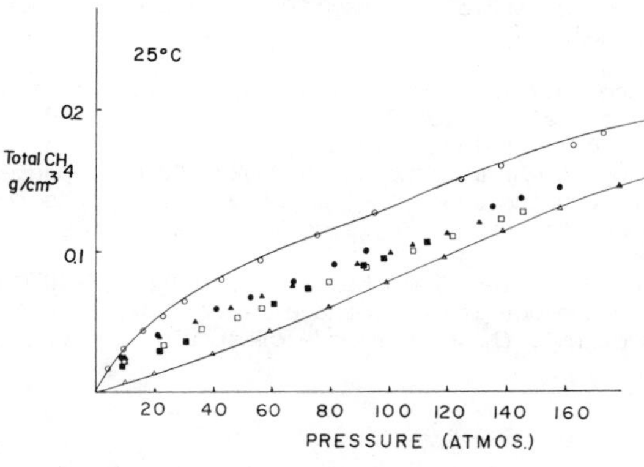

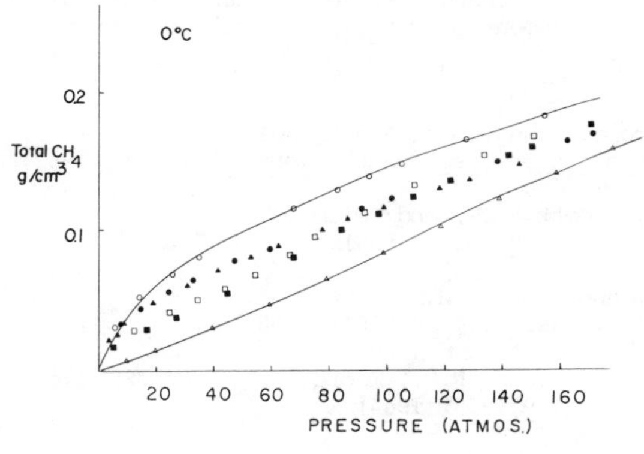

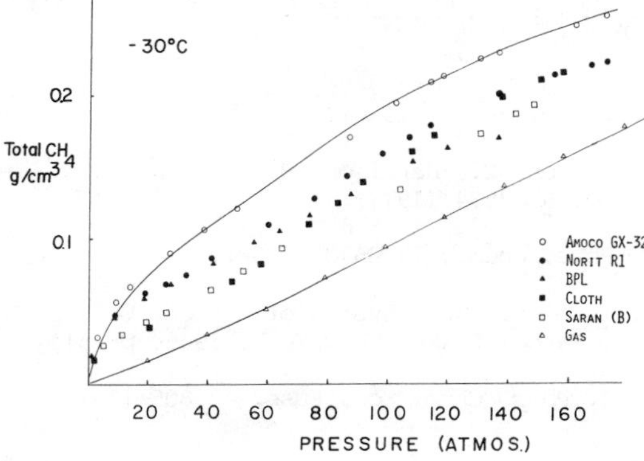

and no adsorbent would contain less than 40 percent the methane in an Amoco GX-32 filled vessel at the same pressure. At higher pressures, however, this increased storage density diminishes considerably. Above 120 atmospheres there is little advantage in using a carbon adsorbent to increase storage density.

In general the following conclusions may be drawn from this work. Microporous carbons are effective adsorbents for methane and are superior in this respect to molecular sieve zeolites. One carbon (Amoco GX-32) adsorbs 22% by weight methane at 25°C and 100 atmospheres. Lowering the temperature increases the amount of methane adsorbed. On some carbons the adsorption at -30°C is double that at 25°C.

Under optimum conditions used here (i.e. 160 atm, -30°C and Amoco GX-32) the storage density of methane ≈0.25 Kg/dm^3 is about 60% of the density of LNG, the common mode of storing and transporting methane.

ACKNOWLEDGEMENT

The authors wish to thank Ms Jayne Holland for technical assistance and the Government of Ontario for financial support.

LITERATURE CITED

1. See, for example, Young, D.M. and Crowell, A.D., Physical Adsorption of Gases, Butterworths, London, 1962, p. 285.

2. Brunauer, S., Emmett, P.M. and Teller, E., J. Amer. Chem. Soc. 60, 309 (1938).

3. Masters, K.J. and Gesser, H.D., J. Phys. E: Sci. Instrum., 14, 1043 (1981).

4. Douslin, D.R., Harrison, R.H., Moore, R.T., and McCullough, J.P., J. Chem. Eng. Data, 9, 358 (1964).

5. Dacey, J.R. and Cadenhead, D.A., Proceedings of the Fourth Conference on Carbon, Pergamon Press, Oxford, 1960, p. 9.

6. Ozawa, S., Kusimi, S. and Ogino, Y., J. Coll. and Int. Sc. 56 83 (1976).

7. Barton, S.S., Boulton, G., Harrison, B.H. and Kemp, W., Trans. Farad. Soc. 67 3534 (1971).

8. Pierce, C., J. Phys. Chem., 72 3673 (1968).

9. Dubinin, M.M., Chemistry and Physics of Carbon, Vol. 2, Edited by Walker, P.L., Marcel Dekker, New York, 1966, p. 51.

10. Gurvitsch, L., J. Phys. Chem. Soc. Russ. 47 805 (1915).

11. Harrison, B.H., Barton, S.S., Dacey, J.R. and Sellors, J., J. Coll. and Int. Sc. 71 367 (1979).

12. Wakasugi, Y., Ozawa, S., and Ogino, Y., J. Coll. and Int. Sc. 79 399 (1981).

13. Menon, P.G., Chem. Rev. 68 277 (1968).

14. Menon, P.G., J. Phys. Chem. 72 2695 (1968).

15. Dubinin, M.M., Chem. Rev. 60 235 (1960).

SELECTIVE ADSORPTION OF ORGANIC HOMOLOGUES ONTO ACTIVATED CARBON FROM DILUTE AQUEOUS SOLUTIONS. SOLVOPHOBIC INTERACTION APPROACH - V THE EFFECT OF SIMPLE STRUCTURAL MODIFICATIONS WITH AROMATICS

Georges Belfort, Gordon L. Altshuler, Kusuma K. Thallam, Charles P. Feerick, Jr., and Karen L. Woodfield
Department of Chemical Engineering and Environmental Engineering
Rensselaer Polytechnic Institute
Troy, New York 12181, USA

ABSTRACT

Preferential adsorption of organic compounds onto activated carbon from dilute aqueous solutions is studied to develop a comprehensive theoretical basis for predicting the adsorption of structurally different isomers for a homologous series. The fundamental multidimensional approach of the solvophobic (cϕ) thermodynamic theory is further refined and used to correlate the extent of adsorption for the comprehensive theory with the overall standard net free energy change ($\Delta G^{net}/RT$) for the association-adsorption reaction in solution, and for the simplified theory with the cavity surface area of the solute (TSA).

Experimental adsorption isotherms of an aromatic homologous series (19 alkyl phenols) were measured and used to test the cϕ-theory. Linear, branched and fragmented isomers showed different adsorption behavior. Surface orientation effects are proposed to explain the unexpectedly higher adsorbability for the fragmented versus linear isomers. In addition, with increasing sorbate molecular weight (and size) a saturated adsorbability is experimentally observed above about 160 Daltons (or equivalent spherical radius of greater than 5.0 Å). Above these values, equilibrium adsorption appears to be independent of both molecular weight and branching for the activated carbon studied.

As part of developing a comprehensive data base to test the theory a new innovative procedure to measure batch adsorption isotherms is presented. This method minimizes the loss of organics due to volatilization and adsorption onto extraneous surfaces.

Comparing the coefficients of linear correlation (r), the results for 8 alkyl phenols give greater than 70% confidence that the r-values are different for $\Delta G^{net}/RT$ and TSA when compared with molecular weight (MW). These results agree with our recently reported correlations for two aliphatic homologous series.

INTRODUCTION

During the past several years a concerted effort has been made in the authors' laboratory to understand the role of the **solvent** during liquid-solid adsorption (1,2). One method to approach this question is to study the adsorption of a homologous series of organic isomers from dilute aqueous solution onto activated carbon. Thus, simple structural modifications of these adsorbate isomers allow one to probe the aqueous-phase adsorption process.

It is well known from the early work of McBain, et al. (3) that the propensity of a solute (gases such as SO_2, NH_3, N_2, H_2) to adsorb from the liquid phase is approximately inversely proportional to its solubility, while from the gas-phase the reverse order is observed. This reversal in the readiness to adsorb has recently been confirmed in the authors' laboratory (4) with a comparison of the adsorption of four alcohols in the gas-and liquid-phase. Clearly, the role of the solvent during the adsorption process is important and must be accounted for if a fundamental understanding of the mechanisms of aqueous-phase adsorption is desired. Within this context a comprehensive thermodynamic formalism called the solvophobic theory ($c\phi$) has been adapted and tested for aqueous phase adsorption of aliphatic compounds (1,2).

In this paper, we extend the testing of the $c\phi$-theory to include the adsorption of a homologous series aromatic compounds (alkyl phenols). In contrast to all our previous publications but one (2), we report here our own experimentally measured isotherms. Besides being more extensive in number of isomers within a homologous group allowing for better statistical predictions, the data presented here are also internally self-consistant, i.e., the isotherms are measured with similar carbon, water, pH, temperature, duration and range of initial concentrations. In addition, we compare correlations for adsorbability ($\ln Q°b$) with the comprehensive $c\phi$-model ($\Delta G^{net}/RT$), with the modified model (TSA), and with molecular weight.

THEORY

The adsorption of a solute onto activated carbon may be represented by the following reversible association reaction

$$S_i + C \rightleftharpoons S_iC \qquad (1)$$

where S_i is the adsorbing solute, C is the activated carbon and S_iC is the adsorbed solute-carbon complex. The equilibrium constant for this reaction, in the presence of the solvent, $K_{solvent,i}$, is related to the overall free energy change, $\Delta G^{assoc}_{solvent}$ under standard conditions by

$$\Delta G^{assoc}_{solvent} = -RT \ln K_{solvent,i} \qquad (2)$$

The solvophobic theory (cϕ) takes into account the effect of the solvent on the above association reaction by considering two hypothetical steps. First, a cavity in the solvent must be created to contain the reacting species. Second, after the gas phase reaction, the "molecules" are placed into the cavities, where each species may interact with the surrounding solvent. This process is depicted in Fig. 1.

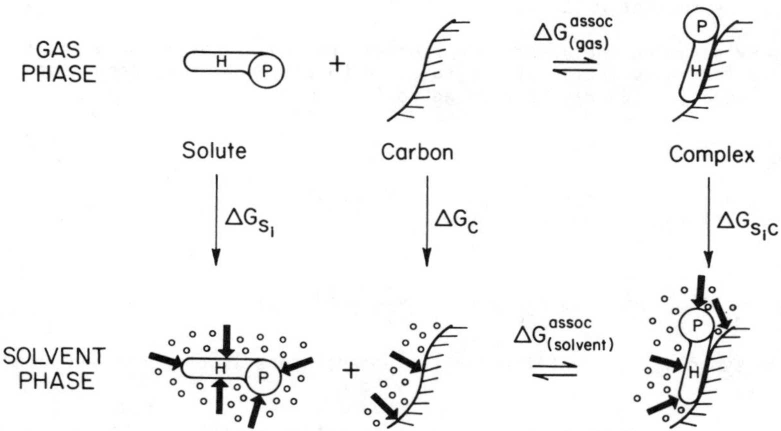

Figure 1. ASSOCIATION-ADSORPTION REACTION.

The effect of the solvent may be calculated from the difference of the free energy changes of the solvent and gas phase reactions

$$\Delta G^{net}_{solvent\ effect} = \Delta G^{assoc}_{solvent} - \Delta G^{assoc}_{gas} \tag{3}$$

In addition, this solvent effect is due to the interaction of each species with the solvent. Therefore,

$$\Delta G^{net}_{solvent\ effect} = \sum_j \{\Delta G^{net}_{j,S_iC} - \Delta G^{net}_{j,S_i} - \Delta G^{net}_{j,C}\} \tag{4}$$

where subscript j represents all possible interactions. In previous applications of the solvophobic theory (1,2,5-7) the interactions considered were cavity formation, van der Waals (and reduction due to presence of a solvent), and electrostatic.

Combining Eqs. (3) and (4) gives

$$\Delta G^{assoc}_{solvent} = \Delta G^{assoc}_{gas} + [\Delta G_{cav} + \Delta G_{vdw} + \Delta G_{es}]^{net}_{S_iC-S_i-C} - RT\ \ln\ (RT/P_\circ V) \tag{5}$$

These terms may now be calculated in terms of the appropriate physical properties of the species. For purposes of brevity, the explicit equations are not presented here, but the important features of the interactions are summarized in Table 1. The reader is referred to (1,2) for further details.

By assuming the adsorption process from the aqueous phase results mainly from association of the solute with a hydrophobic carbon surface, Equation (5) may be reduced to (1)

$$\Delta G^{assoc}_{solvent} = \Delta G^{assoc}_{gas} - \Delta G_{vdw,S_i} + \frac{N(\lambda-1)}{2\lambda}\ \frac{\mu^2_{S_i}}{\nu_{S_i}}\ DP$$

$$- N\gamma\Delta A - 4.836\ N^{1/3}\ \gamma(\kappa^e-1)V^{2/3} - RT\ \ln\ (RT/P_\circ V) \tag{6}$$

where symbols are identified in the nomenclature list at the end of the paper.

The free energy change of the association reaction may be related to its equilibrium constant by Eqn. (2) where

$$\ln K_{solvent,i} = \frac{-\Delta G^{assoc}_{gas}}{RT} + \frac{\Delta G_{vdw,S_i}}{RT}$$

$$- \frac{N(\lambda-1)}{2\lambda}\ \frac{\mu^2_{S_i}}{\nu_{S_i}}\ \frac{DP}{RT} + \frac{\gamma}{RT}\ (N\Delta A + 4.836N^{1/3}(\kappa^e-1)V^{2/3}) + \ell n\ \frac{RT}{P_\circ V} \tag{7}$$

Since this work will focus on the interactions of the dilute solute with the surrounding solvent, the gas-phase term (ΔG^{assoc}_{gas}) will be considered constant. Recent experimental results from the author's laboratory confirm that ΔG^{assoc}_{gas} varies only slightly (<5%)

TABLE I. INTERACTIVE FREE ENERGY CHANGES

Interaction process	Free energy designation	Interaction forces	Important parameters
Cavity formation	$\Delta G^{net}_{cav,i}$	Surface forces	Solute molecular surface area, TSA corrected macroscopic surface tension, γ_i
Solvent-solute interaction			
Van der Waal's	$\Delta G^{net}_{vdw,i}$	Dispersive (attractive) forces	Ionization potential and molecular volume ν_i
Electrostatic	$\Delta G^{net}_{es,i}$	Electrostatic interaction	Inverse molecular volume, ν_i (a) Simple dipole, Solute dipole moment μ_i and dielectric constant of solvent ε (b) Charged species, Ionic charge ze and ionic strength, I
Mixing effects (polymers)	ΔG^{net}_{mix}	Free energy of mixing	
Dispersive reduction as a result of solvent	ΔG^{net}_{red}	Masking of the dispersive forces	
Free volume reduction (cratic)	$RT \ln (RT/PV)$	Entropic reduction	Solvent molar volume V and pressure P

when compared to changes in $\Delta G^{assoc}_{solvent}$ (~30%) for the adsorption of linear isomers of the alcohol homologous series (4). For convenience, the remaining terms will be referred to with respect to their origin, i.e.,

$$VDW = \frac{\Delta G_{vdw,S_i}}{RT} \qquad \text{(van der Waals)} \qquad (8)$$

$$ES = \frac{-N(\lambda-1)}{2\lambda} \frac{\mu^2_{S_i}}{\nu_{S_i}} \frac{DP}{RT} \qquad \text{(electrostatic)} \qquad (9)$$

$$CAV = \frac{\gamma}{RT} (N\Delta A + 4.836N^{1/3} (\kappa^e-1)V^{2/3}) \qquad \text{(cavity)} \qquad (10)$$

$$CRAT = \ln \frac{RT}{P_\circ V} \qquad \text{(cratic)} \qquad (11)$$

At low solute concentrations, $K_{solvent,i}$ is proportional to the initial slope of the Langmuir isotherm, Q°b. Therefore, omitting the relatively constant term, ΔG^{assoc}_{gas}, Eqns. (7)-(11) may be rewritten as:

$$\ln Q°b \; \alpha \; VDW + ES + CAV + CRAT = NET \qquad (12)$$

where $\Delta G^{net}/RT \equiv NET$.

EXPERIMENTAL

Adsorbent Filtrasorb 400 granular activated carbon (Calgon Corp., Pittsburgh, PA) was ground in a jar mill so as to pass through U.S. sieve series #200 (0.074 mm) and to be retained by U.S. sieve series #400 (0.037 mm). The carbon was cleaned with successive washing and supernatant removal. The carbon was then analyzed in a surface area-pore volume analyzer (Model 2100E, Micrometrics Instruments Corp., GA) and the data was fit to a B.E.T. model. The B.E.T. surface area, using N_2 gas, was 1031 m^2/g, the pore volume was 0.95 cm^3/g, the equivalent particle diameter was 37-74 μm and the bulk density was 0.478 g/cm^3.

The carbon was dried in a vacuum oven (Isotemp No. 281, Fisher Scientific, NY) at 100°C and 25 mmHg vacuum for a five day period. Any extraneous adsorbed organics were volatilized by increasing the temperature to 120°C for two days until the weight of the carbon remained constant. The dried carbon was stored in a desiccator until use.

Adsorbates The alkyl phenols were purchased from Aldrich Co. (WI), graded 99% pure. Stock solutions were made with 0.01 M phosphate buffer and stored in amber glass bottles.

Adsorption Isotherm Procedure Several innovations have been introduced into the adsorption isotherm procedures in an attempt to improve reproducibility and accuracy and reduce solute losses. Although there is a widely held view among some water technologists that liquid-phase adsorption is relatively insensitive to temperature, and therefore need not be accurately controlled, temperature was kept constant at 20 ± 0.5°C during this study. The pH was also buffered at 7.0 ± 0.1 with standard 0.01 M phosphate buffer.

To minimize extraneous solute-loss and maximize solute/sorbent contact the following steps and precautions were taken:

a) Adsorption was conducted in completely filled and capped 38-mℓ stainless steel centrifuge tubes (Part No. 301112, Beckman, CA). Thus there was no head-space for possible solute loss to the gas phase, and the solutions were only exposed to stainless steel surfaces for both minimum adsorption loss and to reduce the possibility of introducing extraneous organics.

b) The completely filled tubes were rotated end-over-end 360° at 2 rpm in a constant-temperature bath (20 ± 0.5°C). This allowed the carbon to fall the length of the tube twice per rotation resulting in good mixing and increased effective external film mass transfer coefficient. The time to reach pseudo equilibrium (i.e., >95% of the 14th day saturation value, Q°) varied from two to several hours depending on the kinetics of the particular solute and rotation rate. The mixing period employed in this study was 24 hours for most solutes.

c) Once pseudo equilibrium was attained, the tubes were placed directly into an ultracentrifuge (Model L8-55, Beckman, CA) and the carbon was spun-down at 20,000 rpm for 20 min.

Dissociation Effects Zogorski and Faust (8) have shown that inorganic ions (phosphate buffer) do not affect the adsorption of the undissociated form of several halogenated phenol adsorbates. Adsorption of the anionic form of the compounds, however, is increased in the presence of these ions, possibly through reduction of the repulsive forces between adsorbed molecules and the carbon surface, or between anions adsorbed at the surface. If a large percentage of the solute molecules were present in the dissocated form, the inclusion of phosphate buffer in our procedure may have influenced adsorption considerably.

Of the nineteen alkyl phenols studied, phenol has the greatest tendency to dissociate. It has a dissociation constant $K_A = 1.2 \times 10^{-10}$ (9). At pH = 7.0, less than one percent of phenol exists in the anionic form in aqueous solution (10). Subsequently, we can assume our equilibrium adsorption parameters are characteristic of the undissociated molecular species.

Data Analysis The adsorption data for this study is presented in terms of the Langmuir isotherm equation:

$$Q = \frac{Q^\circ bC}{1+bC} \quad \text{or} \quad \frac{1}{Q} = \frac{1}{Q^\circ b}\,\frac{1}{C} + \frac{1}{Q^\circ} \tag{13}$$

where Q is the amount adsorbed (mmoles/g carbon) and C is the equilibrium solution concentration (mmoles/ℓ), Q° is the asymptotic saturation value of Q, and b is the energy-related constant.

RESULTS AND DISCUSSION

The results of the adsorption isotherm experiments on nineteen alkyl phenols are presented in Table 2. In addition to the Langmuir isotherm parameters, the correlation coefficient, r, for the linearized isotherms is given, as well as the total cavity surface area for each solute, calculated from geometrical considerations by the method of intersecting spheres (11).

The linearized Langmuir isotherms for several representative linear alkyl phenols are shown in Figure 2. Note that as the alkyl portion of the solute is lengthened, both the intercept and slope decrease, corresponding to an increased saturation value Q°, and an increased initial slope for the Langmuir isotherm, $Q^\circ b$, as expected.

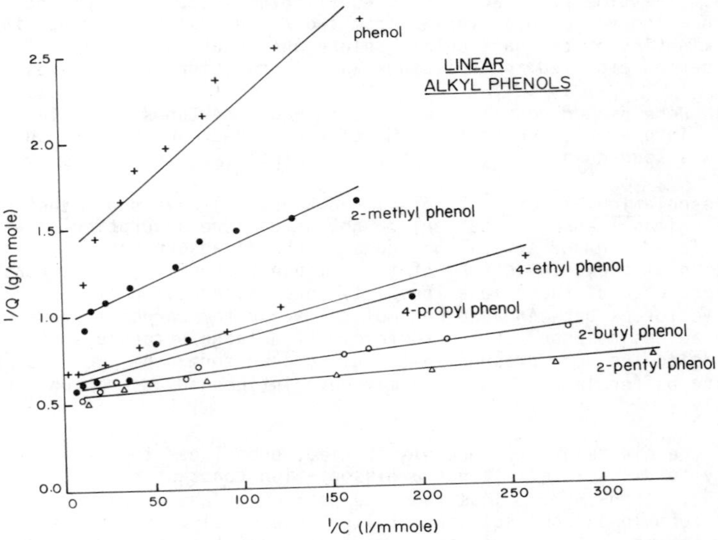

Figure 2. LINEAR LANGMUIR ISOTHERMS FOR SIX NORMAL ALKYL PHENOLS.

TABLE 2. ADSORPTION ISOTHERM MEASUREMENTS FITTED TO THE LANGMUIRIAN MODEL FOR ALKYL PHENOLS

Solute	MW	TSA[a] ($\AA^2$)	Concentration Range (mmole/L)	No. or Points	Q^o (mmole/g)	b $(\text{mmole/L})^{-1}$	ln Q^o (mmole/g)	ln $Q^o b$[b]	r
1. Phenol	94.1	249.1	.132 - .004	11	0.73(8)	150.4	-0.30(4)	4.7(3)	0.93(4)
2. 2-methyl Phenol	108.1	274.3	.174 - .006	10	1.03(5)	213.7	0.03(4)	5.4(0)	0.96(7)
3. 4-ethyl Phenol	122.2	306.3	.202 - .002	9	1.44(5)	266.6	0.36(8)	5.9(5)	0.93(2)
4. 2,6-dimethyl phenol	122.2	296.3	.171 - .006	8	2.27(1)	308.7	0.82(0)	6.5(5)	0.93(5)
5. 2,3-dimethyl phenol	122.2	295.3	.140 - .012	11	1.79(5)	282.4	0.58(5)	6.2(3)	0.97(5)
6. 3,4-dimethyl phenol	122.2	297.9	.158 - .005	10	1.81(3)	266.4	0.59(7)	6.1(8)	0.95(2)
7. 3,5-dimethyl phenol	122.2	306.1	.188 - .007	9	1.61(7)	299.7	0.48(1)	6.1(8)	0.96(4)
8. 2,5-dimethyl phenol	122.2	302.8	.136 - .008	9	1.92(1)	238.2	0.65(3)	6.1(3)	0.98(0)
9. 4-propyl phenol	136.2	336.8	.189 - .005	7	1.62(2)	343.6	0.48(4)	6.3(2)	0.95(8)
10. 4-isopropyl	136.2	321.7	.145 - .011	6	3.77(3)	92.4	1.32(8)	5.8(5)	0.95(8)
11. 2,3,5-trimethyl phl.	136.2	323.5	.230 - .004	11	2.27(2)	302.9	0.82(1)	6.5(3)	0.93(4)
12. 2,3,6-trimethyl phl.	136.2	317.3	.194 - .006	10	2.45(2)	335.0	0.89(7)	6.7(1)	0.92(7)
13. 2-butyl phenol	150.2	363.1	.149 - .002	6	1.70(2)	833.1	0.53(2)	7.2(6)	0.95(6)
14. 4-butyl phenol	150.2	366.9	.057 - .011	6	2.26(5)	613.3	0.81(8)	7.23(7)	0.62(9)[c]
15. 4-tert butyl phenol	150.2	345.1	.133 - .008	8	1.53(9)	473.5	0.43(1)	6.5(9)	0.93(2)
16. 2-pentyl phenol	164.3	394.0	.160 - .002	16	1.78(3)	800.9	0.57(8)	7.2(6)	0.94(1)
17. 4-pentyl phenol	164.3	397.4	.127 - .004	10	2.56(3)	672.3	0.94(1)	7.4(5)	0.89(4)
18. 4-tert pentyl phenol	164.3	362.3	.138 - .015	7	4.93(5)	177.4	1.59(6)	6.7(8)	0.93(3)
19. 2-hexyl phenol	178.3	423.7	.163 - .001	6	1.79(4)	733.3	0.58(4)	7.1(8)	0.91(9)

a Total cavity surface areas (TSA) calculated using the MSD Program (11).

b Experimental error for one standard deviation is about ± 0.10 units for ln $Q^o b$.

c difficult to dissolve

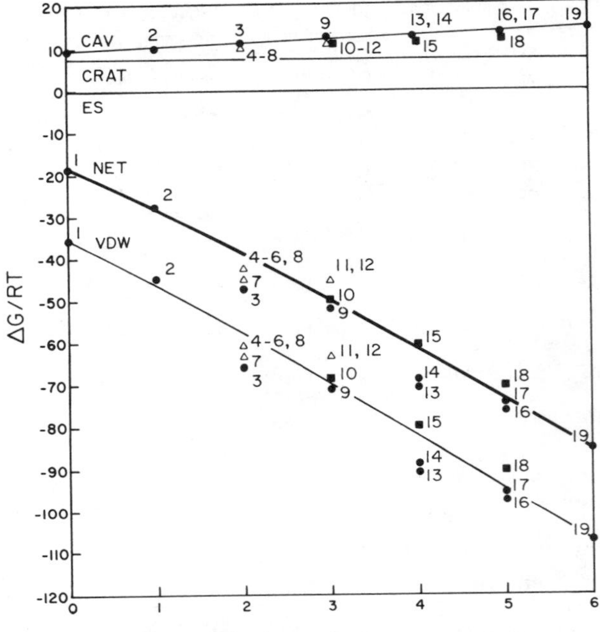

Figure 3. SOLVOPHOBIC CONTRIBUTIONS TO $\Delta G^{net}/RT$ FOR ALKYL PHENOLS.

Figure 4 shows a levelling off of adsorbability starting at the pentyl phenols. This effect has been seen in previous work (2) on alkyl ketones, and can be explained by steric hindrance of the larger solutes in the carbon micro-pores. The equivalent spherical radius of the pentyl phenols is approximately 5.0Å.

Instead of considering the sum of the interactions, it can be informative just to consider the first step of the hypothetical process shown in Figure 1. This is represented by the cavity term, which is proportional to ΔA, the microsurface area change of reaction (6). In this work, we have estimated this value as 25% of the total cavity surface area (TSA). Therefore, the linear dependence of lnQ°b versus TSA can be considered a "simplified" version of the solvophobic theory. Figure 4c shows this dependence. The fragmented isomers lie closer to the correlated line than in Figure 4b, but there is not a large difference in the TSA's of the fragmented isomers. The correlation coefficient is r = 0.88(1) for the entire group, and is r = 0.94(6) for the linear and branched phenols. In addition, the correlation coefficients for the non-fragmented alkyl phenols with an equivalent spherical radius of less than 5.0 Å (#1-3, 9, 10, 13-15) were r = 0.97(6) for both $\Delta G^{net}/RT$ and TSA versus lnQ°b, and r = 0.93(9) for MW versus lnQ°b.

Of the nineteen alkyl phenols studied, nine were linear alkyl groups, three were branched (branched alkyl groups at one position on the phenol), and seven were fragmented (alkyl group split and arranged at more than one position). Figure 4a illustrates the increase of adsorbability ($\ln Q°b$) with molecular weight. However, molecular weight considerations alone cannot distinguish among isomers. The branched isomers all exhibit lower adsorbability than their linear counterparts, as shown in previous work for other homologous series (2). The fragmented isomers all exhibited higher adsorbability. A correlation coefficent of $r = 0.87(9)$ was obtained for the complete set, and $r = 0.92(2)$ omitting the fragmented solutes.

The solute physical properties required to calculate the solvophobic interactions are listed in Table 3. In most cases, data were taken from standard reference sources. Where this was not possible, values were estimated from other compounds in the homologous series. In addition, a value of 8.50 eV was used for the ionization potential of all of the phenols, along with a dipole moment of 1.50 D. Previous work (1,2) has shown these properties to vary slowly among a homologous series, and the $\Delta G^{net}/RT$ results to be fairly insensitive to the exact values used.

The results of the solvophobic calculations are presented graphically in Figure 3. Note that $\Delta G^{net}/RT$ is dominated by the term representing the van der Waal's interaction between the solute and the solvent. Also, for a non-dissociated solute (experiments were run under buffered conditions of pH = 7.0) the electrostatic interaction is practically negligible here.

As seen from Figure 3, the van der Waal's contribution dominates the value of $\Delta G^{net}/RT$. This interaction is represented by an effective Kihara potential (a modified Lennard Jones potential) (6). The term "effective" refers to the fact that it is corrected due to the presence of the surrounding solvent. This potential is integrated over a discrete solvent layer, then an outer solvent continuum. Since this is a non-polar interaction, the accentric factor of a non-polar analog solute is used in the Kihara potential. Therefore, some error could be introduced into the ΔG^{net} calculations by the estimation of the accentric factor, and the neglected interactions due to the aromatic ring of the alkyl phenols.

From Eqn. (12), $\Delta G^{net}/RT$ should be proportional to $\ln Q°b$. This relationship is shown in Figure 4b. The adsorbability of the linear and branched alkyl phenols are fairly well correlated by $\Delta G^{net}/RT$, while the fragmented alkyl phenols show no apparent pattern, other than lying at a higher adsorbability. Although the adsorbability of the isomers can be differentiated here the correlation coefficient of $\Delta G^{net}/RT$ versus $\ln Q°b$ for the entire group of nineteen isomers ($r = 0.89(7)$) is not statistically different from that obtained for MW versus $\ln Q°b$. Excluding the fragmented isomers, the correlation of $\Delta G^{net}/RT$ versus $\ln Q°b$ is better than that of MW, at a greater than 70% confidence limit.

TABLE 3 - SOLUTE PHYSICAL PROPERTIES

	Alkyl Phenols	Molecular Weight MW -	Density[a] ρ g-cm^{-3}	Index of Refraction[a] n_D -	Accentric Factor[b] ω -	Surface Area[c] TSA Å^2	Volume[c] ν Å^3	$\Delta G^{net}/RT$
1.	phenol	94.1	1.0767	1.5496	0.257	249.1	35.8	18.03
2.	2-methyl phenol	108.1	1.0460	1.5442	0.314	274.3	379.4	27.36
3.	4-ethyl phenol	122.2	1.054	1.5239	0.425[d]	306.3	425.4	28.61
4.	2,6-dimethyl phenol	122.2	1.132	1.542[d]	0.39	296.3	426.0	42.59
5.	2,3-dimethyl phenol	122.2	1.164	1.542	0.39	295.3	425.2	42.57
6.	3,4-dimethyl phenol	122.2	1.138	1.542[d]	0.39	297.9	428.0	42.69
7.	3,5-dimethyl phenol	122.2	1.115	1.542[d]	0.398	306.1	442.5	45.00
8.	2,5-dimethyl phenol	122.2	1.189	1.542[d]	0.39	302.8	434.0	42.74
9.	4-n propyl phenol	136.2	1.009	1.5379	0.425[d]	336.8	473.0	51.37
10.	4-isopropyl phenol	136.2	0.990	1.5228	0.425[d]	321.7	471.6	50.08
11.	2,3,5 trimethyl phenol	136.2	0.980[d]	1.52[d]	0.400[d]	323.5	474.8	44.83
12.	2,3,6 trimethyl phenol	136.2	0.980[d]	1.52[d]	0.400[d]	317.3	471.7	44.83
13.	2-n butyl phenol	150.2	0.975	1.5182	0.500[d]	363.1	537.3	70.59
14.	4-n butyl phenol	150.2	0.974	1.5170	0.500[d]	366.9	520.9	68.86
15.	4-tert butyl phenol	150.2	0.908	1.504	0.475[d]	341.5	505.4	60.63
16.	2-n pentyl phenol	164.3	0.950[d]	1.5[d]	0.520[d]	394.0	588.4	76.35
17.	4-n pentyl phenol	164.3	0.950[d]	1.5[d]	0.520[d]	397.4	568.5	74.52
18.	4-tert pentyl phenol	164.3	0.950[d]	1.5[d]	0.510[d]	362.2	539.0	70.51
19.	2-n hexyl phenol	178.3	0.950[d]	1.5[d]	0.540[d]	424.9	639.5	85.30

FOOTNOTES

a. From (Manufacturing Chemists Association (12)
b. For a non-polar homologue, from (Reid, Prausnitz, and Sherwood (13)
c. Calculated from Merk, Sharp & Dohme computer program (11)
d. Estimated

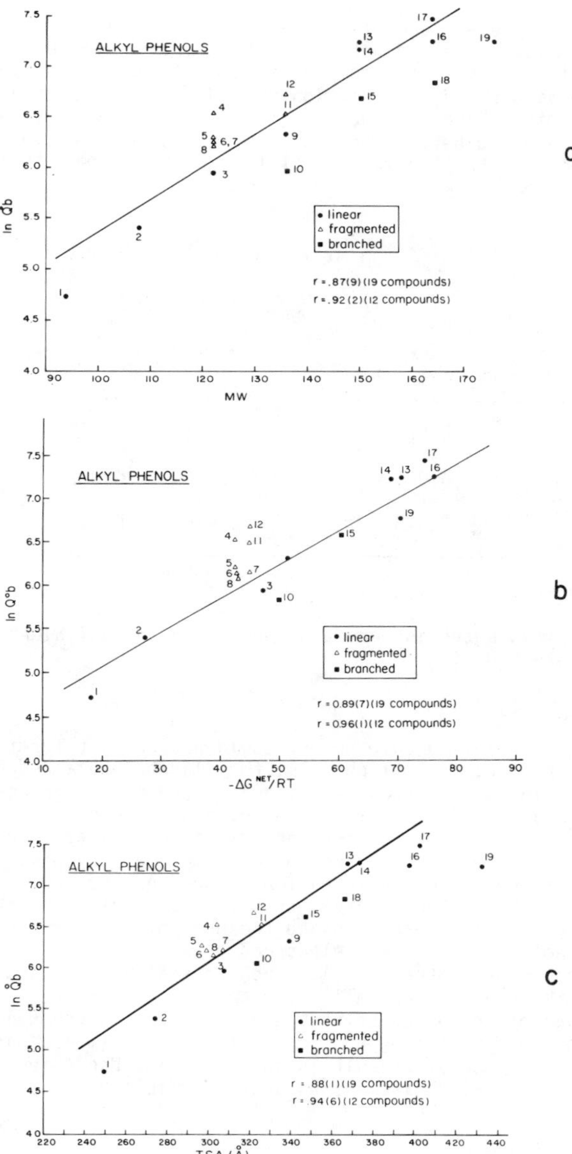

Figure 4. lnQ°b AS A FUNCTION OF MOLECULAR WEIGHT, $\Delta G^{net}/RT$, AND CAVITY SURFACE AREA FOR ALKYL PHENOLS.

For the complete group of solutes the correlations for $\Delta G^{net}/RT$, TSA, or MW versus $\ln Q°b$ did not yield statistically different correlation coefficients. For the group of 12 non-fragmented solutes, $\Delta G^{net}/RT$ yields a better correlation than MW at greater than the 70% confidence limit. Finally, for the group of eight solutes, both TSA and $\Delta G^{net}/RT$ yielded better correlations than MW at greater than the 70% confidence limit. The results of the correlations with $\ln Q°b$ are presented in Table 4.

TABLE 4. STATISTICS - MODEL COMPARISON

No. of Isomers	$\Delta G^{net}/RT$	r-values TSA	r-values MW	$r_{G} \neq r_{MW}$
n	-	Å^2	-	
19 (all)	0.897	0.881	0.879	< 50%
12[a]	0.961	0.946	0.922	> 70%
8[b]	0.976	0.976	0.939	> 70%

[a] Excluding fragmented isomers.

[b] Excluding fragmented isomers and those with an equivalent radius greater than 5.0 Å.

Although the solvophobic theory could not account for the adsorption of the fragmented alkyl phenols, they may be explained qualitatively by shielding of the phenol-OH group by the methyl groups distributed nearby. This would serve to reduce hydrogen bonding with the surrounding water, decreasing the solute solubility, and hence increasing the adsorption. A second and complementary possibility has to do with the geometric configuration of the sorbate during adsorption. With one substituent alkyl group, it is likely that the non-polar alkyl group is oriented toward the non-polar adsorbent surface with the phenol ring and its π-electrons oriented toward the polar solvent (i.e., end-on attachment). When several alkyl substituents are attached to the phenol ring, all the non-polar groups are "squeezed-out" of the polar solvent and drawn toward the non-polar sorbent causing the phenol ring to "hug" this surface (i.e., parallel attachment). This should result in an increased adsorption capacity for the fragmented alkyl isomers as shown in Table 2 and Figure 4.

CONCLUSIONS

The results presented here are part of a larger study to develop and test the solvophobic theory ($c\phi$) with a statistically valid database.

In general, the results indicate that the comprehensive $c\phi$-theory is able to account for the adsorption of linear and branched alkyl phenols and that the simplified version of the $c\phi$-theory provides a useful physical interpretation of hydrophobic adsorption. However, for a complete account of the aqueous phase adsorption of alkyl phenols, an extension of the $c\phi$-theory to account for the adsorption of fragmented alkyl phenols is necessary.

The ability to predict the effects of even simple structural modifications on the adsorption of organic molecules from dilute aqueous solutions could be of great value in the design and operation of large scale commercial water and wastewater treatment plants.

In addition, the solvophobic theory extends beyond that of a purely predictive tool. It is significant in that it provides a rational basis for considering specific contributions to aqueous phase adsorption, leading to an improved understanding of the entire process.

NOTATION

A	Surface area, A^2
b	Langmuir isotherm parameter
C	Solute concentration, mmol/l
D	Electrostatic variable, Eq. (6)
G	Gibbs free energy, kcal/mol
I	Ionization potental, eV
K	Reaction equilibrium constant
MW	Molecular weight
N	Avogadro's constant, mol^{-1}
P	Electrostatic variable, Eq. (6)
P	Pressure, kPa
Q	Langmuir isotherm loading, mmole/g
Q°	Langmuir isotherm saturated loading, mmole/g
R	Universal gas constant

T	Temperature, K
V	Molar volume, cm^3/mol

Greek

γ	Interfacial tension, erg/cm^2
κ^e	Energy correction for curved surface, Eq. (6)
λ	Complex/solute volume ratio
μ	Dipole moment, Debyes
ν	Molecular volume, $cm^3/molecule$
ρ	Density, g/cm^3
ω	Accentric factor

Superscripts

assoc	Association reaction, Eq. (1)
net	Sum of all interactions

Subscripts

cav	Cavity interaction
C	Carbon
es	Electrostatic interaction
gas	Gas phase reaction
i	i^{th} solute
S_i	Solute
S_iC	Carbon-solute complex
solvent	Solvent phase reaction
vdw	van der Waals interaction
w	Water

LITERATURE CITED

1. Altshuler, G.L., and G. Belfort, "Selective Adsorption of Organic Homologues onto Activated Carbon from Dilute Aqueous Solutions. Solvophobic Interaction Approach-Branching and Predictions" in "Treatment of Water by Granular Activated Carbon" Advances in Chemistry Series No. 202, McGuire, M.J. and I.H. Suffet, ed., Am. Chem. Soc., Washington, D.C., 1983.

2. Belfort, G., G.L. Altshuler, K.K. Thallum, C.P. Feerick, Jr., and K.L. Woodfield, "Selective Adsorption of Organic Homologues onto Activated Carbon from Dilute Aqueous Solution: Solvophobic Interaction Approach-The Effect of Simple Structural Modifications with Aliphatics", Submitted to AIChE J. (1983).

3. McBain, J.W., "The Sorption of Gases and Vapors by Solids", George Routledge & Sons, Ltd., London (1932).

4. Neuhaus, D.M., and G. Belfort, "Adsorption of Alcohols onto Activated Carbon: From Vapor and Aqueous Solution", Submitted for publication (1983).

5. Sinanoglu, O., "Solvent Effects on Molecular Associations" in Molecular Associations in Biology, B. Pullman, ed., Academic Press, NY, 427 (1968).

6. Halicioglu, J. and O. Sinanoglu, "Solvent Effects on Cis-Trans Azobenzene Isomerization", Ann. N.Y. Acad. Sci., 158, 308 (1969).

7. Horvath, C., W. Melander, and I. Molnar, "Solvophobic Interactions in Liquid Chromatography with Nonpolar Stationary Phases", J. Chrom., 125, 129 (1976).

8. Zogorski, J. and S.D. Faust, "The Effect of Phosphate Buffer on the Adsorption of 2,4-dichlorophenol, and 2,4-dinitrophenol Onto Activated Carbon", J. Environ. Sci. Health, A11, 501 (1976).

9. Dean, J.A., ed., "Lange's Handbook of Chemistry", 12th Edition, McGraw-Hill, NY (1979).

10. Sawyer, C.N., and P.L. McCarty, "Chemistry for Environmental Engineers", 3rd Ed., McGraw-Hill, NY (1978).

11. Smith, G., Merck-Sharp and Dohme, personal communication (1981).

12. Manufacturing Chemists Association Research Project, Chemical Thermodynamics Properties Center, "Selected Values of Properties of Chemical Compounds", Texas A&M University, College Station, TX (1966).

13. Reid, R.C., J.M. Prausnitz, and T.K. Sherwood, "The Properties of Gases and Liquids", McGraw-Hill, NY (1977).

ADSORPTION RATE OF ORGANICS FROM AQUEOUS SOLUTIONS ONTO GRANULAR ACTIVATED CARBON

W.A. Beverloo, G.M. Pierik, K.Ch.A.M. Luyben

Agricultural University,
Department of Food Science,
Process Engineering Group,
De Dreijen 12,
6703 BC WAGENINGEN,
The Netherlands

ABSTRACT

The described research is aimed at the provision of sound design data for activated carbon adsorbers as applied in water treatment.
Finite bath experiment results, processed by means of a modified 'shrinking core' kinetic model reveal that the dominant transport phenomenon within the adsorbent granules is surface diffusion. Adsorption rate is at least 60 times higher than could be ascribed to mere pore diffusion. The rate controlling surface diffusivity drops to extremely low values for moderately low adsorbent loads.

INTRODUCTION

Increasing amounts of drinking water obeying to more and more demanding quality standards will have to be prepared from polluted water sources. To protect these surface water sources as well as possible, increasingly severe control regulations are imposed on industrial effluents discharged into the surface waters. Adsorption onto activated carbon is necessary for the final removal of dissolved pollutants escaping all preceding biological and physicochemical treatment steps. It is applied in the preparation of drinking water as well as in the obligatory purification of industrial effluents. The described research should contribute to better understanding of the activated carbon adsorption process needed for a sound design of adsorbers.

The at decreasing rate approached and never reached end point of the adsorption process is equilibrium between adsorbate load on the adsorbent and solute concentration in the solution. It would be impossible to leave out consideration of the equilibrium relation, described by means of 'sorption isotherms', from any work on adsorption. Therefore we include a review of some equilibrium correlations that combine convenience with a more or less credible description of measured equilibria.

The main question to be answered by the adsorption process designer is, how much contact time of solution and adsorbent is to be provided for in order to approach equilibrium to a predetermined extent. In other words the designer needs insight in the rate of the adsorption process. Therefore we constructed a series of more or less sophisticated process rate models and determined model parameters for optimal fit to results of 'finite bath' experiments. A 'finite bath' experiment is one in which a known amount of solution - the 'finite bath' -at zero time is suddenly brought into contact with a known amount of virgin activated carbon. The adsorption rate is monitored by continuous or repeated determination of the bath concentration. This is a relatively simple experiment needing a relatively complex data processing program. An 'infinite bath' experiment could be simulated by controlled and measured supply of solute to maintain a constant concentration. This more complicated experiment would produce more easily interpretable results.

SORPTION EQUILIBRIA

Sorption equilibria for single solutes in aqueous solution and adsorbates on activated carbon might conveniently be expressed as a relation between q in [mol/kg carbon] or in [kg/kg carbon] and c in $[mol/m^3]$ or in $[kg/m^3]$. Experimental relations can be classified to the number of parameters in it. Real sorption equilibrium relations will predictably have three features. The first that for low concentrations the relation asymptotically approaches proportionality, the second that the adsorbent load will asymptotically approach a maximum 'saturation value' for very high solution concentrations and the third that the adsorption load has no maximum value in the range of positive concentration values. In fact we consider the presence of the three features as a credibility test for equilibrium correlation formulae. In Table 1 a number of correlation formulae are given classified by the number of parameters. The results of the two asymptote credibility tests are indicated. The third credibility feature is eventually secured by limiting the value range of parameters. Because of their physically realistic behaviour at very low and very high values of c, the exponential, Langmuir-, Volmer- and Toth-equations are to be recommended.

Table 1 Adsorption equilibrium equations (Ref. 2,3,4)

Equation	Name	Parameters	Credibility Low c	Credibility High c
$q = Hc$	Henry	1	+	-
$q = q^*$ for $c > 0$; $c = 0$ for $q < q^*$	saturation	1	-	+
$q = A(c + \frac{1}{B})$	linear	2	-	-
$q = \frac{A}{B}\{1 - \exp(-Bc)\}$	exponential	2	+	+
$q = \frac{Ac}{1 + Bc}$	Langmuir	2	+	+
$c = \frac{q}{A - Bq} \exp\left(\frac{Bq}{A - Bq}\right)$	Volmer	2	+	+
$c = \frac{q}{A} \exp\left(\frac{Bq}{A}\right)$	Myers (reduced)	2	+	-
$q = A_f\, c^n$	Freundlich	2	-	-
$q = \left\{\frac{(Ac)^n}{1 + (Bc)^n}\right\}^{1/n}$	Toth	3	+	+
$c = \frac{q}{A} \exp\left(\frac{Bq}{A}\right)^n$	Myers	3	+	-
$c = \frac{q}{A - Bq} \exp\left(\frac{Dq}{A - Bq}\right)$	Volmer (extended)	3	+	+
$q = \frac{Ac}{1 + Bc}\{1 - \exp(-Dc)\}$	Langmuir (extended)	3	+	+
$q = \frac{Ac}{(1 + Bc)(1 - Dc)}$	B.E.T.	3	+	-
$q = \frac{A(Bc)^m}{B\{1 + (Bc)^n\}}$ $(0 < n < m)$	Fritz--Schluender	4	-	-
$q = \frac{Ac(1 + Dc)}{(1+Bc)(1+Ec)}$ $(B > D > E)$	Langmuir (extended)	4	+	+

ADSORPTION RATE MODELS

Transport to the adsorbent particles

Adsorbable matter is transported from the bulk of the solution to the outer boundary of the adsorbent particles, driven by a concentration difference:

$$\phi_A = k_l a(c_{bA} - c_{iA}) = h \frac{d\bar{c}_A}{dt} \qquad (1)$$

where ϕ_A is the adsorption rate (in [kg/m³s] or [mol/m³s]).

k_l is the liquid-side mass transfer coefficient (in [m/s]).

a is the specific outer particle surface (in [m^{-1}]).

c_{bA} is the bulk solution concentration of "A" (in [kg/m³] or [mol/m³]).

c_{iA} is the solution concentration of "A" at the outer particle boundary (in [kg/m³] or [mol/m³]).

$\bar{c}_A$ is the average concentration of "A" in the particles (in [kg/m³] or in [mol/m³]).

t is time in [s].

h is particle fraction of total volume.

Transport within the adsorbent particles

Within the adsorbent particles the adsorbate molecules are transported from the outer boundary inward to the unoccupied adsorption places on the internal surface by diffusion. The most simple model for the internal diffusion is 'Fick' diffusion with constant 'overall' diffusivity (ref. 5).

$$\phi''_A = -D_{oA} \nabla c_A \qquad (2)$$

where ϕ''_A is the flux of adsorbate "A" in [kg/m²s] or in [mol/m²s], D_{oA} is the 'overall' diffusivity of "A" in [m²/s], c_A is the local concentration of "A" in [kg/m³] or [mol/m³] and ∇ is the "gradient" operator in [m^{-1}].

In case of spherical symmetry Equation (2) becomes:

$$\phi_A = -4\pi D_{oA} r^2 \frac{dc_A}{dr} = 4\pi D_{oA} \frac{dc_A}{d(1/r)} \qquad (3)$$

where ϕ_A is the flux of adsorbate "A" in [kg/s] or in [mol/s] through a spherical surface with radius r in [m] around the centre of spherical symmetry.

The local accumulation of adsorbate 'A' is equal to the negative divergence of the adsorbate flux:

$$\frac{\delta c_A}{\delta t} = -\nabla.\phi''_A = \nabla.(D_{oA} \nabla c_A) = D_{oA} \nabla^2 c_A \qquad (4)$$

where $\frac{\delta c_A}{\delta t}$ is the rate of local accumulation of 'A' in [kg/m³s] or in [mol/m³s], $\nabla.$ is the divergence operator in [m^{-1}] and ∇^2 is the Laplace operator in [m^{-2}].

The last step in Equation (4) is valid only in case the overall diffusivity D_{oA} is really a constant.

In case of spherical symmetry Equation (4) becomes:

$$\frac{\delta c_A}{\delta t} = D_{oA} \{\frac{1}{r^2} \frac{\delta}{\delta r} (r^2 \frac{\delta c_A}{\delta r})\} = D_{oA} \{\frac{2}{r} \frac{\delta c_A}{\delta r} + \frac{\delta^2 c_A}{\delta r^2}\} = \frac{D_{oA}}{r} \frac{\delta^2 (c_A r)}{\delta r^2} \tag{5}$$

A more realistic model for the intraparticle diffusion is described earlier (1). It states two parallel diffusion phenomena. The first is 'Fick' diffusion through tortuous solution-filled pores. The second is 'Fick' diffusion across the internal adsorbent surface. Instantaneous local equilibrium between pore solution concentration and adsorbent surface load is presumed in the model

$$\phi''_A = - D_{app} \nabla C_A \tag{6}$$

where $c_A = \varepsilon c_{pA} + (1 - \varepsilon) \rho_c q_A$

$$\text{and } D_{app} = \frac{\frac{\varepsilon D_{pA}}{\tau_p} + \frac{(1 - \varepsilon) \rho_c D_{sA}}{\tau_s} \frac{dq_A}{dc_{pA}}}{\varepsilon + (1 - \varepsilon)\rho_c \frac{dq_A}{dc_{pA}}}$$

ε is the pore fraction of the particle volume

$D_{p,A}$ is the pore diffusivity of "A" in $[m^2/s]$

τ_p is the pore tortuosity

ρ_c is the density of the solid fraction of the particle in $[kg/m^3]$

$D_{s,A}$ is the surface diffusivity in $[m^2/s]$

τ_s is the surface tortuosity

q_A is the local adsorbate load in [kg A/kg] or in [mol A/kg]

$c_{p,A}$ is the total pore solution concentration in $[kg\ A/m^3]$ or in $[mol\ A/m^3]$

q_A and $c_{p,A}$ are related by the equilibrium relation.

Superficially Equations 2 and 6 are identic, the essential difference is that D_{app} in Equation 6 is not a constant:

$$\frac{d\, Da_pp}{dc_A} = \frac{\varepsilon(1-\varepsilon)\rho_c \{\frac{D_{s,A}}{\tau_s} - \frac{D_{p,A}}{\tau_p}\}}{\{\varepsilon + (1-\varepsilon)\rho \frac{dq_A}{dc_{p,A}}\}^2} \cdot \frac{d^2 q_A}{dc^2_{p,A}} \tag{7}$$

It is to be emphasized, that for an absorbent like activated carbon and an adsorbate like nitrobenzene $\rho_c \frac{dq_A}{dc_{pA}}$ is in the order of 10^4 to 10^5. Surface diffusion will contribute substantially to total diffusion for rather low values of $D_{s,A}$ already. Eventually rate experiments indicating concentration dependence of the overall diffusivity cannot be sufficiently explained by Equation 7 and leave no other conclusion than load dependent surface diffusion.

'Shrinking core' model

In the 'shrinking core' adsorption rate model the equilibrium relation is the 'saturation' equation. The surface diffusivity is considered to be infinitesimally small. The contribution of the pore solution concentration $c_{p,a}$ to the total concentration c_A is considered to be so small that accumulation of "A" in the pore solution has a negligible effect on the radial total concentration profile. The absorbent particles are considered to be spherical. The result of the model presumptions is that a spherical 'shrinking core' of diminishing radius r_f is free of adsorbate and that adsorbable material 'A' is transported by pore diffusion along a quasi-stationary concentration gradient through the saturated outer shell of the adsorbent particle to the progressing 'adsorption front' to be there absorbed. The resulting differential equation is:

$$\frac{d\bar{q}}{dt} = \frac{12\,\varepsilon\, D_{p,A}\, c_{iA}}{(1-\varepsilon)\,\tau_p \rho_c\, d_p^2} \cdot \frac{\sqrt{1-\bar{q}/q^*}}{1-\sqrt{1-\bar{q}/q^*}} \tag{8}$$

Where $\bar{q}$ is the average adsorbent load, d_p is the particle diameter and c_{iA} the pore solution concentration at the outer particle boundary.

The difficulty with application of the 'shrinking core' model is the choice of q^* from some real sorption equilibrium relation. The best choice is to take q^* at the adsorbent load level of the asymptotically approached end point of the experiment.

Modified shrinking core model

For experiments with changing bath concentration, like 'finite-bath' experiments, the shrinking core model was modified (ref. 6). The modification implied a periodical shift to a new value of q^* in the numerical simulation of the adsorption process. In fact q^* was subsequently adapted to values in equilibrium with the surrounding bath, following the real equilibrium relation.

Analytical solutions for 'infinite-bath' experiments

The result of Equation 1 for an 'infinite-bath' experiment with infinitely fast intraparticle diffusion ($\bar{q}_A$ in equilibrium with c_{iA}) reads:

$$\frac{k_\ell\, a}{h\, P_c\, (1-\varepsilon)}\, dt = \frac{d\bar{q}_A}{c_{b,A} - c_{i,A}} \tag{9}$$

The solution for the Langmuir equilibrium relation reads:

$$\frac{k_\ell\, at}{A\, h\, P_c\, (1-\varepsilon)} = \frac{1}{1 + Bc_{bA}} \left[\frac{B\bar{q}_A}{A} - \frac{1}{1+Bc_{bA}} \ln\left\{1 - \frac{(1+Bc_{bA})\,\bar{q}_A}{Ac_{bA}}\right\}\right] \tag{10}$$

The solution of Equation 5 for an 'infinite bath' experiment with infinitely high values of k_1 reads:

$$\bar{q}_A = \frac{24}{d_p^3} \int_0^{d_p/2} q_A\, r^2\, dr = q_{A,\infty} \left[1 - \frac{6}{\pi^2} \sum_{n=1}^{\infty} \frac{1}{n^2} \exp\left\{-(2\pi n)^2 \frac{D_{o,A}\, t}{d_p^{\,2}}\right\}\right] \tag{11}$$

where $\bar{q}_A$ is the average value of q_A and $q_{A,\infty}$ is the asymptotically approached end value of $\bar{q}_A$.

The solution of Equation 8 for an 'infinite bath' experiment with infinitely high value of k_l reads:

$$\bar{q}/_{q*} = 4\sqrt{\frac{3\,\varepsilon\, D_{p,A}\, c_{i,A}\, t}{(1-\varepsilon)\,\tau_p\,\rho_c\, d_p^{\,2}\, q*}}\left(1-\sqrt{\frac{3\,\varepsilon\, D_{p,A}\, c_{i,A}\, t}{(1-\varepsilon)\,\tau_p\,\rho_c\, d_p^{\,2}\, q*}}\right) \quad (12)$$

Note that Equation 12 predicts completion of the adsorption process within a finite time. This is a feature of the 'shrinking-core' model, indicating its intrinsic irreality.

Analytical solutions for finite bath experiments

The solution of equation (1) for a 'finite-bath' experiment and Langmuir equilibrium relation between $\bar{q}$ and $C_{i,A}$ reads:

$$y = \tfrac{1}{2}\Big[\ln\Big\{\frac{\alpha^2+(\alpha+\beta-1)z-1}{2\alpha+\beta-2}\Big\} + \frac{\alpha-3\beta+3}{\sqrt{(\alpha+1)^2+2\beta(\alpha-1)+\beta^2}}$$

$$\ln\Big\{\frac{(3\alpha+\beta-1)z+\alpha+\beta-3+(z-1)\sqrt{(\alpha+1)^2+2\beta(\alpha-1)+\beta^2}}{(3\alpha+\beta-1)z+\alpha+\beta-3-(z-1)\sqrt{(\alpha+1)^2+2\beta(\alpha-1)+\beta^2}}\Big\}\Big]$$

$$\text{where } y = \frac{k_\ell a t}{1-h}\;;\; z = \frac{C_{b,A}}{C_{o,A}} \quad (z=1 \text{ for } y=0)$$

$$\alpha = (Bc_{o,A})^{-1} \quad ; \quad \beta = \frac{A\, h\, \rho_c\,(1-\varepsilon)}{Bc_{oA}\,(1-h)} \quad (13)$$

The solution of Equation 8 for a 'finite-bath' experiment with infinitely high value of k_l (Ref. 1) reads:

$$\frac{12\, h\, \varepsilon\, D_p t}{(1-h)\,\tau_p\, d_p^{\,2}} = \Big\{1-\tfrac{1}{2}\Big(\frac{\beta}{1-\beta}\Big)^{1/3}\Big\}\ln\frac{c_{b,A}}{c_{o,A}} +$$

$$+\Big(\frac{\beta}{1-\beta}\Big)^{1/3}\Big[\tfrac{3}{2}\ln\frac{(1-\beta)^{1/3}+\big(\beta-1+\frac{c_{b,A}}{c_{oA}}\big)^{1/3}}{(1-\beta)^{1/3}+\beta^{1/3}} +$$

$$+\sqrt{3}\Big\{\arctan\frac{2\beta^{1/3}-(1-\beta)^{1/3}}{\sqrt{3}\,(1-\beta)^{1/3}} - \arctan\frac{2(\beta-1+c_{oA})^{1/3}-(1-\beta)^{1/3}}{\sqrt{3}\,(1-\beta)^{1/3}}\Big\}\Big] \quad (14)$$

$$\text{with } \beta = \frac{q*\, h\, \rho_c\,(1-\varepsilon)}{c_{oA}\,(1-h)}$$

for $\beta = 1$ Equation 14 reduces to

$$\frac{12\ h\ \varepsilon\ D_p\ t}{(1-h)\ \tau_p\ d_p^2} = \ln \frac{c_{b,A}}{c_{o,A}} + 3\ \frac{c_{o,A}^{1/3} - c_{b,A}^{1/3}}{c_{b,A}^{1/3}} \tag{15}$$

for $\beta < 1$ the process ends at

$$\frac{c_{b,A}}{c_{o,A}} = 1-\beta \quad \text{in finite time:}$$

$$\frac{12\ h\ \varepsilon\ D_p\ t}{(1-h)\ \tau_p\ d_p^2} = \ln(1-\beta) - \tfrac{3}{2}\left(\frac{\beta}{1-\beta}\right)^{1/3} \ln\{(1-\beta)^{1/3} + \beta^{1/3}\} +$$

$$+ \sqrt{3}\left(\frac{\beta}{1-\beta}\right)^{1/3} \left\{ \arctan \frac{2\beta^{1/3} - (1-\beta)^{1/3}}{\sqrt{3}\ (1-\beta)^{1/3}} - \arctan \frac{-1}{\sqrt{3}} \right\} \tag{16}$$

Again the possible ending in finite time is a feature of the intrinsically irrealistic 'shrinking core' model.

Allowance for finite mass transfer rate from the bath solution to the outer boundary of the particles can be made in an interesting way. The mass transfer boundary layer around the absorbent particle has the same resistance as would have an additional layer of saturated adsorbent material. The virtual additional particle diameter is added to d_p in Equations 14 through 16. The virtual time needed to saturate the virtual outer layer and the virual additional bath concentration needed to provide for the virtual amount of adsorbate to be stored in the virtual outer layer are to be added to t and c_{oA} respectively in Equations 14 through 16 (Ref. 1).

Numerical solutions

Analytical solutions for finite bath experiments are not available for cases even slightly more complex than the few preceding examples. Especially the 'modified shrinking core' model demands numerical solution. It works in this way that Equation 14 is applied with $q^* = q(c_{o,A})_{eq.}$ the value in equilibrium with the initial bath concentration. After a predetermined reduction of $c_{o,A}$ to $c_{1,A}$ is reached, the value of q^* is changed to that in equilibrium with $c_{1,A}$. The zero point of time is corrected, because with the new value of q^* the process would have been slower and Equation 16 is applied again untill the next allowed reduction of $c_{1,A}$ to $c_{2,A}$. And so on (ref. 6).

Experiments

The experimental equipment for the 'finite bath' experiments is schematically shown in Figures 1 and 2. A weighed amount of activated carbon is encaged between two sieves as indicated. The entire circulation flow generated by a propeller pump is passed through the cage with activated carbon. To start the process, the vessel is closed and a measured amount of solution with known concentration is introduced. A minor sample flow is led through a U V spectro-

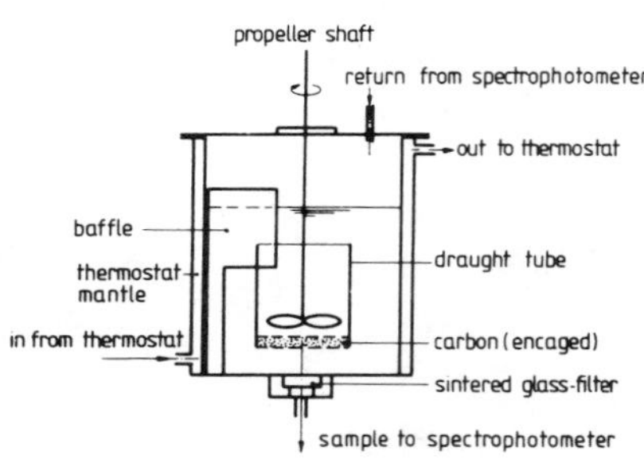

Fig. 1 Vessel for 'finite bath' adsorption experiments

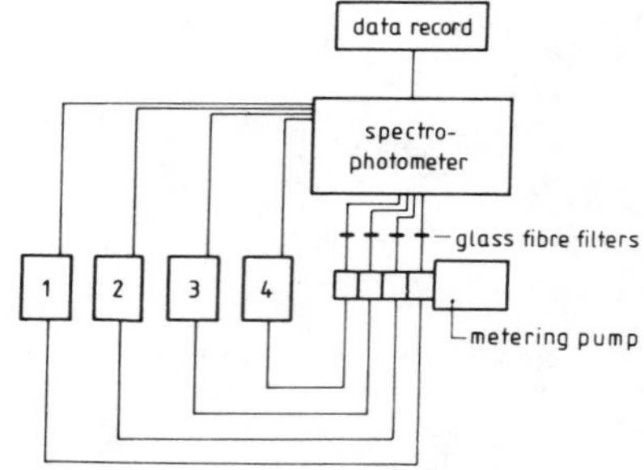

Fig. 2 Flow scheme of fourfold adsorption measuring equipment

photometer cuvette and returned. Temperature is controlled by water circulation between the mantle of the vessel and the thermostat. Four experiments can be made simultaneously.

Experiments were made with nitrobenzene and 2-4-dichlorophenoxy acetic acid (2-4-D) as adsorbates and a granular activated carbon of the type 'Norit ROW 0.8 Supra' as adsorbent. Experiments were made at 20°C and 70°C. In Figures 3 and 4 areas are shown in which all 'operating lines' of the experiments lay. Data on the used activated carbon are given in Table 2.

Experimental data are automatically recorded in digital form on papertape. A numerical minimum searching method was used to find the combination of k_1 and D_p/τ_p resulting in a minimum value of the sum of squares for the differences between measured concentrations and those calculated for one of the theoretical models.

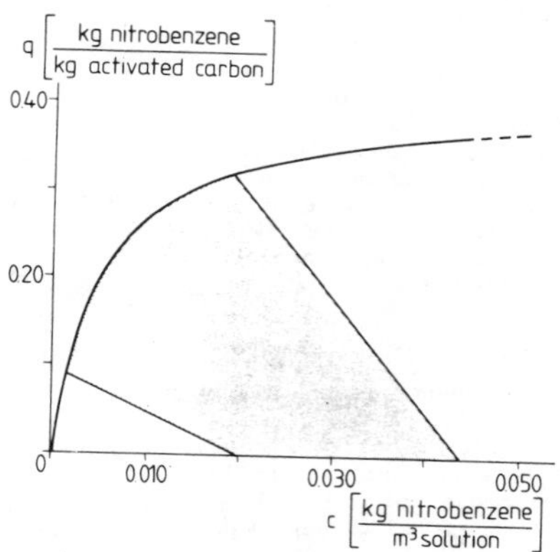

Fig. 3 Concentration-load domain of 'finite-bath' experiments with nitrobenzene

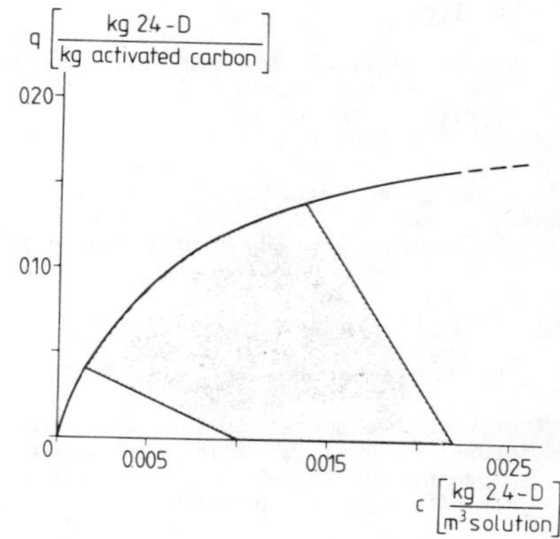

Fig. 4 Concentration-load domain of finite-bath experiments with 2, 4-D

Table 2 Norit ROW 0,8 Supra Activated Carbon

Density in liquid mercury, $\rho_c(1-\varepsilon)$ =	590	kg/m³
Density in helium, ρ_c =	2100	kg/m³
Porosity ε =	0,719	m³/m³
Cilinder diameter	0.8	mm
Cilinder length	1,6	mm

Results

For nitrobenzene at 20°C the results varied around:

$$k_1 = 1.5 \times 10^{-4} \text{ [m/s] and } D_p/\tau_p = 3.0 \times 10^{-8} \text{ [m}^2\text{/s]}$$

and for 2,4-dichlorophenoxy acetic acid on the same carbon at 20°C the results varied around:

$$k_1 = 1.0 \times 10^{-4} \text{ [m/s] and } D_p/\tau_p = 0.4 \times 10^{-8} \text{ [m}^2\text{/s]}$$

both for the 'modified shrinking core' model.
This implies that D_{app} to be applied in eq. 6 would vary from 5×10^{-13} [m²/s] in the initial phase of the process to 2×10^{-12} [m²/s] in the last phase for nitrobenzene. For 2.4.D the apparent diffusivity in eq. 6 would increase during the process from 1.5×10^{-13} [m²/s] to 2×10^{-12} [m²/s].

Discussion

The found mass transfer coefficients have no general value. They are necessarily determined to elimate their influence on the diffusivity value.

The diffusivity of nitrobenzene in water can be estimated by the Wilke and Chang relation (Ref. 5):

$$D_{NB\text{-}H_2O} = K \frac{(\Psi_{H_2O} \cdot M_{H_2O})^{\frac{1}{2}} \rho_{o,NB}^{\ 0,6} \cdot T}{\mu \, M_{NB}^{\ 0,6}} = 4,24 \times 10^{-10} \text{ [m}^2\text{/s]}$$

where $D_{NB\text{-}H_2O}$ is the diffusivity of nitrobenzene in water in [m²/s]

Ψ_{H_2O} is the association parameter for water = 2,6.

M_{H_2O} is the molecular weight of water = 0,018 [kg/mol].

$\rho_{o,NB}$ is the density of pure liquid nitrobenzene = 1204 [kg/m³].

T is the absolute temperature = 293 [K].

μ is the viscosity of the solution = 10^{-3} [kg/m³].

M_{NB} is the molecular weight of nitrobenzene = 0,133 [kg/mol

K is a dimensional factor = $5,88 \times 10^{-17}$ $\left[\dfrac{m^{2,8}\ kg^{0,5}}{mol^{0,1}\ s^2\ K}\right]$

For 2-4-dichlorophenoxy acetic acid the molecular weight is $M_{2\text{-}4\text{-}D}$ = 0,221 [kg/mol], the density of pure liquid is unknown but might be estimated to be 1500 [kg/m³]. The diffusivity would so be:

$$D_{2\text{-}4\text{-}D\text{-}H_2O} = 5{,}88\times10^{-17} \frac{(2{,}6 \times 0{,}018)^{\frac{1}{2}} \times 1500^{0,6} \times 293}{10^{-3} \times 0{,}221^{0,6}} = 7{,}42\times10^{-10}\ [m^2/s]$$

Obviously the values found for the pore diffusivity are orders of magnitude too high, the more so if for the pore diffusivity is taken its probable value of $\tau_p = 6$. This means that pore diffusion does not considerably contribute to the transport within the particles. The mechanism controlling the intraparticle transport appears to be surface diffusion. However if surface diffusion is the main transport mechanism, concentration dependence of D_{app} could not be explained by Equation 7. It was found, that the experiments give more constant values of the 'apparent pore diffusivity' with the 'modified shrinking core' model than of the 'overall diffusivity' D_{OA} of Equations 2 and 3 with the Fick' diffusion model. The only explanation is that the surface diffusivity controlling the intraparticle transport is severely concentration dependent. The 'modified shrinking core' model is an extreme case of a model with a constant overall diffusivity dropping suddenly to zero for some finite value of $c_A = c_{AZ}$. For $c_{AZ} = 0$ the model reduces to mere 'Fick' diffusion.

For $c_{AZ} = \dfrac{q^* \quad C^{\infty}_{bA}}{\rho_c \quad (1-\varepsilon)}$ with $q^*\ (c^{\infty}_{bA})$ = the value of q in equilibrium with the bath concentration $c_{bA} = c^{\infty}_{bA}$ at the end the process, the model reduces to the 'shrinking core' model. Obviously the overall diffusivity behaves approximately in the last way.

Possible extension of the research project could be to find the combination of k_1, D_{AO} and c_{AZ} for optimal description of adsorption experiments.

Another extension could be to complicate the 'finite bath' experiments by automatic control of the bath concentration at a constant value, so obtaining the equivalent of 'infinite bath' experiments. These are easier to calculate. In view of application to column experiments the bath concentration could also be controlled to follow a straight 'operating line' describing coinciding values of $\bar{q}$ and C_{bA} in a 'constant pattern' adsorption column process.

References

1 W.A. Beverloo & S. Bruin, Kinetics of activated carbon adsorption. Ch.19 in G. Mattock (ed.) New Processes of waste water treatment and recovery SCI London (1978)

2 J. Langmuir, J.Am.Chem.Soc. 40 1361-1403 (1981)

3 H. Freundlich, Colloid and Capillary Chemistry, London (1926)

4 L. Jossens, J.M. Prausnitz, W. Fritz, E.U. Schluender & A.L. Myers, Chem. Eng. Sci. 33, 1097-1106 (1978)

5 R.B. Bird, W.E. Stewart & E.N. Lightfoot, Transport phenomena, Wiley International Edition

6 K.Ch.A.M. Luyben, Non-linear diffusion in sorption and desorption processes. Diss. Wageningen (1983) (in press)

ENTHALPY OF ADSORPTION ON SOLIDS FROM SOLUTION

K.S.Birdi

Fysisk-Kemisk Institut, Technical University,
Building 206, DK-2800 Lyngby, Denmark.

ABSTRACT: The enthalpy of adsorption, h_{ads}, of solutes on solids from solution is of much importance in various biological and industrial processes. The thermodynamic aspect of this reaction has been analyzed by various investigators in recent years. The present study reports the various data as found in literature and measured in our own laboratory, which allows us to describe the various forces which are involved in the adsorption process. It will be shown that the magnitude of h_{ads} is dependent on the solid, adsorbate and the solvent.

SCOPE: The adsorption phenomena at a solid-liquid interface is of much interest in many biological (evolution, cell surface, transplants) and industrial (chromatography, soil surface, enhanced oil recovery, pollution) processes. In spite of this observation, one finds that the current understanding of these adsorption phenomena is not satisfactorily analyzed in the current literature. The physico-chemical and the thermodynamic aspects of the adsorption process at the solid-liquid interface has been analyzed, to some degree, on the basis of Gibbs adsorption theory for the liquid interface (1-7).

Furthermore, a great many studies have been reported on the free energy of adsorption, ΔG_{ads}, while the enthalpy of adsorption (or immersion) data are very scarce in literature (7,8,9,10,11,12,13). As is well known from other interfacial systems, like liquid-air or liquid-liquid, the knowledge of entropy and enthalpy is essential in order to be able to understand the molecular forces acting at these interfaces (14).

If we consider a process where a solid (S) is immersed into a liquid (L), whereby a change in enthalpy of immersion, h_{imm}, per unit area of solid surface is given as (4,7,12):

$$h_{imm} = h_{SL} - h_S \tag{1}$$

where h_{SL} is the enthalpy of solid-liquid interface, and h_S is the enthalpy of solid interface. The magnitude of h_{imm} of TiO_2 and Teflon in water is 2300 J/cm^2 (550 cal/cm^2) and 25 J/cm^2 (6 cal/cm^2), respectively (14). It has been found that the magnitude of h_{imm} of a polar solid in a polar liquid is much larger than in a non-polar liquid. The difference in h_{imm} for TiO_2 and Teflon in water thus arises from the polar and non-polar character of solids. On the other hand, the magnitude of h_{imm} of solids in non-polar liquids is small and varies very little.

As has been shown elsewhere (4,15,16,17):

$$\gamma_S = \gamma_S^D + \gamma_S^P , \qquad \text{for solid} \tag{2}$$

$$\gamma_L = \gamma_L^D + \gamma_L^P , \qquad \text{for liquid} \tag{3}$$

that the surface tension, γ, of a solid (γ_S) or a liquid (γ_L) is composed of dispersion (D) and polar (P) forces. Accordingly, it is thus safe to propose that the magnitude of h_{imm} would be given as:

$$h_{imm} = h_{imm}^D + h_{imm}^P \tag{4}$$

In the current literature no detailed analysis of the enthalpy arising from the dispersion or polar forces has been given. This procedure should be carried out, since the molecular analyses of the surface tension terms has been given (15-18).

Another adsorption process where a solute adsorbs from a solution onto a solid, has been studied by using the technique of flow calorimetry (8, 12,13). In this method the heat evolved on displacement of one adsorbate (1) system in contact with solid (i.e. $h_{SL,1}$) by another (2), $h_{SL,2}$, is given as:

$$h_{ads} = h_{SL,2} - h_{SL,1} \tag{5}$$

It is also noticed that the term h_{ads} can be derived from the difference between the h_{imm} terms for the two adsorbates. The purpose of this study is to discuss the various adsorption enthalpy data available in the current literature, which we consider useful in providing some systematic information, as regards the dispersion and polar forces (Equation 4).

EXPERIMENTAL

All chemicals were used as supplied with >99 % purity of analytical grade. The sample of activated charcoal used was supplied by B.D.H., U.K. The surface area was estimated as about 1000 m^2/g, as described herein. The heat of adsorption at different concentrations of adsorbate (n-propanol, 2-propanol, n-butanol, n-pentanol, n-hexanol, n-heptanol, n-octanol) in n-heptane (as solvent) were measured. A flow microcalorimeter (Microscal, Ltd., London) was used for the determination of the heat of adsorption. The adsorbent (Graphon) was saturated with the carrier liquid (n-heptane) until the calorimeter reaches equilibrium. A 5 μL solution of alcohol (in n-heptane) was injected which causes heat to be evolved. The instrument is calibrated by the introduction of a known quantity of heat through a small coil which can be introduced into the adsorbent.

RESULTS

The heat of adsorption, h_{ads}, of n-hexanol at different concentrations in n-heptane on Graphon by flow calorimeter is given in Fig.1. The monolayer heat of adsorption, h^m_{ads}, was obtained from the plateau of the curves of Fig.1, or from the Langmuir plots of integral heats at different concentrations (13):

$$C/q_{ads} = C/q^{m}_{ads} + 1/(k\ q^{m}_{ads}) \tag{6}$$

where C is the concentration of the adsorbate (Mol/Liter) and q_{ads} and q^{m}_{ads} are the integral heats at C or at monolayer coverage, respectively, and k is a constant.

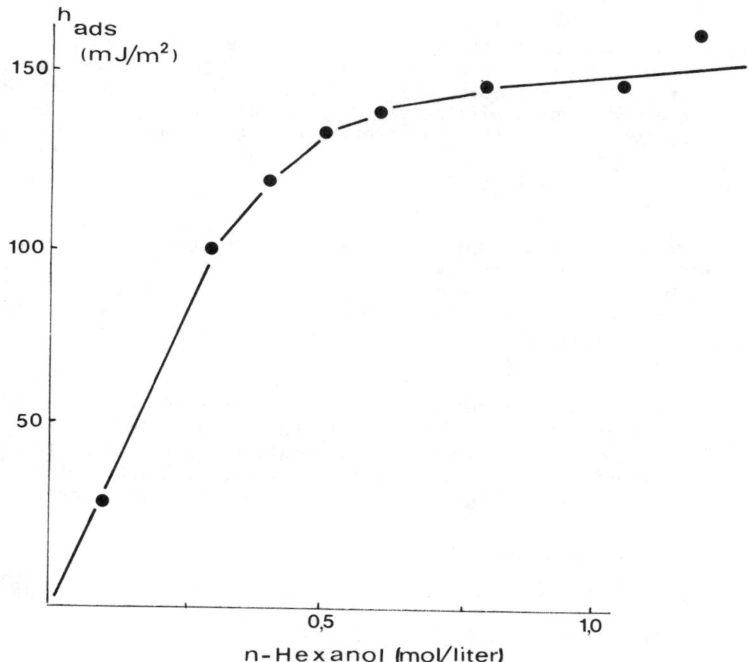

Figure 1 - Enthalpy of Adsorption, h_{ads}, of n-hexanol on carbon from n-heptane by flow calorimeter. Carbon = 20 mg (this study) (25 ^{o}C).

As shown later herein, the magnitude of h^{m}_{ads} is dependent on the surface area of solid. This allows one to estimate the surface area of any solid, provided one has a standard. We found that the ratio of h^{m}_{ads} (Graphon): $h^{m}_{ads}(TiO_2)$ was about 100. Since the BET surface area of TiO_2 was 10 m^2/g, we conclude that the surface area of Graphon used is roughly 1000 m^2/g.

DISCUSSION

Before describing our results, it is worth considering the energetics of an adsorption process. In the adsorption reaction where a solute, C_S, from its solution adsorbs onto a solid, the latter being covered by the solvent molecules, is given by the equilibrium (4):

Initial state = (solute in solution) + (adsorbed solvent)
Final state = (adsorbed solute on solid) + (solvent in solution)

Free energy, entropy, and enthalpy of this reaction is under consideration in this report, as given below:

$$\Delta G_{ads} = \Delta H_{ads} - T\Delta S_{ads} \quad (7)$$

As already mentioned above, it is clear that if one only considers ΔG_{ads}, then the information available can be useful for formulating a model valid for only one temperature. On the other hand, if one desires to formulate a model which will be valid for a range of temperatures, then the entropy and enthalpy of adsorption need to be measured. This point will become more clear, when we consider later, that the effect of temperature on the energetics of solute in solution or in the adsorbed state will be much different.

If we now break down this complicated adsorption process, and analyze a few adsorption equilibria, then we will make our postulate more clear. In order to do this, we shall consider the magnitudes of h_{imm} values of Graphon in different n-alkanes (8), Fig.2. It is seen that the variation of h_{imm} with the alkane chain length (n_C) is non-linear: $h_{imm} = -64.8 + 94.63 \log(n_C)$.

These data thus allow us to conclude that the adsorption of alkane molecules on Graphon is non-linear function of chain length, for linear chains. This becomes more obvious when we consider the effect of temperature on the magnitude of h_{imm}, for two systems, e.g. Graphon plus n-C_7H_{16} or n-$C_{16}H_{34}$, (14), Fig.3. The variation of h_{imm} with temperature is as:

Graphon + n-heptane:

$$h_{imm} = 128.0 - 0.1532\ t \quad (8)$$

Graphon + n-hexadecane:

$$h_{imm} = 325.5 - 4.865 + 0.0342\ t^2 \quad (9)$$

where t is the temperature, C°. These data have been analyzed earlier (14), but we mention these here only to point out that one cannot understand these adsorption data if one has not available the enthalpy and entropy of a process. These data thus allow us to conclude that the differences between h_{imm} of n-heptane and n-hexadecane must arise from some configurational differences after adsorption on Graphon. This, we will argue, arises from the possibility that at low temperatures n-hexadecane is most likely in a "solid" state on adsorption, while n-heptane is still more "liquid-like". This consideration is analogous to the phase transitions as one observed in monolayer studies at air-water interfaces (4, 7,16). At higher temperatures, where the adsorbed alkane molecules become "liquid-like", then the energy, h_{imm}, becomes less dependent on the chain length and temperature. It is of interest to consider the effect of alkyl chain lengths on h_{ads} of different organic compounds on solids from their solutions.

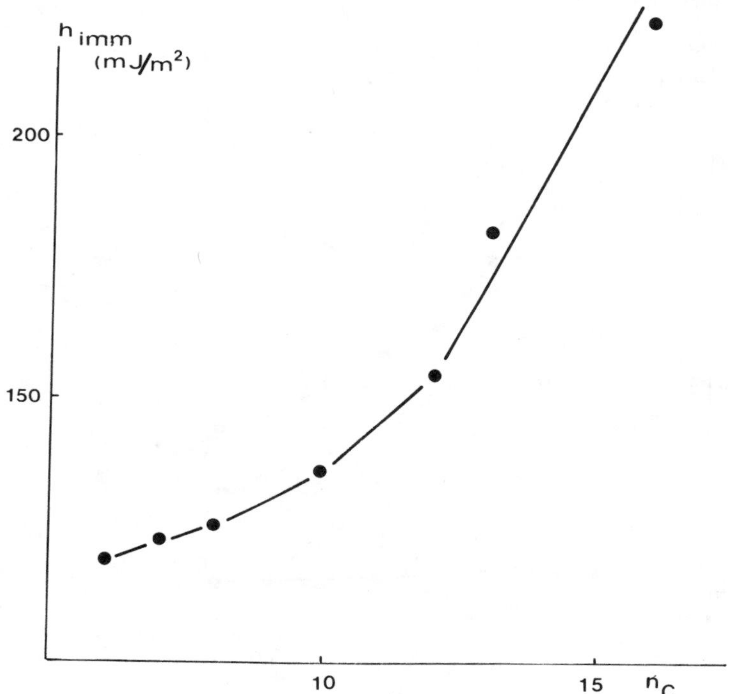

Figure 2 - Heat of Immersion, h_{imm}, of Graphon in n-alkanes (at 25 °C) (Ref.8) as a function of n_C.

In Fig.4 we give the variation of h^m_{ads} for various systems (8). In the current literature, although a few such systematic studies have been reported, we feel though that much more detailed analyses is still needed. It was reported (8) that on many polar solids, e.g. TiO_2, Fe_2O_3, the plots of h^m_{ads} versus number of carbon atoms, n_C, in the linear alkyl chain length of alcohols, one found a maximum around n_C = 6. On the other hand, similar plots of adsorption of n-fatty acids gave a decrease with n_C, and h^m_{ads} was constant after n_C = 6 (8). Our studies carried out on Graphon of n-alcohols, Fig.4, show that these h^m_{ads} values vary with n_C in similar manner as the n-fatty acids.

The data in Fig.4 thus convincingly show that h_{ads} is dependent on the solid and as well as the adsorbate. Let us further consider the dependence of heat of adsorption and the surface area of the solid. In Fig.5 are given the plots where h^m_{ads} versus surface area (from BET) show a linear dependence for two different systems. The slopes are different by by a factor 2.5, i.e. enthalpy/area. This observation has not been thoroughly analyzed in the current literature, and is being pursued in the author's laboratory

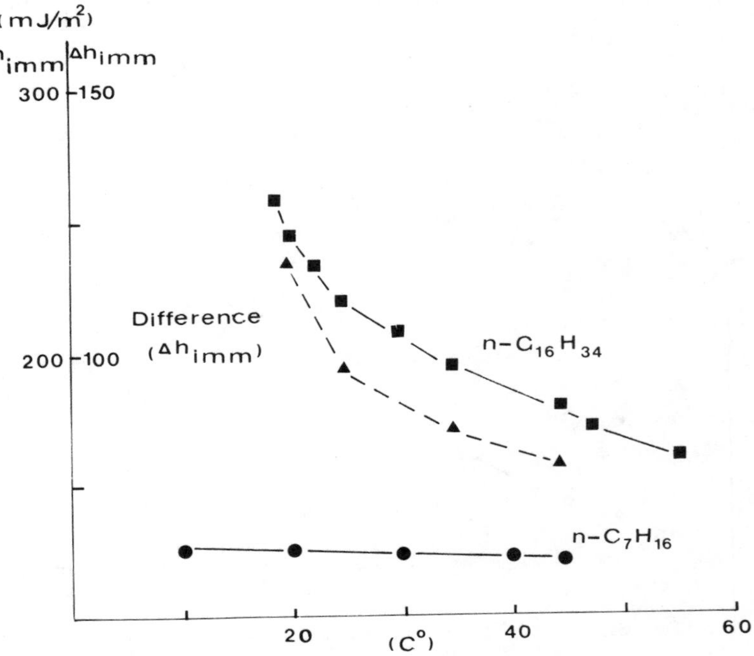

Figure 3 - Variation of h_{imm} of n-alkanes on Graphon with temperature. (Ref. 11).

Let us examine the relation between enthalpy per unit area, as a function of solvent. In Table 1 we find that the enthalpy of adsorption is dependent on the solvent.

The data in Table 2 are given to further show that the ratio h^m_{ads} for two different solids can be used quite successfully to estimate the surface area of different solids. These data further show that the packing of different n-fatty acids on different solids is very much alike. We have not been able to derive a simple dependence of h^m_{ads} on n_C from these data at this stage.

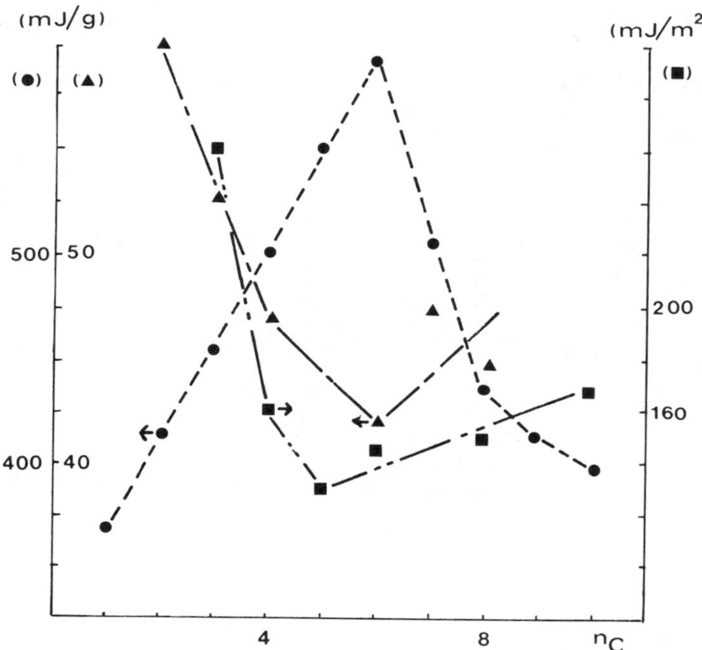

Figure 4 - Variation of h^m_{ads} with n_C in Systems: (▲) Heptane/TiO_2/ n-fatty Acids (8); (●) Heptane/Fe_2O_3/n-Alcohol (8); (■) Heptane/Graphon/n-Alcohols (this study). (Surface area, BET; TiO_2 = 10 m^2/g; Fe_2O_3 = 3.5 m^2/g).

CONCLUSION

It is shown that the enthalpy of adsorption is a linear function of solid surface area (m^2/g from 0.5 to ∿ 1000). In systems n-alcohol Fe_2O_3 and n-alcohol - Graphon, the conformations of adsorbed alcohols are different. On Fe_2O_3. the hydroxyl groups are adsorbed, while on Graphon the alkyl chains are adsorbed. This explains the differences in the effect of n_C.

Figure 5 - Variation of h^m_{ads} with surface area (by BET method) of different solids for two different systems: (■) n-butanol or (●) n-octanoic acid as adsorbent (from n-heptane) (Data from Ref.8)

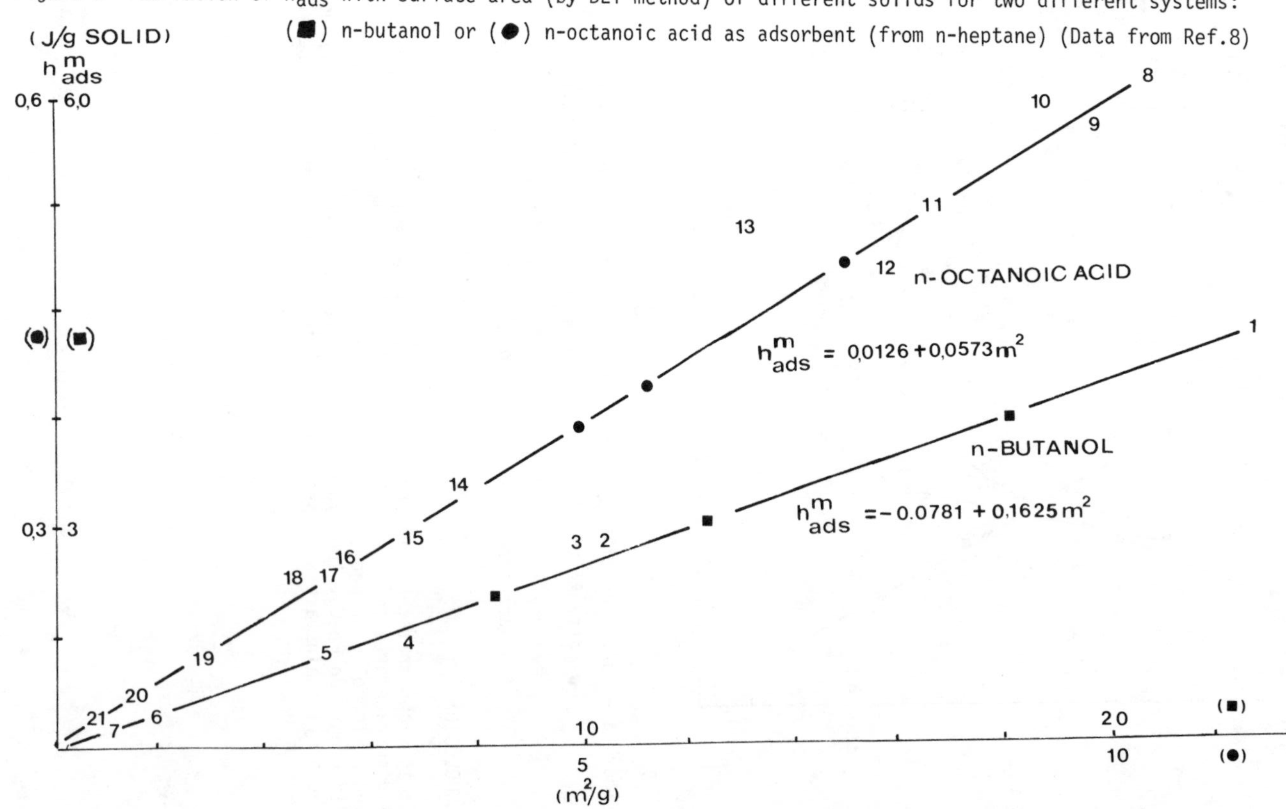

LITERATURE CITED

1. Kipling, J.J., "Adsorption from solutions of Non-electrolytes", Academic Press, London (1965).
2. Everett, D.H., "Colloid Science", vol.1, edited by D.H.Everett, The Chemical Society, London (1973).
3. Broen, C.E. & Everett, D.H., "Colloid Science", vol.2, edited by D.H.Everett, The Chemical Society, London (1975).
4. Adamson, A.W., "Physical Chemistry of Surfaces", 3rd ed., John Wiley & Sons, New York (1976).
5. Mittal, K.L., editor, "Adsorption at Interfaces", ACS Symposium Series, No.8, American Chemical Society, Washington, D.C. (1975).
6. Scamehorn, J.F., Schechter, R.S. & Wade, W.H., J.Colloid Interface Sci. 83 (1982) 463: J.Colloid Interface Sci. 83 (1982) 479.
7. Chattoraj, D.K. & Birdi, K.S., "Adsorption & the Gibbs Surface Excess", in preparation.
8. Grozek, A.J., editor, "British Petroleum Symposium", "Significance of the Heats of Adsorption at the Solid-liquid Interface", (1971).
9. Everett, D.H., Israel J.Chem. 14 (1975) 267.
10. Brown, C.E., Everett, D.H., Powell, A.V. & Throne, P.E., Faraday Discuss.Chem.Soc. 59 (1975) 97.
11. Parfitt, G.D. & Thompson, P.C., Trans.Faraday Soc. 67 (1971) 3372.
12. Razouk, R.I., Saleeb, F.Z. & Said, F.S., J.Colloid Interface Sci. 28 (1968) 487.
13. (a) Allen, T. & Patel, R.M., J.Colloid Interface Sci. 35 (1971) 647. (b) Rahman, M.A., Basu, S.K., Roychoudhry, A. & Ghosh, A.K., J.Colloid Interface Sci. 85 (1982) 452.
14. Parfitt, G.D. & Tideswell, M.W., J.Colloid Interface Sci. 79 (1981) 518.
15. Fowkes, F.M., J.Colloid Interface Sci. 28 (1968) 493.
16. Birdi, K.S., Proceed. VII Scandinavian Surface Chemistry Symposium, editor K.S.Birdi, Holte, Denmark (1981).
17. Birdi, K.S., J.Colloid Interface Sci. 88 (1982) 290.
18. Zettlemoyer, A.C., Ind.Eng.-Chem. 57 (1965) 27: Zettlemoyer, A.C. et al., Can.J.Chem. 33 (1955) 251.

Table 1 - Variation of Enthalpy of Adsorption (at monolayer coverage), h^m_{ads}, on various solids and solvents.

SOLID	SOLVENT	ADSORBATE	h^m_{ads} (mJ/m^2)	BET (m^2/g)	REF.
Cuprix Oxide	C_6H_6	n-$C_{17}H_{35}COOH$	220	9.4	(13,b)
Ferrix Oxide	C_6H_6	n-$C_{17}H_{35}COOH$	220	9.5	(13,b)
	nC_7H_{16}	n-$C_{17}H_{35}COOH$	60	4	(13,a)
Alumina	C_6H_6	n-$C_{17}H_{35}COOH$	220	120	(13,b)

Table 2 - Variation of the ratio of enthalpy of adsorption of n-acids (with alkyl chain length from 1 to 18 carbon atoms) on Fe_2O_3, $BaSO_4$ and TiO_2. The surface area of the solids from BET are given. Data from (13,a).

Number of carbon atoms in the alkyl chain of Fatty Acid	RATIO $h^m_{ads,Fe_2O_3} : h^m_{ads,BaSO_4}$	RATIO $h^m_{ads,Fe_2O_3} : h^m_{ads,TiO_2}$
1	2.43	2.93
2	2.5	3.08
3	3.42	3.4
4	2.37	3.08
5	2.67	3.09
6	2.39	3.04
7	2.55	3.07
8	2.56	3.03
10	2.5	2.96
12	2.51	3.04
14	2.51	3.0
16	2.56	3.09
17	2.55	3.01
18	2.51	3.03
BET (m^2/g)	Fe_2O_3:$BaSO_4$	Fe_2O_3:TiO_2
	3.45:6.63	3.45:10.1
	1:1.92	1:2.93

ADSORPTIVE SEPARATIONS BY SIMULATED MOVING BED TECHNOLOGY: THE SORBEX PROCESS

by

D. B. Broughton and S. A. Gembicki

UOP Process Division
UOP Inc.
Des Plaines, Illinois, U.S.A.

ABSTRACT

Adsorption from the liquid phase has long been used for removal of contaminants present at low concentrations in process streams. In some instances, the objective is removal of specific compounds. In others, the contaminants are not well defined, and the objective is the improvement of a general property such as color, taste or odor. In contrast to the use of adsorption for removal of trace contaminants, its commercial use for large-scale bulk separation from the liquid phase has been both relatively recent and almost entirely accomplished by the use of a flow scheme simulating the continuous countercurrent flow of adsorbent and process liquid, without actual movement of the solid. In conjunction with suitable adsorbents, the use of simulated moving-bed technology has led, in some cases, to an improvement in the economics of separations previously accomplished by other means. In other cases, its application has allowed separations to be made which are not possible by other known methods. This text summarizes the technology and theory of adsorptive separation by the simulated moving-bed processes developed and licensed by UOP Inc. under the general name, Sorbex[SM].

ADSORPTIVE SEPARATIONS

In theory, adsorption offers a more efficient route to a number of separations of commercial interest than do more conventional techniques, such as liquid-liquid extraction or extractive distillation. This stems in part from the fact that adsorbents are known which are much more selective in their affinity for various materials than are any known solvents.

However, another item of comparable importance is the fact that much higher efficiency of transfer of material between liquid and adsorbed phases can be achieved in adsorptive operations than in the conventional equipment used for extraction or extractive distillation. As an example, laboratory-scale chromatographs commonly show separation efficiencies equivalent to many thousands of theoretical equilibrium stages in columns of modest length. This high efficiency results from the use of small particles to give high interfacial area and from the absence of significant axial mixing of either phase.

In contrast, the trays of fractionating columns and liquid-liquid extractors are designed to obtain practically complete axial mixing in each physical element. With this arrangement, regardless of actual transfer rate coefficients, the number of theoretical equilibrium stages is limited substantially to the number of physical mixing stages installed. This limitation could be avoided, in theory, by the use of packed columns. However, if the particle diameter of the packing is small enough to give interfacial areas comparable to those obtainable in adsorptive beds, the capacity for accommodating counterflow of two fluid phases becomes low, and great difficulty is encountered in obtaining uniform unchanneled flow of both fluid phases. These limitations are much less severe in an adsorptive bed, because only one fluid phase is involved.

In spite of these potential advantages, adsorption has not, until rather recently, achieved widespread commercial acceptance, except in cases where only small quantities of material are to be removed from the process stream and regeneration of the adsorbent can therefore be infrequent. One reason for its slow acceptance has been the lack of a design which would permit operation in a continuous fashion. In the usual fixed-bed adsorptive process, feed streams are discontinuous and the product streams vary continuously in composition. It is difficult to integrate such intermittent processes with continuous processes operating upstream and downstream from it, and problems of control are troublesome.

MOVING-BED PROCESSES

Continuity of operation and the process advantages of countercurrency could theoretically be obtained in an adsorptive process by actually conveying the adsorbent through the system counter to the fluid streams. However, the physical movement of solids introduces problems which tend to nullify the potential advantages of adsorption.

One obvious hazard is that of attrition of the adsorbent. Another basic difficulty is that of maintaining uniform plug-wise flow of both phases over the entire cross section of columns of large diameter. It would be most difficult to move a bed of solids uniformly down a column of large diameter, and to do this without creating packing non-uniformity which would allow extensive channeling of the fluid phase.

The significance of the axial mixing that would be induced by non-uniform flow is shown by the following approximate equation:

$$\frac{1}{n_A} = \frac{1}{n_P} + \frac{1}{n_M} \tag{1}$$

where:

n_A = theoretical-tray equivalent of actual bed

n_P = theoretical-tray equivalent of bed in absence of axial mixing

n_M = number of perfect mixes in series which would give axial dispersion equal to that existing

The term n_P is governed by the mass transfer coefficient and n_M by the flow pattern.

Since many adsorptive systems exhibit high mass transfer rates and correspondingly high values of n_P, the potential performance can be seriously degraded by modest degrees of axial mixing. Schemes have been proposed in which adsorbent is conveyed countercurrent to process fluid in a series of fluidized beds. These, however, suffer from the same limitation as trayed fractionators -- namely, that n_M is substantially equal to the number of physical mixing elements in series and n_A therefore cannot exceed this number, regardless of the value of n_P.

Studies reported in the literature show that, for flow of homogeneous fluids through stationary beds of densely-packed, uniform, impermeable spheres, the value of n_M is the order of the bed height divided by the particle diameter. This presumably sets an upper limit on the efficiency of an ideal packed bed. In practice, however, this limit cannot be approached, because additional axial mixing is induced by such items as porosity of the particles, density and viscosity gradients in the fluid, non-uniform packing, fluid distribution, and the effect of obstructions such as bed-supporting structural elements. Great care with respect to all of these items is required to obtain high efficiency in beds of large diameter.

SIMULATED MOVING BED

In view of these difficulties, a flow scheme has been devised which maintains the process features of continuous countercurrent flow of fluid and solid without actual movement of the solid. Although the commercial operations do not use an actual moving bed, the action taking place is most easily visualized in terms of a hypothetical moving-bed operation.

Consider the moving-bed system shown in Figure 1. The adsorbent circulates continuously as a dense bed in a closed cycle and moves up the adsorbent chamber from bottom to top. Liquid streams flow down through the bed, countercurrently to the solid. The feed is assumed to be a binary mixture of A and B, with component A being adsorbed selectively. Feed is introduced to the bed as shown.

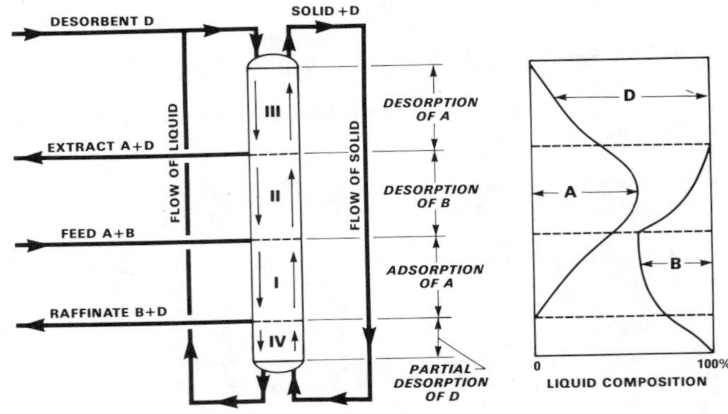

FIGURE 1. ADSORPTIVE SEPARATION WITH MOVING BED

Desorbent D is introduced to the bed at a higher level as shown. It is a liquid of different boiling point from the feed components and can displace feed components from the adsorbent pores. Conversely, feed components can displace desorbent from the pores with proper adjustment of relative flow rates of solid and liquid.

Raffinate product, consisting of the less strongly adsorbed component B mixed with desorbent, is withdrawn from a position below the feed entry. Only a portion of the liquid flowing in the bed is withdrawn at this point; the remainder continues to flow into the next section of the bed. Extract product, consisting of the more strongly adsorbed component A mixed with desorbent, is withdrawn from the bed; again only a portion of the flowing liquid in the bed is withdrawn, the remainder continues to flow into the next bed section.

The positions of introduction and withdrawal of net streams divide the bed into four zones, each of which performs a different function as described below.

Zone 1

The primary function of this zone is to adsorb A from the liquid. The solid entering at the bottom carries only B and D in its pores. As the liquid stream flows downward, countercurrent to this solid, component A is transferred from the liquid stream into the pores of the solid. At the same time, component D is desorbed (transferred from the pores to the liquid stream) to make room for A.

Zone 2

The primary function of this zone is to remove B from the pores of the solid. When the solid arrives at the fresh feed point, the pores contain the quantity of A that was adsorbed in Zone 1.

However, the pores also contain a large quantity of B, because the solid has just been in contact with fresh feed.

The liquid entering the top of Zone 2 contains no B, only A and D. As the solid moves upward, countercurrent to this stream, B is gradually displaced from the pores and is replaced by A and D. Thus, when the solid arrives at the top of Zone 2, the pores contain only A and D.

By proper regulation of the liquid rate in Zone 2, B can be desorbed completely from the pores. This can be accomplished without simultaneously desorbing all of A, because A is more strongly adsorbed than B.

Zone 3

The function of this zone is to desorb A from the pores. The solid entering the zone carries A and D in the pores; the liquid entering the top of the zone consists of pure D. As the solid rises, A in the pores is displaced by D.

Zone 4

The purpose of this zone is to reduce the required circulation rate of fresh desorbent in order to reduce the size and energy consumption of the fractionators that must separate desorbent from the net products.

When the adsorbent leaves Zone 3, the pores are completely filled with desorbent. The liquid entering the top of Zone 4 is of raffinate composition and contains B and D. If the flow rate in Zone 4 is properly regulated, component B will be readsorbed completely from the liquid, and an equal quantity of D will be displaced from the pores and returned to Zone 3 where it functions in the same manner as fresh desorbent.

The liquid-phase composition profile also shown in Figure 1 serves to clarify what is taking place.

In the moving-bed system of Figure 1, solid is moving continuously in a closed circuit past fixed points of introduction and withdrawal of liquid. The same results can be obtained by holding the bed stationary and periodically moving the positions at which the various streams enter and leave. A shift in the positions of liquid feed and withdrawal points in the same direction of internal fluid flow through the bed simulates the movement of solid in the opposite direction.

It is, of course, impractical to move the liquid feed and withdrawal positions continuously. However, approximately the same effect can be produced by providing multiple liquid-access lines to the bed, and periodically switching each stream to the adjacent line. Functionally, the adsorbent bed has no top or bottom and is equivalent to an annular bed. Therefore, the four liquid-access positions can be moved around the bed continually, always maintaining the same distance between the various streams.

The commercial application of this concept is shown in Figure 2 where the adsorbent is a stationary bed. A liquid circulating pump is provided to pump liquid from the bottom outlet to the top inlet of the adsorbent chamber. A fluid-directing device, known as a rotary valve, is provided which functions on the same principle as a multi-port stopcock in directing each of several streams to different lines. At the right-hand face of the valve, the four streams to and from the process are continuously fed and withdrawn. At the left-hand face of the valve, a number of lines are connected that terminate in distributors within the adsorbent bed.

At any particular moment, only four lines from the rotary valve to the adsorbent chamber are active. Figure 2 shows the flows at a time when lines 1, 5, 8 and 11 are active. When the rotating element of the rotary valve is moved to its next position, each net flow is tranferred to the adjacent line. Thus, desorbent would enter line 2 instead of line 1, extract is drawn from 6 instead of 5, feed enters 9 instead of 8, and raffinate is drawn from 12 instead of 11.

Figure 1 shows that in the moving-bed operation the liquid flow rate in each of the four zones is different because of the addition or withdrawal of the various streams. In the simulated moving bed of Figure 2, the liquid flow rate is controlled by the circulating pump. At the position shown in Figure 2, the pump is between the raffinate and desorbent ports and, therefore, should be pumping at a rate appropriate for Zone 4. However, after two switches in position of the rotary valve, the pump will be between the feed and raffinate ports and should, therefore, be pumping at a rate appropriate for Zone 1. Stated briefly, the circulating pump must be programmed to

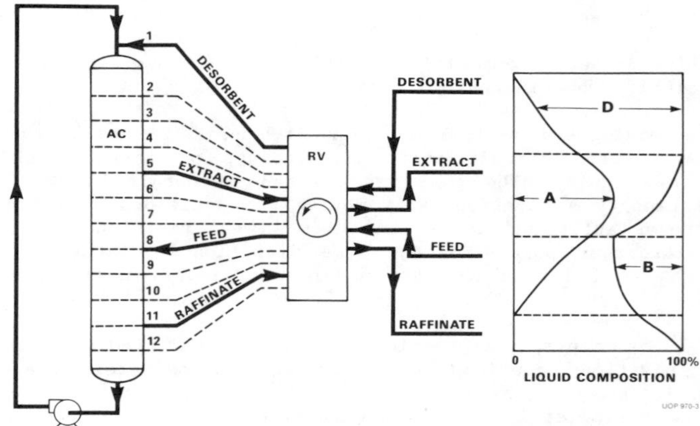

FIGURE 2. SORBEX -- SIMULATED MOVING BED FOR ADSORPTIVE SEPARATION

pump at four different rates. The control point will be altered each time a stream transfers from line 12 to line 1 resulting in the circulating pump crossing a zone boundary.

To complete the simulation, the liquid-flow rate relative to the solid must be the same in both the moving-bed and simulated moving-bed operations. Since the solid is physically stationary in the simulated moving-bed operation, the liquid velocity relative to the vessel wall must be higher than in an actual moving-bed operation.

A minor complication results from the fact that any particular line between the rotary valve and the bed is employed successively in carrying feed to the bed and carrying extract from it. To avoid extract contamination with feed components, the feed has to be flushed from each line before the line is used for withdrawing extract. Flushing is generally accomplished with desorbent, and the function is programmed into the mechanical construction of the rotary valve.

MATHEMATICAL MODELING

It is known that the theoretical performance of this operation, as commercially designed is practically identical to that of a system in which solids flow completely continuously, as a dense bed, countercurrent to the liquid, as depicted in Figure 1. The operation is modeled (1) in terms of theoretical equilibrium trays having the same significance as in fractionating columns. Solid and liquid are assumed to flow continuously through hypothetical well-mixed theoretical trays in which equilibrium is attained. The number of trays is determined by bed height, mass transfer coefficient and flow rates. Axial mixing is generally of much greater significance in liquid systems than in vapor systems because of the greater mass of process fluid in the voids relative to that in the selective pores. In order to allow for axial mixing in the liquid phase, it is necessary to assume that the solid entrains a certain amount of interstitial fluid from tray to tray. This model of equilibrium theoretical trays with entrainment is readily implemented on a computer by methods analogous to those used in the design of fractionating columns.

Two parameters are adjusted in the model to reproduce experimental concentration profiles:

Number of theoretical trays: $n = KkH/Lz$ (2)

Axial mixing ratio: $e = E/L = Kk/DL^2$ (3)

where:

K	=	linear equilibrium constant
k	=	mass transfer coefficient
H	=	bed height
L	=	liquid rate
z	=	$m \ln (m/m-1)$
m	=	adsorption factor = pore circulation rate, K/L

E = entrainment rate
D = axial diffusion coefficient

Each zone is treated separately according to the component whose concentration is changing most rapidly. Values of n and e corresponding to the equilibrium properties of the critical component for each zone are used.

For a simple binary feed, using a desorbent intermediate between the feed components in selectivity, it is possible to calculate the minimum circulation rates of adsorbent and desorbent required to achieve perfect separation when the mass transfer coefficient is infinite, the axial diffusion coefficient is zero, and the equilibrium enrichment factors are constant. For this case:

$$\frac{\text{Pore Circulation Rate}}{\text{Feed Rate}} = \frac{1}{\beta_{AB}-1} + X_{AF} \tag{4}$$

$$\frac{\text{Desorbent Circulation Rate}}{\text{Feed Rate}} = \frac{1 + (\beta_{AB}-1)X_{AF}}{\beta_{DB}} \tag{5}$$

where:

β_{AB} = selectivity for strongly adsorbed feed component A, relative to weakly adsorbed feed component B

β_{DB} = selectivity for desorbent D, relative to weakly adsorbed feed component B

X_{AF} = volume fraction of strongly adsorbed component in feed

$$\beta_{AB} = \frac{Y_A/X_A}{Y_B/X_B} \tag{6}$$

where:

X_A, X_B = volume fraction of components A and B in the liquid phase

Y_A, Y_B = volume fraction of components A and B in the adsorbed phase

COMPARISON WITH FIXED-BED OPERATION

It is of interest to compare the characteristics of the continuous Sorbex operation with those of the conventional liquid chromatography, or batch operation. The batch operation is illustrated in Figure 3. It consists simply of charging increments of feed and desorbent alternately to a fixed bed. As the feed components are eluted through the bed, they gradually separate into bands,

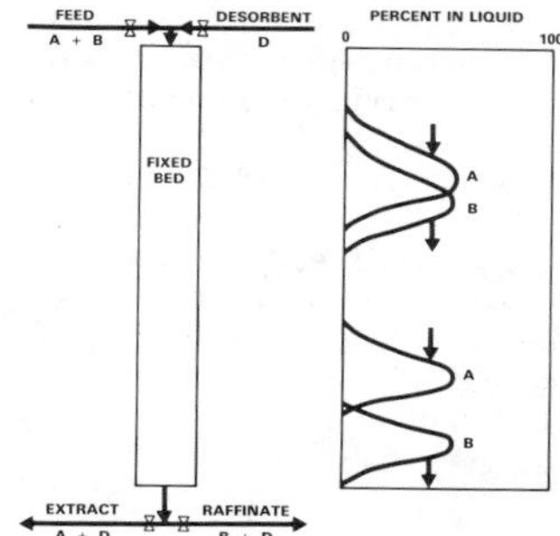

FIGURE 3. BATCH ADSORPTION. CONVENTIONAL CHROMATOGRAPHIC OPERATION

which travel at different rates, and are withdrawn alternately as raffinate and extract. Band separation as the pulse of feed travels through the bed is illustrated by the composition profiles. A second increment of feed must be delayed long enough to ensure that the least strongly adsorbed component does not overtake the most strongly adsorbed component in the first increment.

A comparative mathematical modeling of the two operations has shown that the batch operation requires more adsorbent inventory by a factor of 3-4, and more desorbent circulation by a factor of 2. Without going into the details of these mathematical analyses, it is possible to explain the large difference in adsorbent requirement in physical terms.

In the continuous system, every part of the bed at all times can be identified as performing useful work with respect to the primary function of each zone. In the batch system, however, various parts of the bed at various times can be identified as doing either nothing or something useless.

This is most clearly seen near the entrance of the bed in the batch system. As feed enters, the adsorbent near the inlet rapidly comes to complete equilibrium with the feed; as feed continues to enter, this section performs no further function except that of a pipe, carrying the feed down into the part of the bed where action is occurring. A similar situation exists when desorbent is introduced. Non-useful zones can also be identified further down the bed.

COMMERCIAL SORBEX PROCESSES

Commercial operations include the following specific separations: n-paraffin from naphtha, kerosine, and gas oils (Molex®); olefins from olefin-paraffin mixtures (Olex®); p-xylene from other C_8-aromatics and non-aromatic hydrocarbons (Parex®); p-cresol or m-cresol from mixtures of the cresol isomers (Cresex™); fructose from fructose/glucose mixtures (Sarex®). The extent of commercialization, as of 1982, Table 1, is illustrated by the licensing of 60 units in eight applications which have an aggregate capacity in excess of five million tons per annum of selectively adsorbed product.

TABLE 1. COMMERCIALIZED SORBEX TECHNOLOGY. COMMISSIONED UNITS

PROCESS (YR.)	NO. OF UNITS
PAREX (1971)	17
MOLEX (1962)	12
OLEX (1972)	4
SAREX (1978)	3
CRESEX (1979)	1
	37
IN DESIGN OR CONSTRUCTION	23
TOTAL LICENSED	60

UOP 970-1

Other separations, not yet commercial, have been demonstrated in pilot plants that completely reproduce the commercial simulated moving-bed mode of operation. These include the separation of: p-cymene or m-cymene from an isomer mixture, ethylbenzene from a mixture of C_8-aromatics, 1-butene from a mixture of all C_4-paraffins and olefins, p-diisopropylbenzene from a mixture of the isomers, p-ethyltoluene from other C_9-aromatics, rosin acids from fatty acids, and oleic acid from linoleic acid.

Commercial success has also depended heavily on the development of suitable adsorbents with high capacity, high selectivity and good kinetics. Under conditions of equilibrium control, selectivity is the major determinant of pore circulation rate. The higher the selectivity, the lower the adsorbent inventory required.

Problems associated with large-scale commercial operations have been solved, as evidenced by satisfactory operation of a single-train Parex unit having a capacity of over 130,000 tons of product per year. The largest adsorbent bed in commercial use has a bed diameter of 6.70 meters and is performing as well as a pilot plant with a bed diameter of 0.10 meter. Performance of the rotary valves has been excellent, and they have required only routine maintenance.

Long adsorbent lives have been achieved without any form of adsorbent regeneration. As an example, a n-paraffin unit, operating on hydrotreated feedstock, is still performing satisfactorily with the original adsorbent after twelve years of service.

LITERATURE CITED

(1) D. B. Broughton, R. W. Neuzil, J. M. Pharis, C. S. Brearley, Chem. Eng. Prog., 66 (9), 70-75 (1970).

OPTIMAL SMOOTHING OF SITE-ENERGY DISTRIBUTIONS FROM ADSORPTION ISOTHERMS

Lee F. Brown and Bryan J. Travis
Earth and Space Sciences Division
Los Alamos National Laboratory, Los Alamos, NM 87545, U.S.A.

ABSTRACT

The equation for the adsorption isotherm on a heterogeneous surface is a Fredholm integral equation. In solving it for the site-energy distribution (SED), some sort of smoothing must be carried out. The optimal amount of smoothing will give the most information that is possible without introducing nonexistent structure into the SED. Recently, Butler, Reeds, and Dawson (1) proposed a criterion (the BRD criterion) for choosing the optimal smoothing parameter when using regularization to solve Fredholm equations. The BRD criterion is tested for its suitability in obtaining optimal SED's. This criterion is found to be too conservative. While using it never introduces nonexistent structure into the SED, significant information is often lost. At present, no simple criterion for choosing the optimal smoothing parameter exists, and a modeling approach is recommended.

INTRODUCTION

Single-energy adsorption isotherms, $\theta(p,q)$, such as the Langmuir and Hill-de Boer relationships, normally cannot model experimental isotherms adequately. This has been recognized from the earliest studies of adsorption, and so the heterogeneous nature of an adsorbent or catalytic surface long has been perceived as an important property. The usual equation for the nonuniform-surface adsorption isotherm, relating the amount adsorbed to the pressures and energies of adsorption, has embodied within it both a single-energy adsorption isotherm (local isotherm) and an energy distribution function $\eta(q)$:

$$\theta_{gr}(p) = \int_{q_{min}}^{q_{max}} \theta(p,q)\eta(q)dq \qquad (1)$$

The function $\eta(q)dq$ is defined as the fraction of adsorption sites with energies between q and $q + dq$; $\eta(q)$ is called the site-energy distribution (SED) and has been the principal choice as a device to quantify the heterogeneity of an adsorbent's surface. Beginning with the work of Roginsky (2) almost forty years ago, studies in a continuing stream have examined means of extracting $\eta(q)$ from adsorption isotherms and analyzed the implications of the resulting distributions. Reviews of these efforts (3-5) show the extent of these investigations.

Equation (1) is a first-kind Fredholm integral equation. Because of properties inherent in this equation, some form of smoothing must be carried out before an acceptable solution is obtained (e.g., 6).

Sometimes the experimental isotherm data are smoothed, as when the data are approximated by an analytic function (e.g., 5,7). Other times the solution is smoothed, as in the regularization method for solving first-kind Fredholm equations (cf. 8). Sometimes both data and solution are smoothed (e.g., 3,9).

The smoothing is necessary to prevent nonexistent structure from being introduced into the solution of Eq. (1). An example is presented in Fig. 1. Here a unimodal, gaussian SED was postulated and used with a Langmuir local isotherm to generate an isotherm. Random errors with a standard deviation of 1% were imposed on the isotherm data, and regularization was used to extract the SED. A low amount of solution smoothing created the structured SED, while the best agreement between the postulated and calculated SED required a much greater amount of smoothing. This was true even though the least squares criterion, i.e., the minimum of the sums of the squared differences between the generated isotherm data and the isotherms calculated using the extracted SED's, said that the structured SED was the best possible. Thus such a least squares criterion can be a very poor guide to the optimal smoothing. Some other direction is needed to tell what amount of smoothing will give the maximum amount of information concerning the structure of the SED without introducing nonexisting components.

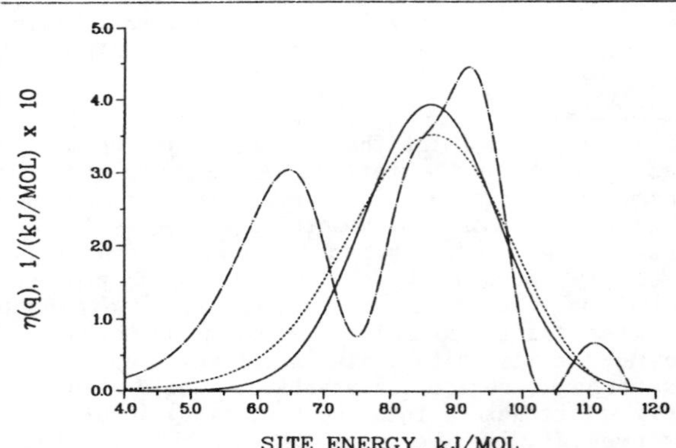

Figure 1. Creation of Nonexistent Structure by Inadequate Smoothing. 1% RMS Error in Isotherm Data.

——— Postulated Distribution.
—·—·— $\alpha = \alpha_{LS}$. Calculated Distribution with Minimum Least-Squares-of-Differences between Original Isotherm and Isotherms from All Calc. Distributions.
········ $\alpha_{opt} = 1000\alpha_{LS}$. Calculated Distribution Closest to Postulated Distrib.

While significant smoothing is needed, too much smoothing obviously can destroy information in the original data. Our purpose is to examine quantitatively this aspect of obtaining SED's from isotherms, and to propose some guidelines for telling how much smoothing is optimal for a level of error in the isotherm data.

Another factor which has the potential of affecting the solution $\eta(q)$ is the choice of local isotherm $\theta(p,q)$ in Eq. (1). We do not examine this factor here. There has been work in this area (e.g., 9), and these efforts indicate that using different, though still reasonable, local isotherms results in qualitatively similar SED's.

Since the particular local isotherm $\theta(p,q)$ does not appear to be a crucial factor in the SED which results from solving Eq. (1), the Langmuir isotherm appears throughout this work; it is the simplest single-energy isotherm which also is physically reasonable.

PRELIMINARY CONSIDERATIONS

The technique of regularization, developed by Phillips (6), Twomey (10), and Tikhonov (8,11-12), is used throughout the present work. Papenhuijzen and Koopal (13) recently have shown that regularization is a superior method for extracting SED's from adsorption isotherms. We employ the formulation of Tikhonov, as it has been found to lead to a straightforward method for solving Eq. (1) numerically. In addition, work by Butler et al. (1) has built upon the Tikhonov formulation to incorporate a nonnegativity constraint upon $\eta(q)$. These latter investigators also proposed a criterion for an optimal value of the regularization smoothing parameter, α. We use the additions proposed by Butler and his coworkers, and the algorithm employed for this work is presented in Appendix A. As was mentioned above, either the data or the solution, or both, must be smoothed in the process of solving Eq. (1). The present work examines directly only the effects of smoothing the solution; it is felt, however, that some of the results may be applicable to situations where instead the data are smoothed.

In spite of the long-recognized need for smoothing of data or solution when obtaining SED's by solving Eq. (1), until very recently no quantitative guideline had been proposed for judging when the optimal amount of smoothing has been carried out. As mentioned above, Butler et al. (1) proposed such a criterion for use with regularization. Their optimal α was supposed to smooth the calculated distribution so as to give the most probable distribution for the existing level of error in the data. They pointed out that their guideline was probably a conservative one, in that it might give some degree of oversmoothing. Our algorithm for implementing the BRD (Butler, Reeds, and Dawson) criterion is included in Appendix A.

In a recent work (14), we calculated isotherms from postulated $\eta(q)$'s, then used regularization to determine the SED using Eq. (1) and the generated isotherm data. Good agreements between postulated and calculated unimodal and bimodal SED's were obtained, and our results suggested that the best agreement between postulated and calculated SED's could be found at an α-value one-tenth that generated by the BRD criterion. In this paper we investigate the optimal smoothing matter more extensively. Amount of error, characteristics of the peaks in the SED, and number and location of data points appear to be aspects which may affect the optimal smoothing.

APPROACH TO PROBLEM

To study the effects of different factors upon the optimal degree of smoothing, a numerical approach was taken. Different situations, each containing characteristics typical of experimental possibilities, were examined. In each situation an SED containing particular properties was postulated, a series of pressures chosen, and Eq. (1) used to generate a vector of accurate isotherm data. Seven significant figures were retained in each of the isotherm points. Random, normally distributed errors, with a specified standard deviation, were then imposed upon the individual points of the generated isotherm. Using the modified isotherm containing the error, regularization was employed to extract a series of calculated SED's, each SED corresponding to a different value of the smoothing parameter α. The best value of α by the BRD criterion, α_{BRD}, was identified.

Trial and error found the optimal value of α. Visual comparison was used to identify the best α, as it was difficult to create analytical criteria which would be satisfactory. Usually, the optimal value of α occurred at the point just before nonexistent structure began to appear in the solution. Once the optimal value of α was found, it was compared with α_{BRD} to see if any consistency occurred. The comparison between α_{opt} and α_{BRD} was also used to see if information was lost using α_{BRD} which otherwise could be extracted without creating nonexisting structure.

RESULTS

The results are in Figs. 2-5, and the parameters used, distributions employed, and ranges covered are presented in Appendix B.

Figure 2 shows the results with a 2-peaked SED and a 1% RMS error. Here the α of the BRD criterion gives a solution with the two peaks, but the solution does not indicate the heights at all well. It is possible, however, to match the peaks quite well by using a more sensitive smoothing parameter. This indicates some loss of information if α_{BRD} is used.

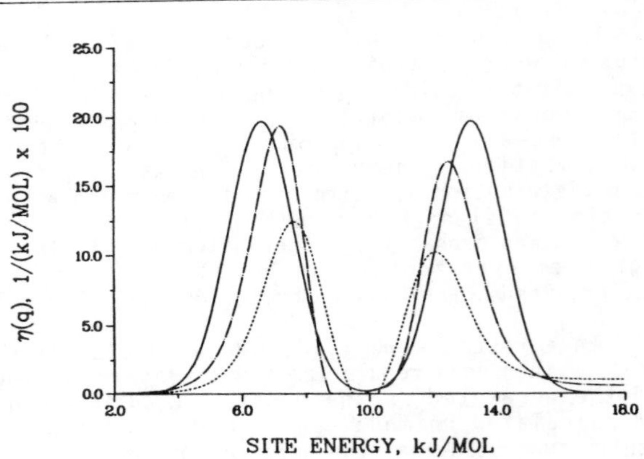

Figure 2. Comparison between Distributions from Optimal α and α_{BRD}. Two-Peaked SED, 1% RMS Error in 51-Point Isotherm.

——— Postulated Distribution.
- - - - - Calculated Distribution Using α_{BRD}.
—·—·— Calculated Optimal Distribution, $\alpha_{opt} = 0.05\alpha_{BRD}$.

It may be noted that the optimal value of α in the 2-peaked, 1% error situation is equal to $0.05\alpha_{BRD}$, while in the 1-peaked, 1% error situation (Fig. 1) it is equal to $0.2\alpha_{BRD}$. This indicates that there may be little consistency between the optimal α and the BRD α, although the optimal α does appear to be always less than α_{BRD}.

In Fig. 3, error in the data is tripled over that in Fig. 2; other factors remain the same. With the higher level of error, use of α_{BRD} destroys the nature of the distribution. Only the approximate range of the distribution is maintained. The optimal calculated distribution does give a good representation of the original SED, although the height is attenuated significantly with the increased error. In this case, $\alpha_{opt} = 0.006\alpha_{BRD}$, confirming the inconsistencies between α_{opt} and α_{BRD}.

Figure 4 presents a 4-peaked distribution. This distribution was so sensitive to error in the data that an error level of 0.01% RMS deviation was necessary to obtain significant results; higher levels of error resulted in distributions that did not reflect the distribution with any realism whatsoever. Even at this level of error, though, important information is lost if α_{BRD} is used as the smoothing parameter. When

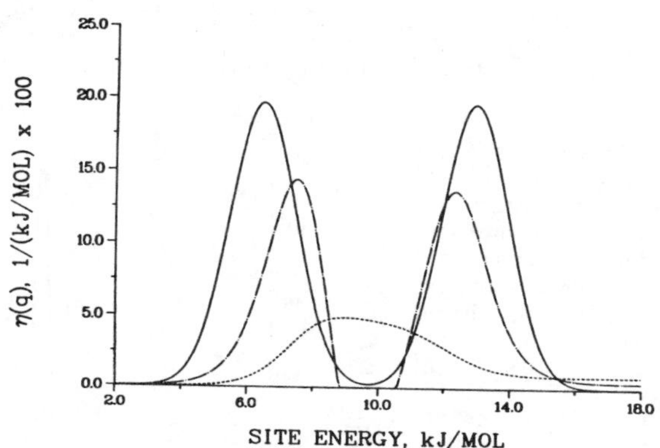

Figure 3. Comparison between Distributions from Optimal α and α_{BRD}. Two-Peaked SED, 3% RMS Error in 51-Point Isotherm.

——— Postulated Distribution.
- - - - - Calculated Distribution Using α_{BRD}.
—·—·— Calculated Optimal Distribution, $\alpha_{opt} = 0.006\alpha_{BRD}$.

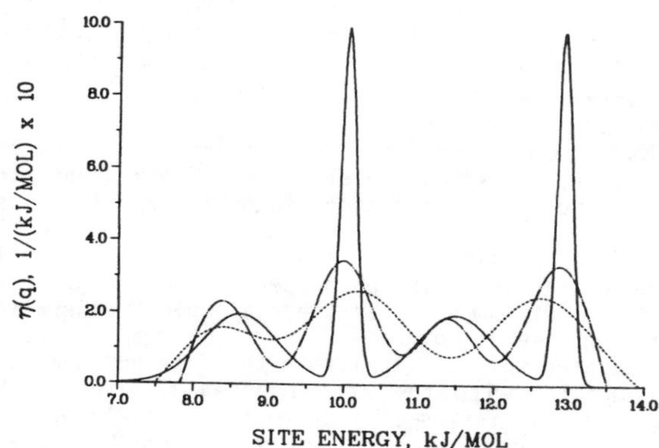

Figure 4. Comparison between Distributions from Optimal α and α_{BRD}. Four-Peaked SED, 0.01% RMS Error in 51-Point Isotherm.

——— Postulated Distribution.
- - - - - Calculated Distribution Using α_{BRD}.
—·—·— Calculated Optimal Distribution, $\alpha_{opt} = 0.006\alpha_{BRD}$.

this is done, only three peaks appear in the resulting distribution, while the optimal SED gives the four peaks. The results reflect the true situation only qualitatively; the true nature of the peaks remains obscured even by this low level of error. Again, the optimal value of the smoothing parameter is far removed from that resulting from the BRD criterion.

A comparison of Figs. 4 and 5 shows the importance of the number of points in the isotherm. Figure 5 results from using one-third the number of points that were used in Fig. 4. The points were spread over the same range, yet one of the peaks has been lost. Here there is little difference in the distributions from α_{BRD} and α_{opt}.

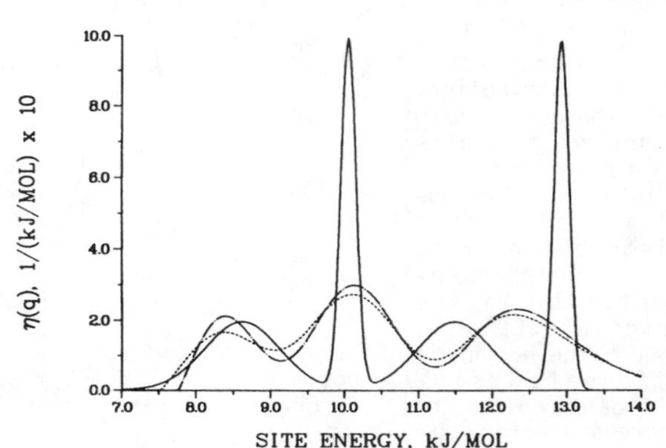

Figure 5. Comparison between Distributions from Optimal α and α_{BRD}. Four-Peaked SED, 0.01% RMS Error in 18-Point Isotherm.

——— Postulated Distribution.
----------- Calculated Distribution Using α_{BRD}.
—·—·— Calculated Optimal Distribution, $\alpha_{opt} = 0.1\alpha_{BRD}$.

DISCUSSION AND CONCLUSIONS

Butler et al. said that their criterion was a conservative one. When extracting SED's from adsorption isotherms, it is too conservative. Nonexistent structure never appears when using α_{BRD}, but significant information loss occurs frequently.

No other simple criterion now exists for choosing the optimal value of the smoothing parameter. Nevertheless, it is important that a guide to optimum smoothing be available. Otherwise, nonexistent structure may be reported if too low an α is used. One way of approaching this difficulty is to carry out numerical calculations in conjunction with the experimental studies. SED's can be postulated with characteristics similar to that calculated, and isotherms generated with the level of error imposed on the data equal to that estimated for the experiments being carried out. The optimal α may then be found by trial and error, as was done for the synthetic situations analyzed here. This may give a reasonable estimate of α_{opt} for a particular situation, albeit with significant effort.

ACKNOWLEDGMENT

The suggestions and comments of Professor John L. Falconer of the University of Colorado were very helpful.

NOTATION

c Nonnegative components of g in the additive smoothing term of Eq. (A3), dimensionless.
D Function of α defined by Eq. (A9).
f Function of experimental variable, various units.
f_e Experimental observations, various units.
g Distribution function, various units.
H Nonlinear differential operator.
I An integral, various units.
I The identity matrix
K Kernel of integral equation, various units.
M Matrix defined by Eq. (A6).
N Number of experimental data points.
p Pressure, Pa.
q Energy of desorption, J/mol (negative of the enthalpy change upon adsorption).
R Gas constant, J/(mol)(K).
T Absolute temperature, K.
T Matrix defined by Eq. (A7).
t Variable over which distribution function g occurs, various units.
w_i Weighting factor, inversely proportional to the variance in the data taken at point i.
x Experimental variable, various units.
α Adjustable smoothing parameter, dimensionless.
η Site-energy distribution function, 1/J.
$\theta(p,q)$ Fraction of sites of energy q covered at pressure p, dimensionless (the local isotherm).
θ_{gr} Gross fractional surface coverage, dimensionless.

LITERATURE CITED

1. Butler, J. P., J. A. Reeds, and S. V. Dawson, SIAM J. Num. Anal., 18, 381 (1981).
2. Roginsky, S. Z., Compt. rend. acad. sci. URSS, 45, 61 (1944)
3. Zolandz, R. R., and A. L. Myers, Prog. Filt. Sep., 1, 1 (1979).
4. Jaroniec, M., A. Patrykiejew, and M. Borowko, Prog. Surf. Membr. Sci., 14, 1 (1981).
5. Jaroniec, M., Adv. Colloid Interface Sci., 18, 149 (1983).
6. Phillips, D. L., J. Assoc. Comput. Mach., 9, 84 (1962).
7. Sips, R., J. Chem. Phys., 16, 490 (1948).
8. Tikhonov, A. N., and V. Y. Arsenin, Solutions of Ill-Posed Problems, W. H. Winston, New York, 1977.
9. Bräuer, P., W. A. House, and M. Jaroniec, Thin Solid Films, 97, 369 (1982).
10. Twomey, S., J. Assoc. Comp. Mach., 10, 97 (1963).
11. Tihonov, A. N., Soviet Math. Dok., 4, 1035 (1963a).
12. Tihonov, A. N., Soviet Math. Dok., 4, 1624 (1963b).
13. Papenhuijzen, J., and L. K. Koopal, in Adsorption from Solution

(R. H. Ottewill, C. H. Rochester, and A. L. Smith, eds.), pp. 211-225. Academic Press, London, 1983.

14. Britten, J. A., B. J. Travis, and L. F. Brown, paper presented at 1982 Annual AIChE Meeting, Los Angeles, CA, November 14-19. AIChE Symp. Ser. (to be published, 1984).

APPENDIX A

ALGORITHM FOR EXTRACTING NONNEGATIVE DISTRIBUTION FUNCTIONS FROM FIRST-KIND FREDHOLM EQUATIONS

The general form of Eq. (1) in the text is

$$f(x) = \int_a^b K(x,t)g(t)dt \quad . \tag{A1}$$

K is representative of the experimental system and procedure. When observations $f(x)$ have been obtained from a system characterized by Eq. (A1), a solution for $g(t)$ would minimize locally the integral

$$I = \int_c^d [f_e(x) - \int_a^b K(x,t)g(t)dt]^2\, dx \quad , \tag{A2}$$

in which $f_e(x)$ represents the experimental observations. In regularization, a smoothing term is added to the RHS of Eq. (A2), and the functional to be minimized is altered so that

$$I = \int_c^d [f_e(x) - \int_a^b K(x,t)g(t)dt]^2\, dx + \alpha \int_a^b [H(g)]dt \quad , \tag{A3}$$

where $H(g)$ is a nonlinear differential operator on g with nonnegative coefficients, and α is a parameter. In the simplest formulation, $H(g) = g^2$, which is what we have used throughout this work. Applying variational calculus gives the necessary condition for a minimum with $H(g) = g^2$ (e.g., 14):

$$\alpha g(t) + \int_a^b \{ \int_c^d K(x,z)K(x,t)dx\} g(z)dz = \int_c^d K(x,t)f_e(x)dx \quad . \tag{A4}$$

Equation (A4) can be put into finite-difference form:

$$(\underline{\underline{K}}^{tr}\underline{\underline{\delta x}}\ \underline{\underline{K}}\ \underline{\underline{\delta t}} + \alpha\underline{\underline{I}})\underline{g} = \underline{\underline{K}}^{tr}\underline{\underline{\delta x}}\ \underline{f}_e \quad , \tag{A5}$$

in which $\underline{\underline{K}}^{tr}$ is the transpose of $\underline{\underline{K}}$, and $\underline{\underline{\delta x}}$ and $\underline{\underline{\delta t}}$ are diagonal matrices whose elements are the weighting factors for the intervals [c,d] and [a,b]. Butler et al. (1) suggested a means by which $g(t)$ can be restricted to nonnegative values, and we have followed their proposal. Equation (A4), using a given α, is solved for $\underline{g}$. The points at which $\underline{g}$ is negative are recorded, and the evaluation of the quantity $(\underline{\underline{K}}^{tr}\underline{\underline{\delta x}}\ \underline{\underline{K}}\ \underline{\underline{\delta t}} + \alpha\underline{\underline{I}})$ is not performed at these points. A new $\underline{g}$ is obtained, and the process is repeated. The iteration is continued until no change in $\underline{g}$ is seen. The entire process is then repeated for all subsequent values of α.

To provide resolution of narrow peaks and near-discontinuities, a variable integration mesh is included in the computer code. For each value of α, the integration mesh is altered to allow finer zoning in regions where the emerging distribution has large gradients.

Butler et al. also offered a criterion for choosing an optimal value of α, based on the estimated error in the calculated $g(t)$ over t between a and b. This in turn depends on the error in the data. To express their criterion, some terms are defined:

$$\underline{\underline{M}} \equiv \underline{\underline{K}}^{tr} \underline{\underline{\delta x}}\, \underline{\underline{K}}\, \underline{\underline{\delta t}} \quad \text{(A6)}$$

$$\underline{\underline{I}} \equiv (\underline{\underline{M}} + \alpha \underline{\underline{I}})^{-1} \quad \text{(A7)}$$

$\underline{c} \equiv$ a vector which satisfies $(\underline{\underline{M}} + \alpha\underline{\underline{I}})\underline{c} = \underline{f}_e$ in which $\underline{\underline{M}}$ is evaluated only for points in t where $g(t) > 0$. (A8)

A function of α is then defined:

$$D(\alpha) \equiv (\underline{f}_e^{tr} \underline{\underline{I}}\, \underline{\underline{M}}\, \underline{\underline{I}}\, \underline{f}_e) - (2\underline{f}_e^{tr} \underline{\underline{I}}\, \underline{f}_e) + [2\sigma(N\, \underline{c}^{tr} \underline{c})^{1/2}] \quad \text{(A9)}$$

The criterion of Butler, Reeds, and Dawson states that the optimal value of α (α_{BRD}) is attained when $D(\alpha)$ is a minimum.

When the levels of error in the various data points are not equal, elements in the $\underline{\underline{K}}$ matrix and $\underline{f}_e$ vector must be weighted. Let the weights w_i^2 be inversely proportional to the variances in the data points f_i. The weights are scaled so that $\sum_{i=1}^{N} w_i^2 = N$. The elements in the $\underline{\underline{K}}$ matrix are then weighted so that k_{ij} becomes $w_i k_{ij}$ and those in the $\underline{f}_e$ vector are weighted so that f_{ei} becomes $w_i f_{ei}$.

APPENDIX B

FACTORS AND PARAMETERS USED IN GENERATION OF FIGURES

Figure 1:

$$\theta(p,q) = (4.464 \cdot 10^{-6})(p)(e^{q/RT})/[1 + (4.464 \cdot 10^{-6})(p)(e^{q/RT})]$$

(p in torr, q in joules)

$$\eta(q) = (3.931 \cdot 10^{-4})[e^{-(4.856 \cdot 10^{-7})(q-8619)^2}] \quad \text{(q in joules)}$$

Range of p: $0 \leqslant p \leqslant 3$ torr. $N = 51$

Range of q: $0 \leqslant q \leqslant 3.5 \cdot 10^4$ joules. $T = 77.5$

Accuracy of isotherm data: Random, normally distributed errors with a standard deviation of 1% were imposed on the isotherm data.

Smoothing parameters: $\alpha = \alpha_{LS}$ {Minimizes $\sum_{i=1}^{N} [f_e(x_i - \int_a^b K(x_i,t)g(t)]^2$ }

$\alpha = 0.2\alpha_{BRD}$ ($\alpha = 1000\alpha_{LS}$)

Figure 2:

$\theta(p,q)$: Same as for Fig. 1.

$$\eta(q) = (1.965 \cdot 10^{-4})[e^{-(4.856 \cdot 10^{-7})(q-6600)^2} + e^{-(4.856 \cdot 10^{-7})(q-13200)^2}]$$

(q in joules)

Range of p: $0 \leq p < 0.01$ torr: 15 points N = 51
$0.01 \leq p < 2$: 25
$2 \leq p \leq 3$: 11

Range of q: $0 \leq q \leq 4 \cdot 10^4$ joules. T: Same as for Fig. 1.

Accuracy of isotherm data: Same as for Fig. 1.

Smoothing parameters: $\alpha_{opt} = 0.05\alpha_{BRD}$

Figure 3:

$\theta(p,q)$: Same as for Figs. 1 and 2. $\eta(q)$: Same as for Fig. 2.

Range of p: Same as for Fig. 3. N = 51

Range of q: $0 \leq q \leq 6 \cdot 10^4$ joules. T: Same as for Figs. 1 and 2.

Accuracy of isotherm data: Random, normally distributed errors with a standard deviation of 3% were imposed on the isotherm data.

Smoothing parameters: $\alpha_{opt} = 0.006\alpha_{BRD}$

Figure 4:

$\theta(p,q)$: Same as for Figs. 1-3.

$$\eta(q) = (1.965 \cdot 10^{-4})[e^{-(1.942 \cdot 10^{-6})(q-8619)^2} + e^{-(1.942 \cdot 10^{-6})(q-11492)^2}]$$
$$+ (9.826 \cdot 10^{-4})[e^{-(4.854 \cdot 10^{-5})(q-10056)^2} + e^{-(4.854 \cdot 10^{-5})(q-12929)^2}]$$

(q in joules)

Range of p: $0 \leq p < 0.01$ torr: 15 points N = 51
$0.01 \leq p < 1$: 30
$1 \leq p \leq 3$: 6

Range of q: $0 \leq q \leq 8 \cdot 10^4$ joules. T: Same as for Figs. 1-3.

Accuracy of isotherm data: Random, normally distributed errors with a standard deviation of 0.01% were imposed on the isotherm data.

Smoothing parameters: $\alpha_{opt} = 0.006\alpha_{BRD}$

Figure 5:

$\theta(p,q)$: Same as for Figs. 1-4. $\eta(q)$: Same as for Fig. 4.

Range of p: $0 \leq p < 0.01$ torr: 5 points N = 18
$0.01 \leq p < 1$: 10
$1 \leq p \leq 3$: 3

Range of q: Same as for Fig. 4. T: Same as for Figs. 1-4.

Accuracy of isotherm data. Same as for Fig. 4.

Smoothing parameters: $\alpha_{opt} = 0.1\alpha_{BRD}$

ON THE KINETICS OF MOLECULAR REARRANGEMENT IN ADSORBED FILMS

Cl. Buess-Herman, L. Gierst
Department of Chemistry, Faculty of Sciences, C.P. 160,
Université Libre de Bruxelles, 50, av. F.D. Roosevelt,
1050 Brussels, Belgium.

ABSTRACT

According to the value of the local electrical field, dipolar surfactants may present at the mercury-water interface various surface states involving the formation of 2D gaseous, liquid or solid phases with different configurations. The kinetics of the associated phase transitions is interpreted on the basis either of the stochastic formation of nuclei (liquid → solid) or by the seeding effect of a preexisting network of defects (solid → liquid).

\- - - - - -

The adsorption of neutral organic surfactants at an electrified interface is usually characterized by the gradual variation of the electrocapillary parameters (double layer capacity, charge density, interfacial tension) and the superficial excess values when the electrode potential is progressively scanned.

Some organic molecules dissolved in an aqueous electrolyte may adsorb at the mercury electrode in the form of compact monolayers which are stabilized by lateral short-range forces. This behaviour can be described by a Frumkin type of isotherm, exhibiting a typical S-shape which presents forbidden regions when the lateral interactions are strong enough. It is not too infrequent to observe well-defined immiscible surface states separated from each other by a sharp 2D phase transition (1-3). These structural changes result in drastic variations of the double layer capacity, charge density and superficial excess as shown in figure 1.

On the basis of a detailed electrocapillary study, supplemented by the evaluation of the inhibition power of the adsorbed films towards faradaic reactions (4, 5) three distinct superficial states have been proposed for the adsorption of isoquinoline (isoQ) : dilute flat oriented molecules, a mixture of molecules flat and standing perpendicularly to the electrode and a monolayer with a variable population of local clusters at less negative potentials.

The domain of prevalence of each film and thus the position of the transitions depend markedly on the surfactant concentration, the electrolyte concentration and the temperature. The shifts of the transition potential (E_T) are interpreted in terms of the relative position of the

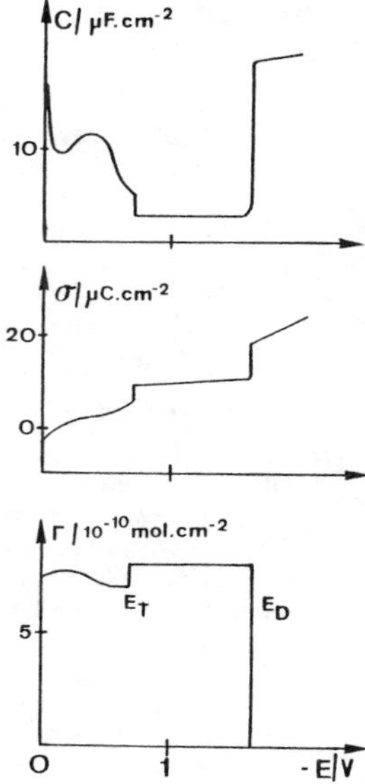

Fig. 1 : Comparison between the capacity, charge and superficial excess plot of isoQ at saturation in 0.5M Na_2SO_4. E_T and E_D are the potentials related to the rearrangement transition and the desorption.

interfacial tension versus potential curves representative of the two structures.
The most compact monolayers stretch out between the phase transition E_T and the desorption discontinuity E_D observed at very negative potentials (Figure 1).
A similar study has shown (6) that in concentrated solutions, where mass transfer is negligible, the major traits of the adsorption behaviour of 3-methyl-isoquinoline (3-MeisoQ) are comparable to those of isoquinoline, at least qualitatively. Rearrangement and desorption transitions, condensed layers and the existence of invariant properties at negative potentials are also observed when the electrode potential or the surfactant concentration is modified.

Figure 2 shows the potential dependence of the double-layer capacity in the presence of 3-MeisoQ at a concentration corresponding to 95% of the saturation value.
For these systems, the study of the kinetic aspects of the 2D phase transitions is made particularly easy because the thermodynamics of the adsorption is known and interferences such as mass transfer, ohmic drop, are much less severe than for similar 2D faradaic processes which involve high faradaic currents.

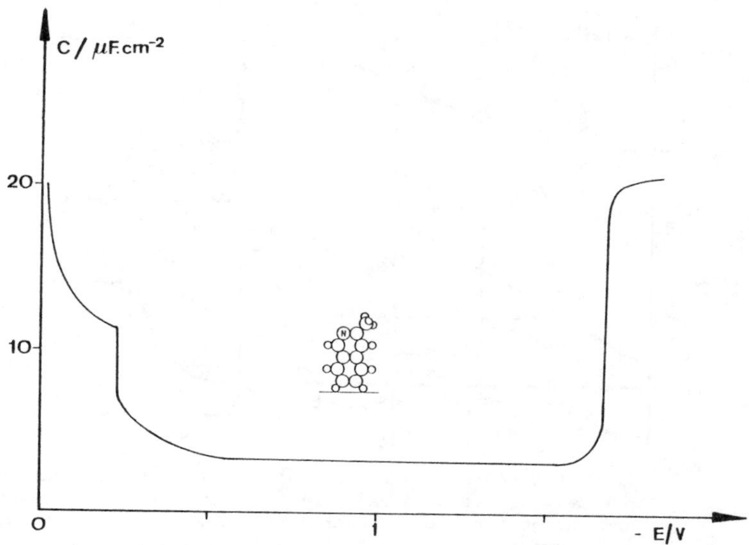

Fig. 2 : 117 Hz differential capacity curve of 3-MeisoQ at 95% of its saturation value in 0.5M Na_2SO_4 at 25°C.

When charge-potential or current-potential curves are recorded in the presence of solutions of 3-MeisoQ under classical polarographic conditions (slow potential ramp) a steep charge step is constantly observed around -0.24V. It is generally proceded by a few erratic peaks which develop at less negative potentials. The occurrence of prior random peaks suggests that the charge step at -0.24V is kinetically controlled by a spontaneous nucleation process similar to the transition displayed by isoquinoline (7).
Accordingly, the extraneous peaks must occur at potentials more negative than the true thermodynamic transition potential. The accurate evaluation of the latter has been performed with double potential step experiments designed with a four seconds prepolarization potential such that the most condensed "solid" like structure is formed, followed by an anodic potential square pulse whose amplitude increases slowly for each successive drop.

Adequate potential step programmes can be easily used in electrochemistry to control the rearrangement process and to elucidate its kinetics by proper analysis of the resulting capacitive transients. When the potential is suddenly stepped from one side of the transition to the other (Figure 3) the resulting current or charge transients are made of two consecutive parts which are in the present case generally well separated. At short times a fast decaying current or charge step is observed resulting from the polarization of the initial film, before the

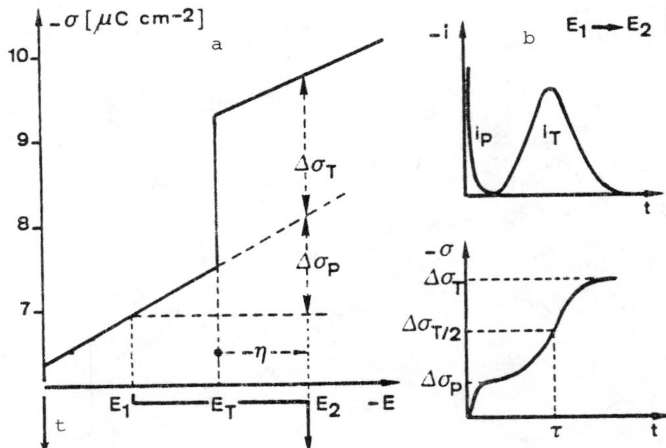

Fig. 3 : Schematic representation of a) the equilibrium charge-potential curve at 4s and b) the corresponding current and charge transients observed after a single potential step performed from E_1 to E_2 (with an overvoltage of η). $\Delta\sigma_p$ and i_p are the fast charging components of the transients while $\Delta\sigma_T$ and i_T are the values associated with the relaxation process.

onset of appreciable restructuration. The second part of the transient unravels the intrinsic characteristics of the phase rearrangement. When the electrode surface is initially covered by a "solid-like" film corresponding to high superficial excesses and a high compactness inducing strong inhibition effects, and if the potential is suddenly stepped to a value less negative than the thermodynamic transition potential, the progressive melting of the film is characterized by current transients of the falling type. Their half-life time decreases rapidly with the overvoltage η (where η is equal to $E - E_T$). The marked dependence of this behaviour on the circumstances of formation of the "solid - like" layer suggests the presence of a network of superficial imperfections acting as seeding sites.

By contrast with the "solid→liquid" process, the 2D "liquid→solid" transitions displayed by 3-MeisoQ as well as isoQ gives typical peaked i-t transients (Figure 4) which are representative of polynucleation and growth mechanisms of the Avrami type (8) as given by equations (1-4).

P.P.N.

$$i = \sigma_T \pi k_N k_G^2 t^2 \exp.(- \frac{\pi}{3} k_N k_G^2 t^3) \qquad (1)$$

$$\sigma = \sigma_T [\ 1 - \exp.(- \frac{\pi}{3} k_N k_G^2 t^3)\ ] \qquad (2)$$

I.P.N.

$$i = \sigma_T \, 2\pi N_N k_G^2 \, t \exp(-\pi N_N k_G^2 t^2) \quad (3)$$

$$\sigma = \sigma_T \left[1 - \exp.(-\pi N_N k_G^2 t^2) \right] \quad (4)$$

where σ_T is the charge corresponding to the full process and N_N the number of pre-existing nuclei.

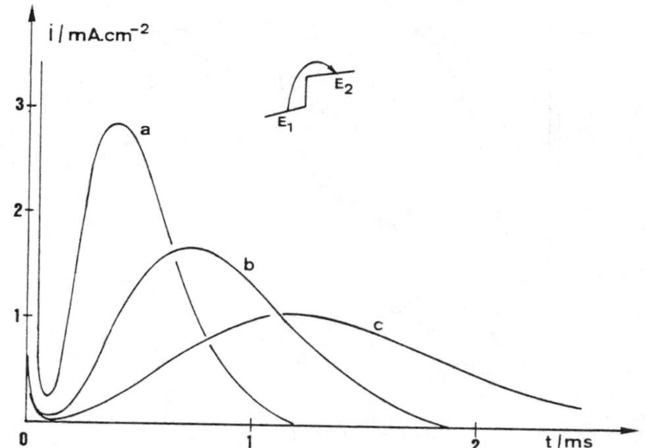

Fig. 4 : Current transients of isoQ observed after a single potential step with an initial potential of -647 mV and variable overpotentials η.
a) 103 mV; b) 96 mV; c) 91 mV.

Unfortunately, the rate constant of nucleation k_N (expressed in nuclei per unit surface and time) and the rate constant of growth k_G (expressed in length per second) appear always as a product and cannot be separated from each other.

By resorting to well-designed double potential step experiments, it is possible to trigger at will the so-called instantaneous (I.P.N.) or progressive (P.P.N.) polynucleation and growth mechanisms. From this kind of measures (Fig. 5), one can easily determine the potential dependence of k_N and k_G.

It is generally assumed (in analogy with the case of 3D phase transformations) that the rate of nucleation is controlled by the probability (9) of formation of circular nuclei having attained the critical size which is determined by the antagonist effects of the overpotential on the difference of the free energy of the two adsorbed layers and the specific edge tension which develops at the periphery of each cluster of new phase.

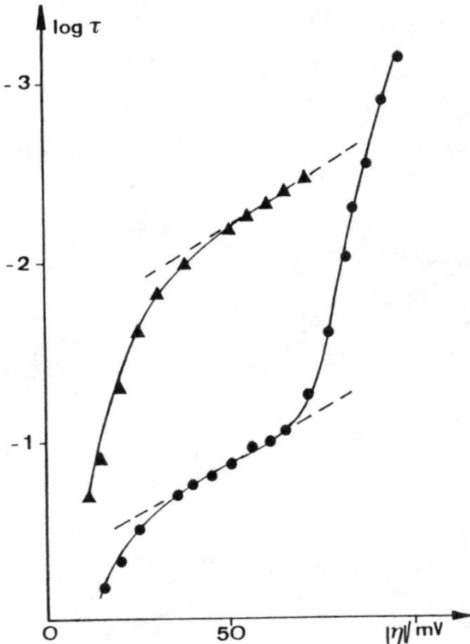

Fig. 5 : Effect of the overpotential on the half-times of the "liquid-solid" transition of 3-MeisoQ. (●) single-potential steps with an initial potential of -185 mV; (▲) double-potential steps (initial potential -185 mV) with a 500 μs pulse at -185 mV before stepping at variable η. The dotted lines correspond to $d \log k_G/d|\eta| = 12\ V^{-1}$.

If experiments can be carried out in such a way that only a few nuclei are allowed to growth, the morphology of the rearrangement transients is no longer given by the Avrami equations, but depends markedly on the geometry and size of the electrode as well as on the position and birth times of the critical nuclei (10, 11). In some cases (highly purified solutions and small electrodes), the full process stems from a single nucleus (10, 12) - a unique circumstance which makes possible the deconvolution of the rate constants of nucleation and growth.

The kinetic analysis of the mononuclear, oligonuclear and polynuclear regimes related to the isoQ and 3-MeisoQ rearrangement transitions gives the overpotential effect on the rate constants k_N and k_G from which the specific edge tensions can be easily derived (Table 1). It appears that the adsorption kinetics of 2D phase transformation does not differ significantly from that observed with isoquinoline although the presence of the methyl group weakens the stability of the flat and tilted orientations at the mercury electrode, as reflected by the appreciable positive shift of the transition potential.

	IsoQ	3-MeisoQ
$\frac{d \ln k_G}{d\lvert\eta\rvert}$	41 V^{-1}	28 V^{-1}
$\frac{d \ln k_N}{d\lvert 1/\eta\rvert}$	-2.58 V	-3.6 V
specific edge tension ε (J/m)	7.6 10^{-12}	1.2 10^{-11}

Table 1 : Comparison of the kinetic parameters for the "liquid→solid" like transition of isoquinoline and 3-methyl-isoquinoline.

As a whole, the kinetics of 2D phase transformations involving adsorption presents remarkable analogies with other monolayer processes such as electrodeposition and anodic passivation.

LITERATURE CITED

1. Armstrong, R.D., J. Electroanal. Chem. 20, 168 (1969).

2. Brabec, V., Kim, M.H., Christian, S.D., Dryhurst, G., J. Electroanal. Chem. 100, 111 (1979).

3. Vanlaethem, N., Lambert, J.P., Gierst, L., Chem. Ing. Technol. 4, 219 (1972).

4. Chevalet, J., Rouelle, F. Gierst, L., Lambert, J.P., J. Electroanal. Chem. 39, 201 (1972).

5. Buess-Herman, Cl., Vanlaethem-Meurée, N., Quarin, G., Gierst, L., J. Electroanal. Chem. 123, 21 (1981).

6. Buess-Herman, Cl., Quarin, G., Gierst, L., J. Electroanal. Chem. 148, 79 and 97 (1983).

7. Quarin, G., Buess-Herman, Cl., Gierst, L., J. Electroanal. Chem. 123, 35 (1981).

8. Avrami, M., J. Chem. Phys. 7, 1103 (1939)
   8, 212 (1940)
   9, 177 (1941).

9. Zettlemoyer, A.C., Nucleation, Marcel Dekker Publications, New York (1969).

10. Gierst, L., Franck, G., Quarin, G., Buess-Herman, Cl., J. Electroanal. Chem. 129, 353 (1981).

11. Smith, A., Fletcher, S., Electrochimica Acta 25, 583 and 889 (1980).

12. Budevski, E., Bostanoff, W., Witanoff, T., Stoinoff, Z., Kotzewa, A., Kaischev, R., Electrochimica Acta 11, 1697 (1966).

EXPERIMENTAL ANALYSIS AND MODELING OF ADSORPTION SEPARATION OF CHLOROTOLUENE ISOMER MIXTURES

S. Carrà, M. Morbidelli, G. Storti and R. Paludetto
Dipartimento di Chimica Fisica Applicata del Politecnico
Piazza Leonardo da Vinci, 32 - 20133 Milano I T A L Y

ABSTRACT

The separation of a mixture of ortho and para-chlorotoluene has been performed through adsorption on 13x zeolite, using toluene as desorbent. A reliable mathematical model of the adsorber dynamics has been developed. The influence of the desorbent characteristics on the separation process efficiency has been investigated in detail.

MATHEMATICAL MODEL OF THE ADSORPTION UNIT

The mathematical model of the adsorption unit is constituted by the mass balances of all the present components in the bulk flowing phase and in the adsorbent particle. With reference to the generic i-th component, the first one can be written as follows

$$\varepsilon \frac{\partial C_i}{\partial t} + \frac{\partial (uC_i)}{\partial z} = \varepsilon D_g \frac{\partial^2 C_i}{\partial z^2} - (1-\varepsilon) N_i \tag{1}$$

where N_i indicates the gas-solid molar rate per unit particle volume. Combining Equation (1) with the equation of state of the mixture, whose behavior is assumed to be close to ideality, the following relationship for the evaluation of the superficial velocity u, is derived

$$\frac{\partial u}{\partial z} = - (1-\varepsilon) \sum_{i=1}^{NC} N_i/\tilde{\rho}_i \tag{2}$$

The formulation of mass balances in the adsorbent particle is complicated by the bidisperse porosity structure which characterizes synthetic zeolites particles (Ruthven and Loughlin, (1)). However as in most cases of practical interest (Ruthven and Lee, (2)), also in the case under examination the mass transfer resistance due to the microporosity in the zeolite crystals is negligible, with respect to those due to the external boundary layer and to the particle macropores. The latter has been described using the lumped model proposed by Glueckauf (3), which has been shown to accurately simulate also multicomponent diffusion processes (Morbidelli et al., (4)). The mass balance of the i-th component in the adsorbent particle is then given by

$$\varepsilon \frac{\partial C_{pi}}{\partial t} = N_i - \rho_s (1-\varepsilon_p) \frac{\partial \Gamma_i}{\partial t} \qquad (3)$$

$$N_i = a_p k_{gi} (C_i - C_{pi}) \qquad (4)$$

where the global mass transfer coefficient k_{gi} is given by

$$(k_{gi})^{-1} = (k_{gi})^{-1} + (k_{pi})^{-1} \quad ; \quad k_{pi} = 5\, D_{pi}/R_p \qquad (5)$$

Equations (1) to (3) are coupled with suitable boundary and initial conditions, which have been reported elsewhere, together with a detailed description of the model (Morbidelli et al., (5)).

The equilibrium conditions between the fluid phase in the macropores and the adsorbate phase on the zeolite crystals are described through the empirical adsorption isotherm proposed by Fritz et al. (6)

$$\Gamma_i/\Gamma_i^{\infty} = K_i C_{pi}^{\beta i} / (1 + \sum_{j=1}^{NC} C_{pj}^{\beta j} K_j) \simeq \alpha_i C_{pi}^{\beta i} / \sum_{j=1}^{NC} C_{pj}^{\beta j} \alpha_j \qquad (6)$$

where $\alpha_i = K_i/K_p$ is the ratio between the equilibrium constants of the i-th component and p-chlorotoluene, and β_i is an empirical parameter, such that for $\beta_i = 1$, Equation (6) reduces to the well known Langmuir adsorption isotherm. The approximation introduced in Equation (6): $\sum_{j=1}^{NC} C_{pj}^{\beta i} K_j >> 1$, can be justified on the basis of the peculiar characteristic of the processes based on displacement chromatography. In it the desorbent exhibits an affinity to the adsorbent which is comparable to the one exhibited by the components to be separated (in contrast to the eluent in elution chromatography), and then the equilibrium ratios of all the compounds in the fluid mixture are of the same order of magnitude, which is usually quite large. It is finally remarkable that due to the large total concentration of adsorbable components, which exhibit very similar saturation concentration Γ_i^{∞}, the total concentration on the adsorbent is always very close to saturation conditions. Therefore, during the adsorption process equimolar counterdiffusion conditions are closely approximated, and, if the adsorption heats of the various involved components are comparable, the temperature inside the unit remains almost constant.

The above described model has been used to simulate the separation process of a (1:1) mixture of o- and p-chlorotoluene through adsorption on commercial 13 zeolite, fully exchanged with calcium, using toluene as desorbent and operating at T = 230°C under atmospheric pressure.

COMPARISON WITH EXPERIMENTAL DATA

All the parameters appearing in the model have been estimated through suitable literature expressions; in particular the axial dispersion coefficient, D_{gi} (Butt, (7)), the macropore diffusion coefficient, D_{pi} (Wheeler, (8)) and the external mass transfer coefficient k_{gi} (Petrovic and Thodos, (9)). The only exception are the equilibrium parameters: the saturation concentration Γ_i^∞ and the equilibrium constants ratio $\alpha_i=K_i/K_p$. The first one has been estimated through independent experimental measurements, leading to $\Gamma_i^\infty = 1.31$, 1.25 and 1.31 x 10^{-3} mol/g for o-chlorotoluene, p-chlorotoluene and toluene, respectively. The parameter α_i has been estimated by comparison of the model results with the experimental breakthrough curves. These were performed in a cyclic mode; i.e. adsorption of a binary mixture up to saturation (stirring with the column saturated with the desorbent), and then desorption with a pure desorbent feedstream. All the examined components have been alternatively used as desorbent in order to investigate in detail all the possible interactions among the various components. The system of partial differential equations has been solved through the orthogonal collocation method, which has been found to be the most efficient among several others (Morbidelli et al., (10)).

In Figure 1 a comparison of the model results with a typical experimental run is shown (Morbidelli et al., (5)). Two models have been considered, in the first one (solid line) the Langmuir adsorption isotherm has been used (i.e., $\alpha_p = 1$, $\alpha_o = 1.5$, $\alpha_t = 2.0$, $\beta_p = \beta_o = \beta_t = 1$ in Equation (6)), while in the second one (dotted line) different values of α and β have been used for toluene ($\alpha_t = 0.5$, $\beta_t = 0.7$). The qualitative difference between the two models is in the prediction of the selectivity $S_{ij} = (\Gamma_i/C_i)/(\Gamma_j/C_j)$, which is constant for the Langmuir model and concentration dependent for the second one. It follows that the intersection between the desorption curves of toluene and o-chlorotoluene shown in Figure 1, which implies an inversion on the selectivity value along the adsorber axis under isothermal conditions, can not be predicted using the Langmuir adsorption isotherm.

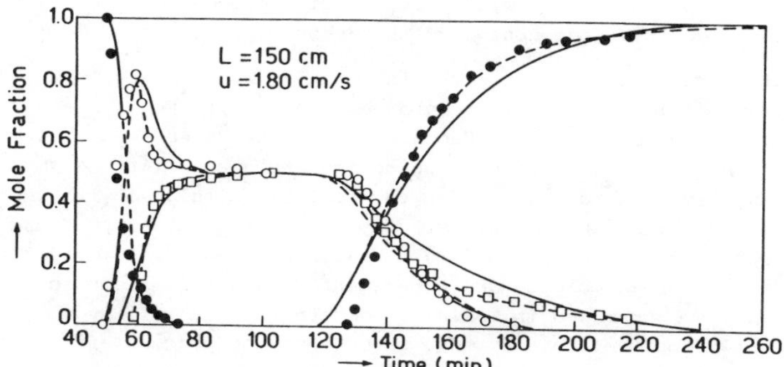

FIG. 1 - ADSORPTION AND DESORPTION OF A 1:1 MIXTURE OF O-CHLOROTOLUENE (o) - TOLUENE (□) WITH P-CHLOROTOLUENE (●) AS DESORBENT.

SELECTION OF THE APPROPRIATE DESORBENT

On the whole the developed model gives a satisfactory representation of the experimental data. Therefore, it constitutes a valuable tool for the optimal design and operation of adsorption units. In particular, the model can be used to identify, with a very limited experimental effort, the most efficient desorbent, which constitutes a key factor in the economy of adsorption separation processes based on displacement chromatography. It is widely recognized (Seko et al., (11)) that the appropriate desorbent must exhibit an affinity value to the adsorbent intermediate between those of the two components to be separated. However, after a preliminary analysis of various components, no such a desorbent could be found. Therefore, attention has been focused on the strongest of the weak desorbents (monochlorobenzene, $\alpha_m = 0.7$) and the weakest of the strong desorbents (toluene, $\alpha_t = 2.0$). The efficiency of each of them has been tested through the analysis of the separation of a given pulse ($\Delta t = 20$ min.)of the (1:1) mixture of o- and p-chlorotoluene. The outlet stream is separated into three regions: the first and third ones are binary mixtures of the separated components and the desorbent, and are fed to the final distillation process, while the second one is a ternary mixture which is then recycled back to the adsorber. The evaluation of the separation performance is based on two overall parameters:

P_i = (pseudocontinuos flowrate of separated i)/(feed flowrate of i) (7)

where the numerator is evaluated as the ratio between the total moles of the i-th component separated in one cycle of the single adsorber and the duration of the cycle itself; P_i is an index of the recycle cost since it is proportional to the fraction of the i-th component separated without need of recycle, and

$\bar{x}_i$ = average mole fraction of i in the separate stream (8)

proportional to the cost of final distillation process, necessary to separate the desired component from the desorbent.

The results obtained are summarized in Table 1, where monochlorobenzene, toluene, and (1:1) mixture of these have been used in run 1,2 and 3, respectively. In run 4 the two desorbents have been fed separately, first toluene ($\Delta t = 15$ min) and then monochlorobenzene ($\Delta t = 43$ min). All the experimental runs have been successfully simulated with the model; for run 4, which is the most complicated since the procedure must be repeated in a cyclic mode up to attain steady state conditions, the comparison of calculated and experimental values is shown in Figure 2. The results summarized in Table 1 indicate the possibility to significantly improve at least some aspects of the process efficiency through the simultaneous use of two different desorbents. However, general conclusions can not be drawn, due to the strong effect of the characteristics of the available desorbents, of the particular adopted

Table 1 - Comparison of the separation process performance using various desorbents.

run	P_p	P_o	$\bar{x}_p$	$\bar{x}_o$
1	0.178	0.150	0.45	0.13
2	0.140	0.198	0.19	0.27
3	0.042	0.102	0.17	0.14
4	0.176	0.134	0.53	0.25

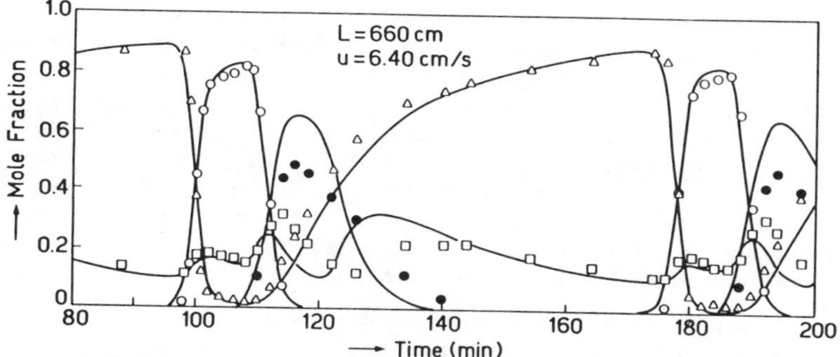

FIG. 2 - ELUTION CURVES FOR RUN 4 OF TABLE 1. EXPERIMENTAL POINTS: (o) O-CHLOROTOLUENE, (●) P-CHLOROTOLUENE, (△) MONOCHLOROBENZENE, (□) TOLUENE. CALCULATED CURVES WITH FRITZ ISOTHERM (—).

separation scheme, of the possible optimization of the composition of the two desorbent mixtures alternatively fed to the unit, and of the distillation cost for the separation of the two desorbents (Morbidelli et al., (12)). Nevertheless, it is worthwhile to investigate the desorbent influence somewhat more in detail possibly using a very simple, although approximate, mathematical model.

ANALYSIS OF THE DESORBENT EFFICIENCY

The aim of this section is to briefly analyse the influence of the desorbent on the efficiency of the separation process, which will be measured through the parameters P_i and $\bar{x}_i$ previously defined. The chromatographic equilibrium theory (using the Langmuir adsorption isotherm) has been used in order to obtain analytical results which greatly simplify the comparison of the various desorbent performances. Such a theory, reported in detail by Rhee et al. (13) and Helfferich and Klein (14), gives the analytical solution of the above reported model, under the assumption of negligible mass transfer resistances, axial dispersion and constant interstitial velocity. Its application retains the essential features of the complete model, while leading to a less accurate approximation of the experimental behavior of the unit.

In order to compare the efficiency of various desorbents, it is first necessary to establish the operating scheme of the adsorption process.

Bailly and Tondeur (15, 16) have analysed several operating schemes: classical, recycle and two way chromatography and defined their relative convenience. Moreover, it was pointed out that the optimal performance in the case of recycle chromatography is closely approximated by using the equilibrium theory in the absence of wave interactions. Therefore, for each examined desorbent, the value of the dimensionless parameter $\Delta = u\,\Delta t_p/L\varepsilon$ is evaluated in order to avoid wave interactions, This allows, for example, for a fixed value of the column length L, to estimate the value of the feed pulse duration Δt_p which closely approximates the optimal performance of the operating scheme with recycle. In Figure 3 the performance of the separation process, expressed in terms of P_i and $\bar{x}_i$ is shown as a function of the desorbent equilibrium parameter α_d. It is remarkable that the comparison among the various desorbents is fully general, since using the equilibrium theory, it can be readily shown that the values of P_i and $\bar{x}_i$ are only functions of the equilibrium parameters α_i of the involved components. From Figure 3, it appears that the best performance is exhibited by the desorbent characterized by a α_d value intermediate between those of the two components to be separated i.e., 1 and 1.5.

The performance of the process with two different desorbent mixtures, d1 and d2, which are successively fed to the unit after the pulse of the mixture to be separated, is investigated. Again the duration of the pulse of each feedstream is evaluated in order to avoid wave interactions, leading to values which closely approximate the optimal operating conditions in the scheme with recycle. Several compositions of the two desorbent mixtures have been examined, and the obtained results are summarized in Figure 4. In it the mixture composition is represented by the parameter

$$\omega_d = \alpha_m \alpha_t / (\alpha_m x_m + \alpha_t x_t) \qquad (9)$$

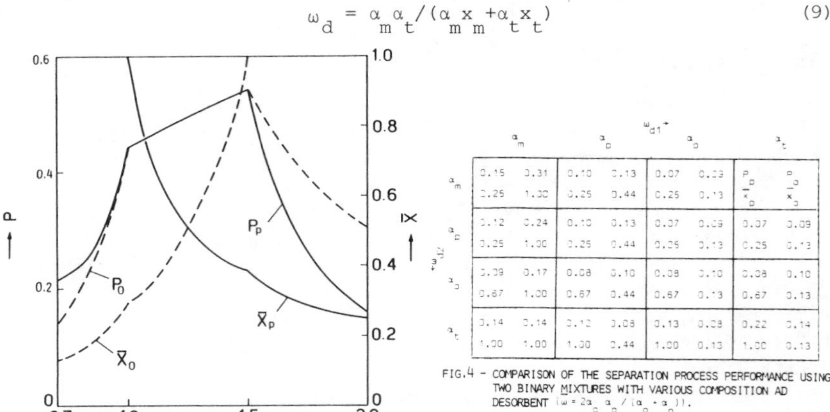

FIG. 3 - PROCESS PARAMETERS P AND X FOR P- AND O-CHLOROTOLUENE AS A FUNCTION OF THE DESORBENT EQUILIBRIUM PARAMETER α_d

$\omega_{d1} \rightarrow$ (columns); $\omega_{d2} \downarrow$ (rows)

	α_m		α_p		α_o		α_t	
α_m	0.15	0.31	0.10	0.13	0.07	0.09	P_p	P_o
	0.25	1.00	0.25	0.44	0.25	0.13	$\bar{x}_p$	$\bar{x}_o$
α_p	0.12	0.24	0.10	0.13	0.07	0.09	0.07	0.09
	0.25	1.00	0.25	0.44	0.25	0.13	0.25	0.13
α_o	0.09	0.17	0.08	0.10	0.08	0.10	0.08	0.10
	0.67	1.00	0.67	0.44	0.67	0.13	0.67	0.13
α_t	0.14	0.14	0.12	0.08	0.13	0.08	0.22	0.14
	1.00	1.00	1.00	0.44	1.00	0.13	1.00	0.13

FIG. 4 - COMPARISON OF THE SEPARATION PROCESS PERFORMANCE USING TWO BINARY MIXTURES WITH VARIOUS COMPOSITION AD DESORBENT $(\omega = 2\alpha_o \alpha_p / (\alpha_o + \alpha_p))$.

where x_m and x_t indicate the mole fraction of monochlorobenzene and toluene in the desorbent mixture. The points on the diagonal represent the case of a unique desorbent, i.e. $\omega_{d1} = \omega_{d2}$. It can be seen that the performance of the mixture is inferior to that of the two single desorbents, for all possible compositions. From inspection of Figure 4 it appears that the two most efficient processes are obtained by using successively pure toluene and pure monochlorobenzene ($\omega_{d1} = \alpha_m$, $\omega_{d2} = \alpha_t$) or using only pure toluene ($\omega_{d1} = \omega_{d2} = \alpha_m$). The first one leads to high purity of the separated stream of the desired component, so reducing distillation cost, while the second reduces recycle cost.

It is apparent that the final decision depends on the particular case under examination; in particular, on the relative cost of the distillation and the recycle processes and on the equilibrium parameter values of the available desorbents. However, it can be concluded that the use of two desorbents, one strong and the other weak, in cases where the appropriate desorbent is not available can lead to significant improvements of the separation process efficiency.

ACKNOWLEDGMENT: The financial support of the Italian Consiglio Nazionale delle Ricerche (Progetto Finalizzato Chimica Fine e Secondaria) is gratefully aknowledged.

NOTATION

a_p	specific external surface of the zeolite crystals, cm^2/cm^3
C, C_p	concentration in external and macroporosity gas phase, respectively mol/L
D_g	axial dispersion coefficient, cm^2/s
D_p	diffusion coefficient into macroporosity, cm^2/s
k_f	external,global and internal mass transfer coefficient, cm/s
K	equilibrium constant of adsorption, L/mol
L	length of the column, cm
N	molar rate of mass transfer, mol/L s
$\dot{P}_i$	ratio between the total moles separated in one cycle and the product of the cycle duration and feed flowrate, for the i-th component
R_p	radius of spherical particle equivalent to the solid pellet, cm
t	time, s
u	superficial fluid velocity, cm/s

$\bar{x}_i$ average mole fraction of the i-th component in the separated stream

z axial length coordinate, cm

α_i equilibrium constants ratio, K_i/K_p

Γ concentration in solid phase, mol/g

Δt_p, Δt_c pulse feed and total cycle duration, min

ε, ε_p external and intraparticle void fraction

$\tilde{\rho}$ molar density, g/mol

ρ effective density of solid, g/L

LITERATURE CITED

1. Ruthven, D.M. and K.F. Loughlin, Can. J. Chem. Eng., 50, 550 (1972).
2. Ruthven, D.M. and L.K. Lee, AIChE J., 81 (1981).
3. Glueckauf, E., Trans. Faraday Soc., 51, 540 (1955).
4. Morbidelli, M., Servida, A., Storti, G. and S. Carrà, Ind. Eng. Chem. Fundam., 21, 123 (1982).
5. Morbidelli, M., Storti, G., Niederjaufner, G., Pontoglio, A. and S. Carrà, Chem. Eng. Sci., in press.
6. Fritz, W. and E.V. Schlünder, Chem. Eng. Sci., 29, 1279 (1974).
7. Butt, J.B., "Reaction Kinetics and Reactor Design", Prentice-Hall, Engelwood Cliffs, N.J. (1978), Ch. 3.
8. Wheeler, A., "Catalysis", P.H. Emmet (Ed.), Reinhold, New York (1955), Vol. 2.
9. Petrovic, L.J. and G. Thodos, Ind. Eng. Chem. Fundam., 7, 274 (1968).
10. Morbidelli, M., Storti, G. and S. Carrà, Proceedings of the 3rd International Conference "Informatique ed Génie Chimique", Vol. II: C-98, Paris (1983).
11. Seko, M., Miyake, T. and K. Inada, Hydrocarbon Process., 59 (1), 133 (1980).
12. Morbidelli, M., Storti, G., Niederjaufner, G., Pontoglio, A. and S. Carrà, to be published.
13. Rhee, H., Aris, R. and N.R. Amundson Phil. Trans. Roy. Soc. London, A267, 419 (1970).

14. Helfferich, F. and G. Klein, "Multicomponent Chromatography", M. Dekker, New York (1970).

15. Bailly, M. and D. Tondeur, Chem. Eng. Sci., 37(8), 1199 (1982).

16. Bailly, M. and D. Tondeur, Chem. Eng. Sci., 36, 455 (1981).

MEDICINAL APPLICATIONS OF ADSORBENTS

David O. Cooney

Chemical Engineering Department
University of Wyoming
Laramie, Wyoming

ABSTRACT

This paper reviews literature on medicinal applications of adsorbents which has appeared since the publication, in early 1980, of the book Activated Charcoal: Antidotal and Other Medical Uses, by the same author. Further impressive reports on the effectiveness of orally-administered activated charcoal in greatly reducing the systemic absorption of various drugs and poisons have continued to appear. Increased studies of resins and clays have shown these adsorbents to be quite effective also in certain cases. Many new evaluations of the effects of combining various salts and cathartics with activated charcoal have appeared. New roles for activated charcoal (e.g., in treating wounds and skin disorders) have been identified.

INTRODUCTION

A very extensive treatise on the use of activated charcoal for antidotal and other medical purposes appeared about three years ago in the book Activated Charcoal: Antidotal and Other Medical Uses by Cooney (1980). This book discusses the various treatment methods that are available for handling drug overdose and poisoning victims (e.g., supportive therapy, gastric lavage, administration of emetics, and the use of sorbents) and argues strongly in favor of the choice of activated charcoal.

This work reviews the extensive history of the use of activated charcoal for antidotal purposes, discusses the manufacture and properties of activated charcoal, and surveys the large number of reports in the medical literature in which activated charcoal was used to treat humans and various animals after they had ingested drugs such as sedatives, hypnotics, analgesics, antidepressants, etc. Over 250 references attest to the general and powerful effectiveness of activated charcoal in binding drugs (and poisons) in the gastrointestinal tract, thereby greatly reducing their systemic absorption.

There is no point in reiterating here the contents of the book cited, and it will be assumed that any reader seriously interested in this subject area will gain access to the book and digest it. We wish to use the limited space available here to update the previous review.

Much new work has appeared in the intervening three years, especially in certain areas that were barely touched-upon previously.

GENERAL REVIEW-TYPE ARTICLES

A few review articles have appeared recently, with one by Neuvonen (1982) on the clinical pharmacokinetics of oral activated charcoal being the most extensive. Brief reviews by Greenscher (1979) and Dipalma (1979) contain only basic information long known; however, a recent editorial by Levy (1982) does discuss some new work.

IN VITRO STUDIES

Three new in vitro studies may be mentioned here. Javaid and El-Mabrouk (1983) studied phenobarbital adsorption on activated charcoal and found no effect of pH, over the pH range 1.2-8.0, on the degree of adsorption--this seems unusual, as phenobarbital is acidic (pKa = 7.6) and should adsorb significantly less strongly near pH 8.0, where it exists 72% in dissociated form. The other two studies involved other adsorbents besides activated charcoal and will be discussed in a later section.

IN VIVO STUDIES

Neuvonen and Elonen (1980a, 1980b) studied the effect of 50 g activated charcoal given orally 5 min after the oral intake of phenobarbital (0.2 g), carbamazepine (0.4 g), or phenylbutazone (0.2 g) in human volunteers. The systemic absorption of all drugs was more than 95% prevented. Charcoal given after several hours delay was much less effective, yet still reduced the serum half-lives of the drugs very significantly, e.g. 118 g charcoal given in five divided doses between 10 and 48 hours after drug ingestion reduced the three drug half-lives by 80, 45, and 30%, respectively. Neuvonen, Elonen, and Mattila (1980) found similar results with dapsone (an antileprotic and anti-inflammatory drug).

Berg et al. (1982) and Goldberg and Berlinger (1982) studied the effects of repeated doses of activated charcoal given over three days on the elimination of phenobarbital from human volunteers. The charcoal decreased the drug serum half-life from 110 hr to 45 hr and increased the total body clearance of the drug from 4.4 to 12.0 mL per kg per hr.

Glab et al. (1982) found that activated charcoal reduced mortality by about 50% when it was given to propoxyphene-dosed rats 30 min after the drug was administered.

Comstock et al. (1982) studied the efficacy of administering a slurry of 100 g activated charcoal to actual patients presenting to an emergency room with sedative-hypnotic or aspirin overdoses. The charcoal was given via a gastric tube following gastric lavage. The charcoal showed significant benefits only in a few patients. It was concluded that the charcoal may have been given too late to be effective and should have been given earlier.

Pond et al. (1981) recount a case of a 66 year-old woman with digitoxin overdose who was treated with activated charcoal given orally every eight hours. The treatment accelerated the drug half-life from 162 hr to 18 hr. This dramatic effect is largely due to the fact that digitoxin undergoes substantial enterohepatic circulation, allowing it to contact the charcoal in the gastrointestinal tract.

North, Thompson, and Peterson (1981) studied ethanol elimination from dogs with and without simultaneously-administered activated charcoal. Although one would not expect ethanol to adsorb well on charcoal, the charcoal treatment reduced blood ethanol concentrations by about 40% in this study.

Tietze and Laass (1979a, 1979b) found that activated charcoal given to rats which had been dosed with various organic solvents (benzene, 1,2-dichloroethane, carbon tetrachloride) decreased blood concentrations significantly _if_ enough charcoal was used.

EFFECT OF ADDED CITRATE SALTS

Easom, Caraccio, and Lovejoy (1982) studied the effects of adding magnesium citrate to activated charcoal given to aspirin-dosed human volunteers. The magnesium citrate had no apparent influence by itself on the urinary excretion patterns which were determined.

In vitro studies of aspirin adsorption by activated charcoal in simulated gastric fluid and simulated intestinal fluid media, both with and without magnesium citrate, were conducted by LaPierre, Algozinne, and Doering (1981). No significant effect of the citrate was noted.

These results contrast markedly with those of Ryan, Spigiel, and Zeldes (1980), who found that sodium salicylate adsorption by activated charcoal in vitro, in a pH 4 medium, was greatly enhanced by the presence of magnesium or sodium citrates, by sodium chloride, and by sodium sulfate. One possible explanation for this effect is that sodium salicylate is more strongly ionized in solution than is aspirin.

EFFECT OF ADDED CATHARTICS

The effects of added saline cathartics (e.g., sodium sulphate) have also been studied by other researchers. Sketris et al. (1982) found that sodium sulphate added to activated charcoal had no incremental effect on patterns observed for aspirin elimination from humans.

Chin, Picchioni, and Gillespie (1981) found that sodium sulphate alone reduced the absorption of aspirin, but not of three other drugs tested, in humans. With three of the four drugs, sodium sulphate enhanced the action of activated charcoal in reducing systemic absorption of the drugs.

Laass (1980) studied the effects of charcoal in organic solvent dosed rats and found that the addition of castor oil or liquid paraffin had, if anything, detrimental effects.

FORMULATIONS OF ACTIVATED CHARCOAL

Further work has continued on attempts to create activated charcoal formulations which are palatable yet effective. A review of previous studies using Medicoal (an effervescent charcoal formulation) has appeared (Anonymous, 1979) and a study of the effectiveness of Medicoal in reducing theophylline absorption in vivo was reported by Helliwell and Berry (1981).

Oppenheim (1980) reported on an effective charcoal mixture flavored with strawberry powder and Chung, Murphy, and Taylor (1982) described a formulation prepared with fructose. Navarro, Navarro, and Krenzelok (1980) mixed bentonite and bentonite plus chocolate syrup with activated charcoal. Bentonite aided palatability, and bentonite plus chocolate syrup was even more highly favored; however, while bentonite did not interfere with the effectiveness of the charcoal in reducing aspirin absorption in vivo, the chocolate syrup interfered significantly.

Cooney (1982) studied the effects of the type and amount of carboxymethylcellulose (CMC) on the in vitro adsorption of sodium salicylate by activated charcoal and recommended that, whatever type of CMC is employed, sufficiently little be used to keep the charcoal mixture "pourable". Picchioni, Chin, and Gillespie (1982) have reported that the addition of substantial amounts of sorbitol to activated charcoal actually somewhat enhanced the effect of charcoal in several in vivo tests with four drugs.

Van de Graff et al. (1982) have found that mannitol and sorbitol had little effect on in vitro acetaminophen adsorption by charcoal, but both "diminished the charcoal inhibition of acetaminophen absorption" in vivo (but not by unduly large amounts).

CONTRAINDICATIONS FOR ACTIVATED CHARCOAL USE

Two chemical antidotes are widely used to prevent liver damage in acetaminophen overdose. These are methionine and N-acetylcysteine.

In vitro tests by Klein-Schwarz and Oderda (1981) have shown that activated charcoal can effectively adsorb both of these agents. Further in vitro work by Chinouth and Czajka (1980) on N-acetylcysteine also showed that this chemical is adsorbed by charcoal. Van de Graff et al. (1982) have reported, however, that charcoal does "not avidly adsorb methionine or acetylcysteine in vitro", and did not decrease the ability of these agents to protect against acetaminophen damage in vivo.

Furthermore, in vivo tests by North, Peterson, and Krenzelok (1981) showed no significant effect of activated charcoal on acetylcysteine blood level curves. Until this discrepancy between in vivo and in vitro behavior is resolved, charcoal should not be administered concurrently with either chemical antidote.

USE OF CLAYS, RESINS, AND OTHER NON-CHARCOAL AGENTS

Interest in evaluating the effectiveness of various clays, synthetic resins, and other non-charcoal sorbents seems to have increased recently. Okonek et al. (1982) evaluated Fuller's earth and bentonite clay, as well as activated charcoal, for treating paraquat poisoning in rats. The charcoal was the best of the three, but all were reasonably effective for moderate paraquat doses.

Browne et al. (1980) conducted an in vitro study of various cation-saturated montmorillonite clays and determined which forms adsorbed a test drug (atrazine) best. Some tests using bentonite clay were also run.

Juhl (1979) compared the effects of a kaolin-pectin suspension versus activated charcoal on aspirin absorption in humans, and found the charcoal to be far superior (70% of the aspirin appeared in the urine following charcoal, whereas 90-95% of the aspirin appeared in the urine after the kaolin-pectin).

Said and Al-Shora (1980) studied the adsorption of oral hypoglycemics (acetohexamide, tolazamide, tolbutamide) on kaolin and on activated charcoal in vitro and found substantial and nearly equal adsorption on both adsorbents (kaolin adsorbed 91-96% as much drug as did the charcoal).

Decker et al. (1981) examined the effectiveness of charcoal, a non-ionic carbonized resin, magnesium silicate, calcium silicate, a commercial gelling agent, and hydrophobic silica in immobilizing kerosene, methanol, and ethylene glycol added to water. The activated charcoal was best, but all agents were similar in effect.

McConnell, Harris, and Moore (1980) studied the effectiveness of activated charcoal and cholestyramine (a strong-base anion exchange resin) in extracting polybrominated biphenyls from the bodies of rats previously fed these chemicals in their diet. Neither sorbent effectively reduced tissue levels of the chemicals; however, the resin did prevent progressive neuropathy from developing.

Cady, Rehder, and Campbell (1979) have observed that cholestyramine given to two digitoxin-overdosed human patients caused a rapid decline in serum digitoxin levels, presumably because this drug is subject to significant enterohepatic cycling.

Scholtens et al. (1982) studied the in vitro adsorption of oxalic acid and glyoxylic acid by activated charcoal, by a series of neutral and ionic resins, and by hydrous zirconium oxide. The zirconium oxide was the best by far. The other sorbents were generally ineffective.

Ganjian, Cutie, and Jochsberger (1980) investigated the in vitro adsorption of cimetidine by activated charcoal, kaolin, talc, and magnesium trisilicate at pH 5. The amounts of cimetidine adsorbed per gram of adsorbent were 26, 0.40, 0.29, and 0.34 mg, respectively. Clearly, activated charcoal was far superior.

OTHER MEDICAL USES FOR ACTIVATED CHARCOAL

Beckett et al. (1980) have tried application of an activated charcoal cloth (made from carbonized and activated rayon cloth) on various infected and malodorous wounds. A striking reduction in wound odor occurred. Further study showed that the cloth strongly adsorbs bacteria (especially gram-negative types).

Pederson et al. (1980) studied the effects of oral activated charcoal on the pruritis (itching) experienced by 11 hemodialysis patients and found that the condition was relieved within eight weeks in all but one patient by a regimen of 6 g charcoal daily.

In further connection with uremic patients, Manis et al. (1980) have reported that oral charcoal reduces serum cholesterol and triglyceride levels in uremic rats (and also in diabetic rats). This effect in uremic humans has previously been reported by this group.

Hall, Thompson, and Strother (1981) report that oral activated charcoal greatly reduced the generation of intestinal gas in test subjects following meals rich in legumes.

Krasopoulos, deBari, and Needle (1980) found that bile salts adsorb well on activated charcoal in vitro and Decker and Corby (1980) similarly found that charcoal adsorbs aflatoxins (toxins occurring in several foods, such as nuts) well in vitro.

Gadgil et al. (1982) administered charcoal to patients receiving methotrexate, an anti-cancer drug, and observed that 25 g charcoal given orally at 12, 18, 24 and 36 hr time reduced serum methotrexate levels significantly after the 18 hr mark.

LITERATURE CITED

Anonymous (1979). Medicoal (effervescent activated charcoal) in the treatment of acute poisoning, Drug Ther. Bull. 17, 7.

Beckett, R., Coombs, T.J., Frost, M.R., McLeish, J., and Thompson, K. (1980). Charcoal cloth and malodorous wounds, Lancet (Sept. 13) 594.

Berg, M.J., Berlinger, W.G., Goldberg, M.J., Spector, R., and Johnson, G.F. (1982). Acceleration of the body clearance of phenobarbital by oral activated charcoal, New Eng. J. Med. 307, 642.

Browne, J.E., Feldkamp, J.R., White, J.L., and Hem, S.L. (1980). Potential of organic cation-saturated montmorillonite as treatment for poisoning by weak bases, J. Pharm. Sci. 69, 1393.

Cady, W.J., Rheder, T.L., and Campbell, J. (1979). Use of cholestyramine resin in the treatment of digitoxin toxicity, Am. J. Hosp. Pharm. 36, 92.

Chin, L., Picchioni, A.L., and Gillespie, T. (1981). Saline cathartics and saline cathartics plus activated charcoal as antidotal treatments, Clin. Toxicol. 18, 865.

Chinouth, R.W., and Czajka, P.A. (1980). N-acetylcysteine adsorption by activated charcoal, Vet. Hum. Toxicol. 22, 392.

Chung, D.C., Murphy, J.E., and Taylor, T.W. (1982). In vivo comparison of the adsorption capacity of "superactive charcoal" and fructose with activated charcoal and fructose, J. Toxicol.-Clin. Toxicol. 19, 219.

Comstock, E.G., Boisaubin, E.V., Comstock, B.S., and Faulkner, T.P. (1982). Assessment of the efficacy of activated charcoal following gastric lavage in acute drug emergencies, J. Toxicol.-Clin. Toxicol. 19, 149.

Cooney, D.O. (1980). "Activated charcoal - Antidotal and Other Medical Uses," Marcel Dekker, N.Y.

Cooney, D.O. (1982). Effect of type and amount of carboxymethylcellulose on in vitro salicylate adsorption by activated charcoal, J. Toxicol. - Clin. Toxicol. 19, 367.

Decker, W.J., and Corby, D.G. (1980). Activated charcoal adsorbs aflatoxin B_1, Vet. Hum. Toxicol. 22, 388.

Decker, W.J., Corby, D.G., Hilburn, R.E., and Lynch, R.E. (1981). Adsorption of solvents by activated charcoal, polymers, and mineral sorbents, Vet. Hum. Toxicol. 23 (Suppl. 1), 44.

Dipalma, J.R. (1979). Activated charcoal - a neglected antidote, Am. Fam. Physician 20, 155.

Easom, J.M., Caraccio, T.R., and Lovejoy, F.H. (1982). Evaluation of activated charcoal and magnesium citrate in the prevention of aspirin absorption in humans, Clin. Pharm. 1, 154.

Gadgil, S.D., Damle, S.R., Advani, S.H., and Vaidya, A.B. (1982). Effect of activated charcoal on the pharmacokinetics of high-dose methotrexate, Cancer Treat. Rep. 66, 1169.

Ganjian, F., Cutie, A.J., and Jochsberger, T. (1980). In vitro adsorption studies of cimetidine, J. Pharm. Sci. 69, 352.

Glab, W.N., Corby, W.G., Decker, W.J., and Coldiron, V.R. (1982). Decreased absorption of propoxyphene by activated charcoal, J. Toxicol. - Clin. Toxicol. 19, 129.

Goldberg, M.J., and Berlinger, W.G. (1982). Treatment of phenobarbital overdose with activated charcoal, JAMA 247, 2400.

Greensher, J., Mofenson, H.C., Picchioni, A.L., and Fallon, P. (1979). Activated charcoal updated, JACEP 8, 261.

Hall, R.G., Jr., Thompson, E., and Strother, A. (1981). Effects of orally administered activated charcoal on intestinal gas, Am. J. Gastroenterol. 75, 192.

Helliwell, M., and Berry, D. (1981). Theophylline absorption by effervescent activated charcoal (Medicoal), J. Int. Med. Res. 9, 222.

Javaid, K.A., and El-Mabrouk, B.H. (1983). In vitro adsorption of phenobarbital onto activated charcoal, J. Pharm. Sci. 72, 82.

Juhl, R.P. (1979). Comparison of kaolin-pectin and activated charcoal for inhibition of aspirin absorption, Amer. J. Hosp. Pharm. 36, 1097.

Klein-Schwartz, W., and Oderda, G. (1981). Adsorption of oral antidotes for acetaminophen poisoning (methionine and N-acetylcysteine) by activated charcoal, Clin. Toxicol. 18, 283.

Krasopoulos, J.C., deBari, V.A., and Needle, M.A. (1980). The adsorption of bile salts on activated carbon, Lipids 15, 365.

Laass, W. (1980). Therapy of acute oral poisonings by organic solvents: treatment by activated charcoal in combination with laxatives, Arch. Toxicol. (Suppl. 4), 406.

LaPierre, G.L., Algozzine, G., and Doering, P.L. (1981). Effect of magnesium citrate on the in vitro adsorption of aspirin by activated charcoal, Clin. Toxicol. 18, 793.

Levy, G. (1982). Gastrointestinal clearance of drugs with activated charcoal, New Engl. J. Med. 307, 676.

McConnell, E.E., Harris, M.W., and Moore, J.A. (1980). Studies on the use of activated charcoal and cholestyramine for reducing the body burden of polybrominated biphenyls, Drug and Chem. Toxicol. 3, 277.

Manis, T., Deutsch, J., Feinstein, E.I., and Lum, G.Y. (1980). Charcoal sorbent-induced hypolipidemia in uremia and diabetes, Amer. J. Clin. Nutr. 33, 1485.

Navarro, R.P., Navarro, K.R., and Krenzelok, E.P. (1980) Relative efficacy and palatability of three activated charcoal mixtures, Vet. Hum. Toxicol. 22, 6.

Neuvonen, P.J. (1982). Clinical pharmacokinetics of oral activated charcoal in acute intoxications, Clin. Pharmacokinet. 7, 465.

Neuvonen, P.J., and Elonen, E. (1980a). Effect of activated charcoal on absorption and elimination of phenobarbitone, carbamazepine, and phenylbutazone in man, Eur. J. Clin. Pharmacol. 17, 51.

Neuvonen, P.J., and Elonen, E. (1980b). Phenobarbitone elimination rate after oral charcoal, Br. Med. J. (Mar. 15) 762.

Neuvonen, P.J., Elonen, E., and Mattila, M.J. (1980). Oral activated charcoal and dapsone elimination, Clin. Pharmacol. Ther. 27, 823.

North, D.S., Peterson, R.G., and Krenzelok, E.P. (1981). Effect of activated charcoal administration on acetylcysteine serum levels in humans, Amer. J. Hosp. Pharm. 38, 1022.

North, D.S., Thompson, J.D., and Peterson, C.D. (1981). Effect of activated charcoal on ethanol blood levels in dogs, Amer. J. Hosp. Pharm. 38, 864.

Okonek, S., Setyadharma, H., Borchert, A., and Krienke, E.G. (1982). Activated charcoal is as effective as Fuller's earth or bentonite in paraquat poisoning, Klin. Wochenschr 60, 207.

Oppenheim, R.C. (1980). Strawberry flavoured activated charcoal, Med. J. Aust. 1, 39.

Pederson, J.A., Matter, B.J., Czerwinski, A.W., and Llach, F. (1980). Relief of idiopathic generalized pruritis in dialysis patients treated with activated oral charcoal, Ann. Intern. Med. 93, 446.

Picchioni, A.L., Chin, L., and Gillespie, T. (1982). Evaluation of activated charcoal-sorbitol suspension as an antidote, J. Toxicol.-Clin. Toxicol. 19, 433.

Pond, S., Jacobs, M., Marks, J., Garner, J., Goldschlager, N., and Hansen, D. (1981). Treatment of digitoxin overdose with oral activated charcoal, Lancet (Nov. 21) 1177.

Ryan, C.F., Spigiel, R.W., and Zeldes, G. (1980). Enhanced adsorptive capacity of activated charcoal in the presence of magnesium citrate N.F., Clin. Toxicol. 17, 457.

Said, S., and Al-Shora, H. (1980). Adsorption of certain oral hypoglycemics on kaolin and charcoal and its relationship to hypoglycemic effects of drugs, Int. J. Pharm. 5, 223.

Scholtens, R., Scholten, J., de Koning, H.W.M., Tijssen, J., ten Hoopen, H.W.M., Olthuis, F.M.F.G., and Feijen, J. (1982). In vitro adsorption of oxalic acid and glyoxylic acid onto activated charcoal, resins, and hydrous zirconium oxide, Int. J. Artif. Organs 5, 33.

Sketris, I.S., Mowry, J.B., Czajka, P.A., Anderson, W.H., and Stafford, D.T. (1982). Saline catharsis: Effect on aspirin bioavailability in combination with activated charcoal, J.Clin. Pharmacol. 22, 59.

Tietze, G., and Laass, W. (1979a). Suitability of carbo medicinalis for the treatment of acute oral poisoning with organic solvents. 3. Adsorption of organic solvents on carbo medicinalis, Pharmazie 34, 253.

Tietze, G., and Laass, W. (1979b). Suitability of carbo medicinalis for the treatment of acute oral poisoning with organic solvents. 4. Blood level studies, Pharmazie 34, 254.

Van de Graaff, W.B., Thompson, W.L., Sunshine, I., Fretthold, D., Leickly, F., and Dayton, H. (1982). Adsorbent and cathartic inhibition of enteral drug absorption, J. Pharmacol. Exper. Therap. 221, 656.

DYNAMICS OF PHENOL ADSORPTION ON POLYMERIC SUPPORTS

C. Costa and A. Rodrigues
Department of Chemical Engineering, University of Porto
4099 Porto Codex, Portugal

ABSTRACT

Effective diffusivities of phenol in a polymeric adsorbent (Duolite ES861) were measured from dynamic experiments carried out in a perfectly mixed basket adsorber operating either in batch or continuous mode and using a pore diffusion model.

Regeneration using sodium hydroxide was carried out in the same equipment. Experimental results were first explained by an "equilibrium model" and then by a "reaction front model".

Adsorption processes for removal of organic compounds (e.g. phenol) are usually carried out in fixed bed of granular activated carbon (Fritz (1,2,3)). Such processes involve three steps: saturation or load, regeneration or desorption (chemical or thermal) and washing (steam); as regenerant agent we often use an organic solvent (e.g. benzene) which is further recovered by distillation.

Macroreticular polymeric adsorbents (e.g. polystyrene matrix cross-linked with divinylbenzene) provide an alternative for phenol removal due to easy regeneration using either sodium hydroxide or methanol (Fox (4)). The porous structure of these adsorbents is such that can be viewed as an ensemble of microspheres with pores among them.

Modelling of adsorption processes involve writing conservation equations (mass, heat and momentum), equilibrium laws at the fluid-solid interfaces, kinetic laws of transport, boundary and initial conditions. Model parameters are then grouped in equilibrium, hydrodynamic and kinetic parameters.

Getting these parameters from independent experiments is not always a simple task, in that, we have to ensure that in a planned experiment for measuring a given parameter the influence of all the others is made negligible.

This paper deals with the determination of the effective diffusivity of phenol in a polymeric adsorbent; furthermore the regeneration step of the saturated resin is studied in detail.

ADSORPTION EQUILIBRIUM ISOTHERMS FROM BATCH EXPERIMENTS

Adsorption equilibrium isotherms for several phenols in Duolite ES861 were determined by contacting different quantities of adsorbent with a given volume of solution (20 or 40 ml,generally) using tubes continuously stirred in a thermostated bath.We first checked the time needed for reaching equilibrium,which was set at 72 hours.After equilibration if c_i^* is the solute concentration in the solution,the adsorbed quantity referred to the mass of dry solid,q_i^*,in equilibrium with c_i^* is

$$q_i^*=(c_o v_f - c_i^*(v_f+v_s))/m_i \tag{1}$$

where m_i is the mass of dry solid,v_f the volume of solution in the tube, v_s the volume of water contained in the solid and c_o the initial solute concentration.

Solute concentrations were measured by UV spectrophotometry in a UNICAM SP-400 at 272 nm for phenol,o-cresol and m-cresol and 320 nm for the p-nitrophenol.The adsorbent characteristics are given in Table 1.

Table 1 - Physical characteristics of Duolite ES861

wet density (kg of wet resin/m^3 of resin)	1.02×10^{-3}
apparent density (kg of dry resin/m^3 of particle volume)	0.537×10^{-3}
true density (kg of dry resin/m^3 of solid)	1.04×10^{-3}
specific BET area (m^2/kg)	5.0×10^{5}
porosity (%)	43.7
mean pore diameter (nm)	7.5

Isotherms were obtained at 20 and 60ºC and various equations(Langmuir, Freundlich,Toth) were used for fitting experimental data:

Langmuir

$$q^*=\frac{k_L Q c^*}{1+k_L c^*} \tag{2a}$$

Freundlich

$$q^*=k_F c^{*n} \tag{2b}$$

Toth

$$q^*=\frac{Qc^*}{(b+c^{*M})^{1/M}} \tag{2c}$$

The parameters were determined by fitting the experimental data (q_i^*,c_i^*) by Equations (2a),(2b) and (2c) using a direct search optimization and an objective function

$$\text{Fobj}=\frac{1}{N}\sqrt{\sum_{i=1}^{N}\left(1-\frac{q^*_{calc}}{q^*_{exp}}\right)^2} \tag{3}$$

N beeing the number of experimental points.

Table 2 summarizes the obtained model parameters using the Langmuir equation.

As an example we show in Figure 1 the experimental and optimized isotherms by using Langmuir equation for the system m-cresol/ES861 at 20 and 60ºC.

Experimental results show an important temperature effect on the adsorption equilibrium isotherms.Assuming Langmuir adsorption and Arrhenius temperature dependence of the equilibrium constant k_L it was shown that the heat of adsorption was of the order of 2 kcal/mole.

Table 2 - Model parameters for the Langmuir equation

SYSTEM	T(ºC)	k_L(l/mg)	Q(mg/g)	Fobj %	Range Conc. (mg/l)
phenol/Duolite ES861	20	4.6×10^{-3}	63.6	4.9	0-200
	60	1.5×10^{-3}	68.6	5.6	0-300
p-nitrophenol/Duolite ES861	20	2.9×10^{-3}	93.4	5.9	0-220
	60	5.0×10^{-3}	49.0	5.5	0-300
m-cresol/Duolite ES861	20	5.4×10^{-3}	160.5	3.7	0-230
	60	5.8×10^{-3}	63.2	5.7	0-200
o-cresol/Duolite ES861	20	6.3×10^{-3}	162.1	3.0	0-220
	60	6.0×10^{-3}	68.8	7.2	0-400

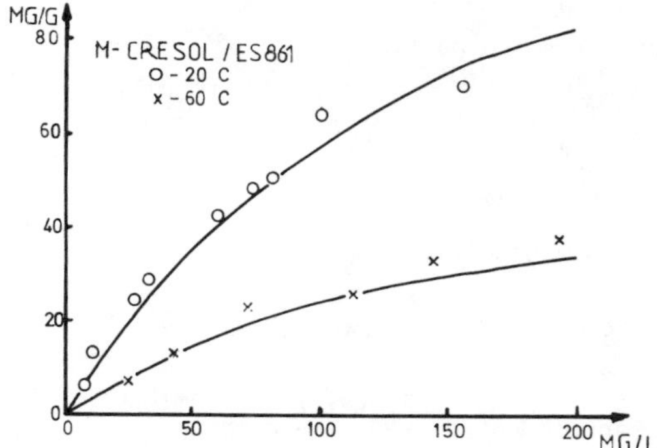

Figure 1 - Experimental and optimized isotherms for m-cresol/Duolite ES861 system;Langmuir equation.

DYNAMICS OF PHENOL ADSORPTION IN BATCH ADSORBERS

Diffusional mechanisms inside the particle of adsorbent are related to its porous structure.Different models should then be considered: homogeneous diffusion model (e.g. gel type resins),pore diffusion model, series model and parallel model (pore diffusion in parallel with surface diffusion as in activated carbon).

Mass balance equation for a volume element of the particle is,generally

$$\frac{\partial J_R}{\partial R} + \frac{2}{R} J_R + \frac{\partial c_{ev}}{\partial t} = 0 \qquad (4)$$

where J_R is the diffusional specific flux of solute normal to the surface located at radius R of the particle and c_{ev} the solute concentration in the volume element.According to the model considered above we have:

Model:	Homogeneous	Pore diffusion	Parallel	Series
$c_{ev}=$	q	χc_p+q	χc_p+q	$\chi c_p+\bar{q}$
$J_R=$	$-D_I\frac{\partial q}{\partial R}$	$-D_{pe}\frac{\partial c_p}{\partial R}$	$-D_{pe}\frac{\partial c_p}{\partial R}-D_I\frac{\partial q}{\partial R}$	$-D_{pe}\frac{\partial c_p}{\partial R}-D_I\frac{\partial q}{\partial R}$

where D_I is the effective diffusivity for the homogeneous model, χ is the internal porosity, D_{pe} the effective diffusivity based on the pore diffusion and c_p is the solute concentration in the pores.

Pore Diffusion Model

As an example we write down the dimensionless model equations for the pore diffusion model in a batch adsorber after introducing reduced variables $R^*=R/Ro$, $x_p=c_p/co$, $u^*=R^{*2}$ and $\theta=t/\tau_p$ ($\tau_p=R_o^2\chi/D_{pe}$ pore diffusion time constant).

Mass conservation for the particle

$$\frac{\partial x_p(u^*,\theta)}{\partial \theta}=\frac{\chi}{\chi+\dfrac{k_L Q\rho_{ap}}{(1+k_L c_o x_p)^2}}\left(6\frac{\partial x_p(u^*,\theta)}{\partial u^*}+\frac{\partial^2 x_p(u^*,\theta)}{\partial u^{*2}}4u^*\right) \tag{5}$$

Boundary and initial conditions

symmetry condition ($u^*=0$) $\quad \left.\dfrac{\partial x_p(u^*,\theta)}{\partial u^*}\right|_{u^*=0}=0$

solid/fluid interface ($u^*=1$) $\quad \dfrac{dx_p(1,\theta)}{d\theta}=-6\chi\xi\dfrac{1+k_L c_o}{k_L Q\rho_{ap}}\left.\dfrac{\partial x_p(u^*,\theta)}{\partial u^*}\right|_{u^*=1}$

initial condition $\quad x_p(u^*,0)=0 \quad (u^*<1)$

$\quad x_p(1,0)=1$

where $\xi=(1-\varepsilon)q_o/(\varepsilon co)$, ε the adsorber porosity and q_o the solid concentration in equilibrium with c_o.

The numerical integration of this nonlinear parabolic partial differential equation was made by using the method of lines with two types of basis functions: Lagrange polynomials and B-splines. In the case of Lagrange polynomials we have

$$S(u^*,\theta)=\sum_{i=0}^{N+1}S_i(\theta)\,L_i(u^*) \tag{6}$$

while in the case of B-splines we get

$$S(u^*,\theta)=\sum_{i=0}^{N+1}d_i(\theta)\,\phi_i(u^*) \tag{7}$$

in which $S(u^*,\theta)$ is the numerical solution, $L_i(u^*)$ are the Lagrange polynomials and $\phi_i(u^*)$ the B-spline functions. By substituting either of Equations (6) or (7) in the model equations we get a residue which should be minimized in order to find the coefficients $S_i(\theta)$ or $d_i(\theta)$. The minimization criterium was the orthogonal collocation method, in which the residue is equated to zero in N interior collocation points; moreover the tentative solution should satisfy the boundary conditions resulting in a total of N+2 equations enabling us to calculate the N+2 coefficients (Villadsen(5) and Boor(6)).

If very sharp profiles are encountered we can use collocation on finite elements by dividing the collocation domain in subintervals; within each subinterval orthogonal collocation is used and continuity of solution and its first derivative are imposed between subintervals.

The pore diffusion model equations have been integrated using this finite element technique with cubic splines and 10 subintervals needed to assure non-oscillating radial profiles in the particle;if we use global collocation more than 20 interior collocation points were needed to get the same result.For the integration we used a package written by Madsen(7).

In Figure 2a we show the influence of model parameter ξ in the response of the batch adsorber in terms of reduced concentration versus a reduced time.Figure 2b shows the evolution of the radial profiles inside the particle at different times.The radial profile is very instructive in the sense it enables us to see whether or not simplified models (e.g. Glueckauf) are valid.

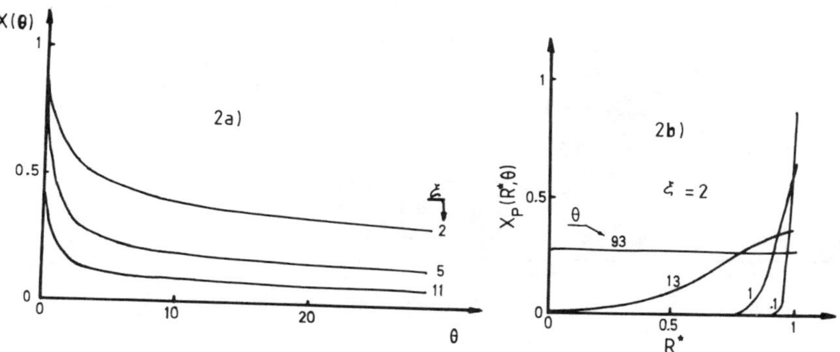

Figure 2a - Influence of ξ on the history of concentration in a batch adsorber using the pore diffusion model;
2b - Radial profiles for the pore diffusion model in a batch adsorber.

Experimental

For the determination of diffusivities we used a "Carberry" type adsorber.This choice has been determined by considering the ease of hydrodynamic characterization,the possibility of elimination of film mass transfer resistance and the conditioning of the adsorbent which prevents breaking.

The adsorber is a cylindrical tank (500ml) with four baffles and a variable speed four blade stirrer;in fact the blades are parallelepipedic baskets (6.45x2.75x1.05 cm) made of a stainless steel net,where the adsorbent is placed.The phenol concentration is continuously measured by circulating the solution through a spectrophotometer;about 2% of the adsorber volume is recirculated at a flowrate of 60 ml/min which gives 8 s of non contact between solution and adsorbent,a small period that prevents perturbations.Figure 3 shows the experimental arrangement used.

As we want that this adsorber behaves as a perfectly mixed tank some tracer experiments were conducted in order to know the speed of the stirrer needed to achieve that goal - we found that 250 rot/min were enough.On the other hand we also want to guarantee that the only kinetic phenomena present are the intraparticle resistances;we conducted saturation experiments at several stirrer speeds (300,500 and 850 rot/min) having concluded that the responses obtained for 500 and 850 rot/min were coincident;we fixed then 500 rot/min as the minimum stirrer speed needed to eliminate film mass transfer resistance.

The start up of the saturation experiments was done by putting on the agitation and the recirculation and then injecting 2 ml of a concentrated phenol solution through a septum located at the adsorber top;this solution dilutes,almost instantaneously,in the adsorber volume,approaching a step input.The initial condition was the absence of solute in the adsorber.Table 3 shows the experimental conditions used.

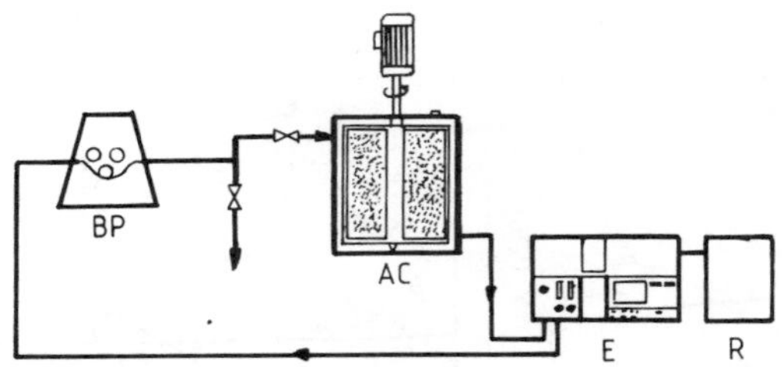

AC-basket adsorber
BP-peristhaltic pump
E -spectrophotometer
R -recorder

Figure 3 - Experimental arrangement for batch runs.

Table 3 - Experimental conditions used in batch runs.

RUN	$2R_o$ (cm)	Stirrer speed (rot/min)	c_o (mg/l)	ε	ξ	c_∞ (mg/l)	T (ºC)
20	0.077	545	96.4	0.951	5.61	11.1	20
21	0.060	546	96.3	0.952	5.49	11.3	20
22	0.034	543	96.0	0.955	5.14	12.0	20

The porosity was calculated from the knowledge of the adsorber volume and the mass of adsorbent.The final equilibrium concentration (c_∞) can be calculated using a global material balance

$$c_\infty = \frac{\sqrt{(1+(1-\varepsilon)k_L Q\rho_{ap}/\varepsilon - k_L)^2 + 4k_L c_o} - (1+(1-\varepsilon)k_L Q\rho_{ap}/\varepsilon - k_L)}{2k_L} \quad (8)$$

To determine the pore diffusivity values that optimize the batch experimental results we note that it is possible to simulate an "universal" history of concentrations curve ($x=f(\theta)$) for each experimental run (ξ,c_o and isotherm parameters known).Once we have those curves they can be compared with the experimental ones and for each concentration value a pair (θ,t) is obtained.Now taking into account that $\theta=D_{pe}t/\chi R_o^2$ it is possible to determine the D_{pe} value that,for each experiment,better fits a straight line to the set of (θ,t) points.Figure 4 shows the experimental and simulated results for the batch experiments (see Table 3) together with the $D_{pe}/\chi R_o^2$ values that produced the best linear fitting to $\theta=f(t)$.

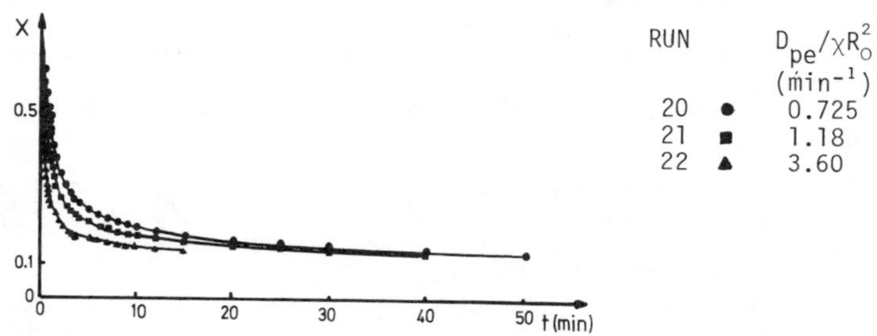

RUN		$D_{pe}/\chi R_o^2$ (min^{-1})
20	●	0.725
21	■	1.18
22	▲	3.60

Figure 4 - Experimental and simulated results for batch runs.

Then runs were carried out in a continuous stirred tank adsorber at different flowrates.Table 4 shows the experimental conditions used,where $t_{st}=\tau(1+\xi)$ is the stoichiometric time,$\tau=V\varepsilon/U$ is the space time,$\beta=\tau_p/\tau$ is the ratio between pore diffusion time constant and space time and $\theta_{st}=t/t_{st}$.

Table 4 - Experimental and simulation conditions for continuous runs.

RUN	U (ml/min)	ξ	c_o (mg/l)	ε	t_{st} (min)	d_p (cm)	$D_{pe}/\chi R_o^2$ (min^{-1})	β
24.1	108.3	5.5	92.6	0.952	25.8	0.060	1.18	0.215
24.2	54.0	5.4	92.6	0.953	50.8	0.060	1.18	0.107
24.3	77.8	5.3	93.8	0.954	34.5	0.060	1.18	0.154

Experimental and simulated results using D_{pe} values obtained from batch runs are shown in Figure 5.As it can be seen the agreement is quite good.

DYNAMICS OF ADSORBENT REGENERATION IN A CONTINUOUS STIRRED ADSORBER

In this work we used sodium hydroxide as regenerant of the polymeric adsorbent saturated with phenol.The acid-base reaction can be written as follows: $C_6H_5OH + OH^- \leftrightarrows C_6H_5O^- + H_2O$
with

$$k_{eq}(20^\circ C)= \frac{c_{FN}}{c_c^F c_c^S}=1.3\times10^4 \; l/mole \tag{9}$$

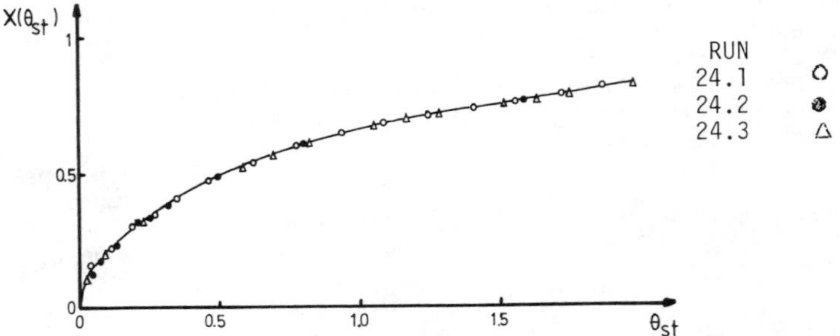

Figure 5 - Experimental and simulated results for continuous runs.

where c^{FN}, c^{F} and c^{S} are molar concentrations of phenol, hydroxide and phenate, respectively.

It is important to calculate the effect of pH on the equilibrium isotherm of phenol on Duolite ES861. If reaction and adsorption equilibria are both instantaneous we get, for a Langmuir type isotherm (Equation (2a) coupled with Equation(9))

$$q^{F}=k_{L}Q\,\frac{c_{T}}{1+k_{eq}10^{pH-14}+P_{M}k_{L}c_{T}10^{3}} \qquad (10)$$

where c_T is the total (phenol+phenate) molar concentration in the liquid phase, q^F is the molar concentration of phenol in the solid and P_M is the phenol molecular weight. The effect of pH on adsorption equilibrium is better illustrated by Figure 6.

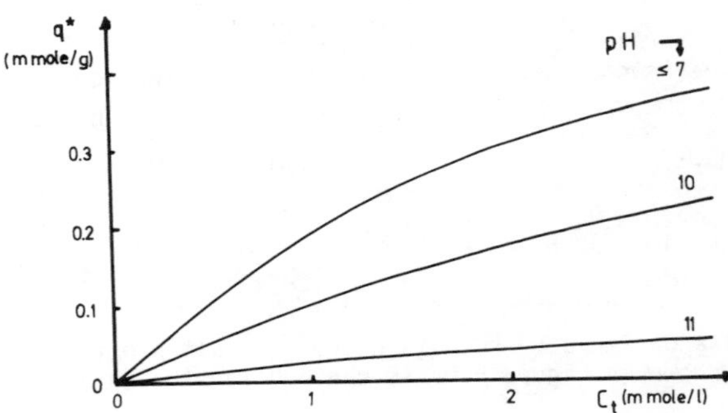

Figure 6 - Influence of pH on the adsorption equilibrium isotherm of phenol on Duolite ES861.

An attempt to modelling of regeneration process was made by using two different models: equilibrium and "reaction front" models. First a simple equilibrium model was developed which assumes instantaneous equilibrium of adsorption and infinitely fast reaction; the reaction takes place in the bulk liquid phase and in the particle pores. The equations for this simple equilibrium model are:

- conservation equations

phenol $$Uc_E^F = Uc^F + V\varepsilon \frac{dc^F}{dt} + V(1-\varepsilon)\frac{d\bar{q}^F}{dt} + \{\text{mass flowrate disappeared by reaction}\} \quad (11)$$

phenate $$Uc_E^{FN} = Uc^{FN} + V\varepsilon \frac{dc^{FN}}{dt} + V(1-\varepsilon)\frac{d\bar{q}^{FN}}{dt} - \{\text{mass flowrate produced by reaction}\} \quad (12)$$

hydroxide $$Uc_E^S = Uc^S + V\varepsilon \frac{dc^S}{dt} + V(1-\varepsilon)\frac{d\bar{q}^S}{dt} + \{\text{mass flowrate disappeared by reaction}\} \quad (13)$$

with $\bar{q}^F = q^F + \chi c^F$, $\bar{q}^{FN} = \chi c^{FN}$, $\bar{q}^S = \chi c^S$; U is the flowrate and subscript E means input concentration.

- equilibrium relationships: Equations (9) and (2a) for reaction and adsorption, respectively.

-initial conditions: $c^F(0) = c_o^F$, $c^{FN}(0) = c_o^{FN}$, $c^S(0) = c_o^S$.

The system of ordinary differential equations was numerically solved by using the GEARB package developed by Hindmarsh(8).

Experimental results obtained in the basket adsorber as described before but operating in a continuous way are shown in Figure 7a together with the simulated results using the described model for the following set of conditions: $c_o^F = 96.3$ mg/l, $c_o^S = 10^{-8}$M, $c_E^F = 0$, $c_E^S = 1.017$M, U=108.7 ml/min, $R_o = 0.0385$ cm and $\varepsilon = 0.951$.

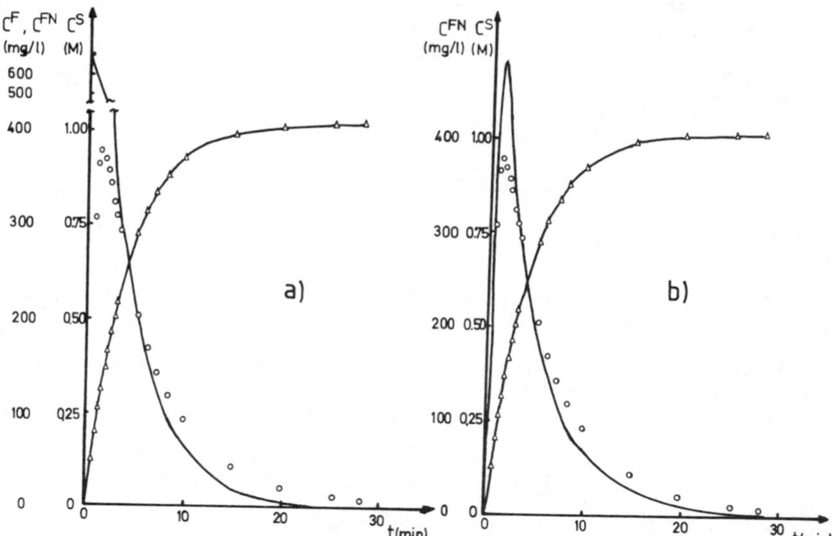

Figure 7 - Experimental and simulated results for the regeneration of phenol with sodium hydroxide in a continuous stirred adsorber; a-equilibrium model, b-reaction front model. (▵ sodium hydroxide, ○ phenate, • phenol)

It is apparent from Figure 7a that the sodium hydroxide history is correctly described by this simple model;however the peak predicted by this model is higher and appears earlier than the experimental one.This is expected since diffusional resistances have been neglected.

Then a more sophisticated model was built based on several assumptions:

- reaction of phenol and sodium hydroxide is instantaneous and irreversible;this assumption can be justified because there is a large excess of sodium hydroxide;
- phenol is stationary inside the particles,i.e.,there is no diffusion of phenol towards the surface of the particle;this is supported by experiments since almost no phenol is detected at the outlet;
- sodium hydroxide diffuses towards the center of the particles;
- sodium phenate,once is formed,is instantaneously distributed in the fluid phase;
- film mass transfer resistances are negligible;this was assured by using enough stirrer speed (more than 500 rot/min).

These assumptions lead to a model in which the reaction between phenol and sodium hydroxide inside the particles takes place in a spherical surface that moves inwards -reaction front (Costa(9)).The model equations were solved by orthogonal collocation for the radial coordinate after transforming the initial moving boundary value problem into a fixed boundary one by making $R^{+}=(R-R_f)/(R_o-R_f)$,where R_f is the coordinate of the reaction front.Simulation and experimental results are shown in Figure 7b for the same set of experimental conditions used in the case of the equilibrium model.In this simulation the value of the effective pore diffusivity for sodium hydroxide ($D_p^S=5.8\times10^{-10}$ m^2/s)was obtained by making simultaneous the simulated and experimental times of phenate peak appearance.As it can be seen in Figure 7b the description of the experimental results has been enhanced by using the "reaction front model".

CONCLUSIONS

When designing a cyclic fixed bed adsorption process it is necessary to have mathematical models that can predict the behavior of the saturation and regeneration steps.The simulation of those models needs the knowledge of equilibrium and kinetic parameters;also numerical techniques must be available.

Relatively to the equilibrium parameters we concluded that a Langmuir type equation applies.To determine intraparticle diffusivities it is important to consider the mechanisms of diffusion into the adsorbent particles;in the present case several models were tried,(Costa(10)),and it has been concluded that a pore diffusion model was sufficient to represent the experimental data in the range of operating conditions used,with $D_{pe}=7.7\times10^{-10}$ m^2/s.Film mass transfer resistance was not treated in this work as various correlations can be used,e.g. Kataoka(13).

For the adsorbent regeneration we used sodium hydroxide.An equilibrium model assuming instantaneous reaction and adsorption equilibrium was tried,giving a qualitatively good representation of the experimental results.Then a more sophisticated model,the reaction front model,was used and a better description was obtained,specially in terms of peak appearance and height prediction.As methanol may be also used as a regenerant it is important to notice that a similar model may be applied using now a distribution equation instead of the reaction equilibrium constant one.

Finally the temperature effect on the isotherms suggested the use of parametric pumping to separate phenol-water solutions (Costa et al(11), Almeida et al(12)).

NOTATION

c_{ev} -solute concentration in volume element of adsorbent particle -M(solute)/L^3(fluid)
c_i^* -equilibrium concentration of phenol in solution-M(solute)/L^3(fluid)
c_p -solute concentration in particle pores-M(solute)/L^3(fluid)
c_o -initial solute concentration in fluid-M(solute)/L^3(fluid)
c_∞ -final equilibrium solute concentration in fluid-M(solute)/L^3(fluid)
c^F,c^{FN},c^S,c_T-solute molar concentration in fluid-mole(sol.)/L^3(fluid)
d_p -particle diameter-L
D_I,D_{pe}-effective diffusivities for the homogeneous model and based in pore diffusion,respectively-L^2T^{-1}
D_p^S -effective sodium hydroxide diffusivity based in pore diffusion-L^2T^{-1}
J_R -diffusional specific flux of solute-$ML^{-2}T^{-1}$
k_{eq} -equilibrium acid-base constant-L^3mole^{-1}
k_L -parameter in Langmuir equation-L^3(fluid)/M(solute)
m_i -mass of dry resin-M(dry solid)
P_M -phenol molecular weight-$Mmole^{-1}$
$\underline{q}$ -concentration of solute in the solid-M(solute)/L^3(solid)
q_F -average microsphere concentration-M(solute)/L^3(solid)
q^F -molar phenol concentration in solid-mole(sol.)/M(dry solid)
$\bar{q}^F$,$\bar{q}^{FN}$,$\bar{q}^S$-average molar concentration of solute in solid-mole(solute)/L^3(solid)
q_i^* -concentration of solute in solid in equilibrium with c_i^*-M(solute)/M(dry solid)
Q -parameter in the Langmuir equation-M(solute)/M(dry solid)
R_o,R,R_f-particle ratius,radial coordinate and coordinate of the moving front,respectively-L
R* -reduced radial coordinate
t,t_{st}-time and stoichiometric time($=\tau(1+\xi)$),respectively-T
u* -modified radial coordinate ($=R^{*2}$)
U -volumetric flowrate-L^3T^{-1}
V,v_f,v_s-adsorber volume,volume of fluid and volume of fluid contained in resin pores-L^3
x_p -reduced concentration of solute in pores
β -ratio between pore diffusion time and space time ($=\tau_p/\tau$)
ε -adsorber porosity
ξ -capacity parameter ($=(1-\varepsilon)q_o/(\varepsilon c_o)$)
ρ -apparent density-M(dry solid)/L^3(solid)
χ -internal porosity
θ,θ_{st}-reduced time,respectively,(t/τ_p) and (t/t_{st})
τ -space time ($=V\varepsilon/U$)-L

τ_p -pore diffusion time constant $(=\chi R_o^2/D_{pe})$-T
superscripts:F-phenol,FN-phenate,S-sodium hydroxide
subscripts:o-initial,E-input,T-total

LITERATURE CITED

(1) Fritz,W. and E.U. Schlunder,Chem.Eng.Sci.,36,721(1981)
(2) Fritz,W.,W. Merck and E.U. Schlunder,Chem. Eng. Sci.,36,731(1981)
(3) Merck,W.,W. Fritz and E.U. Schlunder,Chem.Eng.Sci.,36,743(1981)
(4)Fox,C.R.,Hyd.Proc.,November,269(1978)
(5) Villadsen,J. and M.L. Michelsen,"Solutions of Differential Equations by Polynomial Approximation",Prentice Hall Inc.,New York(1978)
(6) De Boor,C.,SIAM. J. Numer. Anal.,14,441(1977)
(7) Madsen,N.K. and R.F. Sincovec,ACM Trans. Math. Software,5(3), 326(1979)
(8) Hindmarsh,A.C.,"GEAR-Ordinary Differential Equations System Solver" Lawrence Livermore Laboratory,Report UCID-30001,Rev.3,Livermore (1974)
(9) Costa,C. and A. Rodrigues,submitted to the Chem.Eng.Sci.
(10) Costa,C.,"Dynamics of Cyclic Processes:Adsorption and Parametric Pumping",Ph.D.Thesis,University of Porto(1983)
(11) Costa,C.,A.Rodrigues,G.Grevillot and D.Tondeur,AIChE J.,28(1), 73(1982)
(12) Almeida,F.,C.Costa,A.Rodrigues and G.Grevillot,"Removal of Phenol from Wastewater by Recuperative Mode Parametric Pumping",in Physicochemical Methods for Water and Wastewater Treatment ,L. Pawlowski(Ed.),Elsevier Publ. Comp.,Amsterdam(1982)
(13) Kataoka,T.,H. Yoshida and K. Ueyama,J. of Chem. Eng. of Japan,5(2), 132(1972)

ADSORPTION EQUILIBRIUM OF HYDROCARBON GAS MIXTURES ON 5A ZEOLITE

Enrique Costa, Guillermo Calleja and Luis Cabra
Departamento de Ingeniería Química, Facultad de Ciencias Químicas
Universidad Complutense, Madrid (3), Spain

ABSTRACT

Experimental binary and ternary equilibrium data of adsorption of mixtures of ethane, propane and ethylene on 5A zeolite at 20°C and pressures up to 800 torr have been obtained. A real adsorbed solution theory has been applied to correlate the binary data and to predict the equilibria of the ternary mixtures from experimental data of pure components and binary mixtures.

THEORY

The thermodynamic relations between the gas phase and a hypothetical bidimensional adsorbed phase, according to the original idea of Gibbs, lead to equations that relate different thermodynamic properties with the experimental data (Costa et al. (1)). Thus the spreading pressure of a pure adsorbate Π_i^o can be expressed as follows:

$$\frac{\Pi_i^o A}{R\,T} = \int_o^{p_i^o} \frac{n_i^o}{p_i^o}\, dp_i^o = \int_o^{n_i^o} \frac{d \ln p_i^o}{d \ln n_i^o}\, dn_i^o \qquad (1)$$

The spreading pressure of an adsorbed binary mixture Π is given by:

$$\Pi = \Pi_1^o + \Delta\Pi \qquad T,p = \text{const.} \qquad (2)$$

where:

$$\frac{\Delta\Pi A}{R\,T} = \int_{y_1}^{1} n\, \frac{y_1 - x_1}{y_1(1-y_1)}\, dy_1 \qquad T,p = \text{const.} \qquad (3)$$

Generally the integrand of the above equation becomes very large when y_1 approaches unity, making impossible the extrapolation up to $y_1=1$. This difficulty can be solved by changing the variable, $y_1'=1-y_1$ and $x_1'=1-x_1$. Substituting in Equation (3) and operating, it can be obtained:

$$\frac{\Delta\Pi A}{R\,T} = \int_o^{x_1'} n \frac{d\ln \dfrac{y_1'}{1-y_1'}}{d\ln x_1'} dx_1' - \int_o^{y_1'} n \frac{dy_1'}{1-y_1'} \qquad T,p = \text{const.} \tag{4}$$

expression in which both integrands tend to the known value $n_1^o(T,p)$ when x_1', $y_1' = 0$.

When the equilibrium is reached, the chemical potentials in the adsorbed and gas phases become equal. This relation, expressing the chemical potentials as a function of composition, leads to a modified Raoult's law:

$$py_i = \gamma_i p_i^o(\Pi) x_i \tag{5}$$

where the activity coefficient of each component in the adsorbed phase γ_i represents the deviation from ideallity due to mixing and $p_i^o(\Pi)$ is the pressure of the component i in the gas phase in equilibrium with an amount adsorbed that exerts the same spreading pressure of the mixture Π.

According to Myers and Prausnitz (2), the total amount adsorbed of a mixture n can be calculated from the expression:

$$\frac{1}{n} = \sum_i \frac{x_i}{n_i^o(\Pi)} + \sum_i \left[\frac{\partial \ln\gamma_i}{\partial\left(\frac{\Pi A}{RT}\right)}\right]_{T,x_i} \tag{6}$$

<u>The Real Adsorbed Solution Theory</u>

Myers and Prausnitz (2) used the above equations to predict multicomponent equilibria assuming the adsorbed mixtures to behave ideally, γ_i=1 (ideal adsorbed solution theory). This simplification allow the prediction just from pure component experimental isotherms, but it has been shown that a considerable amount of adsorbed mixtures deviate from ideal behaviour.

Costa et al. (1) proposed a method of prediction of multicomponent equilibria from experimental data of pure components and binary mixtures not restricted to ideal behaviour. This method can be summarized in the following steps:

1) From the experimental pure components isotherms, the curves $\Pi_i^o A/RT$ vs. p_i^o are obtained with Equation (1).

2) Select a value of the total pressure p.

3) From the experimental binary data x_i, x_j, y_i, y_j, n at the total pressure p of all the possible binary mixtures i-j in the multicomponent mixture, the spreading pressures of the adsorbed binary mixtures Π are calculated with Equation (2). With these spreading pressures and the curves $\Pi_i^o A/RT$ vs. p_i^o (step 1), the terms $p_i^o(\Pi)$ and $p_j^o(\Pi)$ are determined and substituting them in Equation (5), the isobaric curves γ_i-x_i and

γ_j-x_j are obtained.

4) The obtained curves γ_i-x_i and γ_j-x_j are fitted to Wilson or UNIQUAC equations, and pairs of parameters Λ_{ij}-Λ_{ji} (Wilson) or τ_{ij}-τ_{ji}(UNIQUAC) of each binary mixture are obtained.

5) Select the values of the compositions in the adsorbed multicomponent mixture x_i.

6) The activity coefficients of the components in the multicomponent mixture are calculated from Wilson or UNIQUAC equations, which contain the binary parameters determined in step 4.

7) Assume a value of the spreading pressure of the multicomponent mixture Π_m.

8) The equilibrium molar fractions y_i are calculated from Equation (5). The terms $p_i^o(\Pi)$ are obtained from the curves $\Pi_i^o A/RT$ vs. p_i^o (step 1) for the assumed value of Π_m.

9) The correct value of Π_m is determined by trial and error until the necessary condition $\Sigma y_i=1$ is reached.

10) The total amount adsorbed is calculated from Equation (6).

EXPERIMENTAL SYSTEM AND MATERIALS

The experimental set, consisting of a closed circuit with a bed of adsorbent, an oil-free compressor for circulating the gas and a gas chromatograph, was operated as previously explained (Costa et al. (1)).

The adsorbents were supplied by Union Carbide International Co. in the form of pellets of 1/16 inch diameter, containing about 20% of inert binder. The gases were supplied by Sociedad Española de Oxígeno, S.A., with the following minimum purity: ethane (99.0%), ethylene (99.9%), propane (99.5%).

RESULTS AND DISCUSSION

Experimental adsorption isotherms on 5A zeolite at 20°C and pressures up to 800 torr have been obtained in the following cases:

- Pure components: methane, ethane (also at 5 and 40°C), ethylene, propane, propylene and carbon dioxide.

- Binary mixtures: ethane-propane, ethane-ethylene and propane-ethylene.

- Ternary mixtures: ethane-propane-ethylene.

Adsorption isotherms of ethane, ethylene and propylene and their mixtures on 13X zeolite have been also obtained (3) and will be published in the future due to the limited extension of this paper.

Pure Component Equilibria: Spreading Pressure and Characteristic Equilibrium Curve

The spreading pressures of pure ethane, propane and ethylene on

5A zeolite have been calculated from the experimental isotherms with Equation (1). The experimental isotherms and the calculated spreading pressures are summarized in table 1.

Table 1. Experimental isotherms and spreading pressures of pure components on 5A zeolite at 20°C.

ETHANE			PROPANE			ETHYLENE		
p (torr)	$n \times 10^3$ (mol/g)	$\frac{\pi_i^o A}{RT} \times 10^3$ (mol/g)	p (torr)	$n \times 10^3$ (mol/g)	$\frac{\pi_i^o A}{RT} \times 10^3$ (mol/g)	p (torr)	$n \times 10^3$ (mol/g)	$\frac{\pi_i^o A}{RT} \times 10^3$ (mol/g)
9.18	0.150	0.152	0.44	0.264	0.313	0.89	0.501	0.790
24.27	0.373	0.391	1.45	0.580	0.800	3.69	0.998	1.844
46.45	0.666	0.666	4.56	0.990	1.700	14.85	1.479	3.598
75.70	0.945	1.113	16.23	1.385	2.955	42.85	1.845	5.446
116.39	1.203	1.578	98.37	1.646	5.752	110.47	2.134	7.330
173.93	1.429	2.107	210.48	1.768	7.057	198.07	2.302	8.627
255.80	1.608	2.691	314.40	1.822	7.779	310.92	2.423	9.697
359.32	1.747	3.262	419.88	1.873	8.315	436.79	2.518	10.541
476.81	1.849	3.769	524.67	1.911	8.739	572.18	2.595	11.234
600.54	1.940	4.204	640.30	1.957	9.126	673.23	2.654	11.663
739.34	2.021	4.614	739.10	1.980	9.410	775.34	2.709	12.043

Myers and Sircar (4) demonstrated that the function:

$$F(\Theta) = \frac{mRT \ln p/p_s}{\Delta G} \quad (7)$$

should be a characteristic function of the adsorption on an heterogeneous adsorbent, not depending on the type of adsorbate nor the temperature. All the experimental pure component data have been represented in the form F(Θ) vs. Θ, leading to the characteristic equilibrium curve represented in Figure 1, showing an excellent fit of equilibrium data of different adsorbates at various temperatures to a single curve, as a result of a certain degree of surface heterogeneity.

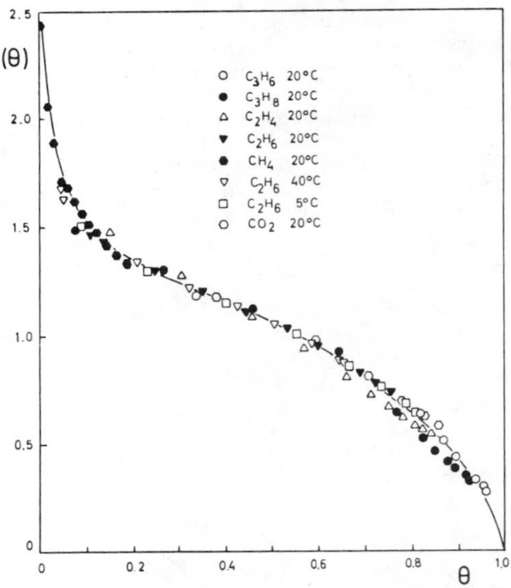

Figure 1. Characteristic equilibrium curve of adsorption on 5A zeolite

Binary Mixture Equilibria

Different isotherms of the binary mixtures have been obtained, each one corresponding to a fixed value of the initial composition in the gas phase. All the mixture isotherms lie between those of the pure components, except for the mixture propane-ethylene on 5A zeolite that shows a certain range of pressures and compositions in which the amount adsorbed is larger than the corresponding to pure ethylene, as can be observed in Figure 2, indicating a clear deviation from ideallity.

The isobaric equilibrium data x_1, y_1, n have been obtained by interpolation from the experimental isotherms. The data corresponding to adsorption on 5A zeolite at 20°C and three total pressures of 100, 500 and 700 torr are presented in Table 2.

From the isobaric data x_1, y_1, n the activity coefficients of the components in the adsorbed mixtures have been evaluated at different total pressures by the method proposed in the theoretical section (RAS theory). The characteristic equilibrium curve has been used to extrapolate the isotherm of ethane,the less adsorbed component, in order to calculate the term $p_i^o(\Pi)$ of the Raoult's law at the high spreading pressures of the mixtures.

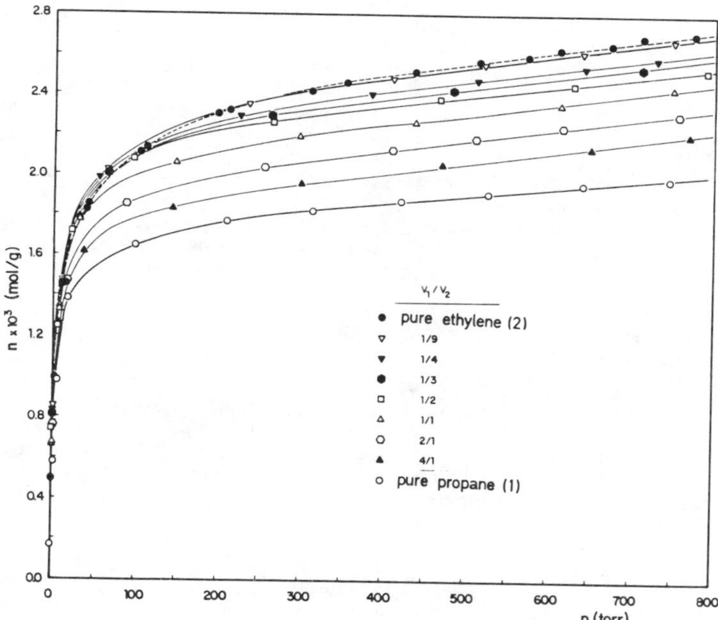

Figure 2. Experimental adsorption isotherms of the binary mixture propane-ethylene on 5A zeolite at 20°C

Table 2. Isobaric equilibrium data of binary mixtures on 5A zeolite at 20°C

	Ethane(1)-Propane(2)			Ethane(1)-Ethylene(2)			Propane(1)-Ethylene(2)		
	x_1	y_1	$nx10^3$	x_1	y_1	$nx10^3$	x_1	y_1	$nx10^3$
p = 100 torr	0.771	0.989	1.220	0.770	0.996	1.250	0.785	0.944	1.780
	0.622	0.972	1.300	0.622	0.988	1.370	0.646	0.873	1.870
	0.440	0.944	1.390	0.444	0.960	1.560	0.483	0.677	2.000
	0.266	0.874	1.462	0.275	0.879	1.710	0.328	0.394	2.090
	0.136	0.735	1.550	0.152	0.685	1.858	0.250	0.241	2.110
	0.057	0.490	1.600	0.070	0.420	1.979	0.202	0.177	2.125
							0.103	0.067	2.131
p = 500 torr	0.712	0.975	1.872	0.712	0.992	2.053	0.734	0.946	2.069
	0.530	0.939	1.860	0.534	0.971	2.161	0.585	0.855	2.181
	0.325	0.847	1.862	0.343	0.882	2.279	0.433	0.664	2.298
	0.166	0.665	1.863	0.194	0.690	2.395	0.303	0.412	2.398
	0.081	0.442	1.898	0.102	0.462	2.471	0.237	0.283	2.435
	0.032	0.273	1.893	0.056	0.204	2.493	0.193	0.218	2.467
							0.104	0.092	2.542
p = 700 torr	0.689	0.967	1.985	0.686	0.990	2.199	0.703	0.935	2.162
	0.498	0.918	1.962	0.499	0.963	2.328	0.566	0.841	2.276
	0.296	0.805	1.968	0.313	0.848	2.443	0.418	0.649	2.404
	0.151	0.603	1.951	0.175	0.639	2.532	0.294	0.408	2.485
	0.076	0.385	1.986	0.091	0.416	2.595	0.232	0.285	2.529
	0.030	0.206	1.987	0.056	0.154	2.600	0.191	0.219	2.569
							0.103	0.094	2.645

In the range of pressures investigated, the mixtures ethane-propane on 5A zeolite and ethylene-propylene on 13X zeolite (3) are ideal, showing the rest of the mixtures paraffin-olefin on both adsorbents non-ideal behaviour, increasing the deviation from ideallity with the total pressure, as it could be expected. The mixture propane-ethylene on 5A zeolite exhibits azeotropic compositions at all the pressures studied. The azeotropic point at each total pressure has a maximum value of the spreading pressure, causing the mixtures with compositions around the azeotropic one to have spreading pressures larger than that of the more adsorbed pure component and leading to the abnormally high amounts adsorbed previously commented.

The obtained isobaric curves γ_i-x_i and γ_j-x_j of each binary mixture i-j have been fitted to Wilson and UNIQUAC equations, the best values of $\Lambda_{ij}-\Lambda_{ji}$ and $\tau_{ij}-\tau_{ji}$ being determined by a least squares criterium using the Marquardt optimization algorithm (5). The parameters obtained with the Wilson and UNIQUAC equations for the binary mixtures adsorbed on 5A zeolite at different pressures are summarized in Table 3. As an example, the experimental and correlated activity coefficients of the same mixtures at a total pressure of 100 torr are presented in Figure 3, showing the ideal behaviour of the mixture ethane-propane ($\gamma \simeq 1$).

With the correlated activity coefficients and the modified Raoult's law, Equation (5), the equilibrium data of the binary mixtures have been reproduced, weighting the obtained values of y_1 and y_2 in order to reach the necessary condition $y_1+y_2=1$. An excellent agreement between experimental and reproduced values has been attained, confirming the consistency of the model. Almost identical results have been obtained with both Wilson and UNIQUAC equations. As an example, the results obtained for the mixture propane-ethylene on 5A zeolite at 100 torr are presented in figure 4a, showing the commented azeotropic composition.

Table 3. Values of binary Wilson parameters Λ_{ij}-Λ_{ji} for the adsorption on 5A zeolite at 20°C. Ethane=1, propane=2, ethylene=3

	p = 100 torr			p = 500 torr			p = 700 torr		
	Component i			Component i			Component i		
Component j	1	2	3	1	2	3	1	2	3
1	--	1.0000	1.6413	--	1.0000	1.6690	--	1.0000	2.4574
2	1.0000	--	2.5925	1.0000	--	2.3211	1.0000	--	2.0780
3	1.7625	1.4275	--	2.4303	1.7758	--	1.9456	2.1478	--

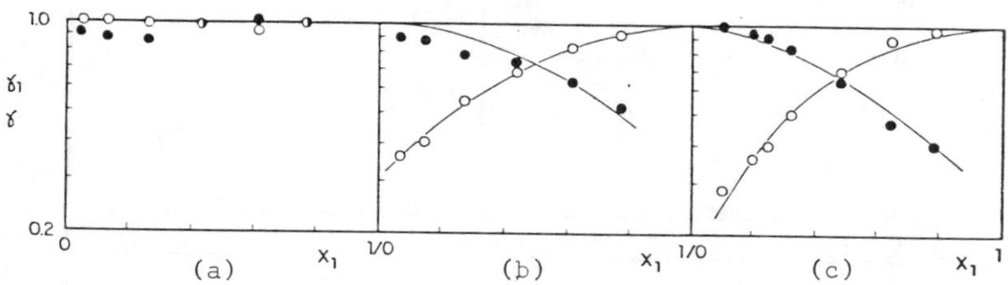

Figure 3. Activity coefficients of the binary mixtures on 5A zeolite at 20°C and 100 torr. White points=experimental γ_1; black points=experimental γ_2; lines= correlated. a) ethane(1)-propane(2); b) ethane(1)-ethylene(2); c) propane(1)-ethylene(2)

The total amount adsorbed can be calculated from Equation (6). However the evaluation of the last term on the right-hand side of this equation is extremely tedious. The influence of this term was checked, showing to correct the total amount adsorbed not more than 2 to 4 % in excess for the less ideal mixtures. Therefore it was preferred to use the simplified expression:

$$\frac{1}{n} = \sum_i \frac{x_i}{n_i^o(\Pi)} \qquad (8)$$

The calculated and experimental values of the total amount adsorbed for the mixture propane-ethylene on 5A zeolite at 100 torr are presented in Figure 4a.

The ideal adsorbed solution, IAS (2), the vacancy solution, VS (6, 7) and the statistical-thermodynamic, ST (8,9) models have been applied to predict the equilibria of the binary mixtures just from pure component experimental data. The two last ones require the fit of these data to theoretical isotherms with four and two parameters respectively, the values of such parameters being obtained with the aid of the Marquardt optimization algorithm. Predicted and experimental data showed not to be in good agreement, specially for the azeotropic mixture propane-ethylene, as can be observed in figure 4b. The IAS and ST models provide a good prediction for the ideal mixture ethane-propane. Although the ST and VS models present the advantage of not needing any extrapolation beyond the experimental pressure, they are limited by the need of a correct fit of the experimental pure component data to the theoretical isotherms.

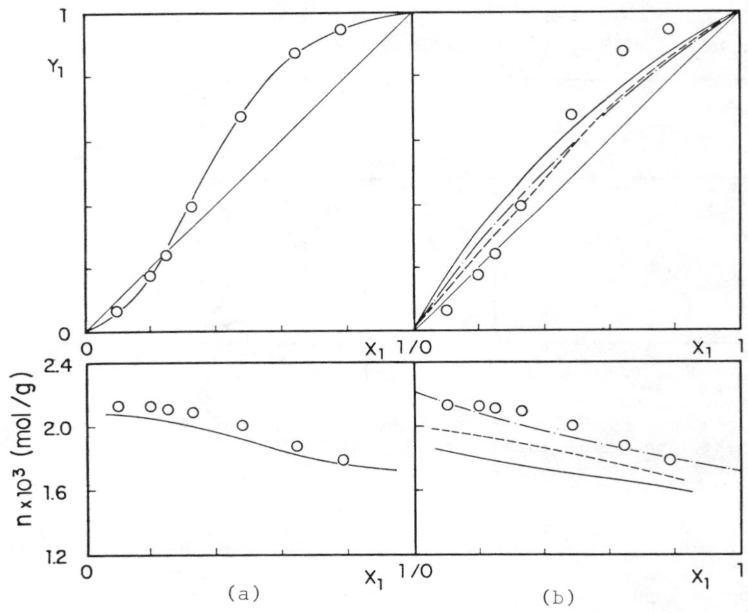

Figure 4. Equilibrium data of the mixture propane(1)-ethylene(2) on 5A zeolite at 20°C and 100 torr. Points=experimental; a) —— reproduced, RAS; b) —— predicted, IAS; —·— predicted, ST; --- predicted, VS.

Ternary Mixture Equilibria

The prediction of the ternary mixture equilibria has been carried out by the method proposed in the theoretical section (RAS theory) from the experimental data of pure components and binary mixtures. The results obtained for the mixtures ethane-propane-ethylene on 5A zeolite at 20°C and three total pressures of 100, 500 and 700 torr are presented in Table 4, together with the values predicted by the IAS theory. The standard deviations of the predicted molar fractions y_i in Table 4 with respect to the experimental ones are: 0.51×10^{-2} (RAS, 100 torr),

Table 4. Experimental and predicted ternary equilibrium data of adsorption on 5A zeolite at 20°C. Ethane=1, propane=2, ethylene=3

	Experimental							RAS(Wilson)				IAS			
	x_1	x_2	x_3	y_1	y_2	y_3	$nx10^3$	y_1	y_2	y_3	$nx10^3$	y_1	y_2	y_3	$nx10^3$
p = 100 torr	0.138	0.426	0.435	0.790	0.144	0.067	1.778	0.795	0.141	0.064	1.643	0.767	0.150	0.083	1.594
	0.061	0.209	0.730	0.517	0.103	0.380	1.985	0.521	0.101	0.378	1.895	0.590	0.142	0.271	1.771
	0.232	0.657	0.112	0.864	0.130	0.006	1.531	0.854	0.144	0.004	1.496	0.842	0.142	0.013	1.488
	0.440	0.223	0.337	0.962	0.023	0.015	1.458	0.964	0.021	0.014	1.379	0.950	0.029	0.020	1.344
	0.140	0.320	0.541	0.790	0.105	0.106	1.791	0.795	0.100	0.105	1.689	0.779	0.115	0.106	1.604
	0.056	0.834	0.110	0.494	0.493	0.013	1.632	0.492	0.496	0.012	1.615	0.482	0.484	0.035	1.621
p = 500 torr	0.075	0.422	0.503	0.499	0.347	0.154	2.245	0.515	0.335	0.150	2.160	0.530	0.315	0.157	2.063
	0.032	0.208	0.760	0.266	0.180	0.553	2.356	0.274	0.182	0.542	2.406	0.360	0.267	0.372	2.267
	0.139	0.718	0.143	0.636	0.354	0.011	1.958	0.654	0.337	0.009	1.893	0.658	0.316	0.028	1.875
	0.319	0.262	0.419	0.904	0.061	0.035	2.100	0.900	0.066	0.038	2.009	0.894	0.059	0.048	1.925
	0.073	0.320	0.607	0.507	0.251	0.241	2.266	0.504	0.249	0.245	2.240	0.542	0.258	0.200	2.120
	0.030	0.830	0.140	0.246	0.738	0.018	1.950	0.237	0.745	0.016	1.949	0.227	0.724	0.049	1.939
p = 700 torr	0.067	0.409	0.523	0.438	0.383	0.179	2.357	0.440	0.377	0.183	2.294	0.483	0.345	0.170	2.167
	0.032	0.207	0.760	0.197	0.192	0.611	2.417	0.193	0.209	0.601	2.516	0.340	0.287	0.375	2.359
	0.124	0.720	0.157	0.574	0.414	0.012	2.056	0.608	0.380	0.010	1.964	0.600	0.367	0.034	1.947
	0.280	0.269	0.451	0.873	0.082	0.045	2.219	0.860	0.090	0.052	2.127	0.864	0.076	0.059	2.034
	0.065	0.314	0.621	0.447	0.272	0.282	2.386	0.410	0.293	0.296	2.382	0.498	0.285	0.216	2.223
	0.029	0.818	0.153	0.210	0.773	0.018	2.044	0.223	0.760	0.016	2.033	0.220	0.729	0.052	2.016

3.55×10^{-2} (IAS, 100 torr), 1.03×10^{-2} (RAS, 500 torr), 5.85×10^{-2} (IAS, 500 torr), 1.90×10^{-2} (RAS, 700 torr), 7.77×10^{-2} (IAS, 700 torr), showing the better prediction provided by the RAS theory. The extrapolation beyond the experimental pressures needed for the pure component isotherms makes the standard deviation to increase with the total pressure, but the larger increase with the IAS theory can only be explained in terms of the deviation from ideallity at high concentrations in the adsorbed phase.

In conclusion, the RAS theory provides an accurate method of prediction of adsorption equilibria of multicomponent mixtures, being needed experimental data of pure components and binary mixtures. Other methods of prediction available in the literature, when applied to non-ideal mixtures, do not lead to satisfactory results. It would be interesting to incorporate experimental binary data for the prediction of multicomponent mixture equilibrium with these last models.

NOTATION

A = adsorption area
F = characteristic function, Equation (7)
ΔG = free energy of immersion
m = moles adsorbed at saturation per unit mass of adsorbent
n = moles adsorbed per unit mass of adsorbent
p = pressure
R = gas constant
T = temperature
V_1/V_2 = pre-adsorption volumetric ratio of components 1 and 2 in a binary mixture
x = molar fraction in the adsorbed phase
y = molar fraction in the gas phase

Greek letters

Δ = increment

Θ = reduced adsorbed amount (=n/m)
Λ = binary parameter in Wilson equation
Π = spreading pressure
τ = binary parameter in UNIQUAC equation

Superscripts

^{o} = pure component

Subscripts

1 = component 1
2 = component 2
3 = component 3
i = component i
j = component j
m = multicomponent mixture
s = saturation (referred to pressure, Equation (7))

LITERATURE CITED

1. Costa, E., Sotelo, J.L., Calleja, G. and Marrón, C., AIChE J., 27, 5 (1981)

2. Myers, A.L. and Prausnitz, J.M., AIChE J., 11, 121 (1965)

3. Cabra, L., "Equilibrio de Adsorción de Mezclas Gaseosas de Hidrocarburos en Zeolitas 5A y 13X", Tesis doctoral, Universidad Complutense, Madrid, Spain (1983)

4. Myers, A.L. and Sircar, S., to be published in AIChE J.

5. Marquardt, D.W., J. Soc. Ind. Appl. Math., 11, 431 (1963)

6. Suwanayuen, S. and Danner, R.P., AIChE J., 26, 68 (1980)

7. Suwanayuen, S. and Danner, R.P., AIChE J., 26, 76 (1980)

8. Ruthven, D.M., Loughlin, K.F. and Holborow, K.A., Chem. Eng. Sci., 28, 701 (1973)

9. Ruthven, D.M., AIChE J., 22, 753 (1976)

MODELING OF ADSORPTION, DESORPTION AND DISPLACEMENT IN FIXED-BED ADSORBERS

John C. Crittenden, Associate Professor
Department of Civil Engineering
David W.Hand, Assistant Research Engineer
Water and Waste Management Programs
Michigan Technological University
Houghton, MI 49931

ABSTRACT

A model which included diffusion resistances for liquid and intraparticle phases was successful in predicting adsorption, desorption and displacement of single- and two-solute fixed-bed data. Complete reversibility was assumed and competitive interactions were fully accounted for by using bisolute equilibrium expressions. The model was also successful in predicting the effluent concentration of known solutes in complex mixtures of unknown composition when significant competitive interactions did not occur. The model was less successful in predicting effluent concentrations of nonspecific measures of organic concentration such as total organic carbon (TOC). However, suggested improvements in the model may allow the prediction of TOC or known adsorbates in competing mixtures of unknown composition in the future.

INTRODUCTION

Granular activated carbon is capable of removing many dissolved organic compounds from waters and wastewaters. Predictive mathematical models can save both time and money by planning the scope of pilot studies and interpreting pilot data for full scale design. In order to be useful, a model must be able to predict the effluent concentrations of adsorbates in complex mixtures and the impact of process variables on process performance so that the least cost operation can be identified. The successes and limitations of the homogeneous surface diffusion model (HSDM) in predicting the effluent concentration history profiles of adsorbates in complex mixtures is presented.

MODEL FRAMEWORK AND JUSTIFICATION

As shown in Figure 1, these diffusion mechanisms are involved in the adsorption rate: (a) liquid-phase mass transfer resistance, (b) diffusion in the fluid contained in the pores known as pore diffusion, and (c) diffusion along the surface of the pores known as surface diffusion. With respect to intraparticle diffusion mechanisms, the diffusion rates can vary spatially within the adsorbent due to steric interactions in macro-, meso- and micropores. Some of these effects have been incorporated into complex mathematical representations and additional experiments have been conceived to estimate the required parameters (1-5). However, this increases the number of parameters that need to be determined and the model complexity especially when the model is extended to multicomponent systems. Model parameter sensitivity analyses have demonstrated that the description of equilibrium surface capacity is usually much more important in predicting fixed-bed adsorber dynamics than mass transfer resistances (6-12). For example, Crittenden et al. (13) were able to describe the data of Peel and Benedek (5) with the HSDM as well as their model which included micropore mass transfer resistance. In addition, the HSDM has been applied with success to a number of adsorbate-adsorbent combinations (3, 4, 6-18).

The assumptions of the HSDM include (See Fig. 1): (a) the rate is diffusion limited, i.e., local equilibrium prevails at the external surface of the adsorbent particle; (b) liquid-phase diffusion resistance occurs at the external surface, and it can be described by a linear concentration gradient; (c) solid or surface diffusion is the principal intraparticle mass transport mechanism and the impact of pore structure and concentration can be simulated with a single effective diffusivity (See references 6 and 19 for justification and criteria.); (d) the adsorbent is spherically shaped and the surface concentration depends only on radial position (homogeneous for a given radial position); (e) plug flow occurs in the bed; (f) for desorption, the model assumes complete reversibility with identical mass transfer parameters; and, (g) in multisolute systems, the adsorbates diffuse independently of one another such that interaction of the solutes is fully accounted for by multisolute equilibrium expressions. Consequently, the adsorption behavior of a solute is characterized by a rather simple model which involves its equilibrium capacity description and liquid and intraparticle mass transfer coefficients.

The HSDM is comprised of the following equations which are written for each adsorbing component. Details are given elsewhere (6-12). The following equations are described by the manner in which they were derived. The liquid-phase mass balance, and its initial and boundary conditions are:

$$- v \frac{\partial C(z,t)}{\partial z} = \frac{\partial C(z,t)}{\partial t} + \frac{3k_f(1-\varepsilon)}{R\,\phi\,\varepsilon} \left[C(z,t) - C_s(z,t)\right] \quad (1)$$

$$C(tv < z \le L, t < \tau) = 0 \quad (2)$$

$$C(z=0, t \ge 0) = C_0 \quad (3)$$

The intraparticle mass balance, and its initial and boundary conditions are:

$$\frac{D_s}{r^2} \frac{\partial}{\partial r} \left[r^2 \frac{\partial q(r,z,t)}{\partial r} \right] = \frac{\partial q(r,z,t)}{\partial t} \quad (4)$$

$$q(0 \le r \le R, 0 \le z \le L, t=0) = 0 \quad (5)$$

$$\frac{\partial}{\partial r} q(r=0, 0 \le z \le L, t \ge 0) = 0 \quad (6)$$

$$\frac{\partial}{\partial r} q(r{=}R,z,t) = \frac{k_f}{\rho_a D_s} \left[C(z,t) - C_s(z,t) \right] \qquad (7)$$

The equation(s) which couples Eqs. 1 and 4 are the equilibrium relationships which are represented by the Freundlich equation for single components or Ideal Adsorbed Solution Theory (IAS) for multicomponents (20). The partial differential equations constituting the HSDM were converted by the technique of orthogonal collocation to ordinary differential equations which were then integrated numerically. The application of the orthogonal collocation technique was presented elsewhere (10, 14).

MODEL PARAMETER ESTIMATION

The flow of information used to verify the HSDM is given in Figure 2. No model parameters were adjusted for comparing model predictions to the data. Sensitivity analyses on the column model were used to help refine the required parameter accuracy which was in turn used in the design of experiments to determine the parameters. Hand et al. (21) have discussed the impact of these parameters on single solute HSDM calculations.

The parameters which define the equilibrium capacity are usually the most important parameters that need to be defined for model predictions. To avoid the pitfalls which are associated with conducting isotherm tests, the tests were conducted with the methods that were discussed by Randtke and Snoeyink (22) and Crittenden and Hand (19).

Hand et al. (23) discuss the pitfalls of conducting rate studies and these problems must be recognized in order to obtain accurate rate data for the determination of intraparticle rate parameters, regardless of the model which is used to describe the intraparticle adsorption rate.

Hand et al. (23) have also provided the solutions to the HSDM for a batch reactor which do not require the use of complex numerical algorithms. The solutions to the HSDM which are given in Figure 3 for various 1/n values assume that the equilibrium liquid-phase concentration is 0.5 of the initial concentration. These solutions may be used to determine the experimental surface diffusivity by comparing elapsed time to the dimensionless time given in Figure 3 (23).

CONSTANT PATTERN SINGLE SOLUTE SOLUTIONS TO THE HSDM

According to Hand et al. (21), a dimensional analysis of Eqs. 1-7 revealed that solutions to the HSDM depend only on three dimensionless parameters, if the Freundlich isotherm was used to describe equilibrium: (a) the Biot number, Bi, (b) the Stanton number, St, and (c) the isotherm slope, 1/n. Constant pattern solutions to the HSDM require that 1/n is less than 1.0 and the fixed-bed is long enough to establish a mass transfer zone (MTZ). Figure 4 displays the effluent concentration history profiles for an adsorber with increasing empty bed contact times (EBCTs) and St numbers. As the EBCT or St increases, the MTZ spreads out and approaches constant pattern. If only constant pattern solutions are provided then only one solution for each Bi and 1/n value needs to be given; i.e., the solution corresponding to the minimum $EBCT_{min}$ or St_{min} value that is required for constant pattern. Predicted profiles for other EBCTs can be calculated from the wave velocity and that single constant pattern solution.

Hand *et al.* (21) have discussed the applicability of the constant pattern solutions and have provided a complete set of constant pattern solutions for 1/n from 0.0 to 0.9. Constant pattern solutions allow easy calculation of the complete concentration history profile because a complex numerical algorithm which has been used to solve the HSDM equations is no longer needed.

HSDM PREDICTIONS FOR KNOWN COMPONENTS

Single Solutes. -- In order for a model to be useful in design, it must be capable of predicting desorption of previously adsorbed components. As a first approximation, the predictive capability of the HSDM was tested assuming complete reversibility. Figure 5 compares HSDM predictions (solid lines) to fixed-bed data (solid circles) for desorption and reintroduction of dichlorophenol (DCP). Other details and excellent model comparisons for other adsorbate-adsorbent systems were given by Thacker *et al*. (10-12). Accordingly, the HSDM may be used for design applications that involve desorption of single solutes.

Bisolutes. -- The predictive capability of the HSDM has been tested for bisolute mixtures assuming that there were no intraparticle diffusion interactions and complete reversibility. Model details are given by Thacker *et al*. (10-12) and Crittenden *et al*. (14). Figure 6 compares HSDM predictions to column data for sequential feeding of two adsorbates. The weaker-adsorbed species, dimethylphenol (DMP), initially was fed alone and then DCP was introduced into the influent. This caused the DMP to be desorbed and greatly exceed its influent concentration due to competitive interactions. The HSDM closely predicts the displacement of DMP. In other bisolute studies (10-12, 14), the HSDM was shown to predict the adsorber dynamics for the displacement of the stronger-adsorbed species and desorption in bisolute mixtures due to reductions in influent concentrations. However, the HSDM has not been extensively tested; e.g., in adsorbate-adsorbent systems in which the competing molecules have different sizes and can compete on only part of the adsorbent surface. Accordingly, the HSDM is suitable for design applications involving bisolute mixtures of molecules of similar size.

HSDM PREDICTIONS FOR MIXTURES OF UNKNOWN COMPOSITION

In most applications involving water or wastewater treatment, adsorbents are used to remove or reduce the concentrations of specific organics in a complex background mixture of unknown composition or the concentration of nonspecific parameters that measure organic concentration such as total organic carbon (TOC). To predict adsorber dynamics for such a complex problem represents a significant step beyond what the HSDM has been shown to predict for known components. However, the current capability of the HSDM to describe this phenomenon and data illustrating future modeling needs will be discussed. Excellent progress has been made through the use of breaking down the TOC into pseudo-components based on competitive equilibrium interactions (24), molecular weight (25),or refined representations of intraparticle mass transfer resistances (1,5), and these efforts will not be discussed in detail.

Fate of Known Adsorbates in the Presence of a Complex Background of Unknown Composition. -- In drinking water applications, the synthetic organic chemicals (SOCs) that are to be removed are often at much lower concentrations than the organic background. However, in some cases, the organic background does not significantly affect the breakthrough of individual SOCs (15) and the competitive interactions with the background TOC can simply be described by performing rate and isotherm studies in the presence of the TOC background (15,18,26). Figure 7 compares the results of two physically identical column runs for the adsorption and desorption of chloroform. Column 1 was conducted in organic free water and the influent to column 2 contained 5 mg/L TOC after 8,000 bed volumes of treatment. The model simulation which is plotted contains several breaks in the curve due to interuptions in the test. (This model calculation is designated a simulation because the surface diffusivity was determined by fitting this data.) These results and others (15,18,26) have demonstrated that no significant competitive interactions were found between several untreated fulvic acids and chloroform adsorbed by carbons which have a variety of pore volume distributions. In addition, it was found that the breakthrough behavior of other more strongly adsorbing SOCs including competitive interactions with chloroform were not affected by the presence of an untreated fulvic acid. Based on these results, it appears that the background TOC can be ignored in some instances with respect to predicting the fate of individual SOCs and competitive interactions between SOCs.

Although no significant competitive interactions appear to occur with untreated TOC, it appears that chlorination which is standard water treatment practice in the US can modify the TOC such that competitive interactions with chloroform and other SOCs have been observed. In one set of tests, it was demonstrated that if a column was fed a chlorinated fulvic acid before the introduction of chloroform then the column capacity for chloroform was significantly reduced. Furthermore, a chromatographic overshoot of chloroform was possible when a chlorinated fulvic acid was present. In another study, significant competitive interactions were observed between the chlorinated fulvic acid and the more strongly adsorbing species, trichloroethylene and bromoform.

Prediction of Nonspecific Measures of Organic Concentration. -- As a first approximation, the effluent concentration history profiles for TOC were predicted using one component to represent TOC. Figure 8 (which is adapted from Ref. 16) compares HSDM predictions (solid lines) and simulations (dashed lines for which D_S was adjusted to fit the TOC data) to TOC data (various symbols) for several granular activated carbons (GACs). Although the relative performance of the GACs can be predicted, it would appear that significant improvements are needed. Several options offer promise with respect to improving the predictive capability of the HSDM and they include using more components to represent the TOC (24), a better resolution of the components which make up the TOC (25) and/or the inclusion of micropore resistance (1,5).

SUMMARY

The HSDM accounts for diffusion resistances in the fluid and intraparticle phases by film resistance and Fick's law for surface diffusion. If complete reversibility of adsorption is assumed, the HSDM can be used to predict adsorption and desorption of single adsorbates. By accounting for competitive interactions using only competitive equilibrium isotherm expressions, the HSDM is capable of predicting the competitive interactions of two adsorbates of similar size. For some drinking water applications, the organic background does not interfere with the adsorption or desorption of synthetic organic chemicals (SOCs) or competitive interactions between SOCs and the HSDM can be used to describe these systems. There are instances when significant competitive interactions with the organic background is found, e.g., after chlorination, and the simpler forms of the HSDM have not been successful in predicting the effluent concentration history profiles of specific SOCs in such mixtures. More work is needed to refine the HSDM so that it can predict the effluent concentration history profiles of nonspecific measures of organic contamination. By using pseudo-components (24) or molecular weight fractionation (25) to describe the organic background and by using more refined representations of intraparticle mass transfer resistances (1,5), progress is being made at describing such complex systems.

ACKNOWLEDGEMENTS

This research was supported, in part, by grants from the National Science Foundation (No. CME 79-24589 from the Water Resources and Environmental Engineering Program) and the Water and Waste Management Programs at Michigan Technological University.

NOTATION

Roman Letters

Bi	=	$k_f R(1-\varepsilon)/(D_s D_g \varepsilon \phi)$, Biot number based on the surface diffusion coefficient (dimensionless);
C_o	=	influent fluid-phase concentration (M/L^3);
$C(z,t)$	=	fluid-phase concentration as a function of time and axial position (M/L^3);
$C_s(z,t)$	=	fluid phase concentration at the exterior particle surface as a function of time and axial position (M/L^3);
D_s	=	surface diffusion coefficient (L^2/t);
D_g	=	$\rho_a q_e(1-\varepsilon)/(\varepsilon C_o)$ or $\rho_b q_e/(\varepsilon C_o)$, solute distribution parameter or partition coefficient (dimensionless);
EBCT	=	τ/ε, V_B/Q or L/V_s, fluid residence time in the bed which is devoid of the adsorbent or empty bed contact time (t);
$EBCT_{min}$	=	minimum empty bed contact time required for constant pattern (t);
k_f	=	film transfer coefficient (L/t);
K	=	Freundlich isotherm capacity constant $(L^3/M)^{1/n}$;
L	=	length of fixed-bed (L);
1/n	=	Freundlich isotherm intensity constant or slope on a

		log-log plot (dimensionless);
Q	=	fluid flow rate (L3/t);
q(r,z,t)	=	adsorbent-phase concentration as a function of radial and axial position, and time (M/M);
q_e	=	adsorbent phase concentration in equilibrium with the influent fluid-phase concentration (M/M);
r	=	radial coordinate (L);
R	=	adsorbent particle radius (L);
St	=	$k_f\tau(1-\varepsilon)/(R\varepsilon\phi)$, Stanton number (dimensionless);
St_{min}	=	minimum Stanton number required to establish constant pattern (dimensionless);
t	=	real or elapsed time (t);
T	=	$t/\tau(D_g+1)$, single solute mass throughput (dimensionless);
v	=	v_s/ε, intersitial fluid velocity (L/t);
V_B	=	volume of the bed (L3);
v_s	=	$v\varepsilon$, hydraulic loading or superficial fluid velocity (L/t);
z	=	axial coordinate (L).

Greek Letters

ε	=	bed void fraction (dimensionless);
τ	=	L/V, or EBCTε, fluid residence time in the pack bed or packed bed contact time;
ρ_a	=	$\rho_b/(1-\varepsilon)$, adsorbent density which includes pore volume (M/L3);
ρ_b	=	bulk density of the adsorbent in the fixed bed (M/L^3);
ϕ	=	sphericity, ratio of the surface area of the equivalent-volume sphere to the actual surface area of the adsorbent particle, (dimensionless).

LITERATURE CITED

1. Weber, W.J., Jr., and S. Liang, Environmental Progress, 2 (3) 167 (1983).
2. Weber, W.J., Jr., and K.T. Liu, Chem. Engrg. Comm., 6 49 (1980).
3. Liu, K.T., and W.J. Weber, Jr., J. Water Poll. Control Fed., 53 (10), 1541 (1981).
4. Van Vliet, B.M., and W.J. Weber, Jr., J. Water Poll. Control Fed., 53, 11, 1585 (1981).
5. Peel, R.G., and A. Benedek, J. Env. Eng. Div. Amer. Soc. Civil Engrs., 106 (EE4) 797.
6. Crittenden, J.C., "Mathematical Modeling of Fixed-Bed Adsorber Dynamics--Single Component and Multiple Component," Ph.D. Thesis, University of Michigan at Ann Arbor, 1976.
7. Crittenden, J.C. and W.J. Weber, Jr., J. Env. Eng. Div Amer. Soc. Civil Engrs., 104 (EE2) 185 (1978).
8. Crittenden, J.C. and W.J. Weber, Jr., J. Env. Eng. Div. Amer. Soc. Civil Engrs., 104 (EE3) 433 (1978).
9. Crittenden, J.C. and W.J. Weber, Jr., J. Env. Eng. Div. Amer. Soc. Civil Engrs., 104 (EE6) 1175 (1978).
10. Thacker, W.E., "Modeling of Activated Carbon and Coal Gasification Char Adsorbents in Single-Solute and Bisolute Systems." Ph.D. Thesis, University of Illinois, Urbana, IL (1981).

11. Thacker, W.E., V.L. Snoeyink, and J.C. Crittenden, J. Amer. Water Works Assoc., 75, 144 (1983).
12. Thacker, W.E., J.C. Crittenden, and V.L. Snoeyink, "Modeling of Fixed-Bed Adsorber Performance: Variable Influent Concentration and Comparison of Adsorbents," J. Water Poll. Control Fed., (in press, 1983).
13. Crittenden, J.C., M. Ari, J.L. Oravitz, and W.E. Thacker, J. Env. Eng. Div. Amer. Soc. Civil Engrs., 107 (EE4) 870 (1981).
14. Crittnden, J.C., B.W.C. Wong, W.E. Thacker, V.L. Snoeyink, and R.L. Heinrichs, J. Water Poll. Control Fed., 52, 2780 (1980).
15. Crittenden, J.C., B.R. Sabin, M. Ghosh, and G. Mueller, "Competitive Interactions Between Humic Substances and Chloroform in Fixed-Bed Adsorbers." 102nd Ann. Conf. AWWA, Miami, FL (1982).
16. Lee, M.C., J.C. Crittenden, V.L. Snoeyink, and M. Ari, J. Env. Eng. Div. Amer. Soc. Civil Engrs., 109, 631 (1983).
17. Merk, W., W. Fritz, and E.U. Schlunder, Chem. Eng. Sci., 36, 743 (1980).
18. Weber, W.J., Jr. and M. Pirbazari, Jour. Am. Water Works Assoc., 74, 203 (1982).
19. Crittenden, J.C. and D.W. Hand, "Design Considerations for GAC of Synthetic Organic Chemicals and TOC," Proceedings of the Am. Water Works Assoc. Seminar "Strategies for Controlling Trihalomethanes," 103rd Ann. Conf., Las Vegas, Nevada (1983).
20. Jossens, L., J.M. Prausnitz, W. Fritz, E.U. Schlunder, and A.L. Myers, Chem. Eng. Sci., 33, 1097 (1978).
21. Hand, D.W., J.C. Crittenden, and W.E. Thacker, "Simplified Models for Design of Fixed-Bed Adsorber," J. Env. Eng. Div. Amer. Soc. Civil Engrs., (in press, 1983).
22. Randtke, S.J. and V.L. Snoeyink, J. Am. Water Works Assoc., 75 (8) 406 (1983).
23. Hand, D.W., J.C. Crittenden, and W.E. Thacker, J. Env. Eng. Div. Amer. Soc. Civil Engrs., 109, 82 (1983).
24. Frick, B.R. and H. Sontheimer, in "Treatment of Water by Granular Activated Carbon." Amer. Chem. Soc. Books, Washington, D.C., Advances in Chemistry Series No. 202, pp247-268, 1983.
25. Summers, R.S. and P.V. Roberts, "Simulations of DOC Removal in Activated Carbon Beds." J. Env. Eng. Div. Amer. Soc. Civil Engrs., (in press, 1983).
26. Sabin, B.R., "Competitive Interactions between Humic Substances and Chloroform in Fixed-Bed Adsorbers," M.S. Thesis, Dept. of Civil Engrg., Mich. Tech. Univ., 1983.
27. Lee, M.C., V.L. Snoeyink, and J.C. Crittenden, "Activated Carbon Adsorption of Humic Substances." J. Am. Water Works Assoc., 73, 440 (1981).
28. Glaze, W.H., C.C. Lin, J.C. Crittenden, and R. Cotton, "An Attempt to Model a GAC Pilot Plant for Removal of Natural Organic Compounds and THM Precursors," Proceedings Amer. Chem. Soc., Seattle, Washington (March, 1983).
29. Lin, C.C., Ph.D. Thesis, Univ. of Texas at Dallas, 1983.

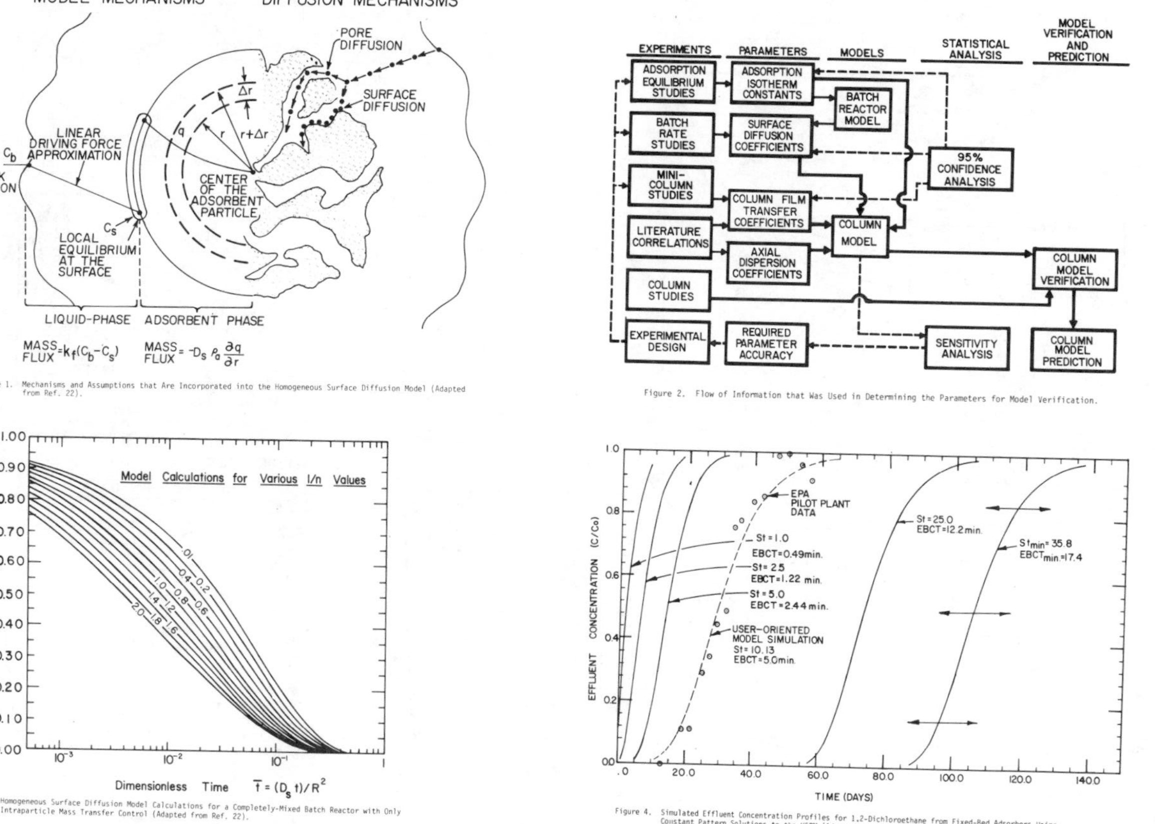

Figure 1. Mechanisms and Assumptions that Are Incorporated into the Homogeneous Surface Diffusion Model (Adapted from Ref. 22).

Figure 2. Flow of Information that Was Used in Determining the Parameters for Model Verification.

Figure 3. Homogeneous Surface Diffusion Model Calculations for a Completely-Mixed Batch Reactor with Only Intraparticle Mass Transfer Control (Adapted from Ref. 22).

Figure 4. Simulated Effluent Concentration Profiles for 1,2-Dichloroethane from Fixed-Bed Adsorbers Using Constant Pattern Solutions to the HSDM (Adapted from Ref. 23).

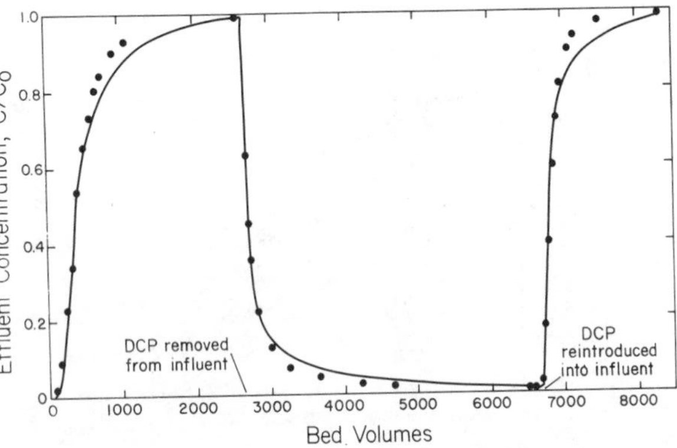

Figure 5. Predicted Effluent Concentration Profile (Solid Line) Using the HSDM as Compared to Data (Solid Circles) for Dichlorophenol (Adapted from Ref. 11 -- C_0 = 0.987 mmole/L, v_s = 9.24 m/hr, EBCT = 21.6 sec).

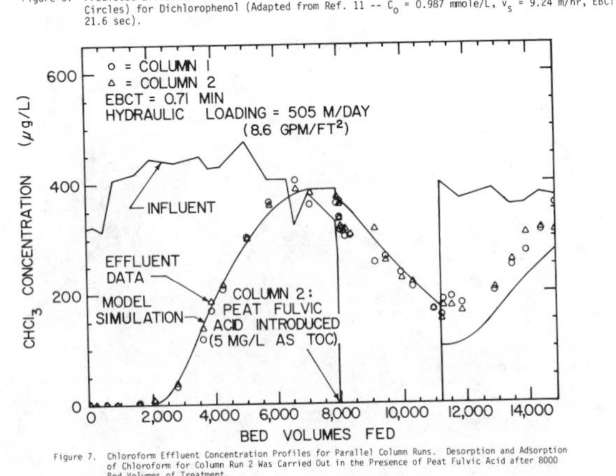

Figure 7. Chloroform Effluent Concentration Profiles for Parallel Column Runs. Desorption and Adsorption of Chloroform for Column Run 2 Was Carried Out in the Presence of Peat Fulvic Acid after 8000 Bed Volumes of Treatment.

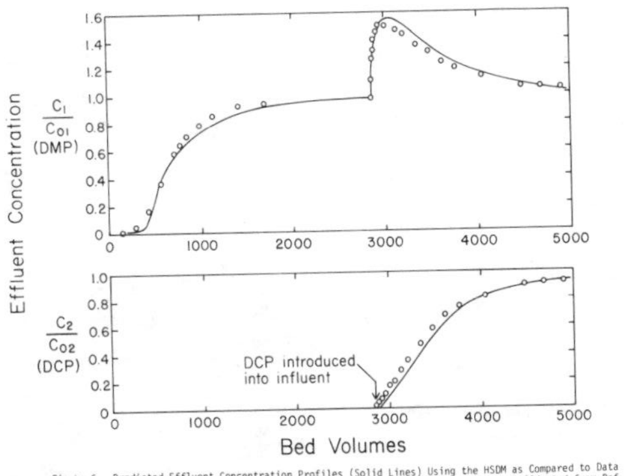

Figure 6. Predicted Effluent Concentration Profiles (Solid Lines) Using the HSDM as Compared to Data (Solid Circles) for the Displacement of Dimethylphenol by Dichlorophenol (Adapted from Ref. 12 -- C_{01} = 0.990 mmole/L, C_{02} = 1.02 mmole/L, v_s = 7.16 m/hr, EBCT = 25.4 sec).

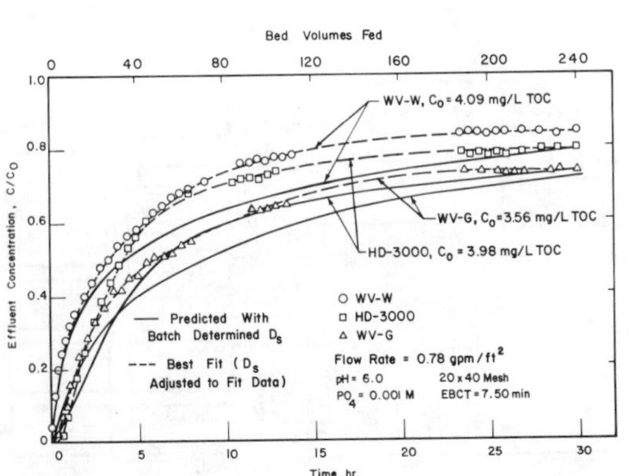

Figure 8. Predicted Effluent Concentration History Profiles (Solid Lines) Using the HSDM as Compared to Data (Solid Circles) for Peat Fulvic Acid Adsorbed by Different Activated Carbons.

APPLICATION OF VACANCY SOLUTION THEORY TO GAS ADSORPTION

Ronald P. Danner
133 Fenske Laboratory
The Pennsylvania State University
University Park, PA 16802

ABSTRACT

Of the available methods for predicting gas-mixture adsorption equilibria, those based on solution thermodynamic theory are the most useful. In particular, the model based on the vacancy solution concept in conjunction with the Flory-Huggins activity coefficient equations provides the simplest and most accurate means of predicting mixture equilibria as a function of composition, temperature, and pressure.

INTRODUCTION

Adsorption systems are considerably more difficult to characterize than vapor-liquid (or liquid-liquid) equilibrium systems. In the latter cases excellent progress has been made in predicting multicomponent equilibria using only binary interaction parameters between the components (Van Laar, Wilson, UNIQUAC, etc.) or between chemical groups in the molecules (e.g., ASOG, UNIFAC). If one is separating two components by adsorption, however, there are actually three species in the system because the solid adsorbent plays an important role. Thus, for separation of two components by adsorption, binary parameters are needed for these components on each adsorbent (and perhaps for each regeneration condition of that adsorbent). For practical use, these parameters should be correlated with, or present in a correlation which accounts for, the temperature and pressure dependencies of the equilibria. Gas-mixture adsorption equilibria data are difficult to obtain. Only in exceptional cases will binary data be available or determined experimentally for all the components in the system on the particular adsorbent at the temperature and pressure of interest. For general purposes then, a prediction scheme for gas-mixture equilibria is required which uses only pure-component data and has temperature and pressure dependencies incorporated into it.

Adsorption models based on solution thermodynamics appear to hold the most promise for predicting gas-mixture equilibria using only pure-gas data. In this case the system is treated in terms of macroscopic thermodynamic properties such as spreading pressure and chemical potential. Details of the structure of the adsorbent do not enter in; therefore, this approach can be applied to most any type of adsorption system. Some of the methods introduce the activity coefficient or osmotic coefficient which can be correlated with composition, temperature and pressure by equation forms analogous to those used in other fluid phase equilibria (VLE,LLE). The extension of these methods to mixtures is, in general, straightforward.

Myers and Prausnitz (1) introduced the concept of the ideal adsorbed solution in 1965. This ideal model is analogous to Raoult's Law. It has the disadvantages that in many cases pure-component adsorption data are required at pressures far in excess of the pressure of the mixed-gas system, that there is no method of directly incorporating temperature dependence, and that it can not predict highly nonideal systems such as adsorption azeotropes. It has been quite successful, however, in predicting adsorption equilibria for a number of more regularly behaved systems.

Lee (2) used lattice solution theory and the concept of volume filling of micropores to develop a correlation that takes into account the adsorbed phase nonidealities. However, one of the model parameters, an interchange energy, has to be calculated from mixture data.

In the mid-1970's Soviet workers (3,4,5) introduced the concept of a vacancy solution. A vacancy solution is composed of adsorbates and vacancies which are the spaces that could be filled by adsorbates. Adsorption equilibrium then is treated as an equilibrium between a vacancy solution in the gas phase and vacancy solution in the adsorbed phase. Standard solution thermodynamic equations relating the equilibrium between these two phases can be applied in order to determine the equilibrium compositions. This has been done in terms of both osmotic coefficients and activity coefficients. The remarkable thing about this approach is that even for pure-component adsorption one has a binary vacancy solution (gas + vacancies); and, thus, for the pure component data one can incorporate expressions for the osmotic or activity coefficients and extend these relationships to multicomponent systems by the standard methods found in other fluid phase equilibria. Bering and coworkers (6,7,8) tended to use a very general osmotic coefficient approach while workers in our laboratory (9,10,11,12) have adopted the activity coefficient approach and examined the applicability of three different activity coefficient equations to vacancy solutions containing one to four gases. The vacancy solution model provides an excellent framework on which to develop correlation and prediction methods for gas adsorption. The theory is thermodynamically sound, the extensive analogous work in vapor-liquid equilibria can be used as a guide, and the extension to multicomponent systems is straightforward.

THEORY

In the vacancy solution model (VSM) both the gas and adsorbed phases are considered to be solutions of moleules in a hypothetical solvent called "vacancy". A vacancy is a vacuum entity occupying spaces that can be filled by adsorbate or gas-phase molecules. The entire system including the solid is in thermal equilibrium, but only the gas and adsorbed phases are in phase equilibrium.

$$\mu_i^g = \mu_i^s \tag{1}$$

One of the i components in Equation (1) is the vacancy. How one now expresses the chemical potentials is a matter of choice. The Soviet workers (4,5) have primarily used the osmotic coefficient approach where the chemical potential of the vacancy in the surface phase is expressed as:

$$\mu_v^s = \mu_v^{os} + gRT \ln x_v^s + \pi a \tag{2}$$

Here the osmotic coefficient, g, characterizes the nonideality of the vacancy solution, x_v^s is the absolute mole fraction of vacancies, and a is the molar surface area.

We have elected to write the chemical potential of the components (including vacancy) in the surface phase in terms of the activity coefficient (11,12).

$$\mu_i^s = \mu_i^{os} + RT \ln \gamma_i^s x_i^{s,e} + \pi \bar{a}_i \tag{3}$$

Here the activity coefficient, γ_i^s, characterizes the nonideality of the vacancy solution, $x_i^{s,e}$ is an **excess** mole fraction, and $\bar{a}_i$ is a **partial** molar area. We have found that the use of activity coefficients has allowed us to examine a number of composition dependencies of the chemical potential easily by using analogs from vapor-liquid equilibria. The appearance of an excess property indicates that a Gibbs dividing surface has been invoked. The incorporation of an appropriately defined dividing surface together with standard activity coefficient equations has allowed us to move directly into multicomponent systems without introducing any other arbitrary assumptions. Thus, all of our work has been on this latter approach, and all further discussion will be limited to this case.

Pure Gas

For the vacancies in the gas phase the chemical potential is written as:

$$\mu_i^g = \mu_i^{og} + RT \ln \gamma_i^g x_i^g \tag{4}$$

The relationship between the standard state potentials can be obtained by considering the limiting case of pure vacancies where $x_v^g = x_v^{s,e} = 1$ and $\gamma_v^g = \gamma_v^s = 1$ and the spreading pressure, π, is zero. Thus, from Equations (1), (3) and (4):

$$\mu_v^{og} = \mu_v^{os} \tag{5}$$

For the vacancy solution in the gas phase one can argue that at reasonable pressures, it is so dilute that $x_v^g \simeq 1$ and $\gamma_v^g \simeq 1$. Thus, the logarithmic term in Equation (4) is negligible for the vacancy component. By substituting Equations (3), (4), and (5) into Equation (1) and rearranging terms, one obtains the equation of state for the adsorbed phase:

$$\pi = \frac{-RT}{\bar{a}_v} \ln \gamma_v^s x_v^{s,e} \tag{6}$$

The excess properties are defined in terms of a dividing surface recommended by Lucassen-Reynders (13). This surface is chosen such that the sum of the excess surface concentrations of the vacancy and of the gas is equal to a constant which is taken to be the limiting excess surface concentration. Cochran (9) has shown that by combining the equation of state, the Gibbs adsorption equation, and the above dividing surface, the following isotherm equation is obtained.

$$P = \left[\frac{n_1^{\infty}}{b_1}\frac{\theta}{1-\theta}\right]\left[\exp - \int \frac{d\ell n\gamma_v^s}{\theta}\right]\left[\lim_{\theta\to 0} \exp\left(\int \frac{d\ell n\gamma_v^s}{\theta}\right)\right] \tag{7}$$

The first bracket is in the form of the well-known Langmuir equation. The exponential terms are corrections which depend on the form of the activity coefficient equation selected. Since the vacancy solvent is an abstract entity, the nonidealities accounted for by the activity coefficient are actually caused by concentration-dependent effects at the gas-solid interface. Such effects include adsorbate-adsorbate and adsorbate-adsorbent interactions.

Note that for the pure component isotherm an activity coefficient for the solution of gas and vacancy has been introduced. The "binary" parameters used in the activity coefficient equations can then be incorporated into multicomponent expressions.

Gas-Mixtures

For a gas mixture Equation (3) is used to express the chemical potential in the adsorbed phase. The chemical potential for an adsorbate in the gas phase is expressed by the usual equation for a gas mixture.

$$\mu_i^g = \mu_i^{og} + RT\ \ell n\ \phi_i\ y_i\ P \tag{8}$$

This equation has the advantage over the equation for a gas solution [Equation (4)] that at high pressures the nonideality of the gas phase can be taken into account through the fugacity coefficient, ϕ_i. The relationship between the mole fractions in the adsorbed phase as observed experimentally and as expressed in terms of the vacancy solution can be written as (see 12):

$$x_i^s = \frac{x_i n_m}{n_m^{\infty}} = x_i\theta \tag{9}$$

$$x_v^s = 1-\theta \tag{10}$$

Equating the chemical potentials and substituting, the equilibrium equation becomes:

$$\phi_i y_i P = \gamma_i^s x_i \frac{n_m}{n_m^{\infty}} \exp\left[\frac{\mu_i^{os}-\mu_i^{og}}{RT}\right] \exp\left[\frac{\pi\bar{a}_i}{RT}\right] \tag{11}$$

The dividing surface is defined such that the limiting adsorption concentration is the average of those of the adsorbates. This is consistent with the definition used for the pure gas adsorption case and it allows description of adsorbed mixtures containing adsorbates which have unequal values of the limiting amount adsorbed. As shown by Cochran (9) the final equilibrium relation can then be written as:

$$\phi_i y_i P = \gamma_i^s x_i \frac{n_m}{n_m^\infty} \frac{n_i^\infty}{b_i} \left[\lim_{x_i^s \to 0} \frac{1}{\gamma_i^s} \right] \exp \left[\frac{\pi \bar{a}_i}{RT} \right] \tag{12}$$

$$\frac{\pi \bar{a}_i}{RT} = - \left[1 + \frac{n_m^\infty - n_i^\infty}{n_m} \right] \ln \gamma_v^s x_v^s \tag{13}$$

For multicomponent adsorption, there is one equilibrium equation, Equation (12), for each gas species. For most cases the errors introduced by setting the fugacity coefficients to unity are insignificant compared to the accuracy of the adsorption equilibrium data. Equation (12) is also written in a general form which can be applied for any selected activity coefficient model.

Activity Coefficient Models

Three different activity coefficient models have been evaluated in terms of their incorporation into the VSM. These are given in Table 1 together with their corresponding pure gas isotherm equations and gas-mixture equilibria equations.

Two-Suffix Margules Equation

Suwanayuen and Danner (11) examined the two-suffix Margules equation and found that the pure gas isotherm derived from the vacancy solution equation is the same as the well-known Fowler equation. It can only be applied to adsorption equilibria which deviate only slightly from the Langmuir isotherm. Therefore, it was not investigated further by these authors. Nakahara et al. (14) applied this equation to the ethylene-propylene-carbon molecular sieve system and confirmed that this form did not give very good fits of the pure component isotherms nor very good predictions of the adsorption equilibria. We have found that the data of Nakahara et al. can be treated with good accuracy by the Cochran et al. form of the vacancy solution equations.

Wilson Equation

Suwanayuen and Danner (11,12) introduced the Wilson equation into their VSM. There are now four parameters for each gas - Henry's law constant, b_i, the limiting amount adsorbed, n_i^∞ , and the two Wilson parameters, Λ_{iv} and Λ_{vi}. Most pure component isotherms are of course fit very well by an equation form containing four parameters. The objective of this type of work is not to correlate pure component data

TABLE 1

Summary of Vacancy Solution Models

Activity Coefficient Equation	Pure Gas Isotherm	Gas-Mixture Equilibria Equation
Two-Suffix Margules		
$\ln \gamma_v^s = \frac{W_{1v}}{RT}(x_1^s)^2$	$P = \left[\frac{n_1^\infty}{b_1}\frac{\theta}{1-\theta}\right] \exp\left[\frac{-2W_{1v}}{RT}\theta\right]$	
Wilson		
$\ln \gamma_k^s = 1 - \ln\left[\sum_{j=1}^{j=N} x_j^s \Lambda_{kj}\right] - \sum_{i=1}^{i=N}\left[\frac{x_i^s \Lambda_{ik}}{\sum_{j=1}^{j=N} x_j^s \Lambda_{ij}}\right]$	$P = \left[\frac{n_1^\infty}{b_1}\frac{\theta}{1-\theta}\right]\left[\Lambda_{1v} \cdot \frac{1-(1-\Lambda_{v1})\theta}{\Lambda_{1v}+(1-\Lambda_{1v})\theta}\right] \cdot \exp\left[\frac{-\Lambda_{v1}(1-\Lambda_{v1})\theta}{1-(1-\Lambda_{v1})} - \frac{(1-\Lambda_{1v})\theta}{\Lambda_{1v}+(1-\Lambda_{1v})\theta}\right]$	$y_i \phi_i P = \gamma_i^s x_i n_m \frac{n_i^\infty \Lambda_{iv}}{n_m^\infty b_i} \cdot \exp(\Lambda_{vi}-1) \exp\left[\frac{\pi \bar{a}_i}{RT}\right]$
Cochran et al. (Based on Flory-Huggins approach)		
$\ln \gamma_i^s = -\ln \sum_{j=1}^{j=N} \frac{x_j^s}{\alpha_{ij}+1} + \left[1 - \left(\sum_{j=1}^{j=N} \frac{x_j^s}{\alpha_{ij}+1}\right)^{-1}\right]$	$P = \left[\frac{n_1^\infty}{b_1}\frac{\theta}{1-\theta}\right] \exp\left(\frac{\alpha_{1v}^2 \theta}{1+\alpha_{1v}\theta}\right)$	$y_i \phi_i P = \gamma_i^s x_i \frac{n_m}{n_m^\infty} \frac{n_i^\infty}{b_i} \cdot \frac{\exp \alpha_{iv}}{(1+\alpha_{iv})} \exp\left[\frac{\pi \bar{a}_i}{RT}\right]$

but to extract information from the pure component data which can be used to predict gas mixture behavior. Suwanayuen and Danner obtained good results for the prediction of binary mixtures of O_2, N_2, and CO on molecular sieves and for light hydrocarbons on activated carbon. In these evaluations they assumed that the interactions between adsorbates were insignificant compared to the interaction between the adsorbate and the adsorbent, i.e., they set $\Lambda_{ij}=1$ when neither i or j were the vacancy. They did, however, provide a means of estimating these interaction parameters from the heat of adsorption and found in several cases that these estimated values improved the prediction of binary equilibria.

Hyun and Danner (15) studied several hydrocarbon-CO_2-molecular sieve systems which exhibit azeotropes. They reported that while the VSM was the only model which could even qualitatively exhibit azeotropic behavior, the Wilson form did not give good accuracy for these systems. They also found, that using adsorbate-adsorbate interaction parameters estimated as suggested by Suwanayuen and Danner (12) gave poorer predictions than assuming that they were unity.

Kaul (16) evaluated the VSM in the Wilson form for predicting binary and ternary gas mixtures on several adsorbents. He concluded that for low surface coverage one can predict multicomponent data from pure component data by either the ideal adsorbed solution model or the VSM.

For high surface coverage he found that binary interaction parameters extracted from the binary-gas data were needed in order to predict three-component equilibria in the CO-N_2-O_2-molecular sieve 10X system. This has also been confirmed by Cochran (9).

Cochran found that there is a strong correlation between the regressed adsorbate binary interaction parameters in the Wilson form of the VSM. He showed that this relationship can be expressed in terms of the pure component parameters as follows:

$$\frac{\Lambda_{12}}{\Lambda_{21}} = \frac{\Lambda_{1v}}{\Lambda_{v1}} \frac{\Lambda_{v2}}{\Lambda_{2v}} \qquad (14)$$

This modification greatly simplifies the computational difficulties in determining the binary adsorbate parameters and no loss in accuracy is encountered. Cochran also showed that the parameters of interest can be correlated with temperature as follows:

$$b_1 = b_{o1} \exp\left(\frac{-q_1}{RT}\right) \qquad (15)$$

$$n_1^{\infty} = n_{o1}^{\infty} \exp\left(\frac{k_1}{T}\right) \qquad (16)$$

$$\Lambda_{v1} = \frac{a_v}{a_1} \exp\left(\frac{\lambda_{vv}-\lambda_{1v}}{RT}\right) \qquad (17)$$

$$\Lambda_{1v} = \frac{a_1}{a_v} \exp\left(\frac{\lambda_{11}-\lambda_{v1}}{RT}\right) \qquad (18)$$

If one has a number of isotherms available for the gases of interest, instead of regressing to obtain four parameters for every temperature, it is recommended that all the data be regressed simultaneously to extract at the most seven parameters - b_{o1}, q_1, n_{o1}^{∞}, k_1, a_v/a_1, $(\lambda_{vv}-\lambda_{1v})$ and $(\lambda_{11}-\lambda_{v1})$. This approach provides a means of interpolating between the experimental temperatures and for cautious extrapolation outside the temperature range of the data. In addition, it was found to give more meaningful physical significance to the values of the parameters b_i, n_i^{∞}, and Λ_{ij}. These temperature correlated values in general gave better predictions for the mixtures than the values extracted from isothermal data.

The VSM in the Wilson form has been applied to adsorption from dilute aqueous solutions by Fukuchi et al. (17). For two dilute solutes these authors found that good agreement with experimental data required interaction parameters between the two solutes.

Cochran et al. Equation

Cochran et al. (10) derived an activity coefficient equation based on an excess Gibbs free energy expression for the vacancy solution which was written by analogy to that derived by Flory and Huggins. This approach was suggested by physical interpretations regarding the relative changes in entropy and enthalpy (ΔH was assumed negligible) during the interchanging or mixing of adsorbate molecules and vacancies as well as by the empirical observation that the correlation found between the two Wilson equation parameters lead essentially to the Flory-Huggins model.

The Cochran et al. form has only three parameters for a pure gas - b_1, n_1^{∞} and α_{1v}. It was found, however, to represent the pure-component isotherms essentially as well as the four-parameter, Wilson form. For binary gas mixtures, Cochran et al. found that the following equation could be used to determine the adsorbate-adsorbate binary parameter.

$$\alpha_{ij} = \frac{\alpha_{iv}+1}{\alpha_{jv}+1} - 1 \qquad (19)$$

After studying numerous examples the authors found that for most cases the value of α_{ij} obtained from this equation gave essentially the same results as regressing a binary parameter from the binary-mixture data. This approach was also, overall, the most accurate means of predicting binary and ternary adsorption equilibria. This is quite fortunate because in most cases only pure-gas data will be available.

Cochran et al. (10) also developed temperature correlations for their form of the VSM. The b_i and n_i^{∞} parameters are given by the relations used for the Wilson form, i.e. Equations (15) and (16). The α_{iv} parameter is given by:

$$\alpha_{1v} = m_1 n_1^{\infty} - 1 \qquad (20)$$

Thus, if a number of isotherms are available for a pure gas, instead of regressing three parameters for each temperature, all the data should be used simultaneously to obtain only five parameters - b_{o1}, q_1, n_{o1}, k_1 and m_1. This approach will allow excellent interpolation within, and reasonable extrapolation outside of, the experimental temperature range. Furthermore, the parameters determined in this way are more consistent and have been found to give better binary and ternary mixture predictions than parameters determined from only isothermal data. For the azeotropic adsorption system of isobutane-ethylene-13X zeolite, the Cochran et al. method predicted an azeotrope, but not in very good quantitative agreement with the experimental data. For most other systems including some ternary and quaternary systems good predictions were obtained. Cochran (9) recommends that if binary data are available, both the Wilson and Cochran et al. forms should be evaluated using binary interaction parameters regressed from the data. Whichever method correlates the binary data better should then be used for predictions in multicomponent systems. If only pure component data are available, the recommended procedure is to use the Cochran et al. form using the temperature correlations for the model parameters, if sufficient data are available.

CONCLUSIONS

As would be expected in such a complex area as adsorption, there is no single approach that will best suit all cases. For the **correlation** of pure component adsorption data, most any of the three or four parameter models will do an acceptable job. For **predicting** adsorption equilibria in systems containing two or more adsorbates, the vacancy solution model seems to be the most promising. If no binary data are available the Cochran et al. form of the vacancy solution model is the most likely to give reasonable predictions. If a number of isotherms are available for the same gas, the temperature dependent functions should be used to determine the model parameters. If binary equilibrium data are available, the Wilson and Cochran et al. forms of the vacancy solution model should be evaluated - in both cases examining the effect of using regressed binary parameters. Whichever method gives the most accurate result for the binary data should be used to predict higher order systems. For systems where extreme nonideality is anticipated such as found with adsorption azeotropes, all methods must be viewed with suspicion. But, only those based on the vacancy solution model are able to predict azeotropic behavior even qualitatively.

ACKNOWLEDGMENT

Much of the work reviewed in this paper is the result of the efforts of my colleague R. L. Kabel and my students, T. W. Cochran, S. H. Hyun, and S. Suwanayuen. We are grateful for the financial support we have received for this work from the National Science Foundation (Grant No. CPE-8012423) and from the Frontiers of Separations program of the Exxon Research and Engineering Company.

NOTATION

a_i Molar area, m^2/kmol

$\bar{a}_i$ Partial molar area, m^2/kmol

b_i Henry's law constant, kmol/(kg)(kPa)

b_{oi} Temperature independent constant, kmol/(kg)(kPa)

g Osmotic coefficient

k_i Temperature independent constant, K

m_i Temperature independent constant, kg/kmol

n_i Moles adsorbed per mass of adsorbent, kmol/kg

n_i^{∞} Limiting amount adsorbed per mass of adsorbent, kmol/kg

n_{oi}^{∞} Temperature independent constant, kmol/kg

P Equilibrium adsorption pressure, kPa

q_i Temperature independent constant (isosteric heat of adsorption at infinite dilution), kJ/kmol

R Gas constant, kJ/(kmol)(K)

T Temperature, K

W_{1v} Constant in two-suffix Margules equation

x_i Mole fraction in adsorbed mixture

x_i^g Mole fraction in vapor phase vacancy solution

x_i^s Mole fraction in adsorbed phase vacancy solution

$x_i^{s,e}$ Excess mole fraction in adsorbed phase vacancy solution

y_i Mole fraction in vapor phase

Greek Letters

α_{ij} Interaction parameter in adsorbed phase

γ_i^g Activity coefficient in vapor phase vacancy solution

γ_i^s Activity coefficient in adsorbed phase vacancy solution

θ Fraction of limiting adsorption

λ_{ij} Potential energy of interaction, kJ/kmol

Λ_{ij} Wilson interaction parameter

μ_i^g Chemical potential in vapor phase vacancy solution, kJ/kmol

μ_i^s Chemical potential in adsorbed phase vacancy solution, kJ/kmol

μ_i^{og} Standard state chemical potential in vapor phase vacancy solution, kJ/kmol

μ_i^{os} Standard state chemical potential in adsorbed phase vacancy solution, kJ/kmol

π Spreading pressure, N/m

ϕ_i Fugacity coefficient

LITERATURE CITED

1. Myers, A. L. and J. M. Prausnitz, **AIChE J.**, 11, 121 (1965).

2. Lee, A. K. K., **Can. J. Chem. Eng.**, 51, 688 (1973).

3. Bering, B. P. and V. V. Serpinskii, **Izv. Akad. Nauk SSSR, Ser. Khim.**, No. 12, 2679 (1973).

4. Bering, B. P. and V. V. Serpinskii, **Izv. Akad. Nauk SSSR, Ser. Khim.**, No. 11, 2427 (1974).

5. Dubinin, M. M., **Amer. Chem. Soc. Symposium Ser.**, 40, 1 (1977).

6. Bering, B. P., V. V. Serpinskii and T. S. Jakubov, **Izv. Akad. Nauk SSSR, Ser. Khim.**, No. 4, 727 (1977).

7. Bering, B. P. and V. V. Serpinskii, **Izv. Akad. Nauk SSSR, Ser. Khim.**, No. 8, 1732 (1978).

8. Jakubov, T. S., B. P. Bering and V. V. Serpinskii, **Izv. Akad. Nauk SSSR, Ser. Khim.**, No. 5, 991 (1977).

9. Cochran, T. W., **Vacancy Solution Models for Gas-Mixture Adsorption**, M.S. Thesis, The Pennsylvania State University, University Park, PA (1982).

10. Cochran, T. W., R. L. Kabel and R. P. Danner, "Vacancy Solution Theory of Adsorption using Flory-Huggins Activity Coefficient Equations," submitted to AIChE J.

11. Suwanayuen, S. and R. P. Danner, **AIChE J.**, 26, 68 (1980).

12. Suwanayuen, S. and R. P. Danner, **AIChE J.**, 26, 76 (1980).

13. Lucassen-Reynders, E. H., **Prog. Surface Membrane Sci.**, 10, 253 (1976).

14. Nakahara, Tomoko, Mitsuho Hirato and Hisashi Mori, **J. Chem. Eng. Data**, 27, 317 (1982).

15. Hyun, S. H. and R. P. Danner, **J. Chem. Eng. Data**, 27, 196 (1982).

16. Kaul, B. K., **Correlation and Prediction of Adsorption Isotherm Data for Pure and Mixed Gases**, presented at AIChE Annual Meeting, Los Angeles, CA (1982).

17. Fukuchi, Kenji, Shigetoshi Kobuchi and Yasuhiko Arai, **J. Chem. Eng. Japan**, 15, 316 (1982.)

HIGH PRESSURE PHYSICAL ADSORPTION OF GASES ON HOMOGENEOUS SURFACES

Gerhard H. Findenegg
Institute of Physical Chemistry, Ruhr-University,
4630 Bochum, West Germany

ABSTRACT

Physical adsorption of several pure gases on graphitized carbon black has been studied at elevated pressures up to 150 bar in the author's laboratory over the past five years. On the basis of these results, the general features of high pressure physical adsorption on homogeneous high energy solid surfaces are presented.

Above the critical temperature T_c the surface excess isotherms generally pass through a maximum at pressures greater than the critical pressure and densities below the critical density of the fluid. At high temperatures (typically $T > 1.5\,T_c$) the surface excess concentration is small and can be accommodated in one or two monolayers. In this temperature range the isotherms can be represented quantitatively by a three-parameter equation.

In the so-called supercritical region ($T_c < T < 1.2\,T_c$) the surface excess of the fluid exhibits a pronounced dependence upon temperature near the maximum of the isotherms, due to multilayer formation ('prewetting' of the solid by the supercritical fluid). Beyond the maximum, the surface excess increases with increasing temperature at fixed pressure, but decreases with increasing temperature at fixed density of the fluid. Thus density is a more natural variable than pressure in supercritical fluid adsorption.

Below the critical temperature the multilayer regime of the surface excess isotherms can be represented by a modified Frenkel-Halsey-Hill equation. For the system studied (propane/graphite) an exponent $m = 2.55 \pm 0.25$ is obtained for the temperature range from $T/T_c = 0.7$ to 1.0. There are indications that, on approaching T_c, the fluid layer next to the solid is in a compressed state, i.e., local density exceeds the density of the saturated liquid.

The paper also deals with the question of how to predict adsorption isotherms of gases at high temperatures ($T > T_c$) for a given substrate, on the basis of a corresponding states conjecture.

Physical adsorption of gases at elevated pressures is arousing at present considerable interest in several fields such as, the separation and purification of the lower hydrocarbons and several other gases (1), storage of fuel gases (2) or radioactive gases (3) in microporous solids, and adsorption from supercritical gases in extraction processes and chromatography (4,5). Furthermore, knowledge of the structure and thermodynamics of compressed gas/solid interfaces is fundamental to an understanding of wetting phenomena and of heterogeneous catalysis.

Most experimental adsorption studies at elevated pressures reported in the literature have been made on adsorbents with heterogeneous surfaces or on microporous solids (6 to 11). In our laboratory we have been studying the adsorption of several gases at elevated pressures on graphitized carbon black (g.c.b.), a non-porous adsorbent of highly homogeneous surface (see Table 1). On the one hand, these data can be used to test modern theories of the structure of the fluid/solid interface (12,13) and thus they help to understand the connection between gas adsorption, wetting phenomena, adsorption from liquid mixtures, and several related topics. On the other hand, knowledge of simple fluid/solid interfaces is also useful for the modelling and prediction of high temperature adsorption in microporous materials (14).

Table 1. Gas adsorption at elevated pressures on graphitized carbon black at pressures up to 150 bar. Systems studied in Bochum 1978 to 1983.

Gas	T_c/K	p_c/bar	T/T_c range	Ref.
Argon	150.7	48.7	1.7 to 2.1	20
Krypton	209.3	54.9	1.2 to 1.8	21
Methane	190.6	45.9	1.3 to 1.7	20
Ethylene	282.3	50.4	0.93 to 1.14	22
Propane	369.8	42.4	0.7 to 1.00	23
SF_6	318.7	37.6	0.97 to 1.07	24

DEFINITIONS

The *surface excess amount* of an adsorbed substance is defined experimentally as

$$n^{\sigma} = n - \rho(V - V_s) \qquad (1)$$

where n is the total amount of gas in the volume V, ρ is the molar density of the bulk gas at given temperature and pressure, and V_s is the volume of the solid adsorbent, i.e., the dead space from which the solid excludes a reference fluid for which n^{σ} is zero by definition. Helium gas is recommended (15) for determination of V_s in low pressure gas adsorption, but liquid nitrogen (16) or liquid cyclohexane (17) have also been used as reference liquids in high pressure gas adsorption and liquid adsorption studies, respectively. For a model adsorbent of perfectly homogeneous surface the surface excess amount is related to the molar density profile $\rho(z)$ perpendicular to the surface by (15)

$$n^{\sigma} = A_s \int_{-\infty}^{z_o} \rho(z)dz + A_s \int_{z_o}^{\infty} [\rho(z) - \rho]dz \qquad (2)$$

where A_s is the total surface area of solid, and z_o represents the position of the Gibbs surface, often defined as the plane above the surface layer of lattice points of the solid at which the solid-gas potential $u_s(z)$ passes through zero (12).

Gas adsorption at elevated pressures ($p > 1$ bar) on homogeneous surfaces falls into three typical ranges of temperature relative to the critical temperature T_c of the adsorptive:

(i) low temperature region ($T < T_c$)

(ii) supercritical region ($T_c < T < 1.2\,T_c$)

(iii) high temperature region ($T > 1.5\,T_c$)

Above the critical temperature (ranges ii and iii) surface excess isotherms $n^{\sigma} = n^{\sigma}(p)$ generally pass through a maximum at some pressure $p_{max}(T) > p_c$. When pressure is increased beyond p_{max} the amount of gas in the adsorption space increases less than the amount of gas in an equal volume of the bulk fluid; thus the surface excess amount n^{σ} decreases. This situation is closely analogous to adsorption from binary liquid mixtures, when density ρ of the gas is replaced by composition of the liquid mixture (15). At sufficiently high temperature (region iii) the adsorption space (i.e., the region in which $\rho(z)$ deviates significantly from the bulk density) is confined essentially to a range of two or three molecular diameters. In the supercritical region, particulary at near-critical densities, the compressibility of a fluid becomes very large, i.e., the density is changed substantially by modest changes of pressure, or, conversely, at fixed pressure by modest changes of temperature. In the interfacial region of a supercritical fluid near an adsorbing surface the density profile $\rho(z)$ becomes long-ranged as the critical point is approached. The combination of these two features causes a complex behaviour of the surface excess amount in the supercritical region. Below the critical temperature (region i) multilayer adsorption will occur on homogeneous high-energy surfaces as p approaches the saturation pressure $p_o(T)$. These data can be treated in the framework of the Frenkel-Halsey-Hill (F.H.H.) theory and some of its refinements which take account of the compressibility of the adsorbed film, and the finite thickness of the interface between a thick adsorbed film and the gas at higher reduced temperatures (18,19).

RESULTS AND DISCUSSION

High Temperature Region

We have studied high temperature adsorption of argon, krypton and methane up to 150 bar at several temperatures (cf. Table 1). Surface excess isotherms of krypton/g.c.b. are shown in Figure 1. A comparison of the adsorption behaviour of these gases can be made in terms of the reduced variables Γ^*, ρ^*, T^* (21):

$$\Gamma^* = \Gamma^\sigma L\sigma^2, \qquad \rho^* = \rho L\sigma^3, \qquad T^* = kT/\varepsilon_s, \tag{3}$$

where $\Gamma^\sigma = n^\sigma/A_s$ is the surface excess concentration, L is Avogadro's constant, σ is the distance parameter in the pair-potential of the gas, and ε_s is the energy parameter of the gas-solid potential. The reduced variables Γ^* and ρ^* depend sensitively on the distance parameter σ. Thus it is important to use a consistent set of pair-potential parameters (σ,ε) for a comparison of different adsorbates. In our analysis (21), the pair potential parameters proposed by McDonald and Singer (obtained by fitting Monte Carlo simulation results to experimental data of several fluids in their liquid state) were used. A self-consistent set of energy parameters of the gas-solid potential (ε_s) of the systems studied was obtained by fitting the low density region ($\rho^* < 0.03$) of the experimental adsorption isotherms. Figure 2 shows a comparison of reduced surface excess isotherms of argon/g.c.b. and krypton/g.c.b. at reduced temperatures $T^* = 0.264$ and 0.285. Agreement is remarkably good over almost the whole density range up to $\rho^* = 0.2$. However, a similar comparison of krypton and methane at reduced temperatures $T^* = 0.206$ to 0.244 exhibits systematic deviations at reduced densities $\rho^* > 0.1$, where Γ^* of krypton exceeds Γ^* of methane. This deviation can be explained qualitatively by the fact that adsorption is favoured not only by strong adsorbate-solid interactions but also by mutual adsorbate-adsorbate attraction. From our analysis the energy parameter of the gas-solid potential of methane/g.c.b. ($\varepsilon_s/k = 1325$ K) is somewhat larger than for krypton/g.c.b. ($\varepsilon_s/k = 1255$ K), whereas the energy parameter of the pair potential of methane ($\varepsilon/k = 149.1$ K) is lower than the value of krypton ($\varepsilon/k = 163.1$ K). Thus, when the two systems are compared at reduced temperatures T^* corresponding to the same strength of gas-solid interaction we may expect, at higher densities ρ^*, a higher adsorption of the molecules with the stronger pair interaction (i.e., krypton), as was indeed found. At lower temperatures the attractive adsorbate-adsorbate interaction causes formation of a second and further adsorbed layers. For example, at the lowest experimental temperature of the krypton/g.c.b. system ($T/T_c = 1.2$) the maximum surface excess concentration amounts to 13.0 μmol m^{-2}, corresponding to a reduced surface excess concentration $\Gamma^* = 1.03$. The minimum number of layers to accommodate this surface excess can be estimated as follows: A close-packed monolayer of Lennard-Jones molecules with a nearest neighbour distance of $2^{1/6}\sigma$ (corresponding to the separation at the minimum of the pair potential) has a reduced surface concentration $N_m^a\sigma^2/A_s = 2^{2/3}\sqrt{3} = 0.916$. A reduced surface excess $\Gamma^* = 1.03$ thus corresponds to the amount accommodated in 1.12 close-packed monolayers. However, as Γ^* is a surface *excess* we have to add the amount of fluid contained in an

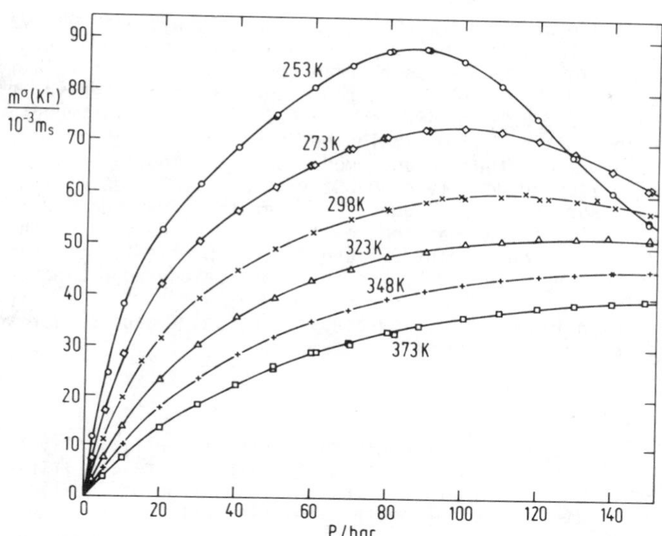

Figure 1. Krypton/graphitized carbon black (Graphon): Isotherms of the specific surface excess mass m^σ/m_s as a function of pressure p for six temperatures (253 to 373 K).

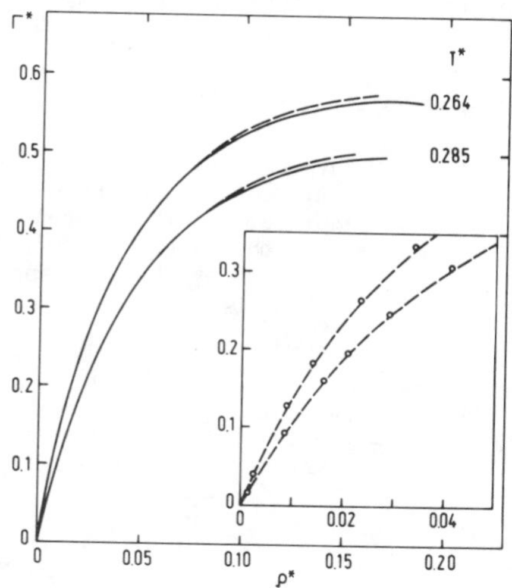

Figure 2. Comparison of reduced surface excess isotherms of argon (full curves) and krypton (dashed curves) on Graphon at reduced temperatures $T^* = 0.264$ and 0.285. Insert shows the fit of the experimental data points of argon with the (interpolated) krypton isotherms at low reduced densities ρ^*.

equivalent volume of the bulk fluid. Assuming the width of a layer of Lennard-Jones molecules to be 1.1σ, such a layer accommodates $N^a/A_s = 1.1\rho L\sigma$ molecules per unit area, corresponding to a relative packing density $N^a/N^a_m = 1.20\rho^*$. In the example mentioned above, Γ^* refers to a reduced bulk density of $\rho^* = 0.20$, i.e. a relative packing density in the bulk gas of 0.24. The surface excess of 1.12 close-packed monolayers may thus be accommodated in two monolayers with a mean relative packing density of 0.80. A theoretical treatment of high temperature adsorption (12) suggests that at the given reduced temperature and density of the gas the relative packing density of the first and second layer is only about 0.7 and 0.5, respectively, so that more than two layers are affected by the influence of the adsorbing wall.

In the high-temperature, low-density region Γ^σ can be represented by a virial expansion in density,

$$\Gamma^\sigma = B_{gs}\rho + C_{gs}\rho^2 + D_{gs}\rho^3 + \ldots \tag{4}$$

where B_{gs}, C_{gs}, D_{gs},... are the gas-solid virial coefficients. For a system of hard spheres in contact with a structureless attractive wall it was shown by Fischer (25) that the expansion of Equation (4) starts approximately like a geometrical series with alternating signs, i.e.,

$$\frac{D}{C} \simeq \frac{C}{B}\,. \tag{5}$$

It was suggested (26) that Γ^σ can then be approximated over a wider density range by the sum of an infinite geometrical series,

$$\Gamma^\sigma \simeq \frac{B\rho}{1 + q\rho} \tag{6}$$

with $q = -C/B$. Equation (6) should be applicable up to gas densities $\rho = 1/q$. For argon/g.c.b. this limit corresponds to a density of $3.5\ \mathrm{mol\ dm^{-3}}$ (90 bar) at 323 K and $2.0\ \mathrm{mol\ dm^{-3}}$ (40 bar) at 253 K. In order to account for the maximum of the experimental surface excess isotherms, Equation (6) may be replaced by the Padé approximation

$$\Gamma^\sigma = \frac{B\rho - k\rho^2}{1 + q\rho}\,, \tag{7}$$

which fits the data for argon up to 150 bar over the entire experimental temperature range. Formally, Equation (7) may be considered as a Langmuir-type isotherm for the *surface excess amount* of adsorbed gas. If we write

$$\Gamma^\sigma = \frac{(B + k/q)\rho}{1 + q\rho} - (k/q)\rho = \Gamma^a - \rho z^a \tag{7a}$$

the first term represents the total amount of gas per unit area of the adsorption space, Γ^a, and the second term gives the amount of gas in this layer of thickness z^a, in the absence of adsorption forces. For practical application of Equations (6) and (7) the density ρ may be replaced by pressure p.

Low Temperature Region

Surface excess isotherms of the system propane/g.c.b. (23) for temperatures ranging from $T/T_c = 0.7$ to 1.0 are shown in Figure 3. The thin vertical lines indicate the saturation pressure $p_o(T)$ at the respective temperatures. The isotherm at 98°C corresponds to a temperature about 1 K above the critical temperature.

The multilayer region of the isotherms has been analysed in terms of the F.H.H. equation

$$\ln(f/f_o) = \frac{u_p(z_i)}{kT} = -\frac{\Delta\varepsilon_1}{kT}\left(\frac{z_i}{z_1}\right)^{-m} \tag{8}$$

where f is fugacity and f_o fugacity of the saturated vapor; $u_p(z)$ is the perturbation energy (or replacement energy) of an adsorbate molecule at a distance z from the surface, z_i is the distance of the th statistical layer of molecules, and $\Delta\varepsilon_1$ is the perturbation energy of a molecule in the first layer (at a distance z_1 from the surface). In order to test the applicability of Equation (8) in the range of elevated pressures, z_i was calculated from the experimental surface excess Γ^σ on the basis of a single-step density profile

$$\rho(z) = \begin{cases} \rho_o^\ell & \text{for } z \le z_i \\ \rho & \text{for } z > z_i \end{cases} \tag{9}$$

which yields

$$z_i = \Gamma^\sigma/\left[\rho_o^\ell - \rho\right] \tag{10}$$

where ρ_o^ℓ is the density of the saturated liquid at the given temperature. When Equation (8) is rewritten in the form

$$z_i^{-m} = (kT/a)x, \tag{8a}$$

where $a = z_1^m\Delta\varepsilon_1$ and $x = \ln(f_o/f)$, a plot of $\ln z_i$ against $\ln x$ should be linear. Figure 4 shows such a plot for the data of propane/g.c.b. for relative fugacities f/f_o from 0.75 to 0.97. Table 2 summarizes the best-fit values of parameter m, and interpolated values of z_i at $f/f_o = 0.9$, for several experimental temperatures. For the exponent m a mean value of 2.55, with a weak positive temperature derivative, is found. When the thickness z_i of the adsorbed film at $f/f_o = 0.9$ is calculated by Equation (10) with the single-step density profile, Equation (9), values of 1.5 to 1.65 nm are obtained for an extended temperature range up to 343 K ($T/T_c = 0.93$), but a significantly higher value of z_i is found at 363 K ($T/T_c = 0.98$). Such an increase in the thickness of the adsorbed layer with increasing temperature, for a given f/f_o, is physically unreasonable and probably an artifact introduced by the single-step density profile. It is likely that the attractive potential u_p causes a compression of the liquid next to the solid surface, proportional to its compressibility. In other words, next to the surface the density of the adsorbate will exceed the

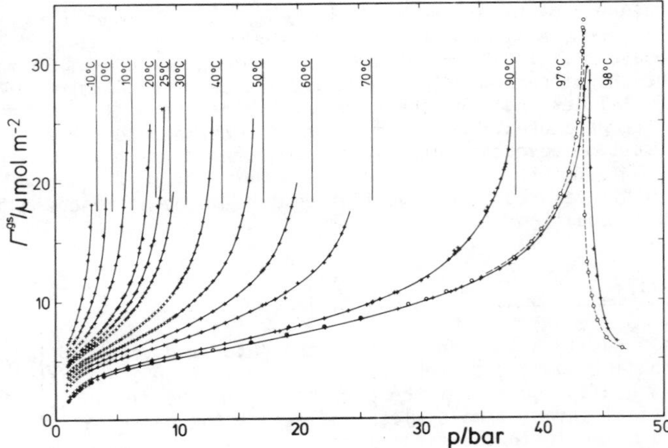

Figure 3. Propane/graphitized carbon black (Graphon): Isotherms of surface excess concentration Γ^{σ} as a function of pressure p at temperatures from 263 K to 371 K. Vertical lines indicate saturation pressure at the respective temperatures.

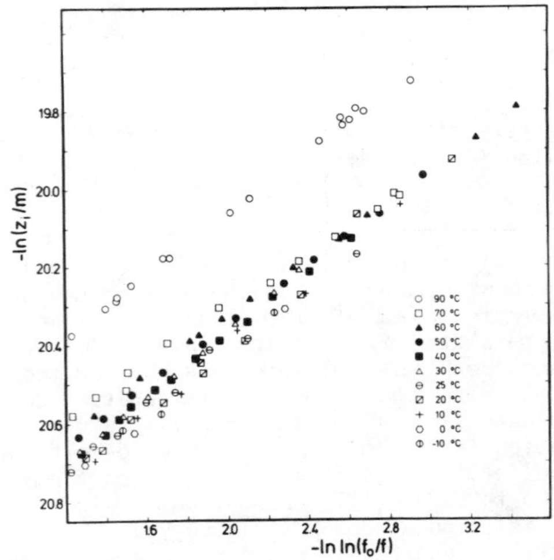

Figure 4. Propane/Graphon: Experimental data of Figure 3 plotted according to the logarithmic form of Equation (8a); the width of the layer, z_i, is calculated from Γ^{σ} by Equation (10).

Table 2. Frenkel-Halsey-Hill analysis of the system propane/g.c.b. at elevated pressures: Exponent m of Equation (8), surface excess Γ^σ and thickness of adsorbed film z_i at relative fugacity $f/f_o = 0.9$, based on the single-step profile Equation (9) and on the two-step profile Equation (11).

T	ρ_o^ℓ	ρ^g	Γ^σ	z_i/nm		m
K	mol dm^{-3}		µmol m^{-2}	Eq.(8)	Eq.(11)	Eq.(8a)
273	11.98	0.21	17.9	1.51	1.45	2.39
283	11.67	0.27	17.3	1.51	1.43	2.34
298	11.16	0.40	16.6	1.54	1.43	2.44
313	10.60	0.57	15.6	1.56	1.41	2.42
323	10.17	0.72	15.1	1.60	1.42	2.58
333	9.70	0.89	14.4	1.63	1.42	2.68
343	9.15	1.09	13.2	1.64	1.38	2.86
363	7.45	1.62	12.4	2.10	1.60	2.50

density of the saturated liquid. As an alternative to the single-step density profile, one may consider a two-step profile,

$$\rho(z) = \begin{cases} \rho_m & \text{for } z \le z_1 \\ \rho_o^\ell & \text{for } z_1 < z \le z_i \\ \rho & \text{for } z > z_i \end{cases}$$

where ρ_m represents the density of the monolayer next to the solid surface. Assuming $\rho_m = 13.8$ mol dm^{-3} ($= \rho_o^\ell$ at the normal boiling point), $z_1 = 0.455$ nm (cross section of propane molecules), one finds, at $f/f_o = 0.9$, a thickness of the adsorbed film z_i which is nearly independent of temperature up to $T/T_c = 0.98$ (see Table 2). Thus, the two-step density profile seems to be more realistic than the single-step profile.

Supercritical Region

Physical adsorption in the supercritical region has been studied for ethylene and sulfur hexafluoride (cf. Table 1). Figure 5 shows surface excess isotherms of SF_6/g.c.b. for pressures up to 100 bar (24). These isotherms exhibit several unusual features:

- The maximum of the isotherms $\Gamma^\sigma = \Gamma^\sigma(p)$ is sharply peaked near $T = T_c$, and the maximum Γ^σ_{max} decreases markedly within a narrow temperature range $T_c < T < T_c + 10$ K.
- Up to nearly p_{max} near-critical isotherms are of Brunauer type II like the subcritical isotherms, indicating multilayer adsorption.
- Beyond p_{max} the supercritical isotherms $\Gamma^\sigma(p)$ cross over; i.e. at given pressures $p > p_c$, Γ^σ *increases* with increasing temperature.

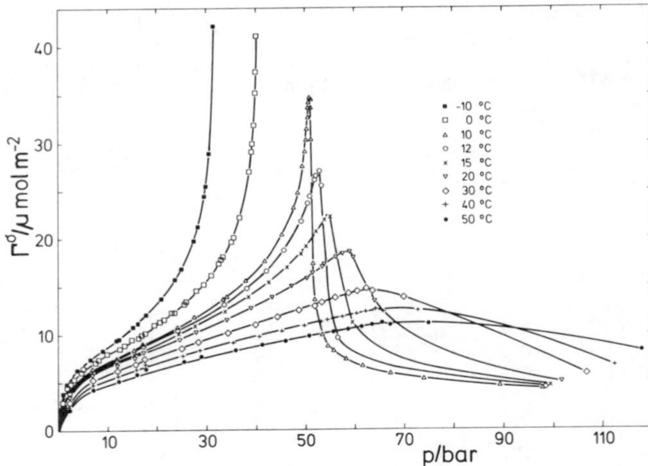

Figure 5. Ethylene/graphitized carbon black (Graphon): Isotherms of surface excess concentration Γ^σ as a function of pressure p at temperatures from 263 K to 323 K. The two isotherms at 263 K and 273 K refer to subcritical temperatures (cf. Figure 3), the other isotherms to supercritical temperatures (cf. Figure 1).

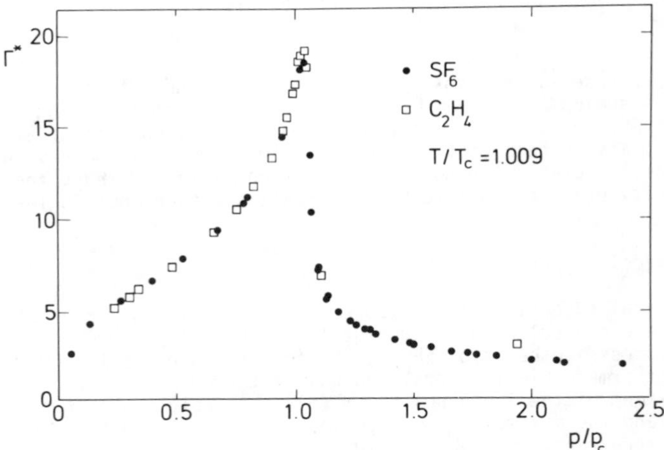

Figure 6. Comparison of reduced surface excess isotherm of SF_6 with the corresponding surface excess isotherm of ethylene, plotted vs. reduced pressure p/p_c; the two isotherms correspond to the same reduced temperature $T/T_c = 1.009$ of the two gases.

The first two features clearly demonstrate that the formation of thick adsorbed layers does not end abruptly at the critical temperature but breaks down over a temperature range of typically 10 K above T_c. This phenomenon may be seen as a *prewetting* of the solid by the supercritical fluid.

The reversed temperature dependence of $\Gamma^\sigma(p)$ beyond p_{max} is a consequence of the fact that at given pressure p the density of a supercritical fluid decreases sharply as temperature is increased and so, by Equation (1), the surface excess increases. Therefore, the surface excess isotherms do not cross over when Γ^σ is plotted as a function of density ρ, as was shown for ethylene/g.c.b. (22). However, accurate p-ρ-T data for the near-critical region are available only for a few gases.

Figure 6 shows a comparison of near-critical adsorption isotherms for SF_6 and ethylene, plotted against the reduced pressure p/p_c, for the same reduced temperature $T/T_c = 1.009$. The Γ^σ values of SF_6 are multiplied by a scaling factor 1.41. This factor agrees reasonably well with the ratio of cross-sectional area $\pi\sigma^2$ of the two molecules (27): $\sigma(SF_6) = 5.13$ Å, $\sigma(C_2H_4) = 4.16$ Å, $\sigma^2(SF_6)/\sigma^2(C_2H_4) = 1.52$. Thus it may be concluded that close to the critical point the reduced adsorption $\Gamma^* = \Gamma^\sigma L\sigma^2$ of different gases should be compared at the same reduced temperature T/T_c (corresponding to the same kT/ε, where ε is the energy parameter of the adsorbate pair potential), rather than at the same reduced temperature $T^* = kT/\varepsilon_s$, as in the high temperature region.

ACKNOWLEDGEMENT

The author thanks Dr. J. Fischer for valuable discussions. Part of this work was supported by the Bundesministerium für Forschung und Technologie (DFVLR-Project No. 01 QV 070-ZA/SN/SLN 7902-9.9).

LITERATURE CITED

1. Jüntgen, H., B. Harder and K. Knoblauch, Chem. Ind. 35, 38 and 87 (1983).

2. Stockmeyer, R. and M. Monkenbusch, Ber. Bunsenges. Phys. Chem. 84, 1072 (1980).

3. Niephaus, D., Ber. Kernforschungsanlage Jülich, No. 1765 (1982).

4. Randall, L.G. and L.M. Bowman, Jr. (Eds.), Special Topics Issue on "Supercritical Gases in Extraction and Chromatography", Separation Sci. and Technol. 17 (1), (1982).

5. Ladner, W.R. (Ed.), Special Issue on "Supercritical Fluids, Their Chemistry and Application", Fluid Phase Equilibria, 10 (2+3), (1983).

6. Menon, P.G., Chem. Rev. 68, 277 (1968); Advances in High Pressure Res. (R.S. Bradley, Ed.) Vol.3, 313 (1969).

7. Gachet, Ch. and Y. Trambouze, J. Chim. phys. 67, 380 (1970).

8. Hori, Y. and R. Kobayashi, J. Chem. Phys. 54, 1226 (1971). Ind. Eng. Chem. Fundam. 12, 26 (1973).

9. Ozawa, S., S. Kusumi and Y. Ogino, J. Colloid Interface Sci. 56, 83 (1976).

10. Wakasugi, Y., S. Ozawa and Y. Ogino, J. Colloid Interface Sci. 79, 399 (1981).

11. Findenegg, G.H., B. Körner, J. Fischer and M. Bohn, Ger. Chem. Eng. 6, 80 (1983).

12. Fischer, J., in J.M. Haile and G.A. Mansoori (Eds.), "Molecular-Based Study of Fluids", Advances in Chemistry Series 204, p. 139, American Chemical Society, Washington (1983).

13. Sokolowski, S., Adv. Colloid Interface Sci. 15, 71 (1981); J. Chem. Soc. Faraday Trans. 2, 78, 255 (1982).

14. Fischer, J., M. Bohn, B. Körner and G.H. Findenegg, Ger. Chem. Eng. 6, 84 (1983).

15. Everett, D.H., Pure Appl. Chem. 31, 579 (1972).

16. Antropoff von, A., Kolloid-Z. 137, 105 (1954).

17. Findenegg, G.H., J. Chem. Soc. Faraday Trans. 1, 68, 1799 (1972).

18. Steele, W.A., "The Interaction of Gases with Solid Surfaces", Chap. 5, Pergamon, Oxford (1974).

19. Steele, W.A., J. Colloid Interface Sci. 75, 13 (1980).

20. Specovius, J. and G.H. Findenegg, Ber. Bunsenges. Phys. Chem. 82, 174 (1978).

21. Blümel, S., F. Köster and G.H. Findenegg, J. Chem. Soc. Faraday 2, 78, 1753 (1982).

22. Specovius, J. and G.H. Findenegg, Ber. Bunsenges. Phys. Chem. 84, 690 (1980).

23. Findenegg, G.H. and R. Löring, submitted to J. Phys. Chem.

24. Lewandowski, H. and G.H. Findenegg, to be published.

25. Blümel, S. and J. Fischer, Molec. Phys. 34, 1237 (1977).

26. Fischer, J., J. Specovius and G.H. Findenegg, Chem.-Ing.-Techn. 50, 41 (1978).

27. Reid, C., J.M. Prausnitz and T.K. Sherwood, "The Properties of Gases and Liquids", McGraw-Hill, New York (1977).

THE EFFECT OF RELATIVE HUMIDITY ON THE ADSORPTION OF WATER-IMMISCIBLE ORGANIC VAPORS ON ACTIVATED CARBON

R. J. Grant, R. S. Joyce and J. E. Urbanic
Calgon Corporation, Calgon Center, Pittsburgh, PA 15230

ABSTRACT: The adsorptive capacities of activated carbon pre-equilibrated with moisture for two ternary mixtures of paraffins, aromatics and halocarbons has been determined and compared with similar results with dry gas and dry activated carbon. With the exception of methylene chloride, an immiscible compound with significant (2 parts/ 100 parts H_2O) solubility, the reduction in adsorptive capacity directly attributable to the effect of relative humidity has been adequately accounted for on the basis of the water isotherm for the carbon and the Polanyi-Dubinin adsorption potential theory.

The adsorption of dry multi-component organic gases and vapors in equilibrium with dry activated carbon has been the subject of several earlier papers.(1-3) Adsorptive capacities of the individual components have been predicted by means of the Polanyi-Dubinin (5,6) adsorption potential theory and the results have been experimentally verified. The only data required were the physical properties of the components and the Polanyi correlation curve of the adsorbent. The equilibrium (non-flow) calculation has been extended (3) to include development of chromatographic bands assuming negligibly small mass transfer zones between them.

Very little attention has been paid to the effect of adsorbed water vapor.(7) Therefore, the purpose of this work is to examine the effect of moisture pre-equilibrated on carbon upon the adsorption of water-immiscible organic vapors at concentrations in the threshold limit value range of approximately 100 ppm. It was also of interest to determine whether the reduction in capacity directly attributable to the effect of relative humidity could be successfully accounted for using the Polanyi adsorption potential theory and a model incorporating the assumption that the adsorbed moisture would simply preempt pore space that would otherwise be available for adsorption. Such a model, if validated, could aid immeasurably in enhancing our understanding of the role played by relative humidity on the adsorption of organic vapors on activated carbon from moist air.

THEORY

Initially, the adsorption space is equilibrated with water at the relative humidity of the influent stream. Consequently, the adsorption potential for adsorption of the organic vapor is reduced by the adsorption potential of an equal volume of water. Manes (8) described an approach which evaluated the net driving force as follows:

$$\varepsilon_{s1} = \varepsilon_s - \frac{\varepsilon_1 V_s}{V_1} \tag{1}$$

Where ε_s and ε_1 are the adsorption potentials of the organic vapor and water vapor, respectively.

Equation (1) can be rewritten as:

$$\frac{\varepsilon_{s1}}{V_s} = \frac{\varepsilon_s}{V_s} - \frac{\varepsilon_1}{V_1} \quad \text{. . . for 100\% RH} \tag{2}$$

Therefore, the correlation curve for an organic vapor in the presence of water vapor at 100% relative humidity, is the difference between the gas-phase correlation curve of the organic vapor and that of water vapor. If however, the relative humidity is less than 100%, a correction must be made as follows:

$$\frac{\varepsilon_{s1}}{V_s} = \frac{\varepsilon_s}{V_s} - \frac{\varepsilon_1}{V_1} - \frac{RT}{V_1} \ln \left(\frac{1}{RH}\right) \quad \text{. . . <100\% RH} \tag{3}$$

At a relative humidity of 100%, the correction vanishes. In Figure 1 are shown the Polanyi curves for organic vapor and for water vapor obtained on the activated carbon sample used in these experiments, from which it can be seen that:

$$\frac{\varepsilon_1}{V_1} = 0.4 \frac{\varepsilon_s}{V_s} \tag{4}$$

Converting to common logarithms and rearranging,

$$\frac{\varepsilon}{4.6V} = \frac{T}{0.6V} \log \left(\frac{P_s}{P}\right) - \frac{T}{0.6V_{H_2O}} \log \left(\frac{1}{RH}\right) \tag{5}$$

which is used for determining the correlation curve abscissa for pure vapors. In the case of multi-component mixtures, the evaluation of component mole fractions is carried out as for the dry case according to the following equation from an earlier paper (1):

$$\frac{1}{V_1} \log x_1\left(\frac{P_s}{P}\right)_1 = \frac{1}{V_2} \log x_2 \left(\frac{P_s}{P}\right)_2 \text{, etc.} \tag{6}$$

The correlation curve abscissa of each component is then determined as:

$$\frac{\varepsilon_i}{4.6V_i} = \frac{T}{0.6V_i} \log x_i\left(\frac{P_s}{P}\right)_i - \frac{T}{0.6V_{H_2O}} \log \left(\frac{1}{RH}\right) \tag{7}$$

for purposes of evaluating the total volume adsorbed from the correlation curve.

Once the adsorptive capacities have been determined in this manner, band lengths and breakthrough times can be calculated.

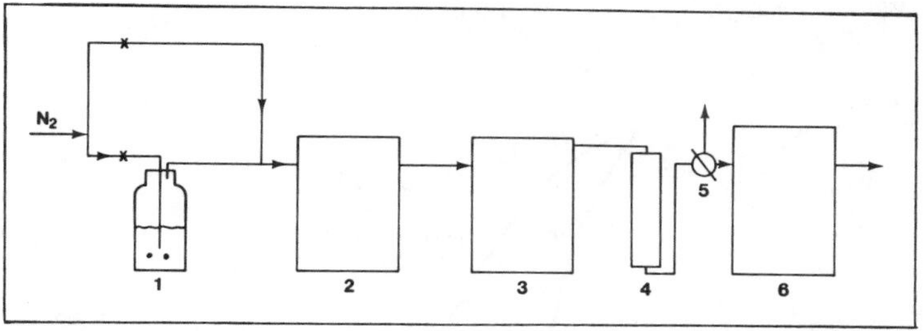

FIGURE 1. Adsorption Test Apparatus Flow System: 1-Humidifier, 2-RH Detector, 3-Standards Generator, 4-Adsorption Tube, 5-Sampling Valve, and 6-Chromatograph with flame ionization detector and electronic integrator with output to a mini-computer.

EXPERIMENTAL

A diagram of the flow system is given in Figure 1. A water bubbler humidifier and a model 350 standards generator (Analytical Instrument Development, Inc., Avondale, PA) are modifications to the conventional flow system previously described.(3) A General Eastern model 400 C/D relative humidity and temperature detector was also a part of the system. The generator used diffusion tubes (one for each solvent) and was calibrated by measuring the change in weight of the diffusion tubes over a period of several days.

The activated carbon used was 30x100 mesh BPL, Audit sample 104, manufactured by the Calgon Corporation. The sample size was 100 mg and the nitrogen flowrate was 100 cc/min at ambient conditions. The n-Hexane, n-Heptane, and Toluene were Fisher, ACS Spectranalyzed grade: the Dichloroethylene was Eastman, 98% minimum grade; and the Methylene Chloride was Burdick and Jackson, B.P. 40-41°. The Polanyi correlation curve for the carbon was determined by the method of Semonian and Manes.(9) The curve was fit to the following polynomial equation in X (= $\mathcal{E}$ /4.6 V):

$$\log V_t = 1.725 + (2.642 \times 10^{-3}) X - (1.231 \times 10^{-2})X^2 + (6.068 \times 10^{-4})X^3 - (1.678 \times 10^{-5})X^4 + (1.847 \times 10^{-7})X^5 \quad (8)$$

V_t is the adsorbate volume in cm^3/100 grams carbon. The following inverse equation in Y (= $\log V_t$) was also used.

$$X = 17.45 - 7.067\,Y - 1.055\,Y^2 + (8.538 \times 10^{-3})Y^3 + 0.6\,Y^4 + 0.115\,Y^5 - 0.17Y^6 - (7.6\times10^{-2})Y^7 - (8.67 \times 10^{-3})Y^8 \quad (9)$$

FIGURE 2

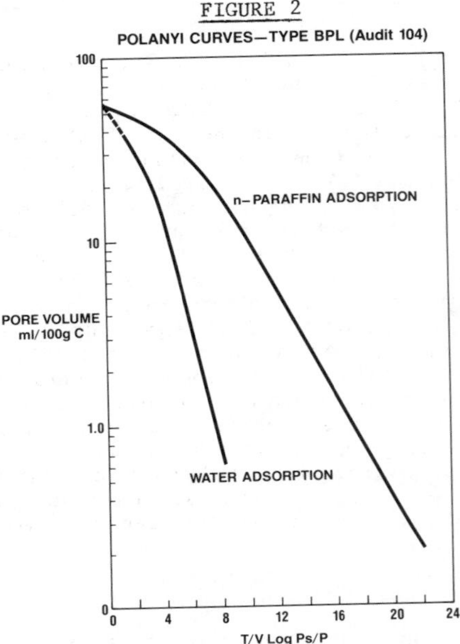

RESULTS AND DISCUSSION

In order to check the validity of the correlation curve of the activated carbon, several runs were made on single component mixtures of Toluene, n-Heptane, dichloroethylene (DCE) and methylene chloride (MECL) in nitrogen as shown in Table 1. Data were collected at both dry conditions and with the carbon pre-equilibrated at 80% RH. In the case of both Toluene and n-Heptane, the agreement between observed and predicted results is excellent. However, in the case of the halocarbons, DCE and MECL, the predicted values were considerably lower than those observed, suggesting that the abscissa scale factors (1.16 and 1.09) used were too low. The scale factors were calculated theoretically (11) from the liquid refractive index relative to a reference compound, namely, n-heptane. A single Polanyi curve has been assumed for the normal paraffins (4). New abscissa scale factors were therefore experimentally evaluated from the dry case data in Table 1 and were used in subsequent predictions for the mixtures described below. The new values were 1.28 and 1.24 for DCE and MECL, respectively. The chlorinated compounds are apparently more strongly adsorbed than would be predicted from the theoretically calculated scale factors.

Dichloroethylene and methylene chloride were chosen as adsorbates since these vapors are encountered in pollution control situations.

TABLE 1. EXPERIMENTAL AND PREDICTED EQUILIBRIUM CAPACITIES OF SINGLE ADSORBATES ON BPL-A104 ACTIVATED CARBON

Compound	Conc. PPMV	Dry Exp.	Dry Pred.	80% RH Exp.	80% RH Pred.
Toluene	106±7	32.0	29.6	29.3	25.9
Toluene	59±1	25.8	26.7	22.4	22.6
n-Heptane	81±0.5	22.7	23.2	19.4	20.2
Dichloroethylene	84±3	13.0	9.6	9.4	3.1
Methylene Chloride	122	5.0	3.0	5.0	1.9

The effluent concentration profiles for the adsorption of the mixture consisting of 138 ppm dichloroethylene, 90 ppm n-hexane and 97 ppm toluene from dry and 80% RH nitrogen are shown in Figures 3 and 4. Dichloroethylene is displaced with "rollover" by toluene. The effluent profiles for the same mixture at 80% RH adsorbed on carbon pre-equilibrated at the same RH show that the components breakthrough earlier due to the presence of water in the pores and this results in reduced adsorptive capacity. The "rollover" is less pronounced than in the dry case since there is less adsorbate to displace.

FIGURE 3

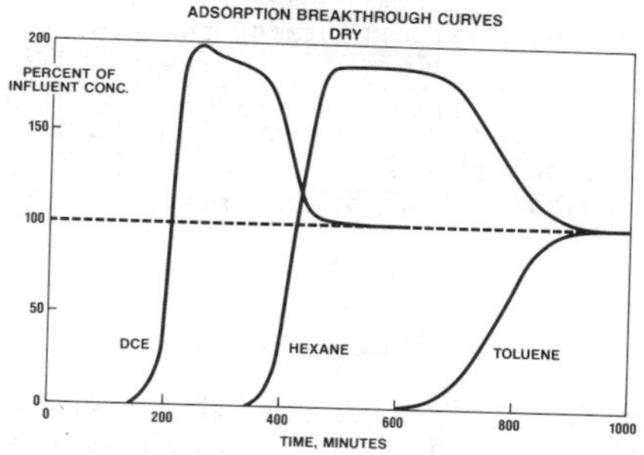

ADSORPTION BREAKTHROUGH CURVES
DRY

FIGURE 4

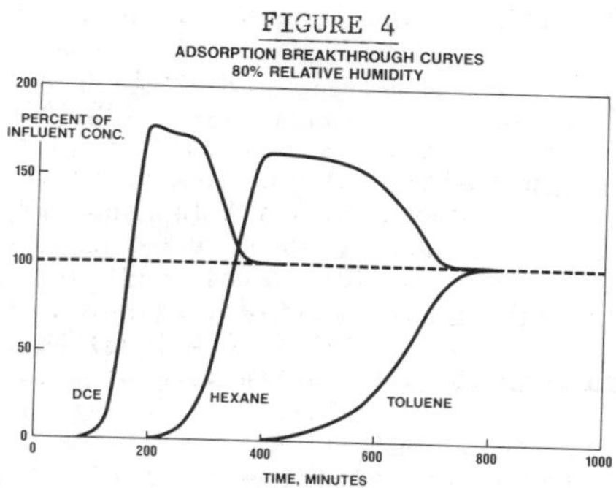

In Tables 2 and 3, the experimental equilibrium and breakthrough adsorptive capacities from Figures 3 and 4 are listed together with the corresponding theoretical values predicted by the "uniform adsorbate" model and the "non-uniform adsorbate" model. The latter values were predicted using the integrated Hansen-Fackler Equation (10). The best agreement was observed in the case of the non-uniform model. Since there is such a large separation between the three components, this result is not totally unexpected, although there is not a great deal of difference between either model. The effect of relative humidity is adequately predicted in both cases.

TABLE 2. EXPERIMENTAL VERSUS PREDICTED EQUILIBRIUM ADSORPTIVE CAPACITIES FOR A MIXTURE IN DRY AND 80% RH NITROGEN OF 138 PPMV DICHLOROETHYLENE (DCE), 90 PPMV n-HEXANE AND 97 PPMV TOLUENE ON BPL-A104 ACTIVATED CARBON AT 25°C

	Dry			80% RH		
		Predicted			Predicted	
Compound	Exp.	Uniform	Non-Uniform	Exp.	Uniform	Non-Uniform
DCE	0.5±0.4	0.8	0.8	0.6±0.04	0.7	0.8
Hex.	4.0±0.2	4.8	3.5	3.6±0.05	4.2	2.9
Tol.	26.7±1.1	23.2	25.7	22.4±0.2	21.5	23.7

TABLE 3. EXPERIMENTAL VERSUS PREDICTED BREAKTHROUGH CAPACITIES FOR DCE, n-HEXANE AND TOLUENE MIXTURES IN TABLE 2

	Dry			80% RH		
		Predicted			Predicted	
Compound	Exp.	Uniform	Non-Uniform	Exp.	Uniform	Non-Uniform
DCE	10.1±0.3	10.0	10.5	7.0±0.1	6.9	7.3
Hex.	12.1±0.5	11.2	12.1	9.4±0.5	9.4	10.3
Tol.	27.1±0.7	23.3	25.1	23.5±0.1	21.5	23.7

The adsorptive capacities in Table 2 are, by definition, the capacities in the first chromatographic band during column adsorption. The experimental breakthrough capacities in Table 3 are calculated from Figures 3 and 4 at an effluent concentration corresponding to 50% of influent. Theoretical breakthrough times are independent of effluent concentration. There is very good agreement between experimental and predicted breakthrough capacities.

The effluent concentration profiles for the mixture consisting of 265 ppm methylene chloride, 75 ppm heptane and 110 ppm toluene are shown in Figures 5 and 6. Methylene chloride is displaced with "rollover" by the other two components. However, heptane and toluene were not separated but eluted essentially together. The breakthrough curves for the same mixture at 80%RH on carbon pre-equilibrated at this relative humidity show only a very slight reduction in breakthrough time and extent of "rollover."

FIGURE 5

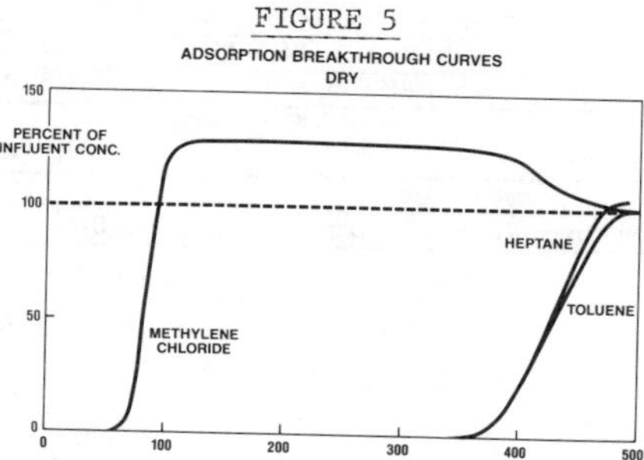

FIGURE 6

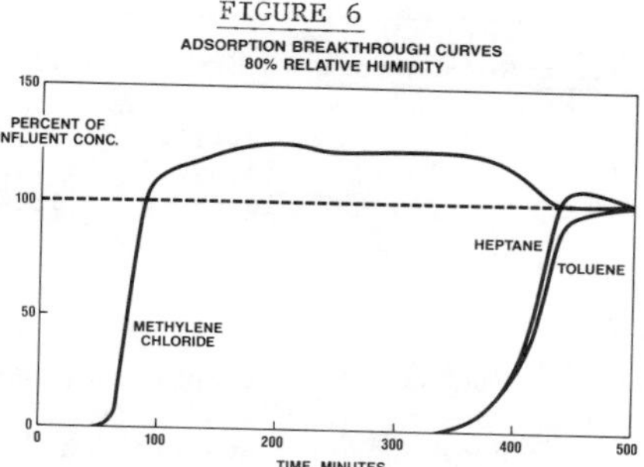

Equilibrium capacities for the methylene chloride-heptane-toluene mixture are shown in Table 4, and were predicted somewhat more accurately using the uniform model. Perhaps this is understandable to some extent since there was no separation of heptane and toluene in the gas phase and, therefore, there would be little likelihood of a concentration gradient in the adsorbed phase. This development suggests that neither model has general applicability. The effect of relative humidity, however, is predicted for hexane and toluene and the displacement of methylene chloride is predicted.

The breakthrough capacity comparisons are shown in Table 5. They indicate that the use of the displacement theory for relative humidity effect did not predict the behavior of methylene chloride. The experimental methylene chloride capacity is significantly higher than predicted for the 80% RH conditions. This could be due to the fact that the more soluble methylene chloride (solubility 2.0 grams per 100 grams H_2O) is able to diffuse through the water layer more readily than the relatively insoluble dichloroethylene.

TABLE 4. EXPERIMENTAL VERSUS PREDICTED EQUILIBRIUM ADSORPTIVE CAPACITIES FOR A MIXTURE IN DRY AND 80% RH NITROGEN OF 265 PPMV METHYLENE CHLORIDE (MECL), 75 PPMV n-HEPTANE AND 105 PPMV TOLUENE

	Dry			80% RH		
		Predicted			Predicted	
Compound	Exp.	Uniform	Non-Uniform	Exp.	Uniform	Non-Uniform
MECL	-0.9±0.05	0.2	0.2	0.5±0.4	0.2	0.2
n-Hep.	14.0±1.3	13.8	9.7	12.7±0.7	12.8	9.7
Tol.	18.1±0.5	16.2	23.6	16.7±0.3	15.4	22.1

TABLE 5. EXPERIMENTAL VERSUS PREDICTED BREAKTHROUGH CAPACITIES FOR MECL, n-HEPTANE AND TOLUENE MIXTURE DESCRIBED IN TABLE 4

	Dry			80% RH		
		Predicted			Predicted	
Compound	Exp.	Uniform	Non-Uniform	Exp.	Uniform	Non-Uniform
MECL	7.9±0.3	7.4	7.5	6.1±1.6	2.7	2.7
n-Hep.	14.0±0.8	13.8	13.7	13.0±0.7	12.8	13.0
Tol.	18.3±0.3	16.2	23.1	16.7±0.4	15.4	22.1

SUMMARY AND CONCLUSIONS

The effect of relative humidity on the adsorption of water-immiscible adsorbates on activated carbons has been studied in the case of two ternary mixtures of halocarbons, normal paraffins, and aromatics in humid and dry nitrogen. The use of Polanyi adsorption potential theory and a model which postulates that adsorbed water reduces the pore space available for adsorption on a one to one volume basis proved adequate to predict the effect of moisture on breakthrough capacities of a mixture containing low concentrations of dichloroethylene, n-hexane and toluene. The model did not predict the breakthrough capacity of the relatively soluble methylene chloride in a ternary mixture which included n-heptane and toluene. The model used here should be studied further for water insoluble adsorbates over a broad range of compound types.

Use of the Hansen-Fackler (10) equation to account for non-uniform mixtures in the adsorbed phase, improved the accuracy of the capacity predictions to some degree for the DCE, hexane and toluene mixture, but was not applicable in the case of heptane and toluene which were not separated in the gas phase.

NOTATION

ε_i Adsorption potential of the i'th component.

ε_s Adsorption potential of organic solvent

ε_1 Adsorption potential of water vapor

ε_{s1} Adsorption potential of organic in the presence of water vapor

V_s Molar volume of organic vapor

V_1 Molar volume of water vapor

P Equilibrium partial pressure

P_s Saturation pressure

X_i Mole fraction of the i'th component

T Absolute temperature, °K

V_t Total adsorbate volume of organic components

LITERATURE CITED

1. Grant, R.J. and Manes, M. Ind. Eng. Chem Fundamentals, 5, 490 (1966).

2. Manes, M. and Grant. R. J., Presented at the Adsorption-Ion Exchange Symposium, AIChE Meeting, New York, N.Y., April 1978.

3. Grant, R.J. and Joyce, R.S., "Multicomponent Gas Adsorption on Activated Carbon Via Computerized Data Acquision", presented at the Adsorption-Ion Exchange Symposium, AIChE Meeting, Houston, Texas, April 1979.

4. Grant, R. J. and Manes, M., A.I.Ch.E. Journal, "Adsorption of Normal Paraffins and Sulfur Compounds on Activated Carbon," 8, 403, (1962).

5. Polanyi, M. Venh. Deut. Phys. Ges., 16, 1012 (1914); 18, 55 (1916); Z. Elektro-Chem. 26, 370 (1920); Trans. Far. Soc. 28, 316 (1932).

6. Dubinin, M. M. and Timofeyev, D. P., Akad. Nauk SSSR, 54, 8, 701 (1946); 55, 2, 137 (1947).

7. Robell, A. J., Arnold, C. R. Wheeler, A. Kersels, G. J. and Merrill, R. P., NASA Report Cr-1582, 1970.

8. Manes, M. Presented at the Engineering Foundation Conference, "Fundamentals of Adsorption", Bavaria, W. Germany, May 1983.

9. Semonian, B. P., and Manes, M., Anal. Chem., 49, 991 (1977).

10. Hansen. R. S. and Fackler, W. V., J. Phys. Chem., 57, 634 (1953).

11. Manes, M. and Hofer, L. J.E., J. Phys. Chem., 73, 584 (1969).

SPECTROSCOPIC CHARACTERIZATION OF ACTIVE CARBON SURFACES AS AN AID IN THE UNDERSTANDING OF ADSORPTION PROCESSES

Chanel Ishizaki
Instituto Venezolano de Investigaciones Científicas
I.V.I.C.
Apartado 1827, Caracas 1010-A, Venezuela

ABSTRACT

An infra-red analytical technique has been developed capable of measuring qualitative as well as quantitative differences on a series of active carbons. The IR spectra of all the carbons investigated look qualitatively similar, the main differences being quantitative. Three broad but distinctive bands are observed, centered around 1735, 1585 and 1240cm^{-1}. In the range of carbon 0.2-0.7mg the background absorbance follows a linear relationship. In this same range the intensity of the bands is linearly but inversaly proportional to the amount of carbon present in the pellet. The intensity of the three bands per unit weight of carbon are related in the form:

$$B_{1240} = K + 0.2\, B_{1585} + 1.7\, B_{1735}$$

for all the carbons investigated. The observed bands and correlation found among them for a carbon that is basic in nature and shows only the 1585 and 1240cm^{-1} bands, is interpreted as a confirmation of the basic oxide structures, α-pyronic like, proposed by Voll and Boehm. The 1735 cm^{-1} band is assigned to acidic carbonyl groups, mainly lactonic, of the α-pyrone like type structures.

The surface properties of active carbons and carbon blacks are strongly influenced by the presence of surface oxides which are formed during the manufacturing process. The actual chemical nature of these oxygen containing groups has drawn the attention of a considerable number of investigators, but up to the present time total agreement about the different functionalities has not been reached. The determination of the chemical identity and structure of the surface functional groups and active sites is extremely necessary when trying to understand, for instance the different behaviour of active carbons towards the removal of particular adsorbates in the treatment of drinking waters, which has been the main interest of the author (1-4). In the past the selection of active carbons for a particular application has been done in a very empirical manner, without understanding the nature of the interaction of the surface with the solvent and the different solutes to be removed. Ishizaki et al. (4) clearly showed that the chemical nature of the carbon plays a very important role in the adsorption process at very low solute concentrations, as is the case in the drinking water processes. Carbons having the same surface area and pore size distributions, but different chemical characteristics behave differently.

In the past, characterization and quantification of surface functional groups on active carbons has been based on indirect methods such as selective neutralization techniques, analysis of thermal decomposition products, reactions with specific reagents, etc. These methods have proved not to give a representative image of the real nature of the unmodified surface, and therefore, failed to fulfill the objective in aiding the understanding of the role of the different groups on the adsorption mechanisms. On the other hand, infra-red spectrum analysis provides direct evidence of functional groups occurring in the non-modified substance: the adsorption bands observed correspond to definite vibration frequencies of certain atomic bands. In principle, the intensity of these absorption bands is also a measure of the number of bands of a certain type present.

In a previous publication by the author (5) some understanding on the nature of the surface functional groups of a commercially available active carbon was obtained combining infra-red direct transmission spectroscopy with the selective neutralization technique outlined by Boehm (6). In the present work, the emphasis is on the development of an infra-red analytical technique capable of measuring qualitative as well as quantitative differences on a series of active carbons. In order to accomplish this objective, the amount of oxygen surface functional groups of a reference carbon was varied by heat treatment in an inert atmosphere and by oxidation with oxygen gas. The reason for taking this approach was based on the very well known facts that the heat treatment at high temperatures in an inert atmosphere eliminates most of the oxygen functional groups from the surface and that surface oxygen groups can be produced again by oxygen oxidation (7).

EXPERIMENTAL

Two different commercial granular active carbons produced from coal by high temperature steam activation were investigated. These carbons were produced by the same manufacturer, under the same commercial identification but obtained in different years. They will be identified as carbon A and carbon B.

Carbon A was subjected to the following treatments: the carbon was heat treated at 1,000°C in a nitrogen atmosphere for 17 hours and allowed to cool down to room temperature and kept under the same atmosphere. This carbon will be identified as A-OG. The surface of carbon A was also modified by heating at 1,000°C under nitrogen atmosphere for 17 hours, and allowed to cool to 400°C, the nitrogen was then displaced by oxygen and the oxidation accomplished at this temperature for 30 minutes with an oxygen flow of 40ml/min. After the oxidation process the carbon was cooled to room temperature under nitrogen atmosphere. This carbon will be labeled A-OGOX400°C.

All these carbons were subject to infra-red investigation using KBr pellets. The KBr pellets of the different carbon samples were prepared in the following manner: a given amount of carbon sample was ground for 4 minutes in a stainless steel grinding capsule with a vibrating mill. After the addition of 200mg of KBr the mixture was ground for an additional half minute and pressed. Extreme precautions were taken to avoid water contamination. Spurious bands are observed due to water contamination (10). The spurious bands were eliminated in our case if the pellets were storage under vacuum in a desiccator with silica gel. The same observation was reported by Robin and Rouxhet (11). Infra-red direct transmission spectra of the dried pellets were obtained using a Perkin-Elmer Model 580 IRS coupled with an Interdata Model 6/16 Computer. The spectra were recorded in a moisture and CO_2-free atmosphere. The analysis was confined to the 1850-1150cm^{-1} region due to the uncertainty of possible water adsorption in the OH stretching region.

The amount of carbon present in the pellets was optimized to obtain the spectra without using attenuation in the reference beam. The working range obtained was up to 0.5mg for the carbon A-OGOX400°C and up to 0.7mg for the other carbons. The average background absorbance, $\frac{1}{2}$ (Absorbance at 1850cm^{-1} + Absorbance at 1150cm^{-1}), and the amount of carbon in each pellet follow a linear direct relationship. The correlations obtained for each carbon were used to calculate the actual amount of carbon present in each pellet. The spectrum of each pellet was accumulated five times. Spectral accumulation means that each spectrum is added to the previously recorded spectrum/spectra to give an overall data enhancement. A given percent transmittance was added to each accumulated spectrum to obtain 99% transmittance at 1150cm^{-1}, converted to absorbance and multiplied by a factor of two. The multiplication command multiplies each point in the spectrum by the selected factor, and is applicable only if the spectra is stored in absorbance. The areas under the peaks were then calculated. All these operations are done with the aid of the computer.

RESULTS AND DISCUSSION

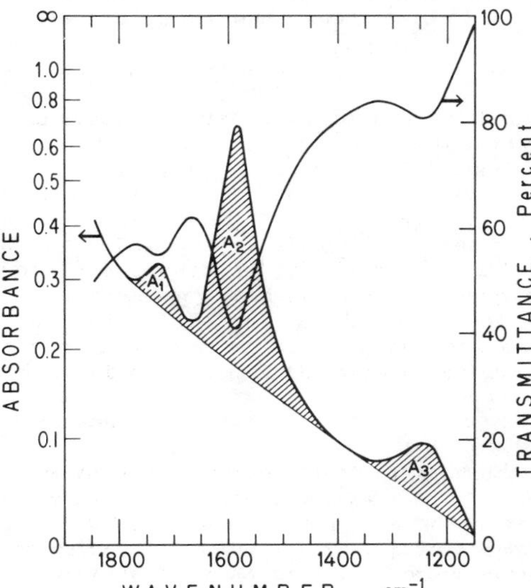

Figure 1. IR-direct transmission spectrum of carbon B.

The spectra of carbons A, A-OGOX400°C and B show three broad but distinctive bands centered at 1735, 1585 and 1240cm^{-1}, and carbon A-OG only presents the 1585 and 1240cm^{-1} bands. Figure 1 shows a typical five times accumulated spectrum in transmittance and the corresponding spectrum in absorbance times a factor of two.

For the A-OG carbon a linear direct correlation was found between the intensity of the absorbance of the A_2 band per unit weight of carbon (B_2) and the corresponding intensity of the A_3 band per unit weight of carbon (B_3) as shown in Figure 2.

The equation that represents this correlation is:

$$B_3 = 0.94\ (\pm 0.67) + 0.20\ (\pm 0.05)\ B_2 \quad (R=1) \tag{1}$$

The slope of this equation was used to calculate the value of B_3 for the other carbons considering $A_1 = 0$. These values were substracted from the measured B_3 and the difference correlated with the corresponding intensities of the band A_1 per unti weight of carbon (B_1). The resulting plots for the A-OGOX400°C and B carbons are presented in Figure 3. The correlations indicate that the intensities of the bands per unit weight of carbon, in the range of weights investigated, can be related in the general form:

$$B_3 = K + 1.7\ B_1 + 0.2\ B_2, \tag{2}$$

the value of K being different for each carbon. Equation 2 holds for the four carbons investigated in the present study, giving an indication of the qualitative similarities of the surface functionality of the oxygen groups on active carbons.

The most controvertial assignment for the observed bands in a series of carbonaceous materials; coals (10-21), carbon blacks, (22-27) graphite (27-28), chars (13-20, 29-34), active carbons (5, 35, 22, 27, 34) has been the $1600cm^{-1}$ band. This band has been assigned to the aromatic structure, hydrogen bonded carbonyl groups, quinone structures and thermally stable carboxyl-carbonate structures. It is generally agreed that the absorption in this region is too intense to be attributed solely to aromatic ring vibrations and it has been suggested that the intensity may be largely due to the aromatic ring vibrations enhanced by polar (oxygenated) substituents or to direct contribution by conjugatively chelated carbonyl groups.

An effort was made by Friedel and Retcofsky (21) to search the nuclear magnetic resonance spectrum for the possible existence of the conjugated chelated carbonyl structures

OH ----- O = C - C = C -

Their results established that appreciable numbers of strongly chelated protons are not present in the same structure that show intense infra-red absorption in the $1600cm^{-1}$ region. The authors gave a possible explanation for this band in coals as metal derivates of chelated, conjugated carbonyl structures

- O - METAL - O = C - C = C -

Friedel et al. (29) using oxygen - 18 labeled chars failed to obtain any appreciable shift in the $1600cm^{-1}$ band due to the isotope effect and consider that this observation appears to suggest that oxygen containing groups may not be contributing, either directly or indirectly to the intensity of the $1600cm^{-1}$ band, eventhough this is not compatible with other investigations of Friedel and co-workers (13, 20) which indicate that the chars prepared from carbohydrates ($\sim$500°C) invariably exhibit coal-like spectra with an intense absorption feature near $1600cm^{-1}$, also chars from other oxygenated compounds and of hydrocarbons chared in presence of oxygen are often coal like. This apparent contradiction was solved by Friedel after obtaining the infra-red spectra of graphite

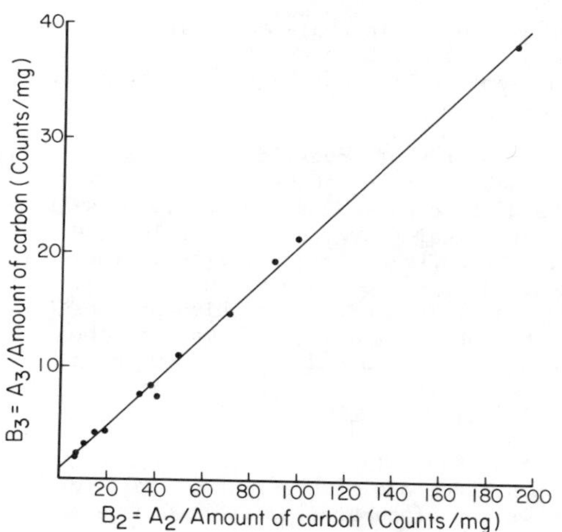

FIGURE 2. Correlation between B_3 and B_2 for the A-0G carbon.

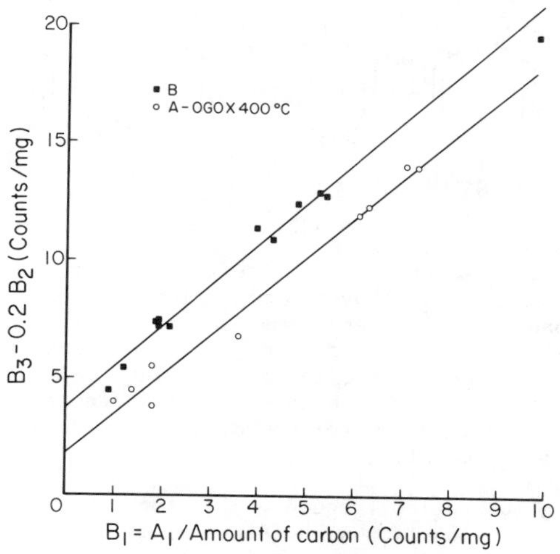

FIGURE 3. Correlation between $B_3 - 0.2\ B_2$ and B_1 for the A-0G0X400°C and B carbons.

(28) reassigning the $1600 cm^{-1}$ in coals and chars to graphitic structures. As a major assignment this may be correct but the enhancement of the band due to oxygen chemisorption has been shown by Zawadsky (30, 31) in a very convincing manner.

The results of the present investigation show for active carbon a very good correlation between the absorption band for the - C - O - vibrations and the band centered around $1585 cm^{-1}$. This is also true for the carbon blacks investigated by Prest and Mosler (24). These two works together with the works of Zawadsky clearly indicate that at least indirectly the oxygen groups are contributing to the intensity of the $1600 cm^{-1}$ band. On the other hand, the results presented by Studebaker and Rinehard (23) show that the assignment made by them to the $1600 cm^{-1}$ band, as quinone groups is supported by the good correlation found between the integrated intensity across the $1675\text{-}1550 cm^{-1}$ and the quinone content from reduction with $NaBH_4$.

It has been shown, Marti (36), that the surface groups on the A-OG carbon are basic in nature, and Figure 2 shows that for each spectrum the areas per unit weight of carbon for the two observed bands follow a linear correlation. Since the acidic nature of this carbon (A-OG) is negligible, it may be considered that the observed absorption bands correspond to basic oxides. Based on chemical evidence Voll and Boehm (33) proposed a pyrone-like structure for the basic oxides with the two oxygen atoms located in different rings of a graphitic layer. For simplicity let's consider the behaviour of the γ-pyrones. These compounds are resonance hybrids of the form

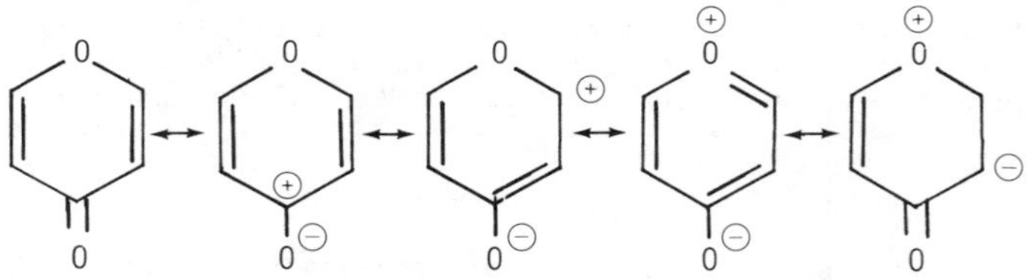

(a) (b) (c) (d)

The polar forms arise partly from the tendency of the carbonyl group to assume the polar state by the migration of a pair of electrons from the double bond of the oxygen atoms, and partially from participation of unshared electrons of the ring-oxygen to a pseudo aromatic system in formula (c). Molecular orbital calculations indicate that formula (a) is the best single representation of the electronic structure for γ-pyrones. Confirmation and refinement of the resonance picture comes from dipole measurements (38).

The stretching absorption frequency of ketonic carbonyls is known to decrease from the conjugation with unsaturated chromophores (i.e. olefinic double bonds) and that conjugation on both sides of the carbonyl group tend to be additive. Furthermore contributions from the resonance forms (b), (c) lead to a decrease in the force constant of the carbonyl group (more single bond character) and a further decrease in the frequency is observed (39). This type of structure will also explain the enhancement of the aromatic bands vibrations. As Bellamy (39) indicates the $1600 cm^{-1}$ doublet arises from vibrations in which the main dipole moment change is produced by the movements of the substituents on opposite sides of the ring acting in opposition. With polar substituents the change is larger and the band intensifies. This occurs independently of whether the substituent is an electron acceptor or an electron donor.

The correlation found between the intensity of the $1670-1400 cm^{-1}$ and the $1370-1190 cm^{-1}$ bands for the A-OG carbon can be explained by structures similar to the γ-pyrones. The ether linkages will give absorption in the $1370-1190 cm^{-1}$ range, and the carbonyl plus the graphitic structure enhanced by the oxygen substituents will give corresponding overlapping absorptions in the $1670-1400 cm^{-1}$ range. The quantity of ether linkages and carbonyl groups appears to have a systematic array, and therefore the absorption intensities correlate for each given amount of carbon sample.

The spectra of the oxidized carbon A-OGOX400°C as well as the B carbon show an additional intense band centered at $1735 cm^{-1}$. It has been shown (Figure 3) that the intensity of the A_1 band per unit weight of carbon, B_1, correlates with the difference, B_3-0.2 B_2, indicating that at least partially the functionality or functionalities that give rise to the A_1 band contribute to the A_3 band in a definite and constant manner. Structures that can explain the observed structural behaviour are the α-pyrones. The carbonyl frequency for these compounds is around $1740-1720 cm^{-1}$, the C=C at $1650-1600 cm^{-1}$, and $1590-1560 cm^{-1}$ and the C-O at $1300-1250 cm^{-1}$, $1100-1180 cm^{-1}$ (40). This type of structure was also postulated by Zawadsky (33) to explain his observed bands in carbonic films at $1760-1720 cm^{-1}$ (C=O) and $1260 cm^{-1}$ (C-O-C).

In view of the results and discussions presented the functionality of the carbon surface is interpreted as a combination of discrete structural units as the ones shown below.

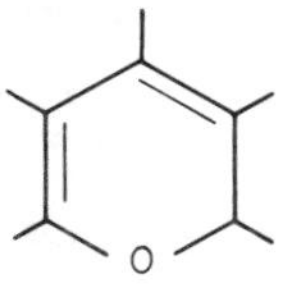

α - PYRAN

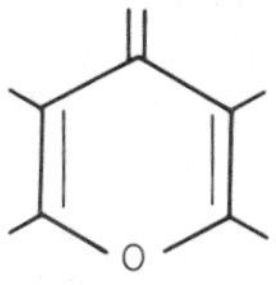

γ - PYRAN

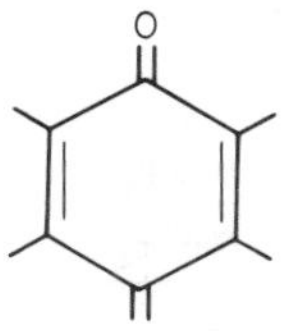

QUINONIC TYPE CARBONYL

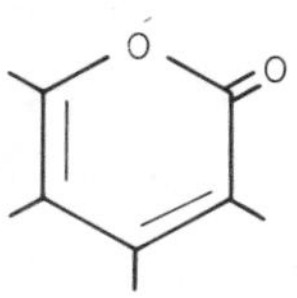

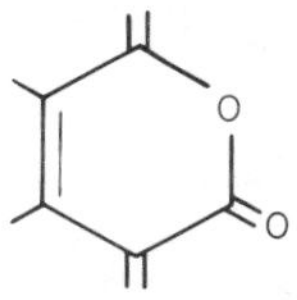

α - PYRONES

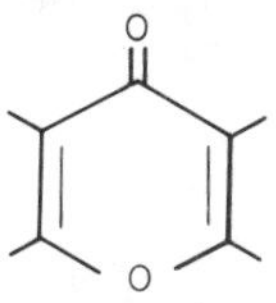

γ - PYRONIC TYPE CARBONYL

CONCLUSIONS AND SIGNIFICANCE

The results of the present study together with previously reported spectra of active carbons indicate that the surface functional groups show distinctive and reproducible features, which are also similar to the spectra obtained for carbon blacks. Three broad but distinctive bands are observed in the range 1150-1500cm^{-1} wavenumbers for carbons with acidic and basic characteristics. Heating at 1,000°C in an inert atmosphere eliminates the 1735cm^{-1} band corresponding to the acidic carbonyl groups.

By the analytical method developed, it is possible to enhance the band intensities by accumulation of spectra without using attenuation in the reference beam and measure the area of bands (Figure 1). A systematic study of the spectra obtained indicated that the intensity of the observed peaks are intimately related to one another, and that the functional groups are not erratically distributed. The intensity of the bands per unit weight of carbon are related in the general form described by Equation (2).

The structural units proposed are in agreement with the observed correlations. A given carbon surface appears to be the result of discreet combinations of these units. The proposed structures can explain the differences found in the acidic and basic character among active carbons depending on the activation temperature and therefore the labilityof the different structures.

Using the technique described, it is possible to distinguish relativedifferences among active carbons surface chemical characteristics, and in the future by the extension and refinement of the method it may be possible to predict the behaviour of a given surface towards the adsorption of classes of chemicals with distinctive characteristics.

The technique may also be used to control the chemical characteristics of active carbons by the manufacturers to produce and reproduce certain required characteristics, and to evaluate the performance of the regeneration processes.

ACKNOWLEDGEMENTS

The present work was partially performed at the Swiss Federal Institute for Water Resources and Water Pollution Control (EAWAG) and the Swiss Federal Laboratories for Materials Testing and Research (EMPA), Switzerland, during the author's sabbatical year. The financial and technical assistance received is greatly appreciated.

LITERATURE CITED

1. Cookson, Jr., J.T., C. Ishizaki and C.R. Jones, AIChE, Symposium Series Water 1971, 68 (124), 157 (1972).

2. Ishizaki, C. and J.T. Cookson, Jr., Wat. Poll. Control Fed. J., 45, 515 (1973).

3. Ishizaki, C. and J.T. Cookson, Jr., In Chemistry of Water Supply Treatment and Distribution (Ed. by A.J. Rubin) p. 201, Ann Arbor Science Pub. Inc., Ann Arbor (1974).

4. Ishizaki, C., I. Martí and M. Ruiz, In Treatment of Water by Granular Activated Carbon, (Ed. by M.J. McGuire and J.H. Suffet) Adv. in Chem. Series Nº 202, p. 95, ACS, Washington D.C., (1983).

5. Ishizaki, C. and I. Martí, Carbon, 19, 409 (1981).

6. Boehm H.P., In Advances in Catalysis (Ed. by D.D. Eley, M. Pines and P.B. Weiss), Vol. 16, p. 179. Academic Press, New York (1966).

7. Puri, B.R., In Chemistry and Physics of Carbon (Ed. by P.L. Walker, Jr.) Vol. 6, p. 191. Marcel Dekker, New York (1970).

8. Friedel, R.A., In Applied Infra-red Spectroscopy (Ed. by D.N. Kendall) p. 312. Reinhold Pub. Corp., New York (1966).

9. Robin P.L. and P.G. Rouxhet, Fuel, 55, 177 (1976).

10. Cannon C. and B. Sutherland, Trans. Farad. Soc., 41, 279 (1945).

11. Gordon, R.R., W.N. Adams and G.J. Jenkis, Nature, 170, 317 (1952).

12. Cannon, C.G. Nature, 171, 308 (1953).

13. Friedel, R.A. and M.G. Pelipetz, J. Opt. Soc. Am., 43, 1051 (1953).

14. Gordon, R.R., W.N. Adams, G.J. Pitt and G.H. Watson, Nature, 174, 1098 (1954).

15. van Vucht, H.A., B.J. Rietveld and D.W. van Krevelen, Fuel, 34, 50 (1955).

16. Brown, J.K. and W.E. Wyss, Chem. and Ind., p. 1118, (1955).

17. Brown, J.K., J. Chem. Soc., p. 744 (1955).

18. Friedel, R.A. and J.A. Queiser, Anal. Chem., 28, 22 (1956).

19. Brooks, J.D., R.A. Durie and S. Sternhell, Aus. J. Appl. Sci., 9, 63 (1958).

20. Friedel, R.A., Proceeding of the Fourth Carbon Conference, p. 321, Pergamon Press (1960).

21. Friedel, R.A. and H. Retcofsky, Proc. 6th Carbon Conference, Vol. 2, p. 149, Pergamon Press, (1963).

22. Garten, V.A., D.E. Weiss and J.B. Willis, Aust. J. Chem., 10, 295 (1957).

23. Studebaker, M.L. and R.W. Rinehard, Rubber Chem. Technol., 45, 106 (1972).

24. Prest, W.M., Jr., and R.A. Mosler, In Colloids and Surfaces in Reprographic Technologies (Ed. by M. Hair and M. Croucher) Chapter 12, ACS Symposium Series, Washington D.C. (1982).

25. Friedel R.A. and G.L. Carlson, Fuel, 51, 194 (1972).

26. Lygin, V.J., N.V. Kovaleva, N.N. Kautaradze and A.V. Kiselev, Kolloidn. Zurnal, 22, 334 (1960).

27. Mattson, J.S. and H.B. Mark, Jr., J. Colloid Interface Sci., 31, 131 (1969).

28. Friedel, R.A. and G.L. Carlson, J. Phys. Chem., 75, 1149 (1971).

29. Friedel, R.A., R.A. Durie and Y. Shewchyk, Carbon, 5, 559 (1967).

30. Zawadzki, J., Carbon, 16, 491 (1978).

31. Zawadzki, J., Polish J. of Chem., 52, 2157 (1978).

32. Zawadzki, J., Carbon, 18, 281 (1980).

33. Zawadzki, J., Carbon, 19, 19 (1981).

34. Mattson, J.S., L. Lee, H.B. Mark Jr. and W.J. Weber, Jr., J. Colloid Interface Sci., 33, 284 (1970).

35. Friedel, R.L. and L.J.E. Hofer, J. Phys. Chem., 74, 2921 (1970).

36. Martí, I.,"Estudio de las Modificaciones Químicas y Físicas en la Superficie de un Carbón Activado," Thesis, Universidad Simón Bolívar, Caracas (1979).

37. Voll M. and H.P. Boehm, Carbon, 9, 481 (1971).

38. Rodd's Chemistry of Carbon Compounds Vol. IV, Heterocyclic Compounds Part E, (Edited by S. Coffey), Elsevier Scientific Pub. Co. The Netherlands. Second Edition (1977).

39. Bellamy, L.J., The Infra-red Spectra of Complex Molecules, Chapman and Hall, London, Third Edition (1975).

40. Nakanishi, N., Infra-red Absorption Spectroscopy, Nankodo Co. Ltd. Tokyo (1962).

PHYSICAL ADSORPTION ON HETEROGENEOUS SOLIDS

Mieczysław Jaroniec
Institute of Chemistry
M. Curie-Skłodowska University
Lublin 20031, Poland

Abstract

Two fundamental isotherm equations, describing single-gas adsorption without lateral interactions on heterogeneous surfaces, are presented with emphasis on their special cases. A simple method is proposed to extend these equations to adsorption with lateral interactions, multilayer adsorption, mixed-gas adsorption and adsorption from dilute and concentrated solutions.

- - - - - -

INTRODUCTION

Considerable progress was certainly made in the 1970's in the theoretical studies of adsorption from gas and liquid mixtures on energetically heterogeneous surfaces of solids. Such an important problem in physical adsorption as the study of correlations between gas and liquid adsorption was considered by some authors. Unfortunately, their studies were concerned mainly with adsorption on homogeneous surfaces. Interest in these studies, whose aim is the elaboration of a unified description of physical adsorption from gaseous and liquid phases on heterogeneous surfaces, remains strong because of applications for proper design and modeling of many extremely important processes of utilitarian significance such as: separation of gas and liquid mixtures by means of solids; purification of water and air; catalytic processes, etc.

The objective of this study was to present a simple method for extending single-gas isotherm equations to those describing adsorption from gas and liquid mixtures on heterogeneous surfaces. This study creates the possibility of developing a thermodynamically consistent and comprehensive description of gas and liquid adsorption on heterogeneous surfaces, and gives the theoretical foundations for predicting the total adsorption from a mixture by means of parameters characterizing the adsorption of single components.

The majority of papers published on physical adsorption on heterogeneous solids concern single-gas adsorption [1-5]. Many equations have been proposed to describe single-gas adsorption isotherms [1,6-10].

These isotherms may be obtained by solving the well-known integral equation for Langmuir local behaviour and different energy distributions:

$$\theta_t(p) = \int_{\Delta} \theta(p,\varepsilon)\chi(\varepsilon)d\varepsilon \tag{1}$$

where

$$\theta(p,\varepsilon) = Kp/(1+Kp) \tag{2}$$

and

$$K = K_0 \exp(\varepsilon/RT) \tag{3}$$

In the above $\theta_t(p)$ is the overall isotherm describing monolayer adsorption on a heterogeneous surface characterized by the energy distribution $\chi(\varepsilon)$, $\theta(p,\varepsilon)$ is the local adsorption isotherm describing adsorption on sites of adsorption energy ε, p is the equilibrium pressure, K is the Langmuir constant, K_0 is the pre-exponential factor of K, and Δ is the region of integration. Eq. (2) describes monolayer single-gas adsorption without lateral interactions in the surface phase. Therefore, the overall isotherms derived by means of Eqs. (1) and (2) refer to the same adsorption model.

In this paper a classification of isotherm equations describing monolayer single-gas adsorption without lateral interactions will be proposed. Moreover, a simple method for extending them to monolayer adsorption with lateral interactions, multilayer single-gas adsorption, adsorption from gas and liquid mixtures, and multi-solute adsorption from dilute solutions will be discussed.

TWO FUNDAMENTAL SINGLE-GAS ADSORPTION ISOTHERMS

It follows from the review [1] that the most popular isotherm equations used for single-gas adsorption on heterogeneous surfaces may be divided into two groups. The adsorption isotherms belonging to the first group become Langmuir's Equation (2) for the heterogeneity parameter equal to unity. The isotherm equations belonging to the second group may be considered as special cases of the exponential adsorption isotherm proposed in 1975 [10].

Quite recently, a new general isotherm equation has been proposed to describe single-gas adsorption on heterogeneous surfaces [11]; its form is analogous to the empirical equation used recently for single-solute adsorption from dilute solutions [12]. This general equation has the following mathematical form [11]:

$$\theta_t(p) = \left[\frac{(\bar{K}p)^n}{1+(\bar{K}p)^n}\right]^{m/n} \tag{4}$$

where $\bar{K}$ is the constant analogous to K, m and n are the heterogeneity parameters assuming values in the interval (0,1). For $n=m=1$, Eq. (4) becomes Langmuir's Equation (2). Moreover, for special values of m and n, Eq. (4) reduces to three popular isotherm equations belonging to the first group. These are: the generalized Freundlich isotherm for $n = 1$ and $0 < m < 1$ [6,8], the Langmuir-Freundlich isotherm for $n = m$ and $0 < n,m < 1$ [6] and the Tóth isotherm for $m = 1$ and $0 < n < 1$ [9]. Generally, Eq. (4) describes single-gas adsorption on heterogeneous surfaces characterized by quasi-Gaussian energy distributions. These distributions are symmetrical for $n = m \in (0,1)$ and asymmetrical for other sets of parameters n and m [11]. If $n > m$ they show a widening to the right-hand side, whereas for $n < m$ this widening appears on the left-hand side.

The isotherm equations belonging to the second group may also be considered as special cases of the Dubinin-Astakhov equation [13]:

$$\theta_t(p) = \exp\left[-B_q\left(\ln \frac{p^0}{p}\right)^q\right] \quad \text{for } p \leq p^0 \tag{5}$$

where p^0 is the parameter with a physical meaning analogous to $\bar{K}$, and B_q and q are the heterogeneity parameters. The energy distribution corresponding to Eq. (5) is asymmetrical with a widening to the right-hand side,and shows a minimum adsorption energy which is connected with the parameter p^0. For $q = 1$, Eq. (5) becomes the classical Freundlich isotherm [14], whereas for $q = 2$ it becomes the Dubinin-Radushkevich equation [15]. Thus Eq. (5) produces two types of adsorption isotherm equations which are not reducible to Langmuir's Eq. (2). It should be noted that Eq. (5) has only an approximate explanation on the basis of Eqs. (1) and (2).

Eqs. (4) and (5) represent two general four-parameter isotherms (the fourth parameter is the monolayer capacity used to define the relative surface coverage θ_t) which describe single-gas adsorption on heterogeneous surfaces characterized by simple energy distributions. These distributions show a quasi-Gaussian shape with a widening to the right-hand side or left-hand side, or they may be decreasing exponential functions. Their course is dependent on the values of the heterogeneity parameters. All special cases of Eqs. (4) and (5), i.e., the isotherm equations belonging to the first and second groups, contain three parameters.

In the case of adsorption on heterogeneous surfaces characterized by complex energy distributions, the overall isotherm may be represented by a linear combination of the simple isotherm equations (special cases of Eqs. (4) and (5)) or by the exponential adsorption isotherm proposed by the author [10].

EXTENSIONS OF EQUATIONS (4) AND (5)

GENERAL CONSIDERATIONS

Many isotherm equations describing different models of adsorption from gaseous and liquid phases on homogeneous surfaces may be presented in the following mathematical form:

$$Y(X) = AX/(1 + AX) \tag{6}$$

In the above

$$A = A_0 \exp(E/RT) \tag{7}$$

is defined similar to the constant K. Table 1 presents the definitions of Y, X and A for different models of physical adsorption on homogeneous surfaces, which are frequently used in adsorption theory. The physical meaning of the parameters and variables appearing in Table 1 is given in the Notation section. The isotherm equations summarized in Table 1 may be used as local isotherms in the integral equation:

$$Y_t(X) = \int Y(X,E)F(E)dE \tag{8}$$

which is formally identical to Eq. (1). In Eq. (8), $Y_t(X)$ and $Y(X,E)$ denote the overall and local adsorption isotherms, respectively, and $F(E)$ is the distribution function of E normalized to unity.

The analytical solutions of Eq. (8) with Eq. (6) as the local isotherm are analogous to those obtained from Eqs. (1) and (2). They may be written in the following forms:

$$Y_t(X) = \left[\frac{(\bar{A}X)^n}{1 + (\bar{A}X)^n}\right]^{m/n} \tag{9}$$

and

$$Y_t(X) = \exp\left[-B_q\left(\ln\frac{X^0}{X}\right)^q\right] \qquad \text{for } X \le X^0 \tag{10}$$

where $\bar{A}$ and X^0 are constants analogous to A; they are connected with the value of $\bar{E}$, which is characteristic of the given distribution function $F(E)$. The constant $\bar{A}$ is proportional to $\exp(\bar{E}/RT)$ (cf. Eq. (7)) whereas X^0 is proportional to $\exp(-\bar{E}/RT)$ [16].

Table 2 contains the specifications of Y_t, $\bar{A}$ and X^0 for adsorption models summarized in Table 1. Below, additional information concerning the adsorption models under consideration will be given.

Model B

As an example of an adsorption model with lateral interactions, we presented in Table 1 the Fowler-Guggenheim (FG) equation in Langmuirian form. Extension of Eqs. (4) and (5) for FG local behaviour is possible when the exponential interaction term in the FG isotherm characterizes the average force field for the entire surface and is independent of the adsorption energy; it may be fulfilled for heterogneous solids showing a random distribution of adsorption sites on the surface. Using the above assumption, other adsorption isotherms with lateral interactions, e.g., the isotherm obtained by the quasi-chemical approximation [17], or isotherms involving molecular association [18], may be presented in Langmuirian form as for the FG equation.

Model C

The general form of the BET equation involving formation of a finite number of adsorbed layers may be presented in Langmuirian form according

Table 1 . Presentation of different local adsorption isotherms in Langmuir's form given by Equation (6)

Code	Adsorption model	Y	X	A
A	Monolayer single-gas adsorption without lateral interactions	θ	p	K
B	Monolayer single-gas adsorption with lateral interactions on "random" heterogeneous surfaces	θ	$p \cdot \exp(\alpha \theta_t)$	K
C	Multilayer single-gas adsorption occuring according to the BET model	$\theta^M / G(h)$	$y(h)$	$C = K \cdot p_s$
D	Monolayer adsorption from multicomponent gas mixtures without lateral interactions	$\theta_{(\underline{r})}$	$p_1 + \sum_{i=2}^{r} K_{i1} p_i$	K_1
E	Monolayer adsorption from multicomponent liquid mixtures on "random" heterogeneous surfaces	$X^s = \sum_{i=1}^{r-1} x_i^s$	$\beta_{1r} + \sum_{i=2}^{r-1} K_{i1} \beta_{ir} x_{ir}$	K_{1r}
F	Monolayer adsorption from binary ideal liquid mixtures	x_1^s	x_{12}	K_{12}
G	Monolayer multi-solute adsorption from dilute ideal solutions	$\theta_{(\underline{r})}$	$c_1 + \sum_{i=2}^{r} K_{i1} c_i$	$K_1^* = K_{1s}/c_s$
H	Monolayer single-solute adsorption from dilute ideal solutions	θ	$c = c_1$	$K^* = K_{1s}/c_s$

Table 2 . Specification of Y_t , $\bar{A}$ and X^o appearing in Equations (9) and (10)

Code of adsorption model	Y_t	$\bar{A}$	X^o
A	θ_t	$\bar{K}$	p^o
B	θ_t	$\bar{K}$	p^o
C	$\theta_t^M/G(h)$	$\bar{C} = \bar{K} p_s$	p^o/p_s
D	$\theta_{(\underline{r})t}$	$\bar{K}_1$	p_1^o
E	x_t^s	$\bar{K}_{1r}$	x_{1r}^o
F	$x_{1,t}^s$	$\bar{K}_{12}$	x_{12}^o
G	$\theta_{(\underline{r})t}$	$\bar{K}_1^* = \bar{K}_{1s}/c_s$	$x_1^* = x_{1s}^o c_s$
H	θ_t	$\bar{K}^* = \bar{K}_1^*$	$x^* = x_1^*$

to Eq. (6) with Y, A and X specified in Table 1. The function G(h) describes the formation of second and higher adsorbed layers, whereas

$$y(h) = h(1 - h^t)/(1 - h) \quad ,$$

where t is the number of adsorbed layers. For the classical BET equation (t tends to infinity) the above functions are defined as follows:

$$G(h) = (1 - h)^{-1} \quad \text{and} \quad y(h) = h/(1 - h) \tag{11}$$

The analytical forms of G(h) and y(h) for other multilayer adsorption isotherms obtained by modification of the classical BET model were discussed previously [19]. Application of the BET equation presented in Langmuirian form to describe local adsorption requires fulfillment of two conditions: (1) Energetic heterogeneity of the adsorbent surface is not transmitted from the first to the second and higher layers; otherwise, heterogeneity effects in the second and higher layers are neglected, and (2) Lateral interactions in each adlayer are neglected, which means that the topography of adsorption sites on the surface is arbitrary.

Model D

The application of a Langmuir-type equation for local adsorption to adsorption from multicomponent gas mixtures is possible when the constants K_{i1} for $i = 1,2,\ldots,r$ are characteristic of the entire surface, i.e., the differences in the adsorption energies between ε_i and ε_1 are constant. Theoretical studies have shown that this assumption is fulfilled in the case of identical heterogeneity parameters characterizing the single-gas adsorption isotherms; then, their energy distributions have identical shapes and are only shifted on the energy axis [20].

Therefore the heterogeneity parameters appearing in the mixed-gas and single-gas isotherms should be identical.

Model E

In contrast to gas adsorption, in which the adsorption energy is used to define the surface heterogeneity, in liquid adsorption this heterogeneity is specified by means of the difference in adsorption energies of both components. Although the same symbols are used to denote the heterogeneity parameters in gas and liquid adsorption isotherms, their values cannot be compared because they characterize distributions for different variables.

The variable β_{ir} appearing in Table 1 is defined as the ratio of the activity coefficients in the surface phase and bulk phase:

$$\beta_{ir} = (f_i^l f_{r,t}^s)/(f_{i,t}^s f_r^l) \tag{12}$$

As in the case of single-gas adsorption with lateral interactions, we assume that the activity coefficients $f_{i,t}^s$ for $i = 1,1,\ldots,r$ are characteristic of the entire surface phase; this is fulfilled for heterogeneous surfaces with a random distribution of adsorption sites. However, for adsorption models involving ideality of the surface phase, the assumption about topography of adsorption sites is useless.

Similarly as in mixed-gas adsorption, we assume that the constants K_{il} for $i = 1,1,\ldots(r-1)$ are characteristic for the entire surface; this means that the heterogeneity exhibited by a given solid surface for adsorption of $(r-1)$ components is similar, but different from that of the r'th component adsorption. Model F is a simple case of Model E.

Model G

Model G is a special case of Model E. Assuming ideality of both phases in Model E, and taking into account the condition that the concentrations of solutes are infinitely low, we obtain the multi-solute adsorption isotherms from those describing liquid adsorption in the whole concentration region. For this purpose we replaced

$$x_i^l/x_s^l \quad \text{by} \quad c_i/c_s$$

and defined

$$K_i^* = K_{is}/c_s \quad .$$

Model H is the simple case of Model G.

CONCLUSIONS

Two fundamental single-gas adsorption isotherms, comprising the most popular equations used in single-gas monolayer adsorption without lateral interactions on heterogeneous surfaces, have been considered. A simple method has been proposed to generalize these isotherm equations to adsorption with lateral interactions, multilayer adsorption, and adsorption from gas and liquid mixtures. The theoretical foundation of this method is a unified treatment of physical adsorption from gaseous and liquid phases on heterogeneous surfaces, which has been elaborated upon by utilizing the mathematical similarities in the description of these phenomena.

The main results of this work have been summarized in tabular form. In the tables we have presented a set of variables and constants which are used to replace the adsorbate pressure and the Langmuir-like constant in single-gas adsorption isotherms to obtain isotherm equations describing specific models of adsorption from gas and liquid mixtures. Such as summary of the isotherm equations used for the theory of physical adsorption on heterogeneous surfaces is interesting, especially for those who would like to utilize these equations for practical purposes.

NOTATION

B_q = heterogeneity parameter in Eqs. (5) and (10)

c = solute concentration

f = activity coefficient

h = relative pressure

K = Langmuir's constant

$\bar{K}$ = Langmuir-type constant connected with the characteristic adsorption energy

K_{ij} = ratio of the constants K_i and K_j

$\bar{K}_{ij}$ = constant K_{ij} analogous to $\bar{K}$

m,n = heterogeneity parameter in Eqs. (4) and (9)

p = equilibrium pressure

p^0 = constant in Eq. (5) connected with the minimum adsorption energy

p_s = saturation pressure

q = heterogeneity parameter in Eqs. (5) and (10)

r = number of components in bulk phase

t = number of adsorbed layers

x = mole fraction

x_{ij} = ratio of the mole fractions x_i^1 and x_j^1

x_{ij}^0 = constant connected with the minimum difference of adsorption energies in liquid adsorption

Greek Letters

α = constant characterizing lateral interactions in FG equation

ε = adsorption energy

$\chi(\varepsilon)$ = energy distribution function

θ, θ^M = relative monolayer and multilayer surface coverages

Superscripts

M = multilayer adsorption

l = bulk phase

s = surface phase

Subscripts

i,j = i'th and j'th components

($\underline{r}$) = total quantity for r components

s = solvent in adsorption from dilute solutions

t = entire surface

Remark

Subscripts relating to the adsorption of individual components are omitted when the reference is obvious.

ACKNOWLEDGMENT

The author wishes to thank Professor A.L. Myers, who suggested the title of this paper and supported financially his participation in the Fundamentals of Adsorption Conference.

LITERATURE CITED

1. Jaroniec. M., "Physical Adsorption on Heterogeneous Solids," *Advances in Colloid and Interface Sci.* 18, 149 (1983).
2. Jaroniec, M., A. Patrykiejew and M. Borówko, "Statistical Thermodynamics of Monolayer Adsorption from Gas and Liquid Mixtures on Homogeneous and Heterogeneous Surfaces," *Progress in Surface and Membrane Sci.* 14, 1 (1981).
3. Zolandz, R.R. and A.L. Myers, "Adsorption on Heterogeneous Surfaces," *Progress in Filtration and Separation Sci.* 1, 1 (1979).
4. House, A.W., Adsorption on Heterogeneous Surfaces," in *Specialist Periodical Reports, Colloid Science,* Everett, D.H. (Ed.), Royal Soc. of Chemistry, London (1983), chap. 1.
5. Cerofolini, G.F., "Localized Adsorption on Heterogeneous Surfaces," *Thin Solid Films* 23, 129 (1974).
6. Sips, J.R., *J. Chem. Phys.* 16, 420 (1948); 18, 1024 (1950).
7. Misra, D.N., *Surface Sci.* 18, 367 (1969).
8. Misra, D.N., *J. Chem. Phys.* 52, 5499 (1970).
9. Tóth, J., W. Rudziński, A. Waksmundzki, M. Jaroniec and S. Sokolowski, *Acta Chim. Hung.* 82, 11 (1974).
10. Jaroniec, M., *Surface Sci.* 50, 553 (1975).
11. Jaroniec, M. et al., to be published.
12. Marczewski, A.W. and M. Jaroniec, *Monatsh. Chem.*, in press.

13. Dubinin, M.M. and V.A. Astakhov, *Izv. Akad. Nauk SSSR Ser. Khim.*, 71, 5 (1971).

14. Freundlich, H., *Trans. Faraday Soc.* 28, 195 (1932).

15. Dubinin, M.M. and L.V. Radushkevich, *Dokl. Akad. Nauk SSSR* 55, 331 (1947).

16. Cerofolini, G.F., *J. Low Temperature Phys.* 6, 473 (1972).

17. House, W.A., *J. Colloid and Interface Sci.* 67, 166 (1978).

18. Garbacz, J.K., M. Jaroniec and A. Deryło, *Thin Solid Films* 75, 307 (1981).

19. Jaroniec, M. and W. Rudziński, *Acta Chim. Hung.* 88, 351 (1976).

20. Jaroniec, M., W. Rudziński and J. Narkiewicz, *J. Colloid and Interface Sci.* 65, 9 (1978).

ADSORPTION EQUILIBRIUM DATA AND MODEL NEEDS

Bal K. Kaul* and Norman H. Sweed
Exxon Research and Engineering Company
Florham Park, NJ 07932

Adsorption equilibrium data are important for selection, design, and simulation of adsorption-based separation processes, yet availability of such data in the literature at realistic process conditions is limited. Reliable methods for correlating and predicting adsorption equilibrium data for mixtures are also limited. This paper discusses the needs for data, reviews available methods for predicting and correlating mixture isotherms, and points out areas where improvements are needed. For example, there are some limited pure component data available in the literature for components important in hydrogen purification yet almost none exist for their mixtures at process conditions. Also, models describing adsorbed phase mixtures at high coverage often assume these mixtures are ideal when in fact non-idealities are more common.

INTRODUCTION

The purpose of this Engineering Foundation Conference on Fundamentals of Adsorption is to "define gaps in engineering-related knowledge, particularly in the interdisciplinary areas among various branches of engineering and technology and then to find ways to fill these gaps". This paper focusses on defining gaps in knowledge related to adsorption-based bulk separation processes in the petroleum, petrochemical, and related industries. In particular, this paper discusses the gap between adsorption equilibrium data in the open literature and the data needed for industrial separation processes. This paper also discusses the need for improved adsorption equilibrium models for correlating and predicting adsorption data for both gases and liquids.

Adsorption equilibrium data provide the capacity and the selectivity of an adsorbent needed for the selection and the design of adsorption processes. Most adsorption processes are cyclic, and require the desorption or regeneration of the adsorbent by temperature or pressure swings or by replacement by another species. Calculations for the time required (cycle time) for adsorption and desorption modes of the process need adsorption equilibrium models. Adsorption equilibrium models are also needed for correlation and prediction of the equilibrium data in order to minimize the number of experiments for measuring such equilibrium data.

---

*To whom correspondence should be addressed

Many pure component adsorption equilibrium isotherm data are available in the literature, but two-component and three-component mixture data are relatively scarce. Most available data are usually taken at 25°C and low pressures (~ 1 atmosphere).

A number of pure gas and mixed-gas isotherm models are available in the literature. Among the frequently cited pure gas isotherms are the Langmuir isotherm (1), the Freundlich isotherm (1), the vacancy solution model (2,3), the Ruthven's Statistical thermodynamic model (4), and others (5,6). These pure gas isotherm models have been also extended to gas mixtures, but in some cases the extension is strictly empirical which makes correlation and prediction of mixture data unreliable. Another gas mixture model is the ideal adsorbed solution model (IASM) proposed by Myers and Prausnitz (7).

Since there are no pure component isotherms available for liquids, the approach to modeling the equilibrium data for liquids has been essentially starting from binary mixtures and then building multicomponent models from these binaries. The ideal adsorbed solution model developed for gas mixtures has been applied in some cases for modeling adsorption from liquids (8). The success of this model is limited as nonideal adsorbed-phases are more common.

LITERATURE DATA ARE USUALLY FAR FROM PROCESS CONDITIONS

Gaps between adsorption equilibrium data and needs for commercial practice are illustrated below for the several widely used separation processes.

Pressure swing adsorption (PSA) process purifies hydrogen from mixtures containing hydrocarbons (e.g., methane, ethane, ethylene), oxides of carbon (e.g., carbon monoxide, carbon dioxide), sulfur compounds (e.g., hydrogen sulfide), nitrogen, and water vapor (9,11). These impurities typically range from 0.5 to 45% by volume depending upon the feed stock. Most PSA units operate from 150 to 600 psia during part of their cycle. At these conditions significant quantities of all the components, including hydrogen, are adsorbed. However, existing literature data were taken at conditions far from where the process operates.

Figure 1 shows the temperature-pressure operating window typical of PSA hydrogen processes together with the conditions at which the literature data for gas mixtures were taken. The mixture data are rarely reported above 50 psig; in fact most of the data are at atmospheric pressure or below. Even pure component adsorption data in the literature do not cover the commercial operating pressure range. Figure 2 shows that there is only a small region of overlap between the literature data and process conditions. Note that methane and ethane adsorption data are available to high pressures, but these data are exceptions.

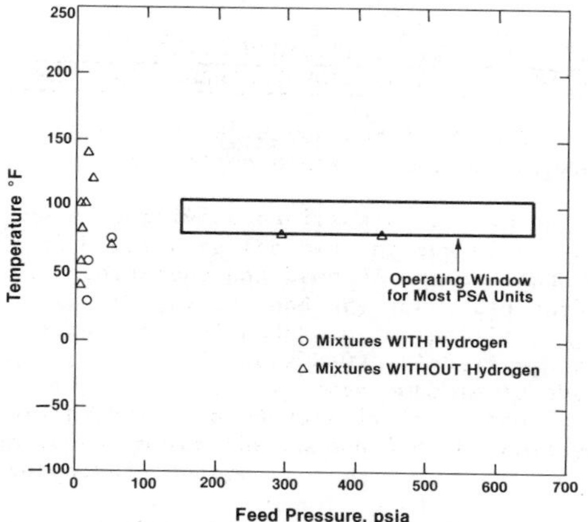

FIG. 1 MIXTURE DATA NOT AVAILABLE AT REALISTIC PROCESS CONDITIONS (Hydrogen Purification by PSA)

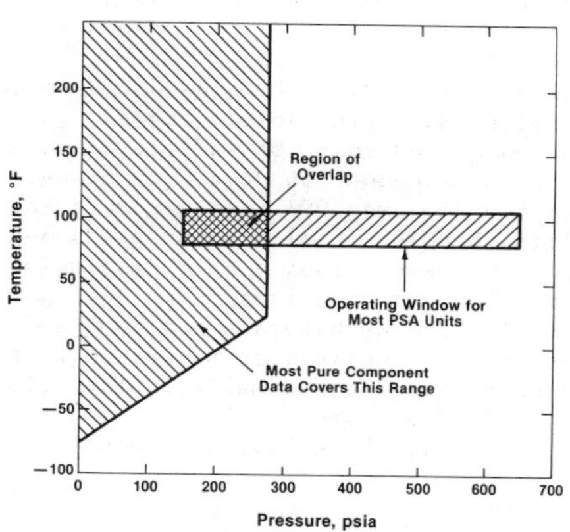

FIG. 2 PURE COMPONENT DATA NOT AVAILABLE OVER ENTIRE PRESSURE RANGE (Hydrogen Purification by PSA)

Another common commercial application of adsorption is paraffin isomer separation (10) where separation is by sieving rather than by equilibrium, but the equilibrium mixture data of n-paraffins is of importance. The process temperature is usually kept high to maintain the feed in the vapor-phase. The data published in open literature for this application are extremely limited and these are at 25°C, far from process conditions.

Separation of aromatics (e.g., paraxylene from C_8 aromatics) is a common liquid-phase commercial adsorption application. Data published in the open literature are sparse and usually at 25°C (12, 13). At 25°C, mass transport of these aromatics are slow, and there are incentives to run such processes hotter. Therefore, the lack of data at high temperatures (actual process conditions) makes simulation of this process more difficult.

From our discussion above it is clear that there is a need for high quality equilibrium data at actual process conditions, i.e. high temperature and pressure. There is also a need for compilation of the critically evaluated adsorption equilibrium data. This compilation could be similar to DECHEMA Series for vapor-liquid equilibrium data used extensively for distillation calculations (14). Penn State University, is making such efforts under the grants from the American Petroleum Institute.

GAS-PHASE ADSORPTION EQUILIBRIUM MODELS

Pure Gas

Successful multicomponent adsorption isotherm calculations rely on an accurate representation of the pure component isotherms. Based on the regression of isotherm data for many components, Kaul (16) concluded that isotherm models having at least three parameters can be successful in correlating the pure gas data. For example, Figure 3 shows an excellent fit by the Langmuir-Freudlich (3-parameters) and the vacancy solution model (4-parameters) to the experimental isotherm data of carbon monoxide on molecular sieve (10x) at 144 K and up to 20 psia (15). However, the commonly used Langmuir isotherm (2-parameters) shows substantial deviations from the experimental data in comparison to other models (12% average absolute deviation by the Langmuir model against <2% for the other models).

There are other three parameter isotherm models (besides those reported in Figure 3) which can fit the data equally well, but these isotherms have limitations. Briefly, the three parameter virial isotherm is not reliable for extrapolation in pressure and its extension to mixtures is not straightforward. However, the virial isotherm is one of the powerful methods for obtaining Henry's law constant from the isotherm data. Ruthven's statistical correlation (4) is applicable for zeolites only and this limits its application. Several authors (6) have shown success with the potential theory for hydrocarbon gases on nonpolar adsorbents such as activated carbon. For polar adsorbents, such as molecular sieves, potential theory is

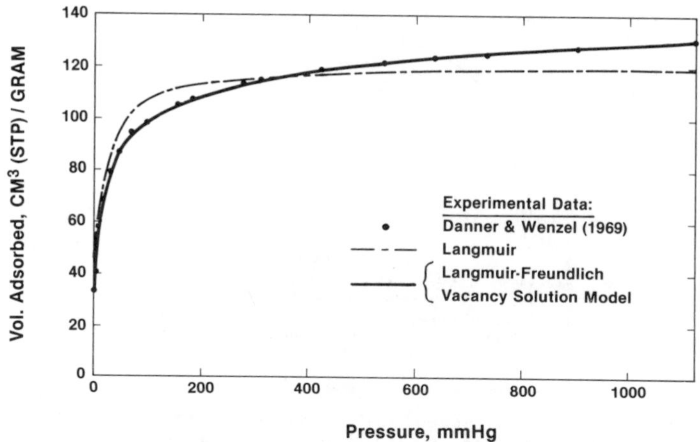

FIG. 3 ADSORPTION ISOTHERM FOR CARBON MONOXIDE ON MOLECULAR SIEVE (10x) AT 144 K (-200 °F)

reported to have only limited success due to lack of characterizing the interaction with the solid adsorbent by a suitable parameter. Recently Myers (17) modified the potential theory by taking into account the surface heterogeneity of an adsorbent; but again, it is reported to have limitations for adsorbents such as molecular sieves.

Gas Mixtures

Adsorptive separations involve mixtures and need mixture equilibrium data for the simulation and the design of the adsorbers. In principle, many pure component isotherm models can be extended to the mixtures. However, in many cases extension is empirical which makes mixture prediction by these models unreliable and thermodynamically inconsistent. Extension of the vacancy solution model (VSM) and the ideal adsorbed solution model (IASM) proposed by Myers and Prausnitz (7) to mixtures is based on solution thermodynamic principles, and therefore these models are promising mixture models.

Low surface coverage mixture data can be predicted from pure component data alone by both IASM and VSM. Figure 4 shows excellent agreement between the predicted and the experimental binary isotherm data of Costa et al. (18) for methane-ethylene on activated carbon (AC-40) at 20°C. The predicted values are by VSM using the pure component data alone. The predictions by IASM are close to the predictions by VSM again using pure component data only. Other successful low coverage binary and ternary isotherm predictions by IASM and VSM using pure component data alone are discussed elsewhere (16).

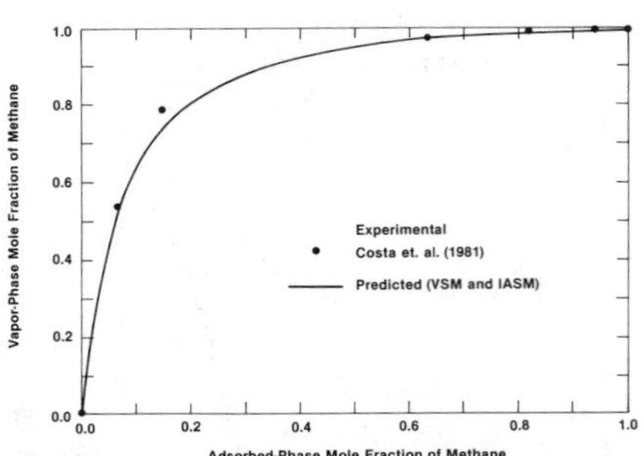

FIG. 4 COMPARISON OF PREDICTED AND EXPERIMENTAL LOW COVERAGE DATA FOR METHANE-ETHYLENE ADSORBED ON ACTIVATED CARBON (AC-40) AT 20°C AND 75 mmHg

High surface coverage multicomponent data can be predicted successfully from pure and binary isotherm data by VSM. Figure 5 shows poor predictions for oxygen-carbon monoxide system on molecular sieve 10x adsorbent at 144K (15) using pure component isotherms and the VSM and IASM models. The surface coverage for this system is so high (>50%) that the adsorbate-adsorbate interactions cannot be ignored. These interactions can be obtained from the binary data. Figure 5 shows excellent fit to the binary data by VSM with adsorbate-adsorbate interactions incorporated into the model. For IASM there is no framework available to incorporate these interactions into the model, and so it cannot be used. High coverage ternary isotherm data can also be predicted by VSM from pure and binary isotherm data as pointed out by Kaul (16).

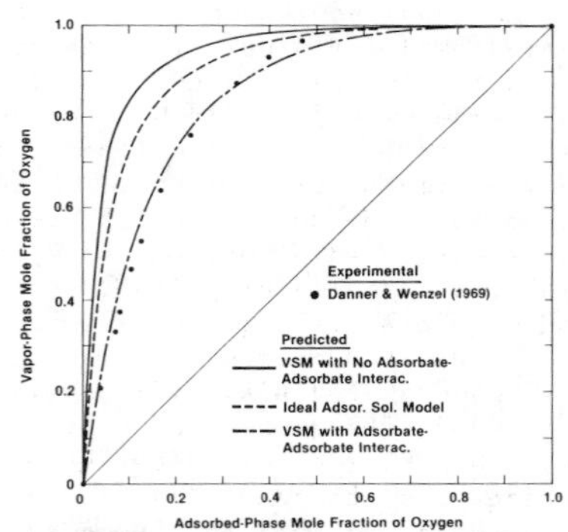

FIG. 5 HIGH COVERAGE ADSORPTION EQUILIBRIUM FOR OXYGEN-CARBON MONOXIDE ON MOLECULAR SIEVE (10X) AT 144 K & 760 mmHg

Thus, the vacancy solution model can be used for the prediction of both low coverage and high coverage mixture isotherm data although the prediction of high coverage data also needs binary data. But, as pointed out in the last section, there are limited data available at actual process conditions. To establish the predictive powers of VSM at these conditions further testing is needed.

LIQUID-PHASE ADSORPTION EQUILIBRIUM MODELS

Gas-phase adsorption equilibrium models are not directly applicable for modeling adsorption from liquids. Unfortunately, most developed gas-phase models are at their best below the saturation limit of the surface whereas adsorption from liquids occurs close to the surface saturation. Another difference is that most gas mixture adsorption isotherm models rely on accurate representation of pure component isotherms; but in the case of adsorption from liquids, pure component isotherms cannot be measured.

There are two promising approaches for modeling adsorption from liquids. Ruthven's Statistical Model is limited to zeolites, and even for such adsorbents this model needs further development in terms of characterizing adsorbate-adsorbate interactions within the zeolite cages. Another attractive model which the authors believe has greater potential is a classical thermodynamic model. This model needs a realistic activity-coefficient model for the adsorbed-phase, but such a model is not available. An activity coefficient model accounts for the nonidealities of the adsorbed-phase, and such nonidealities are common in actual practice. The activity coefficient enters the adsorption equilibrium calculations as shown by the following equilibrium equations.

$$f_i \text{ (Liquid-Phase)} = f_i' \text{ (Adsorbed-Phase)} \tag{1}$$

$$\gamma_i x_i P_i^s \exp\left(\frac{P - P_i^s}{\rho RT}\right) = \gamma_i' x_i' P_i^s \exp\left(\frac{\phi_i^o - \phi}{m_i RT}\right) \tag{2}$$

For an ideal liquid-phase under pressure close to 1 atm, equation (2) simplifies to:

$$x_i = \gamma_i' x_i' \exp\left(\frac{\phi_i^o - \phi}{m_i RT}\right) \tag{3}$$

Equation (3) is the simplified adsorption equilibrium equation but it needs a model for calculating activity coefficient (γ_i') for the adsorbed-phase. The free energy of immersion (ϕ) is related to γ_i' and the compositions in the liquid-phase (x_i) and the adsorbed-phase (x_i') .

Many authors (18) use the activity coefficient models developed for vapor-liquid equilibrium calculations (e.g., Wilson, UNIQUAC), but these models may not be appropriate for adsorption equilibrium calculations. For example, Figure 6, a plot of the adsorbed-phase activity coefficient for a mixture of benzene-cyclohexane on activated carbon at 30°C, shows that the activity coefficient for cyclohexane (8) can be as low as 0.4. A value of 0.4 shows negative deviation from Raoult's Law whereas the same mixture for vapor-liquid equilibrium calculations is nearly ideal, showing slight positive deviations from Raoult's Law. The vapor-liquid equilibrium models may empirically fit the data but problems arise in interpolation and extrapolation of activity coefficients with respect to temperature. Therefore, to model adsorption equilibrium from liquids, there is a need for realistic activity coefficient model for commonly occurring nonideal adsorbed-phases.

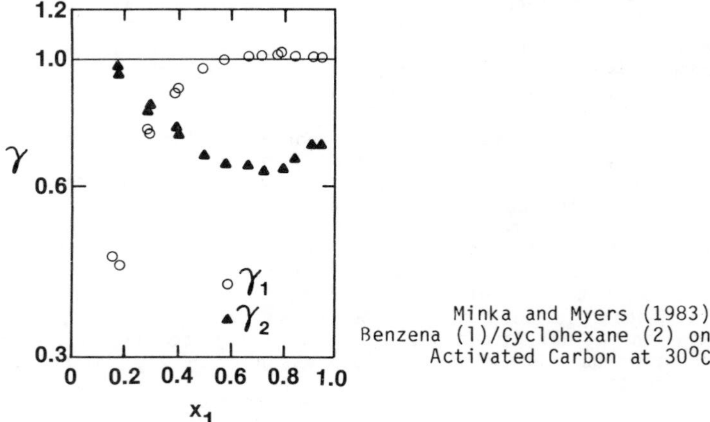

FIG. 6 ADSORBED-PHASE IS NONIDEAL FOR BENZENE-CYCLOHEXANE SYSTEM WHEREAS LIQUID-PHASE IS IDEAL FOR VLE

CONCLUSIONS AND SIGNIFICANCE

Most commercial adsorption processes run hotter than 25°C and/or at pressure above 1 atmosphere yet most of the data published in open literature are at 25°C and 1 atmosphere. Therefore, gaps exist between the data published in the open literature and those needed in actual practice.

The vacancy solution model is successful for correlation and prediction of the adsorption equilibrium from gases but needs testing at realistic process conditions. Low coverage adsorption equilibrium mixture data can be successfully predicted from the pure component data alone by either the vacancy solution model or the ideal adosrbed solution model. High coverage mixture data also needs binary data for successful multicomponent predictions by the vacancy solution model.

There are limited number of equilibrium models available for adsorption from liquids and even these models need further development. There is a pressing need for an activity coefficient model for the adsorbed-phase to model nonideal adsorption systems.

The significance of this work is that it points out gaps in the area of adsorption equilibrium data and models. Hopefully some of these gaps will be bridged in future for better understanding, simulation and design of adsorption processes.

ACKNOWLEDGEMENT

The authors thank Exxon Research and Engineering Company for the permission to present and publish this work.

NOTATION

f	-	fugacity
m	-	saturation capacity of adsorbent
P	-	pressure
R	-	Gas constant
T	-	temperature
x	-	composition
γ	-	activity coefficient
ϕ	-	Free energy of immersion
ρ	-	Molar density

Subscript

i	-	component

Superscript

o	-	pure component
´	-	adsorbed-phase
s	-	saturation condition

Literature Cited

1. Adamson, A. W., "Physical Chemistry of Surfaces", John Wiley and Sons, N.Y. (1976).

2. Suwanayuen, S., Danner, R. P., AIChE J., 26, 68 (1980).

3. Suwanayuen, S., Danner, R. P., AIChE J., 26, 76 (1980).

4. Ruthven, D. M., Loughlin, K. F., Holborow, K. A., Chem. Eng. Sci., 28, 7016 (1973).

5. Barrer, R. M., "Zeolites and Clay Minerals as Sorbents and Molecular Sieves", Academic Press, London, England (1978).

6. Grant, R. J., Manes, M., Ind. Eng. Chem. Fund., 5, 490 (1966).

7. Myers, A. L., and Prausnitz, J. M., AIChE J., 11, 1216 (1965).

8. Minka, C., Myers A. L., AIChE J., 19, 453 (1973).

9. Stewart, H. A., Heck, J. L., Chemical Engineering Progress, 65(9), 78 (1969).

10. "N-Paraffins (ISOSIV Process - Kerosine/Gas Oil Range) - Engineering Products and Processes, Union Carbide Corp.," Hydrocarbon Processing, Nov. 1977.

11. Cassidy, R. T., "POLYBED Pressure-Swing Adsorption Hydrogen Processing", ACS Symposium Series 135, San Francisco, California, Aug. 25-26, 1980.

12. Broughton, D. B., Kirk-Othmer's Encycl. of Chem. Techn., 3rd Ed., vol. 1, pp 563-581, 1978.

13. Santacesaria, E., Morbidell, M., Danise, P., Mercenari, M., Carra, S., Ind. Eng. Chem. Proc. Des. Dev., 21, 440 (1982).

14. Gmehling, J., Onken, U, Arlt, W., "Vapor-Liquid Equilibrium Data Collection", Chemistry Data Series, DECHEMA Publisher, W. Germany (1978).

15. Danner, R. P., Wenzel, L. A., AIChE J., 15, 515 (1969).

16. Kaul, B. K., "Correlation and Prediction of Adsorption Isotherm Data for Pure and Mixed-Gasses", Presented at the AIChE Meeting Los Angeles, California, Nov. 14-19, 1982.

17. Myers, A. L., AIChE National Meeting, Orlando, Fl., Feb. 28-March 4, 1982.

18. Costa, E., Sotelo, J. L., Calleja, G., and Marron, C., AIChE J., 27, 5 (1981).

CONSEQUENCES FROM THE SECOND LAW OF THERMODYNAMICS FOR THE DYNAMICS OF ADSORPTION PROCESSES

J.U. Keller
Inst. Thermodynamik u. Anlagentechnik
Technische Universität Berlin, TK 7
Str. d. 17. Juni 135, D-1000 Berlin 12, West-Germany

ABSTRACT

The Second Law of Thermodynamics is formulated for a fixed-bed adsorption column. From it so-called constitutive equations are derived, which describe exchange processes of heat and mass occurring either within the column or between the column and its surroundings.

The general linear functional form of these equations is given. Besides, it is shown that these equations, supplemented by the thermostatic equations of state and the balance equations of the column, enable one to calculate all sorption processes in the column on a sound thermodynamic basis.

THE ADSORPTION COLUMN

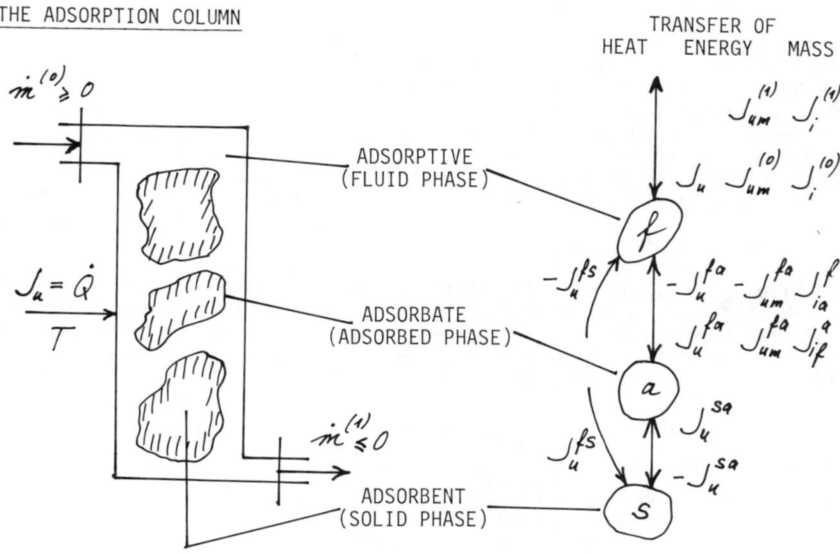

Fig. 1: Adsorption column.
Exchange processes of mass an energy between the fluid (f), adsorbate (a) and (solid) adsorbent (s).

We consider a fixed-bed adsorption column including a fluid, i.e. liquid or gaseous bulk phase (adsorptive (f)) an adsorbed phase (adsorbate (a)) and a solid phase (adsorbent (s)). Moreover heat and mass is supplied to (index: 0) or withdrawn from (index: 1) the column (cp. Figure 1).

Thermodynamic equilibrium and non-equilibrium states of any of the phases (f, a, s) can phenomenologically be described by the usual extensive and intensive variables which are connected by Gibb's equation as follows (1, 2):

$$f: \quad dS^f = \frac{1}{T^f} dU^f \qquad - \sum_i^n \frac{\mu_i^f}{T^f} dm_i^f, \tag{1}$$

$$a: \quad dS^a = \frac{1}{T^a} dU^a + \frac{P^a}{T^a} dA \qquad - \sum_i^n \frac{\mu_i^a}{T^a} dm_i^a, \tag{2}$$

$$s: \quad dS^s = \frac{1}{T^s} dU^s. \tag{3}$$

Here S denotes the entropy, U the internal energy, $m_i = 1...n$, the masses, T the temperature, μ_i, $i = 1...n$ the chemical potentials of the various components in the respective phases (f, a, s). Besides, A is the surface of the adsorbed phase, P^a its spreading pressure (2). We here assumed the volumes of the fluid and solid phase to be constant ($dV^f = 0$, $dV^s = 0$) and moreover the solid adsorbent to be rigid and inert, i.e. to have constant "active" surface. For non-equilibrium states the intensive parameters occurring in Eqs. (1-3) can be interpreted as "accompanying" parameters whose values uniquely are determined by the respective thermostatic equations of state (3,4).

The fluid, adsorbed and solid phase, f, a, s, exchange mass, convective energy and heat as sketched in Figure 1. The respective fluxes $J_x^{\alpha\beta}$ indicate the transfer of mass of component i ($x = i = 1,...n$), convective energy ($x = um$), or heat ($x = u$) from phase α to phase β. Bearing this in mind the conservation laws for mass and energy of the various phases can be written as:

f: mass:

$$\dot{m}_i^f = - J_i^{fa} + J_i, \quad i = 1...n \tag{4}$$

energy:

$$\dot{U}^f = - J_u^{fa} - J_u^{fs} + J_u - J_{um}^{fa} + J_{um}, \tag{5}$$

a: mass: $\dot{m}_i^a = J_i^{fa}, \quad i = 1 \dots n$ (6)

energy: $\dot{U}^a = J_u^{fa} + J_u^{sa}$

$+ J_{um}^{fa} \qquad - P^a \dot{A}$ (7)

s: mass: $\dot{m}_i^s = 0,$ (8)

energy: $\dot{U}^s = - J_u^{sa} + J_u^{fs}.$ (9)

Summation of Eqs. (4, 6, 8) and Eqs. (5, 7, 9) yields the overall balances of the system, namely

mass: $\dot{M}_i = J_i, \quad i = 1 \dots n$ (10)

energy: $\dot{U} = J_u + J_{um},$ (11)

where

$$M_i = m_i^f + m_i^a + m_i^s \tag{12}$$

$$U = U^f + U^a + U^s, \tag{13}$$

indicate the total mass of component i and the internal energy of the column. Besides, the fluxes

$$J_i = \sum_{\alpha=0}^{1} \dot{m}_i^{(\alpha)} = \sum_{\alpha=0}^{1} \gamma_i^{(\alpha)} \dot{m}^{(\alpha)} \tag{14}$$

and

$$\left(\gamma_i = \dot{m}_i^{(\alpha)} / \dot{m}^{(\alpha)}, \quad \sum_{i=1}^{n} \gamma_i = 1 \right)$$

$$J_{um} = \sum_{\alpha=0}^{1} h^{(\alpha)} \dot{m}^{(\alpha)}, \tag{15}$$

describe the exchange of mass and convective energy (neglecting kinetic and potential energy) of the column with its surroundings.

The mass flow between the fluid and the adsorbed phase can be split in two (non-negative) parts

$$J_i^{fa} = J_{if}^{a} - J_{ia}^{f}, \tag{16}$$

indicating the net flow from the fluid to the adsorbed phase and vice versa respectively. The energy flow connected with these mass flows is (3, 4),

$$J_{um}^{fa} = \sum_{i}^{n} \left(\mu_i^{f} J_{if}^{a} - \mu_i^{a} J_{ia}^{f} \right). \tag{17}$$

THE SECOND LAW OF THERMODYNAMICS

Consider an adsorption process of the column which starts at time $t = -\infty$ in any equilibrium state Z^- and ends at $t = \infty$ in another equilibrium state Z^+. For such a process the Second Law of Thermodynamics states (3-5),

$$S^+ - S^- \geq \int_{Z^-}^{Z^+} \left\{ \frac{dQ}{T} + \sum_{\alpha=0}^{1} s^{(\alpha)} dm^{(\alpha)} \right\}. \tag{18}$$

Here $S^\pm$ are the (well defined) entropies of the column in the equilibrium states $Z^\pm$. The inequality sign holds for natural, i.e. irreversible processes, the equality sign is valid only for quasi-static and reversible adsorption processes. Assume now the column and all thermodynamic phases (f, a, s) included to be suddenly isolated from each other at (an arbitrary) time t. Then all phases approach in (t, ∞) certain equilibria states Z^{f+}, Z^{a+}, Z^{s+}. Taking the additivity relations for the entropy

$$Z^\pm : \qquad S^\pm = S^{f\pm} + S^{a\pm} + S^{s\pm} \tag{19}$$

into account, relation (18) can be rewritten as

$$\int_{-\infty}^{t} \left\{ \dot{S}^f + \dot{S}^a + \dot{S}^s - \frac{\dot{Q}}{T} - \sum_{\alpha=0}^{1} s^{(\alpha)} \dot{m}^{(\alpha)} \right\} dt \geq 0. \tag{20}$$

Inserting Gibbs' equations (1-3) and the conservation laws (4-9), we get after some algebra the Fundamental Inequality or Passivity Inequality of the column (4, 5):

$$\int_{-\infty}^{t} \Big\{ \underbrace{\left(\frac{1}{T^f} - \frac{1}{T} \right) J_u}_{\text{external HT}} + \underbrace{\left(\frac{1}{T^a} - \frac{1}{T^f} \right) J_u^{fa}}_{\text{HT } (f \to a)} + \underbrace{\left(\frac{1}{T^s} - \frac{1}{T^f} \right) J_u^{fs}}_{\text{HT } (f \to s)}$$

$$+ \underbrace{\left(\frac{1}{T^a} - \frac{1}{T^s} \right) J_u^{sa}}_{\text{HT } (s \to a)} \tag{21}$$

$$+\sum_{i=1}^{n}(\mu_i^f-\mu_i^a)\left(\frac{J_{if}^a}{T^a}-\frac{J_{ia}^f}{T^f}\right)$$

MT (f → a) (adsorption process!)

$$+\sum_{\alpha=0}^{1}\left(\frac{1}{T^f}-\frac{1}{T^{(\alpha)}}\right)h^{(\alpha)}\dot{m}^{(\alpha)}+\sum_{i=1}^{n}\sum_{\alpha=0}^{1}\left(\frac{\mu_i^{(\alpha)}}{T^{(\alpha)}}-\frac{\mu_i^f}{T^f}\right)\dot{m}_i^{(\alpha)}\Big\}dt\geq 0,$$

external ET external MT

HT ... heat transfer
ET ... convective energy transfer ... all t.
MT ... mass transfer

To each term in this inequality, a thermodynamic process can be assigned as indicated above. The inequality formally can be written as

$$\int_{-\infty}^{t}\sum_k X_k Y_k\,dt\geq 0 \qquad \ldots \text{all } t, \tag{22}$$

with the (internal and external) thermodynamic fluxes

$$Y_k=\{J_u, J_u^{fa}, J_u^{fs}, J_u^{sa}, J_{if}^a, J_{ia}^f, h^{(\alpha)}\dot{m}^{(\alpha)}, \dot{m}_i^{(\alpha)}\} \tag{23}$$
$$\alpha=0,1,\quad i=1\ldots n$$

and (internal and external) thermodynamic forces

$$X_k=\Big\{\left(\frac{1}{T^f}-\frac{1}{T}\right),\left(\frac{1}{T^a}-\frac{1}{T^f}\right),\left(\frac{1}{T^s}-\frac{1}{T^f}\right),\left(\frac{1}{T^a}-\frac{1}{T^s}\right),$$
$$(\mu_i^f-\mu_i^a),\left(\frac{1}{T^f}-\frac{1}{T^{(\alpha)}}\right),\left(\frac{\mu_i^{(\alpha)}}{T^{(\alpha)}}-\frac{\mu_i^f}{T^f}\right),\Big\} \tag{24}$$
$$\alpha=0,1,\quad i=1\ldots n.$$

THE CONSTITUTIVE EQUATIONS OF THE COLUMN

The forces and fluxes are not independent quantities, but rather depend on each other due to inequality (22), i.e. the Second Law. Indeed, one can formulate constitutive equations (C. Eq.) between fluxes and forces as

$$Y_k(t)=\mathfrak{F}_k\{X_l(s),-\infty<s\leq t,\ l=1\ldots\}, \tag{25}$$

where the symbol $\mathfrak{F}_k$ indicates either a function or a functional of its arguments which can be linear or nonlinear (4-6). We here restrict to mention two linear cases, namely

1. Classical thermodynamics of irreversible processes (TIP):
   $\mathfrak{F}_k$... linear function,

$$Y_k(t) = \sum_{\ell} L_{k\ell} \; X_\ell(t) \; ; \tag{26}$$

2. Extended thermodynamics of processes (4, 6)
   $\mathfrak{F}_k$... linear (passive) functional,

$$Y_k(t) = \sum_{\ell} \Big\{ A_{k\ell} \; \dot{X}_\ell(t) + B_{k\ell} \int_{-\infty}^{t} X_\ell(s) \, ds + \int_{-\infty}^{t} P_{k\ell}(s) \, X_\ell(t-s) \, ds + \int_{-\infty}^{t} [P_{k\ell}(0) - P_{k\ell}(s)] \ddot{X}_\ell(t-s) ds \Big\}. \tag{27}$$

The quantities $L_{k\ell}$ in (26) and $A_{k\ell} \ldots P_{k\ell}(s)$ in (27) are phenomenological parameters which are characteristic for the system, i.e. the column considered and which have to be determined experimentally. As far as nonlinear C.Eqs. are concerned we refer to the literature (3-6).

Concerning the external exchange of mass and energy we always assume all parameters of the incoming mass flow (index: 0) to be known. The thermodynamic parameters of the outgoing mass flow (index: 1) on principle can be calculated by an iterative procedure from the equations of state (Eq.s.)

caloric Eq.s. : $$h^{(1)} = h^{(1)}(T^{(1)}, p^{(1)}, \gamma_i^{(1)}, i = 1 \ldots n), \tag{28}$$

thermal Eq.s. : $$\rho^{(1)} = \rho^{(1)}(\text{———} \, '' \, \text{———}), \tag{29}$$

chemical Eq.s.: $$\mu_i^{(1)} = \mu_i^{(1)}(\text{———} \, '' \, \text{———}), \tag{30}$$

and the C.Eqs. of the outgoing flow

$$\dot{m}_i^{(1)} = \mathfrak{F}_i^{(1)} \Big\{ \Big(\frac{1}{T^f} - \frac{1}{T^{(1)}}\Big), \Big(\frac{\mu_k^{(1)}}{T^{(1)}} - \frac{\mu_k^f}{T^f}\Big), k = 1 \ldots \Big\}, \tag{31}$$

$$h^{(1)} \dot{m}^{(1)} = \mathfrak{F}^{(1)} \{ \text{———} \, '' \, \text{———} \}, \tag{32}$$

if all thermodynamic parameters of the fluid phase (f), among them $P^{(f)}$, $P^{(1)}$, T^R, μ_i^f are known.

These parameters can be calculated as follows. Assume the state of the column, i.e. all thermodynamic parameters of all phases (f, a, s) to be given at any (initial) time t. Then, from the thermostatic equations of state of the single phases included in equations (1-3), and properly chosen C.Eqs. (25) inserted in the balance equations (4-9), the extensive quantities $m_i^f \ldots U^s$ of phases f, a, s at time (t + Δ t) and from them the respective intensive parameters μ_i^f, $T^f \ldots T^s$ at the same time can be calculated. This determines the thermodynamic state of the column at (t + Δ t).

Space limitations do not allow us to give more details here. However, we will give a numerical example of the calculation procedure of adsorption processes mentioned above in a future publication. There, also our method will be compared with common phenomenological procedures (7-9).

ACKNOWLEDGEMENTS

The author is grateful to H. Knapp, TU Berlin and M. Streich, Linde AG, Munich for discussions.

LITERATUR CITED

1. Schottky, W., H. Ulrich and C. Wagner, Thermodynamik, reprint of the edition 1929, Springer, Berlin-New York (1973).

2. Myers, A.L., and J.M. Prausnitz, "Thermodynamics of Mixed-Gas Adsorption", AIChE Journal, 11 ( ), 121 (1965).

3. Keller, Jürgen U., Thermodynamik der irreversiblen Prozesse, Teil 1, Thermostatik und Grundbegriffe, de Gruyter, Berlin-New York (1977).

4. Keller, Jürgen U., Über den Zweiten Hauptsatz der Thermodynamik irreversibler Prozesse, de Gruyter, Berlin-New York (1976).

5. Keller, Jürgen U., "The Fundamental Inequality for Thermodynamic Systems with Heat and Mass Transfer", J. Non-Equilib. Thermodyn., 1 (1), 67 (1976).

6. Keller, Jürgen U., Int. J. Engng. Science, 17 (6), 715 (1979).

7. Jüntgen, Harald, "Grundlagen der Adsorption", VDI-report No. 253, VDI-Verlag, Düsseldorf (1876).

8. Kast, Werner, Chem.-Ing.-Tech., 53 (3), 160 (1981).

9. Mersmann, A., U. Münstermann and J. Schadl, Trennen von Gasgemischen durch Adsorption, Chem.-Ing.Tech., 55 (6), 446 (1983).

WATER-SORPTION ISOTHERMS OF BIOPOLYMERS: HYDRATE H_2O AND LIQUID-LIKE H_2O

Hubertus Kleeberg and Werner A.P.Luck
Dept. Phys. Chem., Philipps University, D-3550 Marburg

ABSTRACT

H_2O Sorption Isotherms of biopolymers are similar in shape to typical adsorption Isotherms of solids. Both may be described with analogous mathematical formalism.

Infrared spectroscopic investigations of H_2O bands in biopolymers like collagen, polysaccharides etc. indicate that hydrate-H_2O - i.e. H_2O forming H-bonds with the polar groups of the polymer - is present below 50% relative humidity (r.h.). Above 50% r.h. liquid-like H_2O - i.e. H_2O with an average H-bond structure like water - is present in addition to hydrate-H_2O. There are some indications that a second type of H_2O structure exists between these both types mentioned (2,6,7).

The aim of our investigation is to gain information about the amount and strength of H_2O-polymer interactions. This question seems to be of significance in the processing of leather and cloth or the storage conditions of food stuff for example.

In many cases the properties of polymers change at about 50% r.h., thus reflecting the absence/presence of liquid-like H_2O. The strength of hydrate-H_2O-polymer interactions may be of importance in the development of desalination membranes.

RESULTS AND DISCUSSION

In Figure 1 our results for H_2O-collagenous protein are summarized. In Figure 1c the desorption isotherms of gelatin and collagen (tendon) are shown. The observed shape is typical for biopolymers and several synthetic polymers.

The H-bond state may be investigated by infrared spectroscopy using the H_2O combination band at about 1900 nm. In this spectral region polymer absorption hardly disturbs the H_2O spectrum. Even the corresponding combination band of alcohol OH-groups is well separated from the H_2O band because of the different frequency of the C-O-H deformation vibration.

In Figure 1b the position of the maximum of the H_2O-band (λ_{max}) at different relative humidity (r.h.) is given. Note that λ_{max} changes until 50% r.h. and remains constant below 50% r.h. In the latter region the molecular monolayer H_2O-polar groups of the protein is desorbed. At 50% r.h. one H_2O molecule is present per polar group of the protein. The H-bond strength, which is linearly correlated to the position of the H_2O band (1-4), corresponds to that between H_2O and amides.

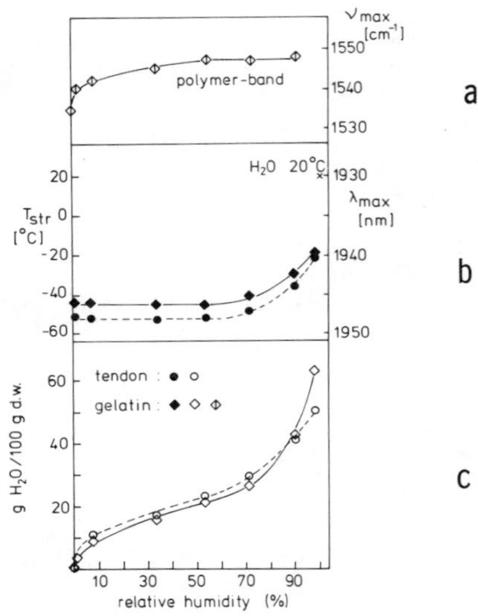

Figure 1. c: Desorption Isotherms of collagen (tendon) and gelatin.
b: Position of the H_2O combination band (right scale) at different relative humidities. The structure temperature (T_{str}) (left scale) indicates the extrapolated temperature which liquid water should have to produce a similar spectrum like the sample.
a: Position of the amide II band of gelatin at different relative humidities.

At higher r.h. liquid-like H_2O is present in addition to hydrate-H_2O. Above 50% r.h. an overlapping of the band of hydrate H_2O and of liquid like H_2O is observed.

In Figure 1a the frequency of an IR-band (amide II) of the protein is given. In contrast to the H_2O-band, the frequency of amide II changes at low H_2O content and asymptotically approaches its limiting value with the presence of all hydrate-H_2O. Above 50% r.h. the position of the polymer band is unchanged. These results indicate the hydrate-H_2O is of considerable importance for the state of the protein; its desorption seems to cause a change in the protein structure.

Especially at low H_2O content (below 25% hydrate-H_2O or approx. one H_2O/tripeptide) the amide II band is shifted.

For several other systems including proteins, polysaccharides (5,6), desalination membranes (3,4), biological tissues like cornea, cartilage or leaves (2,7,8) the results may be interpreted in an analogous manner.

As an example Figure 2 shows the results for three glycosaminoglycans (2,6) at different r.h. λ_{max} of hydrate-H_2O is found at shorter wave length (i.e. weaker H-bonds) with increasing degree of sulphonation. The H_2O band position in hyaluronate is very similar to that in polysaccharides containing OH and C-O-C as polar groups only as well as to the band found in alcohol-mixtures dilute in H_2O (1,2,5,9); the H_2O-uptake at 50% r.h. increases with the degree of sulphonation.

The change in the position of the polymer band (see Figure 2a) indicates that the state of the polymer changes below 50% r.h.

These results indicate that mainly OH-groups of hyaluronate act as H-bond acceptors of hydrate-H_2O; in comparison to this the sulphate groups of chondroitinsulphate and heparin are weaker H-bond acceptors (2,5,6), although more H_2O is attached to them, since more lone electron pairs are present in $-OSO_3^-$ than in -OH.

Investigations of solutions of these and other polysaccharides indicate (2,5,6) that the hydrate-H_2O which is present at 50% r.h. forms H-bonds with the polymers in moderately dilute solutions to the same extent and with comparable H-bond energy.

Our results of these systems support the presence of (mainly; see (2,7,8)) two different states of H_2O which are discussed under "bound" and "free" water in biology and medicine for example. To us it seems that the terms "hydrate-H_2O" and "liquid-like H_2O" are more precise and less confusing (particularly for "non-specialists").

For several biochemical processes the excistence of liquid-like H_2O is a necessary prerequisite (6,8,10).

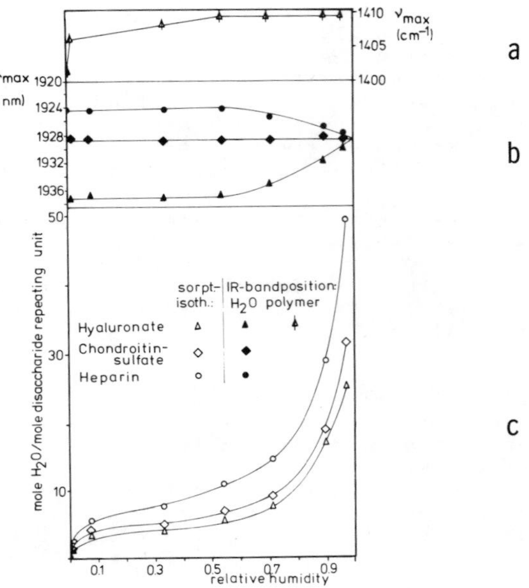

Figure 2. Comparison between desorption Isotherms (c), the position of the H_2O combination band (b) and a polymer band (a) of different glycosamino-glycans at different relative humidities (T=25°C)

To illustrate the order of magnitude of the difference between the spectra of water and aqueous systems we use the structure temperature T_{str}, i.e. T_{str} is that temperature at which pure water produces a similar spectrum (with the same λ_{max} for example) as H_2O in the system under investigation. In the case of collagenous protein hydrate-H_2O would correspond to pure water at (an extrapolated temperature of) -50°C (see Figure 1).

The shifts of the H_2O I.R. bands are relatively small (see Figure 1 and 2). But the heuristic T_{str} scale helps to understand the big effects of small changes in the H-bond structure of H_2O on biochemical reactions.

The final clarification of the mathematical description of H_2O sorption in polymers as well as on surfaces of solids (11,12) is yet to come. Although the number of equations developed (10-12) is very large, none describes the sorption process satisfactorily in the total humidity region (i.e. between 0 and 100% r.h.).

Thus the reasons causing hysteresis are still unclear and up for discussion. In many cases the difference in H_2O uptake between desorption and adsorption of polymers is as large as 30% (at about 50% r.h.). It thus seems to us, that at last one of the reasons for the hysteresis effects is due to the H_2O-dependent state of the polymer.

CONCLUSIONS

During desorption, the H_2O which is evaporated from the polymer sample has an H-bond structure very similar (or identical) to liquid water until approx. 50% r.h. is reached. The polymer structure usually seems to be similar to that in aqueous solutions.

During desorption of hydrate-H_2O below 50% r.h. the polymer structure changes in the cases investigated. Thus infrared spectroscopic investigations at about 50% r.h. are an easy tool to gain information about polymer hydration in solution. The changes of the polymer structure (mainly below 50% r.h.) may be an explanation for the hysteresis between desorption and adsorption.

Since microorganisms grow only if at least small amounts of liquid-like H_2O are present, the storage of food stuff at about 50% r.h. seems to be favourable.

The examples of hydrate and liquid-like H_2O mentioned here are similar in many respects to those discussed in biology or medicine where such words as "bound" and "free" water are often used. Our results indicate that the nomenclature "hydrate-H_2O" and "liquid-like H_2O" should be preferred.

LITERATURE CITED

1. Kleeberg, H., Koçak, Ö. and Luck, W.A.P., J.Solution Chem., 11, 611 (1982).

2. Kleeberg, H., Thesis, Philipps Universität, Marburg (1983).

3. Luck, W.A.P., in: Water in Polymers (S.P.Rowland, ed.) p. 43, ACS Symposium Series, Amer. Chem. Soc., Washington (1980)

4. Luck, W.A.P., Progr. Colloid & Polymer Sci., 65, 6 (1978).

5. Kleeberg, H. and Luck, W.A.P., in: Glycoconjugates (R.Schauer et al., eds.) p.98, G.Thieme, Stuttgart, (1979).

6. Kleeberg, H. and Luck, W.A.P., Proceedings of Symp. "Water and Ions", Bukarest Academic Press, (1982) (accepted for publ.).

7. Luck, W.A.P. and Kleeberg, H., in Photosynthetic Oxygen Evolution (H.Metzner, ed.) P.1, Academic Press, London (1978).

8. Luck, W.A.P. and Kleeberg, H., in: Water Activity: Influences on Food Quality (L.B.Rockland, G.F.Stewart, eds.) p.421 Academic Press, New York (1981).

9. Luck, W.A.P. and Ditter, W., J. Phys. Chem., 74, 3687 (1970).

10. Tome, D. et Bizot, H., Les Aliments a Humidité Intermediaire Serie Sytheses Bibliographiques no. 16, APRIA, Paris (1978).

11. Tagung der deutschen Kolloid-Gesellschaft, Bochum (1980).

12. Berg van den C. and Bruin, S., in loc. cit. 8, p.1.

ANALYSIS OF COMPLEMENTARY PRESSURE SWING ADSORPTION

Kent S. Knaebel
Department of Chemical Engineering
The Ohio State University
Columbus, Ohio

ABSTRACT

Difficult separations, which are characterized by narrowly spaced isotherms, are relatively inefficient in pressure swing adsorption applications because a substantial portion of the purified product is recycled for pressurization and/or purging the column. In addition, a relatively large portion of the feed may be exhausted. To overcome these disadvantages, a process configuration that involves parallel columns containing adsorbents having complementary selectivities for the components is developed, in which the purged product (waste) of one column is supplied as the purging agent for the other column. Results of mathematical analysis of the process are given for a range of selectivities and pressure ratios, based on an equimolar binary feed. The results indicate the purities and flow rates of the products.

INTRODUCTION

Since the mid-1960's several process configurations of pressure swing adsorption have been developed. Beginning with the work of Skarstrom (1), and progressing through those reviewed by Lee and Stahl (2), then Keller and Jones (3), researchers have aimed at better ways to exploit both the adsorbent and operating conditions in order to recover or enrich a portion of the feed. In many cases there are large differences between values of the components in a mixture. For example, if the feed were inexpensive but contained a small percentage of a valuable component, then the desired recovery might be relatively low. Conversely, if the feed were costly and contained a large amount of a valuable component, then the recovery could be high but the exhaust gas might simply be discarded.

The present work is oriented toward a slightly different goal. The basis of the concept is that both components of a binary feed are valuable, and that material and energy conservation may be possible when the desired products are not required to be very pure. Such a situation might arise in removing helium from natural gas, or both oxygen and nitrogen from air.

A significant restriction of the proposed process, however, is that adsorbents exist that are *complementary* for the binary mixture, i.e., they must possess inverse selectivity for the components. The likelihood of this condition being met, however, increases as does the molecular similarity of the components. For example, several synthetic zeolites exist that have greater selectivity for nitrogen than oxygen. Conversely, some activated carbons exhibit greater selectivity for oxygen. It is envisaged that isomers and homologs also may have such properties in selected adsorbents. Note that this selectivity difference may be based on kinetic rather than equilibrium behavior, although the analysis presented below is based upon equilibrium behavior. Taking for granted the existance of such adsorbents, a process scheme is suggested which conserves the purified product by employing the enriched purge product material as the feed during pressurization and purging of the adjacent column. This approach is contrary to normal practice in heatless adsorption and apparently all other multibed processes.

BASIC THEORY

This section describes, in abbreviated form, the mathematical nature of a single adsorbent bed within an arbitrary process. From this general material, performance equations that apply to the specific application are obtained, in the next section. The theory presented here is idealized to the extent that linear adsorption isotherms are assumed, as is ideal gas-phase behavior. All dissipative effects such as pressure gradients; film, intraparticle, and axial diffusion; and latent heat effects are ignored. It should be noted that general results that are essentially identical to these have been presented elsewhere (4,5, 6), and so the theory is merely outlined here.

The continuity equations for the binary mixture and for one component are

$$\varepsilon\left\{\frac{\partial p_A}{\partial t} + \frac{\partial u p_A}{\partial z}\right\} + RT\ (1-\varepsilon)\ \frac{\partial n_A}{\partial t} = 0 \tag{1}$$

$$\varepsilon\left\{\frac{\partial P}{\partial t} + \frac{\partial u P}{\partial z}\right\} + RT\ (1-\varepsilon)\ \frac{\partial n}{\partial t} = 0 \tag{2}$$

in which the subscript A denotes the component, which exhibits the partial pressure, p_A, and the capacity within the adsorbent, n_A. The interstitial velocity is u, while the total pressure is P. The equilibrium isotherms are expressed in a linear form, as follows

$$n = n_A + n_B = (k_A p_A + k_B p_B)/RT \tag{3}$$

It is helpful in the later development to identify the relative affinities of the components. Thus, component A is identified as being selectively adsorbed. Accordingly, for this binary mixture only the mole fraction of that component is used, and the subscript is omitted.

The parameters β_A, β_B and their ratio, β, are used to describe the intensive nature of the adsorbent bed.

$$\beta_A = \frac{1}{1 + (1-\varepsilon)k_A/\varepsilon} \tag{4}$$

$$\beta_B = \frac{1}{1 + (1-\varepsilon)k_B/\varepsilon} \tag{5}$$

$$\beta = \frac{\beta_A}{\beta_B} = \frac{\varepsilon + (1-\varepsilon)\ k_B}{\varepsilon + (1-\varepsilon)\ k_A} \tag{6}$$

According to these definitions and the previous statement concerning the adsorbent selectivity, the range of β is restricted, i.e., $0 \leq \beta < 1$.

The material balance equations yield expressions for the interstitial velocity, as follows for the conditions of

1. Varying pressure; u = 0 at z = 0

$$u = \frac{-z}{\beta_B[1+(\beta-1)y]}\ \frac{1}{P}\ \frac{dP}{dt} \tag{7}$$

2. Constant pressure

$$\frac{u_1}{u_2} = \frac{1+(\beta-1)y_2}{1+(\beta-1)y_1} \tag{8}$$

Replacing the original terms in the balance equations, yields a first order partial differential equation, which may be solved by the method of characteristics, with the following result

$$\frac{dz}{dt} = \frac{\beta_A u}{1+(\beta-1)y} \tag{9}$$

$$\frac{dy}{dP} = \frac{(\beta-1)(1-y)y}{[1+(\beta-1)y]P} \tag{10}$$

The former precribes characteristic directions in the z-t plane, while the latter describes composition shifts due to pressure changes along any characteristic. Upon rearrangement and integration with general initial conditions the following expressions are obtained

$$\frac{y}{y_0} = \left(\frac{1-y}{1-y_0}\right)^{\beta} \mathcal{P}^{\beta-1} \tag{11}$$

$$\frac{z}{z_0} = \left[\frac{y}{y_0}\right]^{\frac{\beta}{1-\beta}} \left[\frac{1-y_0}{1-y}\right]^{\frac{1}{1-\beta}} \left[\frac{1+(\beta-1)y}{1+(\beta-1)y_0}\right] \tag{12}$$

The initial condition y_0 depends on its initial position z_0, and pressure, P_0. The pressure ratio is defined as $\mathcal{P} = P/P_0$

Upon addition of material to the bed that is richer in component A than the bed contents, a shock-wave is formed. It propagates according to

$$u_s = \left.\frac{dz}{dt}\right|_s = \frac{\beta_A\ u_1}{1+(\beta-1)y_2} = \frac{\beta_A\ u_2}{1+(\beta-1)y_1} \tag{13}$$

And composition shifts that accompany pressure changes follow

$$y_2 = \frac{K\, y_1}{1+(K-1)y_1} \qquad (14)$$

$$K = \frac{y_{2_o}\,(1-y_{1_o})}{y_{1_o}\,(1-y_{2_o})} \qquad (15)$$

In eqs. (13)-(15) the subscript 1 refers to material at the leading edge of the shock-wave, while 2 represents the trailing edge.

THE OPERATING CYCLE

Four steps comprise a typical cycle: pressurization, high pressure feed, blowdown, and purge. Earlier work indicated that, in a more standard P.S.A. process, there is uniform incentive to pressurize the column with recycled product rather than fresh feed. Hence that approach is taken here, as well. In this process, however, the flowsheet is designed so that the adsorbent beds are coupled in a different manner than before. Although two column units are conceptually feasible, a four column process has the advantage of being analogous to a pair of dual column P.S.A. units. Therefore, this layout is described in Table 1 and Figure 1. Similarly, Figure 2 depicts a typical operating cycle.

Pressurization

The pressurization step for any cycle involves adding gas to the column from the bottom (cf. Figure 2), while it is closed at the opposite end. Since the feed and initial contents of the column have identical compositions, the characteristics described by eqs. (9) and (10) do not intersect or diverge. Hence, there is no shock-wave or simple-wave created.

The ultimate composition reached during pressurization by the initial column contents is given by eq. (11). During an arbitrary cycle, i, the pressurizing gas composition is y_I, so the ultimate composition, y_U, becomes

$$\left.\frac{y_U}{(1-y_U)^{\beta}}\right|_i = \frac{y_I}{(1-y_I)^{\beta}}\, P^{\beta-1} \qquad (16)$$

The penetration of the feed into the bed is determined from eq. (12), in which $y_o = y_F$ and $y = y_U$. Note that z is the distance from the closed end, so $z_o = 1$. From the penetration it is possible to assess the moles supplied to the column by a material balance. The result is simply

$$N_{PR} = \phi\, \frac{\beta\,(P-1)}{1+(\beta-1)y_I} \qquad (17)$$

where $\phi = \varepsilon A_{CS}\, P_L/(\beta_A\, RT)$

High-Pressure Feed

Following the pressurization step, the process feed is introduced to the column from the top and the product is withdrawn from the bottom (cf. Figure 2). Since the feed concentration of the preferentially adsorbed component is higher than (or equal to) that of the column contents, a shock-wave is formed. When the adsorbent is exploited to the maximum possible extent, the shock-wave propagates to the product end during this step.

As in the pressurization step, the overall results of the high-pressure feed step can be evaluated by material balances. In this case, the initial state is given by

$$dN = [\varepsilon + (1-\varepsilon)(k_A y_A + k_B y_B)] \frac{A_{cs} P}{RT} dz \qquad (18)$$

Rather than integrating numerically, it is more convenient to assess the value by material balances over the sequence of steps. The result is

$$\int_0^1 y dz = \frac{y_I}{P} [1 + \frac{\beta (P-1)}{1 + (\beta-1)y_I}] \qquad (19)$$

Conversely, the final state corresponds to the instant at which the shock-wave reaches the bottom of the column. Thus, the contents of the column are uniform at the composition of the feed. Therefore, the integrated form of eq. (18) becomes

$$N_F = \phi [\beta + (1-\beta)y_F] P \qquad (20)$$

In addition, the molar quantities supplied in the feed and removed in the product may be determined from the relation for the total number of moles supplied during the high-pressure feed step.

$$N_H = \phi [1 + (\beta-1)y_U] P \qquad (21)$$

The corresponding number of moles in the product may be determined from eqs. (18)-(21) through an overall material balance. An equivalent procedure may be used to determine the quantities of species A, i.e., the preferentially adsorbed component.

Blowdown

After the shock-wave has reached the bottom of the bed, the product flow is stopped and the column is depressurized through the top. As this occurs, the shock-wave disintegrates and a simple-wave is formed which is described by the diverging characteristics, as in eqs. (9-12). As the pressure declines the composition within the column becomes increasingly enriched in the selectively adsorbed component. The product from the blowdown step, accordingly, has a time-dependent composition. Typically, the analysis of the net change that occurs during this step is more readily obtained by evaluating the other steps and applying overall material balances. This will be expanded upon in the next section.

Purge

Finally, the column contents that are rich in the preferentially adsorbed component are exhausted by essentially recycling a portion of the purified product. Again, since the material being supplied is leaner than the effluent, a simple-wave is formed. The step is complete when the characteristic that defines the trailing edge of that wave reaches the top of the column. That result may be obtained from eq. (9). The purging material or low-pressure feed is assumed to have a fixed concentration, viz. y_I.

As indicated in Figure 1, the net changes during the blowdown and purge steps are combined. The initial state is identical to the final state of the high-pressure feed step, while the final state corresponds to the initial state of the pressurization step. The latter case is the result of sufficient purified product being recycled to fill the column. This step proceeds according to eqs. (8) and (9). The ultimate molar contents of the bed are

$$N_{PU} = \phi \; [\beta + (1-\beta) y_I] \tag{22}$$

As in the high-pressure feed step, the total number of moles supplied during the low-pressure purge step may be evaluated which may be combined with eq. (9) to give

$$N_L = \phi \; [1 + (\beta - 1) y_I] \tag{23}$$

Again, the quantity expelled during the blowdown and low-pressure purge steps may be found by an overall material balance. Similarly, the net flows of species A may be determined by the same procedure.

PROCESS PERFORMANCE

From the preceding development it is possible to extract relevant parameters for determining the performance of the complementary pressure swing adsorption process. With such information it is possible to make general comparisons between operating conditions and even for alternate process layouts.

There are two primary incentives for developing and applying this process: more complete exploitation of adsorbent and greater recovery or enrichment of the products. These general goals may be ascertained by simply evaluating the elementary material balance equations.

For example, since an ordinary P.S.A. process produces essentially pure high pressure product, it is relevant to determine the composition for this arrangement. In order to generalize the development, it is helpful to consider a symmetric system. The system should be symmetric so that a binary mixture, comprised of components 1 and 2 can be treated without confusion. This can be stated concisely by a number of conditions.

Condition 0: Component 1 = A in Column 1; Component 2 = A in Column 2

Condition 1: β (Column 1) = β (Column 2)

Condition 2: The low pressure product of Column I-1 is the purging agent for Column I-2. Similarly for Columns I-2 to I-1, and II-1 to II-2 and II-2 to II-1. (See Figure 1.)

Condition 3: The blowdown product of Column I-1 is the pressurizing agent for Column II-2. Similarly for Columns I-2 to II-1, and II-1 to I-2 and II-2 to I-1.

This condition can only be actually attained if an intercolumn compressor is used because the pressure at the source declines as the pressure at the destination increases. Ultimately an intermediate pressure would be reached that would be unsatisfactory for both columns.

From the previous equations, one gets

$$\bar{y}_{HP} = \frac{\frac{y_I}{P}\left\{1+\frac{\beta(P-1)}{1+(\beta-1)y_I}\right\}+(\beta-1)y_u y_F}{1+(\beta-1)\left(y_u+y_F-\frac{y_I}{P}\left\{1+\frac{\beta(P-1)}{1+(\beta-1)y_I}\right\}\right)} \tag{24}$$

While the number of moles of this net product reduces to

$$N_{HP} = \phi P\left[1+(\beta-1)\left(y_u+y_F-\frac{y_I}{P}\left\{1+\frac{\beta(P-1)}{1+(\beta-1)y_I}\right\}\right)\right] \tag{25}$$

Similarly, if the net product is combined with the excess "waste" from the alternate column then the economy of operation may improve due to the increased quantity. Nevertheless, purity of this total product is less than that of the net product. It is found that for equimolar mixtures

$$y_{TP} = y_F + \left\{1+\frac{N_{PR}+N_L-N_{HP}}{N_H}\right\}(1-2y_I) \tag{26}$$

Finally, the number of moles is given by eq. (21).

RESULTS AND DISCUSSION

The preceding equations have been solved based on parameters that apply to the purification of an equimolar binary mixture. The performance of the process is determined in terms of the purities of the high-pressure product and the total product (as given by eqs. (24) and (26), shown in Figures 3 and 4. Since the feed is assumed to be equimolar, both components have identical behavior, so only one is shown.

The ratio of the number of moles in the net (high pressure) product to that in the feed is shown in Figure 5. Note that, since the feed is equimolar and the columns have inverse selectivities, the number of moles in the feed and that of the total product are identical.

The results shown in Figures 3 and 4 clearly indicate that a substantial amount of enrichment is possible in the complementary pressure swing adsorption process. The quality of the products increases uniformly as the operating pressure ratio increases. The quality also increases in proportion to the difference between the slopes of the isotherms (i.e., as β decreases). There is a trade-off, however, between the purities of the products and their flow rates, as shown in Figure 5. As a result, if the products need not be very pure, a large throughput can be attained with only a moderately selective pair of adsorbents.

In conclusion, the complementary pressure swing adsorption process is predicted to serve the needs of rough separations (e.g., for products having about 95% or less of the desired component). Advantages of the process are that essentially no waste is generated, and that both components in a binary mixture may be recovered at relatively high concentrations.

NOMENCLATURE

A_{CS}	column cross sectional area
k_i	equilibrium distribution coefficient for species i
n_i	moles of species i adsorbed per unit column volume
p_i	partial pressure of species i
P	total pressure
$\mathcal{P}$	P_H/P_L
t	time
u	interstitial gas velocity
z	axial distance

Greek Letters

β	β_A/β_B
βi	$\varepsilon/[\varepsilon+(1-\varepsilon)k_i]$
ε	bed void fraction
ϕ	$\varepsilon A_{CS}\ P_L/\beta_A\ RT$

Subscripts

A,B	species A or B
F	feed step
H	high pressure feed
HP	high pressure product
I	input of recycled material
L	low pressure feed
PR	pressurization step
PU	purge step
S	shock wave
U	ultimate value reached (during pressurization)
W	low pressure (exhaust) product
0	initial or feed condition
1,2	property or condition before and after shock waves

LITERATURE CITED

1. Skarstrom, C.W., "Heatless Fractionation of Gases over Solid Adsorbents," in Rec. Dev. in Sep. Sci., N.N. Li, Ed., 2, CRC Press, 1972.
2. Lee, H.; Stahl, D.E., AIChE Symp. Ser., 1973, 69 (134) 1.
3. Keller, G.E.; Jones, R.L., ACS Symp. Ser.,
4. Flores-Fernandez, G.; Kenney, C.N., Chem. Eng. Sci., 1983, 38, 827.
5. Chan, Y.N.I.; Hill, F.B.; Wong, Y.W., Chem. Eng. Sci., 1981, 36, 243.
6. Knaebel, K.S.; Hill, F.B., Chem. Eng. Sci. (submitted 1983).

Table 1.

Complementary Pressure Swing Adsorption Operating Sequence

STEP	COLUMN I-1		COLUMN I-2	
	Influent	Effluent	Influent	Effluent
1. High Pressure Feed	Feed	B	Feed	A
2. Blowdown	-	A to II-2	-	B to II-1
3. Purge	B from I-2	A to I-2	A from I-1	B to I-2
4. Pressurization	B from II-2	-	A from II-1	-

Notes: A and B indicate enriched but not pure components.

The corresponding steps for Columns II-1 and 2 may be determined by replacing I with II and vice-versa.

Column pairs I and II operate 180° out-of-phase; refer to Figures 1 and 2 for process configuration.

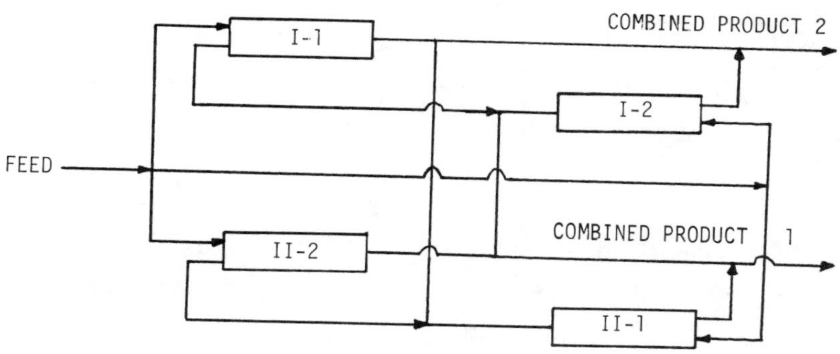

Figure 1. Schematic diagram of a four-column complementary pressure-swing adsorption process. Columns I-1 and II-1 are selective for component 1. Columns I-2 and II-2 are selective for component 2. Streams for which arrows are omitted have flow in alternating directions.

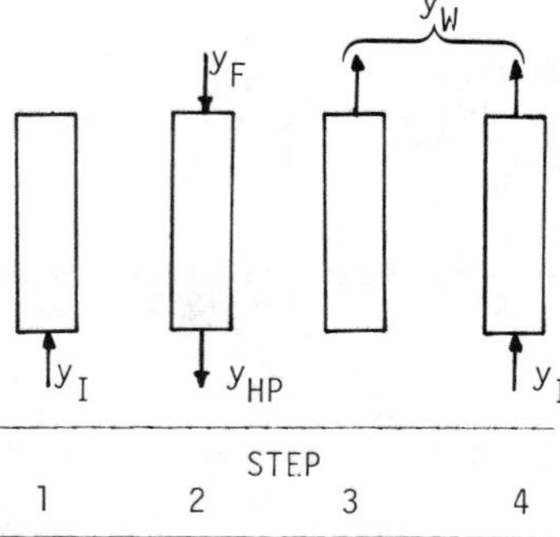

Figure 2. The four steps in a typical PSA cycle. (1) Pressurization, (2) High-pressure feed, (3) Blowdown, (4) Purge.

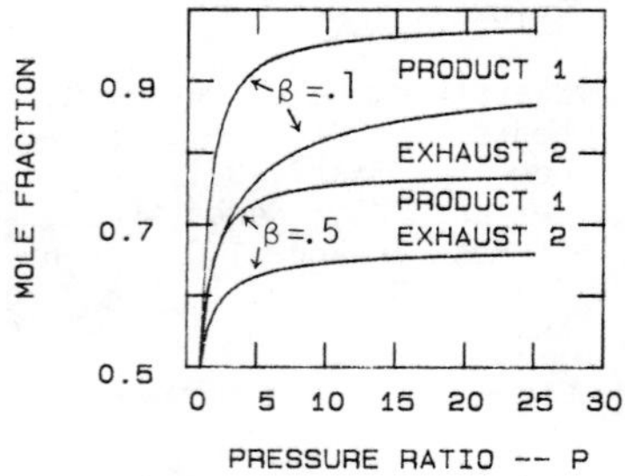

Figure 3. Product compositions of component 2. The high pressure product of column 1 and the exhaust of column 2 are shown.

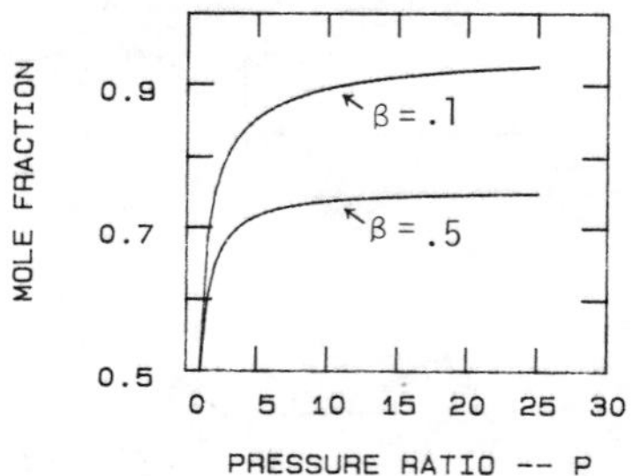

Figure 4. Total product mole fractions of component 2. These represent the combined product of column 1 and the exhaust of column 2.

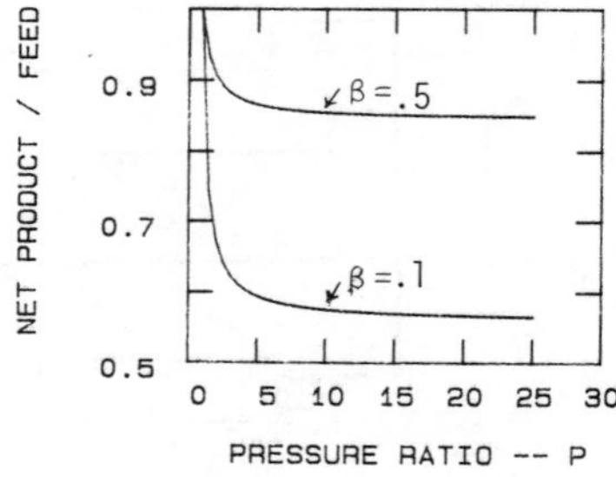

Figure 5. Ratio of the total number of moles in the high pressure product to that of the feed, both to column 1.

ADSORPTION OF LOW MOLECULAR WEIGHT ORGANIC COMPOUNDS FROM AQUEOUS SOLUTIONS ON CHARGED SORBENTS

Luuk K. Koopal

Laboratory for Physical and Colloid Chemistry, Agricultural University, De Dreijen 6, 6703 BC Wageningen, The Netherlands.

ABSTRACT

The adsorption of a weak electrolyte onto a charged surface is described using the regular solution and adsorbed phase model including electrostatic interactions. Comparison of theoretical predictions with experimental results shows the importance of lateral interactions and counter ion association.

\- - - - - -

A model for the adsorption of *weak* organic electrolytes is derived which can be seen as an extension of earlier models (1 to 7). New aspects are that (1) 'chemical' and electrostatic lateral interactions between the adsorbate molecules are taken into account and (2) the association of the organic ion in the adsorbed state with its counter ion is allowed to be different from that in the bulk solution.

ISOTHERM MODELS

Neutral molecules

The adsorption of a neutral organic compound from an aqueous solution is essentially an exchange reaction. In the case of molecules of similar size the equilibrium constant, K^*, of the exchange of water with the organic component, A, can be written as (8)

$$K^* = \frac{\gamma_W^l \gamma_A^\sigma}{\gamma_A^l \gamma_W^\sigma} \frac{1-x}{x} \frac{\theta}{1-\theta} \tag{1}$$

where γ's are the activity coefficients and x and θ are the mole fractions of A in the solution (l) and adsorbed phase (σ) respectively. Substitution of the expressions for γ_i^l and γ_i^σ (8) in (1) leads to

$$\frac{\theta}{1-\theta} = K^* \frac{x}{1-x} \exp\{q^l(1-2x) - q^\sigma(1-2\theta)\}/RT \tag{2}$$

where q^l and q^σ are composite interaction parameters. As the organic component is considered to be uncharged q^l and q^σ contain only 'chemical' energies such as dispersion and hydrophobic interactions. R is the gasconstant and T the absolute temperature. Frequently adsorption occurs from dilute solutions where x << 1 so that (2) reduces to

$$\theta/(1-\theta) = K\,c\,\exp(-b\theta) \tag{3}$$

where c is the concentration of the organic compound, $b = -2q^{\sigma}/RT$ a lateral interaction parameter and $K = \{K^{*}\exp(q^{1}-q^{\sigma})\}/55{,}5$ a measure of the affinity. A negative value for b indicates lateral attraction, a positive value repulsion. The equilibrium constant K^{*} is related to the standard free energy of adsorption, Δg_a^{o}: $K^{*} = \exp(-\Delta g_a^{o}/RT)$. Δg_a^{o} is independent of θ.

For *charged* surfaces Δg_a^{o} depends on the electrical field strength and formally it can be separated into a 'chemical' and an 'electrical' or dipole contribution

$$\Delta g_a^{o} = \Delta g_{a,ch}^{o} + \Delta g_{a,dip}^{o} \tag{4}$$

$\Delta g_{a,dip}^{o}$ contains the obvious electrostatic contributions due to the energies of permanent or induced dipoles at the interface. In the case of aqueous solutions with water dipoles present it may be clear that $\Delta g_{a,dip}^{o}$ is constant only at constant field strength. In the next section, where an isotherm equation will be derived for organic *ions*, changes in $\Delta g_{a,dip}^{o}$ due to changes in the dipole interactions will be neglected, they are considered as a second order effect in comparison with charge-charge interactions.

Charged molecules

Equation (3) can be extended to *specific* adsorption of charged adsorbates by including coulombic interactions. With specific adsorption we indicate (1) that chemical interactions also contribute to Δg_a and (2) that the adsorption of the organic ions in the diffuse part of the electrical double layer is not included. To maintain electrical neutrality it is assumed that the interfacial charge resulting from the adsorption of the organic ions is compensated for in the diffuse part of the double layer. Under these conditions the isotherm equation can be derived as before provided that the electrical work to transport the ionic charges from bulk solution to their position at the interface is included in Δg_a. If the potential difference between the locus of adsorption and the bulk solution is given by ψ_A, then the coulombic part of the adsorption free energy can be defined as $-\, z_A F \psi_A$ where z_A is the valency of the organic ion and F the Faraday. Incorporation of this term in (3) leads to

$$\frac{\theta}{1-\theta} = K\,c\,\exp(-b\theta)\,\exp(-z_A Y_A) \tag{5}$$

where $Y_A = \psi_A F/RT$. As explained above K is only constant if changes in dipolar interactions with θ are negligible.

In order to use (5) ψ_A has to be evaluated. ψ_A will be considered as an average, smeared-out potential. ψ_A is different from the surface potential because the centres of charge are a distance d removed from the surface due to the size of the organic ions. The magnitude of ψ_A depends in general on the surface charge density, σ_S, and on the charge density, $z_AFN_A\theta$, due to adsorbed organic material. N_A is the maximum adsorption of the organic component in moles per unit area. If there are no other specifically adsorbing ions and when the centres of charge of the adsorbed ions are located at the Stern-plane, then Y_A can be calculated from σ_S and $z_AFN_A\theta$ using the Gouy-Chapman theory:

$$z_iY_A = \ln \{Q + (Q^2 + 1)^{0.5}\}^2 \tag{6}$$

where

$$Q = (z_S\sigma_S + z_AFN_A\theta) \;/\; (8\varepsilon\varepsilon_o RTI)^{0.5} \tag{7}$$

where ε is the relative static permittivity and ε_o the permittivity in vacuum, z_S indicates the sign of σ_S and I is the ionic strength. Equation (6) applies for symmetrical indifferent electrolytes with valency z_i and flat double layers. Substitution of (6) in (5) gives the expression for the adsorption isotherm. For low potentials (6) simplifies considerably in which case z_iY_A equals 2Q.

In the above derivation it has been tacitly assumed that I is dominated by the presence of an indifferent electrolyte. In this situation the ionic activity coefficient for the organic ion is constant and incorporated in K. In the surface region the coulombic term in the isotherm equation serves as an activity correction. A second advantage of the presence of the indifferent electrolyte is that the adsorption of the organic component in the diffuse part of the double layer can be safely neglected.

Weak organic electrolytes (ion association)

In practice the dissociation of organic electrolytes is often incomplete and both neutral and ionic components are present in solution. The relative amount of each component is given by the degree of dissociation, α, which in turn depends on the dissociation constant K_d and the concentration, c_M, of the associating inorganic ion. By virtue of having two organic components present, competition for the adsorption sites has to be taken into account.
Assuming that K, b and N_A are not affected by the dissociation, the Following overall adsorption equation can easily be derived

$$\frac{\theta}{1-\theta} = K\,c\,(1-\alpha)\exp(-b\theta) + K\,c\,\alpha\exp(-b\theta - z_AY_A) \tag{8}$$

The first term on the RHS of (8) accounts for the adsorption of the neutral, the second for that of the charged component. As before Y_A can be obtained with (6). However, the relation for Q has to be modified because Y_A is affected only by the charged fraction of molecules adsorbed. Therefore

$$Q = (z_S\sigma_S + z_A F\alpha^{\sigma} N_A\theta) \;/ (8\varepsilon\varepsilon_o RTI)^{0.5} \tag{9}$$

where

$$\alpha^{\sigma} = K_d/(K_d + c_M^{\sigma}) \quad \text{with} \quad c_M^{\sigma} = c_M \exp(-z_M Y_A) \tag{10}$$

α^{σ} is the degree of dissociation of the organic ion in its adsorbed state. It differs from α because the concentration of associating inorganic ions at distance d from the wall, c_M^{σ}, is different from c_M. Substitution of (10) in (9) and the resulting equation in (8) gives the full expression for the overall adsorption of a weak organic electrolyte.

At c_M much larges than K_d α is small and few organic ions are present. However, when the organic ion and the surface are oppositely charged α^{σ} may still be close to 1 and K $\exp(-z_A Y_A)$, the *apparent affinity*, could be much larger than K. Therefore, the charged component may still contribute signifantly to the overall adsorption. For small α and with equal signs for z_A and z_S α^{σ} is smaller than α and the second term on the RHS of (8) can be neglected.

SENSITIVITY OF THE PARAMETERS

To investigate the effect of different parameter values on the adsorption behaviour some calculations have been made for a hypothetical surfactant molecule absorbing at say the air/water interface. Results are shown in Figure 1, where the adsorption is plotted versus the logarithm of the surfactant concentration. K = 10^{+4} M^{-1} indicating a strong preferential adsorption of the surfactant, I = 10^{-3} M. The dashed curves represent the case in which no ion association occurs (pK_d = -4). With b = 0 and c_A = 10^{-2} M, which is close to the critical micelle concentration (CMC) of sodium dodecyl-sulfaat (SDS), it follows that θ equals only about 0.2. A drastic increase in the lateral attraction (b = -4) gives θ = 0.3 at 10^{-2} M. This is still a rather low value, whereas in the absence of lateral electrostatic repulsion b = -4 would lead to surface condensation. In reality surfactants like SDS strongly adsorb at the air/water interface before the CMC is reached. Apparently the electrostatic repulsion is much smaller than accounted for in the dashed curves. We therefore introduced pK_d = 0, which allows ion association to take place in the vicinity of the surface. In the solution the surfactant is fully dissociated. Isotherms are shown in Figure 1, the drawn curves give results for several values of b. We see that ion association strongly affects the adsorption. At b = -4 part of the isotherm is nearly vertical. Increasing the attraction by 1RT leads to surface condensation (hemi-micelle formation) between θ equals 0.3 to 0.7.

The plots of α^{σ} and Y_A against θ in Figure 2 show a strong decrease in α^{σ} with θ and that Y_A reaches about a constant value for θ > 0.2. The latter effect is caused by the fact that the adsorbed charge remains nearly constant with increasing θ because of the decrease in α^{σ}. When no ion association occurs Y_A also levels off, but at much higher value. At high surface charges the Gouy-Chapman theory also leads to effective screening of the charge.

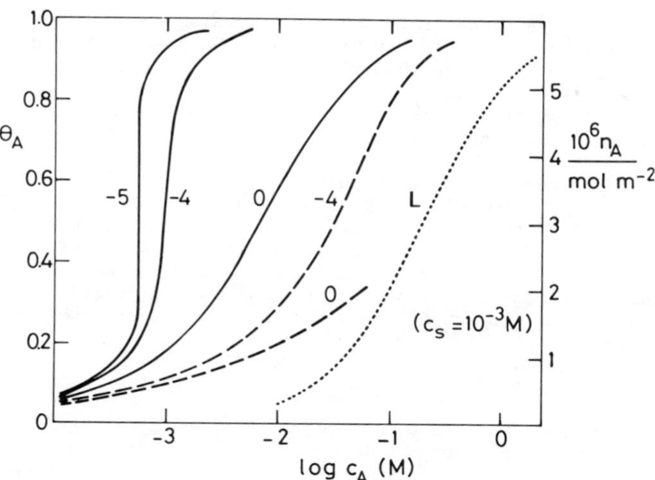

Figure 1. Adsorption of a charged surfactant on an originally uncharged surface. The lateral interaction parameter b is indicated. Dashed curves: complete dissociation in the surface region; full curves: association of the surfactant with counter ions ($pK_d = 0$). The dotted curve (L) shows a Langmuir type isotherm.

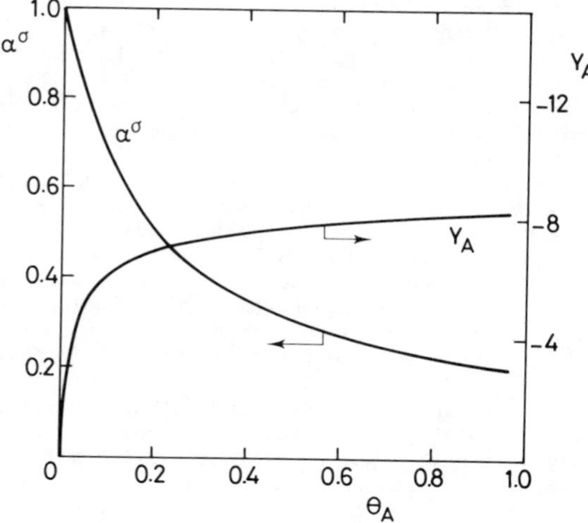

Figure 2. The degree of dissociation, α^σ, of the adsorbed molecules and the reduced potential, Y_A, as a function of the surface coverage. The curves correspond to the full curves in Figure 1.

In Figure 1 a Langmuir isotherm (L) is also shown. Comparing this isotherm with the dashed one at b = -4 and the drawn curve at b = 0 shows that for $\theta > 0.1$ the shapes are similar, indicating nearly *pseudo-ideal* behaviour, i.e. the lateral attractions and repulsions mutually compensate. In practice such *pseudo-ideal* behaviour may lead to the erroneous conclusion that lateral interactions are absent.

Two parameters not illustrated are K and I. Different values of K lead to an approximately parallel shift in the isotherms. A factor of 10 in K shifts the isotherms by about one unit along the log c axis. I is especially effective in suppresssing the electrostatic repulsion in the case of fully dissociated molecules. For less well dissociated molecules the effect is smaller because at low indifferent electrolyte concentration the ion association, which is dependent on c_M, is relatively large.

COMPARISON WITH EXPERIMENTAL RESULTS

A preliminary version of the model in which no lateral chemical interactions were considered was successful in predicting the adsorption behaviour of paranitrophenol (PNP) on an activated carbon (7). In this paper results will be given on the adsorption of (1) PNP on Sterling MT, a non porous carbon black, and (2) dodecyl trimethyl ammonium bromide (DTAB) on a non porous pyrogenic silica, Cab-O-Sil M5.

PNP/Sterling MT. The adsorption of PNP on carbon MT was measured by the depletion method at differnt pH values and salt concentrations. The surface charge of Sterling MT as a function of pH is obtained by potentiometric titration of the carbon. The experimental methods used have been described in more detail elsewhere (7). Adsorption isotherms are plotted in Figures 3 and 4. Sterling MT has a negative surface charge which is very low at pH = 3, in 10^{-3} M KNO_3 it raises to 2.5 $\mu C/cm^2$ at pH = 7 and 6 $\mu C/cm^2$ at pH = 10. In 10^{-1} M KNO_3 these figures are 4.5 and 7.5 $\mu C/cm^2$ respectively. The specific surface area, 10 m^2/g, was obtained from the N_2 adsorption isotherm. At pH = 3 the dissocoation of PNP (pK_d = 7.15) is negligibly small both in solution and near the surface. In these conditions (3) can be applied if surface heterogeneity is neglected. In doing so we found K = 510 M^{-1}, N_A = 3.2 $\mu mol/m^2$ and b = 0 i.e. pseudo-ideal behaviour. The drawn curve at pH = 3 (Figure 3) shows the calculated isotherm. The isotherms at pH = 7.2 and 9.8 where calculated using the same constants and the appropriate values of σ_S. Comparison of the calculated curves with the experimental points shows that at pH = 7.2 the agreement is quite good and that at pH = 9.8 the trend is correctly predicted.

The difference between experimental and calculated results at pH = 9.8 probably has to do with surface heterogeneity. The 10% of sites occupied may well have a somewhat higher affinity than the average value used. Another reason might be that at low salt concentrations and high surface charges the potential ψ_A is somewhat overestimated. Results at high salt concentration are in better agreement. At pH 3 and 7 the salt concentration has no or little effect respectively on the adsorbed amount. This is also found theoretically as can be seen in Figure 4 where the adsorption of PNP at con-

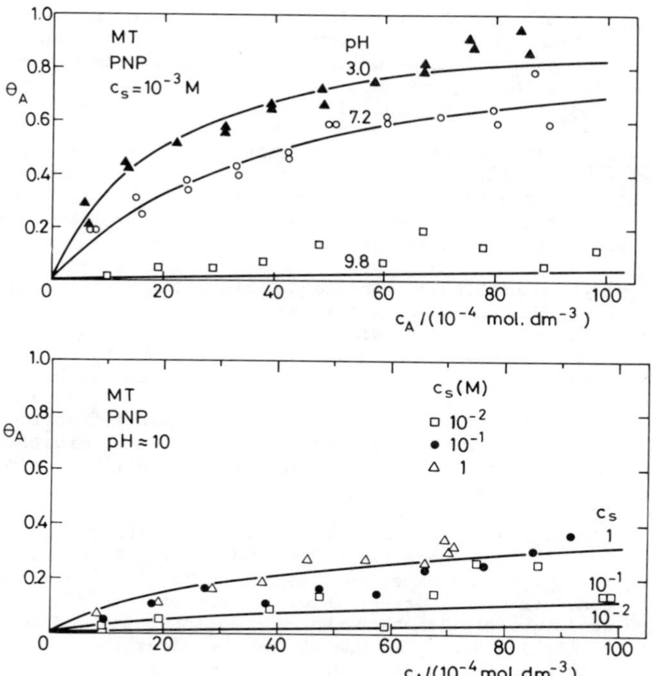

Figure 3. Adsorption of PNP on Sterling-MT carbon. c_s is the KCl concentration. Symbols: experimental results, curves: theoretical predictions (see text).

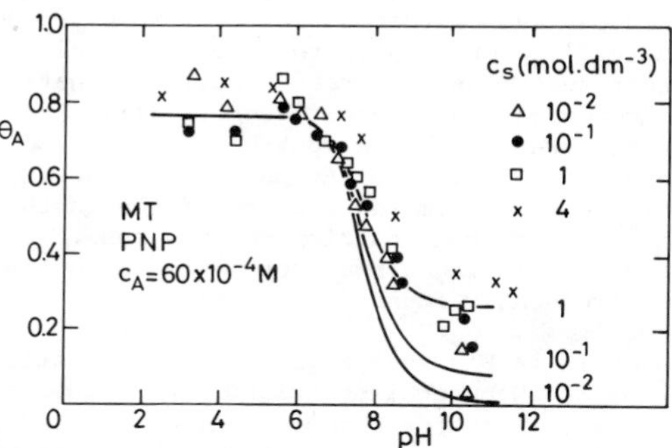

Figure 4. Adsorption of PNP on carbon MT as a function of pH at constant equilibrium concentration PNP and different KCL concentrations. Curves: theoretical predictions.

stant (equilibrium) concentration is shown a function of pH at various salt concentrations. Experimental and theoretical results agree well up to about pH = 8, and the pH at which the decrease in adsorption starts is well predicted. At high pH values and low salt concentrations the agreement is not so good as was discussed before.

DTAB/Cab-O-Sil. Adsorption data of DTAB on Cab-O-Sil at pH 8, 9 and 10 are taken from literature (9). The surface charge of Cab-O-Sil as a function of pH has been studied by Abendroth (10). The surface charge density at any pH value can also be obtained from the maximum amount of DTAB adsorbed in the presence of 1M butanol (9).

The adsorption of DTAB on uncharged Cab-O-Sil (pH < 8) is negligible and it may be concluded that K is rather small, we took $K = 10\,M^{-1}$. Using this value of K together with the other selected parameter values gives at the CMC (15 mM) an insignificant adsorption. N_A is inferred from the maximum adsorption at pH = 10, it amounts to 4.5 $\mu mol/m^2$. A further increase in pH no longer increased the maximum adsorption. DTAB is a long chain molecule and strong hydrophobic attraction between the adsorbed molecules may be expected. From studies on micellar systems it can be inferred that one CH_2-group could contribute 0.7 to 1.1 RT to the lateral attraction (11). In the present study a value of b = -6 is adopted, indicating strong lateral attraction. Values of σ_S at pH 8, 9 and 10, derived from the saturation adsorption values of DTAB in the presence of 1M butanol, were 2.9, 8.1 and 24 $\mu C/cm^2$ respectively. The ionic strenght in the experiment is not constant but given by the surfactant concentration. Correct incorporation of this changing concentration would greatly lengthen the calculations. Instead we assumed $I = 10^{-2}$M and $[Br^-] = 5.10^{-3}$ M. Due to these choices the selected value of $pK_d = 1.6$ is also fairly arbitrary. However, for a reasonable prediction of the results ion association had to be taken into account.

Experimental and calculated isotherms are shown in Figure 5. The agreement between both is worse than in the case of PNP on MT. Trends however, are predicted well. Near the CMC experiment and theory deviate seriously. This is partly due to the fact that we have chosen fixed values for I and $[Br^-]$, a gradual increase of these along the isotherm would improve the results. Another reason could be that close to the CMC some surface condensation occurs. Theoretically at pH = 9 this is predicted at a concentration only slightly above the CMC as indicated by the dashed part of the isotherm. At pH = 10 the strong lateral attraction can still be seen from the S-shape of the isotherm, but charge repulsion prevents surface condensation in this case. The fact that the model is capable of predicting a two step isotherm is encouraging as such behaviour is frequently found in practice.

The filled symbols in Figure 5 indicate the saturation levels of the DTAB adsorption in 1M butanol solutions. The dashed curves, which show the theoretical results for b = 0, predict these saturation levels nicely. Apparently in the presence of butanol the hydrophobic tails of the surfactant molecules associate with the butanol chain. This association is favoured over association with surfactant because with butanol there is no counteracting electrostatic repulsion. Obviously for b = 0 surface condensation will not occur.

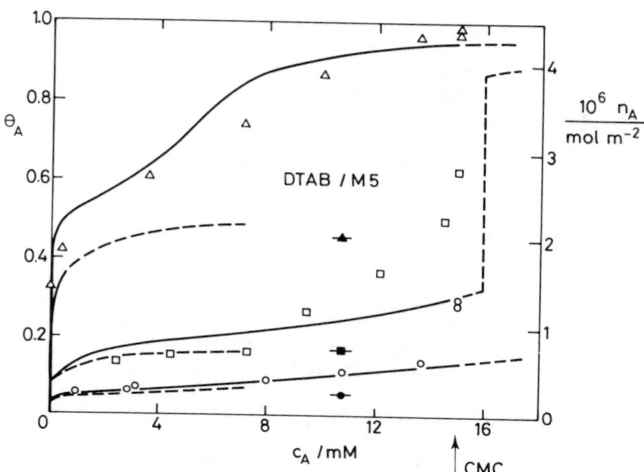

Figure 5. Adsorption isotherms of DTAB on Cab-O-Sil M5 at pH 8 (O), 9 (□) and 10 (Δ) respectively. Curves: theoretical predictions; filled symbols: adsorption level of DTAB in 1M butanol.

DISCUSSION

Although the model predicts the adsorption behaviour reasonably well, two serious omissions are made in its derivation: the size of the molecules and surface heterogeneity are not taken into account. Thus, it is worthwhile to give a short discussion on both factors.

Size effects

More often than not adsorbate and water molecules will differ in size. For flat adsorption of a large rigid molecule, such as for instance PNP, one may assume that the monolayer hypothesis still holds but one molecule exchanges with p water molecules. This case can be worked out relatively easily and introduces p as an extra parameter. For $p > 3$ the form of equation (3) is still adequate (12). However, b should now be considered as a composite parameter containing both lateral interactions and size effects: $b = b' + p$. For the PNP-Sterling MT system we may assume that $p \approx 4$. Hence, $b = 0$ would indicate lateral attraction between the adsorbed PNP molecules.

For flexible chain molecules such as DTAB incorporation of the size is complicated. In principle this can be done using a similar model as that presented by Scheutjens et al. (13) for oliomers and polymers. However, the present situation is more complex; for an accurate description two types of segements (head and tail) and charge effects have to be incorporated. In the model presented here, say only the headgroup of the surfactant is considerd to exchange with a water molecule and the tail, although near the surface, to remain in solution. As long as the chemical affinity is small and the lateral interaction large, this approximation is reasonable. Even at high θ, when the molecules may also adsorb 'upside down' (last tail segment adsorbed), the model may be approximately correct if the lateral interactions dominate. In the case of a strong chemical affinity of the chain for the surface and a large lateral attraction the model breaks down. Up to intermediate values of θ a relatively flat adsorption will occur, whereas at high θ the molecules will adsorb end-on. This leads to a complicated configurational entropy contribution.

Surface heterogeneity

In practice all surfaces are to some extent heterogeneous and hence for a correct description of adsorption surface heterogeneity should be considered. In doing so let us assume that the lateral interactions are not affected by the heterogeneity and that ψ_A may still be considered as an average potential smeared out over the whole surface. Instead of one K value we now have a distribution of affinities. When this distribution is given by a continuous function f(K), the overall adsorption Θ can be given by the following relation (14)

$$\Theta = \int_0^\infty \theta\{K,c,b\theta^*, \psi_A(\Theta)\}f(K)\,dK \tag{11}$$

where $\theta\{K,c,b\theta^*, \psi_A(\Theta)\}$ is the local isotherm function. Both the distribution and the local isotherm function are a priori unknown and their separation is arbitrary. To solve (11) at least an assumption has to be made regarding one of the functionalities. As in general little is known about the heterogeneity we prefer to take a function for θ, in which case equation (3), (5) of (8) can be chosen. In (5) and (8) ψ_A is now a function of Θ instead of θ, because ψ_A is considered to have an average value over the whole surface. For a *patchwise* heterogeneous surface $b\theta^*$ becomes $b\theta$ indicating that the interactions are counted per patch only. For a *random* heterogeneous surface $\theta^* = \Theta$, because all lateral interactions now have to be counted over the whole surface.

Once θ is selected, in principle f(K) can be obtained from the overall isotherm under the condition that both α and α^σ are close to zero, see Papenhuijzen et al. (15). After calculation of f(K) this function can be used to obtain $\overline{\Theta}(c)$ for other conditions. However, the procedure to obtain f(K) from the isotherm requires an accurate set of data and in practice it may be necessary to assume the functionality f(K) as well as a function for θ. Frequently authors go even further and assume rather simplified functions for both θ and f(K). When θ is a stepfunction or a Langmuir isotherm and f(K) is a single peaked function several analytical expressions for Θ can be found (14, 16). In the case of adsorption of organic electrolytes such simplifications of θ can only be made when b is about zero or large and negative, and when the electrostatic component to the lateral interaction is approximately constant. That is to say for net positive or negative charges larger than 10 $\mu C/cm^2$ (see Figure 2).

Concluding, incorporation of surface heterogeneity is rather difficult in practice and although not always justified, there will be some preference for the use of analytical equations for the overall isotherm. In using these it should be realized that the obtained parameters often lose at least part of their physical significance.

ACKNOWLEDGEMENT

The author expresses his thanks to Mr. Leo Keltjens for his help with the computations.

LITERATURE CITED

1. Trasatti, S., Electroanal. Chem., 53, 335 (1974).

2. Bijsterbosch, B.H. and J. Lyklema, Advan. Colloid Interface Sci., 9, 147 (1978.

3. Chander, S., D.W. Fuerstenau and D. Stigter in "Adsorption from Solution", Ottewill, R.H., C.H. Rochester and A.L. Smith (Eds.) Academic Press, 197 (1983).

4. Sigg, L.M. and W. Stumm, Colloids and Surfaces, 2, 101 (1981).

5. Müller, G., C.J. Radke and J.H. Prausnitz, J. Phys. Chem., 84, 369 (1980).

6. Koopal, L.K., Extended Abstr. 15th Biennial Conf. on Carbon, 266 (1981).

7. Koopal, L.K., Z. Wasser Abwasser Forsch., 16, 91 (1983).

8. Everett, D.H., Trans Faraday Soc., 61, 2478 (1965).

9. Bijsterbosch, B.H., J. Colloid Interface Sci., 47, 186 (1974).

10. Abendroth, R.P., J. Colloid Interface Sci., 34, 591 (1970); J. Phys. Chem., 76, 2547 (1972).

11. Wennerström, H. and B. Lindman, Physics Reports, 52 (1), 1 (1979).

12. Dhar, H.P., B.E. Conway and K.M. Joshi, Electrochim. Acta, 18, 789 (1973).

13. Scheutjens, J.M.H.M. and G.J. Fleer, Advan.Colloid Interface Sci., 16, 341 (1982).

14. Jaroniec, M., Advan. Colloid Interface Sci., 18, 149 (1983).

15. Papenhuijzen, J. and L.K. Koopal in "Adsorption from Solution", Ottewill, R.H., C.H. Rochester and A.L. Smith (Eds.), Academic Press, 211 (1983).

16. Myers, A.L. in "Fundamentals of Adsorption", Myers A.L. and G. Belfort (Eds.) Engineering Foundation, ... (1983).

Models for Thermal Regeneration of Adsorption Beds

M. Douglas LeVan* and David K. Friday+
Department of Chemical Engineering
University of Virginia
Charlottesville, Virginia 22901

This paper reviews and compares stage model and equilibrium theory analyses of solute condensation in adsorption beds during thermal regeneration with hot purge gas. Examples are given for regeneration of an activated carbon bed with adsorbed benzene and a 4A molecular sieve bed with adsorbed water.

The economic feasibility of many adsorption-based separation processes depends on efficient removal of the adsorbate from the adsorbent. Fixed-bed adsorbers are regenerated by heating, using either a purge gas or steam, or by depressurizing. Thermal regeneration is used in systems designed for solvent recovery and drying applications. Pressure swing regeneration is applied to separations of gases such as nitrogen and oxygen or light hydrocarbons.

This paper reviews briefly and then compares two models for solute condensation in adsorption beds during thermal regeneration with hot purge gas. While there have been numerous papers published on adiabatic adsorption in fixed beds, the development of a condensed liquid phase in an adsorption bed during thermal regeneration with a purge gas in a new area for fundamental work. The list of published studies of this phenomenon is short. We have carried out analyses using the method of lines-based stage model (1) and the method of characteristics-based equilibrium theory (2). An analysis and discussion of the role that transfer resistances may play on water condensation in gas dehydration beds is contained in a recent and thorough review article by Basmadjian (3). Related work on non-isothermal adsorption is discussed in these three references.

During the course of an adsorption or desorption step transitions develop in and pass through the bed. The design and efficient operation of the bed is based on the accurate prediction of the magnitude and velocity of these transitions. Thus, it is important to be aware of the factors that determine the heights of plateaus (regions of constant concentration and temperature that connect transitions) and the general shapes and locations of the transitions in deep beds. These are: (1) material balances, (2) the energy balance including heat losses, and (3) phase equilibria relations. Additional but secondary

* Author to whom correspondence should be addressed.
+ Present Address: Westvaco Corp., Laurel Research Center, Laurel, MD 20707

factors that affect only the specific shapes of transitions are: (1) mass and heat transfer resistances and dispersion and (2) deviations from plug flow (channeling). This paper is concerned with the primary factors that affect bed behavior. Secondary factors, although discussed briefly, will not be considered in any detail. It should be recognized, however, that in shallow beds (in which transitions are not clearly discernable) all factors referred to above are of equal importance.

MATHEMATICAL MODELS

The material and energy balances that describe the behavior of a non-isothermal fixed-bed adsorber are developed here and then cast into two mathematical models, the stage model and the equilibrium theory. In setting up the balances the bed is divided into two parts, a stationary phase and a vapor phase. The stationary phase consists of the solid porous adsorbent particles, the adsorbate, and condensed liquid (if any). The vapor phase includes the solute and regeneration gas in both the intersticies of the bed and the pores of the particles. In this development only one component adsorption is considered. However, the approach can be readily extended to multicomponent adsorption.

The formulation of the mathematical models neglects any effects of dispersion or channeling and assumes no radial gradients in temperature or concentration in the bed. Heat and mass transfer resistances are considered small and plug flow is assumed. One additional assumption is that the pressure drop through the bed is small relative to the total system pressure.

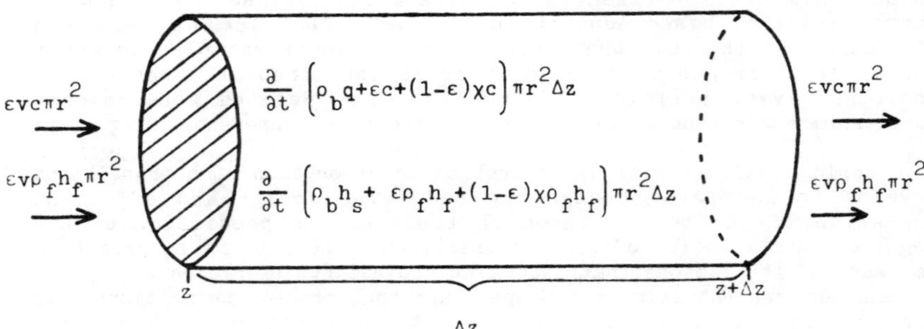

Figure 1. Control Volume for a Cylindrical Fixed-Bed Adsorber.

Figure 1 shows the control volume for an adiabatic fixed-bed adsorber. The void fraction of the packing and the porosity of the particles are denoted by ε and χ respectively. The vapor-phase concentration of the solute is c, the stationary-phase concentration q, and the interstitial velocity v. The solid-phase and fluid-phase enthalpies h_s and h_f are defined by

$$h_s = (c_s + c_a q)(T - T_{ref}) - \int_0^q \lambda dq \quad (1)$$

$$h_f = c_f (T - T_{ref}) \quad (2)$$

where c_s, c_a, and c_f are the heat capacities of the solid, adsorbate, and fluid. The reference states for the enthalpies are adsorbate-free adsorbent at the reference temperature T_{ref} and pure carrier and gaseous adsorbate at T_{ref}.

From Figure 1 the balances are

$$\frac{\partial}{\partial t}\left(\rho_b q + \{(1-\varepsilon)\chi+\varepsilon\}c\right) + \frac{\partial}{\partial z}\left(\varepsilon v c\right) = 0 \quad (3)$$

$$\frac{\partial}{\partial t}\left(\rho_b h_s + \{(1-\varepsilon)\chi+\varepsilon\}\rho_f h_f\right) + \frac{\partial}{\partial z}\left(\varepsilon v \rho_f h_f\right) = 0 \quad (4)$$

Equations (3) and (4) can be further simplified and written in terms of dimensionless independent variables. In adsorption operations involving gases at low and moderate pressures, the rates of accumulation of mass and energy in the gas phase are generally small compared to the rates in the stationary phase. Therefore, these terms in Equations (3) and (4) may be discarded. The following dimensionless variables can be defined

$$v^* = \frac{v}{v_o} \qquad \zeta = \frac{z}{L} \qquad \tau = \frac{\varepsilon v_o t}{L} \quad (5,6,7)$$

The dimensionless time τ is equal to the number of bed volumes of hot regeneration gas that have been passed into the bed, based on an open column. Substituting the above expressions into Equations (3) and (4) gives

$$\rho_b \frac{\partial q}{\partial \tau} + \frac{\partial (v^* c)}{\partial \zeta} = 0 \quad (8)$$

$$\rho_b \frac{\partial h_s}{\partial \tau} + \frac{\partial (v^* \rho_f h_f)}{\partial \zeta} = 0 \quad (9)$$

Stage Model

The equilibrium stage model is obtained by writing the spacial derivatives in the material and energy balances in backward difference form. From Equations (8) and (9) we obtain

$$\rho_b \frac{dq_i}{d\tau} = \frac{\{(v^*c)_{i-1} - (v^*c)_i\}}{\Delta\zeta} \tag{10}$$

$$\rho_b \frac{dh_{si}}{d\tau} = \frac{\{(v^*\rho_f h_f)_{i-1} - (v^*\rho_f h_f)_i\}}{\Delta\zeta} \tag{11}$$

Each stage is well-mixed with the effluent concentration and temperature of stage i constituting the feed concentration and temperature of stage i+1.

Equilibrium Theory

Equations (8) and (9) are of reducible form. This means that for constant initial and boundary conditions the dependent variables are only functions of the combined variable τ/ζ. Rearrangement of the material and energy balances gives for a gradual transition

$$\frac{\tau}{\zeta} = \rho_b \frac{dq}{d(v^*c)} = \rho_b \frac{dh_s}{d(v^*\rho_f h_f)} \tag{12}$$

where the derivatives are directional derivatives. For an abrupt transition Equation (12) is replaced by

$$\frac{\tau}{\zeta} = \rho_b \frac{\Delta q}{\Delta(v^*c)} = \rho_b \frac{\Delta h_s}{\Delta(v^*\rho_f h_f)} \tag{13}$$

Further details concerned with the development of the stage model and equilibrium theory are given elsewhere (1,2).

EXAMPLES

Two example systems, benzene adsorbed on activated carbon regenerated by hot nitrogen and water adsorbed on 4A molecular sieve regenerated by hot methane, are considered here using the models. The physical parameters used in obtaining the solutions are given in our first paper (1). Stage model solutions were obtained using 50 stages.

Benzene on Activated Carbon

We consider the regeneration of an activated carbon bed with adsorbed benzene using nitrogen at a total pressure of 1.0 MPa. Adsorption equilibria are described by the Langmuir isotherm (4,5)

$$q = \frac{QKc}{1 + Kc}$$

$Q = 4.4$ mol/kg
$K = K_o\sqrt{T}\, \exp(\lambda_d/RT)$
$K_o = 3.88 \times 10^{-8}$ $m^3/mol\,°K^{1/2}$
$\lambda_d = 43.5$ kJ/mol

Vapor-liquid equilibria are given by the Antoine equation

$$\log_{10} p = A - \frac{B}{C + T}$$

p in MPa T in °K
A = 3.0305 B = 1211 C = -52.2

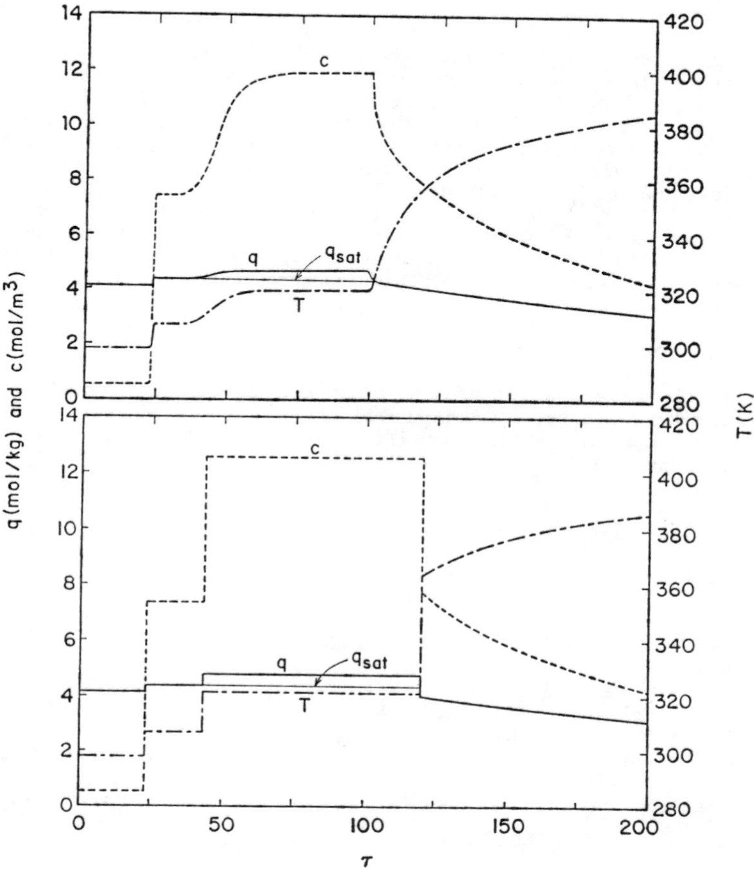

Figure 2. Concentrations and temperature at bed outlet for regeneration of activated carbon with adsorbed benzene. Stage model (top) and equilibrium theory (bottom).

At $\tau = 0$, let the bed have a uniform temperature and loading given by

$$T = 298^{o}K \qquad q = 4.12 \text{ mol/kg}$$

and for $\tau > 0$, let the feed to the bed be

$$T_o = 403^{o}K \qquad c_o = 0$$

This initial loading corresponds to an adsorbed-phase concentration in equilibrium with gas ten percent saturated with solute at the initial bed temperature. Because the isotherm is favorable the initial value of q is large, 94% of q_{sat}.

Stage model and equilibrium theory solutions for these conditions are compared in Figure 2. There, concentrations and temperature at the bed outlet are shown as functions of time. Breakthrough begins at roughly $\tau = 23$ for both models. The first transition to leave the bed is of combined form with an intermediate plateau on which q equals q_{sat}. On the principal plateau, liquid is present at the bed outlet and q exceeds q_{sat} by 8.3% for the stage model and 9.3% for the equilibrium theory. The second transition begins to leave the bed at $\tau = 100$ for the stage model and at $\tau = 120$ for the equilibrium theory, which shows this transition to be of abrupt-gradual combined form. Thereafter, the temperature and fluid-phase concentration approach feed values.

<u>Water on 4A Molecular Sieve</u>

For the regeneration of a 4A molecular sieve bed with adsorbed water using methane let the total system pressure be 4.0 MPa. Adsorption and vapor-liquid equilibria are described by a modified Antoine equation (1)

$$\log_{10} p = A - \frac{B(\theta)}{C(\theta) + T} \qquad \begin{array}{ll} p \text{ in MPa} & T \text{ in } {}^{o}K \\ \theta = q/q_{sat} & q_{sat} = 12.8 \text{ mol/kg} \end{array}$$

$A = 4.092$

$$\begin{array}{ll} B(\theta) = \theta^{-0.372}(2176 - 67.4\,\theta - 8700\,\theta^2) & \text{for } \theta \leq 0.196 \\ B(\theta) = 3845 - 2542\,\theta + 103\,\theta^2 + 262\,\theta^3 & \text{for } 0.196 \leq \theta \leq 1 \\ B(\theta) = 1668 & \text{for } \theta \geq 1 \end{array}$$

$$\begin{array}{ll} C(\theta) = -273 + \theta^{-0.329}(200.3 - 81\,\theta - 163\,\theta^2) & \text{for } \theta \leq 0.196 \\ C(\theta) = 50 - 37\,\theta - 347\,\theta^2 + 289\,\theta^3 & \text{for } 0.196 \leq \theta \leq 1 \\ C(\theta) = -45 & \text{for } \theta \geq 1 \end{array}$$

The initial condition of the bed is

$$T = 298^{o}K \qquad q = 11.35 \text{ mol/kg}$$

and the feed to the bed is

$$T_o = 533^{o}K \qquad c_o = 0$$

As with benzene adsorbed on activated carbon, this initial condition corresponds to a ten percent initial saturation of the vapor phase in the bed. The initial value of q is 89% of q_{sat}.

Figure 3 compares solutions for these conditions. Breakthrough is predicted at τ = 17 by the stage model and at τ = 18 by the equilibrium theory. Along the single plateau liquid exists at the bed outlet with q exceeding q_{sat} by a maximum of 5.8% from the stage model and by 5.9% from the equilibrium theory. The stage model gives 410°K for the temperature of this plateau while the equilibrium theory gives 411°K.

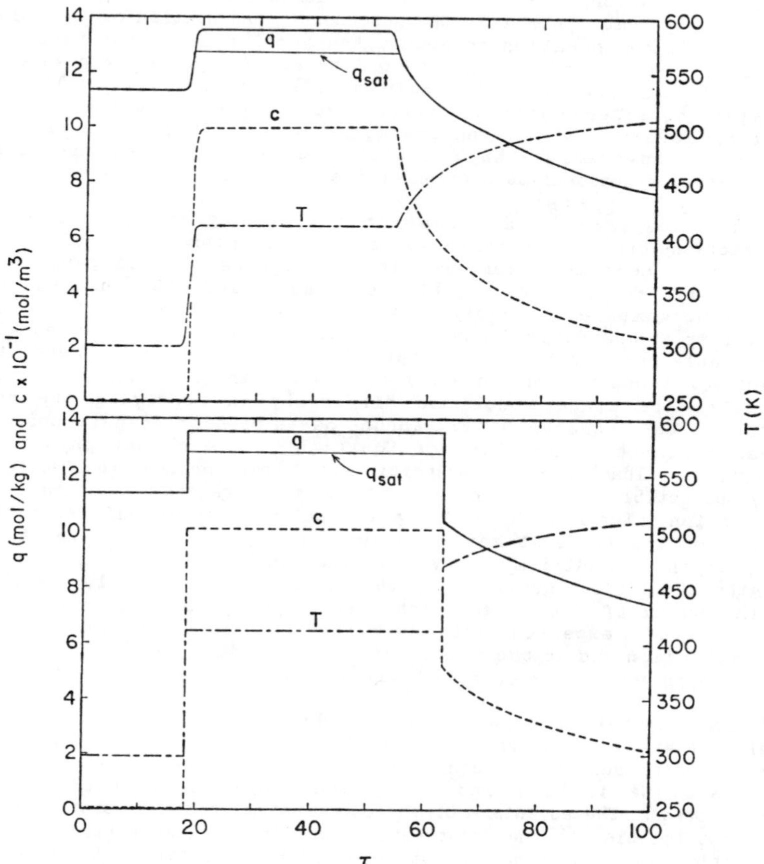

Figure 3. Concentrations and temperature at bed outlet for regeneration of 4A molecular sieve with adsorbed water. Stage model (top) and equilibrium theory (bottom).

The plateau breaks at $\tau = 54$ according to the stage model and at $\tau = 63$ based on the equilibrium theory, after which unsaturated vapor leaves the bed.

DISCUSSION

It has been shown with both models and for both systems that a condensed liquid phase can be developed in and passed through an adsorption bed during thermal regeneration. The liquid results from the passage of hot regeneration gas from the inlet end of the bed, where the adsorbate is stripped from the adsorbent to create a concentrated vapor, towards the cooler outlet end. As this rich vapor is cooled by heat transfer to the adsorbent, condensation of the solute occurs. As regeneration proceeds, the zone of saturated vapor and liquid is transferred through the bed by evaporation and condensation until it is passed from the column as saturated vapor. The two examples considered involve strongly adsorbed components with high initial loadings of the adsorbents. Because the isotherms are favorable, however, the high initial loadings can be obtained with a low initial fractional saturation of the vapor phase.

The stage model and equilibrium theory give qualitative descriptions of bed performance without considering the complicating effects of heat and mass transfer resistances and dispersion or deviations from plug flow. Figures 2 and 3 indicate that the two solutions compare favorably in most respects. General shapes and locations of transitions and heights of plateaus are predicted with reasonable agreement. Given that the equilibrium theory generates exact solutions to Equations (8) and (9) it can be noted that in the examples the stage model had difficulty in resolving the true mathematical character of the abrupt parts of some transitions. In these instances a solution of Equation (12) is almost physically acceptable. These parts of transitions can therefore be expected to be very susceptible to spreading by heat and mass transfer resistances and dispersion. The lengths of the principal plateaus predicted by the stage model with 50 stages are in both examples shorter than those given by the equilibrium theory. In other work we have found that the lengths of plateaus generated by the stage model are fairly sensitive to the number of stages used with more stages giving longer plateaus. This should be expected since the stage model is a finite difference approximation and as the number of stages is increased the solution should approach the exact result given by the equilibrium theory.

In comparing the two methods, advantages of the stage model include that it is not restricted to adiabatic behavior, that heat and mass transfer resistances can be included within the stages, and that non-constant initial and feed conditions can be used without complication. The advantage of the equilibrium theory is that it gives an exact solution to the conservation equations, clearly resolving the transitions and providing insight into the mathematical character of the solution. Also, the stage model can be implemented more quickly, especially for the case of complicated transitions like those shown here, but the equilibrium theory requires significantly less computer time.

In practice, due to the secondary factors, the breakthrough behavior of real beds is more likely to resemble the stage model solution rather than that given by the equilibrium theory. Researchers in the field of chromatography have long recognized the role that channeling plays, even in packed columns of small diameter (6). The stage model is simply a central difference approximation of the equations of the plate theory of chromatography (7), from which the number of plates is determined to fit experimental chromatographic data. By analogous application of the stage model equations, the behavior of a particular adsorption column can be described, lumping the effects of mass and heat transfer, eddy dispersion, and especially channeling into a single parameter, the number of stages.

ACKNOWLEDGEMENT

Acknowledgement is made to the Donors of the Petroleum Research Fund, administered by the American Chemical Society, for the support of this research.

NOTATION

c	= gas phase concentration of solute (mol/m^3)
c_a	= heat capacity of adsorbate and condensate (kJ/mol ^oK)
c_f	= heat capacity of gas phase (kJ/kg ^oK)
c_s	= heat capacity of solid phase (kJ/kg ^oK)
h	= enthalpy (kJ/kg)
L	= bed length (m)
q	= solid phase concentration (mol/kg)
q_{sat}	= adsorbed phase concentration in equilibrium with saturated vapor (mol/kg)
r	= bed radius (m)
t	= time (s)
T	= temperature (^oK)
v	= interstitial velocity (m/s)
v^*	= dimensionless velocity
z	= axial coordinate (m)

Greek Letters

ε	= void fraction of packing
ζ	= dimensionless axial coordinate
λ	= latent heat (kJ/mol)
ρ	= density (kg/m^3)
τ	= dimensionless time

Subscripts

i	= stage index
f	= fluid phase
s	= solid phase
o	= inlet condition

LITERATURE CITED

1. Friday, D. K. and M. D. LeVan, AIChE J., 28, 86 (1982).

2. Friday, D. K. and M. D. LeVan, AIChE J., in press (1983).

3. Basmadjian, D., "The Adsorption Drying of Gases and Liquids," in Advances in Drying, Vol. 3, Mujumdar, A. S. (Ed.), Hemisphere, in press.

4. James, D. H. and G. S. G. Phillips, J. Chem. Soc., 1954, 1066 (1954).

5. Rhee, H. K., E. D. Heerdt and N. R. Amundson, Chem. Eng. J., 1, 279 (1970).

6. Giddings, J. C., Anal. Chem., 34, 1186 (1962).

7. Glueckauf, E., Trans. Faraday Soc., 51, 34 (1955).

EFFECTS OF CHANNELING AND CORRECTIVE RADIAL DIFFUSION IN FIXED-BED ADSORPTION USING ACTIVATED CARBON

M. Douglas LeVan
Department of Chemical Engineering
University of Virginia
Charlottesville, Virginia 22901

Theodore Vermeulen
Department of Chemical Engineering
University of California
Berkeley, California 94720

Rate-controlled adsorption of solvent vapors in fixed beds has been found not to correspond to conventional predictive methods. A previously untreated rate mechanism, involving channeling combined with radial and axial dispersion, is analyzed theoretically in a manner analogous to G. I. Taylor's analysis for dispersion during laminar flow. Constant-pattern conditions, with a mildly favorable isotherm, are assumed. The results are found to correspond to breakthrough data for n-octane or n-decane vapors in a carrier gas.

For the design of fixed-bed adsorbers, it is of interest to be able to account for all factors that affect column performance. Since the breadth of the breakthough curve is determined by the combined effects of adsorption equilibria and mass transfer, a knowledge of the magnitudes of the various mass transfer resistances in relation to the design variables (particle size, fluid velocity, and bed geometry) is essential for scale-up of laboratory runs. If an empirical approach is adopted in place of a fundamental analysis, much more time and money must be invested in pilot-plant investigations.

The mechanisms which are thought to determine rate behavior in fixed-bed adsorbers are external mass transfer, pore diffusion, surface diffusion, and axial dispersion, all of which have been reviewed by Vermeulen, Klein and Hiester (1). Theoretical investigations have been made of breakthrough behavior with each of these mechanisms for both linear and nonlinear isotherms.

In the present study, breakthrough curves are found to be broadened by a mechanism which has not previously been treated explicitly for fixed-bed adsorption - dispersion caused by deviations from uniform velocity across the bed, known as "channeling". Under this flow condition, transverse concentration gradients are formed by the variation of interstitial velocity and are reduced in magnitude by transverse molecular and eddy diffusion. By pursuing the approach of G. I. Taylor (2), the mathematical analysis derived here for constant-pattern behavior of a solute with a slightly favorable isotherm serves to account for the combined effects of channeling, and radial and axial molecular and eddy diffusion (all expressed in terms of a composite apparent axial dispersion coefficient).

In the experimental program, vapors of either n-octane or n-decane were adsorbed in fixed beds of activated carbon, and full-range breakthrough curves were recorded. Bed diameter, particle diameter, and fluid velocity were varied, and their effects on the mid-height breakthrough slopes were analyzed. As indicated above, channeling was found to be the predominant cause of breakthrough-curve spreading for the operating conditions selected.

THEORY

The dispersion model originally proposed by G. I. Taylor (2) for flow in cylindrical unpacked tubes is equally valid for flow in packed tubes, and will be retraced here.

The axial distance relative to a moving plane is

$$z_1 \equiv z - vt \tag{1}$$

The convective diffusion equation for the unpacked tube then is

$$\partial c/\partial t + (v_z - v)\partial c/\partial z_1 = D[(1/r)\partial(r\partial c/\partial r)/\partial r + \partial^2 c/\partial {z_1}^2] \tag{2}$$

Taylor assumed that the time scale for radial gradients to form by convection ($L/\Delta v_z$) is large compared to the time constant for decay of radial gradients by molecular diffusion. For a parabolic velocity profile, he deduced that

$$L/\Delta v_z \gg {r_0}^2/(3.8^2 D) \tag{3}$$

$\partial c/\partial t$ was set equal to zero, $\partial c/\partial z_1$ taken to be independent of r, and axial diffusion neglected, thus converting Equation (2) into an ordinary differential equation for the radial variation of c. The solution, given by direct integration, was used to evaluate a radially averaged concentration. This resulted in a partial differential equation in t and z_1 for the axial distribution of solute, for a moving or stationary origin of z_1 or z:

$$\partial \bar{c}/\partial t = K\partial^2 \bar{c}/\partial {z_1}^2 \qquad \partial \bar{c}/\partial t + v\partial \bar{c}/\partial z = K\partial^2 \bar{c}/\partial z^2. \tag{4,5}$$

For the parabolic velocity profile, K was found to be given by

$$K = {r_o}^2 v^2/(48D). \tag{6}$$

Equation (4), with K = D, is the equation for axial diffusion. Hence, K is the apparent axial diffusion (or dispersion) coefficient. Aris (3) later showed that Equation (4) can be used to describe the combined effects of axial diffusion and of the dispersion mechanism treated by Taylor, if K is given by

$$K = D + \kappa\, {r_o}^2 v^2/D \tag{7}$$

where $\kappa = 1/48$ for the parabolic profile. For other velocity profiles, Aris showed that Equation (4) still applies but with different values of κ. It can be noted that κ decreases as the velocity profile is flattened; $\kappa = 0$ corresponds to plug flow. A similar result has been obtained by Bischoff and Levenspiel (4).

In practice, Inequality (3) can be relaxed at the expense of low-percentage variations in fitting experimental data. Bournia, Coull, and Houghton (5) successfully interpreted dispersion data this way for gases in laminar flow through an unpacked tube; at their highest velocities, the left-hand side of the inequality was only 1.5 times the right-hand side. Similarly, data taken by Bailey and Gogarty (6) for liquid-phase laminar flow through an unpacked tube fit Equation (4) if the left-hand side of the inequality is at least 2.6 times the right-hand side.

In certain cases of adsorption, the rate of breakthrough may be controlled by the combined effects of channeling, radial dispersion, and axial dispersion. Mathematical modeling of such cases, analogous to Taylor's analysis, follows. The model will be applied specifically for a slightly favorable isotherm under constant-pattern conditions.

By assuming angular symmetry and neglecting accumulation in the fluid phase, the continuity equation for fixed-bed adsorption in cylindrical coordinates takes the forms

$$\rho_b \partial q/\partial t + v_z \partial c/\partial z = (E_r/r)\partial(r\partial c/\partial r)/\partial r + E_z \partial^2 c/\partial z^2 \tag{8}$$

$$\Lambda \partial Y/\partial t + v_z \partial X/\partial z = (E_r/r)\partial(r\partial X/\partial r)/\partial r + E_z \partial^2 X/\partial z^2 \tag{9}$$

The dimensionless quantities in the second of these equations are defined as

$$X \equiv (c-c')/(c''-c') \qquad Y \equiv (q-q')/q''-q') \qquad \Lambda \equiv \rho_b(q''-q')/(c''-c') \tag{10,11,12}$$

<u>For linear and slightly nonlinear isotherms</u>, because of the similarity between Equations (5) and (9), axial and radial contributions can again be added, as in Equation (7). Equation (9) becomes

$$\Lambda \partial \overline{Y}/\partial t + v_o \partial \overline{X}/\partial z = (E_z + \kappa r_o^2 v_o^2/E_r)\partial^2 \overline{X}/\partial z^2 \tag{13}$$

The value of κ depends on how v_z varies within a representative cross section; the point-to-point variation of v_z in Equation (9) determines κ in Equation (13), which can be made more compact by using nondimensional variables:

$$\partial \overline{Y}/\partial NT + \partial \overline{X}/\partial N = \partial^2 \overline{X}/\partial N^2 \tag{14}$$

$$T \equiv tv_o/(\Lambda z) \qquad N \equiv zv_o/(E_z + \kappa r_o^2 v_o^2/E_r) \tag{15,16}$$

Following Acrivos (7) and Quilici (8), the constant-pattern form utilizes a "substantive" variable Z analogous to the relative length defined in Equation (1). With this, Equation (14) becomes

$$\partial \overline{Y}/\partial NT - \partial \overline{Y}/\partial Z + \partial \overline{X}/\partial Z = \partial^2 \overline{X}/\partial Z^2 \qquad Z \equiv N - NT \tag{17,18}$$

For <u>constant-pattern</u> profiles, concentrations are functions of Z only, and do not depend on time NT. Thus

$$-\partial \overline{Y}/\partial Z + \partial \overline{X}/\partial Z = \partial^2 \overline{X}/\partial Z^2 \qquad -\overline{Y} + \overline{X} = (\partial \overline{X}/\partial Z)_{NT} \tag{19,20}$$

where Equation (20) is the integrated form. In stationary coordinates, this result becomes analogous to the known constant-pattern rate equation for axial dispersion:

$$(1/N)(\partial \bar{X}/\partial T)_N = \bar{Y} - \bar{X} \tag{21}$$

With concentrations constant across the radius of the bed, $\bar{Y}$ and $\bar{X}$ are related directly by the adsorption isotherm. Where channeling is important, $\bar{Y}$ and $\bar{X}$ are related by a function less favorable than the isotherm Y(X). Expanding Y in a Taylor series around $X = \bar{X}$ yields

$$\bar{Y} = Y\Big|_{\bar{X}} + d^2Y/dX^2\Big|_{\bar{X}} \int_0^1 (X-\bar{X})^2 \xi d\xi + \ldots.. \tag{22}$$

Since $d^2Y/dX^2 < 0$ for a favorable isotherm, it follows that $\bar{Y}(\bar{X}) < Y(\bar{X})$.

EXPERIMENTS

Apparatus and Materials

Adsorption equilibria and rates were measured with a single apparatus, shown schematically in Figure 1. The system provided controlled carrier gas and liquid feed inputs, a small thermostated adsorption bed, and effluent analysis by gas chromatography.

To concentrate the solute vapor, the gas stream leaving one of the flowmeters could be sparged through pure liquid hydrocarbon in a constant-temperature bath. The vapor-rich stream was then mixed with additional pure nitrogen from the other flowmeter. This feed stream was then passed through a fixed bed of activated carbon contained in a second constant-temperature bath, and the leaving vapor was discharged

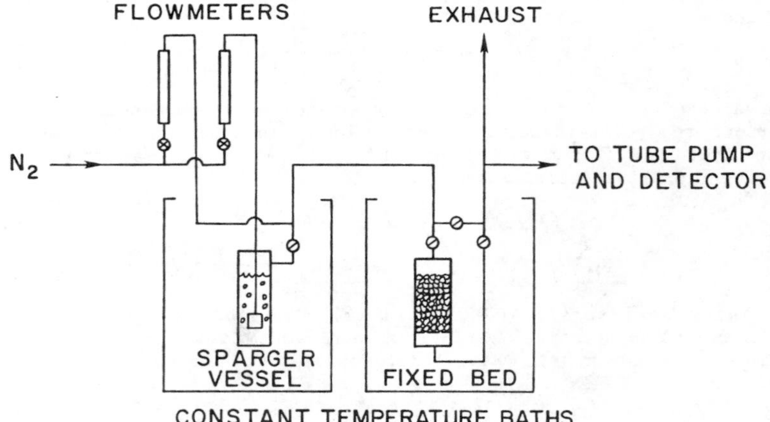

Figure 1. Apparatus.

at atmospheric pressure. A small fraction of the effluent was led into a gas chromatograph to determine the concentration of hydrocarbon with a flame-ionization detector and to monitor it on a strip-chart recorder.

As activated carbon, Pittsburgh Type BPL Granular Carbon, manufactured by Calgon Corporation from bituminous coal and designed for vapor phase applications, was selected. Calgon reports that much of the pore volume is in micropores 18 to 21 Angstroms in diameter, and that the particles are completely permeated by a system of macropores. The carbon granules were randomly shaped, as found by microscopic examination, and were obtained from a single sample lot by grinding and screening. Four mesh-size ranges were used. Expressed as U. S. Sieve-Series values and as average particle diameters, the ranges were 120/140 (0.115 mm), 60/70 (0.230 mm), 30/35 (0.547 mm), and 14/16 (1.30 mm).

Vertical beds of 0.554, 1.10, and 2.67 cm inside diameter were used with downward flow. Fine-mesh wire screens were installed at the foot of each bed to support the particles. Entrance and exit regions in the beds were packed with glass wool. For the tube-and-particle combinations selected, ratios of tube inside diameter to average particle diameter were between 20 and 100. The radio of packing height to inside bed diameter varied from 0.55 to 20. Reynolds numbers, defined by

$$Re \equiv v_o d_p / \nu \tag{23}$$

ranged from 0.01 to 5. In packing the beds, efforts were made to minimize channeling.

Equilibrium Data

Adsorption equilibria for n-octane and n-decane were measured by column runs at known values of bed temperature (50°C and 100°C), partial pressure of hydrocarbon in the entering vapor, vapor-stream flow rate, and quantity of adsorbent in the bed. The number of bed volumes of vapor fed to the bed for stoichiometic breakthrough ranged from 3×10^3 to 1×10^6. The quantities adsorbed were obtained by integrating the breakthrough curves and by weighing the bed before and after adsorption. The results are plotted in Figure 2 as q vs. hydrocarbon partial pressure.

Breakthough Experiments

More than fifty rate runs were made, all under laminar flow conditions. Vapors of either n-octane or n-decane in nitrogen were absorbed onto Type BPL carbon at 50°C. Feed concentrations, bed diameters, particle diameters, packing heights, and fluid velocities were varied. Midheight slopes of the experimental breakthrough curves were measured as being the most characteristic and most useable result.

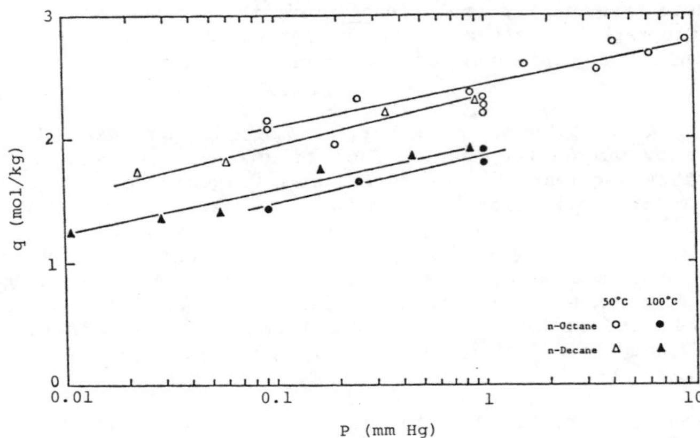

Figure 2. Isotherm Data.

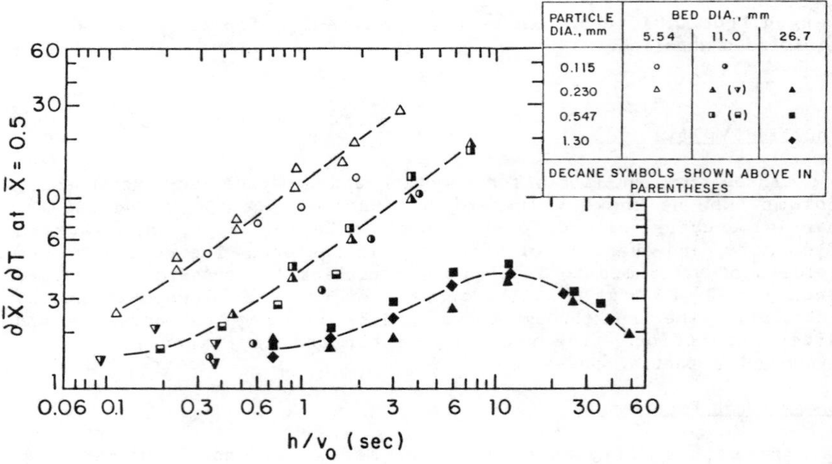

Figure 3. Rate Data.

The feed concentrations for these runs ranged from 2.9×10^{-9} to 2.9×10^{-7} mol/cm^3. The ratio of feed concentation to presaturated vapor-phase concentration was kept constant at 2.72. For this ratio, the nondimensional isotherm is essentially the same for every fed concentration, with both hydrocarbons. The isotherm is slightly favorable, with $Y = 0.62$ at $X = 0.5$; this corresponds to a separation factor R of roughly 0.6, sufficiently far from unity so that many of the rate runs should approach the constant-pattern condition.

RESULTS AND DISCUSSION

A plot of midheight slopes ($\partial X/\partial T$ at $X = 0.5$), determined from the experimental breakthrough curves, versus the superficial residence time h/v_o of fluid in the packing is shown in Figure 3. This residence time or "space time" appears in the number of transfer units for channeling that is offset by radial dispersion, Equation (16). Observations that can be made immediately, are: 1) the data fall onto three curves which depend upon bed diameter; 2) the smallest particles used with a column may give slightly lower midheight slopes than do larger particles; 3) the midheight slopes of breakthrough curves for the large bed, taken by holding h constant and varying v_o, pass through a maximum (h is the value of z at the bed outlet); 4) the dependence of midheight slope on h/v_o, even for the small beds, is somewhat less than first-power.

A dependence of breakthrough slope of bed diameter, like the one shown in Figure 3, is expected to occur either when channeling combined with radial diffusion controls, or when heat released during the adsorption step and lost through the wall of the column produces non-uniform temperature within each cross section. Heat release would serve to broaden the breakthrough curves. If this were to occur, the dilute-vapor curves would be sharper than those of more concentrated vapors. However, in this work, no correlation was found between midheight slopes and feed concentrations, at constant ratios of feed concentration to presaturated concentrations. Hence, thermal effects are not a controlling factor for the data reported here.

We thus conclude that channeling is the controlling factor for the bed-diameter dependence shown in Figure 3. The nonuniform velocity profile in a packing produces concentration gradients perpendicular to the direction of bulk flow, which are reduced by radial dispersion. When the linear velocity through the bed is decreased, radial dispersion has a relatively greater effect, and breakthrough is steeper. The smaller the column diameter is, the shorter the diffusion length; consequently, sharper breakthrough results. A small dependence of midheight slope on particle diameter, shown by the data, indicates that random packings of small particles can give slightly more channeling than random packings of larger particles.

In Figure 3, the breakthrough curves for the large-diameter bed become shallower as v_o decreases, pointing to axial molecular diffusion, for which the increased residence time in the bed increases the spreading of the axial gradients. Axial diffusion is known to arise for Peclet numbers ($d_p v_o/D$) below 1. For the data presented here, the minimum value of the Peclet group is 0.03.

As indicated above, constant-pattern breakthrough is approached for a favorable-equilibrium system under plug flow in a comparatively long column. For constant pattern, a plot of midheight slope against the number of transfer units will have a slope of unity [Vermeulen et al. (1)]. As this slope is approximately equal to the ratio of height h of packing to the length of packing in which appreciable changes in concentration occur, if h were doubled, $\partial X/\partial T$ at X = 0.5 would also double. With the slopes of the dashed curves in Figure 3 being somewhat less than unity, further analysis is needed. It is necessary to specify the two dispersion coefficients E_z and E_r, and the parameter κ, which depends only upon the channeled velocity profile within the packing.

In the present experiments, the dispersion coefficients are determined by molecular diffusion:

$$E_z = E_r = \varepsilon D/\tau \tag{24}$$

We assume $\varepsilon = 0.43$ and $\tau = \sqrt{2}$. Values of D given by Perry (9) were adjusted to 50°C, using the approximation that diffusion coefficients vary with the 3/2 power of absolute temperature. The D values at 50°C, 0.0781 cm^2/sec for n-octane and 0.0706 cm^2/sec for n-decane (in nitrogen), are substituted into Equation (24) to give $E_z = E_r = 0.0237$ and 0.0215 cm^2/sec for n-octane and n-decane respectively.

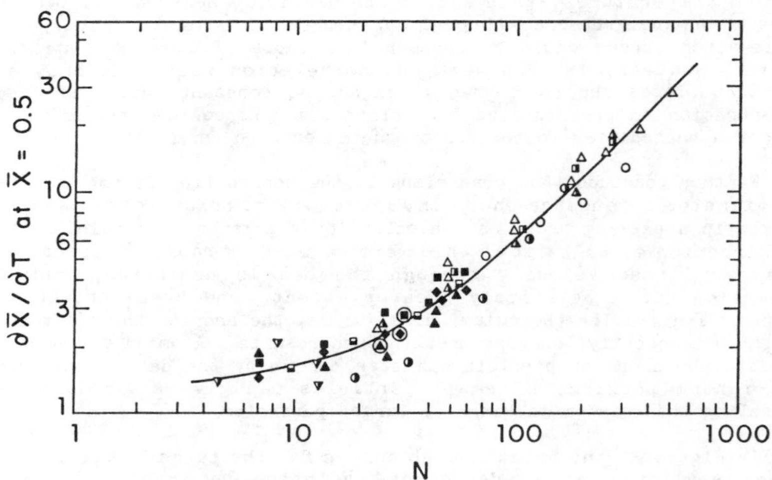

Figure 4. Rate Behavior. For key to symbols, see Figure 3.

The needed value of κ was determined in the following manner. It was for the largest-diameter bed that the maximum number of transfer units occurs where the midheight slope of the breakthrough curve is a maximum, i.e. where the two denominator terms in Equation (16) are equal. Thus,

$$\kappa = E_z E_r / (r_o^2 v_o^2) \tag{25}$$

Here v_o is the superficial velocity at the maximum midheight slope. By substituting into this equation the values $E_z = E_r = 0.0237$ cm^2/sec, r_o = 1.34 cm, and v_o = 0.478 cm/sec, we obtain κ = 1/730. (κ could perhaps be smaller for the larger, and larger for the smaller, particles.) Using κ = 1/730, the data are replotted on logarithic scales in Figure 4, to show midheight slope versus number of transfer units calculated from Equation (16). A consistent correlation is obtained.

CONCLUSIONS

Conforming to G. I. Taylor's treatment of dispersion in an unpacked tube, a model has been developed here to treat fixed-bed adsorption with rates of breakthrough controlled by the combined effects of channeling, radial dispersion, and axial dispersion. For slightly favorable isotherms, relations have been obtained for the constant-pattern profile. An inequality criterion, which limits the range of validity of the theoretical model, also has been derived.

Midheight slopes of experimental breakthrough curves for vapors of either n-octane or n-decane adsorbed on activated carbon show rates controlled by channeling and dispersion, and correlate with a new number of transfer units, derived for the combined mechanism.

Causes of channeling have also been reviewed. Under the conditions of the present experiments, channeling probably results from packing inhomogeneities. In the calculation of the number of transfer units, the bed radius (or diameter) is the representative length in the term that describes the effect of channeling. The extent to which this definition of the number of transfer units will apply to larger-scale columns is a subject for future study.

NOTATION

- c concentration in fluid phase
- c' c in equilibrium with presaturated adsorbent
- c'' c in feed
- d_p particle diameter
- D molecular diffusion coefficient
- E_r radial dispersion coefficient
- E_z axial dispersion coefficient
- h bed depth
- K apparent axial diffusion or dispersion coefficient
- L axial length over which appreciable changes in concentration occur
- N dimensionless number of transfer or dispersion units
- NT dimensionless time
- q concentration in adsorbent

q'	presaturation value of q
q''	q in equilibrium with feed
r	radial distance
r_o	tube radius
R	equilibrium parameter
Re	Reynolds number
t	time
T	dimensionless throughput parameter
v	velocity of a plane normal to axis
v_o	superficial velocity of feed stream
v_z	axial velocity dependent on r
X	dimensionless concentration in gas phase
Y	dimensionless concentration in adsorbent
z	axial distance (z_1, z relative to plane moving with velocity v)
Z	substantive variable analogous to relative length

Greek symbols:

Δv_z	difference between maximum and minimum velocities
ε	void fraction of packing
κ	constant in Equation (7)
Λ	dimensionless partition ratio
ν	kinematic viscosity
τ	tortuosity of packing
ρ_b	bulk density of adsorbent

Superscript

$\bar{\ }$	radially averaged value

LITERATURE CITED

1. Vermeulen, T., G. Klein, and N. K. Hiester, "Adsorption and Ion Exchange," R. H. Perry and C. H. Chilton (Eds.), Chemical Engineers' Handbook (5th Ed.), Sect. 16, McGraw-Hill, New York (1973).

2. Taylor, G. I., Proc. Roy. Soc., A219, 186 (1953); Ibid., A225, 473 (1954).

3. Aris, R., Proc. Roy. Soc., A235, 67 (1956).

4. Bischoff, K. B. and O. Levenspiel, Chem. Eng. Sci., 17, 245 (1962).

5. Bournia, A., J. Coull, and G. Houghton, Proc. Roy. Soc., A261, 277 (1961).

6. Bailey, H. R. and W. B. Gogarthy, Proc. Roy. Soc., A261, 227 (1961).

7. Acrivos, A., Chem. Eng. Sci., 15, 1 (1960).

8. Quilici, R., "Axial-Dispersion Constant-Pattern Kinetics of Ion-Exchange and Adsorption Columns," M.S. Thesis in Chemical Engineering, University of California, Berkeley (1969).

9. Perry, R. H. and C. H. Chilton (eds.), Chemical Engineers' Handbook (5th Ed.), p. 3-222, McGraw-Hill, New York (1973).

ADSORPTION AND DIFFUSION OF GASES IN SHAPE SELECTIVE ZEOLITES

Yi Hua Ma
Department of Chemical Engineering
Worcester Polytechnic Institute
Worcester, Massachusetts 01609

ABSTRACT

Recent work on adsorption and diffusion of C_6 and C_8 hydrocarbons in shape selective zeolites such as ZSM5 and silicalite is reported. Simultaneous measurements of diffusion and heat effects during methanol sorption in ZSM5 show rapid temperature rises (up to 80°C in certain cases) followed by slow radiative cooling. The effect of shape selective property on the adsorption equilibria and diffusion of C_6 and C_8 hydrocarbons are discussed in terms of the size and shape of the adsorbing molecules. Adsorption data on cation exchanged ZSM5 show that the effect of cation size on the reduction of adsorption capacity can be interpreted by correlating the size and the locations of the cations with the shape and size of the diffusing molecules.

INTRODUCTION

The current wave of advances in zeolite synthesis has created a brilliant prospect for catalyst and separation in both basic research and industrial applications. In particular, synthetic silica-rich porotectosilicates such as ZSM5 and silicalite show unique shape-selective properties, high thermal stability and low rate of deactivation.

ZSM5 is a high silica zeolite with SiO_2/Al_2O_3 ratio ranging from about 20 to 2000 (1). The crystal has an orthorhombic symmetry Pnma with unit cell parameters a=20.07 Å, b=19.92 Å and c=13.42 Å. The channel system consists of straight channels running parallel to [100]. Straight channels have openings defined by 10-membered rings of size 5.4x5.6 Å based on oxygen radii of 1.35 Å. Also defined by 10-membered rings are sinusoidal channels of size 5.1x5.4 Å (2,3). Silicalite is a microporous crystalline silica isostructural to ZSM5.

Sorption measurements on n-hexane, 3-methylpentane, 2,3-dimethylbutane, toluene, p-xylene, o-xylenes and mesitylene on HZSM5, NaZSM5 and silicalite were carried out by Anderson et al. (4) at 293 K. Their results show that the presence of sodium ions caused the pores of NaZSM5 to be more constricted than in HZSM5 and silicalite. This constriction is sufficient to affect the sorption of large molecules such as 3-methylpentane, toluene and p-xylene although no appreciable effects were observed for sorption of the straight chain hydrocarbon, n-hexane. Systematic measurements of the adsorption isotherms of N_2, O_2, H_2S, CO_2, pentane through n-decane, isobutane, isopropane, neopentane, xylenes and monomethylnonanes on HZSM5 and HZSM11 were made by Jacob et al. (5). Derouane and Galelica (6) measured adsorption capacities of several hydrocarbons on ZSM5 and suggested that aliphatics can diffuse and be accommodated in both channel systems of ZSM5, while iso-aliphatics may prefer to diffuse in the straight channels. Aromatics and methyl-substituted aliphatics, have a strong preference for diffusion and adsorption in the straight channels. The adsorption of methanol, dimethyl ether and various hydrocarbons on two HZSM5 samples prepared by two different methods were studied by Doelle et al (7). Their results indicate that sorption rates were dependent on the method of zeolite preparation. The diffusion and equilibrium adsorption of n-hexane, benzene, 3-methylpentane, p-xylene and o-xylene on HZSM5 were reported by Olson et al. (2). Their results were interpreted in terms of molecular pore filling as well as interaction forces between diffusing molecules and zeolitic pores.

Most diffusion and sorption studies assumed isothermal conditions within the zeolite sample. However, non-isothermal conditions were recently observed. A 25°C temperature difference between a Calsit 5A molecular sieve pellet and ambient was observed by Ilavsky et al. (8) while adsorbing n-heptane. A 50°C temperature rise was measured by Eagan et al. (9). A mathematical model to account for the thermal effects was presented by Brunovska et al. (10,11) and Chichara et al. (12).

The present paper summarizes our recent work on adsorption and diffusion of C_6 and C_8 hydrocarbons on silicalite (13) and cation exchanged ZSM5 (14). Furthermore, simultaneous measurements of diffusion and heat effects during sorption of methanol in ZSM5 are also reported (15).

EXPERIMENTAL

All the adsorption and diffusion measurements were made gravimetrically in a constant volume and constant pressure system. The equipment and detailed experimental procedure can be found in Wu et al. (13). A unique feature of the set-up worthy of mentioning is that a specially designed sample pan equipped with a thermocouple was used to measure simultaneously the uptake and the temperature during sorption. A schematic of the equipment is shown in Figure 1. As the temperature of the zeolite sample can be continuously monitored, it is possible to control the temperature rise to be less than 1 to 2°C by using small pressure steps to minimize the temperature effect on diffusion measurements.

The ZSM5 used in the study was synthesized in Na form which was then treated with 8% NH_4NO_3 followed by heating at 520°C for 7 hours to give HZSM5 (16). The Li, K, Rb, and Cs forms of the ZSM5 were then obtained through cation exchange with HZSM5. The chemical composition of the NaZSM5 is listed in Table 1 and the calculated unit cell formula is $Na_{3.8}Al_{3.8}Si_{92.9}O_{192}$. The silicalite sample was obtained from Union Carbide Corporation and was fractionated by sedimentation. The fraction mainly composed of 2μm was collected for adsorption and diffusion measurements. ZSM5 sample obtained from Mobil Oil Corporation was used for the measurements of temperature rise during methanol sorption.

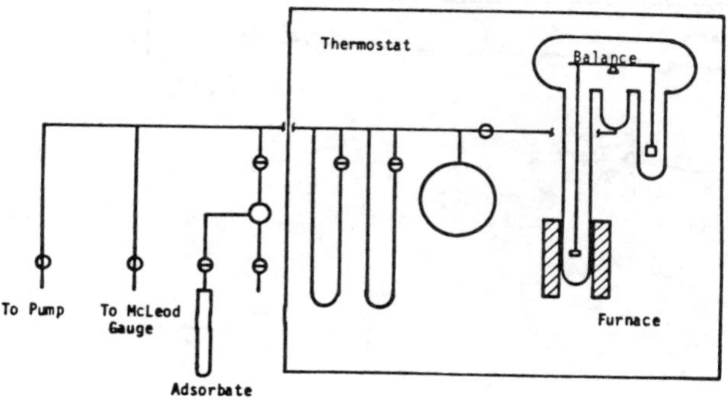

Figure 1. Schematic of the Adsorption Equipment

Table 1
Chemical Analysis of NaZSM5

	% wt.
SiO_2	94.3
Al_2O_3	3.2
Na_2O	2.0
CaO	0.2
Fe_2O_3	0.3

DISCUSSION

Adsorption in Silicalite

As silicalite is hydrophobic and organophilic, it appears reasonable to expect that adsorption of organic molecules occurs through the volume filling of micropores by physical adsorption at low relative pressures. From the adsorption isotherms measured for benzene, cyclohexane, hexane, p-xylene, o-xylene, m-xylene and ethylbenzene at 20°C, 100°C and 200°C (13), it appears that they can be divided into four groups: (1) benzene and hexane, (2) p-xylene and ethylbenzene, (3) m-xylene and o-xylene and (4) cyclohexane. A typical adsorption isotherm is shown in Figure 2 and the measured saturation adsorption capacities at 20°C are summarized in Table 2.

As the size of the adsorbed molecules is comparable to that of the channel, it appears reasonable to expect that the difference in adsorption capacities is primarily caused by steric effect. For example, if we assume that each sinusoidal channel accommodates one benzene molecule and each straight channel accommodates two, the total number of benzene molecules would be 8 in each unit cell which is consistent

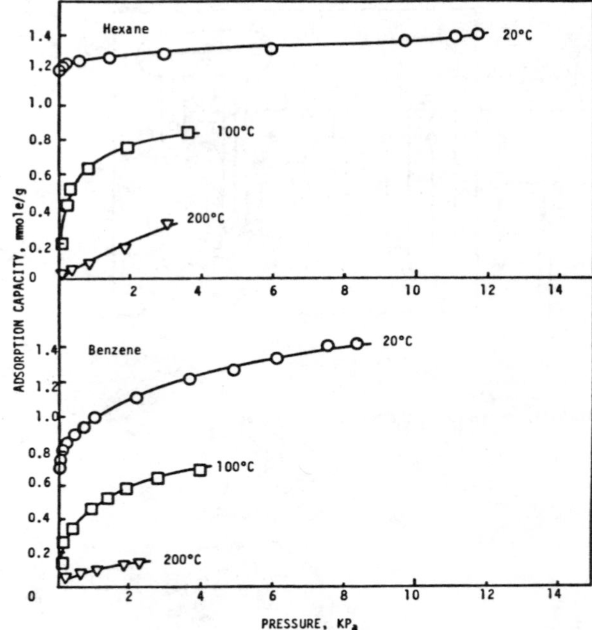

Figure 2. Adsorption Isotherms of Hexane and Benzene on Silicalite

Table 2. Adsorption Capacity and Heat of Adsorption of Hydrocarbons on Silicalite (T=20°C)

Adsorbate	Kinetic Diameter (2)	Adsorption Capacity mmole/g	molecules/unit cell	V_p cm^3/g	q_{st} KJ/mole
Benzene	5.85	1.42	8.2	0.126	80
Cyclohexane	6.0	----	---	----	86*
Hexane	4.3	1.42	8.2	0.185	87
p-xylene	5.85†	1.06	6.1	0.13	41
ethylbenzene	5.85†	1.04	6.0	0.14	44
m-xylene	6.8	0.69	4.0	0.085	--
o-xylene	6.8††	0.51	2.9	0.062	48

*Estimated by isotherms at 100 and 200°C, adsorption amount = 0.16 mmole/g.

†Taken as the same as benzene.

††Taken as the same as m-xylene.

with the measured value of 8.2 molecules/unit cell. Furthermore, calculations based on the length of a benzene molecule and the length of channels in silicalite substantiates such an arrangement. Hexane molecules are adsorbed in a somewhat coiled configuration which is substantiated by the fact that the calculated hexane pore volume as normal liquid is essentially equal to that of the calculated silicalite pore volume. On the other hand, the relatively longer and rigid molecules of ethylbenzene and p-xylene would either occupy a sinusoidal channel and block the cross section of a straight channel or align in the straight channel but block the sinusoidal channel. This gives six molecules per unit cell, again consistent with the experimental results. Although m-xylene and o-xylene molecules have much larger kinetic diameters than the size of the channel, measurable adsorption capacities can still be obtained due to the possibility of the molecules being situated at the intersects of the channels, with one methyl group in the straight channel and the other in the sinusoidal channel. The rigid cyclohexane molecules have such low rate of adsorption at 20°C that it was not possible to determine its isotherms.

Diffusion of these hydrocarbons appears to be closely related to their kinetic diameter. Molecules with a kinetic diameter up to 5.85 Å can rapidly diffuse into silicalite with a diffusion coefficient in the order of 1×10^{-11} cm^2/sec at 20°C. Although m-xylene and o-xylene have a kinetic diameter of 6.8 Å, they could still diffuse into silicalite with a measurable rate due, probably, to the fact that their 2- or 3-position methyl group could be bent to align with the axis of the channel during diffusion.

Adsorption in Ion Exchanged ZSM5

Relatively few studies were performed to investigate the effects of cations on the adsorption in ZSM5. Recent work by Wu and Ma (14) appears to show that the adsorption capacity decreases as the radius of the cation increases for adsorption of several hydrocarbons in Li, Na, K, Rb, and Cs forms of ZSM5. This trend is quite obvious as shown in Table 3. Typical adsorption isotherms for benzene are shown in Figure 3.

Table 3. The Effect of Cation on Adsorption Capacity (t=30°C)

Adsorbate	Adsorption Capacity, molecules/unit cell					
	Silicalite	LiZSM5	NaZSM5	KZSM5	RbZSM5	CsZSM5
Benzene	8.2*	8.3	7.2	5.9	5.0	4.1
Hexane	8.2*	7.9	7.7	7.2	6.1	3.9
p-Xyelene	6.1*	6.9	6.6	5.5	4.3	2.8
m-Xylene	4.0*	vs+	---	---	---	vs+
Methanol	22.6	20.8	18.9	18.2	16.1	16.2
Dimethyl Ether	12.1	11.8	11.4	10.9	10.6	10.6

*at 20°C, refer to (8)

+adsorbed very slowly, equilibrium capacity not determined.

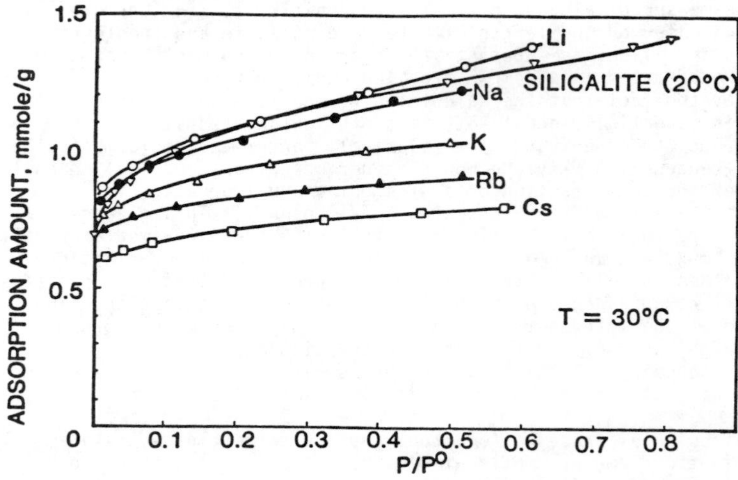

Figure 3. Adsorption Isotherms of Benzene on ZSM5

The effect of cation on adsorption depends primarily on (1) the size and shape of the guest molecule, (2) the size of the cation and its location in the channel, and (3) the interaction between the cation and the guest molecule. Based on the assumption that negligible exchangeable cations were present in silicalite and the measured adsorption capacities in silicalite, Wu and Ma (14) presented a relation to correlate the adsorption capacities of the cation exchanged ZSM5 zeolites. The correlation simply takes the difference between the total space occupied by the cations in a unit cell as free length accessible to adsorption. The simple correlation works rather well for most cases investigated as shown in Table 4. However, deviation from the correlation was observed for adsorption of rigid structure molecules such as benzene in ZSM5 with larger cations K, Rb and Cs. This is due, in part, to the fact that local blocking caused by large cations may make some sections of the channel inaccessible for adsorption. This blockage may very likely occur in the sinusoidal channel as its unblocked length is very close to the length of a benzene molecule. Further indication of this blockage is evidenced by the fact that values of a/a_c shown in Table 4 decrease with increasing cation radius for both benzene and p-xylene adsorption (a and a_c are measured and calculated adsorption capacities respectively).

The effect of cation on diffusion appears to be small due, in part, to the fact that diffusion into the crystal may be preferred via the straight channels as they offer less resistance than that of the sinusoidal channels. If the locations of the cations are at the channel intersections, then one would not expect the cation to affect the diffusion through straight channels in a significant way.

Table 4. Calculation of Adsorption Capacity on ZSM5 with Cation

Cation Form	Li	Na	K	Rb	Cs
Radius of Cation, Å	0.74	1.02	1.38	1.49	1.70
$(L_T\text{-}nd)/L_T$	0.916	0.884	0.843	0.831	0.807
Adsorbate: Methanol					
a_c	20.7	20.0	19.0	18.8	18.2
a/a_c	1.00	0.95	0.96	0.86	0.89
Adsorbate Dimethyl Ether					
a_c	11.1	10.7	10.2	10.1	9.8
a/a_c	1.06	1.07	1.07	1.05	1.09
Adsorbate n-Hexane					
a_c	7.5	7.2	6.9	6.8	6.6
a/a_c	1.05	1.07	1.04	0.90	0.59
Adsorbate Benzene					
a_c	7.5	7.2	6.9	6.8	6.6
a/a_c	1.11	1.00	0.86	0.74	0.62
Adsorbate p-Xylene					
a_c	5.6	5.4	5.1	5.1	4.9
a/a_c	1.23	1.22	1.08	0.84	0.57

L_T: total channel length in a unit cell; n: number of cations in a unit cell; d: diameter of a cation; a_c: calculated adsorption capacity; a: measured adsorption capacity.

Thermal Effects During Sorption

Recent work by Kmiotek et al. (15) showed that a substantial temperature rise (as high as 80°C in certain cases) could result during sorption. Furthermore, this larger temperature rise generally occurred in a very short period of time. With a large temperature rise in a short period of time (within 10-15 seconds), the thermal effect on the diffusion process would be rather severe. The assumption of a constant temperature normally employed in the evaluation of diffusion coefficients would most likely be in error.

Simultaneous measurements of sorption rates and temperature histories were made for methanol vapor in ZSM5 zeolites at three ambient temperatures (30°, 45° and 60°C) for pressures in the range of 1.7 mmHg to 266.8 mmHg (15). A typical uptake curve and temperature history curve are shown in Figure 4 for adsorption of methanol in ZSM5.

Kmiotek et al. (15) indicated that the maximum temperature rise can be correlated as a function of one of the dimensionless parameters characteristic of adsorbate-adsorbent system. The dimensionless maximum temperature rise $(\Delta T)_{max}/T_0$ can be correlated with the parameter $(\Delta H)(M_\infty)/M_0 C_p T_0$ as shown in Figure 5 for two different bed depths. The slopes of the line are very close to unity as expected as the system is essentially adiabatic as a result of reaching maximum temperature almost instantaneously with negligible heat loss due to the slow radiative cooling.

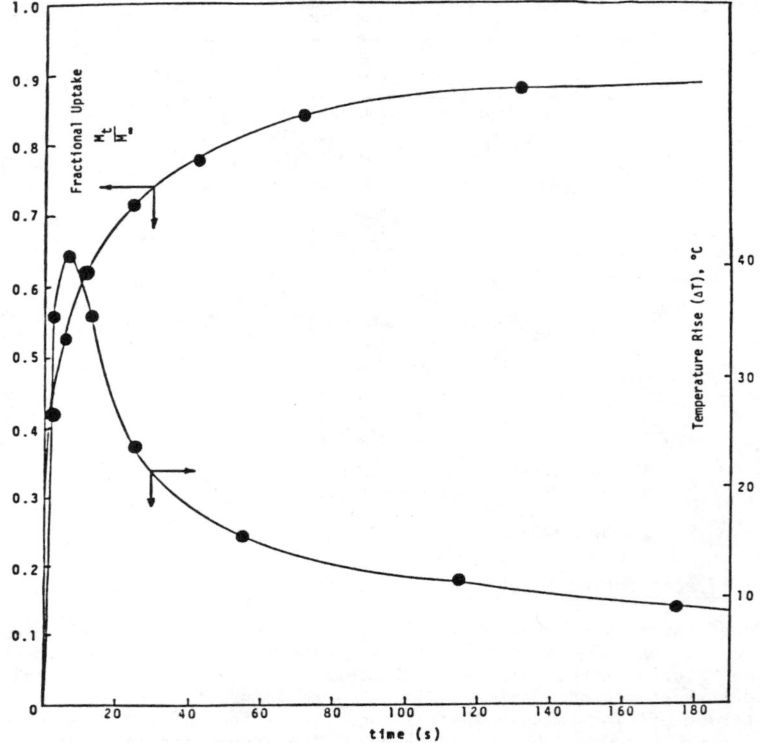

Figure 4 Fraction Uptake and Temperature History of Methanol Adsorption in ZSM5
System Temperature = 45°C, P = 51.0 Torr

The rapid initial temperature rise is the result of the high initial sorption rate and low rate of heat dissipation. The maximum temperature rise is, therefore, a function of initial pressure as the rate of sorption generally increases as the pressure increases. This is shown in Table 5. Temperature rises ranging from a few degrees to 85°C are observed for the system studied. These results demonstrate the importance of proper temperature control during sorption experiments. Furthermore, the need to have small pressure increments especially with a freshly prepared sample appears to be imperative.

CONCLUSIONS

The molecular sieving effect of silicalite depends mainly on the relative size of the channel and the shape and size of the adsorbate. The effects of cation size on the reduction of adsorption capacity are correlated with the shape and size of the diffusing molecules taking into account the location of the cations. No significant effect of cation on diffusion rate is observed.

Large temperature rises may be experienced during sorption if heats of adsorption are high and heat transfer is slow compared to mass transfer. The initial heating is approximately adiabatic which results in a steep temperature rise. The maximum temperature rise can be correlated as a function of the ratio of the total heat generated during sorption to the heat capacity of the zeolite.

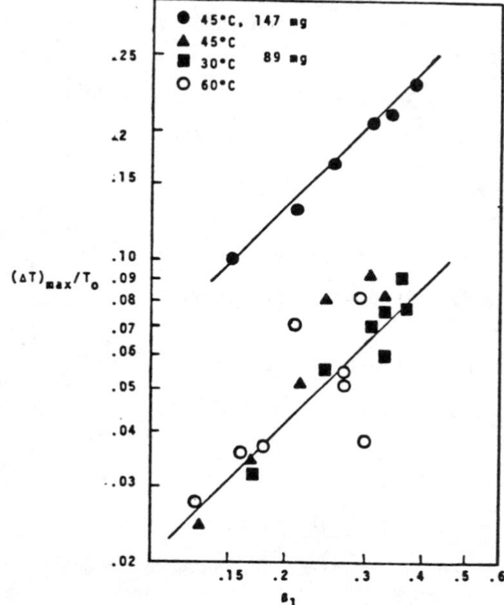

Figure 5

Maximum temperature rise as a function of β_1.

$$\beta_1 = \frac{M_\infty(\Delta H)}{M_o C p T_o}$$

M_∞: Equilibrium adsorption capacity

M_o: Mass of zeolite

ΔH: Heat of adsorption

Cp: Heat capacity of zeolite

T_o: Initial temperature

Table 5. Maximum Temperature Increase During Sorption

INITIAL TEMPERATURE (°C)	INITIAL WEIGHT (mg)	PRESSURE (mmHg)	(ΔT) MAX (°C)
30	89	1.7	9.8
		9.6	17.1
		18.3	21.5
		29.0	18.3
		53.4	24.6
		70.3	20.3
		82.4	23.3
45	89	2.9	7.7
		8.7	10.7
		17.5	16.5
		29.6	25.9
		63.6	29.4
		116.6	26.1
45	147	4.7	32.4
		11.5	43.2
		17.0	54.1
		31.2	68.0
		56.9	70.5
		127.5	84.7
60	89	2.0	0.4
		10.0	9.2
		19.2	17.1
		31.2	11.9
		49.3	24.0
		102.2	18.3
		110.1	17.1
		131.1	12.5
		266.8	26.8

REFERENCES

1. Olson, D.H., W.O. Haag and R.M. Lago, J. Catalysis, 61, 390 (1980).

2. Olson, D.H., G.T. Kokotailo, S.L. Lawton, and W.M. Meier, J. Phys. Chem., 85, 2238 (1981).

3. Kokotailo, G.T., S.L. Lawton, D.H. Olson and W.M. Meier, Nature (London), 272, 437 (1978).

4. Anderson, J.R., K. Foger, T. Mole, R.A. Rajadh and J.V. Sanders, J. Catalysis, 58, 114 (1979).

5. Jacobs, P.A., H.K. Beyer and J. Valyon, Zeolite, 1, 161 (1981).

6. Derouane, E.G. and J. Galelica, J. Catalysis, 65, 486 (1980).

7. Doelle, H.-J., J. Heering and L. Riekert, J. Catalysis, 71, 27 (1980).

8. Ilavsky, J., A. Brunovska and V. Hlavacek, CES, 35, 2475 (1980).

9. Eagan, J.D., B. Kindl, and R.B. Anderson, Adv. Chem. Ser., 102 164 (1971).

10. Brunovska, A., J. Ilavsky and V. Hlavacek, CES, 36, 123 (1980).

11. Brunovska, A., V. Hlavacek, J. Ilavsky and J. Vallyni, CES, 33, 1385 (1978).

12. Chihara, K., M. Suzuki, and K. Kawazol, CES, 31, 505, (1976).

13. Wu, Pingdong, A. Debebe and Y.H. Ma, Zeolite, 3, 118 (1983).

14. Wu, Pingdong and Y.H. Ma, "The Effect of Cation on Adsorption and Diffusion in ZSM5" presented at the Sixth International Zeolite Conference, Reno, Nevada (1983).

15. Kmiotek, S.J., Pingdong Wu and Y.H. Ma, "Recent Advances in Adsorption and Ion Exchange," AIChE Symposium Series, 78, No. 219, 83 (1982).

16. Xu, Qinha, Yan Aizhen, and Yan Yichun, Chinese Journal of Catalysts, 2, 308 (1981).

OZONE/GAC FOR DRINKING WATER TREATMENT

S. W. Maloney
K. Bancroft
I. H. Suffet

Environmental Studies Institute
Drexel University
Philadelphia, PA 19104

H. Neukrug

Philadelphia Water Department
Municipal Services Building
Philadelphia, PA 19102

ABSTRACT

Ozonation was shown to reduce the applied total organic carbon (TOC) to GAC contactors by stripping and/or conversion to more biodegradable forms. However, adsorption efficiency for chloroform was shown to decrease despite less applied TOC. Ozonation appears to produce lightweight organic compounds which increase competition for adsorption sites with chloroform.

INTRODUCTION

The combination of ozonation followed by granular activated carbon (GAC), the so-called BAC process, has been proposed as a method to increase the cost-effectiveness of the GAC process for organic removal in drinking water treatment, based on European experience (1,2). Ozonation has been shown to increase biodegradability of various waters (3,4,5) but may reduce adsorbability (4), as measured by TOC. This paper will examine the effects encountered during implementation of the combination of ozone and GAC processes within conventional U.S. potable water treatment. One expected effect is greater total organic carbon (TOC) removal for GAC contactors receiving ozonated water. This greater TOC removal may induce longer bedlife to breakthrough for specific organics (6,7) by reducing competition for adsorption sites. Conversely, the ozonation unit process may produce less adsorbable compounds (4), causing more rapid breakthrough for specific organics.

Conventional U.S. water treatment is designed to meet two goals: 1) removal of suspended and colloidal material from raw water, and 2) removal of pathogenic microorganisms. The treatment processes of chlorination, coagulation, flocculation, sedimentation and rapid sand filtration have proven very effective for these purposes, and the combination of these processes is commonly referred to as conventional water treatment. Conventional treatment does not provide significant removal of most trace organic compounds (8). This paper addresses additions to and modifications within conventional treatment designed to control trace organic chemicals in the finished drinking water. This research was conducted at the Torresdale water treatment plant on the

Delaware River in Philadelphia, PA.

The objective of this study is to examine the effects of preozonation on the adsorption process. Three parameters are used in this presentation to characterize the ozone/GAC process and to judge its effectiveness within the framework of conventional treatment. These are: microbial organic carbon removal, TOC removal and chloroform removal.

Microbial organic carbon removal has been proposed to increase GAC bedlife by reducing competition for adsorption sites (6,7). The results of the microbial studies have been presented elsewhere (9,10) and will be used here to interpret the breakthrough characteristics of TOC and specific organic chemcals.

TOC removal is presented as a measure of removal of organics which include natural humic substances as well as specific organics of health concern and as a comparison between the results of the Philadelphia study and others. TOC is not a water quality factor of health concern, but it has been linked to trihalomethane formation (11), has been shown to compete for adsorption sites with organics of health concern (6,7), and has been implicated in "carrying" associated compounds through GAC contactors (12,13).

Chloroform was selected as a model compound representing volatile organics which tend to break through GAC adsorbers rapidly. Chloroform is a compound of health concern for which regulations (14) have been developed. Chloroform has not shown competition with background organics in laboratory (7) and pilot scale (15) studies in the absence of ozonation. Displacement of chloroform was observed in full scale studies which included ozonation (16). This study evaluates displacement of chloroform in the presence and absence of preozonation using a common influent.

EXPERIMENTAL

Four carbon contactors were evaluated in this study to determine their relative effectiveness for removing trace organics. Each contactor was preceded by a different treatment train. Each treatment train included the conventional unit operations of raw water sedimentation, flocculation/sedimentation, and rapid sand filtration. Chlorination and ozonation were evaluated separately, in combination, and without either predisinfectant. Water from the Philadelphia Torresdale water treatment plant was used for the chlorinated plant (CP) studies (with and without ozonation) and water from a pilot scale plant operated in parallel to the Torresdale plant was used for nonchlorinated (NCP) studies.

The carbon contactors were 6-inch diameter glass columns with an overall height of 6 feet. Carbon was added to the contactors to a depth of 4 feet. The flowrate was set at approximately 2140 liters per day, based on a surface loading rate of 2 gallons per minute per square foot. This results in an empty bed contact time of 15 minutes. A complete description is available elsewhere, (17).

Organic

Organic constituents were measured by the group parameter of total organic carbon and specific volatile halogenated organic (VHO) compounds. Samples were collected once per week, dechlorinated and stored at 4 degrees C. Samples for TOC were acidified with phosphoric acid and pressure filtered through a 0.2 micron surfactant free polycarbonate membrane to remove carbon fines. TOC analysis was performed on a Dohrman DC-54 Low Level Analyzer using procedures as detailed by Dohrman (18).

Volatile halogenated organic analysis was performed using a Tekmar LSC Model 1, Purge and Trap and a Hall detector according to EPA method 601 (19) with two minor modifications. A 10 foot glass column was used in lieu of an 8 foot column, and the temperature program was 60 degrees C for four minutes and 6 degrees per minute to 215 degrees C. The concentrations of organic constituents are summarized in Table 1.

RESULTS

Figure 1 shows influent and effluent TOC data for four GAC contactors. The least efficient treatment train (in terms of residual organic carbon concentration) had prechlorination without ozonation while the most efficient had ozonation without prechlorination. It must be stressed that a large portion of the overall TOC removal in the plant operating without prechlorination occurred prior to GAC adsorption. This figure represents combined removal including all the processes.

Figure 2 presents mass removal data for chloroform for the two GAC contactors which received prechlorinated influent. Note that although the ozone/GAC train had less TOC applied, it was less effective in removing chloroform. This figure must be interpretted carefully because the ozonation unit process stripped a large portion of the chloroform from the process stream applied to the ozone/GAC train. However, in the long term, much of the chloroform applied to the ozone/GAC train eluted from the contactor. The GAC train retained chloroform at greater than that amount predicted by equilibrium isotherms (20), but the ozone/GAC train was well below the predicted surface concentration. Table 2 shows the observed and predicted mass loading at the peak load and after one year of operation. These predictions were made based on the different influent chloroform concentrations. The influent concentration used for prediction was the average of the three weeks prior to the observed surface concentration.

DISCUSSION

Three parameters were used to judge the GAC contactors performance Microbial growth was used to assess TOC removal by biodegradation. TOC removal was used to assess overall organic removal by adsorption and biodegradation. Chloroform removal was used to assess the interaction between adsorption and ozone induced biodegradation on GAC.

Microbial Growth

The detailed results have been presented elsewhere (9,10). Ozonation was shown to increase biodegradability, however, accelerated bioactivity was not observed. The data indicate that the preozonated GAC contactors had substantially less TOC applied due to biodegradation. Thus, competition for adsorption sites should have theoretically been reduced, but competition may actually have increased for chloroform.

TOC Removal

The use of ozone prior to application of process water to GAC has been shown to increase TOC removal in the carbon contactors in this study, both for the chlorinated and nonchlorinated plants. The steady state removal was greater for ozone/GAC contactors than for GAC contactors by approximately 10 to 15 per cent. More efficient TOC removal in the nonchlorinated plant resulted in final TOC concentrations of smaller magnitude in the nonchlorinated plant than in the chlorinated plant.

This result demonstrates that direct comparison of a plant using ozone/GAC technology with a conventional U.S. treatment plant is inappropriate. TOC removals were greatly affected by chlorination as a pretreatment for disinfection. Chlorine appears to do more than inhibit biological growth in the conventional treatment processes. It may react with the background organics to make them less biodegradable, because the biological TOC removal in the chlorinated ozone/GAC train is less than the sum of the TOC removal in the nonchlorinated plant conventional unit treatment processes plus the biological TOC removal in the nonchlorinated plant ozone/GAC train. The data in this study cannot conclusively prove reduced biodegradability because of the possibility of improved floc/sed in the nonchlorinated plant, as discussed elsewhere (17).

Chloroform Removal

Ozonation preceding GAC adsorption had two effects on the removal of chloroform. First, it reduced the peak mass removed of a specific compound in comparison to its equilibrium surface concentration calculated from literature isotherm values. Second, it tended to displace previously adsorbed chloroform (i.e., a chromatographic effect where the instantaneous effluent concentration is greater than the influent concentration).

Due to variable influent concentration of several of the compounds investigated, it is difficult to distinguish between the mechanism of displacement (desorption) of a compound through re-equilibration after a peak load of the compound, and displacement of the compound by competing adsorbable solutes. However, the chloroform data in this study appear to show re-equilibration as the mechanism behind declining surface concentrations in the absence of preozonation, while re-equilibration and displacement both appear to be operating on the ozone/GAC treatment train. This can be seen in Table 2 where the ozone/GAC contactor retained chloroform at a surface concentration which was only 55 percent of the predicted value, whereas GAC alone retained chloroform at greater than 100 percent of the predicted value. This suggests that the chloroform was being displaced by competing substances. Ozonation is the most likely source for production of competing compounds because it is the only difference between the two treatment trains.

Competition between the specific organics appears to be the most important mechanism in determining removal (21). Chloroform breaks through relatively rapidly, and there are few compounds in the water which appear to compete with it (due to the fact that it achieves 100 + % of the surface concentration predicted by isotherms). All other compounds fail to reach 100%, and preozonation forces the value as low as 55%. Ozonation appears to breakup background organics such that the products can compete with chloroform. Thus, competition must be a strong consideration in removal of mixtures of organic compounds from water, and the production of low molecular weight organics in the ozonation process appears to increase the competition for adsorption sites, and consequently decrease removal. This decreased effectiveness of GAC can be masked if the organics being analyzed are effectively stripped in the ozone contactor.

CONCLUSIONS

The results of the study indicate that the arrangement of the unit processes is an important consideration and may change depending on the desired treatment goal. The most effective of the treatment trains investigated (cost not considered) for TOC removal is conventional treatment without prechlorination, followed by ozone/GAC. However, a substantial portion of the TOC removal occurs in the conventional treatment train when prechlorination is not practiced. Economic considerations, discussed elsewhere (17), indicate that the cost of the ozonation unit operation is greater than the benefit received from extended bedlife.

Eliminating prechlorination may not be acceptable to many U.S. water treatment plant operators because of the potential for bacterial problems in the plant and associated fouling of sand filters. In this case, adding ozone/GAC to the treatment train will produce lower effluent TOC values than GAC alone. Other data have shown that ozone reduces adsorbability (4), so that improvement for TOC removal may not always occur with preozonation. It should be noted that parallel studies (17) indicated that post chlorination was effective in reducing bacteria densities to acceptable levels for all of the treatment trains presented here.

The addition of ozone followed by GAC to conventional water treatment proved to be less effective for chloroform removal (in the GAC contactor) than GAC alone. This suggests that the ozone unit process may produce light weight organics which compete with chloroform for adsorption sites more effectively than natural background organics.

If ozonation is producing light weight organics, then GAC following ozonation may not protect "vulnerable" water supplies (14) from spills of lightweight organics as well as GAC alone, unless the organic is efficiently stripped in the ozonation process.

Table 1. Concentrations of Organic Parameters in the Influent of the GAC Contactors for the Four Treatment Trains - All Values ug/L

	Chlorinated Plant				Non-Chlorinated Plant			
	GAC		Ozone/GAC		GAC		Ozone/GAC	
	Ave.	Range	Ave.	Range	Ave.	Range	Ave.	Range
TOC	2559	-	2452	-	1826	-	1622	-
CHLOR	73	20-163	46	16-110	4.3	0.3-43	2.8	0.2-26

TABLE 2. Observed and Predicted Mass Removals for Chloroform Chlorinated Plant

	PEAK LOAD		ONE YEAR	
	GAC	O3/GAC	GAC	O3/GAC
Observed Mass Removal, mg/g	1.36	0.83	0.93	0.26
Influent Concentration, mg/L	99	65	38	28
Predicted Mass Removal, mg/g	1.18	0.87	0.59	0.47
Percent of Predicted Value (Obs./Pre.)	115	95	150	55
Day on which Peak Occurred	73	66	354	354

LITERATURE CITED

1. Rice, R. G., et al, "A Review of the Status of Preozonation of Granular Activated Carbon for the Removal of Dissolved Organics and Ammonia from Water and Wastewater", in Carbon Adsorption Handbook, Cheremisinoff, P.N., and Ellerbusch, F., (Eds.) Ann Arbor Science, Ann Arbor, MI, (1978) p 485

2. Constantine, T. A., "Advanced Water Treatment for Color and Organics Removal", JAWWA, 74, (6), 310, (1982)

3. Mallevialle, J., "Transformation of Humic Acids by Ozone", in Oxidation Techniques in Drinking Water Treatment, EPA-570/9-79-020, (1979), p 291

4. Benedek, A., et al, "Mechanistic Analysis of Water Treatment Data", Ozonews, 6, (1), (1979)

5. Narkis, N. and M. Schneider-Rotel, "Evaluation of Ozone Induced Biodegradability of Wastewater Treatment Plant Effluent", Water Research, 14, 929, (1980)

6. Frick, B., et al, "Predicting Competitive Adsorption Effects in Granular Activated Carbon Filters", in Activated Carbon Adsorption of Organics from the Aqueous Phase, Volume 1, eds. McGuire, M.J., and Suffet, I.H., (Eds.) Ann Arbor Science, Ann Arbor, MI, (1980), p. 507

7. Snoeyink, V. L., et al, Activated Carbon Adsorption of Trace Organic Compounds, EPA-600/2-77/223, 1977

8. National Academy of Science, "Drinking Water and Health, Volume 2, National Academy Press, Washington, D.C., (1980)

9. Maloney, S. W., et al, "TOC Removal on Sand and GAC" in Proceedings of the 1982 ASCE National Conference on Environmental Engineering, ASCE, New York, NY, (1982), p 600

10. Bancroft, K., et al, "Assessment of Bacterial Growth and Total Organic Carbon Removal on Granular Activated Carbon Contactors", Applied and Environmental Microbiology, In Press (1983)

11. Singer, P. C., et al, "Trihalomethane Formation in North Carolina Drinking Waters", JAWWA, 73, (8), 392, (1981)

12. Carter, C. W, and I. H. Suffet, "Binding of DDT to Dissolved Humic Materials", Environ. Sci. and Tech., 16, (11), 735, (1982)

13. Pirbazari, M, and W. J. Weber, Jr., "Removal of Dieldrin from Water by Activated Carbon" in Proceedings of the 1982 ASCE National Conference on Env. Eng. ASCE, New York, NY, (1982) p 506

14. USEPA, National Interim Primary Drinking Water Regulations, 44 Federal Register 68624 (Nov. 29, 1979)

15. Yohe, T. L., I. H. Suffet, and P. R. Cairo, "Specific Organic Removals by Granular Activated Carbon", JAWWA, 73, (8), 402, (1981)

16. Summers,R. S., and P. V. Roberts,"Dynamic Behavior of Organics in Full Scale Granular Activated Carbon Columns", in Treatment of Water by Granular Activated Carbon, eds. McGuire, M.J., and Suffet, I.H., (Eds.) Advances in Chemistry Series: 202, American Chemical Society, Washington, D.C., (1983), p 503

17. Neukrug, H.M., et al, "Removing Organics from Philadelphia Drinking Water by Combined Ozonation and Adsorption" Final report to EPA, Project # 806256-02 In press (1983)

18. Dohrman,"Dohrman DC-54 Ultra Low Level Total Organic Carbon Analyzer System Equipment Manual," Dohrman Division of Envirotech Corporation, Santa Clara, CA, (1977)

19. Environmental Protection Agency, "Guidelines Establishing Test Procedures for the Analyses of Volatile Halogenated Organic Compounds," Federal Register 44 (233), Dec. 3, (1979)

20. Weber, W. J., Jr., et al, "Effectiveness of Activated Carbon for Removal of Volatile Halogenated Hydrocarbons from Drinking Water" in Viruses and Trace Contaminants in Water and Wastewater, eds. Borchardt, J.A., (Eds.) et al, Ann Arbor Science, Ann Arbor, MI, (1977), p 125

21. Yohe, T. L.,"Specific Organic Removals and The Significance of Competitive Adsorption in Granular Activated Carbon Treatment of Drinking Water," Ph.D. Thesis, Drexel University, Philadelphia, (1982)

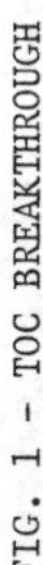

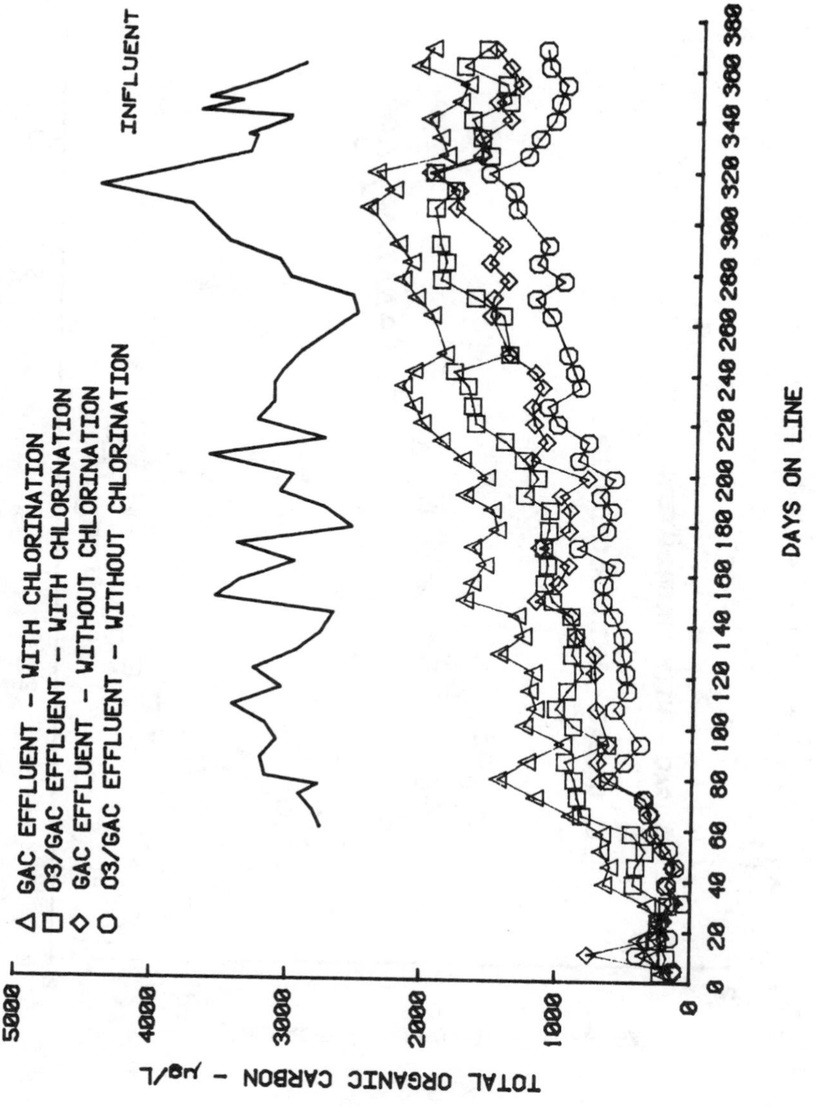

FIG. 1 - TOC BREAKTHROUGH

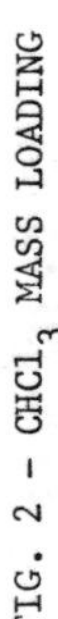

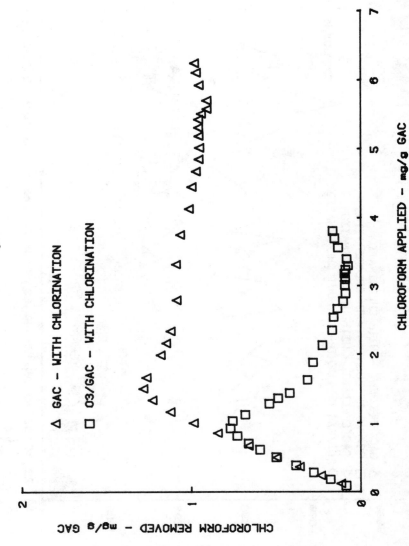

FIG. 2 - $CHCl_3$ MASS LOADING

ESTIMATION OF THE EFFECTS OF HUMIDITY ON THE ADSORPTION ONTO ACTIVATED CARBON OF THE VAPORS OF WATER-IMMISCIBLE ORGANIC LIQUIDS,

Milton Manes, Chemistry Department, Kent State University, Kent, OH 44242

ABSTRACT

A Polanyi-based model is presented that assumes a two-phase adsorbate and predicts zero adsorption of water vapor wherever the (liquid) volume of (water-free) organic adsorbate exceeds the adsorbate volume of (organic-free) water vapor; it makes quantitative predictions for the adsorption of water vapor and single or multiple organic vapors from the individual isotherms, which may in turn be experimental or estimated. It is well suited for estimating the conditions under which water vapor will significantly affect the adsorption of the organic vapors; for example, the effect of humidity tends to be negligible at relatively high capacities and at all but the highest relative humidities, and is most pronounced at very low loadings, where humidity may make as much as a tenfold difference in capacities. These factors may be expected to be of considerable importance in the removal of trace atmospheric contaminants. Conditions are also given that are best suited for distinguishing between alternative adsorption models. Finally, the calculations are readily adaptable to earlier work on the equilibrium adsorption of multicomponent organic vapors as chromatographic bands in carbon columns. (Data are presented in a companion article by Grant, Joyce, and Urbanic (9)).

Consider first the problem of estimating the effect of atmospheric humidity on the adsorptive removal of trace contaminants from air by activated carbon. Most such contaminants are the vapors of organic liquids (e.g., hydrocarbons) that are mutually miscible and relatively insoluble in water. The Polanyi adsorption potential model has already been applied to closely related systems, some of which would appear to be of greater complexity. For example, Greenbank and Manes (1) have extended the Polanyi-based treatment of Wohleber and Manes (2) to the adsorption of multiple (partially water-miscible) organic liquids from water solution; these treatments should become equivalent to a vapor-phase approach as the relative humidity approaches saturation. An alternative approach, which turns out to be exactly equivalent, is to start with a somewhat modified Grant-Manes treatment (3) for multiple organic vapors and to find an appropriate method for incorporating the adsorption of (immiscible) water vapor.

In addition to the obvious practical interest in the effect of humidity on the adsorptive removal of atmospheric contaminants, the coadsorption of water presents an interesting challenge to existing models for multicomponent vapors (e.g., the Myers-Prausnitz (4) and the Polanyi-based Grant-Manes (3) approaches), which have yet to come to grips with the possibility of a two-phase adsorbate (although it does appear in the Wohleber-Manes (2) treatment of liquid-phase adsorption). We shall later consider the experimental conditions for which alternative models give strongly divergent results.

We shall first consider a Polanyi-based model that incorporates the concept of a two-phase adsorbate, and then consider some alternatives and their experimental consequences.

THEORETICAL

As in earlier publications (1), we assume the basic Polanyi model, which postulates an "adsorption space" (corresponding to pores of different sizes and shapes in activated carbon), and an adsorption potential, ε_A (for adsorbate A) that is equal to the loss of potential energy per mole relative to the bulk phase, and that varies within the adsorption space from some maximum value to zero, depending on proximity to the carbon surface. In this model, a vapor in the adsorption space concentrates, relative to the bulk phase, to an extent that depends on the local value of ε and is given by the Boltzmann equation,

$$p_A^*/p_A = \exp(\varepsilon_A/RT) \qquad (1)$$

where p_A^* and p_A are the local and bulk values of the partial pressure of component A, and ε_A is the local value of the adsorption potential. Condensible vapors will liquefy wherever the local partial pressure exceeds the saturation pressure $(p_S)_A$, or wherever

$$\varepsilon_A \geq RT \ln(p_S/p)_A \quad . \qquad (2)$$

Adsorption (here defined as the amount of component A in excess of what the adsorption space would contain if filled with bulk gas) results from the excess concentration of vapor and of the enhanced liquefaction; although both can be explicitly dealt with in the Polanyi model, the enhanced liquefaction usually accounts for practically all of the adsorption, and is the only effect we here consider.

A plot of the total (liquid) volume adsorbed against the corresponding value of ε_A at the presumed liquid-vapor interface (i.e., where the equality holds in Equation (1)) is referred to as the "characteristic curve" for the adsorbate-carbon system; its shape is determined by the structure of the carbon, and it is quite analogous to a cumulative pore-size distribution, where the pores are characterized by energy rather than dimensions. In the Polanyi model, the ratio $\varepsilon_A/\varepsilon_B$ of two components in the same location remains constant for all locations, with the result that the characteristic curves for all adsorbates on a given carbon are the same, except for an abscissa scale factor that can either be determined directly from individual isotherms, or else estimated by various methods. (This assumption is useful but not strictly necessary for calculating multicomponent adsorption.)

An adsorption isotherm (mass adsorbed vs. relative pressure) is readily transformed into a characteristic curve; the adsorbed mass is converted to an adsorbate (liquid) volume by using the bulk liquid density (or an adjusted adsorbate density), and the equilibrium partial pressure (or fugacity) converted to the equilibrium (or interface) adsorption potential by using Equation (2). The reverse calculation is similarly carried out. As shown elsewhere (5) the model is quite analogous to the behavior of liquefiable vapors in an extremely powerful gravitational field.

We now consider the adsorption of the vapor of a single water-immiscible liquid in the presence of water vapor, and we start with saturated water vapor. This situation is exactly equivalent to the adsorption of a vapor from water solution. In this case the net adsorption potential for the liquefied vapor is its normal adsorption potential, diminished by the adsorption potential of an equal volume of water. The resulting equation is as given by Polanyi (6), namely

$$RT \ln(p_s/p)_A = \varepsilon'_A = \varepsilon_A - \varepsilon_W V_A/V_W \quad , \qquad (4)$$

where ε'_A is the net (organic-water) interface adsorption potential corresponding to any given (organic liquid) adsorbate volume, ε_W and ε_A are the corresponding adsorption potentials read from the individual characteristic curves (estimated or experimenal) for water and for component A on the same carbon, and V_A and V_W are the respective molar volumes. Given the individual adsorption isotherms, one can use Equation (4) to calculate the adsorption isotherm in saturated vapor. Again, the calculation is analogous to the gravitational model, and the correction for water is quite analogous to Archimedes' principle. Moreover, the problem becomes exactly equivalent to the Wohleber-Manes (1) treatment for liquid-phase adsorption of immiscible liquids if one substitutes concentrations for partial pressures. We now consider adsorption in unsaturated water vapor. Here (following Hansen and Fackler (7)) we consider ΔG_A for adsorption of a liquid at any point in the adsorption space as

$$\Delta G_A = -\varepsilon_A + RT \ln(p_s/p)_A \, , \tag{5}$$

so that ΔG for adsorption is zero at the (single adsorbate) interface, positive above the interface, and negative below it.

Then for the adsorption of water vapor,

$$\Delta G_W = -\varepsilon_W + RT \ln(p_s/p)_W \tag{6}$$

$$= -\varepsilon_W - RT \ln H \, , \tag{7}$$

where H is the (fractional) relative humidity. For the condensation of an organic liquid phase below the presumed water-vapor interface, the two conditions (noted by Hansen and Fackler (7)) are:

$$dG/dn_A = \Delta G_A + \Delta G_W \, (dn_W/dn_A) \tag{8}$$

and

$$V_A dn_A + V_W dn_W = 0 \, , \tag{9}$$

whence

$$\frac{RT}{V_A} \ln \frac{p_s}{p}_A = \frac{\varepsilon_A}{V_A} - \frac{\varepsilon_W}{V_W} - \frac{RT}{V_W} \ln H \, . \tag{10}$$

One can now find the adsorbate volume of any immiscible organic adsorbate as a function of its partial pressure and the relative humidity, wherever the organic adsorbate volume is less than the (liquid) volume of adsorbed water. If the single-component organic adsorbate volume is greater than the volume of adsorbed water, then there is zero interference by water vapor. One can imagine a situation in which the relative humidity is fixed and the partial pressure of the organic adsorbate is systematically increased from some initial low value. At low adsorptions the competitive effect of the water vapor is maximal, since it depends on the difference between the local and the interface adsorption potential of water. As the organic partial pressure increases, the effect of water displacement also decreases, becoming zero at the original water-vapor interface. The model therefore predicts significant interference by relative humidity with trace components.

Equation (10) may be readily related to the individual isotherms for the water and the organic adsorbate. At any given volume of the organic adsorbate the value of ε_A/V_A is readily read from the adsorption isotherm of the pure adsorbate vapor, and ε_W/V_W from the pure water isotherm. The final term is readily calculated from the relative humidity, and represents the effect of relative humidity on the displacing power of the water. If $p_A(V)$ is the equilibrium partial pressure of pure component A at total volume V, and if $p_W(V)$ is the partial pressure read from the water isotherm at the same volume, and if p_A^* and p_W are the equilibrium partial pressures of the bulk organic component and of water, then the effect of water is given by Equation (11) as

$$p_A^*(V) = p_A(V)(p/p(V))_W^{V_A/V_W}; \quad (p/p(V)) > 1 \qquad (11)$$

where again V_A and V_W are molar volumes.

The analogy between the Polanyi-based model and a gravitational model is illustrated in Figure 1. Assume a vessel of undetermined shape, in which a powerful gravitational field can concentrate the vapors of water and of organic liquid to condensation. Where both liquids can condense, their competition depends on the relative values of "adsorption potential density" (ordinary mass density in a gravitational field). At 100% RH the liquid water is at level "A" and has maximum competitive power; it raises the partial pressure of organic vapor that can condense at any given location. If the water level is at "C" (in absence of organic vapor), then the water vapor will have no effect on organic adsorption and its own adsorption will be zero. Equation (4) applies for water at "A" and Equation (10) for level "B".

The calculation of adsorbate volumes from experimental conditions is illustrated in Figure 2 for the adsorption of butane at zero, 60%, and 100% RH. The curve to the right is the correlation curve for butane on the carbon. Also shown is the estimated curve for water (2). The curve to the extreme left shows the adjustment for 60% relative humidity which is (T/18)log(1/0.6). The abscissa of this curve is subtracted from the butane correlation curve. For butane volumes above 60 cc/100g (the limit of adsorption of water), the water adsorption is zero. For lower butane volumes, the volume of water adsorbed is the difference between 60 cc and the butane volume. (The same calculation can be readily carried out with isotherms rather than correlation curves.)

The generalization to multiple organic adsorbates in the presence of water vapor is quite straightforward and not especially novel. We first consider the adsorbate volume and composition at the presumed adsorbate-water

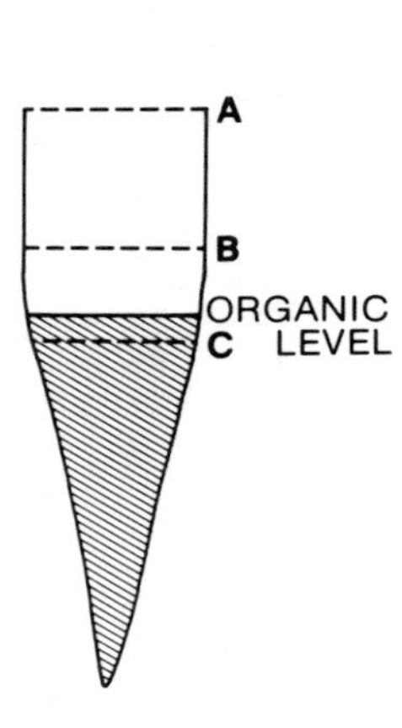

Figure 1. Schematic of two-phase model.

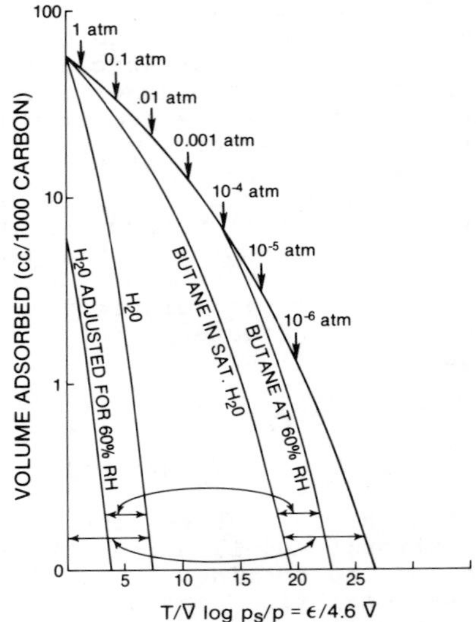

Figure 2. Illustrative graphic calculation for adsorption of butane and water at 60% and 100% R.H.

interface. We can then either assume the adsorbate to be uniform (which suffices for those adsorbates that do not exhibit significant adsorptive selectivity from the corresponding mixed liquids), or apply the more general Hansen-Fackler (7) approach to adsorbate nonuniformity, which becomes equivalent to adsorption from mixed liquids in the limit of saturated vapors.

Although the Grant-Manes (3) treatment is expressed in terms of correlation curves, we can express it equally well in terms of adsorption isotherms, where adsorbate volume is plotted against relative pressure. We start with the assumption that the partial pressure of each organic component at the organic-water (or organic-vapor) interface is the same as for a bulk liquid mixture of equal composition, where the vapor pressure of each component is the pressure p_i^* that is read from the isotherm of the corresponding component, this time as a single adsorbate in the presence of water vapor. The mole fraction, x_i, of each component at the interface (assuming ideal solution for simplicity) is given by

$$x_i = p_i/p_i^* \qquad (12)$$

For any assumed total adsorbate volume one reads the set of p_i^* from the individual isotherms, calculates the x_i, and determines whether they sum to 1. If not, the value of V is correspondingly adjusted. From here on the calculation is exactly the same as in Grant-Manes (3) for vapors or in Greenbank-Manes (1) (for mixed liquids from water solution).

Suppose now that the components are not of equal adsorption potential per unit volume, so that the adsorbate composition is not uniform. Here we use the Greenbank-Manes (1) modification of the Hansen-Fackler (7) calculation. (A similar derivation to the Greenbank-Manes modification appears in Schenz (8).) The only difference between the two is in the reference solution; in the original Hansen-Fackler treatment it is the bulk solution, whereas here it is the solution at the organic side of the organic-water interface. The appropriate equations are:

$$-\frac{\varepsilon_i-\varepsilon_i^+}{V_i}+\frac{RT}{V_i}\ln\frac{x_i}{x_i^+}=-\frac{\varepsilon_j-\varepsilon_j^+}{V_j}+\frac{RT}{V_j}\ln\frac{x_j}{x_j^+} \qquad (13)$$

for all i and j, where the values of ε_i, ε_j and ε_i^+ and ε_j^+ corresponding to each value of V (the local cumulative adsorbate volume) may be read directly from the corresponding isotherms, either with or without the water-vapor correction; both alternatives used consistently give the same answer.

The next step is to determine the mole numbers, n_i, in the total adsorbate by integration over the adsorbate volume, this time to the organic-water interface. This is carried out in the same fashion as in Greenbank-Manes (1), and is readily programmed for a relatively small computer.

DISCUSSION

The utility of the basic model for predicting the effects of humidity on the adsorption of trace contaminants will be illustrated by the data to be presented by Grant et al. (9). The model has the additional advantages of being straightforward and thermodynamically consistent; as noted earlier, the mathematical treatment is wholly equivalent to the behavior in a gravitational field of the vapors of immiscible liquids of different densities. Moreover, the model is essentially the same for both gas-phase and liquid-phase adsorption, and applies without difficulty from trace concentrations to saturation. Furthermore, it is readily incorporated into a computer program that calculates equilibrium adsorption of multicomponent adsorbates in columns, following the procedures of Manes and Grant (10).

However, the agreement of the model with at least some experimental data does not necessarily exclude alternative models, which we here consider. We first consider the basic Polanyi model with the assumption of ideal miscibility of the adsorbate, and we consider the alternative possibilities of a uniform (interface composition) and a nonuniform (Hansen-Fackler (7)) adsorbate. The first alternative may be ruled out because it leads to the unrealistic expectation of no selective adsorption from a water solution (11). However, Greenbank and Manes (1) found that the miscible (nonuniform) and the immiscible (Polanyi-based) models gave similar results for solute mixtures of components with widely differing adsorption potential densities. They also found quite frequently similar results with the Radke-Prausnitz (12) model, which is essentially the same as the Myers-Prausnitz (4) model for vapors. By analogy, one would expect the latter model to work at least reasonably well for trace vapor adsorbents in water. However, it may be expected to break down at relatively high adsorbate loadings of organic adsorbate, where the calculated "spreading pressures" exceed that of saturated water vapor. Therefore, although the immiscible adsorbate model presented here has the advantage of utility and generality, other models may be expected to work, at least for selected systems, although they may break down for others.

One should note that a critical test between alternative models would be provided by the determination of the adsorption of water vapor in the presence of immiscible organic adsorbates. The immiscible model is the only one of the preceding that predicts complete exclusion of adsorbed moisture under prescribed conditions that are readily derivable from the model. However, the experimental test would have to be carried out on ash-free carbons in order to avoid specific interactions with water vapor.

The difference in behavior between these alternative models is illustrated in Figures 3 and 4, which give estimated calculations for butane adsorption on a typical gas-phase carbon from 1 to 10^{-6} atm. Figure 3 shows the behavior of the immiscible model for butane adsorption, illustrating the pronounced effect of humidity at low butane capacities, and zero effect at higher loadings. Figure 4 compares the predictions of the miscible and the Myers-Prausnitz models at 100% RH with the predictions of the immiscible model. The alternative models give similar results at some partial pressures and diverge more sharply at others; the Myers-Prausnitz model is out of bounds at high butane capacities. As Figure 5 shows, the most striking differences between the alternative models are in water adsorption, where the immiscible model predicts zero water adsorption, whereas the other models predict significant water adsorption under similar conditions. The fact that the models diverge sharply under some conditions but agree reasonably well under others suggests the kind of critical experiments that should best be done if one wishes to distinguish between the alternative models.

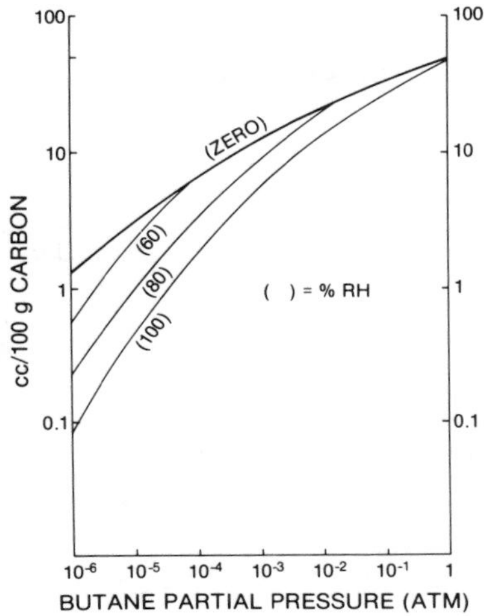

Figure 3. Calculated adsorbate volumes of butane on a typical carbon (two-phase model).

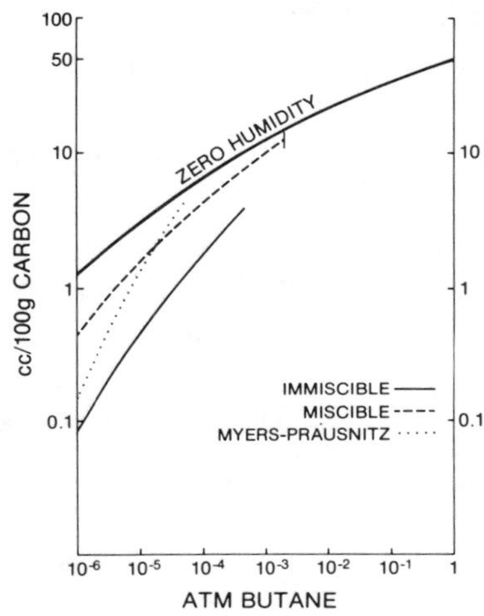

Figure 4. Calculated adsorbate volumes for butane at 100% R.H. by alternative models.

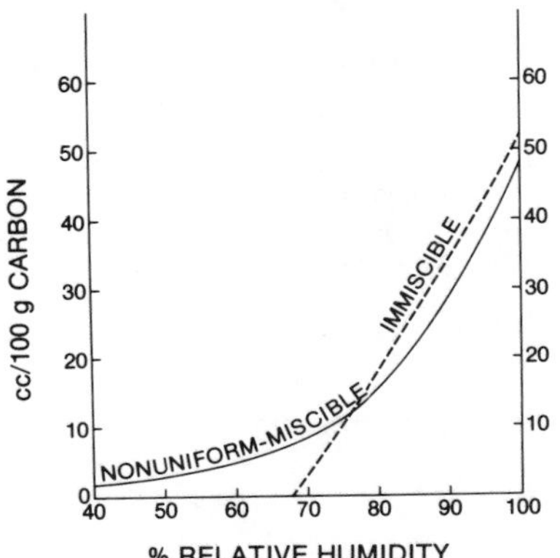

Figure 5. Comparison of prediction of water adsorption by the two-phase and the miscible (nonuniform) models.

The utility of the Polanyi-based immiscible model will be illustrated in a companion article by Grant et al. (9). One may anticipate that the evidence thus far tends to support the idea of a two-phase adsorbate.

LITERATURE CITED

1. Greenbank, M. and M. Manes, J. Phys. Chem. 85, 3050 (1981).

2. Wohleber, D. A. and M. Manes, J. Phys. Chem. 75, 61 (1971).

3. Grant, R. J. and M. Manes, Ind. Eng. Chem. Fundamenals 5, 490 (1966).

4. Myers, A. L. and J. M. Prausnitz, A.I.Ch.E.J. 11, 121 (1965).

5. Manes, M. and M. Greenbank, "Treatment of Water by Granular Activated Carbon," in Advances in Chemistry Series #202, McGuire, M. J. and I. H. Suffet, (eds.) (Am. Chem. Soc.), p. 9.

6. Polanyi, M., Z. Elekrochem. 26, 370 (1920); Z. Physik 2, 111 (1920).

7. Hansen, R. S. and W. V. Fackler, J. Phys. Chem. 57, 634 (1953).

8. Schenz, T. W., Ph.D. Dissertation, Kent State University, Dec. 1973 (p. 29).

9. Grant, R. J., R. S. Joyce, and J. E. Urbanic (this volume).

10. Manes, M. and R. J. Grant, "Estimation of Multicomponent (Chromatographic) Gas Adsorption on Activated Carbon Columns," presented at 10th Annual Meeting of A.I.Ch.E., New York, Nov. 15, 1977.

11. Myers, A. L. and S. Sircar, J. Phys. Chem. 76, 3412 (1972).

12. Radke, C. J. and J. M. Prausnitz, A.I.Ch.E.J. 18, 761 (1972).

Dynamics of Adsorption in a Fixed Bed of Polydisperse Particles

Alexander P. Mathews, Assistant Professor
Department of Civil Engineering
Kansas State University
Manhattan, Kansas 66506, USA

ABSTRACT

The effect of a distribution of adsorbent particle sizes on the breakthrough profiles from fixed-bed adsorbers, is examined using the homogeneous solid phase adsorption model. The bed is assumed to be stratified with the smaller particles at the top and the larger particles at the bottom of the bed. A three layer model, with the logarithmic mean diameter representing each layer, provides predictions in good agreement with experimental data for the adsorption of p-chlorophenol on activated carbon. A dispersion of particle sizes results in earlier breakthrough in the downflow mode, and as shown by theoretical analysis, in delayed breakthrough in the upflow mode, when compared to a uniform particle size.

Adsorption from the liquid phase in batch and fixed-bed reactors has been studied extensively from both experimental and modeling standpoints over the last three decades. The importance of liquid-phase adsorption in the process industries, water purification, wastewater renovation, and the control of toxic material in the environment, has accelerated theoretical and experimental developments in this area. The prediction of breakthrough profiles from fixed beds is of importance in both design and operation of adsorption systems. Several models have been proposed to this end, incorporating transport resistances in both fluid and solid phases coupled with nonlinear isotherm (Morton and Murril, 1967; Stuart and Camp, 1973), and bimodal pore distributions (Ruckenstein, et al., 1971; Kawazoe and Takeuchi, 1974; Peel and Benedek, 1980). However, none of these models have accounted for the effects of a distribution of particle sizes that may be found in the adsorbers.

Commercial adsorbents generally contain a range of particle sizes and shapes. Activated carbon used in water and wastewater treatment contain a range of particle sizes from U. S. Standard Sieve No. 8 to Sieve No. 40. The effect of a Gaussian particle size distribution on adsorption kinetic curves has been theoretically studied by Ruthven and Longlin (1971) for molecular sieves, where intraparticle resistance through the micropores is the controlling rate mechanism. A linear isotherm was assumed for the solute. Sensitivity analysis indicated that both the shape and particle size distribution can significantly affect the adsorption kinetic curves and diffusivities estimated from these curves. Mathews (1983) presented a model for batch reactors with an adsorbent size distribution, for solutes with nonlinear equilibria, and where fluid and solid phase transport resistances are important. Cooney, et al. (1983), studied the effects of particle size distributions on adsorption in finite bath and infinite bath reactors.

The effect of particle size distribution on adsorption in fixed beds has been evaluated theoretically by Moharir, et al. (1980). The homogeneous solid phase diffusion model (Rosen, 1952) was used, and the predominant mass transfer resistance was assumed to be in the micropore diffusion in the particle. The isotherm was assumed to be linear.

In the present work, a general fixed-bed adsorption model is developed that includes unsteady state transport within the particle, external mass transfer from the fluid phase to the solid phase, and a general nonlinear isotherm. The effect of a log-normal particle size distribution on effluent breakthrough profiles is presented. Several types of mean diameters are examined to represent the size distribution, and the prediction using a single mean diameter are compared with experimental data.

MATHEMATICAL MODEL

The homogeneous solid phase adsorption model proposed by Rosen (1952) for solutes with linear isotherms is adapted here for adsorption in a fixed bed with a range of particle sizes. The bed is assumed to be stratified, with the smallest particles at the top and the largest particles at the bottom. If there are P layers of discrete size fractions, each with bed length L_i, the following rate, equilibrium and bed material balance equations apply for i^{th} size fraction with particle radius R_i.

The external mass transfer rate is equal to the average particle uptake rate as shown in Equation (1).

$$\frac{k_i R_i^2 (C - C_s)}{\rho} = \frac{\partial}{\partial t} \int_0^{R_i} q_i \, r^2 dr, \quad i = 1, P \tag{1}$$

Intraparticle transport for a spherical particle assuming concentration independent diffusivity and symmetry in two directions is given by

$$\frac{\partial q_i}{\partial t} = \frac{D}{r^2} \frac{\partial}{\partial r} \left(r^2 \frac{\partial q_i}{\partial r}\right), \quad i = 1, P \tag{2}$$

A further subdivision of intraparticle diffusion into micropore and macropore diffusion is not made, since this increases parameter estimation requirements without increasing model prediction accuracy (Thacker, et al., 1981). The initial and boundary conditions for particle size fractions 1 to P are

$$q_i\ (r,0) = 0 \tag{3}$$

$$q_i\ (R,t) = q_{si}(t)$$

$$\frac{\partial q_i(0,t)}{\partial r} = 0 \tag{4}$$

Adsorption equilibrium at the surface is given by

$$q_s = \frac{aC_s}{1 + bC_s^\beta}, \quad \beta \leq 1 \tag{5}$$

The fixed-bed adsorber material balance equation and the initial and boundary conditions for the bed are:

$$\frac{\partial C}{\partial t} + v\,\frac{\partial C}{\partial z} + \frac{3(1 - \varepsilon)k_i\ (C - C_s)}{\varepsilon R_i} = 0 \tag{6}$$

$$C(z,0) = 0 \tag{7}$$

$$C_1(0,t) = C_0 \tag{8}$$

$$C_i(0,t) = C_{i-1}\ (L_{i-1},\ t), \qquad i = 2,\ P \tag{9}$$

The boundary conditions represented by Equations (8) and (9) state that the influent to the first layer is at constant concentration C_0, and the influent to any subsequent layer is the effluent from the preceding layer. The model equations were solved using a modification of the orthogonal collocation procedure developed by Thacker, et al. (1981).

EXPERIMENTAL METHODS

Reagent grade p-chlorophenol was obtained from the J. T. Baker Chemical Company. Bituminous coal-based activated carbon supplied by the carborundum company was used as the adsorbent. The adsorbent was sieved into eight size fractions, washed with high purity deionized water and dried to constant weight for use in fixed-bed studies. Powdered carbon of size 200 μm to 230 μm was used for the equilibrium studies.

Adsorption equilibrium studies were conducted by the bottle point method at a constant temperature of 17°C, using tap water passed through a pre-adsorption column. Approximately six days were required for achievement of equilibrium following which, the solutions were filtered and analyzed using ultraviolet/visible spectrophotometry.

Fixed-bed studies were conducted in a plexiglass column 5.08 cm in diameter. The water was supplied to the column by a Masterflex pump fed from a constant head tank filled with tap water filtered through a pre-adsorption column. A metering pump discharged p-chlorophenol to the discharge side of the pump. The column was operated downflow. Inlet flow rates and concentrations were monitored to assure constant input conditions. Outlet concentrations were measured to obtain the breakthrough profiles.

RESULTS AND DISCUSSION

The parameters required for the prediction of fixed-bed breakthrough profiles using the model Equations (1) to (9) are the isotherm constants, and the external mass transfer and internal diffusion coefficients. The adsorption isotherm for p-chlorophenol in tap water at a temperature of 17°C is shown in Figure 1. The data points are fitted with the three-parameter isotherm over a wide concentration range. The isotherm constants a, b and β are 212.3, 123.4 and 0.872 respectively, when q and C are reported in mmoles/gm and mmoles/ℓ, respectively. The external mass transfer coefficient for the bed operating conditions is obtained from the correlation of Dwivedi and Upadhyay (1977). The surface diffusion coefficient for p-chlorophenol is estimated from batch reactor studies (Mathews, 1983) to be 1.11 x 10^{-8} cm^2/sec.

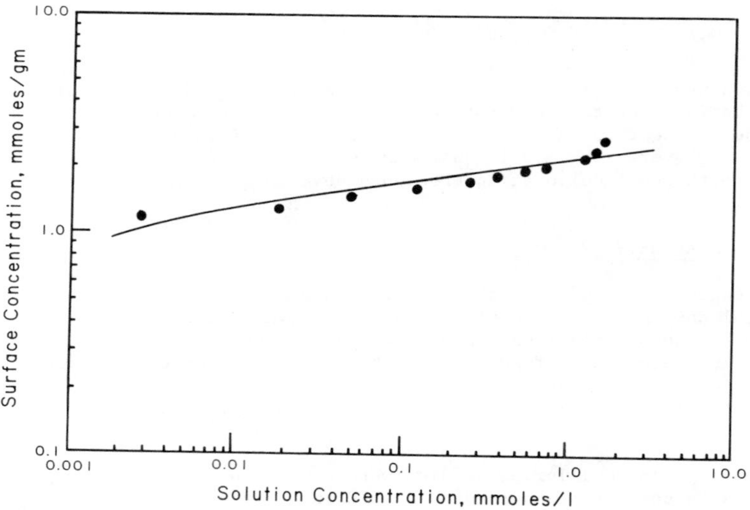

Figure 1: p-Chlorophenol isotherm at 17°C: (●) experimental data, (-) three-parameter isotherm

Fixed-Bed Adsorption With Particles of Uniform Size

The first experiment was conducted using uniformly sized particles of size 18/20 (passing Sieve No. 18 and retained on Sieve No. 20), with a p-chlorophenol influent concentration of 59.77 mg/ℓ, and a flow rate of 800 mℓ/min. The amount of carbon used in the column was 250 gms, and the bed height was 27.5 cm. The effluent breakthrough profile and the model predictions are shown in Figure 2. The excellent agreement between the two assures that the model can be reliably used to predict adsorption with mixed particle sizes.

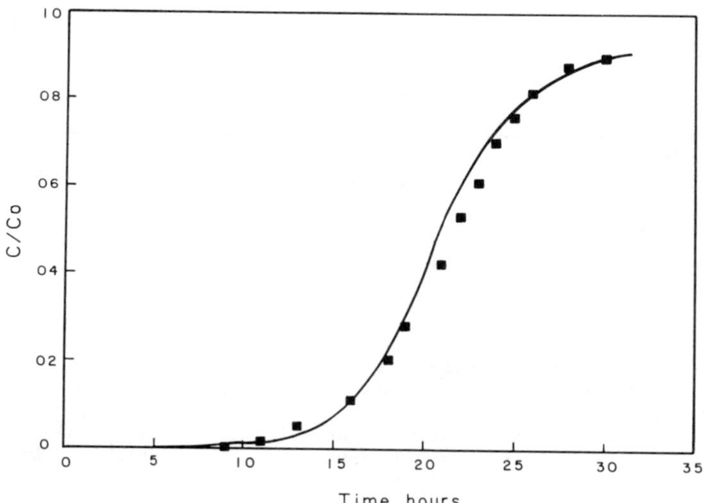

Figure 2: p-Chlorophenol breakthough profile with uniform particle size: (■) experimental and (-) predicted data

Fixed-Bed Adsorption With Particles of Varying Size

The second experiment was conducted using a particle size distribution comprising of eight different size fractions. The size distribution is log-normal as shown in Figure 3, and this is generally the distribution obtained when size reduction processes are employed during manufacturing. The experimental conditions were nearly identical to the previous run, with 250 gms of carbon, influent concentration 59.37 mg/ℓ, and flow rate of 800 mℓ/min. The bed height was measured to be 26.2 cm.

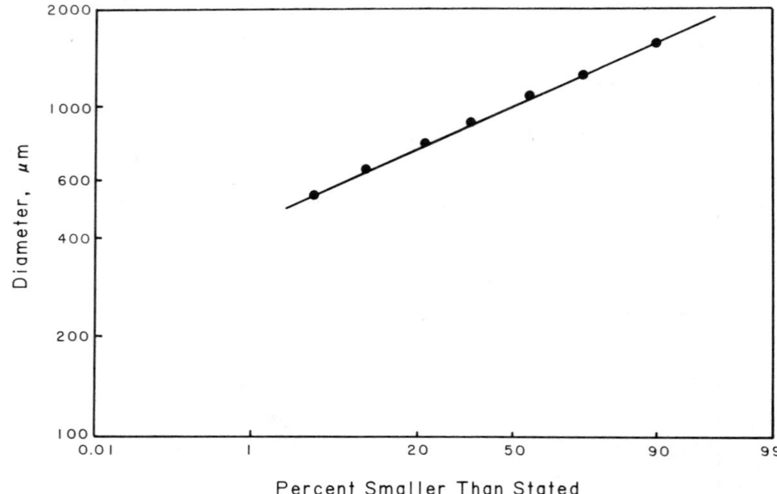

Figure 3: Particle size distribution:

Three types of mean diameters were used to characterize the particle size distribution. These are: (i) the log-mean (Smith and Jordan, 1964), (ii) length mean, and (iii) Sauter mean (Foust, et al., 1960). These diameters are listed in Table 1. The predictions using a single mean diameter are shown in Figures 4 and 5. Table 1 shows a comparison of the percent deviations between experimental and predicted data for the three means. As can be seen, the log-mean diameter gives a better representation if a single mean diameter is to be used. Other mean diameters, such as the surface mean, and volume mean were tried, but did not provide any better results and are not reported here.

The effect of using three distinct layers, as opposed to a single layer are also shown in Figures 4 and 5. In Figure 4, each of the three layers is represented by the length mean, and in Figure 5, each of the three layers is represented by the log-mean diameter. Table 1 lists these diameters and size fractions. In both cases predictions are much better than when a single layer is used. The log-mean diameter provides somewhat better results than the length-mean diameter. As the number of layers in the model is increased,

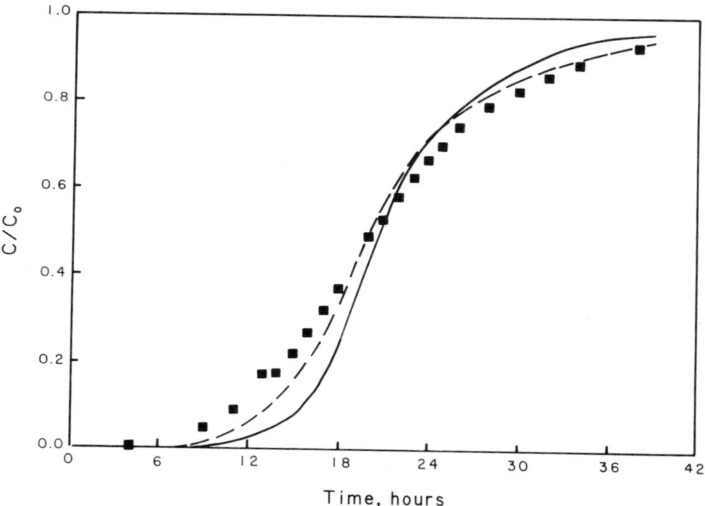

Figure 4: p-Chlorophenol breakthrough profile with size distribution:(■) experimental data, and predictions using (-) single length mean and (--) three length means.

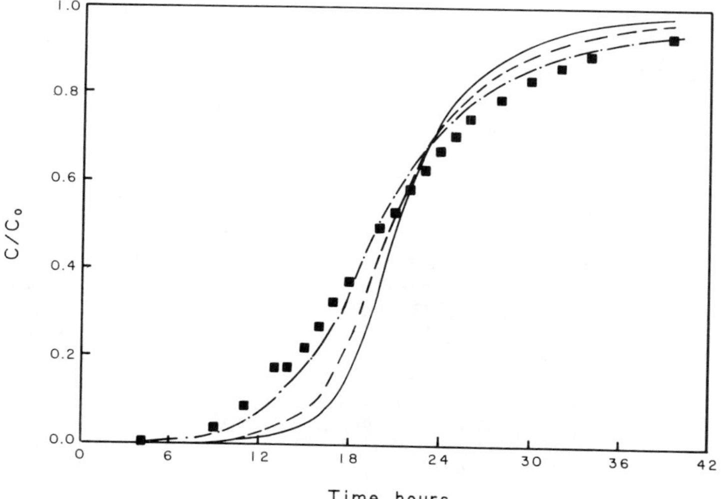

Figure 5: p-Chlorophenol breakthrough profile with size distribution: (■) experimental data, and predictions using (-) Sauter mean, (--) log-mean, and (-·-) three log-mean

predictions can be expected to become closer in agreement to the experimental data. However, computational time can also be expected to increase as the number of layers is increased.

Table 1. Comparison of Model Predictions for Various Particle Diameters

Type of diameter	Weight fraction	Diameter (cm)	Mass Transfer Coefficient (10^3 cm/sec)	Deviation experimental vs predicted*
Length mean	1	0.0977	4.9715	0.2889
Sauter mean	1	0.0880	5.3604	0.3567
Log-mean	1	0.1000	4.8879	0.2593
Three length means	0.26	0.1386	3.8641	
	0.38	0.0999	4.8900	0.1662
	0.36	0.0657	6.6119	
Three log-means	0.25	0.1500	3.6503	
	0.50	0.1000	4.8900	0.1211
	0.25	0.0670	6.5215	

*Deviation $= \frac{1}{N} \Sigma \, |(C_{ei} - C_{ci})/C_{ei}|$

Upflow Versus Downflow Operation

The experimental conditions for data reported in Figures 2 and 4 are identical except for the particle size distribution. However, as can be seen from the data, breakthrough occurs earlier in the downflow mode when a particle size distribution is present. This can also be seen from the theoretical analysis presented in Figures 4 and 5.

If the adsorber is operated upflow, the effect of a particle size distribution is to delay the breakthrough, compared to uniformly sized particles. This is shown by theoretical analysis in Figure 6. The extent to which breakthrough will be delayed is determined by the dispersion in the particle size distribution.

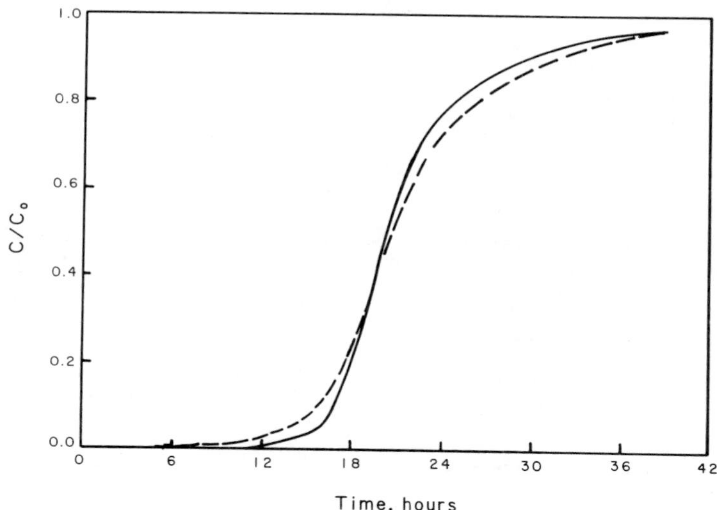

Figure 6: Compurison of breakthrough profiles for upflow operation: (--) uniform particle size, (-) size distribution

NOTATION

a, b	isotherm constants
C	liquid phase solute concentration
C_0	inlet concentration to first layer
C_s	equilibrium liquid phase concentration at particle surface
D	internal diffusion coefficient
k	external mass transfer coefficient
L	length of each layer
P	number of particle size fractions
q	pointwise concentration inside the adsorbent
q_s	equilibrium solid phase concentration at the particle surface
r	radial coordinate
R	particle radius
t	time
v	interstitial velocity
z	axial distance

Greek Letters

β isotherm constant

ε bed porosity

ρ apparent density of particle

LITERATURE CITED

1. Cooney, D. O., B. A. Adesanya, and A. L. Hines, "Effect of Particle Size Distribution on Adsorption Kinetics in Stirred Batch Systems," presented at the Spring National AIChE Meeting, Houston (1983).

2. Dwivedi, P. N., and S. N. Upadhyay, Ind. Eng. Chem. Proc. Des. and Dev., 16, No. 2, 157 (1977).

3. Foust, A. S., L. A. Wenzel, C. W. Clump, L. Maus, and L. B. Anderson, "Principles of Unit Operations," John Wiley and Sons, Inc., N.Y. (1960).

4. Kawazoe, K., and Y. Takeuchi, J. of Chem. Eng. of Japan, 7, No. 6, 431 (1974).

5. Mathews, A. P., "Adsorption in an Agitated Slurry of Polydisperse Particles," presented at the Spring National AIChE Meeting, Houston, (1983).

6. Moharir, A. S., D. Kunzru, and D. N. Saraf, Chem. Eng. Science, 35, 1795 (1980).

7. Morton, E. L., and P. W. Murrill, AIChE J., 13, 965 (1967).

8. Peel, R. G., and A. Benedek, J. Env. Eng. Div., ASCE, 106, 797 (1980).

9. Rosen, J. B., J. Phys. Chem., 20, 387 (1952).

10. Ruthven, D. M., and K. F. Loughlin, Chem. Eng. Science, 26, 577 (1971).

11. Ruckenstein, D., A. S. Vaidyanathan and G. R. Youngquist, Chem. Eng. Sci. 26, 1305 (1971).

12. Smith, J. E., and M. L. Jordan, J. Coll. Science, 19, 549 (1964).

13. Stuart, F. Y., and D. T. Camp, AIChE Symp. Ser., 33 (1973).

14. Thacker, W. E., V. L. Snoeyink, and J. C. Crittenden, "Modeling of Activated Carbon and Coal Gasification Char Adsorbents in Single-Solute and Bisolute Systems," Res. Report 161, University of Illinois Water Resources Center, July 1981.

ADSORPTION OF CATIONIC POLYACRYLAMIDE ON HEMATITE AND SILICA: EFFECT OF POLYMER-SODIUM DODECYLSULFONATE INTERACTION

Brij M. Moudgil* and P. Somasundaran
Henry Krumb School of Mines
Columbia University
New York, NY 10027

ABSTRACT

Effect of interactions between cationic polyacrylamide (PAMD) and sodium dodecylsulfonate on the polymer adsorption on hematite and silica was studied. PAMD adsorption on hematite increased or decreased depending on the order of reagent addition and the solution pH. Adsorption of the cationic polyacrylamide on silica was found not to be affected by the surfactant. Polymer adsorption on hematite and silica is discussed in terms of nature of polymer-surfactant interactions, surface charge characteristics of solids and solution chemistry of cationic polyacrylamides.

INTRODUCTION

Surfactant-polymer interactions are of importance in systems such as enhanced oil recovery by micellar flooding and mineral processing. However, the effect of such interactions on polymers and surfactant adsorption at the solid/liquid interface has been investigated only to a limited extent in the past. Tadros (1) has reported an increase in the adsorption of polyvinyl alcohol (PVA) on silica in the presence of cetyltrimethylammonium bromide (CTAB) and sodium dodecylbenzenesulfonate (NaDBS). Polymer and surfactants were premixed before contacting the silica particles with the solution. Increase in the PVA adsorption at pH 9.1 in the presence of CTA^+ ions was attributed to the CTA^+ ions acting as hydrophobic anchors between $-SiO^-$ on the surface and the PVA chains. Improvement in the PVA adsorption on silica at pH 3.6 in the presence of NaDBS was ascribed to the formation of a cationic PVA-NaDBS "polyelectrolyte complex." In the above study, adsorption of CTA^+ ions at pH 3.3 was determined to be higher in the presence of PVA. This was explained on the basis of PVA molecules acting as anchors between the undissociated SiOH groups and CTA^+ ions. Adsorption of DBS^- ions on silica with preadsorbed PVA at low pH was stated to be complex and no explanation was provided by the author for the observed increase in the DBS^- adsorption.

*Present Address: Department of Materials Science and Engineering
University of Florida, Gainesville, FL 32611

In the polyacrylamide/kaolin/sodium dodecylbenzenesulfonate system, Hollander (2), has reported a decrease in the surfactant adsorption in the presence of nonionic polyacrylamide. Polyacrylamide adsorption was also reported to decrease when surfactant was added to the system. The results have been explained on the basis of competitive adsorption of the two species on kaolin.

It should be noted that in none of the past studies effect of order of reagent addition has been investigated.

EXPERIMENTAL

Materials

Synthetic hematite (99.2-99.5% Fe_2O_3) of particle size between 0.1 and 10 microns was supplied by Pfizer Inc. Specific surface area of hematite using nitrogen as the adsorbate was determined by the Quantasorb to be 8.7 m^2/g.

Silica (Bio-Sil A) was purchased from Bio-Rad, Inc. Particle size of the material was quoted by the manufacturer to be more than 90% less than 10 microns. Surface area of silica using nitrogen gas was determined by Quantasorb to be 280 m^2/g.

C^{14} tagged cationic polyacrylamide was synthesized by the radiation induced precipitation technique using 3 mol% dimethylaminopropyl methacrylamide (DMAPMA) as the comonomer.(3) Viscosity average molecular weight of the polymer used in this study was estimated to be 1.9 million.

Sodium dodecylsulfonate (+99.9% pure) was obtained from the Aldrich Chemical Co and was used without any further purification.

Fisher certified NaOH and HCl were used for pH modification. ACS reagent grade NaCl was used for adjusting the ionic strength. Triple distilled water of specific conductivity of 10^{-6} mho was used in this study.

Techniques

Adsorption tests. 0.4 g of material was equilibrated with 8 cm^3 of 3×10^{-2} $kmol/m^3$ NaCl solution in a screw cap glass vial for 4 hours by shaking at the desired temperature in a wrist action shaker. After 4 hours of equlibration, pH was adjusted using NaOH or HCl and the suspension was further equilibrated for 2 hours. Required amount of the polymer solution was then added and the suspension was agitated for an additional 12 hours before adding the surfactant solution. After adding the surfactant, 2 hours of further agitation was required to reach equilibrium. After equilibration, pH of the suspension was measured using a thin glass electrode attached to a digital Corning pH meter (Model 125). The contents of the vial were transferred to a centrifuge tube and were centrifuged in a IEC Model B20-A centrifuge at 15,000RPM for 10 minutes. Residual amounts of the polymer and dodecylsulfonate were determined in the supernatant and adsorption densities were calculated using the respective calibration curves.

Analytical. Amount of C^{14} labelled polymers was determined using a Beckman Model LS 100C spectrophotometer. It was confirmed that the polymer employed in this investigation did not interfere with the surfactant analysis and vice versa.

RESULTS AND DISCUSSION

Nature of PAMD-Dodecylsulfonate Interaction

Interactions between polymer and surfactant species can be categorized mostly as those arising from (i), the changes in the solvent power of the medium and (ii), associative interactions in the bulk or at interfaces as a result of electrostatic, hydrogen and hydrophobic bonding between the polymer and the surfactant species. (4-10)

Results of bulk interaction studies between PAMD and dodecylsulfonate using conductivity, relative viscosity and precipitation techniques have been published elsewhere (6). It was concluded from these results that significant bulk complexation leading to precipitation occurs in this system above a sulfonate concentration of 10^{-3} kmol/m^3.

In the present investigation, effect of charge neutralization of the polymer by surfactant will alter its conformation towards coiling. Also, pH will have a major role in determining the nature of interaction between the dodecylsulfonate and the cationic polyacrylamide, since the amine (from DMAPMA) and amide functional groups attached to the polymer are hydrolyzable.

Adsorption of PAMD and Sodium Dodecylsulfonate on Hematite and Silica

Adsorption of a polymer or a surfactant on an oxide mineral is attributed to one or more of the following: electrostatic charge attraction, hydrogen bonding, covalent bonding and hydrophobic interactions. Charge characteristics of the solid surface as well as that of the polymer and surfactant play an important role in determining the extent of adsorption of a particular specie on the mineral particles. Modifications in variables such as pH and ionic strength can result in (i), alterations in surface charge characteristics of hematite or silica since H^+ and OH^- are potential determining ions, and (ii), changes in polymer structure due to hydrolysis, especially under basic pH conditions.

A brief description of the role of pH in determining the surface chemical properties of hematite and silica, and in the solution chemistry of cationic polyacrylamide is presented below.

Surface Chemical Properties of Hematite and Silica. The isoelectric point (IEP), of hematite and silica was determined using a Zeta Meter to be pH 8.1 and pH 3.1, respectively. At pH values higher than the IEP the minerals are negatively charged and below it they exhibit a net positive charge. Addition of an indifferent electrolyte such as NaCl in the present study will result only in compression of the electrical double layer, thereby, reducing the zeta potential values without affecting the isoelectric point.

Solution Chemistry of Cationic Polyacrylamide. Hydrolysis of polyacrylamide type polymers containing amide ($-CONH_2$) groups has been reported to occur under both acidic and basic pH conditions. (11-14) Since hydrolysis of amide groups by acids has been reported to be significant only at temperatures higher than 100°C, it can be assumed that under experimental conditions of 25°C polymer structure did not change in the acidic pH range. Kinetics of hydrolysis of polyacrylamides under basic pH conditions are fast enough to result in a conversion of a significant number of $-CONH_2$ groups to $-COOH$ groups in the following manner. (15)

$$RCONH_2 + HOH \rightleftarrows RCOOH + NH_3$$

Amine functional groups in the comonomer also are hydrolizable and will go through a transition from RNH_3^+ to ionomolecular complex ($RNH_3^+ \cdot RNH_2$), to neutral RNH_2 groups as the pH is varied from the acidic to basic range. (16) Thus, depending on the pH which governs the dissociation of $-COOH$ groups, negatively charged sites can develop on a cationic polyacrylamide also and affect its conformation.(17)

Adsorption of Cationic Polyacrylamide on Hematite. Adsorption of the cationic polyacrylamide (PAMD) on hematite as a function of pH is plotted in Figure 1. Hematite is positively charged at pH 2.7 and pH 7.2 with the charge density being higher at the lower pH values. Cationic nature of the polymer charge under these pH conditions remains unaffected. Because of the similar nature of the charges electrostatic repulsion will oppose the adsorption of cationic polyacrylamide on hematite. It has, however, been reported that polyacrylamide type polymers adsorb on the oxide mineral surfaces by hydrogen bonding (18) and therefore, charge repulsion can only decrease the amount of the cationic polyacrylamide adsorbing on the hematite surface. A tendency towards lower PAMD adsorption on hematite at pH 2.7 as compared to that at pH 7.2 is observed in the data presented.

At pH 10.9, hematite is negatively charged, and consequently, a greater uptake of an oppositely charged cationic polyacrylamide is expected. It has, however, been discussed earlier that under such pH conditions, cationic polyacrylamide will acquire negative charges ($RCOO^-$) due to hydrolysis, thus, resulting in an electrostatic charge repulsion between hematite and the polymer. In the present investigation polymer adsorption at pH 10.9 was determined to be lower than that at pH 2.7 and pH 7.2.

Adsorption of dodecylsulfonate on Hematite. Adsorption of dodecylsulfonate on hematite occurs mostly due to electrostatic charge attraction. It was therefore, determined to be maximum at pH 3 and decreased at higher pH values.

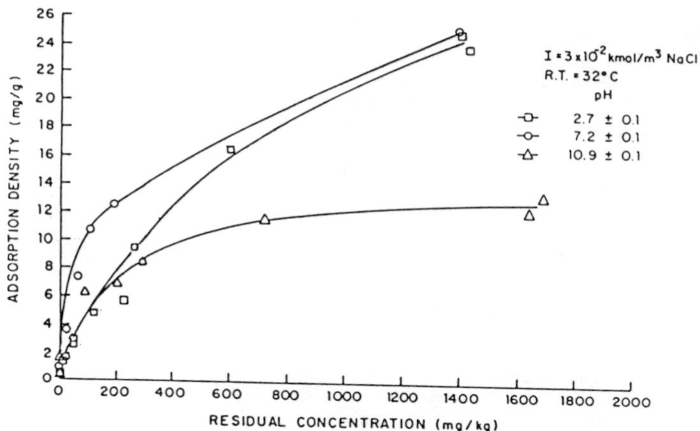

Figure 1. Adsorption Isotherms of Cationic Polyacrylamide (PAMD) on Hematite.

Effect of Dodecylsulfonate on PAMD adsorption on Hematite.

Effect of dodecylsulfonate on cationic polyacrylamide adsorption on hematite at three pH levels, is presented in Table 1 as a function of the order of reagent addition.

Table 1

Effect of Dodecyl Sulfonate on Cationic Polyacrylamide (PAMD) Adsorption on Hematite

NaCl Conc. = 3×10^{-2} kmol/m^3
Polymer Conc. = 500 mg/kg
Dodecylsulfonate Conc. = 6.25×10^{-4} kmol/m^3

pH	Polymer Adsorption Density,mg/kg (P)	(P+S)	(S+P)	($\overline{P+S}$) (Premixed)
2.3	8.3, 8.2	12.1, 12.0	7.6	11.2
7.1	11.1, 11.6	11.6	8.4	11.1
11.5	7.5, 7.9	6.3	6.2	6.9

Presence of dodecylsulfonate can affect polymer adsorption by (i), adsorbing on the substrate, thus, leaving less vacant sites for polymer molecules to adsorb (ii), altering the adsorption behavior of a polymer by forming an associative complex with it or, in an extreme case, forming a precipitate with the polymer, thereby reducing the effective concentration of the polymer (iii), modifying the bulk solution properties such as ionic strength, and (iv), desorbing the adsorbed polymer molecules. In the present case, depletion of the polymer by precipitatioin can be ruled out since precipiation studies did not reveal formation of an insoluble complex in the sulfonate concentration range employed . Also, the changes in ionic strength due to dodecylsulfonate addition will not be significant because adsorption measurements were conducted in the presence of 3 x 10^{-2} kmol/m^3 NaCl.

At pH 2.3, hematite and PAMD are positively charged, therefore, an anionic surfactant can interact electrostatically with both of them. The amount of cationic polyacrylamide adsorbed on hematite should be positively infuenced by polymer-surfactant interactions, since this will neutralize some of the positive charged sites on the polymer backbone, thereby reducing the electrostatic repulsion between the similarly charged hematite and PAMD molecules. Also, polymer conformation as a result of such interaction will tend towards more coiling and occupying less surface area upon adsorption. However, polymer adsorption will be adversely affected by the adsorption of dodecylsulfonate on hematite, because it will reduce the number of anchoring sites available for polymeric species. Overall adsorption should be governed by the relative significance of these factors.

Polymer adsorption increased when dodecylsulfonate followed the cationic polyacrylamide addition to the system(P+S) (see Table 1). It indicates that, possibly because of reduced repulsion between hematite and cationic polyacrylamide as a result of its interaction with the anionic dodecylsulfonate, additional PAMD molecules were adsorbed on the solid surface . In this case, therefore, modifications in the surface charge characteristics seem to have overcome the adverse effect of the reduction in vacant sites available for polymer adsorption.

When cationic polyacrylamide addition followed the dodecylsulfonate addition (S+P), polymer adsorption was found to be lower than in the presence of polymer alone. Adsorption of the polymer in this case can be considered to be essentially taking place on a hematite surface which is already covered with dodecylsulfonate molecules with their non-polar hydrocarbon chains oriented towards the bulk solution. Uptake of the polymer, therefore, can occur by adsorption at vacant sites and also by hydrophobic interactions with the surfactant chains. Bulk solution studies, however, indicated that hydrophobic interactions are not significant in this system, therefore, adsorption of the polymer is governed mostly by the availability of vacant sites on the surface.

When PAMD and dodecylsulfonate were premixed ($\overline{\text{P+S}}$) , polymer adsorption was determined to be slightly lower than when the polymer was added prior to the surfactant addition (P+S). In this case, interaction between oppositely charged species can reduce positive charges on the polymer and result in a higher tendency towards coiling. Both of these factors would favor increased uptake of the polymer on the hematite surface. Some of the residual surfactant molecules, however, will also adsorb, thus reducing the number of vacant sites available for polymer-surfactant complex to adsorb on the solid substrate. Overall, adsorption of the polymer, would be a result of these two opposing tendencies.

At pH 7.1, since the nature of charges associated with hematite, cationic polyacrylamide, and dodecylsulfonate are similar to that at pH 2.3, the effect of order of reagent addition can be expected to be like that observed at pH 2.3. It should, however, be noted that the amount of polymer adsorbed is higher at pH 7.1 because of reduced charge density of the hematite particles and therefore, less electrostatic repulsion between the polymer and the solid. Additionally, as a result of reduced charges on hematite, adsorption of the dodecylsulfonate molecules will also decrease thus, reducing their influence on polymer adsorption. Polymer adsorption at pH 7.1 in the presence of dodecylsulfonate reflected these changes in the adsorption behavior.

At pH 11.5, hematite particles are highly negatively charged, and also the cationic polyacrylamide as discussed earlier, is hydrolyzed. This will decrease the electrostatic charge attraction between PAMD and hematite particles, thus, causing a reduction in the adsorption. Presence of dodecylsulfonate molecules is not expected to have a major influence on the polymer adsorption because no significant interactions of the solid substrate and of the polymer with the dodecylsulfonate molecules are expected. Polymer adsorption therefore, is not expected to be influenced by the order of reagent addition as was experimentally observed.

Adsorption of Cationic Polyacrylamide (PAMD) on Silica.

It was determined that out of the three polymers tested, namely nonionic, anionic, and cationic polyacrylamide, only the cationic polyacrylamide (PAMD) adsorbed on silica (3). This suggested that the major mechanism of polyacrylamide adsorption on silica is electrostatic in nature.

Adsorption isotherms of PAMD on silica at three pH levels are presented in Figure 2. Silica particles at pH 2.6 (<IEP = pH 3.1) are positively charged, and at pH 4.4 and 9.8, exhibit a net negative charge. However at pH 9.8, cationic polyacrylamide will be hydrolyzed resulting in a decrease in the positive charge density of the polymer.

Based on the charge characteristics, it is expected that PAMD adsorption on silica will be maximum at pH 4.4, since the solid and the polymer at this pH will exhibit an opposite charge. Adsorption should be lower at pH 2.6 and pH 9.8, because of positive charge repulsion in the first case and negative charge repulsion at pH 9.8. Experimentally, however, adsorption of PAMD at pH 9.8 was found

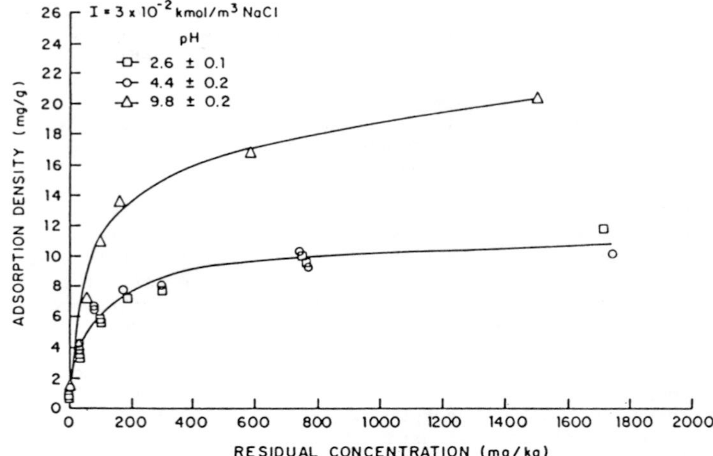

Figure 2. Adsorption Isotherms of Cationic Polyacrylamide (PAMD) on Silica.

to be maximum. One possible explanation for this observation could be the enhanced dispersion of silica particles at pH 9.8, thus resulting in an increased surface area available for polymer adsorption. Conformational changes in the hydrolyzed PAMD can also favor an uptake of the polymer molecules into silica pores . Furthermore, under basic pH conditions, specific interactions between the amine functional groups attached to the polymer backbone and the silica surface cannot be ruled out.

Although, adsorption of PAMD at pH 4.4 and at pH 2.6 according to the nature of silica and polymer charges were expected to be different, they were determined to be similar at both pH levels. At present, the reason for this observation is not clear.

Dodecylsulfonate Adsorption on Silica. According to the electrostatic charge considerations, significant adsorption of sodium dodecylsulfonate on silica is expected to occur only at pH values below its IEP.

Effect of Dodecylsulfonate on PAMD Adsorption on Silica.

Effect of order of reagent addition on adsorption of PAMD on silica was studied at pH 2.2, 4.2 and at pH 10.0. In all three cases, as shown in Figure 3, no significant change in polymer adsorption was observed as a result of surfactant addition. Charge considerations dictate that interaction between dodecylsulfonate and PAMD can occur only at pH 2.2 and at 4.2. As a result of such interactions, adsorption of PAMD at 2.2 should increase because of a decrease in charge repulsion between the silica surface and the

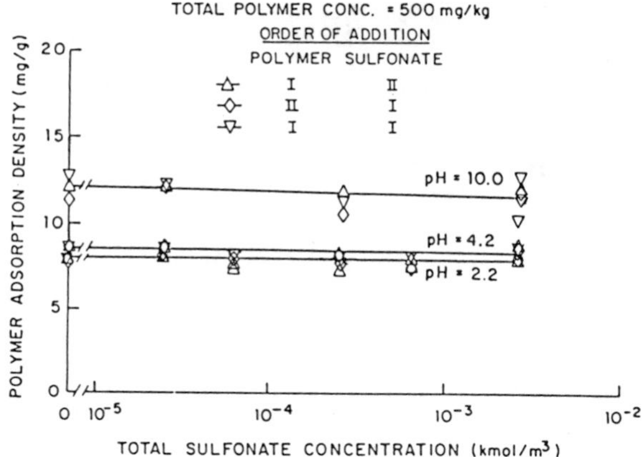

Figure 3. Effect of Dodecylsulfonate on Adsorption of Cationic Polyacrylamide (PAMD) on Silica.

cationic polyacrylamide. However, as discussed earlier, an opposite effect on adsorption can take place because of adsorption of surfactant molecules, thereby leaving less number of vacant sites for PAMD molecules to anchor on the solid. It is, therefore, possible that the two factors will cancel out each other leaving the PAMD adsorption on silica unchanged.

At pH 4.2, no significant adsorption of dodecylsulfonate on silica is expected because of charge repulsion. However, attraction between dodecylsulfonate and PAMD is possible, which will neutralize the positive charges on the polymer backbone. This should result in lowering of the polymer adsorption on silica in the presence of dodecylsulfonate. However, no such reduction was experimentally observed suggesting that interaction of the cationic polyacrylamide with silica is stronger than its interaction with the dodecylsulfonate molecules.

SUMMARY

It is evident from the above discussion that although electrostatic charge attraction between the cationic polyacrylamide (PAMD) and hematite influence the polymer adsorption, the major adsorption mechanism is the hydrogen bonding. In the case of silica, adsorption of polymer was found to occur mainly due to electrostatic interaction between the cationic polyacrylamide and the silica surface.

PAMD adsorption on hematite was found to be affected

significantly by the solution pH and by the order in which the polymer and surfactant were added to the system. Adsorption of PAMD on silica also was dependent on pH, however, the effect of dodecylsulfonate was not significant. Modifications in the cationic polyacrylamide adsorption on hematite, as a function of the order of dodecylsulfonate addition are attributed to (i), the competition between the polymer and surfactant molecules and (ii), the adsorption behavior of the complex formed as a result of associative interactions between the oppositely charged polymeric and surfactant species.

ACKNOWLEDGEMENTS

The authors would like to thank Dr. K.P. Ananthapadmanabhan for helpful discussions and Mrs. Regina Gorelik for help in experimentation. The National Science Foundation (Grant No. DAR-79-09295) and the INCO Inc. are acknowledged for financial assistance. One of the authors (BMM) would also like to thank the Occidental Research Corp. for granting him education leave at Columbia University.

LITERATURE CITED

1. Tadros, Th. F., J. Colloid Interface Sci., 46, 528 (1974).
2. Hollander, A., "A Study of the Interactions of Poly (Acrylamide) and its copolymers with Sodium Kaolinite," M.S. Thesis, Columbia University, New York (1979).
3. Moudgil, B.M., "The Role of Polymer-Surfactant Interactions in Interfacial Processes," Eng. Sc. D. Thesis, Columbia University, New York (1981).
4. Goddard, E.D., T.S. Phillips and R.B. Hannan, J. Soc. Cosmet Chem., 26, 461 (1975).
5. Jones, M.N., J. Colloid Interface Sci., 23, 36 (1967).
6. Somasundaran, P. and B.M. Moudgil, "Effect of Polymer-Surfactant Interactions on Polymer Solution Properties," in Macromolecular Solutions, Seymour, R.B. and G.A. Stahl (Eds.), Pargamon Press, New York (1982).
7. Cabane, B., J.Phys. Chem. 81, 1639 (1977).
8. Schwuger, M.J., J. Colloid Interface Sci., 43, 491 (1973).
9. Lewis, K.E. and C.P. Robinson, J. Colloid Interface Sci., 32, 539 (1970).
10. Cabane, B., J. Physique, 43, 1529 (1982).
11. Pinner, S.H., J. Polymer Sci., 10, 376 (1953).
12. Moens, J. and G. Smets, J. Polymer Sci., 29, 931 (1957).
12. Smets, G. and A.M. Hesbain, J.Polymer Sci., 40, 217 (1959).
14. Higuchi, M. and R. Senju, Polymer J. 3, 370 (1972).
15. Kulicke, W. and J. Klein, Angwandte Makromol. Chem. 69, 189 (1978). Quoted in reference 2.
16. Ananthapadmanabhan, K.P., "Associative Interactions in Surfactant Solutions and Their Role in Flotation," Eng.Sc.D. Thesis Columbia University, New York (1979).
17. Muller, G., J.P. Laine and J.C. Fenyo, Polymer Sci. (Polym. Chem.) 17, 659 (1979).
18. Linke, W.F. and R.B. Booth, Trans. AIME, 217, 364 (1959).

ADSORPTION OF PURE GASES AND THEIR MIXTURES ON HETEROGENEOUS SURFACES

Alan L. Myers
Chemical Engineering Department
University of Pennsylvania
Philadelphia, Pennsylvania, 19104
USA

Abstract

The thermodynamics of adsorption of mixtures is guided by the Gibbs adsorption isotherm. In addition, boundary conditions place mathematical constraints on allowable forms for the adsorption isotherm of the pure gas. There exists a family of three-constant equations which incorporate surface heterogeneity in the model but ignore adsorbate-adsorbate interactions. Any of these three-constant equations provides a basis for predicting multicomponent equilibria by means of ideal-adsorbed-solution (IAS) theory. Here a specific algorithm for the Langmuir-uniform-distribution (LUD) equation is provided. Agreement with experimental mixture data is excellent up to 50% of saturation, but at higher coverage negative deviations from Raoult's law and enhanced loading are observed. These deviations are not due to adsorbate-adsorbate intermolecular forces. Sites of different energies possess different selectivities, and the resultant segregation in composition is primarily responsible for the nonideal behavior in the adsorbed phase.

INTRODUCTION

The theory of ideal-adsorbed-solutions (IAS), which is based upon the thermodynamics of mixed-gas adsorption, was proposed in 1965 [1]. At surface coverages up to about 50% of saturation the theory is in excellent agreement with experiment, but at higher coverage negative deviations from Raoult's law and enhanced loadings (i.e., total amount adsorbed greater than that predicted by IAS) have been observed [2,3,4]. This phenomenon can be explained by heterogeneity in the energies of adsorption [5], as well as the more obvious explanation of increased affinity of unlike molecular interactions.

A cell model of adsorption from liquid mixtures for an ideal adsorbed phase and a nonideal bulk phase provides an analytical adsorption isotherm [6]. This theory was later extended [8] to mixtures of molecules of unequal size, but the resultant equation for surface excess is an implicit function of composition. In the case of ideal bulk phases, the equation simplifies to the phase-exchange reaction model of Everett [7]. For dilute, strongly adsorbed solutes, the isotherm reduces to the Langmuir equation in which the concentration of the solute is replaced by its activity [9]. IAS theory can be applied directly to multi-solute adsorption from dilute solution [10]. The solvent adsorbs competitively because of its relatively high concentration in the bulk phase, but its activity is fixed and therefore provides a constant potential environment for competitive adsorption of mixed solutes.

Liquid mixture pairs A-B, A-C and B-C on the same adsorbent obey an integral thermodynamic consistency test [11]. The same principle applies to ion exchange for three pairs of counter-ions [12]. A binary liquid mixture A-B must follow an integral consistency test involving the pure-vapor isotherms [13]. The redundant information which allows the consistency test suggests several practical applications. For example, adsorption from binary liquids can be predicted in terms of the pure-vapor isotherms [4]. For binary gas adsorption, compositions in the adsorbed phase can be calculated rigorously from a differential equation, provided that extensive experimental data is taken for several isotherms at different vapor compositions [14,30].

For adsorption on non-heterogeneous surfaces like graphitized carbon, two-dimensional equations of state combined with realistic gas-solid intermolecular potentials provide the basis for theories which agree with experiment for adsorption of monatomic gases [16,17]. Thus, adsorption of simple molecules on non-heterogeneous surfaces is understood in principle. Vapor-adsorption equilibria on graphitized carbon approach the corresponding vapor-liquid equilibria asymptotically at saturation, but no such correspondence is observed for adsorption on microporous, heterogeneous surfaces [14].

For non-heterogeneous surfaces, adsorbate-adsorbate intermolecular forces determine the shape of the adsorption isotherm, which can be predicted by the application of a two-dimensional principle of corresponding states to the adsorbed layer [18]. For heterogeneous surfaces, the dominant factor is the character of the energy distribution, which itself provides a basis for a dimensionless correlation of adsorption of different gases on the same solid [19]. The principle has been extended to adsorption from dilute liquid solutions [20].

A better understanding of physical adsorption on high-area microporous solids has emerged as a result of numerous recent publications on heterogeneity of the energy of adsorption. In addition, Monte Carlo calculations have helped in the development of more realistic theories of adsorption based on specific gas-gas and gas-solid intermolecular potentials [21]. Most research on the statistical mechanics [22] of adsorption has focused upon pure-gas adsorption. Here the effect of heterogeneity [24,25] on mixed-gas adsorption is considered.

THERMODYNAMICS OF ADSORPTION

The Gibbs adsorption isotherm is the starting point for thermodynamic treatments of adsorption of pure gases and their mixtures:

$$A d\Pi = \sum_{i=1}^{N} n_i d\mu_i \tag{1}$$

This equation can be integrated for several cases, the simplest of which is that of the pure-gas isotherm:

$$\Pi A/RT = \int_{P=0}^{P} (n/P)dP \tag{2}$$

Here and in the following it is assumed that the gas phase follows the perfect gas law, which is a satisfactory approximation in most cases, even at pressures as high as 2 MPa (see discussion below).

Eq. (1) can also be integrated for a binary mixture at constant pressure:

$$\frac{(\Pi_1 - \Pi_2)A}{RT} = \int_{y_1=0}^{1} \frac{n(s-1)}{1+(s-1)y_1} dy_1 \tag{3}$$

where s, the relative separation factor or selectivity, is defined by:

$$s = \frac{x_1/y_1}{x_2/y_2} \tag{4}$$

This powerful equation can be used to test isobaric mixture data for thermodynamic consistency because the left-hand-side of Eq. (3) is calculable from the pure-gas isotherms by Eq. (2). It also provides an interesting insight into the selectivity, which should be at least 3 if adsorption is to provide the means for a practical separation process. Although both n and s vary with composition along an isobar, Eq. (3) reduces to the following simple expression under the approximation of constant n and s:

$$\bar{s} = \exp\left[A(\Pi_1 - \Pi_2)/\bar{n}RT\right] \tag{5}$$

Recognizing that the values of $\bar{n}$ and $\bar{s}$ in Eq. (5) are mean values which satisfy Eq. (3), it is apparent that selectivity is an exponential function of spreading pressure. A small error in either Π_1 or Π_2 would cause a greatly magnified error in the selectivity.

This point is so important that it merits additional discussion. The calculation of spreading pressure by Eq. (2) is sensitive to the low pressure region of the isotherm, where both n and P approach zero but have a finite ratio which is proportional to the adsorption second virial coefficient B_{1s}:

$$B_{1s} = \lim_{P \to 0} (nRT/P) \tag{6}$$

Some adsorption equations have singularities at the origin; an example is the Freundlich isotherm:

$$n = C\,P^t \tag{7}$$

where $0 < t < 1$. This equation incorrectly predicts that the limit of (n/P) is infinite at the origin and gives large errors in the spreading pressure from Eq. (2). As a result, the Freundlich equation gives unsatisfactory estimates of selectivity.

The Gibbs adsorption isotherm is particularly useful for the prediction and analysis of mixture equilibria. Equality of fugacities in the gas and adsorbed phases is written:

$$Py_i\phi_i = P_i^\circ \phi_i^\circ \gamma_i x_i \tag{8}$$

for each component. ϕ_i and ϕ_i° are the gas-phase fugacity coefficients of component i in the mixture and in the standard state, respectively. Under the conditions mentioned above, it is usually adequate to assume that $\phi_i = \phi_i^\circ = 1$. Adsorbed-phase activity coefficients are conveniently correlated in terms of the excess Gibbs free energy function:

$$\Delta g^e/RT = \sum_{i=1}^{N} x_i \ln \gamma_i \tag{9}$$

All of the properties of the Gibbs function familiar in vapor-liquid equilibrium (VLE) apply with appropriate change of variable names, especially the restricted form of the Gibbs-Duhem equation:

$$\sum_{i=1}^{N} x_i d \ln \gamma_i = 0 \qquad \text{(constant } T, \Pi) \tag{10}$$

Thus any of the usual equations for liquid-phase activity coefficients (quadratic, Margules, van Laar, Wilson, etc.) may be used for adsorption, provided that the spreading pressure is held constant. Unfortunately activity coefficients taken at constant pressure (P) do not obey Eq. (10). The problem is analogous, but much more serious, than that of fitting isobaric (i.e., nonisothermal) activity coefficients in VLE. For adsorption, activity coefficients must be written as functions of both composition and spreading pressure [14]. In addition, the following boundary conditions must be satisfied:

$$\lim_{x_i \to 1} \gamma_i = \lim_{P \to 0} \gamma_i = 1 \tag{11}$$

In adsorption the total loading (n) as well as the selectivity is needed. Analogous to the excess volume in VLE is the excess area, or, since the specific surface area is a constant, the excess reciprocal loading:

$$(1/n)^e = (1/n) - (1/n)^{id} \tag{12}$$

where the ideal loading is given by:

$$(1/n)^{id} = \sum_{i=1}^{N} x_i (1/n_i^\circ) \tag{13}$$

In VLE the excess volume is so small that activity coefficients are considered to be independent of pressure, but in adsorption the situation is different. It can be shown from Eq. (1) that

$$(1/n)^e = \frac{\partial(\Delta g^e/RT)}{\partial(\Pi A/RT)} \tag{14}$$

Deviations from Raoult's law in the adsorbed phase ($\Delta g^e \neq 0$) are accompanied by deviations from ideality of the loading curve. Since Δg^e must be zero at the limit of zero pressure in accordance with Eq. (11), the slope of Δg^e versus spreading pressure can be large (see below).

The special case where Raoult's law is obeyed in the adsorbed phase ($\gamma_i = 1$ and $\Delta g^e = 0$) is called the theory of ideal adsorbed solutions (IAS). Obviously IAS obeys the differential consistency requirement of Eq. (10) and satisfies the boundary conditions in Eq. (11). IAS provides a useful means of predicting mixture adsorption entirely in terms of the pure gas isotherms. It is exact in the limit of zero surface coverage, and is accurate for coverages up to about 50% of saturation. At high loadings approaching saturation, errors become appreciable as shown below.

Various empirical equations have been proposed which are pure-gas adsorption isotherms extended to mixtures by mixing rules for the constants contained in the equation. Such theories do not satisfy the Gibbs adsorption isotherm. A theory which satisfies all the rules of thermodynamics is not necessarily correct, but there is certainty that an inconsistent theory is wrong.

In summary, the most important step toward prediction of mixture adsorption equilibria is the selection of an equation for the pure gases which fits the experimental data and has no singularities at the origin.

Mixture theories must satisfy Eqs. (3), (10) and (14), all of which are special forms of the Gibbs adsorption isotherm.

ADSORPTION EQUATIONS FOR PURE GASES

The simplest and most familiar adsorption equation is the two-constant Langmuir isotherm [23]:

$$n = \frac{mKP}{1 + KP} \tag{15}$$

This equation has the required mathematical properties and correct qualitative behavior for adsorption on microporous solids, but with only two constants (m,K) does not have sufficient flexibility to fit experimental data from zero pressure to saturation. Quantitative agreement with experiment requires an equation containing at least three constants.

The Langmuir equation ignores adsorbate-adsorbate molecular interactions and surface heterogeneity. Both are important, but in order to obtain a relatively simple equation containing only three constants, only the effect of heterogeneity is considered here. The adsorption integral equation for a patchwise-heterogeneous surface is [24]:

$$\frac{n}{m} = \theta(P,T) = \int_{E=E_{min}}^{E_{max}} \theta(P,T,E)f(E)dE \tag{16}$$

For the adsorption energy distribution, the first choice would be a Gaussian form but the same effect can be achieved by using the uniform distribution:

$$\begin{cases} f(E) = 1/(2\sqrt{3}\sigma) & (E_{min} \le E \le E_{max}) \\ f(E) = 0 & (\text{otherwise}) \end{cases} \tag{17}$$

where

$$\mu = (E_{max} + E_{min})/2$$
$$\sigma = (E_{max} - E_{min})/(2\sqrt{3})$$

μ is the mean energy and σ is the square root of the variance. E, a positive number, is the decrease in energy accompanying adsorption.

Integration of Eq. (16) using Eq. (17) for the distribution and Eq. (15) for the local adsorption isotherm yields the Langmuir-Uniform-Distribution (LUD) equation:

$$n = \frac{m}{2s} \ln\left[\frac{1 + CPe^{s}}{1 + CPe^{-s}}\right] \tag{18}$$

where

$$C = K_0 \exp(\mu/RT)$$
$$s = \sqrt{3}\sigma/RT$$

The LUD equation has three constants: m, C and s. Eq. (18) predicts that C increases exponentially with 1/T, and s increases linearly with 1/T. The saturation loading m decreases with T in approximate agreement with the Gurvitsch rule [26] which is that m = V/v°, where V is the (constant) pore volume of the solid and v° is the molar volume of adsorbate in the state of saturated liquid.

Constants of the LUD equation (and several other three-constant adsorption isotherms) have been determined by an optimization program which minimizes the sum of the squares of the differences between calculated and experimental points. For example, values are given in Table 1 for the experimental data of Reich, Ziegler and Rogers [27].

Table 1. Constants of the LUD Isotherm for Adsorption on Pittsburgh BPL Activated Carbon, 988 m^2/g.

Gas	T, K	m mmol/g	C $(kPa)^{-1}$	s
Methane	212.7	8.3229	0.00543	2.8244
"	260.2	7.1056	.00156	2.0965
"	301.4	5.8140	.00086	1.4303
Ethylene	212.7	9.4343	.05297	4.8430
"	260.2	7.3329	.01429	3.2457
"	301.4	6.5455	.00449	2.6867
Ethane	212.7	7.8000	.19231	4.0182
"	260.2	6.8747	.02646	3.4180
"	301.4	5.9514	.00838	2.5155

An adsorption data base [28] contains procedures for comparing calculated and experimental data for various three-constant isotherms. Computer graphics are especially helpful. For example, Fig. 1 is for the adsorption of ethylene on BPL activated carbon at 260.2 K. The average absolute error is 0.3% and the maximum error is less than 1%. The solid line on Fig. 1 is the LUD equation with constants from Table 1, but other three-constant equations such as that of Toth [29] fit these data [27] with equal precision.

Fig. 1

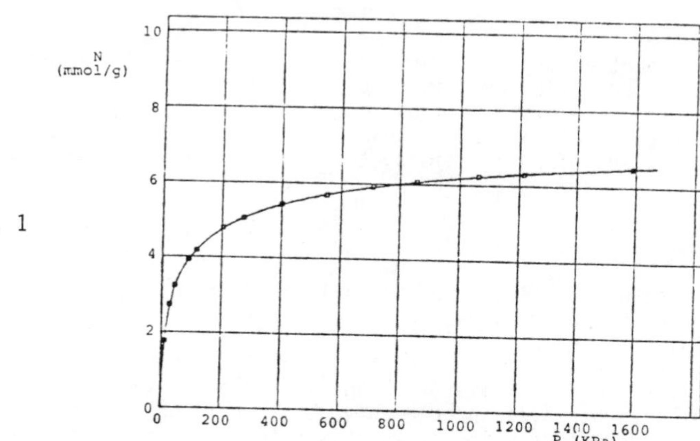

PREDICTION OF MIXTURE EQUILIBRIA

Given adsorption isotherms for the pure gases in terms of the LUD or any suitable equation, the mixture equilibria for an N-component mixture can be predicted from Eq. (8) by setting $\gamma_i = 1$ for each component:

$$Py_i = P_i^\circ x_i \tag{19}$$

where

$$\sum_{i=1}^{N} x_i = 1 \tag{20}$$

Integration of the pure-gas isotherms for spreading pressure by Eq. (2) gives:

$$\Pi A/RT = \frac{m}{2s} \int_{P=0}^{P} \frac{1}{P} \ln\left|\frac{1 + CPe^{s}}{1 + CPe^{-s}}\right| dP \tag{21}$$

The integration may be performed numerically but converges slowly. It can be shown that, instead of pressure, the integration may be performed with respect to the energy distribution:

$$\Pi A/RT = \frac{m}{2s} \int_{z=-s}^{s} \ln(1 + CPe^{z})dz \tag{22}$$

and this converges rapidly. Since the standard-state pressures (P_i°) are evaluated at equal values of spreading pressure, we have for each gas Eq. (22) which is of the form:

$$\Pi = f_i(P_i^\circ) \tag{23}$$

so there are (N - 1) equations needed to fix the common spreading pressure which defines the standard state:

$$f_1(P_1^\circ) = f_2(P_2^\circ) = \cdots = f_N(P_N^\circ) \tag{24}$$

For N-component adsorption there are (N + 1) degrees of freedom which are conveniently specified by temperature, pressure and (N - 1) vapor compositions $\{y_i\}$. There are 2N equations: N from Eq. (19), (N - 1) from Eq. (24), plus Eq. (20), and 2N unknowns: $x_1 \cdots x_N$ and $P_1^\circ \cdots P_N^\circ$. After solving this set of 2N equations, the total loading is given by Eq. (13):

$$n = 1/\sum_{i=1}^{N} (x_i/n_i^\circ) \tag{25}$$

where the standard state loadings (n_i°) are from Eq. (18) evaluated at the standard-state pressures (P_i°).

The problem is to solve for the total loading (n) and adsorbed phase compositions $\{x_i\}$ given T, P and vapor compositions $\{y_i\}$. For the design of adsorber breakthrough curves, it would be convenient to have a computer subroutine which makes this calculation for the general non-isothermal case. Eq. (19) is straightforward but its solution is complicated by the fact that the reference pressures $\{P_i^\circ\}$ are calculated at fixed Π. Therefore the inverse function of Eq. (22), itself a numerical integration, is needed for the simultaneous solution of the set of 2N equations.

An algorithm for solving the N-component equilibrium problem is given in the Appendix. It requires as input data the constants of two adsorption isotherms. Values of the constants at other temperatures are determined by interpolation: $m \propto T$, $s \propto T$, and $\ln C \propto (1/T)$. A severe test of the accuracy of this interpolation is to calculate an adsorption isotherm at 212.7 K by extrapolation from the isotherms measured at 260.2 and 301.4 K. Results for ethylene are given in Table 2 for a few points. High temperature constants were taken from Table 1. The good agreement of the extrapolated points with the experimental data at 212.7 K indicates that the interpolation procedure is sound.

Table 2. Comparison of Experimental Points for Adsorption of Ethylene on Activated Carbon at 212.7 K with Values Calculated by Extrapolation from 260.2 and 301.4 K.

	n, mmol/g	
P, kPa	Exper.	Calc.
4.14	3.010	3.174
19.3	4.736	4.732
106.18	6.346	6.369
599.84	7.884	7.583

Integration of Eq. (22) is by Simpson's rule and the inversion of this equation is accomplished by Newton's method using the subroutine INVSP in the Appendix. The main solution algorithm in the IASLUD routine for the unknown pressures $P_i^\circ(\Pi)$ is similar to that [15] for finding vapor pressure $P_i^\circ(T)$ in the dew-point temperature algorithm for VLE. For faster convergence, the reference component is automatically selected as the main constituent of the vapor phase (largest y_i).

A program for the N-component equilibrium problem described in the Appendix was written in interpreted BASIC language supplemented by software for subprograms with the ability to pass arguments like FORTRAN. Since the BASIC compiler was not used, it takes about one minute on an IBM Personal Computer to solve for a ternary mixture. Therefore, if compiled and run on a faster computer, a similar non-interactive program could be written as a subroutine for the calculation of adsorption column breakthrough curves under realistic nonisothermal conditions.

In summary, given the constants of the adsorption isotherms for the pure gases at two temperatures, the algorithm described in the Appendix solves for loading and adsorbed-phase composition as a function of temperature, pressure and vapor composition. Isothermal calculations require three constants for each gas. The IASLUD algorithm is written specifically for the LUD equation, but the solution can easily be modified for other three-constant adsorption isotherms such as that of Toth.

COMPARISON OF IAS WITH EXPERIMENT

Reich, Ziegler and Rogers [27] measured adsorption of binary and ternary mixtures of methane, ethane and ethylene on BPL activated carbon with a BET surface area of 988 m^2/g. 100 mixture points were taken for various vapor compositions at three temperatures: 212.7, 260.2 and 301.4 K. The pure-gas isotherms were measured at the same temperatures (see, for example, Fig. 1) and on the same sample of carbon.

Most of the loadings were high, but 9 points are for moderate loading less than 50% of saturation and these are reproduced in Table 3. Values predicted by IAS using the LUD constants in Table 1 and calculated using the program described in the Appendix are shown for comparison in parentheses. Table 3 indicates that IAS provides excellent predictions of the adsorbed-phase composition, but the loadings are about 5% less than the experimental values. Similar results were obtained for the ternary data.

Table 3. Experimental Mixture Adsorption Equilibria Below 50% of Saturation Loading for Activated Carbon. Values in Parentheses Calculated by IAS.

Comp. #1	Comp. #2	T, K	P kPa	y_1	n, mmol/g	x_1	$\Pi A/RT$ mmol/g
C_2H_4	CH_4	301.4	129	.235	1.716(*1.583*)	.754(*.760*)	(2.133)
"	"	"	122	.464	2.053(*1.994*)	.894(*.896*)	(2.945)
"	"	"	131	.740	2.453(*2.445*)	.964(*.964*)	(3.967)
C_2H_6	"	"	130	.255	1.994(*1.923*)	.833(*.837*)	(2.818)
"	"	"	132	.501	2.596(*2.467*)	.929(*.934*)	(4.078)
"	"	"	130	.733	2.893(*2.780*)	.974(*.974*)	(4.921)
"	"	260.2	125	.255	3.488(*3.435*)	.875(*.893*)	(7.262)
"	C_2H_4	301.4	144	.682	3.267(*3.069*)	.762(*.760*)	(5.744)
"	"	"	138	.240	2.988(*2.866*)	.281(*.318*)	(5.098)

At higher loadings approaching saturation of the adsorbent, results similar to those given in Table 4 were obtained for the binary and ternary mixtures studied by Reich, Ziegler and Rogers. Again, values in parentheses are from IAS theory, which does not predict the observed drop in selectivity (decrease in x_1) with pressure. Moreover, the predicted values for total loading (n) are as much as 12% less than the measured points. Similar results were found for the rest of the data.

Table 4. Experimental Equilibria at High Loading for Mixtures of Ethane(1) and Methane(2) Adsorbed at 301.4 K on Activated Carbon. Vapor Composition y_1 = 0.733. Values in Parentheses Predicted by IAS.

P, kPa	n, mmol/g	x_1	$\Pi A/RT$ mmol/g	$(mmol/g)^{-1}$ $(1/n)^e$	$\Delta g^e/RT$
130	2.893(*2.780*)	.974(*.974*)	(4.92)	-.014	0
351	3.908(*3.775*)	.942(*.972*)	(8.19)	-.009	-.02
685	4.520(*4.387*)	.925(*.970*)	(10.92)	-.007	-.06
1234	5.263(*4.849*)	.894(*.969*)	(13.65)	-.016	-.10
2006	5.811(*5.161*)	.872(*.968*)	(16.08)	-.022	-.14

Several reasons for these discrepancies can be postulated:

(1) The experimental data are incorrect.

(2) The disagreement may be due to imperfections in the vapor phase, since the perfect gas law was used for the calculations and the pressure is as high as 2 MPa.

(3) IAS does not take into account adsorbate-adsorbate interactions.

(4) Heterogeneity was ignored in the mixture calculations.

The first possibility is highly improbable because these data satisfy the most stringent kind of differential consistency test. For example, for the points in Table 4, activity coefficients were calculated from the experimental data and the excess Gibbs free energy obtained from Eq. (9) is tabulated. These are large negative deviations from Raoult's law; activity coefficients for methane are as small as 0.25 . The excess reciprocal loading $(1/n)^e$ was calculated by Eq. (12) and this excess function is also given in Table 4. Finally, the fact that the values of Δg^e and $(1/n)^e$ obey Eq. (14) indicates a high degree of thermodynamic consistency.

The second possibility was studied by recalculating the spreading pressure in the case of an imperfect gas, which leads to the addition of an additional term in Eq. (2). Fugacities were also recalculated according to Eq. (8). It was found that the effect of ignoring gas-phase imperfections is small because of a cancellation of errors, and cannot account for the discrepancies noted above.

The third postulate cannot be so easily dismissed, but it does seem extraordinary that all of these data indicate the existence of large negative deviations from Raoult's law in the adsorbed phase, even though the same mixtures in the bulk liquid would be expected to be nearly ideal or, if anything, to exhibit positive deviations from Raoult's law. Moreover, the effect is not limited to activated carbon, or to these particular mixtures of methane, ethane and ethylene. The phenomenon of negative deviations from Raoult's law was reported as early as 1969 for mixtures of ethane and carbon dioxide adsorbed in 5A molecular sieves [2]. Since then, numerous laboratories have reported negative deviations from Raoult's law for various mixtures adsorbed on microporous adsorbents such as activated carbon and zeolites.

The key to the explanation is that the adsorbed mixture is being treated as if it were uniform in composition, even though the surface is in fact heterogeneous with respect to energy. Different selectivities on different parts of the surface would produce a segregation in the composition of the adsorbed phase. It was shown recently [5] that surface heterogeneity, if ignored by assuming the adsorbed phase to have a fixed composition, accounts qualitatively for the apparent deviations. This effect is easily large enough to account for the magnitude of the deviations. It is concluded that heterogeneity is primarily responsible for the negative deviations from Raoult's law and the accompanying enhancement in loading observed in Table 4 and elsewhere.

Calculations for various heterogeneous models of physical adsorption have been started at the University of Pennsylvania. The IAS theory discussed here assumes that the adsorbed phase is homogeneous even though the LUD equation upon which the mixture calculations are based is for sites of various energies. Different sites would be expected to have different selectivities, and therefore different compositions, depending upon the ratio of energies of adsorption of the adsorbed species. IAS theory can still be used in the framework of a heterogeneous model, but is applied at the local level instead of for the surface phase as a whole. The heterogeneous mixture calculation is actually simpler than the homogeneous algorithm in the Appendix because only one explicit integral over the various sites is needed.

CONCLUSIONS

1. Adsorption isotherms containing three or more constants are needed to describe adsorption of single gases on heterogeneous surfaces. If thermodynamic properties and mixture calculations are desired, these equations must satisfy certain boundary conditions.
2. Agreement of IAS theory with experiment is excellent up to about 50% of saturation, but at higher loadings large negative deviations from Raoult's law are observed with accompanying enhancement of the loading curve.
3. Negative deviations from Raoult's law calculated for the average composition of adsorbed mixtures are not due to stronger attractive forces between unlike molecules. Sites of different energies have different selectivities, and the resultant segregation in the composition is primarily responsible for the apparent negative deviations.
4. The algorithm in the Appendix for computer calculations of nonisothermal N-component adsorption equilibria by IAS theory is fast enough to be used as a subroutine in process design calculations.

NOMENCLATURE

A	specific surface area of solid
B_{1s}	adsorption second virial coefficient, Eq. (6)
C	constant, Eq. (18)
E	decrease in energy due to adsorption
f	a function, Eq. (23)
Δg^e	excess Gibbs free energy function, Eq. (9)
K	constant, Eq. (15)
K_0	constant, Eq. (18)
m	saturation capacity, mmol/g
n	loading or amount adsorbed, mmol/g
N	number of components
P	pressure

NOMENCLATURE (Cont'd)

R	gas constant
s	selectivity, Eq. (4); dimensionless std. deviation, Eq. (18)
T	temperature
x	mole fraction in adsorbed phase
y	mole fraction in gas phase
z	dummy variable, Eq. (22)
γ	activity coefficient in adsorbed phase
θ	dimensionless amount adsorbed, n/m
μ	chemical potential; mean value of E, Eq. (17)
σ	standard deviation of E, Eq. (17)
Π	spreading pressure
ϕ	gas-phase fugacity coefficient, Eq. (8)

Subscripts

i	refers to component i

Superscripts

e	denotes excess function
o	refers to standard state
id	ideal value

REFERENCES

1. Myers, A.L. and J.M. Prausnitz, *A.I.Ch.E. Journal* 11, 121 (1965).
2. Glessner, A.J. and A.L. Myers, *Chem. Engrng. Progress Symp. Series* 65(96), 73 (1969).
3. Minka, C. and A.L. Myers, *A.I.Ch.E. Journal* 19, 453 (1973).
4. Sircar, S. and A.L. Myers, *A.I.Ch.E. Journal* 19, 159 (1973).
5. Myers, A.L., *A.I.Ch.E. Journal*, 29, 691 (1983).
6. Sircar, S. and A.L. Myers, *J. Phys. Chem.* 74, 2828 (1970).
7. Everett, D.H., *Trans. Faraday Soc.* 60, 1803 (1964).
8. Larionov, O.G. and A.L. Myers, *Chem. Engrng. Science* 26, 1025 (1971).
9. Sircar, S., A.L. Myers and M.C. Molstad, *Trans. Faraday Soc.* 66, 2354 (1970).
10. Jossens, L., J.M. Prausnitz, W. Fritz, E.U. Schlünder and A.L. Myers, *Chem. Engrng. Science* 33, 1097 (1978).
11. Sircar, S. and A.L. Myers, *A.I.Ch.E. Journal* 17, 186 (1971).
12. Novosad, J. and A.L. Myers, *Can. J. Chem. Engrng.* 60 500 (1982).
13. Myers, A.L. and S. Sircar, *J. Phys. Chem.* 76, 3415 (1972).

REFERENCES (Cont'd)

14. Myers, A.L., C. Minka and D.Y. Ou, *A.I.Ch.E. Journal* 28, 97 (1982).

15. Van Ness, H.C. and M.M. Abbott, *Classical Thermodynamics of Non-Electrolyte Solutions*, McGraw-Hill Book Co., New York (1982), p. 355.

16. Glandt, E.D. and A.L. Myers, *J. Phys. Chem.* 70, 4243 (1979).

17. Glandt, E.D., D.D. Fitts and A.L. Myers, *Chem. Engrng. Science* 33, 1659 (1978).

18. Myers, A.L. and J.M. Prausnitz, *Chem. Engrng. Science* 20, 549 (1965).

19. Sircar, S. and A.L. Myers, "Principle of Correspondence for Adsorption of Vapors on Heterogeneous Adsorbents," *A.I.Ch.E. Journal*, in press.

20. Myers, A.L. and S. Sircar, *Adv. in Chem. Series 202*, Amer. Chem. Soc., Wash., D.C. (1983), p. 63.

21. Nicholson, D. and N.G. Parsonage, *Computer Simulation and the Statistical Mechanics of Adsorption*, Academic Press, London (1982).

22. Steele, W.A., *The Interaction of Gases with Solid Surfaces*, Pergamon Press, Oxford (1974).

23. Langmuir, I., *J. Amer. Chem. Soc.* 40, 1361 (1918).

24. Zolandz, R.R. and A.L. Myers, "Adsorption on Heterogeneous Surfaces," in *Progress in Filtration and Separation*, Elsevier Scient. Pub. Co., New York (1979).

25. House, W.A., "Adsorption on Heterogeneous Surfaces" in *Specialist Periodical Reports, Colloid Sci., Vol. 4*, Roy. Soc. Chem. (London).

26. Gurvitsch, L., *J. Phys. Chem. Soc. Russ.* 47, 805 (1915).

27. Reich, R., W.T. Ziegler and K.A. Rogers, *Ind. Eng. Chem. Process Des. Dev.* 19, 336 (1980).

28. Valenzuela, D., "Adsorption Data Base," University of Pennsylvania, in preparation.

29. Toth, J., *Acta Chim. Acad. Sci. Hung.* 30, 1 (1962).

30. Van Ness, H.C., *Ind. Eng. Chem. Fund.* 8, 464 (1969).

ACKNOWLEDGMENT

This work was supported by National Science Foundation Grant CPE-8117188.

APPENDIX

Program IASLUD

N-component vapor adsorption equilibrium
Uses subprogram INVSP

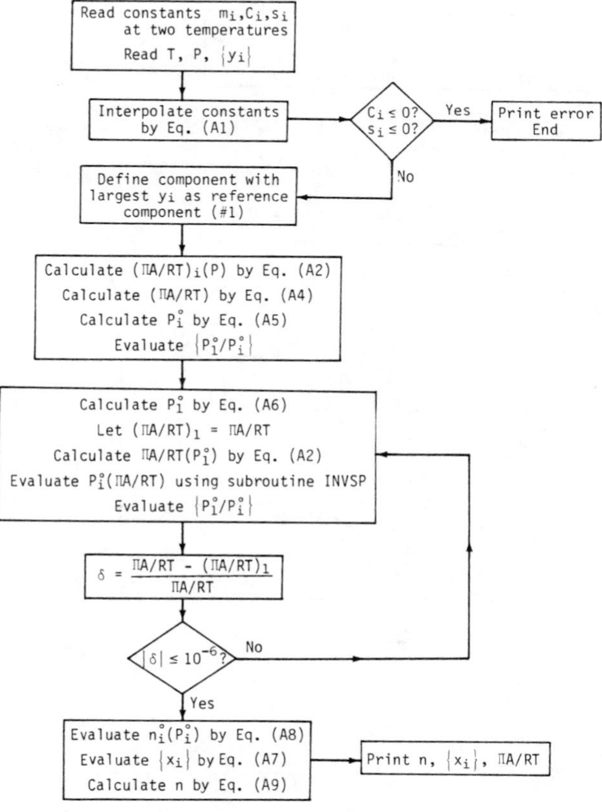

APPENDIX (Cont'd)

Subroutine INVSP

Inverse function of Eq. (A-2) by Newton's method.

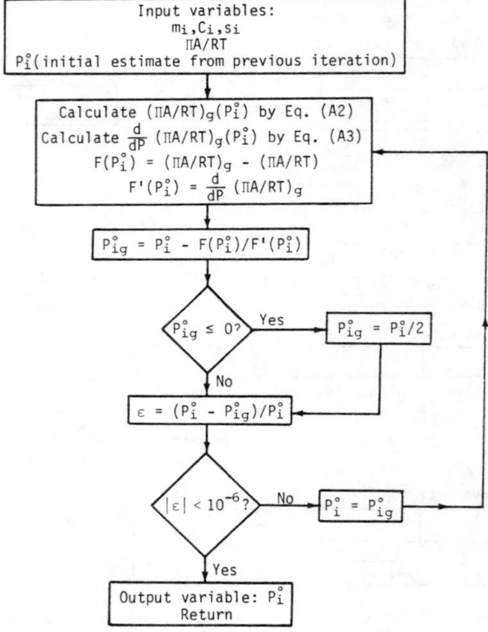

Appendix (Cont'd)

EQUATIONS

$$m_i \propto T$$
$$s_i \propto T \qquad \text{(A1)}$$
$$\ln C_i \propto 1/T$$

$$(\Pi A/RT)(P) = \frac{m_i}{2s_i} \int_{z=-s_i}^{s_i} \ln(1 + C_i P\, e^z)\, dz \qquad \text{(A2)}$$

$$\frac{d}{dP}(\Pi A/RT)(P) = \frac{m_i}{2s_i} \int_{z=-s_i}^{s_i} \frac{C_i e^z}{\left[1 + C_i P\, e^z\right]}\, dz \qquad \text{(A3)}$$

$$\Pi A/RT = \sum_{i=1}^{N} y_i (\Pi A/RT)_i \qquad \text{(A4)}$$

$$P_i^\circ = \frac{1}{C_i}\left(\exp\left[(\Pi A/RT)_i/m_i\right] - 1\right) \qquad \text{(A5)}$$

$$P_1^\circ = P \sum_{i=1}^{N} y_i \left\{\frac{P_1^\circ}{P_i^\circ}\right\} \qquad \text{(A6)}$$

$$x_i = P y_i / P_i^\circ \qquad \text{(A7)}$$

$$n_i^\circ(P) = \frac{m_i}{2s_i} \ln\left[\frac{1 + C_i P\, e^s}{1 + C_i P\, e^{-s}}\right] \qquad \text{(A8)}$$

$$n = 1/\sum_{i=1}^{N} (x_i/n_i^\circ) \qquad \text{(A9)}$$

Comments:

(A1) are interpolation formulas

(A2) and (A3) are integrated numerically by Simpson's rule.

(A4) and (A5) are initial estimates.

(A6) is $\sum x_i = 1$, and is used to recalculate the reference pressure in terms of pressure ratios.

THE IMPORTANCE OF SORPTION PROCESSES IN RETARDING RADIONUCLIDE MIGRATION IN THE GEOSPHERE

Ivars Neretnieks
Department of Chemical Engineering
Royal Institute of Technology
Stockholm, Sweden

ABSTRACT

In most industrialized countries with nuclear power plants, studies have been started to investigate how the very highly active spent nuclear fuel can be disposed of. One of the disposal options is to bury the radionuclides in deep geologic repositories and to surround the waste with a backfill with high sorption power for the important nuclides.

This paper discusses the retardation in a backfill based on sorption and diffusion measurements. There are backfill materials available which will retard the shorter lived nuclides so long that they will have decayed to insignificance before the breakthrough curve emerges at the outside of the barrier. So far there have not yet been devised sorbents strong enough to retard the very long lived nuclides.

The bedrock surrounding a repository can also retard the nuclides considerably if the flowrate is low and the sorption capacity of the rock is high. All investigated crystalline rocks are porous to some extent, and species dissolved in the water moving in the larger fissures will diffuse into the micropores of the rock. The vast majority of the radionuclides of importance in this context are cations. They will adsorb or be ion exchanged on the inner surfaces of the rock and thus be considerably retarded and given time to decay on their way to the geosphere.

Radionuclide migration in the bedrock is treated as an adsorption column in this paper. Breakthrough calculations based on flow, sorption and diffusion measurements are presented for a repository in crystalline rock.

— — —

1 BACKGROUND

In many countries spent fuel and reprocessed waste from nuclear power plants will be disposed of by emplacement in deep geologic repositories under the water table. The waste canisters will be surrounded by some suitable backfill material which may act as an adsorbent or ion exchanger for many of the radionuclides. The sorption processes may retard some of the important radionuclides for a time long enough to let them decay considerably during their migration through the backfill. Backfill materials with high sorptive properties have been proposed and are being investigated.

Those nuclides which pass the backfill move with the flowing water in the bedrock. In some countries e.g. Sweden and Canada crystalline rocks such as granites and gneisses are predominant and are considered as potential repository sites. The water moves in fissures, but there is also stagnant water in intercrystalline voids. The latter exchange dissolved species with the flowing water by diffusion. The crystal surfaces in the micropores have a considerable sorptive capacity for many of the important radionuclides. The radionuclides which migrate into the micropores by diffusion, may thus be considerably retarded on their way through the bedrock.

Figure 1 shows the potential hazard index of spent fuel compared to the amount of uranium from which the fuel was obtained. The figure indicates that it will take nearly a million years for the spent fuel to decay by the necessary orders of magnitude to the same hazard as uranium ore. The repository should isolate the radionuclides for at least this time. A possible repository concept is shown in figure 2.

2 BACKFILL MATERIALS

Several natural backfill materials have been proposed and investigated. They include zeolites (9), natural clays such as bentonite (15) without or with special admixtures such as ferrous phosphates (24). At present bentonite clay seems to be a very strong candidate because of its extremely low hydraulic conductivity (16) and good swelling properties which ensures that the fuel canister will not be contacted by flowing water. All exchange of dissolved species takes place by diffusion.

2.2 Properties of bentonite

The major constituent of bentonite clays is the Alumino-silicate mineral Montmorillonite. When compacted to a wet density of 1.9 - 2 g/cm^3 the hydraulic conductivity is less than 10^{-13} m/s (16). Several investigations of the sorptive and diffusional properties of this material have been performed (10,4,5,23,24). An analysis of the various diffusion investigations and a compilation of results are given by Neretnieks (13).

2.3 Migration through the backfill

If the matrix of the spent fuel (95 % UO_2) dissolves at a constant rate, the first nuclides in a chain will be released at a rate which decreases in time due to radioactive decay.

$$N_i = N_o \cdot e^{-\lambda t} \qquad (1)$$

N_i is the rate of inflow to the backfill barrier. N_o is the release rate at time 0 . The nuclide will migrate in the backfill barrier by diffusion only. For a barrier with plane geometry we have

$$\frac{\partial c}{\partial t} + \frac{1-\varepsilon}{\varepsilon} \cdot \frac{\partial c_s}{\partial t} = D_p \frac{\partial^2 c}{\partial z^2} - \lambda(c + \frac{1-\varepsilon}{\varepsilon} c_s) \tag{2}$$

For linear equilibrium

$$c_s = K' \cdot c \tag{3}$$

and Equation (2) becomes

$$\frac{\partial c}{\partial t} = \frac{D_p \varepsilon}{K} \frac{\partial^2 c}{\partial z^2} - \lambda c \tag{4}$$

where

$$K = \varepsilon + (1-\varepsilon)K' \tag{5}$$

For a semiinfinite medium ($c = 0 \quad z \to \infty \quad t > 0$) Equation (4) with the boundary condition Equation (1) for $t > 0$ at $z = 0$ and initial condition $c = 0, t > 0, z > 0$ the solution is (13)

$$N_{z_o}/N_o = e^{-\lambda t} \operatorname{erfc} \frac{z_o}{2\sqrt{\frac{D_p \varepsilon}{K} t}} = e^{-\lambda t} \operatorname{erfc} \frac{A}{\sqrt{t}} \tag{6}$$

Equation (6) gives the flux N_{z_o} past a plane z_o at time t related to the leach rate i.e. the flux N_o into the backfill at time 0. The release from the backfill will rise at first as the breakthrough curve progresses through the buffer, and then it will decline as the nuclide decays. Figure 3 shows the maximum value of Equation (6) as a function of λA^2. It also shows at what dimensionless times λt the maximum occurs. It is seen that for λA^2 values between 10 and 100 the maximum release falls from 10^{-3} to 10^{-9} of the release that would occur at time 0 and if the buffer would have no retardation for the nuclides.

Table 1 shows the apparent diffusivity D_a ($= D_p\varepsilon/K$), the sorption equilibrium constant K, the decay constant λ, and λA^2 for a buffer thickness of 0.5 m for some of the most important nuclides in spent fuel. The data are taken from (13).

The nuclides are also plotted in Figure 3 on the line for maximum flux. It is seen that five of the nuclides fall in a range where a considerable decrease of the maximum release occurs. The fairly shortlived Am^{241} will decay to insignificance. I^{129} and U^{238} will be practically unaffected by the buffer. Tc^{99} and U^{238} have very long halflives and will not decay much. Iodide is an anion and does not interact with the mildly negative natural minerals. Technetium is also present as a negative pertechnetate ion in oxidizing conditions, but can be reduced to the tetravalent state in reducing waters. In the tetravalent state it will have a considerably smaller solubility and higher sorption. Neptunium may also be reduced and sorb better (1).

The thickness of the barrier probably cannot be increased much. The diffusivity in the pore water of the backfill also cannot be much changed. The remaining important entity which might be influenced is the sorption capacity. This could be increased by making the environment reducing by addition of some reducing agent, but radiolysis will tend to change it to an oxidizing environment again. No sorbents have as yet been proposed which will have a high sorption capacity for Np and Tc in oxidizing conditions. The five retarded nuclides might also be retarded to give off insignificant amounts, if sorbents with a 1 to 2 order higher sorption capacity or lower diffusivity could be found. Cs and Sr have been found to diffuse faster in bentonite clay than pore diffusion can account for (13) and probably moves by surface diffusion. The latter process is as yet not well understood for ionic species.

3 CRYSTALLINE ROCK

Crystalline rock e.g. granites and gneisses are fractured even at large depths. The distance between the flow fractures are from 0.1 to 10 m or even more in "good" rock at large depths. The fissures are sometimes coated with rock alteration materials or precipitated materials such as calcite. The rock matrix consists of crystals of quartz, feldspars, micas, etc. Crystal sizes may vary from 0.1 mm or less and up to many mm. Microfissures between the crystals make up a connected pore structure. The fissures and micro-fissures contain water. All the rock-forming minerals sorb cations on their surfaces.

3.1 Radionuclide interaction with the rock

The radionuclides which escape from the repository migrate with the flowing water in the fissures. Many of them sorb on the fissure coating materials and migrate through the coating and into the connected pore space of the micro-fissures. The process is illustrated in figure 4. Porosities and pore diffusivities have been measured for crystalline rocks from many locations (20,21). Field tests in a mine at a depth of 360 m have shown that the micro-pores are open and connected in rock which is compressed by the rock over burden (3) in its natural state. The fissure coating materials are porous enough to permit the radionuclides to penetrate to the rock (21).

The uptake of strontium and cesium in rock tablets by diffusion into the micro-fissures and sorption on the inner surfaces have been measured by finite bath techniques (21) and by profile analysis of the sorbed species in the rock tablets (25). Cesium and strontium have a considerable surface migration component. The actinides sorb so strongly that it has not been possible as yet to determine if the migration is enhanced by surface diffusion or not. Rock cores with natural fissures have been used in the laboratory for "column" experiments with nonsorbing as well as sorbing tracers (12). Tracer tests in the field with nonsorbing and sorbing tracers have been performed and analysed (6,8). Diffusion and sorption data for some of the important nuclides are given in Table 2.

Sorption on the surfaces of the fissures with flowing water and in the thin layers of fissure coating or filling material is negligible compared to the sorption in the rock matrix at the contact times of interest here. The penetration depth of a nuclide into the rock matrix can be assessed by solving the instationary diffusion equation. For the classical constant concentration boundary at $z = 0$ the penetration depth of a nondecaying species into a medium with concentration 0 originally, we have (2)

$$n_{0.01} \approx 4 \sqrt{D_a t} \tag{7}$$

The penetration depths for 10^3 and 10^5 years contact time are also given in Table 2. They vary considerably from a few mm to several meters for the shorter contact time for different nuclides.

4 MODELS AND PREDICTIONS

The movement of the radionuclides with the flowing water is modelled by the advection dispersion equation in a single fissure or in a porous medium with blocks of a given size. For the case of modelling, the blocks are assumed to be spherical or infinite plates.

The model is formulated for a decay chain where i is the number of the nuclide in the chain.

For the flowing water we have

$$\frac{\partial c_f^i}{\partial t} + U_f \frac{\partial c_f^i}{\partial z} - D_L \frac{\partial^2 c_f^i}{\partial z^2} = aD_p^i \varepsilon_p \left. \frac{\partial c_p^i}{\partial x} \right|_{x=0} - \lambda^i c_f^i + \lambda^{i-1} c_f^{i-1} \tag{8}$$

and for the rock matrix (formulated for the planar fissure)

$$\frac{\partial c_p^i}{\partial t} = D_a^i \frac{\partial^2 c_p^i}{\partial x^2} - \lambda^i c_p^i + \frac{K^{i-1}}{K^i} \lambda^{i-1} c_p^{i-1} \tag{9}$$

The initial conditions used indicate zero concentration everywhere. The boundary conditions may be flux or concentration conditions at $z = 0$ and $c_p = c_f$ at $x = 0$. For blocks of finite thickness the symmetry condition is used at the centre of the slab.

The Equations (8) and (9) have been solved analytically for single nuclides for a single fissure with no dispersion (11), for a single fissure with dispersion (22) and, for porous beds with spherical particles (17). These solutions have been used to test the numeric solutions (18,19), which must be used for decay chains and complex geometries (14).

The models and data have been used in a safety analysis performed for the Swedish nuclear fuel supply company (7). In the study the water flux in the rock is less than 0.1 l/m^2 year, the fissure spacing (block size) is 5 m. Although the repository is at 500 m depth, the travel distance for the nuclides until they reach the biosphere is taken to be 100 m because the bedrock is intersected by major crushed zones with much higher water flowrates than in the "good" rock (see Figure 5). The leach time of the spent fuel runs into many million years because of the low solubility of the uranium oxide of the fuel matrix. With these data, the data in table 2 and data on the release from the "near field", the breakthrough behaviour of the nuclides to the biosphere have been calculated. They are shown in Figures 6 and 7. Figure 6 shows the release from the near field.

That makes up the influx boundary condition to the "far field" calculations. The breakthrough curves after 100 m travel in the rock are shown in Figure 7.

A comparison of the figures shows that I^{129} does not decrease in concentration, Cs^{135} decreases by a factor of 30, U^{238} decreases by a factor of 2, and Tc^{99} by several hundred times during the passage through the rock. All other nuclides decrease by many more orders of magnitude. The decay chain U^{238} - U^{234} - Th^{230} - Ra^{226} leaves Ra^{226} as an important species. In a closed system Ra^{226} would attain equilibrium with its immediate parent Th^{230}. Ra has an equilibrium constant K 50 times smaller than its parent and grandparents and moves faster than these. It overtakes its parent and is reconcentrated in the effluent. Radium becomes the most important nuclide.

5 DISCUSSION AND CONCLUSIONS

The naturally occuring clay Bentonite has mechanical and hydraulic properties which makes it well suited as a backfill material. Its inherent sorptive and ion exchange properties retard some of the important nuclides, giving them time to decay considerably. Some other important nuclides are not retarded enough in the expected oxidizing conditions. Admixtures have been proposed (ferrous phospate) which would ensure reducing conditions. α-radiolysis may, however, oxidize the ferrous iron and nullify the effect. Sorbents with specific affinity for these species would add to the value of the buffer.

The rock has proved to be a very strong barrier to radionuclide migration due to its high sorption capacity on the micropore surfaces. The porosity, although low, is sufficient to allow the nuclides to diffuse into rock matrix and sorb there.

If these mechanisms were not active, there would only be a decrease due to sorption on the surfaces of the flow fissures and due to dispersion. These effects would in the present scenario hardly modify the effluent from the near field. Under such conditions the outlet of the radionuclides would become uncomfortably high.

6 NOTATION

A	$z_o/(2D_p\varepsilon/K)^{\frac{1}{2}}$	$s^{\frac{1}{2}}$
a	surface area per volume of water in fissure	m^2/m^3
c	concentration in water	mol/m^3

c_f	concentration in water in fissure	mol/m^3
c_p	concentration in pore water	mol/m^3
c_s	concentration in solid	mol/m^3
D_a	"apparent diffusivity" $D_p\varepsilon/K$	m^2/s
D_p	diffusivity in pore water	m^2/s
D_L	longitudinal dispersivity	m^2/s
K	volume equilibrium constant for bulk phase (solid + liquid)	m^3/m^3
K'	volume equilibrium constant for solid phase	m^3/m^3
N	nuclide flux	mol/s
N_o	nuclide flux at time 0 at inlet side of buffer	mol/s
N_{z_o}	nuclide flux in buffer at distance z_o	mol/s
t	time	s
U_f	water velocity	m/s
x	distance into rock	m
z	distance into buffer or along flow path in fissure	m
z_o	distance to observation point in buffer or in flow path	m
ε	porosity	-
$\eta_{0.01}$	penetration depth	m
λ	decay constant	$s^{-1}(year^{-1})$

7 LITERATURE CITED

1. Allard B., Olofsson U., Torstenfelt B., Kipatsi H., Andersson K., Sorption of actinides in well-defined oxidation states in geologic media. In "Scientific basis for nuclear waste management" vol. 5, Elsevier North Holland N.Y., 1982, p 775-764.

2. Bird B.R., Stewart W.E., Lightfoot E.N., Transport Phenomena. Wiley 1960, p 354.

3. Birgersson I., Neretnieks I., Diffusion in the matrix of granitic rock. Field test in the Stripa mine. Paper presented at the 5th international symposium on the scientific basis for nuclear waste management. Berlin, June 1982. Proceedings North-Holland 1982, p 519-528.

4. Eriksen T., Jakobsson A., Ion diffusion through highly compacted bentonite. KBS TR 81-06.* 1981.

5. Eriksen T., Jakobsson A., Ion diffusion in compacted sodium and calcium bentonites. KBS TR 81-12.* 1982.

6. Hodgkinson D.P., Lever D.A., Interpretation of a field experiment on the transport of sorbed and nonsorbed tracers through a fracture in crystalline rock. Theoretical physics division, Atomic Energy Research Establishment, Harwell, England, October 1982. AERE-R-10702.

7. Final Storage of Spent Nuclear Fuel. KBS 3, Stockholm 1983.

8. Moreno L., Neretnieks I., Klockars K., Evaluation of some tracer tests in the granitic rock at Finnsjön. April 1983. KBS TR 83-49.*

9. Neretnieks I., Retardation of escaping nuclides from a final repository. 1977. KBS TR 30.*

10. Neretnieks I., Skagius C., Diffusivitetsmätningar av metan och väte i våt lera. 1978. KBS TR 86.*

11. Neretnieks I., Diffusion in the rock matrix: An important factor in radionuclide retardation? J Geophys Res 85, 1980, p 4379.

12. Neretnieks I., Eriksen T., Tähtinen P., Tracer movement in a single fissure in granitic rock – Some experimental results and their interpretation. Water Resources Res, 1982, p 849.

13. Neretnieks I., Diffusivities of some dissolved constituents in compact wet bentonite clay MX80 and the impact on radionuclide migration in the buffer. KBS TR 83-33.* 1983.

14. Neretnieks I., Rasmuson A., An approach to modelling radionuclide migration in a medium with strongly varying velocity and block sizes along the flow path. KBS TR 83-69.*

15. Pusch R., Highly compacted Na bentonite as buffer substance. KBS TR 74.* 1978.

16. Pusch R., Permeability of highly compacted bentonite. KBS TR 80-16.* 1980.

17. Rasmuson A., Neretnieks I., Exact solution of a model for diffusion in particles and longitudinal dispersion in packed beds. AIChE J 26, 1980, p 686.

18. Rasmuson A., Narasimhan T.N., Neretnieks I., Chemical transport in a fissured rock. Verification of a numerical model. Water Resources Res 18, 1982 a, p 1479.

19. Rasmuson A., Bengtsson A., Grundfelt B., Neretnieks I., Radionuclide chain migration in fissured rock – The influence of matrix diffusion. April 1982 b. KBS TR 82-04.*

20. Skagius K., Svedberg G., Neretnieks I., A study of strontium and cesium sorption on granite. Nuclear Technology, 1982, p 302.

21. Skagius K., Neretnieks I., Diffusion measurements in crystalline rock. KBS TR 83-15.* 1983.

22. Tang D.H., Frind E.O., Sudicky E.A., Contaminant transport in fractured porous media: An analytical solution for a single fracture. Work done by University of Waterloo for Atomic Energy of Canada Limited, TR 132, Nov 1980.

23. Torstenfelt B., Andersson R., Kipatsi H., Allard B., Olofsson U., Diffusion measurements in compacted bentonite. In "Scientific basis for nuclear waste management" vol. 4, Elsevier N.Y., 1982 a.

24. Torstenfelt B., Kipatsi H., Andersson K., Allard B., Olofsson U., Transport of actinides through a bentonite backfill. In "Scientific basis for nuclear waste management" vol. 5, Elsevier North Holland N.Y., 1982 b, p 659-668.

25. Torstenfelt B., Ittner T., Allard B., Andersson K., Olofsson U., Mobilities of radionuclides in fresh and fractured crystalline rock. KBS TR 82-26.* 1982 c.

---

*) KBS Technical reports may be obtained from
INIS Clearing House
International Atomic Energy Agency (IAEA)
P O Box 590
A-1011 Vienna

Nuclide	$D_a \cdot 10^{12}$ m^2/s	λ years^{-1}	$K^{a)}$ m^3/m^3	λA^2 -
I^{129}	9	$3.5 \cdot 10^{-8}$	0.4	$7.7 \cdot 10^{-6}$
Tc^{99}	53	$3.5 \cdot 10^{-6}$	0.4*	$1.3 \cdot 10^{-4}$
Cs^{137}	8	$2.3 \cdot 10^{-2}$	3000	5.7
Sr^{90}	25	$2.5 \cdot 10^{-2}$	5000	2.0
U^{238}	1	$1.6 \cdot 10^{-10}$	300*	$3.2 \cdot 10^{-7}$
Np^{237}	0.4	$3.5 \cdot 10^{-7}$	300*	$1.7 \cdot 10^{-3}$
Pu^{239}	$3 \cdot 10^{-2}$	$2.9 \cdot 10^{-5}$	10000	1.9
Pu^{240}	$3 \cdot 10^{-2}$	$1.1 \cdot 10^{-4}$	10 000	7.3
Am 241	$1.5 \cdot 10^{-2}$	$1.5 \cdot 10^{-3}$	20000	198
Am^{243}	$1.5 \cdot 10^{-2}$	$9.4 \cdot 10^{-5}$	20000	12.4

a) The K-values may vary between various bentonites.

*) Under oxidizing conditions.

Table 1. Apparent diffusivity D_a, sorption equilibrium constant K, decay constant λ and λA^2 for a 0.5 m thick buffer.

Nuclide	Da* m^2/s	K m^3/m^3	$\eta_{0.01}$ mm $t=10^3$	10^5 years
I^{129}	$2.5 \cdot 10^{-11}$	$2 \cdot 10^{-3}$**	3 500	$35 \cdot 10^3$
Tc^{99}	$4 \cdot 10^{-16}$	130	$(14)^b$	$(140)^b$
Cs^{137}	$\sim 10^{-12}/K$	$(130)^a$	$(60)^b$	$(600)^b$
Sr^{90}	$\sim 10^{-13}$	10	220	2200
U^{238}, Np^{237}, Pu^{239}, Am^{243}	$4 \cdot 10^{-18}$	> 13000	1.4	14

*) $D_a = D_p \varepsilon_p / K$

**) Nonsorbing species, sorbing capacity only due to porosity of rock.

a) Nonlinear Freundlich isotherm. The value given is for reasonably high but still trace concentrations.

b) These nuclides have decayed to insignificance at times > 500 years.

Table 2. Diffusion and sorption data in reducing waters for some of the important nuclides. The penetration depth (equation (2)) for 1000 and 10^5 years contact time is also given.

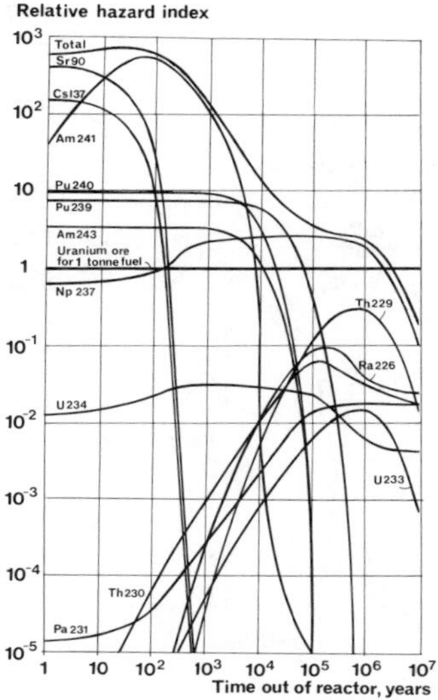

Figure 1. Potential hazard index for spent fuel. The hazard is related to that of an original amount of uranium from which the fuel was obtained.

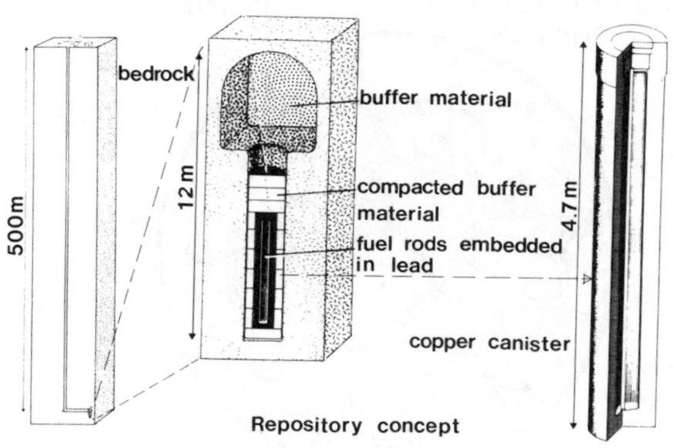

Figure 2. Repository as proposed in the Swedish Nuclear Fuel Safety study (KBS 1983).

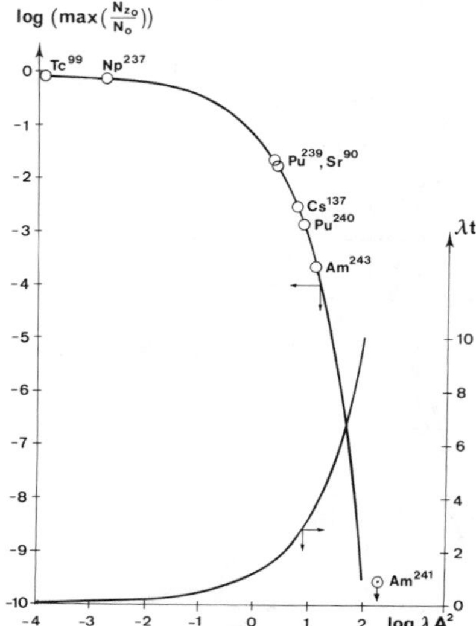

Figure 3.
The maximum release N_{z_o} ever from the outside of a backfill buffer of thickness z_o as a function of λA^2.
Some important nuclides are indicated.

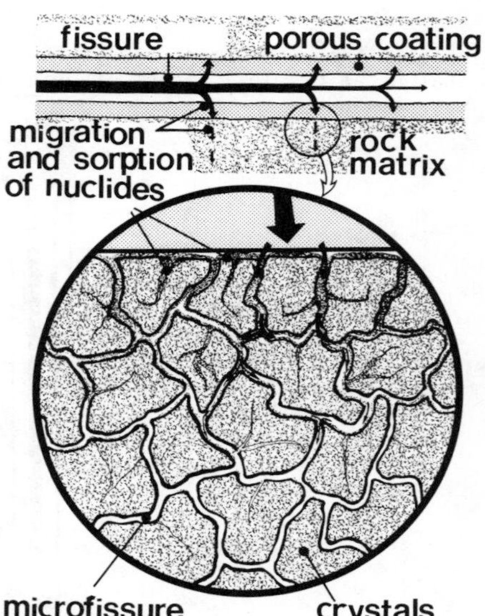

Figure 4. Illustrates how the nuclides diffuse into the micropores of the rock and sorb onto the micropore surface. The flowing water in the fissure is depleted accordingly.

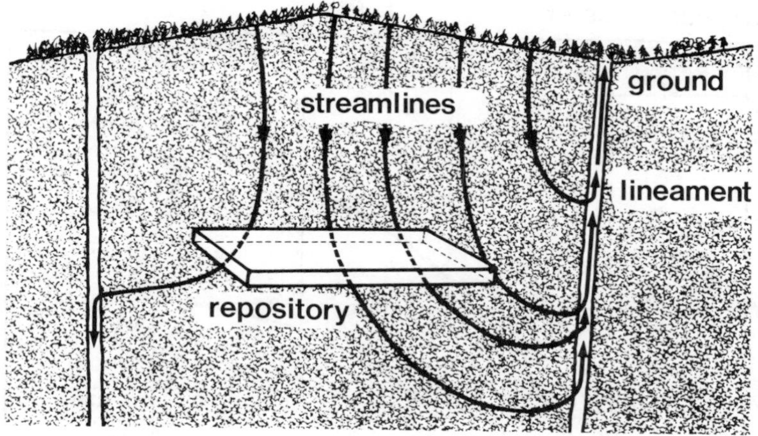

Figure 5. Artist's view of a repository in crystalline rock with major fissure zones.

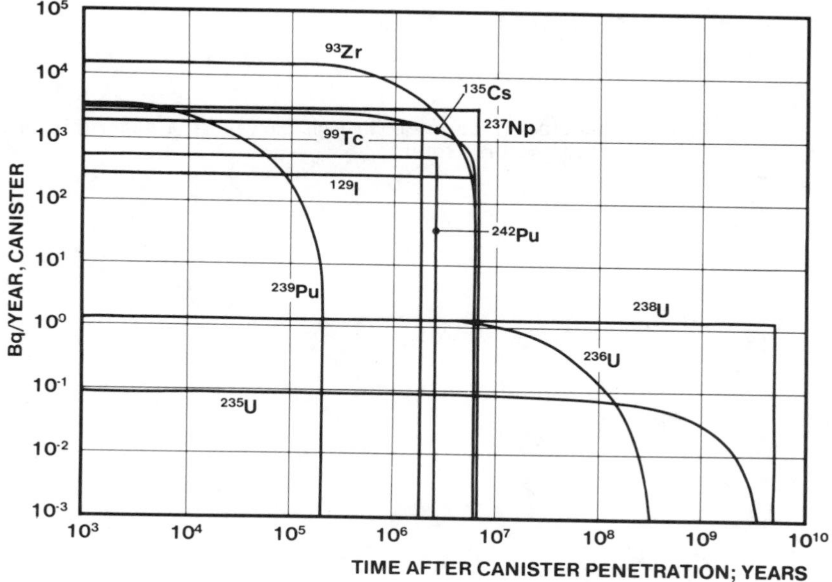

Figure 6. Radionuclide release from the near field.

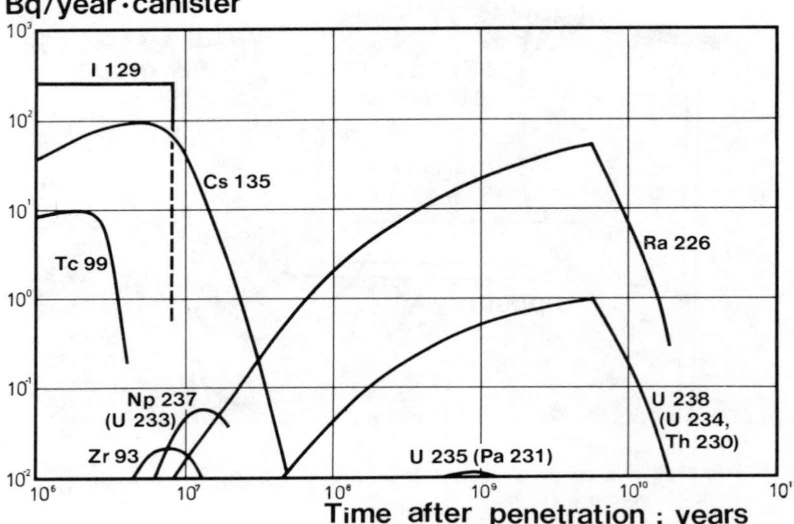

Figure 7. Radionuclide breakthrough curves to a major fissure zone 100 m distant.

INTERMOLECULAR FORCES AND ADSORPTION ON GRAPHITE

DAVID NICHOLSON

Department of Chemistry, Imperial College, London SW7 2AY

ABSTRACT

Experimental results relating to structures and transitions in adsorbed phases are briefly reviewed. Intermolecular potentials can be constructed from pairwise and other contributions; these contain electrostatic, induced and dispersion terms. The anisotropy of polarizability as well as higher order terms are important in the last two. Computer simulation studies which link experimental results directly to potentials reveal shortcomings in existing functions.

A SURVEY OF ADSORPTION ON GRAPHITE

The importance of carbon as an industrial material provides a strong motivation towards gaining a deeper understanding of adsorption on its surface, whilst the availability of well characterized graphites offers the opportunity of studying the process at a fundamental level. The recognition that graphitization, at a sufficiently high temperature, can produce an energetically homogeneous surface for adsorption goes back to the fifties (1). However the more recent work dates from the development of exfoliated graphites in the late sixties.

Thermodynamic Measurements

Measurements of isotherms, adsorption heats and heat capacities on such substrates have been made with high precision. These have established the existence of submonolayer phase transitions in the rare gas adsorbates (2),(3),(4), and enabled detailed phase diagrams to be constructed (5), (6).

Improved techniques and adsorbents have frequently revealed significant detail absent from earlier work. For example, although it was well known that the $q_{st}(\theta)$ curve passed through a maximum its fine structure only became apparent when automatic recording was employed (4) (Figure 1).

A large number of molecular adsorbates have been investigated on exfoliated substrates including nitrogn (7), oxygen (8), nitric oxide (9), and several hydrocarbons (10) (including CH_4(11), C_2H_2(12), C_2H_6(14), C_6H_6(15) and n-C_4H_{10} (16)).

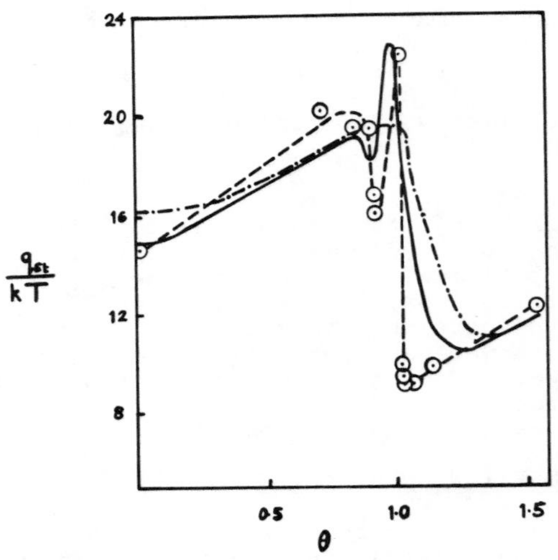

Fig.1. Isosteric heats of adsorption versus coverage. Experimental data from Grillet et al., reference 4:———, Argon on Sterling 2700, —·—·—·— , Argon on Spheron (1). MC calculations, — — O — — with 12-6 potentials (52).

Scattering Measurements

For most of the adsorbates mentioned, scattering techniques, especially LEED and neutron scattering and diffraction, have played a vital, if not exclusive part in establishing the structure of the various phases. In early work on rare gases such methods demonstrated the existence of localised $\sqrt{3}\times\sqrt{3}$ commensurate layers for Kr (17) and Xe (6) and have shown that incommensurate monolayers of Kr, Ar and Ne are rotated in relation to the graphite substrates by angles of 0.5° (17), 4° (18) and 17° (19) respectively. As well as commensurate and incommensurate solid phases, liquid and gaseous surface phases have also been characterized. In view of the complexity of the rare gas/graphite phase diagrams it is not surprising to find that polyatomic molecules exhibit even richer behaviour, including orientational transitions (10).

To take one example, ethane (14), has been found to possess 3 solid phases below 60K. As the coverage is increased from 0.8 to 1.2 of a monolayer the structure passes from a herringbone pattern, with molecules flat or slightly tilted to the surface, through a second phase with an angle of 50° between adjacent molecular axes,to a third one in which the molecules stand upright on CH_3 tripods. The herringbone structure with molecules lying flat on the surface has also been found for N_2(20) and a similar set of orientational transitions to those observed for ethane have been proposed for nitric oxide (9).

Adsorbent Properties

In applying theoretical descriptions to such systems it is important to have reliable information about the nature of the substrate. Adsorption measurements have been used (21) to compare different graphites and the relationship between temperature of "graphitization" and adsorption has long been recognized and exploited (1),(4).

Some recent studies have sought to place the characterization of the graphite on still firmer footing. It has been shown for example (22), using LEED intensity/energy plots, that the stacking is ABAB, and that there is a small contraction (0.005 nm) of the outer layer. Cleaved graphite was found to be almost devoid of surface steps, however EXAFS studies of Kr on exfoliated graphite (23) indicate that high energy sites contribute significantly to the adsorption suggesting that,as well as (0001) planes, the edges normal to these may be important.

INTERMOLECULAR POTENTIALS

Calculation of the Adsorption Potential

The transition between commensurate, incommensurate-rotated and different orientational phases over small ranges of coverage point to a subtle balance between the various intermolecular potentials which control them. It is well known that the adsorbent potential $u^{(1)}$ is periodic in phase with the substrate, and the height of barriers between C-hexagons is important in determining epitaxy and the rotation of monatomic phases. When orientational dependence of the intermolecular potential $u^{(2)}$ is introduced for polyatomic molecules, it is to be expected that a more complicated phase diagram will result.

The main avenue of approach to the calculation of $u^{(1)}$ has been pairwise summation over atoms in the adsorbent. Contributions from triple body terms have also received some attention (24).

When molecular adsorbates are considered a similar decomposition of the molecule into separate atomic interaction centres is an attractive way to procede, and calculations of this kind have been made for a large number of species on continuum model surfaces (25) and of periodic potentials for N_2 (26) and hydrocarbons (27,28). For smaller molecules at least, there may be some advantages to be gained from treating the molecule as a whole, since orientational dependence can be displayed explicitly as a closed equation, and this alternative method could help to resolve some doubts about atom-atom summations. A further advantage is possible if expressions can be obtained which can be rapidly evaluated in a computer simulation.

These objectives are made more attractive because it was shown some time ago by Steele (29) that the pairwise summation could be expressed as a Fourier expansion, and $u^{(1)}$ written in the general form

$$u^{(1)} = u_o^{(1)} + \sum_i u_i^{(1)} f_i \qquad (1)$$

The first and most important term on the RHS can be evaluated by treating each graphite layer as a continuum, integrating to find its interaction with the adsorbate, and then summing over all the layers. Since the summation can, to a good approximation, also be put into a simple closed form (30) an easily evaluated expression results. For an inverse nth power dependence of the pair potential function on the distance between the adsorbate and a carbon atom in the substrate, the sum over layers can be written,

$$S_n(z) = \frac{1}{z^{n-2}} + \frac{1}{(n-3)\ d\ (z + k_n d)^{(n-3)}} \qquad (2)$$

where z is the distance of the adsorbate from the first layer, d is the spacing between the graphite layers and k_n ranges from 0.62(n=4) to 0.71 (n=12).

Only the first term of the summation over periodic terms in Equation (1) has been found to be important in studies made so far. This gives the periodic part of the potential as product of $f_1(x,y)$ (29) and a rapidly decreasing term $u_1^{(1)}(z)$. For an inverse nth power dependence of the pair potential function, $u_1^{(1)}(z)$ depends on $(g/2z_\lambda)^{n-2/2} K_{n-2/2}(gz_\lambda)$ where g is a wave vector and K_m is an mth order modified Bessel function.

To obtain an expression for $u^{(1)}$ therefore it is necessary to have an expression for the potential acting between the molecule (a) or its atomic components and a substrate atom.

Pair Potentials

For the separations of interest in physical adsorption, the interaction between the components of a dimer is weak compared to the total energy of the system. A perturbation treatment of the wave equation is therefore appropriate over much of the range and we shall concentrate attention on this aspect of the problem.

Exchange and coulombic repulsion terms do of course become very important in the vicinity of the potential minimum and ab initio methods are the only rigorous way of calculating the latter. Considerable advances which have recently been made in this area (31), (32), (33), encourage the view that a reliable semi-empirical treatment of the total potential may be feasible in which these two terms are represented by a function such as f(R) exp(-bR) of the separation R, where f(R) is a polynomial. Terms from the perturbation expansion can then be added to this to account for the long range part of the interaction.

In the perturbation treatment of the intermolecular potentials the interaction can be divided into three parts due to electrostatic, induction and dispersion contributions. Within this framework, several points can be considered.

Contributions to $u^{(1)}$

Electrostatic terms. Graphite can be pictured as a layer structure in which planar arrays of localised ion cores alternate with non-localised negative charges from the π electrons. It is reasonable to approximate this model as an array of linear quadrupoles of magnitude $\Theta^{(c)}$. Spurling and Lane (34) have shown that such an array produces no external field in the continuum approximation, i.e. the electrostatic interaction with a polar molecule produces no contribution to $u_o^{(1)}$ in Equation (1). However the periodic part of this interaction does not vanish. When the interacting molecule (a) possesses a dipole $\mu^{(a)}$ or a normalized linear quadrupole $\Theta^{(a)}$ it can be shown that (35),

$$u^{(1)}_{EQDi}(z_\lambda) = \frac{\Theta^{(c)} \mu^{(a)}}{2\sqrt{2}\, a_s} g_i^2 \cos {}_a^{(s)} e^{-\kappa_i} \qquad (3)$$

$$u^{(1)}_{EQQi}(z_\lambda) = \frac{\Theta^{(c)} \Theta^{(a)}}{16\, a_s} g_i^3 (3 \cos^2 {}_a^{(s)} - 1) e^{-\kappa_i} \times$$

$$\times\left[\frac{1}{4}-\frac{3}{2\kappa_i}-\frac{9}{4\kappa_i^2}-\frac{9}{4\kappa_i^3}\right] \quad (4)$$

where a_s is the area of a unit cell in the graphite layer, $\kappa_i = g_i z_\lambda$ and $\theta_a(s)$ is the angle made by the axis of symmetry through the molecule with a normal through the surface. The periodic contributions also vanish for dipoles lying parallel to the surface and for quadrupoles inclined at an angle of 35.2° to it. They would therefore be expected to be of significance in determining the barrier heights to dipolar species (since these also gives rise to quadrupolar terms) and, for example, to N_2 which has been found to lie flat on the surface.

Anisotropy of polarizability. The dipole-dipole polarizability of carbon in graphite is anistropic. If the polarizability is assumed to have cylindrical symmetry then $\alpha_\perp(c)$ (normal to the z-axis) is considerably greater than $\alpha_{11}(c)$. Anisotropy of polarizability for species with cylindrical symmetry is conveniently expressed by the parameter

$$\gamma = (\alpha_{11} - \alpha_\perp)/(\alpha_{11} + 2\alpha_\perp) = (\alpha_{11} - \alpha_\perp)/3\bar{\alpha} \quad (5)$$

Estimates of this quantity for carbon in graphite lie between -0.50(36) and -0.31 (37). The dipole-dipole polarizability appears in the first order (inverse 6th power) dispersion term, and also in the low order induction terms. Evidence from He-scattering (38) suggests that this anisotropy can have an important effect on surface potential barriers, increasing their height by perhaps a factor of two.

Molecules of low symmetry are also anisotropically polarizable with γ values typically of the order of 0.1 to 0.3.In cylindrical molecules, the inverse 6th power dispersion interaction in the pair potential gives rise to a term contributing to u_0 in Equation (1), which can be written (35).

$$u_{DDO}^{(1)} = -\frac{2\pi A\,\bar{\alpha}^{(a)}\bar{\alpha}^{(c)}\,S_4(z)}{a_s}$$

$$\times\ (1-\tfrac{1}{2}\ \gamma^{(c)})[3 + \tfrac{3}{4}\ \gamma^{(a)}(3\ \cos^2\theta_a^{(s)}\ -1)] \quad (6)$$

in which $A=U_a U_c/4(U_a + U_c)$ and U is an excitation energy. Accurate estimates of A are possible if frequency dependent dynamic polarizabilities are available, and lower and upper bounds are given by the London and Slater-Kirkwood formulae (39).

The orientational dependence of Equation (6) very nearly vanishes for the typical $\gamma^{(a)}$ values given above which is not inconsistent, for example, with N_2 lying flat on the surface, since its interaction centre is closer in this configuration (26). On the other hand, if $\gamma^{(c)} = 0$, the vertical configuration would be quite strongly favoured. It is also readily deduced that anistropy in $\alpha(c)$ causes some 15% reduction in $u^{(1)}_{DDo}$, in comparison to its value when the polarizability is isotropic.

Induced interactions. The induced interaction is usually a small contribution to the overall potential, typically of the order of 5% of the total. Induced interactions between permanent multipoles and the substrate do however have a strong orientational dependency (35), and for some molecules such as NH_3 or H_2O, they may contribute as much as 20% of the overall interaction $u^{(1)}$ in certain orientations. An angle-averaging of these terms may not therefore always be justified in considering the data obtained from the sensitive experiments now possible.

Higher order dispersion terms. Inverse 8th and 10th power contributions to the dispersion energy were included in the calculations of Avgul and Kiselev (25),who gave expressions analogous to the Kirkwood-Muller formula for the interaction coefficients. When considered in more detail it is seen that these terms depend on quadrupole-quadrupole and octupole-octupole polarizabilities (40) which will be anisotropic like their lower order counterparts. Since the inverse 10th power term was found (25) to contribute less than 1% of the overall energy it would appear safe to neglect this complication here, if not the term itself.

Of interest in the present context is the inverse 7th power term, first discovered by Buckingham (40). This contains the dipole-quadrupole polarizability,and for a dimer interaction between molecules of $C_{\infty v}$ symmetry depends on odd powers of the cosine of the angle between them. If both molecules rotate freely then this interaction term averages to a negligible inverse 14th power contribution (41).However, in a localised monolayer at low temperature, the adsorbate molecules are likely to be held fairly rigidly in particular orientations (42) so that inverse 7th (and indeed inverse 9th) power dispersion terms might be expected to contribute to the interaction of CO with graphite, but not to that of N_2. Unfortunately experimental values for higher order polarizabilities are hard to obtain and no estimates have been made for carbon in in graphite.

Three body terms. It was shown many years ago by Sinanoğlu and Pitzer (43) that the triple dipole dispersion interaction involving two adsorbates and one substrate atom would contribute an inverse z^3 term to the pairwise interaction between adsorbate atoms. This term is repulsive and of the order of 10% of the total potential. This and various later estimates have been well reviewed (24). Little attention has been paid to the C-C-a triple dipole term (44). It is by no means certain that triple dipole terms make the most important contribution to three-body interactions (45).

Surface heterogeneity. The evidence for step-free basal planes on graphites (23) which can give rise to reproducible transition phenomena (21) encourages the view that a fundamental understanding of adsorption may be achieved in a context free of the complications of surface heterogeneity. However,it is inevitable that some proportion of edge planes must be present (23) and $u^{(1)}$ for these will probably be quite different from that for the basal planes. This is readily appreciated from Equation (6) since $\gamma^{(c)}$ must be positive at edge planes if it is negative for the basal plane. Some effects due to this have been investigated (46).

A second source of non-uniformity in the adsorption potential is the high energy sites formed by the close spacing of adsorption planes when adsorbent particles are packed together. In crevices or pores, formed in this way, the adsorbate molecule interacts with more than one adsorbent surface and $u^{(1)}$ consequently has a high value.

Obviously this difficulty will be serious for high area adsorbents, and absent in single crystal studies.

Intermolecular Forces and Computer Simulation

One criterion by which a potential can be judged is the zero coverage heat,although strong adsorption sites make reliable extrapolation difficult (4).

Away from the low coverage region it is more probable that most of the adsorbate behaviour on low area graphites can be attributed to molecules on the basal plane but here interactions between molecules are also important in determining the properties observed and statistical mechanical models may not be reliable. Ideally, computer simulation bypasses this difficulty in providing a direct link between intermolecular potentials and experiment.In the grand ensemble Monte Carlo (GEMC) method (47), (48), (49), coverage and structure can be determined at chosen values of μ, V, T and the distinction between metastable and stable states can be made using free energy criteria.

As an illustration of the use of this method we return to the simple monatomic adsorbates and consider krypton on graphite. In this system the existence of a commensurate phase is well established (5),(18). Simulations on the 90.2K isotherm have been reported in detail elsewhere (50), (51). In the more recent work (51) a graphite plane with 648 epitaxial sites was chosen such that 722 close packed atoms could also be accommodated without disturbing the continuity of the periodic boundaries. Runs of several million configurations, using potentials based on the pairwise summation of Lennard-Jones functions (29) for the Kr-graphite interactions, and standard 12-6 functions for the Kr-Kr pairs, failed to reproduce the commensurate phase. A reasonable conjecture is that a graphite modulated triple dipole term (43), (24), added to the pair potential, may stabilize the commensurate phase, because it is repulsive. In fact a barely perceptible shift towards commensurate behaviour was found in those runs started from close packed configurations as can be seen from the density plots shown in Figure 2 (51), and this only occured at the expense of severe depletion in the coverage, so it cannot be claimed that the experimental data have been represented. In a similar way, although the qualitative features of the liquid-solid transition in adsorbed argon have been reproduced, quantitative agreement between experiment and simulation is lacking (52). It should be mentioned that in simulations where the number of molecules is kept constant,and where an estimation of free energy is difficult metastability of certain phases may go undetected. The GEMC method offers certain advantages in this respect.

The evidence from this work is supported by potential energy calculations (54) and points to the need for potential functions giving higher barriers, such as may be obtained when anisotropy of polarizability is accounted for (53),but it is not known how high the barriers should be, and no suitable potential function exists as present (5).

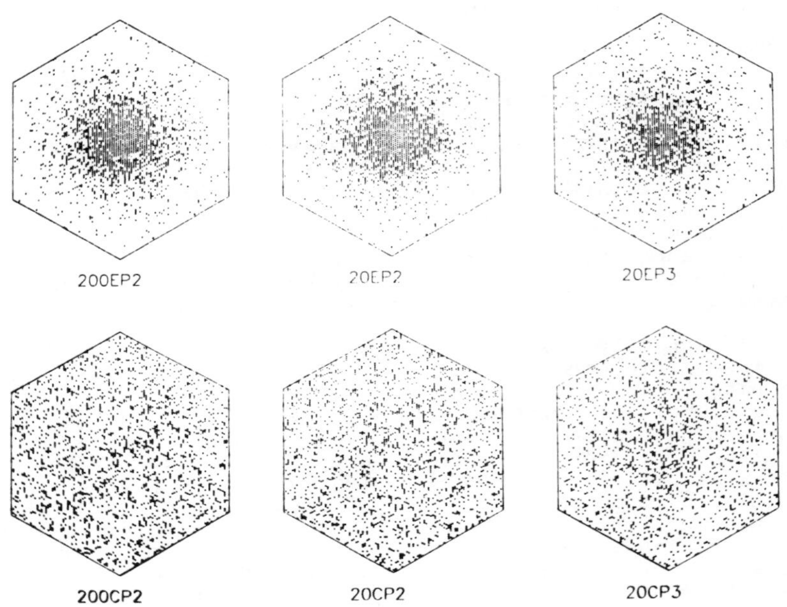

Fig.2. Kr densities, averaged over a graphite hexagon from GEMC simulations on the 90.2K isotherm, are represented by the density of points. CP and EP refer to close packed and epitaxial starting configurations respectively, 200 and 20 signify pressures above and below the experimental transition from commensurate to incommensurate, (p = 2.1mB and 0.21mB respectively) and the final digit (2 or 3 respectively) indicates 2-body only or 2 + graphite modulated 3 body interactions between Kr atoms. In all cases the CP runs had significantly lower free energies than the EP runs, (from ref. 51).

CONCLUSION

The detail and reproducibility of many of the newer adsorption data indicate that theories more accurate than those used hitherto would be appropriate.

The new results summarized here in the sections on electrostatic and dispersion terms suggest that the former could play a part in determining barrier heights between sites and illustrate the importance of anisotropy of the dipole-dipole polarizability of carbon in graphite in relation to orientational preference of molecular species.

Computer simulations for monatomic systems have shown that conventional potentials are incapable of reproducing phase transitions with quantitative accuracy, nor does the introduction of three body terms involving C-Kr-Kr interactions resolve the problem for this adsorbate. It is concluded that a reconstruction of potential functions would be needed if satisfactory agreement is to be reached

LITERATURE CITED

1. Beebe, R.A. and Young, D.M., J.Phys.Chem., 58, 95, (1954).

2. Thomy,A. and Duval, X.,J.Chim Phys., 66, 1966, (1969).

3. Larher, Y., J.C.S. Faraday Trans. I, 70, 320, (1974).

4. Grillet, Y., Rouquerol, J.and Rouquerol,F.,J. Colloid and Interf. Sci.,70,239,(1979).

5. Butler, D.M., Litzinger, J.A. and Stewart, G.A., Phys.Rev.Lett., 44,466,(1980).

6. Venables, J.A.,Kramer, H.M. and Price, G.L,.Surface Sci., 55, 373, (1976).

7. Larher, Y., J. Chem. Phys. 68, 2257, (1978).

8. Pan, R.P.,Etters, R.D. Kobashi, K., Chandresekhran, S., J. Chem. Phys., 77, 1035 (1982).

9. Matecki, M., Thomy, A. and Duval, X., J.Chim.Phys., 71, 1484 (1974). Suzanne, J., Coulomb, J.P.,Bienfait, M., Matecki, A., Croset, B., and Marti, C., Phys. Rev. Letts., 41, 760 (1978).

10. Suzanne, J. "Phase transitions, dynamics and orientational ordering in hydrocarbon molecules adsorbed on graphite" in Adsorption at the Gas-Solid and Liquid-Solid Interface, Rouquerol, J.and Sing, K.S.W. (Eds.) Elsevier, Amsterdam, 1982.

11. Thomy,A. and Duval, X., J.Chim Phys. 67, 1101,(1970).

12. Menancourt, J., Thomy, A and Duval, X.,J.Chim.Phys. 77, 954, (1980).

13. Menancourt, J., Thomy, A. and Duval, X.,J.Phys.,Paris, 38, C4-195, (1977).

14. Suzanne, J., Sequin, J.L., Taub, H., and Biberian, J.P., Surf. Sci., 125, 153 (1983).

15. Khatrir, Y., Coulon, M. and Bonnetain, L, J. Chim. Phys., 75, 789, (1978).

16. Trott, G.J., Taub, H., Hansen, F.Y. and Danner, H.R., Chem.Pohys. Letts. 78, 5045, (1981).

17. Shaw, C.G., Fain, Jr., S.C. and Chinn, M.D.,Phys. Rev. Letts., 41, 955, (1978).

18. Chinn, M.D. and Fain, Jr. S.C., Phys. Rev. Letts. 39,146, (1977).

19. Calisti,S., Suzanne, J.and Venables, J.A. Surface Sci., 116, 135, (1982).

20. Diehl,R.D., Toney, M.F. and Fain, Jr.S.C., Phys.Rev. Letts., 48, 177, (1982).

21. Duval, X. and Thomy, A., Carbon, 13, 242, (1975).

22. Wu, N.J. and Ignatiev, A., Phys Rev. B., 25, 2983 (1982).

23. Bouldin, C. and Stern, E.A., Phys. Rev.B. 25, 3462 (1982).

24. Takaishi, T. Prog. Surf.Sci. 6, 43 (1975).

25. Avgul, N.N. and Kiselev, A.V., "Physical adsorption of gases and vapours on graphitized carbon blacks" in Chemistry and Physics of Carbon,6, Walker Jr., P.L. (Ed), Marcel Dekker, 1970.

26. Steele, W.A. J. Phys., Paris, 38, C4-61 (1977).

27. Hansen, F.Y. and Taub, H. Phys. Rev.B.19, 6542 (1979).

28. Battezzati, C., Pisani, C. and Ricca, F., J.C.S. Faraday Trans II, 71, 1629, (1975).

29. Steele, W.A., Surface Sci., 36, 317 (1973).

30. Steele, W.A., J. Phys. Chem. 82, 817, (1978).

31. Claverie, P. "Elaboration of Approximate Formulas for the interactions between large molecules" in Intermolecular Interactions: From Diatomics to Biopolymers, Pullman, B. (ed.), Wiley, 1978.

32. Koide, A., Meath, W.J. and Allnatt, A.R.,Mol. Phys., 39, 895 (1980).

33. Gerratt, J. and Papadopoulos, M., Mol. Phys., 41, 1071 (1980).

34. Spurling, T.H. and Lane, J.E., Aust. J. Chem., 33, 1967, (1980).

35. Nicholson, D. and Steele, W.A. to be published.

36. Lippincott, E.R. and Stutman, J.M., J.Phys.Chem., 68, 2926, (1964).

37. Carlos, W.E. and Cole, M.W., Phys.Rev.Letts., 43, 697 (1979).

38. Carlos, W.E. and Cole, M.W., Surf.Sci., 91, 339, (1980).

39. Amos, A.T. and Crispin, R.J. "Calculations of Intermolecular Interaction energies" in Theoretical Chemistry, Advances and Perspectives, Vol. 2, Eyring, H., and Henderson, D. (eds), Academic Press (1976).

40. Buckingham, A.D., Disc. Faraday Soc., 40, 232, (1965).

41. Coulson, C.A., Disc. Faraday Soc. 40, 278, (1965).

42. Severin, E.S. and Tildesley, D.J., Mol. Phys., 41, 4101, (1980).

43. Sinanoğlu, O. and Pitzer, K.S., J.Chem.Phys., 32, 1279 (1960).

44. Schmit, J.W., Surface Sci., 55, 589, (1976).

45. O'Shea, S.F. and Meath, W.J., Mol. Phys. 31, 515 (1976).

46. Meyer, E.F., J.Phys. Chem., 71, 4416, (1967).

47. Rowley, L.A., Nicholson, D. and Parsonage, N.G., Mol. Phys., 31, 365, (1976).

48. Lane, J.E. and Spurling, T.H., Aust. J. Chem., 29, 2103 (1976).

49. Nicholson, D. and Parsonage, N.G. "Computer Simulation and the Statistical Mechanics of Adsorption", Academic Press, 1982.

50. Spurling, T.H. and Lane, J.E., Aust. J. Chem. 31, 465, (1978).

51. Whitehouse, J.S., Nicholson, D. and Parsonage, N.G., Mol. Phys. 49, 829, (1983).

52. Nicholson,D.,Rowley,L.A. and Parsonage, N.G.,J.Phys., Paris, 38, C4-69,(1977).

53. Bonino, G., Pisani, C., Ricca, F, and Roetti, C.,Surf. Sci., 50, 379, (1975).

54. Gooding, R.J., Joos, B and Bergersen, B, Phys.Rev.B., 27, 7669 (1983).

55. Pisani, C., Ricca, F. and Roetti, C., J.Phys.Chem., 77, 657 (1973)

DIRECT DIFFERENTIAL REACTOR STUDIES ON ADSORPTION FROM CONCENTRATED LIQUID AND GASEOUS SOLUTIONS

by

K.E. Noll, Ph.D., C.N. Haas, Ph.D., J-P Menez,
A.A. Aguwa, M. Satoh Dr. of Eng., A. Belalia and P.S. Bartholomew

Pritzker Department of Environmental Engineering
Illinois Institute of Technology
Chicago, IL 60616
U.S.A.

ABSTRACT

Equilibria and dynamics of various solute/sorbent combinations have been conducted in the gaseous and liquid systems. In the gaseous system (toluene on activated carbon), the sorption capacity was determined by correlation of the extension of a quartz spring hung from the basket containing the sorbent. In the liquid systems, bottle technique was used for the equilibrium studies and differential reactor columns were used for the dynamics (p-chlorophenol on resins). The most important results obtained from the dynamic studies was that there was a strong dependency of the intraparticle diffusivity on the sorbate concentration. The intraparticle diffusivities (liquid system) are successfully correlated to the equilibrium isotherm.

INTRODUCTION

Whether used in gaseous or liquid systems, adsorption processes are all similar in that separation of components is achieved by the migration of one component from fluid phase to solid phase (the adsorbent surface). Much of the prior work on adsorption systems has been performed using outlet breakthrough curves and mass transfer zone concepts with the application of the unsteady-state multidimensional continuity equations for data analysis. The approach adopted in this study has been to directly measure the saturation history of the adsorbent particles within a differential-reactor and analyzing the data with the resultant, simplified forms of the continuity equation. This project also emphasizes the use of resins instead of activated carbon as adsorbents. Pollutants are more readily desorbed, and thus recovered, from the former, which may be of greater practical application. Furthermore, the geometry of resins is simpler, and this facilitates adsorption investigations.

THEORETICAL CONSIDERATIONS

Adsorption is a surface phenomenon but it also involves transfer of a molecule from one phase to another. To derive the fundamental equations of the process, it is assumed that the system is isothermal, that no resistance is due to adsorption/desorption at the active sites, that surface diffusion is the main internal transfer mechanism and that the velocity is sufficiently high so as to render external mass transfer resistance negligible. The differential reactor assumption means that the external liquid concentration stays constant througout the bed. Assuming monodisperse spherical adsorbent particles (which is closely approximated by most resins) and that equilibrium can be represented by a Freundlich isotherm, from mass balances, the following system of equations are applicable:

- $$\frac{\partial q}{\partial r} = \frac{1}{r^2} \frac{\partial}{\partial r} \left(D_s\, r^2 \frac{\partial q}{\partial r}\right) \qquad (1)$$

- $$q = K\, C_o^{1/n} \qquad @\ r = r_p \qquad (2)$$

- $$\frac{\partial q}{\partial r} = 0 \qquad @\ r = 0 \qquad (3)$$

- $$q = 0 \qquad @\ t = 0 \qquad (4)$$

where

D_s = Effective surface diffusion coefficient
r = Radial coordinate
C_o = External liquid concentration
t = Time
q = Solid-phase concentration
K = Freundlich isotherm parameter
n = Freundlich isotherm parameter

If D_s is constant, then the system of Equation (1), (2), (3) and (4) can be integrated to give (Crank, 1956):

- $$\frac{q_t}{q_e} = 1 - \frac{6}{\pi^2} \sum_{m=1}^{\infty} \frac{1}{m^2} \exp\left(-\pi^2\, D_s\, m^2 t / r_p^2\right) \qquad (5)$$

EXPERIMENTAL APPARATUS AND PROCEDURES

Kinetic Studies. Four different laboratory experiments have been developed to provide information on the kinetics of sorption. Two differential reactor apparatus, one for gases and one for liquids (Figure 1), have been fabricated. These allow solid phase sorption to be measured as a function of time. Studies conducted at various concentrations and flow rates provide information on the effective diffusivity for use in the development of sorption models.

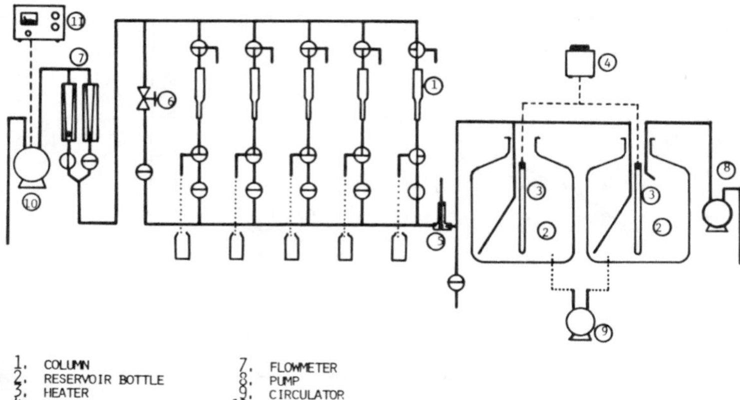

Figure 1. Schematic Diagram of Experimental Apparatus for Aqueous Solutions

The differential reactor concept eliminates attrition of the adsorbent and fluid dynamics are very close to those of the long-column. The differential column is 1 cm high and 1.25 cm in diameter. Rate studies have been conducted at various superficial velocities to determine the effect of fluid - film resistance on adsorption kinetics. Figure 2 shows that the velocity, if high enough, has an almost negligible effect on the adsorption kinetics. At the chosen interstitial velocity (V_z = 13.0 cm s^{-1}), it was shown by sensitivity analysis that there was no significant difference with the results obtained assuming a negligible mass transfer resistance. For an example, typical rate data is shown in Figure 3 where the solid line represents the model prediction. Figure 4 shows sensitivity analysis done on the predicted profile (Figure 3). The predicted profile (Figure 3) was unaffected by variation in K_F hence, it is very reasonable to assume that the main resistance is within the particle. Notice that Equation 5 assumes that the external mass transfer coefficient is negligible hence Equations 6.41 and 6.43 (Crank, 1956) which incorporate the two resistances were used.

In the air experiment, Barnebey-Cheney Co. activated carbon (8 x 14 mesh coconut shell base) was used as adsorbent and carbon dioxide (CO_2) as adsorbate. A gravimetric method was used to determine the adsorption capacity. This method was feasible due to the high affinity of CO_2 for activated carbon. Nitrogen was used as the carrier.

In the water experiments, run at a temperature of 25 $\pm$ 1°C and a pH of 6.7, data have been collected using XAD-2 and XAD-4 as the sorbents and p-chlorophenol (PCP) as the adsorbate. The amount adsorbed was determined by chemical extraction of PCP by sodium hydroxide (0.1N). The efficiencies have been shown to be consistently very high (around 98%) (Noll, 1983). Analysis of PCP was done by ultraviolet (UV) spectrophotometry at wavelength of 300 nm.

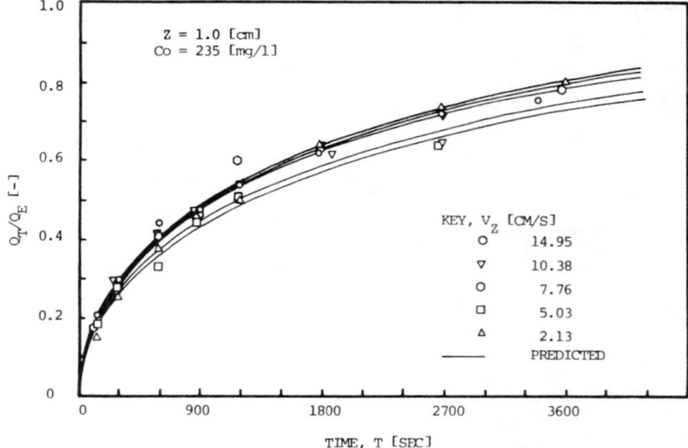

Figure 2. Changes in Adsorption Kinetics for Varying Linear Velocities

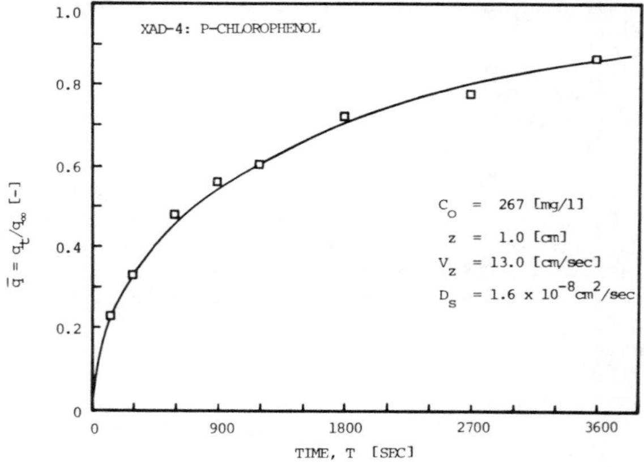

Figure 3. Rate Data: Solute Uptake as a Function of Time (Equation 7).

Equilibrium Studies. A gas phase sorption apparatus has been designed to produce equilibrium isotherms of organic air pollutants (Figure 5). In the experiment, sorption capacity is being correlated to the extension of a quartz spring. A mesh basket, containing the sorbent, is hung from the quartz spring. The spring expansion caused by the weight increase of the sorbent due to adsorption of the pollutant is detected with a cathetometer.

The spring and adsorbent are housed within two pyrex tubes of 14 and 25 mm diameter. In this manner, the outer tube provides a layer which serves as a water jacket monitored by a thermistor. A second hot water bath is used to generate organic vapors.

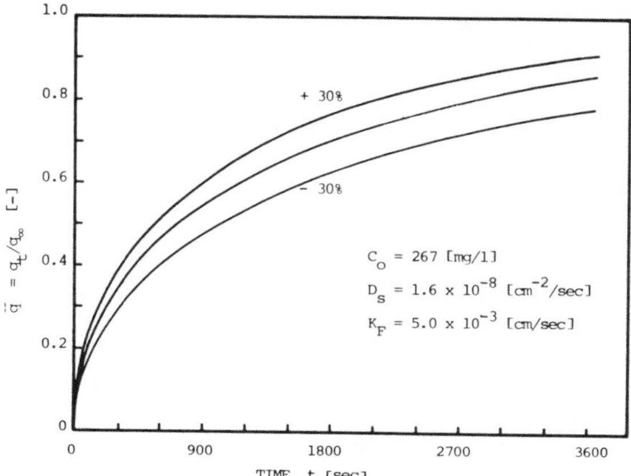

Figure 4. Sensitivity of Predicted Profile to Variation in the Intra-particle Mass Transfer Coefficient, D_s

This method has been verified as reliable by adsorbing toluene on Kureha G-BAC, which is a oil pitch base activated carbon. The re-results obtained (Figure 6) match those of Urano et al. (1981). Also, results of the adsorption of toluene on the Davison Chemical Type 13X molecular sieve are presented on the same figure.

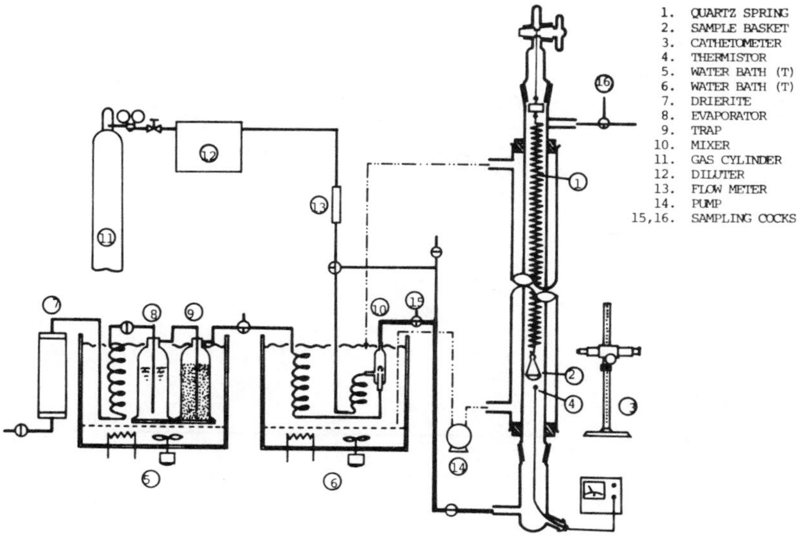

Figure 5. Experimental Apparatus for Organic Air Pollutant Isotherms

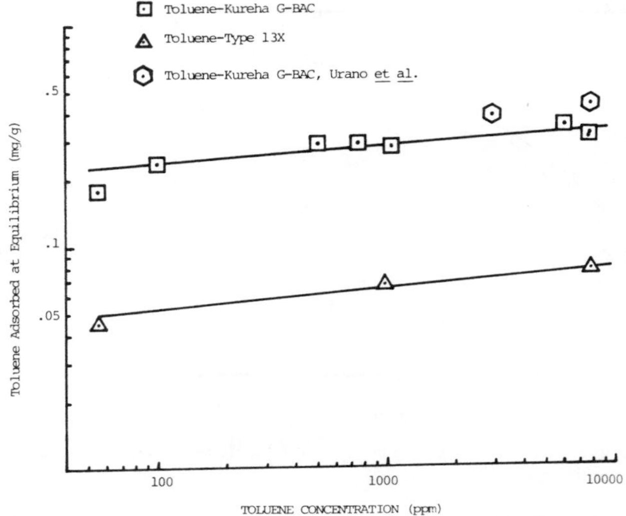

Figure 6. Isotherms for Toluene-Kuruha G-BAC and Toluene-Davison Chemical Type 13X Systems at 25°C.

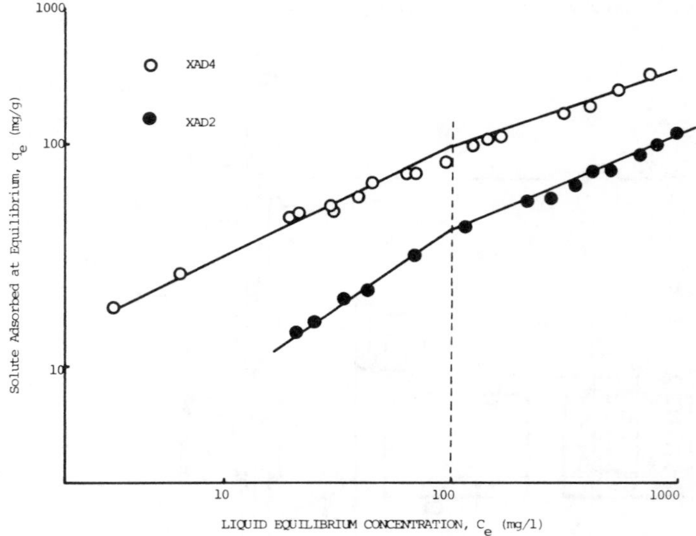

Figure 7. Freundlich Isotherms for PCP - XAD4 and PCP - XAD2 Systems at 25°C and pH = 6.7.

Liquid-phase adsorption equilibrium studies with PCP were conducted using the bottle technique at 25 ± 0.5°C. The equilibrium data (Figure 7) correlated well with the Freundlich isotherm but a better fit was obtained if the liquid concentration range was divided into two parts. The values of K and n obtained are presented in Table 1.

Table 1. Constants for the Freundlich Isotherm for PCP.

Resin	C_e < 100 mg/l K	n	C_e > 100 mg/l K	n
XAD2	1.98	1.54	4.53	2.18
XAD4	10.5	2.04	15.6	2.51

DATA EVALUATION

As stated previously, one of the objectives was the study of the effect of concentration on the effective diffusivity. Figure 8 (for water) and Figure 9 (for air), show a strong dependency is apparent.

A semi-empirical relationship was developed to try to correlate kinetic parameters with equilibrium parameters since concentration diffusivity dependency has been attributed to the nature of the isotherm (Ruthven and Loughlin, 1971, Gard and Ruthven, 1972). The plot of log D_s vs. q_e/Q_o (where Q_o is the maximum capacity of the resin obtained by fitting a Langmuir isotherm to the equilibrium data points)

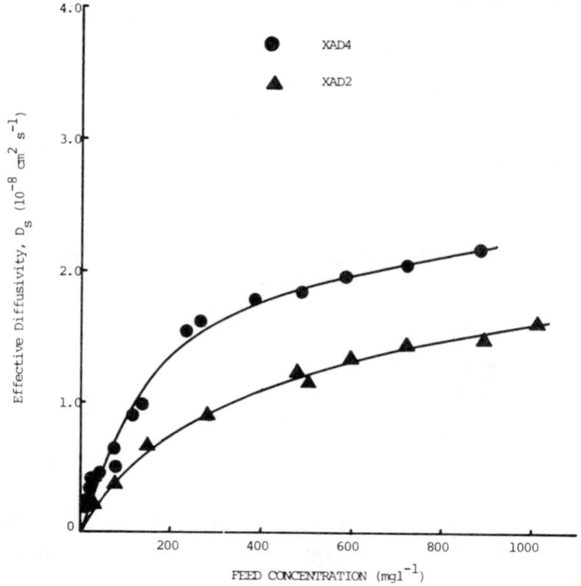

Figure 8. Effective Diffusivity of PCP in XAD Resins as a Function of Feed Concentrations.

gave, as shown in Figure 10, two straight lines with very high correlation coefficients (more than 0.99). The slopes of these two lines are not statistically different but the intercepts (defined as D°_s) are quite different. In the same fashion data reported by Fritz et al. (1981) were analysed and identical relationship was found for PCP and the activated carbon B 10 I, and the same slope was found as with PCP and the XAD resins in our study. A preliminary working hypothesis would be that the magnitude of the slope of this line may be due to the extent of interactions between solute molecules.

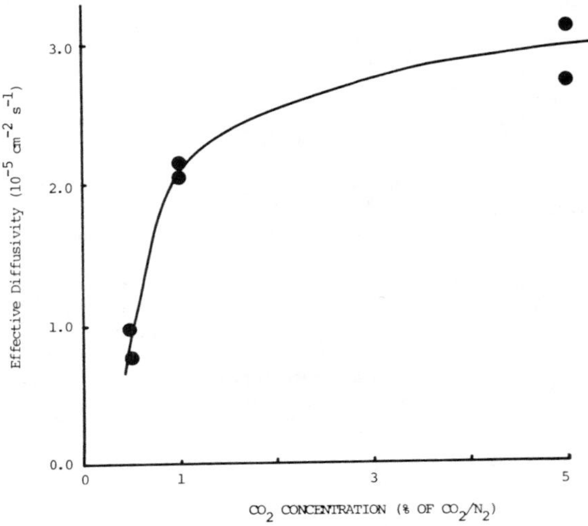

Figure 9. Plot, For a Linear Velocity of 15 cm/s, of the Effective Diffusivity vs CO_2 Concentration.

CONCLUSIONS

This study has demonstrated that the differential reactor operated on the solid-side is a simple and useful technique to obtain reliable values of the effective intra-particle diffusivity. Solid-side studies are facilitated by resins from where organics are readily desorbed.

The initial results, obtained over a broad range of liquid concentrations (10 mg/l - 1000 mg/l), show that the effective diffusivity cannot be considered concentration independent. A similar conclusion can be drawn from the air data collected. Some uncertainty remains as to the exact reasons for this phenomenon.

This concentration dependency of the effective diffusivity has already been noted by several investigators who studied broad concentration ranges. The effective diffusivity can be satisfactorily correlated with the solid-phase equilibrium concentration, in agreement with work of other investigators.

ACKNOWLEDGEMENTS

This research has been supported by the U.S. Environmental Protection Agency through the Industrial Waste Elimination Research Center (Chicago, Illinois, U.S.A.).

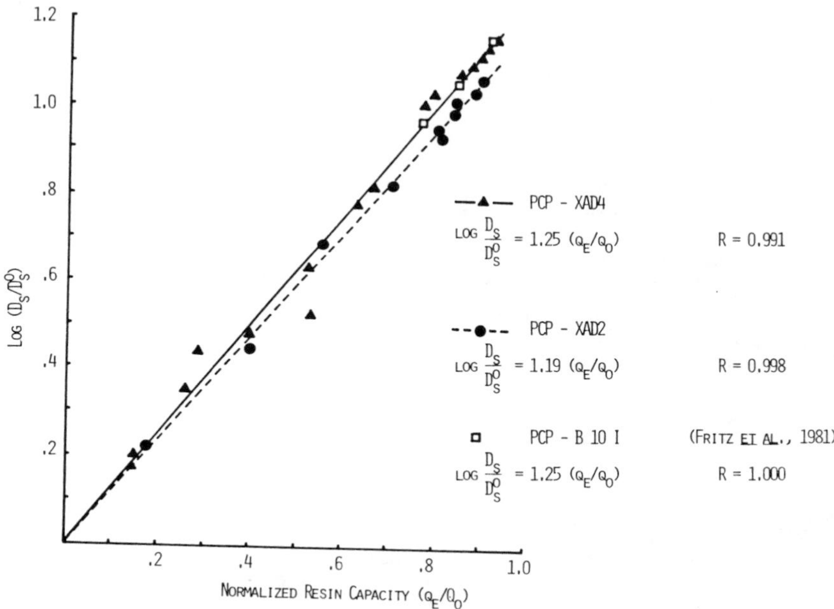

Figure 10. Surface Diffusion vs. Normalized Resin Capacity.

REFERENCES

Crank, J. 1956. The Mathematics of Diffusion. Oxford University Press. London.

Fritz, W., Merk W. and Schlunder E.U. 1981. Competitative Adsorption of Two Dissolved Organics onto Activated Carbon II. Chem Eng Sci V36:731.

Noll, K.E. 1983. Evaluation of the Dynamics of Multicomponents Sorption/Desorption Processes with Differential Reactor Columns Progress Report to the Industrial Waste Elimination Research Center.

Urano, K., Omori, S., and Yamamoto, E. 1982. Prediction Methods for Adsorption Capacities of Commerical Activated Carbons in Removal of Organic Vapors Env. Sci. Tech. V16:10.

Ruthven, D.M., and Loughlin, D.F., 1971. Correlation and Interpretation of Zeolite Diffusion Coefficients. Tran. Fara. Soc. V67.

Gard, D.R. and D.M. Ruthven, 1972. The Effect of the Concentration Dependence of Diffusivity on Zeolite Sorption Curves, Chem. Eng. Sci. V27.

NOTATION

a = constant
C_o = external liquid concentration
C_e = liquid equilibrium concentration
D_s = effective diffusion coefficient
D_s° = constant surface diffusion coefficient
K = Freundlich isotherm coefficient

n = Freundlich isotherm coefficient
Q_o = Saturation sorbate concentration in Langmuir isotherm
q = Solid-phase concentration
q_e = solid equilibrium concentration
q_t = average solid-phase concentration
r = radial coordinate
t = time
V_z = interstitial velocity

SURFACTANT RETENTION IN HETEROGENEOUS CORES

by J. Novosad
Petroleum Recovery Institute
3512 - 33rd Street N.W.
Calgary, Alberta, Canada

ABSTRACT

Experimental procedures allowing differentiation between surfactant losses resulting from trapping of surfactants in the immobile oil phase and surfactant losses caused by adsorption at the solid-liquid interface are applied to heterogeneous cores. The experimental results indicate that the mechanism responsible for surfactant loss within heterogeneous cores may be different in each core material.

INTRODUCTION

One of the most difficult problems facing enhanced oil recovery applications of surfactant flooding is the large amount of surfactants retained in the reservoir during a flood. Several physico-chemical mechanisms may cause surfactant retention, the most important being adsorption at the solid-liquid interface, trapping in the immobile oil phase, and precipitation.

Surfactant losses resulting from trapping and precipitation can be drastically reduced by the proper design of the surfactant solution for a specific rock-fluid system (1). However, there is no effective way to eliminate losses resulting from surfactant adsorption at the solid-liquid interface. It is therefore desirable to differentiate the surfactant loss caused by adsorption from other types of surfactant retention. The experimental procedure that allows determination of surfactant loss resulting from a single mechanism (trapping) was developed previously, and its use was demonstrated in studies of the temperature dependence of surfactant retention (2).

The trapping of surfactant in the immobile oil phase can have a severe effect on the performance of a surfactant flood. It has been shown previously that the best oil recoveries are achieved if the surfactant systems form middle phase microemulsions which usually produce the lowest interfacial tensions between the microemulsion phase and oil and also between the microemulsion phase and brine (3). However, the range of conditions at which such a phase arrangement takes place is a very narrow one. If conditions change, either a lower or upper phase microemulsion is formed. In the former case, surfactant dissolves in the aqueous phase together with some solubilized oil; and in the latter case, surfactant dissolves in the hydrocarbon phase together with some solubilized water (Figure 1). These changes are usually accompanied by a rapid increase in interfacial tension between the excess phase and the microemulsion phase. In some situations, the formation of an upper phase microemulsion can lead to the complete loss of surfactant as a result of its being trapped in the immobile oil phase.

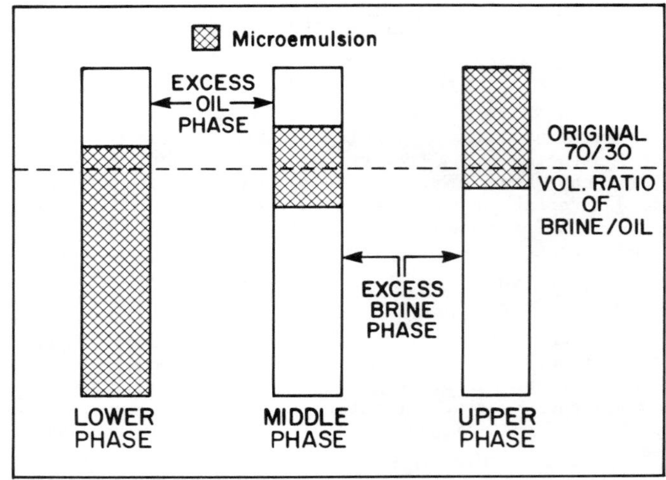

Figure 1. Volumetric Phase Behavior of Injected Surfactant Solution

Middle Phase Microemulsion: (3% 1/1 Petrostep 420/465 + 0.75% SBA) in 70/30 3% NaCl/octane
Lower Phase Microemulsion: Same system at changed conditions (for example, higher temperature)
Upper Phase Microemulsion: Same system at changed conditions (for example, in brine containing divalent ions)

In this paper, application of the experimental method for differentiating surfactant trapping from other types of surfactant retention is extended to studies of surfactant flooding in composite cores. The incentive for this work lies in the many observations that laboratory studies of surfactant retention in reservoir cores seem to underestimate the losses experienced in field projects. One possible explanation for such discrepancies could be that reservoir heterogeneities are very difficult, if not impossible, to model in laboratory-scale experiments, and therefore the proportions of the surfactant flow through different reservoir layers have to be estimated. (In this paper, the term heterogeneity is used in the petroleum engineering context, i.e. heterogeneity is a measure of differences in permeabilities of each reservoir layer and the permeabilities are defined by Darcy's Law.) A different rate of surfactant solution propagation in each reservoir layer may result in different surfactant compositions flowing through each layer even in the presence of cross-flow between the layers. Since the formation of the upper phase microemulsion is affected by surfactant concentration, by the cosurfactant/surfactant ratio, and by the oil/water ratio, surfactant trapping could be occurring in some layers of heterogeneous reservoirs and not in others. Therefore, it may not be observed when the retention experiments are performed with small-sized homogeneous reservoir cores in the laboratory.

Composite cores are used as one possible means of modelling heterogeneous systems. Such cores are prepared by assembling two different types of core materials in parallel while maintaining capillary contact between the fluids in the core components. The materials used in this study were Berea sandstone, having an absolute brine permeability range of 0.25 to 0.5 μm^2, and silicon carbide, ranging in permeability from 3 to 5 μm^2.

The main objective of this paper was to determine if it is possible to observe simultaneously the different mechanisms responsible for surfactant losses in different parts of the composite core.

EXPERIMENTAL

Chemicals

Petrostep 420 and 465	- Petroleum sulfonate manufactured by Stepan Chemical Co. According to the supplier the surfactants are 50% active with equivalent molecular weight of 420 and 465, respectively. Both surfactants, used as received, were combined in a 50/50 weight mixture in all experiments. A 3% (active) solution produced the middle-phase type of microemulsion when mixed 70/30 (volume) with 3% NaCl brine and octane.
Octane	- Technical grade, supplied by Phillips Petroleum Co., Bartlesville, Oklahoma.
Secondary Butyl Alcohol (SBA)	- Boiling point range 98 to 100°C, supplied by Eastman Kodak Co., Rochester, N.Y.
Brine	- NaCl, analytical reagent grade, supplied by BDH Chemicals Ltd., and dissolved in deionized water.
Solid	- Berea sandstone cores with air permeability of approximately 0.5 μm^2, supplied by Cleveland Quarries, Inc.
Silicon Carbide	- Supplied by Kennametal Ltd. under the trade name "Norton Dressing Slicks", 100 grit in 2.5 x 2.5 x 15 cm dimensions.
Flocon Bio-polymer 4800	- Xanthan broth produced by Pfizer Inc. for enhanced oil recovery applications. A 1000 ppm solution in

3% NaCl brine was used (viscosity, μ = 20.5 cp at 5 sec^{-1}; screen factor = 2.1).

Procedures

Composite cores were prepared by assembling 2.5 x 2.5 x 15 cm rectangular component pieces side by side with two layers of tissue paper between them to assure capillary contact between the fluids. While the pieces were held together by three "C" clamps, the entire assembly was wrapped in Teflon tape and cast in epoxy resin. Figure 2 shows the assembled core with eight Swagelok fittings attached. Injection ports A and B were used for water and surfactant injection (C and D ports plugged) so that the two materials were in parallel. The C and D injection ports were used for oil saturation (ports A and B plugged) to assure sufficient flow of oil through both core materials in order to achieve irreducible water saturation in both parts of the core. The following flooding sequence was used in all experiments with the composite cores:

1. Brine injection into the evacuated core until constant core weight was achieved.

2. Oil injection into port C and collection from port D until no further brine was produced.

3. Waterflooding at the rate of 2 ml/hour through port A and effluent collection from port B until no more oil was produced. This flow rate corresponds to an average frontal velocity of less than 1 cm/hour which is the range of a typical oil field operation.

4. A 50% pore volume (PV) injection of the surfactant solution at a 2 ml/hour flow rate, followed by the brine or the Flocon Biopolymer 4800 solution until complete cessation of oil, surfactant, or cosurfactant production.

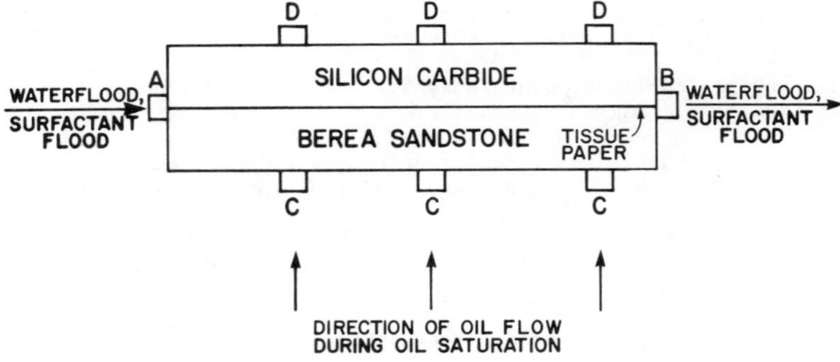

Figure 2. Assembled Composite Core

At the end of the flooding sequence, the two materials comprising the composite core were separated and coated separately with epoxy resin. The previously developed procedure for differentiating surfactant losses resulting from adsorption and losses resulting from trapping in the immobile hydrocarbon phase (2) was then applied to each core material separately. First, the bulk hydrocarbon phase was miscibly displaced by octane or nonane and analyzed for surfactant content. This miscible flood recovered only that surfactant trapped in the remaining immobile oil phase. A strong surfactant solvent (50/50 ethanol/brine mixture) was then injected. The amount of surfactant recovered in this flood is equivalent to the sum of the precipitated and adsorbed surfactant. This flooding sequence allows completion of the surfactant material balance and determination of residual oil, retained surfactant, and the type of retention in each material separately. Procedures used in flooding experiments with Berea cores are described elsewhere (2).

RESULTS AND DISCUSSION

Surfactant Flooding in Composite Cores

A series of four core floods was performed using the same surfactant system, a 3% 50/50 (weight ratio) Petrostep 420/465 and 0.75% SBA in 3% NaCl brine. The first two floods were run in a single silicon carbide core and in a single Berea sandstone core. The next two floods were performed in composite cores comprised of both materials. In one of the latter two floods, the surfactant slug was followed by a brine, and in the other by a 1000 ppm solution of Flocon 4800 biopolymer. Floods 1, 2, and 3 were run at flow rates resulting from the same pressure drop across the core. Flood 4 was run at the same flow rate, 4 ml/hour, as in Core 3. Since core permeabilities varied by a factor of as much as ten, the flow rates varied considerably, and the oil recovery results cannot be compared since they are influenced by the flow rate. Core characteristics and flooding results are shown in Table 1.

Examination of the core characteristics listed in Table 1 shows that the permeabilities and fluid saturations of the composite cores are consistent with single core characteristics. Incremental oil recoveries listed in the last column of Table 1 indicate the benefits of surfactant flooding. The importance of mobility control in floods in heterogeneous cores is demonstrated by a 50% improvement in the oil recovery (observed in Core 4 vs. Core 3) when the polymer solution was driving the surfactant slug.

Examples of effluent analysis for floods in Core 3 are shown in Figures 3 and 4, with Figure 3 depicting oil recovery and Figure 4 showing surfactant and cosurfactant breakthrough curves. Table 2 summarizes the surfactant retention data.

Surfactant retention data listed in Table 2 provide a good example of the useful information obtained from determining not only the amount of retention but also its type in a specific material in the composite core. Retention in Core 3 is predominantly in the silicon carbide, and most of it is caused either by surfactant adsorption or by

Table 1. Core Characteristics and Flooding Results

CORE No.	MATERIAL DIMENSIONS	$(k_a)_{brine}$	$(k_o)_{S_{wi}}$	$(k_w)_{S_{OR}}$	S_{wi}	S_{OR} AFTER WATER FLOOD	S_{OR} AFTER SURFAC. FLOOD	OIL RECOVERY
	cm	md	md	md	% PV	% PV	% PV	ROIP
1	BEREA 2.5X2.5X15.2	270	260	21	36.6	36.8	30.0	14
2	SILICON CARBIDE 2.5X2.5X15.4	4200	1260		25.3	16.5	16.0	3
3	50/50 BEREA SILICON CARBIDE 2.5X5.0X15.3	5200	1700	200	30.5	26.8	15.6	41
4	50/50 BEREA SILICON CARBIDE 2.5X5.0X15.5	4800	1200	170	29.2	28.3	11.2	60

Table 2. Surfactant Retention

CORE NUMBER	TOTAL RETENTION	SILICON CARBIDE		BEREA SANDSTONE	
		TRAPPING	ADSORPTION AND OTHER	TRAPPING	ADSORPTION AND OTHER
	mg/g				
1	1.3	N/A	N/A	0	1.3
2	1.4	0.1	1.3	N/A	N/A
3	2.1	0.2	3.7	0	0.6
4	1.2	0.8	0.5	0.1	0.5

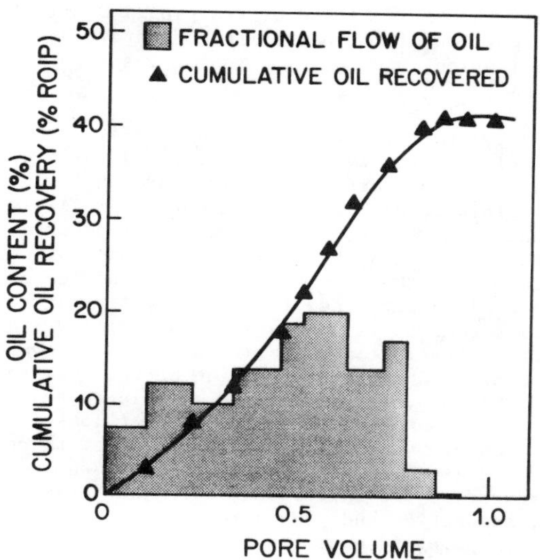

Figure 3. Oil Recovery in Composite Core #3 (0.5 PV surfactant solution followed by brine injection)

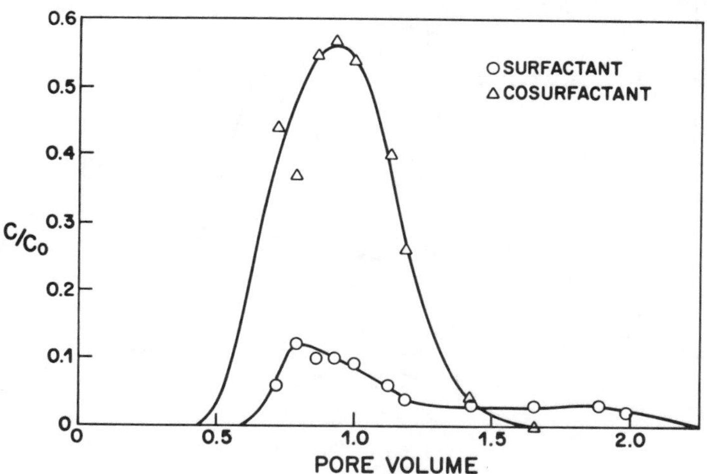

Figure 4. Surfactant and Cosurfactant Concentration in the Effluent - Core #3

precipitation. Less than 5% is caused by trapping. Similarly, the losses in the Berea portion of the composite core are caused by adsorption or precipitation, and no surfactant appears to be trapped in the immobile oil phase.

The retention data are quite different in Core 4. Here the polymer solution provides effective mobility control which results in a more uniform flow of the surfactant solution through both parts of the composite core. This results in a different "overall dispersion" of the surfactant and cosurfactant as indicated by the surfactant and cosurfactant breakthrough curves shown in Figures 4 and 5. The peak in surfactant concentration at the Core 4 outlet is significantly delayed. However, the most striking difference is in the surfactant retention levels. The overall retention in Core 4 is substantially reduced as the adsorption or precipitation in the silicon carbide drops by almost one order of magnitude. Since the distinction between adsorption and precipitation cannot be made, it can only be speculated that precipitation was the primary cause of the large surfactant loss in the silicon carbide portion of Core 3. However, surfactant trapping in the silicon carbide has increased, so the overall loss of surfactant in this material is reduced by two-thirds only. In the Berea portion of the composite core, adsorption and precipitation loss is effectively unchanged, but some trapping is observed in Core 4 while none was determined in Core 3.

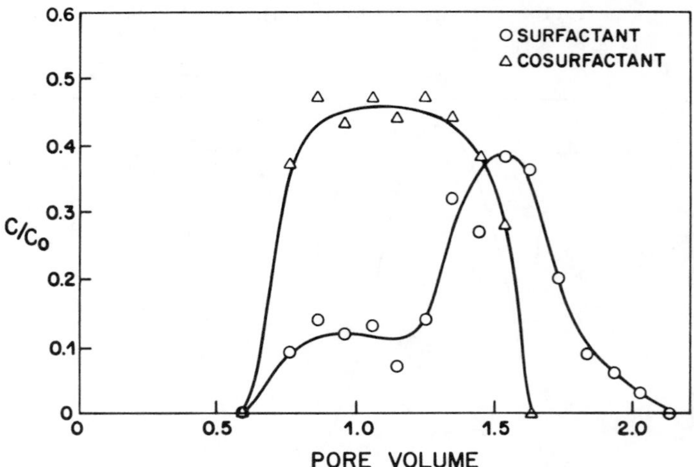

Figure 5. Surfactant and Cosurfactant Concentration in the Effluent - Core #4

CONCLUSIONS

This paper describes applications of previously developed experimental procedures for the assessment of the level and type of surfactant retention to flooding experiments in composite cores.

It is shown that these techniques are suitable for studies of surfactant retention in heterogeneous cores, and the experimental results indicate that different types of surfactant retention can coexist in the component parts of the core. This phenomenon could explain why laboratory evaluation of surfactant retention may underestimate the subsequent retention of surfactants in the field.

ACKNOWLEDGEMENTS

The author expresses his thanks to L. Baxter, D. Moon, and V. Masata who performed most of the experiments reported in this paper.

LITERATURE CITED

1. Nelson, R.C., "The Salinity Requirement Diagram - A Useful Tool in Chemical Flooding Research and Development," SPEJ 22, 259-270 (April 1982).

2. Novosad, J., "Surfactant Retention in Berea Sandstone - Effects of Phase Behavior and Temperature," SPEJ, 22, 962-970 (December 1982).

3. Healy, R.N., R.L. Reed, and D.G. Stenmark, "Multiphase Microemulsion Systems," SPEJ, 16, 147-160 (1976).

HEAT TRANSFER IN AN ADSORBENT FIXED-BED

by Patrick OZIL
Laboratoire d'Adsorption et Réaction de Gaz sur Solides
ENSEEG - BP 75 - Domaine Universitaire
38402 Saint Martin d'Hères - FRANCE.

This work is a contribution to a better comprehension of the thermal phenomena owing to adsorption and the parameters acting on them. A simplified theoretical study of the heat transfer during the dynamical adsorption of a single diluted solute in a fixed-bed allows us to predict the thermal wave as an analytical function of time and length. This analysis is limited to the case of an equilibrium corresponding to an irreversible isotherm and assume that the heat transfer does not affect the mass transfer.

These results are applied to the systems C_3H_6 or C_3H_8 - 13 X zeolite. For the first one, empirical kinetics of cooling allow to take into account the influence of the thermal phenomena on the breakthrough curves.

The prediction of the heat transfer is important for understanding the dynamical behaviour of an adsorbent fixed-bed. Only a few works concern the dynamical adsorption of a single solute in non-isothermal conditions without assuming equilibrium theory [1]-[7].

We propose here such a study for which analytical equations can be obtained for the thermal wave.

THEORETICAL STUDY.

Let us consider a binary mixture (linear velocity u - adsorbate concentration C_o - density ρ_f - heat capacity C_f) entering a fixed-bed (length H - radius R - void fraction ε - adsorptive capacity q_o^* - bulk density ρ_b - heat capacity C_s).

Mass balance.

We shall assume the following hypotheses : (a) the fluid is incompressible, (b) mass diffusion in axial and radial directions is negligible, (c) adsorbate concentrations and densities do not depend on temperature, (d) the sorption equilibrium is described by an irreversible isotherm.

Then the mass balance can be classically expressed as :

$$x = y \qquad (1)$$

after introducing the dimensionless variables $x = C/C_o$ and $y = q/q_o^*$ defined from the concentrations C, q in the fluid and solid phases.

Moreover the sorption rate equation is supposed to be written as :

$$\frac{\partial x}{\partial \tau} = \frac{\partial y}{\partial \tau} = g(\tau - \tau_o [h]) \quad (2)$$

function g being characteristic of the step(s) limiting the mass transfer and τ_o being the breakthrough time defined as :

$$\tau_o (h) = (1 + \Lambda/\varepsilon) h/u - b ; x[\tau_o, h] = 0 - h \quad (3)$$

Here the dimensionless parameter Λ is the one characterizing the mass distribution between the solid and fluid phase, as defined by Vermeulen [8] :

$$\Lambda = q_o^* \rho_{b/C_o} \quad (4)$$

Thermal balance.

Its modelling needs additionnal assumptions :
(e) axial and radial dispersions of heat are negligible,
(f) the thermal accumulation in the fluid phase is of no account
(g) heat capacities do not depend on temperature nor concentration
(h) the temperatures of the solid and fluid phases are equal [9].
Then the thermal balance [5] [7] is :

$$\varepsilon \rho_f C_f u \frac{\partial t}{\partial h} + \rho_b C_s \frac{\partial t}{\partial \tau} + \frac{2H_B}{R} t = \rho_b q_o^* \Delta H \frac{\partial y}{\partial \tau} \quad (5)$$

ΔH being the sorption heat and H_B the over-all heat transfer coefficient between the bed and the outside [10].

Resolution.

Let us define the following variables :

$$T = C_s t / (\Delta H q_o^*) ; \theta = \tau - \tau_o[h] ; L = h/u \quad (6)$$

and the parameters :

$$\alpha = \rho_b C_s / (\varepsilon \rho_f C_f) ; \beta = 2 H_B/(R \rho_b C_s)$$

$$A = \alpha - 1 - \Lambda \varepsilon^{-1} \quad (7)$$

It is interesting to note that the dimensionless parameter α is characteristic of the heat distribution between the solid and fluid phases while β is representative of the heat losses.

Hence the Equation (5) can be rewritten as :

$$A \frac{\partial T}{\partial \theta} + \frac{\partial T}{\partial L} + \alpha\beta T = \alpha g (\theta) \quad (8)$$

This latter equation may be solved by using the Laplace transformation and then the method of the integral factor when taking into account the initial conditions : T $(\theta, 0) = 0$ for any value of θ [7]. Calculations lead to the equation :

$$T(\theta,L) = z(\theta)\ \gamma(\theta) - z(\theta - AL)\ \exp(-\alpha\beta L)\ \gamma(\theta - AL) \quad (9)$$

in which $z(\theta)$ is the inverse Laplace transform of the function $\bar{z}$ (p) defined from the Laplace transform $\bar{g}$ (p) of function $g(\theta)$ as

$$\bar{z}(p) = \alpha\bar{g}(p) \,/\, [A_p + \alpha\beta] \quad (10)$$

$\gamma(X)$ being the step function ($\gamma(X) = 0 \ \forall X < 0$; $\gamma(X)=1 \ \forall X>0$)

The Equation (9) allows to predict the temperature profile as a function of time and length in adiabatic or isothermal cooling conditions whatever the steps controlling the mass transfer may be.

Before studying each of these limiting steps on which depends the knowledge of the function $g(\theta)$, we can present some general informations deduced from Eqn (9).

* First two cases are to be considered according to the only operative conditions :

· 1st case : $A > 0$ ($\alpha > 1 + \Lambda/\varepsilon$)

The temperature increases from the breakthrough time at any length [$T(0,L) = 0$; $\forall L$]

. 2nd case : $A < 0$ ($\alpha < 1 + \Lambda/\varepsilon$)

The temperature increases before the breakthrough time [$T(AL, L) = 0$; $\forall L$] and the maximum temperature rise is reached for $\theta = 0$. This important conclusion had been pointed out by Convers from a very simple model [11].

* Secondly, the thermal losses have such a large influence that the thermal wave tends to a quasi-steady state if they are great enough. However, the shape of temperature profiles remains characte-tic of the limiting steps as it can be shown latter for external and (or) intraparticular diffusion(s) control and for macroporous diffusion control (Figure 1).

Complete calculations leading to the final equations of the ther-mal waves for each mechanism would be too long [7]. So we shall treat here in details the one case of intraparticular diffusion control. Moreover, we shall give some information about the other cases.

Mass transfer under intraparticular diffusion control.

This case is the most frequent one for dynamical adsorption in zeolitic beds.The linear driving force approximation proposed by Glueckauf and Coates [12] for the rate equation is :

$$\frac{\partial y}{\partial \tau} = k\,(1 - y) = k \exp(-k\,[\tau - \tau_o(h)]) \quad (11)$$

with $k = 0.894\ D_p/(60\ d'^2_p)$ for an irreversible equilibrium.

From Eqns (10), (11), we deduce :

$$z(\theta) = \frac{\alpha\, k}{\alpha\beta - Ak}\left[\exp(-k\theta) - \exp\left(-\frac{\alpha\beta}{A}\theta\right)\right] \quad (12)$$

and then the Equation (9) leads to the expression of the temperature evolution.

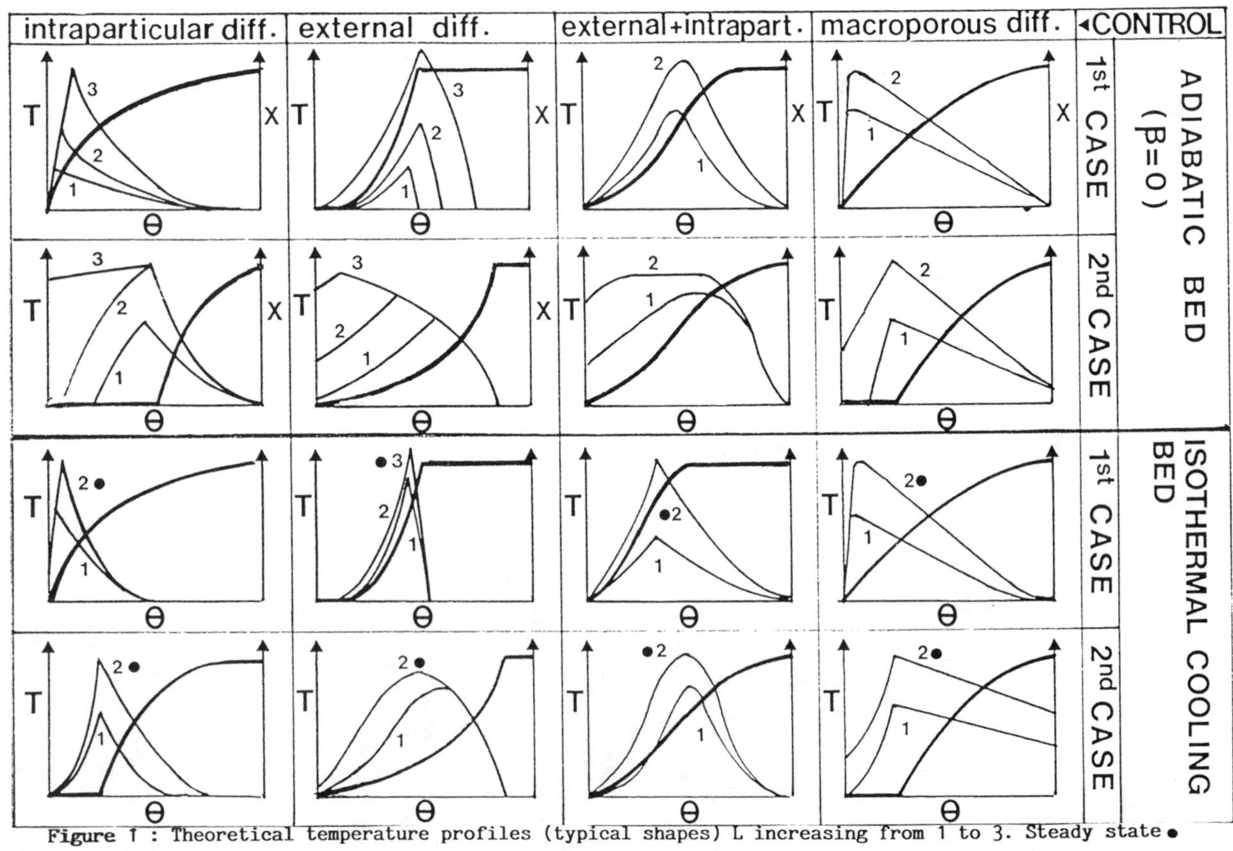

Figure 1 : Theoretical temperature profiles (typical shapes) L increasing from 1 to 3. Steady state ●

Mass transfer under external diffusion control or under external and intraparticular diffusions control.

Then the problem is more complex because of the initial conditions on the mass-transfer due to the external diffusion ($x \to 0$ for $\tau \to -\infty$ which is inconvenient with our assumption given by Eqn (3).

So we have to define a fictitious breakthrough time corresponding to a very small value of $x = \exp(-\Psi)$ in order to use our general mathematical method. After obtaining the $T(\theta,L)$ expression, we have only to put $\Psi \to \infty$ to eliminate this arbitrary value. Hence the final solution is given as $T(\tau,L)$.

Mass transfer under macroporous diffusion.

Solving this case does not present any difficulty. The rate equation proposed by Hall [13] leads to :

$$g(\theta) = 0.1544\ \omega^2 \left[\theta - \frac{3.604}{\omega}\right] \gamma(\theta) - \gamma\left(\theta - \frac{3.604}{\omega}\right) \qquad (13)$$

with : $$\omega = 32.88\ \frac{D_{pore}}{d_p^2}\ \frac{(1-\varepsilon)}{\Lambda} \qquad (14)$$

and our mathematical treatment can be easily applied.

EXPERIMENTAL STUDY.

Our experiments concern the dynamical adsorption of propylene or propane diluted in helium at 25°C under the atmospheric pressure on 1/16" pellets of 13 X zeolite.

Ten thermocouples allow to study the progression of the thermal wave all along the bed and the gas composition is analyzed near the inlet, the outlet and at the middle of the column with a katharometer

Mass transfer study.

For both the systems C_3H_6 - 13 X and C_3H_8 - 13 X, the main steps limiting the sorption are successively [7] [14] :

- intraparticular diffusion, characterized by a mean mass transfer coefficient in the particles K_1 (= 0.66 mn^{-1} for C_3H_6 - 13 X).
- empirical kinetics of cooling for the bed, characterized by a rate equation formally identical to the one for intraparticular diffusion control :

$$\frac{\partial x}{\partial \tau} = K_2\ (1 - x) \qquad (15)$$

this step corresponds to the increase of the adsorptive capacity of the saturated bed when temperature decreases (K_2 = 0.030 mn^{-1} for C_3H_6 - 13 X). So it allows to consider the influence of the thermal phenomena on the mass transfer.

Heat transfer study.

First these experiments prove that the two theoretical cases according to the A value exist truly (Figures 2 and 3).

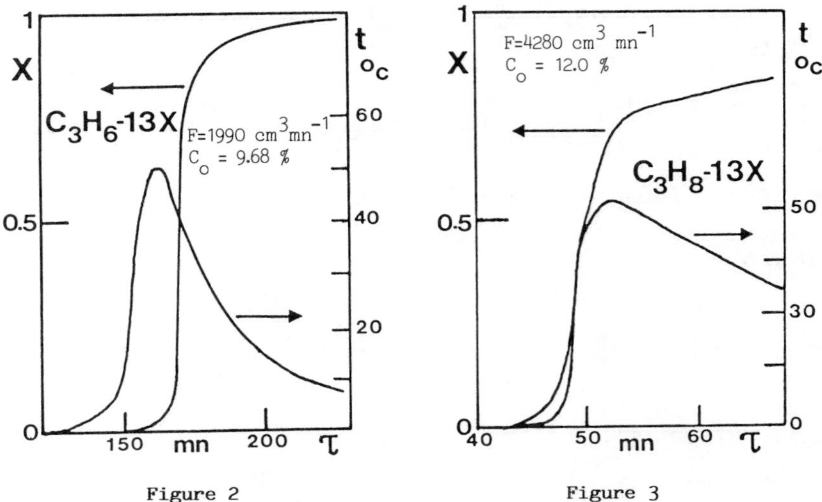

Figure 2

Figure 3

In fact, for experiments on C_3H_6 - 13 X corresponding to such operative conditions as required by the second case (A < 0), the thermal wave grows before the breakthrough time for which the maximum temperature is reached (Figure 4). On the other hand, the system C_3H_8 - 13 X is studied for the conditions of the first case (A> 0) and then the temperature profile and the concentration wave start at the same time (Figure 5).

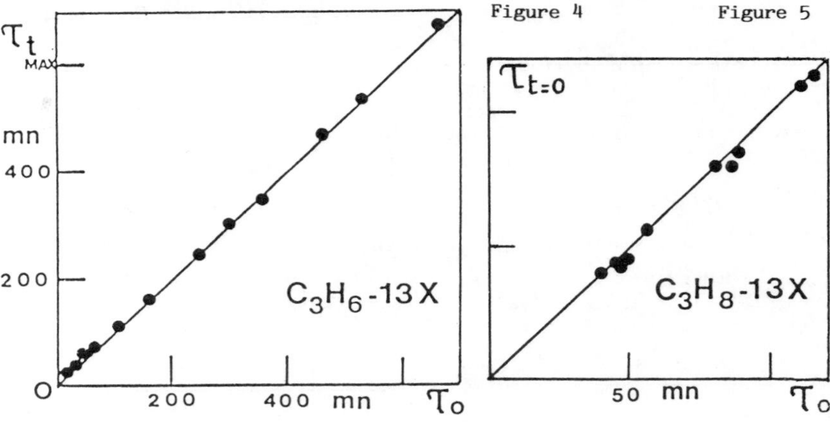

Figure 4

Figure 5

For both cases the thermal wave tends to a stationnary shape as shown for C_3H_6 - 13 X on Figure 6.

Now we shall examine in more details the results about C_3H_6. The inverse of the experimental steady state maximum of temperature rise follows a linear law with the mass distribution parameter Λ as predicted by theory from Eqns (9) (12) for $\theta = 0$: (Figure 7).

$$\frac{1}{t_{max}} = \frac{C_s}{\Delta Hq_o^*}[\beta'/K-1+1/\alpha + \frac{1}{\varepsilon\alpha}\Lambda] \quad (16)$$

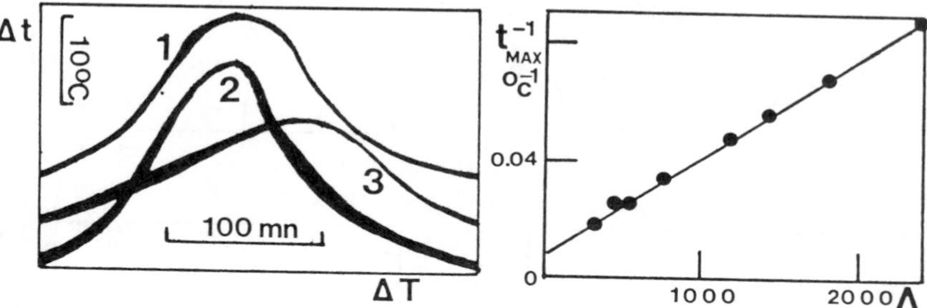

Figure 6 : Deformation of the thermal wave through the bed
1 C_o = 2.60 %. F = 2550 cm^3mn^{-1}
2 C_o = 4.17 %. F = 2590 cm^3mn^{-1}
3 C_o = 1.60 %. F = 2520 cm^3mn^{-1}

Figure 7
Plotting of t^{-1}_{max} vs. Λ

Moreover at the steady state the Equation (12) can be rewritten as :

$$Ln(t) = Ln\ (t_{max}) - \frac{\alpha\beta}{A}\theta + [\frac{\alpha\beta}{A}\theta - K_2\theta]\ \gamma(\theta) \quad (17)$$

and the fitting of our experimental thermal waves as $Ln(t)=f(\theta)$ gives two linear parts (Figure 8) ; the second parts are characterized by a constant slope for all experiments and the first ones have slopes depending only on Λ. This plotting allows us to deduce : α = 630 and β' = 0.2 mn^{-1}. These values are physically acceptable as it can be shown [6] [7].

Then the very simple Equation (17) supplies a good representation of all our temperature profiles (Figure 9). It is important to point out that this model needs the estimation of only three parameters α, β', K_2, the third one having been determinated from the study of the mass transfer.

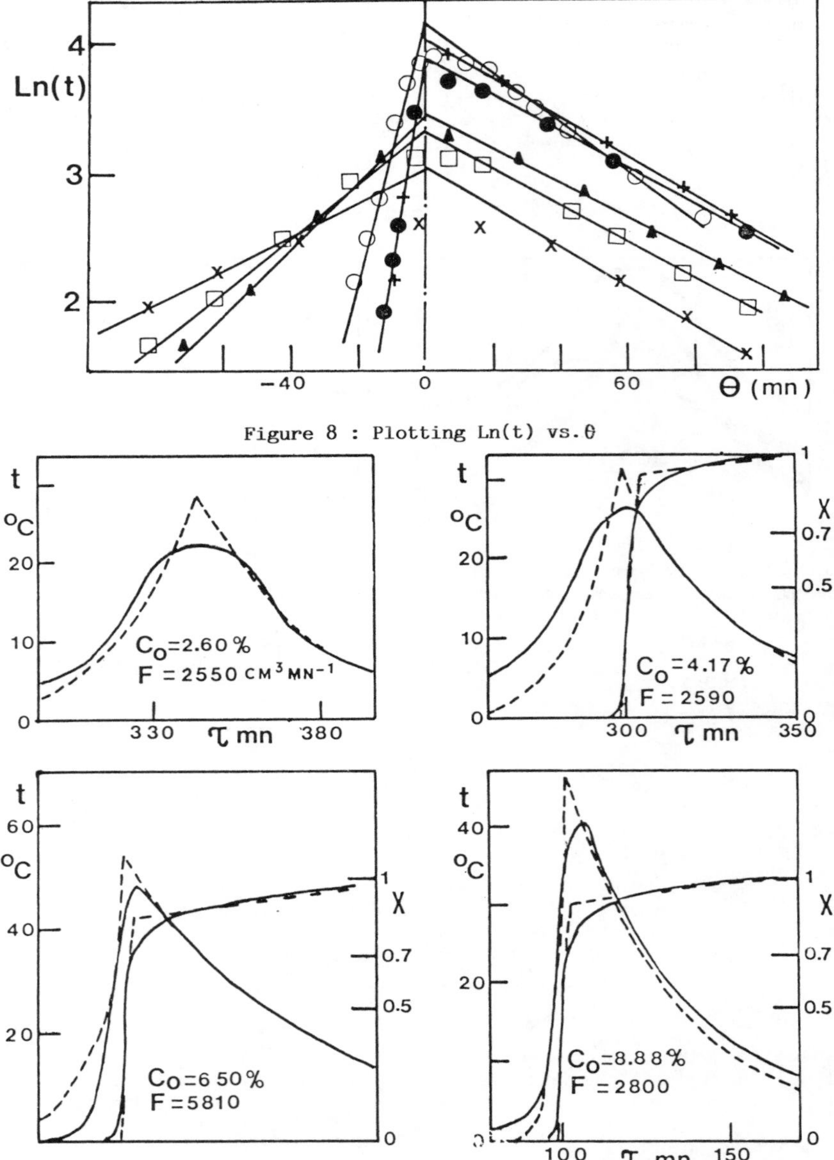

Figure 8 : Plotting Ln(t) vs. θ

Figure 9 : Comparison between experiments on C_3H_6-13X and theory

CONCLUSIONS.

This work shows that a model based on certain reasonable assumptions can predicts experimental thermal phenomena in a adsorbent fixed-bed.

The major interest of such an approach is an analytical treatment which permits a better comprehension of the influence of each parameter.

Now it will be necessary to test the validity of theory on adsorbate-adsorbent systems for which the equilibrium is governed by other isotherms than the rectangular one.

NOTATION.

A	Parameter defined by eqn (7)
C	Adsorbate concentration in the fluid phase
C_o	Initial adsorbate concentration in the fluid phase
C_f, C_s	Heat capacities of the fluid and solid phases
d'_p, d_p	Diameters of the particles and of the pellets.
D_p	Coefficient of intraparticular diffusion
D_{pore}	Coefficient of macroporous diffusion
h	Length in the bed.
H_B	Over-all heat-transfer coefficient between the bed and the outside.
k	Mass transfer coefficient in the particles
L	Reduced length (=h/u)
q	Adsorbate concentration in the solid phase.
q_o^*	q value in equilibrium with C_o
R	Bed radius
t	Temperature rise in the bed
T	Reduced temperature ($= C_s t/[\,Hq_o^*\,]$)
u	Linear velocity of the fluid
x	Dimensionless ratio C/C_o
y	Dimensionless ratio q/q_o^*

Greak symbols.

α	Dimensionless parameter of heat distribution ($=\rho_b C_s/[\varepsilon\ \rho_f C_f]$)
β	Parameter characteristic of heat losses ($=2H_B/[R\rho_b C_s]$)
ΔH	Sorption heat
ε	Void fraction in the bed
θ	Reduced time ($= \tau - \tau_o(h)$)
Λ	Dimensionless parameter of mass distribution ($= \rho_b\ q_o^*/C_o$)
ρ_f, ρ_b	Densities of the fluid and solid phases.
τ	Time
τ_o	Breakthrough time
ω	Parameter defined by eqn (14).

LITERATURE CITED.

1. Meyer O. and Weber T.W., AICHE J., 13, 457 (1967)

2. Lee R.G. and Weber T.W., Can. J. Chem. Engng. 47, 60 (1967)

3. Ruthven D.M., Garg D.R. and Crawford R.W., Chem. Engng. Sci. 30, 803 (1975)

4. Ikeda K., Fujino C. and Katoh K., "Heat and mass transfer in the non isothermal fixed bed adsorption column with non linear equilibrium", presented at the "contribution of computers to the development of Chemical Engineering". International Congress, Paris (1978).

5. Ozil P. and Bonnetain L., Chem. Engng. Sci. 33, 1233 (1978)

6. Ozil P. and Bonnetain L., Proc. of the 2nd World Congress of Chemical Engineering (Montréal) III, 241 (1981)

7. Ozil P., "Transferts de matière et de chaleur dans un lit fixe adsorbant", Thesis Grenoble (1981)

8. Perry J.H., and Chilton C., Chemical Engineers' Handbook, 5th Ed., p. 1617, Mc Graw-Hill, New-York (1973)

9. Denbigh k. Chemical Reactor Theory, p. 144, Cambridge University Press, Cambridge (1965)

10. Johnson B.M., Froment J.C., Watson C.C., Chem. Engng. Sci., 17, 835 (1962)

11. Convers A., "Adsorption des hydrocarbures en lit fixe sur tamis moléculaire", Thesis, Nancy (1964)

12. Glueckauf E. and Coates J.I., J. Chem. Soc., 1315 (1947)

13. Hall, Eagleton L., Acrivos A., Vermeulen T., Ind. Engng. Chem. Fundls, 5, 2, 212 (1966)

14. Caire J.P., Ozil P., Unpublished works.

PILOT-PLANT INVESTIGATIONS OF THE ADSORPTION OF 1,2-DICHLOROETHANE AND 1,2-DIBROMO-3-CHLOROPROPANE FROM WATER SUPPLIES

Massoud Pirbazari, Assistant Professor of Civil and Environmental Engineering, University of Southern California, Los Angeles, CA 90089.

Leown A. Moore, Chemist, Drinking Water Research Division, Municipal Environmental Research Laboratory, U.S. Environmental Protection Agency, Cincinnati, OH 45268.

Walter J. Weber, Jr., Professor of Environmental Engineering and Chairman, University Water Resources Program, University of Michigan, Ann Arbor, MI 48109.

Alan A. Stevens, Chief, Organics Control, Chemical Studies Section, Drinking Water Research Division, Municipal Environmental Research Laboratory, USEPA, Cincinnati, OH 45268.

ABSTRACT

This research considers the results of granular activated carbon (GAC) pilot-plant studies related to the adsorption of an industrial solvent and a pesticide, namely, 1-2-dichloreothane (DCE), and 1,2-dibromo-3-chloropropane (DBCP), respectively, from river water. A predictive mathematical model [Michigan Adsorption Design and Application Model-MADAM-film/particle diffusion version (1-4)] which has demonstrated potential for use in the design and operation of GAC for wastewater treatment, was used in simulation and prediction of the dynamic performance of fixed-bed pilot-plant adsorbers. Good agreement between the predicted breakthrough profiles and the pilot-plant verification data was observed.

EXPERIMENTAL APPROACH

Materials and Methods

Carbon Type. 12 x 40 US standard sieve size Filtrasorb® 400 (Calgon Corp., Pittsburgh, PA)*. Carbon particle sizes larger than 12 mesh were screened out and discarded; less than 5% of the particles were smaller than 40 mesh. The carbon that was used for small scale testing was washed with organic free water, dried at 105°C and stored in air-tight screw cap bottles.

*Mention of commercial products does not imply endorsement by the U.S. Environmental Protection Agency.

Background Solution. Cincinnati Water Works unchlorinated effluent from a 3-day detention reservoir which is fed with "alum dosed" Ohio River water. [Average total organic carbon (TOC) was 1.45 mg/L].

Analysis. Analyses for DCE were performed using a purge and trap method similar to that described by Bellar and Lichtenberg (5); 1,2-dichloropropane was used as an internal standard. For DBCP analysis, a modification of the purge and trap (Tekmar Model LSC-1 Liquid Sample Concentrator; Cincinnati, OH) was employed. The gas chromatograph detector response was recorded graphically, and peak area printouts were used to measure concentrations.

Adsorption Procedures

Batch Studies. Completely mixed batch reactor (CMBR) equilibrium studies for DCE were conducted using the static bottle-point technique. Different quantities of unpulverized activated carbon were placed in 250-mL bottles. The bottles were filled with spiked settled river water (SRW), sealed with teflon-lined caps, and tumbled until equilibrium was achieved.

CMBR rate studies for DCE were conducted in 1.7 liter (total volume) Celstir® bottles (Wheaton Scientific; Millville, NJ) equipped with suspended magnetic stirrers. The bottles were filled, headspace free, with spiked SRW. Weighed quantities of carbon were transferred to the reactors and the solutions stirred at 500 rpm. Samples of 5-mL volume were withdrawn periodically for analysis and unspiked SRW was added to reestablish a headspace-free condition.

Pilot-Column Studies. Column studies were performed for DCE and DBCP in 4-cm ID glass columns containing 400 grams of 12 x 14 mesh carbons. All tubing and connections were constructed of glass, stainless steel, or Teflon® . Glass sampling ports, extending into the center of the column, were provided at various depths and were equipped with Teflon® stop-cocks.

Settled Ohio River Water (SRW) was transported from the Cincinnati Water Works and stored in a large tank outside the building. The water in the storage tank was continuously recirculated with pump to prevent stagnation. The flow from the storage tank to a 100-gallon stainless steel tank inside the laboratory was controlled by the needle value to deliver a slightly greater volume than was required by the column experiments; the excess was overflowed to waste.

A concentrated solution of the test compound was contained in an air-tight 20-L glass bottle. To prevent loss by volatilization, a Teflon® bag was attached to the bottle. The bag would expand as fresh spike solution was added, or contract as solution was withdrawn. The spike solution was injected into the discharge tubing of the SRW pump. Mixing was provided by a small glass column filled with beryl saddles. The pressure drop across the adsorber was measured using a mercury manometer. Figure 1 illustrates the experimental apparatus and arrangement. The salient features of the pilot column investigations are summarized in Table 1.

RESULTS AND DISCUSSION

Adsorption Equilibrium

Several theoretical and empirical equations, including the Freundlich, Langmuir, and three-parameter isotherms, were investigated for the mathematical description and quantification of the DCE adsorption equilibrium data. The experimental data correlated best with the Freundlich and three-parameter isotherms. However, the Freundlich isotherm was found to be superior to the three-parameter model because of simpler and more efficient parameter estimation. The Freundlich equation relates the amount of solute adsorbed on carbon, $q_e = \log K_f + 1/n \log C_e$, where K_f and $1/n$ are indicators of adsorption capacity and adsorption intensity, respectively (6). Figure 2 demonstrates the equilibrium data and Freundlich isotherm for DCE.

Adsorption Rates

The MADAM model (1-4) was used for the estimation of mass transfer coefficients for DCE simultaneously from the CMBR rate data. To minimize the program search for the appropriate values of the film transfer coefficient, $k_{f,b}$, and the intraparticle diffusion coefficient, D_s, a close control value for $k_{f,b}$ was estimated by conducting a separate CMBR rate experiment using a large amount of carbon to nearly deplete the solute within the first few minutes of the experiment (7). The $k_{f,b}$ value was then calculated from the slope of the regression line for the initial rate data. Figure 3 illustrates the results of simulation analysis for a two-parameter search. The estimated coefficients, $k_{f,b} = 4.05 \times 10^{-3}$ cm/sec and $D_s = 1.77 \times 10^{-9}$ cm^2/sec, were used to predict the experimental profile represented by the solid line.

Adsorption Columns

Simulation modeling analysis for DCE were conducted for three sets of experimental conditions. Table 2 summarizes values for the physical and chemical parameters associated with these studies, including values for the surface diffusion coefficient, D_s, and the column film transfer coefficient, $k_{f,c}$. The surface diffusion coefficient was estimated from the CMBR rate tests, and the film transfer coefficients by a correlation technique proposed by Williamson et al (8).

Experimental data and the predicted breakthrough profile for a typical fixed-bed pilot-plant adsorber are presented in Figure 4. The MADAM model was generally able to simulate and predict the performance of fixed-bed adsorbers for removal of DCE.

Biodegradation Studies

All column experiments in this study exhibited biological activity. Biological growth was apparent in each column within a few days from the onset of operation.

Adsorber run 1 was the first to be conducted in an attempt to investigate the characteristic breakthrough profiles. After reaching exhaustion, the adsorber column was emptied and recharged with fresh

TABLE 1

Adsorber Run#	Test Compound	Flowrate (mL/min) [hydraulic loading (liter/min/m^2)]	Avg. Soln. Conc. ($\mu g/L$)	EBCT* (min)	Total run time (days)
1	DCE	100[80]	112	10	82
2	DCE	100[80]	89	10	88
3	DCE	50[40]	196	20	88
4	DCE	200[160]	21	5	88
5	DBCP†	100[80]	18	10	166

†SRW was used for the first 40 days of operation and tap water was used for the remainder of the run.

* EBCT: Empty Bed Contact Time

TABLE 2. PARAMETERS USED FOR FIXED-BED ADSORBER PREDICTIVE BREAKTHROUGH PROFILES FOR DCE

Adsorber Run #	$\overline{C}_{in}$	d	L	γ	v	$k_{f,c}$	D_s
1	112	0.106	80	0.4	0.33	2.6×10^{-3}	1.77×10^{-9}
3 (Figure 4)	200	0.106	80	0.4	0.17	2.05×10^{-3}	1.77×10^{-9}
4	20	0.106	80	0.4	0.66	3.29×10^{-3}	1.77×10^{-9}

MERL Reported Parameters

$\overline{C}_{in}$: average influent concentration ($\mu g/L$)

d : carbon particle diameter (cm)

L : bed length (cm)

γ : bed porosity (dimensionless)

v : interstitial velocity (cm/sec)

EBCT: empty bed contact time (min)

MADAM Determined Parameters

$k_{f,c}$: column film transfer coefficient (cm/sec)

D_s : intraparticle diffusion coefficient (cm^2/sec)

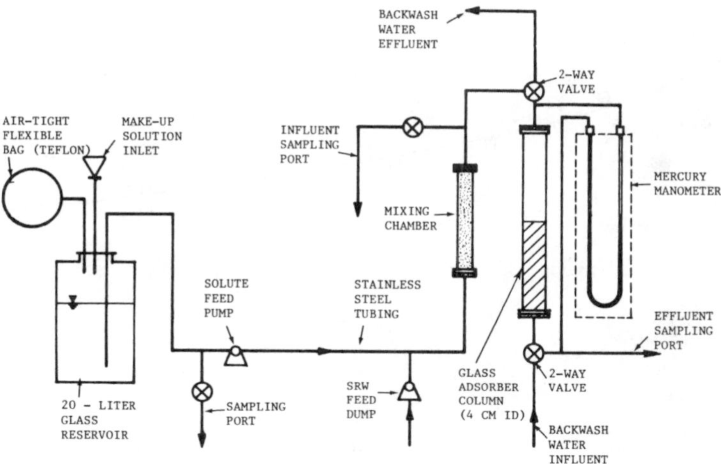

Figure 1. experimental apparatus for fixed-bed adsorber studies.

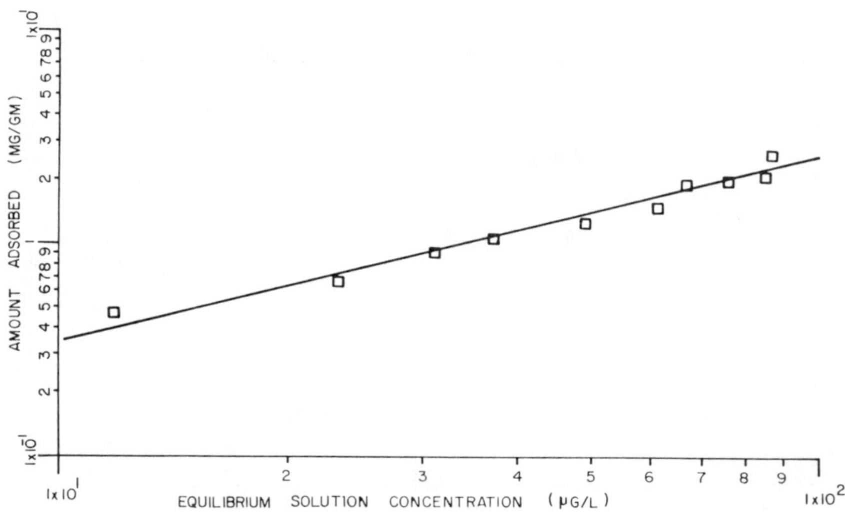

Figure 2. experimental data and Freundlich isotherm for DCE.

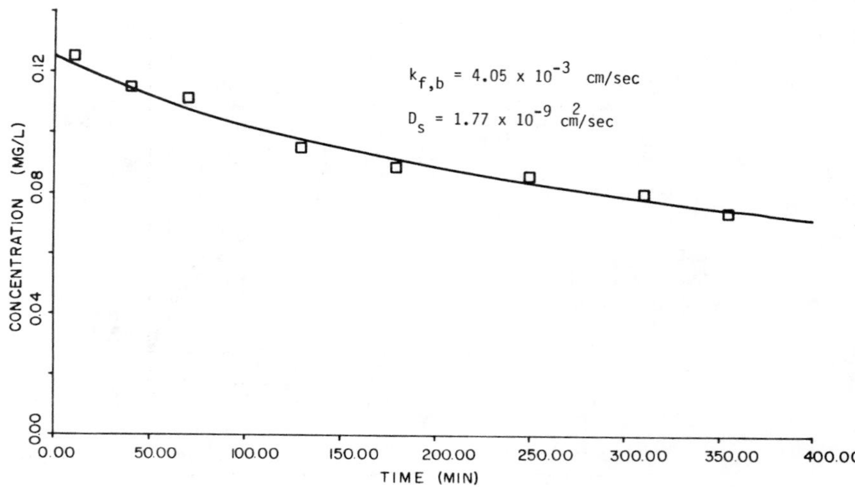

Figure 3. experimental data and model profile for rate of adsorption of DCE in a CMBR (C_o = 124 µg/ℓ; 0.118 g/ℓ of 12 x 40 carbon).

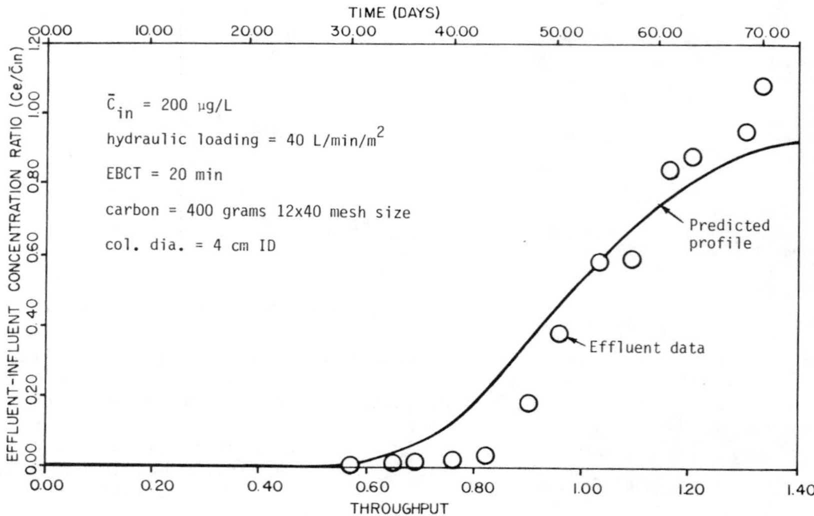

Figure 4. effluent concentration profiles for DCE (fixed-bed adsorber run 3).

activated carbon and was run (absorber run 2) in parallel with adsorber runs 3 and 4. It is of interest to note that, although adsorber runs 1 and 2 were conducted under similar conditions, no characteristic breakthrough profile was observed for adsorber run 2, as shown in Figure 5. The most logical explanation for this behavior is the unintentional seeding of adsorber run 2 by the bacteria which had been acclimated in the stainless steel column feeder tubing during an earlier run (adsorber run 1).

Figure 5 also presents the results of biodegration studies for adsorber run 4, demonstrating a decline in DCE effluent concentrations after breakthrough. It is postulated that bacterial acclimation during the initial stages of breakthrough may have been responsible for this decline. Roberts (9) reported observations for a number of trace contaminants for reclaimed water during aquifer passage. This phenomenon, however, was not experienced with adsorber runs 1 and 3. Furthermore, no breakthrough was observed for DBCP after 166 days of operation (adsorber run 5), a duration far beyond the theoretical carbon capacity (10).

That biodegradation may be responsible for removal of DCE and DBCP is not a surprising phenomenon. In fact, a recent comprehensive study reveals that almost every class of anthropogenic compounds can be degraded by some microorganisms (11). More specifically, biodegradation of DCE has been reported by Patterson and Kodukala (12). These investigators conducted static flask tests whereby the test solution was dosed with bacterial innocula from settled domestic sewage. The extent of biodegradation for DCE in these experiments was about 60%.

Evidence is available (13,14) indicating that biodegradation of trace organic materials in biofilm applications can occur in general by secondary utilization or cometabolism of secondary substrate provided that abundant primary organic substrate is available and bacteria are capable of decomposing both the primary and secondary substrates. In the present context, DCE and DBCP may be regarded as a secondary substrate while the TOC of the background solution acts as an aggregate primary substrate providing energy needed to sustain the biofilm. Figure 6 represents the TOC concentration-history profiles for adsorber runs 2 and 4, indicating a substantial decrease in the so-called "primary substrate" with an increase of EBCT during the entire course of column operation.

To investigate the extent of biodegradation of DCE in the adsorber run 2, samples were taken at various column depths and analyzed for DCE. Figure 7 shows the DCE profile after 68 days of operation, indicating extensive biodegradation within the first 2-3 cm of the column depth.

Similarly, to investigate biodegradation phenomenon for DBCP, samples were taken from various column depths at three different run times of 17, 76, and 166 days, as shown in Figure 8. The sharp initial drop in concentration within the first few centimeters of column depth would be expected for short operating times, but not after 166 days. It is thus postulated that biodegradation may be operative in the entire column depth.

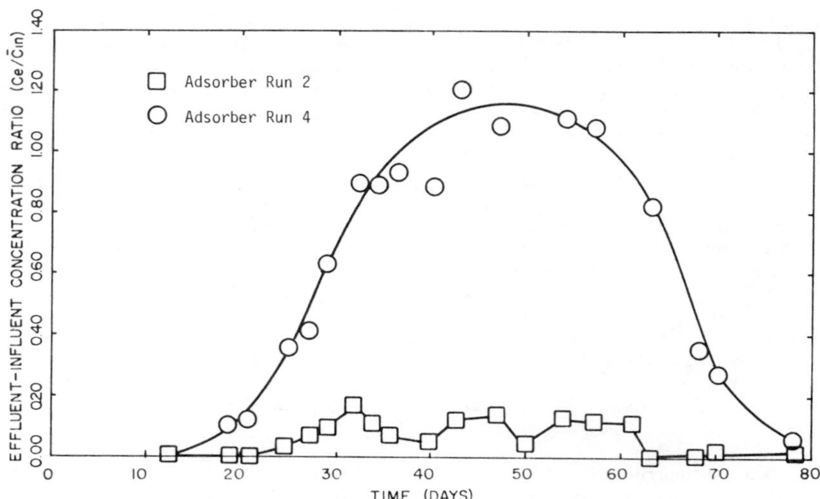

Figure 5. DCE breakthrough profiles for fixed-bed adsorber runs 2 and 4 (effect of biodegradation).

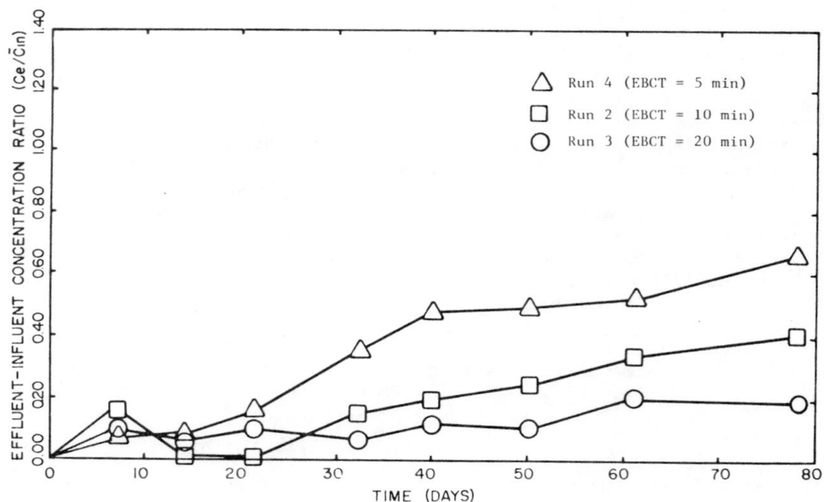

Figure 6. TOC breakthrough profiles for fixed-bed adsorber runs 2 to 4.

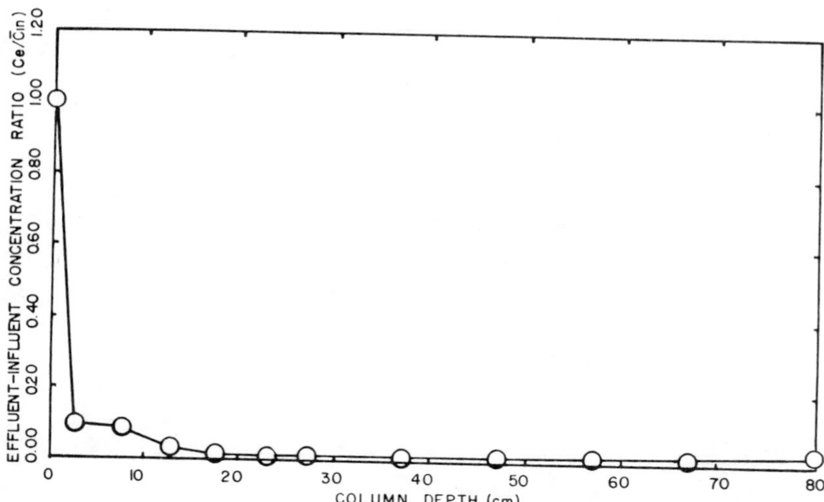

Figure 7. DCE concentration at various depths in fixed-bed adsorber run 2 after 68 days of operation.

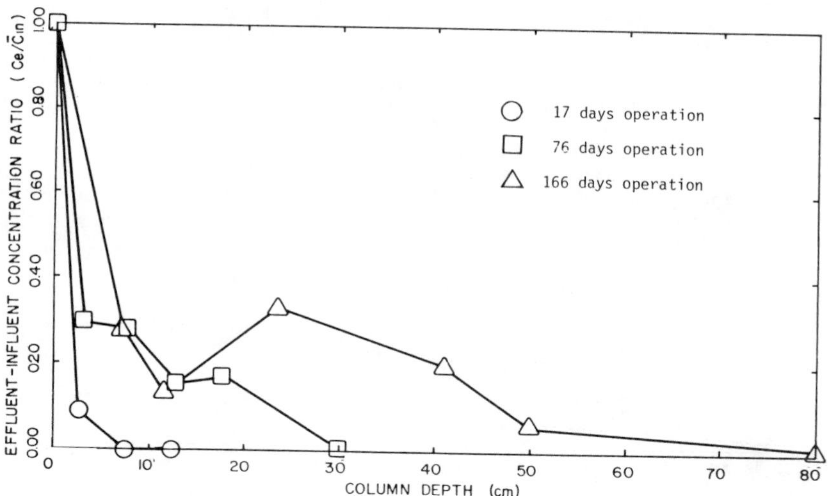

Figure 8. DBCP concentration at various depths in fixed-bed adsorber run 5 after different run times.

ACKNOWLEDGEMENTS

The research reported here was sponsored in part by Research Grant No. R-804369 from the Drinking Water Research Division, MERL, USEPA, Cincinnati, Ohio. MERL conducted the pilot test and provided the experimental data used in these analyses.

LITERATURE CITED

1. Weber, W. J., Jr., and J. C. Crittenden, J. Water Poll. Control Fed. 47:5 (1975).

2. Crittenden, J. C., and W. J. Weber, Jr., J. Environ. Eng. Div., ASCE 104 (EE2) (1978).

3. Crittenden, J. C., and W. J. Weber, Jr., J. Environ. Eng. Div., ASCE 104 (EE3) (1978).

4. Crittenden, J. C., and W. J. Weber, Jr., J. Environ. Eng. Div., ASCE 104 (EE6) (1978).

5. Bellar, T. A., and J. J. Lichtenberg, USEPA, 670/4-74-009, (1974).

6. Weber, W. J., Jr., Physicochemical Processes for Water Quality Control, Wiley-Interscience, New York (1972).

7. Pirbazari, M., "Performance Predications for Removal of Toxic and Carcinogenic Compounds from Water Supplies," Ph.D. Dissertation, The University of Michigan, Ann Arbor, MI (1980).

8. Williamson, J. E., K. E. Bazaire and C. J. Geankopolis, Ind. Eng. Chem. Fund., 2(2), 126 (1963).

9. Roberts, P. V., "Removal of Trace Contaminants from Reclaimed Water During Aquifer Passage," In: Oxidation Techniques in Drinking Water Treatment, W. Khun, and H. Sontheimer (eds.), Karlsruhe, FRG, 1978, U.S. Environmental Protection Agency, EPA-570/9-79-020 (1979).

10. Dobbs, R. A., and J. M. Cohen, USEPA, 600/9-80-023 (1980).

11. Kobayashi, H., and B. E. Rittman, "Microbial Removal of Hazardous Organic Compounds," Environ. Sci. Technol., 16:3:170A (1982).

12. Patterson, J. W., and P. S. Kodukala, Chem. Eng. Prog., 77 (4), 48 (1981).

13. McCarty, P. L., B. E. Rittman, and M. Reinhard, Environ Sci. Tech. 15, 40 (1981).

14. Rittman, B. E., and P. L. McCarty, Biotech. Bioeng., 22, 2343 (1980).

ADSORPTION EQUILIBRIA ON ACTIVATED CARBON OF MIXTURES OF SOLVENT VAPOURS

Åke C. Rasmuson
Department of Chemical Engineering
Royal Institute of Technology
S-10044 Stockholm, Sweden

ABSTRACT

The objective of this work is to find a model or method for prediction of equilibria of adsorption of solvent vapour mixtures, in low concentrations in air, on activated carbon. A very promising model is the Ideal Adsorbed Solution Theory (IAST). This model and some closely related models have been evaluated with experimental data. The results show that the IAST can predict the binary data reasonably well. Single component isotherms for toluene and butanol are correlated well by the Dubinin-Radushkevich (DR), the Kisarov and the Freundlich equations. A Langmuir equation cannot represent the isotherms. The possible capacity of the DR and the Kisarov equations, for prediction of single component isotherms, is not completely supported. However, due to experimental difficulties, the amount of experimental data is so far too limited for any final conclusions.

INTRODUCTION

It has been found (Olin(1)) that chemists who continued with laboratory work, for at least a few years after graduation, and specifically worked with organic compounds, display an increased frequency of death from certain forms of cancer. One way to reduce the exposure to vaporized organic chemicals would be to clean the ambient air using filters with activated carbon. There are very few experimental studies regarding adsorption equilibria on activated carbon of mixtures of solvent vapours. None of them are concerned with partial pressures below $5 \cdot 10^{-5}$ atm.

THEORETICAL

Models and methods for prediction of multicomponent gas adsorption equilibria have been (Sircar and Myers (2)) classified into three categories: 1) models based on single component isotherm equations; 2) models based on potential theory; 3) thermodynamic models.

Older models of the first category are reviewed by Young and Crowell (3). Among more recently developed models are the one of Hoory and Prausnitz (4) based on the two-dimensional analogue of van der Waals' equation. Another interesting model is the Vacancy Solution Model of Suwanayuen and Danner (5). In this model the single component adsorption is treated as binary adsorption by introduction of the imaginary substance "vacancies". Some of the models in the first category are only derived for binary mixtures. For mixtures of a number of components some models will demand complex calculations. Some of the models are based on single component isotherm equations which do not correlate single component data well.

Among models based on potential theory are those of Lewis et al. (6), Grant and Manes (7) and Maslan, Altman and Aberth (8). The model of Grant and Manes (7) is similar to the one of Lewis et al. (7) in that the total adsorbate volume is used together with the pure component characteristic curve to get the adsorption potential of the substance in the mixture. Grant and Manes (7) explicitly assume that Raoult's law, using the adsorbate volume as the intensive variable (see below), is obeyed. Lewis et al. (6) introduce an empirical correlation in the calculations. Both the model of Lewis et al. (6) and the model of Maslan, Altman and Aberth (8) violate Gibbs phase rule. The model of Lewis et al. (6) contains one unnecessary equation. Maslan, Altman and Aberth (8) specify, for a binary mixture, four variables instead of the permitted three. Bering, Serpinskii and Surinova (9) have extended the Dubinin-Radushkevich equation (see below) to mixtures. However, by this extention only the total amount adsorbed may be calculated. By application of the empirical correlation of Lewis et al. (6), they are able to calculate partial isotherms for binary mixtures.

A very interesting thermodynamic model is the Ideal Adsorbed Solution Theory (IAST) of Myers and Prausnitz (10):

$$P_i = x_i \cdot P_i^o(\pi, T) \qquad (1)$$

Equation (1) may be looked upon as a Raoult's law of adsorption. The pure component variable, P_i^o, is the equilibrium pressure for pure component adsorption at the same spreading pressure as that of the mixture. As discussed by Sircar and Myers (2), other versions of Raoult's law with, instead of π, the total pressure, the adsorbate volume or the adsorbed number of moles as the intensive variable, have been suggested. Only the use of the spreading pressure is thermodynamically consistent in the sense that Gibbs adsorption isotherm is satisfied. Equation (1) is very attractive for mixtures with a large number of components, because of its apparent simplicity. However, the appearance of the spreading pressure complicates the calculations. Integrations of pure component adsorption data from an equilibrium pressure of zero have to be done. Activated carbon has a high capacity for solvent vapours even at one ppm in air; that is, partial pressure of 10^{-6} atm. Consequently, a considerable amount of the integral value is due to pressures below 10^{-6} atm. However, experiments at pressures much lower than 10^{-6} atm. are difficult to perform. Therefore a test of the IAST, for the adsorption of solvent vapours at low concentrations on activated carbon, may also be a test of the chosen method of extrapolation of the single component isotherms.

In the literature a number of single component isotherm equations are proposed. Among these may be mentioned the Dubinin-Radushkevich equation (11) and the Kisarov equation (12).

Dubinin-Radushkevich

$$a = \frac{W_o}{V_a} \exp\left[-B_D\left(\frac{T}{\beta} \ln\left(\frac{P_s}{P}\right)\right)^2\right] \tag{2}$$

Kisarov

$$A = \frac{W_o}{V_a} \text{ according to Kisarov (12)}$$

$$a = \frac{AB(P/P_s)^n}{1 + B(P/P_s)^n} \qquad A = \frac{W_o}{V_a}\left(\frac{1+B}{B}\right) \text{ according to Begun et al. (13)} \tag{3}$$

$$n = \frac{k \cdot T}{\beta}$$

These two equations are able to correlate adsorption isotherms of solvent vapours on activated carbon very well. They are interesting also because of their potential capacity of prediction of pure component isotherms. The regression constants in the Dubinin-Radushkevich equation and the Kisarov equation are related to parameters of the adsorbent. These parameters are supposed to be independent of the adsorbate. In the Dubinin-Radushkevich equation the parameters are W_o, the volume of micropores, and B_D. In the Kisarov equation the parameters are W_o, the limiting sorption volume minus the part in which capillary condensation can occur, B, which is a dimensionless quantity and k . β is the affinity coefficient and may be calculated from physical properties of the adsorbate substances. β relates the adsorption of one component to the adsorption of a chosen reference substance for which β, by definition, equals unity. V_a is usually approximated by the liquid molar volume. Two other equations of interest, because of their simplicity, are the Freundlich equation and the Langmuir equation (3). Of these four equations only the Langmuir equation conforms to Henry's law at low pressures.

EXPERIMENTAL

The experimental apparatus consists of three, nearly identical, separate static adsorption systems, a gas analyzing system and a standard gas generation system for calibration of the analyzing system. In each adsorption system, Figure 1, the gas phase is circulated through the activated carbon. At first, the weighed carbon sample is placed in a specially made glas bottle. The bottle is immersed in a temperature controlled water bath and connected to the rest of the system. Then the complete system is evacuated and synthetic air is introduced to approximately 1 atm. After that, a chosen amount of solvent or mixture of solvents is injected through a membrane and allowed to vaporize in the system. The exact amount injected is determined by weighing the syringe before and after the injection. The equilibration process may be followed and the actual equilibrium gas phase concentration may be determined by sampling and analyzing on a gas chromatograph. The gas chromatograph is equipped with a gas sampling valve. The gas sampling loop is filled by a vacuum technique. The glas bottle of about 2 litres increases the gas volume in the system, thus making the pressure changes at sampling negligibly small. The response of the chromatograph is calibrated with samples of standard gas from a dynamic standard gas generation system. The standard gas is generated by flowing nitrogen over temperature controlled diffusion tubes (Nelson (14) and Blacker and Brief (15)). By the steady-state diffusion of vapour from the surface of pure liquid, through a narrow tube, to the by-passing nitrogen, the diffusion tubes generate a constant mass flow of solvents to the nitrogen stream. This nitrogen stream may be diluted with synthetic air up to one hundred times.

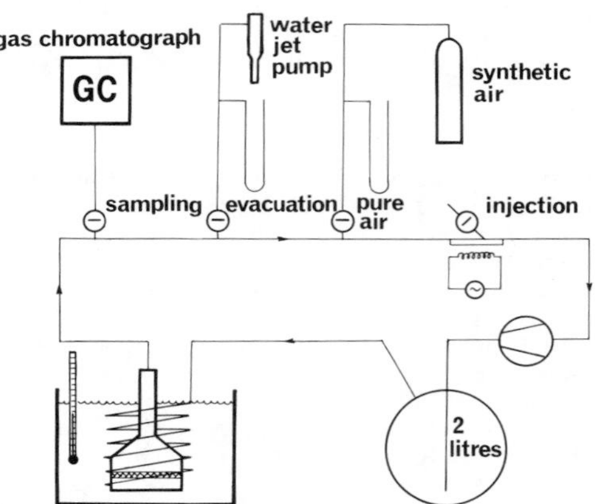

Figure 1:

An adsorption system connected to the gas chromatograph.

The activated carbon, Pittsburgh BPL 6x16 mesh (Calgon Corporation), is crushed and sieved. The selected fraction is then washed in distilled water several times. The purpose is to clean the particles of carbon dust. The particles are then dried and desorbed at 200°C, for seven days, in a continuously flowing stream of nitrogen.

The very long times of equilibrium have been a major problem in the experiments. Figure 2 shows the equilibration process in a single component experiment. In this experiment the system gas volume was circulated approximately 0.3 times per hour. One explanation for the long times of equilibrium is the very high equilibrium constants K (in the order of 10^7-10^8 m^3/m^3). With a reasonable amount of carbon and volume of gas, the amount of solvents will always be distributed so that the adsorbed amount will dominate over the amount still in the gas. The high equilibrium constants will, during the equilibration process, cause situations where comparatively large amounts of solvents have to be spread out in the carbon bed using a gas phase with a rather low concentration. The process is equilibrium controlled. In Figure 3 is shown the equilibration process in a three component mixture experiment. In this case the system gas volume was circulated approximately 90 times per hour.

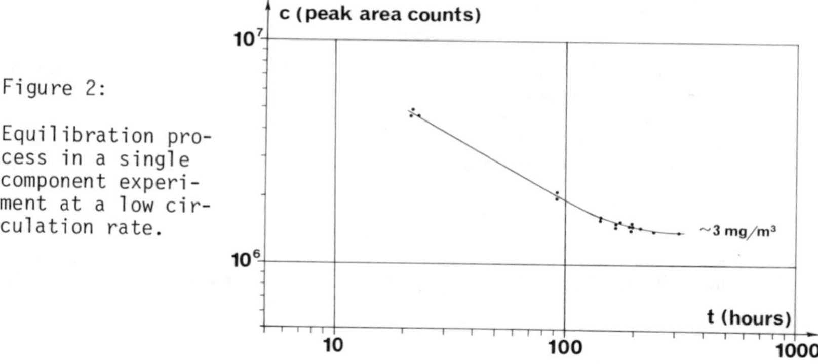

Figure 2:

Equilibration process in a single component experiment at a low circulation rate.

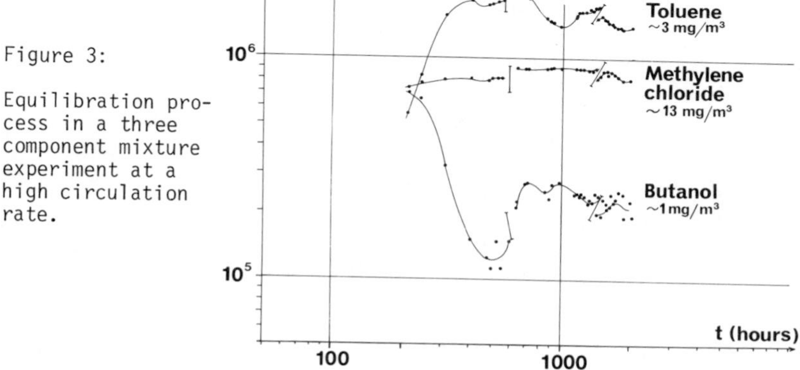

Figure 3:

Equilibration process in a three component mixture experiment at a high circulation rate.

The figure shows that even after several hundred hours, the competition between butanol and toluene results in an exchange of the two. Methylene chloride is only affected to a small extent by this exchange. Methylene chloride is displaced readily by the other two and in this case obviously to approximately the same extent. It is still unclear whether there is also, besides the effect of the high equilibrium constants, a very slow equilibration process in the very narrow micropores.

EVALUATION OF EXPERIMENTAL DATA

Equilibrium data of adsorption of solvents, in low concentrations in air, on activated carbon have been determined at 293.1 K. Figure 4 shows the experimental single component adsorption isotherms of toluene and butanol. This diagram does not contain all data for these substances. There are some more scattered data for butanol. However, their accuracy are in doubt and they are therefore not included at this time. Four different equations have been tried for correlation. The Langmuir equation gives a very unsatisfactory description of the experimental data. The Freundlich equation and particularly the Dubinin-Radushkevich equation and the Kisarov equation correlate the data well. The solid drawn lines in Figure 4 represent correlations by the Kisarov equation.

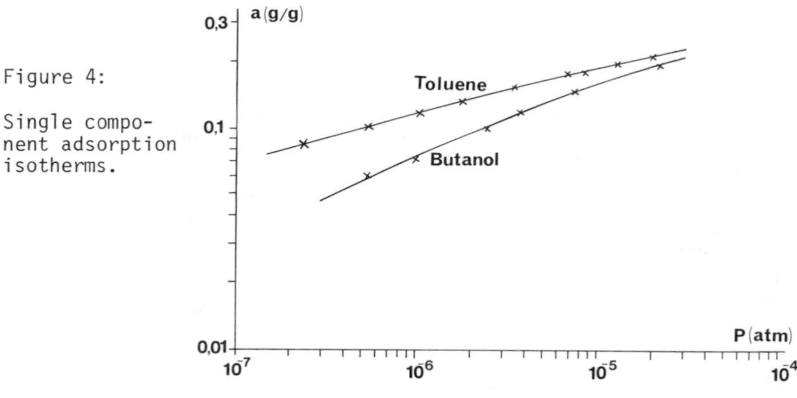

Figure 4:

Single component adsorption isotherms.

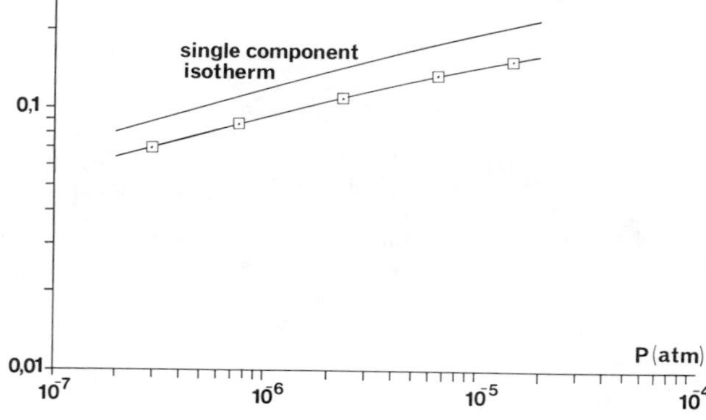

Figure 5:

Partial adsorption isotherm of toluene in a mixture with butanol.

In Table 1 are the experimental values of the constants in the Dubinin-Radushkevich equation and in the Kisarov equation. V_a is calculated as the liquid molar volume at 20°C. β for butanol, using toluene as the reference substance, is calculated as the ratio of the electronic polarizations. Molar volumes and parachors were also tried but gave poorer agreement between the B_D values and the k values respectively.

	Toluene	Butanol
Dubinin-Radushkevich:		
W_o(cc/g)	0.46	0.49
$B_D(10^{-7}\ K^{-2})$	1.33	1.35
Kisarov:		
W_o(cc/g) (Kisarov)	0.71	0.56
W_o(cc/g) (Begun)	0.58	0.51
B	4.27	9.65
$k(10^{-3}\cdot K^{-1})$	0.96	1.11

Table 1: Experimental values of constants in the Dubinin-Radushkevich equation and in the Kisarov equation.

In Figures 5 and 6 are shown binary equilibrium data of adsorption of toluene and butanol. The equilibrium pressures of each substance have been determined for increasing total amount adsorbed at constant mole fraction in the adsorbed phase ($x_{Toluene}$ = 0.7393 kmol/kmol). The points in the two figures originate from the same experiments and are pairwise coupled with each other. These partial isotherms are also well represented by the Dubinin-Radushkevich, Kisarov and Freundlich equations. The Langmuir equation gives very poor correlations. The diagrams show that even in the low-ppm-range, the competition is significant.

In Figure 7 is shown a comparison of experimental binary data with predictions based on the IAST. For each substance there are three solid lines drawn. In the calculation of each line, the single component isotherms have been correlated by the isotherm equation in question. These correlations have been used in the whole integration interval in the calculations of the spreading pressures. The diagram shows that the predictions are fairly good for both components. However, for both components the model tends to overestimate the carbon capacity.

By exchanging the intensive variable in the IAST from the spreading pressure to the total volume of adsorbate or the total number of moles adsorbed, the model is simplified. In these cases it is not necessary to know the single component isotherms from zero pressure. However, as is shown by Sircar and Myers (2), the model is now no longer thermodynamically consistent. Figure 8 shows predictions of the binary data by the simplified version of the IAST, using the total volume of adsorbate as the intensive variable. This model does not give as good prediction of the butanol adsorption as the IAST. Application of the total number of moles adsorbed, results in slightly better predictions for toluene but worse predictions for butanol. A Raoult's law type of equation, using the total number of moles adsorbed as the intensive variable, is thermodynamically consistent (Kidnay and Myers (16)) if logarithmic plots of the pure component isotherms are capable of superposition, by means of lateral translation parallel to the pressure axis. As is shown in Figure 4 a good superposition is not possible which in part may explain the deviations in the predictions of the binary data.

Figure 6:

Partial adsorption isotherm of butanol in a mixture with toluene.

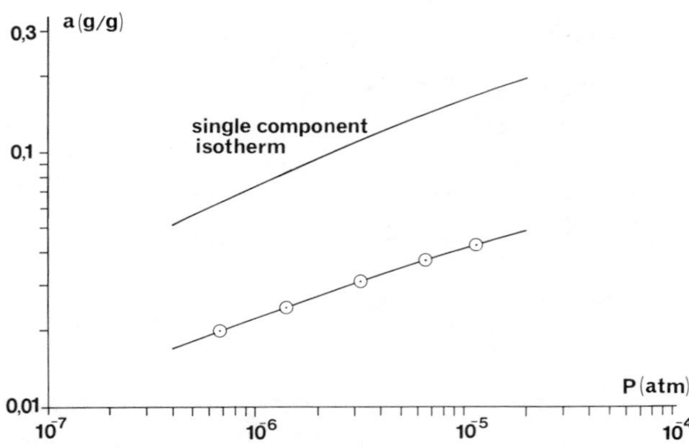

Figure 7:

Prediction of binary data by the IAST.

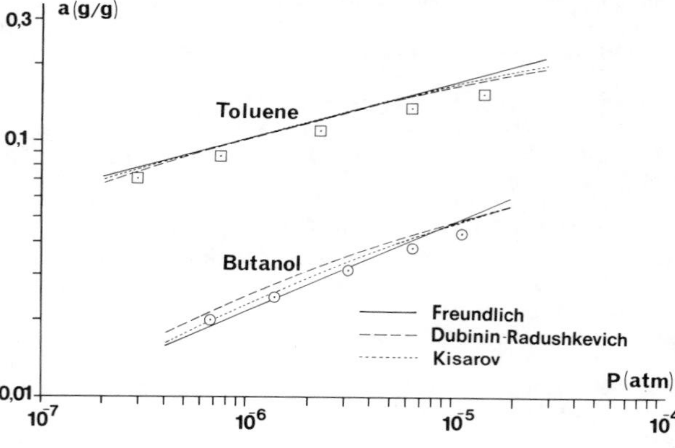

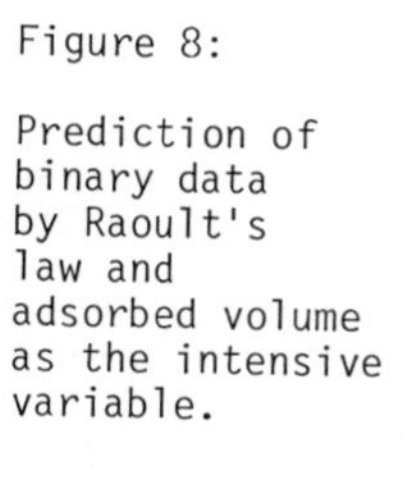

Figure 8:

Prediction of binary data by Raoult's law and adsorbed volume as the intensive variable.

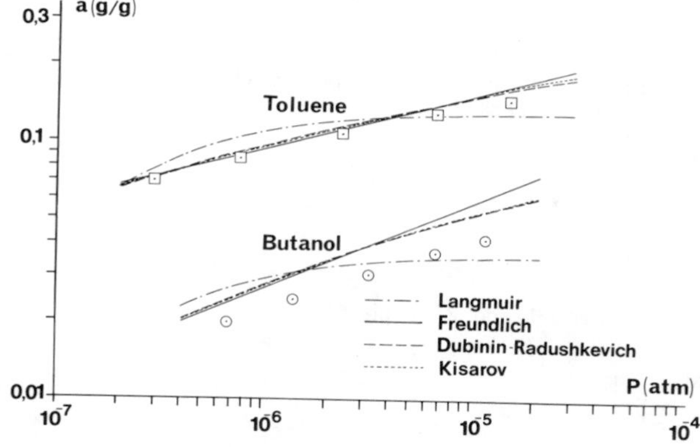

ACKNOWLEDGEMENT

This work has been financially supported by the Swedish Council for Building Research and the Swedish Work Environment Fund.

NOTATION

- A constant
- a number of moles adsorbed, kmol/kg (g/g in figures only)
- a_{vol} adsorption per volume of adsorbent, $kmol/m^3$
- B constant
- B_D constant, K^{-2}
- c concentration of the substance in the gas phase, $kmol/m^3$
- K adsorption equilibrium constant = a_{vol}/c, m^3/m^3
- k constant, K^{-1}
- n constant
- P equilibrium pressure, atm
- P_s vapour pressure of liquid, atm
- P_t total pressure, atm
- R gas constant
- T temperature, K
- t time, hours
- V_a molar volume of adsorbate, $m^3/kmol$
- W_o constant, m^3/kg (cc/g in table only)
- x mole fraction in the adsorbed phase, kmol/kmol

Greek symbols

β affinity coefficient
γ activity coefficient in the adsorbed phase
π spreading pressure, N/m

Superscripts

o single component

Subscripts

i component i
1,2 component 1 and 2 respectively

\- - - - -

LITERATURE CITED

1 Olin, G.R., Amer. Ind. Hyg. Assoc. J. 39 (7), 557 (1978)

2 Sircar, Shivaji and A.L.Myers, Chem. Eng. Sci. 28 (2), 489 (1973)

3 Young, D.M. and A.D. Crowell, Physical Adsorption of Gases, Butterworths, London (1962)

4 Hoory, S.E. and J.M. Prausnitz, Chem. Eng. Sci. 22 (7), 1025 (1967)

5 Suwanayuen, Solot and R.P. Danner, AIChE J. 26 (1), 76 (1980)

6 Lewis, W.K., E.R. Gilliland, B. Chertow and W.P. Cadogan, Ind. Eng. Chem. 42 (7), 1319 (1950)

7 Grant, R.J. and Milton Manes, Ind. Eng. Chem. Fundam. 5 (4), 490 (1966)

8 Maslan, F.D., M. Altman and E.R. Aberth, J. Phys. Chem. 57 (jan), 106 (1953)

9 Bering, B.P., V.V. Serpinskii and S.I. Surinova, Acad. Sci. USSR Bull., Div. Chem. Sci. (5), 753 (1965)

10 Myers, A.L. and J.M. Prausnitz, AIChE J. 11 (1), 121 (1965)

11 Dubinin, M.M., E.D. Zaverina and L.V. Radushkevich, Zhur. Fiz. Khim. 21, 1351 (1947)

12 Kisarov, V.M., Russ. J. Phys. Chem. 43 (4), 580 (1969)

13 Begun, L.B., V.M. Kisarov, A.I. Subbotin and V.I. Trachenko, Sov. Chem. Ind. 5 (3), 162 (1973)

14 Nelson, G.O., Controlled Test Atmospheres, Ann Arbor Science Publ., Inc., Ann Arbor (1971)

15 Blacker, J.H. and R.S. Brief, Amer. Ind. Hyg. Assoc. J. 32 (10), 668 (1971)

16 Kidnay, A.J. and A.L. Myers, AIChE J. 12 (5), 981 (1966)

TIME DOMAIN SOLUTIONS OF SOME MODELS FOR DIFFUSION AND ADSORPTION IN PARTICLES AND FLOW IN PACKED OR SLURRY ADSORBERS

Anders Rasmuson
Department of Chemical Engineering
Royal Institute of Technology
S-100 44 Stockholm
Sweden

ABSTRACT

Transient mass transfer in fixed-bed and slurry adsorbers are studied mathematically. Analytical solutions are derived in the form of infinite integrals. The solutions include the effects of fluid-to-particle diffusion, intraparticle diffusion and reversible first-order adsorption for a first-order reaction on the intrapore surface. Efficient numerical schemes for evaluating the infinite integrals are given. Several examples are presented.

INTRODUCTION

The problem of predicting transient mass and heat transfer during flow through packed beds or agitated vessels has numerous applications in the chemical process industries. In the following we use the language of mass transfer although it is recognized that the analysis is readily applicable to the analogous heat transfer problems. The scope of this study is to derive analytical solutions of some linear models for fixed-bed and batch adsorption. The following effects are taken into account:

- o Diffusion of the component from the main body of the flowing phase to the external surface of the adsorbent particle (external diffusion)
- o Diffusion through the porous network of the particle (internal diffusion)
- o The adsorption process itself

In the general case, all three steps can contribute to the overall rate of adsorption. The adsorption rate (third mass transfer resistance) is usually rapid. However, it may be of importance in certain systems (1). For generality we also include the effect of a first-order chemical reaction at the interior particle surface.

In the case of a fixed-bed, mathematical solutions for the breakthrough curve have been presented for special cases where one of the three processes controls the rate (see literature survey in Rasmuson (2)). Previous theoretical treatments which include more than one of the three rate-controlling resistances are limited to those of Rosen

(3, 4) and Masamune and Smith (5, 6). Rosen considered the combined effects of intraparticle and external diffusion. Masamune and Smith developed a solution of the three-resistance case. However, they assumed in addition, negligible accumulation of the adsorbing component in the void space of bed and particles. These results have supposed that the effect of longitudinal dispersion in the flowing phase in the bed is insignificant and that there is no chemical reaction. Longitudinal dispersion is included in the paper by Lapidus and Amundson (7), who considered the effect of linear surface adsorption kinetics. In the present paper we give a solution of the general three-resistance case including longitudinal dispersion and chemical reaction.

Smith and Amundson (8) derived the steady-state solution for the continuous-flow agitated reactor using a somewhat more complicated reaction model. They also derived the transient solution for a batch reactor, using the same reaction model as in the present paper but neglecting film diffusion and adsorption rate resistance. Huang and Li (9) derived an analytical solution for a batch adsorber (no chemical reaction) including film diffusion but assuming infinite adsorption rate. When the resistance in the film becomes negligible this solution simplifies to the classical one for intraparticle diffusion only (10). Recently Datta et al. (11) derived time domain solutions for the transient response of an isothermal CFSR. Irreversible first-order reaction, internal and external diffusion, but no adsorption rate resistance, is included in their model. The solutions were obtained in the form of infinite sums via the inversion of the Laplace transform using the calculus of residues. Hatton et al. (12) considered the same problem for the limiting case of diffusion and reaction in a body suddenly immersed in an isolated volume of stirred fluid. They used a generalized Sturm-Liouville approach to solve the problem.

The solutions in the present paper are obtained using the Laplace transform technique and inversion in the complex plane. The solutions are obtained in the form of infinite integrals. Efficient numerical schemes have been written to evaluate these integrals.

MATHEMATICAL MODEL AND SOLUTION

The particles are assumed to be spherical and of equal size. In the case of a fixed-bed, the effect of longitudinal dispersion is included and the particle diameter is small in comparison with the overall bed length. The adsorption equilibrium relationship is linear. The temperature is constant. The concentration C of the adsorbable component may now be obtained by solving the following system of equations:

mass balance of the adsorbable component in the fluid phase:

fixed-bed:

$$\frac{\partial C}{\partial t} + V\frac{\partial C}{\partial z} - D_L\frac{\partial^2 C}{\partial z^2} = -\frac{3N_o}{mb} \tag{1a}$$

slurry:

$$\frac{dC}{dt} = \frac{1}{\tau}(C_o - C) - \frac{3N_o}{mb} \tag{1b}$$

batch: $(\tau \to \infty)$

$$\frac{dC}{dt} = -\frac{3N_o}{mb} \tag{1c}$$

mass balance of this component in the particle:

$$\varepsilon_p \frac{\partial C_p}{\partial t} + N_i = \varepsilon_p D_p \left(\frac{\partial^2 C_p}{\partial r^2} + \frac{2}{r}\frac{\partial C_p}{\partial r}\right) \tag{2}$$

where

$$N_i = k_{ads}\left(C_p - \frac{C_s}{K_A}\right) = \frac{\partial C_s}{\partial t} + r_c \tag{3}$$

$$r_c = k_r C_s. \tag{4}$$

The boundary conditions used are:

fixed-bed:

$$C(0,t) = C_o \tag{5}$$

$$C(\infty,t) = 0 \tag{6}$$

$$C(z,0) = 0 \tag{7}$$

$$\frac{\partial C_p}{\partial r}(0,z,t) = 0 \tag{8}$$

$C_p(b,z,t) = C_p|_{r=b}$ given by

$$N_o = \varepsilon_p D_p \left(\frac{\partial C_p}{\partial r}\right)_{r=b} = k_f (C - C_p|_{r=b}) \tag{9}$$

$$C_p(r,z,0) = C_s(r,z,0) = 0 \tag{10}$$

For a slurry adsorber essentially the same boundary conditions are used except that (5) and (6) are not needed (no z-coordinate) and (7) is modified to:

$$C(0) = C_i \tag{7a}$$

N_o is the mass flux from the flowing phase to the outer surface of the particles (couples Equations (1) and (2)).

Equations (1) - (4) subject to the boundary conditions (5) - (10) have been solved analytically using the Laplace transform technique and inversion in the complex plane (2, 13-15). The solutions are obtained in the form of infinite integrals:

$$\frac{C}{C_o} = \frac{1}{2}\frac{C_\infty}{C_o} + \int_0^\infty G(p_i,\lambda)\ H\ [h(p_i,y,\lambda)]\ d\lambda \tag{11}$$

where

C_o inlet (or initial) concentration of C

C_∞ steady-state value of C

G positive decreasing function of λ

H sine or cosine function

h complicated function of p_i, y and λ

The entity p_i stands for the following dimensionless parameters:
fixed bed:

$\delta = \frac{3D_p\varepsilon_p}{b^2}\frac{z}{mV}$ residence time/characteristic time of internal diffusion

$Pe = \frac{zV}{D_L}$ Peclet number

$R_1 = \frac{K}{m}$ distribution ratio

$R_2 = \frac{K_A}{\varepsilon_p}$ distribution ratio

$A = \frac{b^2 k_{ads}}{D_p\varepsilon_p}$ internal diffusion resistance/adsorption rate resistance

$\nu = \frac{D_p\varepsilon_p}{k_f b}$ external diffusion resistance/internal diffusion resistance

$B = \frac{K_A k_r}{k_{ads}}$ rate of reaction/rate of adsorption

For a slurry adsorber the residence time z/V in δ is given by τ and the Peclet number disappears. In the batch situation $\delta \to \infty$ (as $\tau \to \infty$) also. The quantity y is the dimensionless time and is given by $y = (k_{ads}t)/K_A$. If the rate of adsorption is infinite y is given by $y = (2D_p\varepsilon_p t)/(Kb^2)$.

NUMERICAL EVALUATION

G is a positive decreasing function of λ and H is either a sine or a cosine function. The total integrand is thus a decaying sine wave, in which both the period of oscillation and the degree of decay are functions of the system parameters. In some instances, especially for longer times y, the magnitude of the integrand is still considerable after a thousand oscillations of the sine wave. With ordinary integration methods one must choose a step size which is small with respect to the wave-length. Thus some special integration method must be devised for these cases.

In the method used here the integration is performed over each half-period of the oscillatory integrand. The convergence of the alternating series obtained (which may be very slow) is then accelerated by repeated averaging of the partial sums (15-17). Convergence is usually very rapid. The main problem in this method is to find the roots of the trigonometric function (where the integrand changes sign). This is done by a combination of explicit expressions and Newton-Raphson iteration.

In practice a standard (library) integration routine is used up to a certain n = NMAX, where n is the n:th root of

$$H\left[h(p_i,\ y,\ \lambda_n)\right] = 0 \tag{12}$$

The magnitude of the integrand at the point half-way between λ_{NMAX-1} and λ_{NMAX} is then calculated. If ABS (integrand) < 10^{-10} the integration is stopped. If not, the averaging procedure is entered. Approximately 10-20 terms are calculated up to n = NMAX + (10-20). The integration between the zeros is also done with a standard integration routine.

So far this procedure has been implemented in the case of a fixed-bed for $k_{ads} \to \infty$ and $k_r = 0$ (17) and in the cases of slurry and batch adsorbers for $k_{ads} \to \infty$ (15). Typical computing times on a CDC Cyber 170-720 are in the range of 10-40 seconds for 15-40 points on the concentration-time curve. The operation time for one multiplication on this computer is 4.0 μs. Faster running times could be obtained by relaxing the very high degree of accuracy requested.

SOME RESULTS

Using the analytical solutions a large number of calculations have been performed (15-17). Some examples are given in Figures 1 - 6. Figures 1 - 4 are for a fixed-bed, Figure 5 for a continuous slurry and Figure 6 for a batch case. Note that $k_{ads} \to \infty$ and $y = (2D_p\varepsilon_p t)/(Kb^2)$. For the fixed-bed AB = 0.0 and R = R1.

In Figure 1 the influence of the Peclet number is shown. Low Peclet numbers have a pronounced effect on the breakthrough curves. For example the time of first arrival in the case shown, is 100 times earlier for Pe = 1.0 than for the same case without dispersion. This effect could be of great importance for the migration of radionuclides from a repository in fissured rock (16). Figure 2 differs from Figure 1 in the value of δ only. Note the steep breakthrough curve for Pe = ∞ in this long contact-time case. The influence of the film-resistance ν is shown in Figure 3. This parameter is, of course, most important for short bed lengths (δ small). In Figure 4 the impact of R_l is demonstrated. It is seen that the breakthrough curves for $R_l = \infty$ and $R_l = 10^6$ are nearly identical. For increasing δ this situation occurs for lower values of R_l. For example, for $\delta = 1.0$ the curves for $R_l = \infty$ and $R_l = 10^3$ coincide. Figure 5 gives an example for a continuous-flow reactor ($C_i = 0.0$). The influence of ν at high reaction rates ($AB = 10^4$) is clearly demonstrated. Figure 6 shows some results for a batch case.

CONCLUDING REMARKS

The analytical solutions developed provide viable tools for use in the prediction of breakthrough curves and in parameter estimation studies. The codes developed have been tested over a large range of the input parameters and no special difficulties have been encountered. A high degree of accuracy is obtained. This is also the case at low relative concentrations and the evaluation is actually somewhat simpler at lower values of y. Typical computing times on a CDC Cyber 170-720 are in the range of 10-40 seconds for 15-40 points on the concentration-time curve. Apart from making quantitative predictions, the analytical expressions developed should be of use in benchmark calculations for more general numerical codes.

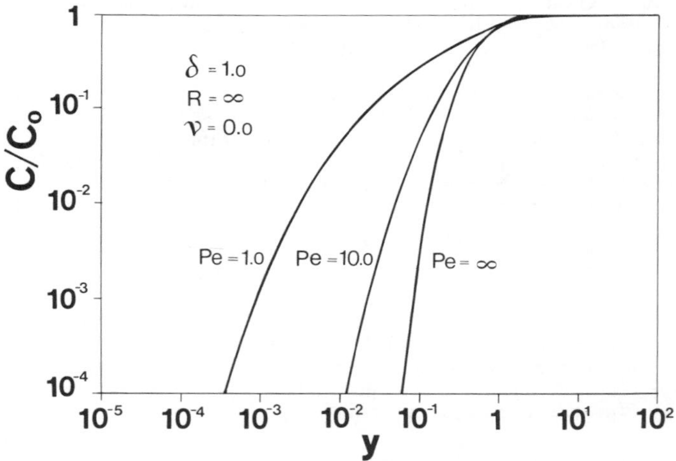

Figure 1: Breakthrough curves for fixed-bed.

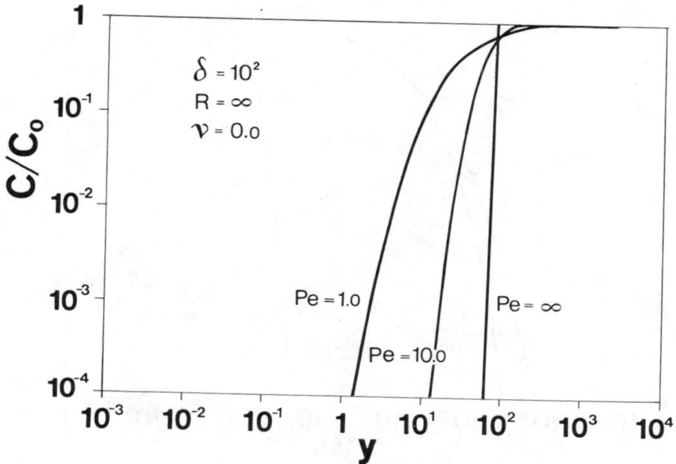

Figure 2: Breakthrough curves for fixed-bed.

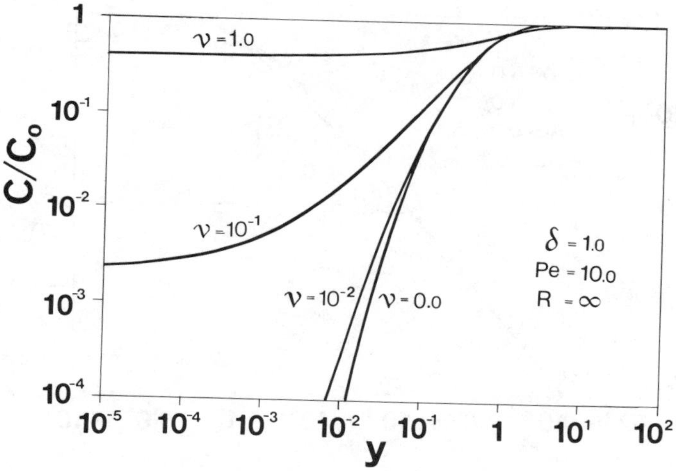

Figure 3: Breakthrough curves for fixed-bed.

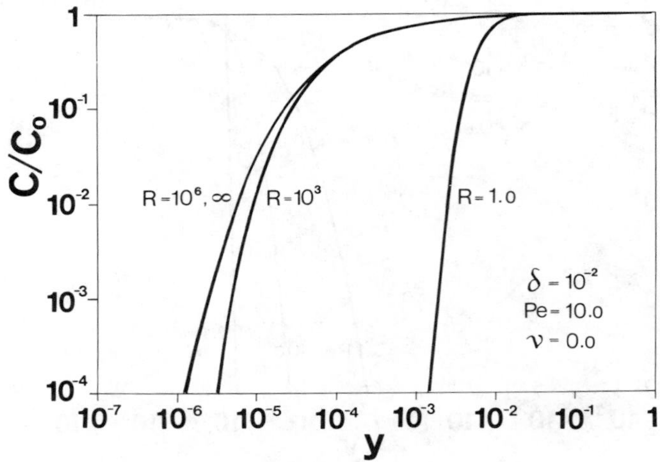

Figure 4: Breakthrough curves for fixed-bed.

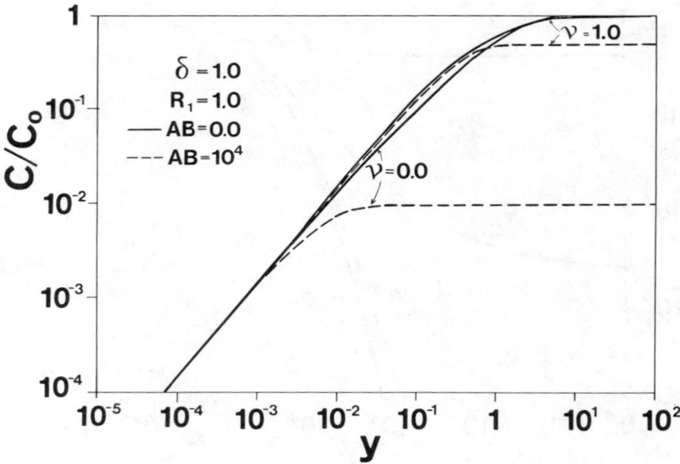

Figure 5: Breakthrough curves for continuous adsorber.

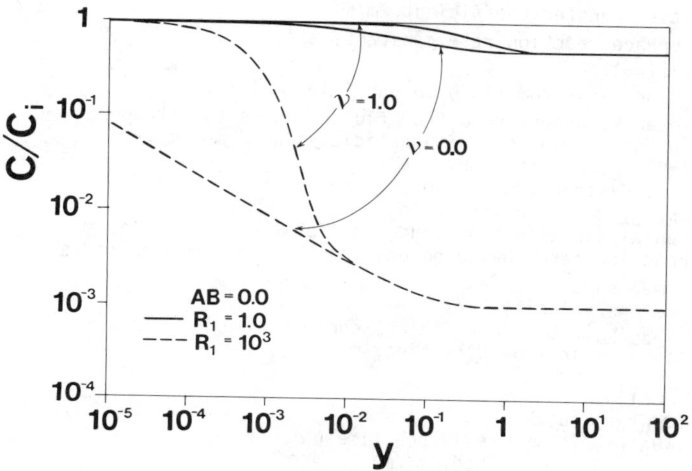

Figure 6: Concentration decay in batch adsorber.

NOTATION

A = $(b^2 k_{ads})/(D_p \varepsilon_p)$
B = $(K_A k_r)/k_{ads}$
b = particle radius, m
C = concentration in fluid, mol/m^3
C_i = initial concentration in fluid, mol/m^3
C_p = concentration in fluid in intrapores, mol/m^3
C_s = concentration in solid material, mol/m^3
C_o = inlet concentration in fluid, mol/m^3
C_∞ = steady-state value of C, mol/m^3
D_L = longitudinal dispersion coefficient, m^2/s
D_p = diffusivity in fluid in intrapores, m^2/s
G = see Equation (11)
H = see Equation (11)
h = see Equation (11)
K = $K_A + \varepsilon_p$, volume equilibrium constant, m^3/m^3
K_A = adsorption equilibrium constant, m^3/m^3
k_{ads} = adsorption rate constant, s^{-1}

k_f = mass transfer coefficient, m/s
k_r = surface reaction rate constant, s^{-1}
m = $\varepsilon/(1-\varepsilon)$
N_o = molar flux from fluid to particle, $mol/m^2,s$
N_i = molar transport rate from fluid in pore to intrapore surface, based on unit particle volume, $mol/m^3,s$
Pe = $(zV)/D_L$, Peclet number
R_1 = K/m, distribution ratio
R_2 = K_A/ε_p, distribution ratio
r = radial distance from center of spherical particle, m
r_c = reaction rate, based on unit particle volume, $mol/m^3,s$
t = time, s
V = average velocity, m/s
y = $(k_{ads}/K_A)t$ or $(2D_p\varepsilon_p/Kb^2)t$, contact time parameter
z = distance in flow direction, m

Greek symbols
δ = $(3D_p\varepsilon_p z)/(b^2 mV)$ or $(3D_p\varepsilon_p\tau)/(b^2 m)$, residence time parameter
ε = void fraction of bed, m^3/m^3
ε_p = void fraction of particle, m^3/m^3
λ = variable of integration
τ = residence time in slurry adsorber without particles, s
ν = $(D_p\varepsilon_p)/(k_f b)$, film resistance parameter

LITERATURE CITED

1. Ramachandran, P.A. and J.M. Smith, Ind. Engng. Chem. Fundam. 17, 148 (1978).
2. Rasmuson, A., AIChE J. 27, 1032 (1981).
3. Rosen, J.B., J. Chem. Phys. 20, 387 (1952).
4. Rosen, J.B., Ind. Engng. Chem. 46, 1590, (1954).
5. Masamune, S. and J.M. Smith, Ind. Engng. Chem. Fundam. 3, 179 (1964).
6. Masamune, S. and J.M. Smith, AIChE J. 11, 34 (1965).
7. Lapidus, L. and N.R. Amundson, J. Phys. Chem. 56, 984 (1952).
8. Smith, N.L. and N.R. Amundson, Ind. Engng. Chem. 43, 2156 (1951).
9. Huang, T.-C. and K.-Y. Li, Ind. Engng. Chem. Fundam. 12, 50 (1973).
10. Crank, J., The Mathematics of Diffusion, 2nd ed. Oxford University Press (1975).
11. Datta, R., B. Croes and R.G. Rinker, Chem. Engng. Sci. 38, 885 (1983).
12. Hatton, T.A., A.S. Chiang, P.T. Noble and E.N. Lightfoot, Chem. Engng. Sci. 34, 1339 (1979).
13. Rasmuson, A. and I. Neretnieks, AIChE J. 26, 686 (1980).
14. Rasmuson, A., Chem. Engng. Sci. 37, 411 (1982).
15. Rasmuson, A., Submitted to Chem. Engng. Sci. (February 1983).
16. Rasmuson, A. and I. Neretnieks, J. Geophys. Res. 86, 3749 (1981)
17. Rasmuson, A., Accepted for publication in AIChE J. (August 1983).

Mixed Vapor Adsorption on Activated Carbon

P. J. Reucroft
Department of Metallurgical Engineering and Materials Science
University of Kentucky

ABSTRACT

Adsorption of mixed vapors by an activated carbon has been investigated by different procedures. Vapor mixtures include water-hydrogen cyanide, water-cyanogen chloride and cyanogen chloride-hydrogen cyanide. The results indicate that adsorption of the component with the higher affinity coefficient tends to dominate in mixed vapor situations.

- - - - -

Most treatments of mixed component adsorption have dealt with low sorption coverage, or low pore volume filling, at low relative pressure, P/P_0, where P_0 is the saturation vapor pressure. However, it is often necessary to deal with mixed component adsorption in situations where the relative pressure of one or both components is high. Assessing the effect of high humidity (high water vapor relative pressure) on the adsorption of various organic and inorganic vapor species is an example of such a situation. At low relative pressure (<0.4), water vapor adsorption by activated carbon adsorbents is low and less than 4% of the available pore volume is occupied. At high relative pressure (~0.8) more than 80% of the pore volume becomes occupied by water adsorbate. Studies have been carried out on the adsorption of vapor mixtures such as water-cyanogen chloride, water-hydrogen cyanide and cyanogen chloride-hydrogen cyanide by activated carbon in order to evaluate the effect of water vapor at high relative pressures on the single vapor adsorption characteristics of cyanogen chloride and hydrogen cyanide. Another objective has been to assess the role of the single vapor isotherm determined affinity coefficient in mixed vapor adsorption. Three different procedures were utilized in the investigation.

In the first procedure, a vapor mixture consisting of hydrogen cyanide and water was employed as an adsorbate and the carbon was simultaneously exposed to the two components and allowed to equilibrate. The ratio of the two components was kept constant and the total equilibrium pressure was varied until the hydrogen cyanide equilibrium relative pressure was 0.15 and the water equilibrium relative pressure was 0.8. In a second set of experiments vapor mixtures were employed in which the equilibrium water vapor pressure was kept constant (at P/P_0 = 0.73) and the equilibrium hydrogen cyanide vapor pressure was varied (up to P/P_0 = 0.25). The adsorbent was again simultaneously exposed to the two components in the second procedure.

The third procedure consisted of pre-adsorbing the first component and introducing the second adsorbate component without disturbing the equilibrium established between the adsorbent and the first component.

EXPERIMENTAL

Adsorption isotherms were determined with a Cahn RG electrobalance/gas handling system which has been described previously (1-3). The three procedures employed in the mixed vapor adsorption studies are outlined in the foregoing and described in more detail in (2). The activated carbon adsorbent was a Calgon activated carbon, Type BPL, 12-30 mesh, which has been described previously (1,4). The carbon has an internal surface area of ~1000 m^2/g with approximately 70-75% of the internal surface area associated with pores less than 20Å in diameter. Hydrogen cyanide (HCN) was obtained from Formica, Inc. with a purity of 99.5%. Cyanogen Chloride (CNCl) was obtained from Scientific Gas Products with a minimum purity of 99%. The vapors were further purified by vacuum distillation until the vapor pressure at dry-ice temperature agreed with the literature values (5). All adsorption measurements were carried out at room temperature (298°K).

RESULTS

Single Vapor Adsorption

Single vapor adsorption isotherm data are shown in Figure 1 for hydrogen cyanide, cyanogen chloride and water vapor on BPL activated carbon (2,6). The isotherm data are plotted as log W versus ε^2, where W is the volume of adsorbate per gram of adsorbent at equilibrium vapor pressure P and ε is given by:

$$\varepsilon = RT\,\ell n(P/P_o) \tag{1}$$

where R is the gas constant and P_o is the saturated vapor pressure of adsorbate liquid at T°K. Linear (high pressure) regions of the single vapor isotherm can be analyzed to yield an affinity coefficient (β) by employing a reference isotherm and applying the Dubinin-Polanyi (Dubinin-Radushkevich) equation (1,6):

$$\log W = \log W_o - k\varepsilon^2 \tag{2}$$

where W_o is the maximum pore volume per gram of adsorbent and:

$$\beta = (k_{REF}/k)^{1/2} \tag{3}$$

k and k_{REF} refer to the isotherm slope and reference isotherm slope, respectively. β values determined from the isotherms shown in Figure 1 are in the sequence $\beta_{CNCL} > \beta_{HCN} > \beta_{H_2O}$ and are listed in Table 1 (reference vapor chloroform).

Mixed Vapor Adsorption (Simultaneous Exposure)

Figure 2 shows isotherm data obtained by exposing the carbon to hydrogen cyanide/water constant ratio mixtures employing procedure 1. The results are shown as grams per gram of total adsorbate as a function of hydrogen cyanide relative pressure and are compared in the same figure with the single vapor hydrogen cyanide isotherm. The single

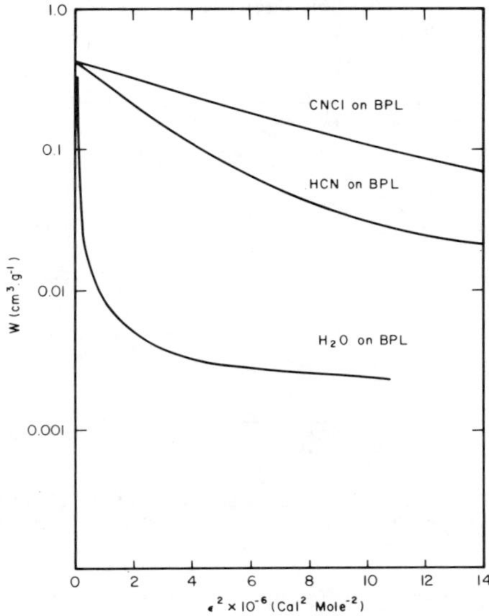

Fig. 1 Single vapor isotherms on BPL activated carbons.

Table 1. Adsorption Parameters for Single and Mixed Vapor Adsorption on BPL Activated Carbon.

	W_o $(cm^3.g.^{-1})$	W_{total} $(cm^3.g.^{-1})$	k $(cal.^{-2}mole^2x10^6)$	β^*
HCN on BPL	0.42	0.42	0.15	0.41
HCN on H_2O/BPL	0.14	0.50	0.07	0.61
HCN on CNCl/BPL	0.052	0.41	0.14	0.38
H_2O on HCN/BPL	0.016	0.35	0.20	0.35
CNCl on BPL	0.42	0.42	0.06	0.64
CNCl on H_2O/BPL	0.20	0.53	0.07	0.60
CNCl on HCN/BPL	0.25	0.55	0.19	0.36
H_2O on CNCl/BPL	0.008	0.34	0.30	0.29
H_2O on BPL	0.42	0.42	5.10	0.07

*Reference vapor $CHCl_3$; $k_{REF} = 2.5 \times 10^{-8}$ cal.$^{-2}$ mole^2

vapor isotherm is also plotted as grams per gm of adsorbate as a function of hydrogen cyanide relative pressure. The adsorption level for the mixed vapor isotherm is only slightly greater than the adsorption level for the single vapor hydrogen cyanide isotherm, indicating that the adsorption of water under these conditions represents a small fraction of the total amount adsorbed.

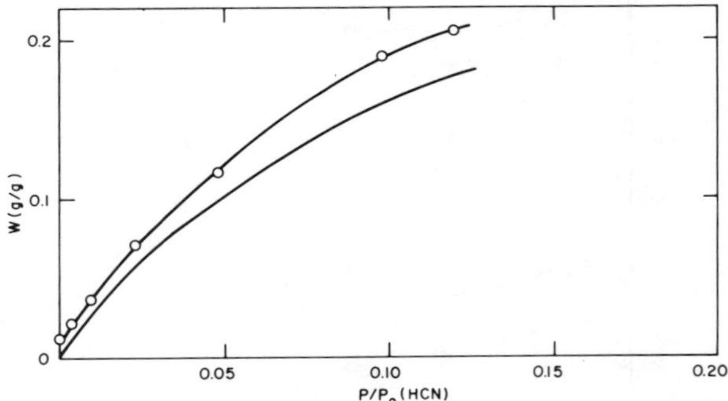

Fig. 2 HCN/H_2O adsorption on BPL activated carbon with a constant H_2O to HCN relative pressure ratio. -O-, HCN/H_2O on BPL; ———, HCN on BPL.

Isotherm data obtained by procedure 2 are shown in Figure 3. At zero hydrogen cyanide relative pressure, the adsorbate is pure water vapor at P/P_o = 0.73 and 0.32 grams per gram is adsorbed. As the hydrogen cyanide relative pressure increases, the adsorption level first decreases and then increases as the relative pressures is further increased. At P/P_o (HCN) 0.05, the mixed vapor isotherm approximates to the single vapor hydrogen cyanide isotherm which is also shown in Figure 3 for comparison. The results again indicate that adsorption of water vapor accounts for only a small fraction of the total adsorbate at the high hydrogen cyanide relative pressures even though the relative pressure of water vapor is relatively high.

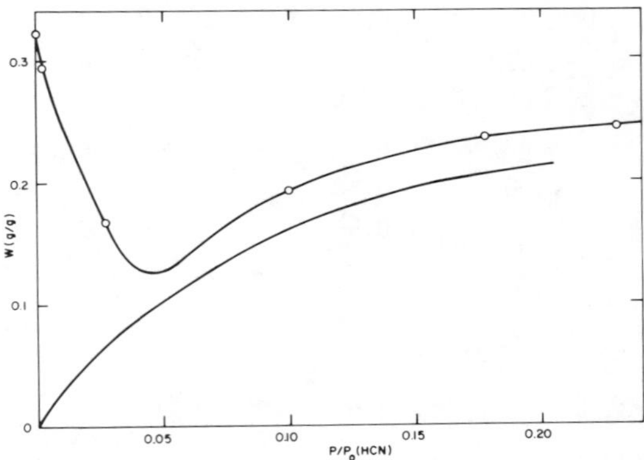

Fig. 3 HCN/H_2O adsorption on BPL activated carbon with a constant H_2O relative pressure. -O-, HCN/H_2O on BPL; ———, HCN on BPL.

Mixed Vapor Adsorption (Consecutive Exposure)

In the third procedure, the adsorbent was initially equilibrated with the first component at a vapor pressure which caused approximately three quarters of the pore volume to be filled. By suitable manipulation of the gas storage flasks and various valves, the vapor pressure of the first component was kept constant while the second vapor was introduced into the system. The sample mass and adsorbate pressure were then allowed to equilibrate. Additional doses of the second component and successive equilibrations allowed compilation of an isotherm. At equilibrium, the increase in pressure was taken to be the equilibrium pressure of the second vapor and the increase in the amount adsorbed was assumed to be associated with the second adsorbate.

Isotherms showing the adsorption of hydrogen cyanide by BPL activated carbon in equilibrium with cyanogen chloride or water are compared with the single vapor hydrogen cyanide isotherm in Figure 4. The same isotherms plotted as log W versus ε^2 are shown in Figure 5. Figure 5 also includes an isotherm which shows the adsorption of water vapor on BPL activated carbon which was first equilibrated with 0.33 $cm^3 g^{-1}$ of hydrogen cyanide. The results show that the adsorption of hydrogen cyanide vapor is generally reduced at all equilibrium pressures from the single vapor isotherm values when the adsorbent is first equilibrated with another adsorbate. Pre-adsorption of cyanogen chloride has a greater effect then pre-adsorption of water in reducing subsequent adsorption of hydrogen cyanide. Pre-adsorption of hydrogen cyanide reduces subsequent adsorption of water vapor to negligible levels.

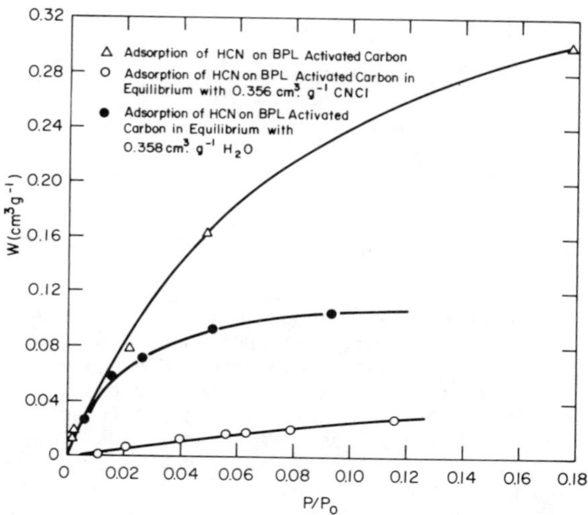

Fig. 4 Adsorption of HCN on BPL activated carbon systems.

Isotherms showing the adsorption of cyanogen chloride by BPL activated carbon in equilibrium with hydrogen cyanide or water are compared with the single vapor cyanogen chloride isotherm in Figure 6. The isotherms plotted as Log W versus ε^2 are shown in Figure 7. Figure 7 again includes an isotherm which shows the adsorption of water vapor on BPL activated carbon which was first equilibrated with 0.33 $cm^3 g^{-1}$ of cyanogen chloride. The results are similar to those obtained for hydrogen cyanide adsorption (Figures 4 and 5) and indicate that pre-adsorption generally reduces subsequent adsorption of cyanogen chloride below the single vapor isotherm value. Pre-adsorption of hydrogen cyanide has a greater effect than pre-adsorption of water in reducing subsequent adsorption of cyanogen chloride. Pre-adsorption of cyanogen chloride reduced subsequent adsorption of water vapor to negligible levels.

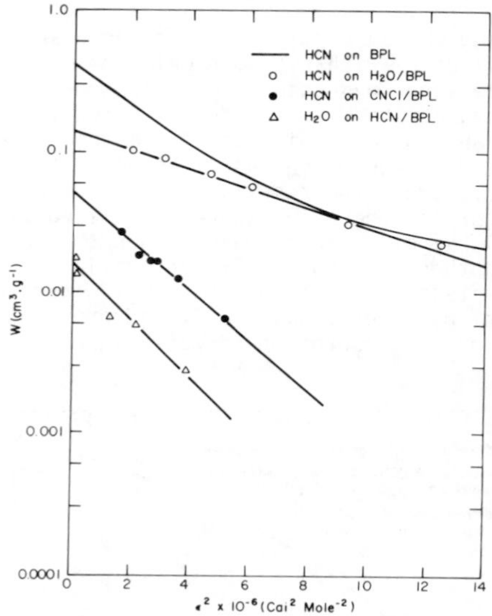

Fig. 5 Adsorption of HCN and H_2O on BPL activated carbon systems (DP plots).

The equilibrium adsorption of cyanogen chloride on HCN/BPL is generally greater than the equilibrium adsorption of hydrogen cyanide on CNCl/BPL (Figure 8). The amount of pre-adsorbed cyanogen chloride (0.356 $cm^3 g^{-1}$) is slightly greater than the amount of pre-adsorbed hydrogen cyanide (0.306 $cm^3 g^{-1}$), and this may account for part of this difference, since presumably less pore space is available for the subsequent adsorption of hydrogen cyanide. However, the increased adsorption of cyanogen chloride cannot be completely accounted for in these terms. The greater level of cyanogen chloride adsorption is most likely due to the higher affinity coefficient of cyanogen chloride compared to hydrogen cyanide when chloroform is employed as a reference vapor (6).

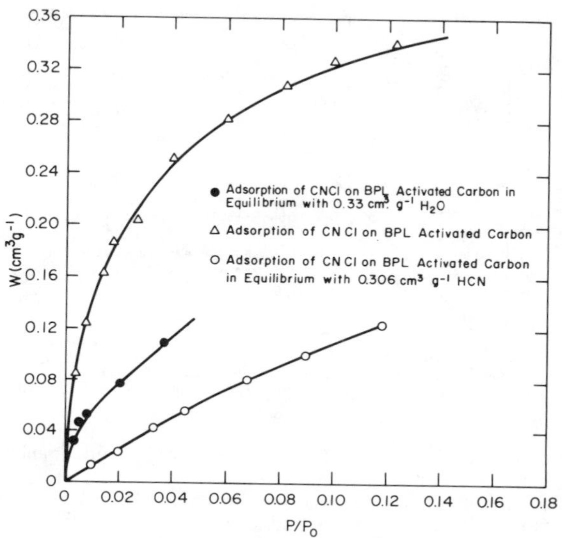

Fig. 6 Adsorption of CNCl on BPL activated carbon systems.

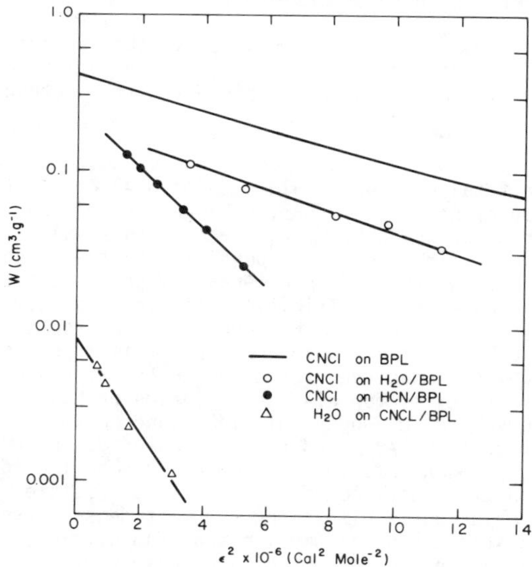

Fig. 7 Adsorption of CNCl and H_2O on BPL activated carbon systems (DP plots).

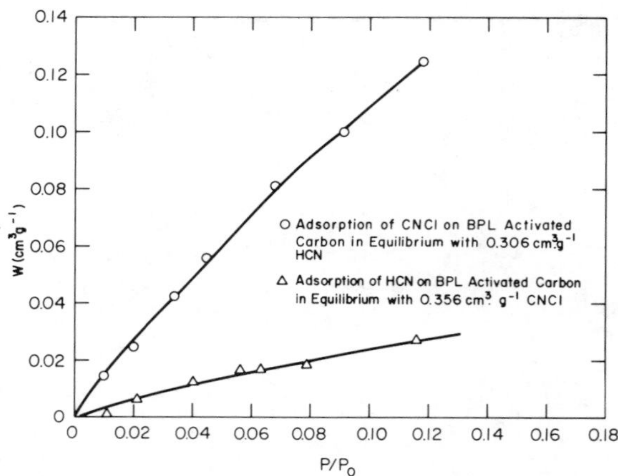

Fig. 8 Adsorption of HCN and CNCl on BPL activated carbon systems.

It is of interest to determine the parameters W_0 and k from Equation 2 for hydrogen cyanide and cyanogen chloride vapors adsorbed on carbons which have been pre-exposed to a second vapor. The results are summarized in Table 1. Also included are experimental affinity coefficients referenced to chloroform vapor.

k_{REF} was taken from previous studies on the adsorption of chloroform on BPL activated carbon (1). W_{total} is the volume of pre-adsorbed vapor plus the W_0 value obtained from the second vapor isotherm.

The adsorption of hydrogen cyanide and cyanogen chloride on BPL determines the total micropore volume to be 0.42 cm^3g^{-1} (6). In three cases, hydrogen cyanide on H_2O/BPL, cyanogen chloride on H_2O/BPL and cyanogen chloride on HCN/BPL, W_{total} exceeds this value by approximately 0.1 cm^3g^{-1}. This suggests, possibly, that enhanced affinity for the adsorbate is produced when certain vapors are pre-adsorbed. However, an increase in the affinity coefficient is only observed in the case of hydrogen cyanide on H_2O/BPL. In the case of cyanogen chloride on H_2O/BPL, the affinity coefficient is approximately the same as that which is observed for cyanogen chloride alone on BPL. The affinity coefficient for cyanogen chloride on HCN/BPL is lower than that observed for cyanogen chloride alone on BPL. In the case of cyanogen chloride on H_2O/BPL and cyanogen chloride on HCN/BPL, the second vapor, cyanogen chloride, has both a higher liquid state density and a higher affinity coefficient. It is possible, therefore, that cyanogen chloride is displacing the less dense first adsorbate. In these circumstances the assumption that an increase in adsorption on exposure to the second adsorbate is a measure of the amount of cyanogen chloride adsorbed is not valid and could lead to a value of W_{total} which is greater than W_0. In the case of hydrogen cyanide on

H_2O/BPL, however, the second vapor has a lower liquid density and the higher W_{total} value cannot be explained in this way. A more detailed chemical analysis of the adsorbate is required in order to resolve these questions.

When hydrogen cyanide is adsorbed on CNCl/BPL, W_{total} is similar to the W_0 value which is obtained when hydrogen cyanide is adsorbed alone. This suggests that the second adsorbate is filling the pore volume which remains after first pre-adsorbing cyanogen chloride. The affinity coefficient in this case is very similar to the affinity coefficient which is obtained for hydrogen cyanide alone on BPL.

The observations on the W_{total} values obtained for the two systems, cyanogen chloride on HCN/BPL and hydrogen cyanide on CNCl/BPL, suggest that cyanogen chloride is displacing hydrogen chloride but that hydrogen cyanide is not displacing cyanogen chloride. Such behavior is consistant with the greater affinity coefficient observed for cyanogen chloride compared to hydrogen cyanide in single vapor adsorption studies (6) and is thus not unexpected. The results which indicate that cyanogen chloride is displacing water are also consistent with the greater affinity coefficient of cyanogen chloride compared to the affinity coefficient (0.07) which can be estimated from the high pressure region of the single vapor water adsorption isotherm obtained previously (2).

When water is adsorbed on HCN/BPL or CNCl/BPL, the adsorption level is low in comparison to single vapor H_2O adsorption (2). W_{total} is much lower than the W_0 value which can be obtained from the single vapor adsorption isotherms. The affinity coefficient appears to be higher than the affinity coefficient which can be estimated from the high pressure region of the single vapor water adsorption isotherm (see above). However, the results indicate that previous adsorption of cyanogen chloride or hydrogen cyanide in BPL strongly discourages additional adsorption of water vapor.

The affinity coefficients are generally lower in the mixed vapor systems in comparison with the single vapor affinity coefficients. Hydrogen cyanide on H_2O/BPL appears to be an exception in that the affinity coefficient is higher than the value observed for hydrogen cyanide alone on BPL activated carbon. The higher affinity coefficient taken with the high W_{total} value may thus be evidence for an added interaction between the hydrogen cyanide adsorbate and the pre-adsorbed water in which the hydrogen cyanide dissolves in the adsorbate water phase. The affinity coefficients for water on CNCl/BPL and HCN/BPL are apparently higher than the affinity coefficient for water on BPL activated carbon alone. The single vapor isotherm for water vapor shows significant departures from a linear plot, however, and the determination of k is more uncertain.

In summary, interpretation and analysis of multicomponent vapor adsorption is generally more complicated than adsorption of single component vapors. The results obtained on several mixed vapor systems indicate that the Dubinin-Polanyi concepts of pore volume filling and affinity coefficient have some utility in analyzing mixed vapor adsorption. However, more detailed analysis of the mixed adsorbate and the vapor mixture in equilibrium with the adsorbate is needed in order to refine these concepts and apply them more effectively to mixed vapor adsorption.

CONCLUSIONS AND SIGNIFICANCE

Mixed vapor adsorption studies on an activated carbon adsorbent by several procedures have shown that (a) appreciable adsorption of hydrogen cyanide or cyanogen chloride occurs when the adsorbent is in equilibrium with relatively high levels of water vapor, (b) pre-adsorbed hydrogen cyanide or cyanogen chloride inhibits additional adsorption of water vapor, (c) adsorption of hydrogen cyanide strongly dominates in any process involving the simultaneous adsorption of hydrogen cyanide/water vapor mixtures, (d) adsorption of cyanogen chloride by an adsorbent which has been pre-equilibrated with hydrogen cyanide is generally greater than the adsorption of hydrogen cyanide by an adsorbent which has been pre-equilibrated with cyanogen chloride.

The overall results indicate that in mixed vapor adsorption situations the component with the higher affinity coefficient tends to dominate. This conclusion is less clear-cut in the case of water vapor since an affinity coefficient cannot be uniquely defined from the single vapor isotherm. Single vapor isotherm results indicate, however, that affinity coefficients obtained from high pressure regions of the isotherm and referenced to chloroform decrease in the sequence cyanogen chloride > hydrogen cyanide > water. The results also indicate that cyanogen chloride displaces pre-adsorbed hydrogen cyanide and water to some extent but that hydrogen cyanide fills up the pore volume which remains after first pre-adsorbing cyanogen chloride.

LITERATURE CITED

1. C. T. Chiou and P. J. Reucroft, Carbon, 15, 49 (1977).

2. G. B. Freeman and P. J. Reucroft, Carbon, 17, 313 (1977).

3. P. J. Reucroft, P. B. Rao and G. B. Freeman, Carbon, 21, 171 (1983).

4. P. J. Reucroft, W. H. Simpson and L. A. Jonas, J. Phys. Chem. 75, 3526 (1971).

5. Handbook of Chemistry and Physics, 56th Edn., Chemical Rubber Co. (1975).

6. P. J. Reucroft and C. T. Chiou, Carbon, 15, 285 (1977).

ADSORPTIVE FRACTIONATION: CYCLIC RECYCLE CASCADES

by

Richard G. Rice
Department of Chemical Engineering
Louisiana State University
Baton Rouge, Louisiana 70803
United States of America

ABSTRACT

Countercurrent behavior can be accomplished in sorption cascades by manipulating internal recycle loops, without recourse to attrition-prone moving bed operation. By incorporating internal recycle along with a pseudo (non-linear) equilibrium curve we show that simple and classical design procedures evolve which are strictly analogous to the well known McCabe-Thiele method. The structure and important idealizations for the binary process called Adsorptive Distillation are extended for multi-component systems.

BACKGROUND AND QUALITATIVE ARGUEMENTS

In the middle sixties, Wilhelm and co-workers (1) published details of a cyclic separation process called Parametric Pumping (PP). Indeed, parameters were not found to be pumpable, as the unfortunate name implied. The gist of the early work on PP showed how the difference between two adsorption isotherms at two slightly different temperatures could be exploited for the separation of chemical species. The early work dealt almost exclusively with the transient nature of batch, closed systems. Sometime later (2), rigorous analysis led to prediction of the ultimate or limiting separation, which is the final steady-state for batch systems. In this latter review article, and the papers described therein, the physical requirements for a separation to exist are delineated, and these include firstly, selectivity of the species to be separated on a given sorbent, and secondly, on sensitivity to some intensive thermodyanmic variable (such as temperature) so that a retard-release (adsorption-desorption) mechanism can be triggered by manipulating this intensive variable. The sensitivity requirement is often difficult to achieve in practical systems owing to very small shifts in the equilibrium isotherm as temperature is varied. We refer to such systems as insensitive. Enhanced sensitivity usually occurs for sorbate-sorbent systems with large heats of adsorption, since Gregg and Sing (3) show that Henry's law constant for gas adsorption depends as $\sim (1/T) \exp(E/RT)$ where E is the differential heat of adsorption. In fact, it is the differential change in the Henry parameter that gave rise to the name Parametric Pumping, since the cyclic variation of the Henry parameter causes the species to separate when coupled with a synchronous velocity variation. In short, species is retarded (adsorbed) when flow is upward, and released (desorbed) when flow is downward with the net effect of species migration toward a reservoir attached to the bottom of the traditional packed column arrangement.

The probability of the occurrence of both selectivity and sensitivity for industrial processing would seem to be quite low. In the present work, we propose a three phase separation scheme, whereby liquid phase adsorption with high selectivity (but low sensitivity in the liquid phase) is followed by regeneration into a vapor phase (or alternately, regeneration by supercritical gas solvents such as carbon dioxide). Thus the sensitivity problem can be overcome by simply vaporizing the sorbed species by application of heat, vacuum or extraction via supercritical gas solvents. This leads to the condition that sorbent choice can be based mainly on species selectivity, with sensitivity playing a minor role.

The basis of the separation mode we have called Adsorptive Distillation (4) builds on the four "mini-cyles" depicted in Figure 1a. Focusing on any stage "n" of the cascade, one sees the countercurrent nature of the process as follows:

(1) Liquid equilibration cycle: liquid charged to stage "n" is equilibrated with a certain mass of solid, M_n, which thereby extracts the selective species from the bulk liquid.

(2) Drain Cycle: lean solution is drained away from the stationary solid and passed downstream as part of the feed to stage "n-1".

(3) Vaporize Cycle: the rich "adsorbed" solution remaining is next removed by vaporization or supercritical gas extraction thereby producing a rich "overflow" which is passed upstream as partial feed to stage "n+1".

(4) Charge Cycle: the total cycle sequence begins anew after combining lean liquid from stage "n+1" with rich condensed vapor from stage "n-1" to produce the feed to stage "n". This specific sequence of course leads to internal recycle. The four "mini-cycles" comprise a total cycle for stage "n". All other stages are thus required to follow the same total cycle time, unless additional off-line storage tanks of variable capacity are added to the process.

One sees by inspection of Figure 1b and 1c that the overall effect of the above sequence of events is to produce a countercurrent recycle battery with the unique feature that, generally, departing streams are not in thermal equilibrium. In the idealized model that follows, we shall take the solid phase to be completely voided at the end of the "vaporize" cycle.

Idealized Model Predictions

A major objective of the exercise is to produce a simplified conceptual framework which can be used to delineate the main effects owing to design variables such as stage-to-stage solid loading, reflux ratio and feed composition and location. As we shall see, under not too restrictive conditions, an exact analog of the classical McCabe-Thiele analysis results.

The material balance around stage "n" under steady cyclic conditions is seen from Figure 1b to be:

$$V_{n-1} + L_{n+1} = L_n + V_n \quad (1)$$

and for any component in a multi-component system is:

$$V_{n-1}\, y_{n-1} + L_{n+1}\, x_{n+1} = L_n x_n + V_n y_n \quad (2)$$

where as usual x_n, y_n represent liquid and vapor material fractions, respectively, departing stage "n".

The residual or accumulation which remains following the drain cycle is given by, for total material

$$V_{n-1} + L_{n+1} - L_n = M_n\,(H_s + H_p) \quad (3)$$

and for any component:

$$V_{n-1}\, y_{n-1} + L_{n+1}\, x_{n+1} - L_n x_n = M_n\,(H_s z_n + H_p x_n) \quad (4)$$

where M_n represents mass (dry) of solids on stage "n" and z_n is the adsorbed phase material fraction for any species (usually the desired or selective species) on stage "n". The sorbent "capacity" or holdup is represented by H_s (e.g. moles per gram of dry solid) and H_p is the unadsorbed pore capacity which can also account for drained solids fluid; both of these latter quantities represent the liquid phase which is, ultimately, in equilibrium with the solid phase at the fixed (low) equilibration temperature.

Next, the residual or accumulated liquid undergoes a thermodynamic phase change, by raising the temperature and lowering the chamber pressure for example, until all material is removed from the solid and interstices, hence

$$V_n = M_n\,(H_s + H_p) \quad (5)$$

and for any component

$$V_n\, y_n = M_n\,(H_s z_n + H_p x_n) \quad (6)$$

We have taken the capacity or holdup terms (H_s, H_p) to be invariant from stage-to-stage. This restriction could be relaxed, for instance, by allowing different values in sections (stripping and enriching).

The equilibration cycle allows sufficient time so that thermodynamic phase equilibrium is approached between components in the liquid and solid phases. If the equilibrium isotherms can be taken to be mutually independent, or in any case if one considers only binary systems, then

$$z_n = f(x_n) \tag{7}$$

which implies that the species solid phase composition depends only on the species liquid phase composition. A specific structure used by Shelstad and Wong (5) for binary liquid adsorption of C_8 isomers gave

$$z_n = \frac{\alpha x_n}{1+(\alpha-1)x_n} \tag{8}$$

which is of the Langmuir type. Here, α represents the selectivity for the strongest species, and for binary systems is simply the ratio of the pure component Henry constants.

Combining the regeneration balances, equations (5) and (6) and replacing z_n with the equilibrium relationship given by equation (7), gives

$$y_n = \varepsilon_s\ f(x_n) + \varepsilon_p x_n \tag{9}$$

which is the sought after connection between departing streams; this is the pseudo equilibrium relationship alluded to earlier. We have denoted the dimensionless holdup ratios as $\varepsilon_s = H_s/(H_s + H_p)$ and $\varepsilon_p = H_p/(H_s + H_p)$. For binary systems, one can see the well-behaved properties $y_n \to 0$ as $x_n \to 0$ and $y_n \to 1$ as $x_n \to 1$. Selectivity enters via the isotherm function $f(x_n)$, according to structures such as equation (8) for instance. Note here that holdup plays an important role in determining the maximum driving force for mass transfer and this role enters the calculations via the pseudo equilibrium curve (eqn 9) through the hold-up parameters ε_s, ε_p. Large driving forces are exerted for large α and large ε_s.

Further simplifications can result if we rearrange equation (1), coupled with the voiding or regeneration statement given by equation (5), hence

$$L_{n+1} - L_n = (M_n - M_{n-1})[H_s + H_p] \tag{10}$$

The equivalent to "equimolal overflow" can be realized by taking the assumption of invariant solids loading, so that $M_n = M_{n-1} = M_{n-2}$ etc. for a given section of the cascade. Thus, for the stripping section we take

$$M_n = M_{n-1} = \bar{M} \tag{11}$$

and later for the enriching section, take

$$M_m = M_{m+1} = M \tag{12}$$

Any number of sections may be specified and each will contribute a unique operating line, as we shall see.

For the usual case of a two-section cascade separated by the feed stage, an overall balance between any stripping stage "n" and the lean end of the cascade (Figure 1c) gives

$$L_{n+1} = V_n + B \tag{13}$$

$$L_{n+1}\, x_{n+1} = V_n y_n + Bx_B \tag{14}$$

and since vapor flows are invariant if solids loading does not change, then a straight operating line results, since L_{n+1} (= $\bar{L}$) is invariant, then according to equation (13), V_n is also constant

$$y_n = (\bar{L}/\bar{V})\, x_{n+1} - (B/\bar{V})\, x_B \tag{15}$$

and similarly for the enriching section

$$y_{m-1} = (L/V)\, x_m + (D/V)\, x_D \tag{16}$$

where D denotes overhead or distillate product rate. The material balance at the top of the cascade is simply V = L + D, hence the slope in equation (16) can expressed, in the usual fashion, as $R_T/(1+R_T)$ where $R_T = L/D$ is the top reflux ratio.

q-line Determination So far, we see that the structure of the system equations follows the McCabe-Thiele analysis exactly, and as usual the locus of intersections of the enriching and stripping lines is given by

$$y = \frac{q}{q-1}\, x - \frac{1}{q-1}\, x_F \tag{17}$$

where

$$q = \frac{\bar{L}-L}{F} \tag{18}$$

In the McCabe-Thiele method, q is related to the thermal condition of the feed. The parallel for the present system is not so obvious. However, we take note that the solids ratio is a system constant

$$\frac{\bar{M}}{M} = \frac{\bar{V}}{V} = R_M \tag{19}$$

and we exploit this by writing a material balance around the feed stage to give

$$\frac{\bar{V}-V}{F} = q - 1$$

and combining this with the top separator balance, V = L + D, leads to

$$q = \frac{(R_M-1)(R_T+1)}{(F/D)} + 1 \tag{20}$$

If we specify equal solids loading throughout the cascade (i.e. R_M=1), then the q-line is vertical; in the classical distillation sense, this would be equivalent to saturated liquid feed.

For the simple binary system, calculations can proceed stage-to-stage using simple graphical methods as shown in Figure 2, which applies to the separation of a 50:50 isomeric mixture of xylenes (ortho is the most selective in a mixture with meta) into a 95:5 split. Selectivity and holdup parameters were estimated using the data of Shelstand and Wong (5) who gave $\alpha \simeq 5$, sorbed phase holdup of 0.362 mℓ/g and total pore volume of 0.586 mℓ/gram. From these we compute $\varepsilon_s = .362/.586 = 0.618$ hence $\varepsilon_d = 0.382$. The graphical results indicate the number of stages and the feed stage location for a reflux ratio twice the minimum, taking equal solids loading throughout (vertical q-line) with the "pinch" occuring, as usual, at the feed stage. The parameters apply only for the Columbia grade activated carbon used by Shelstad and Wong (5). We have ignored drained solids wetting, which usually amounts to around 0.1 mℓ/g. It is remarkable that only nine stages are necessary for this separation. Several hundred would be needed if traditional distillation were used. We note that if the values of ε_s, ε_d are reversed (4) nearly twice as many stages would be needed.

Extension to Multicomponent Systems. The multicomponent problem is invariably of the trial-and-error type and cannot be expressed in simple graphical terms. In the traditional distillation calculation, one begins computation from the reboiler by performing a material balances around the first stage and these are coupled with the equilibrium relationships to solve for y_o and x_1. The pressure is specified and usually fixed (but can vary to account for pressure drop) and one guesses the stage temperature until $\Sigma y_i = 1$ is satisfied. Here, the relative volatility, α_i, changes strongly with temperature and the simplest equilibrium connection between departing streams is given by Raoult's law, hence $y_i = P_i^o(T)x_i/P_T$ where the pure component vapor pressure, $P_i^o(T)$, must be known as a function of temperature, and the changing selectivity is related to $P_i^o(T)/P_T$. Calculations proceed from stage-to-stage in the usual fashion, by trial-and-error.

Computations proceed in an analogous manner for the multicomponent Adsorptive Fractionation problem. However, the usual "handle", temperature, is fixed (isothermal operation) and does not vary from stage-to-stage, but selectivity must change nonetheless. Obviously, a species selectivity must depend on the fraction of solid surface covered by other molecules. For example, in a three component system, one may use binary Henry constants, H_{ij}, to relate sorbate and solution equilibria along with an overall balance and known temperature.

The equivalence in the multicomponent trial-error adsorption problem thus requires three selectivity relations which in turn are connected to fraction of surface covered by other unlike molecules. As pointed out by Myers (6) for vapor adsorption, as coverage increases, selectivity decreases rapidly as lower energy sites become filled. However, for liquids the adsorbent is nearly always saturated or near saturation. Hence, selectivity is strictly definable only for binary mixtures, where it is simply the ratio of species Henry constants, $\alpha = (z/x)/(1-z)/(1-x) = H_{12}/H_{21}$. Here we denote the most

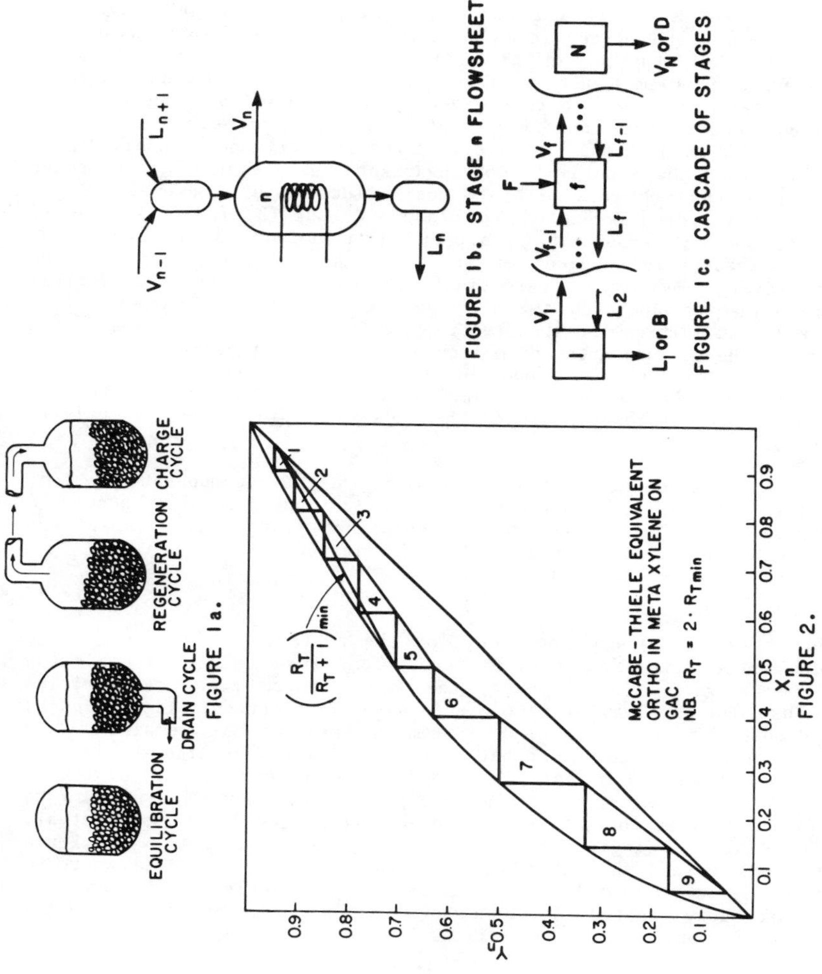

FIGURE 1a.

FIGURE 1b. STAGE n FLOWSHEET

FIGURE 1c. CASCADE OF STAGES

FIGURE 2.

strongly adsorbed with the subscript unity. For the multicomponent problem, H_{ij} should be viewed as more akin to the K-factors used in distillation. Ruthven et al. (7) have treated the multicomponent adsorption problem based on first principles and arrived at Langmuir-type relation similar to that given by equation (8) and these results suggest under what conditions the ideal-solution sorbate model of Myers and Prausnitz (8) is valid. In the present model, we have taken total sorbed phase holdup to be invariant (hence ε_s and ε_d are proper constants) and this is probably not strictly true. However, in the light of the experiments of Shelstand and Wong (5) for binary liquid adsorption on carbon of C_8 isomer mixtures and similar organics, holdup apparently changed very little with changing composition and selectivity remained essentially constant. The statistical thermodynamics approach taken by Ruthven et al. (7) suggests (for gases) that multicomponent selectivity may, in principle, be estimated from the Henry constants and molecular volumes of the pure components. However, for liquids one would expect that sorbate-sorbate interactions to be just as important as the interaction between sorbate and sorbent, since the adsorbent will be essentially saturated.

Thus we conclude that to solve the multicomponent problem, species selectivity is not a simple invariant but depends on adsorbate compositions. Trial and error calculations proceed just as for the distillation example, given the binary relations

$$z_i = H_{ij}\, x_i \tag{21}$$

and the binary Henry constants for the system. While there is no fundamental difference between adsorption from the liquid phase and from the gas phase, however, the liquid phase poses difficulties because one is always very close to saturation concentration. Hence, one must determine binary H_{ij} from experiment. It remains to be determined the extent of sorbate-sorbate interaction which are ignored in using binary data only, but the problem becomes quite complex on the theoretical level. A light key/heavy key approach may also be fruitful in this context, such as used recently by Santacesaria et al. (10) wherein a single xylene in n-octane was compared to two xylenes in n-octance and the ratio of the Langmuir constants was shown to be quite similar. Finally, we note that the variable solids case has been recently treated (10) and shows that a condition of zero reflux results for optimum solids profiles.

CONCLUSIONS

The author suggested sometime ago (2) that a McCabe-Thiele equivalent should exist for continuous or semi-continuous parapump-type processes. To accomplish this idealization exactly, we show here that each stage must be completely regenerated in a cyclic manner. Following our original suggestion, Grevillot and Tondeur (11) published a staged model of parapumping which led to a McCabe-Thiele type of graphical procedure for batch separations. At the time, however, there existed a more realistic approach (which included dissipative effects) and this was clearly pointed out to these authors by way of an exchange of letters to the editor (12). Finally, we note that Grevillot (13) published a McCabe-Thiele equivalent for open systems which gave rise to three different "partition curves" each depending on reflux ratio. While the approach taken by Grevillot was tedious, nonetheless he showed that open systems could be designed by stepping off stages in a graphical format. However, the exact analogy with the method of McCabe-Thiele was somewhat elusive, and may need further clarification. In the present effort, the door has been opened for non-uniform solids loading and this leads to some surprising results (10). Finally, we take note that the work presented here is strictly analogous to the classical McCabe-Thiele method.

SYMBOLS

B bottoms product rate per cycle

D top product rate per cycle

F feed rate per cycle

H_{ij} Henry constants for binary system

H_p pore holdup

H_s sorbed solution holdup

L_n n^{th} stage liquid underflow

M_n n^{th} stage solids loading

$\bar{M}$ solid loading in stripping stages

M solid loading in enriching stages

R_M solids ratio, $\bar{M}/M$

R_T top reflux ratio

V_n n^{th} stage vapor rate per cycle

x_n species liquid fraction for n^{th} stage

y_n species vapor fraction for n^{th} stage

z_n species sorbate fraction for n^{th} stage

Greek

α selectivity (eqn 8)

ε_p fraction pore holdup, $H_d/(H_s+H_p)$

ε_s fraction sorbate holdup, $H_s/(H_s+H_p)$

LITERATURE CITED

(1) Wilhelm, R.A., Rice, A.W., and A.R. Bendelius, Ind. Eng. Chem. Fundam. 5, 141 (1966)

(2) Rice, R.G., "Progress in Parametric Pumping", Separation and Purification Methods, 5(i), 139-188, Marcel-Dekker (1976).

(3) Gregg, S.J. and K.S.W. Sing, "Adsorption, Surface Area and Porosity", Academic Press, London (1967).

(4) Rice, R.G., Chem. Eng. Comm., 10, 111 (1981).

(5) Shelstand, K.A. and S.W. Wong, Can. Jour. Chem. Eng., 47, 66 (1969).

(6) Myers, A.L., AIChE Annual Meeting, Symp. #40, New Orleans (1981).

(7) Ruthven, D.M., Loughlin, K.F. and K.A. Holborow, Chem. Eng. Sci., 28, 701 (1973).

(8) Myers, A.L. and J.M. Prausnitz, AIChE Jour., 11 121 (1965).

(9) Santacesaria, Ello et al., Ind. Eng. Chem. Proc. Des. Dev., 21, 400 (1982).

(10) Knopf, F.C. and R.G. Rice, Chem. Eng. Comm., 15, 109 (1982).

(11) Grevillot, G. and D. Tondeur, AIChE Jour., 22, 1055 (1976).

(12) Rice, R.G., AIChE Journ., 25, 734 (1979).

(13) Grevillot, G., AIChE Journ., 26, 120 (1980).

Criteria for the Evaluation of GAC Saturation

F. Riera, P. Charles, F. Fiessinger, and J. Mallevialle

ABSTRACT

Several tests were performed on GAC using samples from various lengths of service time as well as regenerated GAC. These tests were evaluated for their ability to detect a gradually increasing level of GAC saturation. The two most promising tests were adsorption isotherms with phenol and pollutant spreading factor.

INTRODUCTION

Activated carbon treatment of drinking water has been practiced in France for several years to control taste and odor problems. The evaluation of these compounds, although subjective, is relatively rapid. The identification of undesirable organic compounds, for which GAC is also an effective treatment, at trace levels has led to costly and more time consuming analyses. This paper presents results of research performed to establish a test which could be used as a criteria for regenerating activated carbon.

Potable water treatment by activated carbon is particularly complex and effects a broad spectrum of organic compounds. These undesirable compounds are at low concentration and are accompanied by a largely uncharacterized background matrix of natural organic materials. For the purposes of this paper organics will be placed in two general concentration categories (1,2):

$$\text{Organics} > 10 \text{ ug/L}$$

$$\text{Organics} < 10 \text{ ug/L}$$

Compounds with concentrations $>$ 10 ug/L are the natural background organics, major pollutants and byproducts of the treatment process. They are often evaluated by non-specific parameters such as total organic carbon (TOC), fluorescence, ultra violet (UV) absorption and total organic halogen (TOX).

The compounds below 10 ug/L are often of synthetic origin and exhibit great variability in raw water supplies, both qualitatively and quantitatively. Their identification and quantification (3,4,5) present greater problems than materials at higher concentration, usually involving one or more concentration steps followed by chromatographic analysis.

Generally, the state of saturation of a GAC filter has been determined by the breakthrough characteristics of a non-specific parameter such as TOC. There are two successive phases for TOC breakthrough normally observed on a GAC filter. The first phase exhibits a large reduction in TOC but has a short duration. The second phase has a much lower efficiency but is maintained for a longer time. This long term low efficiency removal has been attributed to a combination of biodegradation and slow adsorption (6). This type of breakthrough curve poses many problems in explanation and planned operation of a full scale GAC filter in potable water treatment. For example:

* The breakthrough curve is a manifestation of the overall interaction of all the organics present and the GAC. Thus, the specific efficacy of a GAC for organic removal is a function of the organic matrix as well as the characteristics of the GAC.

* The extreme variability (qualitatively and quantitatively) of the organics has a definite but poorly understood effect on the phenomena of competition and reequilibration.

These two points lead to a most important question for operators of GAC filters: when should the GAC be regenerated? Choosing the point at which high efficiency removal for an easily analyzable parameter such as TOC is exhausted is not entirely satisfactory, because this parameter does not quantify the compounds of health and organoleptic concern. Conversely, the identification and quantification of the broad spectrum of undesirable compounds is costly and time consuming. The purpose of this paper is to report on several physical chemical tests which were evaluated for their potential as criteria for carbon regeneration. The goal of this research was to identify analyses of the carbon surface which exhibit a measurable change as the carbon surface becomes exhausted. A criteria of this type would have the advantage of predicting exhaustion prior to actual breakthrough of undesirable compounds.

MATERIALS AND METHODS

The carbon source in this study was a full scale GAC filter at the SLEE Morsang sur Seine treatment plant. The filter was preceded by prechlorination, clarification, sand filtration and ozonation. The GAC filter was 1.3m of Chemviron F400 operated with a contact time of 10 minutes. Carbon was sampled just prior to regeneration, just after, and at one month intervals for the next seven months. Samples from the operating filter were taken in a hollow tube and separated into three depths: top, middle and bottom.

Adsorbed TOC

A sample of carbon (15 to 20 mg) was dried to constant weight at 110 C and then introduced into an apparatus shown in Figure 1. The adsorbed total organic carbon (TOC) is desorbed into an inert atmosphere at 850 C and then burned in a second oven. The carbon dioxide liberated is measured by microcoulometry.

Adsorbed TOX

The method for total organic halogen (TOX) is similar to that for TOC. A diagram of the apparatus is shown in Figure 2. Complete details are available elsewhere (7).

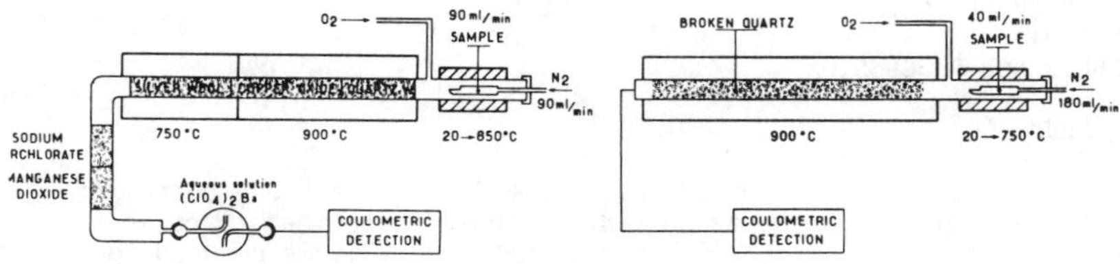

Figure 1. Apparatus to Measure Adsorbed TOC

Figure 2. Apparatus to Measure Adsorbed TOX

Broad Spectrum Analysis of Adsorbed Organics

Samples (500 mL) of wet carbon were extracted for 72 hours in a Soxhlet apparatus using two liters of dichloromethane. The dichloromethane was then concentrated to 10 mL. Identification and quantification of the extracted compounds was accomplished by coupled GC-MS. (GC conditions: splitless injection, 50m fused silica column of CP SIL 5, 50 to 250 C at 2 C/min, identification on a RIBERMAG R-10-10 at 70eV).

Adsorption Isotherms

Adsorption isotherms were developed using the frontal analysis method. The method is described in detail elsewhere (8,9). A solution of the pollutant is pumped through a minicolumn containing the carbon sample. When breakthrough is obtained, on point on the isotherm is determined. The pollutant concentration is then increased and another point on the isotherm is determined. Two compounds, representing two polarities, were used.

The method is automatic using a Waters HPLC system (two 6000A pumps, a M720 solvent programmer, a WISP injector, and an M440 UV detector). Results are obtained in a few hours. A stainless steel column is filled with the carbon sample (45 mg) after it is powdered and sieved to less than 200 mesh.

Pollutant Spreading Factor

The same mini-column system described above was used to simulate an accidental spill. The system was modified such that a spike of pollutant was injected and then eluted by the mobile phase water (instead of continuously feeding). A sample (2 mL) of 10-3 M pyridine solution was injected at 0.8 mL/min. The spike was then eluted and the shape of the breakthrough curve was used to estimate the spreading factor.

RESULTS AND DISCUSSION

Adsorbed TOC

Figure 3 shows the TOC adsorbed on GAC at the top and bottom of the filter. Data from the exhausted carbon indicate that there is approximately 35 to 40 mg TOC per g of GAC, and this is reduced to approximately 11 by regeneration, for a reduction of about 70%. Adsorption is very great during the first month after regeneration, then, desorption occurs, first in the upper layer but eventually in the bottom also. A comparison between the chemical oxygen demand (COD-KMnO4 method) of the influent water and the TOC loading shows that the reduction in TOC loading on the carbon is following a decrease in influent COD, especially during the first few months after regeneration. It seems the carbon reacts rapidly to the influent variation.

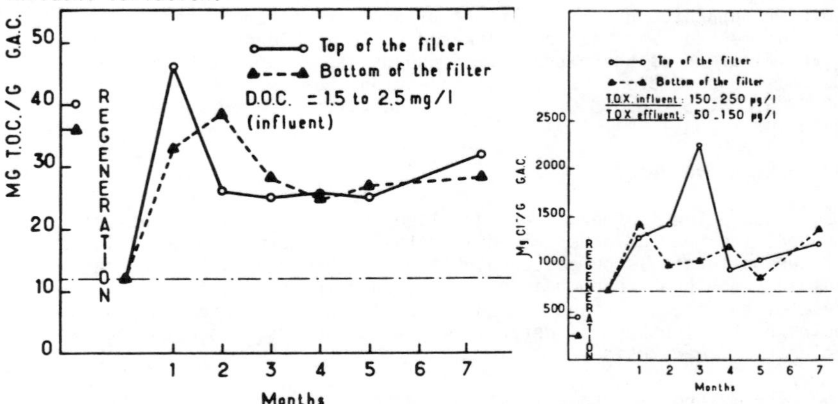

Figure 3. TOC Adsorbed on GAC

Figure 4. TOX Adsorbed on GAC

The TOC loading demonstrates a chromatographic effect when the high organic load, noted in the influent by COD, is eluted from the column. First, the load on the upper layers is reduced, then the bottom layers. During months 2 and 3, the load on the bottom actually appears higher than on the top.

It is difficult to explain the apparent constant carbon loading in the third through fifth months, during which a consistent COD removal was observed on the filters. Two hypotheses are offered:

* Organics are oxidized to CO2 by bacterial growth. (There were 10 to 10 bacteria per gram of carbon.)

* Organic compounds on the surface are exchanged with less oxidizable compounds during this period of time. This exchange could combine with bacterial oxidation by exchanging nonbiodegradable compounds in the liquid phase for biodegradable materials on the surface. Once in the liquid phase, the biodegradable compounds would be oxidized by the attached bacteria.

Adsorbed TOX

Figure 4 shows that no progressive pattern was observed for TOX, although the difference in influent and effluent indicates that partial removal consistantly occurred. The quantity of the organochlorine compounds adsorbed on GAC fluctuated during the months following regeneration. The chromatographic effect was observed for this parameter in that the values determined at the top of the filter were sometimes less than values at the bottom.

It is curious that the concentration of adsorbed TOX on the exhausted carbon is less than the amount on the regenerated carbon of relatively short usage. This may be due to displacement of the chlorinated organics by less polar compounds. Supporting evidence for greater TOX on regenerated than spent carbon is provided by GC-MS analysis. Some chlorinated compounds were found only on the regenerated carbon (e.g., trichloroacetic methylester) or in greater concentration (e.g., trichloroethylene).

Broad Spectrum Analysis

Extracts of carbon samples were analyzed by GC-MS. An example of the results is presented in Figure 5. About 50 compounds were identified including about 10 known toxic compounds (e.g., tetrachloromethylene, dichlorobenzene). The concentrations varied from the nanogram to microgram (per gram dry carbon) level. The most abundant substances were the phthalates (10 to 20 ug/L). Regeneration of the carbon eliminated most of these adsorbed organics. Unidentified high molecular weight substances were also observed in the extracts. These materials were not necessarily removed by regeneration. There are compounds present on the virgin and regenerated carbons which disappear (assumed to be displaced) from the carbon surface after a few months of operation (e.g., trichloroacetic acid methylester, triamide phosphoric hexamethyl). After one month of operation, the same compounds and same concentrations were found at the top and the bottom of the column. After two months, a reduction (by a factor of 10) was seen for a number of compounds (e.g., xylene, C3 alkylbenzene, dichlorobenzene). The greatest reduction occurred at the top. In the following months,

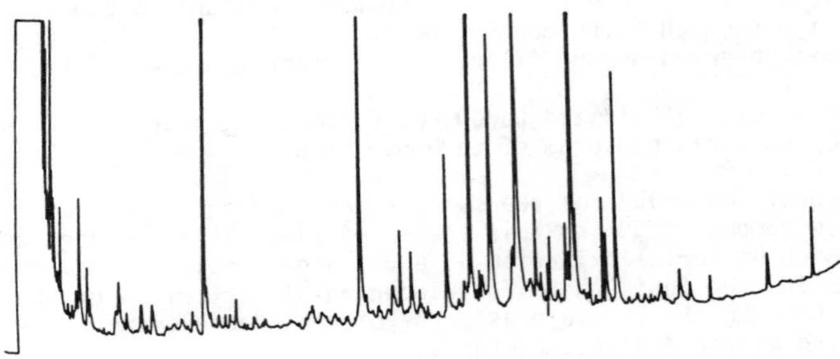

2 months after regeneration

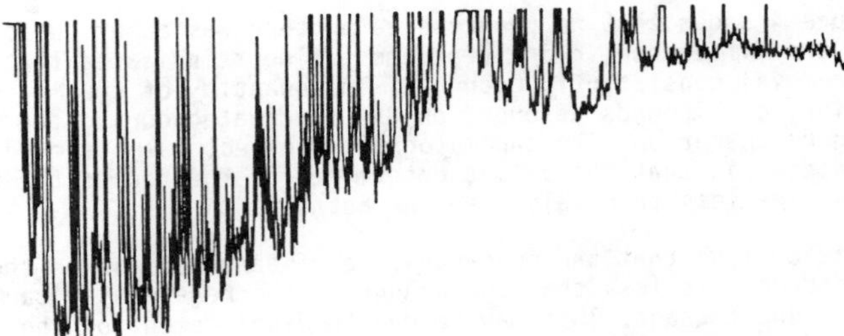

7 months after reg.

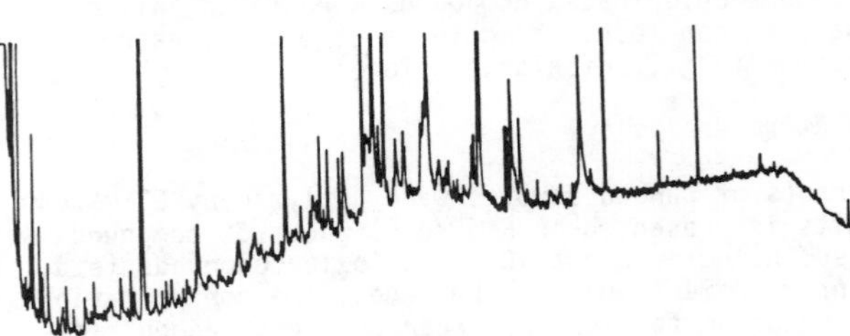

10 months after reg.

Figure 5. Broad Spectrum Chromatographic Analysis of Organic Compounds Extracted from the GAC surface after Various Lengths of Service (as shown)

adsorption and desorption phenomena were continuously observed. The adsorption and desorption observed can be attributed to the mechanisms of reequilibration, competition, influent variability and the chromatographic effect. Most of the adsorbed compounds are affected by one or more of these mechanisms, with the possible exception of compounds similar to phthalates.

Adsorption Isotherms

Adsorption isotherms were developed for two compounds: pyridine and phenol. These compounds represent polar and non-polar organics, respectively. The isotherms (Figure 6) for pyridine show a high initial capacity on regenerated carbon which rapidly declines to the capacity of the spent carbon, and not in a continuous manner. This may indicate that the compounds which compete with pyridine are appearing and disappearing over time by mechanisms of adsorption desorption described previously.

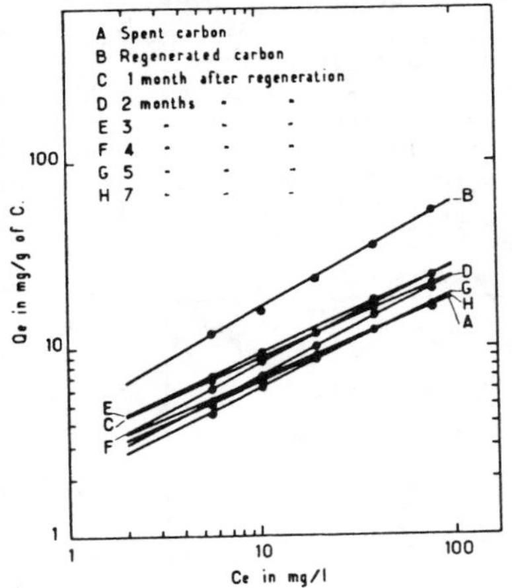

Figure 6. Pyridine Adsorption Isothermes on GAC

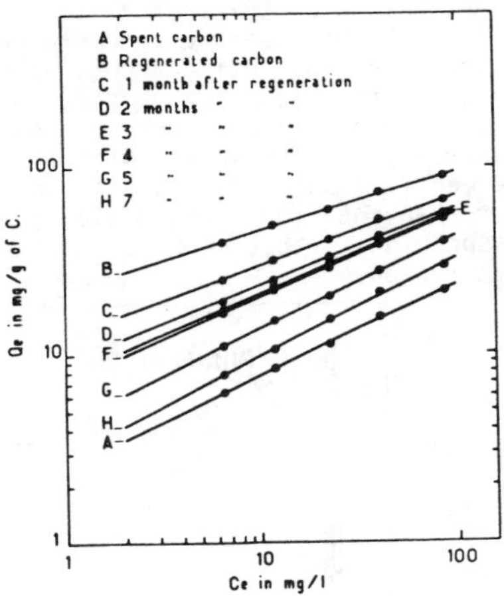

Figure 7. Phenol Adsorption Isothermes on GAC

Conversely, the phenol isotherms (Figure 7) show a continuous pattern of reduced capacity with time and bed depth. The sample of carbon after 6 months operation exhibits greater capacity than the spent carbon (6 years operation) although its capacity is less than all the preceding months. The progression is more evident at the top of the filter than at the bottom.

These results are interpreted as follows: At the onset of use, there are a large number of active sites on the carbon. Adsorption occurs for both polar and non-polar compounds. Over time, there is a gradual replacement on the GAC surface of polar molecules by non-polar molecules, due to the effects of competition. A group parameter such as adsorbed TOC does not detect this pattern because it does not distinguish between polar and non-polar organic molecules. The GC-MS procedure also does not determine the polar molecules.

Pyridine, a polar molecule, is ineffective in displacing less polar molecules from the carbon's surface. Thus, a rapid decrease in carbon's capacity for pyridine is observed. Phenol, a less polar molecule with higher affinity for carbon than pyridine, is capable of displacing some of the adsorbed molecules, but as the composition of the adsorbed organics changes, the capacity for phenol also declines. The change in phenol capacity is more detectable than that for pyridine, and does present a progession of measurements which indirectly indicates that the carbon's surface is becoming more saturated.

Pollutant Spreading Factor

An accidental spill of pyridine was simulated on two different samples of carbon (before regeneration and 2 months after - Figure 8). Over time, the entire amount of the compound elutes from the column, but the peak is spread out of a large broad peak in which the greatest concentration is only a few percent of the initial concentration of the spill. With the 6 year old carbon, the pyridine elutes earlier and the greatest concentration observed in the peak is higher than that for 2 month old carbon. However, the concentration corresponding to that peak is still only 5% of the initial concentration.

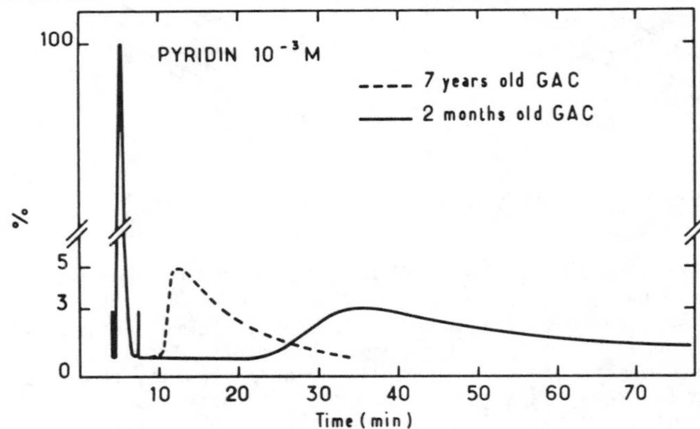

Figure 8. Breakthrough Profile for a Spile Concentration of Pyridine, breakthrough shown as a percent of original spike concentration.

These results are very important in the drinking water field, because, in the case of an accidental spill, the problems of short term toxicity are more inportant than the long term toxicity. Thus, it is particularly interesting to transform a very sharp peak of pollution to a large broad peak with a maximum as low as possible, even if all of the contaminant eventually elutes from the column. The test with the exhausted carbon indicates that the ability of carbon to spread out a peak is still effective after several years of operation.

CONCLUSIONS

Analysis of non-specific parameters (e.g., TOX, TOC) on the carbon's surface show an apparent saturation after one or two months of operation. This occurs despite the fact that these parameters continue to be partially removed by the GAC. These parameters would not show displacement of weakly adsorbed species by more strongly adsorbed species. Because of the great variability of the influent, such parameters do not allow observation of the slow adsorption kinetic phenomenon. These parameters, when measured on the carbon's surface, are not good criteria for regeneration of GAC.

Identification of specific organic compounds show that the same organic compounds are present on the activated carbon and in the influent water (phthalates, pesticides, alkylbenzenes, etc.). Some oxygenated organics (alcohols, ketones, etc.) were identified on the GAC but not in the influent. These specific organics may indicate that the carbon was concentrating them from the influent and/or they were displacing more weakly adsorbed compounds (i.e., the chromatographic effect). Due to the variability of the organic matrix in the influent water, the results obtained for specific organics in this study do not allow description of the slow adsorption kinetic phenomenon.

Adsorption isotherms with compounds of little or no polarity demonstrated a steady decline in carbon capacity over service time. The capacity reduction continued into the period when slow adsorption kinetics were most important. This type of measurement indicates that the carbon's surface was becoming increasingly saturated with competing species, and the mini-column frontal chromatographic technique used here is much less costly and time consuming than specific organic analysis. The reduced capacity results from the saturation of the surface by all competing species, thus this test appears to be a good measurement of surface saturation and therefore a reasonable parameter to be developed into a criteria for carbon regeneration.

Simulations of accidental spills with a small column of activated carbon show that, even after several years of operation, the carbon filter is able to attenuate a peak organic concentration. The compound eventually elutes from the column, but the concentration in the effluent is spread out over time. This is of practical importance to a water utility which must meet maximum contaminant level standards and indicates that the carbon may serve a useful purpose for several years of operation.

Acknowledgements - We wish to thank the Agence Financiere de Bassin Seine Normandie for financial help.

REFERENCES

1. Suffet, I. H., P. R. Cairo and M. J. McGuire, "New Developments in Removal of Organics", Special Subject, 13th IWSA Congress, Paris, France, Sept., 1980

2. Suffet, I. H., and M. J. McGuire, Activated Carbon Adsorption of Organics from the Aqueous Phase - Volume 1, Ann Arbor Science, Ann Arbor, MI, p. 508, 1980

3. Mallevialle, J., F. Fiessinger, P. Semet, and M. Bigoit, "Capteurs industriels pour la mesure en continu des parametres organiques", 2nd World Congress of Chemical Engineering, IX Inter-American Congress of Chemical Engineering, Montreal, Oct. 1981

4. Suffet, I. H., and J. Mallevialle, "Broad Spectrum Analysis for Trace Organic Chemicals in Water Treatment Processes", 2nd International Congress on Analytical Techniques in Environmental Chemistry, Barcelona, Spain, Nov., 1981

5. Mazet, H., A. Bruchet, C. Rousseau, and J. Mallevialle, "Identification par GC-MS des composes organiques dans les eaux de la Region Parisienne et Etude de leur evolution dans les usines de production d'eau potable", Congres de AGHTM, La Baule, France, 1982

6. Fiessinger, F., J. Mallevialle, and A. Benedek, "Interaction of Adsorption and Bioactivity in Full Scale Activated Carbon Filters: the Mont Valerien Experiment", in Treatment of Water by GAC, eds., M. J. McGuire and I. H. Suffet, Advances in Chemistry Series 202, ACS Books, Washington, D.C., 1983, p. 319

7. Charriaux, B., "Mise au point du dosage du chlore organique total dans les eaux, dans les charbons ou dans tout autre materiel de filtration et dans les sediments", DEA Hydrolie Appliquee, Faculte des Sciences Pharmaceutiques et Biologiques de Paris, Luxembourg, 1979

8. Rosene, M. R., M. Ozcan., M. J. Manes, Jour. Phys. Chem., 23, 80, 1976

9. Bilello, L. J., and B. A. Beaudet, "Evaluation of Activated Carbon by the Dynamic Minicolumn Adsorption Technique", in Treatment of Water by GAC, eds., M. J. McGuire and I. H. Suffet, Advances in Chemistry Series 202, ACS Books, Washington, D.C., 1983, p. 213

INFLUENCE OF THE ORIENTATION OF THE NITROGEN MOLECULE UPON ITS ACTUAL CROSS-SECTIONAL AREA IN THE ADSORBED MONOLAYER

J. Rouquerol, F. Rouquerol, Y. Grillet and M.J. Torralvo[+]

Centre de Thermodynamique et de Microcalorimétrie du CNRS
26, rue du 141ème R.I.A. - 13003 Marseille (France)

ABSTRACT : To understand the discrepancy between nitrogen and argon BET surface areas, the authors analyse a set of microcalorimetric and volumetric adsorption data obtained with various divided oxides. They conclude to a variable orientation of the nitrogen molecule (cross-sectional area between 0.112 and 0.162 nm^2) which cannot of course take place in the case of argon (0.138 nm^2).

INTRODUCTION

The starting point of this work is the well kown discrepancy between BET surface areas obtained with different adsorbates. When choosing for instance nitrogen and argon, the discrepancy often amounts to more than 20 %, but no general rule is presently available to explain it. A widely accepted way to cope with this problem is to consider nitrogen as a kind of reference adsorptive, giving good results with the BET method, with a molecular cross-sectional area taken equal to 0.162 nm^2. The latter value had been derived by Brunauer (1) from the density of three-dimensional liquid nitrogen at 77 K with the implicit asumption of random orientation and free rotation of the adsorbed nitrogen molecule (i.e. like in the three-dimensional liquid state). The choice of this cross-sectional area was strongly supported by the unexpectedly good agreement found, when using it, between the specific surface area measured by the BET method and that measured by Harkins and Jura's "absolute" calorimetric method in the case of *one* titania sample (2). One now knows that this calorimetric method may be trusted but at the condition of being used with a procedure (pre-adsorption of about only two layers of water molecules to "screen" the chemical functions of the surface, prior to immersion in water (3) rather different from that used by Harkins and Jura in their reference experiment (pre-adsorption in saturating water vapour, giving rise to capillary condensation and to thick adsorbed layers, so that the area of the liquid/vapour interface could differ from the surface area of the solid adsorbent). In these conditions, this "historical" agreement may be considered as fortuitous. Nevertheless, another reason to select the nitrogen mole-

----

[+] Present address : Instituto de Quimica Inorganica, Universidad Complutense, Ciudad Universitaria, Madrid, Spain.

cule is that it often leads to adsorption isotherms with a "knee" which is likely to make easier the location of the statistical monolayer by the BET method (4).

On the other hand, the use of nitrogen is sometimes questioned : according to Kiselev et al. (5) and Kaganer (6), it is likely that the quadrupole moment of the nitrogen molecule makes it sensitive to the chemical state of the adsorbent surface, so that argon could well be a better choice.

Nevertheless, during the last two decades, experimental evidence has been lacking for a definitive choice between nitrogen and argon. We therefore decided to compare systematically the behaviour of these adsorptives with respect to :

(i) their *relative surface concentration* as obtained from the respective apparent contents of the statistical monolayer, when calculated from the BET equation ;

(ii) their *energetical interaction with the surface* as measured in terms of enthalpy of adsorption.

A number of systems were considered, involving nitrogen and argon as adsorptives, at 77 K, and a number of oxides (silica gel, beryllia, alumina, zirconia, titania, at various stages of surface dehydroxylation and of surface cation content) as adsorbents.

The two non-conventional techniques used in a large part of this work will be now briefly described.

EXPERIMENTAL

Quasi-Equilibrium Adsorption Volumetry

Most adsorption isotherms were obtained with a small, all-metal and high-resolution equipment whose principle is represented Figure 1. The constriction, directly connected on one side to the bottle of compressed adsorptive, allows the latter to enter *at a constant and very slow rate* (0.1 to 5 cm^3 STP for gas per hour) and to reach the adsorbent enclosed in a pyrex bulb, itself immersed into a constant-level cryogenic bath (7), specially designed for the quality of its level-control ($\pm$ 0.5 mm), for its autonomy (several days for a 10 liters container) and for its isothermicity (no disturbances from the operation of the level control). The quasi-equilibrium pressure is continuously recorded *vs*. time, i.e. *vs*. the amount of adsorptive introduced into the volumetric system. A conventional dead-volume correction allows to get (automatically in the case of a computerised data-storage) *a continuous adsorption isotherm*. The validity of the method lies on the quality of the "quasi-equilibrium". The latter may be easily checked from two experiments carried out with two different inlet flow rates : the adsorption isotherms must be superimposed. Otherwise, a slower specific flow-rate must be chosen (either by decreasing the flow-rate or by increasing the mass of adsorbent). In our experience, this method (confined to *adsorption* isotherms up to a pressure of about 300 torrs and therefore designed *for the determination of specific surface areas*) has the interest of high resolution and accuracy.

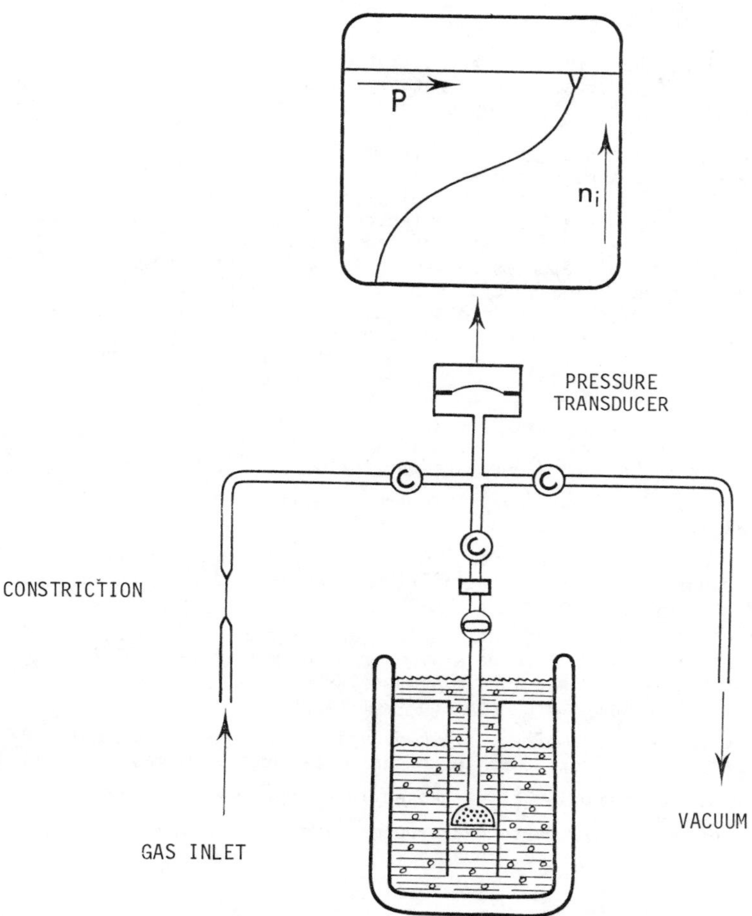

Figure 1 - Principle of quasi-equilibrium adsorption volumetry

Adsorption microcalorimetry

Figure 2 gives the principle of the isothermal microcalorimeter used for our determinations of derivative enthalpies of adsorption of argon and nitrogen at 77 K. Tube 1 is connected to a volumetric equipment of the type described above. The microcalorimeter, immersed into a liquid nitrogen bath 2, is made of an aluminium block 3 containing two cylindrical heat-flowmeters ("thermopiles" of the Tian-Calvet type, made of $\sim$ 1000 chromel-constantan thermocouples each), differentially connected. The sample is in a pyrex bulb 6. After being outgassed (in an appropriate furnace) it may be easily introduced through opening 7, never closed, since the microcalorimeter is continuously filled with an over-pressure of helium. The heat-flow generated by gas adsorption is continuously recorded *vs*. time (together with the quasi-equilibrium pressure), i.e., here again, *vs*. the amount of adsorptive introduced. After appropriate dead-volume correction, the heat curve is converted into a curve of *derivative enthalpy of adsorption vs*. amount adsorbed.

Adsorbents

The adsorbents considered here are the following :

- o A mesoporous silica gel available from the National Physical Laboratory, Teddington, G.B., as "Gasil II".
- o A microporous silica gel from the same manufacturer (Crossfield, U.K.) and known as "Gasil I" in the SCI/IUPAC/NPL Project on Surface Area Standards (8).
- o A set of alumina samples obtained by carefully controlled thermal decomposition (controlled water vapour of 60 mtorrs up to 207° C, following a method described elsewhere (9) of an industrial gibbsite sample up to temperatures ranging from 183 to 1092° C.
- o A set of beryllia samples, obtained by a similar heat treatment (10) of a beryllium hydroxide up to temperatures ranging from 20 to 1075°C.
- o A microporous zirconia gel outgassed in a similar way as above (11).
- o A non-porous rutile, whose exact outgassing conditions may be found in (12).

RESULTS

The microcalorimetric curves, previously published separately in (13) for silica, in (11) for zirconia and in (12) for rutile are redrawn all together in Figure 3.

The results of adsorption volumetry will be given in the form of tables in the course of the discussion.

DISCUSSION

Interpretation in the case of Silicas

The enthalpy curves for the *mesoporous* silica (i.e. those represented in Figure 3 under the heading "Silica") show a striking difference between argon and nitrogen : it is only with nitrogen that the extent of dehy-

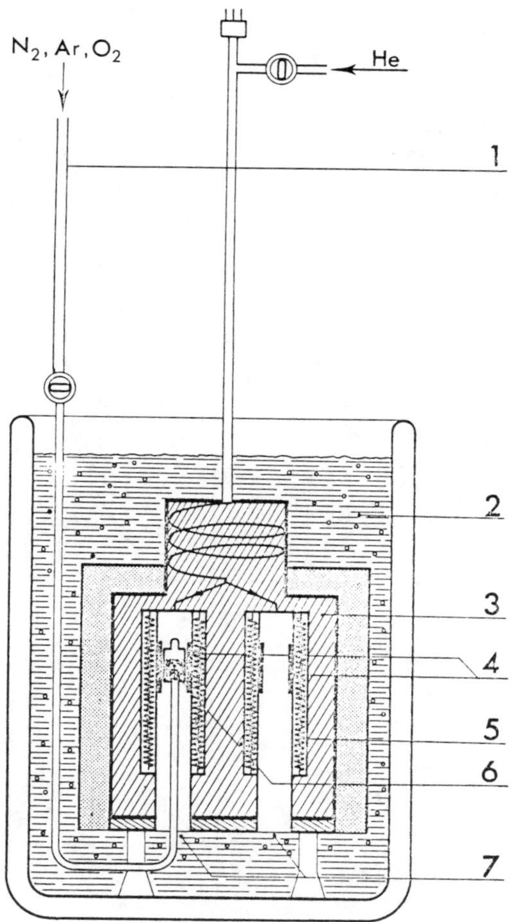

Figure 2 - Principle of the low temperature and isothermal adsorption microcalorimeter

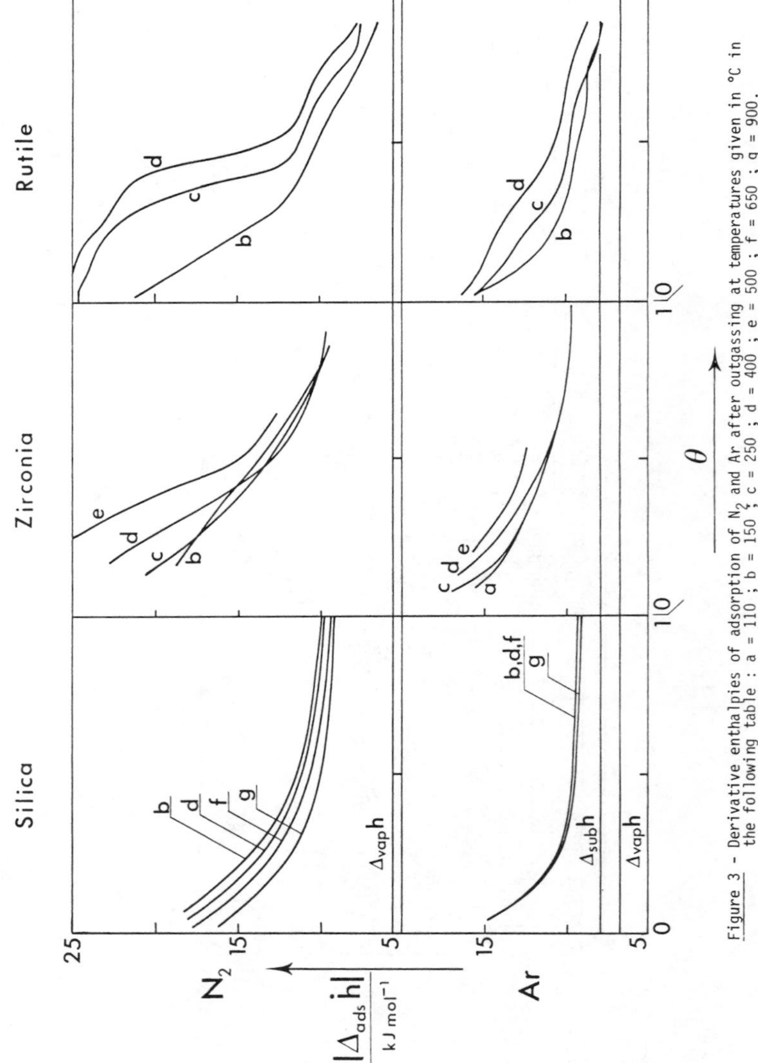

Figure 3 - Derivative enthalpies of adsorption of N_2 and Ar after outgassing at temperatures given in °C in the following table : a = 110 ; b = 150 ; c = 250 ; d = 400 ; e = 500 ; f = 650 ; g = 900.

droxylation of the surface plays a part which allows us to speak of "specific" interaction of the nitrogen molecule with the surface hydroxyles. The more dehydroxylated the surface, the lower the derivative enthalpy of adsorption of nitrogen, which tends to approach that of argon. Between a fully hydroxylated surface (sample b, outgassed at 150° C) and a nearly completely dehydroxylated one (sample g, outgassed at 900° C), the drop is of about *3 kJ mol^{-1}* at a coverage of 0.1 : this drop may figure out the *energy of specific interaction* (i.e. involving the permanent quadrupole moment of the nitrogen molecule) *between nitrogen and hydroxyles*.

The volumetric results are given in table 1.

Table 1 : *Meso*porous silica

t/°C	150	400	650	900
$(V_M)_{N_2}$/cm^{3}g^{-1}	62.6	63.7	67.0	61.6
$(V_M)_{Ar}$/cm^{3}g^{-1}	56.5	57.2	58.4	56.0
R	1.11	1.11	1.15	1.10

In this table, like in the following ones, the data reported are, from the top line to the bottom one, the following : (i) final temperature of outgassing or heat treatment, (ii) volume of nitrogen (S.T.P.) necessary to build, following the BET equation, a statistical monolayer, (iii) the same for argon, (iv) the ratio R of the two volumes above, i.e. the "ratio of apparent surface concentrations" of nitrogen and argon in their monolayer.

Let us keep in mind that, for this non-microporous sample, R remains steadily *above 1.1* .

For the *microporous* silica, the volumetric results are given in table 2.

Table 2 : *Micro*porous silica

t/°C	150	400	650	900
$(V_M)_{N_2}$/cm^{3}g^{-1}	99.6	104.2	101.2	72.6
$(V_M)_{Ar}$/cm^{3}g^{-1}	111.5	117.6	115.1	86.2
R	0.89	0.89	0.88	0.84

In contrast with what happens with the former silica, here R remains always *under 0.9*. Our interpretation is the following : on the "open" surface of the mesoporous silica, the nitrogen molecules are able to orient themselves.

A 1.1 value for R corresponds to a mean cross-sectional area of 0.138/1.1 = 0.125 nm^2 for the nitrogen molecule, so that taking 0.162 nm^2 as its standard cross-sectional area would introduce an error -in excess- of 30 %. On the other hand, in the micropores, whatever the orientation of the nitrogen molecule, the content of argon or nitrogen tends to approach the ratio of the molar volumes of these adsorbates in the three-dimensional liquid state (i.e. 0.82). Let us see now whether this conducting thread is able to account for the results obtained with the other series of adsorbents.

Interpretation in the case of beryllia and alumina

The volumetric results for both types of solids are given in tables 3 and 4.

Table 3 : Microporous (200-503) and mesoporous (916-1075) *beryllias*

$t/°C$	200	298	389	1075
$(V_M)_{N_2}/cm^3g^{-1}$	133.3	131	114	39.3
$(V_M)_{Ar}/cm^3g^{-1}$	148	156.3	139	37.3
R	0.90	0.84	0.82	1.05

Table 4 : Microporous (183-686) and mesoporous (1092) aluminas

$t/°C$	183	207	606	686	1075
$(V_M)_{N_2}/cm^3g^{-1}$	3.7	58.5	79	72.5	15.7
$(V_M)_{Ar}/cm^3g^{-1}$	3.8	64.3	93.7	88.5	14.8
R	0.97	0.91	0.84	0.82	1.05

We know, from another study (10), that the heat treatment of beryllium hydroxide first produces micropores whose diameter progressively increases up to the mesopore range. We interprete the slow decrease of R from 0.90 (sample heated to 200° C) down to 0.82 (sample heated up to 389° C) by the fact that the nitrogen molecule, due to its shape and to its stronger energy of interaction with the surface, is able to penetrate the micropores -when they are of molecular diameter- more easily than the argon molecule ; nevertheless, as the micropore diameter increases, the penetration of the argon molecule tends to be the same, so that R tends to reach the limiting theoretical value of 0.82. Above 389° C, sintering occurs (as seen from the strong decrease in surface area) and the available surface is progressively "opened", so that R is able to increase, indicating the possibility of a monolayer with oriented nitrogen molecules.

The case of alumina is extremely similar : here again, the progressive widening of micropores (initially only able to accomodate water molecules) causes R to drop down to 0.82 (alumina treated at 686° C) whereas the subsequent sintering makes it jumping up to 1.05 because of a possible orientation of the nitrogen molecule on the free surface.

Interpretation in the case of zirconia

We see on Figure 3 that outgassing at 500° C produces a significant increase of the nitrogen derivative enthalpy of adsorption whereas it also influences the corresponding curve for argon, but at a much lower level (about 2 kJ instead of 7.5 kJ at a coverage of about 0.2). The corresponding volumetric results are given in table 5. We know, from a previous study (11) that it is likely that this zirconia gel sinters by

Table 5 : Microporous zirconia

$t/°C$	110	250	400	500
$(V_M)_{N_2}/cm^3g^{-1}$	58.5	39.0	26.2	5.4
$(V_M)_{Ar}/cm^3g^{-1}$	66.8	44.1	27.9	6.9
R	0.88	0.88	0.94	0.78

progressive reduction of the micropore size. This would explain a provisional increase in the R-value (sample heated up to 400° C), due to a more difficult penetration of argon. The decrease (down to ≃ 0.78, the accuracy being here limited by the smaller surface area) which follows the recrystallisation of the zirconia gel (at 430° C) may be explained by an increase in size of the residual microporosity. At the same time, the crystalline field (and possibly also Zr^{3+} cations) increase significantly the interaction between nitrogen and the surface (cf. enthalpy curve "d" for N_2 adsorbed on zirconia). If zirconia was not microporous, this strong interaction would orientate the nitrogen molecule and would increase the value of R instead of lowering it to a minimum.

Interpretation of the case of rutile

The enthalpy curve shows a strong increase in the case of nitrogen when the outgassing temperature is raised from 150 (curve b) to 250° C (curve c). The interpretation already given is that dehydroxylation produces a number of highly polar T^{4+} sites, themselves surrounded with microgaps resulting from the departure of co-ordinated water molecules (12). The corresponding volumetric data are in the following table :

Table 6 : non-porous rutile

t/°C	150	250	400
$(V_M)_{N_2}/cm^3g^{-1}$	5.07	5.75	5.74
$(V_M)_{Ar}/cm^3g^{-1}$	4.98	5.50	5.56
R	1.02	1.04	1.03

We may notice, between 150 and 250° C, the increase in V_M (both for N_2 and Ar) which may be due to the above-mentioned micro-gaps. In the meanwhile, the ratio remains slightly above 1, indicating that an orientation of the nitrogen molecule is taking place (under the action of the strong interactions shown on the enthalpy curves) over an open surface.

CONCLUSION

Concerning the energetical interaction with the surface, it is shown here that, for the three types of adsorbents examined from this point of view (silica, zirconia and rutile), the *derivative enthalpy of adsorption of nitrogen* was always (11 samples examined) *significantly higher than that of argon* (whereas the enthalpies of condensation of both vapours are of the same order of magnitude). Also, its decrease during the progressive formation of the monolayer is always larger (more than two-fold in the case of rutile) than that of argon. Furthermore, dehydroxylation of silica has no measurable influence on the derivative enthalpies of adsorption of argon, whereas an interaction of about 3 kJ mol^{-1} may be measured between the nitrogen molecules and the surface hydroxyl groups of the silica.

Finally, the adsorption enthalpy data clearly shows that the enthalpy of adsorption is ca. 7 kJ/mol as θ approaches unity, in all systems. Especially, the enthalpy values at $\theta = 1$ for Ar-Silica and N_2-Silica are *exactly* the same. This indicates that the energetics at $\theta = 1$ are the same for Ar and N_2 on same solid.

Concerning the ratio of apparent surface concentrations of nitrogen and argon, it is shown that in the case of *microporous* adsorbents, this ratio tends to go *down to about 0.8* (the ratio of the molar volumes of these adsorptives at the three-dimensional liquid state) whereas on *"open" surfaces* (i.e. mesoporous, macroporous or non porous) the ratio *may reach 1.15* to be compared with 1.23, the theoretical ratio when the nitrogen molecule is supposed to be standing with its main axis normal to the surface, in a close-hexagonal arrangement, with a smaller molecular diameter of 0.36 nm).

The general conclusion is that *the specific interaction of the nitrogen molecule with the surface* (as shown by direct microcalorimetric determinations, and involving mainly a quadrupole/field gradient interaction) *is enough to explain a statistical orientation of the nitrogen molecule in the monolayer,* so that, finally, the spherical argon molecule seems to lend itself (with a unique possible cross-sectional area of 0.138 nm^2) to a safer determination of the specific surface area of oxides.

LITERATURE CITED

1. Brunauer, S., "The adsorption of gases and vapours", vol. I, Physical Adsorption, Oxford Univ. Press (1977).

2. Harkins, W.D. and Jura, G., J. Amer. Chem. Soc., 66, 1362 (1944).

3. Partyka, S., Rouquerol, F. and Rouquerol, J., J. Colloid and Interf. Science, 68, (1), 21 (1979).

4. Gregg, S.J. and Sing, K.S.W., "Adsorption, Surface Area and Porosity", 2d Edition, Academic Press, London (1982).

5. Aristov, B.G. and Kiselev, A.V., Russian J. Phys. Chem., 37, 1359 (1963) ; idem, 38, 1077 (1964).

6. Kaganer, M.G., Dokl. Akad. Nauk SSSR, 138, 405 (1961).

7. Rouquerol, J. and Davy, L., Thermochimica Acta, 24, 391 (1978)

8. Everett, D.H., Parfitt, G.D., Sing, K.S.W. and Wilson, R., J.Appl. Chem. Biotechnol., 24, 199 (1974) ; Harvard, D.C. and Wilson, R., J. Colloid Interface Sci., 57, 276 (1976).

9. Rouquerol, J., "Thermal Analysis", vol. I, A. Schwenker ed., p. 281, Academic Press, New-York (1969) ; Rouquerol, J. and Ganteaume, M., J. Thermal Anal., 11, 201 (1977).

10. Rouquerol, F., Rouquerol, J. and Imelk, B., "RILEM Symposium on Porous structure", M. Haynes ed., Bristol (1983).

11. Torralvo, M.J., Grillet, Y., Rouquerol, F. and Rouquerol, J., J. Chim. Phys. Fr., 77(2), 125 (1980).

12. Furlong, D.N., Rouquerol, F., Rouquerol, J. and Sing, K.S.W., _J. C. S. Faraday I_, 76, 774 (1980).

13. Rouquerol, J., Rouquerol, F., Pérès, C., Grillet, Y. and Boudellal, M., "Characterization of porous solids"; Gregg, S.J., Sing, K.S.W. and Stoeckli, H.F. ed. London, Society of Chemical Industry (1979).

EXCESS ISOTHERMS AND HEATS OF IMMERSION IN MONOLAYER ADSORPTION FROM BINARY LIQUID MIXTURES ON SLIGHTLY HETEROGENEOUS SOLID SURFACES

W.Rudzinski*,J.Michalek,and J.Zajac
Department of Theoretical Chemistry
Institute of Chemistry UMCS
20-031 Lublin, Nowotki 12
P o l a n d

S.Partyka
Laboratoire de Physico-Chimie des Systemes Polyphases
Université des Sciences et Techniques du Languedoc
34060 Montpelier Cedex
F r a n c e

ABSTRACT

The theoretical approach to the problem of excess isotherms and heats of immersion in monolayer adsorption from binary liquid mixtures on heterogeneous surfaces of actual solids, developed recently by Rudzinski and Partyka,is extended here for the case when the molecules of the liquid mixture occupy different surface areas. Equations for excess isotherms and heats of immersion are developed, and next illustrated by appropriate computer model investigation

* To whom the correspondence should be adressed

THEORY

The monolayer adsorption from binaries composed of molecules of different sizes onto solid surfaces may be represented as the following phase exchange equilibrium between an adsorbed phase and a bulk liquid,

$$(A)^{(s)} + r(B)^{(b)} \rightleftharpoons (A)^{(b)} + r(B)^{(s)} \tag{1}$$

where superscripts (s) and (b) refer to the surface and the equilibrium bulk phases respectively, whereas r represents the ratio of the surface areas occupied by the molecules A and B. The equilibrium constant K for this process is expressed in the form (1),

$$K = \frac{\varphi_A^{(s)}\gamma_A^{(s)}}{\varphi_A^{(b)}\gamma_A^{(b)}}\left(\frac{\varphi_B^{(b)}\gamma_B^{(b)}}{\varphi_B^{(s)}\gamma_B^{(s)}}\right)^r \tag{2}$$

where the symbols "φ" and "γ" are used to denote the volume fractions and the activity coefficients of each component in both phases. The excess of the energy of mixing β due to the formation of the solid/liquid interface, obtained by means of the Bragg-Williams approximation, reads

$$\beta = n_A^o\left[a_p\varepsilon_p\varphi_A^{(s)}\varphi_B^{(s)} + a_v\varepsilon_v\left(\varphi_A^{(s)}\varphi_B^{(b)} + \varphi_A^{(b)}\varphi_B^{(s)} - \varphi_A^{(b)}\varphi_B^{(b)}\right)\right] \tag{3}$$

where n_A^o is the surface capacity for the component whose molecules are smaller "monomer". Further a_p and a_v are the numbers of the nearest neighbours-adsorption sites in the direction parallel (p) to the solid surface and the direction vertical (v) to the surface. ε_p and ε_v denote appropriate "exchange energy" terms, which may differ from their bulk counterparts because of possible perturbations in the adsorbate-adsorbate interactions due to the presence of the solid phase. Similarly, the lattice parameters a_p and a_v should be treated as different from their bulk counterparts. We shall further replace the products $a_p\varepsilon_p$ and $a_v\varepsilon_v$ by symbols A_p and A_v respectively.

By an appropriate differentiation of β, one obtains the formulas for the surface activity coefficients,

$$\gamma_A^{(s)} = \exp\left\{\frac{1}{RT}\left[A_p\left(\varphi_B^{(s)}\right)^2 + A_v\left(\varphi_B^{(b)}\right)^2\right]\right\} \tag{4a}$$

$$\gamma_B^{(s)} = \exp\left\{\frac{1}{RT}\left[\frac{A}{r}p\left(\varphi_A^{(s)}\right)^2 + \frac{A}{r}v\left(\varphi_A^{(b)}\right)^2\right]\right\} \tag{4b}$$

We assume the monolayer surface solution as a quadratic mixture in the sense of Flory-Huggins approach here, but such an assumption is not necessary for the equilibrium bulk solution. That allows us to replace the products $\varphi_A^{(b)}\gamma_A^{(b)}$ and $\varphi_B^{(b)}\gamma_B^{(b)}$ by appropriate bulk activities of the components A and B, $a_A^{(b)}$ and $a_B^{(b)}$.

Now, we consider the detailed meaning of the equilibrium constant K. Following Rudziński and Partyka (2),

$$K = \exp\left(\frac{\varepsilon_{As} - r\,\varepsilon_{Bs}}{RT}\right) \tag{5}$$

where

$$\varepsilon_{As} = \varepsilon_A + RT\ \ln\left(q_{As}/q_{Ab}\right) \tag{6a}$$

$$\varepsilon_{Bs} = \varepsilon_B + RT\ \ln\left(q_{Bs}/q_{Bb}\right) \tag{6b}$$

and where ε_A and ε_B are the adsorption energies of the molecules A and B. Further q_b and q_s denote their molecular partition functions in the bulk and adsorbed state respectively. To make further consideration easier, we introduce a new quantity ε,

$$\varepsilon = \varepsilon_{As} - r\,\varepsilon_{Bs} \tag{6c}$$

We shall further call it the "adsorption energy". Because of the local variations in the stoichiometry and crystallography of the outermost part of the actual solid surfaces, the adsorption energy will be different on different adsorption sites.

Let $\chi(\varepsilon)$ denote the differential distribution of the adsorption sites among the related values of ε, normalized to unity,

$$\int_{\Omega} \chi(\varepsilon)\,d\varepsilon = 1 \tag{7}$$

where Ω is the physical domain of the adsorption energy ε. Rudziński et al.(3) have shown that the problem of the integration limits becomes essential only when the concentration of one of the components is very small. So, for the purpose of mathematical convenience we shall assume further that ε varies between $(-\infty)$ and $(+\infty)$.

Now, let us consider the two most important adsorption observables: the excess adsorption isotherm and the heat of wetting (immersion).

For a homogeneous surface, the excess adsorption isotherm, n_A^e, is related to the individual adsorption isotherm $\varphi_A^{(s)}$ by the equation,

$$n_A^e = n_A^o\left[\varphi_A^{(s)} - \left(\varphi_A^{(s)} + r\varphi_B^{(s)}\right)x_A^{(b)}\right] \tag{8}$$

In the case of a heterogeneous surface, $\varphi_A^{(s)}$ has to be replaced by its averaged value $\varphi_{At}^{(s)}$,

$$\varphi_{At}^{(s)} = \int_{\Omega} \varphi_A^{(s)}(\varepsilon)\, \chi(\varepsilon)\, d\varepsilon \tag{9}$$

The other observable, i.e., the heat of immersion Q_w, may be a source of a very interesting information about the properties of an adsorption system. For a homogeneous solid surface the equation for Q_w takes the form,

$$Q_w = n_A^o \left(\varphi_A^{(s)} Q_{wA} + r\, \varphi_B^{(s)} Q_{wB} \right) + \beta \tag{10}$$

where Q_{wA} and Q_{wB} are the molar heats of wetting by pure liquids A and B. The variables Q_{wA} and Q_{wB} may be identified with the adsorption energies ε_A and ε_B to a good approximation. Thence, Equation (10) can be written in the following alternative form,

$$Q_w = Q_B + n_A^o\, \varphi_A^{(s)} \varepsilon - n_A^o\, \varphi_A^{(s)} \lambda + \beta \tag{11}$$

where Q_B denotes the heat of immersion of the whole solid sample in the pure component B, and λ is defined by

$$\lambda = RT \ln \left[\frac{q_{As}}{q_{Ab}} \left(\frac{q_{Bb}}{q_{Bs}} \right)^r \right] \tag{12}$$

The form of the equation for the heat of immersion of a whole heterogeneous surface Q_{wt} will depend on the "topography of surface". Similarly as in our previous publications (4,5) two extreme models of the surface topography are taken into account.

The first one is the "patchwise" model. It assumes that the adsorption sites having the same adsorption energy are grouped into large patches and the interactions between admolecules on two different patches are neglected in considering the state of the adsorption system. In other words, the adsorption system is considered as a collection of independent subsystems being only in thermal and material contact. The potential of the average force acting on a molecule adsorbed on certain patch will depend only on the surface concentration on this patch. Consequently, the surface activity coefficient $\gamma_{Ap}^{(s)}$, associated with this patch, is given by the formula,

$$\gamma_{Ap}^{(s)} = \exp \left[A_p \left(1 - \varphi_A^{(s)}(\varepsilon)\right)^2 + A_v \left(\varphi_B^{(b)}\right)^2 \right] \tag{13}$$

where the subscript "p" denotes the patchwise topographical model. Also, the excess of the energy of mixing β is strictly related to the local surface concentration $\varphi_A^{(s)}(\varepsilon)$ on this patch. Taking this fact into account, we write the equation for Q_{wt} in the form,

$$Q_{wp} = Q_B + n_A^o \int_{\Omega} \varphi_A^{(s)} \varepsilon \chi \, d\varepsilon - n_A^o \varphi_{At}^{(s)} \lambda + \int_{\Omega} \beta \, d\varepsilon \quad (14)$$

where the dependence of q_{As} and q_{Bs} on ε_A and ε_B is neglected.

The random model of the surface topography assumes that the adsorption sites having various adsorption energies are distributed on the heterogeneous surface completely at random. For this reason any local concentration on such a surface will be equal to the average surface concentration. In this case, the surface activity coefficient $\gamma_{Ar}^{(s)}$ is expressed in the following form,

$$\gamma_{Ar}^{(s)} = \exp\left[A_p\left(\varphi_{Bt}^{(s)}\right)^2 + A_v\left(\varphi_B^{(b)}\right)^2\right] \quad (15)$$

where subscript "r" refers to the random topographical model. For the same reason, β takes the form,

$$\beta_r = n_A^o \left[A_p \varphi_{At}^{(s)} \varphi_{Bt}^{(s)} + A_v\left(\varphi_{At}^{(s)} \varphi_B^{(b)} + \varphi_A^{(b)} \varphi_{Bt}^{(s)} - \varphi_A^{(b)} \varphi_B^{(b)}\right)\right] \quad (16)$$

and is independent of ε. Therefore, the expression describing the heat of immersion of the surfaces with random topography has the form,

$$Q_{wr} = Q_B + n_A^o \int_{\Omega} \varphi_A^{(s)} \varepsilon \chi \, d\varepsilon - n_A^o \varphi_{At}^{(s)} \lambda + \beta\left(\varphi_{At}^{(s)}\right) \quad (17)$$

Similarly as in our previous works we assume here the following gaussian-like energy distribution,

$$\chi(\varepsilon) = \frac{\frac{1}{c}\exp\left[\frac{1}{c}(\varepsilon - \varepsilon_o)\right]}{\left\{1 + \exp\left[\frac{1}{c}(\varepsilon - \varepsilon_o)\right]\right\}^2} \quad (18)$$

centered about $\varepsilon = \varepsilon_o$. The spread of the function (18) is described by the heterogeneity parameter c. When $c \rightarrow 0$, this function degenerates into Dirac delta distribution $\delta(\varepsilon - \varepsilon_o)$. In such a hypothetical case, we would have to deal with an ideally homogeneous surface, characterized by the adsorption energy ε_o.

Now, we shall repeat shortly the principes of the mathematical approach that would let $\varphi_{At}^{(s)}$ and Q_{wt} be expressed by simple analytical formulas, in the case of slightly and moderately heterogeneous solid/solution interfaces.

The partial derivative of the adsorption isotherm $(\partial \varphi_A^{(s)}/\partial \varepsilon)$ takes then the following explicite form,

$$\frac{\partial \varphi_A^{(s)}}{\partial \varepsilon} = \frac{\frac{1}{RT}\exp\left[\frac{1}{RT}(\varepsilon_c^o - \varepsilon)\right]}{\left\{1 + \exp\left[\frac{1}{RT}(\varepsilon_c^o - \varepsilon)\right]\right\}^2} \tag{19}$$

where

$$\varepsilon_c^o = -RT \ln \frac{a_A^{(b)}}{a_B^{(b)}} \tag{20}$$

The derivative $(\partial \varphi_A^{(s)}/\partial \varepsilon)$ has essentially the same form as the adsorption energy distribution in Equation (18). The gaussian-like derivative (19) is centered about $\varepsilon = \varepsilon_c^o$, and its spread is given by the value of RT. When RT >> c, the function $\varphi_A^{(s)}(\varepsilon)$ may be considered as a slowly-varying function of ε, compared to $\chi(\varepsilon)$. In such a case it is to be expected that the expansion of $\varphi_A^{(s)}(\varepsilon)$ around $\varepsilon = \varepsilon_o$ should be an effective way of evaluating the integral (7).

$$\varphi_{At}^{(s)} = \varphi_{Ao}^{(s)} + 2\sum_{n=0}^{+\infty} \frac{(\pi c)^{2n}}{(2n)!}\left(2^{2n-1} - 1\right) B_n D_o^{(n)} \tag{21}$$

where

$$D_o^{(n)} = \left(\frac{\partial^n \varphi_A^{(s)}}{\partial \varepsilon^n}\right)_{\varepsilon = \varepsilon_o} \tag{22}$$

Neglecting the terms of order higher than $(c/RT)^2$ in the expansion (21), we arrive at the following equation for $\varphi_{At}^{(s)}$,

$$\varphi_{At}^{(s)} = \varphi_{Ao}^{(s)} + \frac{(\pi c)^2}{6} D_o^{(2)} \tag{23}$$

where the second term on right hand of Equation (23) is a correction term, describing the perturbation in the behaviour of the adsorption isotherm caused by surface heterogeneity.

In a similar way we evaluate the integral,

$$\int_{-\infty}^{+\infty} \varphi_A^{(s)} \varepsilon \chi \, d\varepsilon = \varphi_{Ao}^{(s)} \varepsilon_o + \frac{(\pi c)^2}{6}\left(\varepsilon_o D_o^{(2)} + 2D_o^{(1)}\right) \tag{24}$$

appearing in Equations (14) and (17). Equations (21) and (24) are of a general validity, i.e., they apply to the case when $\gamma_{A,B}^{(s)} \neq 1$ and $r \neq 1$. The influence of the surface topography is reflected in the explicite form of the derivative $D^{(2)}$.

After performing appropriate differentiations, we arrive at the following explicite equations:

For the patchwise surface topography we have,

$$D_p^{(1)} = \frac{1}{RT}\left[\frac{1 + (r-1)\varphi_A^{(s)}}{\varphi_A^{(s)}(1-\varphi_A^{(s)})} - \frac{2A_p}{RT}\right]^{-1} \tag{25}$$

$$D_p^{(2)} = \left(\frac{1}{RT}\right)^2\left[\frac{1 + (r-1)\varphi_A^{(s)}}{\varphi_A^{(s)}(1-\varphi_A^{(s)})} - \frac{2A_p}{RT}\right]^{-3} \frac{1 - 2\varphi_A^{(s)} + (1-r)(\varphi_A^{(s)})^2}{[\varphi_A^{(s)}(1-\varphi_A^{(s)})]^2} \tag{26}$$

whereas for the random surface topography, we obtain,

$$D_r^{(1)} = \frac{1}{RT}\,\frac{\varphi_A^{(s)}(1-\varphi_A^{(s)})}{1 + (r-1)\varphi_A^{(s)}} \tag{27}$$

$$D_r^{(2)} = \left(\frac{1}{RT}\right)^2 \frac{\varphi_A^{(s)}(1-\varphi_A^{(s)})\left[1 - 2\varphi_A^{(s)} + (1-r)(\varphi_A^{(s)})^2\right]}{\left[1 + (r-1)\varphi_A^{(s)}\right]^3} \tag{28}$$

According to Equations (14), (23) and (24), the heat of immersion Q_{wt} for the heterogeneous surfaces with the patchwise topography, should be expressed as follows,

$$Q_{wp} = Q_B + n_A^o\varphi_{Ao}^{(s)}\varepsilon_o + n_A^o\frac{(\pi c)^2}{6}\left(D_{po}^{(2)}\varepsilon_o + 2D_{po}^{(1)}\right) - n_A^o\varphi_{At}^{(s)}\lambda + \beta_p \tag{29}$$

where

$$\beta_p = n_A^o\Big\{A_p\Big[\varphi_{Ao}^{(s)}\varphi_{Bo}^{(s)} + \frac{(\pi c)^2}{6}\Big(D_{po}^{(2)}(\varphi_{Bo}^{(s)} - \varphi_{Ao}^{(s)}) - 2\big(D_{po}^{(1)}\big)^2\Big)\Big] + A_v\left(\varphi_{At}^{(s)}\varphi_B^{(b)} + \varphi_A^{(b)}\varphi_{Bt}^{(s)} - \varphi_A^{(b)}\varphi_B^{(b)}\right)\Big\} \tag{30}$$

For the random surface topography Q_{wt} takes the following form,

$$Q_{wr} = Q_B + n_A^o\varphi_{Ao}^{(s)}\varepsilon_o + n_A^o\frac{(\pi c)^2}{6}\left(D_{ro}^{(2)}\varepsilon_o + 2D_{ro}^{(1)}\right) - n_A^o\varphi_{At}^{(s)}\lambda + \beta_r \tag{31}$$

where

$$\beta_r = \beta(\varphi_{At}^{(s)}) \tag{35}$$

The papers where both experimental excess isotherms, and heats of immersion are reported are very rare. Thus, similarly as in our two previous publications (2,4), we have decided to carry out appropriate model investigation rather, based on the equations for $\phi_{At}^{(s)}$ and Q_{wt} developed here by us. This time, we focused our attention on the effect of different surface areas occupied by the molecules of a liquid mixture.

Figure 1 shows the behaviour of the excess isotherm n_{At}^{e} of the component A on the heterogeneous surfaces exhibiting a random surface topography.

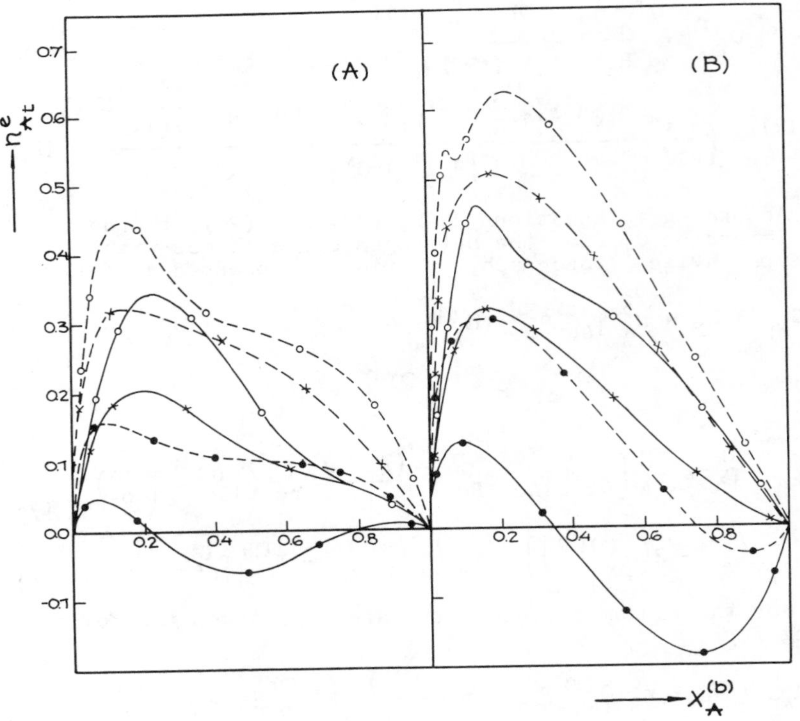

FIGURE 1

There the behaviour of the excess isotherm n_{At}^{e} is shown for various values of the adsorption parameters appearing in our theoretical treatment,but still for the same value of the heterogeneity parameter (c/RT)=0.75 .The solid lines correspond to the situation when the most probable value of the adsorption energy $(\varepsilon_o/RT)=1.0$,whereas the broken ones were evaluated by assuming that $(\varepsilon_o/RT)=2.0$ .Three values of the geometrical parameter r were accepted in our model investigation; r=1/2 (empty circles), r=1 (crosses), and r=2 (shaded circles).Part A of Figure 1 shows the behaviour of n_{At}^{e} when both the bulk and surface solution exhibit negative deviations from Raoult's Law,characterized by the parameters $A_p=2A_v=-0.5RT$ .Part B shows a reverse situation when $A_p=2A_v=+0.5RT$. Of course,many other combinations of these interaction parameters could also be considered in our model investigation.This is because the presence of a solid surface causes in general perturbations in the structure of the surface solution compared to its structure in the equilibrium bulk phase.However,a limited volume of the present publication does not allow us to show many other interesting physical situations.

We have shown it in our previous publication that the behaviour of both the excess isotherms n_{At}^{e} and the heats of immersion Q_{wt} in adsorption on patchwise heterogeneous surfaces is essentially similar to that in adsorption on the surfaces exhibiting a random surface topography,except that the heterogeneity effects are demonstrated much more strongly in the case of the surfaces exhibiting the patchwise topography.For this reason,we have abandomed the related model investigation for the situation when our adsorption systems would exhibit the patchwise topography.

The next Figure 2 shows the related behaviour of Q_{wt} , for the same physical situations /adsorption parameters/, which were considered in the model investigation presented in Figure 1. The additional parameter λ was assumed to be equal to zero,and the scale for Q_{wt} was chosen so that $Q_B=0$

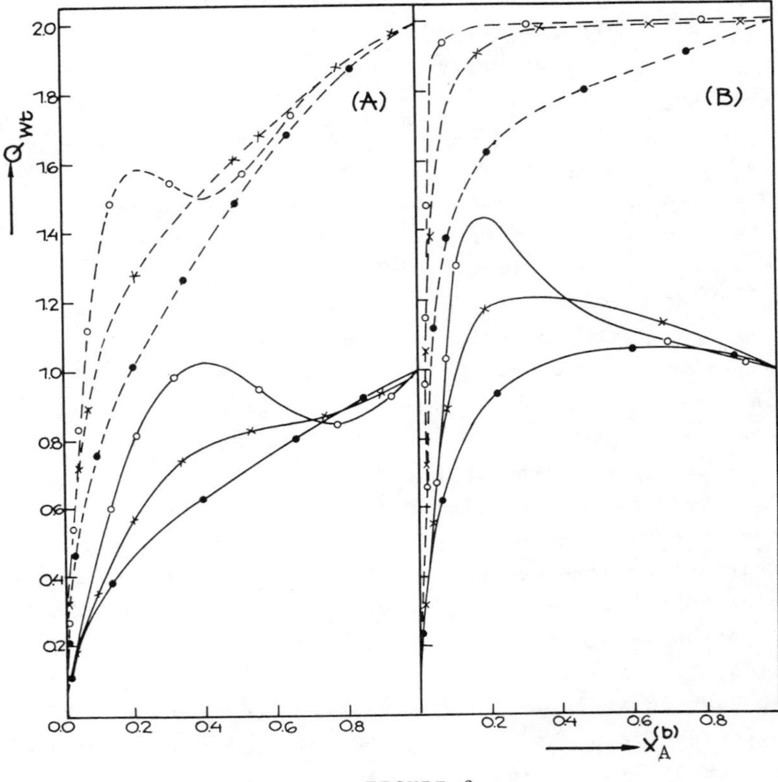

FIGURE 2

LITERATURE

1. Everett, D.H., Trans. Faraday Soc., 61, 2478 (1965)

2. Rudzinski, W. and Partyka, S., J. Chem. Soc. Faraday Trans. 1 77, 2577 (1981)

3. Rudzinski, W., Michalek, J., Schöllner, R., Herden, H., and Einicke, D.W., Acta Chim. Acad. Sci. Hungaricae (in press)

4. Rudzinski, W., Michalek, J., and Partyka, S., J. Chem. Soc. Faraday Trans. 1, 78, 2361 (1982)

5. Rudzinski, W., and Partyka, S., J. Coll. Interface Sci., 89, 25 (1982)

ADSORPTION STUDIES OF ETHYLENE AND PROPYLENE MIXTURES ON CARBON - COMPARISON OF THEORETICAL PREDICTIONS WITH EXPERIMENTAL DATA

Steve Russell and E. R. Haering
Department of Chemical Engineering
The Ohio State University
Columbus, Ohio

ABSTRACT

Pure and binary adsorption isotherms were determined on activated carbon at 300°K and over a pressure range from 3.3 to 100.0 kPa. The experimental data were compared with predictions based on well-known theories. It was necessary to use the concept of activity coefficients in order to correlate the data.

INTRODUCTION

Multicomponent adsorption plays an important part in several areas of chemical engineering ranging from separation processes to heterogeneous catalysis. While numerous theories and equations have been presented for single component adsorption, it is still necessary, for the majority of systems, to determine the isotherm data experimentally. In the case of multicomponent adsorption, the need for such studies is even greater.

Significant contributions to multicomponent, physical adsorption theory have been made by the extension of the Gibbs spreading pressure theory by Myers and Prausnitz (1), Myers (2) and Van Ness (3) as well as the extension of the Polanyi potential theory to multicomponent adsorption by Grant and Manes (4). While these theories have been well documented, their applicability has not been tested extensively due to lack of published data dealing with multicomponent adsorption. The purpose of this work was to determine adsorption isotherm data for a binary system of gases over a common adsorbent and to compare the experimental results with predictions based on the above theories.

THEORY

The concept of the adsorbed film behaving as an ideal adsorbed phase has been well developed in the literature. A basic premise of these developments is that the adsorbed film behaves thermodynamically as a two dimensional solution. Two basic approaches have been used to develop the necessary theoretical equations used to describe multicomponent physical adsorption equation.

The first method developed by Myers and Prausnitz (1) is based on the Gibbs adsorption equation. In this approach it is possible, based on pure component data, to calculate the spreading pressure for the adsorbed phase as shown in Equation (1).

$$\frac{\pi A}{RT} = \int_0^P \frac{N}{P}\, dP \qquad (1)$$

With this information and the assumption that an equation analogous to Raoult's law is applicable, it is possible to predict multicomponent adsorption equilibria. Several modified techniques for calculating the spreading pressure have been proposed (Hoory (5), Van Ness (3), Kidnay (6)) primarily to overcome the difficulty associated with evaluating the integral of Equation (1) at low pressure.

The second method for estimating multicomponent adsorption equilibria is based on the Polanyi potential theory of adsorption as developed by Grant and Manes (4). In this development, the adsorption potential for pure components is given by:

$$\varepsilon_i = RT\, \ell n\, (Ps/p^*)_i \qquad (2)$$

By making use of Dubinin affinity coefficients, it is possible to prepare a single characteristic adsorption curve for a given adsorbent. With this characteristic curve and again the assumption that the adsorbed film behaves as an ideal solution, it is possible to predict multicomponent adsorption equilibria. It has been shown by Grant and Manes that although the two concepts are different physically, that they are related mathematically and have indeed been shown capable to correlating the same data.

Both Myers and Van Ness have extended the spreading pressure theory to non-ideal adsorbed film, through the use of the activity coefficient concept or

$$P_T y_i = Pi^o\, (\pi)\, \gamma_i\, x_i \qquad (3)$$

Myers has also developed Equation (4) for calculating the spreading pressure, π, for a given adsorbed mixture at constant temperature and total pressure over a range of gas compositions.

$$\frac{\pi A}{RT} = \int_o^{y_1} N_1 \frac{dy_1}{y_1} + \int_o^{y_2} N_2 \frac{dy_2}{y_2} \qquad (4)$$

Equations (3) and (4) provide sufficient information for the calculation of the activity coefficients.

EXPERIMENTAL

The need for accurate experimental data for the evaluation of the above theories has been presented by both Myers (1) and Van Ness (3). In order to decrease the time required to reach equilibrium, an apparatus similar to that proposed by Lewis (7) was assembled and is shown in Figure 1. This apparatus allows for stepwise adsorption studies to be undertaken. A stainless steel U-tube contains the adsorbent and the temperature controlled by a circulating water jacket. Quantities of gas are measured and added to the system by means of calibrated gas burets. Gas compositions can be determined by taking small samples

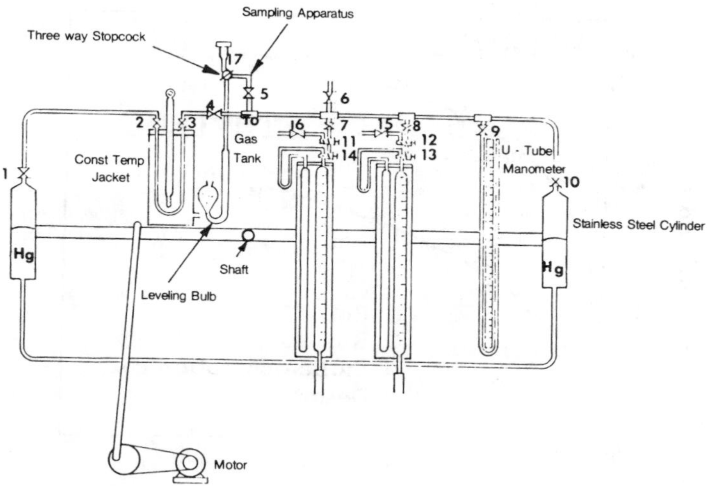

Figure 1. "Mixing" Adsorption Apparatus.

followed by analysis by gas chromatography. Good gas mixing and gas contact with the adsorbent is provided by means of a reciprocally operated mercury pump. In order to check the accuracy of the data obtained from this device, separate pure component adsorption data were obtained in a static adsorption apparatus patterned after a similar device of Harris and Emmett (8) and shown to be capable of providing very accurate data. Details concerning these apparati and their operation is available elsewhere (Russell (9) and Svanks (10)).

Pure component adsorption isotherms for both ethylene and propylene on Barnebey-Cheney KE760 activated carbon at 303°K are shown in Figure 2 with the data given in Table 1. None of the common isotherm equations were found capable of representing these data. Multicomponent adsorption data are given in Table 2 for this system.

Results of the predictions of the multicomponent equilibria based both on the spreading pressure concepts and the Polanyi potential are shown in Figures 3A, 3B, 3C, and 3D for the four total pressures investigated. As might be expected, the calculated results from both theories are similar. The greatest discrepancy, particularly at the lower pressures, was noted with the use of the Van Ness suggestion. Much of the difference noted with this technique is thought to be due to errors associated in manipulating the data rather than being associated with the theory itself. The Van Ness technique requires data to be taken at constant temperature and gas composition. The latter restriction was not strictly met in this study but by means of a series of cross plots it was possible to estimate the necessary information. This approximation along with errors associated in evaluating $\partial(\pi A/RT)\partial y$, are believed to account for most of the differences noted.

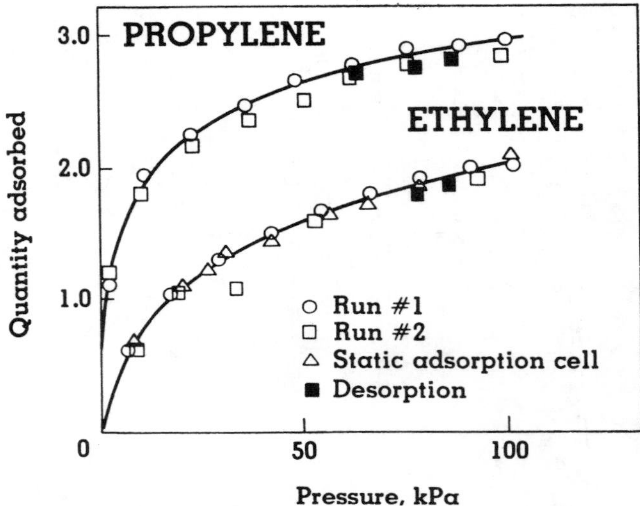

Figure 2. Pure Component Isotherms.

Table 1. Pure Component Adsorption Data on Activated Carbon at 303°K

ETHYLENE		PROPYLENE	
Pressure kPa	Quantity Adsorbed mmol/g	Pressure kPa	Quantity Adsorbed mmol/g
6.4	0.6142	2.0	1.1314
16.8	1.0370	10.9	1.9504
28.3	1.2965	22.4	2.2599
41.1	1.4574	36.0	2.4775
53.6	1.6508	48.7	2.6451
65.7	1.7868	62.3	2.7662
77.6	1.8954	75.7	2.8765
89.7	1.9649	89.0	2.8876
100.0	1.9760	100.5	2.9411
6.8	0.6134	2.8	1.1996
18.8	1.0291	10.4	1.8100
32.1	1.0748	22.9	2.1611
41.1	1.4233	36.9	2.3484
51.9	1.5735	50.4	2.4136
64.9	1.6893	61.6	2.6858
78.5	1.8146	75.3	2.7547
91.6	1.8923	88.6	2.7757
84.0[1]	1.8398	98.6	2.8234
79.6[1]	1.7832	87.3[1]	2.8200
		77.6[1]	2.7552
		68.8[1]	2.6974
6.7[2]	0.6236		
19.9[2]	1.0802		
25.6[2]	1.2064		
30.1[2]	1.3196		
55.2[2]	1.6433		
77.0[2]	1.8547		
99.2[2]	2.0670		

[1] Stepwise desorption
[2] "Static" Adsorption cell

Table 2. Binary Adsorption Data for Ethylene-Propylene Mixtures on Activated Carbon at 303°K.

Pressure kPa	Total Adsorption mmol/g	Propylene Adsorption mmol/g	Ethylene Adsorption mmol/g	Composition	
				$y_{propylene}$	$x_{propylene}$
14.7	1.3703	1.1215	0.2488	0.265	0.818
12.7	1.0091	0.4606	0.5485	0.145	0.456
15.7	1.3372	1.0761	0.2611	0.110	0.805
15.6	0.9263	0.2935	0.6338	0.205	0.317
14.4	1.4908	1.3588	0.1320	0.355	0.911
40.1	2.0482	2.0444	0.0038	0.300	0.999
38.3	1.7127	0.9806	0.7321	0.149	0.573
41.1	1.6474	1.0045	0.6027	0.083	0.634
39.2	2.2183	2.1542	0.0641	0.545	0.955
39.1	2.1514	1.9857	0.1657	0.450	0.530
41.3	1.7621	1.3163	0.4458	0.126	0.747
66.0	2.4428	2.2815	0.1613	0.365	0.934
68.0	2.0744	1.4805	0.5939	0.141	0.714
66.3	1.7253	1.0021	0.7231	0.055	0.582
68.2	2.4576	2.7081	-	0.715	1.10
68.1	2.4335	2.4031	0.0304	0.700	0.988
65.7	1.8968	1.4055	0.4913	0.137	0.741
65.3	2.3905	2.3865	0.0040	0.382	0.998
91.8	2.6861	2.7381	-	0.380	1.021
92.4	2.2634	1.7601	0.5033	0.180	0.778
91.8	1.9407	1.2197	0.7210	0.135	0.692
93.0	2.8456	2.7821	0.0635	0.495	0.978
94.1	2.5284	2.5834	-	0.775	1.020
93.7	1.9576	1.3677	0.5898	0.095	0.699
94.5	2.6742	2.6958	-	0.622	1.008

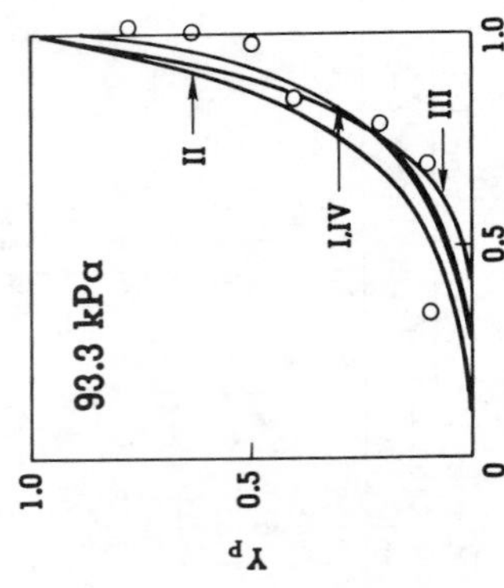

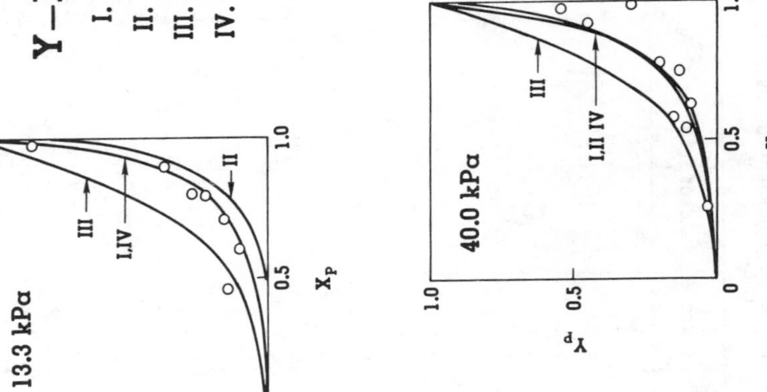

Figure 3.

Y – X DIAGRAMS

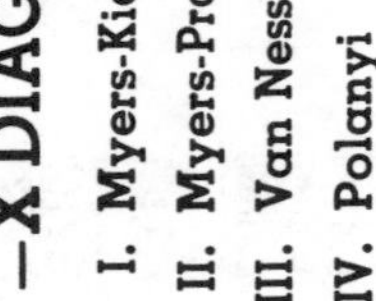

I. Myers-Kidnay
II. Myers-Prausnitz
III. Van Ness
IV. Polanyi

66.7 kPa

For most of the data, the predicted values for the phase equilibria concentrations were generally lower than the experimental values. This appears to indicate that the adsorbed film may not be behaving as an ideal phase. Using the method of Myers and Prausnitz (1), activity coefficients were calculated. The spreading pressure curves evaluated by Equation (4) are shown in Figure 4. The activity coefficients thus calculated by Equation (3) are shown in Figure 5.

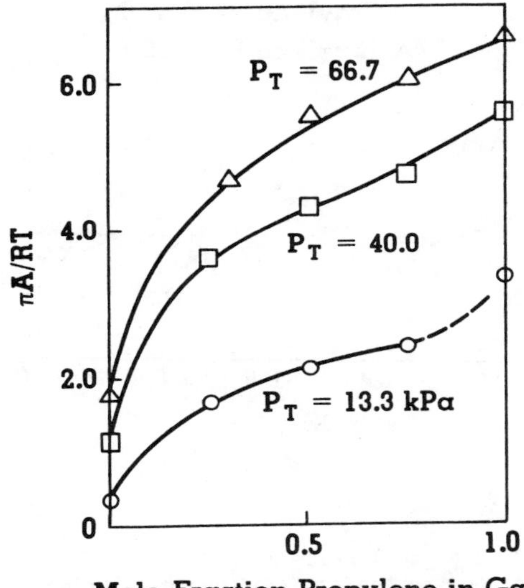

Figure 4. Mixture Spreading Pressures.

DISCUSSION

The pure component data obtained with the "mixing" apparatus shown in Figure 1 compared favorably (±3%) with the data generated with the more accurate "static" equipment. Most of the error (∿2%) is associated with the determination of the quantity of gases added to the system during an experiment. For most studies this level of precision is probably adequate but for the careful evaluation of the various thermodynamic theories the level of precision needs improvement. In the case of binary adsorption, the average deviation in the composition of the two phases is estimated to be 15%, while the difference between individual replicate samples averages 3%. Most of this observed error is associated with accumulative errors in material balances relating total material added and that remaining in the gas phase. The greatest relative error in the total quantity adsorbed was observed, as expected, at the lowest pressures and is estimated to be less than 10%.

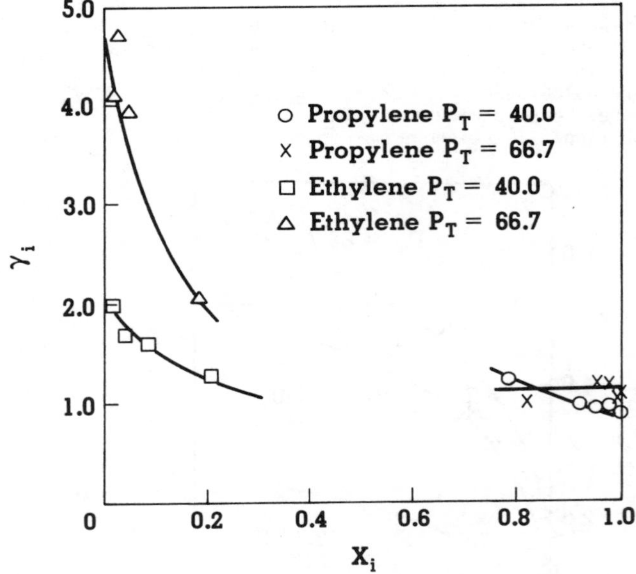

Figure 5. Activity Coefficient.

As noted earlier, the equilibrium compositions for the binary adsorption data as predicted by the ideal adsorbed solution theory are on the order of 20% lower than those observed. This is true for both the predictions based on the Gibbs spreading pressure theory or on the Polanyi Potential theory. Both theories also predicted a greater total quantity adsorbed than was measured, with the greatest deviation being at the lower propylene compositions. It was necessary in the case of the Polyani Potential method to use an arbitrary correction factor in order to obtain a single characteristic curve. The value chosen was 0.903 V for the affinity factor for ethylene. Ethylene was chosen because it is above its critical temperature at the conditions studied here. The Polanyi method gave a somewhat unusual (∿) curve for the total quantity adsorbed as a function of gas composition at a constant total pressure. The meaning of this observation is not clear but may be an artifact of the calculation. Neither the observed data nor the Gibbs technique gave any indication of this behavior.

By means of the concept of activity coefficients it was possible to achieve better agreement between the predicted and observed phase equilibria. The behavior of the calculated activity coefficients is observed to be at least qualitatively correct. Insufficient data for this system is available at present to check the validity of these coefficients or to predict them by some independent means such as that presented by Hoory (5).

Further experimental studies are required for the complete evaluation of the various theories and modifications thereof. In particular for this and similar type systems, more data are required at the low mole fraction range (0-0.25) and in the low pressure region for both components. Additional studies at different temperatures would aid in the evaluation of the applicability of the Polanyi theory. Greater precision in the data needs to be obtained in order to more fully evaluate the applicability and accuracy of the various techniques. A constant volume adsorption cell with the capability or providing in-situ gas analyses similar to one developed in our laboratories for the study of adsorption with reaction (Stolk (11)) may prove helpful in generating the required data.

CONCLUSIONS

Both pure component and binary adsorption isotherms were determined on activated carbon at 303.0°K and over a pressure range from 3.3 to 100.0 kPa. None of the usual isotherm equations were found capable of representing the experimental data over the entire pressure range.

Binary adsorption data and equilibrium concentration relationships between vapor and adsorbed phases were determined. Multicomponent adsorption equilibria were predicted using the Myers-Prausnitz and Grant-Manes theories for an ideal solution type adsorbed phase. In general, these theories were capable of predicting the total amount adsorbed to within 15% of the experimentally determined values. Predicted phase compositions deviated from experimental data by as much as 30%. This analysis indicates that ethylene and propylene mixtures may form a non-ideal adsorbed phase, particularly at low concentrations.

The results of this study indicate the need for more extensive and precise data, particularly in the low pressure region, in order that the various theories, procedures and their modifications can be more fully evaluated. This need is most important if the thermodynamic consistency of the results is to be determined.

NOTATION

A = adsorbent surface area (m^2/kg)
N = moles adsorbed (mmol/g)
P = pressure (kPa)
R = gas constant
T = temperature (°K)
V = molar volume (cc/gmole)
x = mole fraction adsorbed phase
y = mole fraction gas phase

Greek

ε = adsorption potential
γ = activity coefficient
$(\frac{\pi A}{RT})$ = spreading pressure function (mmole/g)

Subscript

i = component i
s = saturated conditions
t = total

LITERATURE CITED

(1) Myers, A.L. and Prausnitz, J.M., "Thermodynamics of Mixed-Gas Adsorption," AIChE J., 11, 121 (1965).

(2) Myers, Alan L., "Adsorption of Gas Mixtures," I & EC, 60, 45 (1968).

(3) Van Ness, H.C., "Adsorption of Gases on Solids," I & EC Fund., 8, 464 (1969).

(4) Grant, R.J. and Manes, Milton, "Adsorption of Binary Hydrocarbon Gas Mixtures on Activated Carbon," I & EC Fund., 5, 490 (1966).

(5) Hoory, S.E. and Prausnitz, J.M., "Monolayer Adsorption of Gas Mixtures on Homogeneous and Heterogeneous Solids," Chem. Eng. Sci., 22, 1025 (1967).

(6) Kidnay, A.J. and Myers, A.L., "A Simplified Method for the Prediction of Multicomponent Adsorption Equilibria from Single Gas Isotherm," AIChE J., 12, 981 (1966).

(7) Lewis, W., Gilliland, E., Chartow, B. and Cadogan, W., "Adsorption Equilibria-Hydrocarbon Gas Mixtures," I & EC, 42, 1319 (1950).

(8) Harris, B.L. and Emmett, P.M., "Adsorption Studies - Physical Adsorption of Nitrogen, Toluene, Benzene, Ethyl Iodate, Hydrogen Sulfide Water Vapor, Carbon Disulfide, and Pentane on Various Porous and Nonporous Solids," J. Phys. Chem., 53, 811 (1949).

(9) Russell, S., "Binary Adsorption of Various Mixtures of Ethylene and Propylene on Activated Carbon," M.S. Thesis, The Ohio State University (1970).

(10) Svanks, Karlis, "Flow Characteristics of Adsorbed Gas Through Vycon," Ph.D. Dissertation, The Ohio State University (1966).

(11) Stolk, R.D. and Syverson, A., "Adsorption Studies at Reaction Conditions - Reactor Development and Evaluation for Transient Studies and Millisecond Rates," ACS Sympos. 65, 50 (1978).

Correlation and Analysis of Equilibrium Isotherms for Hydrocarbons on Zeolites

Douglas M. Ruthven and Marcus Goddard

Department of Chemical Engineering
University of New Brunswick
P.O. Box 4400
Fredericton, N.B., Canada
E3B 5A3

ABSTRACT

The development of the statistical approach to the analysis and correlation of adsorption equilibrium data for zeolites is reviewed. A generalized statistical model isotherm is proposed and applied to the analysis of experimental equilibrium data for several hydrocarbons on zeolites NaX and NaY, including both single component and binary systems.

SCOPE

The methods of statistical thermodynamics have been applied by many authors to both the *a priori* prediction of Henry constants and the correlation and analysis of equilibrium data at higher concentration levels. Earlier attempts to develop a suitable general isotherm depended on approximating the configuration integral by a suitable simplified expression. Although such models have proved successful for several systems there are many other systems which do not conform. In order to increase the generality we have suggested the use of a general form of isotherm equation in which the non-ideal parts of the configuration integrals are retained as the model parameters. The equilibrium isotherms for several systems are interpreted in terms of this model and the relationship with classical ideal adsorbed solution theory is defined.

CONCLUSIONS AND SIGNIFICANCE

The general model provides a good representation of the single component isotherms for all the systems investigated. Sorbate-sorbate interaction effects are generally found to be insignificant in cages containing only two molecules and for some systems such effects are still small in cages containing three sorbate molecules. In this respect benzene appears anomalous since interaction effects are significant even at low concentrations. It is shown that the assumption that the configuration integrals for the binary subsystems may be approximated by the geometric means of the values for the corresponding single components is equivalent to the ideal adsorbed solution model of Myers and Prausnitz (20). This approximation provides a good prediction of the binary equilibrium isotherm for n-heptane-cyclohexane on 13X from the single component isotherm constants.

Because of their structural regularity and the existence of more or less discrete cages which can accommodate several guest molecules, zeolitic adsorbents are well suited for the application of the statistical approach to the correlation and analysis of adsorption equilibrium data. The basic theory, originally given by Hill (1), has been applied to a wide range of zeolitic systems by numerous author (2-6). The approaches vary from more or less exact *a priori* calculations of configuration integrals to the development of simplified model isotherms which use the general form of the statistical relationships as a basis for the correlation of experimental data. In this brief review some of the more important results of these investigations are summarized, a simple generalization of the statistical model isotherm is suggested and the relationship between the statistical approach and classical thermodynamic theory is clarified.

Calculation of Henry Constants and Limiting Heat of Sorption

The ratio of equilibrium concentrations between adsorbed and fluid phases is given, according to the basic principle of statistical thermodynamics (1), by:

$$c/c_g = (f_s/f_g)e^{-u/kT} \qquad (1)$$

In the low concentration limit f_s becomes independent of concentration and Equation (1) reduces to Henry's Law. If there is no change in rotational or internal freedom on sorption the ratio of partition functions reduces simply to the configuration integral:

$$\frac{c}{c_g} = K = \int_V \exp[-u(\underline{r})/kT]\cdot d\underline{r} \qquad (2)$$

which is evaluated over all possible positions of the probe molecule within the zeolite cage. For the inert gases, which have no rotational or internal freedom, Equation (2) provides, in principle, an exact method for the *a priori* calculation of the Henry constant. The potential energy may be calculated as a function of position by summing the pairwise contributions from each atom in the zeolite lattice. For non-polar sorbates the main contribution to the potential arises from dispersion-repulsion energy and the contribution from polarization is relatively small. Such calculations have been carried out by many authors with varying success. A useful summary has recently been given by Kiselev et al. (7-9). The main difficulty is that the semi-empirical expressions (Slater-Kirkwood, London and Kirkwood-Müeller) used to predict the dispersion constants are not very accurate (10). Kiselev recommends the Kirkwood-Müeller expression but this generally predicts dispersion constants which are somewhat too large and an empirical correction factor was therefore introduced which, for Ar, Kr and Xe on KX, NaX and NaY, varies from 0.7 to 0.95.

An alternative approach was adopted by Soto et al. (11) who evaluated the Henry constant in generalized form as a function of the dispersion constant and showed that the values required to match the experimental data were of the same order as the values estimated from the Slater-Kirkwood and Kirkwood-Müeller expressions.

Similar calculations may be extended to polyatomic and polar molecules but it then becomes necessary to calculate also the electrostatic contributions to the potential energy and to take account of molecular rotation in computing the configuration integral. For ethane and propane in NaX and KX zeolites Kiselev and Du (9) found good agreement between theory and experiment with correction factors of the same order as for the inert gases.

To extend this approach to higher concentration levels outside the Henry's Law region requires the calculation of configuration integrals for cages containing more than one sorbate molecule as attempted by Sargent and Whitford (12). Although the numerical intergrations can be carried out, the increased importance of sorbate-sorbate interaction energy and the limited accuracy with which the relevant force constants and the electric field within the cage are knownleads to large errors in the calculated results. It has therefore proved more useful to adopt a less fundamental approach.

A Simple Model Isotherm

Consider a macro-system composed of M sub-systems (cages) each of which is statistically independent and can accommodate a maximum of m sorbate molecules. The grand partition function for such a system is (1):

$$\Xi = \xi^M = [1 + Z_1 a + Z_2 a^2 + \dots Z_m a^m]^M \qquad (3)$$

and the equation for the equilibrium isotherm is given by:

$$c = \frac{\partial \ell n \Xi}{\partial \ell n a} = \frac{Z_1 a + 2Z_2 a^2 + \dots + MZ_m a^m}{1 + Z_1 a + Z_2 a^2 + \dots Z_m a^m} \qquad (4)$$

where $Z_1, Z_2 \dots Z_m$ represent the configuration integrals for cages containing 1, 2 ... m molecules. If we assume that the molecules within a given cage are freely mobile or that all sites are equivalent, we may write:

$$Z_s a^s = \frac{(Kp)^s}{s!} \cdot A_s \qquad (5)$$

where A_s is a temperature dependent constant representing the non-ideal portion of the configuration integral. This yields for the isotherm equation:

$$c = \frac{Kp + (Kp)^2 A_2 + \frac{1}{2}(Kp)^3 A_3 + \dots + \frac{1}{(m-1)!} \cdot (Kp)^m A_m}{1 + Kp + \dots + \frac{1}{2}(Kp)^2 A_2 + \frac{1}{3!}(Kp)^3 A_3 + \dots \frac{1}{m!}(Kp)^m A_m} \qquad (6)$$

If there is no interaction between sorbate molecules, even when more than one molecule is present within the same cage, $A_1 = A_2 = \dots = A_m = 1.0$ and Equation (6) reduces to the isotherm equation suggested by Riekert (3). Such an assumption is clearly unrealistic except at low loading since, apart from any attractive interaction, the freedom of movement of a given molecule will be restricted by the finite size of the other molecules present within the cage.

If only the reduction in the free volume of the cage is considered, one may write, as rough approximation:

$$A_s = (1-s\beta/v)^s \;;\; Z_s a^s = \frac{(Kp)^s}{s!} \cdot (1-s\beta/v)^s \tag{7}$$

where β is the effective molecular volume of the sorbate which is given, approximately, by the van der Waals 'b'. This leads to the isotherm equation suggested by the present author (1,13,14):

$$c = \frac{Kp + (Kp)^2(1-2\beta/v)^2 + \ldots + (Kp)^m(1-m\beta/v)^m/(m-1)!}{1 + Kp + \frac{1}{2}(Kp)^2(1-2\beta/v)^2 + \ldots + (Kp)^m(1-m\beta/v)^m/m!} \tag{8}$$

where $m \leq v/\beta$. This expression has been shown to provide a useful representation of the isotherms for several non-polar sorbates in various cationic forms of zeolite A (15-17). With the reasonable assumption that the effective molecular volume increases somewhat with temperature, the increase in isosteric heat of sorption with coverage, which is generally observed for non-polar sorbates, is correctly predicted, thus providing an alternative to the more common hypothesis that such an increase in isosteric heat results from sorbate-sorbate attraction.(14)

It is somewhat more logical to write the Henry constant also as the product of an ideal Henry constant for a point molecule (K*) and a free volume reduction factor:

$$K = K^*(1-\beta/v) \;;\; Z_s a^s = \frac{(K^*p)^s}{s!} \cdot (1-s\beta/v)^s \tag{9}$$

This leads to a slightly modified form of isotherm equation(18):

$$c = \frac{Kp + (Kp)^2\left[\frac{1-2\beta/v}{1-\beta/v}\right]^2 + \ldots + \frac{(Kp)^m}{(m-1)!}\left[\frac{1-m\beta/v}{1-\beta/v}\right]^m}{1 + Kp + \frac{1}{2}(Kp)^2\left[\frac{1-2\beta/v}{1-\beta/v}\right]^2 + \ldots + \frac{(Kp)^m}{m!}\left[\frac{1-m\beta/v}{1-\beta/v}\right]^m} \tag{10}$$

In both Equations (8) and (10) the saturation limit is given by $c_s = m-1$.

Theoretical isotherms, calculated from Equation (10) are shown in Figure (1) for a range of values of c_s (or m). For $c_s = 1.0$ the expression reduces to the Langmuir isotherm while for large values of c_s the curves become numerically identical to the Volmer isotherm with $b = K/c_s$:

$$bp = \frac{\Theta}{1-\Theta}\exp\left(\frac{\Theta}{1-\Theta}\right) \tag{11}$$

The numerical difference between Equations (8) and (10) becomes significant only when m is small. The behaviour of Equation (10), is somewhat more realistic since it gives a smooth transition while Equation (8) gives an abrupt change between $v/\beta = 2.0$ and $v/\beta = 3.0$.

A General Statistical Model

The assumption that the main effect of sorbate-sorbate molecular interaction is simply to reduce the free volume of the cage is clearly inappropriate for many systems. In order to increase the generality of the statistical model we may consider Equation (6) as the fundamental expression for the isotherm, retaining the constants A_2, A_3 ... A_m as empirical parameters. This formulation is similar to the virial isotherm in that sufficient parameters may be included to fit most experimental isotherms.

The spreading pressure (ϕ) may be calculated from a single component isotherm directly by integration according to the Gibbs equation:

$$\ell n\xi = \frac{\phi_o}{RT} = \int_0^{p_o} c\frac{dp}{p} = \ell n\left\{1+Kp_o+\frac{A_2}{2}(Kp_o)^2+\frac{A_3}{3!}(Kp_o)^3 + \dots\right\} \qquad (12)$$

This provides a convenient means of calculating the parameters $A_2, A_3 \dots$ from the equilibrium data as well as giving the link between the statistical and classical approaches to the theory of adsorption equilibrium. In order to obtain reliable values for the higher order parameters the Henry constant must first be determined accurately and the coefficients must then be evaluated in sequence. Values determined directly by multivariable optimization of the fit of the experimental isotherm to Equation (10) are less reliable due to the tendency for compensation between terms of different order.

As examples of the application of this model experimental equilibrium isotherms for several hydrocarbons on faujasite type zeolites are shown in Figure (2) as plots of $(\xi-1-Kp)/(Kp)^2$ vs Kp. For these sorbates the saturation capacity is three molecules per cage so that, according to Equation (12), such a plot should be linear with intercept $A_2/2$ and slope $A_3/6$. For smaller molecules such plots show an upward curvature but the limiting slope may still be determined quite easily.

The isotherms for a number of zeolitic systems have been analyzed in this way and representative values of the coefficients so derived are summarized in Table (1). For different patterns of behaviour may be identified as follows:

1. $A_2 \approx A_3 \approx 1.0$ — No significant intermolecular interaction even with 3 molecules/cage. (C_6H_{12}, C_2H_6, C_2H_4, o and m-xylene)
2. $A_2 \approx 1.0, A_3 << 1.0$ — No interaction with two molecules/cage but significant repulsion or restriction of rotation with 3 molecules/cage. (nC_2H_{16}, toluene, ethylbenzene)
3. $A_2 \approx 1.0, A_3 >> 1.0$ — No interaction with two molecules/cage but significant attraction for three molecules/cage. (p-xylene)
4. $A_2 << 1.0, A_3 << A_2$ — Repulsive interaction or restriction of rotation even at two molecules/cage. (benzene)

The most striking feature is that intermolecular attraction appears to be of minor importance in most systems; in only one system (p-xylene-NaY) were the A_3 coefficients significantly greater than A_2. The behaviour of benzene also appears anomalous since this is the only system for which

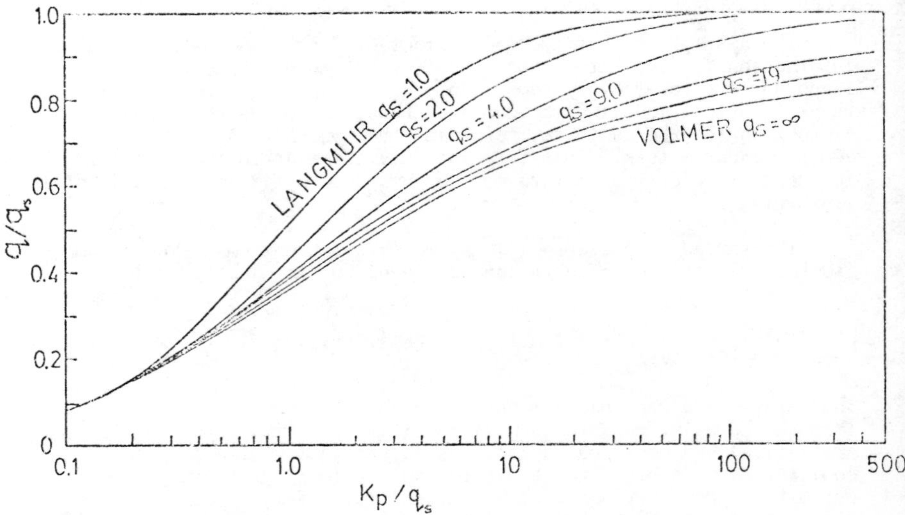

Figure 1. Theoretical Isotherms from Equation (10). ($q \equiv c$; $q_S \equiv c_S$)

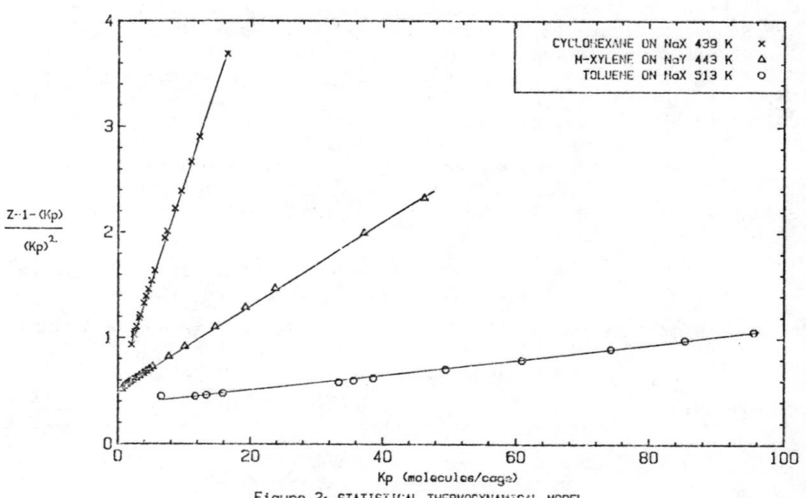

Figure 2: STATISTICAL THERMODYNAMICAL MODEL

interaction effects are important even when only two molecules are present within the cage.

If the parameters A_2,A_3 etc. are not significantly temperature dependent it follows from Equation (6) that a plot fo c vs Kp will provide a generalized isotherm giving a concise correlation of the equilibrium data at all temperatures as illustrated in Figure (3) for n-heptane-13X.

Extension to Binary Systems

The extension of Equation (6) to a binary system (A,B) may be written in the general form:

$$c_A = \frac{K_A p_A + \sum_j \sum_i \frac{(K_A p_A)^i (K_B p_B)^j}{(i-1)!j!} \cdot A_{ij}}{1 + K_A p_A + K_B p_B + \sum_j \sum_i \frac{(K_A p_A)^i (K_B p_B)^j}{i!j!} A_{ij}} \tag{13}$$

with a similar expression for c_B. The summations are carried out over all values of i and j satisfying the requirement $i\beta_A + j\beta_B \leq v$. If there are no non-ideal interactions in the mixed adsorbed phase one may write, as an initial approximation:

$$A_{ij} = (A_{An}^{\ i} \cdot A_{Bn}^{\ j})^{\frac{1}{i+j}} \tag{14}$$

where A_{An},A_{Bn} represent the coefficients for subsystems with n molecules of A and B respectively in the pure component systems. The expression for the spreading pressure for the pure components (Equation (12)) for either A or B may be written:

$$\exp(\phi_o/RT) = 1 + Kp_o + \frac{(K'p_o)^2}{2} + \frac{(K''p_o)^3}{3!} + \ldots \tag{15}$$

where $K' = KA_2^{\frac{1}{2}}$, $K'' = KA_3^{1/3}$ .... while, with the approximation of Equation (14) the expression for the spreading pressure of the mixture may be shown by simple algebra to reduce:

$$\xi = \exp(\phi/RT) = 1 + K_A p_A + K_B p_B + \tfrac{1}{2}(K_A' p_A + K_B' p_B)^2 + \frac{1}{3!}(K_A'' p_A + K_B'' p_B)^3 + \ldots \tag{16}$$

The concentrations are given by $c_A = (\partial \ell n \xi / \partial \ell n p_A)_{p_B}$ whence:

$$X_A = \frac{c_A}{c_A + c_B} = \frac{K_A p_A + K_A' p_A (K_A' p_A + K_B' p_B) + \tfrac{1}{2} K_A'' p_A (K_A'' p_A + K_B'' p_B)^2 + \ldots}{(K_A p_A + K_B p_B) + (K_A' p_A + K_B' p_B)^2 + \tfrac{1}{2}(K_A'' p_A + K_B'' p_B)^3 + \ldots} \tag{17}$$

Clearly if $K_A' p_{oA} = K_A' p_A = K' p_{oB}$ and $K'' p^o_A = K_A'' p_A + K_B'' p_B = K_B'' p^o_B$ then:

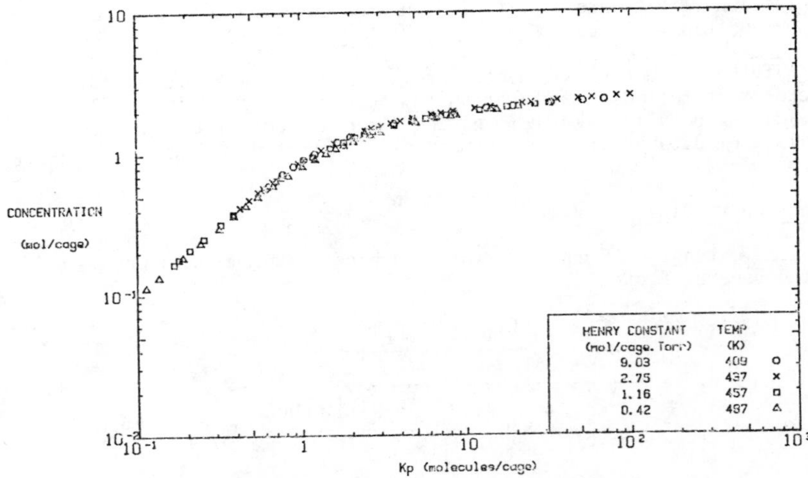

Figure 3: GENERALIZED ISOTHERM FOR N-HEPTANE ON NaX

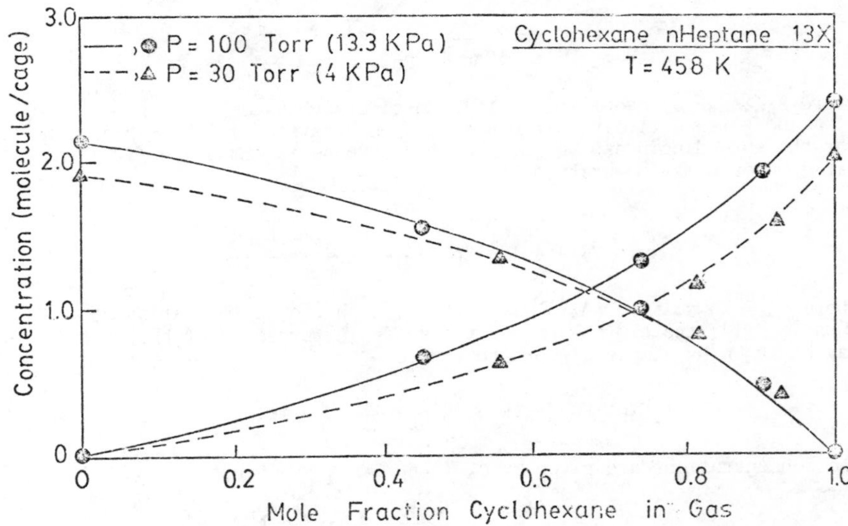

Figure 4. Binary Isotherms for C_6H_{12}-nC_7H_{16}-13X Showing Comparison of Experimental Points with Theoretical Curves from Equations (13) and (14).

$$\phi_{oA} = \phi_{oB} = \phi \; ; \; X_A = p_A/p^o_A \; ; \; X_B = p_B/p^o_B \qquad (18)$$

corresponding to an ideal adsorbed solution as defined by Myers and Prausnitz (20).

Representative binary equilibrium data for the system n-heptane-cyclohexane on 13X zeolite (25) at 409K are shown in Figure (4) together with the theoretical curves calculated from Equations (13) and (14) using the parameters derived from the single component isotherms (Equation (6)). It is evident that the theoretical curves provide a reasonable representation of the experimental data showing the system to be approximately ideal in accordance with the Myers-Prausnitz definition. Similarly good agreement was found for the system C_2H_6-C_2H_4-13X, which is not surprising since the data for this system have been shown previously to conform to the IAST model (19).

Ideal adsorbed solution behaviour appears to be quite common for adsorbed mixtures of simple molecules and on the basis of the present analysis one may suggest that such behaviour is likely up to moderate concentrations for any system in which the A_2 coefficients for both components are close to unity. However ideal behaviour is by no means universal and systems such as CO_2-C_2H_6-5A (21), C_2H_4-C_3H_8-5A (22), C_2H_4-cycloC_3H_6-5A (23) and C_2H_4-CO_2-5A (24) have been shown to be highly non-ideal. The hypothesis that ideal mixture behaviour can be expected only if $A_2 \sim 1.0$ for both components could in principle, be tested by examining the single component isotherm data for these systems but this has not yet been done.

It is possible by appropriate choice of the coefficients A_{ij} to correlate highly non ideal binary data by Equation (13) but since there is no obvious way to estimate these coefficients *a priori* such an approach reduces the model simply to a method of data correlation and is equivalent to the introduction of activity coefficients in the classical approach.

Table 1

Correlation of Isotherm Data According to Equation (6)

Sorbate	Sorbent	C_s	T(K)	$K^{\dagger}$	A_2	A_3	A_4
C_6H_{12} *	NaX	3	439	0.38	0.99	1.45	-
			458	0.18	1.12	1.25	-
			488	0.078	1.10	1.59	-
C_2H_6 +	NaX	6	298	0.018	0.97	1.1	1.1
			323	0.009	1.01	1.02	0.54
C_2H_4 +	NaX	6	298	0.16	0.99	1.17	0.54
nC_7H_{16} *	NaX	3	409	9.0	1.1	0.2	-
			457	1.2	0.95	0.053	-
C_6H_6 *	NaX	4	458	10.6	0.65	0.096	0.0032
			488	3.6	0.37	0.11	0.0035
			513	1.25	0.23	0.104	0.003
C_6H_6	NaY	4	405	34	0.37	0.047	
			448	5	0.75	0.036	
$C_6H_5 \cdot CH_3$ *	NaX	3	513	4.95	0.81	0.001	
$C_6H_5 \cdot C_2H_5$	NaY	3	477	15.9	0.97	0.007	
$o\text{-}C_6H_4(CH_3)_2$	NaY	3	477	8.7	1.04	1.01	
$m\text{-}C_6H_4(CH_3)_2$	NaY	3	477	5.9	0.98	2.3	
$p\text{-}C_6H_4(CH_3)_2$	NaY	3	477	5.7	1.16	4.8	

† In molecules/cage.Torr. * Data of Doetsch (26). + Data of Danner and Choi (19). Other systems from Goddard (unpublished).

Notation

a	activity of sorbate
A_s	coefficient defined according to Equation (5) for cage containing s sorbate molecules
A_{ij}	coefficient corresponding to A_s for cage containing i molecules of type A, j molecules of type B where s=i+j
b	Langmuir equilibrium constant = K/c_s
c	sorbate concentration in adsorbed phase (molecules/cage)
c_s	saturation limit
c_g	sorbate concentration in vapour phase
f_s	partition function for adsorbed molecule
f_g'	partition function per unit volume for vapour phase
k	Boltzmann constant (=R/N)
K	equilibrium (Henry's Law) constant
K',K"	modified values of K defined by $K'=KA_2^{\frac{1}{2}}$, $K''=KA_3^{1/3}$ etc.
K*	modified value of K (Equation (9))
m	maximum number of molecules in one cage
p, p_o	equilibrium pressure for single component system
p_A, p_B	partial pressures of A and B in binary system at equilibrium
R	gas constant
T	absolute temperature

u potential energy
v volume of zeolite cage
X_A mole fraction of A in adsorbed phase
Z_s configuration integral for cage containing s sorbate molecules
β effective molecular volume of sorbate molecule
Ξ grand partition function (Equation (3))
ϕ, ϕ_o spreading pressure (binary, single component)
ξ defined by $\ell n\xi=\phi_o/RT$
Θ fractional saturation = c/c_s

Literature Cited

1. Hill, T.L., Introduction to Statistical Thermodynamics, Addison Wesley, Reading, Mass (1960).
2. Bakaev, V.A. Doklady Akad. Nauk SSSR 167, (2), 369 (1966).
3. Riekert, L., Adv. Catalysis 21, 287 (1970).
4. Kiselev, A.V., Adv. Chem. 102, 37 (1971).
5. Brauer, P., A.A. Lopatkin and G. Ph. Stephanez, Adv. Chem. 102, 97 (1971).
6. Ruthven, D.M., Nature Phys. Sci. 232, (29), 70 (1971).
7. Kiselev, A.V., A.G. Bezus, A.A. Lopatkin and P.Q. Du, J. Chem. Soc. Faraday Trans II, 74, 367 (1978).
8. Kiselev, A.V. and P.Q. Du, Ibid 77, 1 (1981).
9. Kiselev, A.V. and P.Q. Du, Ibid 77, 17 (1981).
10. Derrah, R.I. and D.M. Ruthven, Can. J. Chem. 53, 996 (1975).
11. Soto, J.L., P.W. Fisher, A.J. Glessner and A.L. Myers, J. Chem. Soc. Faraday Trans I, 77, 157 (1981).
12. Sargent, R.W.H. and C.J. Whitford, Adv. Chem. 102, 144 (1971).
13. Ruthven, D.M., A.I.Ch.E. J. 22, 753 (1976).
14. Ruthven, D.M. and K.F. Loughlin, J. Chem. Soc. Faraday Trans I 68, 696 (1975).
15. Coughlin, B. and S. Kilmartin, J. Chem. Soc. Faraday Trans I 71, 1809 (1975).
16. Coughlin, B., J. McEntee and R.G. Shaw, J. Colloid Interface Sci. 52, 386 (1975).
17. Zuech, J.L., A.L. Hines and E.D. Sloan, Ind. Eng. Chem. Process Design Develop 22, 172 (1983).
18. Ruthven, D.M., Zeolites 2, 242 (1982).
19. Danner, R.P. and E.C.F. Choi, Ind. Eng. Chem. Fund. 17, 248 (1978).
20. Myers, A.L. and J.M. Prausnitz, A.I.Ch.E. Jl. 11, 121 (1965).
21. Glessner, A.J. and A.L. Myers, C.E.P. Symp. Series 65, 73 (1969).
22. Holborow, K.A. and K.F. Loughlin, A.C.S. Symp. Series 40, 379 (1977).
23. Loughlin, K.F., K.A. Holborow and D.M. Ruthven, A.I.Ch.E. Symp. Series (152), 71, 24 (1976).
24. Holborow, K.A., Ph.D. Thesis, Univ. of New Brunswick, Fredericton (1975).
25. Wong, F., M.Sc. Thesis, Univ. of New Brunswick, Fredericton (1979).
26. Ruthven, D.M. and Doetsch, I.H., A.I.Ch.E. Jl. 22, 882 (1976).

INFLUENCE OF ISOTHERM SHAPE AND HEAT OF ADSORPTION ON THE SORPTION KINETICS OF THE SYSTEM CO_2/H_2O/MOLECULAR SIEVE

Josef Schadl and Alfons Mersmann

Institute B for Chemical Engineering,
Techn. Univ. München
Arcisstr. 21, 8000 München 2, FRG

ABSTRACT

Fundamental studies on sorption kinetics were carried out with spherical single pellets in a flow system to determine the rate limiting step (external mass transfer, macro- or micropore diffusion or heat transfer). Good agreement with experimental results was obtained with a numerically calculated macropore diffusion-model taking into account the isotherm shape. Heat effects here are only of secondary influence.

In contrast, experimental results obtained from a nearly adiabatic fixed-bed adsorber show for high adsorptive concentrations a first mass transfer and a second heat transfer controlled zone.

Molecular sieves (ms) are widely used in chemical plants. By drying of air or purification of buffer gas, for instance, CO_2 and H_2O are removed by ms-adsorbers. Often the expenditures for desorption are decisive for the economy of adsorption processes. For the design of an adsorption process, both equilibrium and kinetic data are required. Contrary to adsorption, only a few studies for desorption were carried out. In principle three different procedures exist: temperature swing, pressure swing and displacement desorption. These processes in practice are mostly overlapped and often it is difficult to distinguish between them. The objective of our experiments was to obtain fundamental results for adiabatic adsorption as well as the purging step of pressure-swing desorption and thermal regeneration.
For adsorption, diffusion in the macropores of the porous adsorbents as the rate controlling step is commonly confirmed (1,2). The adsorption step resistance itself is negligible. For desorption however the findings are contradictory. Also the desorption step itself as the rate controlling factor is found (3,4). Thermal-swing processes are calculated only with a heat balance (5). With the "Equilibrium Theory", no

mass- and heat-transfer resistances are considered (6,7). Existent studies often postulate isothermal conditions and/ or linear adsorption isotherms. For adsorption, the influence of heat effects and isotherm curvature are shown by Ruthven (8).
For fundamental research, gravimetrical experiments were carried out with a single pellet in a flow system, because by this way it is possible to investigate separately the different parameters, such as pellet size, temperature, gas velocity and partial pressure jump. Simultaneously, the temperature profiles within a single pellet were measured. With regard to technical adsorbers, experiments in an adiabatic fixed-bed adsorber were carried out.

Equilibrium Isotherms

Adsorption systems are characterized at equilibrium by the isotherms. Besides the primary information about the adsorptive capacity $X = f(p,\vartheta)$ the curvature of the isotherm reveals the forces of attraction.

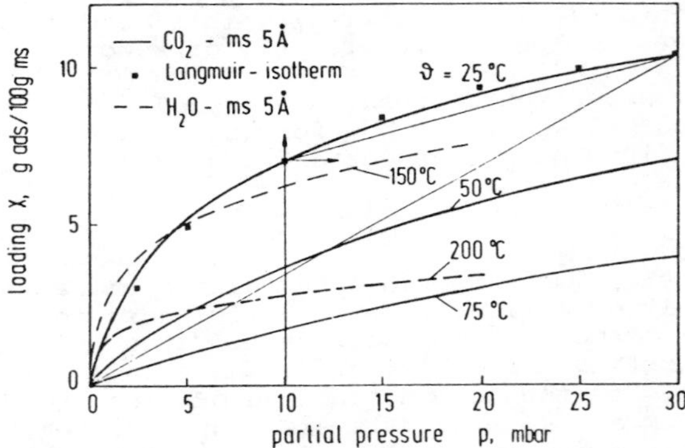

Figure 1. Adsorption isotherms of investigated systems continuous and broken lines: measured points: correlated with Langmuir-equation thin lines: influence of pressure jump on the deviation of the isotherm from a linear one

On account of the polarity of water, the H_2O-isotherms are highly convex. Because at ambient temperature, molecular sieves have an almost constant adsorptive capacity irrespective of the vapor pressure, with this system the purge gas desorption experiments must be carried out at higher temperatures. The very different driving force for desorption is revealed by the isosteres: for example, at $\vartheta = 50\ ^{\circ}C$ and $X = 0.02$ kg ads/kg ms, the equilibrium partial pressure is about 10^{-3} mbar for H_2O whereas it is 3.5 mbar for CO_2. In the special case of H_2O-preadsorption the CO_2-isotherm (reduced CO_2-capacity) and therefore the sorption rates are influenced by preadsorption (10). In practice, this is important for thermal regeneration with humid air.

Structure of Molecular Sieve Pellets

For the technical use of molecular sieves, the very small zeolite crystals (1 - 10 μm) are pelletized by means of a clay binder. By this production process, molecular sieves obtain a very characteristic bimodal pore structure (9). The adsorbate molecules diffuse through this macropore system. In figure 2 the macropore size distribution of two fractions is shown, measured by mercury penetration. The average pore diameter and free length of path are in the same order of magnitude. In this case, the pore diffusion coefficient is in the transition region between Knudsen (D_K) and free gas diffusion (D_G) (12).

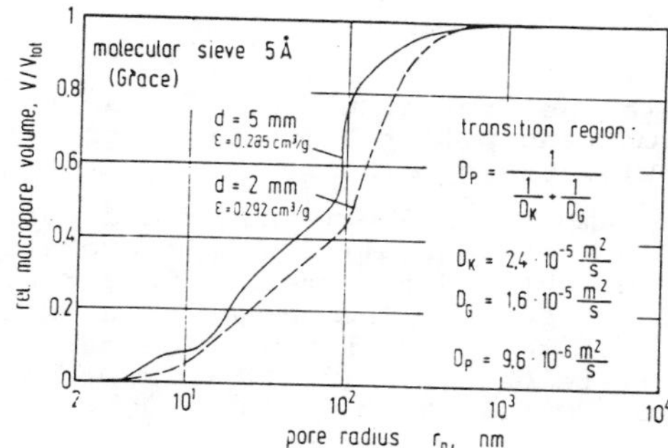

Figure 2. Macropore radius distribution and diffusivity (values: CO_2 in air, 25 °C, 1 bar, $\bar{r}_p$ = 80 nm)

SINGLE PELLET STUDIES

Apparatus and Procedure

By the experiments with single pellets in a flow system at ambient pressure the sorption and temperature profiles were measured simultaneously. Commercial spherical ms-pellets (Grace) were used. The change of loading was measured by means of an electronic microbalance; the temperature change was measured with very thin thermocouples, imbedded within the area of fracture of split pellets recemented after catious wetting with the original mixture of molecular sieve powder and alumina binder. Very thin NiCr-Ni-wires (30 μm in diameter) with small mass compared to pellet mass and with low thermal conductivity were used. The preparation influenced the sorption kinetics of the pellet only very slightly (10).

After regeneration of the adsorbent with dry air (300 °C, 3 h), the shell was thermostated. By switching over with solenoid valves different partial pressure jumps in the flow system are achieved. For the system CO_2-ms test gas flasks were available. For H_2O-ms, a part of the gas stream was wetted, mixed with the main stream, and afterwards cooled

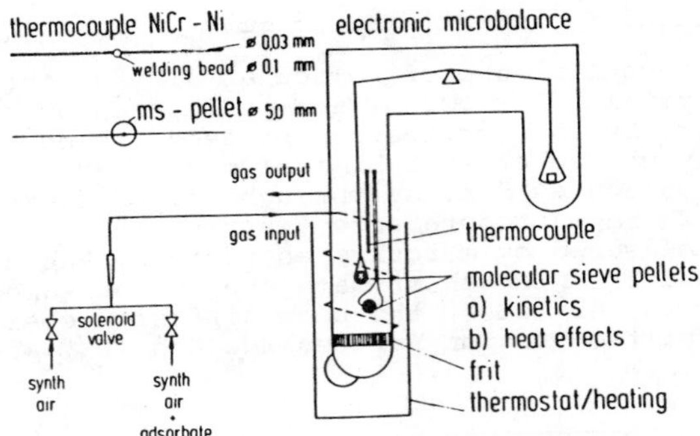

Figure 3. Experimental set-up: simultaneous measurement of sorption- and pellet temperature profiles after partial pressure jumps

down to the dewpoint temperature by means of a cryostat. Additionally the dewpoint was controlled by a hygrometer (Panametrics). Changes in weight and temperature were recorded with a continuous-line-recorder. The variables studied were: pellet diameter d_p (2.5 to 5 mm); superficial gas velocity, v_G (0.05 to 0.25 m/s); temperature (25 °C to 200 °C); and partial pressure jump (CO_2: 0 or 10 to 10 or 30 mbar), H_2O: dry gas up to 20 °C dew point, reciprocal for desorption)

Nonsteady Transport Processes

Adsorption and desorption processes are caused by the non-equilibrium of the system. If a partial pressure jump takes place in the ambient atmosphere of a pellet, the adsorbate diffuses in the pores, respectively to the surface of the pellet for desorption. The following steps can limit the sorption rate:

a) external mass transfer through the boundary layer,
b) diffusion in the macropore system,
c) diffusion in the micropores,
d) the sorption step itself,
e) internal or external heat transfer.

The effective molecular diameters of the two gas components are nearly the same (11) $d_{m,CO_2} = 2.8$ Å, $d_{m,H_2O} = 2.6$ Å compared with a zeolite pore diameter of about 5 Å. Therefore the activation energy for zeolitic diffusion is very low.

Assuming macropore diffusion after a partial pressure jump, the equation for nonsteady diffusion in a sphere is valid (13).

$$\frac{\partial X}{\partial t} = \frac{D_p/\mu_p}{1 + K} \left[\frac{\partial^2 X}{\partial r^2} + \frac{2}{r}\left(\frac{\partial X}{\partial r}\right) \right] ; \quad K = \frac{\rho_{ad}}{\varepsilon} \frac{\mathcal{R}T}{M_{ad}} \frac{\partial X}{\Delta p} \qquad (1)$$

An approximate solution exists only for the special case of a linear isotherm. However, for technical systems, the isotherm curvature is mostly very favourable for adsorption, but unfavourable for desorption. For these nonlinear systems Equation (2) is valid (2).

$$\frac{\partial X}{\partial t} = \frac{D_p/\mu_p}{1 + \alpha(X)} \left[\frac{\partial^2 X}{\partial r^2} + \frac{2}{r}\left(\frac{\partial X}{\partial r}\right) + \beta(X)\left(\frac{\partial X}{\partial r}\right)\right]^2 \qquad (2)$$

$$\alpha(X) = \frac{\rho_{ad}\ \mathcal{R}\ T}{\varepsilon\ M_{ad}} \cdot \frac{dX}{dp} \quad ; \qquad \beta(X) = -\frac{d^2X}{dp^2} \bigg/ \left(\frac{dX}{dp}\right)^2$$

Instead of the constant slope K (Eq. (1)) here $\alpha(X)$ is a function of the loading. $\beta(X)$ takes into account the curvature of the isotherm. For convex isotherms, caused by the term $\beta(X)$, the adsorption is faster and the desorption is slower, as for a linear one. This differential equation must be solved numerically.

Therefore the adsorbent pellet was divided into spherical shell elements. For each element, a mass balance was established. In this way, a system of n coupled differential equations resulted, which was solved numerically with the Runge-Kutta-Merson-method. The number of elements was varied from n = 5 up to 80 (computing time was proportional n^3). For n = 20 or greater no more influence on the relative loading X/X_{max} up to a three-digit accuracy was determined.

The measured isotherm was approximated by a Langmuir-equation (see figure 1, CO_2-ms 5 Å, ϑ = 25 °C) to obtain a differentiable function for the isotherm for reason of computation.

In general, transport by gas diffusion and in the adsorbed phase is possible. Here only gas phase diffusion was considered. Caused by the production processes, the walls of the macropores are mainly composed of the clay binder, where only small quantities of CO_2 and H_2O are adsorbed. Transport in the adsorbed phase, however, occurs only with higher degrees of loading. Also no capillary condensation was assumed to occur.

Results and Discussion

As shown in figure 4, good agreement between experimental and theoretical results was obtained.

The tortuosity factor μ_p = 5 was also found by other authors (2) to be of the same order of magnitude. Only in the beginning there are deviations between calculation and measurement. Firstly, at very high mass transfer rates an influence of external mass transfer in the boundary layer is observed, which slows down the sorption process. This was also confirmed by calculating the film transport resistance.

Secondly, expecially for the adsorption process, a part of the preadsorbed nitrogen from the carrier gas air is displaced by CO_2 or H_2O. Therefore, no change in weight can be measured.

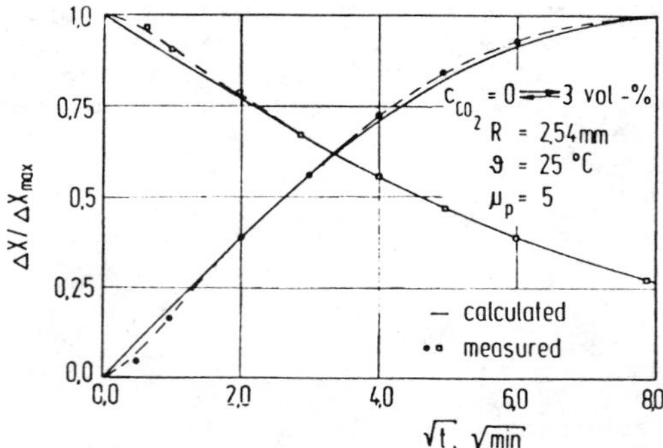

Figure 4. Sorption profiles with spherical single pellets (calculated with macropore diffusion model)

The remarkable slower desorption rate after identical partial pressure jumps in the ambient atmosphere, in comparison to adsorption, is explicable, if the numerically calculated local profiles of partial pressure and loading within the pellet are compared (figure 5). Assuming diffusion in the macropores to be the limiting factor, very different profiles result, caused by the unfavourable isotherm curvature for desorption.

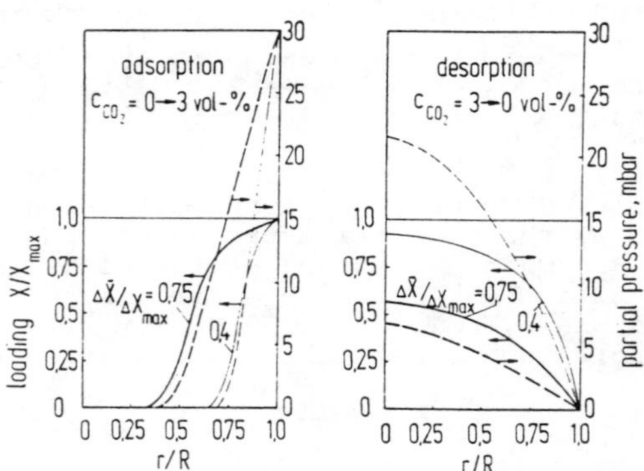

Figure 5. Calculated local partial pressure and adsorbent loading profiles inside a single pellet, showing the influence of shape of adsorption isotherm

At the same arbitrary chosen degree of loading change, the partial pressure profiles within the pellet illustrate that the pressure gradient is remarkably greater for adsorption in comparison to desorption. For example, at $\Delta X/\Delta X_{max}$ = 0.75 the driving partial pressure gradient between the center of the pellet and the surface for desorption is only about 6 mbar compared to 30 mbar for adsorption.

In contrast, for linear isotherms at the same degree of

change in loading homologous profiles for local partial pressure and loading within the pellet are obtained (10).

Experimental Verification

If macropore diffusion controls the sorption kinetics, the following proportionality is valid for short times:

$$\Delta X/\Delta X_{max} \sim \sqrt{t}/R^n$$

By varying the pellet diameter, the measurements can be fitted as shown in figure 6, with an exponent n = 0.9 instead of the theoretical value n = 1 for the pellet radius R.
The small deviation also confirms that the sorption processes are mainly governed by macropore diffusion. Other effects are only of secondary influence. Analogous results are obtained with measurements carried out with the system H_2O-ms 5 Å.

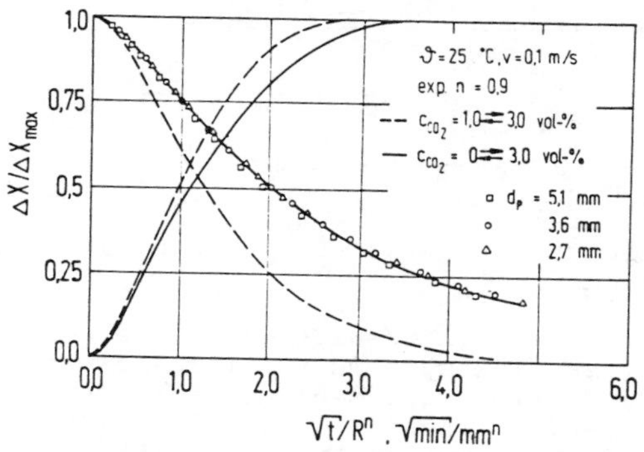

Figure 6. Measured sorption profiles with spherical single pellets
a) influence of pellet diameter
b) influence of isotherm curvature (pressure jump)

By varying the partial pressure jump, the influence of the isotherm curvature on sorption kinetics can be demonstrated. As shown in figure 1, the deviation from linearity for the system CO_2-ms at a partial pressure jump 10 ⇄ 30 mbar, corresponding to 1 ⇄ 3 vol-% CO_2, is much smaller compared with deviations for partial pressure jumps 0 ⇄ 30 mbar. For linear isotherms there should occur identical adsorption and desorption profiles. In figure 6 these theoretical considerations are confirmed. For smaller deviation from linearity the adsorption and desorption profiles are more similar.

Temperature Effects

As shown in figure 7 sorption and temperature profiles of single pellets are measured simultaneously. The change of temperature as a function of time is dependent on the sorption rate and heat of adsorption, heat transfer by convection (predominant in flow systems), and radiation.

$$m_p \cdot c_p \frac{dT}{dt} = q_{ad} \cdot m_p \frac{dX}{dt} - \alpha \cdot d_p^2 \cdot \pi (T - T_O) - \varepsilon_p \cdot c_S \cdot d_p^2 \cdot \pi (T^4 - T_O{}^4) \quad (3)$$

By approximating the measured sorption profiles with a polynomial (broken lines in figure 4) and under the following simplifying assumptions:

a) no internal heat transfer resistance in the pellet
b) constant heat of adsorption
c) ideal flow of gas around the single pellet
d) constant temperature of the carrier gas

the temperature profile of the pellet was calculated numerically by the Runge-Kutta-method.

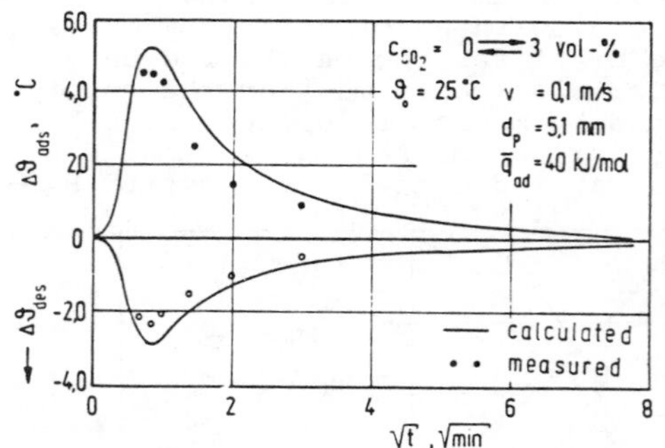

Figure 7. Temperature profiles of single pellets during sorption

In contrast to experimental results in a pure gas atmosphere (vacuum systems), in flow systems the temperature effects, especially for desorption, are negligible because of the better heat transfer. An essential hindrance (10) or even limitation (14) of the sorption by the heat transfer, as in a pure gas atmosphere, didn't occur with single pellets in the flow system. Figure 1 shows that for the adsorption of CO_2 (0 to 3 % step change), a 5 ^oC change in the adsorbent temperature changes ΔX_{max} from 10 to 9.7 g/100 g only. To a first approximation, an isothermal calculation is permissible.

ADIABATIC FIXED-BED STUDIES

Apparatus and Procedure

With regard to technical adsorbers an adiabatic fixed-bed was chosen for the experiments. By measuring the axial and radial temperature profiles with thermocouples placed at 0.25 m intervals along the bed the adiabaticity of two different constructions was checked. On one hand a very thin stainless steel tube with small heat capacity and heat conductivity (thickness of wall 0.25 mm, diameter 100 mm, length 1.5 m) was insulated with different insulation materials (thickness of layer 100 mm). Secondly a modified Dewar-glass double jacket was used as shown in figure 8.

The purified and dried air stream was adjusted with a mass flow controller. For adsorption, CO_2 was fed in and before entering the column the gas stream was thermostated. Alternately the feed concentration was controlled while the concentration profiles at the different working planes were measured continuously. Axial and radial temperature profiles were recorded with a 12-point-recorder, and concentration profiles were recorded with a continuous-line recorder.

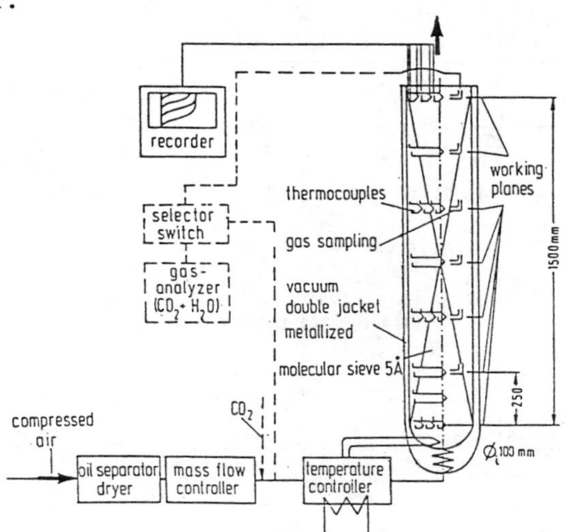

Figure 8. Experimental set-up to measure concentrations and temperature profiles (axial and radial) at different working planes

The comparison of the two different constructions showed that the modified Dewar-flask was advantageous. Firstly, the measured heat losses, especially during thermal regeneration, were smaller. Secondly, within the layer of insulation material a temperature profile occured. Concerning the storage term in the heat balance it is necessary to integrate over the thickness of the layer. With the Dewar-flask it can assumed that the inner jacket is at the fixed-bed temperature, whereas the outer jacket is nearly cold.
The variables studied were: pellet diameter $\bar{d}_P$ (2.0 and 3.8 mm), superficial gas velocity (0.05 to 0.4 m/s) and feed concentration (CO_2: 0.3 to 10 vol-%).

Results and Discussions

The concentration and the temperature profiles of an adiabatic column differ very strongly from earlier results with a "non-isothermal"* column (10) (diameter 50 mm, jacket tempered by means of a thermostat). For example, at a feed concentration c_{CO_2} = 3 vol-% for the non-isothermal case, a steady S-shaped break-through curve was obtained. For the adiabatic column measurements show a break in the concentration profiles.

* "non-isothermal" measurements in cooperation with U. Münstermann at our Institute.

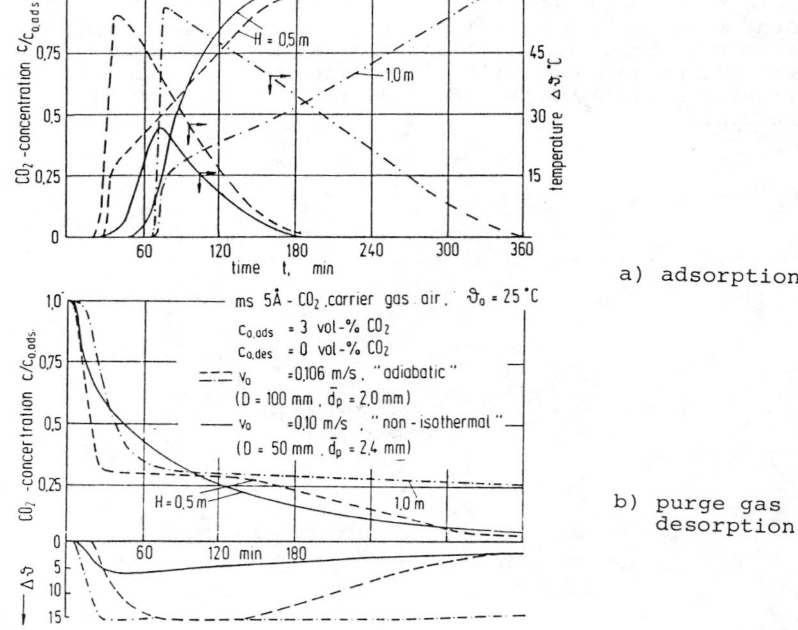

Figure 9. Concentration and temperature profiles in an adiabatic fixed-bed adsorber, showing a comparison of "non-isothermal" and adiabatic sorption

Here a first mass transfer and a second heat transfer controlled zone occur. This is confirmed by the temperature measurements: for adsorption in the adiabatic column the temperature rise is about 55 °C. Regarding the isotherms (figure 1), it can be noted that the loading capacity is decreased to about one third of the amount of the isothermal case. Further adsorption is limited by heat transfer. For adsorption, at feed concentrations of 3 vol-% CO_2, because of the high temperature effects heat losses through the column jackets occur. As shown by Sircar et al. (15) in this case no marked temperature plateaus occur, in contrast to results with 1 vol-% CO_2 feed concentrations ($\Delta\vartheta_{max}$ = 21 °C).

Thermal regeneration is necessary to obtain a fast and a high degree of desorption in the case of convex isotherms. After complete saturation, the adiabatic fixed-bed was heated with a hot gas stream. Within a short time the bed temperature rose slightly. This temperature remained nearly constant until, caused by complete desorption at the actual measuring points the temperature rose very steeply. Analogous results are reported by Basmadijan et al. (7) for the system CO_2-ms and Chi (5) for the system H_2O-ms.

Acknowledgement

Financial support of this project by the Deutsche Forschungsgemeinschaft, SFB 153, is gratefully acknowledged.

NOTATION

c	concentration
c_p	specific heat capacity
c_s	radiation constant
d	diameter
D	diffusivity
m	mass
M	molecular weight
p	pressure
q	heat of adsorption
r, R	radius
$\mathfrak{R}$	universal gas constant
t	time
T, ϑ	temperature (^oK, ^oC)
v	superficial gas velocity
X	adsorbate loading
α	heat transfer coefficient
ε	void volume fraction
μ_p	tortuosity factor
ρ	density

Indices

ad	adsorbate
ads	adsorption
des	desorption
p	pellet
O	ambient

LITERATURE CITED

1 Wicke, E.: Kolloid-Z., 93 (2), 129-157, (1940)

2 Jokisch, F.: Dissertation, TU Darmstadt (1975)

3 Fukunaka, P., Hwang, K.C. et al.: Ind. Eng. Chem. Proc. Des. Dev., 7 (2), 269-275, (1968)

4 Seewald, H., Jüntgen, H.: Ber. Bunsenges. phys. Chem., 81 (7), 638-645, (1977)

5 Chi, C.W.: AIChE Symp. Ser., 74 (179), 42-46, (1978)

6 Pan, C.Y., Basmadjian, D.: Chem. Eng. Sci., 25, 1653-1664, (1970)

7 Basmadjian, D., Ha, K.D., Proulx, D.: Ind. Eng. Chem. Proc. Des. Dev., 14 (13), 340-347, (1975)

8 Ruthven, D.M., Garg, D.R., Crawford, R.M.: Chem. Eng. Sci., 30, 803-810, (1975)

9 Ullmanns Enzyklopädie techn. Chemie, Vol. 2, Verlag Chemie, Weinheim (1972)

10 Münstermann, U., Mersmann, A., Schadl, J.: Ger. Chem. Eng. 6, 1-8, (1983)

11 Grubner, O., Jiru, P., Ralek, M.: Molekularsiebe, VEB Verlag, Berlin (1968)

12 Satterfield, O.M.: Mass Transfer in Heterogeneous Catalysis, M.I.T. Press, Cambridge, Mass. (1970)

13 Crank, J.: The Mathematics of Diffusion, Clarendon Press, Oxford (1975)

14 Doelle, H.J.: Dissertation, TU Karlsruhe, (1978)

15 Sircar, S., Kumar, R., Anselmo, K.H.: Ind. Eng. Chem. Proc. Des. Dev., 22, 10-15, (1983)

CORRELATION BETWEEN DESORPTION BEHAVIOUR AND ADSORPTION CHARACTERISTICS OF ORGANIC SUBSTANCES ON ACTIVATED CARBON

H. Seewald, J. Klein and H. Jüntgen
Bergbau-Forschung GmbH, BRD

ABSTRACT

The knowledge of desorption behaviour of loaded adsorbents is of great importance for characterization of the adsorbate and for design of adsorption plants. Whereas adsorption processes have been studied extensively desorption kinetics are not yet sufficiently known and accordingly the necessary data for adequate desorption models are not yet at hand.

Therefore, experiments are carried out to study the correlation between adsorption and desorption of organic substances on activated carbon with defined porous structure. Adsorption from gas phase and aqueous solution is followed by temperature programmed desorption in a thermobalance combined with a mass spectrometer. Isothermally and non-isothermally determined desorption curves can be described by means of a superposition model based on hypothetical load-dependent reaction parameters. In the case of physisorption a relation between the activation energy of desorption and the adsorption heat is established so that the model used enables the calculation of desorption behaviour from adsorption data.

It is shown, that surface reactions can take place during adsorption, which involve a change of the adsorbates, consisting of reversibly bonded compounds or reaction products and irreversibly bonded residues. Identifying the desorption products and evaluating the desorption curves it is possible to establish the type of reaction and to derive the binding energies of the adsorbate components. With regard to the quantity of desorbable substance an appropriate description of porous structure influenced by the deposit is developed.

INTRODUCTION

In recent years increased efforts were made for expanding the application range of carbonaceous adsorbents for adsorptive separation and cleaning processes.[1)] Particular attention is given in this context to the possibilities of controlled variation of micropore structures which in turn result from variation of individual process steps during manufacture as well as from the use of different raw materials and blends.

The characterization of micropore structure frequently is confined to the evaluation of adsorption isotherms according to suitable theories. However, this adsorption characteristic gives little information about transport phenomena, which essentially control the velocity of adsorption and desorption processes as well as heterogeneous reactions. If an adsorbate or reactant comes in contact with an adsorbent or porous solid particle, film and pore diffusion take place before reaching active sites at the inner surface and adsorption equilibrium. On the other hand desorption begins with a separation step. The predominant mechanism controlling the reaction velocity depends on flow conditions, pore width, heating rate etc. In the following a concise view on structural properties of commercial activated carbon is given, and some particular aspects of desorption kinetics in relation to equilibrium data shall be discussed.

ADSORPTION EQUILIBRIUM

The adsorption equilibrium in the gaseous phase is expressed by the equation as set up by Dubinin and Radushkevich[2]:

$$V = V_o \exp\left[-\left(\frac{\varepsilon}{E}\right)^2\right] \tag{1}$$

This equation stands for a relation between the adsorbate volume V - as calculated from the molar volume of the liquid for the range below the critical temperature and by the Van der Waals-constant for the range above the critical temperature - and the adsorption potential $\varepsilon = RT \ln (p_s/p)$, where p_s is the saturation pressure of the liquid or of a hypothetical liquid (R = gas constant, T = temperature in K, p = partial pressure of the adsorption). The adsorption energy E can be written as $E = \varepsilon_o \beta$. ε_o is the characteristic value for a reference adsorbate - very frequently benzene - and is an inverse measure for the average pore width. The affinity coefficient β describes the influence of the physical properties of the adsorbate relative to those of the reference adsorbate (where $\beta = 1$), and may be calculated from a relation between adsorption potential und pore width.[3]

For temperature invariance of the characteristic curve $V(\varepsilon)$ the differential heat of adsorption, Q, can be derived as

$$Q = \Delta H_v + \varepsilon_o \beta \left[\left(\ln \frac{V}{V_o}\right)^{\frac{1}{2}} + \frac{\alpha T}{2}\left(\ln \frac{V_o}{V}\right)^{-\frac{1}{2}}\right] \tag{2}$$

where ΔH_v is the heat of vaporization and α the thermal expansion coefficient of the adsorbate at temperature T.

The adsorption isotherms established at 25° C for a variety of gases and vapours on activated carbon made from hard coal are shown in linear mode in figure 1 according to equation (1). At low load values ($V/V_o < 0.1$) the plots for methane, ethylene and methanol show a break thus resulting in two straight sections. This probably is an indication for the existence of two different pore systems. With high load volumes or after extrapolation to $\varepsilon \rightarrow 0$ the isotherm intersection point is at V_o, except for the methane isotherm. This graph shows clearly that a standard description of adsorption equilibrium as per equation (1) is subject to some limitations. The calculation of the volume adsorbed in the range above the critical temperature as well as the description of the submicropore system where adsorption interaction might be different, as may be concluded from the flat methanol isotherm in the latter case, is somewhat difficult. It has been stated repeatedly that at high adsorption energy the postulate of the characteristic curve being temperature-invariant is no longer met.[4,5,6] Fig. 2 shows as an example the methane isotherms for 30, 40 and 50° C on an activated carbon. With increasing adsorption temperature the second straight section is shifted to lower values for the volume adsorbed.

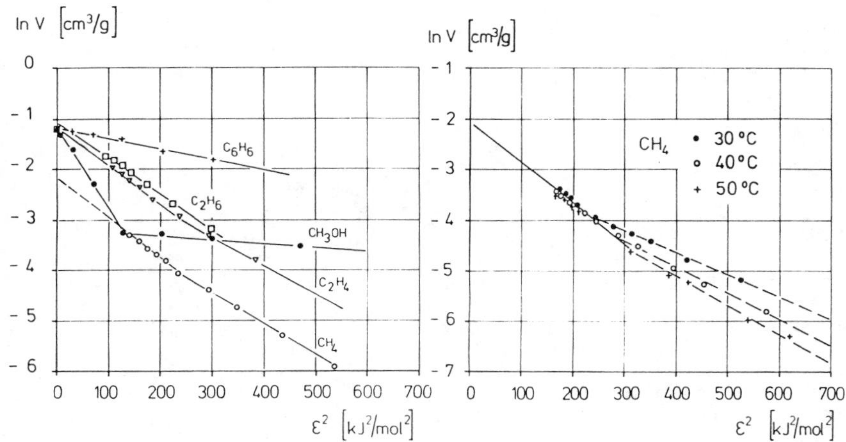

Figure 1. Adsorption isotherms of various adsorbates on activated carbon

Figure 2. Variation of adsorption isotherm (methane) with temperature

With regard to higher loadings only, it is sufficient for characterisation of the pore structure to have the micropore volume V_o and the adsorption energy ε_o as obtained from the benzene isotherm. Figure 3 shows a correlation between these structure data for granular commercial activated carbon. For activated carbon made from hard coal and differing only in the degree of activation - marked here by light points -, we obtain a curve showing clearly that with increase of micropore volume a decrease of adsorption energy i.e. a widening of pores is implied. Activated carbon made up by other production processes and from other raw materials - marked here by black dots - may also be plotted within narrow limits.

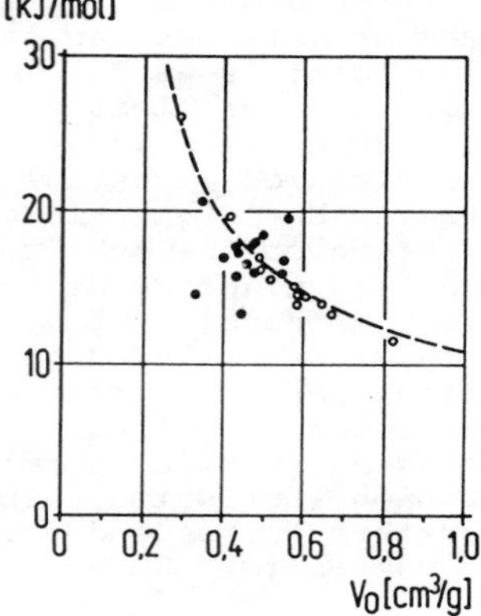

Figure 3. Adsorption energy as a function of micropore volume for commercial activated carbon

DESORPTION KINETICS

Unlike the exothermic adsorption, desorption is always endothermic. The desorption mechanism functions, depending on the bonding energy of adsorbate molecules, either by deconcentration viz. pressure diminution or by temperature, increase. As far as the desorption mechanism from heterogeneous surfaces viz. porous solids is concerned, the literature reflects contradictory opinions. This causes difficulties when having to select appropriate evaluation methods for non-isothermal and isothermaldesorption alternatives and when having to translate fundamental model concepts into large-scale separation. Thermo-gravimetric investigations into the desorption of organic vapours were carried out using adsorbents of varying pore structures[7,8]. The systems that were examined as well as the type of reaction control were selected in a way as to preclude diffusion as a rate-determining factor. Initial measurements were non-isothermal with a time-proportional temperature function. The non-isothermal method is superior to isothermal method in that wider load ranges are covered. This is due to the fact that any isothermal measurement is preceded by a heating-up period where certain reactions already take place. Furthermore the non-isothermal method simulates the technical heating conditions.

Desorption is considered a first order reaction, i.e. the desorption rate dn/dt is proportional to the number n of the molecules adsorbed:

$$\frac{dn}{dt} = -k\ n \qquad (3)$$

Considering the temperature/time function $T = T_o + mt$ as well as the Arrhenius' approach for the rate constant k, we obtain:

$$\frac{dn}{dt} = - \frac{k_o}{m} \exp \left[- \frac{E_D}{RT} \right] n \tag{4}$$

where k_o is the frequency factor and E_D the activation energy of the desorption. The desorption rate as a function of temperature shows a maximum. The initial curve rise is due to the Arrhenius' law, whereas the subsequent drop is caused by the decrease in concentration of adsorbed molecules. The example in figure 4 shows the desorption curves of benzene for varying initial loads n_o until equilibrium is reached. The load- or initial temperature was T_o = 25° C and the heating rate m = 2 K/min. With increasing load there is a temperature shift of the maximum desorption rate towards the lower temperature range. The sloping branches of the desorption curves finally merge. This desorption behaviour can be explained by the presence of load-dependent kinetic parameters. By evaluating the initial curve behaviour it is possible to associate both activation energy as well as frequency factor to the load status at any given moment. Figure 5 represents the activation energies obtained for different organic vapours with activated carbon, as a function of load. Compared with the adsorption heats calculated according to the potential theory there are slight deviations.

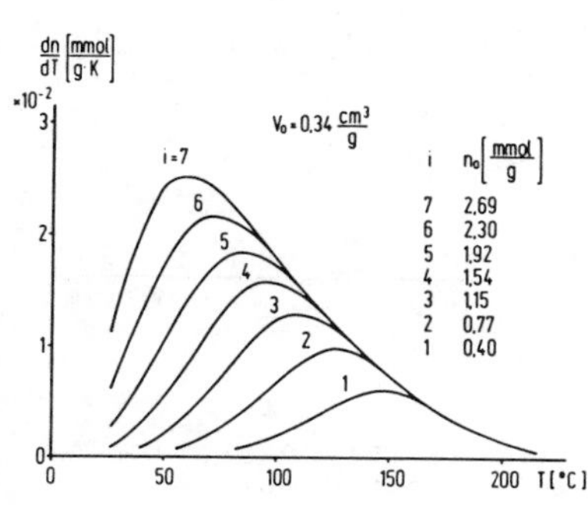

Figure 4. Non-isothermal desorption of benzene

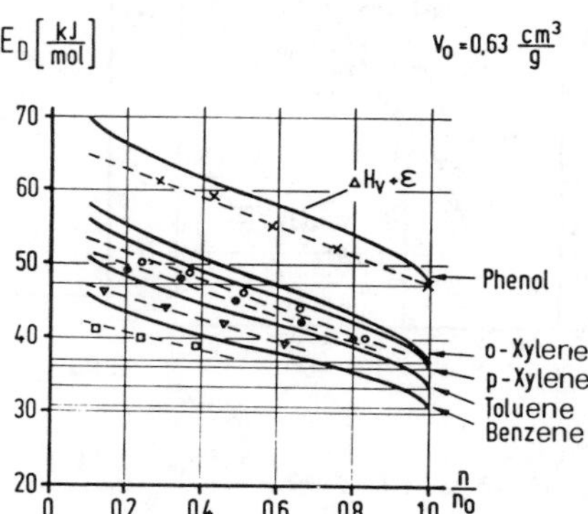

Figure 5. Comparison between activation energy of desorption and adsorption heat

The relationship between activation energies and derived frequency factors for desorption is shown on figure 6. Such a relationship corresponds to the compensating effects known from catalysis. Since adsorption heat is indicative of the amount of activation energy, the kinetic parameters needed to precalculate desorption processes can be derived from the aforementioned relationship.

It remains to be seen whether the reaction parameters obtained by the non-isothermal approach are also appropriate for describing isothermal processes and, if so, which type of model needs to be considered. On the one hand it is possible that the bonding characteristics of all the adsorbate changes with progressive desorption. On the other hand desorption may happen simultaneously everywhere - and independently of single aggregates of active sites - with in any case constant activation energies and frequency factors.

Figure 7 showing isothermal desorption of the toluene from an activated carbon is given as an example of the desorption mechanism calculated according to the two different models described above. From comparison with the measured data it becomes obvious that the desorption behaviour can be calculated and the second model is applicable.

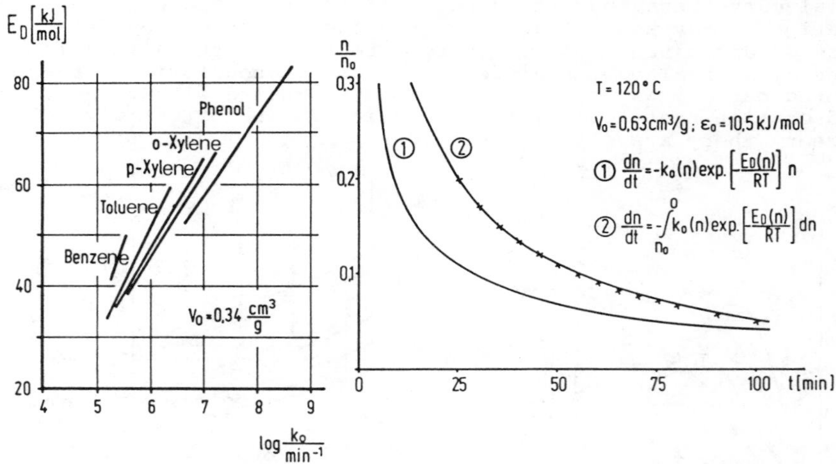

Figure 6. Correlation between activation energy and frequency factor of desorption

Figure 7. Isothermal desorption of toluene from activated carbon

SURFACE REACTIONS

The adsorption of organic substances on activated carbon from the gaseous phase and aqueous solution frequently is combined with surface reactions. These reactions involve a change of the adsorbates, consisting of reversibly bonded compounds or reactions products and irreversibly bonded residues. The knowledge of the nature of such reactions is of great importance to the perfomance of regeneration of carbon and the recovery of valuable compounds.

On different qualities of activated carbon desorption behaviour of some solvents, e.g. cyclohexanone and phenols, has been examined. After isothermal adsorption up to equilibrium at different temperatures above 25° and in various media, helium, air and aqueous solution, desorption measurements are carried out non-isothermally with a constant heating rate in a thermobalance combined with a mass spectrometer, using helium as a flow gas. Identifying the desorption products and evaluating the desorption curves, it is possible to establish the type of reaction and to derive the bonding energies of adsorbate components. By means of this method we are not able to clarify the nature of the active sites on the activated carbon surface, on which a small part of mineral matter is present.

As a first example of the measured desorption curves the figure 8 shows the influence of different adsorption atmospheres on desorption behaviour of cyclohexanone. We see a second and a third peak, if the activated carbon is loaded in the presence of oxygen.

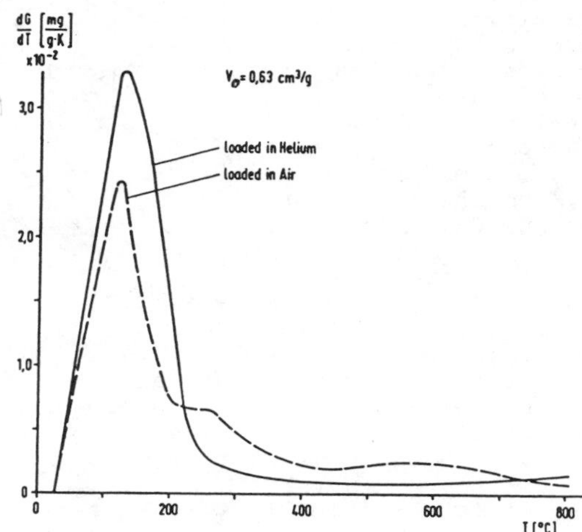

Figure 8. Desorption of cyclohexanone from activated carbon

Another example shows the influence of adsorption temperature (see figure 9). The desorption of phenol, which at room temperature is reversibly bonded, leads to two additional peaks, if the adsorption temperature is increased. These examples suggest that the reason for the different desorption behaviour is not the desorption reaction itself, but the adsorption condition. So we are able to analyse the state of adsorbate by desorption measurements.

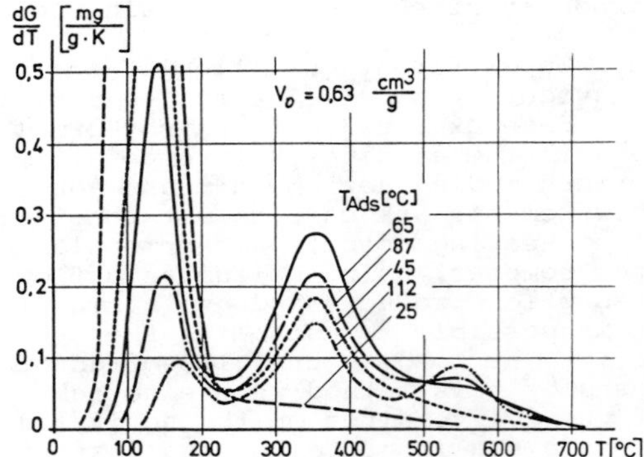

Figure 9. Influence of adsorption temperature on desorption of phenol

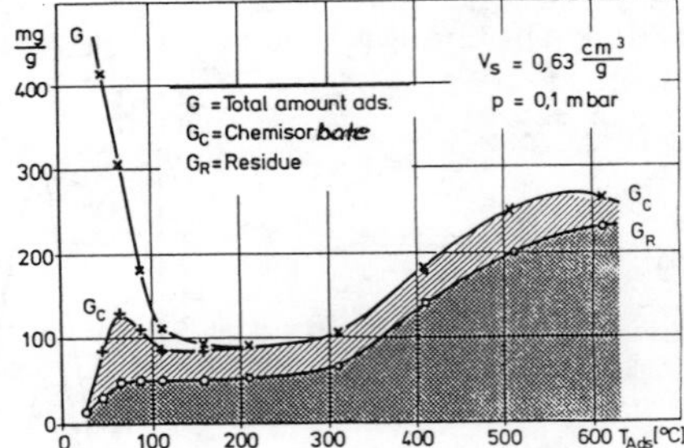

Figure 10. Phenol adsorption isobar

The thermogravimetric measurement enables a classification of the adsorbate, which distinguishes between the total amount adsorbed, G, the portion of adsorbate influenced by surface reactions, the "chemisorbate" G_C, and the non-desorbable residue G_R. G_C is determined by substracting the amount desorbed in the first peak up to a temperature of 250° C. Figure 10 shows the phenol adsorption isobar at a partial pressure of 0.1 mbar broken down by results of desorption measurements. Thus the typical curve of combined physisorption and activated adsorption with a minimum value of G is obtained.

It is possible, to take these results into account in description of the foregoing adsorption isotherms. Without the knowledge of the state of adsorbate the volume adsorbed has to be calculated from the liquid density. The correction of the adsorbate weight by the chemisorbate G_C leads to straight lines, which attach reduced volumina and wider pores to the reversibly bonded portion of adsorbate (see figure 11). This means that the chemisorbate can be considered as a part of the adsorbent, which is located in narrower pore regions.

In the temperature range above 300° C the kinetics of the deposit formation are studied, assuming a saturation

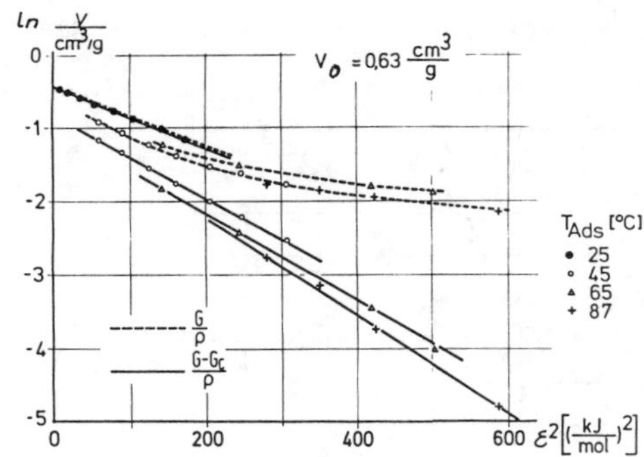

Figure 11. Influence of adsorption temperature on phenol isotherm

covering and a first order reaction related to the concentration in the gaseous phase. The determination of activation energy from weight increase amounts to the relatively low value of 46 kJ/mol[8]. The change of pore structure depending upon the amount of residue is given by adsorption isotherms. We find a reduction of pore volume, whereas the pore size distribution is influenced slightly. From the reduction of micro pore volume the density of the residue could be derived as 0.8 g/cm^3 in comparison to 2.0 g/cm^3 of the original activated carbon. Studies on the reactivation of carbon with steam have shown that the reactivity of the residue is higher than the reactivity of the carbon material[9].

Analysing the desorption products of G_C, mainly a mixture of hydroxy-diphenylether and dibenzofuran in the second peak and dibenzofuran in the third peak were found. Figure 12 shows the desorption curves of the identified components. The distribution of the desorbable reaction products is dependent on adsorption temperature, adsorption atmosphere and pore structure. In the case of adsorption from air (or from aqueous solution) the reaction is intensified. The characteristic phenomenon of the influence of oxygen is the increase of the non-desorbable residue.

The reaction scheme derived from the present results is shown in figure 13. In the proposed mechanism dehydrogenation is the predominant step of the surface reaction. It should be mentioned that this concept is confirmed by the desorption behaviour of cyclohexanone, which shows similar desorption curves like phenol. The additional peaks of which, however, have to be related to the phenol reaction. Adsorption of cresols leads to chemisorbate with a very small portion of desorbable products.

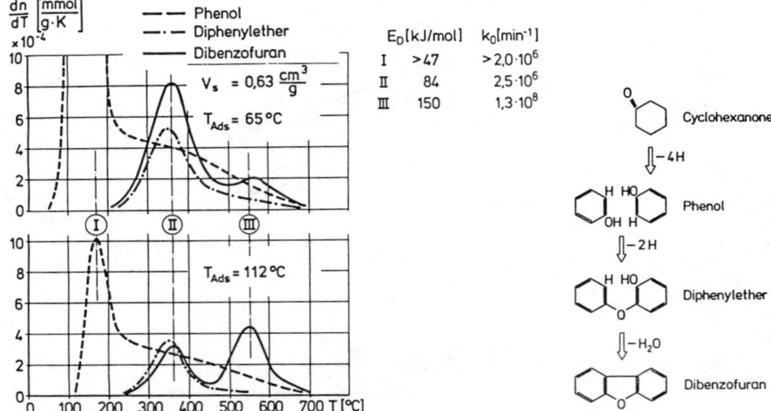

Figure 12. Desorption of reaction product

Figure 13. Reaction scheme

REFERENCES

1) Jüntgen, H., Carbon 15, 273 (1977)

2) Dubinin, M.M., Advan. Colloid Interface Sci., 2, 217 (1968)

3) Jüntgen, H. and H. Seewald, Ber. Bunsenges. phys. Chemie, 79, 734 (1975)

4) Halblaub, H. and K. Schäfer, Ber. Bunsenges. phys. Chemie, 80, 487 (1976)

5) Stoeckli, H.F. and J.Ph. Houriet, Carbon, 14, 253 (1976)

6) Hey, W. Thesis TH Aachen 1977

7) Seewald, H. and H. Jüntgen, Ber. Bunsenges. phys. Chemie, 81, 638 (1977)

8) Seewald, H., Thesis TH Aachen 1974

9) Klein, J. and H. Jüntgen, Paper presented at the EPA-Conference "Adsorption Techniques", Reston, Virg., 30. 4. - 2. 5. 1979

REPORTING PHYSISORPTION DATA FOR GAS/SOLID SYSTEMS

By Kenneth S. W. Sing

Department of Chemistry, Brunel University, Uxbridge, Middlesex, UB8 3PH, England.

ABSTRACT

A Subcommittee of the IUPAC Commission on Colloid and Surface Chemistry including Catalysis was formed in 1979 to consider the reporting of gas adsorption data and with the membership: K.S.W. Sing (Chairman), D.H. Everett, R. Haul, L. Moscou, R.A. Pierotti, J. Rouquerol, T. Siemieniewska. In view of the widespread use of physisorption measurements for the determination of surface area and pore size distribution, the Subcommittee decided to focus attention on this area of application of gas adsorption and to attempt to produce some notes of guidance in the form of an IUPAC manual. The present interim report of the Subcommittee has already appeared in the form of a 'provisional' publication (Pure and Applied Chemistry, 54, pp. 2201-2218, 1982) and is reproduced here by kind permission of the IUPAC.

The main objectives of the Subcommittee are: (1) to draw attention to the problems and ambiguities which have arisen in the reporting of physisorption data; (2) to formulate tentative proposals for the standardisation of procedures and terminology. The provisional manual is essentially a discussion document but it contains a set of general conclusions and recommendations including a check list to assist authors in the measurement and presentation of physisorption data.

SECTION 1. INTRODUCTION

Gas adsorption measurements are widely used for determining the surface area and pore size distribution of a variety of different solid materials, e.g. industrial adsorbents, catalysts, pigments, ceramics and building materials. The measurement of adsorption at the gas/solid interface also forms an essential part of most fundamental and applied investigations of the nature and behaviour of solid surfaces.

Although the role of gas adsorption in the characterisation of solid surfaces is firmly established, there is still a lack of general agreement on the evaluation, presentation and interpretation of adsorption data. Unfortunately, the complexity of most solid surfaces - especially those of industrial importance - makes it difficult to obtain any independent assessment of the physical significance of the derived quantities (e.g. the absolute magnitude of the surface area and pore size).

A number of attempts have been made (see Note a), at a national level, to establish standard procedures for the determination of surface area by the BET-nitrogen adsorption method. In addition, the results have been published (see Note b) of an SCI/IUPAC/NPL project on surface area standards. This project brought to light a number of potential sources of error in the determination of surface area by the gas adsorption method.

The purpose of the present Manual is two-fold: first to draw attention to the problems and ambiguities which have arisen in connection with the reporting of gas adsorption (physisorption) data and secondly to formulate tentative proposals for the standardisation of procedures and terminology which, through further discussion, will lead to a generally accepted code of practice. It is not the purpose of this Manual to provide detailed operational instructions or to give a comprehensive account of the theoretical aspects of physisorption. Determination of the surface area of supported metals is of special importance in the context of heterogeneous catalysis, but this topic is not dealt with in this Manual since chemisorption processes are necessarily involved.

The present proposals are based on, and are in general accordance with, the *Manual of Symbols and Terminology for Physicochemical Quantities and Units* (see Note c) and Parts I and II of Appendix II (see Note d). Although it has been necessary to extend the terminology, the principles are essentially those developed in Part I.

Note a. British Standard 4359: Part 1: 1969. Nitrogen adsorption (BET method).

Deutsche Normen DIN 66131, 1973. Bestimmung der spezifischen Oberfläche von Feststoffen durch Gasadsorption nach Brunauer, Emmett und Teller (BET).

Norme Française 11-621, 1975. Détermination de l'aire massique (surface spécifique) des poudres par adsorption de gaz.

American National Standard, ASTM D 3663-78. Standard test method for surface area of catalysts.

Note b. D.H. Everett, G.D. Parfitt, K.S.W. Sing and R. Wilson, *J. appl. Chem. Biotech.*, 24, 199 (1974).

Note c. *Manual of Symbols and Terminology for Physicochemical Quantities and Units* prepared for publication by D.H. Whiffen, *Pure Applied Chem.*, 51, 1-41 (1979).

Note d. *Part I of Appendix II, Definitions, Terminology and Symbols in Colloid and Surface Chemistry*, prepared by D.H. Everett, *Pure Applied Chem.*, 31, 579-638 (1972).

Part II of Appendix II, Terminology in Heterogeneous Catalysis, prepared for publication by R.L. Burwell, Jr., *Pure Applied Chem.*, 45, 71-90 (1976).

SECTION 2. GENERAL DEFINITIONS AND TERMINOLOGY

The definitions given here are essentially those put forward in Appendix II, Part I, §1.1 and Part II, §1.2. Where a caveat is added, it is intended to draw attention to a conceptual difficulty or to a particular aspect which requires further consideration.

Adsorption (in the present context, positive adsorption at the gas/solid interface) is the enrichment of one or more components in an interfacial layer. *Physisorption* (as distinct from *chemisorption*) is a general phenomenon: it occurs whenever an adsorbable gas (the *adsorptive*) is brought into contact with the surface of a solid (the *adsorbent*). The intermolecular forces involved are of the same kind as those responsible for the imperfection of real gases and the condensation of vapours. With some adsorption systems, certain specific molecular interactions occur (e.g. polarisation, field-dipole, field gradient-quadrupole), arising from particular geometric and electronic properties of the adsorbent and adsorptive.

It is convenient to regard the interfacial layer as comprising two regions: the *surface layer* of the adsorbent (often simply called the *adsorbent surface*) and the *adsorption space* in which enrichment of the adsorptive can occur. The material in the adsorbed state is known as the *adsorbate*, i.e. as distinct from the adsorptive, the substance in the fluid phase which is capable of being adsorbed.

When the molecules of the adsorptive penetrate the surface layer and the structure of the bulk solid, the term *absorption* is used. It is sometimes difficult or impossible to distinguish between adsorption and absorption. In such cases it is convenient to use the wider term *sorption* which embraces both phenomena and to use the derived terms *sorbent*, *sorbate* and *sorptive*.

The term *adsorption* may also be used to denote the process in which adsorptive molecules are transferred to, and accumulate in, the interfacial layer. Its counterpart, *desorption*, denotes the converse process, i.e. the decrease in the amount of adsorbed substance. Adsorption and desorption are often used adjectivally to indicate the direction from which experimentally determined adsorption values have been approached, e.g. the adsorption curve (or point) and the desorption curve (or point). *Adsorption hysteresis* arises when the adsorption and desorption curves deviate from one another.

The relation, at constant temperature, between the quantity adsorbed (properly defined in Section 3.2) and the equilibrium pressure of the gas is known as the *adsorption isotherm*.

Many adsorbents of high surface area are porous and with such materials it is often useful to distinguish between the *external* and *internal* surface. The *external surface* is usually regarded as the geometrical envelope of discrete particles or agglomerates, but a difficulty arises in defining it because solid surfaces are rarely smooth on an atomic scale. A suggested convention is that the external surface be taken to include all the prominences and also the surface of those cracks which are wider than they are deep; the internal surface then comprises the walls of all cracks, pores and cavities which are deeper than they are wide and which are accessible to the adsorptive. In practice, the demarcation is likely to depend on the methods of assessment and the nature of the pore size distribution. Because the accessibility of pores may depend on the size and shape of the gas molecules, the area of, and the volume enclosed by, the internal surface as determined by gas adsorption may be controlled by the dimensions of the adsorptive molecules (*molecular sieve* effect). On a molecular scale the roughness of a solid surface may be characterized by a *roughness factor*, i.e. the ratio of the external surface to the chosen geometric surface.

It is expedient to classify pores according to their sizes:

(i) pores with widths exceeding about 50 nm (0.05μm) are called *macropores;*

(ii) pores with widths not exceeding about 2nm are called *micropores;*

(iii) pores of intermediate size are called *mesopores*.

These limits are to some extent arbitrary since the pore filling mechanisms are dependent on the pore shape and are influenced by the properties of the adsorptive and the adsorbent-adsorbate interactions. The whole of the accessible volume in micropores may be regarded as adsorption space and the process of *micropore filling* thus occurs as distinct from coverage of the external surface and the walls of open macropores or mesopores. Micropore filling may be regarded as a primary physisorption process (see Section 8); on the other hand, physisorption in mesopores takes place in two more or less distinct stages (monolayer-multilayer adsorption and capillary condensation).

In *monolayer adsorption* all the adsorbed molecules are in contact with the surface layer of the adsorbent. In *multilayer adsorption* the adsorption space accommodates more than one layer of molecules and not all adsorbed molecules are in contact with the surface layer of the adsorbent. In *capillary condensation* the residual pore space which is left after multilayer adsorption has occurred is filled with liquid-like material separated from the gas phase by menisci. Capillary condensation is often accompanied by hysteresis. The term capillary condensation should not be used to describe micropore filling.

For physisorption, the *monolayer capacity* (n_m) is usually defined as the amount of adsorbate (expressed in appropriate units) needed to cover the surface with a complete monolayer of molecules (Appendix II, Part I, §1.1.7). In some cases this may be a close-packed array but in others the adsorbate may adopt a different structure. Quantities relating to monolayer capacity may be denoted by the subscript m. The *surface coverage* (θ) for both monolayer and multilayer adsorption is defined as the ratio of the amount of adsorbed substance to the monolayer capacity.

The *surface area* (A_s) of the adsorbent may be calculated from the monolayer capacity (n_m in moles), provided that the area effectively occupied by an adsorbed molecule in the complete monolayer (a_m) is known.

Thus,

$$A_s = n_m . L . a_m$$

where L is the Avogadro constant. The *specific surface area* (a_s) refers to unit mass (m) of adsorbent:

$$a_s = \frac{A_s}{m} .$$

Appendix II, Part I recommends the symbols A, A_s or S and a, a_s or s for area and specific area, respectively, but A_s and a_s are preferred to avoid confusion with Helmholtz energy A or entropy S.

In the case of micropore filling, the interpretation of the adsorption isotherm in terms of surface coverage may lose its physical significance. In that event, it may be convenient to define a *monolayer equivalent area* as the area, or specific area, respectively, which would result if the amount of adsorbate required to fill the micropores were spread in a close-packed monolayer of molecules (see Section 8).

SECTION 3. METHODOLOGY

3.1 Methods for the determination of adsorption isotherms

The many different procedures which have been devised for the determination of the amount of gas adsorbed may be divided into two groups: (a) those which depend on the measurement of the amount of gas leaving the gas phase (i.e. gas volumetric methods) and (b) those which involve the measurement of the uptake of the gas by the adsorbent (e.g. direct determination of increase in mass by gravimetric methods). Many other properties of the adsorption system may be related to the amount adsorbed, but since they require calibration they will not be discussed here. In practice, a static or a flow technique may be used in the application of volumetric or gravimetric methods.

In the static volumetric determination a known quantity of gas is usually admitted to a confined volume containing the adsorbent, maintained at constant temperature. As adsorption takes place, the pressure in the confined volume falls until equilibrium is established. The amount of gas adsorbed at the equilibrium pressure is given as the difference between the amount of gas admitted and the amount of gas required to fill the space around the adsorbent, i.e. the *dead space*, at the equilibrium pressure. The adsorption isotherm is constructed point-by-point by the admission to the adsorbent of successive charges of gas with the aid of a volumetric dosing technique and application of the gas laws. The volume of the dead space must, of course, be known accurately: it is obtained (see Section 3.2) either by pre-calibration of the confined volume and subtracting the volume of the adsorbent (calculated from its density), or by the admission of a gas which is adsorbed to a negligible extent (see Section 3.2). Nitrogen adsorption isotherms at the temperature of the boiling point of nitrogen at ambient atmospheric pressure are generally determined by the volumetric method and provide the basis for the various standard procedures which have been proposed for the determination of surface area (see references in Section 1).

Volumetric measurements may be conducted with the aid of conventional gas chromatographic equipment for the measurement of the change in composition of a flowing gas stream (a mixture of carrier gas and adsorptive gas) to obtain the amount adsorbed at a given partial pressure when the composition of the exit gas stream has returned to the inlet composition.

If the establishment of equilibrium is sufficiently fast and the adsorption of the carrier gas negligible, this method may be regarded as equivalent to a 'static' procedure. Other types of flow methods involve the introduction of the pure adsorptive, e.g. at a slow and constant rate under quasi-equilibrium conditions. The validity of flow techniques should be checked by changing the flow-rate.

Recent developments in vacuum microbalance techniques have revived the interest in gravimetric methods for the determination of adsorption isotherms. With the aid of an *adsorption balance* the change in weight of the adsorbent may be followed directly during the outgassing and adsorption stages. A gravimetric procedure is especially convenient for measurements with vapours at temperatures not too far removed from ambient. At both high and low temperatures, however, it becomes difficult to control and measure the exact temperature of the adsorbent, which is particularly important in the determination of mesopore size distribution.

3.2 Operational definitions of adsorption

To examine the fundamental basis on which experimental definitions depend, consider an adsorption experiment incorporating both volumetric and gravimetric measurements (see Fig. 1). A measured amount, n, of a specified gas (see Note e), is introduced into the system whose total volume, V, can be varied at constant temperature, T. Measurements are made of V, p (the equilibrium pressure) and w (the apparent weight of a mass, m, of adsorbent).

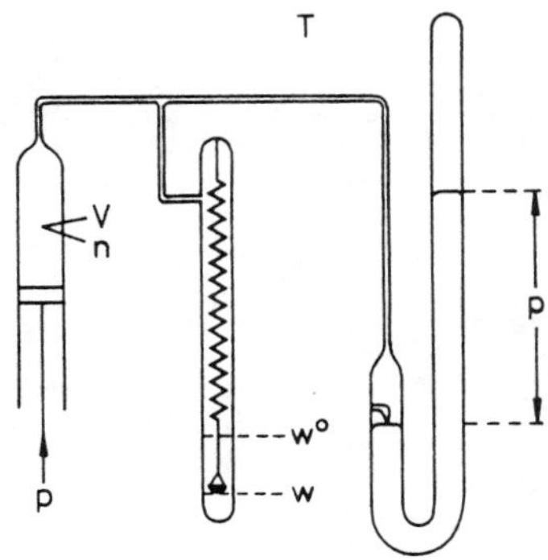

Fig. 1. Schematic arrangement of a simultaneous volumetric and gravimetric adsorption experiment.

In a calibration experiment the balance pan contains no adsorbent. The total volume, V^o, of the system is now simply related to the amount, n^o, of gas admitted:

$$V^o = n^o v^g(T,p),$$

where $v^g(T,p)$ is the molar volume of the gas at T and p, and is known from its equation of state. If the buoyancy effect arising from the balance itself is negligible, the apparent weight will remain constant. Since the gas concentration, $c^o = 1/v^g$, is constant throughout the volume V^o,

Note e. for simplicity, adsorption of a single gas is considered here (cf. Appendix II, Part 1, §1.1.11).

$$n^{o} = \int_{\text{all } V^{o}} c^{o} dV = c^{o}V^{o} = \frac{V^{o}}{v^{g}} .$$

A mass, m, of adsorbent (weighing w^{o} in vacuum) is now introduced and the experiment repeated using the same amount of gas. If adsorption is detectable at the given T, p, the volume V will usually be less than V^{o} and the apparent weight of the adsorbent will increase from w^{o} to w.

V/v^{g} is the amount of gas which would be contained in the volume V if the gas concentration were uniform throughout the volume. That the amount actually present is n^{o} means that local variations in gas concentration must occur: the gas concentration within the bulk of the solid is zero, but is greater than c^{o} in the interfacial layer. The difference between n^{o} and V/v^{g} may be called the *apparent adsorption*

$$n^{a}\ (\text{app}) = n^{o} - V/v^{g} ,$$

and is a directly observable quantity. The precision with which it can be measured is controlled only by the experimental precision in T, p and V, and by the reliability with which v^{g} (and c^{o}) can be calculated from the equation of state of the gas.

The apparent adsorption may, alternatively, be defined by measuring the amount of gas which has to be added to the system at constant T, p to increase the volume V back to V^{o}. The apparent adsorption is then equal to the extra amount of gas which can be accommodated in a volume V^{o} at a given T, p when the solid is introduced. It can, therefore, be expressed in terms of the local deviations of the concentration, c, of adsorptive molecules, from the bulk concentration c^{o}:

$$n^{a}(\text{app}) = \int_{V^{o}} (c-c^{o})\, dV.$$

If the gas does not penetrate into the bulk solid (i.e. is not *absorbed*), the above integral consists of two parts, that over the volume occupied by the solid (V^{s}) within which c = o, and that over the adsorption space plus the gas phase volume, which taken together is denoted by V^{g}:

$$n^{a}(\text{app}) = -c^{o}V^{s} + \int_{V^{g}} (c-c^{o})\, dV.$$

The first term represents the amount of gas excluded by the solid, while the second is the extra amount of gas accommodated because of the accummulation of gas on the neighbourhood of the solid surface. If the adsorption is very weak the first term may exceed the second and the apparent adsorption may be negative ($V > V^{o}$).

The quantity

$$n^{a} = n^{a}(\text{app}) + c^{o}V^{s} = \int_{V^{g}} (c-c^{o})\, dV$$

is thus equal to the *Gibbs adsorption* (see note f) (surface excess amount of adsorbed substance - see Part I, §1.1.8) when the surface of the solid is taken as the Gibbs dividing surface: it is the difference between the amount of substance actually present in the interfacial layer and that which would be present at the same equilibrium gas pressure in a reference system in which the gas phase composition is constant up to the Gibbs surface, and in which no adsorptive penetrates into the surface layer or the bulk of the solid.

The operational definition of n^{a} is thus

$$n^{a} = \int_{V^{g}} (c-c^{o})\, dV = n - c^{o}V^{g},$$

where n is the total amount of gas admitted.

Note f. n_{i}^{σ} was used to denote the Gibbs adsorption in Part I, §1.1.8.

The precision with which n^a can be determined depends, not only on the precision of T, p, V and v^g but also on the precision with which V^s (and hence v^g) is known.

The volume of the solid (i.e. the volume enclosed by the Gibbs surface) is often defined experimentally as that volume which is not accessible to a non-sorbable gas (e.g. helium - leading to the *helium dead-space*). In making this identification it is assumed that the volume available to He atoms is the same as that for molecules of the gas under investigation (which is not true, for example, if the solid acts as a molecular sieve, or if the molecules of the gas are significantly larger than the He atom), and that the solid does not swell under the influence of the adsorbate. Helium adsorption may occur if the solid contains very fine pores (or pore entrances) and the only proof that the adsorption is zero is that the apparent value of V^s is independent of temperature. It is usual to take the high temperature limit of V^s as being the correct value, but if V^s is determined at a temperature widely different from that used in an adsorption experiment, a correction for the thermal expansion of the solid may be required.

Alternatively, V^s may be estimated from the known density of the bulk solid with the implied assumption that this is the same as that of the material of the adsorbent.

The above discussion in terms of the volumetric technique when applied to gravimetric measurements gives for the apparent change in weight

$$\Delta w = w - w_o = \left[n^a - \frac{V^s}{v^g} \right] M$$

where M is the molar mass of the adsorptive.

Thus $$n^a = \frac{\Delta w}{M} + \frac{V^s}{v^g} .$$

The second term on the right-hand-side is the buoyancy correction which has the same origin as the dead-space correction in a volumetric determination.

An alternative but less useful definition of adsorption is

$$n^s = \int_{V_a} c\mathrm{d}V ,$$

where $V_a = \tau A_s$ is the volume of the interfacial layer (see Note g) and c is the local concentration. V_a has to be defined on the basis of some appropriate model of gas adsorption which gives a value of τ the layer thickness. Provided that the equilibrium pressure is sufficiently low and the adsorption not too weak, then

$$n^s \approx n^a ;$$

the surface excess amount (n^a) and total amount (n^s) of substance in the adsorbed layer become indistinguishable and the general term *amount adsorbed* is applicable to both quantities.

SECTION 4. EXPERIMENTAL PROCEDURES

4.1 Outgassing the adsorbent

Prior to the determination of an adsorption isotherm most if not all of the physisorbed species must be removed from the surface of the adsorbent. This may be achieved by outgassing, i.e. exposure of the surface to a high vacuum - usually at elevated temperature. The outgassing conditions (temperature programme, change in pressure over the adsorbent and the residual pressure) required to attain reproducible isotherms must be controlled to within limits which are dependent on the adsorption system. Instead of exposing the adsorbent to a high vacuum, it is sometimes expedient to achieve adequate cleanliness of the surface by flushing the adsorbent with an inert gas (which may be the adsorptive) at elevated temperature. With certain microporous solids reproducible isotherms are only obtained after one or more adsorption-desorption cycles. This problem can be overcome by flushing with the adsorptive and subsequent heating in vacuum.

Note g. In part I, §1.1.11, the volume of the interfacial layer was denoted by V^s

Contrary to chemisorption studies where more rigorous surface cleanliness is required, outgassing to a residual pressure of ~ 10 mPa is usually considered sufficient if physisorption measurements are to be employed for the determination of surface area and/or porosity. Such conditions are readily achieved with the aid of conventional vacuum equipment - usually a combination of a rotary and diffusion pump. The rate of desorption is strongly temperature dependent and to minimize the outgassing time, the temperature should be the maximum consistent with the avoidance of changes in the nature of the adsorbent and with the achievement of reproducible isotherms. Outgassing at too high a temperature or under high vacuum conditions (residual pressure < 1 μPa), as well as flushing with certain gases may lead to changes in the surface composition, e.g. decomposition of hydroxides or carbonates, formation of surface defects or irreversible changes in texture.

For most purposes the outgassing temperature may be conveniently selected as within the range over which the thermal gravimetric curve obtained in vacuo exhibits a minimum slope. To monitor the progress of outgassing it is useful to follow the change in gas pressure by means of suitable vacuum gauges and, if the experimental technique permits, the change in weight of the adsorbent. Further information on the effect of outgassing may be obtained by the application of temperature programmed desorption (thermal desorption spectroscopy) in association with mass spectrometric analysis.

4.2 Determination of the adsorption isotherm

It is essential to take into account a number of potential sources of experimental error in the determination of an adsorption isotherm. In the application of a volumetric technique involving a dosing procedure it must be kept in mind that any errors in the measured doses of gas are cumulative and that the amount remaining unadsorbed in the dead space becomes increasingly important as the pressure increases. In particular, the accuracy of nitrogen adsorption measurements at temperatures of about 77K will depend on the control of the following factors:-

(i) Gas burettes and other parts of the apparatus containing appreciable volumes of gas must be thermostatted, preferably to ± 0.1°C. If possible the whole apparatus should be maintained at reasonably constant temperature.

(ii) The pressure must be measured accurately (to ± 10 Pa). If a mercury manometer is used the tubes should be sufficiently wide - preferably ~1 cm in diameter.

(iii) The level of liquid nitrogen in the cryostat bath must be kept constant to within a few millimetres, preferably by means of an automatic device.

(iv) The sample bulb must be immersed to a depth of at least 5cm below the liquid nitrogen level.

(v) The temperature of the liquid nitrogen must be monitored, e.g. by using a suitably calibrated nitrogen or oxygen vapour pressure manometer or a suitable electrical device.

(vi) The nitrogen used as adsorptive must be of purity not less than 99.9%.

(vii) The conditions chosen for pretreatment of the adsorbent must be carefully controlled and monitored (i.e. the outgassing time and temperature and the residual pressure, or conditions of flushing with adsorptive).

(viii) It is recommended that the *outgassed weight* of the adsorbent should be determined either before or after the adsorption measurements. In routine work it may be convenient to admit dry air or nitrogen to the sample after a final evacuation under the same conditions as those used for the pretreatment.

SECTION 5. EVALUATION OF ADSORPTION DATA

5.1 Presentation of primary data

The quantity of gas adsorbed may be measured in any convenient units: moles, grams and cubic centimetres at s.t.p. have all been used. For the presentation of the data it is recommended that the amount adsorbed should be expressed in moles per gram of the *outgassed* adsorbent. The mode of outgassing and if possible the composition of the adsorbent should be specified and its surface characterised. To facilitate the comparison of adsorption data it is recommended that adsorption isotherms be displayed in graphical form with the amount adsorbed (n^a in mol g^{-1}) plotted against the equilibrium relative pressure (p/p^o), where p^o is the saturation pressure of the pure adsorptive at the temperature of the measurement, or against p when the temperature is above the critical temperature of the adsorptive. If the surface area of the adsorbent is known the amount adsorbed may be expressed as number of molecules, or moles per unit area, (i.e. N^a molecules m^{-2} or n^a mol m^{-2}). Adsorption data obtained on well-defined surfaces or in model pore systems should be given in tabular form.

5.2 Classification of adsorption isotherms

The majority of physisorption isotherms may be grouped into the six types shown in Figure 2. In most cases at sufficiently low surface coverage the isotherm reduces to a linear form (i.e. $n^a \propto p$), which is often referred to as the Henry's Law region (see Note h).

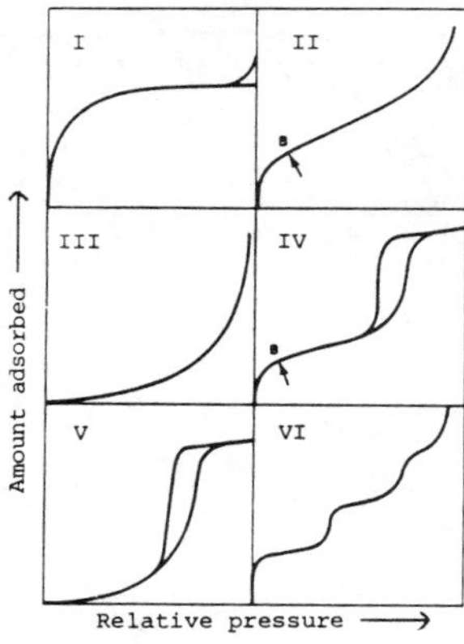

Fig. 2. Types of physisorption isotherms.

The reversible *Type I* isotherm (see Note i) is concave to the p/p^o axis and n^a approaches a limiting value as $p/p^o \rightarrow 1$. Type I isotherms are given by microporous solids having relatively small external surfaces (e.g. activated carbons, molecular sieve zeolites and certain porous oxides), the limiting uptake being governed by the accessible micropore volume rather than by the internal surface area.

The reversible *Type II* isotherm is the normal form of isotherm obtained with a non-porous or macroporous adsorbent. The Type II isotherm represents unrestricted monolayer-multilayer adsorption. Point B, the beginning of the almost linear middle section of the isotherm, is often taken to indicate the stage at which monolayer coverage is complete and multilayer adsorption about to begin.

The reversible *Type III* isotherm is convex to the p/p^o axis over its entire range and therefore does not exhibit a Point B. Isotherms of this type are not common; the best known examples are found with water vapour adsorption on pure non-porous carbons. However, there are a number of systems (e.g. nitrogen on polyethylene) which give isotherms with gradual curvature and an indistinct Point B. In such cases, the adsorbent-adsorbate interaction is weak as compared with the adsorbate-adsorbate interactions.

Characteristic features of the *Type IV* isotherm are its hysteresis loop, which is associated with capillary condensation taking place in mesopores, and the limiting uptake over a range of high p/p^o. The initial part of the Type IV isotherm is attributed to monolayer-multilayer adsorption since it follows the same path as the corresponding part of a Type II isotherm obtained with the given adsorptive on the same surface area of the adsorbent in a non-porous form. Type IV isotherms are given by many mesoporous industrial adsorbents.

Note h. On heterogeneous surfaces this linear region may fall below the lowest experimentally measurable pressure.

Note i. Type I isotherms are sometimes referred to as *Langmuir isotherms*, but this nomenclature is not recommended.

The *Type V* isotherm is uncommon; it is related to the Type III isotherm in that the adsorbent-adsorbate interaction is weak, but is obtained with certain porous adsorbents.

The *Type VI* isotherm represents stepwise multilayer adsorption on a uniform non-porous surface. The step-height now represents the monolayer capacity for each adsorbed layer and, in the simplest case, remains nearly constant for two or three adsorbed layers. Amongst the best examples of Type VI isotherms are those obtained with argon or krypton on graphitised carbon blacks at liquid nitrogen temperature.

5.3 Adsorption hysteresis

Hysteresis appearing in the multilayer range of physisorption isotherms is usually associated with capillary condensation in mesopore structures. Such hysteresis loops may exhibit a wide variety of shapes. Two extreme types are shown as *H1* (formerly Type A) and *H4* in Figure 3. In the former the two branches are almost vertical and nearly parallel over

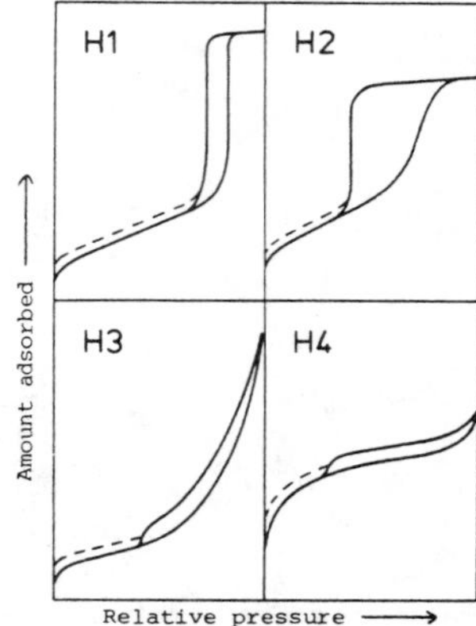

Fig. 3. Types of hysteresis loops.

an appreciable range of gas uptake, whereas in the latter they remain nearly horizontal and parallel over a wide range of p/p^{o}. In certain respects *Types H2* and *H3* (formerly termed Types E and B, respectively) may be regarded as intermediate between these two extremes. A feature common to many hysteresis loops is that the steep region of the desorption branch leading to the lower closure point occurs (for a given adsorptive at a given temperature) at a relative pressure which is almost independent of the nature of the porous adsorbent (e.g. for nitrogen at its boiling point at $p/p^{o} \sim 0.42$ and for benzene at 25°C at $p/p^{o} \sim 0.28$).

The shapes of hysteresis loops have often been identified with specific pore structures. Thus, *Type H1* is often associated with porous materials known, from other evidence, to consist of agglomerates or compacts of approximately uniform spheres in fairly regular array, and hence to have narrow distributions of pore size. Some corpuscular systems (e.g. silica gels) tend to give Type H2 loops, but in these cases the distribution of pore size and shape is not well-defined. Indeed, the H2 loop is especially difficult to interpret: in the past it was attributed to a difference in mechanism between condensation and evaporation processes occurring in pores with narrow necks and wide bodies (often referred to as 'ink bottle' pores), but it is now recognised that this provides an over-simplified picture.

The Type H3 loop, which does not exhibit any limiting adsorption at high p/p^{o}, is observed with aggregates of plate-like particles giving rise to slit-shaped pores. Similarly, the *Type H4* loop appears to be associated with narrow slit-like pores, but in this case the Type I isotherm character is indicative of microporosity.

With many systems, especially those containing micropores, *low pressure hysteresis* (indicated by the dashed lines in Figure 3), may be observed extending to the lowest attainable pressures. Removal of the residual adsorbed material is then possible only if the adsorbent is outgassed at higher temperatures. This phenomenon is thought to be associated with the swelling of a non-rigid porous structure or with the irreversible uptake of molecules in pores (or through pore entrances) of about the same width as that of the adsorbate molecule.

SECTION 6. DETERMINATION OF SURFACE AREA

6.1 Application of the BET method

The Brunauer-Emmett-Teller (BET) gas adsorption method has become the most widely used standard procedure for the determination of the surface area of finely-divided and porous materials, in spite of the oversimplification of the model on which the theory is based.

It is customary to apply the BET equation in the linear form

$$\frac{p}{n^a \cdot (p^o - p)} = \frac{1}{n^a_m \cdot C} + \frac{(C-1)}{n^a_m \cdot C} \cdot \frac{p}{p^o}$$

where n^a is the amount adsorbed at the relative pressure p/p^o and n^a_m is the monolayer capacity.

According to the BET theory C is related exponentially to the enthalpy (heat) of adsorption in the first adsorbed layer. It is now generally recognised, however, that although the value of C may be used to characterise the shape of the isotherm in the BET range it does not provide a quantitative measure of enthalpy of adsorption but merely gives an indication of the order of magnitude of the adsorbent-adsorbate interaction energy. Thus, in reporting BET data it is recommended that C values are stated, but not converted to enthalpies of adsorption.

A high value of C (~ 100) is associated with a sharp knee in the isotherm, thus making it possible to obtain by visual inspection the uptake at Point B, which usually agrees with n^a_m derived from the above equation to within a few per cent. On the other hand, if C is low (<20) Point B cannot be identified as a single point on the isotherm. Unfortunately Point B is not itself amenable to any precise mathematical description and the theoretical significance of the amount adsorbed at Point B is therefore questionable.

The BET equation requires a linear relation between $p/n^a(p^o-p)$ and p/p^o (i.e. the *BET* plot). The range of linearity is, however, restricted to a limited part of the isotherm - usually not outside the p/p^o range of 0.05-0.30. This range is shifted to lower relative pressures in cases of the energetically more homogeneous surfaces, e.g. for nitrogen or argon adsorption on graphitised carbon or xenon on clean metal films under ultra high vacuum conditions.

The second stage in the application of the BET method is the calculation of the surface area (often termed *BET area)*. This requires a knowledge of the average area, a_m *(molecular cross-sectional area)*, occupied by the adsorbate molecule in the complete monolayer. Thus

$$A_s(\text{BET}) = n^a_m \cdot L \cdot a_m$$

and $$a_s(\text{BET}) = A_s(\text{BET})/m$$

where $A_s(\text{BET})$ and $a_s(\text{BET})$ are the total and specific surface areas, respectively, of the adsorbent (of mass m) and L is the Avogadro constant.

For the close-packed nitrogen monolayer at 77K, $a_m(N_2) = 0.162\ nm^2$, as calculated from the density of liquid nitrogen at 77K by assuming hexagonal close packing. This value appears to be satisfactory to within about ± 10% for the adsorption of nitrogen on a wide number of different surfaces. With other adsorptives, arbitrary adjustments of the a_m value is generally required to bring the BET area into agreement with the nitrogen value. The adjusted values of a_m for a particular adsorptive are dependent on temperature and the adsorbent surface. They may also differ appreciably from the value calculated for the close-packed monolayer on the basis of the density of the liquid or solid adsorptive. In view of this situation and the fact that full nitrogen isotherms may be conveniently measured at temperatures ~77K, it is recommended that nitrogen should continue to be used for the determination of both surface area and mesopore size distribution (Section 7.3).

The standard BET procedure requires the measurement of at least three and preferably five or more points in the appropriate pressure range on the N_2 adsorption isotherm at the normal boiling point of liquid nitrogen.

For routine measurements of surface areas, e.g. of finely divided or porous industrial products, a simplified procedure may be applied using only a single point on the adsorption isotherm, lying within the linear range of the BET plot. For N_2, the C value is usually sufficiently large ($\geqslant 100$) to warrant the assumption that the BET straight line passes through the origin of the coordinate system. Thus

$$n^a_m = n^a(1 - p/p^o) .$$

The validity of the simplifying assumption is usually within the variance of surface area determinations (about ± 10%) for different materials and may be checked by calibration against the standard BET procedure or by using surface area reference samples (see Section 6.2).

It is strongly recommended that in reporting a_s(BET) values, the conditions of outgassing (see Section 4.1), the temperature of the measurements, the range of linearity of the BET plot, the values of n^a_m and C and the value taken for the cross-sectional area a_m should all be stated.

If the standard BET procedure is to be used, it should be established that monolayer-multilayer formation is operative and is not accompanied by micropore filling (Section 8.3), which is usually associated with an increase in the value of C (>200, say). It should be appreciated that the BET analysis does not take into account the possibility of micropore filling or penetration into cavities of molecular size. These effects can thus falsify the BET surface areas and in case of doubt their absence should be checked by means of an empirical method of isotherm analysis or by using surface area reference samples (see Section 6.2).

For the determination of *small specific surface areas* ($<5 m^2 g^{-1}$, say) adsorptives with relatively low vapour pressure are used in order to minimise the dead space correction, preferably krypton or xenon at liquid nitrogen temperature. In the case of krypton, use of the extrapolated p^o value for the supercooled liquid tends to lead to a wider range of the linear BET plot and larger monolayer capacities (the difference being <10%) as compared with p^o for the solid. Evaluation of surface areas is further complicated by the choice of the a_m value which is found to vary from solid to solid if compared with the BET nitrogen areas (a_m(Kr) = 0.17-0.23 nm^2; a_m(Xe) = 0.17-0.27 nm^2). Since no generally valid recommendations can be made, it is essential to state the chosen p^o and a_m values.

Ultrahigh vacuum techniques (basic pressure ~10^2 μPa) enable adsorption studies to be made on stringently *clean solid surfaces* whereas degassing under moderate vacuum conditions, as normally applied in surface area determinations, leave the adsorbent covered with a preadsorbed layer of impurities and/or the adsorbate. On subsequent adsorption (e.g. N_2 or noble gases) the complete physisorbed monolayer is usually reached at $p/p^o \sim 0.1$ while with clean surfaces this state occurs at p/p^o values which are smaller by orders of magnitude. In the latter case it is found that straight line BET plots may be obtained in various ranges of p/p^o, leading to larger monolayer capacities as the relative pressure increases. It should, however, be realised that it is not the range of p/p^o which is the determining factor for the relevant BET plot, but rather the region of θ around $\theta = 1$.

Noble gas adsorption is often assumed to be the least complicated form of physisorption. However, on clean solid surfaces additional Coulombic forces may contribute to the adsorption bond, e.g. Xe on clean metal surfaces. Such specific interactions may lead to ordered structures of the adsorbate in registry with the substrate lattice and thus affect the molecular cross sectional area (e.g. a_m (Xe) = 0.172 nm^2 on Pd(110)).

6.2 Empirical procedures for isotherm analysis

In view of the complexity of real solid/gas interfaces and the different mechanisms which may contribute to physisorption, it is hardly surprising to find that none of the current theories of adsorption is capable of providing a mathematical description of an experimental isotherm over its entire range of relative pressure. In practice, two different procedures have been used to overcome this problem. The first, and older, approach involves the application of various semi-empirical equations (e.g. the Langmuir, Dubinin-Radushkevich, Frenkel-Halsey-Hill equations), the particular mathematical form depending on the range of the isotherm to be fitted and also on the nature of the system. The second procedure (e.g. the t-method, α_s-method, comparison plot) makes use of standard adsorption isotherms obtained with selected non-porous reference materials and attempts to explain differences in the isotherm shape in terms of the three different mechanisms of physisorption, i.e. monolayer-multilayer coverage, capillary condensation and micropore filling. Much discussion has surrounded the choice of the standard isotherm, but it now seems generally accepted that it should be one obtained on a chemically similar adsorbent rather than one having the same value of C as the isotherm to be analysed. In favourable cases, this approach can provide an independent assessment of the *total* surface area (for mesoporous, macroporous or non-porous solids) and an assessment of the *external* area for microporous solids (see Section 8.3).

SECTION 7. ASSESSMENT OF MESOPOROSITY

7.1 Properties of porous materials

Most solids of high surface area are to some extent porous. The *texture* of such materials is defined by the detailed geometry of the void and pore space. Porosity, ε, is a concept related to texture and refers to the pore space in a material. An *open pore* is a cavity or channel communicating with the surface of a particle, as opposed to a *closed pore*. *Void* is the space or interstice between particles. In the context of adsorption and fluid penetration *powder porosity* is the ratio of the volume of voids plus the volume of open pores to the total volume occupied by the powder. Similarly, *particle porosity* is the ratio of the volume of open pores to the total volume of the particle. It should be noted that these definitions place the emphasis on the accessibility of pore space to the adsorptive.

The *total pore volume*, V_p, is often derived from the amount of vapour adsorbed at a relative pressure close to unity - by assuming that the pores are then filled with liquid adsorptive.

If the solid contains no macropores, the isotherm remains nearly horizontal over a range of p/p^o approaching unity and the total pore volume is well-defined. In the presence of macropores the isotherm rises rapidly near $p/p^o = 1$ and in the limit of large macropores may exhibit an essentially vertical rise. In this case the limiting adsorption at the top of the vertical rise can be identified reliably with the total pore volume only if the temperature on the sample is very carefully controlled and there are no 'cold spots' in the apparatus (which lead to bulk condensation of the gas and a false measure of adsorption in the volumetric method).

The *mean hydraulic radius*, r_h , is defined as

$$r_h = \left(\frac{V}{A_s}\right)_p$$

where $(V/A_s)_p$ is the ratio of the volume to the area of walls of a group of mesopores.

If the pores have a well-defined shape there is a simple relationship between r_h and the *mean pore radius*, r_p. Thus, in the case of non-intersecting cylindrical capillaries

$$r_p = 2\,r_h \quad .$$

For a parallel-sided slit-shaped pore, r_h is half the slit width.

The *pore size distribution* is the distribution of pore volume with respect to pore size. The computation of pore size distribution involves a number of assumptions (pore shape, mechanism of pore filling, validity of Kelvin equation etc).

7.2 Application of the Kelvin equation

Mesopore size calculations are usually made with the aid of the Kelvin equation in the form

$$\frac{1}{r_1} + \frac{1}{r_2} = -\frac{RT}{\sigma^{lg} v^l} \ln\left(\frac{p}{p^o}\right)$$

which relates the principal radii, r_1 and r_2, of curvature of the liquid meniscus in the pore to the relative pressure, p/p^o, at which condensation occurs. Hence σ^{lg} is the surface tension of the liquid condensate and v^l is its molar volume. It is generally assumed that this equation can be applied locally to each element of liquid surface.

In using this approach to obtain the *pore radius* or *pore width*, it is necessary to assume: (i) a model for the pore shape and (ii) that the curvature of the meniscus is directly related to the pore width. The pore shape is generally assumed to be either cylindrical or slit-shaped: in the former case, the meniscus is hemispherical and $r_1 = r_2$; in the latter case, the meniscus is hemicylindrical and $r_2 = \infty$.

Rearrangement of the Kelvin equation and replacement of the principal radii of curvature terms by $2/r_K$ gives

$$r_K = \frac{2\sigma^{lg} v^l}{RT \ln (p^o/p)}$$

(r_K is often termed the *Kelvin radius*).

If the pore radius of a cylindrical pore is r_p and a correction is made for the thickness of a layer already adsorbed on the pore walls, i.e. the *multilayer thickness*, t,

$$r_p = r_K + 2t.$$

Correspondingly, for a parallel-sided slit, the slit width, d_p, is given by

$$d_p = r_K + t.$$

7.3 Computation of mesopore size distribution

Many attempts have been made to calculate the mesopore size distribution from physisorption isotherms. In such calculations it is generally assumed (often tacitly): (a) that the pores are rigid and of a regular shape (e.g. cylindrical capillaries or parallel-sided slits), (b) that micropores are absent, and (c) that the size distribution does not extend *continuously* from the mesopore into the macropore range. Furthermore, to obtain the pore size distribution, which is usually expressed in the graphical form $\Delta V_p/\Delta r_p$ vs. r_p, allowance must be made for the effect of multilayer adsorption in progressively reducing the dimensions of the free pore space available for capillary condensation.

The location and shape of the distribution curve is, of course, dependent on which branch of the hysteresis loop is used to compute the pore size. In spite of the considerable attention given to this problem, it is still not possible to provide unequivocal general recommendations. In principle, the regions of metastability and instability should be established for the liquid/vapour meniscus in the various parts of a given pore structure, but in practice this would be extremely difficult to undertake in any but the simplest types of pore systems.

Recent work has drawn attention to the complexity of capillary condensation in pore networks and has indicated that a pore size distribution curve derived from the desorption branch of the loop is likely to be unreliable if pore blocking effects occur. It is significant that a very steep desorption branch is usually found if the lower closure point of the loop is located at the limiting p/p^o (see Section 5.3). In particular, the desorption branch of a Type H2 loop is one that should not be used for the computation of pore size distribution.

It is evident from the above considerations that the use of the physisorption method for the determination of mesopore size distribution is subject to a number of uncertainties arising from the assumptions made and the complexities of most real pore structures. It should be recognised that derived pore size distribution curves may often give a misleading picture of the pore structure. On the other hand, there are certain features of physisorption isotherms (and hence of the derived pore distribution curves) which are highly characteristic of particular types of pore structures. Physisorption is one of the few non-destructive methods available for investigating mesoporosity, but it is to be hoped that future work will lead to refinements in the application of the method - especially through the study of model pore systems and the application of modern computer techniques.

SECTION 8. ASSESSMENT OF MICROPOROSITY

8.1 Terminology

It is generally recognised that the mechanism of physisorption is modified in very fine pores (i.e. pores of molecular dimensions) since the close proximity of the pore walls gives rise to an increase in the strength of the adsorbent-adsorbate interactions. As a result of the enhanced adsorption energy, the pores are filled with physisorbed molecules at low p/p^o. Adsorbents with such fine pores are usually referred to as *microporous*.

The limiting dimensions of micropores are difficult to specify exactly, but the concept of *micropore filling* is especially useful when it is applied to the primary filling of pore space as distinct from the secondary process of capillary condensation in mesopores.

The terminology of pore size has become somewhat confused because of the attempts made to designate the different categories of pores in terms of exact dimensions rather than by reference to the gas-solid system (the size, shape and electronic nature of the adsorptive molecules and the surface structure of the adsorbent) as well as to the pore size and shape.

The upper limit of 2.0 nm for the micropore width was put forward as part of the IUPAC classification of pore size (see Appendix II, Part I). It now seems likely that there are two different micropore filling mechanisms, which may take place at p/p^o below the onset of capillary condensation: the first, occurring at low p/p^o, involves the entry of isolated adsorbate molecules into very narrow pores; the second, at a somewhat higher p/p^o, is a *cooperative* process in that the interaction between adsorbate molecules is involved.

It is recommended that, on the basis of this approach, attention should be directed towards the *mechanism* of pore filling rather than to the specification of the necessarily rather arbitrary limits of pore size. Until further progress has been made it is undesirable to modify the original IUPAC classification or to introduce any new terms (e.g. ultrapores or ultramicropores).

8.2 Concept of surface area

In recent years a radical change has been taking place in the interpretation of the Type I isotherm. According to the classical Langmuir theory, the limiting adsorption n^a_p (at the plateau) represents completion of the monolayer and may therefore be used for the calculation of the surface area. The alternative view, which is now widely accepted, is that the initial (steep) part of the Type I isotherm represents micropore filling (rather than surface coverage) and that the low slope at the plateau is due to multilayer adsorption being confined to a small external area.

If the latter explanation is correct, it follows that the value of A_s (as derived by either BET or Langmuir analysis (see Note j)) cannot be accepted as the true surface area of a microporous adsorbent. On the other hand, if the slope of the isotherm is not too low at higher p/p^o and provided that capillary condensation is absent, it should (in principle) be possible to assess the *external* surface area from the multilayer region.

In view of the above difficulties, it has been suggested that the term *monolayer equivalent area* should be applied to microporous solids. However, the exact meaning of this term may not always be clear and it is recommended that the terms *Langmuir area* or *BET area* be used where appropriate, with a clear indication of the range of linearity of the Langmuir, or BET, plot, the magnitude of C etc. (see Section 6.1).

8.3 Assessment of micropore volume

No current theory is capable of providing a general mathematical description of micropore filling and caution should be exercised in the interpretation of derived quantities (e.g. micropore volume) obtained by the application of a relatively simple equation (e.g. the Dubinin-Radushkevich equation) to adsorption isotherm data over a limited range of p/p^o and at a single temperature. The fact that a particular equation gives a reasonably good fit over a certain range of an isotherm does not in itself provide sufficient evidence for a particular mechanism of adsorption.

The t-method and its extensions provide a simple means of comparing the shape of a given isotherm with that of a standard on a non-porous solid. In the original t-method, the amount adsorbed is plotted against t, the corresponding multilayer thickness calculated from the standard isotherm obtained with a non-porous reference solid. Any deviation in shape of the given isotherm from that of the standard is detected as a departure of the 't-plot' from linearity. For the assessment of microporosity, the thickness of the multilayer is irrelevant and it is preferable to replace t by the 'reduced' adsorption, α_s, defined as $(n^a/n^a_s)_{ref}$ where n^a_s is the amount adsorbed by the reference solid at a fixed relative pressure, $p/p^o = s$. An advantage of this method is that it can be used even when the standard isotherm does not exhibit a well-defined Point B, i.e. when the value of C is low. Once the standard α_s-curve has been obtained for the adsorption system at the given temperature, the α_s-method can be applied in an analogous manner to the t-method. It should be recognised that such methods can only be expected to give an understanding of the true nature of the micropore filling process (and hence of the micropore structure) if the standard isotherms (or t-curves) are obtained with non-porous reference solids of known surface structure. It is strongly recommended that the appropriate standard isotherm is selected for the given adsorption *system* and not by arbitrarily choosing a Type II isotherm having the same C value as the isotherm on a particular microporous solid.

Another procedure which may be used for the assessment of microporosity is the *pre-adsorption method*. In this approach the micropores are filled with large molecules (e.g. nonane), which are not removed by pumping the adsorbent at ambient temperature. In

Note j. In fact, many microporous solids do not give linear BET plots although their Langmuir plots may be linear over an appreciable range of p/p^o.

the most straightforward case, this procedure can provide an effective way of isolating the micropores and leaving the external surface available for the adsorption of nitrogen, or some other adsorptive.

SECTION 9. GENERAL CONCLUSIONS AND RECOMMENDATIONS

9.1 For evaluation of both the surface area and the pore size distribution from a single adsorption isotherm, nitrogen (at ~77K) is the recommended adsorptive. If the surface area is relatively low (a_s <5 $m^2 g^{-1}$, say), krypton or xenon, also at ~77K, offer the possibility of higher precision in the actual measurement of the adsorption, but not necessarily higher accuracy than could be obtained with nitrogen in the resultant value of the surface area.

9.2 For a given system at a given temperature, the adsorption isotherm should be reproducible, but the possibility of ageing of the adsorbent - e.g. through the addition or removal of water - must always be borne in mind. The reproducibility of the adsorption should be checked whenever possible by measurement of an isotherm on a second sample (of different mass) of the given adsorbent.

9.3 The first stage in the interpretation of a physisorption isotherm is to identify the isotherm type and hence the nature of the adsorption process(es): monolayer-multilayer adsorption, capillary condensation or micropore filling. If the isotherm exhibits low pressure hysteresis (i.e. at $p/p^o < 0.4$, with nitrogen at 77K) the technique should be checked to establish the degree of accuracy and reproducibility of the measurements.

9.4 The BET method is unlikely to yield a value of the actual surface area if the isotherm is of either Type I or Type III; on the other hand both Type II and Type IV isotherms are, in general, amenable to the BET analysis, provided that the value of C is not too high and that the BET plot is linear for the region of the isotherm containing Point B. It is recommended that both the value of C and the range of linearity of the BET plot be recorded. If the value of C is found to be higher than normal for the particular gas-solid system, the presence of microporosity is to be suspected even if the isotherm is of Type II or Type IV; the validity of the BET area then needs checking, e.g. by the α_s-method, in order to ascertain how closely the shape of the isotherm conforms to that of the standard isotherm in the monolayer range.

9.5 The computation of mesopore size distribution is valid only if the isotherm is of Type IV. In view of the uncertainties inherent in the application of the Kelvin equation and the complexity of most pore systems, little is to be gained by recourse to an elaborate method of computation. The decision as to which branch of the hysteresis loop to use in the calculation remains largely arbitrary. If the desorption branch is adopted (as appears to be favoured by most workers), it should be appreciated that neither a Type H2 nor a Type H3 hysteresis loop is likely to yield a reliable estimate of pore size distribution, even for comparative purposes.

9.6 If a Type I isotherm exhibits a nearly constant adsorption at high relative pressure, the micropore volume is given by the amount adsorbed (converted to a *liquid* volume) in the plateau region, since the mesopore volume and the external surface are both relatively small. In the more usual case where the Type I isotherm has a finite slope at high relative pressures, both the external area and the micropore volume can be evaluated provided that a standard isotherm on a suitable non-porous reference solid is available. At present, however, there is no reliable procedure for the computation of micropore size *distribution* from a single isotherm; but if the size extends down to micropores of molecular dimensions, adsorptive molecules of selected size can be employed as molecular probes.

9.7 The following *check list* is recommended to assist authors in the measurement of adsorption isotherms and the presentation of the data in the primary literature. The reporting of results along generally accepted lines would considerably facilitate the compilation of data in the secondary literature and would thus promote interdisciplinary scientific cooperation (see Note k).

Note k. see "Guide for the Presentation in the Primary Literature of Numerical Data Derived from Experiments". Report of the CODATA Task Group on Presentation of Data in the Primary Literature, CODATA Bulletin No. 9 (1973).

It is suggested that the following items be checked and the relevant experimental conditions and results reported:-

(i) Characterisation of the sample (e.g. source, chemical composition, purity, physical state).

(ii) Pretreatment and outgassing conditions (e.g. temperature, residual pressure/partial pressures, duration of outgassing, flushing with adsorptive).

(iii) Mass of outgassed sample (m in g).

(iv) Adsorptive (e.g. chemical nature, purity, drying).

(v) Experimental procedure for isotherm determination: method (e.g. static volumetric, adsorption balance, chromatographic; calibration of dead space or buoyancy). Measurement and accuracy of pressure (p in Pa (or mbar) or p/p^o), equilibration times.

(vi) Reproducibility (a) second run, (b) with fresh sample of adsorbent.

(vii) Adsorption isotherm: plot of amount adsorbed (n^a in mol g^{-1} or N^a in molecules m^{-2}) versus pressure (p in Pa (or mbar) or p/p^o), statement of measured/calculated p^o value at T.

(viii) Type of isotherm, type of hysteresis, nature of adsorption (monolayer-multilayer adsorption, capillary condensation, micropore filling).

(ix) BET data: adsorptive, temperature (T in K), region of p/p^o and θ in linear BET plot, single point method, monolayer capacity (n^a_m in mol g^{-1} or N^a_m in molecules m^{-2}), C value, molecular cross-sectional area (a_m in nm^2 per molecule), specific surface area (a_s in $m^2 g^{-1}$).

(x) Porosity (ε with reference to powder porosity or particle porosity indicating in the latter case whether open pores or open plus closed pores are considered).

(xi) Assessment of mesoporosity (pore width ~ 2-50 nm), method of computation, choice of ad- or desorption branch, p^o value at T and region of p/p^o used, surface tension of liquid adsorptive (σ^{lg} in Nm^{-1} at T), model for pore shape. Correction for multilayer thickness, t-curve: plot of t in nm vs p/p^o (indication whether a standard curve is assumed or an adsorption isotherm determined on a non-porous sample of the adsorbent). Pore size distribution: plot of $\Delta V_p/\Delta r_p$ vs r_p (pore volume V_p per unit mass of adsorbent in cm^3 g^{-1} as calculated with the density ρ in g cm^{-3} of the liquid adsorptive, mean pore radius r_p in nm), total pore volume at saturation.

(xii) Assessment of microporosity (pore width < ca 2 nm), method of evaluation, t-plot: amount adsorbed n^a in mol g^{-1} vs multilayer thickness t in nm, α_s-plot: n^a vs $(n^a/n^a_s)_{ref}$, where the index refers to a chosen value $s = p/p^o$, Dubinin-Radushkevich plot or pre-adsorption method. Micropore volume per unit mass of adsorbent in $cm^3 g^{-1}$ as calculated with the density ρ^{ℓ} of the liquid adsorptive, monolayer equivalent area of microporous solid, external surface area.

Dynamics of Sorption in Adiabatic Columns

Shivaji Sircar
Air Products & Chemicals Inc.,
Allentown, PA, U.S.A., 18105

ABSTRACT

Instant thermal equilibrium between the gas and the adsorbent may not be attained in a column at low gas flow rates. Consequently, the interactions between the gas-solid mass and heat transfer and their effects on the sorbate mass transfer zone need to be considered in order to estimate the transfer coefficients from the experimental breakthrough data. The present work describes a parametric evaluation of these effects by numerical solution of the simultaneous mass and heat balance equations. It is shown that for a Type I adsorption system where two pairs of mass and heat transfer zones are formed, the front transfer zone is primarily governed by mass transfer and the rear transfer zone is controlled by heat transfer. Thus, these transfer coefficients can be independently estimated from the front and the rear zone breakthrough data. On the other hand, for a Type II adsorption system where a pure thermal wave and a single pair of mass and heat transfer zones are formed, the mass transfer coefficient cannot be estimated without a prior knowledge of the heat transfer coefficient.

Design and optimization of commercial adsorbent columns for gas separation and purification require good equilibrium adsorption and mass transport data for the sorbates. Accurate equilibrium data can be independently measured using a batch sorption apparatus[1] without any specific knowledge about the structure of the adsorbent. The sorbate mass transfer characteristics in a column, on the other hand, depend on the transfer resistances imposed by the external gas film, the anisotropic skin at the surface of the adsorbent particle, and the internal macro and microporosity of the adsorbent particle etc.[2,3]. These resistances are determined by the nature of the packing in the column, the shape and the size distribution of the particles, the physical properties of the adsorbate, the gas flow rate, the adsorbent surface porosity, the internal pore structure and distribution, the adsorbent crystal structure and size distribution (for pelleted zeolites), etc. One or more of the resistances may be the controlling transport mechanism which in conjunction with the equilibrium properties, determine the shape and the size of the sorbate mass transfer zone (MTZ) in the column.

The length of MTZ for isothermal adsorption of a single sorbate has been shown to be inversely proportional to the over-all mass transfer coefficient (k) for Langmuir-type adsorption equilibrium[4] and inversely proportional to $\sqrt{k}$ for linear equilibrium[5]. Thus an accurate value of k is needed for design.

Theoretical methods for estimation of various transport resistances provide only the order of magnitudes of these variables[2,3] since detailed information about the column packing and the adsorbent structure are seldom known. Analysis of sorption kinetic data from a batch apparatus may provide an estimation of the overall internal resistances but the experiments are often tedious and time consuming[6,7]. One still lacks the information about the flow characteristics in the column and its effect on MTZ. A more practical and direct approach is to measure the actual dynamics of adsorption in a laboratory column for the system of interest and estimate k from the experimental MTZ data using an appropriate mathematical model to describe the dynamics. Numerous works based on this approach have been published. Most of the earlier works assume that the adsorbent columns remain isothermal during the sorption process but this may lead to erroneous data interpretation even for small temperature changes[8]. Models to take account of the adsorbent non-isothermality in an adiabatic column have been formulated but they often assume instantaneous thermal equilibrium between the gas and the solid phases[9,10]. Others include gas-solid heat transfer resistance in the model but the transfer coefficients are large due to high flow rates[11,12]. Instant thermal equilibrium may not be achieved under some typical conditions of operation of the laboratory columns and the interactions between the mass and the heat transfer resistances and their effects on the characteristics of the MTZ need to be considered for evaluation of k. This article is a brief review of some of our recent findings[13,14].

We chose adsorption of single sorbates such as C_2H_6 and n-C_4H_{10} from an inert carrier gas (He) on 5Å zeolite as test cases. Equilibrium adsorption data for these systems[4,8] can be adequately described by the Langmuir-type adsorption equation:

$$n = \frac{mb\bar{y}}{1+b\bar{y}} \qquad (1)$$

$$b = b_o P_o e^{-q/RT_s} \qquad (2)$$

Table 1

System Properties used in Calculations

	C_2H_6-He-5Å	C_4H_{10}-He-5Å
System Pressure, P_o	1 atm	1 atm
Initial Column and Feed Gas Temperature, T_o	323°K	320°K
Carrier Gas Flow rate, Q_o	$1.62X10^{-4}$ $\frac{moles}{cm^2\text{-}sec}$	$1.25X10^{-3}$ $\frac{moles}{cm^2\text{-}sec}$
Adsorbate mole fraction in feed gas, y_o	0.10	0.001
Adsorbent density, ρ_s	0.70 gm/cm^3	0.70 gm/cm^3
Adsorbent heat capacity, C_s	0.22 cal/gm/°K	0.22 cal/gm/°K
Helium void fraction, ϵ	0.72	0.72
External void fraction, $\bar{\epsilon}$	0.40	0.40
Gas heat capacity, Cg	5.0 cal/mole/°K	5.0 cal/mole°K
Particle diameter, dp	0.32 cm	0.32 cm
Column Length, L	91.5 cm	100 cm
Langmuir Parameters		
m	1.4 mmoles/g	1.6 mmoles/g
b_o	$5.9X10^{-6}$ atm^{-1}	$3.3X10^{-7}$ atm^{-1}
q	8.8 kcal/mole	13.0 kcal/mole

Table 1 summarizes the Langmuir parameters and the conditions of operation of the test cases. The gas flow rates used in the tests (Q_o = $1X10^{-4}$ - $1X10^{-3}$ moles/cm^2/sec) represent typical orders of magnitude for Q_o in laboratory columns as well as for some commercial columns[9,15,17]. The external gas-solid heat transfer coefficient (h_e) at these flow rates may be significantly small due to channelling of fluid and mutual interactions of the neighboring particles in the column[18]. Table 2 gives the range of experimental h_e values under the conditions of the test cases. The corresponding h_e values for the ideal 'no channelling' case as well as those calculated by the well known Ranz correlation are also given in Table 2. The effective external gas film mass transfer coefficient (k_e) can also be significantly low at low gas flow rates due to axial dispersion[18,19] as shown in Table 2. The table also shows orders of magnitude of macropore diffusivities (D_p) for the sorbates in various typical pore sizes calculated by the Satterfield correlations[2] and the overall mass transfer coefficients (k) assuming that the external film and the macropore resistances are the controlling transport mechanism[5]. It may be seen that h_e may vary between ~$1X10^{-4}$ - $1X10^{-2}$ cal/cm^{3}bed/sec/°K and k may vary between ~$1X10^{-5}$ - $1X10^{-2}$ moles/gm/sec under the conditions of the test cases. Thus these transfer coefficients can be significantly lower than those estimated by the well-known Ranz equation.

Table 2

Order of Magnitudes for Transport Coefficients

Heat Transfer:

Q_o moles/cm^2/sec	h_e, Cal/cm^{3}bed/sec/°K			h_i, Cal/cm^{3}bed/sec/°K
	Experiment*	No Channelling*	Ranz Eqn.*	
1.6×10^{-4}	1×10^{-4} - 5×10^{-4}	2.5×10^{-3}	4.6×10^{-2}	3.5×10^{-2} - 0.35
1.6×10^{-3}	1×10^{-3} - 1×10^{-2}	2.5×10^{-2}	8.9×10^{-2}	

Mass Transfer:

System	Q_o moles/cm^2/sec	k_e* moles/gm/sec	Average Pore Radius, micron 1.0		0.10		0.05	
			Dp(a)	k(b)	Dp	k	Dp	k
C_2H_6-He	1.6×10^{-4}	1×10^{-4} - 1×10^{-3}	2.9×10^{-2}	$0.6\text{-}2 \times 10^{-4}$	1.4×10^{-2}	$4\text{-}7 \times 10^{-5}$	8×10^{-3}	$3\text{-}5 \times 10^{-5}$
C_4H_{10}-He	1.6×10^{-3}	1×10^{-3} - 1×10^{-2}	2.4×10^{-2}	$4\text{-}8 \times 10^{-4}$	1.0×10^{-2}	$5\text{-}6 \times 10^{-5}$	6×10^{-3}	3×10^{-5}

*Reference [18], (a) - cm^2/sec, calculated using $\tau = 4$, $\bar{\epsilon} = 0.3$ (b) moles/gm/sec

The mathematical model used for simulation of the column dynamics is that developed for adsorption of a single sorbate from an inert carrier gas in an isobaric and adiabatic column[13]:

Adsorbate Mass Balance

$$\epsilon \rho_g \left(\frac{\delta y}{\delta t}\right)_x = -\left[\frac{\delta}{\delta x}(Qy)\right]_t - \rho_s \left(\frac{\partial n}{\partial t}\right)_x \quad (3)$$

Carrier Gas Mass Balance

$$Q = Q^c/(1-y) \quad (4)$$

Gas Phase Heat Balance

$$\epsilon \rho_g C_g \left[\frac{\partial}{\partial t}(T_g - T_o)\right]_x = -C_g \frac{\partial}{\partial x}\left[Q(T_g - T_o)\right]_t + h(T_s - T_g) \quad (5)$$

Solid Phase Heat Balance

$$\rho_s C_s \left[\frac{\partial}{\partial t}(T_s - T_o)\right]_x = -q\rho_s \left(\frac{\partial n}{\partial t}\right)_x - h(T_s - T_g) \quad (6)$$

Local Rate of Adsorption

$$\left(\frac{\partial n}{\partial t}\right)_x = k\,[y - \bar{y}] \quad (7)$$

Equation (7) assumes a linear driving force model for the mass transfer. k is the lumped-up mass transfer coefficient accounting for all the external and the internal resistances. Similarly h in Equations (5) and (6) can be treated as a lumped-up heat transfer coefficient accounting for the internal and the external resistances. The internal heat transfer coefficient ($h_i = 60(1-\epsilon)\eta/d_p^2$) in the adsorbent particle depends on the magnitude of the adsorbent thermal conductivity (η). Table 2 gives the value of h_i for a typical η of 1×10^{-4} - 1×10^{-3} cal/cm/sec/°K which shows that h_i may be comparable with h_e only at high flow rates. The external heat transfer resistance is controlling at low flow rates.

Equations (1)-(7) were solved simultaneously by the method of lines[20] for the step input boundary conditions:

$$y(x,t) = o \qquad x \geq o,\ t < o$$

$$= y_o \qquad x = o,\ t \geq o$$

The solutions were obtained in terms of dimensionless gas phase concentration ($\alpha = y/y_o$), adsorbent loading ($\beta = n/n_o$), gas phase temperature [$\Psi = (T_g - T_o)/(T_s^* - T_o)$] and adsorbent temperature [$\lambda = (T_s - T_o)/(T_s^* - T_o)$] as functions of dimensionless time ($\theta = t/t^c$) and distance in the column ($z = x/L$). It should be noted that α and Ψ are the only two experimental variables.

RESULTS

Type I Adsorption

The test conditions of the C_2H_6-He system were chosen in such a way $[n_0/y_0<C_s/C_g]$ that Type I column dynamics is exhibited[21,22]. This is characterized by the formation of two pairs of mass and heat transfer zones in the column separated by an equilibrium section. Figure 1 shows the α and β profiles as a function of z at $\theta = 0.6$ for a given value of k and various values of h. Type I behavior is exhibited. The gas phase composition of the middle equilibrium section is $\alpha^* = 0.975$. It may be seen from Figure 1 that the α and β profiles in the front transfer zones are weak functions of h. The profiles for very low values of h are almost identical to those for $h \geq 10^{-2}$ which corresponds to instantaneous thermal equilibrium between the gas and the solid (see Figure 2) under the test condition.

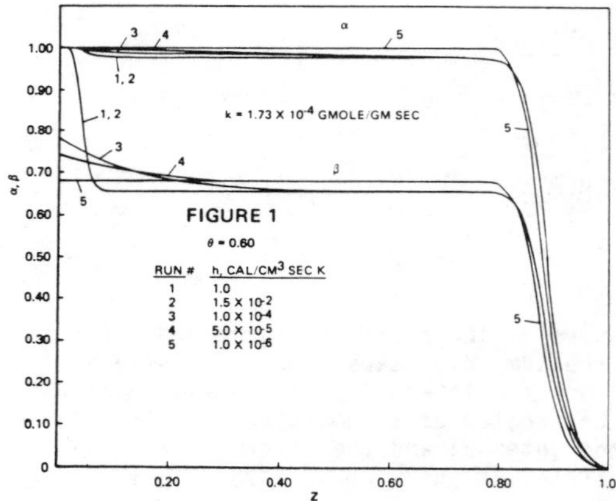

FIGURE 1

The rear zone β profiles, on the other hand, are strong functions of h. For sufficiently high values of h ($\geq 10^{-2}$), a complete rear transfer zone is formed at $\theta = 0.6$ as indicated by $\beta = 1$ at $z = 0$. For low values of h, the rear zone is still forming at $\theta = 0.6$. The Figure also shows that the rear zone is stretched as h decreases. This can be more clearly seen from the β profiles than the α profiles because α^* is close to unity.

Figures 2 and 3 show the corresponding λ and Ψ profiles, respectively, as functions of z. The solid phase temperature of the middle equilibrium section is 336.9°K ($\lambda^* = 0.71$). Figure 2 shows that the front zone adsorbent temperature profiles are also practically independent of h but they are strong functions of h in the rear transfer zones. Figure 3, on the other hand, shows the gas phase temperature profiles are strong functions of h both in the front and the rear transfer zones. The front zone gas phase temperature at lower values of h is higher than that for instant thermal equilibrium case ($h \geq 10^{-2}$) as expected. The adsorbent is heated in this zone. The pattern is reversed in the rear zone where the adsorbent is cooled by the incoming cold feed gas.

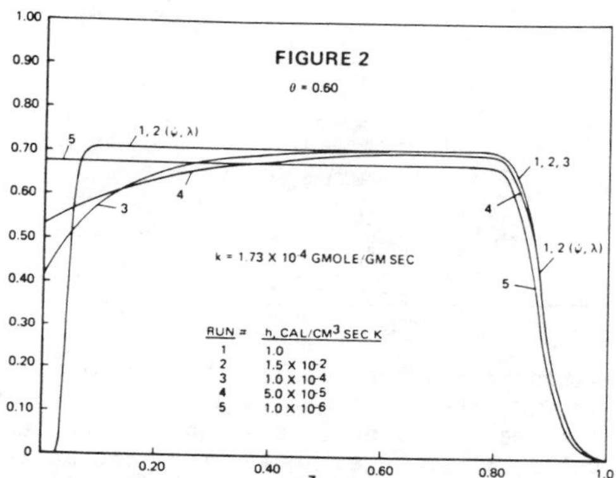

Figures 2 and 3 also show that for sufficiently low values of h, the rear zone may stretch all the way to the front zone and over-lap on it. This may distort the front zone and no meaningful estimation of transport properties can be made in that case. The experimental gas phase temperature profile at the column exit in such a case will not exhibit a plateau after the initial rise from T_0 but will slowly decrease back to T_0 after attaining a maximum. This may be an indication of the overlap. Note from Figure 1, that such overlap may not be recognizable from the experimental α profile.

Figure 4 shows the α and β profiles at $\theta = 0.6$ for a given value of h and different values of k. It shows that the front transfer zone is very strongly influenced by k. The lower the value of k, larger is the zone as expected. The rear mass transfer zone, on the other hand, is a relatively weaker function of k. For sufficiently low value of k, the front zone can stretch all the way to the rear zone and distort it. No meaningful estimation of transport coefficients can be made in such a case.

Figure 5 shows the corresponding temperature profiles in the column. $\Psi = \lambda$ in this Figure due to high value of chosen h. As for the α profiles, the front zone temperature profiles are functions of k but they are practically independent of k in the rear zones. Again overlap of the zones can be recognized by the absence of the plateau in the experimental Ψ profiles.

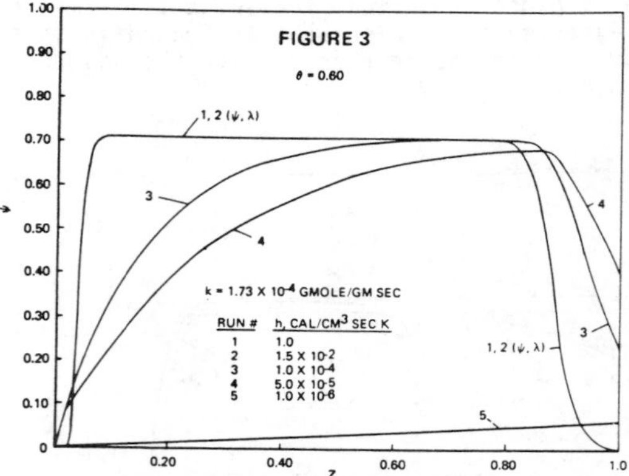

It may be concluded from the above results that for a fully formed Type I adsorption system, the front MTZ is primarily governed by the mass transfer and the rear transfer zone is governed by the heat transfer. Consequently, k can be estimated from the analysis of the front zone data by assuming instantaneous thermal equilibrium in the model and h may be estimated from the rear zone data by assuming instant mass equilibrium in the model. In other words, k and h can be estimated, respectively, from the experimental α profile in the front zone and the experimental Ψ profile in the rear zone alone. This may also allow use of simplified models for data analysis. For sufficiently low values of k or h the two zones may not be fully formed and actually over-lap on each other. Thus no meaningful estimation of these parameters can be made by simpler models. Earlier Cooney had briefly addressed this problem for high pressure adsorption systems.[23]

Type II Adsorption

The test conditions of the C_4H_{10}-He system were chosen in such a way ($n_o/y_o > C_s/C_g$) that they satisfy the conditions for the formation of the Type II behavior[21,22]. This is characterized by the formation of a pure heat transfer zone (front zone) that moves ahead of a combined mass and heat transfer zone (rear zone) which is followed by an equilibrium section at the feed gas conditions. Figure 6 shows typical α and Ψ profiles for this case in the column at different values of θ. Type II behavior is exhibited and the maximum temperature rise in the column is 3.1°C.

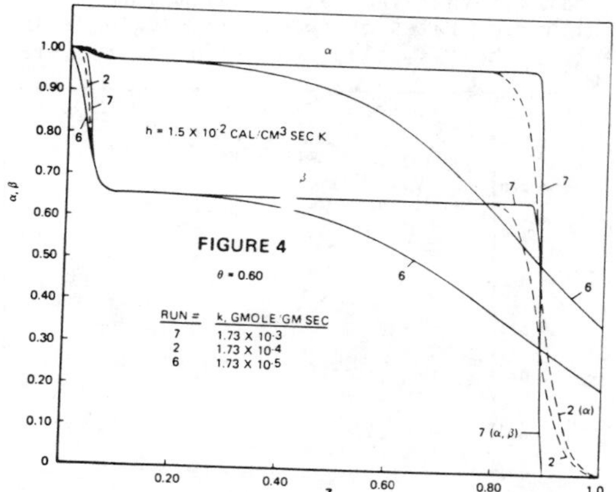

The effects of h and k on the size of the mass transfer zone is demonstrated in Figure 7. The length of the mass transfer zone (LMTZ) is defined by the section of the zone bounded by α = 0.05 to 0.95. LMTZ/L is plotted in the ordinate as a function of k in the abcissa with h as a parameter. The Figure shows that LMTZ is dependent on both k and h. LMTZ increases both for decreasing h and k. However, for sufficiently low values of h, the LMTZ is practically controlled by h. On the other hand, if k is very small, the MTZ is practically independent of h. Neverthe- less, Figure 7 shows that unambiguous estimation of k may not be possible from the analysis of experimental MTZ data for Type II systems unless h is known a priori. Assumption of instant thermal equilibrium in the model for low flow rate experiments may yield a wrong (low) value for k.

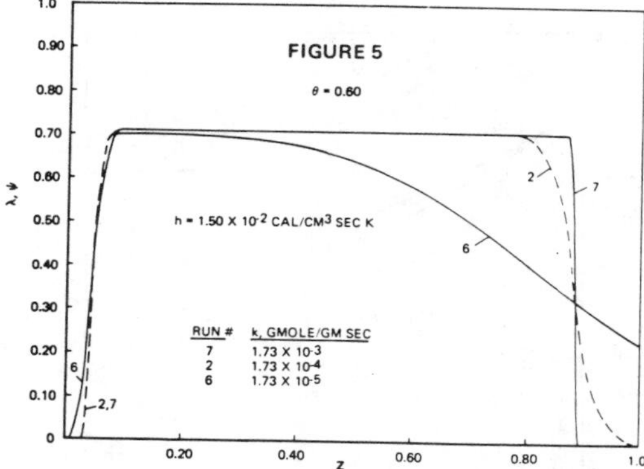

The above results indicate that laboratory columns for type II systems need to be run at high flow rates of commercial scale design to eliminate the effect of the gas-solid heat transfer resistance in order to estimate k.

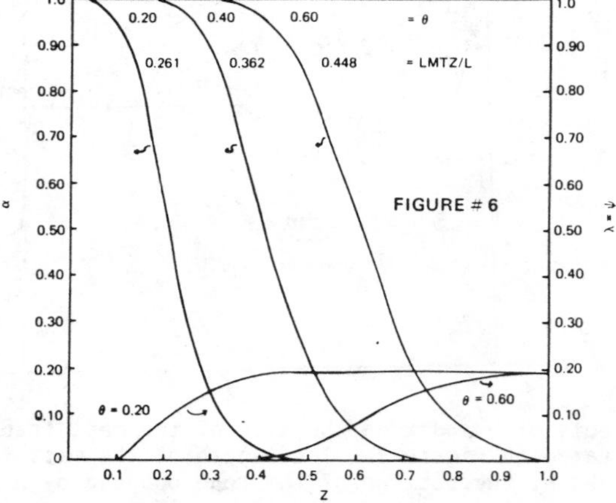

In general, it may be concluded that instant thermal equilibrium between the gas and the solid may not be attained in laboratory adsorbent columns at low gas flow rates. Consequently, the interaction between the mass and the heat transfer coefficients and their effects on the shape of the sorbate mass transfer zone need to be considered during analysis of the breakthrough data for estimation of the mass transfer coefficient. It may be possible to estimate k and h independently from experimental column dynamics for Type I adsorption systems. A priori knowledge of h is needed for estimation of k from column dynamics of Type II systems.

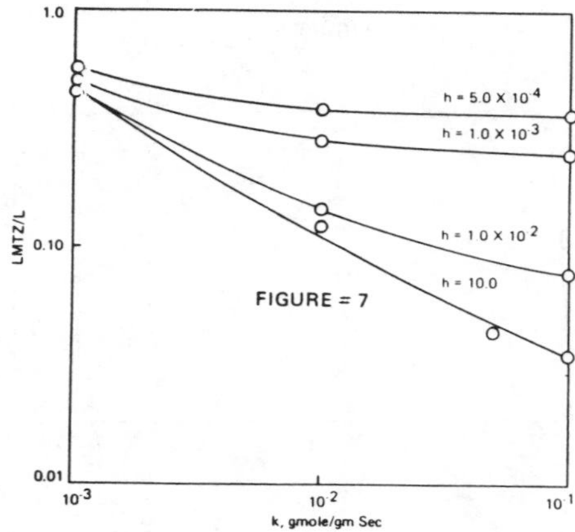

NOMENCLATURE

b	=	constant in Langmuir Equation
b_o	=	b at infinite T
C	=	Heat capacity
d_p	=	Particle diameter
D_p	=	Macropore diffusivity
h	=	over-all heat transfer coefficient
h_e	=	External heat transfer coefficient
h_i	=	Internal heat transfer coefficient
k	=	over-all mass transfer coefficient
k_e	=	External mass transfer coefficient
L	=	column length
LMTZ	=	Length of mass transfer zone
m	=	Saturation capacity (Langmuir)
n	=	Amount adsorbed
P	=	Gas pressure
q	=	Heat of adsorption
Q	=	Flowrate of gas mixture
Q^c	=	Flow rate of carrier gas
R	=	Gas constant
t	=	time
t_c	=	$\dfrac{L\rho_s n_o(1-y_o)}{Q^c y_c}\left[1+\dfrac{\epsilon\rho_g y_o}{\rho_s n_o}\right]$
T	=	Temperature
$(T_s^*-T_o)$	=	$\dfrac{qn_o}{C_s}$
x	=	Axial column distance
y	=	Adsorbate mole fraction
$\bar{y}$	=	mole fraction in equilibrium at P_o, n, T_s
ϵ	=	Helium void fraction
$\bar{\epsilon}$	=	Macropore void fraction
ρ	=	Density
τ	=	Tortuosity
η	=	Adsorbent thermal conductivity
o	=	Feed and initial column conditions (subscript)
*	=	Equilibrium section between the zones (superscript)
g	=	Gas phase (subscript)
s	=	Solid phase (subscript)

REFERENCES

1. D. M. Young and A. D. Crowell, "Physical Adsorption of Gases," Butterworths, Washington, D.C., (1962).
2. C. N. Satterfield, "Mass Transfer in Heterogeneous Catalysis," M.I.T. Press, Mass. (1970).

<u>REFERENCES</u> (cont'd)

3. R. H. Perry and H. Chilton (Eds)., "Chemical Engineer's Handbook (5th ed.) McGraw-Hill, New York (1974).
4. S. Sircar and R. Kumar, I.&E.C. Proc. Des. Dev. <u>22</u> 271, (1983).
5. K. Kawazoe and Y. Takeuchi, J. Chem. Eng. Japan, <u>1</u>, 431 (1974).
6. S. Sircar, J.C.S. Faraday Trans. I., <u>79</u> 785, (1983).
7. K. Chilhara, M. Suzuki and K. Kawazoe, Chem. Eng. Sci., <u>31</u>, 305 (1976).
8. S. Sircar and R. Kumar, I.&E.C. Proc. Des. Dev., <u>22</u>, 280 (1983).
9. F. W. Leavitt, Chem. Eng. Prog. <u>58</u>, 54 (1962).
10. C. Y. Pan and D. Basmadjian, Chem. Eng. Sci., <u>22</u>, 285 (1967).
11. O. A. Meyer and T. W. Weber, A.I.Ch.E. J., <u>13</u>, 457 (1967).
12. J. W. Carter, Trans. Inst. Chem. Engrs., <u>44</u> T253 (1966).
13. R. Kumar and S. Sircar, Chem. Eng. Comm. In press (1983).
14. R. Kumar and S. Sircar, Chem. Eng. Comm. In press (1983).
15. D. R. Garg, Ph.D. Thesis, University of New Brunswick, N.B. Canada (1972).
16. O. P. Mahajan, M. Morishita, P. L. Walker, Carbon, <u>8</u>, L67 (1970).
17. R. L. Gariepy and I. Zwiebel, AIChE Symp. Ser. 117 <u>67</u>, L7 (1971).
18. D. Kunii and M. Suzuki, J. Int. Heat Mass Transfer, <u>10</u>, 845 (1967).
19. N. Wakao and T. Funazkri, Chem. Eng. Sci., <u>33</u>, L375 (1978).
20. S. Sircar, R. Kumar and K. J. Anselmo, I&EC Proc. Des. Dev., <u>22</u> 10 (1983).
21. C. Y. Pan and D. Basmadjian, Chem. Eng. Sci., <u>25</u> 1563 (1970).
22. S. Sircar and R. Kumar, I&EC Proc. Des. Dev. (1983).
23 D. O. Cooney, I&EC Proc. Des. Dev. <u>13</u>, 368 (1974).

COMPUTER SIMULATION OF PHYSISORPTION

William A. Steele
Department of Chemistry
The Pennsylvania State University
University Park, Pennsylvania 16802
USA

ABSTRACT

The principal techniques used to generate simulations of dense phases made up of simple model molecules are briefly described. Several such studies of interfacial systems are reviewed: these include Monte Carlo simulations of density profiles of bulk liquids in contact with a rigid wall; the changes in these profiles for a liquid between two walls of variable separation, and the liquid-mediated wall-wall force that results. Similar simulations of the ionic densities for an electrolyte solution in contact with a charged electrode are reviewed. Finally, molecular dynamics simulations of a monolayer of nitrogen adsorbed on graphite at low temperature are reviewed; it is shown that both the equilibrium and the dynamical results indicate the presence of a sharp change in the orientational properties in this system at $33^{o}K$.

The determination of both molecular and thermodynamic properties of model interfacial systems by computer simulation has emerged as a practical and powerful method. This article comprises a brief description of some of the techniques that have been of particular utility in this respect, followed by a discussion of a few of the many interesting simulation studies reported over the past few years. It should be noted at the outset that some overlap between this paper and the recently published book on this subject by Nicholson and Parsonage (1) is almost inevitable. Indeed, the reader is directed to this monograph for a more extensive review of work in this field.

BASIC ASSUMPTIONS AND COMPUTATIONAL TECHNIQUES

In general, simulations of molecular systems are based on either the Monte Carlo or the molecular dynamics algorithm. There are now a variety of options within these general categories, some of which are particularly well adapted to the fluid-solid case, as we will see below. Both techniques involve the assumption of classical statistical mechanics, so that the kinetic and configurational factors in the partition functions and probability distributions can be separated.

In either scheme, a system containing a few hundred to a few thousand molecules is typically considered; in the computer memory, this small sample of molecules is surrounded by an (infinite) set of images of itself, and molecules and their images are allowed to pass in and out of the central system. Complications arise when the range of the intermolecular potentials is comparable to or larger than the sample dimensions.

However, when this problem is either avoided or solved, the sample behaves as if it were drawn from a much larger bulk fluid. Of course, the infinite set of images extends only in two dimensions for a fluid in contact with a solid surface. The third dimension is frequently closed off by studying the adsorbed phase enclosed between two parallel solid surfaces; these can either be set at a large separation so that two independent adsorption regions form separated by bulk material, or the surfaces can be moved together to simulate adsorption in a parallel-walled pore or (approximately) between two colloidal particles.

The next step in characterizing the system is to choose the potential energies of interaction between the fluid molecules and between a fluid molecule and the solid. As far as fluid-fluid molecular interactions go, a great variety of pair-wise additive functions have been used in bulk simulations and can be used in solid-fluid studies. Starting from the simplest hard sphere potentials, one can work up to the Lennard-Jones inverse (12-6) power potential for intermolecular energy. Purely repulsive potentials, both hard and soft, have also been studied. Non-spherical molecules can also be modeled by altering the atomic potentials, first by adding electrostatic energies of dipoles, quadrupoles, etc. Secondly, one represents molecular shape by site-site potentials in which (for example) diatomics are assumed to be two fused spherical sites separated by the bond distance L; the pair-wise molecular potential is then given by the sum of the four site-site energies. This approach has proven surprisingly effective in describing bulk fluids (2) and has been extended to simulations of relatively large molecules, such as tetrahedra (3) and benzene (4) (either as a six-site or with more difficulty, a twelve-site model).

Simulations of physisorption start from the premise that the solid is an inert, rigid boundary whose role is to provide an external potential for the adsorbate molecules. The simplest potential is the planar hard-wall model; various soft repulsive potentials have also been treated. More realistic functions are based on the assumption that the energy is due to a pair-wise sum of the adsorbate-solid atom interactions. A simple representation of the resulting potential is given by a Fourier expansion (5):

$$u_s(\underset{\sim}{r}) = E_0(z) + \sum_{\underset{\sim}{g}\neq 0} E_{\underset{\sim}{g}}(z)\cos(\underset{\sim}{g}\cdot\underset{\sim}{\tau}) \tag{1}$$

where $\underset{\sim}{\tau}$ is the vector position of the gas atom in a plane parallel to the surface, $\underset{\sim}{g}$ is a multiple of the reciprocal lattice vector of the surface which ensures that Equation (1) conforms to the periodicity of the solid surface. Analytic expressions have been derived for the coefficient E_n; in particular, the leading term, which in isolation gives the interaction of an atom with an array of planes each containing continuous a distribution of atoms, is

$$E_o = 2\pi\rho_s \epsilon_{s\alpha} \sum \left\{ \frac{2}{10}\left(\frac{\sigma_s}{z_\alpha}\right)^{10} - \left(\frac{\sigma_s}{z_\alpha}\right)^4 \right\} \tag{2}$$

where z_α is the distance of the gas atom from the α'th plane, ρ_s is the density of solid atoms in the plane and ϵ_s, σ_s are the parameters of the gas-solid pair-wise potential. Accurate approximations to the sum in Equation (2) have been suggested (6). Site-wise models for non-spherical molecules have been extended to gas-solid interactions by utilizing Lennard-Jones functions for the gas-solid atom energy for each site in a molecule. Of course, electrostatic interactions can play an extremely significant role in many adsorption systems of interest; these may be included either by summing over the Coulomb energies for discrete charge models of the fluid molecule and the solid, or by evaluating the surface field-gas molecule multipole energies.

Having defined the interactions present in the system of interest, one must now select an appropriate simulation technique. The choice is not merely between Monte Carlo and molecular dynamics, but also involves the specification of the statistical ensemble to be simulated. In molecular dynamics, the energy conservation property of the Newtonian equations of motion means that the natural choice is constant N,V,E (a microcanonical ensemble) with computed averages basically chosen to be pressure, including spreading pressure in a fluid-solid system, and the separate potential and kinetic energies. The temperature of the system is then obtained from the average kinetic energy. The simulation consists in solving the equations of motion for the molecular positions and velocities (and angular variables as well, if the molecules are nonspherical) by numerical integration over a series of finite time intervals equal to $\sim 10^{-14}$ sec. Changes in all variables are evaluated in each time step; starting from an arbitrary (but reasonable) initial configuration, statistical equilibrium is generally achieved in $\sim$ 2000 time steps. Subsequently, the time-evolution is followed for another 5000-50000 (or more) time steps, accompanied by evaluation of averages or data-saves or both. It is important to realize that the resulting data file of coordinates and velocities can then be utilized to calculate time-dependent as well as equilibrium average properties.

In the Monte Carlo technique, ensemble averages are explicitly achieved by taking the molecular system through a series of coordinate displacements chosen to make it approach and then fluctuate around the most probable configuration. In the best-known scheme, a canonical ensemble (constant N,V,T) is simulated by evaluating the probability p_i at each displacement, with p_i defined to be proportional to $\exp[-U_i/kT]$, where U_i is the total potential energy of the system in the i'th configuration. Since one wishes to locate the most probable distribution, one accepts the new configuration if $p_i > p_{i-1}$. If this is not so, the acceptance is determined by asking whether another random variable is greater than some arbitrarily chosen value. In this way, fluctuations from the most probable configuration and, more importantly, transitions over potential barriers are allowed.

Ensembles other than canonical can also be simulated by the Monte Carlo technique; in particular, a grand canonical ensemble (constant μ,V,T) are generated if the probability p_i is chosen appropriately and if the number of molecules in the system is randomly varied. This ensemble is particularly suited to adsorption studies, since an isotherm can be viewed as a plot of the total number of molecules on or near the surface as a function of the chemical potential $\mu^{(ad)}$ of the adsorbed phase in equilibrium with bulk ideal gas at a pressure p. Thus, the

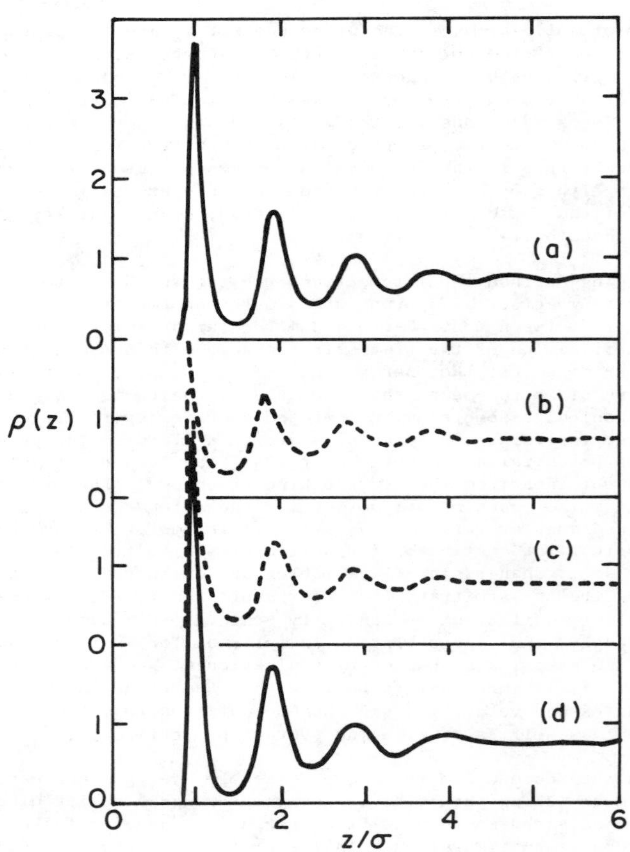

Figure 1. Density as a function of distance for a dense ($\rho\sigma^3 = 0.75$) fluid near several model walls. a) Lennard-Jones fluid, attractive summed 10-4 wall potential; b) Lennard-Jones fluid, hard repulsive wall; c) Lennard-Jones fluid, soft repulsive wall; d) hard spheres against a wall with the attractive interaction of (a).

variables fixed are $\mu^{(ad)} = \mu^{(gas)}$, V and T, where V = volume available to the adsorbate.

With this rather sketchy description, we can now discuss a few of the numerous interfacial studies that have been reported over the past few years.

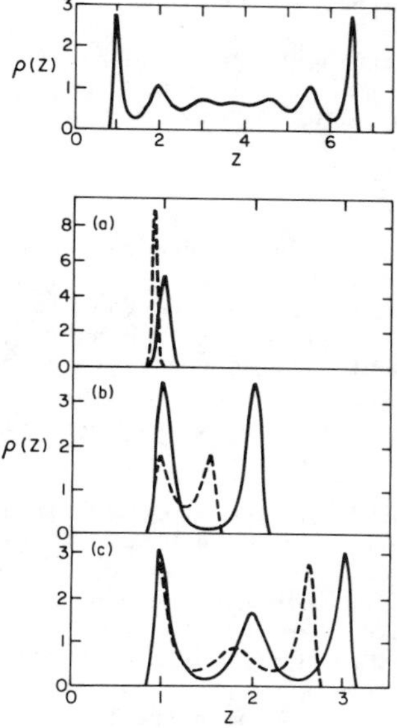

Figure 2. Densities for a Lennard-Jones fluid between two walls with wall-fluid interactions given by a sum of Lennard-Jones energies over the solid. Densities are plotted as a function of position Z = distance/σ, where σ is the Lennard-Jones parameter, for various values of the wall separation. The top figure is for $h/\sigma = 7.5$; the dashed and solid curves in the other figures are for: a) $h/\sigma = 1.8$, 2.0 respectively; b) $h/\sigma = 2.5$, 3.0 respectively; c) $h/\sigma = 3.6$ and 4.0 respectively.

RESULTS

We first discuss a few of the simulation results obtained for dense fluids in contact with structureless planar walls (7,8). Figure 1 shows that a wall induces pronounced structure in the liquid density $\rho(z)$. Evidently, roughly three distinguishable layers of "adsorbed" atoms form at the wall. The most interesting feature of these calculations is that these layers form equally well for repulsive and attractive wall interactions and for attractive and repulsive fluid-fluid potentials. Evidently, the structure is a consequence of packing (or entropic) requirements rather than energy minimization.

Figure 2 (8) shows the same density plots for a LJ fluid between two walls with summed (10-4) interaction potentials for different wall separations h (given in units of distance/σ). Temperature $T^* = kT/\epsilon$ and bulk density $\rho_b^* = \rho_b\sigma^3$ were 1.2 and 0.5925 in all cases. For the largest separation ($h/\sigma = 7.5$), the density profiles on each side of the container are essentially those for the one-wall problem, but considerable alterations in these profiles occur as the walls approach each other. The bulk fluid region is absent for h = 4 or less; the three layers of fluid atoms present at h = 4 are progressively squeezed down into two and then one and eventually vanish altogether. An important feature of these simulations is that it allows one to calculate the solvation force f_s (9), defined to be the excess force per unit area between the solid walls due to the presence of liquid in the intervening space. One has (8,10)

$$f_s = \int_0^h \rho(z) \frac{\partial u_s(z)}{\partial z} dz \tag{3}$$

Because of the large density of fluid $\rho(z)$, this force can be much larger than the direct force due to the interaction of the solids with each other. The solvation force calculated as a function of h from Equation (3) can be quite large compared to the direct wall-wall force; it goes through a series of maxima as h decreases and eventually decreases to zero as <u>all</u> the fluid is squeezed out at very small separations. Calculations of this force have a variety of practical applications, including especially problems in colloid coagulation.

The calculation ion distributions in an electrolyte solution near a charged electrode is of course a classic problem in physical chemistry. The solution of the Poisson-Boltzmann equation leading to the well-known Gouy-Chapman theory gives expressions for both the ion densities and for the electrostatic potential in the fluid near the electrode surface. This theory can be compared with the results of Monte Carlo simulations of fluids of singly charged electrolytes near a surface with a uniform charge density (11). Figurs 3 shows that the Gouy-Chapman theory for the ion densities is reasonably accurate. The solution simulated was reasonably concentrated (1 molar in the bulk) and surface charge densities were selected which give considerable enhancement of the bulk counterion density near the surface (i.e., considerable adsorption). In contrast to the densities shown for the uncharged systems in Figure 1, no layering of the adsorbed ions occurs - one concludes that the Coulombic energies are the dominant factor, at least for this value of the bulk solution concentration. In these simulations, the ions in solution were taken to be charged hard spheres of diameter 4.25 Å, immersed in a continuum of dielectric constant $\epsilon = 78$ at room temperature. Comparisons between the simulations and the theoretical electrostatic potential near the charged wall indicate that the theory is quite successful in this case as well, considering the solution and electrode charge densities taken.

As an example of the molecular dynamics simulation, we now present some recent results obtained for model nitrogen adsorbed on a graphite basal plane at very low temperatures (12). The inclusion of the non-spherical molecular shape and the periodicity in the molecule-solid potential causes a dense monolayer to form at 0° K which consists of molecules lying parallel to the surface with their axes in a herringbone pattern and with their centers over the centers of non-nearest neighbor graphite hexagons in a $\sqrt{3} \times \sqrt{3}$ registered array. In addition to the energy and heat capacity, various translational and orientational order parameters of this dense layer were simulated as a function of temperature. Figure 4 shows the distribution of the in-plane orientation angles for a 96-molecule system; the peaks at low temperature are at the angles for the herringbone structure; at $\sim 31^o$K., these peaks are replaced by a weakly sixfold distribution that reflects preferential orientations in the 6-coordinate registered lattice. Time-dependence in statistical mechanics

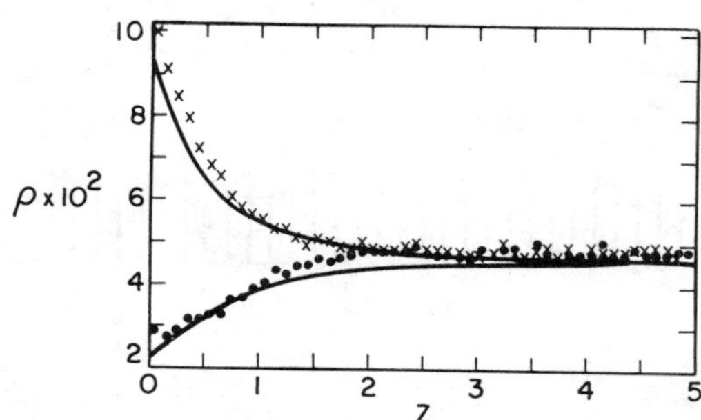

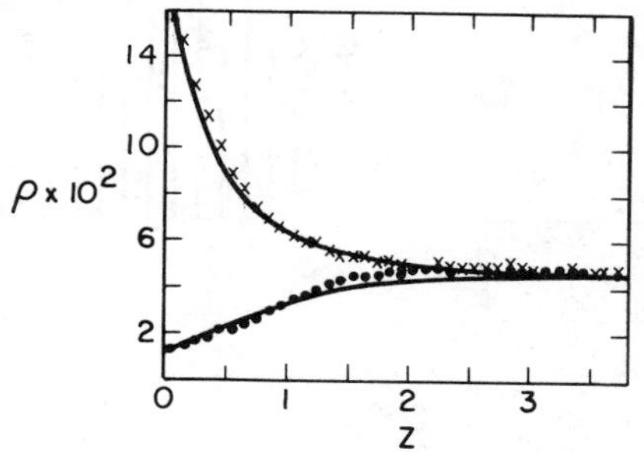

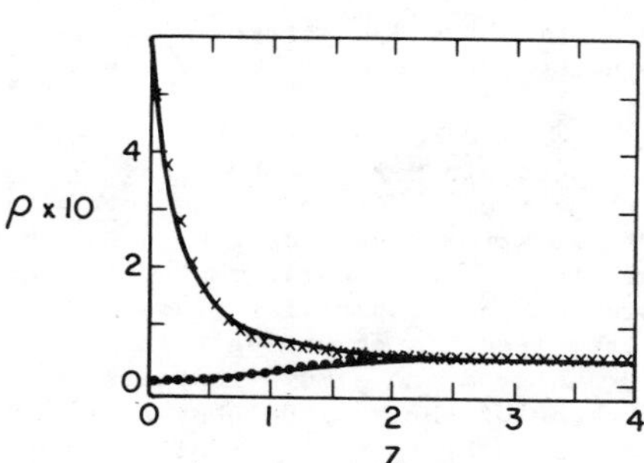

Figure 3. Ion densities for a 1 molar (1-1) electrolyte are plotted as a function of the distance from a charged wall. From the top down, the wall change densities are $4.38 \cdot 10^{-2}$ $8.76 \cdot 10^{-2}$, and $19.72 \cdot 10^{-2}$ coulomb/m^2.

is typically characterized by correlation functions and Figure 5 shows a few of the possibilities for an adsorbed monolayer of diatomic molecules. Reorientation in the surface plane is shown by the $C_n(t)$, which are defined to be $< \cos n\delta\phi(t) >$, where $\delta\phi$ is the change in the in-plane projection of the orientation of the molecular axes. At the low temperature, the $C_n(t)$ approach constant values indicating that only libration

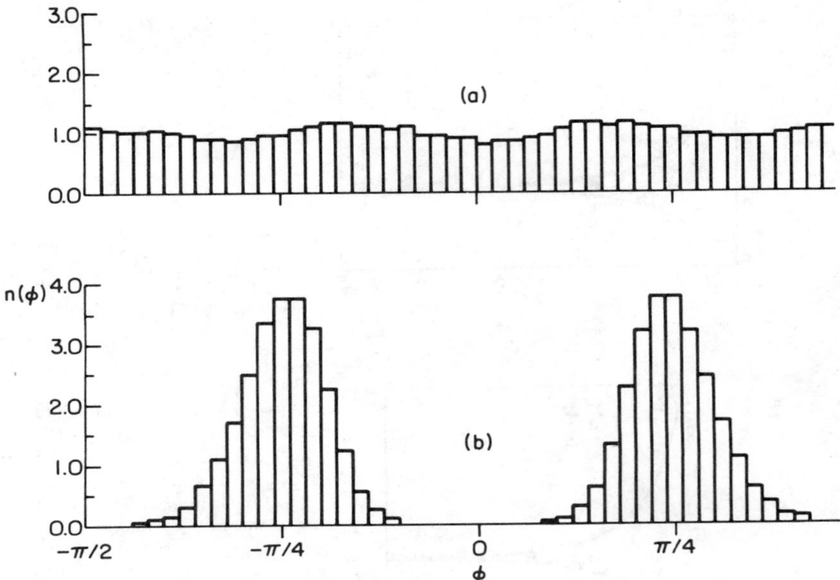

Figure 4. Distributions of the in-plane orientation angle ϕ for a (registered) layer of N_2 or graphite at two temperatures. Histograms above ((a), $T = 42^{o}K$) and below ((b), $T = 19^{o}K$) the orientational disordering temperature. The herringbone sublattices correspond to $\phi = \pm\ \pi/4$; the six neighbors in the registered film are located at $\phi = n\pi/3$, $n = 0$-5.

is occurring, whereas the gradual decay to zero at the higher temperature signifies rotation under the influence of the fluctuating intermolecular torques. The angular velocity correlation functions show a damped harmonic oscillation for libration in the plane perpendicular to the surface; the damping is much stronger for rotation in the parallel plane and again shows that rotation, while far from free, is occurring rather than libration. Some of the conclusions drawn from the simulations are that significant molecular rotation in the plane of the surface begins at $\sim 31^{o}K.$; that onset of this rotation is assisted by librational displacements perpendicular to the surface; that the electrostatic quadrupole-quadrupole energy plays an extremely important role in determining the in-plane angular correlations. Finally, the formation of a registered phase at these very low temperatures (compared to the two-dimensional triple point for N_2 on graphite, which is $\sim 50^{o}K.$) requires only a small periodic variation in gas-solid energy. There are indications that the layer loses registry at a temperature significantly lower than the experimental value; this suggests that improved potential functions are needed for an accurate

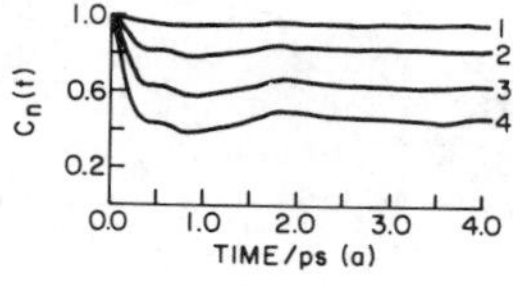

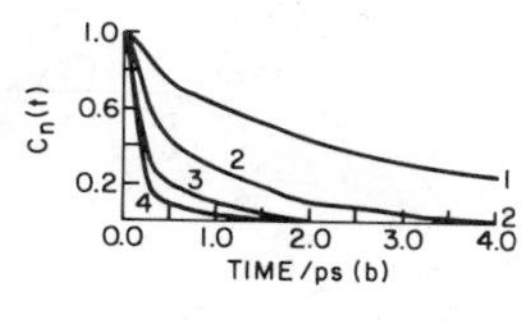

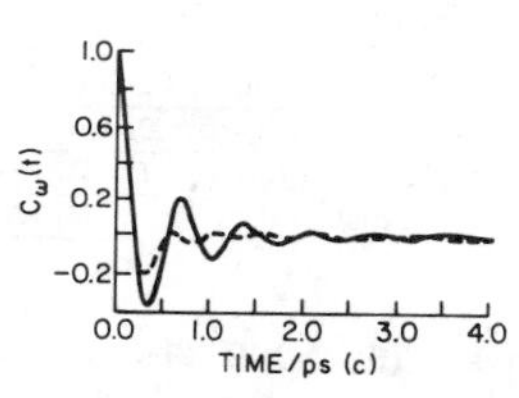

Figure 5. Time correlation functions for the in-plane orientation angle of a N_2 molecule in a (registered) monolayer on graphite. The top two plots are for one temperature below the disordering point (T = 19°K) and one above (T = 41°K.) The bottom plot shows the in-plane angular velocity correlation at T = 19°K (dashed curve) and at 41°K (solid curve).

description of the properties of this (and other gases) on graphite at higher temperatures.

SUMMARY

As in the case of bulk phases, computer simulations of interfacial systems can provide us with quite detailed "experimental" information. One can use these results to test theories and to obtain physical insights concerning molecular behavior. Of course, the systems studied tend to be idealized representations of reality, partly for mathematical convenience and partly because the reality of interfacial systems (involving heterogeneous surfaces, for example) can be so varied and complex as to defeat attempts at comprehensive and detailed modelling. Nevertheless, many more interesting and useful results than those touched upon here have been reported already. Computer power and availability are still increasing rapidly and thus the potential for future progress using this technique seems almost unlimited.

LITERATURE CITED

1. Nicholson, D. and N.G. Parsonage, Computer Simulation and the Statistical Mechanics of Adsorption, Academic Press, New York (1982).

2. Cheung, P.S.Y. and J.G. Powles, Molec. Phys. 32, 1383 (1976); Raich, J.C., N.S. Gillis and T.R. Koehler, J. Chem. Phys. 61, 1311 (1974).

3. Murad, S., D.J. Evans, K.E. Gubbins, W.B. Streett and J.J. Tildesley, *Molec. Phys*. 37, 725 (1979); Steinhauser, O. and M. Neumann, *Molec. Phys*. 40, 115 (1980); MacDonald, I.R., D.G. Bounds and M.L. Klein, *Molec. Phys*. 45, 521 (1982); Murad, S. and K.E. Goubbins, in "Computer Modelling of Matter," Ed. P. Lykos ACS Symposium Series 86, 62 (1978).

4. Evans, D.J. and R.O. Watts, *Molec.Phys*. 32, 93 (1976); 31, 83 (1976); Taddei, G., H. Bonadeo, M.P. Marzocchi and S. Califano, *J. Chem. Phys*. 58, 966 (1973).

5. Steele, W.A., *Surf. Sci*. 36, 317 (1973).

6. Steele, W.A., *J. Phys. Chem*. 82, 817 (1978).

7. Rao, M., B.J. Berne, J.K. Pereus and M.H. Kalos, *J. Chem. Phys*. 71, 3802 (1979); Snook, I,K. and D. Henderson, *J. Chem. Phys*. 68, 2134 (1978); Sullivan, D.E., D. Levesque and J.J. Weis, *J. Chem. Phys*. 72, 1170 (1980); Abraham, F.F. and Y. Singh, *J. Chem. Phys*. 67, 2384 (1977).

8. Snook, I.K. and W. van Megen, *J. Chem. Phys*. 70, 3099 (1979); 72, 2907 (1980); 75, 4738 (1981); 74, 1409 (1981); van Megen, W. and I.K. Snook, *J. Chem. Soc. Faraday Trans. 2* 75, 1095 (1975).

9. Snook, I.K. and W. van Megen, *Phys. Lett*. 74A, 332 (1979); *J. Chem. Soc. Faraday Trans. 2* 77, 181 (1981); Lane, J.E. and T.H. Spurling, *Aust. J. Chem*. 33, 231 (1980).

10. Ash, S.G., D.H. Everett and C. Radke, *J. Chem. Soc. Faraday Trans. 2*, 69, 1256 (1973); Mitchell, D.J., B.W. Ninham and B.A. Pailthorpe, *J. Chem. Soc. Faraday Trans. 2*, 74, 1116 (1978).

11. Torrie, G.M. and J.P. Valleau, *J. Chem. Phys*. 73, 5807 (1980); van Megen, V. and I.K. Snook; *J. Chem. Phys*. 73, 4656 (1980); Snook, I.K. and W. van Megen, *J. Chem. Phys*. 75, 4104 (1981).

12. Talbot, J., D.J. Tildesley and W.A. Steele, *Molec. Phys*., in press. Migone, A.D., H.K. Kim, M.H.W. Chan, J. Talbot, D.J. Tildesley and W.A. Steele, *Phys. Rev. Lett*., 52, 192 (1983).

BEHAVIOR OF DISSOLVED ORGANIC CARBON MOLECULAR WEIGHT FRACTIONS DURING ACTIVATED CARBON ADSORPTION

R. Scott Summers and Paul V. Roberts
Research Assistant and Associate Professor
Civil Engineering Department
Stanford University, Stanford, CA 94305

ABSTRACT

The equilibrium and kinetic adsorptive behavior of dissolved organic carbon molecular weight (MW) fractions in treated wastewater was characterized using ultrafiltration. The applied fixed-bed model adequately simulates column breakthrough behavior of the MW fractions, but the predicted values overestimate both the mass-transfer resistance and the partition parameter inferred from column behavior. The composite behavior of the MW fractions provides a good representation of the unfractionated organic matter with regard to the rate and equilibrium.

For activated carbon to be effectively used for the removal of specific organic compounds, the effects of the solute properties on adsorption equilibrium and rate must be understood. Specific compounds can be well characterized according to structure and properties, but the properties of the collective mass of organic matter, e.g., humic substances, are difficult to elucidate. A proven method of characterizing the collective organic matter is the MW distribution. Ultrafiltration has been used by previous investigators (1-7) and in this laboratory (8) to fractionate water and wastewater organic matter into MW ranges. Ultrafiltration MW fractions inherently are operationally defined parameters, because of the variable chemical nature of the solution matrix of different water and wastewater systems.

The objective of this study was to characterize the adsorption of organic matter in treated wastewater, measured collectively as dissolved organic carbon (DOC), by investigating the molecular weight distribution. The approach taken was to: measure the kinetic and equilibrium behavior of DOC MW fractions during activated-carbon treatment; apply a fixed-bed mass transfer model to simulate the granular activated-carbon (GAC) column breakthrough data; and relate the composite behavior of the MW fractions to that of the unfractionated DOC.

BACKGROUND

Previous Studies

The MW characterization of the adsorption of organic matter has been investigated using treated wastewater (1-4,9,10), as well as humic substances (5-7,11). Most studies have focused on the adsorption of different MW fractions after isolation of those fractions from the original matrix solution by ultrafiltration, gel chromatography, or reverse osmosis (1,3,5-7,9). The most extensive research with isolated MW fractions has been performed by DeWalle and Chian (3) for wastewater and by Snoeyink and coworkers (5-7) for humic substances. These investigations indicated that the highest MW fraction was least strongly sorbed.

The other approach taken was to investigate the matrix solution behavior by characterizing the organic MW distribution before and after activated-carbon treatment using batch or column contactors (2-4,10,11). These matrix studies indicated that the highest MW fraction had the lowest percent removal, with one exception (4), where the the lowest MW fraction was the least strongly sorbed.

A few studies have investigated the mass transfer rate. Kim et al. (4) monitored the column breakthrough response of the organic MW distribution using three temporal samples. Lee and Snoeyink (5) simulated the uptake rate of isolated MW fractions of humic substances in batch experiments. Arbuckle (1) evaluated the batch kinetics of isolated MW fractions of wastewater. Both batch studies (1,5) indicate an increasing rate of mass transfer with decreasing MW. Goto et al. (9) found that the mass transfer rate and equilibrium adsorption increased when the high MW material was removed from the wastewater solution.

Mass Transfer Model

The model utilized in this study was developed by Hiester and Vermeulen (12) for mass transfer in fixed-bed adsorbers. Extensions of this model as well as reviews of fixed-bed adsorber theory, have been presented by Vermeulen (13) and Vermeulen et al. (14). Applications of this model to adsorption of DOC have previously been presented by the authors (15,16). The essential equations are summarized below.

The material balance in a column of constant cross-sectional area yields the following continuity of mass equation in dimensionless form:

$$-\left(\frac{\partial X}{\partial N}\right)_{NT} = \left(\frac{\partial Y}{\partial NT}\right)_{N} \quad (1)$$

where $X = C/C_o$ = fractional fluid-phase concentration, and $Y = q/q_o$ = fractional solid-phase concentration. The number of mass transfer units, N, is defined by

$$N = \overline{t}/R \quad (2)$$

where R = resistance to mass transfer. The throughput parameter T is defined by

$$T = C_e(V - v\varepsilon_b)/q_e \, \rho \, v \quad (3)$$

where V = cumulative volume of water treated at a given time, v = volume of the bed, ε_b = bed porosity, and ρ = the particle density. When the volume of water treated is significantly larger than the volume of the bed ($V \gg v$) and the linear equilibrium relationship applies, the throughput parameter can be simplified to

$$T \simeq \theta/\Omega \quad (4)$$

where $\Omega = \rho K_L$ = partition parameter, K_L = linear equilibrium constant, and $\theta = V/v$ = throughput.

In the special case of linear equilibrium (K_L independent of concentration), the form of the mass-transfer-rate equation is independent of the rate-determining mechanism. Thus the model can be used for film diffusion, pore diffusion, surface diffusion, or second-order reaction kinetics expressed as

$$\left(\frac{\partial Y}{\partial NT}\right)_N = X - Y \text{ .} \tag{5}$$

Combining Equations (1) and (5) and integrating yields a solution in the form of a J-function, which has been summarized in tabular and graphical form (13,14,17).

EXPERIMENTAL PROCEDURE

Reclamation Facility

The GAC system investigated in this study was part of the Palo Alto Reclamation Facility (18). The reclamation facility received chlorinated, nitrified activated sludge effluent and provided sedimentation, filtration and ozonation at 4 g/m^3 (0.5 g O_3/g DOC) prior to adsorptive treatment in two columns filled with Filtrasorb-300 (Calgon Corp.) GAC. The upflow GAC columns were operated in parallel at a superficial linear velocity (u) of 10.8 m/hr (4.4 gpm/ft^2), providing an empty-bed contact time (t) of 34 min. The flow through each of the columns was 0.022 m^3/s (0.5 MGD). On-site regeneration was accomplished with a multiple hearth furnace. The GAC system was monitored by refrigerated (4°C) composite samplers (18) with a sampling interval of 30 min over a 7 to 10 day composite period. A 4-ml aliquot of phosphoric acid was added prior to sampling, to retard biodegradation in the composite sampler, thus lowering the pH range from 6 to 7 to 2.5 to 3.0. A 0.10-l aliquot of a saturated solution of sodium thiosulfate was added prior to sampling to minimize the formation of chlorinated organics as the GAC removal of these compounds was also being studied.

Equilibrium Isotherm Experiments

The isotherm experiments were conducted in one-liter glass bottles with 0.800-l aliquots of a 24-hr composite sample of the influent to the GAC adsorbers. The influent was first filtered with a 0.45-μm Millipore membrane filter. Into each bottle was placed an appropriate amount of activated carbon from a 200/400 mesh size fraction. The regenerated GAC was ground and sieved before use. After a measured amount of influent was added to each bottle, they were capped with Teflon-lined caps and shaken on a shaker table at 275 RPM at room temperature (21°C). After 24 hr, the samples were filtered through a 0.45-μm and 0.20-μm Millipore membrane filters and acidified to pH 2.5 to 3.0 prior to ultrafiltration.

Ultrafiltration

Ultrafiltration (UF) analysis was made with one- or two-liter composite samples of the influent and effluent of the GAC system and with one-liter equilibrium isotherm samples. Immediately after sampling, the samples were filtered sequentially through 0.45-μm and 0.20-μm Millipore membrane filters. The UF membranes utilized were XM 50, YM 10, UM 2, and UM 05 (Amicon, Lexington, MA) with nominal molecular size cut-off levels of 50,000, 10,000, 1,000, and 500 amu, respectively. A cascade procedure was used and the MW distribution was calculated based on mass-balance equations (8). The amount of material unaccounted for (i.e., the loss) was determined for each sample. Samples were filtered at pH 2.5 to 3.0. For organic carbon analysis, a Dohrmann DC-80 instrument (Envirotech, Santa Clara, CA) was used.

RESULTS AND DISCUSSION

Equilibrium Adsorption Capacity

Estimates of the adsorbent's equilibrium capacity based on the laboratory results were expressed as equilibrium isotherms. A modified linear isotherm was found to fit adequately the equilibrium data (19). This model accounts for a positive x-intercept with the explanation that the intercept represents the nonadsorbable portion of the organic matter in that MW fraction. The model is of the form

$$q_e = K_L(C_e - C_n) \tag{6}$$

where q_e is the solid-phase concentration, C_e is the equilibrium concentration, and C_n is the nonadsorbable concentration.

The results of these experiments and the DOC MW distribution of the ozonated water prior to adsorption isotherm experiments are presented in Table 1. The MW distribution of DOC is bimodal: the 1,000-10,000 and < 500 MW fractions together contained two-thirds of the total DOC, whereas the 10,000-50,000 MW fraction contained only 2.2% of the total DOC. A mass balance of the UF systems indicated a 7.7% DOC loss. The linear equilibrium coefficient, K_L, increased with increasing MW except for the largest fraction (> 50,000 amu), which showed a decrease in K_L. A possible explanation of this reversal of trend is the size exclusion of large MW material from surface area available for adsorption. Using an average molecular radius of 2.6 nm, only 3% of the sorbent's total surface area would be available for adsorption of the > 50,000 fraction, assuming a one-to-one ratio of pore-to-solute radius. This available surface area represents 40% of that presumably available to the 10,000-50,000 MW fraction. DeWalle and Chian (3) also reported low adsorption capacity for the highest and lowest MW fractions of treated wastewaters.

Table 1. DOC MW Distribution and Equilibrium Isotherm Results

DOC MW Range (amu)	Percent of Total DOC	Linear Equilibrium Coefficient K_L (1/g)	Percent Nonadsorbable $(C_n/C_o) \times 100$	Correlation Coefficient r
Unfractionated	8.5 g/m^3 *	13.	8.5	0.97
> 50,000	12.	11.	1.0	0.95
10,000-50,000	2.2	24.	2.8	0.89
1,000-10,000	38.	19.	0.	0.93
500-1,000	12.	13.	14.	0.85
< 500	29.	2.9	43.	0.68

*Initial concentration.

In the lowest MW fraction, 43% of the organic matter was estimated to be nonadsorbable, whereas all of the DOC in the 1,000-10,000 MW fraction appears to be adsorbable. Infrared spectroscopy of the > 50,000, 1,000-10,000 and 500-1,000 MW fractions revealed no major differences between the spectra of the fractions.

Column Studies

In simulating the breakthrough behavior of DOC, both internal and external transport were considered. We hypothesized that the pore-diffusion mechanism could adequately account for internal transport of organic matter measured as DOC. This hypothesis was substantiated in earlier studies (15,16,20). Accordingly, the overall resistance to mass transfer was calculated as the sum of an external resistance due to film diffusion and an internal resistance due to pore diffusion. To calculate either of these parameters, the bulk (aqueous-phase) diffusivity is needed. If the heterogeneous mixture of compounds as measured by the MW fractions of DOC are treated as a pseudo-single compound, their bulk diffusivities can be estimated based on a mid-range MW with the following equation proposed by Polson (21):

$$D = 2.74 \times 10^{-9}\ (MW)^{-1/3}\ (m^2/s) \tag{7}$$

or based on a molecular radius with the Stokes-Einstein equation:

$$D = BT/6\pi\ r_m \mu \quad (m^2/s) \tag{8}$$

in which B is the Boltzmann constant, T is the absolute temperature, μ is the fluid viscosity and r_m the median molecular radius. Using MW and r_m for the different MW fractions and applying Equations (7) and (8), the diffusion coefficients were calculated. The results are shown in Table 2 along with the average of the two diffusivities, $\overline{D}$.

To account for the porous structure of GAC, the bulk diffusion coefficient is modified to obtain an effective pore diffusion coefficient

$$D_p = D\ \varepsilon_i/\chi \tag{9}$$

Table 2. Predicted Mass Transfer Coefficients

DOC MW Range (amu)	$\overline{MW}$ (amu)	Molecular Radius r_m (nm)	Diffusivity, $D \times 10^{10}$ (m^2/s) Polson	Stokes-Einstein	$\overline{D}$	Effective Internal Porosity ε_i	Number of Mass Transfer Units $\overline{N}$
Unfractionated	1,000	0.60	2.7	3.2	3.0	0.53	2.0
> 50,000	100,000	2.6	0.59	0.74	0.66	0.49	0.5
10,000-50,000	30,000	1.1	0.88	1.7	1.3	0.51	1.0
1,000-10,000	5,500	0.70	1.5	2.8	2.2	0.52	1.5
500-1,000	750	0.55	3.0	3.5	3.2	0.53	2.0
< 500	250	0.25	4.4	7.7	6.0	0.61	3.0

where ε_i = internal porosity and χ = tortuosity factor (22). The total internal porosity of this carbon was measured to be 0.64 (m^3 total pore vol/m^3 particle vol). It is likely that the pore volume accessible to large organic molecules is less than the total pore volume of the adsorbent. Hence, the pore volume associated with pores of a radius less than or equal to the molecular radius were not included in the calculation of an effective internal porosity for each of the MW fractions shown in Table 2. The tortuosity accounts for the deviations in the pore diffusional path of a molecule and is often reported in the range 2 to 4 (22). Since the micropore volume is excluded from the diffusional path, the tortuosity is likely to be at the low end of the usual range; a tortuosity value of 2.5 was used in this work.

The internal resistance to mass transfer, R_p, was calculated using the particle diameter, d, as suggested by Vermeulen (13):

$$R_p^{-1} = k_p a = 60\ D_p (1 - \varepsilon_b)\ d^{-2}\ . \tag{10}$$

The external resistance was calculated from

$$R_f = d/6k_f \tag{11}$$

where the film mass transfer coefficient k_f = Sh D/d and (23)

$$Sh = 2 + 1.1\ Re^{0.6}\ Sc^{1/3} \tag{12}$$

with $Re = du/\nu$ and $Sc = \nu/D$.

Using values of $d = 1.5 \times 10^{-3}$ (m), $\varepsilon_b = 0.45$, $\nu = 1 \times 10^{-6}$ (m^2/s) and Equations (7) to (12), the internal and external resistivities were calculated for each MW fraction. The external resistance was found to be less than 4% of the internal resistance in all MW fractions. Under conditions of the linear equilibrium, overall resistance can be computed by simply summing the external and internal resistances. Using Equation (2) the total number of mass transfer units were calculated for both sets of predicted diffusivities. Because differences were found between the values of N for the different methods of predicting diffusivities, an average N value was calculated using $\bar{D}$ and rounded to the nearest half N unit (in accord with the availability of published J-function values (13,14,17)). This N value was used for the predictive breakthrough curves.

The breakthrough curves of the unfractionated DOC and the five MW fractions are shown in Figures 1 to 6. The dotted horizontal line with the error bars represents the average fractional concentration and 95% confidence interval of the long-term GAC column, operated parallel to the adsorbing column. This column had served 34,000 bed volumes prior to the start of the present study. The breakthrough curves based upon predictive partition parameters and number of mass transfer units are shown as dashed lines. These breakthrough curve calculations include a nonadsorbable portion as estimated by the isotherm experiments and are bounded above by the removal attributed to biodegradation as estimated by the long-term column. The exceptions to this approach were the unfractionated DOC and > 50,000 MW fraction which showed plateaus in their fractional concentration values, which were less than that of the long-term column, and hence were forced in the simulation to be asymptotic to that value.

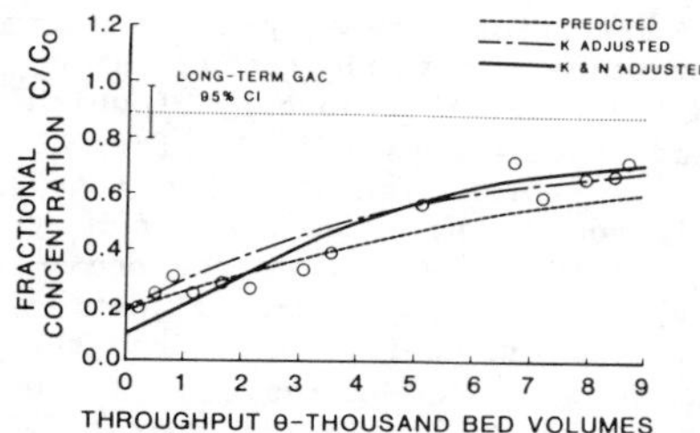

Figure 1. Breakthrough behavior of whole (unfractionated) DOC. Predicted parameters: $K_L = 13$ 1/g; $\bar{D} = 3 \times 10^{-10}$ m^2/s. Adjusted parameters: $K_B = 8.5$ 1/g; $D_B = 5.6 \times 10^{-10}$ m^2/s.

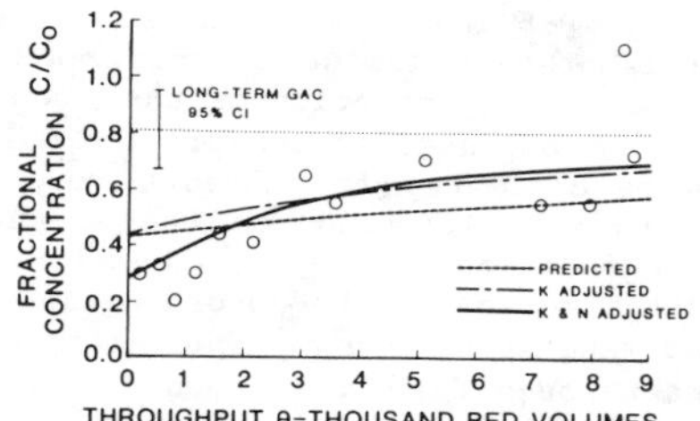

Figure 2. Breakthrough behavior of fraction with MW > 50,000. Predicted parameters: $K_L = 11$ 1/g; $\bar{D} = 0.66 \times 10^{-10}$ m^2/s. Adjusted parameters: $K_B = 4.2$ 1/g; $D_B = 1.7 \times 10^{-10}$ m^2/s.

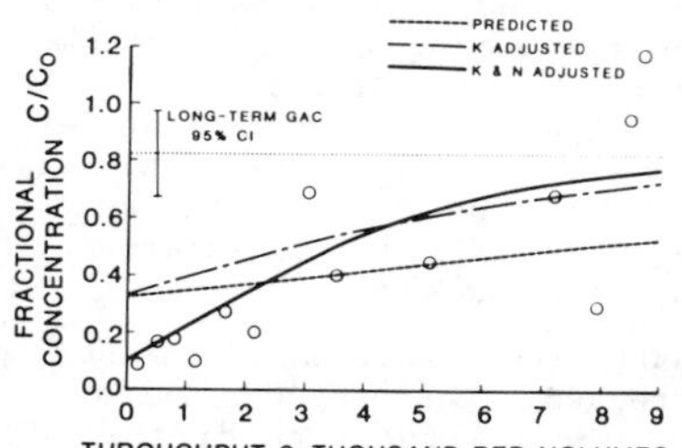

Figure 3. Breakthrough behavior of fraction with 10,000 < MW < 50,000. Predicted parameters: $K_L = 24$ 1/g; $\bar{D} = 1.3 \times 10^{-10}$ m^2/s. Adjusted parameters: $K_B = 8.3$ 1/g; $D_B = 4.2 \times 10^{-10}$ m^2/s.

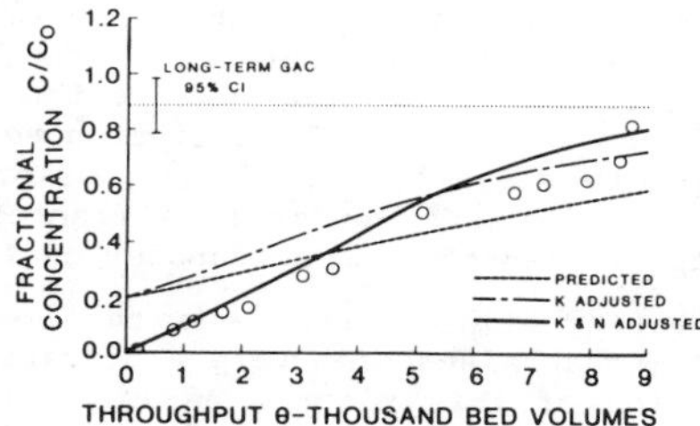

Figure 4. Breakthrough behavior of fraction with 1,000 < MW < 10,000. Predicted parameters: $K_L = 19$ 1/g; $\bar{D} = 2.2 \times 10^{-10}$ m^2/s. Adjusted parameters: $K_B = 11$ 1/g; $D_B = 6.7 \times 10^{-10}$ m^2/s.

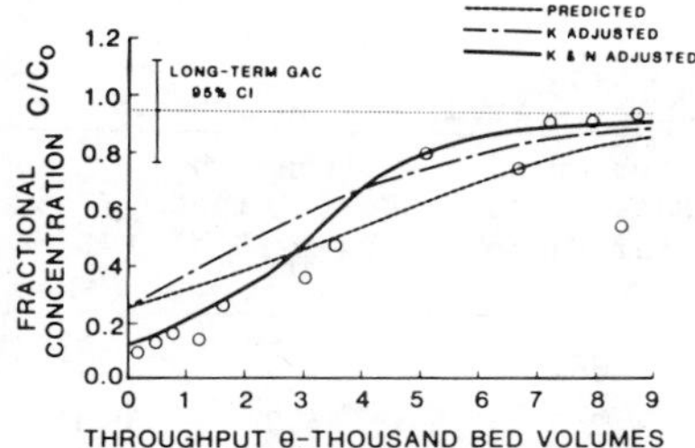

Figure 5. Breakthrough behavior of fraction with 500 < MW < 1,000. Predicted parameters: $K_L = 13$ 1/g; $\bar{D} = 3.2 \times 10^{-10}$ m^2/s. Adjusted parameters: $K_B = 8.0$ 1/g; $D_B = 10 \times 10^{-10}$ m^2/s.

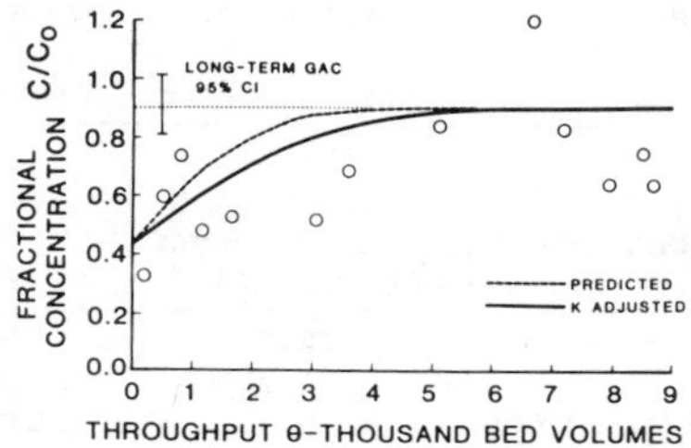

Figure 6. Breakthrough behavior of fraction with MW < 500. Predicted parameters: $K_L = 2.9$ 1/g; $\bar{D} = 6.0 \times 10^{-10}$ m^2/s. Adjusted parameters: $K_B = 4.5$ 1/g; $D_B = 4.3 \times 10^{-10}$ m^2/s.

Analysis of the curves indicate that the partition parameter based upon the laboratory isotherm experiments overestimate the adsorption capacity, with the exception of the lowest MW fraction, the capacity of which seems to be underestimated. The column adsorption capacity, K_B, for DOC can be estimated from breakthrough data by subtracting from the total DOC removed in a column the amount believed to have been removed by biodegradation. This is accomplished by integrating between the breakthrough curves of the adsorbing column and that of the long-term adsorption-exhausted column. The estimated equilibrium coefficients from the breakthrough data are shown in Table 3, as well as the ratio of K_L to K_B. K_L may be viewed as a predicted value of K_B, based on laboratory data; accordingly, the expected value of the ratio $K_L:K_B$ is unity. However, the observed $K_L:K_B$ ratio ranged from 2.9 to 0.6. Several factors have been proposed to account for this difference between isotherm and column adsorption capacity (3-5,16): the principal factors are believed to be the effect of biodegradation on adsorbability, different concentration histories in the batch and column contactors; and the different pore structure of ground carbon versus unsieved granular carbon.

Breakthrough curves were calculated using K_B and the predicted N values and are shown in Figures 1 to 6 as dash-dot lines. The comparatively smooth shape of these curves and the poor fit during immediate breakthrough and the mid-portion of the breakthrough curve indicate that the mass transfer correlations predict greater resistance to mass transfer than was observed. The exception to this was the < 500 MW fraction, for which the transfer resistance was predicted acceptably well. The scatter in the data render this fraction the most difficult to interpret, but the predicted mass transfer coefficients seem to fit.

To better simulate the data, the mass transfer resistance was decreased, i.e., the values of N were arbitrarily increased. The best overall simulation of this kinetic parameter-fitting approach for each of the MW fractions are shown as solid lines in Figures 1 to 6, and the corresponding values of N_B in Table 3. The model most closely simulates the behavior of the unfractionated, > 50,000 and 1,000-10,000 MW fractions. The scatter in the data for the 10,000-50,000 MW fraction, which represents only 2% of the total DOC, as well as for the < 500 MW fraction, which contains 43% nonadsorbable organic matter, makes it difficult to determine the applicability of the model to these fractions. The immediate breakthrough of organics in the two highest MW fractions is attributable predominantly to the mass transfer resistances. With the two lowest MW fractions and the unfractionated DOC, the immediate breakthrough is dominated by the nonadsorbable fraction.

Table 3. Equilibrium and Kinetic Parameters: Predicted and Observed

DOC MW Range (amu)	Column Adsorption Capacity K_B (1 water/g AC)	K_L/K_B (-)	Number of Mass Transfer Units N_B (-)	Diffusion Coefficient $D_B \times 10^{10}$ (m^2/s)	$\bar{D}/D_B$ (-)
Unfractionated	8.5	1.6	3.5	5.6	0.52
> 50,000	4.2	2.6	1.0	1.7	0.38
10,000-50,000	8.3	2.9	2.5	4.2	0.31
1,000-10,000	11.	1.8	4.0	6.7	0.32
500-1,000	8.0	1.6	6.0	10.	0.32
< 500	4.5	0.6	3.0	4.3	1.41

The diffusivities, D_B, associated with the numbers of mass transfer units which best characterize the breakthrough behaviors of the different MW fractions, are listed in Table 3 along with the ratio of predicted D to D_B. The values of D_B are approximately one-third as great as the predicted D values for all fractions but the lowest, and were found to be related to MW by the relationship

$$D = 1.1 \times 10^{-8} (MW)^{\alpha} \tag{13}$$

with $\alpha = -0.34$. An α value of -0.33 results when a molar volume term is substituted for the r_m in the Stokes-Einstein equation, and is also reported by Polson and other investigators for macromolecules in aqueous solution (21,24).

The mean fractional concentration of the long-term adsorption exhausted column and the DOC MW distribution of the effluent from that column are shown in Table 4. The mean C/C_o values indicate that long-term removals of DOC range from 6 to 19%, increasing with increasing MW. The long-term removal is thought to result mainly from biodegradation, as 34,000 bed volumes had been treated over a 2-year period. However, the unfractionated and highest MW fraction long-term removal show trends towards increasing fractional concentration over this study period, indicating a slow adsorption process (25,26). The adsorbing column's asymptotic approach to a C/C_o, less than the long-term average C/C_o, of these two fractions is additional evidence of the slow adsorption phenomena. The breakthrough behavior of all other MW fractions indicated that adsorption capacity had been exhausted by the end of the study period, except the 1,000-10,000 MW fraction which was steadily approaching the exhausted state.

The effect of chemical treatment of the composite sample on the MW distribution can be seen in Table 4. The unpreserved samples' pH was 6.6 and ionic strength 0.01; the preserved samples' pH was reduced to 2.5 with H_3PO_4 and ionic strength was increased to 0.03 with $Na_2S_2O_3$. The results indicate the chemical preservation reduces the median MW and can change the percent total DOC in any one fraction by as much as 9% of the total, e.g. in the < 500 MW fraction.

To assess the composite behavior of the MW fractions in relationship to the unfractionated sample behavior, the values of four parameters of each MW fraction were weighted by the DOC MW distribution and a composite value was calculated. Table 5 shows the results of these calculations.

Table 4. Long-Term Fractional Concentration and MW Distributions

DOC MW Range (amu)	Long-Term Column C/C_o ± 95% CI		Percent DOC in Long-Term Column Effluent ± 95% CI		Percent Total DOC	
					Unpreserved	Preserved
Unfractionated	88.	8.5	–	–	–	–
> 50,000	81.	14.	8.0	1.4	22.	14.
10,000-50,000	82.	15.	2.0	0.4	3.3	2.2
1,000-10,000	88.	10.	32.	3.1	37.	33.
500-1,000	94.	18.	14.	1.5	11.	18.
< 500	91.	9.8	39.	5	25.	34.
Losses	–	–	6.0	1.5	1.8	7.5

Table 5. Relationship of Composite Behavior of Fractions to Unfractionated DOC

	Composite	Unfractionated	Percent Difference
Isotherm K_L (1/g)	12.5	13.2	+ 5.
Column Adsorption Capacity K_B (1/g)	7.24	8.53	+ 15.
Diffusivity D_B (m^2/s)	5.75×10^{-10}	5.6×10^{-10}	- 3.
Long-Term C/C_o (-)	90.0	87.7	- 3.

The composite calculation slightly underestimates the isotherm and column adsorption behavior, and slightly overestimates the diffusivity and long-term fractional concentration. The good agreement between composite and unfractionated sample behavior, as well as the small (less than 10%) loss of DOC during UF, indicate that the UF procedure adequately represents the MW behavior, within the limits imposed by the technique's operationally defined nature.

SUMMARY

The results of the equilibrium isotherm and GAC column studies indicate several tendencies in the behavior of the different MW fractions. The adsorption equilibrium capacity increases with MW except for the highest MW fraction, which exhibits a diminished value. Most of the nonadsorbable organic matter is in the lowest MW fraction. The observed diffusivity is related to MW by a power function with the exponent equal to -0.34, similar to theoretical and other experimental studies. The extent of long-term removal, presumably due to biodegradation, increases with MW, although some evidence exists for a slow adsorption phenomenon for large molecules. The fixed-bed model adequately simulates the column breakthrough behavior but the predicted values overestimated both the mass-transfer resistance and the partition parameter inferred from column behavior. The lowest MW fraction was the most difficult to quantify in both column and isotherm studies, and in several respects behaved anomalously compared to the tendencies displayed by other fractions. The composite behavior of the MW fractions provides a good representation of the unfractionated DOC, with regard to the rate, equilibrium, and biodegradation phenomena.

ACKNOWLEDGEMENTS

The authors thank Ellen Kaastrup for the equilibrium isotherm analysis, Martin Reinhard for assistance with the ultrafiltration analysis and interpretation of the MW distribution data, and Gary Hopkins for assistance in sampling and fabrication of the sampling system. The cooperation of the Santa Clara Valley Water District is appreciated.

LITERATURE CITED

1. Arbuckle, W.B., in Activated Carbon Adsorption of Organics from the Aqueous Phase--Vol. II, M. McGuire and I.H. Suffet, Eds, Ann Arbor Science, Ann Arbor, MI, p. 237 (1980).

2. Chow, D.K., and David, M.M. J. Am. Water Works Assoc., 73 (8), 555 (1981).

3. DeWalle, F.B., and Chian, E.S.K., Jour. Environmental Eng. Div., ASCE, 100 (EE5), 1089 (1974).

4. Kim, B.R., Snoeyink, V.L., and Saunders, F.M., "Jour. Environmental Eng. Div., ASCE, 102 (EE1), 55 (1976).

5. Lee, M.C., and Snoeyink, V.L., "Humic Substances Removal by Activated Carbon," Univ. of Illinois, Water Resources Control Report No. 153 (1980).

6. Lee, M.C., Snoeyink, V.L., and Crittenden, J.C., J. Am. Water Works Assoc., 73 (8), 440 (1981).

7. Snoeyink, V.L., McCreary, J., and Murin, C., "Activated Carbon Adsorption of Trace Organic Compounds," EPA-600/2-77-223, U.S. Environmental Protection Agency, Washington, D.C. (1977).

8. Reinhard, M., submitted to Environ. Sci. Technol. (1982).

9. Goto, C., Tsuchiya, M., and Misaka, Y., Nippon Kagaku Kaishi, 11, 2005 (1975). Chem. Abs., 84, 155185 (1976).

10. Keller, J.V., Leckie, J.O., and McCarty, P.L., J. WPCF, 50 (11), 2522 (1978).

11. Manos, G.P., and Tsai, C.E., Water, Air and Soil Pol. 14, 419 (1980).

12. Hiester, N.K., and Vermeulen, T., Chem. Eng. Prog., 48, 505 (1952).

13. Vermeulen, T., Advances in Chemical Enineering, Vol. II, Drew, T.B., and Hoopes, J.W., Eds., Academic Press, New York, NY (1958).

14. Vermeulen, T., Klein, G., and Hiester, N.K., "Adsorption and Ion Exchange," Chemical Engineer's Handbook, Perry, R.H., and Chilton, C.H., Eds., McGraw-Hill Book Co., Inc., New York, NY, (1973).

15. Roberts, P.V., and Summers, R.S., "Simulation of TOC Removal in Fixed-Bed Activated Carbon Treatment," AIChE National Meeting, Aug. 29-Sept. 1, 1982.

16. Summers, R.S., and Roberts, P.V., accepted by Jour. Environmental Eng. Div., ASCE, (1983).

17. Helfferich, F., Ion Exchange, McGraw-Hill, New York, NY (1962).

18. Summers, R.S., and Roberts, P.V., in Treatment of Water by Granular Activated Carbon, Advances in Chemistry Series No. 202, M. McGuire and I.H. Suffet, Eds., American Chemical Society, Washington, D.C., (1982).

19. Kaastrup, E., Summers, R.S., and Roberts, P.V., submitted to Jour. Environmental Eng. Div., ASCE (1983).

20. Cannon, F.S., and Roberts, P.V., Jour. Environmental Eng. Div., ASCE, 108 (EE4), 766 (1982).

21. Polson, A., J. Phys. and Col. Chem., 54, 649 (1950).

22. Satterfield, C.N., Mass Transfer in Heterogeneous Catalysis, M.I.T. Press, Cambridge, MA (1970).

23. Wakao, N., and Funazkri, T., Chem. Eng. Sci., 33, 1375 (1978).

24. Young, M.E., Carroad, P.A., and Bell, R.L., Biotech. and Bioeng., 22, 947 (1980).

25. Peel, R.G., "The Role of Slow Diffusion and Bioactivity in Adsorption Column Modelling," Doctoral Thesis, McMaster University, at Hamilton, Ontario, Canada (1979).

26. Peel, R.G., and Benedek, A., Jour. Environmental Eng. Div., ASCE, 106 (EE4), 797 (1980).

SURFACE DIFFUSION OF TWO-COMPONENT ORGANIC GASES ON ACTIVATED CARBON

Motoyuki Suzuki, Masafumi Hori and Kunitaro Kawazoe
Institute of Industrial Science, University of Tokyo
Roppongi, Tokyo, JAPAN

ABSTRACT

Intraparticle diffusion rates of propane and butane were measured in an activated carbon particle. First, surface diffusion coefficient for single component systems is determined as a function of amount adsorbed of each component. Then from diffusion fluxes of the two components in coexistent system, four diffusion coefficients were defined and interpreted by taking spreading pressure as a driving force of surface diffusion of the both components.

\- - - - - - - - - - -

Intraparticle diffusion of organic gases in activated carbon particles plays an important role in designing adsorption columns for separation of gas mixtures as well as removal of dilute organic pollutants for air pollution control. As a diffusion mechanism, surface migration of adsorbed species is often a controlling mechanism. Then the surface diffusion coefficient is defined by applying Fickian equation to the diffusion flux and the gradient of the amount adsorbed in the particle. Thus defined the surface diffusion coefficient itself shows dependency on the amount adsorbed, which has been interpreted by several models. Hill (1), Higashi, Ito and Oishi (2) and Smith and Metzner (3) tried to interpret their data by increase of hopping distance or/and rate of adsorbed molecules with increase of amount adsorbed. Gilliland (4) considered that surface migration takes place by the gradient of spreading pressure. In this idea, surface diffusion coefficient increases because of non-linear characteristics of adsorption isotherm. Ash and Barrer (5) applied the idea of chemical potential driving force which is popular in explanation of diffusion in ultra micro pores such as those in zeolite. Sladek, Gilliland and Baddour (6) and Suzuki and Fujii (7) attributed the change of surface diffusion coefficient to the existence of energetic heterogeneity on the adsorption surface. In addition to this complexity, in most cases of the practical adsorption operations, number of adsorbed species is more than one and then it is anticipated that diffusion or migration of different adsorbed species may interact on the adsorption surface.

The purpose of the present work is to clarify how diffusion of two component gases is described. The work consists of three phases. First, adsorption isotherms of propane and butane in coexistent system as well as single component systems are determined from breakthrough experiment. Secondly, diffusion flux through a single activated carbon pellet is measured for each component and surface diffusion coefficient is determined as a function of amount adsorbed for each component. Finally, diffusion of two component gases are measured in the same apparatus and the result is compared with the prediction from the single component results.

EXPERIMENT

Isotherm Measurement

Breakthrough measurement of propane and butane in helium carrier were done with a small column in which crushed and sieved activated carbon particle, TAKEDA HGR-513, were packed. Density of the particle is 0.72×10^{-3}kg/m^3 and surface area is reported to be 1.1×10^6m^2/kg. About 1×10^{-3}kg of the particle was used in the copper column of 3×10^{-3}m in diameter. Propane and butane gases used were of research grade. Adjusted flow rates of these gases were mixed with helium carrier and introduced to the column which was kept at a constant temperature of 273 K for a two component run and 273, 303 and 323 K for single component runs. Effluent concentration changes were followed by sampling with an automatic sampling device connected to a gas chromatograph. The amount adsorbed was obtained by graphical integration of the effluent concentration curve of each component. For two component runs, partial pressure of propane was kept constant at three levels, namely, 76, 114 and 228 Torr, and partial pressure of butane was varied in each set of measurements.

Diffusion Experiment

Wicke-Kallenbach type single pellet reactor was employed. The schematic diagram of the diffusion experimental apparatus is shown in Figure 1. Diffusion cell is made of stainless steel with 2×10^{-2}m in diameter and 3×10^{-2}m in depth. A stainless steel plate of 5.2×10^{-3} m thickness which separates the two diffusion cells has a hall of 4.2×10^{-4}m^2 in diameter which holds an activated carbon pellet of 4×10^{-3}m in diameter. Epoxy cement was used to fix the pellet in the hall. The effective area for diffusion was 0.126×10^{-4}m^2. The concentration of organic gas at one of the cells was kept constant by holding the flow rates of the organic gas and helium constant and the concentration in helium stream flowing out from the other cell was measured by gas chromatography. Special care was taken so that pressure difference across the pellet was negligible. In the case of single component systems, steady state diffusion fluxes were measured by keeping the concentration at the higher concentration cell at several levels. Then surface diffusion coefficient was defined as a function of amount adsorbed by differentiating

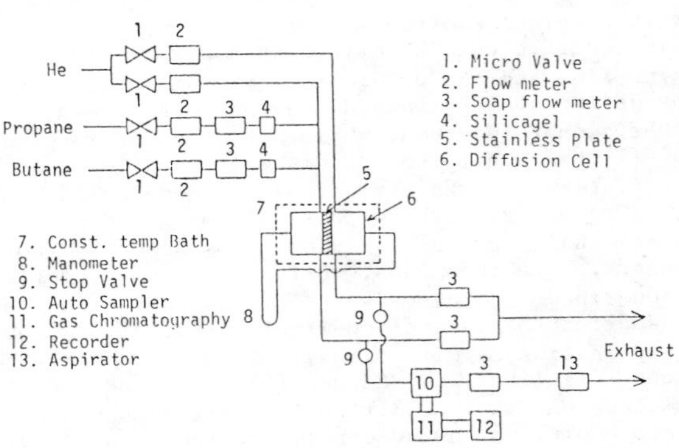

Figure 1. A Schematic Diagram of Diffusion Experiment Apparatus

diffusion flux with respect to the amount adsorbed which is in equilibrium with the concentration at the higher concentration cell. In the case of two component systems, diffusion fluxes of the two components were measured separately. Diffusion runs were made at 252, 273 and 303 K for the single component systems and at 273 K for the two component system.

METHOD OF ANALYSIS OF DIFFUSION DATA

Single Component System

Similar method to the previous one (7) was adopted. Total flux of diffusion of component i through activated carbon pellet is expressed in terms of amount adsorbed, q, gas concentration, c, and flow rate of gas, Q, at the high concentration cell (subscript H) and at the lower concentration side cell (subscript L).

$$N_T = c_L \frac{Q_L}{s} = D_e \frac{dc}{dx} + \rho D_s \frac{dq}{dx} = \frac{1}{L} \int_{q_L}^{q_H} \{D_e \frac{\partial c}{\partial q} + \rho D_s\} dq \qquad (1)$$

Since $q_{L,i}$ is nearly equal to zero, $D_{s,i}$ is obtained as a function of q_H from the following equation.

$$D_{s,i}(q_H) = \frac{\{L (\frac{dN_T}{dq})_{q_H} - D_e (\frac{dc}{dq})_{q_H}\}}{\rho} \qquad (2)$$

$D_{e,i}$ was estimated by using a tortuosity factor, k^2=4.2, which was determined by conducting a separate diffusion study on nitrogen and helium system for the same pellet. Then from the above equation concentration dependence of the surface diffusion coefficient of propane and butane is obtained.

Two Component System

By changing butane concentration for fixed propane concentrations at the higher concentration cell, fluxes of the two gases through the pellet was measured. Diffusion flux of each gas is considered to be a function of the amounts adsorbed of the two gases, qp and qb. Then the flux of component i, for instance, is expressed as

$$N_i = \rho D_{s,ii} \frac{dq_i}{dx} + \rho D_{s,ij} \frac{dq_j}{dx} = \frac{\rho}{L} \{\int_0^{q_i} D_{s,ii} dq_i + \int_0^{q_j} D_{s,ij} dq_j\} \qquad (3)$$

For the analysis of the two component data, correction of pore diffusion was not made since the correction in single component systems suggested that only a small fraction of the total flux was due to pore diffusion. From the above equation, $D_{s,ii}$ and $D_{s,ij}$ where i represents propane or butane and j is the other component, were obtained as

$$D_{s,ii} = \frac{L}{\rho} \frac{\partial N_i}{\partial q_i} \qquad (4)$$

$$D_{s,ij} = \frac{L}{\rho} \frac{\partial N_i}{\partial q_j} \qquad (5)$$

These equations correspond to Equation (2) for a single component system.

RESULTS AND DISCUSSION

Adsorption Isotherms

Adsorption isotherms for single component systems of propane and butane are shown in Figure 2. By applying Myers and Prausnitz method to these isotherms, the two component isotherm at 273 K was calculated and compared with the experimental results for the two component system in Figure 3. Apparently good coincidence is obtained.

Diffusion of Single Component Gases

Figure 4 and 5, respectively show diffusion fluxes of propane and butane plotted against the amounts adsorbed which are in equilibrium with the concentrations at the hither concentration cell. From these plots $D_{s,i}$ is determined as a function of q_i by applying Equation (2).

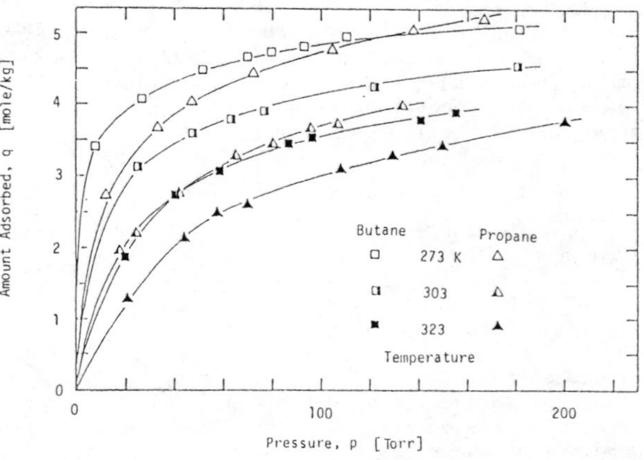

Figure 2. Adsorption Isotherms of Propane and Butane on Activated Carbon; Single Component at 273, 303 and 323 K on Takeda HGR-513

When surface diffusion is governed by the gradient of spreading pressure, then flux of component i, N_i, is written in the following form

$$N_i = \frac{\rho}{c_R} q_i \frac{d\pi}{dx}$$

$$= \frac{\rho}{c_R L} \int_0^{\pi} q_i \, d\pi \qquad (6)$$

Integration is obtained from the isotherms for single component systems and then N_i experimentally obtained are compared with the integrations in Figure 6 and 7

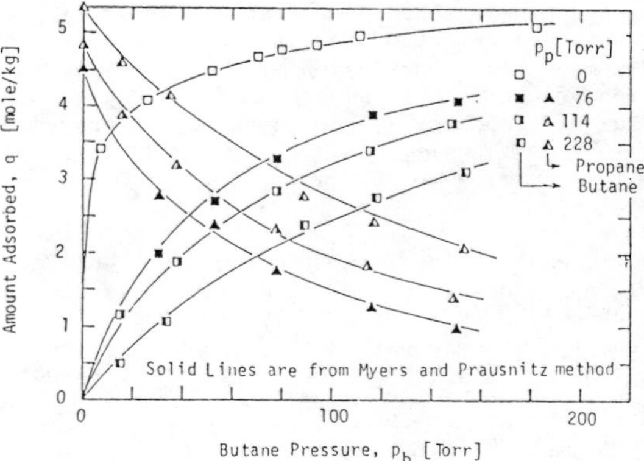

Figure 3. Adsorption Isotherms of Propane and Butane Mixture on Activated Carbon; Amount Adsorbed of Propane and Butane for Propane Pressures of 76, 114 and 228 Torr.

for the case of 273 K. The solid keys in the Figures show that there are linear relationships for each of the two gases and from the slopes of the regression lines coefficients of resistance to surface flow, c_R, are determined for the two gases as $1.92 \times 10^6 kg/m^2 \cdot s$ for propane and $3.26 \times 10^6 kg/m^2 \cdot s$ for butane. When spreading pressure describes surface diffusion phenomena, surface diffusion coefficient, $D_{s,i}$ is written as,

$$D_{s,i} = \frac{RT}{A\ C_{Ri}} \left\{ q_i \frac{d \ln p_i}{d \ln q_i} \right\} \tag{7}$$

This is checked in Figure 8 and 9. These figures include the results obtained at 252 and 303 K.

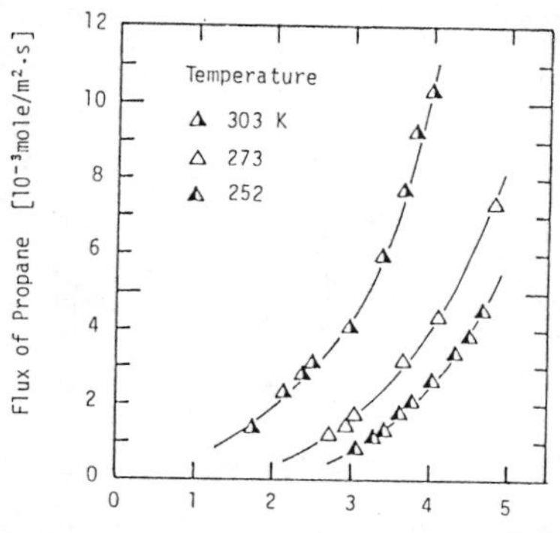

Figure 4. Diffusion Flux of Propane against the Amount Adsorbed in Equilibrium with the Concentration at the Higher Concentration Cell; Single Component Runs

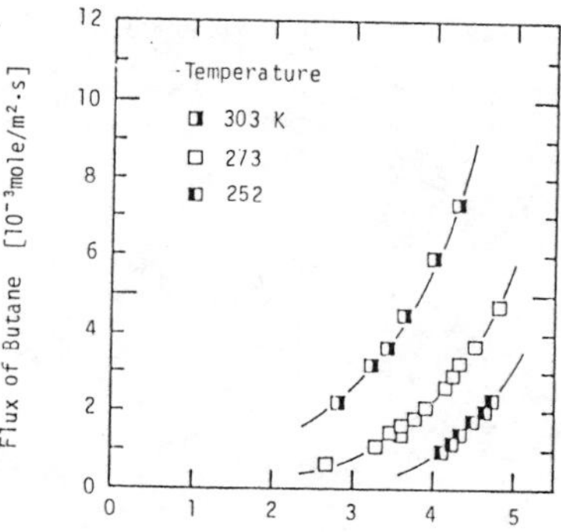

Figure 5. Diffusion Flux of Butane against the Amount Adsorbed in equilibrium with the Concentration at the Higher Concentration Cell; Single Component Runs

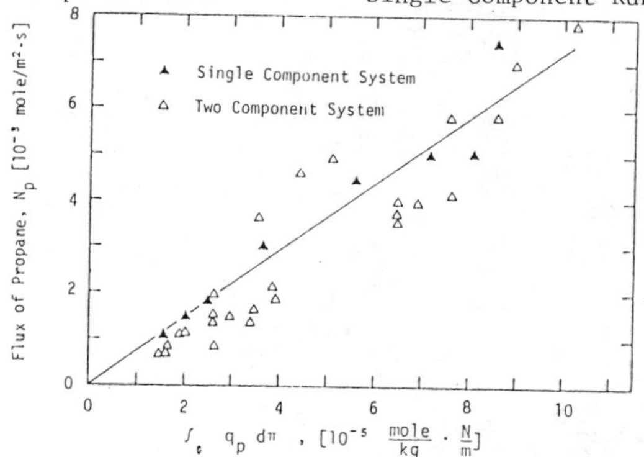

Figure 6. Diffusion Flux of Propane Plotted against $\int_0^{\pi} q_p \, d\pi$; Single and Two Component Runs

Diffusion of Two Component Gases

For various combination of concentrations of propane and butane in flowing stream at the high concentration side cell, fluxes of the both gases were obtained. The results are shown against the amount adsorbed of each species which is determined by applying the two component adsorption isotherm determined in the previous part to the composition at the higher concentration side in Figure 10. In the figure, each point corresponds to one run and first number shows the flux of propane while second number gives the flux of butane. From the figure N_p versus q_p for constant value of q_d and N_p versus q_d for constant q_p are obtained and shown in Figure 11 and 12. Similar procedure is adopted for N_d.

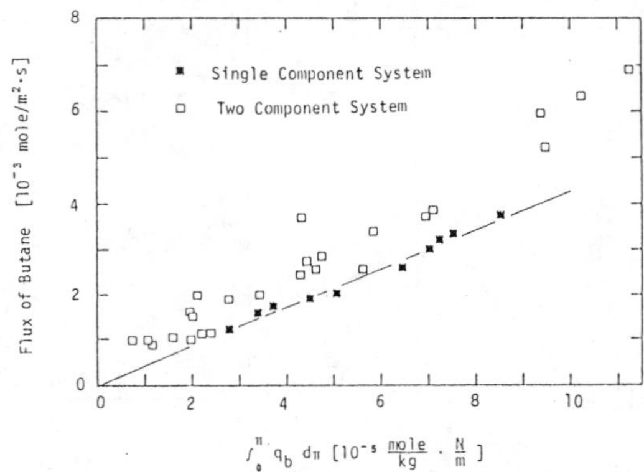

Figure 7. Diffusion Flux of Butane Plotted againt $\int_0^\Pi q_b \, d\pi$; Single and Two Component Runs

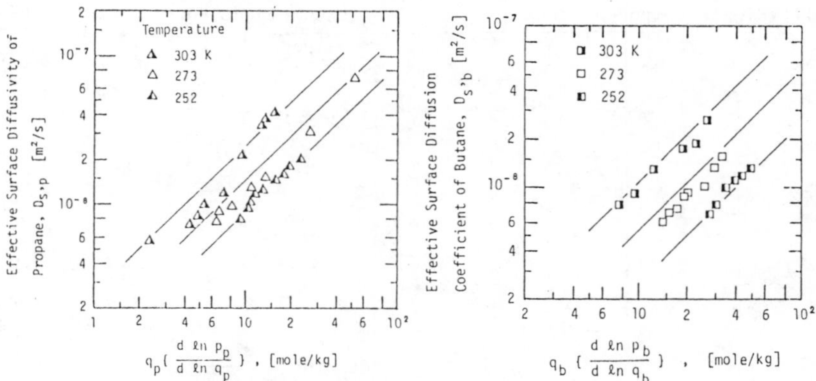

Figure 8. Effective Surface Diffusion Coefficient of Propane in Single Component Runs

Figure 9. Effective Surface Diffusion Coefficient of Butane in Single Component Runs

From the slope of the curves given in the figure, D_s's defined by Equation (5) can be determined.

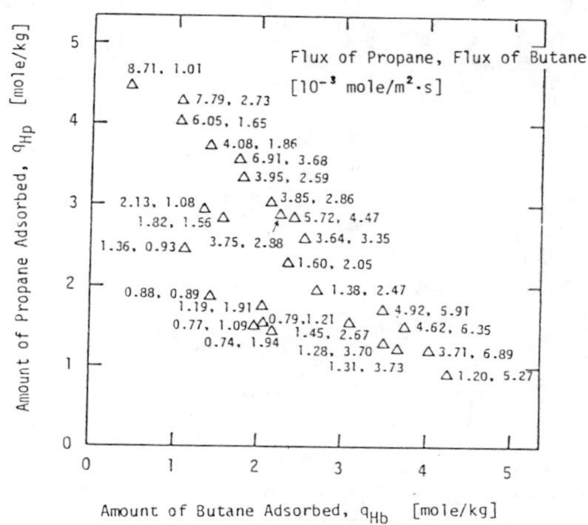

Figure 10. Diffusion Fluxes of Propane and Butane in Two Component Coexistent Systems; Numbers at the Point Show Flux of Propane and Butane in This Order, 10^{-4} mole/m^2·s

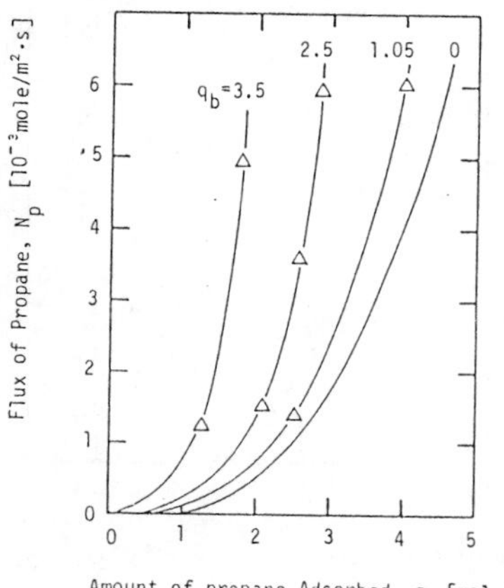

Figure 11. Flux of Propane Plotted against Amount Adsorbed of Propane for Constant Amount Adsorbed of Butane

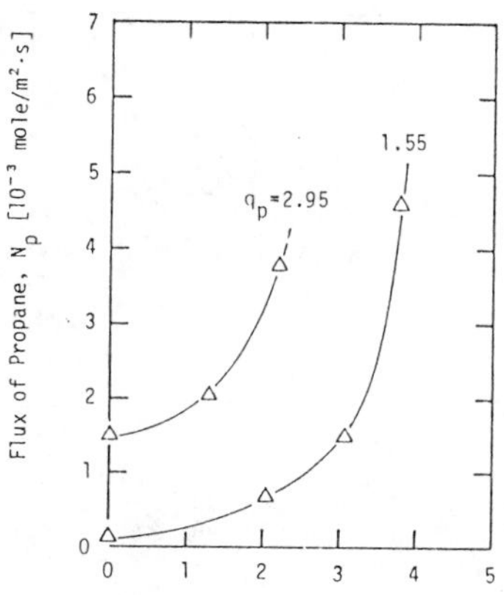

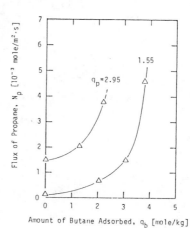

Figure 12. Flux of Propane Plotted against Amount Adsorbed of Butane for Constant Amount Adsorbed of Propane

By assuming that the diffusion flux of each component is determined by the gradient of the spreading pressure which is determined from the two component composition and the surface concentration of the component, flux of component i in the two component system is expected to be written as

$$N_i = \frac{\rho}{C_{Ri}} q_i \frac{d\pi}{dx}$$

$$= \frac{\rho}{C_{Ri}\, L} \frac{RT}{A} \left\{\int_0^{q_i} q_i \frac{d\,\ell n\, p_i}{d\,\ell n\, p_i} dp_i + \int_0^{q_j} q_i \frac{d\,\ell n\, p_j}{d\,\ell n\, q_j} dq_j\right\} \tag{8}$$

Then $D_{s,ii} = \frac{L}{\rho} \frac{dN_i}{dq_i}$

$$= \frac{RT}{A\, C_{Ri}} \left\{q_i \frac{d\,\ell n\, p_i}{d\,\ell n\, q_i} + \int_0^{q_j} \frac{d\,\ell n\, p_j}{d\,\ell n\, q_j} dq_j\right\} \tag{9}$$

and

$$D_{s,ij} = \frac{L}{\rho} \frac{dN_i}{dq_j} = \frac{RT}{A\, C_{Ri}} \left\{q_i \frac{d\,\ell n\, p_j}{d\,\ell n\, q_j}\right\} \tag{10}$$

By comparing Equations (7), (9) and (10), plots of $D_{s,p}$, $D_{s,pp}$ and $D_{s,pb}$ against

$$q_p \frac{d\,\ell n\, p_p}{d\,\ell n\, q_p}, \quad q_p \frac{d\,\ell n\, p_p}{d\,\ell n\, q_p} + \int_0^{q_b} \frac{d\,\ell n\, p_b}{d\,\ell n\, p_b} dq_b \text{ and } q_p \frac{d\,\ell n\, p_b}{d\,\ell n\, q_b},$$

respectively, should give straight lines of the same slope, $RT/A\, C_{Ri}$. This was done in Figure 13 for propane and in Figure 14 for butane.

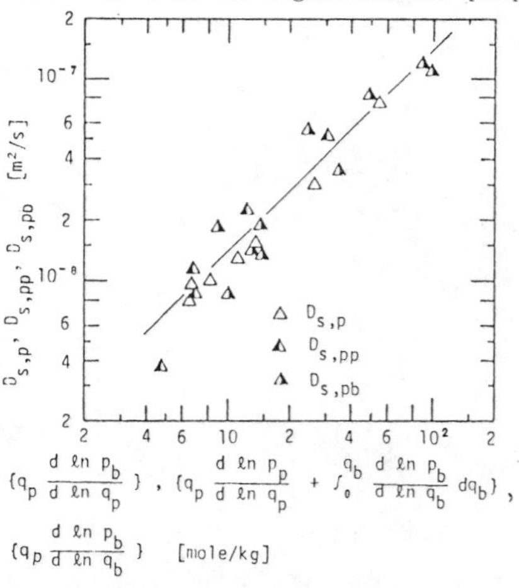

Figure 13. Comparison of Effective Surface Diffusion Coefficients of Propane, $D_{s,p}$ for Single Component System and $D_{s,pp}$ and $D_{s,pd}$ for Two Component System

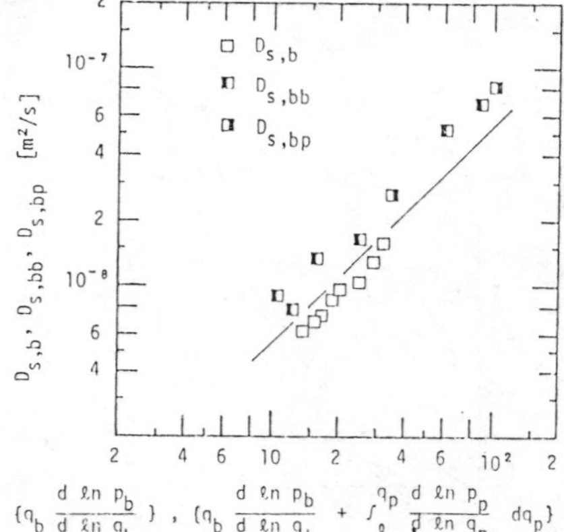

Figure 14. Comparison of Effective Surface Diffusion Coefficients of Butane, $D_{s,b}$ for Single Component System and $D_{s,bb}$ and $D_{s,bp}$ for Two Component System

The data are a little scattered probably due to complexed steps for deriving D_s's. C_R's determined from two component results are C_{Rp} = 1.96 × 10^6 kg/m^2·s and C_{Rd} = 3.56 × 10^6 kg/m^2·s for propane and butane molecules. These figures are in reasonable agreement with those determined from single component runs for the each gas. Also from Equation (8), it is clear that N_i is expressed as

$$N_i = \frac{\rho}{L\ C_{Ri}} \{\int_0^{\pi} q_i\ d\pi\} \tag{11}$$

Plots of N_p against $\int_0^{\pi} q_p\ d\pi$ for two component cases are included in Figure 6, which naturally is in good agreement with the same plot for the single component system. Also the same check was done for butane in Figure 7. Surface migration of propane and butane in coexistent system is explained by taking spreading pressure gradient as a driving force of flow of the both components and coefficient of resistance of the each gas can be considered as a specific value to the component.

NOTATION

- A : effective area for diffusion, m^2
- C : concentration, mole/m^3
- C_{Ri} : resistance to surface flow of component i, kg/m^2·s
- D_e : effective pore diffusion coefficient, m^2/s
- $D_{s,i}$: effective surface diffusion coefficient of a single component, m^2/s
- $D_{s,ii}$, $D_{s,ij}$: effective surface diffusion coefficient in two component system, defined by Equation (5), m^2/s
- N_i : flux of component i through a pellet, mole/m^2·s
- p_i : pressure of component i, Torr
- q_i : amount adsorbed of component i, mole/kg
- R : gas constant
- T : temperature, K
- x : distance of diffusion, m

Greek Letters

- ρ : density of pellet, kg/m^3
- π : spreading pressure, N/m

Subscript

- i : propane or butane
- j : propane or butane, other than i
- p : propane
- b : butane
- H : at the hith concentration side cell
- L : at the lower concentration side cell

LITERATURE CITED

1. Hill, T.L., J. Chem. Phys., 25, 730 (1956).

2. Higashi, K., H. Ito and J. Oishi, J. Atomic Energy Soc. Japan, 5, 846 (1963).

3. Smith, R. K. and A. B. Metzner, J. Phys. Chem., 68, 2741 (1964).

4. Gilliland, E. R., R. F. Baddour and J. L. Russell, A. I. Ch. E. Journal, 4, 90 (1958).

5. Ash, R. and R. M. Barrer, Surface Sci., 8, 461 (1967).

6. Sladek, K. J., E. R. Gilliland and R. F. Baddour, Ind. Eng. Chem., Fundamentals, 13, 100 (1974).

7. Suzuki, M. and T. Fujii, A. I. Ch. E. Journal, 28, 380 (1982).

8. Myers, A. L. and J. M. Prausnitz, A. I. Ch. E. Journal, 11, 121 (1965).

PREDICTION OF THE BREAKTHROUGH IN SOLVENT RECOVERY IN AN ACTIVATED CARBON BED

Yasushi TAKEUCHI and Eiji FURUYA

Department of Industrial Chemistry
Meiji University, Higashi-Mita
Tama-Ku, Kawasaki-Shi 214, Japan

ABSTRACT

Starting from the measurement of a fixed-bed single component adsorption for various solvent vapor — activated carbon systems, binary adsorption was studied in order to compare the observed breakthrough curves with the calculated ones. It became clear that the accuracy of the prediction of breakthrough curves was dependent more strongly on the choice of equilibrium parameters than on that of transport properties, i.e. intraparticle diffusivity of the solvent vapers. Breakthrough behavior of water-soluble solvents changed greatly with the change of moisture content of the carbon bed.

Activated carbon has been used widely in chemical industry to recover solvent vapors entrained in the effluent air. Carbon is used generally in the form of a fixed-bed of particles and the vapor-laden air is fed to the bed until breakthrough occurs. Then desorption of the vapors adsorbed is performed with steam. This whole process is called "solvent recovery" and is an essential application of adsorption in industry.

Since solvents of higher boiling point and of better performance, e.g., toluene, methyl-ethyl-ketone (MEK) and cyclohexanone, are being used more often than before, the adsorption equilibria data as well as transport properties of these solvent-vapors for carbon beds are necessary to design adsorbers properly.

The main purpose of this paper is to present a method of estimating the equilibrium parmeters for the extended Langmuir equation and to show experimentally the validity of the model which describes the processes of single- and bi-component adsorption of solvent vapors from air by a fixed-bed of activated carbons. By comparing the experimental breakthrough curves with calculated ones, the values of intraparticle diffusivity for binary component adsorption were determined.

Furthermore, an additional study was carried out for single component adsorption onto moisture-laden carbon beds. The purpose of these experiments were to determine how the water remaining in the bed affects the subsequent breakthrough curves of the solvent vapors.

EXPERIMENTAL

For single vapor adsorption, commercial activated carbon pellets

(Takeda HGW 750) were used as the adsorbent and methyl acetate, ethyl acetate, acetone, methyl ethyl ketone (MEK), iso-buthyl methyl ketone (MIBK), tetrahydrofuran (THF) and toluene of extra pure grade, respectively, were used to mix with air as solvent-vapor at 298 K. Both pellets, originally 4 mm O.D., were pretreated by crushing and screening to obtain 8/12# particles, i.e., mean particle size of 0.00183 m, and by drying with a nitrogen stream for 2 hrs. or more at 413 K. The particles were packed in glass columns of 0.022 m I.D. up to about 0.050 m in bed length.

Using a well known flow method, the concentration change at the exit of the carbon bed was measured. The experimental procedure is presented elsewhere (1). The flow rate of the vapor-laden air at the bed was chosen to be 0.30 m/s, in order to minimize axial dispersion.

For binary vapor adsorption, acetone-toluene and MEK-toluene systems were studied. Vapor-laden air was supplied to the column packed with HGW 750 pellets at 298 K. The column was 0.040 m I.D. and the bed length was chosen to be 0.10 and 0.20 m in length. Two evaporators were used to obtain two vapor-laden streams before mixing with fresh air. The concentrations of influent and effluent air, respectively, were determined by a gas-chromatograph.

RESULTS AND DISCUSSION

Single Component Adsorption

The equilibrium amount adsorbed determined from the weight increase of the bed was well described by a Langmuir-type equation (Eq.(1)) and the constants in the equation K' and q'_∞ were obtained for each system as summerized in Table 1.

$$q = q'_\infty K'c/(1+K'c) \qquad (1)$$

According to the derivation of the Langmuir equation, the constant q'_∞ is related to the number of adsorption sites per unit weight of adsorbent. Therefore, when the amount adsorbed of various solvents is expressed in terms of the mol per unit weight of sdsorbent, i.e., q_∞ values showed to be constant irrespective of adsorbates, provided the molecular size of

Table 1. Langmuir constants and surface diffusivity for single component adsorption

adsorbate	adsorbent	q'_∞ [g/g]	K' [m³/g]	q_∞ [mmol/g]	Ds × 10¹⁰ [m²/s]
methyl acetate	HGW 750	0.357	0.490	4.82	1.4
ethyl acetate	HGW 750	0.369	1.36	4.19	1.2
acetone	HGW 750	0.280	0.392	4.82	3.1
MEK	HGW 750	0.353	1.02	4.90	1.6
MIBK	HGW 750	0.391	1.85	3.90	0.87
THF	HGW 750	0.368	0.616	5.10	3.0
toluene	HGW 750	0.420	3.32	4.56	1.8

the adsorbate is approximately the same. As shown in Table 1, this assumption is approximately valid for systems studied.

Fixed-bed breakthrough curves were measured for the system as shown in Fig. 1 and values of the effective intraparticle diffusivity were obtained as listed in Table 1. Details of the analysis of the breakthrough curves are presented elsewhere (1,2,3,4).

Binary Component Adsorption (1)

In this case, the amount adsorbed of each component was determined from the graphical integration of breakthrough curves observed. Under the conditions used in the study, the exit concentrations of components c_j (j=1 or 2) change with time as shown in Fig. 2 and the hatched areas correspond to the amount adsorbed of each component, where the component which appears at the exit earlier than the other is expressed as the first component and the other is named the second component, respectively. Total amount adsorbed is clearly the sum of those for each component .

The result was compared with the amount calculated from the weight increase of the carbon bed (ΔW). When the difference between the total amount adsorbed, determined from both methods, was less than 5 % of ΔW, the data were used for further calculation.

It was pointed out that K_i in the extended Langmuir equation described by Eq.(2), was different from those for single component system.

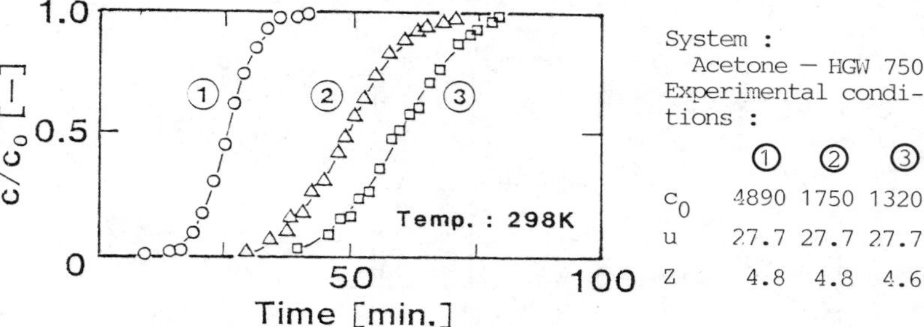

Fig. 1. Observed fixed-bed breakthrough curves for Acetone adsorption

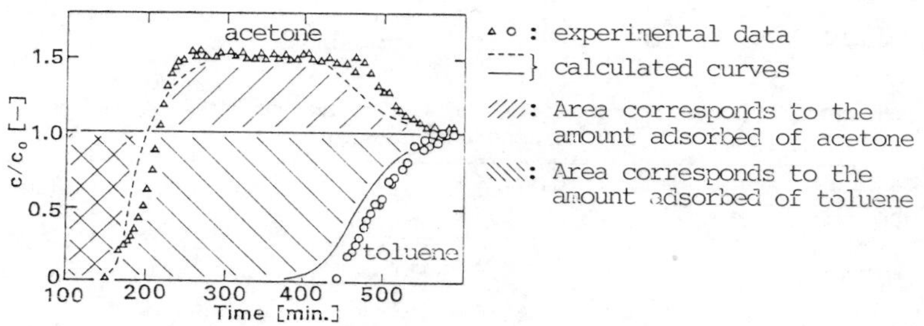

Fig. 2. Breakthrough curves for acetone and toluene adsorptions

$$q_i = q_{\infty i} K_i c_i / (1+\sum K_j c_j) \tag{2}$$

It is considered from the results for single component adsorption that the values of $q_{\infty i}$ are almost the same to each other. Therefore, for binary component adsorption, the ratio of amount adsorbed is expressed by Eq.(3).

$$q_1/q_2 = (q_{\infty 1} K_1 / q_{\infty 2} K_2)(c_1/c_2)$$
$$\simeq (K_1/K_2)(c_1/c_2) \tag{3}$$

That is, when q_1/q_2 is plotted against c_1/c_2, a straight line will be obtained and the slope will give (K_1/K_2), as shown by the solid line in Fig. 3. The values of (K_1/K_2) for MEK – toluene system was 0.231. The broken-line in Fig.3 shows the calculated breakthrough curves using the values of K' listed in Table 1. The lines differ greatly. Langmuir constants, K_1 and K_2 may be determined by the following method in order to get a minimum error in the calculation of amount adsorbed. i) Assume a set of K_1 and K_2 which satisfy the conditions K_1/K_2=0.231. ii) Assume the constant q_∞ to be a value between 4.93 mmol/g (for toluene) and 5.13 (for MEK). iii) Calculate the amount adsorbed $q_{calc.1}$ and $q_{calc.2}$ from concentration c_1, c_2 used in the study by use of the above-assumed constants K_1, K_2 and q_∞. iv) Examine the dependence of errors described by Eq.(4) in the calculation of amount adsorbed on equilibrium constants to find a set of K_1 and K_2 which give the minimum error. (see Fig. 4)

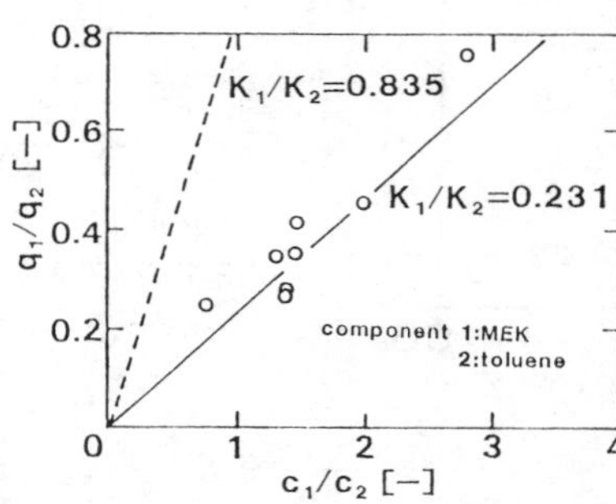

Fig. 3. Plots of q_1/q_2 against c_1/c_2 for MEK – toluene system

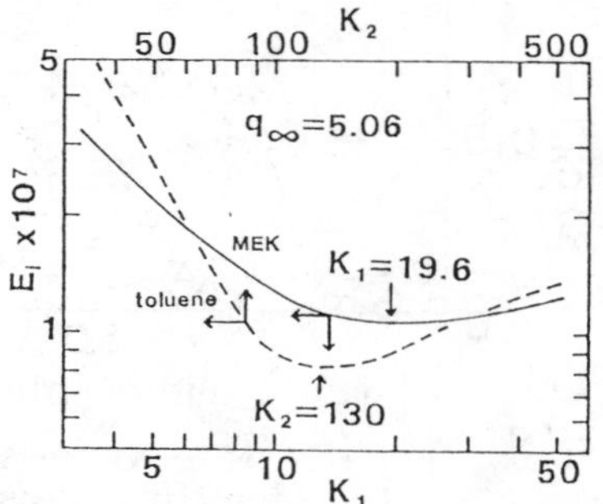

Fig. 4. Dependence of errors in the calculation of amount adsorbed on equilibrium constants

Table 2. Langmuir constants and the values of surface diffusivity for MEK and toluene adsorbed at 298[K]

adsorbate	K [m³/mol] a)*	b)*	c)**	Ds × 10¹⁰ [m²/s] A)	B)	C)	D)
MEK	30.0	19.6	151	1.6	0.8	1.6	0.8
toluene	130	85.0	654	1.8	1.8	0.9	0.9

* q_∞=5.06[mmol/g] ** q_∞=5.01[mmol/g]

$$E_i = \sum(q_{calc.i} - q_{obs.i})^2/n \tag{4}$$

where $q_{obs.}$ is the empirical amount adsorbed and n is the number of measurment. A typical set of Langmuir constants determined by the method mentioned above is listed in Table 2. The set of K, shown in column a) in Table 2 minimized the error E_i for MEK and column b) gave a minimum error for toluene with respect to q_∞=5.06. When q_∞ is 5.01 (see column c) in Table 2), the amount adsorbed, q_1, q_2 calculated from Eq.(2) by use of K_1=151 and K_2=654 m³/mol, respectively, agreed well with the experimental results.

Regarding the estimation of breakthrough curves for the binary system, a set of q_∞ and K_1 which give the best agreement with the experimental ones for less-adsorbable solvent must be used. It is clear that separation of two adsorbates takes place in the bed due to the difference in adsorptivity, and the first adsorption zone is established in the absence of more-adsorbable solvent. The results in Fig. 5 also suggest that equilibrium parameters are the major factor to estimate breakthrough time in this case.

Another factor to determine the shape of concentration profiles is the surface diffusivity. Breakthrough curves for MEK and toluene system were computed using the values listed in Table 2. The shape of breakthrough curves did not change much irrespective of the values Ds_1, Ds_2 used in the calculation. (see Fig. 6)

For aceton — toluene — HGW 750 system at 298 K, the breakthrough curves were computed by use of the method mentioned above and are already shown in Fig. 2.

Single Solvent Vapor Adsorption onto Wet Carbon Bed (5)

Additional experimental studies were performed to make clear how the moisture content of the carbon bed affects the amount adsorbed and the breakthrough of organic vapors. Experiments were carried out using almost the same appartus as above. The bed was moistened with steaming and the moisture content was held to be 0, 20 and 40 %, respectively, of the weight of dry carbon used. The moisture content was determined for every run by measuring the weight of carbon bed before and after steaming.

For the dry or wet carbon bed, the vapor-laden air, i.e., single component vapor with water vapor was introduced and the concentration

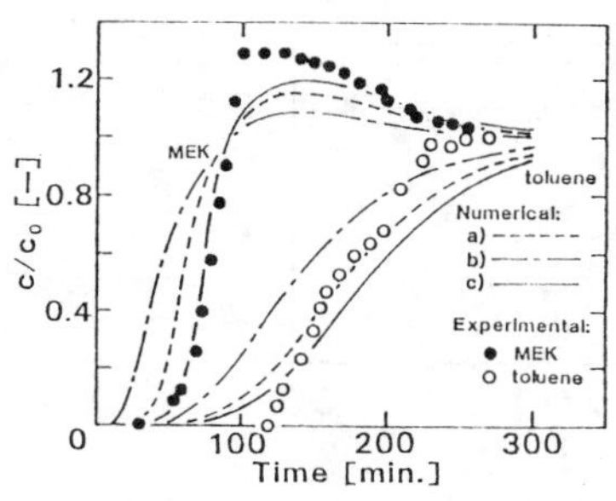

Fig. 5. Breakthrough curves for MEK and toluene with respect to various K_1 and K_2

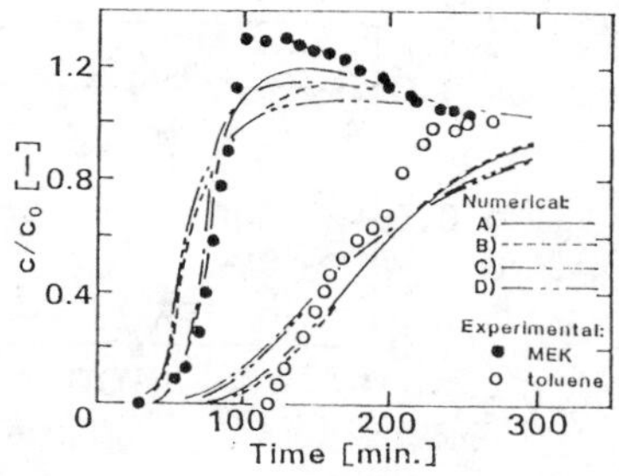

Fig. 6. Breakthrough curves for MEK and toluene with respect to various surface diffusivity

(for a), b), c), A), B), C) and D) refer to Table 2)

changes of organic vapors and water vapor was detected.

Methyl alcohol vapor, MEK vapor and toluene vapor, respectively, were used as representatives of highly water-soluble, water-soluble and water-insoluble solvents, respectively. Concentrations of organic and water vapor were determined by a gas-chromatograph. Activated carbon of smaller size (8/12 #) was chosen simply to speed up the kinetics. Results are shown in Figs. 7 and 8.

As can be seen, the breakthrough curves for toluene did not change with the change in moisture content. For MEK, though the results are not shown, they were almost the same as those for toluene. For methyl alcohol, a highly water-soluble solvent, the breakthrough curves changed greatly with moisture existed in the bed. The curves for methyl alcohol shows a shape similar to those for binary organic adsorption.

This means that methyl alcohol vapor fed to the carbon bed is adsorbed much more in the presence of water. In other words, it is partly absorbed into water and held in the bed. Later, it is desorbed by air which contains less water vapor and evaporation of water in the bed occurs.

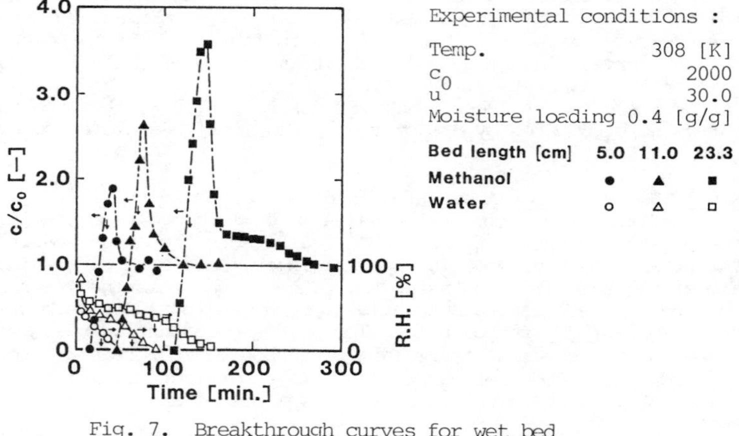

Fig. 7. Breakthrough curves for wet bed

1.0
0.5
0
c/c₀ [—]
100
50
0
R.H. [%]
0
100
200
300
Time [min.]

Experimental conditions:
Temp. = 308 [K], c_0 = 2000, u = 30.0,
Moisture loading 0.40 [g/g]

Bed length [cm]	5.0	11.0	23.3
Toluene	●	▲	■
Water	○	△	□

Fig. 8. Breakthrough curves for wet bed

Adsorption of vapor and evaporation of water from the bed also take place even for the cases of toluene and methyl ethyl ketone adsorption, yet these vapors will almost always be adsorbed, displacing water previously adsorbed. As a result, they did not show any change in the breakthrough curves and in the amount adsorbed. Therefore, one can conclude as follows:

For methyl alcohol adsorption, adsorption capacity of the bed will become larger in the presence of a certain amount of water, however, other solvent vapors will not be affected in their breakthrough behavior by the presence of water in the bed, if the amount of water is not so large as to fill all the pores in carbon granules.

CONCLUSION

Isotherms for single component adsorption on dry carbon were well represented by the Langmuir equation and two parameters were determined for each system.

Breakthrough curves for single component adsorption on dry carbon beds were measured using an experimental apparatus simulating practical carbon beds and were analysed to obtain values of intraparticle diffusivity.

For two solvent vapors adsorption, the following results were obtained. (1) Two equilibrium parameters, i.e., so-called Langmuir constants differ slightly from those for single component adsorption. (2) A comparison of the experimental and theoretical breakthrough curves gave the best set of equilibrium parameters and intraparticle diffusivity for several systems.

The influence of moisture (water) retained in carbon beds on the breakthrough of solvent-vapors was evaluated with single component adsorption on wet carbon beds. Water-soluble solvents, e.g., alcohols were affected greatly in their adsorption capacity and, as a result, in their breakthrough time, while toluene and MEK were not affected by the presence of water in the beds.

Since it is usual to use mixed solvents and to consume less steam in the desorption step, the present results will bring understanding closer to the needs of practical technology.

NOTATION

- c : adsorbate concentration (in respective figures, expressed in terms of ppm)
- Ds : surface diffusivity
- K, K' : Langmuir constant
- q, q' : amount adsorbed
- q_∞, q'_∞ : Langmuir constant
- u : linear flow rate (cm/s, in respective figures)

SUBSCRIPT

- i : component
- cals. : calculated value
- obs. : observed value
- 0 : value at inlet of bed

LITERATURE CITED

1. Furuya, E.; "Design Method for Single- and Multi-component Separation by Fixed-bed Adsorption", Dissertation, Meiji Univ. (1982)

2. Furuya, E. & Y. Takeuchi; "On the Fixed-bed Adsorption Breakthrough Curves and Effective Intraparticle Diffusivity in the Adsorption of Single Component", Research Reports of the Faculty of Engineering, Meiji Univ., No.35, 95 (1978)

3. Takeuchi, Y. & e. Furuya; "A Simple Method to Estimate Effective Surface Diffusivity for Freundrich Isotherm Systems from Breakthrough Curves Based on Numerical Results", J. Chem. Eng. Japan, Vol.13 No.6, 500 (1980)

4. Kawazoe, K. & Y. Fukuda, "Studies on Solvent Recovery by Activated Carbon", Kagaku Kogaku (Chem. Eng. Japan) Abridged Edition in English, Vol.3 No.2, 250 (1965)

5. Takeuchi, Y. & E. Furuya; "Effect of Water Remained in the Carbon Bed on Amount Adsorbed of Solvent", presented at the Annual Meeting of Soc. of Chem. Engrs. Japan (1978, Okayama)

INTRAPARTICLE DIFFUSIVITY OF BICOMPONENT ORGANIC COMPOUNDS IN ADSORPTION ON ACTIVATED CARBON FROM AQUEOUS SOLUTIONS

Yasushi TAKEUCHI and Yoshitake SUZUKI
Department of Industrial Chemistry
Meiji University, Higashi-Mita
Tama-Ku, Kawasaki-Shi, Japan

ABSTRACT

Co-adsorption and displacemental adsorption processes were studied for several pairs of organic compounds adsorbed onto activated carbon particles from their aqueous solutions. Comparing the observed concentrations change of each solute as a function of time (decay curve) with the theoretical predictions, the values of effective intraparticle (surface) diffusivity were determined. Adsorption isotherms for each single component were well expressed by Freundlich equation and bi-component equilibria were expressed by Radke-Prausnitz's method. The values of the diffusivity changed depending on the two cases, i.e., (1)Both components are adsorbed onto virgin carbon simultaneously, (2)The more adsorbable component is adsorbed onto the carbon in which the less adsorbable component is adsorbed beforehand, then the latter component is desorbed partly. However, the values with respect to the more adsorbable component seems to be more important than the other for the proper design of multicomponent adsorption processes.

Regarding multicomponent fixed-bed adsorption, there are many problems to be solved. Among them, the equilibrium and the intraparticle diffusivity of each component are most important things to be known. Since the single component adsorption and displacemental adsorption take place in multicomponent adsorption, one should know how the adsorption proceeds in adsorbent beds, beforehand. When the so-called "constant pattern" occurs due to the difference in adsorbability of each component, the analysis of breakthrough curves and the prediction of break-times are rather easy, if the equilibria are known precisely (1,2). Numerical calculations of fixed-bed adsorption breakthrough curves have been studied by many authors for single and binary systems. It became clear that the adsorption equilibria is more important than the intraparticle diffusivity for the accurate prediction of breakthrough curves and the break-time (3).

Since only a little is known for the prediction of intraparticle diffusivity when two or more solutes are adsorbed on activated carbon particles, the present study was performed to clarify this matter especially for binary systems of alcohols, carboxlic acids and aromatic compounds, respectively.

NUMERICAL CALCULATION OF THE CHANGE OF CONCENTRATION WITH TIME (4)

When two solutes are adsorbed on spherical adsorbent particles from their aqueous solutions in a fully-agitated tank under isothermal and intraparticle (surface) diffusion controlling conditions, the rate process can be expressed by the following equations similar to the case for single solute adsorption described by Suzuki (5). Namely, for intraparticle diffusion:

$$\frac{\partial^2 Y_i}{\partial \rho^2} + \frac{2}{\rho}\cdot\frac{\partial Y_i}{\partial \rho} = \frac{\partial Y_i}{\partial \tau_{si}} \qquad (i=1,\ 2) \qquad \text{--- (1)}$$

The dimensionless time τ_{si} can be chosen so as to correspond to real time elapsed (t) through Equation (2).*

$$t = R^2\cdot(\tau_{s1}/D_{s1}) = R^2\cdot(\tau_{s2}/D_{s2}) \qquad \text{--- (2)}$$

Boundary conditions are expressed as follows,

$$\left.\frac{\partial Y_i}{\partial \rho}\right|_{\rho=1} = -\frac{1}{3\alpha_i}\cdot\frac{\partial X_{Li}}{\partial \tau_{si}} \qquad (i=1,\ 2) \qquad \text{--- (3)}$$

$$\left.\frac{\partial Y_i}{\partial \rho}\right|_{\rho=0} = 0 \qquad (i=1,\ 2) \qquad \text{--- (4)}$$

where α_i denotes as follows.

$$\begin{aligned}\alpha_i &= (W_s\cdot q_{0i})/(V_L\cdot C_{0i}) \\ &= -(1-C_{ei}/C_{0i})/(1-q_{ei}/q_{0i}) \qquad (i=1,\ 2) \qquad \text{--- (5)}\end{aligned}$$

The liquid-to-solid ratio (V_L/W_S) is represented by the straight lines in Figures 1 and 2 for each component.

Initial conditions are expressed as follows, depending on Case 1 and Case 2.

Case 1 : When both components are adsorbed, in other words, the amounts adsorbed of both components increase with time.

$$X_{Li} = 1\ ,\quad q_{0i} = 0 \qquad (i=1,\ 2) \qquad \text{--- (6)}$$

Case 2 : When one component is adsorbed by displaceing the other.

---

* From the real time $t_{\tau si=1}$ which corresponds to $\tau_{si}=1$, the values of intraparticle diffusivity for each component can be calculated by Equation (2').

$$D_{s2} = R^2/t_{\tau_{s2}=1}\ ,\quad D_{s1} = R^2/t_{\tau_{s1}=1} = D_{s2}\,(\tau_{s1}/\tau_{s2}) \qquad \text{--- (2')}$$

$$X_{Li} = 1 \; , \quad q_{01} = k \cdot C_{01}^{1/n_1} \; , \quad q_{02} = 0 \qquad (i=1,\ 2) \qquad \text{--- (7)}$$

Isotherms for the present systems studied were found to be represented by Freundlich equations, respectively, and also binary equilibria were well estimated by Radke-Prausnitz's method (6). Therefore, assuming that equilibria exist always at solid-liquid interface, q_{0i} can be described by Equation (8).

$$q_{0i}Y_{i,N+1,m} = \frac{(C_{0i}X_{Li,m})(k_i n_i/P)^{n_i}}{\sum_{j=1}^{2}((C_{0j}X_{Lj,m})n_j(k_j n_j/P)^{n_j}/P)} \qquad (i=1,\ 2) \qquad \text{--- (8)}$$

Where $P = \pi A/RT$ and P is the value which satisfies the following equation.

$$\sum_{i=1}^{2}((C_{0i}X_{Li,m})(k_i n_i/P)^{n_i}) = 1 \qquad \text{--- (9)}$$

Using the differences of above equations, the following set of simultaneous equations are obtained. In the calculation, the forward finite difference was used for time, while the central finte difference was used with respect to the radial direction of particle.

$$\left\{\begin{array}{l}
(1+1/r_i)Y_{i,1,m+1} - Y_{i,2,m+1} = -(1-1/r_i)Y_{i,1,m} + Y_{i,2,m} \\
(1+1/r_i)Y_{i,2,m+1} - Y_{i,3,m+1} = -(1-1/r_i)Y_{i,2,m} + Y_{i,3,m} \\
\qquad \text{----------} \\
\qquad\quad \text{----------} \\
\qquad\qquad \text{----------} \\
-(1-1/(j-1))Y_{i,j-1,m+1} + 2(1+1/r_i)Y_{i,j,m+1} - (1+1/(j-1))Y_{i,j+1,m+1} \\
\quad = (1-1/(j-1))Y_{i,j-1,m} - 2(1-1/r_i)Y_{i,j,m} + (1=1/(j-1))Y_{i,j+1,m} \\
\qquad \text{----------} \\
\qquad\quad \text{----------} \\
\qquad\qquad \text{----------} \\
-(1-1/(N-1))Y_{i,N-1,m+1} + 2(1+1/r_i)Y_{i,N,m+1} - (1+1/(N-1))Y_{i,N+1,m+1} \\
\quad = (1-1/(N-1))Y_{i,N-1,m} - 2(1-1/r_i)Y_{i,N,m} + (1+1/(N-1))Y_{i,N+1,m} \\
-Y_{i,N,m+1} + (1+1/r_i)Y_{i,N+1,m+1} + (2/3)((N+1)/\alpha_i r_i)X_{Li,m+1} \\
\quad = Y_{i,N,m} - (1-1/r_i)Y_{i,N+1,m} + (2/3)((N+1)/\alpha_i r_i)X_{Li,m}
\end{array}\right.$$

$$(i=1,\ 2) \qquad \text{--- (10)}$$

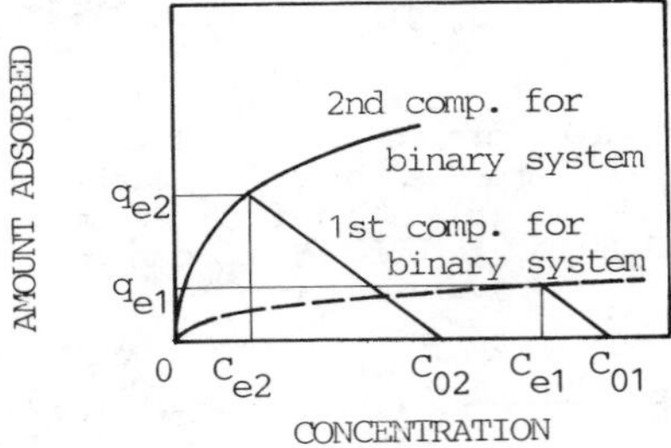

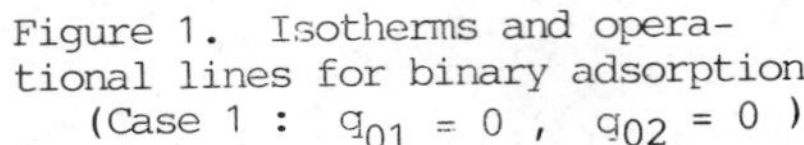

Figure 1. Isotherms and operational lines for binary adsorption (Case 1 : $q_{01} = 0$, $q_{02} = 0$)

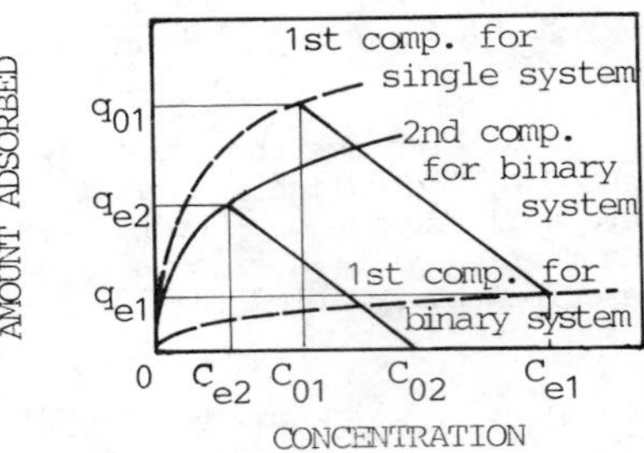

Figure 2. Isotherms and operational lines for binary adsorption (Case 2 : $q_{01} > 0$, $q_{02} = 0$)

Among the above equations, $Y_{i,1,m+1} \sim Y_{i,N,m+1}$ were obtained by the successive over relaxation method and $Y_{i,N+1,m+1}$ and $X_{Li,m+1}$ were calculated by Newton-Raphson's method.

Changing parameters, concentrations of two components at time τ_{si}, in the other words, so-called "decay curves" were calculated.

Comparison of the observed concentration change with the calculated ones gives the disired diffusivity values.

EXPERIMENTAL METHOD AND APPARTUS

Experiments were performed using an agitated tank adsorber, as shown in Figure 3. The tank was made of glass and containd a certain amount (V_L, about 2dm³) of aqueous solution of various compounds. About 0.6g (W_S) of carbon particles (CAL, 16/20mesh) were packed in a rotating double-blade which was set at the center of the tank. To avoid the change in temperature, the tank was placed in a thermostat. A small amount (about 1cm³ or less) of liquid was taken through the hole of the top to analyze the concentration change with time.

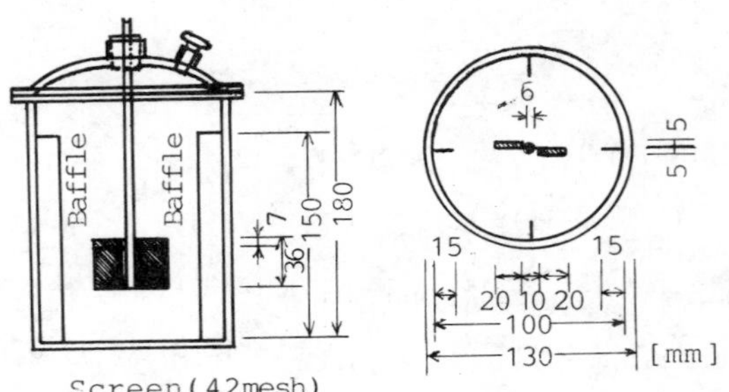

Figure 3. Experimental appartus (Batch agitated tank adsorber)

In Case 1, fresh carbon particles were used and in Case 2, the carbon particles were kept to be in equilibrium with a certain concentration of a component (C_{01}). The solutions were prepared by the addition of a certain amount of compounds (extra pure grade) into deionized water. Though pH of the solution may affect to the adsorption equilibria, experiments were performed under ambient conditions.

Analysis of solutes remaining in water was done by either ultraviolet absorption or total organic carbon detection after separating both solutes by a high speed liquid chromatograph.

EXPERIMENTAL RESULTS AND DISCUSSION

Some examples of decay curves observed for single component adsorption and binary adsorption (Case 1) are shown in Figure 4. The ratio of

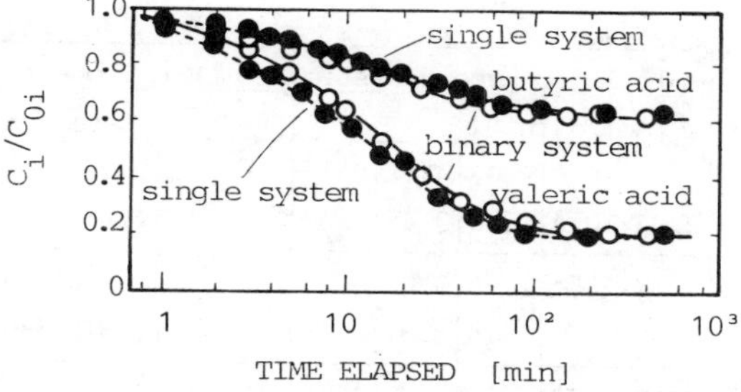

Figure 4. Comparison of calculated and observed changes of concentration eith time (single system; C_{01}=0.556, C_{02}=0.332 [mmol/dm³], binary system; C_{01}=1.49, C_{02}=0.786[mmol/dm³])

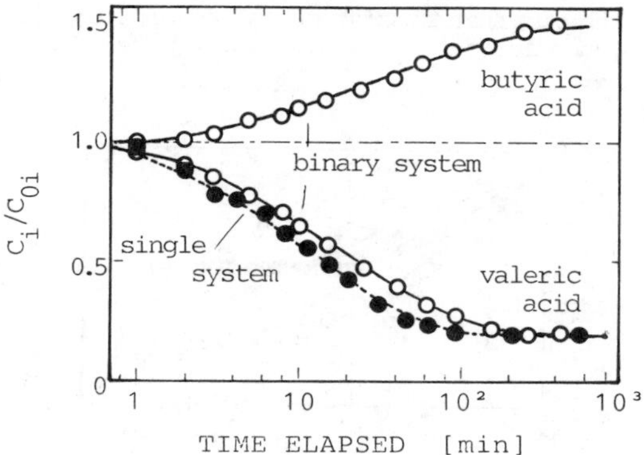

Figure 5. Comparison of calculated and observed changes of concentration with time (single system; C_{01}=0.332[mmol /dm³], binary system; C_{01}=0.614, C_{02}=0.925[mmol/dm³])

concentration at equilibria (C_e) to the initial concentration (C_0) was taken to be the same for single and binary adsorption. Figure 5 shows an example of observed and calculated decay curves for the Case 2, i.e., valeric acid is adsorbed while butyric acid is desorbed. Theoretical decay curves were calculated for the same conditions as above, i.e., C_e/C_0 = constant, and the values of intraparticle effective diffusivity were determined by curve-fitting.

Results are summarized in Table 1. It is clear that the values of intraparticle effective diffusivity (D_s) for binary system are different from those for single component systems. For benzoic acid — p-nitrophenol system in which both components were strongly adsorbed on activated carbon, the values of D_s is much lower than that observed for respective single component adsorption.

Table 1. Comparison of surface diffusivity D_s

$[m^2/s]$ at 298[K]

Adsorbate	Single component system			Bicomponent system		
	k [mol/g]	n[-]	D_s		Case 1	Case 2
A: Benzoic acid	3.00×10^{-3}	3.03	4.1×10^{-12}	A — B	2.7×10^{-12}	2.6×10^{-13}
B: p-Nitrophenol	2.83×10^{-3}	6.40	3.4×10^{-13}		7.7×10^{-13}	5.2×10^{-13}
C: Propionic acid	4.67×10^{-4}	2.80	9.0×10^{-12}	C — D	4.3×10^{-11}	6.2×10^{-12}
D: Valeric acid	1.49×10^{-3}	2.74	8.6×10^{-12}		1.1×10^{-11}	6.2×10^{-12}
E: n-Butanol	4.87×10^{-4}	1.92	8.6×10^{-12}	E — F	1.0×10^{-10}	—
F: n-Pentanol	1.13×10^{-3}	2.32	6.2×10^{-12}		1.0×10^{-11}	—

Furthermore, it is evident that D_s for Case 2 is much smaller than that for Case 1, showing that displacemental adsorption brings about much larger resistance for the transport of the more adsorbable component (7,8).

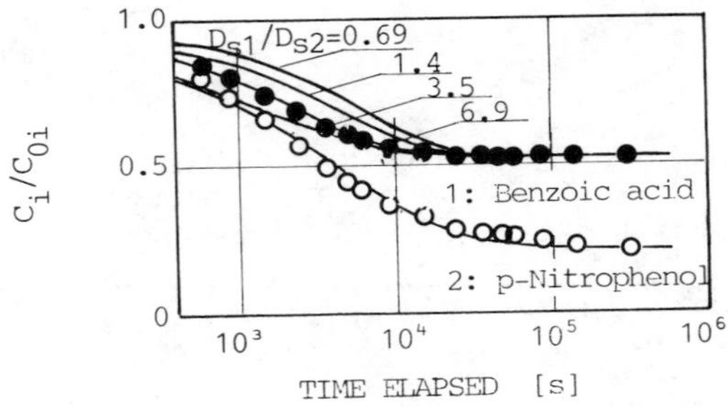

Figure 6. Comparison of calculated and observed changes of concentration with time (C_{01}=0.666, C_{02}=0.463[$mmol/dm^3$])

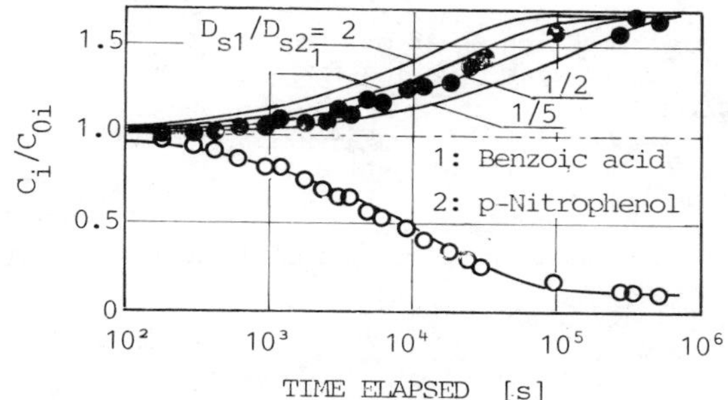

Figure 7. Comparison of calculated and observed changes of concentration with time (C_{01}=0.185, C_{02}=0.356[mmol/dm³])

Figures 6 (Case 1) and 7 (Case 2) show other two examples of comparison between observed data and calculated curves for benzoic acid — p-nitrophenol system.

CONCLUSION

The values of the surface diffusivity were measured for several pairs of organic compounds adsorbed from their aqueous solutions onto activated carbon particles. The values changed depending on two cases particularly, i.e., co-adsorption and displacemental adsorption processes. The values for binary system were different from those for single component systems. However, the values with respect to the more adsorbable component seems to be more important than that for less adsorbable component.

The design of fixed-bed adsorber in which multicomponent adsorption takes place will become easier when one knows the intraparticle diffusivity, if intraparticle diffusion is significant.

Therefore, the present paper will be useful to obtain such information as above for the adsorption between liquid and solid particles.

ACKNOWLEDGEMENT

The authors would like to express their deepest thanks to the Ministry of Education, Japan, for supporting their study through Grant in Aid for Scientific Ressearch (No. 56470093).

They also thank to Mr. H. Motomura for assisting them in experimental works.

NOTATION

A : surface area per unit weight of adsorbent
C : concentration of solute in liquid, [mol/dm³]
D_s : intraparticle diffusivity based on the difference in amount adsorb-

ed, [m^2/s]
k : a coefficient in Freundlich equation, [mol/g]
n : a coefficient in Freundlich equation (expressed as the power of concentration), [-]
N : number of division with respect to radius of particles
P : = $\pi A/RT$, value which satisfies Equation (9)
q : amount adsorbed, [mol/g]
r : = (radial increment)/(time increment)^2
R : gas constant in definition of P or radius of adsorbent particles in the other equations, [m]
t : time elapsed, [s]
T : absolute temperature, [K]
V_L : total volume of liquid in the tank, [dm^3]
W_s : total amount of adsorbent, [g]
X_L : dimensionless concentration of liquid (C/C_0), [-]
Y : dimensionless amount adsorbed (q/q_0), [-]
α : distribution ratio, defined by Equation (5), [-]
π : spreading pressure
ρ : dimensionless length in the radial direction, [-]
τ : dimensionless time elapsed, defined by Equation (2), [-]

SUPERSCRIPT AND SUBSCRIPT

e : value at equilibrium
i : component (1 or 2)
j : component (1 or 2) in Equation (8) or number corresponded to radial position (mesh point) within particle (j=1 and j=N+1 correspond to center and surface of spherical particle, respectively)
m : number corresponded to mesh point with respect to time
0 : initial value

LITERATURE CITED

1. Takeuchi, Y., T. Wasai & S. Suginaka; "Fixed-Bed Breakthrough Curves in Binary and Ternary Trace Component Adsorption from Aqueous Solutions", J. Chem. Eng. Japan, Vol.11 No.6, 458 (1978)

2. Takeuchi, Y., Y. Suzuki & E. Furuya; "On the Break Time and Concentration Distribution in Multicomponent Fixed-Bed Adsorption When Constant Pattern is Established", ibid., Vol.12 No.6, 486 (1979)

3. Furuya, E. & Y. Takeuchi; "On the Fixed-Bed Adsorption Breakthrough Curves and Effective Intraparticle Diffusivity in the Adsorption of Single Component", Research Reports of the Faculty of Engineering, Meiji Univ., No.35, 95 (1978)

4. Takeuchi, Y. & Y. Suzuki; "Study on Binary Solute Adsorption Processes onto Activated Carbon from Aqueous Solutions by Use of Batch Agitated-Tank Adsorber", presented at the Annual Meeting of Soc. of Chem. Engrs. Japan (1981, Kanazawa)

5. Suzuki, M. & K. Kawazoe; "Effective Surface Diffusion Coefficients of Volatile Organics on Activated Carbon during Adsorption from Aqueous

Solution", J. Chem. Eng. Japan, Vol.8 No.5, 379 (1975)

6. Radke, C. J. & J, M. Prausnitz; "Thermodynamics of Multi-Solute Adsorption from Dilute Liquid Solutions" A.I.Ch.E. J., Vol.18 No.4, 761 (1972)

7. Takeuchi, Y., E. Furuya & Y. Suzuki; "On the Measurement and a Simplified Analysis of Fixed-Bed Adsorption Breakthrough Curves for Two Components Aqueous Solution", Kōgyō-Yōsui (J. of Industrial Water), No. 233, 30 (1978)

8. Takeuchi, Y.; "On the Multicomponent Adsorption by Activated Carbon", ibid., No.233, 4 (1978)

SPECIES-GROUPING IN FIXED-BED MULTICOMPONENT ADSORPTION CALCULATIONS

by

Chi Tien

Department of Chemical Engineering and Materials Science
Syracuse University, Syracuse, NY 13210 USA

ABSTRACT

The application of species-grouping for simplifying fixed-bed multicomponent adsorption calculations was explored on heuristic arguments. Procedures were established for grouping adsorbates and subsequently, for assigning physical properties to pseudo-species. Using species grouping, sample calculations were done of effluent concentration histories of fixed bed adsorption, and the results were compared with exact calculations.

\- - -

Important applications of adsorption often involve solutions of multiple adsorbates. The presence of more than a single adsorbate in an adsorption process gives rise to a number of complications manifested both in competition among adsorbates in mass transfer and in the possible displacement of adsorbates with lower adsorption affinity by those with higher affinity. The effort to develop a predictive capability regarding the dynamics of multicomponent adsorption—especially that of fixed-bed processes—which takes into account these complications, has evolved into a major topic of investigation in recent years.

A general algorithm of multicomponent adsorption calculations requires that the base data necessary for the calculations be ascribable to the individual adsorbates. Two such algorithms have been developed of late (1,2). With these algorithms, one can make detailed calculations of fixed-bed adsorption involving an arbitrary large number of adsorbates. On the other hand, of course, there is always a practical limit to the number of adsorbates which can be considered in a given problem.

A logical way to simplify multicomponent adsorption calculations is to group adsorbates with similar adsorption affinity into pseudo-species, thus reducing the number of species to be considered and, consequently, the computational effort. Such a procedure would be extremely useful in a number of instances, for example, the use of granular activated carbon for water and waste treatment. The feasibility of this concept was explored recently by both Calligaris and Tien (3) and Mehrotra (4). The former study was concerned with calculating equilibrium concentrations attained in the batch contacting process. It was found possible to combine into a single group adsorbates whose single-species adsorption data were within the same order of magnitude. On the other hand, Mehrotra considered the use of species-grouping in fixed bed calculations and found that for his study a more restricted criterion was needed than that used by Calligaris and Tien.

The results of these two studies demonstrated the feasibility and utility of species-grouping in multicomponent adsorption calculations. There is, however, an element of ambiguity in the criteria these investigators used. Assuming that the single-species adsorption isotherm data can be approximated by the Freundlich expression, it would be desirable to be able to define the grouping procedure in terms of the magnitude of the Freundlich coefficient and exponent of the individual adsorbates. The establishment of such a criterion and its application in simplifying multicomponent adsorption calculations are the main objectives of the present study.

SPECIES-GROUPING CRITERIA

A simple and yet sufficiently realistic model of fixed-bed adsorption is to consider the adsorption process to be controlled by the external mass transfer and the intraparticle diffusion. Consider a solution of N adsorbates with specific concentrations which flows through a bed packed with fresh adsorbents. The governing equations for such a solution are as follows:

$$u \frac{\partial c_i}{\partial z} + \rho_b \frac{\partial \bar{q}_i}{\partial \theta} = 0 \tag{1a}$$

$$\frac{\partial \bar{q}_i}{\partial \theta} = \frac{3k_{\ell i}}{a_p \rho_p} (c_i - c_{s_i}) = k_{s_i}(q_{s_i} - \bar{q}_i) \tag{1b}$$

$$q_{s_i} = f(c_{s_i}, c_{s_2} \ldots c_{s_n}) \quad i = 1, 2, \ldots N \tag{1c}$$

$$c_i = c_{i_o} \qquad z = 0 \qquad \theta > 0 \quad \text{(1d)} \qquad q_i = 0, \qquad z > 0 \qquad \theta < 0 \quad \text{(1e)}$$

The above set of equations were written with the customary assumption of steady state, one-dimensional plug flow, negligible axial dispersion, etc. The rate equation [i.e., Equation (1b)] assumes that the intraparticle diffusion can be described by the linear driving force model. The surface concentrations, c_{s_i} and q_{s_i} are considered to be in equilibrium. In other words, Equation (1c) represents the multicomponent adsorption isotherm expression.

By applying species-grouping to include all N adsorbates, the analogous expressions of Equation (1a), (1b), and (1c) are

$$u \frac{\partial c_t}{\partial z} + \rho_b \frac{\partial \bar{q}_t}{\partial \theta} = 0 \tag{2a}$$

$$\frac{\partial \bar{q}_t}{\partial \theta} = \frac{3k_\ell}{a_p \rho_p} (c_t - c_{s_t}) = k_s(q_{s_t} - \bar{q}_t) \tag{2b}$$

$$q_{s_t} = g(c_{s_t}) \tag{2c}$$

where the subscript t refers to the summation of the relevant quantities of all N adsorbates. For example, c_t is given as

$$c_t = \sum_{i-1}^{N} c_i \tag{3}$$

On the other hand, by summing Equation (1b) for all N adsorbates one has

$$\frac{\partial \bar{q}_t}{\partial \theta} = \frac{3}{a_p \rho_p} \sum_{i=1}^{N} k_{\ell_i} (c_i - c_{s_i}) = \sum_{i=1}^{N} k_{s_i} (q_{s_i} - \bar{q}_i) \tag{4}$$

Strictly speaking, an exact species-grouping requires that Equations (2a) to (2c) be consistent with Equations (1a) to (1c); in other words, the parameters, k_ℓ and k_s should be defined in such a way that Equations (2b) and (4) are equivalent. The condition under which such requirements can be met provides the criterion for species-grouping.

To establish a species-grouping criterion and to develop procedures for characterizing the pseudo-species formed from species-grouping strictly in accordance with the requirements stated above is, of course, impractical. One may even argue that such an undertaking is impossible. What follows below is an attempt to determine the grouping criterion in an approximate manner under certain limited conditions. The validity of the criterion will then be tested through sample calculations.

Characterization of Adsorption Affinity of Pseudo-Species. Previous studies have demonstrated that the ideal adsorption theory (IAS) provides a practical way of estimating multicomponent adsorption equilibrium. Furthermore, it has been shown that incorporation of the IAS theory for fixed bed adsorption calculations is facilitated if the single-species adsorption isotherm data of the individual adsorbates are represented by the Freundlich expression (2). Accordingly, if an aqueous solution containing a number of adsorbates is to be approximated by a solution of a fewer number of pseudo-species, it is necessary to assign Freundlich coefficients and exponents to the pseudo-species.

To obtain the grouping criterion which defines the Freundlich coefficient and exponent for the pseudo-species, consider a solution containing N adsorbates with concentrations $\{c_i\}$. The equilibrium concentrations in the adsorbed phase are denoted by $\{q_i\}$. The IAS theory enables the calculation of $\{q_i\}$ from the following system of equations:

$$q_i = z_i q_t \tag{5a}$$

$$z_i = c_i \left(\frac{A_i n_i}{\Pi}\right)^{n_i} \tag{5b}$$

$$\sum_{i=1}^{N} z_i = 1 \tag{5c}$$

$$q_t = \left[\sum_{i=1}^{N} \frac{n_i z_i}{\Pi}\right]^{-1} \tag{5d}$$

In other words, Equation (1c) takes the form of the above set of expressions. On the other hand, if the N adsorbates exhibit sufficiently similar adsorption affinity that they can be combined to form a pseudo-species with Freundlich coefficient A and exponent n, q_t becomes

$$q_t = A\, c_t^{1/n} \tag{6}$$

A similar condition of equivalence can be found between Equation (6) and Equation (5d). First, Equation (5d) can be written explicitly as

$$q_t = \left[\frac{n_1 z_1}{\Pi} + \frac{n_2 z_2}{\Pi} + \cdots \frac{n_N z_N}{\Pi}\right]^{-1} \tag{7}$$

If one interprets similarity in adsorption affinity to mean that adsorbates have approximately the same Freundlich expression, namely

$$n_1 \simeq n_2 \simeq n_3 \ldots \simeq n_N \simeq n \tag{8}$$

and

$$n = \sum_{i=1}^{N} n_i/N \tag{9}$$

then substituting Equation (8) into (7) one has

$$q_t \simeq \left[\frac{n}{\Pi}\right]^{-1} \tag{10}$$

Similarly from Equations (5b) and (8), one has

$$z_i \simeq c_i A_i^{n_i} \left(\frac{n}{\Pi}\right)^n \tag{11}$$

Substituting Equation (11) into (5c), one has

$$\sum_{i=1}^{N} A_i^{n_i} c_i \left(\frac{n}{\Pi}\right)^n = 1 \tag{12}$$

Combining Equations (12) and (10) yields

$$q_t = \left[\sum_{i=1}^{N} A_i^{n_i} c_i\right]^{1/n} = A\left[\sum_{i=1}^{N} c_i\right]^{1/n} \tag{13}$$

where

$$A = \frac{\left[\sum_{i=1}^{N} A_i^{n_i} c_i\right]^{1/n}}{\left[\sum_{i=1}^{N} c_i\right]^{1/n}} = \left[\frac{\sum_{i=1}^{N} A_i^{n_i} c_i}{\sum_{i=1}^{N} c_i}\right]^{1/n} \tag{14}$$

Alternately, if one assumes that the adsorbed phase concentrations, $\{q_i\}$, are known, then the total adsorbate equilibrium concentration in the solution phase, Σc_i, becomes

$$c_t = \sum_{i=1}^{N} c_i = \frac{\sum_{i=1}^{N} q_i \left(\frac{\Pi}{A_i n_i}\right)^{n_i}}{q_t} \tag{15}$$

Again, if the n adsorbates are combined on account of the approximate similarity of their Freundlich exponent, n_i, then Equation (15) can be approximated by the following expression:

$$c_t \quad (\frac{\Pi}{n})^n \frac{\sum_{i=1}^{N} q_i \, (\frac{1}{A_i})^{n_i}}{\sum_{i=1}^{N} q_i} = \frac{\sum_{i=1}^{N} q_i \, (\frac{1}{A_i})^{n_i}}{\sum_{i=1}^{N} q_i} (q_t)^n \tag{16}$$

where n is the average value of n_i's as given by Equation (9).

Equation (16) is analogous to Equation (6):

$$c_t = (\frac{1}{A})^n \; (q_t)^n \tag{17}$$

By comparing Equation (17) with Equation (16) one sees that the Freundlich coefficient of the pseudo-species becomes

$$A = [\frac{\sum_{i=1}^{N} q_i (\frac{1}{A_i})^{n_i}}{\sum_{i=1}^{N} q_i}] - \frac{1}{n} \tag{18}$$

Accordingly, with n given by Equation (9), the Freundlich coefficient of the pseudo-species can be calculated either from Equations (14) or (18).

Characterization of Mass Transfer Coefficients of Pseudo-Species. To complete the grouping procedures, one must develop relations which define the liquid and particle phase mass transfer coefficients for the pseudo-species. Generally speaking, one may expect the value of k_{ℓ_i}'s to be quite similar for all the adsorbates. That the k_ℓ's have approximately the same value results from the fact that the diffusivity of dissolved organics in water is often of the same order of magnitude (approximately 0.7×10^{-5} cm^2/sec.). Thus, one may assume k_ℓ to be

$$k_\ell = \sum_{i=1}^{N} k_{\ell_i} / N \tag{19}$$

For particle phase mass transfer coefficient, consider the case where the intraparticle diffusion is the rate controlling step. Equation (16) reduces to

$$\frac{\partial \bar{q}_i}{\partial \theta} = k_{s_i} \, (q_i^* - \bar{q}_i) \tag{20}$$

Application of species-grouping requires

$$\frac{\partial}{\partial \theta} (\sum_i \bar{q}_i) = \sum_i k_{s_i} \; (q_i^* - \bar{q}_i) = k_s \; [\sum_i q_i^* - \sum_i \bar{q}_i] \tag{21}$$

where the superscript * denotes the equilibrium adsorbed phase concentration.

The above equality requirement obviously cannot be satisfied. As an approximation, if it is required that the equality be satisfied initially (i.e., $\bar{q}_i=0$), on the grounds this requirement will be the important factor in determining the initial part of the breakthrough curve, one has

$$\sum_i k_{s_i} q^*_{s_i} = k_s \sum_i q^*_i \quad (22) \quad \text{or} \quad k_s = \frac{\sum_i k_{s_i} q^*_i}{\sum_i q^*_i} = \sum_i k_{s_i} z_i \quad (23)$$

Substituting Equations (5a),(5b), and (10) into (23), one has

$$k_s = \sum_i k_{s_i} A_s^{n_i} c_i \left(\frac{n_i}{\Pi}\right)^{n_i} \simeq \left(\frac{n}{\Pi}\right)^n \sum_i k_{s_i} A_i^{n_i} c_i \quad (24)$$

The quantity (Π/n) according to Equation (10) equals the total equilibrium concentration in the adsorbed phase. With the application of species grouping, one has (Π/n) equals $A(\sum_i c^*_i)^{1/n}$. Equation (24), therefore becomes

$$k_s = \sum_i k_{s_i} \frac{A_i^{n_i} c_i}{A^n(\sum_i c_i)} \quad (25)$$

SAMPLE CALCULATIONS

Systems The multispecies systems considered in this work are aqueous solutions containing up to seven different organic adsorbates. A total of 27 organics taken from a list of toxic substances compiled by the Environmental Protection Agency (5) was used. These substances are exactly those used in Mehrotra's work and correspond to groups II, III, IV, and V considered by Calligaris and Tien.

Since the purpose of the sample calculations is to test the approximate grouping criteria outlined earlier, it is convenient to arrange the substances according to the magnitude of the Freundlich exponent of their single-species carbon adsorption isotherms as shown in Table 1. The 25 organics are arranged into five groups such that the Freundlich exponents of the adsorbate in each group do not vary by more than 20%. Also included in the tables are the values of the Freundlich coefficients and exponents of the single-species carbon adsorption isotherm expressions of these substances and their estimated values of k_{s_i} and k_{ℓ_i}.

Method of Solution. The history of the total adsorbate concentration of the effluent from a fixed carbon bed was predicted by solving the appropriate governing equations for the fixed-bed process [i.e., Equations (1a) through (1e)]. The algorithm used for the numerical solution of these equations was that developed earlier (2). Predictions were made both from exact calculations as well as with the use of species-grouping simplifications. When species grouping was applied, the pseudo-species was characterized according to Equations (9), (14), and (25), with c_i's taken to be the influent values. The conditions used in these sample calculations are given in Table 2.

Results. The sample calculation results and the inferences

Table 1 Properties of Adsorbates used in Sample Calculations

GROUP	COMPOUNDS	A	n	$D_L \times 10^{-6}$ [cm^2/sec]	$k_L \times 10^3$ [cm/s]	k_s [s^{-1}]	C_{inlet} mg TOC/lit
I	Carbon Tetrachloride	12.2	1.2048	9.18	1.726	3×10^{-6}	1.0
	Cyclohexone	9.73	1.333	8.5	1.662	5.35×10^{-6}	1.0
	Average values	10.836	1.2689	8.833	1.694	4.185×10^{-6}	
II	Bis (2 choloroethoxy) methane	12.87	1.5385	6.72	1.477	2.5×10^{-6}	0.5
	Uracil	13.63	1.5873	8.05	1.617	2.32×10^{-6}	0.5
	5 Choloruracil	27.88	1.7241	7.41	1.551	8.31×10^{-7}	0.5
	Bis (2 Chloroisopropyl) ether	28.05	1.7544	6.48	1.451	8.53×10^{-7}	0.5
	Average values	24.21	1.6511	7.149	1.52	1.07×10^{-6}	
	Acrolein	1.742	1.5385	9.46	1.753	5.07×10^{-5}	0.5
	1,1,2 Trichloroethane	4.96	1.6667	8. 2	1.693	9×10^{-6}	0.5
	Average values	3.77	1.6026	9.138	1.723	1.3979×10^{-5}	
III	Tetrachloroethylene	36.79	1.7857	8.90	1.7	2.93×10^{-7}	0.25
	Phenol	31.48	1.8519	10.32	1.831	9.70×10^{-7}	0.25
	Average values	33.99	1.8188	9.594	1.766	6.23×10^{-7}	
	Bromoform	7.7	1.9231	8.69	1.679	1.52×10^{-6}	0.25
	1,1 Dichloroethane	1.56	1.8868	10.0	1.802	4.02×10^{-5}	0.25
	Average values	5.5856	1.905	9.32	1.74	3.209×10^{-6}	
IV	DDT	375.86	2.0	4.64	1.228	2.56×10^{-7}	0.5
	Naphthalene	215.44	2.3810	7.52	1.39	7.64×10^{-7}	0.5
	Average values	291.81	2.19	5.2741	1.31	6.207×10^{-7}	
	1.3 Dichlorobenzene	135.13	2.222	7.61	1.5724	12.31×10^{-7}	0.5
	Dimethyl Phthlate	123.83	2.439	5.4	1.324	23.13×10^{-7}	0.5
	Average values	134.06	2.3305	6.455	1.448	19.89×10^{-7}	
	Toluene	42.02	2.2727	8.43	1.654	2.65×10^{-7}	0.5
	2-Chlorophenol	61.43	2.439	8.5	1.662	0.61×10^{-7}	0.5
	Nitrobenzene	84.93	2.3256	7.52	1.563	0.34×10^{-7}	0.5
	Average values	67.45	2.3457	8.14	1.626	0.64×10^{-7}	
V	DDE	263.18	2.7027	4.9	1.26	5.15×10^{-7}	0.5
	α-Napthol	269.42	3.125	7.15	1.524	2.98×10^{-7}	0.5
	2,4 Dinitrotoluene	145.16	3.2258	5.95	1.39	7.64×10^{-7}	0.5
	1,2,4 Trichlorobenzene	142.28	3.2258	7.2	1.529	7.25×10^{-7}	0.5
	Average values	218.78	3.07	6.26	1.426	4.438×10^{-7}	
	1,1,1 Trichloroethane	1.356	2.9412	9.0	1.71	1.85×10^{-5}	0.5
	DiBromochloro Methane	1.24	2.9412	9.1	1.7195	1.85×10^{-5}	0.5
	Average values	1.3	2.9412	9.05	1.714	1.85×10^{-5}	

Table 2 Operating Condition of Fixed-Bed in Sample Calculation

Bed Height	= 10 cm	Bed Density	= 0.39 gm/cm^3
Particle Radius	= 6.6 x 10^{-2}cm	Particle Density	= 0.59 gm/cm^3
Superficial Velocity	= 0.612 cm/sec.		

drawn from these results are presented below. In all these calculations aqueous solutions containing adsorbates listed in each of the five groups were passed through fixed beds packed with carbon free of adsorbates and under the conditions listed in Table 2.

The degree of accuracy which can be achieved with the use of species grouping simplification can be seen from the sample calculations in which an aqueous solution of carbon tetrachloride and cyclohexanone was considered (i.e., Group I on Table 1). Both of these substances exhibit similar adsorption affinity to the degree that their respective Freundlich coefficients and exponents are within 10%. Furthermore, their particle mass transfer coefficients are within a factor of two. Fig. 1 shows the results of the effluent concentration history (total adsorbate in dimensionless form) obtained by combining carbon tetrachloride and cyclohexanone into a pseudo=species and that from the exact calculation. In terms of the effluent concentration, the difference is well under 0.5 mg TOC per liter and also cannot be detected with common analytical instruments, such as the total organic carbon analyzer.

The accuracy of the species-grouping procedure developed in this work and the extent to which it may be applied can be seen from the results shown in figures 2 through 5. In Figure 2 aqueous solutions containing all the organics of Group II were considered. In carrying out the sample calculations three different groupings were attempted: one pseudospecies was formed by combining all six adsorbates together; two pseudo-species were formed, one combining acrolein and 1,1,2, trichloroethane and the other including the other four adsorbates; or three species were formed with acrolein and 1,1,2 trichloroethane retaining their separate identities and the other four combining to form the third species. The effluent concentration (total adsorbate) histories obtained with these grouping arrangements and that from exact calculation are shown in Figure 2. The great discrepancy between the results obtained with the use of species-grouping of the first type and that from exact calculations can be attributed largely to the differences in the mass transfer coefficient of these adsorbates (for example, the k_s values of acrolein is almost two orders of magnitude larger than that of either 5 chlororuacil or Bis-ether). As Figure 2 also shows, as the differences in mass transfer characteristics are taken into account by grouping the six adsorbates in two or three pseudo-species, a significantly improved agreement was observed. In these two later groupings, the values of k_s of the various adsorbates which were grouped to form a pseudo-species were required to be within the same order of magnitude. It also appears that there was no significant improvement in the accuracy of the results with further restrictions on the k_s values (for example, see the comparisons of the results obtained with the second and third types of grouping).

The importance of considering the difference in the mass transfer characterization of adsorbates in carrying our species-grouping was further demonstrated through the sample calculations involving adsorbates

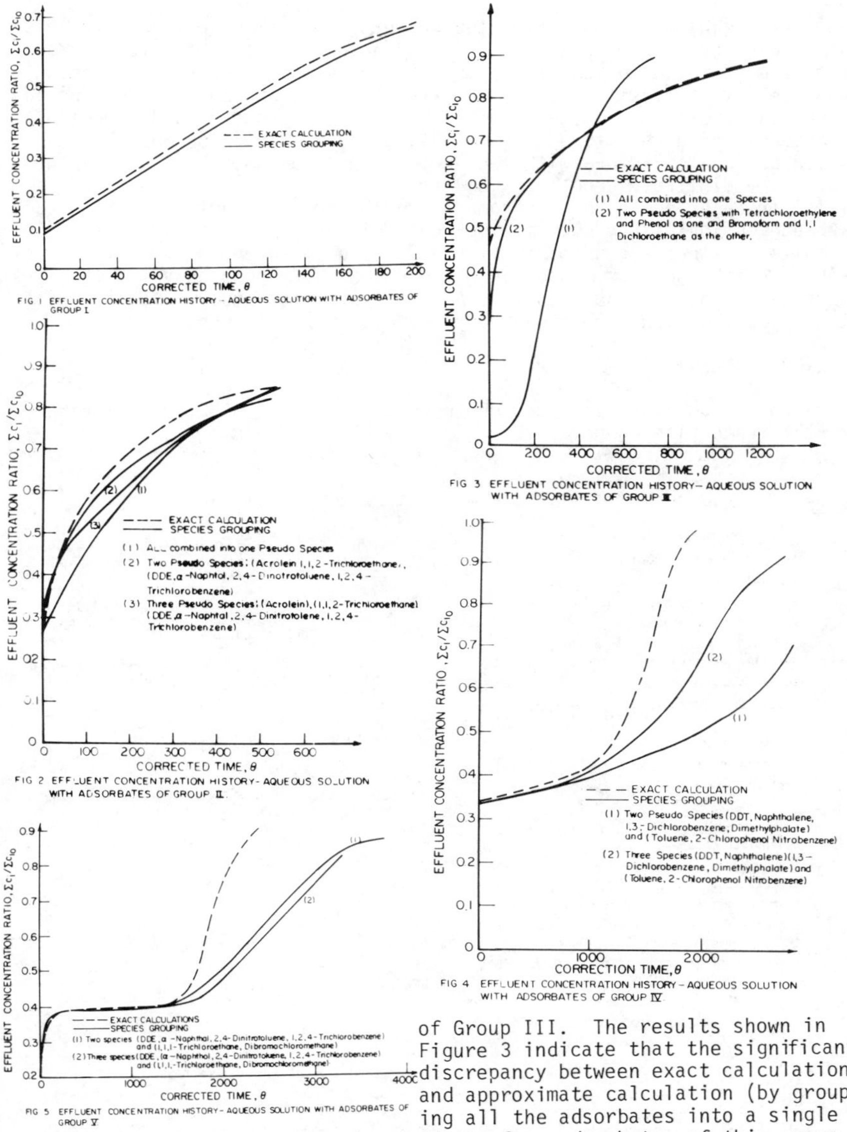

FIG 1 EFFLUENT CONCENTRATION HISTORY - AQUEOUS SOLUTION WITH ADSORBATES OF GROUP I

FIG 2 EFFLUENT CONCENTRATION HISTORY - AQUEOUS SOLUTION WITH ADSORBATES OF GROUP II

FIG 3 EFFLUENT CONCENTRATION HISTORY - AQUEOUS SOLUTION WITH ADSORBATES OF GROUP III

FIG 4 EFFLUENT CONCENTRATION HISTORY - AQUEOUS SOLUTION WITH ADSORBATES OF GROUP IV

FIG 5 EFFLUENT CONCENTRATION HISTORY - AQUEOUS SOLUTION WITH ADSORBATES OF GROUP V

of Group III. The results shown in Figure 3 indicate that the significant discrepancy between exact calculation and approximate calculation (by grouping all the adsorbates into a single pseudo-species) is greatly reduced if the four adsorbates of this group were combined into two pseudo-species (tetrachlorethylene and phenol as one and bromoform and 1,2, dichloroethane as the other). The same conclusion can also be reacted with the sample calculations made with adsorbates from group IV as shown in Fig. 4.

Although, as shown above, the grouping criterion developed in this study was found to be reasonably accurate in estimating the total adsorbate breakthrough curves, there are instances when the criterion does not yield totally satisfactory results. As an example, consider the adsorbates of group V. According to the criterion, one should expect reasonably good results by approximating aqueous solutions of these seven adsorbates as one with two or three pseudo-species (e.g., napthol, DDE, 2,2, dinitrotoluene and 1,2,4 trichloroethane as one group and 1,1,1 trichloroethane and dibromochloromethane as the other, or the same arrangement described but with DDE retaining its separate identity). The results obtained are shown in Figure 5. The agreement between the exact results with those from species-grouping was very good only for the first part of the breakthrough curve. A more systematic study aimed at obtaining an explanation for this discrepancy is currently under way.

ACKNOWLEDGEMENT

This study was performed under Contract No. R809482, U.S. Environmental Protection Agency.

NOTATION

A	Freundlich coefficient
A_i	Freundlich coefficient of the i-th adsorbate
a_p	particle radius
c_i	concentration of the i-th adsorbate in solution
c_{i_o}	value of c_i at the inlet
c_{s_i}	concentration of the i-th adsorbate in the solution phase at the fluid-particle interface
c_t	total adsorbate concentration in solution
k_ℓ	liquid phase mass transfer coefficient
k_i	liquid phase mass transfer coefficient of the i-th adsorbate
k_s	particle phase mass transfer coefficient
k_{s_i}	particle phase mass transfer coefficient of the i-th adsorbate
N	number of adsorbates present in solution
n	Fruendlich exponent
n_i	Freundlich exponent of the i-th adsorbate
q_i	concentration of the i-th adsorbate in the adsorbed phase in equilibrium with $\{c_i\}$
q_{s_i}	concentration of the i-th adsorbate in the adsorbed phase at the fluid-particle interface.
$\bar{q}_i$	average concentration of the i-th adsorbate in the adsorbed phase
$\bar{q}_t$	defined as $\Sigma\, \bar{q}_i$
q_i^*	adsorbed phase concentration in equilibrium with c_i
u	supervelocity
z	axial distance
ρ_b	bulk density
ρ_b	particle density
Π	dimensionless spreading pressure
θ	corrected time

LITERATURE CITED

(1) Hsieh, J.S.C., R.M. Turian & C. Tien, "Multicomponent Liquid-Phase Adsorption in Fixed Beds," AIChE J., 23, 363 (1977).

(2) Wang, S-C. & C. Tien, "Further Work in Multicomponent Liquid Pgase Adsorption in Fixed Beds," AIChE J., 28, 565 (1982).

(3) Calligaris, M.B. & C. Tien, "Species Grouping in Multicomponent Adsorption Calculations," Can. J. Chem. Eng., 60, 772 (1982).

(4) Mehrotra, A.K., "Application of Species-Grouping in Multicomponent Adsorption Calculations in Fixed Beds," M.S. Thesis (Chem. Eng.) Syracuse University, Syracuse, NY.

(5) Dobbs & Cohen, "Carbon Adsorption Isotherm for Toxic Organics," EPA-60018-80-02 (1980).

ISOTHERM EQUATIONS FOR MONOLAYER ADSORPTION OF GASES ON HETEROGENEOUS SOLID SURFACES

J. Tóth
Chemical Research Laboratory
for Mining
of the
Hungarian Academy of Sciences
Miskolc, Hungary

ABSTRACT: A new, semi-empirical isotherm equation to describe and interpret monolayer gas adsorption (isotherms Type I and V) on heterogeneous solid surfaces has been derived. The equation contains two dimensionless constants characterizing the heterogeneity of the surface and/or the lateral interactions between the adsorbed molecules. The equation is of a general character because it comprises all the relationships of the Langmuir type known so far. Three kinds of energy distribution functions can be assigned to the equation: normal, right-hand widened and left-hand widened Gaussian distributions.

\- - - - - -

DERIVATION AND PHYSICAL INTERPRETATION OF NEW ISOTHERM EQUATION

Derivation Based on Mathematical Formalism

From any experimentally determined isotherm, the following differential expression can be calculated numerically:

$$\psi(p) = \frac{\theta}{p}\frac{dp}{d\theta} \tag{1}$$

where p is the equilibrium pressure and θ is the relative coverage.

The character of the isotherm can be defined accurately by Eq. (1), viz:

$$\text{domain convex from below where } \psi(p) < 1 \tag{2}$$

$$\text{domain concave from below where } \psi(p) > 1 \tag{3}$$

$$\text{linear isotherm where } \psi(p) = 1 \tag{4}$$

The differential equation for any isotherm can be written on the basis of Eq. (1). Thus for the Freundlich (F) equation:

$$\theta = (\text{const.})\, p^N$$

we have

$$\psi_F(p) = 1/N \quad , \tag{5}$$

for the Langmuir (L) equation of the form

$$\theta = p/(1/K_1 + p) \quad ,$$

$$\psi_L(p) = K_1 p + 1 \quad , \tag{6}$$

and for the equation (T) derived by the author [1],

$$\theta = p/(1/K_1' + p^m)^{1/m} \quad ,$$

$$\psi_T(p) = K_1' p^m + 1 \quad . \tag{7}$$

It can be shown, without presenting any details here, that mathematical synthesis of any two isotherm equations is possible if their respective differential equations, as defined by Eq. (1), are multiplied with each other and the differential equation obtained in this manner is solved (the so-called rule of multiplication). Thus, using Eqs. (5) and (6), for example, the differential equation for the generalized Freundlich (FL) equation may be written as:

$$\psi_{FL}(p) = \psi_F(p) \times \psi_L(p) = (K_1 p + 1)/N \tag{8}$$

Upon integration, we obtain the FL equation:

$$\theta = \left(\frac{p}{1/K_1 + p}\right)^N \tag{9}$$

In like manner, multiplication of Eqs. (5) and (7) with each other yields the following differential equation:

$$\psi_{FT}(p) = \psi_F(p) \times \psi_T(p) = (K_1' p^m + 1)/N \tag{10}$$

The new isotherm equation (FT) may be obtained from Eq. (10) after separation of variables and integration between the limits $(\theta,1)$ and (p,∞):

$$\theta = \frac{p^N}{(1/K_1' + p^m)^{N/m}} \tag{11}$$

It is apparent that Eqn. (11) comprises all the equations of the Langmuir type, because for $N = 1$ we obtain the T equation, for $m = 1$ the FL equation, for $m = N$ the Sips (S) equation, for $m = N = 1$ the L equation, for $m = 1$ and $p \ll 1/K_1'$ the F equation, and finally for $m = N = 1$ and $p \ll 1/K_1'$ the linear (Henry's law) isotherm equation is obtained.

Physical Interpretation of the Mathematical Formalism

The basis of the physical interpretation is the relationship between the function $\psi(p)$ defined by Eq. (1) and the differential adsorptive potential defined by Ross and Olivier [2]. As is well-known, the following relationship may be written, in general form:

$$q^{diff}(p) = U_0^{diff}(p) + U_K(p) \tag{12}$$

where $q^{diff}(p)$ is the differential heat of adsorption, $U_0^{diff}(p)$ is the differential adsorptive potential and $U_K(p)$ is a function which includes all kinds of energy differences which are *not* due to any solid - gas interaction but rather to the translational and vibrational energies of the adsorbed molecules and, primarily, to the lateral interaction energies. If the condition

$$U_K(p) << U_0^{diff}(p) \tag{13}$$

is satisfied, then the relationships are

$$q^{diff}(p) = U_0^{diff}(p) \tag{14}$$

for a heterogeneous surface and

$$q^{diff} = U_0 = \text{const.} \tag{15}$$

for a homogeneous surface. Again, it is well-known that the constant K_1 appearing in Eqs. (6) and (9) has the following interpretation:

$$K_1 = \frac{1}{k_B} \exp(U_0/RT) \quad , \tag{16}$$

where k_B is the deBoer-Hobson constant [3] which may be written approximately as

$$k_B = 2.346(MT)^{1/2} \times 10^5$$

if p is expressed in kPa and M is the molecular mass of the adsorbate and T is in degrees Kelvin.

With this introduction, the following equations are postulated:

$$\psi_L(p) = \psi_T(p) \quad \text{and} \quad \psi_{FL}(p) = \psi_{FT}(p) \quad . \tag{17}$$

The physical meaning of Eq. (17) is that the heterogeneous solid surfaces characterized by the T or FT equations are divided into n patches (infinitesimally small elements of the surface) and to each of these patches the differential equation

$$\psi_L(p) \quad \text{and} \quad \psi_{FL}(p) \quad ,$$

respectively, is applied, using of course different constants $K_1 \ldots K_i \ldots K_n$ for each patch. According to this interpretation,

Eq. (16) can be written:

$$K_i = \frac{1}{k_B} \exp(U_{0i}/RT) \qquad (i = 1 \ldots n) \tag{18}$$

Eqs. (17) require the following relation:

$$K_i p = K_1' p^m \qquad (i = 1 \ldots n) \tag{19}$$

Eqs. (17) and (19) are explained on Figure 1. If we take into consideration that, based on the definitions of Ross and Olivier [2],

$$U_{0i} = U_0^{diff} \quad , \tag{20}$$

then it may be judged from Figure 1 that the constants K_i in Eq. (19) have been used in calculating the differential adsorptive potential. Specifically we have from Eqs. (18)-(20) that

$$U_0^{diff}(p) = RT \ln (k_B K_1' p^{m-1}) \quad . \tag{21}$$

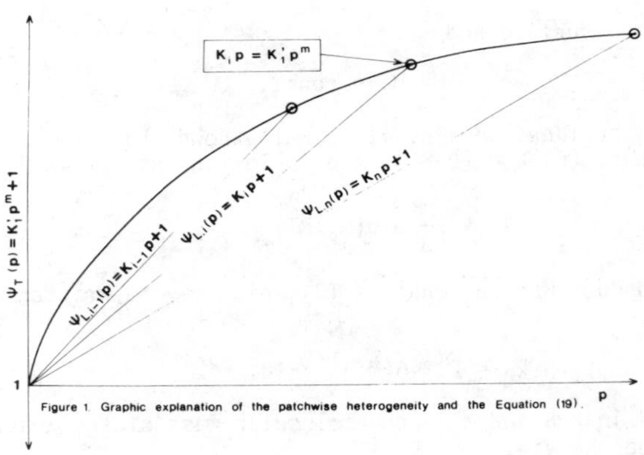

Figure 1. Graphic explanation of the patchwise heterogeneity and the Equation (19).

Interpretation of the Energy Distribution Function and the Constants N and m

If Eq. (11) is solved for p, the following relationship is obtained:

$$p = (1/K_1')^{1/m} \frac{\theta^{1/N}}{\left(1 - \theta^{m/N}\right)^{1/m}} \quad . \tag{22}$$

Upon substitution of Eq. (22) in Eq. (21), the following function of the differential adsorptive potential is found:

$$U_0^{diff}(\theta) = RT\ln\left[k_B(K_1')^{1/m}(\theta^{-m/N} - 1)^{\frac{1}{m}-1}\right] \quad . \tag{23}$$

As is well-known, the energy distribution function based on the condensation approximation [4] can be defined as follows:

$$f_E(U_0^{diff}) = - d\theta/dU_0^{diff} \tag{24}$$

The energy distribution function may be calculated from Eq. (23) by carrying out the differentiation prescribed by Eq. (24). It is however simpler, from a computational point of view, to differentiate Eq. (23) by θ to obtain the function

$$dU_0^{diff}/d\theta \quad ,$$

the negative reciprocal of which gives the simple form:

$$f_E(\theta) = \frac{N}{RT(1-m)}\theta(1 - \theta^{m/N}) \quad . \tag{25}$$

The energy distribution function is determined jointly by Eqs. (23) and (25), in such a manner that pairs of values

$$U_0^{diff}(\theta) \quad \text{and} \quad f_E(\theta)$$

corresponding to the same value of θ determine one point of the energy distribution function.

The form of the energy distribution function defined by Eqs. (23) and (25) is basically determined by the magnitude of the constants m and N relative to one another, namely

if $N \approx m$, the energy distribution is approximately a normal Gaussian function,

if $N < m$, a right-hand widened Gaussian function, and

if $N > m$, or $N = 1$, a left-hand widened Gaussian function

is obtained. If, besides this role of the values of N and m, we also taken into consideration that

$$\lim_{p=0} \psi_{FT}(p) = 1/N \quad ,$$

then we can make the following statements:

→ At very low coverages (equilibrium pressures) the heterogeneity of the surface is characterized by the value of N such that we obtain for

$$N = 1 \quad \text{and} \quad 1/K_1' >> p^m$$

a linear isotherm, for $N < 1$ a Type I isotherm, and for $N > 1$ a Type V isotherm.

→ At higher coverage the heterogeneity of the surface is characterized by the value of m according to Eq. (23), e.g. if m = 1, then

$$U_0^{diff}(\theta) = U_0 = \text{const.}$$

PRACTICAL APPLICATION OF THE FT EQUATION

Mathematical Method of Application

In view of the fact that

$$\theta = n^s/n_u^s \quad ,$$

Eq. (11) can be written in the following linear form:

$$\frac{p^m}{(n^s)^{m/N}} = \frac{1}{(n_u^s)^{m/N}} p^m + \frac{1}{(n_u^s)^{m/N} K_1'} \tag{26}$$

where n_u^s is the total capacity at monolayer adsorption.

Values of N and m are determined iteratively using a two-step fitting procedure. The basic principle of the procedure is that the pair of values (N,m) is found for which the difference between the data calculated according to Eq. (26), n_c^s, and the measured data, n^s, is a minimum. This difference can be defined in various ways; we used the following principle:

$$\Delta R_{min} = 1 - r$$

where r is equal to the correlation coefficient between the measured (n^s) and the calculated data (n_c^s). The iteration minimized the value of ΔR as a function of m and N. The procedure can be completed in less than one minute using a computer of average capacity.

The above method was used in applying Eq. (26) to the Type I isotherms of propane measured on Nuxit-AL microporous charcoal at 40 and 90 C [5]. The measured and calculated data are summarized in Table 1. For purposes of comparison, this table also contains data calculated by the T, S and DR equations. The DR equation was used in the following linear form:

$$\ln(n^s) = B(RT)^2\left(\ln\frac{p_m}{p}\right)^2 + \ln(n_u^s) \tag{27}$$

The value of p_m in Eq. (27) was determined by the same fitting procedure used for determining the values for m and N.

The data in Table 1 are completed by Figure 2, in which the energy distribution functions corresponding to FT, T, S and DR equations are shown for the 40 C isotherm of propane.

The energy distribution function for the DR equation was calculated

Table 1

Comparison of the measured and the calculated data of the isotherm for propane adsorbed on Nuxit-Al charcoal at 40 °C and 90 °C

p MEASURED kPa	n^s MEASURED mmol/g	n^s CALCULATED BY FT EQUATIONS	n^s CALCULATED BY T EQUATIONS	n^s CALCULATED BY S EQUATIONS	n^s CALCULATED BY DR EQUATIONS	PARAMETERS OF ISOTHERM EQUATIONS
At 40 °C						
2.20	1.004	0.996	1.005	1.018	1.026	
8.00	1.834	1.853	1.835	1.800	1.796	
24.53	2.713	2.695	2.685	2.659	2.636	
91.49	3.636	3.632	3.643	3.663	3.666	
307.04	4.323	4.345	4.355	4.378	4.426	
692.08	4.738	4.727	4.722	4.713	4.720	
		0.368	0.441	0.594	-	m
		2.200	1.000	0.594	-	N
		2.121	0.6089	0.1464	-	K'_1
		-	-	-	1321.9	p_m
		-	-	-	0.03766	$B(RT)^2$
		6.062	5.762	5.375	4.795	n^s_u
		62.8	99.2	379.3	808.8	$\Delta R_{min}\ 10^6$
At 90 °C						
8.20	0.812	0.811	0.816	0.827	0.821	
23.60	1.410	1.411	1.403	1.387	1.392	
55.68	2.003	2.006	2.000	1.985	1.987	
161.45	2.815	2.807	2.811	2.819	2.818	
351.57	3.377	3.386	3.391	3.405	3.411	
660.35	3.832	3.828	3.825	3.819	3.827	
		0.344	0.422	0.621	-	m
		1.790	1.000	0.621	-	N
		0.9799	0.3129	0.0522	-	K'_1
		-	-	-	3700.4	p_m
		-	-	-	0.0448	$B(RT)^2$
		6.569	5.967	5.115	4.372	n^s_u
		12.7	24.5	148.8	117.6	$\Delta R_{min}\ 10^6$

by the well-known relationship

$$f_E(U_0^{diff}) = 2B\, U_0^{diff} \exp\left[-B(U_0^{diff})^2\right]$$

where

$$U_0^{diff}(p) = RT \ln(p_m/p) \tag{28}$$

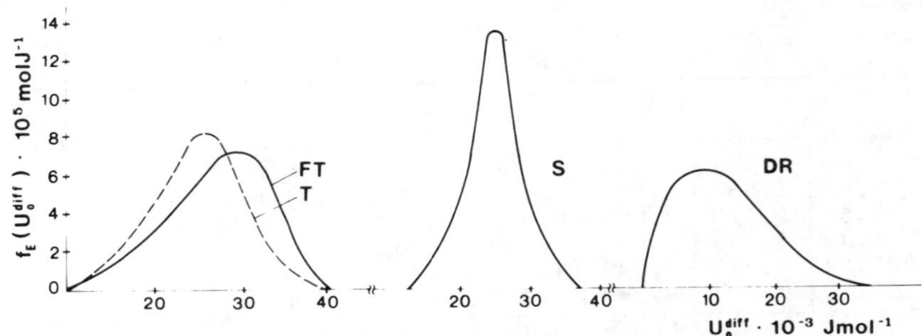

Figure 2. Energy distribution functions corresponding to the FT, T, S and DR equations applied to the isotherm for propane adsorbed on Nuxit charcoal at 40 °C.

Conclusions

The following conclusions may be drawn from the data shown in Table 1 and Figure 2:

The FT equation, with the smallest value of ΔR_{min}, provides the best agreement with the experimental data, but the accuracy of the T equation differs only slightly from this. The less favorable applicability of the S and DR equations lies possibly in the fact that the adsorption energy of the Nuxit charcoal has a left-hand widened Gaussian distribution, while the S equation corresponds to an approximately normal Gaussian distribution and the DR equation to a right-hand widened one.

With regard to the FT and T equations, the energy distribution functions computed by means of the condensation approximation are practically the same as those calculated by the integral representation for the overall adsorption isotherm [6].

It may be of interest to mention the finding that the DR equation lends itself only to Type V isotherms, and the FT equation also uses Type V isotherms to describe the data of Table 1 because, as a result of the iteration, $N > 1$. The statement concerning the DR equation is borne out by the relationship:

$$\psi_{DR}(p) = 2B\left[(RT)^2 \ln(p_m/p)\right]^{-1}$$

according to which, with decreasing p, the condition

$$\psi_{DR}(p) < 1$$

will be satisfied at some value of p. The fact that the convex form of the Type V isotherm passes unnoticed is due to very low equilibrium pressures (coverage) at which the inequality is realized for both the DR and FT equations. For technical reasons, data is rarely measured is this region. For example, for propane at 90 C, the convex section of the isotherm is in the range $0 < p < 52$ Pa for the DR equation, and in the range $0 < p < 500$ Pa for the FT equation. The fact that a section convex from below is found by extrapolation of the extremely accurate FT equation, and a Type V isotherm is described by the DR equation, poses the question of whether we are faced with a physical phenomenon or the problem is simply one of mathematical formalism. Our theory renders the first alternative probable, yet this could be confirmed only with sufficient experimental data.

The FT equation was also applied to Type V isotherms measured at 77.5 and 90.1 K for argon and nitrogen adsorbed on carbon black graphitized at 2700 C [2]. In each case we obtained a very good value of ΔR_{min}, and $m > 1$, which may be explained by the presence of the interaction energies already mentioned in connection with Eq. (12).

LITERATURE CITED

1. Tóth, J., *Acta Chim. Acad. Sci. Hung. (Budapest)* 69, 311 (1971).
2. Ross, S. and J.P. Olivier, *On Physical Adsorption*, Interscience Publishers, New York (1964), p. 135.
3. Hobson, I.P., *Canad. J. Phys.* 43, 1934 (1965).
4. Harris, L.B., *Surface Sci.* 10, 129 (1968).
5. Szepesy, L. and V. Illés, *Acta Chim. Acad. Sci. Hung. (Budapest)* 35, 37 (1963).
6. Tóth, J., W. Rudzinski, A. Waksmundzki, M. Jaroniec and S. Sokolowski, *Acta Chim. Acad. Sci. Hung. (Budapest)* 82, 11 (1974).

ETHANOL SEPARATION FROM FERMENTORS BY STRIPPING AND TWO-STAGE ADSORPTION

P. K. Walsh, C. P. Liu, M. E. Findley, and A. I. Liapis

Department of Chemical Engineering
University of Missouri-Rolla
Rolla, Missouri 65401
USA

ABSTRACT

A process for separating ethanol from dilute aqueous fermentation media has been studied experimentally in parts. Activated carbon and cellulose were used as adsorbents in a two-stage adsorption scheme suitable for use with fermentation media. The separation system involves the stripping of ethanol from a fermentor with recirculated carbon dioxide on a continuous basis. The ethanol and some water are adsorbed onto activated carbon. A carrier gas is then used for desorption of the adsorbed ethanol and water, and the desorbed vapors are then passed through a cellulose adsorbent to remove water and obtain nearly pure ethanol.

Experiments on a cyclic glucose fermentation indicate the feasibility of ethanol removal through the use of the above separation process. The concentration of ethanol in the fermentor was up to 6% by weight. Adsorption experiments with carbon dioxide saturated with 6% ethanol and 94% water by weight, produced an adsorbed phase on activated carbon containing approximately 50% by weight ethanol. The desorbed vapors from the activated carbon were passed through dry cellulose, and were condensed. This condensate contained more than 95% ethanol. The experimental studies indicate that it should be desirable to have a two-stage fermentation process in conjunction with this separation method.

The stripping with a two-stage adsorption separation process appears to be technically feasible, but further studies are needed to establish economic feasibility.

INTRODUCTION

A major economic constraint on the fermentation of carbohydrates to ethanol for industrial or fuel purposes is the product inhibition of the fermentation and the resulting separation costs (1,2). In order to prevent inhibition it is necessary to ferment to only about 7 to 10% ethanol using a dilute substrate. Thus 9 or more kg of water must be carried through the fermentor and into the separation process for each kg of ethanol produced. This water reduces the efficiency of fermentation by dilution of both substrate and yeast, and increases the cost of ethanol separation. Distillation at atmospheric pressure destroys any yeast carried with the large quantity of product solution.

Distillation of the ethanol by vacuum (3,4) at low temperatures permits ethanol removal as it is formed, without damaging the yeast being used for fermentation. It also permits the effective use of higher sugar concentrations and results in higher fermentor productivities (5). Vacuum fermentation requires mechanical energy to remove ethanol and water vapor and also the CO_2 by-product and any air present. However, a relatively high concentration vapor condensate reduces further distillation cost (6,7). It has been proposed that an inert volatile component, such as hexane, can improve these methods (8).

Preferential adsorption is another method of separating ethanol from fermentor media without disturbing the fermentation. In adsorption it is generally desirable to adsorb the dilute component which requires less adsorbent capacity and lower costs of desorption or regeneration (9). Because the numerous components and the yeast suspension present in fermentor liquids could cause problems in adsorbers, a preferred adsorption method would involve stripping vapors from the fermentor with a carrier gas, adsorbing the desired vapor components, and recycling the gas and unadsorbed vapors. In a fermentor, recycled CO_2 by-product would be a convenient carrier gas which would be available and would not change the saturated condition in conventional fermentors.

Desorption of ethanol and water vapor from an ethanol adsorbent should give a relatively high ethanol to water ratio, and one way of purifying the ethanol would be to adsorb the water vapor in a second adsorber and then condense the ethanol. In the adsorption of either ethanol or water, the occurrence of condensation in the adsorbent interstices would reduce the efficiency of the separation, and thus should be avoided.

The adsorbents studied in this investigation were activated carbon to preferentially adsorb ethanol and cellulose to adsorb water. Some data exists on activated carbon adsorption of ethanol (10) and water adsorption on cellulosic type material (11) but not of the type desired for study of the 2-stage process suggested.

This paper presents the results of research on CO_2 stripping of a pulse fed fermentor, on the adsorptivities of ethanol and water on activated carbon and cellulose, and on the effluent compositions from dynamic adsorptions and desorptions. The effects of temperature were only briefly considered, due to the number of possible combinations, but the conditions of the study were considered to be similar to those of a practical process.

FERMENTATION STUDIES

Cells of *Saccharomyces cerevisiae* (ATCC 38637) grown in shake flasks were used to innoculate a 2 l New Brunswick fermentor operated with 1 liter of fermentor media at 37°C (12). Carbon dioxide was circulated by means of a diaphragm pump through a submerged nozzle in the fermentor, a condenser with 0°C cooling water, a condensate trap and back to the pump, as shown in Figure 1. Initially carbon dioxide was supplied from a cylinder. Air (0.1 l/min) was supplied to improve growth (4).

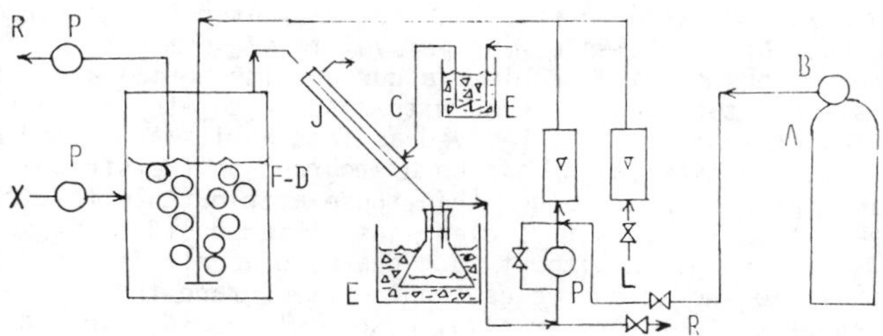

Figure 1. Fermentation Apparatus

Legend
A CO_2 Cylinder
B Pressure Regulator
C Coolant (Heat Transfer Fluid)
D Stripper
E Water or Oil Bath
F Fermentor
G Activated Carbon Column
H Cellulose Column
J Condenser
K Heat Exchanger
L Air
M Settling Tank
P Pump or Blower
R Exhaust or Waste
S Steam
X Fermentor Feed
Y Ethanol Product

At 12 hour intervals, a sample was taken, and feed addition and volume corrections (removal of media or addition of distilled water) were made to maintain constant volume. Condensate and fermentor samples were analyzed, and the runs were continued until results were equivalent to steady-state conditions.

Analyses were made on ethanol content of the fermentor and condensate by gas chromatography, on cell concentration by light scattering, and on glucose in the fermentor by the DNS method (13,14). A Poropak Q column at 115°C was used in the gas chromatograph.

Compositions from one fermentation run are shown in Figure 2.

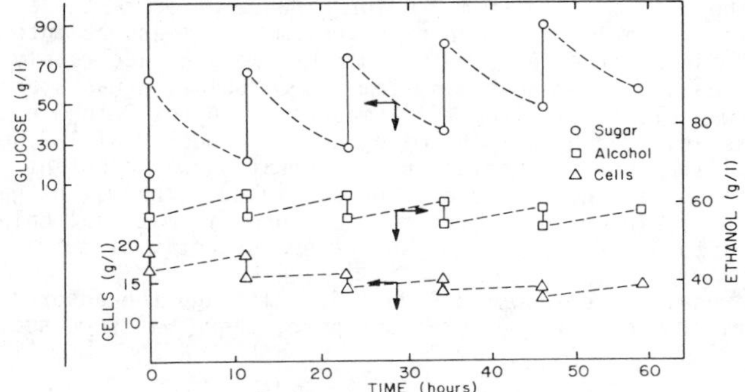

Figure 2. Time Variation of the Concentrations of Glucose, Ethanol and Cells in the Fermentor; Run No. 7.

In this run the glucose did not reach a steady-state cycle, but it was assumed that smaller pulses of glucose feed would have produced steady-state conditions similar to final conditions.

The fermentation study indicated that few problems should occur in recirculating carbon dioxide to strip ethanol and water from a fermentor. Although up to 20 volumes of CO_2 per volume of fermentor were circulated, no serious foaming problems occurred with normal amounts of antifoam. A steady state was achieved with respect to ethanol in all cases indicating ethanol could be removed as fast as it was formed by the fermentation. Tests with the gas chromatograph on the stripping gas outflow indicated no peaks other than those attributable to air, CO_2, water, and ethanol. Thus this clean gas stream could be passed through an adsorption column without much chance of contaminating the surface or pores, regeneration cycles should not degrade the adsorbent appreciably, and a suitable preferential adsorbent should give a good separation. The ethanol concentrations in the 0°C condensate were as follows for various ethanol concentrations in the fermentor

Fermentor Concentration (g/l = $\frac{kg}{M^3}$)	10	12	20	25	32	58
Condensate Concentration g/l	55	90	131	165	190	324

These concentrations represent approximately the total condensable vapors in the stripping gas, and do not represent a condensate that could be obtained by conventional condensation without refrigeration. A suitable adsorbent can adsorb these condensables up to the saturation point.

ADSORPTION STUDIES ON ACTIVATED CARBON

The adsorption studies were carried out by stripping liquid solutions containing pure ethanol and water only (15). Carbon dioxide or air was bubbled through the solutions and then through the desired adsorbent column. In order to prevent internal condensation, the adsorbent was normally maintained 4°C or more higher in temperature than the solution being stripped.

Steady-state adsorption was carried out on the apparatus shown in Figure 3, where the carrier gas was recirculated continuously from adsorbent column to pump, to stripper, to adsorbent column. Activated carbon was held in an oven at 110°C for 6 hours prior to these tests. After analyzing the solution, an exact volume was measured into the stripping flask and the adsorbent column was weighed accurately. Carbon dioxide was added to the system and circulation was started, and the system was closed off from the atmosphere. After 2 to 3 hours the solution was analyzed by gas chromatograph and checked at 1/2 hour periods until the stripper solution concentration was constant. The adsorber column was then wiped dry and reweighed. The weight gain of adsorbent and stripper concentrations were used in material balance calculations to determine ethanol and water adsorptivities.

These results are shown in Figure 4. Although the adsorber was 4°C higher in temperature than the stripper, there was some scatter in

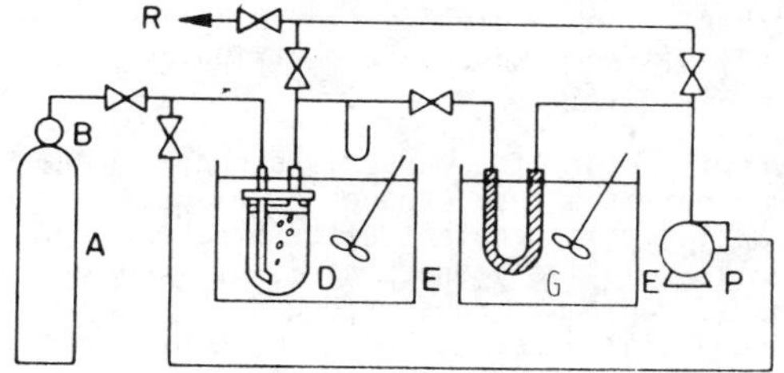

Figure 3. Steady-State Adsorption Apparatus
Legend see Figure 1

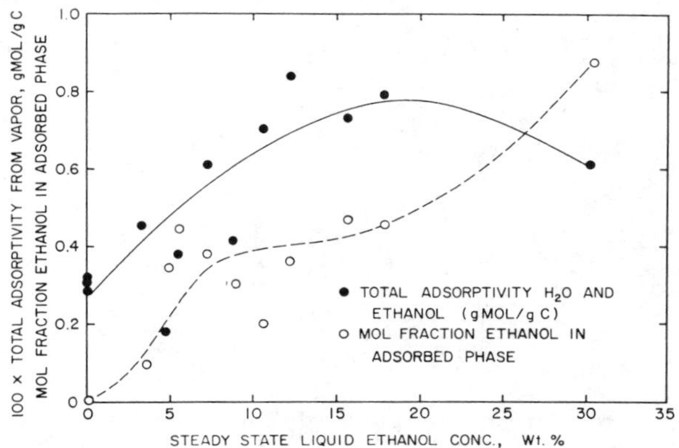

Figure 4. Steady-State Adsorptivity on Activated Carbon. Liquid at 40°C Stripped with CO_2 and Activated Carbon Adsorption at 44°C

the data, probably due to internal condensation. This can occur as a result of vapor pressure lowering due to surface tension effects in small pores. The results show an almost constant composition of the adsorbed phase from about 6% by weight solutions to 15% solutions even though water concentration is decreasing and ethanol concentration is increasing in the stripper. Thus ethanol adsorption seems to enhance the adsorption of water (or perhaps condensation of water), which is similar to results found by Ozaki et al. (16) on methanol and water.

The steady-state results indicate that about 5 to 6% ethanol gives higher adsorptivities and higher ethanol concentration in the adsorbed phase, but higher concentrations are not as beneficial, at least in adsorbed phase concentration. Since concentrations above about 7% are inhibiting in fermentors, the logical choice for fermentor concentration is around 6% by weight ethanol.

Dynamic adsorption tests were carried out in apparatus and conditions similar to Figure 2 except that the gas passed once—through the stripper and adsorber. Carbon dioxide from a cylinder was used with no recirculation. Analyses were made on the stripping solution and the gas phase inflow and effluent.

A typical dynamic adsorption curve is shown in Figure 5. The water breakthrough curve indicates that ethanol displaces previously adsorbed water to produce a high water content after breakthrough. The ethanol breakthrough curve is relatively sharp as is desired in a practical separation system. The dynamic runs produced slightly higher adsorptivities than the steady-state results, which were used for calculations.

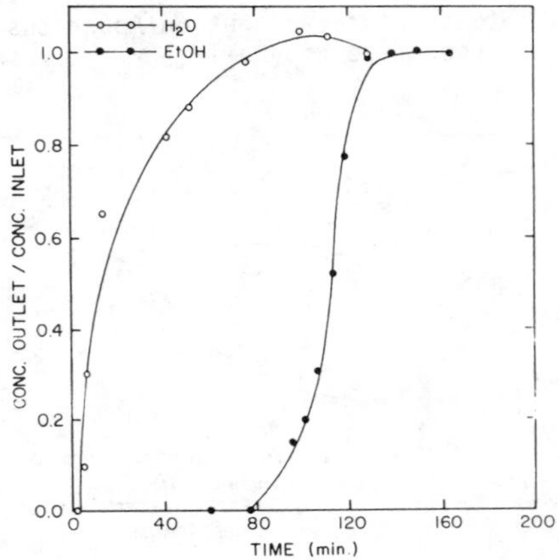

Figure 5. Activated Carbon Adsorption of Ethanol and Water Vapor, Breakthrough Curve at 40°C, CO_2 Flow = 2.3 l/min., Ethanol Vapor = 2.20 to 1.99 mol%, H_2O Vapor = 5.18 to 5.5%, Activated Carbon Wt. = 50g

Desorption was studied by passing CO_2 through a previously saturated activated carbon column in an oil bath at around 110°C. Ethanol and water in the outlet vapor were analyzed with gas chromatography.

Desorption curves are shown in Figure 6, in which very little separation is achieved on desorption. However, one explanation of these curves is that initially the thermal desorption throughout the column produces a high concentration of both water and ethanol with ethanol predominating, but after initial thermal desorption the effect of the carrier gas tends to cause desorption from the inlet end, moving both water and ethanol towards thr outlet and allowing ethanol to displace adsorbed water as it moves downstream. Finally after much of the water is removed, both ethanol and water are gradually desorbed. This interpretation suggests that desorption primarily by carrier gas from inlet to outlet could allow desorbed ethanol (or perhaps ethanol in the carrier gas) to displace adsorbed water and force much of the water out ahead of and separated from the bulk of the ethanol, thus obtaining additional separation on desorption.

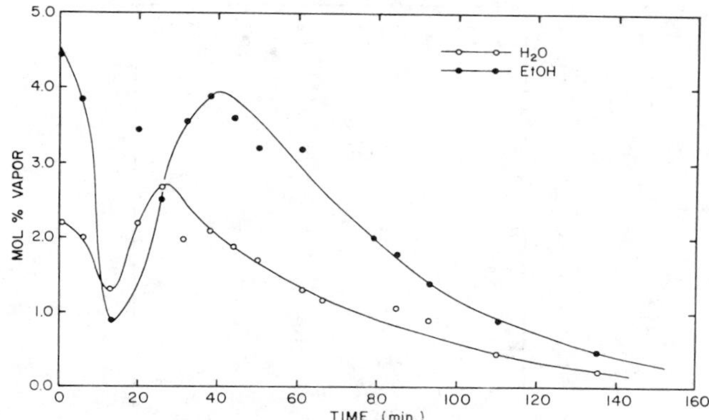

Figure 6. Desorption of Ethanol and Water from Activated Carbon (Saturated by CO_2 Carrier from 6% Ethanol Liquid). CO_2 Flow Rate = 0.9 l/min, T = 110°C, Wt. Activated Carbon = 50g

These studies show that stripping of ethanol and water, followed by adsorption on activated carbon produces an adsorbed phase with more than 50% by weight ethanol. This compares with 35 weight per cent in the approximately 0°C condensate from a 6% original solution. After stripping and adsorption of ethanol, the adsorbed components could be desorbed thermally and directly fed to a distillation column for further purification. However, adsorption of water might be more promising.

Adsorption Studies in Cellulose

A cellulose effluent breakthrough curve is shown in Figure 7. A cellulose desorption curve gave similar adsorptivities of 0.04 g H_2O/g cellulose and 0.001 g ethanol/g cellulose at 80°C. Total alcohol water condensable concentrations obtained by desorption of activated carbon, followed by cellulose adsorption are given in Figure 8.

In the desorption of ethanol and water from carbon followed by cellulose adsorption of water (Fig. 8), initially there is a very high concentration of ethanol in the condensate. In this run the amount of cellulose was small compared to the amount of activated carbon in order to have enough vapors for a cellulose breakthrough curve. The breakthrough of water is at about 10 minutes and the rest of the curve is mainly a function of activated carbon desorption. The use of larger amounts of cellulose adsorbent would produce increased quantities of condensate with greater than 95% concentration of ethanol. The adsorptivity of ethanol was almost negligible as shown in Figure 7. Multiple units in series may be desirable so that saturated units can be regenerated before water breaks through in the downstream units, and so that a periodic-countercurrent type of contacting can be maintained (17).

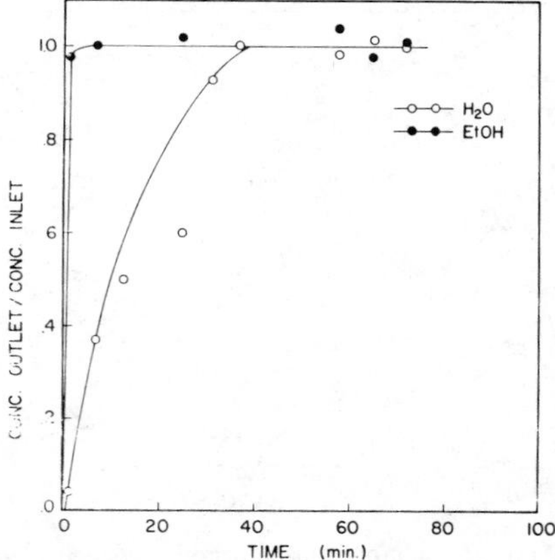

Figure 7. Cellulose Adsorption of EtOH and H_2O Vapors, Breakthrough Curve at 80°C, CO_2 Flow = 1.0 l/min., EtOH Vapor = 1.2 mol%, H_2O Vapor = 2.2 mol%, Cellulose Wt. = 16.7 g

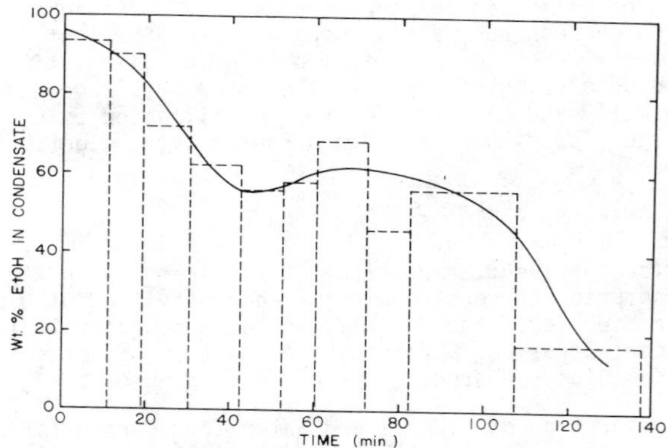

Figure 8. Desorption of Activated Carbon Followed by Adsorption on Cellulose and Condensation of the Effluent. Activated Carbon Saturated by CO_2 from 6% ethanol liquid, Wt. Activated Carbon=50g, Wt. Cellulose = 15.3 g, CO_2 Flow = 0.9 l/min., T = 80°C.

Ladisch and Dyck (11) obtained data on cornstarch adsorption that indicated approximately 0.07 g H_2O/g cornstarch could be adsorbed. It seems likely that cellulose should adsorb similar amounts under more suitable conditions. There are numerous water adsorbents available but their adsorptivity of ethanol, heats of desorption, and temperature requirements need further study.

PROPOSED FERMENTATION-SEPARATION PROCESS

Based on the fermentation and adsorption results, a proposed fermentation-separation system is shown in Figure 9. It contains: 1) two fermentors in series with CO_2 stripping, 2) activated carbon adsorbers to adsorb ethanol and some water, and 3) cellulose adsorbers to remove water from vapors desorbed from the activated carbon.

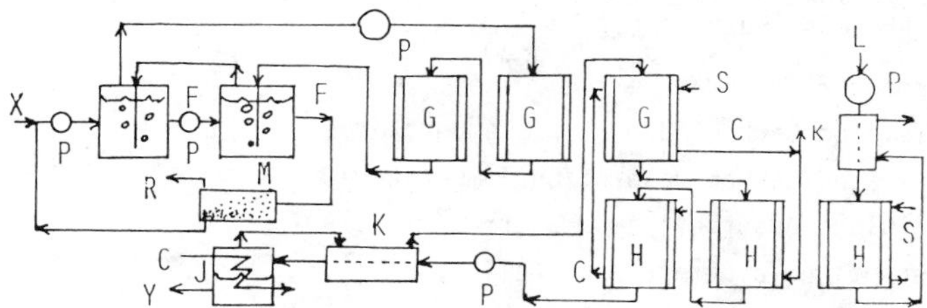

Figure 9. Proposed Fermentation-Stripping-Two-Stage Adsorption Process Legend see Figure 1.

Two stages of continuous fermentation are suggested. The first stage is kept at a high ethanol concentration which is more suitable for stripping and adsorption. The second stage is at a lower concentration of both ethanol and substrate in order to reduce losses with the effluent. Recycle of settled yeast would be desirable to increase yield and rates. It might also be desirable to supply small quantities of air or oxygen to increase cell growth and activity (4).

The CO_2, containing ethanol and water vapor from the fermentor, would be circulated to the activated carbon adsorbers at a temperature slightly higher than the fermentor, to the low concentration fermentor, and then through the high concentration (about 6% ethanol) fermentor. If some channeling occurred in adsorption, no ethanol would be lost because it would be essentially all recycled. For the stripping and adsorption the major energy consumption would be for circulating the CO_2 through the fermentors and activated carbon adsorber.

The desorption or regeneration would take place into circulated gas which would pass through the cellulose adsorber to a condenser. The heat of desorption of ethanol and water would be required for carbon desorption because the gas would be cooled to condense the ethanol product before recirculating. The temperature of desorption is important since appreciable amounts of heat would be required to heat the carbon adsorber to a high regeneration temperature, or to pump and heat the recirculating air if lower temperatures with lower ethanol vapor concentrations were used. Heat removal from cellulose adsorption is necessary, and could supply the majority of the activated carbon desorption heat requirement.

In the cellulose regeneration, the heat required for water desorption would be required, and the temperature level would probably be important.

Of course a number of other adsorbents for both ethanol and water are possible for use in this type of system and eventually other adsorbents should be investigated to determine adsorptivities at various temperatures.

ESTIMATES OF ENERGY REQUIREMENTS

The following assumptions were used to calculate energy requirements for the process shown in Figure 9.

Fermentor Temperature	40°C
Activated Carbon Adsorption Temperature	45°C
Activated Carbon Desorption Temperature	70°C
Cellulose Adsorber Temperature	80°C
Desorption Temperature of Cellulose	90°C
ΔT of Recovery Heat Exchangers on Recycled Gases	10°C
Overall Compressor Efficiency	75%
Compressor Drive Efficiency	95%
Power Generation Efficiency from High Pressure Steam	30%
Waste Heat Recoverable from Power Generation Steam	60%
Reflux Ratio for Distillation	1.5

Heat of Desorption of Components = Heat of Vaporization of Pure Liquids. Ratios of adsorbed component vapor pressures at different temperatures = ratios of pure component vapor pressures at the same temperatures.

On the basis of 1 g mole of ethanol it was estimated that 34 moles of CO_2 would be required for stripping ethanol from the fermentor, 8 moles of gas (air or CO_2) would be required for desorption of the activated carbon, and 1 mole of air would be needed for cellulose drying. For each g mole of ethanol 680 g of activated carbon would be used and regenerated, and 909 g of cellulose would be used and dried, based on experimental results.

Energy requirements (not including cooling water, pumping liquids, etc.) were calculated for the adsorption system shown in Figure 9. The major requirements are for compressing CO_2 an estimated 10 psi, and for desorption of the activated carbon and desorption or drying of the cellulose. Power generation for the 3 compressors is calculated to require as heat for power generation 306810 Joules (73,277 calories) per g mol of ethanol or 2,867 Btu/lb ethanol (1.053 lb at 95%). The waste heat portion of this will more than meet the heat requirements of the system if heat is transferred from water adsorption to carbon desorption. There is about 1200 J/g ethanol (500 Btu/lb) waste heat based on cellulose data and about 1900 J/g (800 Btu/lb) excess if estimates are made from the data of Ladisch and Dyck (11) on cornstarch. This is appreciably less than the requirement of approximately 7700 J/g (3,330 Btu/lb) for distillation given by Kirk and Othmer (18). Fermentation at higher temperatures or ethanol content could reduce the CO_2 requirements and balance the power-heat relationship.

A logical alternative is to go from activated carbon thermal desorption into a distillation column to concentrate the approximately 52% ethanol desorbed to 95%. Approximate calculations indicate that this could provide a system where the waste heat from the power generation from the compressor could supply all the required reboiler heat. The total heat required for power generation would be approximately 4735 J/g ethanol (2,040 Btu/lb). Such a system might require a higher reflux for distillation but this was not considered. It it were necessary to desorb and condense the adsorbed components prior to distillation, the total heat requirement would be about 7190 J/g (3,100 Btu/lb).

In comparison calculations on the same basis indicate that vacuum distillation of water, ethanol, and CO_2 from a 6% fermentation medium followed by condensation of water and ethanol and then distillation should require about 6153 J/g ethanol (2,651 Btu/lb).

In all the described calculations a 95% product is assumed. However one advantage of the use of adsorption for the final separation is that a product composition of over 95% can be obtained as shown by Ladisch and Dyck (11) and by this work. Perhaps the least consumption of energy for over 95% ethanol would be obtained by a combination of stripping, adsorption on activated carbon, distillation, and then adsorption on cornstarch, cellulose, or similar material to remove water.

The energy calculations do not include any consideration of the effects on the fermentor operation. The stripping method of ethanol removal can probably supply all the agitation needed in the fermentor. Also it may be conveniently used in multi-stage fermentors. The stripping plus 2-stage adsorption method, the stripping, adsorption, and distillation method, and the vacuum and distillation methods would all provide for the removal of ethanol from the fermentor, as it is formed, without disturbing the fermentation process. All of these factors affecting the fermentation would be important in considering alternate separation systems.

LITERATURE CITED

1. C. Bazua and C. R. Wilke, "Effect of Alcohol Concentration on Kinetics of Ethanol Production by *S. cerevisiae*", presented at First Chemical Congress of the North American Continent, Mexico City, Mexico, (December 1-5, 1976).

2. S. Aiba, M. Shoda, and M. Nagatave, *Biotechnol. Bioeng.*, 10, 845 (1968).

3. B. C. Boeckeler, U.S. Patent #2,440,925 (1948).

4. A. Ramalingham and R. K. Finn, *Biotechnol. Bioeng.*, 19, 583 (1977).

5. G. R. Cysewski and C. R. Wilke, *Biotechnol. Bioeng.*, 19, 1125 (1977).

6. T. K. Ghose and R. D. Tyagi, *Biotechnol. Bioeng.*, 21, 1387 (1979).

7. E. Maiorella and C. R. Wilke, Biotechnol. Bioeng., 22, 1749 (1980).

8. R. K. Finn and R. A. Boyajion, in Proceedings of the Fifth International Fermentation Symposium, Berlin, (1976).

9. Perry, R. H. and Chilton, C. H., Chem. Engineers Handbook, 5th Ed., 16-2, McGraw-Hill, N.Y., (1973).

10. Kipling, J. J., "Adsorption from Solutions of Non-Electrolytes", 76, Acad. Press, N.Y. (1965).

11. Ladisch, M. R., Dyck, K., Science, 205 898 (1979).

12. Walsh, P., M.S. Thesis, Chemical Engineering, University of Missouri-Rolla (1983).

13. Sumner, J. B., Somers, G. F., "Laboratory Experiments in Biological Chemistry, Acad. Press, N.Y. (1944).

14. Miller, G. L., Analytical Chem., 31, No. 3, 427 (1959).

15. Liu, C. P., M.S. Thesis, Chemical Engineering, University of Missouri-Rolla (1983).

16. Okazaki, M., Tamon, H., and Toel, R., J. of Ch.E. of Japan, 11, No. 3, 209 (1978).

17. Liapis, A. I., and Rippin, D. W. T., AIChE J., 25, No. 3, 455 (May, 1979).

18. Kirk, R. E. and Othmer, O. F., Encyclopedia of Chemical Technology, 1, 270, Wiley, N.Y. (1947).

MODELING OF ADSORPTION AND MASS TRANSPORT PROCESSES IN FIXED-BED ADSORBERS

Walter J. Weber, Jr.
Professor of Civil and Environmental Engineering
and Chairman, University Program in Water Resources
The University of Michigan, Ann Arbor, MI 48109

ABSTRACT

Rate and mass transport processes associated with adsorption from liquid phase by porous adsorbents are described. Parameters affecting these processes are characterized, and methods for their quantification discussed. Interrelationships among critical rate and mass transport steps and the dependence of these interrelationships on system variables are considered.

INTRODUCTION

Adsorption and ion exchange processes have a broad range of potential industrial and environmental applications. Given the diversity of conditions represented by these applications, and the variety of adsorbents available, it is virtually impossible to generalize or extrapolate system design parameters from one situation to another. It is thus usually necessary to independently determine process applicability and specific design criteria for each case to ensure successful and cost effective application. This paper considers the use of modeling techniques as aids to design and operation of fixed-bed adsorbers. The purpose is to demonstrate that, although performance can be specific to each application, predictive mathematical models can prove valuable in reducing the uncertainty otherwise inherent to the translation of bench or pilot scale measurements to full scale design.

FIXED-BED BREAKTHROUGH CURVES

One critical aspect of the design of fixed-bed adsorption systems involves characterization of the effluent concentration profile as a function of throughput (i.e., volume processed or time of operation). This profile, commonly referred to as the breakthrough curve, represents the specific combination of equilibrium and rate factors that control process performance in a particular application. The breakthrough curve may be visualized as a trace generated by a hypothetical "mass transfer zone" which moves through the bed as a function of mass loading and eventual exhaustion of the adsorbent (1). This is illustrated in Figure 1 by the schematic representation of idealized concentration/throughput patterns at different bed depths in a fixed-bed reactor.

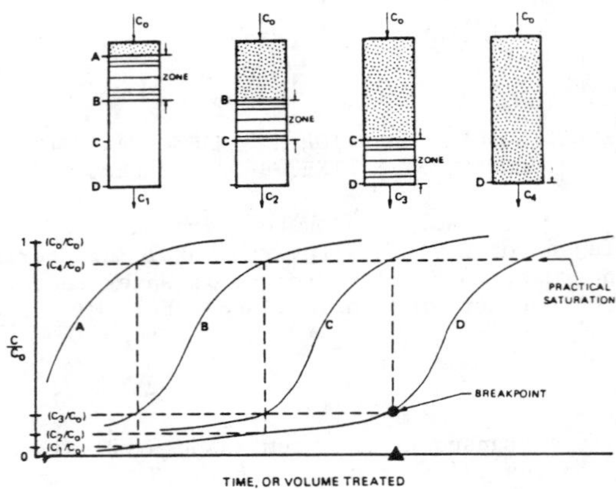

Figure 1. Idealized Breakthrough Curves for a Fixed-Bed Reactor (adapted from Weber, Reference 1)

The effluent breakthrough curve D shown in Figure 1 is the concentration/throughput pattern or profile corresponding to the bed discharge, the "breakpoint" is that point where the effluent concentration no longer meets process objectives, and the point of "practical saturation" is that where the adsorbent is ready for regeneration and beyond which such regeneration is more difficult or less efficient. This conceptualization is useful for defining the boundaries of that region of the reactor in which the majority of removal occurs at any given time, a "zone" which moves through the bed at a velocity and with a length specific to a particular application (2). For a given set of operating conditions, the velocity of the mass transfer front correlates the breakpoint with bed depth (since the breakpoint occurs when the effective zone begins to exit the bed) while its length represents the minimum contact period required to reach the process objectives (since the objectives are met at the end of the zone).

The position and shape of a breakthrough profile depend on several factors, including: the physical and chemical properties of both the sorbate and sorbent, the influent concentration of sorbate, solution pH, particular rate-limiting mechanism, nature of the equilibrium conditions, depth of column, and velocity of flow. The relative effects of these factors are highly specific to a particular application, which constitutes one of the major difficulties for the design of such processes. Mathematical models properly calibrated to a specific system facilitate analysis and evaluation of the effects of design and control variables on adsorber performance for that application.

MODELING OF FIXED-BED ADSORBERS

The first step in developing a mathematical model for description or prediction of adsorption dynamics is to provide a representation of associated rate phenomena. In general, the rate of uptake of a solute by a porous adsorbent is controlled by a resistance to mass transport rather than by a reaction velocity. Figure 2 is a schematic representation of a porous adsorbent comprised of "macropores" and "micropores" and surrounded by a boundary layer or film depicted by the dashed line (3). The macropores are essentially large enough that diffusion in them is unhindered by the pore walls, while the micropores have radii comparable to the size of diffusing species so that diffusion is hindered by the pore walls. This conceptual view translates into the series resistance rate model shown schematically in Figure 2b (4). The overall rate of adsorption is controlled by the step(s) providing the greatest resistance to mass transport.

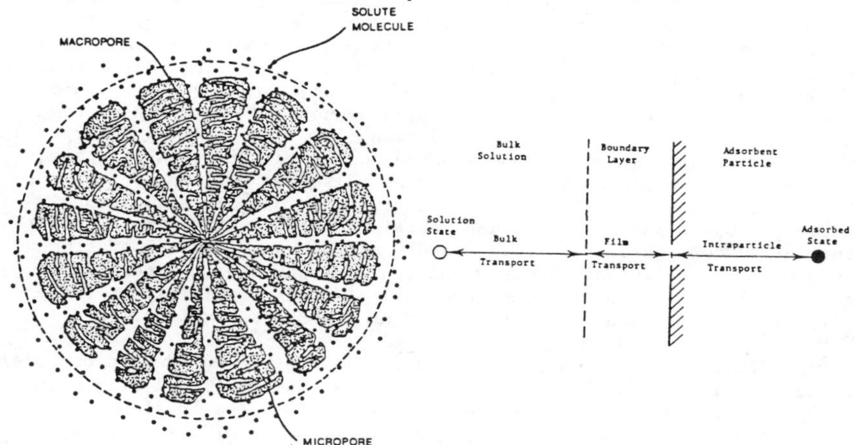

Figure 2a. Conceptual Representation of a Porous Adsorbent Particle (after Weber and Liang, Reference 3)

Figure 2b. Schematic Representation of the Series Resistance Rate Model for Adsorption (after Weber, Reference 4)

A second important aspect of any model is the thermodynamic principle of conservation of mass, or mass continuity. Implementation of this principle results in a material balance for each component of interest within a particular control volume; that is:

$$\begin{bmatrix}\text{NET RATE OF}\\ \text{ACCUMULATION}\\ \text{OR DEPLETION}\end{bmatrix} = \begin{bmatrix}\text{FLUX}\\ \text{IN}\end{bmatrix} - \begin{bmatrix}\text{FLUX}\\ \text{OUT}\end{bmatrix} \pm \begin{bmatrix}\text{NET RATE OF GENERATION}\\ \text{OR REMOVAL BY REACTION}\end{bmatrix} \qquad (1)$$

where the flux of a component entering or leaving the control volume can include bulk flow, dispersive flow, and molecular diffusion.

The normal procedure for applying the continuity principle is to consider a small but finite control volume representative of the overall system. After the appropriate relationships describing the localized mass balance are developed, the control volume is assumed to become infinitesimally small so as to generate the continuity relations on a differential scale. Since development of a usable form for any model requires integration of the differential equations comprising that model, boundary conditions and initial conditions are also required to specify the constants of integration. Thus, the complete model consists of partial differential equations for continuity of each component in each phase as well as the specific boundary and initial conditions corresponding to the particular application. A method for performing the required integration is also required.

It is important to consider the nature of the reaction referred to in Equation 1. If the phase change is assumed to occur over the entire control volume, the reaction is termed "homogeneous" and appears as a source/sink term in the continuity equation. If the phase change is assumed to occur within a restricted region of the control volume, however, the reaction is termed "heterogeneous" and appears as a boundary condition to the continuity equation (5). For modeling adsorption processes an assumption of monodisperse solid phase results in a homogeneous reaction and an assumption of polydisperse solid phase results in a heterogeneous reaction. These definitions of homogeneous and heterogeneous types of adsorption reactions should not be confused with the terms "homogeneous diffusion" and "heterogeneous diffusion" commonly used to differentiate between intraparticle transport models which assume solid diffusion and combined pore and surface diffusion, respectively, for migration of solute within an adsorbent particle.

A third important aspect of the development of an adsorption model is description of the equilibrium behavior or maximum level of adsorption possible for a particular adsorbent/adsorbate combination as a function of the residual liquid phase concentration of the adsorbate. Many relationships have been proposed to represent phase distributions governed by adsorption equilibria, including the Gibbs, Freundlich, Langmuir, Dubinin, and Brunauer-Emmett-Teller equations to cite a few (1). Consideration of the advantages and disadvantages of these several isotherm equations lies outside the scope of this paper. Perhaps the most generally applicable and widely used is the semi-empirical Freundlich isotherm equation:

$$q_e = K_F \, C_e^{\,n} \tag{2}$$

where q_e is the equilibrium amount of solute in the solid phase per unit weight of adsorbent (e.g., mole/kg) and C_e is the solution concentration remaining at equilibrium (e.g., mole/l). The coefficients K_F and n are characteristic parameters relating to the specific equilibrium adsorption capacity and intensity, respectively. Because a variety of system-specific factors influence the magnitude of K_F and n, particularly the adsorbent type and the presence of competing species, it is normally necessary to carry out laboratory-scale batch evaluations for determination of these parameters.

Space does not permit an exhaustive review of the general literature regarding adsorption and ion exchange models predicated on the above principles. For purposes of illustration, three specific approaches involving different governing mass transport phenomena will be described, namely the film, film/particle, and film/dual-particle, resistance models.

The Film Resistance Model

When mass transport is controlled only by the resistance of the hydrodynamic boundary layer pictured in Figure 2, the "film" model can be used to define the overall rate of the adsorption process. In this approach, shown schematically in Figure 3, the liquid phase and solid phases are considered internally uniform and separated by a boundary layer or film. Since the solid phase is considered monodisperse, the adsorption reaction is of the homogeneous type and enters the model as a sink/source term in the continuity relationship. This technique has its roots in early attempts at dynamic modeling of ion exchange and adsorption systems (2).

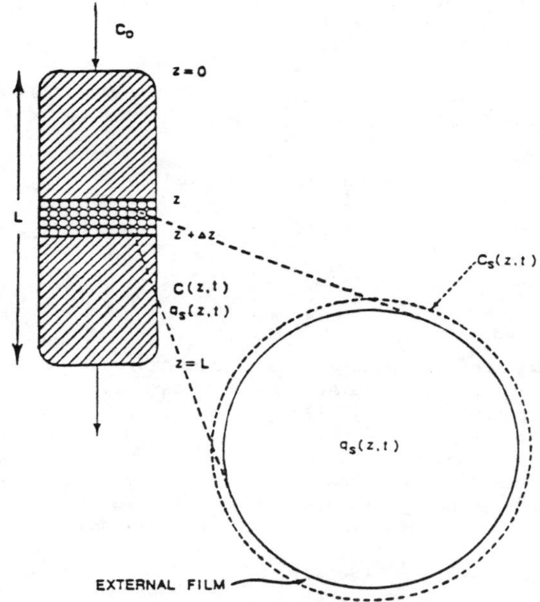

Figure 3. Schematic Representation of the Film Model (adapted from Weber and Liang, Reference 3)

The general version of the film model for a fixed-bed reactor consists of the following continuity relations for the liquid phase bulk concentration, C, and the solid phase surface concentration q_s:

$$\frac{\partial C}{\partial t} = - v_z \frac{\partial C}{\partial z} + D_d \frac{\partial^2 C}{\partial z^2} - \rho_p \frac{(1-\varepsilon)}{\varepsilon} \frac{\partial q_s}{\partial t} \qquad (3)$$

and,

$$\frac{\partial q_s}{\partial t} = \frac{k_f \alpha (C-C_s)}{\rho_p} \qquad (4)$$

Where $C(z,t)$ and $q_s(z,t)$ are the respective liquid phase and particle surface concentrations at depth z and time t, C_s is the equilibrium liquid phase concentration corresponding to a solid phase concentration of q_s, k_f is the specific column-based film transfer coefficient (e.g., cm/sec), and α represents the external surface area of the particle per unit volume (e.g., cm^2/cm^3). The other more generalized constants are defined in Appendix I. Associated with these continuity relationships are boundary conditions of the following form for a fixed-bed adsorber:

$$C = C_o + \frac{D_d}{v_z}\frac{\partial C}{\partial z} \qquad \text{at } z = 0 \qquad (5)$$

and,

$$\frac{\partial C}{\partial z} = 0 \qquad \text{at } z = L \qquad (6)$$

In addition, the initial conditions are specified to be:

$$C = 0 \qquad \text{at } t = 0 \qquad (7)$$

and,

$$q_s = 0 \qquad \text{at } t = 0 \qquad (8)$$

The above equations represent the set of relationships that can be integrated simultaneously to generate the breakthrough curves representing mass transport controlled by film resistance alone.

Successful application of the model requires proper calibration of the specific film transfer coefficient. If the rate is truly controlled by film transport, k_f can be estimated by one of several literature correlations, such as that of Wilson and Geankopolis (6):

$$k_f = 1.09 \frac{D_\ell}{\varepsilon\, d_p} (Sc\ Re)^{1/3} \qquad (9)$$

Where Sc is the dimensionless Schmidt number ($\mu/\rho_\ell D_\ell$), Re is the dimensionless Reynolds number ($d_p v_z \rho_\ell/\mu$), and the other constants are as defined in Appendix I. An alternative calibration method which can be used to empirically account for deviations from strictly film-controlled rate phenomena involves fitting the model to data obtained from a pilot column operating with similar hydrodynamic conditions and using regression analysis to solve for k_f.

The single-resistance film model has been applied successfully for describing ion exchange processes. It also has been used with moderate success for description of activated carbon adsorption of the general class represented by weakly adsorbed solutes, but has proven less satisfactory for systems involving strongly adsorbed substances.

The Film/Particle Diffusion Model

When mass transport is controlled by a combination of film resistance and single (e.g., macroporous) intraparticle resistance, the "film/particle" diffusion model can be used to describe the overall rate of adsorption (7-10). In this approach, shown schematically in Figure 4, the liquid phase is again considered uniform and the external film is again present. However, the particle is now assumed to have a radial (macropore-based) concentration distribution. Because the solid phase is no longer considered uniform, the adsorptive reaction becomes one of the heterogeneous type and enters the model as a boundary condition, although the intraparticle transport component can be described by a homogeneous diffusion model.

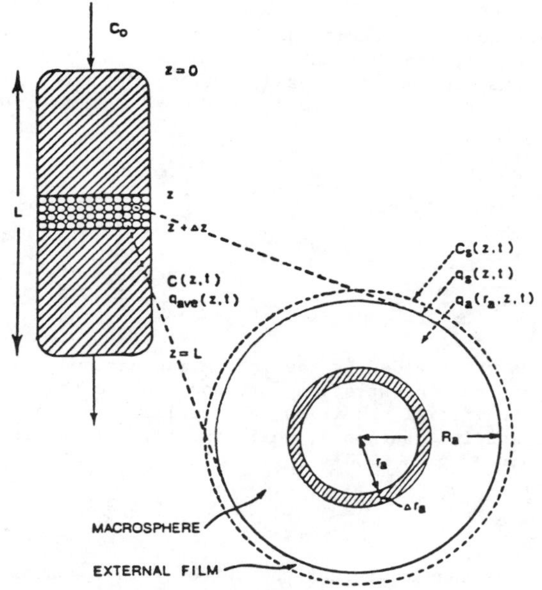

Figure 4. Schematic Representation of Film/Particle Diffusion Model (adapted from Weber and Liang, Reference 3)

The general version of the film/particle diffusion model for a fixed-bed adsorber consists of a liquid phase continuity relationship similar to that of the film model, and a solid phase continuity relationship that reflects Fick's second law of diffusion; for spherical coordinates:

$$\frac{\partial C}{\partial t} = - v_z \frac{\partial C}{\partial z} + D_d \frac{\partial^2 C}{\partial z^2} - \rho_p \frac{(1-\varepsilon)}{\varepsilon} \frac{\partial q_{ave}}{\partial t} \tag{10}$$

and,

$$\frac{\partial q_a}{\partial t} = D_{s,a} \frac{1}{r_a^2} \frac{\partial}{\partial r_a} (r_a^2 \frac{\partial q_a}{\partial r_a}) \tag{11}$$

Here $q_a(r_a,z,t)$ represents the solid phase concentration distribution, and $D_{s,a}$ is the specific surface diffusivity of the solute (e.g., cm^2/sec) in the solid phase. In addition, $q_{ave}(z,t)$ is taken to be the volumetric average over the particle of radius R_a:

$$q_{ave} = \frac{3}{R_a^3} \int_0^{R_a} q_a r_a^2 dr_a \tag{12}$$

The corresponding boundary conditions for the liquid phase are the same as for the film model, namely Equations 5 and 6. The solid phase boundary conditions, on the other hand, reflect the heterogeneous surface-based adsorption reaction and the radial symmetry of the solid phase concentration:

$$k_f (C-C_s) = D_{s,a} \rho_p \frac{\partial q_a}{\partial r_a} \qquad \text{at } r_a = R_a \tag{13}$$

and,

$$\frac{\partial q_a}{\partial r_a} = 0 \qquad \text{at } r_a = 0 \tag{14}$$

In this model, C_s is provided by the isotherm equation (e.g., Eq. 2) used at the macropore concentration corresponding to the surface of the particle. The initial condition for the liquid phase is the same as for the film model, Equation 7, while the initial condition for the solid phase is given by:

$$q_a = 0 \qquad \text{at } t = 0 \tag{15}$$

The above equations thus represent the set of relationships that can be integrated simultaneously to generate the breakthrough curves representing mass transport controlled by the combination of film and a single (macropore) intraparticle resistance.

Application of this model requires the proper calibration of two transport parameters; the film transfer coefficient k_f and the specific intraparticle diffusivity, $D_{s,a}$ which is normally assumed to relate to diffusional transport along the macropore surfaces. This has traditionally been done by subjecting rate data from completely-mixed batch reactor (CMBR) studies to a statistical parameter search to estimate the k_f' (i.e., k_f for CMBR conditions) and the $D_{s,a}$ values. Because the hydrodynamic characteristics of a fixed-bed reactor are significantly different, however, correlation techniques such as Equation 9 are still required to convert k_f' into k_f. An alternative calibration approach was developed by Weber and Liu (11) using specially designed bench-scale "microcolumns" for more accurate and reliable simultaneous determination of both k_f and $D_{s,a}$. The method of Weber and Liu has the further advantage that it inherently includes any miscellaneous particle and solution factors that may be operative but otherwise difficult to describe or correlate in a general way.

Figure 5 presents results that illustrate the general performance of the film/particle diffusion model for predicting p-chlorophenol (PCP) adsorption by several different adsorbents in fixed-bed reactors (12). Good agreement between predicted and measured breakthrough profiles is observed for C/C_o values up to 0.5 and greater in all cases, although some deviation may be noted for several systems at higher relative concentrations. The lattery observation suggests that a mechanism other than either film or macropore transport might control at higher saturation levels.

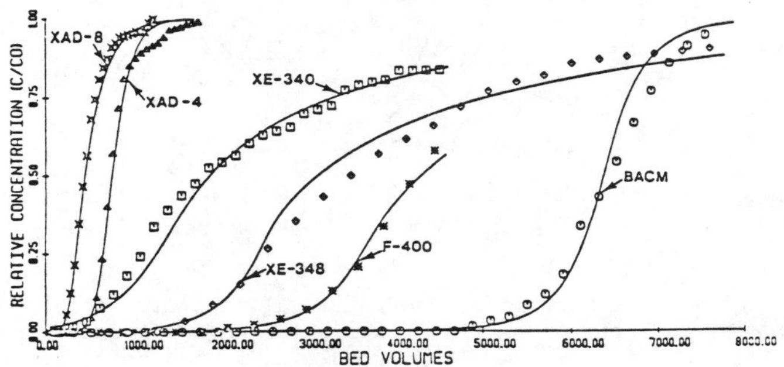

Figure 5. Film/Particle Diffusion Model Predictions and Experimental Data for p-Chlorophenol in Fixed-Bed Reactors of Different Adsorbents (after Van Vliet and Weber, Reference 12)

Film/Dual-Particle Diffusion Model

The "film/dual-particle" diffusion model (3) provides good description of overall rates of adsorption for systems in which mass transport involves a combination of film resistance, macroporous intraparticle resistance, and microporous intraparticle resistance. This approach, shown schematically in Figure 6, is similar to the film/particle diffusion model in that the liquid phase remains uniform, the external film remains present, and the particle macropore concentration is again considered to be radially distributed. However, the particle micropore is now assumed to also have a localized radial concentration distribution. Because the adsorptive reaction remains heterogeneous, it again enters the model as a boundary condition.

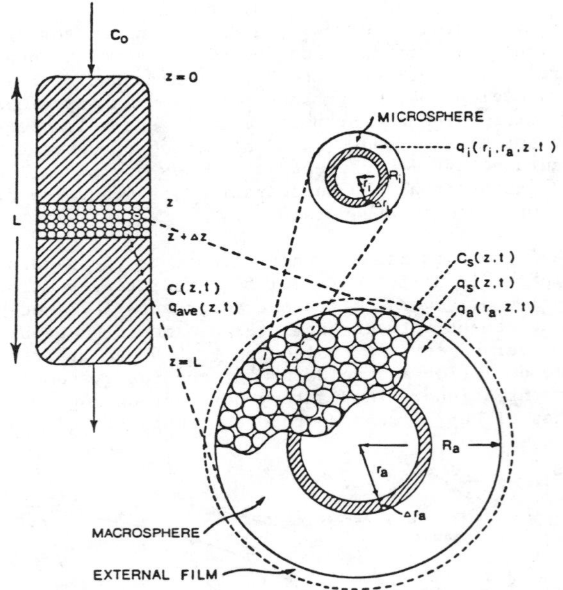

Figure 6. Schematic Representation of the Film/Dual-Particle Diffusion Model (after Weber and Liang, Reference 3)

The general version of the film-dual-particle diffusion model for a fixed-bed adsorber consists of a liquid phase continuity relationship identical to that for the film/particle diffusion model, Equation 10. However, the macropore continuity relationship now employs a sink term to account for the homogeneous "reaction" where the solute changes from macropore phase, $q_a(r_a,a,t)$, to micropore phase, $q_i(r_i,r_a,z,t)$:

$$\frac{\partial q_a}{\partial t} = D_{s,a} \frac{1}{r_a^2} \frac{\partial}{\partial r_a} \left(r_a^2 \frac{\partial q_a}{\partial r_a}\right) - \left(\frac{1-f}{f}\right) \frac{\partial q_{i,ave}}{\partial t} \tag{16}$$

Here f is considered to be the fraction of total solid phase capacity represented by the macropore surface, while $q_{i,ave}$ is determined in a manner similar to that given in Equation 12:

$$q_{i,ave} = \frac{3}{R_i^3} \int_0^{R_i} q_i \, r_i^2 \, dr_i \tag{17}$$

The micropore solid phase continuity relationship is selected to reflect Fick's second law of diffusion; for spherical coordinates, similar to Equation 11:

$$\frac{\partial q_i}{\partial t} = D_{s,i} \frac{1}{r_i^2} \frac{\partial}{\partial r_i} (r_i^2 \frac{\partial q_i}{\partial r_i}) \tag{18}$$

The corresponding boundary conditions for the liquid phase are still the same as for the film model, namely Equations 5 and 6, while the macropore phase boundary conditions again reflect the heterogeneous surface-based adsorptive reaction and the radial symmetry of the macropore concentration distribution.

Although Equation 14 is again applicable for the symmetry condition, the surface reaction condition is slightly modified to reflect the previously defined macropore capacity fraction parameter:

$$k_f (C-C_s) = f D_{s,a} \rho_p \frac{\partial q_a}{\partial r_a} \quad \text{at } r_a = R_a \tag{19}$$

For the micropore phase boundary conditions, the continuous nature of the macropore/micropore interface and the radial symmetry of the micropore concentration distribution are utilized:

$$q_i = q_a \quad \text{at } r_i = R_i \tag{20}$$

and,

$$\frac{\partial q_i}{\partial r_i} = 0 \quad \text{at } r_i = 0 \tag{21}$$

The initial conditions corresponding to a fixed-bed adsorber also remain similar to the previous models, with Equation 7 applying to the liquid phase and Equation 15 to the macropore phase. In addition, the initial condition for the micropore solid phase is:

$$q_i = 0 \quad \text{at } t = 0 \tag{22}$$

The above equations represent the set of relationships that can be integrated simultaneously to generate the breakthrough curves representing mass transport controlled by the combination of film, macropore intraparticle, and micropore intraparticle resistances.

Application of this model requires calibration of five distinct and specific parameters: the film transfer coefficient (k_f), the macropore and micropore surface diffusivities ($D_{s,a}$ and $D_{s,i}$ respectively), the macropore capacity fraction (f), and the micropore radius (R_i). Although this is a significantly more complex task than for the film and film/particle diffusion models, Weber and Liang (3) were able to extend the microcolumn technique of Weber and Liu (11) to simultaneously determine values for these parameters based on bench-scale data.

Figure 7 presents results that demonstrate the general performance of the film/dual-particle diffusion model for describing the PCP solute and BACM adsorbent system for which film/particle diffusion model projections were given earlier in Figure 5. Comparison of Figures 5 and 7 illustrate that the three-resistance model provides a better breakthrough profile prediction than the dual-resistance model for this case, as it has been shown to do for several other systems as well (3).

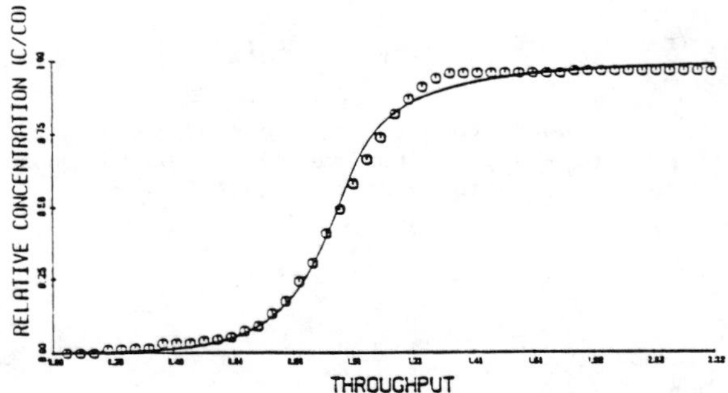

Figure 7. Film/Dual-Particle Diffusion Model Predictions and Experimental Data for Adsorption of p-Chlorophenol by BACM Activated Carbon in a Fixed-Bed Reactor (after Weber and Liang, Reference 3)

SUMMARY

Increasingly sophisticated models capable of predicting breakthrough curves for a variety of applications of adsorption in fixed-bed reactors have evolved over the years. While further developments are required to achieve universal predictive capability, it is evident that the technique of modeling is sufficiently refined to play a useful role in translating bench-scale measurements of certain critical sorption properties for a given sorbent/sorbate/solution combination into forecasts of fixed-bed adsorber dynamics, analysis of performance sensitivity to system variables, and identification of critical design and operational criteria.

APPENDIX I - NOTATION

The general symbols used in this paper and not defined elsewhere include the following:

C_o = Initial liquid-phase feed concentration (e.g., moles/lit)

D_p = diameter of adsorbent particle (e.g., cm)

D_d = liquid-phase dispersion coefficient (e.g., cm^2/sec)

D_ℓ = liquid-phase diffusivity (e.g., cm^2/sec)

v_z = superficial liquid-phase velocity (e.g., cm/sec)

ε = liquid-phase volume fraction (dimensionless)

ρ_ℓ = liquid density (e.g., gm/cm^3)

ρ = apparent particle density (gm/cm^3)

μ = liquid viscosity (e.g., gm/cm sec)

LITERATURE CITED

1. Weber, W.J., Jr., Physicochemical Processes for Water Quality Control, Wiley-Interscience, John Wiley & Sons, Inc., New York, NY (1972).

2. Michaels, A.S., Ind. & Eng. Chem., 44, 1922 (1952).

3. Weber, W.J., Jr., and S. Liang, Environmental Progress, 2, 3, 167 (1983).

4. Weber, W.J., Jr., "Concepts and Principles of Carbon Applications in Wastewater Treatment," Proc. International Conference on Applications of Adsorption to Wastewater Treatment, (W.W. Eckenfelder, Jr., Ed.) pp. 5-28 CBI Publishing Co. Inc., Boston, MA, 1981.

5. Bird, R.B., W.E. Stewart, and E.N. Lightfoot, Transport Phenomena, John Wiley & Sons, Inc., New York, NY (1960).

6. Wilson, E.J., and C.J. Geankopolis, Ind. & Eng. Chem. Fund., 5, 9 (1966).

7. Rosen, J.B., J. Chem. Phys., 20, 387 (1952).

8. Tien, C. and G. Thodos, AIChE J., 11, 945 (1965).

9. Weber, W.J., Jr., and J.C. Crittenden, Jour. Water Poll. Cont. Fed., 47, (5) 924 (1975).

10. Crittenden, J.C. and W.J. Weber, Jr., Jour. Env. Eng. Div., Amer. Soc. Civ. Engrs., 104, (EE6), 1175 (1978).

11. Weber, W.J., Jr., and K.T. Liu, Chem. Engrg. Comm., 6, 49 (1980).

12. Van Vliet, B.M. and W.J. Weber, Jr., Jour. Water Poll. Cont. Fed., 53 (11), 1585 (1981).

KINETICS OF ADSORPTION ON ACTIVATED CARBON IN FIXED BEDS

Franz Josef Weissenhorn
Veehstrasse 43, D-4000 Duesseldorf 1, GERMANY

ABSTRACT

Capacity and adsorption rate are the deciding quantities in fixed bed adsorption. Together they constitute the basis for the selection of suitable adsorbents, the modelling of technical plants and their economical operation. Extensive investigation has shown that the capacity of the adsorbents can be described according to FREUNDLICH, and the adsorption rate calculated by a heterogeneous-kinetic arrangement (1,2,3,4,5,6). For the development of the mathematical basis of fixed bed adsorbers with different adsorption systems, measurings by laboratory, pilot and technical standards have been analyzed. The object of the investigation was to find out the general theoretical basis of fixed bed adsorption and especially that of adsorption on activated carbon in drinking water treatment, because the dispositions made so far, which declared different kinds of diffusion responsible the rate of adsorption in drinking water treatment in the region of the Lower Rhine had not produced satisfactory results (7,8,9). The essential result of these investigations was the development of a system of equations that is applicable for practical operation, too, and that shows the influence of the substantial parameters important for fixed bed adsorption enabling also the demands for optimizing the fixed bed adsorption to be met to all intents. The basis of this system of equations is the saturation function for the loading of the adsorbent depending on time. This function has stood the test not only with regard to the fixed bed adsorption with single-component solutions under constant conditions, but under always changing conditions and with multi-component-solutions of changing composition and concentration, too, it proved extraordinary stable. The ability to quantify the processes of adsorption on actvated carbon in fixed bed rendered possible the optimization of thermal reactivation of spent activated carbon which had been used in drinking water treatment (10).

DERIVATION OF THE EQUATIONS OF FIXED BED ADSORPTION

Single Solutions

The Mass-Transfer-Zone will be defined first of all for the derivation of the equations of the fixed bed adsorption. In this range of the fixed bed the size of which depends on the adsorption system and the process conditions at the time $t = 0$ the adsorptive concentration decreases from $c = c_0$ at the beginning of the Mass-Transfer-Zone to $c = 0$ at the end of it. With the adsorbent mass m_0 and the bed depth z_0 of the Mass-Transfer-Zone result the initial and boundary conditions (1):

$$0 \leq m/m_0 \leq 1, \; 1 \geq c/c_0 \geq 0, \; 0 \leq z/z_0 \leq 1 \qquad (1)$$

For the range of the fixed bed adsorber exceeding the Mass-Transfer-Zone the equations must be slightly modified, as will be shown later on. The capacity q_0 of the adsorbent is connected with the adsorptive concentration c_0 by the FREUNDLICH isotherm (2):

$$q_0 = k_F \, c_0^n \tag{2}$$

This is the result of many measurements of fixed bed adsorption. That means that loadings which can be arithmetically reached, can be examined by batch experiments and vice-versa.
Similar to the derivation of the LANGMUIR isotherm with monomolecular occupation of energetically homogeneous nonporous adsorbents, one can make the disposition for the rate of adsorption of porous adsorbents in the Mass-Transfer-Zone of fixed bed adsorption whereupon the rate is proportional to the capacity still available:

$$\frac{dq}{dt} = k_i \, (q_0 - q(t)) \tag{3}$$

The integration of Equation (3) does not include the parameters of the local equilibrium isotherm.
With the saturation function (4):

$$\frac{q(t)}{q_0} = 1 - \exp\,(-\,k_i \, t) \tag{4}$$

there results the Differential Equation (5):

$$\frac{dq}{dt} = k_i \, q_0 \, \exp\,(-\,k_i \, t) \tag{5}$$

that is needed later in this form for the calculation of breakthrough curves.
The adsorption constant k_i is in addition to the capacity q_0 the second characteristicconstant for the fixed bed adsorption, and it must be evaluated experimentally from breakthrough curves, because in batch tests entirely different kinetics result (2,6). According to the Transition State Theory results the Equation (6):

$$k_i = \frac{kT}{h} \exp\,(-\,\Delta G_A/RT) \tag{6}$$

That means that the adsorption constant depends exponentially on the Free Activation Energy of adsorption, ΔG_A. In the Equation (7), according to GIBBS:

$$\Delta G_A = \Delta H_A - T\,\Delta S_A \tag{7}$$

the Activation Entropie ΔS_A must be calculated with a negative sign, because the adsorptive molecules gain order during the transition state, it is particularly relevant for the size of the adsorption constant. Activation energies of about 90 - 100 kJ/mol were found for different adsorption systems. Batch tests with decreasing adsorptive concentration show that the size of the adsorption constant depends on the order of the adsorption reaction (2).The quantity of the adsorption constant is responsible for the steepness of the breakthrough curves. It decreases with increasing size of the adsorbent granules, and it is obviously depending also of the ratio of volume per unit of time/adsorbent mass. Specific researches are planned for the explanation of these circum-

stances.
For the mass balance in the Mass Transfer Zone the Differential Arrangement (8) is valid:

$$\frac{dq}{dt} = \frac{\dot{V}\, c_o}{m} \left(1 - \frac{c(t)}{c_o}\right) \tag{8}$$

which can be combined with the Differential Equation (5). Thereupon one receives the Equation (9) for the breakthrough curve:

$$\frac{c(t)}{c_o} = 1 - \frac{k_i\, m\, q_o}{\dot{V}\, c_o} \exp\,(-\, k_i\, t) \tag{9}$$

By inverting Equation (9) for the efficiency of the adsorber one gets the subsequent Equation (10):

$$\left(1 - \frac{c(t)}{c_o}\right) = \frac{k_i\, m\, q_o}{\dot{V}\, c_o} \exp\,(-\, k_i\, t) \tag{10}$$

With the abbreviations:

$$\left(1 - \frac{c(t)}{c_o}\right) = \eta \tag{11}$$

and

$$\frac{k_i\, m\, q_o}{\dot{V}\, c_o} = \eta_o \leq 1;\; t_o = 0 \tag{12}$$

the Regression Equation (13) results from Equation (10):

$$\ln \eta = \ln \eta_o - k_i\, t \tag{13}$$

which also can be applied beyond the Mass-Transfer-Zone with delayed breakthrough and the condition (14):

$$\frac{k_i\, m\, q(0)}{\dot{V}\, c_o} = \eta_o > 1;\; t_o > 0 \tag{14}$$

for the evaluation of breakthrough curves.
The loading in the Mass-Transfer-Zone, dependent on concentration, does not correspond to the Equation of LANGMUIR nor to that of FREUNDLICH, but to the linear Equation (15) which results from the combination of the Equations (4) and (9):

$$\frac{q(t)}{q_o} = \left(1 - \frac{1}{\eta_o}\right) + \frac{1}{\eta_o}\,\frac{c(t)}{c_o};\quad \eta_o < 1$$

$$\frac{q(t)}{q_o} = \frac{c(t)}{c_o};\qquad \eta_o = 1 \tag{15}$$

The graph of Equation (15) is shown in Figure 1.
Frequently so-called "sigmoidal" curves have been observed for delayed breakthrough with the condition (14) for the representation of breakthrough curves according to Equation (9), which are distinguished by deviations caused by precocious breakthrough from the calculated straight line according to the Regression Equation (13). These deviations at fixed bed adsorption are caused by longitudinal diffusion. They can be described by the Differential Equation (16):

$$\frac{\partial c}{\partial t} = D\, \frac{\partial^2 c}{\partial z^2} - v\, \frac{\partial c}{\partial z} \tag{16}$$

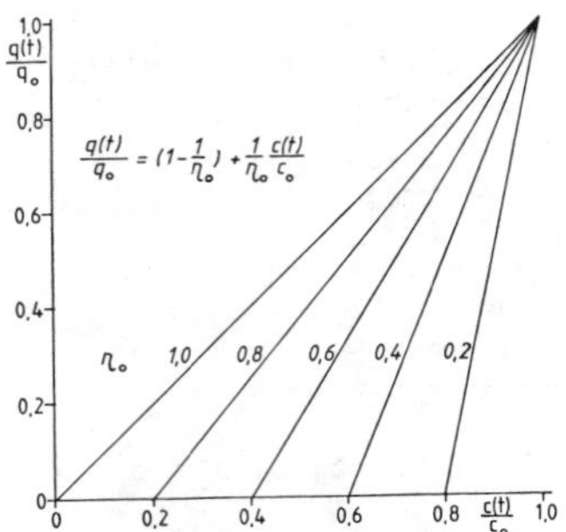

Figure 1. Loading and concentration in the Mass-Transfer-Zone.

with the longitudinal diffusion coefficient D and the front rate v of the adsorption front. The analytical solution for the Differential Equation (16) is:

$$\frac{c(t,z)}{c_o} = \frac{1}{2}\left(1 - \text{erf}\left(\frac{z - v\,t}{2\sqrt{D\,t}}\right)\right) \qquad (17)$$

with the initial and boundary conditions (11,12):

$$t = 0 \qquad z > 0 \qquad \frac{c(0,z)}{c_o} = 0$$

$$t > 0 \qquad z = 0 \qquad \frac{c(t,0)}{c_o} = 1 \qquad (18)$$

$$t = t_{1/2} \qquad z > 0 \qquad \frac{c(t_{1/2},z)}{c_o} = 1/2$$

The precocious breakthrough caused by longitudinal diffusion can amount a great time and reduce considerably the efficiency and economy of fixed bed adsorption. According to the submitted experience, the influence of the longitudinal diffusion is not essentially diminished by faster flow of the adsorptive solution, but is favoured very much by higher front speed v due to higher influent concentration c_o.

ACTIVATED CARBON FIXED BED ADSORPTION AT DRINKING WATER TREATMENT

Multicomponent Solutions

Activated carbon fixed bed adsorption is used with great success at drinkingwater treatment for years (13). The temporal requirement for activated carbon m/t results theoretically from the volume per unit of time of the treated water $\dot{V}$, the influent concentration c_o and the capacity q_o of the adsorbent according to Equation (19):

$$\frac{m}{t} = \frac{\dot{V}\,c_o}{q_o} \qquad (19)$$

In practical water treatment operation the capacity of activated carbons used in fixed bed adsorption cannot be fully utilized on account of safety reasons and because of the tedious breakthrough curves. Therefore a greater temporal demand results as calculated by Equation (19). Under the conditions of drinkingwater treatment where a complete separation of organics is not necessary, one calculates a medium efficiency that results by integration Equation (13):

$$\bar{\eta} = \frac{\eta_o}{k_i\ t} (1 - \exp(-k_i\ t));\ t > 0 \qquad (20)$$

for the conditions (12), and for delayed breakthrough with the conditions (14) one finds Equation (21):

$$\bar{\eta} = \frac{1}{k_i\ (t - t_o)} (1 - \exp(-k_i\ (t - t_o))) \qquad (21)$$

$$t > t_o$$

The measurements of the total organics charge in water is made by the general parameters DOC (Dissolved Organic Carbon) and UV absorption. Under conditions of drinking water treatment in the Region of Lower Rhine with changing concentrations of the raw water, which can be seen as a multicomponent system, and with changing process conditions one finds a great scattering of the measured values, so that an evaluation is rendered very difficult or impossible. The diagram according to UV absorption measurements is shown in Figure 2. This diagram shows very clearly the scattering of the calculated efficiency.

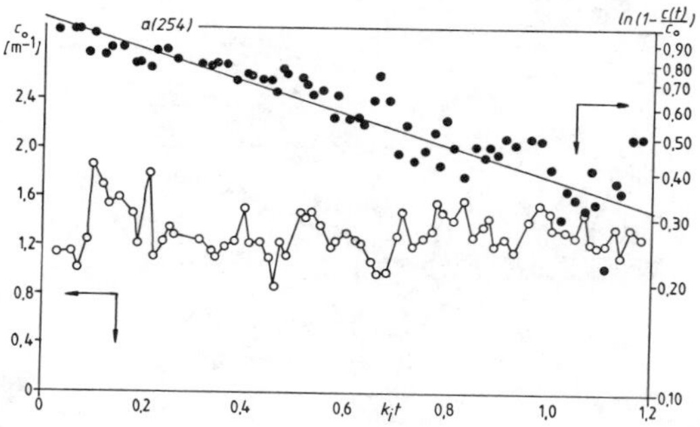

Figure 2. Influent concentration c_o and efficiency $(1 - c(t)/c_o)$ (UV absorption).

Under these circumstances a method has been proved by testing in which one can calculate the loading according to Equation (22):

$$q(t) = \frac{1}{m} \sum_{0}^{V} \Delta V\ (c_o - c(t)) \qquad (22)$$

These calculated values of loading q(t) allow, under the conditions (12) and (14), for instantaneous and delayed breakthrough according to the following Regression Equations (23) and (24), the calculation of the capacity q_o and the adsorption constant k_i. These constants can be used for modelling fixed bed adsorbers:

$$\ln \left(\frac{q_o - q(t)}{q_o}\right) = - k_i \ t; \ t_o = 0 \tag{23}$$

$$\ln \left(\frac{q_o - q(t)}{q_o}\right) = \ln \left(\frac{q(0)}{q_o}\right) - k_i \ t; \ t_o > 0 \tag{24}$$

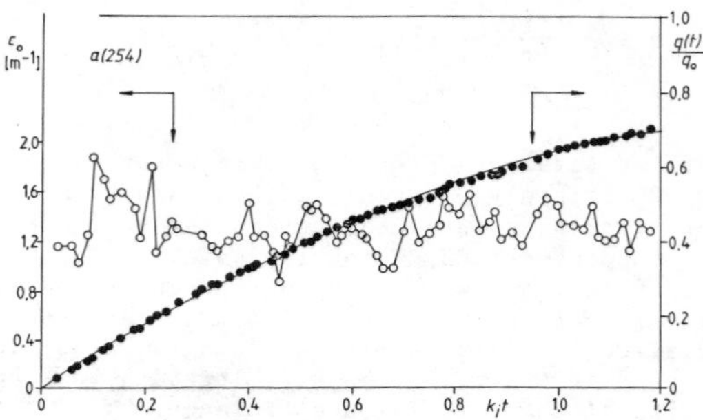

Figure 3. Influent concentration c_o and loading $q(t)/q_o$ (UV absorption).

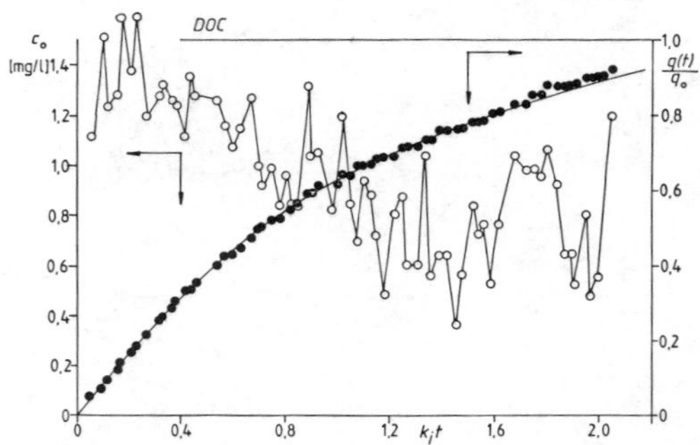

Figure 4. Influent concentration c_o and loading $q(t)/q_o$ (DOC).

Here the value for the capacity q_o will be altered until the best agreement is attained for measurement and calculation. The evaluation of UV- and DOC-measurements according to the Equations (23) and (24) is shown in the Figures 3 and 4. One can see that the measured values

agree very well although the influent concentrations are submitted to continuous and considerable variations.
Many experiments have shown that the adsorption constant k_i and the longitudinal diffusion cannot be calculated by batch tests. Therefore the modelling of technical adsorbers must be made with columns in laboratory or pilot scale under actual conditions.
Recent researches have shown that a calculation is possible taking into consideration that in technical activated carbon adsorbers a classifying of actvated carbon takes place depending on the bed depth regarding apparent density, grain size, adsorption-kinetics and capacity. The system of equations mentioned in this article allows, in spite of these difficulties, an exact calculation.

LITERATURE CITED

1. F. J. Weissenhorn, Chem.-Ztg. 104 (9), 272 (1980).
2. F. J. Weissenhorn, Chem.-Ztg. 104 (12), 358 (1980).
3. F. J. Weissenhorn, Chem.-Ztg. 105 (9), 263 (1981).
4. F. J. Weissenhorn, Chem.-Ztg. 105 (11), 334 (1981).
5. F. J. Weissenhorn, Chem.-Ztg. 106 (2), 93 (1982).
6. F. J. Weissenhorn, Chem.-Ztg. 106 (9), 337 (1982).
7. W. J. Weber and J. C. Crittenden, J. Water Pollut. Control Fed. 47 (5), 924 (1975).
8. J. C. Crittenden et al., J. Water Pollut. Control Fed. 52 (11), 2780 (1980).
9. W. J. Weber and M. Pirbazari, "Effectiveness of Activated Carbon for Removal of Toxic and/or Carcinogenic Compounds from Water Supplies," US EPA Project Summary, EPA-600/S2-81-057, June 1981.
10. F. J. Weissenhorn et al., "Optimierung der Aktivkohleanwendung bei der Trinkwasseraufbereitung," Heft 12, Veröffentlichungen Bereich Wasserchemie Universität Karlsruhe (TH), Karlsruhe (1979).
11. E. Wicke, Kolloid-Z. 86 (3), 295 (1939).
12. J. Crank, "The Mathematics of Diffusion," Clarendon Press, Oxford (1979).
13. P. Schenk, Das Gas- und Wasserfach 103 (30), 791 (1962).

A CONTINUOUS, AUTOMATED TECHNIQUE TO MEASURE MONOLAYER AND MULTILAYER ADSORPTION OF PURE GASES

Sheldon P. Wesson
Owens-Corning Fiberglas
Technical Center
Granville, Ohio 43023

ABSTRACT

Numerical analysis of substrate energetics requires accurate data of high precision. These are best obtained with a volumetric technique that features continuous flow of adsorbate gas and automated data collection. By these means an adsorption isotherm consisting of between 1,000 and 2,000 data points, all at equilibrium, can be measured within a reasonable time. The continuous mode can also be used to improve the gravimetric technique. Experimental data are adduced to demonstrate the equivalence of continuous and manual operation with the volumetric technique; and to demonstrate that volumetric techniques are better suited to determine monolayer adsorption and gravimetric techniques are better suited for multilayer adsorption.

EXPERIMENTAL

Volumetric measurements. Volumetric data were obtained manually using a manifold constructed of Nupro bellows seal valves and 1/4" Swagelok pressure fittings. This apparatus was described previously(1). The experiment consisted of admitting small increments of adsorbate to the sample chamber from the dosing section and recording dosing and sample pressures after equilibration.

Volumetric data acquisition was automated with the manifold shown in Figure 1. Dosing and sample sections consisted of Nupro bellows seal toggle valves connected with 1/4" pipe weld fittings. The dosing section consisted of two MKS Baratron Series 170M pressure sensors, responsive in the pressure ranges 13 kPa and 130 kPa respectively. The sample chamber featured three sensors, operating in the ranges 0.13 kPa, 1.3 kPa, and 130 kPa respectively.

Dosing and sample chambers were connected with a toggle valve that was opened for rapid evacuation and outgassing. This valve was closed during adsorption measurement. The chambers were also connected through a Granville-Phillips Series 203 variable leak valve. Isotherms were obtained by backfilling the dosing section with adsorbate and partially opening the leak valve to admit a slow flux of adsorbate into the sample section. Flow rates in the range 10^{-6} kPa liter sec.$^{-1}$ ensured the system to be near equilibrium throughout the experiment. Data were collected over the pressure range of 10^{-6} to 10 kPa.

Pressure sensors and related instruments were controlled and monitored with a Rockwell AIM-65 microprocessor. Assembly language subroutines called by a FORTH program controlled pressure sensor and range selection for the Baratron manometers. Each datum point consisted of elapsed time from an internal clock; dosing and sample pressures, and ambient temperature from a Fluke 2180A RTD digital thermometer. These readings were collected according to a schedule placed in the microprocessor memory. Data were encoded onto a tape cassette and read into the central computer for storage, analysis and display. Isotherm calculations included correction for thermal transpiration (2).

Gravimetric measurements. The gravimetric apparatus (Figure 2) consisted of two Leybold/Heraeus TMP 360 turbomolecular pumping stations connected with a 3/4" o.d. stainless steel manifold. The manifold consisted of a gas handling section and an adsorption measuring sector, which were connected with a Granville-Phillips Series 216 servo-driven leak valve.

The adsorption section featured a Cahn 1000 recording ultramicrobalance sensitive to 0.1 µg. Adsorbate pressure was monitored with three MKS Baratron sensors operating in the ranges of 1.3 kPa, 13 kPa and 130 kPa respectively. Vapor purity was monitored with an Inficon IQ200 quadrupole mass spectrometer.

Samples were placed on the working arm of the microbalance, counterweighted, and outgassed. The arms of the microbalance were thermostatted symmetrically. The pumps were closed off after outgassing; the dosing and adsorption sectors of the manifold were then isolated. The dosing section was backfilled to atmospheric pressure and kept at constant pressure throughout the experiment. The leak valve was partially opened to admit adsorbate to the sample. Very low flow rates ensured the system was near equilibrium at every pressure. Data were obtained from 10^{-6} kPa to condensation pressure. Desorption isotherms were obtained by closing off the gas supply, pumping on the dosing section and bleeding off adsorbate very slowly through the leak valve.

Output from the microbalance and related instruments was collected periodically with an IBM 7406 device coupler, a programmable A/D converter. Each datum point consisted of readings from an internal clock, followed by mass, pressure and assorted temperature readings, all collected according to a schedule placed in the memory of a microprocessor that controlled the device coupler. Data were encoded onto magnetic tape, and read into the central computer for subsequent processing.

Materials. Volumetric measurements were obtained on untreated Cabosil M5 (fumed silica) manufactured by Cabot Corp., Billerica, Mass. Gravimetric data were obtained on compressed Cabosil M5, consolidated into translucent windows at applied loads of 5 metric tons for 10 seconds in a pellet press. All samples were outgassed at 10^{-6} kPa at ambient temperature for at least 48 hours prior to adsorption measurement. UHP grade nitrogen and argon from Matheson Co. were used without further purification.

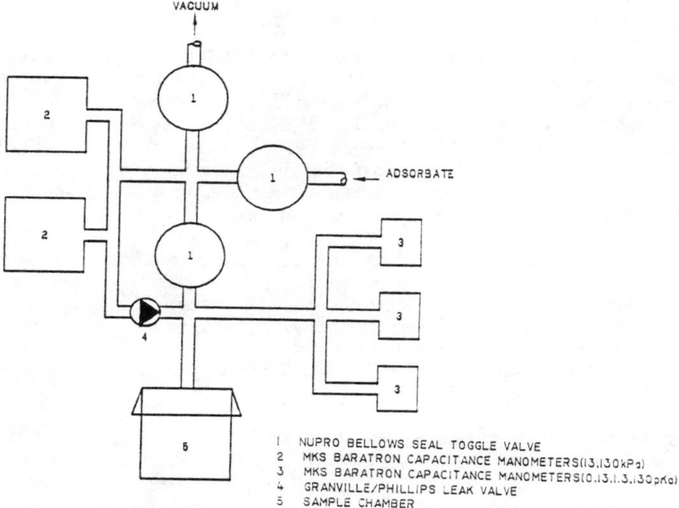

Figure 1. Volumetric adsorption apparatus.

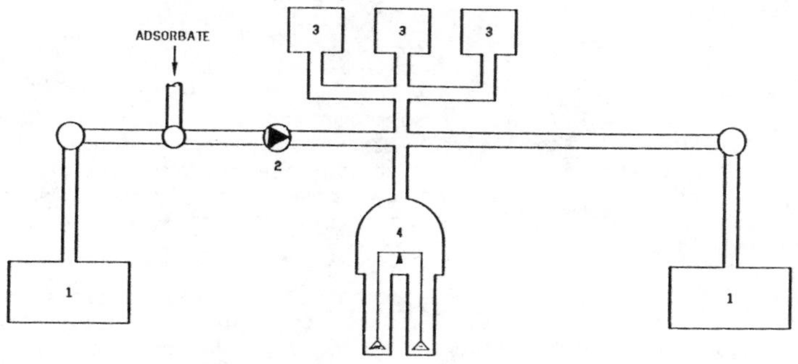

Figure 2. Gravimetric adsorption apparatus.

RESULTS

Figures 3 and 4 compare monolayer isotherms on untreated Cabosil obtained with the manually operated volumetric system (circles) with data produced by constant flow of adsorbate from dosing volume to sample volume (lines). The lines consist of 1,000 to 2,000 points collected by the microprocessor over 48 hours. Registry of data from both methods demonstrates the feasibility of increasing adsorbate pressure in the sample chamber slowly enough so that the amount adsorbed is independent of leak rate (the system is always near equilibrium).

The utility of data obtained by the continuous method is evident in Figures 3 and 4. The smooth curves display differences in adsorption very clearly, especially in the low pressure sector sector (10^{-5} to 10^{-3} kPa) where small differences are often obscured by scatter in conventionally obtained volumetric data. Adsorption at these pressures is used to generate the high energy section of adsorptive distributions by the CAEDMON analysis (3).

Adsorption isotherms were measured on compressed silica with the gravimetric apparatus. Figure 5 shows adsorption (lower line) and desorption (upper line) in good agreement with conventionally obtained volumetric data (circles). The discrepancy between the adsorption and desorption legs above 10^{-3} kPa is attributable to experimental error (drift in the microbalance and presure sensors over the four days during which this isotherm was collected). Gravimetric data below 10^{-3} kPa is subject to non-correctible error caused by thermomolecular flow (4), a phenomenon that obviates the utility of this technique for analysis of adsorption energetics. The data may be used for evaluation of specific surface, however, because the CAEDMON analysis averages out fluctuations caused by thermomolecular flow.

Figure 6 shows the high pressure sector of the gravimetric isotherm that was started in Figure 5. Adsorption and desorption legs are in registry up to 80 kPa (0.8 relative pressure). This shows the system was near equilibrium at all times and that the surface was free of micropores. Hystersis at high relative pressures may be indicative of mesopores, but was also caused in part by slow adsorption kinetics as the system approached condensation pressure.

Volumetric and gravimetric measurement of argon multilayers on the same sample of compressed Cabosil is shown in Figure 7. Both isotherms were obtained using the continuous technique. The data are in good agreement up to 13 kPa, or about 0.5 relative pressure. The volumetric isotherm is systematically higher than the gravimetric data at higher pressures.

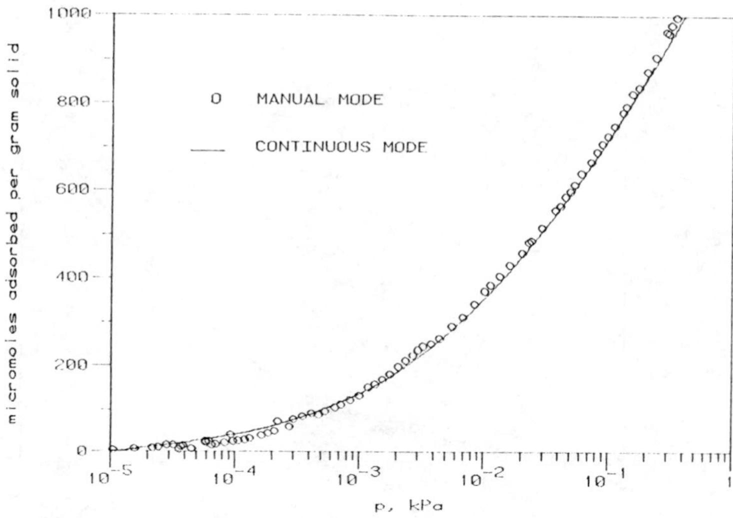

Figure 3. Nitrogen on Cabosil M5 at 77 K.

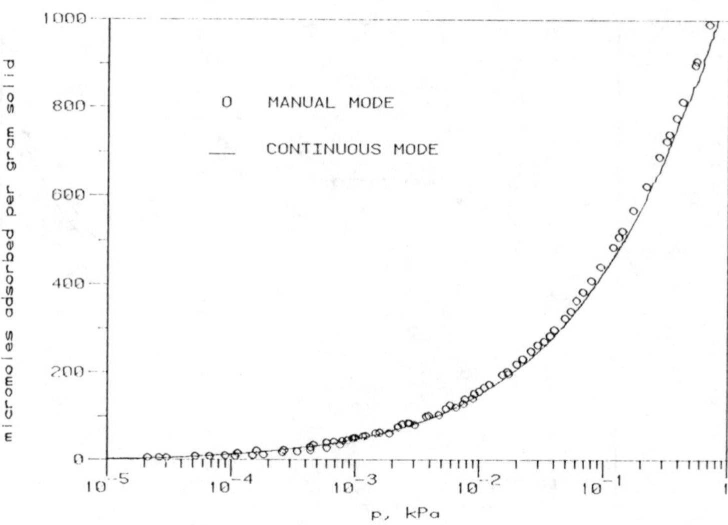

Figure 4. Argon on Cabosil M5 at 77 K.

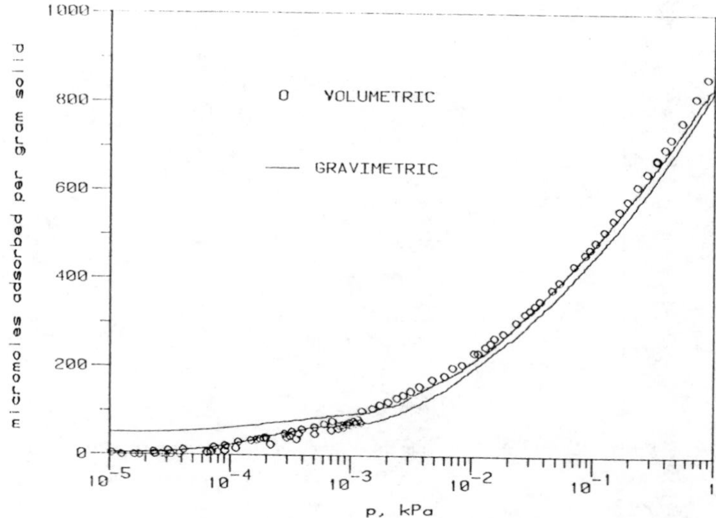

Figure 5. Nitrogen monolayer on compressed Cabosil M5 at 77 K.

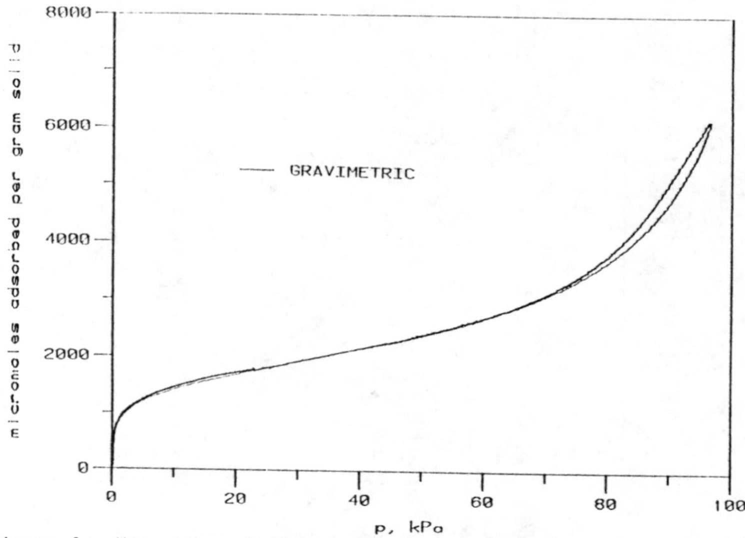

Figure 6. Nitrogen multilayers on compressed Cabosil M5 at 77 K.

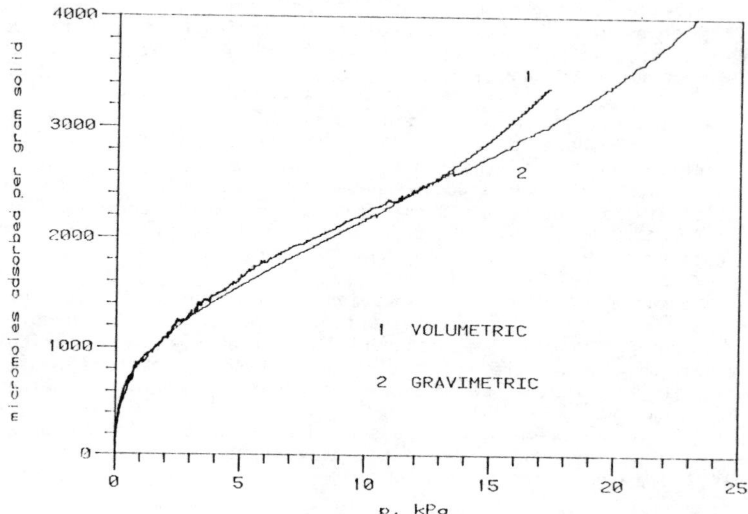

Figure 7. Argon multilayers on compressed Cabosil M5 at 77 K.

DISCUSSION

Volumetric isotherms obtained manually require several days of operator time to manipulate the adsorbate and record pressure changes in the course of multiple runs. The most obvious benefit conferred by our method of automation is that no operator intervention is required after startup except for occasional adjustment of leak rate.

Automated instruments that mimic the manual operating sequence are commercially available. These instruments feature an intricate manifold containing solenoid valves and other electromechanical devices interfaced to a computer that actively drives the data gathering process. Our continuous method requires only that the microprocessor collect data at appointed intervals, and thus benefits from simplicity of design and operation.

The continuous volumetric technique produces more precise data than does discrete operation, as is readily apparent in Figures 3 and 4. Capacitance manometers exhibit hysteresis when subjected to sudden pressure changes; manual operation moves relatively large quantities of gas between chambers. The slow, smooth flux employed by the continuous technique is ideal for mitigating sensor hysteresis and for rapid equilibration of adsorbate on the sample.

The primary objective of monolayer adsorption measurement is to display adsorptive energy distributions obtained by processing data from the low pressure sector of the isotherm. This region presents the greatest difficulty to manual data collection, as evinced by the scatter in points below 10^{-3} kPa in Figure 3. The continuous method produces smooth and precise data that will enable resolution of detail previously obscured by experimental error.

A main concern about the continuous technique is that a leak rate sufficient to produce an isotherm in a reasonable time still be significantly slower than the kinetics of adsorption. The registry of discrete points with continuous curves (Figures 3 and 4) and the reproducibility of adsorption/desorption curves in the continuous mode (Figures 5 and 6) show this to be true for both volumetric and gravimetric determinations.

Volumetric and gravimetric adsorption capabilities are complementary rather than redundant: each operates better in a different sector of the total isotherm. The monolayer isotherm is best obtained volumetrically because of the sensitivity of modern capacitance manometers to small pressure changes. In the low pressure sector of the monolayer isotherm, the determination of n(ads), the total moles adsorbed, is essentially decoupled from measurement of pressure in the sample chamber because virtually all of the gas leaving the dose chamber is taken up on the substrate. The expression for n(ads) is

$$n(ads) = \Delta n(dose) - n(dead) \tag{1}$$

where n(dose) is the moles of gas in the dosing chamber and n(dead) is the gas above the sample. At low equilibrium pressures n(dead) is very small: Δp in the sample chamber is accurately measured, however, giving reproducible determinations of n(ads) that are not affected by small experimental errors.

The sensitivity of the volumetric method declines drastically in the multilayer region, however. The dramatic decrease in $\Delta n(ads)/\Delta p$ exhibited by all adsorbates at the monolayer/multilayer transition causes n(ads) to become a small difference between the two much greater quantities Δn(dose) and n(dead), and scatter increases accordingly.

A more serious problem with volumetric determination of multilayer data is the sensitivity of n(ads) to correction for adsorbate nonideality. The conventional approach is to compensate for nonideality by increasing the V(dead), the effective volume of the thermostatted sample chamber (5):

$$V(dead)' = V(dead)\,(1 + F\alpha p) \tag{2}$$

in which α is the nonideality constant derived from the critical tables or from virial coefficients, p is the equilibrium pressure and F is the fraction of V(dead) at system temperature T(s). If T(a) is ambient temperature,

$$F = (V(dead)_{T(s)} - V(dead)_{T(a)}) \,/\, (V(dead)_{T(a)}\,(T(a)/T(s) - 1)) \tag{3}$$

Typical values of α cause a 3 percent increase in V(dead) at high pressures. This seemingly small adjustment in a calibration constant can modify n(ads) by 50 percent to 100 percent in the multilayer region, however, because there n(ads) is the small difference between two large numbers.

The gravimetric technique cannot be used for low temperature measurements over the pressure region in which thermomolecular flow occurs. This obviates its use for analysis of surface heterogeneity, although values of specific surface will still be returned by the CAEDMON analysis. Gravimetric measurements are most useful in the multilayer region, the sector where volumetric determinations are subject to large experimental and systematic error. Adsorbate nonideality poses no problem for the gravimetric technique, in which the determination of n(ads) is completely decoupled from pressure measurements. Bouyancy effects were eliminated by thermostatting both microbalance arms identically. Gravimetric multilayer data are thus the ideal complement to monolayer isotherms obtained volumetrically.

Figure 7 shows volumetric and gravimetric multilayer isotherms on the same silica sample. Volumetric data were corrected for nonideality as described above. It is apparent that another term is required to eliminate nonideality effects at high relative pressure. An empirical correction could be constructed by zeroing apparent adsorption in an empty sample chamber. The more complicated the correction, however, the less reliable the resulting isotherm.

Gravimetric methods are not conventionally used to measure low temperature adsorption because of controversy about the sample temperature. Congruence of volumetric and gravimetric isotherms in the monolayer region (where correction for nonideality have negligible effect on volumetric data) suggest proper temperature equilibration can indeed be achieved in a gravimetric system. Improper correction of high pressure volumetric data usually causes systematically high values for n(ads). This effect is more likely to cause differences between gravimetric and volumetric measurements than is sample temperature inhomogeneity.

LITERATURE CITED

1. Wesson, S. P., Vajo, J. and Ross, S., J. Colloid Interface Sci., 94(2), 552 (1983).

2. Takaishi, T. and Sensui, Y., Trans. Faraday Soc., 59, 2503 (1963).

3. Sacher, R. S. and Morrison, I. D. J. Colloid Interface Sci., 70 (1), 153 (1979).

4. Garcia Fierro, J. L. and Alvarez Garcia, A. M., Vacuum, 31 (2), 79 (1981).

5. Emmett, P. H. and Teller, E. , J. Am. Chem. Soc., 59, 1553 (1937).

COMPARISON OF ENERGETICS AND SURFACE ARE
OF VARIOUS SILICA SUBSTRATES

By Sheldon P. Wesson
Owens/Corning Fiberglas Corporation, Granville, Ohi
and
Sydney Ross
Rensselaer Polytechnic Institute, Troy, N.Y. 12181

ABSTRACT

CAEDMON analyses of nitrogen and argon adsorption isotherms at 77K on various silica samples before and after compression in a pellet press at an applied load of five metric tons for ten seconds, reveal the effects of the treatment on the surface areas and surface energetics of the materials. This study is part of an ongoing effort to show the application of the CAEDMON analysis of adsorption isotherms to materials before and after physical or chemical treatments that may affect the specific surface area and/or the surface energetics. CAEDMON analysis, as these examples of its application demonstrate, is a powerful tool to investigate treatments in terms of their effects on fundamental surface properties. Previous examples are still too few to convince the profession that the method indeed has a wide range of application.

INTRODUCTION

CAEDMON is the acronym for Computed Adsorptive Energy Distribution in the MONOlayer. As the name implies the method provides an analysis of an adsorption isotherm, in the range of pressures corresponding to the adsorption of a monomolecular layer, to obtain the range and magnitudes of the adsorptive attractions between substrate and adsorbate. The problem is to determine the distribution of adsorptive energies of the substrate from the experimental observations, using the concept that the whole substrate is composed of a finite number of energetically uniform (or homotattic) patches, each with its characteristic adsorptive potential; and that the adsorbed gas is not so perturbed by the substrate as to prevent its behaving essentially as if merely transferred from the three to the two-dimensional state (1), (2), (3). The method also provides a determination of the specific area of the adsorbent, as a summation of the areas of all the adsorptive patches comprising the substrate. Comparisons of results for specific surface area of pristine glass filaments by the CAEDMON method, the BET method, and by mercury porosimetry, show that the three methods agree with each other and also with the result calculated from the geometry and density of the glass filaments (4).

The present paper reports the application of the CAEDMON method to determine the adsorptive-energy distributions and the specific surface areas of two different silicas, one more hydroxylated than the other, each of them before and after compression in a pellet press at an applied load of five metric tons for ten seconds. The two adsorbents are fumed silica and precipitated silica, which have different surface areas and surface energetics. The effect of compression on the Cabosil M5 is to reduce the surface area by about thirty percentum, while the distribution of adsorptive potentials is preserved as what it had been originally. The precipitated silica, which is more contaminated with organic and inorganic matter, undergoes rather less relative reduction of the specific surface area, and the surface energetics are also somewhat affected.

MATERIALS AND METHODS

Adsorption isotherms in the monolayer range of coverage were measured by the volumetric technique on Cabosil M5 (fumed silica) manufactured by Cabot Corporation, Billerica, Mass. Precipitated silica is an amorphous powder prepared from Ludox As-40 (Dupont,) which is a silica sol containing particles of about 220 A.U. diameter, stabilized with ammonium hydroxide to pH = 9. The Ludox is de-ionized with mixed ion-exchange resin, warmed to 50°C, when it gels. It is mixed with an equal volume of n-propyl alcohol, washed on a filter with alcohol, air dried, and dried in an oven at 150°C. The BET area is 124 m^2/g.

THEORY OF THE CAEDMON ANALYSIS

All solid substrates are intrinsically heterogeneous. Adsorption on a heterogeneous substrate may be expressed as

$$a(p_i) = \sum_{j=1}^{n} [\Sigma_j \ f(p_i, K_j)] \text{ for } i = 1,2....n \quad (1)$$

where Σj is the surface area of the jth patch, n is the number of patches and $f(p_i, K_j)$ is the number of moles adsorbed per unit area of patch j at pressure p_i. The function f(p,K) is the two-dimensional virial equation of state of the gaseous adsorbate:

$$\ln p_i = \ln(1/A_j) + 2B_{2D}/A_j + 3C_{2D}/2\, A_j^2 + \ldots -\ln K_j \quad (2)$$

where A_j is the variable area per adsorbate molecule; B_{2D}, C_{2D}, etc. are the two-dimensional reduced virial coefficients, which are listed in reference (1). With the correct choice of units, $A_j^{-1} = f(p_i, K_j)$.

The problem is posed as the solution of n simultaneous equations with 2n unknowns, i.e., the values of K_j and Σ_j for each patch. Values of K_j are determined from the inflexion point of the local adsorption isotherm, Equation (2). This leaves n unknowns, i.e., Σ_j , with n simultaneous equations. The problem now is to determine the set of non-negative values of Σ_j that produces the best agreement between the experimental data and the computed isotherm. Sacher and Morrison (3) describe the optimization procedure.

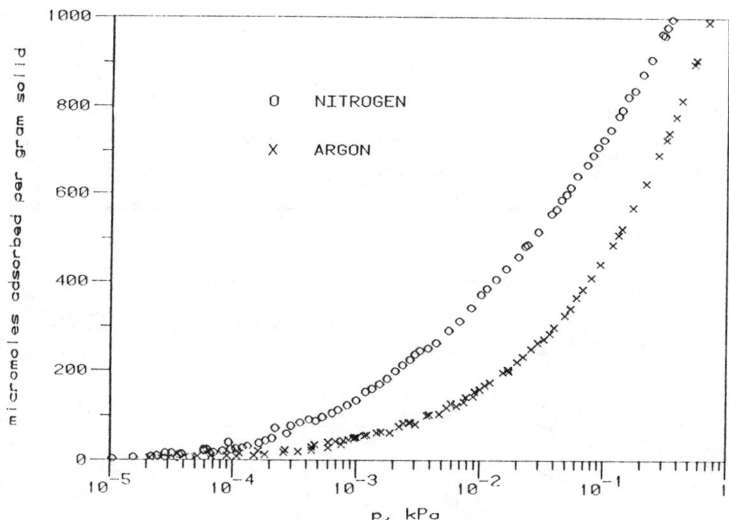

Figure 1. Comparison of monolayer adsorption of nitrogen and of argon on Cabosil M5 ("as is") at 77K, measured volumetrically.

RESULTS

Adsorption isotherms of nitrogen and argon at 77K on Cabosil M5 in the monolayer region, obtained with the volumetric apparatus manually operated, are reported in Figure 1. These are the data that, when subjected to the CAEDMON mechanism, yield the energy-distribution results shown in Figure 2. These data show a larger adsorptive capacity of this silica for nitrogen than for argon. Silica presents a more heterogeneous surface to nitrogen than it does to argon. The quadrupole of the nitrogen molecule interacts with polar constituents of the silica surface, whereas the spherically symmetric argon atom has no permanent polarization and so does not have a similar interaction.

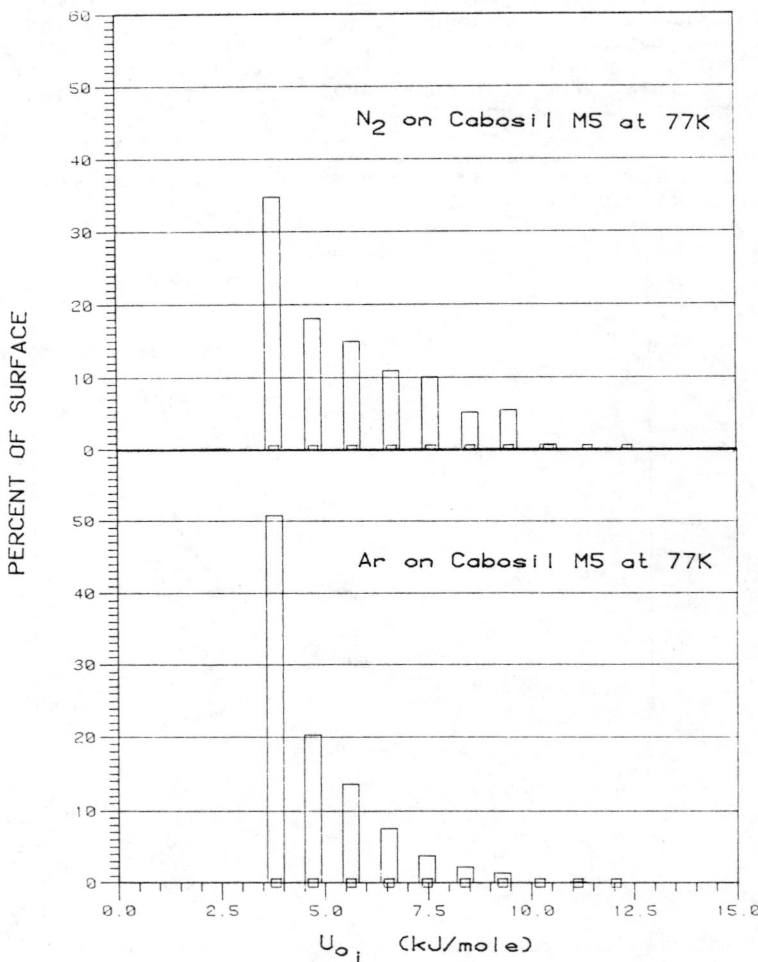

Figure 2. Adsorptive energy distributions of Cabosil M5, based on CAEDMON analyses of the nitrogen and of the argon adsorption isotherms at 77K.

The determination of the specific surface areas of adsorbents by the CAEDMON process is subject to the following difficulty: the theoretical model is confined to monolayer adsorption; but, in practice, multilayers are produced at pressures above one kiloPascal for nitrogen and argon at 77K. The computer program interprets data at that and higher pressures as monolayer formation on patches of lower and lower adsorptive energies; and, indeed, if such patches were actually present they could not be distinguished from multilayer formation on patches of high energy. Inspection of the adsorption isotherm does not reveal exactly where multilayers begin to form. One way to determine the transition point is to subject the data to the CAEDMON process, using each of a series of equilibrium pressures in succession as the terminating datum point, to obtain a plot of specific surface *versus* equilibrium pressure. Even with a relatively small personal computer, CAEDMON analysis is a rapid process, so that this method is not onerous. The selection of a transition point is less certain when the monolayer to multilayer transition is gradual, as is shown in Figure 3 for the "as-is" Cabosil M5; but even there the range of possible values of the specific surface area is narrow, so that any value within that range is still reasonably valid.

The procedure outlined above may appear arbitrary or empirical; it is rationally based, however, on the theoretical model of adsorption, which entails that low-energy sites of the substrate, if present, are going to be hard to distinguish in their effects from second-layer adsorption on high-energy sites.

The result of the CAEDMON analysis of "as-is" and of treated (i.e., compressed) Cabosil M5 shows that the effect of compression of the powder is to diminish its specific surface area by about 30%, without affecting the nature of its surface energetics. This conclusion is brought out clearly in Figure 4, in which the adsorption isotherms of untreated and treated Cabosil M5 are normalized for surface area, that is, the data are presented as the amount adsorbed per unit of substrate area *versus* equilibrium pressure. The coincidence of the two sets of data shows that compression has had a profound effect on the specific surface area but has had no observable effect on the adsorptive-energy profile of the silica substrate.

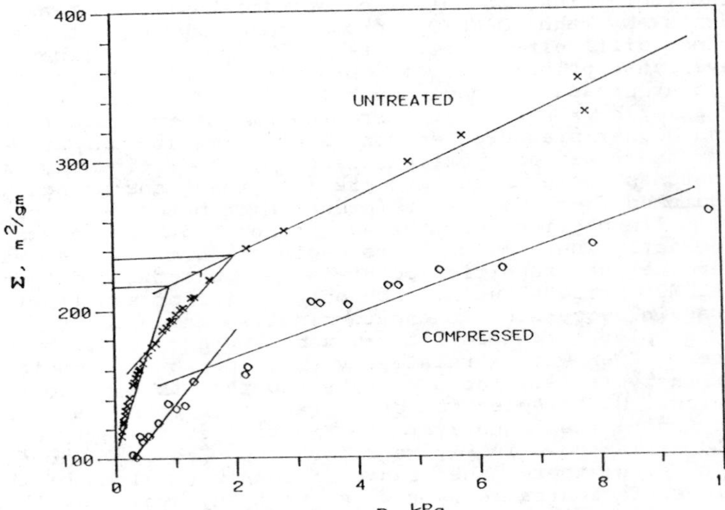

Figure 3. Specific surface areas, computed by CAEDMON program, from the adsorption isotherms of nitrogen at 77K on uncompressed and on compressed Cabosil M5, based on terminating the input data at each datum point in turn in the range of pressures from 0.10 to 10 kPa. The transition between monolayer and multilayer adsorption is taken as at (or near) the point of discontinuity in the results.

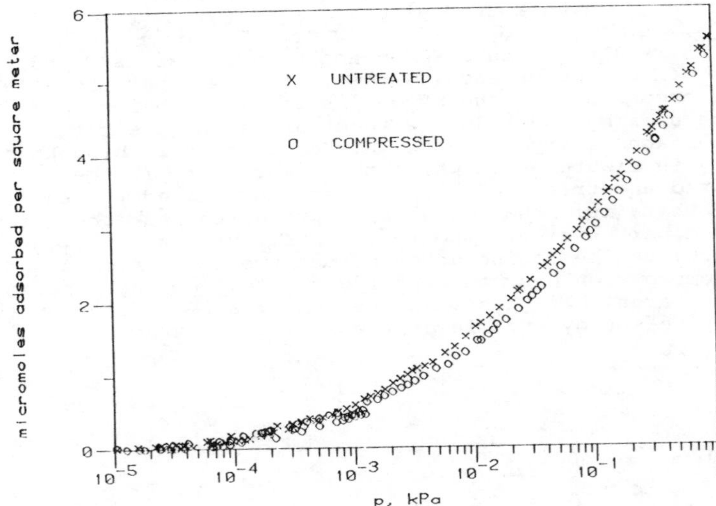

Figure 4. Normalized adsorption isotherms (monolayer) of nitrogen on Cabosil M5 at 77K, showing coincidence of the uncompressed and the compressed samples.

Table 1. Specific Surface from CAEDMON for Cabosil M5 and Precipitated Silica.

Treatment	Apparatus	Adsorbate/ 77 K	Σ (m^2/gm)
CABOSIL M5			
as received	volumetric	N_2	227
		Ar	224
compressed	volumetric	N_2	161
	gravimetric	N_2	164
		Ar	159
PRECIPITATED SILICA			
as received	volumetric	N_2	127
		Ar	119
compressed	gravimetric	N_2	102

Our second silica adsorbent is a sample of precipitated silica. The nitrogen adsorption isotherm (monolayer) at 77K is reported in Figure 5, along with that of untreated Cabosil M5 for comparison. Both the isotherms shown in Figure 5 are normalized with respect to the surface area. The lack of coincidence shows the different energetics of the two substrates. Independent studies (5) have shown that precipitated silica has a surface hydroxyl concentration of 8 to 10 per nm^2 compared with 3.5 to 4.5 per nm^2 for fumed silica. Precipitated silica also has a high concentration of surface sodium ions, whereas the surface of fumed silica is essentially free of alkali. These features of precipitated silica account for its greater normalized adsorptive capacity; nitrogen adsorption in the monolayer is thus seen to respond to differences in surface heterogeneity caused by variations of concentration and composition of surface polar groups.

Compression of precipitated silica does not leave the substrate energetics as undisturbed as was observed for Cabosil M5. Unlike the latter, which is relatively pure, the precipitated silica has both organic and inorganic surface contaminants, which are altered either in concentration or in composition by the compression. The isotherms do indeed coincide at lower pressures, where the higher adsorptive energies characteristic of the silica substrate are operative, but depart from coincidence at higher pressures.

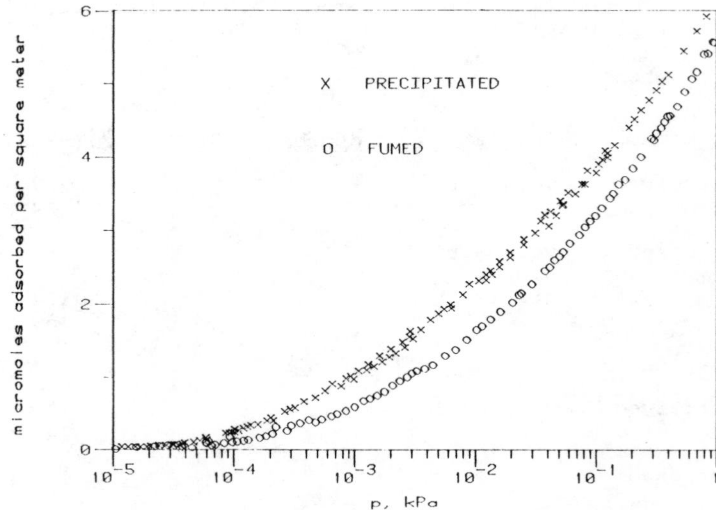

Figure 5. Normalized adsorption isotherms (monolayer) of nitrogen at 77K on precipitated and fumed silicas, showing different adsorptive capacities of the two samples.

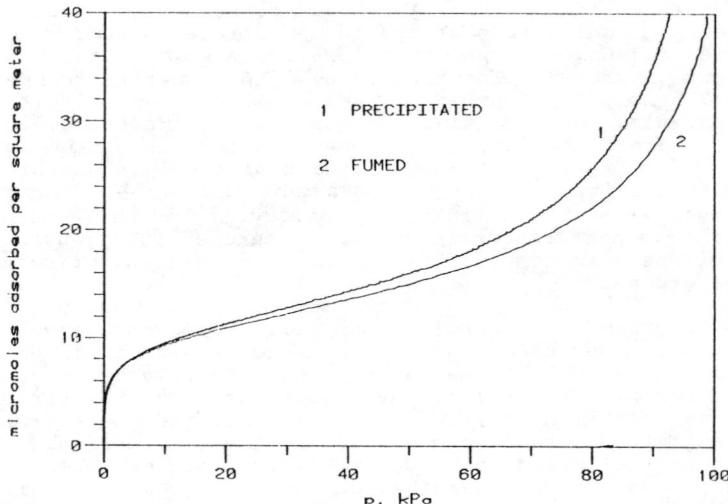

Figure 6. Normalized adsorption isotherms (multilayers) of nitrogen at 77K on Cabosil M5 and on precipitated silica, measured gravimetrically.

In Figure 6 we report the normalized nitrogen-adsorption isotherms (77K) of compressed Cabosil M5 and compressed precipitated silica at higher pressures (multilayers.) These data were obtained by the gravimetric technique: as was pointed out by Wesson (6) the volumetric technique gives considerable scatter of data at pressures greater than 60 kPa. The significant differences between the two silica substrates that were evident in monolayer adsorption, see Figure 5, are not distinguishable in the condensed scale of pressures used in Figure 6; and, except at pressures so great that variations in porosity between the two silicas begins to influence the observations, the two normalized multilayer- adsorption isotherms coincide. This result confirms a widely accepted interpretation (7) that such isotherms reflect the property of the adsorbate, the characteristics of the substrate being overlaid and obscured by a blanket of multilayers of adsorbed gas.

CONCLUSIONS AND SIGNIFICANCE

Comparisons of CAEDMON analyses of treated and untreated samples show that silica, if sufficiently pure, can have its specific surface area reduced by compression without altering the adsorptive-energy distribution of the surface. In accordance with this finding, the normalized monolayer-adsorption isotherms of the unmodified and the compressed adsorbents are in near-perfect registry with each other for the purer silica, but not equally so for the less pure silica.

LITERATURE CITED

(1) Morrison, I.D. and Sydney Ross, Surface Sci., 39, 21 (1973).

(2) Ross, Sydney and I.D. Morrison, Surface Sci., 52, 103 (1975).

(3) Sacher, R.S. and I.D. Morrison, J. Colloid Interface Sci., 70, 153 (1979).

(4) Wesson, S.P., J.J. Vajo, and Sydney Ross, J. Colloid Interface Sci., 94, 552 (1983).

(5) Nishioka, G.M. and J.A. Schramke, J. Colloid Interface Sci., In the press, (1983).

(6) Wesson, S.P., This Symposium, (1983).

(7) Gregg, S.J. and K.S.W. Sing, Adsorption, Surface Area and Porosity, 2nd ed., pp 90 to 94, Academic Press, London and New York, (1982).

ADIABATIC ADSORPTION WITH CONDENSATION: PREDICTION OF COLUMN BREAKTHROUGH BEHAVIOR USING THE BET MODEL

Scott M. Wheelwright, James M. Vislocky,
and Theodore Vermeulen
University of California, Berkeley

ABSTRACT

An equilibrium model based on the isotherm of Brunauer, Emmett and Teller, for the adiabatic desorption of benzene in a nitrogen carrier from activated carbon sorbent, predicts condensation may occur in the bed when the feed and bed temperatures and concentrations lie within the range described by the following inequality

$$1.8 \times 10^{-4}\ (T_B - T_F) + (C_B - C_F) < 0$$

where T_B and T_F are the initial temperatures of the bed and feed, respectively, and C_B and C_F are the initial fluid phase benzene concentrations in equilibrium with the bed and in the feed, respectively, in mol/l, and the initial benzene concentration on the bed is near the saturation level.

Intensive modeling of adsorption and ion exchange has led to an algorithm which quickly and easily computes the column profile for a number of fixed-bed operations. The purpose of this work was to produce a simple model of fixed-bed adsorption under adiabatic conditions and to predict the column performance using this algorithm. Particularly, we wished to investigate the conditions under which condensation would occur within the bed in the course of elution.

THEORY

As the model case, we chose elution of benzene in nitrogen as carrier, with activated carbon as the sorbent. In principle, it appeared that, as heat is taken up by the desorbing species when the bed is eluted to remove the benzene, the temperature of the bed might decrease to the point where condensation occurs. If so, it would be important to know under what conditions condensation occurs, so as to avoid them in operation.

Several workers have approached the problem of modeling adiabatic adsorption on a fixed bed. For example, Rhee, Heerdt, and Amundson (1) applied a Langmuir isotherm in an equilibrium solution to obtain column profiles for the adsorption of benzene on a charcoal bed. Unfortunately, their use of the Langmuir isotherm prevented description of the region near the saturation temperature. Friday and LeVan (2) treated cases in which condensation occurs during thermal regeneration using a Langmuir model, with a discontinuity at the saturation boundary where they set the isotherm slope equal to infinity.

Instead of a Langmuir model, we have used the isotherm of Brunauer, Emmett, and Teller (3), which presumes that multiple layers adsorb as the saturation concentration or saturation temperature is approached. The Langmuir and BET equations and isotherms are compared for a simple isothermal case in Figure 1, and are seen to differ substantially near the saturation boundary. In the Langmuir model the solid phase concentration remains constant, whereas in the BET model the solid phase concentration increases due to gradual adsorption of several layers. For the adiabatic case the equilibrium is described not by a two-dimensional line as shown here, but by a three-dimensional surface with temperature as the additional parameter.

The temperature and concentration of benzene in the gas phase at any time and any location in the column are found by solving a material balance and an enthalpy balance together. For the adsorbed species, the differential material balance is

$$\frac{\partial c_2}{\partial v} + \frac{\partial q_2}{\partial V'} = 0 \qquad (1)$$

where c_2 is the benzene concentration in the fluid phase, q_2 is the benzene concentration in the solid phase, v is the column volume measured from the feed end, and V' is the total volume of feed less the column void volume. The differential enthalpy balance is

$$\frac{\partial T}{\partial v} + \frac{\partial q_1}{\partial V'} = 0 \qquad (2)$$

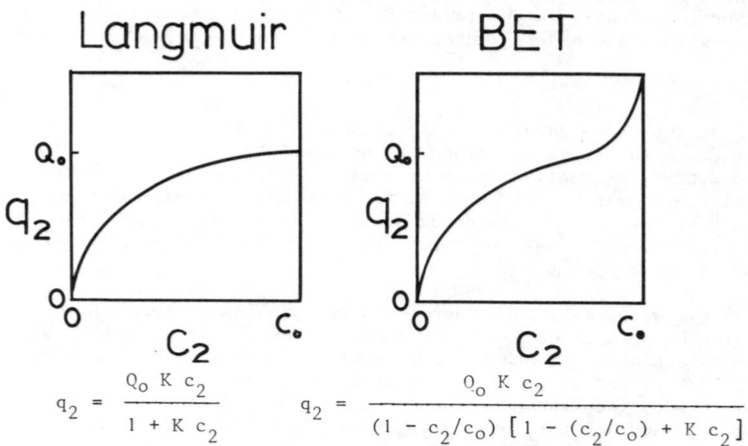

$$q_2 = \frac{Q_o K c_2}{1 + K c_2} \qquad q_2 = \frac{O_o K c_2}{(1 - c_2/c_o)\,[1 - (c_2/c_o) + K c_2]}$$

Figure 1. Comparison of equilibrium models of Langmuir and Brunauer, Emmett, and Teller.

where T is absolute temperature and q_1 is given by

$$q_1 = T \frac{c_s}{c_f} - \frac{(-\Delta H)\, q_2}{c_f} \qquad (3)$$

where c_s is the heat capacity of the solid phase, c_f is the heat capacity of the fluid phase, and ΔH is the heat of adsorption. In this solution we neglected the effects of finite heat and mass transfer rates, and assumed that the bed is in equilibrium with the gas at each point in the column. Therefore Equations (1) and (2) are related by an equilibrium relation, in this case the BET model

$$q_2 = \frac{Q_o\, K\, c_2}{(1 - c_2/c_o)\,[1 - (c_2/c_o) + K\, c_2]} \qquad (4)$$

where Q_o is the monolayer sorbent capacity, and K is the Langmuir parameter,

$$K = K_o\, T^{1/2} \exp(-\Delta H/RT) \qquad (5)$$

where K_o is a constant. These equations were solved using an algorithm written by one of us (4). The adsorbent capacity, calculated from data of James and Phillips (5), is given in Table 1, with other parameters given by Rhee, et al (1). The Antoine equation was fitted to saturation data from Perry and Chilton (6) to give the saturation concentration as a function of temperature. The variables v and V', corresponding to distance and time, can be combined into a single variable, the composition velocity u or relative column length, defined in terms of the monolayer sorption capacity and the saturation concentration:

$$u = \frac{Q_o\, v}{c_o\, V'} \qquad (6)$$

This combination allows the use of general diagrams to describe the column performance, under most conditions.

Q_o = monolayer sorbent capacity = 2.75 mol/l
K_o = constant = 3.88×10^{-5} l/mol $K^{1/2}$
ΔH = heat of adsorption = 10,400 cal/mol
R = gas constant = 1.987 cal/mol K
c_s = heat capacity of solid phase = 405 cal/l K
c_f = heat capacity of fluid phase = 2.7 cal/l K
c_o = saturation concentration = 12.18 exp[9.6 - 2890/(T-43)], mol/l, T in Kelvin
Total pressure = 10 atmospheres

Table 1. Data for benzene adsorption on charcoal.

RESULTS AND DISCUSSION

The two families of curves shown in the composition path diagram in Figure 2 were obtained by integrating the differential material and enthalpy balances under equilibrium conditions, and correspond to the possible routes which the composition may follow as it changes from a given feed to a given initial condition, in terms of the fluid phase benzene concentration and the temperature. The column profile will have two transitions in the case of the adiabatic adsorption or elution of one solute: a downstream or "fast" transition which emerges from the column first, indicated by the dashed lines; and an upstream or "slow" transition which emerges from the column last, indicated by the solid lines. These two transitions will adjoin an intermediate plateau. By inspection of the directions in which the velocities change from point to point along a composition path one determines whether a transition will be abrupt or gradual; the arrows in Figure 2 indicate the direction in which the path is gradual.

For example, consider the elution of a heavily loaded sorbent bed in equilibrium with gas containing 0.008 moles of benzene per liter of bed at 331 K (point A in Figure 2) with a hot nitrogen stream at 360 K and a benzene concentration of 0.006 mol/l (point C). The intermediate plateau will be near B, the intersection of the slow path through the feed C and the fast path through the initial condition A. Both

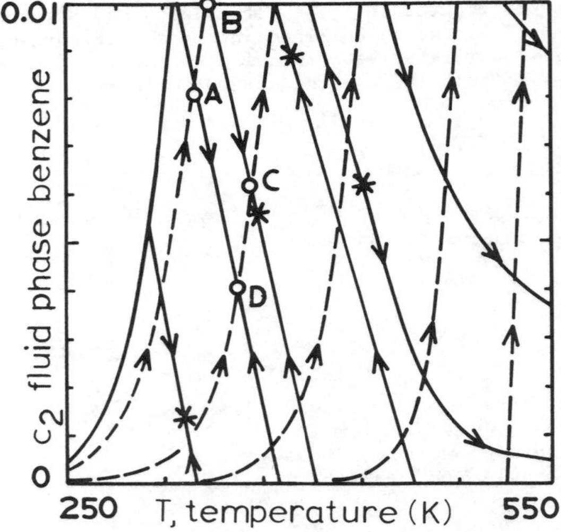

Figure 2. Possible paths followed by the composition as it changes from the feed condition to the intermediate plateau condition (solid lines) and then to the initial condition (dashed lines), in terms of the fluid phase benzene concentration and the temperature. Arrows indicate the direction of increasing velocity and asterisks denote velocity extrema. Lettered points refer to examples.

transitions, from C to B and from B to A, will be abrupt, as indicated by the direction of the arrows. The fluid phase benzene concentration at the intermediate plateau is 0.01 mol/l, which is greater than the initial concentration of 0.008 mol/l. The equilibrium solid phase compositions can be found from the generalized plot shown in Figure 3, in which the solid phase benzene concentration contours are plotted as functions of the fluid phase benzene concentration and the temperature. The solid phase concentration at the intermediate plateau B is 5.15 mol/l, somewhat greater than the initial solid phase concentration at A of 5.02 mol/l.

The column profile for this case has been plotted in Figure 4, with concentrations and temperature as functions of the composition velocity. The velocity of an abrupt transition is simply the ratio of the fluid phase concentration change and the solid phase concentration change between the feed or the initial condition and the intermediate plateau. The feed plateau is on the left, followed to the right by the upstream transition, the intermediate plateau, the downstream transition, and the initial plateau.

As a second example, consider the elution of a bed initially in equilibrium with gas containing 0.01 mole of benzene per liter at a temperature of 338 K (point B) with a feed of 0.004 mol/l and 355 K, point D. The first transition, from D to A, will be abrupt, and the second transition, from A to B, will be gradual as indicated by the arrows. The velocity, or dimensionless time and distance, at each point along the gradual transition can be found from the generalized diagram of Figure 5 which shows contours of constant velocity computed

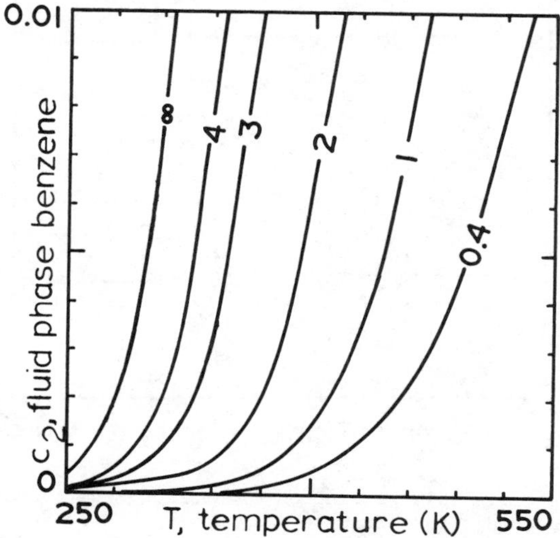

Figure 3. Contours of constant solid phase benzene concentration as function of fluid phase benzene concentration and temperature.

using the partial derivatives of the equilibrium relation, as functions of the fluid phase benzene concentration and the temperature. The solid lines correspond to velocities of the slow paths (or upstream transitions), and the dashed lines correspond to the velocities of the fast paths (or downstream transitions).

The column profile for this case is shown in Figure 6, again with concentrations and temperature as functions of the composition velocity. The solid phase benzene concentration of 4.88 mol/l at the intermediate plateau is less than the 5.00 mol/l at the initial condition.

These two examples of elution of a heavily loaded bed with a warm feed stream illustrate how the intermediate plateau composition depends on the relative positions of the feed and the initial condition. When the slow path through the feed lies to the right of the slow path through the initial condition as in the first example, the solid phase concentration at the intermediate plateau will be greater than at the initial condition. Conversely, when the slow path through the feed lies to the left of the slow path through the initial condition as in the second example, the solid phase concentration at the intermediate plateau will be less than at the initial condition.

Thus condensation occurs only in the first case, where the solid phase concentration at the intermediate plateau exceeds the initial concentration. The range of possible feed conditions under which condensation may occur during elution is shown in the composition path

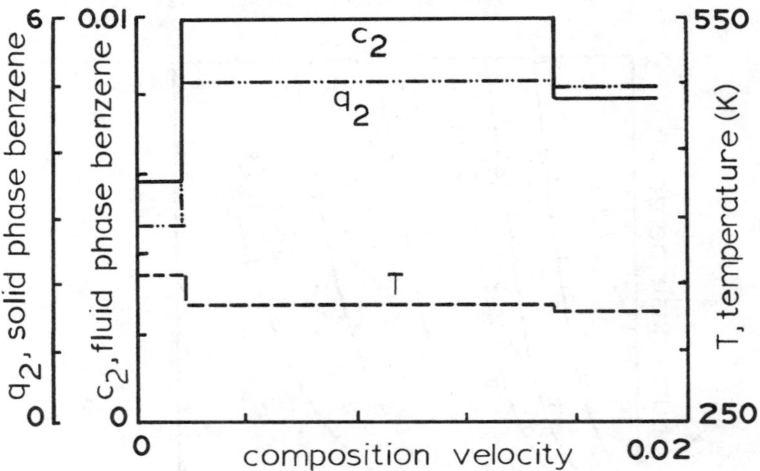

Figure 4. Column profile for the elution of benzene from a charcoal bed near saturation conditions (Example 1), showing the benzene concentration in both phases and the temperature as functions of the composition velocity. Feed plateau is on the left, initial on right.

diagram, Figure 2, as the area to the right of the slow path through the initial condition. This condensation range is described approximately by the following inequality, for the conditions in Table 1:

$$1.8 \times 10^{-4}\ (T_B - T_F) + (C_B - C_F) < 0 \tag{7}$$

where T_B and T_F are the initial temperatures of the bed and feed respectively, and C_B and C_F are the initial fluid phase benzene concentrations in equilibrium with the bed and in the feed, respectively.

While condensation is limited to conditions within this range, its occurance depends on the initial loading of the bed. Typically, condensation within sorbent pores begins to take place at loadings of about twice the monomolecular loading (7). Thus if saturation is specified as an average concentration greater than, say, 5.0 mol/l, then condensation occurs in the first example, but not in the second. Because the solid phase concentration contours in Figure 3 are nearly parallel the fast paths in Figure 2, a lower initial bed concentration will have a correspondingly lower intermediate plateau concentration.

CONCLUSIONS

Adiabatic elution operations including the area near the saturation boundary have been described by the model of Brunauer, Emmett, and Teller. The composition path diagram, or the map of possible composition changes in a transition, was extremely helpful in determining the

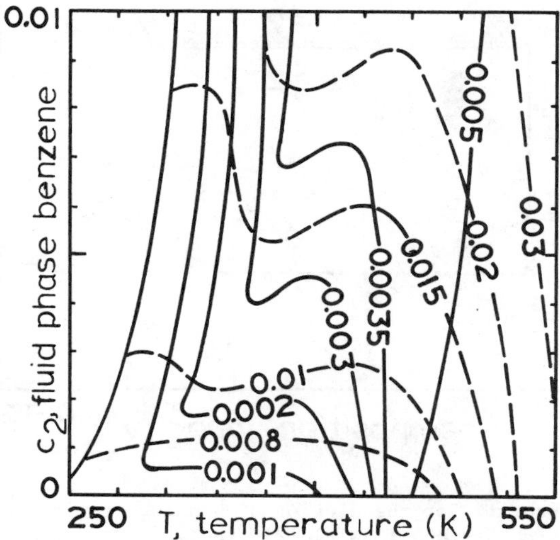

Figure 5. Contours of constant composition velocity as functions of fluid phase benzene concentration and temperature. Solid lines are velocities of upstream transitions and dashed lines are velocities of downstream transitions.

nature of particular transitions and provided general information about types of operations.

Condensation may occur during the elution of a heavily loaded bed with warm feed if the slow path through the feed lies to the right of the slow path through the initial condition, and the initial concentration of benzene on the bed is near the saturation value.

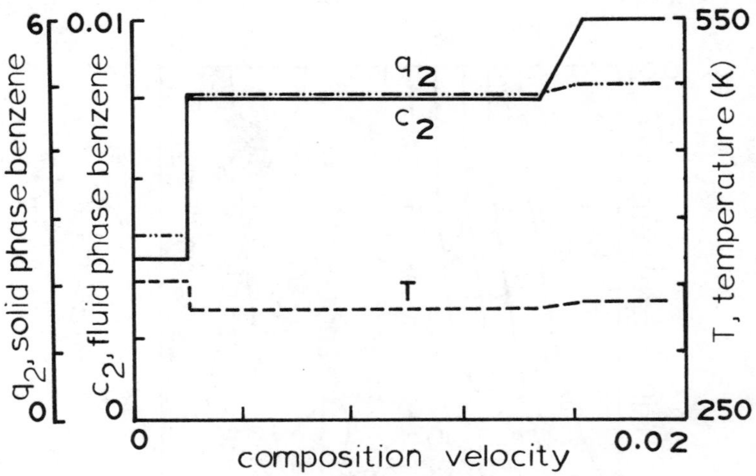

Figure 6. Column profile for the elution of benzene from a charcoal bed near saturation conditions (Example 2), showing the benzene concentration in both phases and temperature as functions of the composition velocity. Feed plateau is on the left, initial on the right.

NOTATION

c_2 = fluid phase benzene concentration, mol/l
c_o = saturation concentration of fluid phase benzene, mol/l
c_f = fluid phase heat capacity, cal/l K
c_s = solid phase heat capacity, cal/l K
C_B = initial fluid phase benzene concentration in equilibrium with bed, mol/l
C_F = initial fluid phase benzene concentration in feed, mol/l
ΔH = heat of adsorption, cal/mol
K = Langmuir parameter, defined by Equation (5)
K_o = empirical constant
q_1 = variable defined by Equation (3)
q_2 = solid phase benzene concentration, mol/l
Q_o = monolayer sorbent capacity, mol/l
R = gas constant
T = temperature, Kelvin
T_B = initial bed temperature
T_F = initial feed temperature
u = composition velocity
v = column volume measured from feed end
V' = total volume of feed less column void volume

LITERATURE CITED

1. Rhee, H.K., E.D. Heerdt, and N.R. Amundson, Chem. Eng. J., 1, 279 (1970).

2. Friday, D.K. and M.D. LeVan, AICHE J., 28, 86 (1982).

3. Brunauer, S., P.H. Emmett, and E. Teller, J. Am. Chem. Soc., 60, 309 (1938).

4. Vislocky, J.J., "Local Equilibrium Theory for Bivariant Fixed-Bed Sorption Systems", PhD dissertation, Univ. of Calif., Berkeley (1982).

5. James, D.H. and C.S.G. Phillips, J. Chem. Soc., 1954, 1066 (1954).

6. Perry, J.H. and C.H. Chilton, Chemical Engineers' Handbook, 5th ed., pp. 3-49, 3-61, McGraw-Hill, New York (1973).

7. Zsigmondy, R., Z. Anorg. Chem., 71, 356 (1911).

CONFERENCE SUMMARY

REPORTER	SUBJECT
G. H. Findenegg	Experimental Aspects
D. M. Ruthven	Modeling and Design of Fixed-Bed Adsorbers
W. A. Steele	Theory and Models
W. J. Weber, Jr.	Environmental Applications

GERHARD H. FINDENEGG was born in Klagenfurt, Austria, in 1938. Following undergraduate study at Vienna Technical University, he did his doctorate work in physical chemistry under Fritz Kohler at the University of Vienna, receiving a Ph.D. in 1965. In 1966-68 he was a postdoctoral fellow at the University of Bristol, UK, working with Douglas H. Everett. After lecturing at Vienna University, he joined the Ruhr-University Bochum, West Germany, in 1973 as a Professor of Physical Chemistry.

His research work has focused on fundamental studies of the physisorption of gases, gas mixtures and the adsorption from solution onto homogeneous surfaces like graphitized carbon. For instance, the influence of molecular shape on the strength of adsorbate - substrate and adsorbate - adsorbate interactions has been studied for a variety of systems. Molecular - statistical theories of the adsorption of chain molecules have been developed and tested experimentally by several experimental techniques.

\- - - - - -

SUMMARY

EXPERIMENTAL ASPECTS

Single Gas Adsorption Isotherms

Single gas physisorption measurements are widely used for determining the specific surface area, the energy distribution of adsorption sites, and the pore size distribution of adsorbents, catalysts, and various other solid materials. K.S.W. Sing discussed several problems and ambiguities which still exist in connection with the evaluation of surface area, and with the assessment of meso- and microporosity, and he introduced a recent recommendation prepared by the IUPAC Subcommittee on Reporting Gas Adsorption Data [1]. Nitrogen (at ≈ 77 K) is still recommended as the adsorbate, in spite of the fact that nitrogen molecules may assume different mean orientations (with different effective cross-sectional areas) on polar and nonpolar surfaces, as was demonstrated by J. Rouquerol and his coworkers.

The three types of methods for measuring single gas adsorption, viz., static gas volumetric methods, gravimetric methods, and flow methods (chromatographic techniques), have been developed to a high standard of accuracy in recent years. Gas volumetric techniques seem to be preferable at low pressures and low temperatures. Adsorption isotherms are obtained either point-by-point, or by pressure scanning, i.e. continuous flow of adsorbate into the system at a low known flow rate and automatic pressure measurement. Such methods were discussed in the contributions of J. Rouquerol et al., and S.P. Wesson. A study of high pressure adsorption (up to 160 bars) by a static gas volumetric method was reported by S.S. Barton et al.

Gravimetric techniques using an adsorption microbalance are most successful at temperatures not too far removed from ambient and can also be improved by operation in the pressure scanning mode (Wesson). At higher gas pressure and/or samples of low specific surface area buoyancy corrections become large, but these corrections can be minimized when a symmetrical-beam balance is used: in this case the buoyancy of the adsorbent can be compensated by a combination of taring materials of equal volume and thus equal buoyancy as the sample. With this volume compensation technique, the gravimetric method yields directly the surface excess amount of adsorbate. I reported a study of high pressure physisorption of several gases on graphitized carbon black using this technique.

Gravimetric techniques also offer a practicable method for studying the kinetics of mass transport in mesoporous and microporous adsorbents. This method was used, among others, by Seewald, Klein and Jüntgen for studying the desorption behaviour of organic substances from activated carbon (AC) of well-defined pore structure; by Schadl and Mersmann for investigating the sorption kinetics of CO_2 + H_2O mixtures in molecular sieves; and by Y.H. Ma for studying the adsorption and diffusion of C_6 and C_8 hydrocarbons in shape selective zeolites. In this case it was possible to measure simultaneously the gas uptake and the temperature of the adsorbent pellet by using a specially designed sample pan equipped with a thermocouple.

Mixed Gas Adsorption Isotherms

Physical adsorption of gas mixtures is of great practical importance for gas separation and purification processes. For such applications it is necessary to know the equilibrium amounts of all adsorbed components as a function of the concentrations of these components in the gas phase. Thus, a number of independent variables has to be controlled and measured simultaneously. For this reason systematic studies of mixed gas adsorption require much more elaborate techniques than single gas adsorption studies. Adsorption equilibrium may be established by passing a gas mixture of known composition at a pre-adjusted pressure over the adsorbent contained in a (small) adsorption chamber; subsequently, the adsorbate may be eluted and transferred to a gas chromatograph for quantitative analysis [2]. Alternatively, a known total amount of a gas mixture of known initial composition may be circulated within a closed system containing the adsorption chamber until equilibrium is attained; the final concentration of the individual components in the gas phase may again be determined by a gas chromatograph, and the adsorption of the individual components is calculated from their depletion in the gas phase. This method was employed by Costa, Calleja and Cabra in their study of the adsorption equilibrium of two- and three-component mixtures of low-molecular-weight hydrocarbons in zeolites. A.C. Rasmuson used a similar method for studying the adsorption of solvent mixtures on AC. In this case the vapour pressures of the sorbable components were so low that almost the entire amounts of these components were contained in the adsorbed phase; hence, the amount and composition of the adsorbed phase could be adjusted externally, and the partial pressures of the sorbable components were determined as a function of the variables of the adsorbed phase.

Mixed gas adsorption can also be measured by determining the breakthrough curves of the individual components when a mixture of known composition is passed through a fixed bed containing the adsorbent. In the terminology of chromatography, this method is called frontal analysis. Compared with the methods described above it has the advantage that no absolute calibration of the detector is required, but the detector must be linear over the experimental concentration step of each component. In single-gas adsorption studies, continuous monitoring of the breakthrough curve is easily achieved by a suitable gas chromatographic detector. In the case of gas mixtures, continuous monitoring of all components is usually not possible; instead, the composition of the gas mixture leaving the bed is analysed in regular intervals of time. For a quantitative analysis of the breakthrough curves it is important to have a sufficiently large number of data points within the mass transfer zone of each component. In favourable cases such measurements yield not only the equilibrium isotherms but also the mass transfer coefficients for the adsorption/desorption process. An example of such as analysis was presented by Takeuchi and Furuya, who studied the equilibrium isotherms and intraparticle diffusivities of binary solvent mixtures on AC. These authors used gas chromatography to analyze the composition of the gas arriving at the column outlet. S. Carrà et al. investigated the separation of chlorotoluene isomers by adsorption on K-exchanged X and Y zeolites, using a related technique called vapour phase displacement chromatography.

P.J. Reucroft reported a combined gravimetric and volumetric method for mixed vapour adsorption studies on AC. Although gravimetric measurements yield only the total mass of the mixed adsorbate, it is possible to derive the composition of the adsorbed phase from a set of such adsorbate-mass isotherms, each obtained at constant composition of the vapour phase, on the basis of the Gibbs adsorption isotherm [3].

An interesting method for measuring surface diffusion coefficients of two-component gases on AC was presented by Suzuki, Hori and Kawazoe. In this work the dependence of the surface diffusion coefficients on the surface concentration of the two components was measured and analyzed in terms of the so-called surface pressure gradient driving force concept.

Adsorption from Solution

Despite the great importance of adsorption from solution by solids in a wide range of industrial processes, fundamental understanding of this phenomenon has developed relatively slowly. Many of the recent developments in this field were initiated by D.H. Everett. In his keynote address, Everett reviewed the format thermodynamic description of adsorption from solution and the formal links with adsorption at the vapour/solid interface, and with wetting phenomena. He also mentioned several experimental methods for studying adsorption from solution. In principle, these methods are similar to those for studying mixed gas adsorption: conventional batch techniques; closed-system type apparatus; and dynamic (frontal analysis) techniques.

The equipment developed in Everett's laboratory consists of two closed systems, one of them containing the adsorbent, which are filled with liquid mixture of the same composition. The liquids in the two systems are circulated through the arms of a differential refractometer and the adsorption is calculated from the change in refractive index of the liquid in contact with adsorbent. This method yields the surface excess isotherm of binary liquid mixtures over the entire composition range, and over a wide temperature range up to nearly the boiling point of the liquid. Less sophisticated apparatus of this type are frequently used to measure adsorption isotherms and the rate of adsorption from dilute solution, as in the work of Beverloo and Pierik, and by Takeuchi and Suzuki.

Adsorption Calorimetry

Calorimetry may be used in several ways to measure the enthalpy changes connected with adsorption and wetting processes. The most important calorimetric methods are:

- Gas adsorption calorimetry
- Immersion calorimetry
- Adsorption from solution calorimetry

New microcalorimeters have been developed in recent years, some of them now being commercially available.

Rouquerol et al. developed an isothermal microcalorimeter to measure the enthalpy of adsorption of simple gases on the surface of various solid materials. Interesting details of the interaction of the adsorbed molecules with the surface are revealed by such measurements.

The immersion calorimetric method of Harkins and Jura represents a reliable method for determination of specific surface area, as was also shown by the Rouquerols and coworkers. The useful role of immersion calorimetry in the field of adsorption from binary liquid systems was demonstrated by Everett. On the one hand, the difference of the enthalpies of immersion in the two pure liquids can be used to test the consistency of the temperature dependence of the surface excess isotherm of the binary system; on the other hand, the dependence of the enthalpy of immersion on the composition of the binary system may give further insights about molecular interactions in the interfacial region. However, such details appear only as small deviations from the expected concentration dependence of the enthalpy of immersion. A much more sensitive measurement of the enthalpy effects caused by solute - substrate or solute-solvent interactions in the adsorbed layer is offered by flow microcalorimetry [4].

Spectroscopic Methods

Several spectroscopic methods, including infrared studies, nuclear magnetic resonance, electron spin resonance, Raman and electron spectroscopy play an important role in the characterization of the structure and configuration of molecules adsorbed at the gas/solid interface [5]. At this Conference, Luck's paper gives an example for the application of infrared spectroscopy for an investigation of the state of water molecules sorbed in biopolymers.

References

1. Sing, K.S.W., *Pure & Appl. Chem.* 54 (11), 2201 (1982).

2. Albrecht, M., P. Glanz and G.H. Findenegg, *Adsorption at the Gas-Solid and Liquid-Solid Interface*, eds. J. Rouquerol and K.S.W. Sing, Elsevier, Amsterdam (1982), p. 75.

3. Van Ness, H.C., *Ind. Eng. Chem. Fundamentals* 8, 464 (1969).

4. Kern, H. and G.H. Findenegg, *J. Coll. & Interf. Sci.* 75, 346 (1980).

5. Cosgrove, T., *Colloid Science*, ed. D.H. Everett, Specialist Periodical Reports, Vol. 3, The Chemical Society, London (1979), p. 293.

G. H. Findenegg

DOUGLAS M. RUTHVEN is an expatriate Scot, born in India, educated in England and a citizen and long-time resident of Canada. He gained his formal education (B.A., M.A., and Ph.D. in Chemical Engineering) at the University of Cambridge. Following a brief spell in industry with Davy-Power-Gas, he joined the University of New Brunswick in 1966 and, apart from a leave of absence spent with EXXON Research and Engineering, he has remained there ever since. He is currently Professor of Chemical Engineering. He has published widely in the area of molecular sieves and adsorption separation processes and has recently completed a monograph on this subject entitled *Principles of Adsorption and Adsorption Processes*, which is to be published by John Wiley in February, 1984.

\- - - - - -

SUMMARY

MODELLING AND DESIGN OF FIXED BED ADSORBERS

Prediction of the dynamic behaviour of an adsorption column requires the simultaneous solution of the differential heat and mass balance equations for a column element together with the appropriate equilibrium and rate expressions for each component. It is therefore convenient to consider the design implications of the conference papers in terms of fluid dynamics, adsorption equilibrium, adsorption/desorption kinetics and column dynamics.

In the first category may be mentioned the paper of LeVan and Vermeulen in which the effect of channeling, which leads to increased axial mixing, was analyzed in terms of a "Taylor Dispersion" model.

Many papers dealt with adsorption equilibrium for both single and multicomponent systems on a range of adsorbents. The commonly used two-parameter models such as Langmuir or Freundlich can often fit individual isotherms, at least over a limited range of concentration, but the physical significance of the parameters is questionable. Extrapolations beyond the range of experimental measurements should therefore be treated with caution and the extension of these simplified models to binary and multicomponent systems is generally unsatisfactory. Nevertheless, the Langmuir equation shows the correct asymptotic behaviour for a type I isotherm and has proved valuable in the analysis of column dynamics since it is equivalent to the constant separation factor approximation. Other more sophisticated models which have been developed to account for sorbate-sorbate interaction and/or energetic heterogeneity of the adsorbent were also discussed by several authors. However, from the practical point of view the problem of predicting binary and multicomponent equilibria from single component isotherms is more important than the precise modelling of single component isotherms which are easily measured experimentally. Experimental binary equilibrium data for several light hydrocarbons on 5A zeolite were presented by Costa, Calleja and Cabra while the commonly used methods of predicting binary and multicomponent equilibria were reviewed by Danner. The need for more extensive experimental data under conditions of high pressure typical of many commercial sorption processes (e.g. PSA hydrogen) were stressed by Kaul and Sweed while problems of irreversibility, which may easily complicate the analysis and interpretation of experimental data were mentioned by several authors.

Results of experimental studies of intraparticle diffusion in silicalite, activated carbon and carbon molecular sieves were reported by Ma; Suzuki, Hori and Kawazoe; and Seewald, Klein and Jüntgen. The influence of the heat of adsorption in retarding adsorption rates and broadening the experimental breakthrough response were discussed by Sircar, and by Schadl and Mersmann. The influence of nonisothermal conditions is greatest at low sorbate concentrations. In the extreme limit of an irreversible isotherm the heat and mass transfer equations become decoupled, permitting a simple analytic solution for the adiabatic breakthrough curve. On the basis of such a solution, the variation in the form of the response curve depending on whether the thermal front lags or leads the mass transfer zone was discussed by Ozil.

In modelling the dynamic behaviour of an adsorption column it is common practice to use a simple linearized rate expression, first introduced by Glueckauf, to approximate the more complex diffusion model. This approach has been shown to work well in many instances but it breaks down in the case of a liquid phase system with two diffusion resistances (micropore-macropore) and significant intraparticle holdup within the macropores. For such systems there appears to be no adequate alternative to the full diffusion equation formulation (Weber).

For modelling the behaviour of many practical systems, particularly in pollution control where the system often contains many adsorbable species, some simplification is required in order to reduce the modelling problem to a tractable level. The possibility of lumping together several components and treating the system as containing a smaller number of pseudo species was examined by Tien. However, in view of the results obtained by Balzi, Liapis and Rippin (*Trans. Inst. Chem. Engr.* 56, 145, (1978)) it is evident that this approach must be treated with caution. In principle each component introduces an additional transition into the column response and lumping is only justified for components for which both the kinetics and equilibria are similar. The correct groupings of the individual species may therefore be difficult to decide *a priori*. Nevertheless lumping seems to be the only feasible approach and further work to decide the precise conditions under which such an approximation is useful seems justified.

Given adequate kinetic and equilibrium data, a reliable numerical prediction of the column response may now be obtained, even for moderately complex multicomponent nonisothermal systems (see, for example, Sircar). No doubt in the future such computational methods will be extended to even more complex "real" systems and to situations such as pressure swing adsorption where the boundary conditions are more complex. However, in many ways the most important barrier to the widespread use of sophisticated mathematical models remains the lack of reliable kinetic and equilibrium data. The problem of predicting multicomponent equilibria from single component isotherms and the study of counter-diffusion in microporous adsorbents are perhaps the two most challenging and important areas for future work.

D. M. Ruthven

WILLIAM STEELE had his first serious contact with surface science and physical adsorption while doing his Ph.D. research under the direction of Professor George Halsey at the University of Washington. He had completed his B.A. degree at Wesleyan University in Connecticut in 1951, and in 1954 he went to Penn State University as a postdoctoral fellow to study the adsorption of simple gases on prepared surfaces in the Cryogenic Laboratory there. He has remained at that university ever since. While retaining an active interest in the theory and more recently, in the computer simulation of physisorbed gases, his interests have broadened to include the theory and simulation of bulk, three-dimensional molecular fluids. He is currently serving as an Associate Editor of the Journal of Physical Chemistry as well as being a present or past member of the editorial boards of Molecular Physics, Advances in Colloid and Interface Science, Journal of Physical and Chemical Reference Data, Journal of Molecular Liquids, Journal of Chemical Physics, Journal of Colloid and Interface Science and Annual Review of Physical Chemistry.

\- - - - - -

SUMMARY

THEORY AND MODELS

As far as physisorption of monolayers of pure gases on homogeneous solids is concerned, the theory is now at a fairly rigorous molecular - statistical mechanics level. Lateral interactions play a significant role in these systems and much of the current effort is in the direction of refining the models of these interactions for more complex adsorbate molecules. For those cases where the gas - solid energy can also be modelled accurately (such as graphitic substrates), the resulting energies can be used either in computer simulations or in integral equation theories to produce numerical results which are in reasonably quantitative agreement with experiment.

Computer simulations now indicate that the layered structure of multilayers plays an important role in theories of thick-film formation, both on free surfaces or in the more restricted geometries encountered in porous solids. Theories are not as detailed as for monolayer adsorption, but Polanyi (or Frenkel-Halsey-Hill) theory is reasonably successful and the modification of these theories is a promising avenue for future progress.

On the descriptive level, the distinction between microporous and mesoporous systems is by now fairly clear, and techniques based on analysis of nitrogen adsorption isotherm data with the aid of the t-plot seem to be successful in characterizing a wide range of porous materials. In addition to complex pore geometries, part of the problem in developing quantitative theories for these systems is the presence of significant surface heterogeneity.

A recent development which is both fundamentally interesting and practically important is the extensive and precise measurements of physical adsorption of various substances on graphitized carbon black at temperatures and pressures in the bulk adsorbate critical region. The experimental data have been obtained mainly by Findenegg and coworkers and are in reasonable agreement with limited integral equation theory and/or simulation studies reported to date. Part of the interest here arises from the fact that layering near the surface is less than at temperatures near the normal boiling point of the adsorbate and, thus, the distinction between multilayer and monolayer adsorption is less.

In spite of significant progress, physical adsorption of pure gases and mixtures on heterogeneous solids remains an unsolved problem. Professor Jaroniec presented an admirable summary of the present theoretical situation with regard to models based on local isotherm equations with negligible lateral interactions; in this much-studied case, only the number distribution of adsorption site energies need be specified to obtain thermodynamic properties. The neglect of adsorbate - adsorbate energies continues to be an unsatisfactory feature of the theory; however, explicit inclusion of the lateral interaction requires additional assumptions which are at least as arbitrary as those made when these energies are omitted. On several occasions during this meeting, it was pointed out that analysis of heats of adsorption (or the temperature dependence of the isotherms) should accompany the analysis of the isotherms themselves; this is an important point and cannot be reiterated too often.

Adsorption in zeolites was reported in a number of papers. These solids provide an interesting set of heterogeneous solid systems in which the geometric distribution of heterogeneity is reasonably well known; Ruthven's data analyses are encouraging and it would seem that additional theoretical work on these systems at the molecular level might be productive.

The mixture adsorption data presented were impressive in their quantity and diversity. It is now becoming apparent that ideal adsorbed solution theory, which quite successful, should often be replaced by more detailed theories. Several alternatives to ideal adsorbed solution theory were suggested: vacancy solution theory, for example. It is not yet clear what the ranges of validity of these various approaches are; one might guess that this would depend upon porosity and surface heterogeneity as well as the nature of the components in the adsorbed mixture, but sorting this out remains a problem for future study.

W. A. Steele

WALTER J. WEBER, JR. is Professor of Environmental Engineering and Chairman of the University Program in Water Resources at the University of Michigan. He holds an Sc.B. (1956) from Brown University, an M.S.E. (1959) from Rutgers University, and A.M. (1961) and Ph.D. (1962) degrees from Harvard University. He is a member of a number of professional and honorary societies and is a registered professional engineer. Weber joined the faculty of the College of Engineering at Michigan in 1963. In the ensuing twenty years he has conducted and directed research and published over 200 papers and three books relating to physicochemical processes and mathematical process modeling in water quality engineering, with particular emphasis on adsorption fundamentals and technology. He has received a number of awards and recognitions for his contributions to research and development.

\- - - - - -

SUMMARY

ENVIRONMENTAL APPLICATIONS

I have been asked to summarize the significance and implications of these proceedings for environmental engineering applications. The task is complicated by the diversity of sorption phenomena operative in environmental systems. Indeed sorption reactions play some role in virtually all environmental transformations and treatment or control operations. To cite a few examples:

→ Atmospheric transport and fallout of gaseous contaminants is affected by their sorptive association with airborne particulates;

→ Sorption of dissolved pollutants by suspended solids and sediments markedly impacts their transport and fate in aquatic systems;

→ Sorption of substrate by microorganisms is an essential step in the biological transformation of substances in the environment;

→ Sorption of harmful chemicals by microorganisms and subsequently larger biological forms plays an important role in the bioconcentration of such chemicals in environmental systems; and,

→ Many of the treatment processes used for separation of contaminants from air, water, and wastewater -- such as coagulation, filtration, membrane separations, solvent extraction, ion exchange, and adsorption by activated carbon -- are based upon sorption phenomena.

Several of these examples involve sorption phenomena coincidentally rather than as primary transport or transformation processes. In these cases the sorption reactions, although integral to a particular process, are usually not delineated and quantified in studies of that process. In other cases sorption is the primary separation process, and is more directly examined and better quantified. The environmentally-related papers given at this Conference have focused on processes in the latter category; most particularly, separation processes involving microporous adsorbents such as activated carbon, synthetic resins, and zeolites. For this type of process, however, it is clear that the circumstances surrounding environmental applications introduce complexities that set these apart from other applications of sorption technology.

Sorbate/sorbent/solution heterogeneities and transient conditions are persistent themes underlying the environmental applications discussed at this conference. In attempting to categorize the papers dealing with these applications, I would identify three major areas of emphasis and concern:

- → The mass transport aspects of sorption;
- → The role of biological activity on adsorbent surfaces; and,
- → The grouping of species of similar sorption characteristics to facilitate the description and prediction of behavior of multicomponent systems.

Mass Transport

A number of papers concerning both theoretical and phenomenological aspects of sorption thermodynamics and equilibria have addressed systems of particular environmental interest, including those by Belfort et al.; Ishizaki; Grant and Urbanic; Koopal; Manes; Ake Rasmuson; Reucroft; and Seewald et al. The transient conditions so commonly associated with environmental applications, however, usually result in systems which are driven by sets of fluctuating thermodynamic conditions, and which therefore shift uncertainly among multiple equilibrium positions. Such fluctuations effectively lead to non-equilibrium or rate control of associated phase separation processes, particularly for sorption from aqueous phase. Because boundary layers at adsorbent/solution interfaces cannot be eliminated in practical reactor systems and because the vast majority of sorbents utilized in separation processes are microporous in nature to provide large surface areas for sorption, rates of separation are frequently controlled by extraparticle and/or intraparticle diffusional mass transport processes. A number of facets of these mass transport processes have been discussed here by Beverloo and Pierik; Costa and Rodrigues; Crittenden et al.; Levan and Vermuelen; Mathews, Noll et al.; Takeuchi and coworkers; Suzuki et al.; Weber; and Weissenhorn. While models for accurate description of mass transport phenomena have been developed for relatively simple systems involving only one or two primary sorbates, it is clear from the discussions at this Conference that the reliability of such models for prediction of contaminant sorption dynamics in environmental systems is adversely affected by such complications as:

- → The multicomponent or multiple-sorbate nature of environmental systems;
- → The typically transient character of such systems with respect to both composition and concentration;
- → The hysteretic behavior which commonly characterizes desorption in such systems; and
- → The transformation of sorbed species at sorbent surfaces.

Not only do these factors complicate predictions of adsorption but, as importantly, they make it difficult to forecast the subsequent displacement of previously adsorbed species. The latter phenomenon is of particular import because of the potential for reintroduction to treated or effluent streams of certain contaminants at levels higher than originally present in the influent streams.

Biological Activity

Microorganisms are present in virtually every environmental system to which sorptive separation operations might be applied. This is particularly true of aqueous systems such as waters and wastewaters treated by adsorption for removal of organic contaminants. The microporous adsorbents used for such separations, particularly activated carbon, provide ideal surfaces for colonization by bacteria and other microbial forms because:

- → Substrate is concentrated at such surfaces;
- → Molecular oxygen is commonly adsorbed by materials such as activated carbon; and,
- → The crevices and pore openings at the external surface of porous adsorbents provide shelter from fluid shear forces.

Biological growth on the external surfaces of a microporous adsorbent may function to reduce the effectiveness of the sorption process by blocking pore openings or by otherwise presenting resistance to the mass transport processes controlling rates of sorption. Conversely, biological degradation of sorbed or sorbing substrate may reduce the overall loading and rate of exhaustion of adsorbents in certain situations. As indicated by the discussions of Maloney et al.; Pirbazari et al.; and Riera et al., it is difficult to make general predictions of the manner in which biological activity will influence the sorption process, and thus difficult to incorporate such effects in the design of sorption separation operations. There is clearly a need to further define factors which control such bioactivity and to develop a more quantitative means for predicting its effects in specific types of sorbate/sorbent/solution systems to facilitate more effective design of such systems.

Species Grouping

A third major aspect of sorption applications in environmental systems addressed at this Conference is the grouping of species of similar sorption characteristics to facilitate description and prediction of multicomponent system behavior. In this approach each group of species is characterized by certain "average" adsorption parameters (thermodynamic and kinetic properties and coefficients). This is the principal thrust of Tien's paper, and is considered as well by Summers and Roberts, and by Pirbazari et al. Although a promising approach for simplification of certain types of complex systems, many compounds may exhibit similar characteristics with respect to sorptive equilibria (i.e., energies and capacities) but dissimilar characteristics with respect to sorption rates (diffusional mass transport). Moreover, sorption operations commonly lead to selective separation of the components of multiple sorbate systems. This, coupled with the variability of the composition of typical feed streams in environmental applications, can markedly and unpredictably change the "average" characteristics of component groups.

In summary, the complexity of environmental systems renders direct application of the theoretical aspects of sorption discussed at this Conference difficult and tenuous. As evidenced by the environmental applications papers, however, theory is slowly but progressively being infused into practice. The substantial research and development challenge is one of adjusting and refining theories and concepts developed for more simple systems to accomodate the heterogeneity and variability of systems as complex as those encountered in environmental applications.

W. J. Weber, Jr.

Author Index

Subject Index

W9-CTJ-893

TEACHER'S EDITION

THE ROBERTS ENGLISH SERIES

A LINGUISTICS PROGRAM

Paul Roberts

BOOK

4

Harcourt, Brace & World, Inc.

NEW YORK CHICAGO SAN FRANCISCO ATLANTA DALLAS

We do not include a teacher's edition automatically with each shipment of a classroom set of textbooks. We prefer to send a teacher's edition only when requested by the teacher or administrator concerned or by one of our representatives. A teacher's edition can be easily mislaid when it arrives as part of a shipment delivered to a school stockroom and, since it contains answer materials, we want to be sure it is sent *directly* to the person who will use it or to someone concerned with the use or selection of textbooks.

If your class assignment changes and you no longer are using or examining this Teacher's Edition, you may wish to pass it on to a teacher who may have use for it.

The Publishers

Contents

Introduction

Aims of the Program

This series aims to improve children's writing by teaching, in a thorough and sequential way, the main features of the writing system—in particular the sound and spelling relationship—and the nature of the syntax.

Though this might seem an obvious plan for an English series to adopt, it has, for several reasons, not been undertaken before. One reason is that until recently not very much was known about either the sound system or the syntactic system of English. Another is that the systems are quite complicated. The rules that govern them are more than can be imparted in a year or two. Because of the difficulties of establishing a real sequence across the years, we have failed to make a frontal attack on the problem and have had to fall back on skirmishing that has sometimes been effective and sometimes not.

But anything less than a frontal attack—a patient and deliberate explanation of the system—has the effect of putting most of the burden of learning to write English on the child. It is he who must figure out the central rules for spelling words. It is he who must, from observation of many sentences, come to unconscious conclusions about what makes some sentences well formed and others not. This is true even though he be given numerous hints: "In certain situations change *y* to *i* and add *es*." "Use *seen*, not *saw*, after a helping verb." "Be sure that each sentence has a subject and a predicate." But such bits of rule, no matter how abundantly supplied, only nibble at the edges. The system itself remains undetailed, unexplained, and, for most children, unacquired.

Of course, many do acquire it, a surprisingly large number. From the specimens of English writing that they encounter, in school and out of it, they somehow work out the rules for writing English. They profit, of course, from the hints and nudges, bits of rules and corrections that they receive from their teachers and their texts. But for the most part they do the job themselves, and some might do it just about as well with no instruction at all. These people are not our prime worry. We hope that they will learn faster and more pleasurably with the teaching provided by this series, but in any case they will arrive.

The ones that concern us more are those of the large segment just below. The pressing current problem for the schools is to develop ways of producing enough young people competent to man the proliferating posts of a society steadily increasing in sophistication—in which there are fewer and fewer jobs for those who merely haul and carry and more and more for those who work with their brains and therefore, necessarily, with their language. These are the many who, for some reason, cannot do it by themselves, who can memorize the spelling of any number of words still without learning how to spell words they haven't memorized, who can puzzle about complete thoughts until Antarctica melts without ever being quite sure which thoughts are complete and which are not. It is hoped and believed that a long and systematic study of English writing will bring such children to a much greater capability in the writing of English than they could otherwise possibly achieve.

To try to reach this goal, we devote seven years, from the third grade through the ninth, to explicating the main features of English writing. In order to make headway, we must necessarily assume progression. That is, we must build each year on the materials taught in the years that went before.

This is not to say that the study of English—any more than the study of mathematics—can move ever forward and be always new. Even the pupil who begins with the third-grade book of the series and proceeds to that of Grade Nine will need much review and repetition. Between every June and September there is inevitably a large loss, and it will often be November before forward movement begins again in earnest. Nor can we altogether ignore the child who encounters the series for the first time in the fourth grade or the seventh or ninth. We must provide for such children such catching-up materials as we can devise. Some of these materials for review, reteaching, or catching up are described on pages 2 and 3 of this introduction.

But we must in the end construct the study first of all for those who go through it all the way. We must suppose that those who enter in the middle, though they should obtain worthwhile results, will work harder for less reward than those who begin at the beginning.

Necessity of Keeping the Sequence

Teachers often pride themselves on not keeping to the order of presentation that they find in the textbook. They work like doctors curing the sick. Noting certain ailments in the class, they look through the text for the appropriate remedies and then apply them.

This procedure is unquestionably the proper one if we conceive the act of learning English as that which was described earlier—in which the chief burden of deducing the system falls on the child and in which the teacher's role is essentially to guide and direct, correct and repair. But if we think of teaching English as the business of teaching a system, then skipping about is clearly impossible. The order of presentation of this series is certainly not the only possible one, but to abandon it without establishing another (i.e., without in effect composing another series) would be to introduce intolerable confusion.

The teacher must therefore take the lessons in the order given. Each one presupposes what went before. One often must go back for reteaching of parts that have not been well enough learned, but one can never go forward, skipping intermediate steps, without breeding certain bewilderment. This is not to say that the teacher should never provide therapy for matters not yet reached in the series. Often this will be necessary to deal with particular points of usage, spelling, punctuation, sentence formation, and the like. But such repairs should be made apart from the text, and for the most part should be made with some caution, the pitfalls of facile generalization being borne in mind. The teacher who does not happen to be also a grammarian should be cautious about formulating generalizations. They may turn out to be invalid and may have to be unlearned in later years.

The General Plan of the Series

The series consists of seven basic books, for Grades Three through Nine. Each book is divided into ten parts and each part into three or four sections—four sections for Grades Three through Five and three thereafter. Each part is composed of two main strands: a reading passage, with notes and questions, and a grammar strand. The grammar includes syntax—the study of the rules that govern the structures that make up sentences, and phonology—the study of the set of rules for the pronunciation of the morphemes that make up sentences. Earlier in the program, in Books 3 and 4, it is possible to identify separately the part of the total grammar program which deals with the writing system. This is called "Sounds and Letters." Later, the components of the grammar are so interwoven in every grammar lesson that it is no longer possible to identify them as separate substrands.

All sections contain other matters as well. There are numerous assignments in composition, letter writing, reports, and the like. Various features of the language and its writing are treated incidentally as the occasion presents itself. Notes of general interest, on such points as etymology and dialect difference, are now and then introduced. However, the bulk of the books is contained in the two strands mentioned.

Some Special Learning Aids in the Textbook

The problem of progression has been mentioned earlier. Pupils in Grade Four who have not studied the Grade Three textbook will need more than the review of Grade Three materials included as reteaching in the early sections of the textbook. This problem grows more acute every year from the fourth grade on because of the accumulating body of concepts and vocabulary.

To help solve this problem, and to provide for systematic review material to be used as needed, each textbook beginning with Grade Four provides at the back a summary of the syntax and the sound and spelling programs of previous grades.

A glossary of terms previously introduced is included also.

In the back of the textbooks for Grades Three and Four, extra word lists illustrating the sounds taught in those grades are provided as a means of reinforcing both sound and spelling. From Grade Five on, the sound and spelling program has passed the stage at which extra words illustrating sounds are useful, and this feature, therefore, is discontinued.

There are ten "Test and Review" sections in each textbook. These are unusually comprehensive and provide a means of diagnosis for immediate reteaching. Page references are given so that the material needed for reteaching may be quickly located.

Publications Other than the Textbook

A complete teacher's edition like this one, with page-by-page teaching suggestions and technical information beside reproductions of pages from the pupil's textbook, is provided for each grade.

A workbook providing extensive further practice and review on the sound and spelling program and grammar program of the textbook is available for each grade.

Record Album

An album composed of two LP vinyl 12-inch 33 rpm records is available for each grade.

These records cover those parts of an English program which textbooks cannot adequately provide for. For example, choral reading, which can only be talked about in a textbook, can be demonstrated on a record.

Here are the types of material presented by the records:

1. Professional recordings of the poems in the textbook
2. Choral reading lessons and demonstrations, for class participation
3. A demonstration of dramatization
4. Models of conversation, introductions, telephone conversations, reports, book reviews, and other oral aspects of some English programs which are more appropriate for a record than for the printed page

Literature Selections

Each part of each section is introduced by a passage to be read and studied. In these, poetry predominates over prose. In Grades Three, Four, and Five, the ratio is about four-fifths poetry to one-fifth prose. In later years the proportion of prose increases somewhat. But in general there is more poetry because it is easier to meet the requirements of the text with poems than with prose passages.

There are several requirements that must be met. One, of course, is that the reading be within the child's presumable capability at a particular grade level. This does not at all mean that it must deal with matter familiar to him. Indeed, we want him to be constantly extending his horizons, not to be forever roaming within them. We want to acquaint him with other times, other places, other views and to let the English lesson play its full role in bringing him into touch with the heritage of western civilization. But obviously one does not go in a bound from Mother Goose to *King Lear.* The attempt has been to find selections that the children, with effort and aid, can at their various ages read and understand.

A further requirement on the reading selections is that they be good literature. They are to be very carefully studied, line by line and word by word. (The teacher may often ask that they be memorized.) They therefore ought to be good enough to merit such close attention, valuable in themselves, whatever other values are wrung from them. The teacher will find among them many well-known items, and some less well known, from eminent writers of English—such as Stevenson, Frost,

Dickinson, Carroll, Eliot, Poe, Churchill, Shelley, Keats, Lincoln, Hardy, Herrick. The pupil who works through the series will embrace intimately a very considerable body of literature—English as well as American—though of course he will not engage in a systematic study or survey of literature.

There is no attempt to make the selections topical—e.g., to have a Halloween poem for around Halloween time, a Christmas tale for Christmas. The other requirements for the selections make this one impractical, and in any case there are available many anthologies of verse for special occasions from which the teacher may draw if the need is felt. Naturally some of the selections do refer to particular times of the year, but these are not integrated with the school calendar: a poem on autumn may be studied in April. The integration is of a different sort, and on no account should the selections be studied out of order.

The Purpose of the Reading Passages

The reading selections serve several ends in the study of English. One justification for them has been mentioned already—that they are for the most part valuable in themselves and need no justification. They are sweet and good for one, like milk. However, they serve in practical ways as well.

For one thing, they are useful as a kind of corpus, as a demonstration of real English being used. From them we can often draw examples as desired and not be wholly dependent on made-up examples. They illustrate English in varied dress, from the Scottish of Burns to the picturesque talk of nineteenth-century American railroaders. They give us a chance to bring the children—some children—to a sensibility of the potential sweep and power of English, to an understanding of what can be done with language. Beyond this, they frequently provide topics for work in composition, for reports, discussions, dramatizations, and the like. This is not to say that the child is required to write appreciations and critiques of the reading passages. Literary criticism can wait until college, or at least until the later years of high school. But it is clearly useful for the class to have a shared body of experience and information and, insofar as possible, to let assignments spring from this rather than from arbitrarily selected topics extraneous to the English lesson.

But the most important purpose of the reading selections is a very simple one: to teach the child to read more accurately and more sensitively. The aim is not to teach him to read with appreciation or to read rapidly—however desirable these goals may be—but simply to be able to discover consistently and accurately what is on the printed page. A large number of even those students who reach college are unable to do this—presumably, in part at least, because they have never been taught to do it, have never been kept steadily to the task. They often get the gist of an argument by reacting to key words here and there, but have no skill in arriving at precise meanings in writing of any difficulty.

It is, of course, hard to learn to read accurately. That is why we must devote so much time to the subject in the school years. Most children beginning this series in the third grade will know how to read in only the simplest sense. They will usually have a general notion of the significance of combinations of letters and an ability to read rather simple sentences by themselves. The reading selections of the third-grade book, though they have been chosen for simplicity, should for the most part be read to the children by the teacher, and played on the phonograph if the poetry recording is available in the classroom. But even in the third grade, the child should not be absolved from the responsibility of working through the passage by himself too and answering the questions on it. Whenever he strays from the poem or story and answers not from the words but from the illustration or fancy's flight, he must be brought back: Is that what the line says? Who does what to whom?

In the fourth grade and later, the child can do much more by himself and can work with material of steadily increasing complexity. Still the teacher should settle for nothing less than an accurate report on what is written there. What is the verb? What is the subject and what the object? How are they doing whatever it is they are doing?

It is to be noted particularly that this series takes

the point of view that the child's attitude toward the prose and poetry is his private affair. He probably will like the selections, but he should not be required to like them or be made to feel a sense of failure if he doesn't. Failure resides in not understanding what is written, not in inability to work up certain emotions about it. It is a reasonable assumption that at least a large proportion of children will appreciate good literature if they understand it, and, conversely, that they haven't much hope of appreciating it if they don't understand it. But appreciation is in some sense a private matter, between the individual and the poem.

Part of the problem of understanding is, of course, the matter of vocabulary. As in other features, the attempt is to provide selections of vocabulary difficulty somewhat greater than what the child is used to, but not so much greater that he cannot arrive at comprehension with the help of the text and the teacher. He is encouraged to arrive at meanings from familiar contexts, from analogies, and, beginning in the fourth grade, through the use of dictionaries.

The Writing System

An important part of each book of the series is the study of the relationship of sounds and letters. In Books 3 and 4 this study is treated as a separate substrand of the grammar, called "Sounds and Letters."

The English spelling system, despite its notorious vagaries, is essentially systematic. It is not an ideographic system like Chinese, in which the character for each word must be learned as a special item unrelated to others. English is indeed quite regular. There is only one possible way to spell such words as **club** or **spat**. There is only one possible way to pronounce such made-up words as **bedge** or **cran**. It is simply not true that good spellers memorize individually the orthography of all words that they may need to spell and then write them correctly when the occasion calls for them. Good spellers somehow come to encompass the system—its rules as well as its exceptions. The exceptions—insofar as they truly are exceptions—they do indeed learn one by one. There is no way to arrive at the spelling of **myth** or **laugh** except to discover that these particular words are spelled in these particular ways. But the vast majority of words are spelled by rule, whether the rule is consciously known or not.

We have tended to let the specter of the exception frighten us away from the rule. How, we ask, can we give rules when there are so many exceptions? But the point is that any rule, no matter how many the exceptions, is better than no rule at all. For if there is no rule, then everything is an exception—every word must be learned as a separate datum, unrelated to any other. The task would put an intolerable strain on all but a very few memories.

In this series we therefore begin with the assumption that English is essentially systematic—has rules—and that the system must be learned by those desirous of spelling the words of the language correctly. We could establish the rules in either of two ways: by beginning with the letters and showing the sounds that they regularly signify; or by beginning with the sounds and showing the letters regularly used to signify them. The first of these directions is that normally pursued by the child learning to read; the second is of more importance to the child learning to spell. Since we suppose that by the time he enters the third grade, the child has normally had a certain amount of work going from letter to sound, we now reverse the direction and go for the most part from sound to letter. We explain, for example, the conventional ways of spelling the vowel sound heard in the words **fine, high**, and **my**. Often we can point out a usual way and some exceptional ways. (The **fine** method is usual, the **high** one exceptional.) Sometimes we can make restricted rules, saying that you do such and so always or only in a particular situation. (The vowel sound of **fine** is represented by the letter **y** at the end of words.)

If we were doing Spanish or Italian or Turkish, we could get through with the spelling system in a relatively short time. In English it takes longer. In the third-grade text we consider the spellings of somewhat less than half of the English sounds, and these only in monosyllabic words. In the fourth grade, we nearly, but not quite, complete the list. In the fifth grade we finish the spelling of monosyllabic words and begin the complicated matter of spelling the reduced vowel schwa—for example, the second vowel sound in such words as **letter** and **grammar**.

From Book 5 on, the spelling substrand is not identified as a separate component. For many of the rules of sound and spelling depend specifically on the syntactic, as well as the phonological, component of the grammar. For instance, the pronunciation of a word—and therefore, perhaps, an essential clue to its spelling—may depend upon whether the word is a noun or a verb. Certain suffixes establish the stress of a word, thereby changing its pronunciation and often making clear its spelling. For example, there is nothing in the pronunciation of the proper noun **Byron** to suggest that the second syllable has the vowel letter **o**. There is, however, in the adjective **Byronic**, where the suffix **-ic**, according to a very regular system, moves the stress to the penultimate syllable. By the sixth grade and later, the child's vocabulary is such as to make it profitable to study systematically the composition and spelling of words of Latin and Greek origin, and this is done.

In order to discuss sounds, it is necessary to have a system of referring to them. One cannot simply use the letters of the alphabet because, among other reasons, there are quite a few more sounds than letters. Various systems are available—such as the International Phonetic Alphabet or the Trager-Smith convention. It is with one of these transcriptions that the linguist would be happiest, but all of them were rejected because of the pull of the dictionary. The child's other experience with the rendering of speech sounds must necessarily be with the dictionary, and in order to avoid putting an additional burden on him, we go along with dictionaries here, insofar as possible. Dictionaries are of course not identical in their usage, but we conform here to the more general practices with only a few modifications.

Thus where the dictionaries generally use two letters for a single sound—as for the **sh** of **shall**—we do too, but we tie the letters together with a ligature as a reminder of the singleness: /sh/. We follow the general dictionary trend away from diacritical marks, using only the macron for the so-called long vowels, like that of **fine** and **high**: /ī/. The symbol /ə/ is used here for the reduced vowel schwa, as it is now in many dictionaries. We adopt the device of slant lines for symbols that indicate sounds and boldface type for those that indicate letters. Thus, the word **cat** has the sounds /k/, /a/, /t/. It has the letters **c, a, t**.

We must also have a convention for naming sounds in oral discussion. The letters already have conventional names—ay, bee, see, dee, ee, eff, gee, aitch, etc. Most of these can be adopted for naming **consonant** sounds. Thus /b/ is called "the sound bee," /h/ "the sound aitch." For the few consonant sounds not symbolized by single letters of the alphabet, devices readily suggest themselves: we call /ch/ "chee" by analogy with "bee" and "dee," /sh/ "esh" by analogy with "ess."

The vowels present a knottier problem. To speak of "short" and "long" vowels is not only misleading but inadequate: there are more vowels than those conventionally called short and long. Neither can all the vowels be named by simply pronouncing the sound. The vowels of **fine** and **moan** can be; one can say "I" and "oh." But those of **fin** and **man** cannot be, or at least not naturally, because these, along with four other vowel sounds, never occur in English except when followed by a consonant sound. We therefore adopt the following convention. We name the vowel sounds by making the sound followed by the consonant sound /k/. The vowel sounds of **fin, man, ten, mine, main, town** are called respectively "ick," "ack," "eck," "ike," "ache," and "owk."

In the course of the study, the pupil is taught something of how some of the sounds are made and some of their distinctive features—such as the difference between voiced sounds and voiceless sounds. But the purpose of this is simply to impress on him the difference between sounds and letters. We do not expect him to learn to describe sound production in any serious way but only to be able to identify the different sounds that he uses and to be aware of them. In general, this series does not attempt to teach the child linguistics; it merely uses linguistic knowledge to pursue the traditional aim of the English class—the competent use of the language.

Syntax

It is in the teaching of syntax that this text leans most heavily on linguistic knowledge, because it is in syntax that this knowledge is most fully developed and most readily applicable. The study of sentence structure (as also, though to a much lesser extent, the study of sound structure) in this series is based on what is called generative transform grammar. In this expression, the term **grammar** has a wider application than is usually given to it. Generative transform grammar contains both a syntactic component and a phonological component. The syntactic component contains such matters as words and parts of words and their arrangement in sentences. It contains what is conventionally thought of as the content of grammar—nouns and verbs, subjects and objects, number and tense, prepositional phrases and relative clauses. The phonological component describes the sound structure of the language; it is composed of the rules for pronouncing English.

Generative transform grammar in America is now fairly well developed so far as its syntactic component is concerned. It has been deeply studied and widely taught, and it is being used more and more for school purposes. There seems to be little doubt that it will establish itself as the central theoretical background for the grammar taught in schools at all levels.

Because of widespread confusion on the subject, it will perhaps be useful here to sort out various "grammars" that have been used in or proposed for the classroom. We shall distinguish three main types: traditional grammar, structural grammar, and generative transform (or, simply, transformational) grammar.

The first of these is what we are all familiar with from our own school days. Its roots go back over two thousand years, and it was most thoroughly worked out as a description of Latin. When, in the seventeenth and eighteenth centuries, attention was turned to the "new" languages—English, French, Italian, etc.—the tendency was to describe these according to the Latin plan, even though they were very different from Latin in structure. Traditional grammar had other drawbacks apart from this tendency to latinize. Even for Latin it was essentially superficial. It made little attempt to grapple with the deep questions, such as how people learn languages, including their native languages, what the nature of correctness is, what is meant by grammaticalness, what the problems of definition are, what the (often hidden) mechanisms are that tie a language together and make it workable.

Essentially, traditional grammar was a description of language that rested on no theory of language. At its best, it was an interesting collection of facts of the language and observations on them. But even at its best it had little explanatory power and largely put on the reader the burden of figuring out the essential systems. At its worst—in the form in which it usually filtered into school texts—it was largely contradictory and frequently absurd. Pupils often could not understand grammar for the very good reason that the grammar presented to them was not understandable.

Traditional grammar was thus a fairly easy prey to the linguistic scientists who attacked it in the first half of the present century. Linguistic science also has roots that go very far back, but until 1910 or so, linguists were for the most part engaged in historical and comparative studies—figuring out, for example, the development and relationships of the Indo-European family of languages and of other language groups. Early in this century, attention turned increasingly to present-day languages and their structure. Many traditional ideas were shown to be untenable and others were proposed to take their place. The grammar that the linguists worked out at this period is usually called structural grammar.

The collision between the traditional grammarians and the structural linguists wasn't as violent as it might have been because it wasn't precisely head on. Traditional grammar was primarily interested in syntax, whereas structural grammar focused, at least in the beginning, on sound structure. Linguistic science was initially absorbed in working out the nature and description of the phoneme (the unit of sound) and the morpheme (the unit of meaning) and paid relatively little attention to nouns and verbs, subjects and predicates, and the like. Gradually, however, linguists came to criticize the traditional presentations of syntax and eventually to put forth syntactic proposals of their own.

Books by linguists like Archibald Hill, Charles C. Fries, George Trager, and Henry Lee Smith presented the new kind of grammar now generally called structural. In these, it was recognized that speech is the underlying fact which writing symbolizes; grammar is therefore essentially a description of speech sounds and of sound combinations and collocations, of course represented by letters of some sort, either those of the traditional orthography or those of a phonetic alphabet designed for the purpose. The structural linguists, further, had an entirely different notion about correctness than did the traditionalists. The latter seemed to have a platonic view of the matter: correctness was an absolute of some sort, discoverable by logic or reason. The structuralists, while they never said (as they have often been misrepresented to have done) that anything goes, did say that anything goes that goes. That is, any structure generally accepted as proper by a given speech community is correct for that community. If the community in question is the one that sets general national standards, then the structure is correct nationally. Logic, in any event, has little to do with it.

The structuralists also pointed out the weakness and contradictions in the categories and definitions of traditional grammar. It was shown that it was impossible to base definitions on meaning or presumed meaning and that all definitions so based reduced ultimately to absurdity.

Suppose we consider such a meaning as "time." We might try to begin by asking how many "times" there are and then discovering the language forms that express these. Probably we would start by supposing three "times"—past, present, and future. But a brief examination of the language (as of philosophy) shows that there are many more time distinctions than three. We have present action ("I am writing"), repetitive present action ("I write every day"), durative present action ("I write for hours"), past action ("I wrote"), unspecified past action ("I have written"), future ("I may write"), near future ("I am about to write"), past near future ("I was about to write"), and many more. If we change the verb to **seem** or **know**, all the meanings shift somewhat and the categories multiply. It is simply impossible, at least in the present state of knowledge, to catalogue ideas and then find the language forms that express them. It is, however, possible to detail the language forms and say something about the meanings they express. It was a serious weakness of traditional grammar that it tried to go from meaning to form. It was a great contribution of structural grammar that it went from form to meaning.

Structural grammar made other contributions too, more than we have space to consider here. However, it also had its fallings short. Though it was much sounder than traditional grammar, it was also much narrower. It confused theory with method and set up methodological rules that imprisoned it and kept it from coming to grips with the broad problems of language and language learning. It also, in its quite proper reaction against the faults of traditional grammar, often failed to recognize that traditional grammarians had certain virtues too and were often on the right track.

There had inevitably to be a second reaction, this time against structural linguistics, and this reaction came in the middle 1950's in the work of the scholars usually called transformationalists. In the often bitter disputes that took place between structuralists and transformationalists, it was not infrequently overlooked that they shared a large extent of common ground. Transformationalists would support the criticisms of traditional grammars mentioned above: that they were too latinized, that they did not perceive that the base of language is speech and not writing, that their definitions were absurd, that their categorization was illogical. Transformationalists looked on such matters as obvious and therefore not worthy of discussion.

What they did instead was to turn attention to deeper problems, such as this one: What is the nature of language such that human beings can learn it? Or, conversely, what is the nature of human beings such that they can learn language? They began by demonstrating that we do not learn language in the sense that we learn facts of some sort. We do not memorize a certain number of sentences which we then use when the occasion arises. Most of the sentences we say and hear, write and read, have never occurred before. It is quite unlikely that any sentence on this page ever occurred before. Yet you can understand these new sentences, as you will understand the new sentences spoken to you in your next conversation. How do you do it?

Obviously, what we learn when we learn English is not a set of sentences but a sentence-making machine. We learn a mechanism for generating sentences according to the requirements of the circumstances through which we move, and for understanding such sentences. This mechanism is what grammar is.

We cannot stop here for a close examination of the nature of the beast. It is unfolded, insofar as space and pupil capability and present knowledge permit, in the syntax strand of this series. But it may be said that the grammar of English (or of any language) is now thought to be best viewed as a small set of sentences, called kernel sentences, plus a set of rules for transforming these into more complicated structures. Given a finite set of kernel sentence structures plus a finite set of transformational rules, we can generate infinitely many sentences, including infinitely many never produced before. The working out of the precise nature of the kernel structure and the precise nature of the transformational rules is the task of the grammarian.

Though the pronouncements of the transformationalists came as a shock to the linguistic world and generated not only sentences but also that homicidal rage that so often characterizes the gentle scholar, the dispute is now simmering down a good deal. Most structuralists will now accept, if not all, still a good deal of transformational theory, and transformationalists will eventually have to agree that they owe much to their structural predecessors. Not only that, but the developments of the last few years have even rehabilitated traditional grammar to some extent. It is now seen that traditional grammarians were not quite so wrong as they were sometimes said to be, that many of their intuitions were right and many of their directions worth pursuing. As a result, the teacher whose only grammatical experience is with the traditional will find transformational grammar much more familiar than structural grammar would have been.

The Presentation of the Grammar

When we present the grammar of a language to people who already speak the language, we run into a very serious initial problem. It is that, however backward the learner may be and however young, he is capable of using and understanding sentences enormously more complicated than those he studies in his grammar lesson. To a foreigner learning English, the sentence "This is a pencil" is a wholly new thing, a collection of grammatical facts to be absorbed and then put to use. To the American third-grader it is as familiar as rain and presents no particular problem. He is very likely already capable of writing sentences of much greater complexity: "When I get home from school, my mommy gives me some milk and cookies."

This gap between the sentences whose structure we can explain to the child and those which he is capable of using probably accounts as much as anything for the general failure to teach grammar in a systematic way. One feels that the grammatical explanation must be what is called functional. That is, it must have an immediate bearing on anything the child may write. But consider how much grammar there is in so simple a sentence as "When I get home from school, my mommy gives me some milk and cookies." It contains an introductory dependent clause with a rather special adverbial (**home**) and a rather special verb (**get**), in addition to a prepositional phrase and a personal pronoun. The main clause has a possessive, an indirect object, and a compound direct object. The sentence displays tense and number, subjects and predicates, different types of nouns. It would take quite a long time to teach the grammar that the sentence contains (and to demonstrate that it is indeed a sentence).

Yet if we want the grammar to function, the grammar must be described. It does no good to talk vaguely about complete thoughts and agreement, things and actions and descriptions. We have to deal precisely and carefully with the grammar itself. This means that we must begin with the structure of very simple sentences—"He is here," "This is a pencil"—sentences way below the level of complexity that the child is actually using. Only on this base can we come in time to the complicated structures in which the conscious knowledge of the grammar can truly function, where we can deal seriously with problems of sentence structure and punctuation and where the pupil can arrive through study, not intuition, at what is called sentence sense. We begin with a description of simple sentences, in which the child is hardly likely to make mistakes, in order to reach, years later, the complicated structures in which the problems of writing abound.

In the third-grade text, we discuss the structure of kernel sentences only. (The term **kernel** is not used here; we speak of these sentences as simple sentences. But the teacher should bear in mind that something quite precise is meant. Thus "Are you hungry?" is not a kernel—simple—sentence. Neither is "Yes, I am." These are transforms.) In this grade we introduce the learner to the noun phrase that functions as the subject of the kernel sentence and the verb phrase that functions as its predicate. Number is introduced in this grade and agreement of subjects with forms of verbs and **be**. In general, spelling problems—such as spelling the plurals of nouns—are taken up and practiced as they arise.

In Grade Four, the structures of noun phrases and verb phrases are studied further. Adjectives are discussed thoroughly. In this grade the first transformation—the possessive—is described. Past tense is studied in Grade Four, but the concept of tense is more thoroughly considered in Five, where expanded forms of the verb phrase—e.g., "is talking," "has talked," "should talk" are introduced. In Five, we introduce the concept of the morpheme and begin to use simple formulas showing the structures that make up sentences. In Grade Six we add to the description of the kernel structures and move more seriously into transformation, taking up, in particular, compounding and the construction of the relative clause.

In Grades Seven through Nine we consider the more complicated structures of which mature writing is composed. We are now in the area where the usual difficulties obtain: punctuation, sentence structure, grammaticalness. On the foundation that has been built, we are able to deal with these in a precise way. At the end of Grade Nine, the pupil doesn't know all there is to know about English grammar, but he should be in firm and conscious control of its central features.

Terminology

The intent in this series is to use the most familiar and traditional terms possible. We avoid when we can terms used only among linguists. We avoid entirely the children's terminology that has sometimes been adopted—"naming word," "picture word," "helping verb," and the like. At some point, the child must come to the adult terms—noun, adjective, auxiliary—and he might as well begin with them. If he needs help in pronouncing them, we help him.

The teacher will therefore find in this series the old familiar terminology—noun, verb, subject, predicate, modifier, prepositional phrase, relative clause. Particular care must be taken, however, not to jump from these to the old familiar definitions. Thus **adjective** does not mean here "any word that modifies a noun." Adjectives do not always modify nouns, and many words that modify nouns are not adjectives. The teacher should follow the text closely in identification and description and ways of talking about particular structures, however familiar the term used to name them, and not try to second guess. Casual embroidery might not hurt the child at the particular grade level at which he is working, might even help him, but it is likely to present serious difficulties in later years and necessitate painful unlearning.

This series is built on the assumption that learning English is a serious business, like learning mathematics. One would be chary in a mathematics lesson about introducing procedures and theories different from those of the text, and so should one be chary in a grammar lesson.

A number of terms are familiar enough but are used in new and unfamiliar ways. This has been allowed to happen only when there was no alternative, but the occasions are not infrequent. Thus the term **noun phrase** includes not only groups of words (**some of the boys, all those little red schoolhouses**) but also single words like **he, Jim, nobody. Noun phrase** includes structures—like **he**—which are not nouns and do not contain nouns. The only alternative to this seems to be to use a wholly arbitrary designation for this structure—"structure X"—and this does not seem desirable. The same thing happens with **verb phrase**, which includes groups of words (**waited on the corner, hailed a taxi**) but also single words—**slept, protested. Verb phrase** comprises also structures that do not contain verbs, like **may be hungry**. The term **verb** is not applied in this text to either modals, like **may**, or forms of **be**.

All of these departures from normal usage are signaled prominently in this teacher's edition in the pages corresponding to those on which the terms are introduced. The teacher will do the child, and the child's teachers of ensuing years, a favor by sticking scrupulously to the procedure of the text, even when it may be thought (perhaps quite justly) that the procedure can be improved upon.

Usage

A prime function of an English class or an English series is to bring the child to familiarity with and easy use of those forms of the language that are approved in polite and prosperous society. There are several problems, however. One is to know just what those forms are, for not everyone agrees on them. Is it, for example, incorrect to say "It was her"? Should we take a stand against "Everybody handed in their papers"?

The position taken in this series is a fairly conservative one. The forms illustrated and explained are for the most part those of a rather formal English, much more characteristic of writing than of speech. We are not interested in teaching the child how to chat with his comrades. This he can do without our assistance. We are interested in teaching him the forms of written English, some of which may not be known to him previously.

We are encouraged in the conservative position by the nature of the grammar we are using. In its simplest applications, the grammar will ordinarily generate the more literary forms. Consider, for example, the transformation which produces comparisons:

a. I am tall.
b. John is tall. } →
c. John is tall + er + than + I am tall.
d. John is taller than I am.

If we now delete **am**, we get

e. John is taller than I.

Thus "John is taller than I" is produced by exactly the same set of rules that produces "John is taller than Jim." If we want to get "John is taller than me," we must introduce a special rule: "If the subject of the second sentence is a personal pronoun and the verb or form of **be** is deleted, change the pronoun (or change it optionally) to the object form."

It is usually simpler to omit this last step and be content with "John is taller than I." We need not tell the child that it is wrong to say "John is taller than me." It is not wrong in any very important sense. But we just don't teach him how to do it.

There is an assumption—implicit in the early books of the series, and explicit later—that grammatical is not necessarily the same thing as **good** and ungrammatical is not the same thing as **bad**. Grammatical means simply generated by the rules of the grammar. Though people generally try to use sentences generated by such rules, they sometimes deviate from them for good and sufficient reasons. It is not at all hard to think of situations in which it would be better to say "John is taller than me" than to say "John is taller than I." There is some sense in which the former is a respectable part of English. But since we cannot teach all varieties of English, and, even if we could, would not be able to detail just those social situations in which one is preferable to another, we content ourselves with a somewhat literary sort, which coincides pretty well with the production of the simplest rules of the grammar.

Another problem in teaching usage is that classes and individuals in classes differ enormously in their backgrounds. In one class the students may come from homes in which the kind of English spoken and written approximates fairly closely the forms recommended in school. Another class may come predominantly from homes in which a very different kind of English is spoken and in which the language is never written at all.

The burden of adjusting to these differences falls necessarily on the teacher. The text can only push along illustrating the kind of English used in colleges, business, government, and similar circles. This differs somewhat from the playground English of all children. No matter how refined the home environments from which they come, the children will use, outside of class, sentence structures not described here: "John is taller than me," "It was her." Since the teacher cannot control this extracurricular speech anyway, it is best not to worry about it. One doesn't say flatly that it is wrong to use such structures, but simply that one doesn't use them in class. In papers written for English and other subjects, one uses the structures set forth in the English text.

At the other end of the scale, we have the pupil who speaks an English different in its fundamental rules—not only in its peripheral ones—from the rules of standard English. This is the child who says, "I knows it but I just can't remember it." Here the problem is akin to that of learning a foreign language, perhaps even more difficult. The text gives what help it can, in the presentation of the standard forms and exercises in their use, these being supplemented in the workbooks. Even these may not be sufficient in extreme cases, however, and more practice must be devised. The teacher should in these cases be aware that there are delicate psychological problems involved, as well as linguistic ones. The child who says "I knows it" doesn't do so because he is stupid. He isn't even, in the largest sense, being ungrammatical. He is simply operating with a different grammar, one whose rules are different from that which generates standard English.

There are no particular usage sections in this series. Usage is demonstrated, and sometimes taught, in all grammar lessons. Variations are, of course, most common in the readings. In these we necessarily encounter dialect differences, such as, in a selection from Grade Five, the sentence, "Why, he ain't no trouble." Here we simply explain to the child something about how language differs from place to place and time to time and social class to social class.

More commonly, the reading selections display what we might call poetic ungrammaticalness—a deliberate departure from the normal forms. Thus, in Grade Three, we have the sentence "Great is the palace," departing from the strictly grammatical "The palace is great." We tell the child simply that when we write poetry, we sometimes use the language in special ways.

In the grammar, the language illustrated is for the most part strictly conventional. In particular, we avoid the persistent illustration of substandard forms. Thus, there are no exercises of the type in which the learner is told to select the correct form in such a sentence as "I saw/seen it." This has merely the effect of encouraging the child to see the wrong form as a possibility. Even the youngster who previously would never have said "I seen it" may become uncertain, after working many such exercises, about whether this is not a possibly correct usage. When we teach the past tense of **see**, we just don't introduce **seen** as a conceivable choice. When we teach the participle, we don't bring up the possibility of **saw**.

Ungrammatical sentences are used in exercises only when they are so far removed from any habit that they present no danger. Thus we might give the pupil the sentence "John hurt John" and direct him to transform it to "John hurt himself," in order to practice the reflexive. There is no possibility that he will be trained by such exercises to use "John hurt John" or "I hurt me."

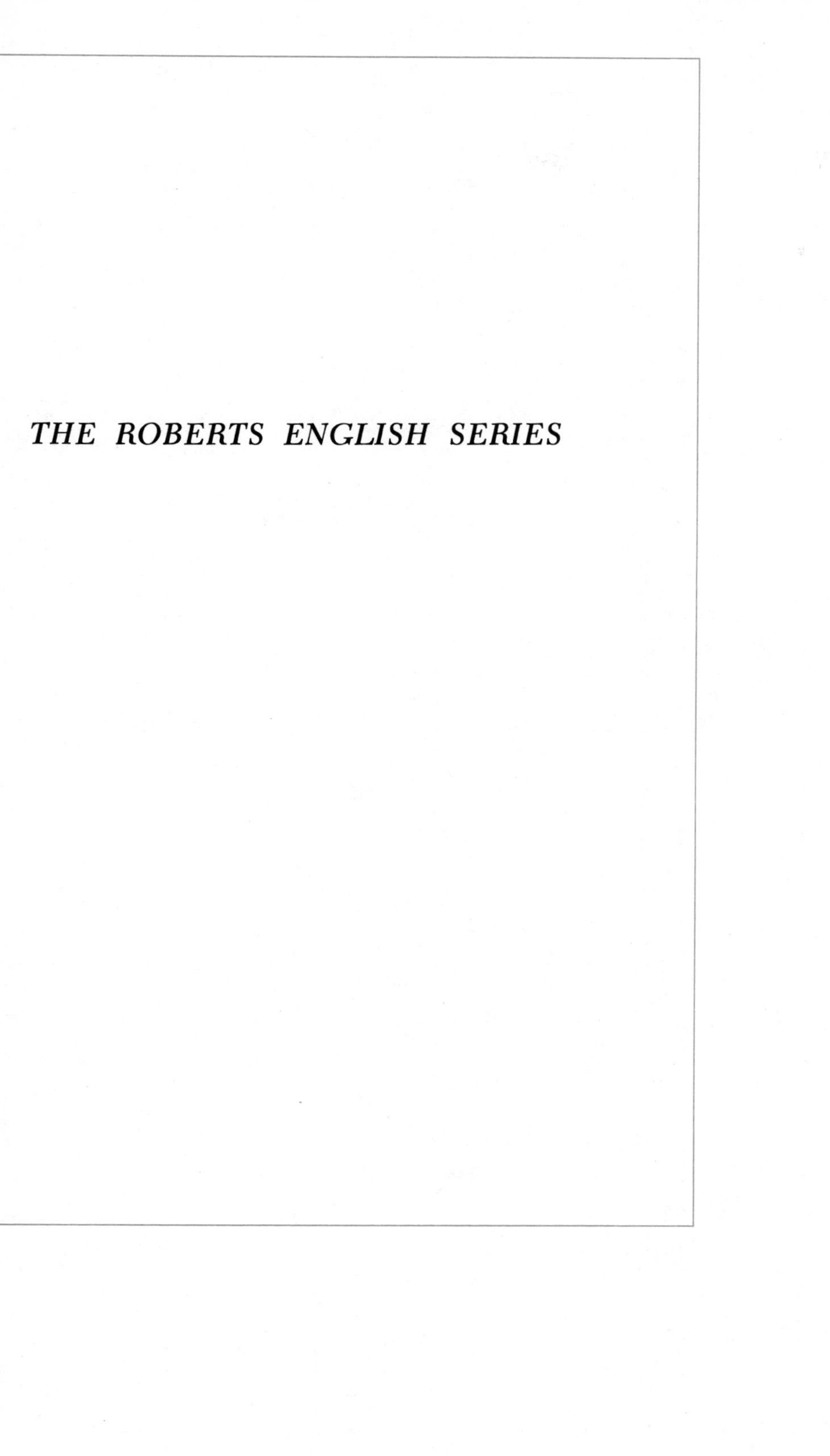

THE ROBERTS ENGLISH SERIES

THE ROBERTS ENGLISH SERIES

A LINGUISTICS PROGRAM

Paul Roberts

BOOK
4

Harcourt, Brace & World, Inc.

NEW YORK CHICAGO SAN FRANCISCO ATLANTA DALLAS

ACKNOWLEDGMENTS For permission to reprint copyrighted material, grateful acknowledgment is made to the following sources:

Basil Blackwell & Mott Ltd.: "The Wind" by Elizabeth Rendall.

E. P. Dutton & Co., Inc. and *William Heinemann Ltd.:* "The Doze" from *Prefabulous Animiles* by James Reeves, copyright © 1957 by James Reeves.

Harcourt, Brace & World, Inc. and *Faber & Faber Ltd.:* "The Ad-dressing of Cats" from *Old Possum's Book of Practical Cats* by T. S. Eliot, copyright 1939 by T. S. Eliot.

Holt, Rinehart and Winston, Inc., Jonathan Cape Limited, and Laurence Pollinger Limited: "Stopping by Woods on a Snowy Evening" from *You Come Too* by Robert Frost, copyright 1923 by Holt, Rinehart and Winston, Inc.; copyright renewed 1951 by Robert Frost.

Houghton Mifflin Company: "Sea Shell" from *A Dome of Many-Colored Glass* by Amy Lowell. "The Plaint of the Camel" from *The Admiral's Caravan* by Charles Edward Carryl.

Arthur S. Johnson: "Travels of a Fox" by Clifton Johnson.

Alfred A. Knopf, Inc. and *Gerald Duckworth & Co. Ltd.:* "Rebecca, Who Slammed Doors for Fun and Perished Miserably" and "The Frog" from *Cautionary Verses* by Hilaire Belloc. U.S. edition published in 1941 by Alfred A. Knopf, Inc.

Lothrop, Lee and Shepard Division of Scott, Foresman and Company: "The Calf Path" from *Whiffs from Wild Meadows* by Sam Walter Foss.

Harold Ober Associates, Incorporated: "W Is for Witch" from *The Children's Bells* by Eleanor Farjeon, © 1960 by Eleanor Farjeon.

The Ben Roth Agency, Inc.: "Noise" by Jessie Pope, © Punch, London.

The Society of Authors, representing The Literary Trustees of Walter de la Mare: "The Old Stone House" by Walter de la Mare.

Illustrations in this book are by the following artists:
Greta Elgaard, pp. 16, 22, 64, 72, 78, 94, 288, 295; Mamoru Funai, pp. 32, 274, 280; Manny Haller, pp. 5, 38, 46, 52, 54, 184, 190, 196, 202, 266; Don Madden, pp. 10, 84, 145, 174, 246, 252; Herb Mott, pp. 128, 136; Rick Schreiter, pp. 100, 106, 112, 214, 222–23, 228–29, 234, 258, 264; Symeon Shimin, pp. xvi–l, 122, 154, 160, 168; David K. Stone, pp. 76, 131.

Contents

Section 1 PART 2

Section 2

Section 3

Section 4

Section 1 PART 4

Section 1 PART 6

Section 2

Section 3

Section 4

Section 1 PART 7

Section 2

Section 3

Section 4

Section 1 PART 8

Section 2

Section 3

Section 4

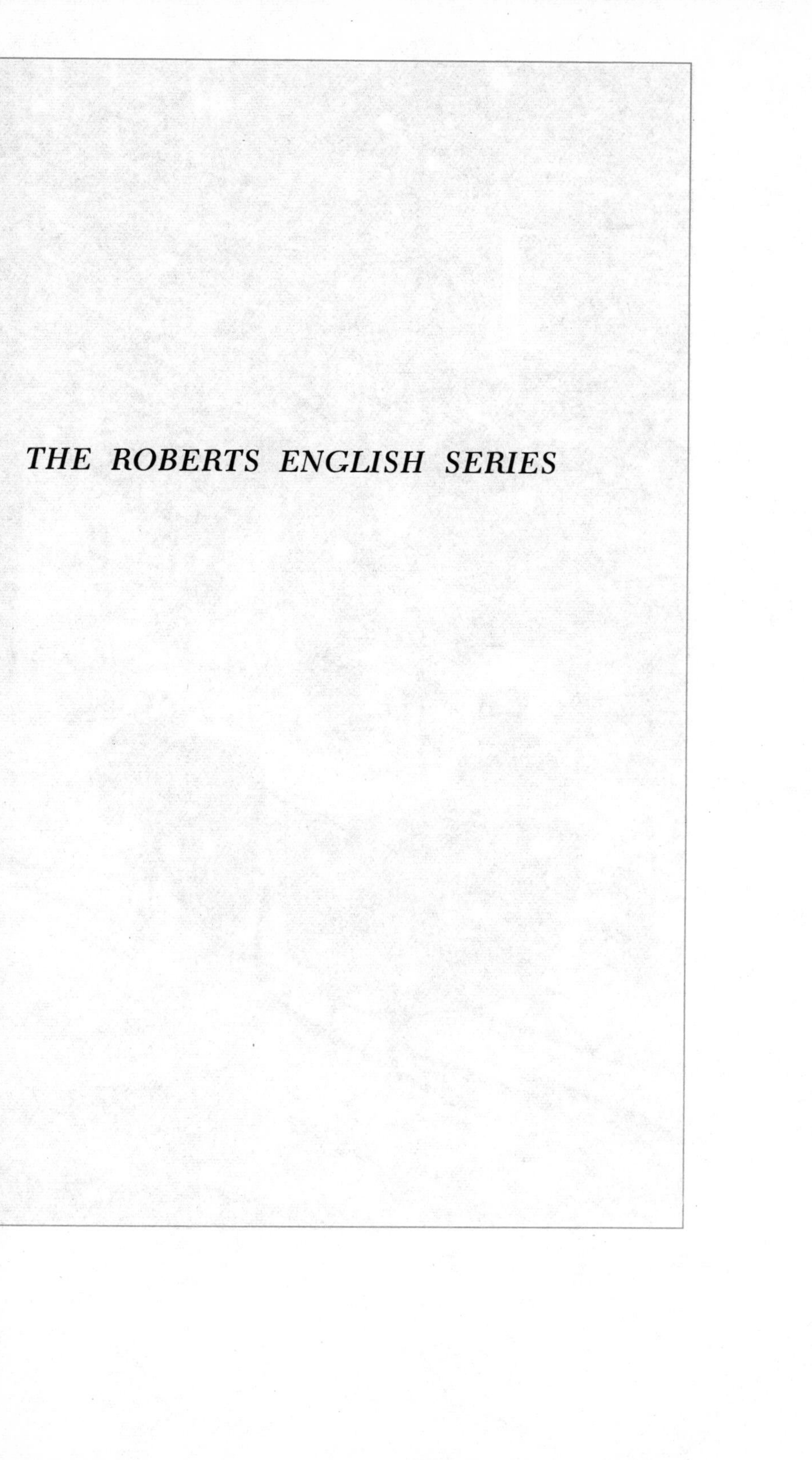

THE ROBERTS ENGLISH SERIES

PART
1

SECTION 1

A Poem

Here is a poem about a man who stops on a winter evening just to look at some woods. Read it, or listen as your teacher reads it.

Stopping by Woods on a Snowy Evening

Whose woods these are I think I know.
His house is in the village though;
He will not see me stopping here
To watch his woods fill up with snow.

My little horse must think it queer
To stop without a farmhouse near
Between the woods and frozen lake
The darkest evening of the year.

He gives his harness bells a shake
To ask if there is some mistake.
The only other sound's the sweep
Of easy wind and downy flake.

The woods are lovely, dark and deep,
But I have promises to keep,
And miles to go before I sleep,
And miles to go before I sleep.

ROBERT FROST

"Stopping by Woods on a Snowy Evening"

The literary selections that begin the sections of this book require thoughtful reading. They are intentionally chosen for their suitability for study and discussion.

Introduction of the Poem

It is usually wise for the teacher to begin by reading the poem to the class or playing the recording of it, or both. Ordinarily, the reading should be repeated several times. Don't be afraid of repetition. The literary pieces are good enough to stand it.

The children should, in general, follow the reading by looking at their texts. Some people find it difficult to comprehend orally and will make better progress if they can simultaneously see what they are hearing.

Teaching Suggestions

When the poem is first introduced, the aim should be a general familiarity, an awareness of the main subject of the poem, an acquaintance with the rhymes and rhythms, rather than a knowledge of particulars. These latter will be taken care of in the teaching of page 2. Do not stop now to explain the meaning of words and structures, unless pressing questions develop.

After the class has heard the poem several times from the teacher or the recording, pupils may be asked to take turns reading the several stanzas. This depends somewhat on whether the class is advanced or not. With a slow class, it might be better to postpone or omit individual reading.

We are not of course engaged here in a history of literature. Nevertheless, when the poet is a famous one, like Frost, the teacher may sometimes want to say something about the poet—when he lived, where, what he wrote about. One might explain that Frost was a New Englander who wrote in the first half of this century about rural New England.

VOCABULARY AND MEANING

"Stopping by Woods on a Snowy Evening"

As explained in the general introduction to this Teacher's Edition, the poems and stories included in the text serve several purposes. One of the more important is to provide training in accurate reading. Poetry predominates over prose partly because poetry gives more to chew on. It is compact, every word and arrangement counts, and only close attention will yield comprehension. The questions that follow each selection require such attention.

Note that the child is rarely asked to express an attitude toward the poem, e.g., to say whether he likes something in it or doesn't. Liking and disliking are in some sense a personal matter and will vary among individuals. What should be asked of all the pupils is not that they have some particular attitude toward the poem but that they thoroughly understand its details. Pleasure will more often than not follow automatically upon understanding and is meaningless in its absence.

Teaching Through Discussion

The material in this text is presented, insofar as possible, in the form of questions for the teacher to work through with the children. The answers yield the knowledge that is to be conveyed. This method ensures active pupil participation and interest.

Teaching Suggestions

Take the class through the questions on this page. They are mostly of a sort that have precise answers rather than subjective ones. When the children cannot answer or answer wrongly, have them return to the text of the poem to find the necessary information.

Do not be content with "imaginary" answers—what might have happened or might have been true. One purpose of these reading passages is to train the child to find out what is actually on the printed page.

Stopping by Woods on a Snowy Evening

In the poem on page 1, the poet used a few words to express many thoughts. It is important to read a good poem like this carefully and think just what it is that the poet says.

• In the first line, the poet does what poets often do. He changes the order of words in a sentence from the order we would usually use. We might say, "I think I know whose woods these are." How does the poet say it? See poem.

• What is the last word in the second line of the poem? What word in the first line rhymes with it? Can you give one reason why the poet didn't simply say, "I think I know whose woods these are"? discuss

• Where is the house of the owner of the woods? What is the poet watching? Do you think he would have stopped to watch if the owner of the woods lived nearby? Blue dots before paragraphs suggest discussion.

• What does the poet say his horse must think queer?

• The poem is divided into four parts, or **stanzas.** The second stanza tells you what day of the year it is. These clues will help you figure it out.

Summer begins on June 21. This is the longest day of the year.

The first day of winter is December 21.

Days grow shorter and shorter from the beginning of summer to the beginning of winter. Then they begin to grow longer again.

What day of the year was it? Dec. 21

• How does the horse ask if there is some mistake? What are the **downy flakes?** See third stanza; snowflakes.

• Do you think the poet would like to stay longer by the woods? Why must he drive on? Yes; see poem.

Rhyme

(Answers may be printed or underlined.)

• You know that lines of a poem rhyme if their last words end with the same sound. What three lines in the first stanza of the poem on page 1 rhyme? 1, 2, 4

The three rhyming words in the first stanza are **know, though,** and **snow.** Do these words all end with the same sound? Is the sound spelled the same in all three words? Which word ends with a different spelling than the other two? yes; no; though

• What line in the first stanza does not rhyme with the other three? Find three lines in the second stanza that rhyme with this line in the first stanza. What are the four rhyming words at the end of these four lines? These words end in the same sound. Is the sound spelled the same way in all four words? What are three different ways this sound is spelled? 3; 1, 2, 4; here, queer, near, year; no; ere, eer, ear

• Does rhyme go by sound or by spelling? sound

• In the second stanza, what is the word that doesn't rhyme with the others? What words does it rhyme with in the third stanza? lake; shake, mistake, flake

• What words rhyme in the last stanza? What word of the third stanza do they rhyme with? deep, keep, sleep; sweep

□ Write the rhyming words you find in each row of words below.

1. go see blow men dough
2. sew who two through few
3. beet bet seat red Pete
4. pin fed bread laugh said
5. queen when mean pan scene
6. main rain mine lane own
7. done fun son run won
8. peep sleep ship leap slap

Notice that some paragraphs are marked (●) and others are marked (□). The former means that the material is intended for oral response; the latter marks a written exercise. Sometimes, of course, depending on the needs of a particular class, the teacher will decide to have the written work done orally or the oral work done also in writing.

For example, when the material of a written exercise is particularly difficult or the class a slow one, it is best to do some of the items on the chalkboard, with class participation, and then have the pupils finish the exercise by themselves. Similarly, after items marked for oral response have been worked through orally, they may often be assigned to some or all of the pupils as written work.

Teaching Suggestions

The children have of course had a good deal of work with rhyme in earlier grades. We review it here, however, partly because of its importance in the poetry, but more essentially because of what it reveals about the connection between sound and spelling. It thus reinforces the work to be begun (or continued for those who have had the third-grade book of this series) in the study of the spelling system.

Go through the questions, which direct attention to all of the rhymes in the Frost poem. Rhyming words may be written on the chalkboard. Establish the fact that words often rhyme even though they are spelled differently, which is a way of saying that we often spell the same sound in different ways.

Do not, at this time, attempt to identify the sounds by name—as to speak, for example, of the "oh" sound of **snow**. Naming the sounds of English, particularly the vowels, is a complicated matter. A scheme for naming them is given as the sounds themselves are studied.

COMPOSITION

In this grade the writing assignments are kept quite simple. They mostly call for single paragraphs on fairly objective subjects. The emphasis should be kept on correctness in mechanical matters—such as indentation, capitalization, end punctuation, spelling—and on the development of fluency. Largely we learn to write through much practice in writing, and the children should do as much as they can.

The aim of correctness sometimes contends with that of fluency. That is, a child who is over-corrected sometimes becomes so cramped and tight that he can't write at all. It is important to keep the pupil from developing the feeling that whenever he puts pencil to paper he is pretty sure to make a mess of it. The teacher must judge how much correction a particular child can take without becoming overly discouraged. Often it is best just to point out obvious mistakes and let others go. With some children it is best not to correct at all until they get used to writing and begin to loosen up.

Teaching Suggestions

Pages 4 and 5 make up a single lesson, but more than one period may be needed for it. Page 4 reviews some matters about capitalization and indentation which most pupils will already have studied. With many classes it will first be necessary to do on the chalkboard one or both of the titles to be capitalized, before having the children write them. Note that we avoid saying that important words in the titles are capitalized and unimportant ones are not. In some sense all the words in the title are important. Until the actual word groups have been identified—as prepositions, determiners, etc.—it is best to say just that "little words within the title" are written with small letters, these being exemplified.

The second written exercise on page 4 reviews capitalization of proper nouns as well as **I** and the initial word of the sen-

Writing

Here is a review of some of the things you have learned about writing.

1. You will often need to write a title for your paragraph or report. Titles are written like this:

The Day I Set Fire to the Kitchen
A Quiet Place in the Woods

Notice that most words in a title are written with capital letters at the beginning. The first word always is. Little words like **to, in, the** are not written with capital letters within a title.

□ Write these groups of words as titles: A Day at the Beach; An Interesting Town in California

a day at the beach
an interesting town in California

2. The first line in a paragraph must be **indented.** This means that it must begin farther to the right than the other lines.

● Find a paragraph on this page. Is the first line of it indented? line 1; yes

3. Each sentence must begin with a capital letter. Each one must end with a period.

● Of course, if your sentence is a question, it doesn't end with a period. What mark do you use at the end of a question? question mark

4. Remember what you have learned about writing some words with capital letters. The word **I** is always written with a capital letter. Names like **Jack** and **Jane** are too.

□ Study these sentences. Notice the capital letters. Then your teacher will dictate the sentences.

a. My brother Bill and I are twins.
b. Where did Mr. and Mrs. Smith go?
c. Mary saw Miss White last Monday.

5. Have you ever been to an amusement park? If you have, write a paragraph about one of the rides or games you remember. If you have never been to one, write a paragraph telling about one or more things you see in the picture.

Write a title for your paragraph, indent the first line, and use what you have learned about capitals.

tence. The children may have to be reminded that the abbreviations **Mr.** and **Mrs.** are written with periods. Note that dictation is called for.

Dictation is the most effective way to provide practice on capitalization and punctuation, but if the children are not used to writing from dictation, considerable help must be provided. For example, read the first sentence aloud: "My brother Bill and I are twins." Call on a child to repeat it exactly. If he fails, read it again and call on another child to repeat it. Continue until all of the children are listening and can repeat the sentence. Then dictate it to be written. This procedure should be followed until the children have developed the skill of listening to and remembering what they have heard.

Use the questions at the bottom of page 5 to stimulate discussion of amusement parks in preparation for the writing assignment. Some children will probably want to tell about an amusement park they have visited and some of the things they have seen. Use the picture to point out items mentioned. On the basis of the discussion, decide which alternative assignment should be used or whether some children should do one and some another.

In this and most of the writing assignments, all but exceptional classes will probably need quite a lot of preparation—what might be called oral warm-up. Pupils often need to be shown the possibilities and ways of using them. For example, children doing the description of the picture will usually need a preliminary class discussion of it, in which the various items are pointed out and named and some conjecture is made about what the various people are doing.

SOUNDS AND LETTERS—/i/, /e/, /a/

Pupils who have used the third-grade book of this series will already have covered most, not all, of the material on phonology and syntax presented in the first few parts of this book. The teacher working with such children in the fourth grade will want to become familiar with this material, in order to know what is new and what is review. The third-grade material is summarized in the back of this book, pages 304–23.

Although it can confidently be assumed that people who have studied the material in the third grade will have forgotten a good deal of it during the summer, they should nevertheless be able to go on faster and more easily than those who begin it in the fourth grade. The latter will need special help and practice. For the study of sounds and letters, more practice can be provided from the lists of words given on pages 324–30.

Teaching Suggestions

Pupils coming afresh to this material will also need special help with certain fundamental concepts and conventions. One very important one is the distinction between sound and letter. The child's previous experience may be such as to blur this distinction; if so, he is likely to suppose that **rips** and **ribs** end in the same sound because they end in the same letter, or that **ss** stands for the same sound in **press** as it does in **pressure**. He may even be unaware, or only vaguely aware, that **bed** and **head** have the same vowel sound.

Similarly, it will take the child some time to get on to the practice of talking about sounds and letters separately so that we may say, for example, that **rips** ends with the sound /s/ and the letter **s**, but **ribs** with the sound /z/ and the letter **s**.

Therefore, though these two pages ideally represent a single lesson, they may for many classes require more time. It may be necessary to return to them on another day before going on for further work. In the third-

The Vowel Sounds /i/, /e/, and /a/

Sounds and letters are different things. When we say words, we make noises, or sounds. When we write words, we make marks on paper, called letters. The letters show what sounds the spoken words have.

• When we say **pick** and **pit**, do we make the same sound at the beginning of both words? When we write **pick** and **pit**, what letter do we use to show the sound that both words begin with? What letter comes after **p** in both words? yes; p; i

To study sounds and letters, we must be able to tell when we are talking about letters and when we are talking about sounds. Here is the way we will do it. When we mean letters, we will use heavy black type: **p, t, i.** When we mean sounds, we will use slanting lines: /p/, /t/, /i/.

• Does /t/ mean a letter or a sound? sound
Does **a** mean a letter or a sound? letter
Does **c** mean a letter or a sound? letter
Does /k/ mean a letter or a sound? sound

Words have two kinds of sounds: **vowel sounds** and **consonant sounds.** We use two kinds of letters to stand for these sounds: **vowel letters** and **consonant letters.**

□ There are five vowel letters: **a, e, i, o, u.** All the others are consonant letters. Name the consonant letters. Write them. How many did you write? Add the 5 vowel letters to this number. Do you get 26 letters in all? b, c, d, f, g, h, j, k, l, m, n, p, q, r, s, t, v, w, x, y, z; 21; +5 26; ye

There are more sounds in English than there are letters in the English alphabet. There are 15 vowel sounds. There are 24 consonant sounds. So we have to use our 26 letters in many special ways to show all of these 39 sounds.

Pit, pet, and **pat** begin with the same consonant sound. They end with the same consonant sound. But each word has a simple vowel sound in the middle.

• Say **pit, pet, pat.** Are the vowel sounds different? yes; What vowel letters are used to spell these sounds? i, e, a

• To name these vowel sounds when you talk about them, make the sound with a /k/ sound after it. For the vowel sound of **pit,** say "ick." For the vowel sound of **pet,** say "eck." What would you say for the vowel sound of **pat?** ack

You have learned that when we mean letters we use heavy black type: **i, e, a.** When we mean sounds we use slanting lines, / /, with a letter between them. The letter we use between the lines is the letter we most often use for writing that sound.

For the vowel sound "ick" we use /i/, for "eck" we use /e/, and for "ack" we use /a/.

□ Write as headings /i/, /e/, and /a/. Look at the words below. Under /i/, write those that have the vowel sound "ick," under /e/, those with "eck," and under /a/, those with "ack."

pit /i/ pat /a/ pet /e/ tell /e/ tack /a/ can /a/
Fred /e/ stiff /i/ brick /i/ bed /e/ tip /i/ damp /a/

Sounds are not always spelled the same way in different words. Remember that we put between slanting lines the letter we **most often use** for writing the sound.

• In the words below, one of the simple vowel sounds is spelled in a special way. Under which heading would you put these words, /i/, /e/, or /a/? /e/

bread head dead

• Under which heading would you put **laugh?** /a/

• Under which heading would you put **said?** /e/

grade book the material presented on pages 6 and 7 occupied several lessons.

Be sure that the class is quite clear on the convention explained on page 6 for distinguishing sounds from letters in writing: slant lines for symbols referring to sounds and boldfaced type for those referring to letters. Give more practice on the chalkboard if needed, but be careful not to use symbols not employed in the system. (For example, the letter **c** is not used to represent a sound.)

Use these symbols for extra examples of sounds: /s/, /z/, /f/, /v/, /l/, /m/, /n/, /r/. Use any letters of the alphabet for extra examples of letters.

Establish or review the distinction between vowel sounds and consonant sounds, vowel letters and consonant letters.

Naming Sounds

We have, of course, a regular convention for naming the letters of the alphabet: ay, bee, see, dee, ee, eff, gee, aitch, etc. There is no such universal convention for naming sounds. It is therefore necessary to establish one.

We can ordinarily use the simple alphabetic names for the consonant sounds. Thus /k/ is "the sound kay," /d/ is "the sound dee." Sounds not represented in the alphabet are given analogical names as they are introduced. Thus /sh/, the first sound of **shall,** is called "esh" by analogy with "ess" (s).

The vowels are much more a problem. The letters of the alphabet don't suffice as names of the vowels, even if we double them by speaking of "long" and "short" vowels, a practice which is to be avoided on other grounds anyway. This series introduces what might be called the /k/ convention. Each vowel is named by making the vowel sound before the consonant sound /k/. We use /k/ as a kind of stopper because many vowel sounds do not occur in English except before consonant sounds.

GRAMMAR—Subjects and Predicates

In the early lessons on syntax, as in those on phonology, the material will be review for some pupils and new for others. The problem is somewhat less acute in syntax, however, since at least some of the concepts are likely to have been previously encountered, whatever text was used in the third grade.

Sometimes the child may have known a category under a different name: "picture word" for **adjective**. Only the regular grammatical terms are used in this series.

Definitions

It will be noticed that the text does not contain definitions of the usual sort. For example, subjects are not defined as "a word or words that tell what the sentence is about." All such definitions lead ultimately to confusion. (In "Jack saw Bill," the sentence is certainly about Bill as well as Jack.) We often say something about the notional content of a category: "The subject names someone or something." But we don't use this definitionally, and we say it only after we have identified the concept in some other way—e.g., by position ("In a simple sentence the subject always comes first") or by example (in "That girl walks to school," the subject is **that girl**).

The Meaning of "Simple Sentence"

What are referred to as simple sentences here are called **kernel** sentences in transformational grammar. They are all active, declarative sentences with little modification. From this base we build by transformation the more complicated sentences that we use in ordinary speech and writing.

Sticking to kernel sentences for a long time enables us to establish the basic structures of the grammar, which will then be easily recognizable in transforms.

Subjects and Predicates

Every simple English sentence has two main parts: the **subject** and the **predicate.**

In a simple sentence the subject always comes first and the predicate comes second. The subject names someone or something. The predicate tells something that the subject **does** or **did** or **is** or **was.**

Here is an example of a simple sentence:

That girl walks to school.

• The subject is **that girl.** What is the predicate?

• Everything in a simple sentence that isn't the subject is the predicate. Everything that isn't the predicate is the subject. In the sentence "The teacher writes on the chalkboard," the predicate is **writes on the chalkboard.** What is the subject? the teacher

In "My brother reads poems," the subject is **my brother.** What is the subject of "My sister reads poems"? my sister

What are the subjects of these sentences?

1. His mother reads poems.
2. The teacher reads poems.
3. Mary reads poems.
4. She reads poems.
5. The class reads poems.
6. Mr. Wilson reads poems.
7. The girl reads poems.
8. The girls read poems.
9. I read poems.
10. You read poems.
11. People read poems.
12. They read poems.
13. Miss Thompson reads poems.
14. Everybody reads poems.

Study the subjects of these sentences:

15. His house is in the village.
16. The woods are lovely.
17. My horse shakes the bells.
18. That man watches the woods.

The subjects are **his house, the woods, my horse,** and **that man.** Each subject is made up of two words. In subjects like this, a word like **his, the, my, that** comes first. Then there is a word like **house, woods, horse, man.**

Words like **house, woods, horse, man** are called **nouns.** Usually nouns name people or things. Here are some other examples of nouns: **boy, teacher, desk, chalkboard, bus, mother.**

Words like **his, the, my, that** are called **determiners.** Determiners come before nouns and they tell, or determine, the meaning of the noun. For instance, "*the* boy" means some particular boy, but "*a* boy" means just some boy. "*My* hat" means that the hat is not yours or Mr. Wilson's, but that it belongs to me. Some other words that are used as determiners are **an, some, this, these, your, their.**

Tell the subjects of these sentences. Tell what word in the subject is a determiner and what is a noun.

	D:	N:
19. The boy walks to school.	the	boy
20. Our teacher explains the poem.	our	teacher
21. A book is on the desk.	a	book
22. His horse shakes the bells.	his	horse
23. This man is the principal.	this	man
24. My dog is little.	my	dog
25. An apple hangs on that branch.	an	apple
26. Your kitten is hungry.	your	kitten
27. Their house looks empty.	their	house
28. Her father works here.	her	father

Teaching Suggestions

The class should not have much difficulty with the sentences given for practice on page 8. Since a simple sentence consists only of a subject and a predicate, everything that is not the predicate must be the subject, and in these sentences the predicate remains constant.

It might be useful to write the various subjects on the chalkboard—**his mother, the teacher, Mary**, etc.—so that the children begin to recognize the *sort* of structure that may function as subject. All the subjects in these early lessons, as indeed in most of the lessons in the fourth-grade book, consist of one or two words.

Page 9 begins to identify the different structures that may function as subjects, beginning with the structure "determiner + noun." Notice again that we say something about the function of determiners: "they tell, or determine, the meaning of the noun." But we do this largely to make the term plausible, not to define the category. That is, we don't expect the child to look to see what determines the meaning of nouns in order to recognize determiners. He recognizes them by position and by simple familiarity with the members of this small group of words: **the, a, some, this, that, these, those, much, many, several, both**, etc.

The possessives—**their, your, his**, etc.—are included here as determiners. As explained later, they result from a transformation in which a definite article is replaced by the possessive.

Notice that although we remark that nouns usually name people or things, we do not use this as a definition. One does not note that a word names something and is therefore a noun. One notes, from syntactic signals, that a word is a noun and therefore names something. The children are probably already familiar with nouns, though possibly under the term "naming words."

LITERATURE – "Noise"

Teaching Suggestions

Read the poem once or twice to the class or play the recording of it, or both. Then let the children take turns reading parts of it. The teacher might begin with "I like noise," and then call on a child to read the first two rhyming lines, then on another for the next two, another for the next two, and so on. When the poem is finished, begin again and continue until each child has had a chance to read.

In the oral reading, the first goal should be accuracy, in the very simple sense of saying just the words that are on the page, not substituting others, not repeating or omitting. Fourth-graders will still have trouble accomplishing this. (Many tenth-graders do too.) When accuracy has been attained, more attention can be given to clear and proper pronunciation, to making the line scan, and so on. It is probably not useful at this grade level to put any stress on "reading with expression." This is likely to be interpreted as an invitation to react emotionally to the poem, and the results are seldom happy ones.

A clear, accurate, rhythmic reading, without singsong, is about all that can be sought of a young child, and the teacher should be pleased indeed when such a reading is attained.

SECTION 2

A Poem

Do you like noise? Here is a poem by someone who does.

Noise

I like noise.
The whoop of a boy, the thud of a hoof,
The rattle of rain on a galvanized roof,
The hubbub of traffic, the roar of a train,
The throb of machinery numbing the brain,
The switching of wires in an overhead tram,
The rush of the wind, a door on the slam,
The boom of the thunder, the crash of the waves,
The din of a river that races and raves,
The crack of a rifle, the clank of a pail,
The strident tattoo of a swift-slapping sail.
From any old sound that the silence destroys
Arises a gamut of soul-stirring joys.
I like noise.

J. POPE

Noise (Use poem as source of answers; discuss)

• One of the kinds of noise is the whoop of a boy. What animal might make the noise the poet calls "the thud of a hoof"? horse

• Some roofs are made of **galvanized iron.** The dictionary says that **galvanized** means:

1. shocked by electricity
2. coated with zinc

Which meaning do you think fits this line? 2

• What is there a hubbub of? a roar of? See poem.

• Have your fingers ever been so cold that they had no feeling in them? If so, they were **numb.** What noise does the poet say makes the brain numb? See poem.

• What is there a switching of? wires A **tram** is a kind of bus or streetcar. Streetcars run on tracks, and buses usually run as cars do. But some buses are connected to wires above them. The wires are electrical, and these are what give the buses their power. An overhead tram is a bus that runs on wires like these.

• What is there a **rush** of, a **boom** of, a **crash** of? How does the door make its noise? See poem.

• What is there a **din** of, a **crack** of, a **clank** of? See poem.

• A **strident** sound is a high or harsh one. A **tattoo** is a beating or drumming sound. You may have heard the expression, "to beat a tattoo on a drum." What is it that the poet thinks of as making a drumming sound? a swift-slapping sail

• The poet likes any old sound at all that destroys something. What does noise destroy? silence

• A **gamut** is a whole range or stretch of something, all the different kinds of it. What kind of gamut does the poet speak of here? soul-stirring joys

11

VOCABULARY AND MEANING

"Noise"

The questions on page 11 should be answered in class discussion. Their purpose is to ensure that the details of the poem are not overlooked, that the children think about what the lines are saying and do not simply accept them as pleasant sound.

The intention in the pages that follow the literary selections is to clear up all possible difficulties in vocabulary and structure or to point to ways of removing difficulties. But often the limitation of space requires omissions. The teacher should be on the alert to see that the children understand words not mentioned in the questions. For example, does everyone in the class know what a rifle is?

When we say that the reader should know what the words mean, we don't of course expect him to be able to define or give synonyms for all of them. Some words are quite hard to define in the dictionary sense: **din, crack, clank**. It is sufficient here to say that a din is the sound a river makes, a crack the sound a rifle makes, and so on.

It should be observed also that the great teacher of the meaning of new words is the context in which the words occur. Children learn relatively few words by looking them up in the dictionary or by having someone tell them what they mean. Mostly they learn from context, the words already known delimiting, if not defining, the new ones. Thus in "The strident tattoo of a swift-slapping sail," the word **strident** will probably be new to the average fourth-grader. But if he knows the rest of the words, he can probably get at least a general notion of its meaning. It is an adjective modifying **tattoo** – a kind of tattoo, then. What kind? Well, the kind that a swift-slapping sail makes. Even before the word is defined for him in the notes as high or harsh, he may have perceived the general nature of the word, the general area in which it applies.

Regularity and Exception

In teaching the spelling of the different sounds, we point out the common, or regular, way, draw attention to one or more less common, though perhaps frequent, ways, and finally point out some of the odd, wholly exceptional ways. We want the learner to think of the spelling of English as essentially systematic, though of course we do not hide the fact that there are many, many items quite outside the system. We want him to accept naturally the idea of system plus exception and not to be terrified by exceptions.

For some sounds the system is largely operative. Thus the sound /i/ in monosyllabic words is almost invariably spelled **i**. The only other possibility (apart from very odd words like **been**) is **y**, and this will occur in only a word or two known to the child: **myth, rhythm**.

The sound /e/ is more complicated. The regular spelling is **e**, as in **red**. But a considerable number of words spell this sound with **ea**. Examples beyond those on this page are **wealth, meant, dread, bread, leather, spread, tread, breast, threat**. Finally there are the real exceptions, belonging to sets of one or two vowels: **friend, any, many**.

Presumably the child who learns to spell learns the exceptions as single items. There is no way to know how to spell **friend** except to know that in this particular word /e/ is spelled **ie**.

Secondary spellings, like **ea** for /e/, are learned partly as items, partly as system: the spelling is more likely before certain sounds than others. The rest is system, applied in all other cases. Thus one spells /e/ as **e** unless otherwise informed. We try here, so far as space permits, to supplement and strengthen this learning process.

The Vowel Sounds /u/ and /o/

- You have learned that the simple vowel sounds in **pit, pet,** and **pat** are called "ick," "eck," and "ack" when we speak of them. What letter do you put between slanting lines to show that you mean the sound "ick"? the sound "eck"? the sound "ack"? i, e, a
- We generally have one or two usual ways of spelling sounds and some special ways. The usual way of spelling the sound /i/ is with the letter **i.** We write **pit, tip, kick, miss, hit.** Can you think of other words in which /i/ is spelled this way? See p. 324.
- The sound /e/ is usually spelled with the letter **e,** as in **pet, get, red, melt, neck.** Can you think of others? See p. 324.
- But in some words we spell the /e/ sound with the letters **ea. Head** is one. **Thread, dead, feather, health** are others. Can you name still more? See p. 324.
- We usually spell the vowel sound /a/ with the letter **a,** as in **pat, tap, tack, fast, match.** Can you think of other words like these? See p. 324.

Most people say the word **laugh** with the vowel sound /a/. What two letters are used to spell /a/ in this word? This is a very special way of spelling the /a/ sound. au

Another simple vowel sound is the one in **stuck** or **mud.** It is usually spelled with the letter **u,** so we write **u** between slanting lines to show the sound this way: /u/. In speaking of the sound, call it "uck," as if you were saying the last part of **stuck** or **duck.**

- Here are some words in which the /u/ sound is spelled with the letter **u:**

stuck mud fun rub fudge

Can you think of other words like these? See p. 324.

In this sentence, there are two words that sound exactly alike:

The sun is shining on his son.

● What two words sound exactly alike? What vowel sound do both words have? sun, son; /u/

We usually spell the sound /u/ with the letter **u,** but in some words we use the letter **o,** as in **son.** Say these words that have the letter **o** in them, but listen to the sound /u/ in them. See p. 324 for more.

some won love shove dozen month

The fifth simple vowel sound is the one in **top.** We usually spell this sound with the letter **o,** so we write o between slanting lines to show the sound, this way: /o/. In speaking of the sound, call it "ock," as if you were saying the last part of **rock** or **lock.**

Here are some other words in which the /o/ sound is spelled with the letter **o.**

Bob rob shot hot not flock

Can you think of other words like these? See p. 324.

□ Write the following sounds as headings for columns across a sheet of paper:

/i/ /e/ /a/ /u/ /o/

Look at the words in the list below. Write all the words with the sound /i/ under that heading, all those with the sound /e/ under the heading /e/ on your paper, and so on. Remember, not every simple vowel sound is always spelled with the letter we put between slanting lines.

rat /a/	**scrub** /u/	**met** /e/	**trip** /i/	**won** /u/
knock /o/	**sweat** /e/	**miss** /i/	**shucks** /u/	**laugh** /a/
quit /i/	**Ted** /e/	**sob** /o/	**hand** /a/	**rich** /i/
health /e/	**stop** /o/	**lick** /i/	**ranch** /a/	**run** /u/

The Vowel Sounds /u/ and /o/

These two sounds, /u/ and /o/, which occupy several lessons in the third-grade book, are here taken care of in a single lesson. For pupils new to the study, more than one day may be required, however.

The spelling of the sound /u/ resembles in some ways the spelling of /e/. The regular spelling of /u/ is **u,** as the regular spelling of /e/ is **e.** But there is an important secondary spelling, **o,** as **ea** is an important secondary spelling of /e/. And as **ea** for /e/ is more likely before certain sounds than others, so **o** for /u/ is largely restricted. It occurs for the most part before a following **m, n,** or **v.** There is a historical reason for this. In early manuscript writing, **m, n, v,** and **u** were all made with very similar up and down strokes. So a **u** followed by an **m** would simply be five strokes and would be hard to read. Scribes therefore closed the top of the **u,** making it an **o.**

Beyond these two ways of spelling /u/, there are, as with /e/, some exceptional exceptions—sets of one or two vowels: **blood, flood, touch.**

At this stage the only spelling given for /o/ is **o.** However, the teacher will want to be aware of certain complications with /o/ that we confront to a much lesser degree with most of the other vowels. For most Americans the vowel of **cot** is a low central vowel. But for many, particularly in New England, it is a low back vowel, like that of **caught**. For some speakers **cot** and **caught** are homophones. Most people have the same vowel sound, approximately, in **cot** that they do in **car,** with the sound spelled **a** when **r** follows: **car, star, bar, jar,** etc. But for those for whom the vowel of **cot** is a back vowel, the vowel of **car** is different.

Before the sound /g/ there is much variation. **Hog, dog, log, bog** have the sound of **cot**—the low central vowel /o/—for some speakers, the low back vowel for others. See page 324 for words for extra practice.

GRAMMAR—The Noun Phrase

Though we do not yet make the distinction explicitly, we are in effect separating here two quite different and very fundamental concepts in grammar: structure and function. The term **noun phrase** refers to a structure. It means a particular kind of word or group of words. The terms **subject** and **object** refer to functions. They mean particular uses in a sentence to which noun phrases may be put.

We may often refer equivalently to either term: What is the object of that verb? What is the noun phrase in that predicate? However, we should be alert to the fundamental difference: a structure is a particular kind of word or group of words; a function is a use of a structure. The confusion of the concepts accounts for much of the muddiness of many school grammars.

The Term *Phrase*

Pupils will hardly be bothered by the rather special use of the word **phrase** in the term **noun phrase**, but teachers may be. Ordinarily the term is used to mean a group of words. Here it also includes single words. Thus **he, Bill, everybody, roses** are all noun phrases, along with **the boy, a golf club, several of the men in the bank**. Some single term is needed for the structures ordinarily used in the functions of subject, object, complement, and none better than **noun phrase** has been put forth.

Teaching Suggestions

Use the first part of this lesson to review and reinforce the concepts taught on pages 8 and 9. If the class is shaky, return to those pages for review.

The class of determiners can of course be subdivided: articles, demonstratives, etc. But we postpone these refinements for later years.

At this point we introduce only those noun phrases which are also phrases in the

The Noun Phrase

Do you remember what the two main parts of a simple sentence are called? Which of these main parts comes first in a simple sentence? subject, predicate; subject

What is the subject of this sentence?

The house is in the village.

We have a name for words like **house.** Can you remember what it is? Other words which are like **house** are **village, noise, bell, mother, garden.** noun

We also have a name for words like **the.** We call them **determiners.** They occur before nouns, and they tell, or determine, something of the meaning of the noun that follows them. Other words that we use as determiners are **a, an, some, this, that, these, those, my, your, her, his, our, their.**

Tell what the subject is in each of these sentences. Tell which word in the subject is a determiner and which word is a noun.

1. The house is in the village. D: the N: house
2. This valley is pretty. this valley
3. Our parents live in a city. our parents

Now look at these sentences.

4. I like that house.
5. That house is small.
6. He lives in that house.

What two words occur in each sentence from 4 to 6? In which sentence are these words the subject? In which sentences are they part of the predicate? that house; 5; 4, 6

We need a term for words like **that house,** whether they are used as subjects or as parts of predicates. We will use the term **noun phrase. That house, the village, this valley, our parents, a city** are noun phrases, no matter how they are used.

There are several kinds of noun phrases, but now we will study just the kind that is made up of a determiner and a noun.

In these noun phrases, which are the nouns? Which are the determiners?

that house **the village** **a city**
D N D N D N

Look at these sentences. Each one has just one noun phrase in it. Each noun phrase is made up of a determiner and a noun. Tell the determiner of each noun phrase. Tell the noun.

7. The girl is pretty. D: the N: girl
8. The children play. the children
9. My hands are dirty. my hands
10. A man is here. a man
11. This cat is little. this cat
12. His sisters are big. his sisters

Look at these sentences. Each one has two noun phrases in it. One noun phrase is the subject. The other is used in the predicate. Each noun phrase is made up of a determiner and a noun. Tell what the noun phrases are in each sentence.

13. The horse shakes the bells.
14. A boy made a noise.
15. A kitten drank the milk.
16. These girls are his sisters.
17. The house is near a village.
18. My desk is near the chalkboard.
19. A pen is on his desk.
20. That man is our principal.
21. The giraffe ate the hay.
22. Some children found the car.
23. This cat got into their cellar.
24. An eagle frightens those birds.

common use of the term **phrase**. They are all composed of two words: determiner noun.

Sentences 1–3 on page 14 check on the understanding of the terms determiner and noun in the subject position. The teacher may want to give further sentences on the chalkboard for more practice but should take care to use noun phrases of the same type. If in doubt, one might be safer to return to pages 8 and 9 for additional practice.

Sentences 4–6 make the point that noun phrases may occur in functions other than that of subject. The other functions are not named however, and will not be until the structure of the predicate has been described. Here it is sufficient for the pupil to observe that a structure like **that house** is not always a subject and that therefore the function term **subject** will not always identify it.

Sentences 7–12 on page 15 give further practice in identifying noun phrases in the subject function and their component parts —determiner and noun in all cases here.

Sentences 13–24 extend this now to noun phrases also in the predicate. Each sentence in 13 through 24 has two noun phrases. One is subject; another is in the predicate. The noun phrases in the predicate have any of three functions: object of a verb, object of a preposition, complement. We do not, however, distinguish these different functions of noun phrases in the predicate at this point. It is here sufficient to be able to identify a simple noun phrase in any function and to be able to say that it is a subject or is not.

SECTION 3

A Poem

In this poem, the poet writes as though the wind were a person with a mind of his own. Read the poem or listen to the recording of it.

The Wind

Why does the wind so want to be
Here in my little room with me?
He's all the world to blow about,
But just because I keep him out
He cannot be a moment still,
But frets upon my window-sill,
And sometimes brings a noisy rain
To help him batter at the pane.

Upon my door he comes to knock.
He rattles, rattles at the lock
And lifts the latch and stirs the key —
Then waits a moment breathlessly,
And soon, more fiercely than before,
He shakes my little trembling door,
And though "Come in, Come in!" I say,
He neither comes nor goes away.

Barefoot across the chilly floor
I run and open wide the door;
He rushes in and back again
He goes to batter door and pane,
Pleased to have blown my candle out.
He's all the world to blow about,
Why does he want so much to be
Here in my little room with me?

E. RENDALL

It is probably best to avoid such a learned term as **personification** in explaining what is going on in this poem. But it should be pointed out that the person speaking thinks of the wind as if it were another person, with intents and desires. This will hardly be the pupil's first experience with personification, which is a familiar device in children's poetry.

Teaching Suggestions

Read the poem to the class once or twice or play the recording of it. Then let the children take turns reading parts of it. Since this poem, like the last, is composed of rhyming couplets, one child may read the first two lines, another the next two, and so on. Continue through the poem and repeat until each child has had a chance to read. By this time the poem will have become quite familiar to everyone.

It is useful to encourage the children to memorize poetry. Poems got by heart become real to the child and a part of his language. In a poem of this length, one might divide the task, having one third of the class memorize the first stanza, another the second, and another the third. Then on another day trios can cooperate in reciting the whole poem.

VOCABULARY AND MEANING

"The Wind"

Go through the questions, eliciting the answers. With slower groups, especially, there will have to be constant reference to the text to find the answers.

Old English had two words for **eat**: **fretan** and **etan**. The latter was used when the subject was a person, the former when it wasn't. Modern German makes the same distinction with the verbs **fressen** and **essen**. We have retained **fretan (fret)**, however, only in the figurative sense explained here.

The Dictionary

Be sure that everyone understands the word **entry**—that is, any one of the words that the dictionary lists and defines. Point out that most words have more than one meaning and therefore more than one definition. In using a dictionary to find out what a word means, emphasize the importance of selecting the definition that suits the way in which the word has been used. Use the example to make the point. The wind seems to gnaw at the window-sill, just as it batters at the pane and rattles at the lock. Besides, this meaning is followed by **on** or **upon**, so definition **2** must be the appropriate one.

Though we are concerned here primarily with meanings, the teacher may wish to point out the other features of the entry. It may be said that the form in parentheses—(fret)—tells how to pronounce the words, using symbols to show sounds, as we do in this text between slanting lines. It may be said that letter *v* stands for the word **verb**, the form **fret-ted** for the past tense of a verb, and **fret-ting** for the form used after a word like **is** or **was** as in "He was fretting." Don't try to define or explain these terms, but say that the class will be meeting them all in the course of the year.

The Wind

- Why does it seem strange to the person speaking in the poem on page 17 that the wind wants to be in the room with him? He has all the world to blow about in.
- What are the different ways in which the wind shows he wants to get into the room? frets, batters the window, rattles the do[or], etc.
- Does the person in the poem invite the wind to come in? What does the wind do? yes; nothing
- Can you tell from the poem how the person in the poem is dressed when he runs to open the door? no; How do you think he is dressed? pajamas
- What does the wind do when the door is opened? What is the wind pleased to have done? See last stanza.
- Why does the person in the poem think the wind is unreasonable? See last three lines.

Dictionary Definitions

You know that the dictionary lists words in alphabetical order. These words are called dictionary **entries.** The dictionary tells the meaning or meanings of each entry — it gives **definitions.**

Here is a word used in the poem. It might appear this way as an entry in a dictionary:

> **fret** (fret) *v.* **fret-ted, fret-ting** 1. To be vexed or troubled. 2. To gnaw: with *on* or *upon.*

- Here just two definitions are given for **fret,** though a regular dictionary gives more. Which of the definitions given tells how **fret** is used in the poem, **1** or **2?** 2
- Find **fret** in a dictionary. Read the other definitions of **fret.**

18

The Simple Vowels

You have studied these five sounds: /i/, /e/, /a/, /u/, /o/. How do you name these sounds when you say them? Are these sounds vowels or consonants? ick, etc.; vowels

These five vowels are simple in the way they are said. Other vowels are said in more complicated ways. Also, they are simple in the way we write them. There are many ways of writing other vowel sounds. But there are just a few ways of writing the vowels /i/, /e/, /a/, /u/, and /o/. So we will call this group of vowels the **simple vowels.**

We usually write the vowel /i/ with the letter **i: pit, tip, kick, sip, bin.** Give some other examples. See p. 324.

We most often write the vowel /e/ with the letter **e: ten, red, help, pet, men.** Give other examples. See p. 324.

In some words we write /e/ with the two letters **ea,** as in **head.** Give some other examples. See p. 324.

We usually write /a/ with the letter **a: rag, had, stand.** Give other examples. Tell what two letters spell this sound in the word **laugh.** See p. 324; au

We usually spell /u/ with the letter **u: tub, truck, pup.** Give other examples. See p. 324.

In some words we write /u/ with the letter **o: one, some, love.** Give other examples. See p. 324.

We usually spell /o/ with the letter **o: lock, pop, rob.** Can you think of others? See p. 324.

□ Write /i/, /e/, /a/, /u/, /o/ as headings. Under each sound, write the words from this list which contain that sound.

far /o/	dread /e/	some /u/	laugh /a/	win /i/
not /o/	tuck /u/	lamb /a/	Bob /o/	sell /e/
pill /i/	rag /a/	mush /u/	trim /i/	shrub /u/
peck /e/	thread /e/	drum /u/	stack /a/	stiff /i/

19

The Simple Vowels

We use the term **simple vowel** as a way of referring collectively to /i/, /e/, /a/, /u/, and /o/. This is the group sometimes called the short vowels, but since there are other vowels that are short, and since these are longer before some consonants than others, **simple** seems a better term. They are simple in the way they are produced; most other vowel sounds are actually diphthongs. They are relatively simple in the way they are spelled.

We need to consider these five vowels as a group because they determine the spelling of following consonants in roughly the same way, in contrast with other vowels. Compare **cap/cape, duck/duke.**

Teaching Suggestions

Go through the questions on page 19 with the class, thus reviewing the ordinary ways of spelling the simple vowels. Although any individual might find it hard to think, for example, of another **ea** spelling for /e/, the class as a whole should be able to provide examples. These may be written on the chalkboard.

The written exercise probably should usually be started on the chalkboard too. Write the headings on the board and have the children write them on their papers. Get a consensus on the proper placement of the first few words, and write them under the proper headings. A good class may then continue on its own. In a slow class the exercise may be completed on the board and later done as an individual writing assignment.

GRAMMAR—The Word *be*

The working out of transformational grammar is considerably simplified if **be** is not treated as a verb. It is quite special in its forms and in the rules that apply to it. If we call it a verb, we are constantly having to add that it is a very unusual one and that what goes for other verbs doesn't go for it.

This dilemma has been faced of course by traditional grammar too, which customarily speaks of "verbs" on the one hand and "the verb **be**" on the other. But it is simpler still to speak just of **verbs** and **be**. This usage will seem natural enough to the children and will be entirely familiar to those who have studied the third-grade book of this series.

Calling **be** a verb was motivated, at least in part, by the desire to classify all words into a small number of word classes, or parts of speech. The number was usually six or seven or eight. However, there is nothing in the nature of the grammar itself which compels such a classification, and there are many things that contend with it.

English is in fact composed first of all of four large word classes: nouns, verbs, adjectives, and adverbs of manner. In addition it is composed of many small word classes, each of limited membership, some consisting of just a single word. It is these numerous small classes that form the skeleton of the syntax, and their several forms, functions, and meanings must be learned separately by anyone learning the language. Thus it is unenlightening to a learner of English to say that **the** is an adjective, since it doesn't behave at all like **happy, angry, tame**. Similarly, **be** doesn't behave like **go, swim, understand**. Its place in English syntax is a special one and must be specially learned.

The important point is that we are not interested here in classification as a goal in itself. If classification were itself the goal, many different classifications would be pos-

The Word *be*

You have learned that every simple sentence has two main parts. One part is the subject. What is the other part? predicate

The subject of a simple sentence is a noun phrase. In "The boy is happy," the subject is the noun phrase **the boy.** This is made up of the determiner **the** and the noun **boy.**

• Tell what noun phrases are the subjects of these sentences. For each noun phrase, tell what word is a determiner and what word is a noun.

		D:	N:
1.	The girl is quiet.	the	girl
2.	My dog is sleepy.	my	dog
3.	This poem is interesting.	this	poem
4.	Our books are new.	our	books
5.	The trail is dangerous.	the	trail
6.	Those animals are tame.	those	animals
7.	Her father is angry.	her	father
8.	Their chickens are white.	their	chickens

• Now look at Sentences 1–8 again. Tell what the predicate is in each sentence. (everything not underlined)

• Look at them once more. Each predicate begins with one of two words. What are those two words? is, are

The words **is** and **are** are two forms of the word **be.** Can you find the word **be** in this sentence?

The boy can be happy.

Now suppose we take away the word **can** in "The boy can be happy." We can't say "The boy be happy." We must change **be** to **is** and say this:

The boy is happy.

So we say that **is** is a form of **be.** It is a form we may use when we do not have a word like **can** before the **be.**

20

Now look at this sentence. What is the subject? Find the word **be** in the predicate.

The boys can be happy.

Take away the **can.** You can't say "The boys be happy." What do you say instead? are

Now that you have learned two forms of the word **be, is** and **are,** you can name all the parts of some sentences. Look at this one.

The boy is a pupil.

Like all simple sentences, this has two parts, a subject and a predicate. The subject is **the boy.** The predicate is **is a pupil.** The subject **the boy** is a noun phrase. It is made up of the determiner **the** and the noun **boy.**

The predicate of "The boy is a pupil" is **is a pupil.** This has two parts. One is the word **is,** which is a form of **be.** The other is **a pupil,** a noun phrase.

• Look at Sentences 9–18. Each has a subject and a predicate. The subject of each sentence is a noun phrase. The predicate is made up of a form of **be** and another noun phrase. Tell what noun phrase is the subject in each sentence. Tell what the form of **be** is and what the noun phrase is that follows the form of **be.**

9. The girl is my friend. is
10. The teacher is a lady. is
11. This building is a school. is
12. These ladies are our mothers. are
13. The boys are the leaders. are
14. The soup is my lunch. is
15. Those mice are my pets. are
16. His teacher is a man. is
17. Our books are our teachers. are
18. These boys are the musicians. are

sible, each defensible. Classification is important only insofar as it contributes to an understanding of the structure of the language.

Teaching Suggestions

The introduction to this lesson on page 20 is a review of the main points established earlier. Sentences 1–8 are a review practice. All the noun phrases are still of the type "determiner + noun."

Sentences 1–8 are now used again to point out the **is** and **are** forms of **be**. (With **be**, as initially with verbs, we work now only in the present tense.) To link these words up with **be**, with which they have no formal similarity, we use a sentence with a modal before **be**: "The boy can be happy." Dropping the **can** of course automatically changes **be** to **is**. Similarly, dropping **can** in "The boys can be happy" changes **be** to **are**. Thus the connection is made, and **is** and **are** are established as forms of **be**.

On page 21, noun phrases are identified in another function—in the predicate, after a form of **be**. Eventually this function will be given the term **complement**, but here it suffices to say merely that the noun phrase follows a form of **be**.

After making sure that the explanation on page 21 is clear, work through Sentences 9–18, identifying in each sentence the two noun phrases and the form of **be**.

LITERATURE – "White Butterflies"

Read the poem to the class or play the recording. Let the children take turns reading, allotting a stanza to each child.

Because it is so short, this poem is an easy one to memorize, and all of the pupils might be asked to do so.

The two rhyming vowel sounds of the poem are those which will be studied in the next lesson on sounds and letters, pages 24–25.

Those of the children who develop literary interests will encounter Swinburne again in later years. It may be said that he lived in England and wrote nearly a hundred years ago. His characteristic mellifluousness and play with sound are present in this little poem.

SECTION 4

A Poem

In this poem, the poet describes a strange sight — great numbers of butterflies flying out to sea.

White Butterflies

Fly, white butterflies, out to sea,
Frail, pale wings for the wind to try,
Small white wings that we scarce can see,
Fly!

Some fly light as a laugh of glee,
Some fly soft as a long, low sigh;
All to the haven where each would be,
Fly!

ALGERNON CHARLES SWINBURNE

22

White Butterflies

- The poet talks to the white butterflies as though they could hear him as persons could. Do you remember a poem in which the poet talks about something as though it were a person? "The Wind," p. 17
- Do the butterflies have strong, sturdy wings? What word in the poem answers this question? no; frail
- One meaning of the word **try** is test. What will test the frail wings of the butterflies? Find another meaning of **try** in a dictionary. the wind
- **Scarce can see** means **can hardly see.** What is it that we can hardly see? small white wings
- The word **glee** means merriment. **A laugh of glee** is a merry, joyful laugh. Do you think "light as a laugh of glee" is a good way to describe the light, rapid flight of tiny wings? discuss
- How does the poet describe the flight of the butterflies that fly softly? What is a **long, low sigh?** discuss
- A **haven** is a shelter. Where does the poet think the butterflies are going? to a safe resting place.
- What three words at the ends of lines rhyme with the word **glee?** Which two of these words sound exactly the same? Which words rhyme with **fly?** sea, see, be; sea, see; try, sigh

A Paragraph to Write

The poet wrote about butterflies flying out to sea. Write a paragraph telling what you see in the picture on the opposite page. If you wish to, tell why you think the butterflies are flying out to sea and where you think they might be going.

Give your paragraph a title. What else must you remember to do in writing a paragraph? indent

VOCABULARY AND MEANING

"White Butterflies"

From the first question, get a recall of the use of personification in the poem "The Wind." The class might now discuss other poems and stories that they know in which things and animals are addressed as people or made to talk or act like people. All the children will know some story of a personified bear or rabbit or chicken. Children who studied the third-grade book of this series should recall Tennyson's "The Brook," in which the brook tells the story of its wanderings to join the brimming river.

Go through the rest of the questions. The questions in the last paragraph foreshadow the phonology lesson that follows.

COMPOSITION – A Paragraph to Write

Again we use a picture to stimulate writing, though in this case the picture is an illustration of a poem that has just been read. The assignment is thus in some sense to write a prose version of the poem, leaving out of course all of the poetic characteristics.

Prepare for the writing by a class discussion of the illustration. Have pupils volunteer to tell things that they see in the picture. When a sufficient substance to write about has thus built up, have the children write their paragraphs.

SOUNDS AND LETTERS—/ī/, /ē/

These are the last of the seven vowels that are studied in the third-grade book of this series. They are studied there at a much more leisurely pace, of course.

The Symbols for the Sounds

In virtually all phonetic or phonemic transcriptions, the symbols employed are different from those used here. Thus the vowel of **my** would usually be transcribed /ay/ or /ai/, not /ī/. The vowel of **see** would be /i/ or /i:/ or /iy/. All such systems have more symmetry and plausibility than the one used here. We would use one of them if we could, but we can't.

We can't because of the weight of the dictionary practice. The child's other large experience with the transcription of sounds is the portrayal of pronunciation provided by the dictionary. It seems too much to ask him to learn (1) regular spelling, (2) a dictionary transcription, (3) a third transcription. We have therefore been compelled to accommodate as closely as possible to ordinary dictionary usage.

The Vowel Sound /ī/

By far the most common way of spelling /ī/ when a consonant sound follows it is with the combination "letter **i** + consonant letter + **e**." Exceptions are few and scattered—for example, **rhyme, aisle, height**. This function of "silent **e**" is of course very important in English spelling. We introduce it here at the beginning and return to it frequently.

The sound /ī/ could be named by just saying the sound, as in **I** or **eye**. Many other vowel sounds could too: /ō/ could be called **oh**. But the five simple vowels and two others cannot be named in this way, since they never occur except with a following consonant. Since we must use something like the /k/ convention to name these vowels, it seems simplest to extend it to the others also.

Two Complex Vowels — /ī/ and /ē/

• You have studied five simple vowels: /i/, /e/, /a/, /u/, and /o/. Name these vowel sounds. ick, etc.

• What is the usual way of spelling the vowel /i/? What are two common ways of spelling /e/? What is the usual way of spelling /a/? What are two ways of spelling /u/? What is the usual way of spelling /o/? i; e, ea; a; u, o

There are ten other vowel sounds in English. We will call them the **complex** vowels. We usually say them in more complicated ways than we do the simple vowels. Also, we spell the complex vowel sounds in several different ways.

In the poem on page 22 did you notice the rhyming words at the ends of lines? The words **try, fly,** and **sigh** rhyme. So do **sea, be,** and **glee.** The vowels in these rhyming words are complex vowels.

The first of the complex vowels that we will study is the vowel of **try, sigh, ride,** and **like.**

We write this complex vowel this way: /ī/. We usually spell this vowel the way we do in **ride** and **like.** We use the letter **i.** But we also put the letter **e** at the end of the word.

• If you didn't put **e** at the end of **ride,** how would you pronounce the word? What vowel sound would it have? rid; /i/

• These words have the sound /ī/; say them: **dime, pine, ripe, stripe, hide.** Now write each word, but leave off the letter **e** at the end. Say them now. What vowel sound do they have? dim, pin, rip, strip, hid; /i/

The letter **e** at the end of a word like **ripe** or **ride** means that the vowel sound is /ī/, not /i/.

To name the sound /ī/ when you speak of it, call it "ike," much as you call /i/ "ick."

The next vowel sound to notice is the sound of **Pete** and **seem** and **treat.** All of these words have the same vowel sound, but you see that it is spelled in different ways.

We will show this vowel sound between slanting lines in this way: /ē/. When we want to speak of it, we will call it "eek."

We now have mentioned the vowels "ike" and "eek" and "ack" and "ock" and three others. What are the three others? eck, ick, uck

Pete, seem, and **treat** show three different ways of spelling /ē/. In **Pete,** we spell it much the way that we spell /ī/ in **bite** and **ride.** We use a letter **e** in the middle and then put another letter **e** at the end. The letter **e** at the end shows that the other **e** means "eek" and not "eck." How would you pronounce **Pete** if the **e** at the end were left off? pet

More often, we spell /ē/ with two **e**'s in the middle, as in **seem. Sweet, queen, deer, tree, feed** are all spelled in this way. Can you think of any other words in which /ē/ is spelled in this way? See p. 324.

We saw that we sometimes use the letters **ea** to spell /e/ as in **head** or **bread.** More often, we use **ea** to spell /ē/ as in **treat** or **bead.** Other words in which **ea** is used to spell the vowel sound /ē/ are **real, steal, meat, tea, reach.** Can you think of more examples? See p. 324.

Write these headings on a piece of paper: /i/, /ī/, /e/, /ē/. Look at the words in the list below. Write each one under the heading that shows its vowel sound.

each /ē/	trim /i/	queen /ē/	like /ī/
seen /ē/	head /e/	rich /i/	well /e/
ride /ī/	lick /i/	sea /ē/	bit /i/
quite /ī/	pet /e/	time /ī/	sled /e/

The Vowel /ē/

In contrast to /ī/, the vowel /ē/ is spelled rather rarely in the system "letter **e** + consonant letter + **e**." In addition to **Pete,** one might cite such words as **Steve, scheme, serene, precede, plebe, impede, mete, Swede, extreme, supreme, Irene, scene, Japanese**. There are others, but not many. There are few monosyllabics and few words within the fourth-grader's vocabulary. We omit here monosyllabics in which /ē/ is followed by /r/—as in **here, mere, sere**—because /r/ always introduces special problems.

Despite the paucity of the spelling "letter **e** + consonant letter + **e**," we introduce the spelling of /ē/ in this way, with the word **Pete**, in order to keep this generally very common system at the forefront. **Pete** seems to be the only word in /ē/ which will yield a word in /e/ by the removal of the final **e**.

The two most common ways of spelling /ē/ before a consonant are with **ee** and **ea**. The former is much the more usual. However, **ea** is more commonly employed to spell /ē/ than to spell /e/.

Teaching Suggestions

Elicit answers to the questions for oral response on page 25. It may be wise to begin the writing assignment at the bottom of the page with class participation also. The headings may be written on the chalkboard, and the first few items written under them at class direction. The pupils may then be able to go on and finish the exercise by themselves. Many are likely to need help, however, particularly those who did not encounter this material in the third grade.

Extra words for practice on these sounds are given on page 324.

GRAMMAR – Adjectives

Any of three structures typically follow **be** in a predicate: (1) a noun phrase, (2) an adjective, (3) an adverbial of place. In this lesson we consider the second of these.

Pupils who have studied the third-grade book of this series will find this review. Others will also probably have had acquaintance with the concept **adjective**, though perhaps under the term "picture word." They may also have had adjectives defined, perhaps as words that describe things.

There is no harm in saying that adjectives describe things. They do. But there is danger in using this observation definitionally. What precisely does **describe** mean, one must ask, and the answer is not a simple one. There is only certain confusion to be derived from saying that adjectives are words that describe or modify nouns. Adjectives do not always modify nouns, and words that modify nouns are not always adjectives.

Here we identify adjectives initially as certain forms that may occur in the predicate after a form of **be**. (It will be pointed out later that they may occur also after certain verbs, such as **seem, look, sound, taste**.) As we go along, we will distinguish them in various ways from other structures that occur in the predicate position after **be**. With practice the pupils should come to recognize adjectives clearly as a word class and be able to follow them through transformations that put them into other positions.

Teaching Suggestions

Page 26 is largely a review of the noun phrase, one of the other functions that share the position after **be** with adjectives. Go through Sentences 1–6 to be sure that no one is still having trouble recognizing noun phrases that consist of determiner + noun and function as subjects or complements of **be**.

Adjectives

You know all the parts of a sentence like "The boy is a pupil." You know that the subject is **the boy** and the predicate is **is a pupil.** The subject **the boy** is a noun phrase. What noun phrase is part of the predicate? a pupil

The other part of the predicate is the word **is. Is** is a form of the word **be.** Can you think of another word that is a form of the word **be?** yes; are

Noun phrases are expressions like **the boy** and **a pupil.** Noun phrases of this sort are made up of a determiner and a noun. The word **the** is a determiner, and the word **boy** is a noun. The word **a** is a determiner and **pupil** is a noun.

Each of the sentences below has two noun phrases in it. Tell what the noun phrases are in each sentence. Each noun phrase is made up of a determiner and a noun. Tell the determiner and the noun of each noun phrase.

Sentence	D:	N:
1. The boy is a pupil.	the, a	boy, pupil
2. The teacher is a lady.	the, a	teacher, lady
3. My dog is a puppy.	my, a	dog, puppy
4. This man is the principal.	this, the	man, principal
5. Some boys are our friends.	some, our	boys, friends
6. These girls are the winners.	these, the	girls, winners

Look back over the sentences above. Name the form of **be** in each predicate. is, is, is, is, are, are

Predicates have to have certain things in them. One thing they may have in them is a form of **be,** like **is** and **are.** But if a simple sentence has **is** or **are** in it, it must have something else too. We have seen that one thing that may follow **is** or **are** in the predicate is a noun phrase, like **a pupil** or **the winners.**

Another kind of word that we may have in a predicate after **be** is what is called an **adjective.** Adjectives are words like **hungry, quiet, good, clean.** Words like these can appear in the predicate after such forms of **be** as **is** and **are.**

Each of the following sentences has an adjective in it after a form of **be.** For each sentence, tell what the form of **be** is and what the adjective is.

7. The boy is hungry. is
8. The girls are quiet. are
9. The poem is short. is
10. The days are long. are
11. The man is angry. is
12. The books are old. are
13. The moon is orange. is
14. The weather is stormy. is
15. The sea is rough. is

Look at Sentences 16–26. Each predicate has a form of **be** in it. In some there are adjectives after the form of **be.** In others there are noun phrases. Tell which predicates have adjectives after **be** and which have noun phrases.

16. The child is thirsty. adj.
17. This girl is my sister. n. p.
18. The boys are noisy. adj.
19. His dog is a collie. n. p.
20. Those girls are the winners. n. p.
21. The girls are happy. adj.
22. The lady is lonely. adj.
23. The prize is a cake. n. p.
24. The days are short. adj.
25. The winners are your friends. n. p.
26. The bread is stale. adj.

Introduce the term **adjective** by discussing paragraph 1 on page 27 with the class. Some pupils will know of the word class either under this name or under the child's term "picture word." Be sure that everyone can pronounce it without difficulty. Point out the words in the first paragraph of this page given as examples of adjectives. If it is felt necessary to give more examples, be sure that they are genuine adjectives. Perhaps it is safer not to add to the list; the children will meet many more as they work through the exercises in this and subsequent lessons.

Do not offer any definition of adjective or ask for one. If a child insists on offering a definition, receive it pleasantly without seeming to accept or support it. The definition of **adjective** is ultimately the sum of what can be said about the forms and behavior of this group of words. We begin by noting one positional characteristic: appearance in predicates after forms of **be**.

Work through Sentences 7–15. The children should have no trouble with this since the word after **is** or **are** in each sentence is an adjective.

Sentences 16–26 are a little more difficult because some sentences have noun phrases after **be**, and others adjectives. Still the children should easily pick out the adjectives. All the single words after **be** are adjectives. All the noun phrases are two words, with a determiner marking a following noun. If there is any trouble, repeat the exercise, this time calling on different children to respond.

TESTS AND REVIEW

Each of the ten parts of this book is followed by three pages of tests and one page of review. The tests cover the essentials taught in the part they test. They are intended not so much for grading as for diagnosis. Weaknesses they reveal should be overcome by reteaching before going on. Sometimes only a few pupils will need extra help, but it may now and then happen that the whole class should review some or all of the matters tested before proceeding to new material.

It is not assumed that the children will often go back on their own initiative to restudy material in which they have made mistakes. The page numbers following each set of directions are intended for the teacher, to permit easy location of material that needs to be retaught.

Teaching Suggestions

The tests should ordinarily be done as written work. Read the directions for the first part of Test I. Be sure that everyone understands them. (It may help to write headings on the chalkboard.) Then let the class do the test. When enough time has been allowed, proceed to the second part of Test I, and so on through the tests.

After each test is completed, the papers may be checked orally in class, the pupils taking turns reading their answers to the different items. However, the teacher should examine the papers later, keep records on the individual children, and note which individuals or groups need further practice.

TEST I Sounds and Letters

1. Write the headings /i/, /e/, /a/, /u/, and /o/ on your paper. Look at the words in the list below. Write each word under the heading that shows the vowel sound it has.

If you make mistakes, study pages 6–7, 12–13.

sit	/i/	tell	/e/	cut	/u/	stop	/o/
bread	/e/	crack	/a/	lift	/i/	tap	/a/
love	/u/	mush	/u/	lad	/a/	wind	/i/
west	/e/	rock	/o/	wealth	/e/	knob	/o/

2. Write the headings /i/, /e/, /ī/, and /ē/ on a piece of paper. Look at the words in the list below. Write each word under the heading that shows the vowel sound that it has.

If you make mistakes, study pages 24–25.

lean	/ē/	Mike	/ī/	teach	/ē/	thread	/e/
wish	/i/	weep	/ē/	pet	/e/	witch	/i/
smile	/ī/	wreck	/e/	strip	/i/	stripe	/ī/
theme	/ē/	ride	/ī/	death	/e/	trick	/i/

3. Each word in the list below has the simple vowel sound /i/ in it. Change each word so that it has the complex vowel sound /ī/ in it.

If you make mistakes, study page 24.

rid strip rip bit kit

ride, stripe, ripe, bite, kite

4. Look at the words below and notice how the vowel sounds are spelled. Then close your books and write the words as your teacher dictates them to you.

bed	sweet	reach	mud	love
real	sled	bread	seem	month
luck	tea	queen	won	head

□ TEST II Grammar

1. Copy each of the following sentences and draw a line between the subject and the predicate, as shown in this example:

The train | crosses that bridge.

If you make mistakes, study pages 8–9.

a. The class | reads poems.
b. Jean | likes poems.
c. She | is a pupil.
d. The boys | clean the board.
e. His book | is on the desk.
f. They | are hungry.
g. The girls | are quiet.
h. Bob | is my friend.
i. The teacher | explains the poem.
j. Her pet | is a kitten.
k. Mr. Wilson | drives the bus.

2. Each of these sentences has two noun phrases in it. Copy the two noun phrases you find in each sentence.

If you make mistakes, study pages 14–15.

a. The class likes the poem.
b. Our teacher explains the lesson.
c. My dog is a collie.
d. Our house is near the supermarket.
e. The prize is a game.
f. A boy is at the desk.
g. The girls clean the chalkboard.
h. His mother is my aunt.
i. These girls are the winners.
j. The car is a Dodge.
k. Her desk is by the chalkboard.

Extra Test

Dictate these words and have the class spell them. Say each word clearly. Use it in the sentence. Say it again and have the children write it.

1. **damp** Dew made the grass **damp**. **damp**
2. **stiff** Her arm was **stiff** and sore. **stiff**
3. **son** His oldest **son** is Bob. **son**
4. **sun** The **sun** went behind a cloud. **sun**
5. **head** His hat is on his **head**. **head**
6. **ride** We **ride** on the school bus. **ride**
7. **try** You can spell if you **try**. **try**
8. **pie** Mother made a lemon **pie**. **pie**
9. **feed** They **feed** sea lions on fish. **feed**
10. **meat** But lions eat **meat**. **meat**

☐ **TEST II Grammar** (CONTINUED)

3. Each of these sentences has just one noun phrase that is made up of a determiner and a noun. Write the determiner and the noun.

If you make mistakes, study pages 14–15.

		D:	N:
a.	The boy is happy.	the	boy
b.	We like the girl.	the	girl
c.	This desk is new.	this	desk
d.	He saw a movie.	a	movie
e.	She has an uncle.	an	uncle
f.	Our class is large.	our	class
g.	They ride the bus.	the	bus
h.	Their mother is pretty.	their	mother
i.	The lesson is easy.	the	lesson

4. Each of these sentences has a predicate that begins with a form of **be**. Some have an adjective after the form of **be**, and some have a noun phrase. Copy what comes after the form of **be** in each predicate. Write **adj.** after it, if it is an adjective. Write **n. p.** after it, if it is a noun phrase.

If you make mistakes, study pages 20–21, 26–27.

a.	The child is happy.	adj.
b.	The horses are wild.	adj.
c.	That animal is a giraffe.	n. p.
d.	Those girls are our friends.	n. p.
e.	This book is short.	adj.
f.	The pictures are funny.	adj.
g.	His father is my uncle.	n. p.
h.	Susan is quiet.	adj.
i.	This man is our principal.	n. p.
j.	Those are our papers.	n. p.
k.	The animals are noisy.	adj.

□ **TEST III Writing and Vocabulary**

Write five sentences. In each sentence, use one of these words.

1. stanza 2. strident 3. hubbub 4. numb 5. din

• **REVIEW Ideas and Information**

1. Who wrote the poem "Stopping by Woods"? Frost
2. What was the day of the year when the poet stopped to watch the woods? Dec. 21
3. How was the poet traveling? horse and buggy
4. Why did the person who wrote the poem "Noise" like noise? It gives him "soul-stirring joys."
5. In the poem about wind, what was the wind trying to do? get in
6. Does **b** mean a sound or a letter? letter
7. Does / e / mean a sound or a letter? sound
8. How do you name these when you say them aloud: / i /, / a /, / ē /? ick, ack, eek
9. What do we call the vowels / i /, / e /, / a /, / u /, and / o / when we think of them together, as a group? simple vowels
10. What does the letter **e** in the word **ripe** show us? that the i stands for /ī/
11. What are the two main parts that each simple sentence has? subject, predicate
12. What is the term for expressions like **the boy, a doll, my desk?** noun phrase
13. What is the term for words like **the, a, some, my, your?** determiner
14. What is the term for words like **boy, doll, desk, cat, woods?** noun
15. What are words like **small, pretty, hungry, quiet** called? adjectives
16. We say that **is** and **are** are forms of what word? be

Writing and Vocabulary

The third test in each set is a group of words taken from the poems or stories that have been studied. They give a little more writing practice and check on whether an acquisition of vocabulary from the literature is going on. Any sentences that use the words in a reasonable way, including the original sentences of the poems, should be accepted gratefully.

Ideas and Information

This review is a means of recalling specific points about the readings, as well as the sound and syntax lessons, not easily tested in a more formal way. It should be done as an oral class exercise.

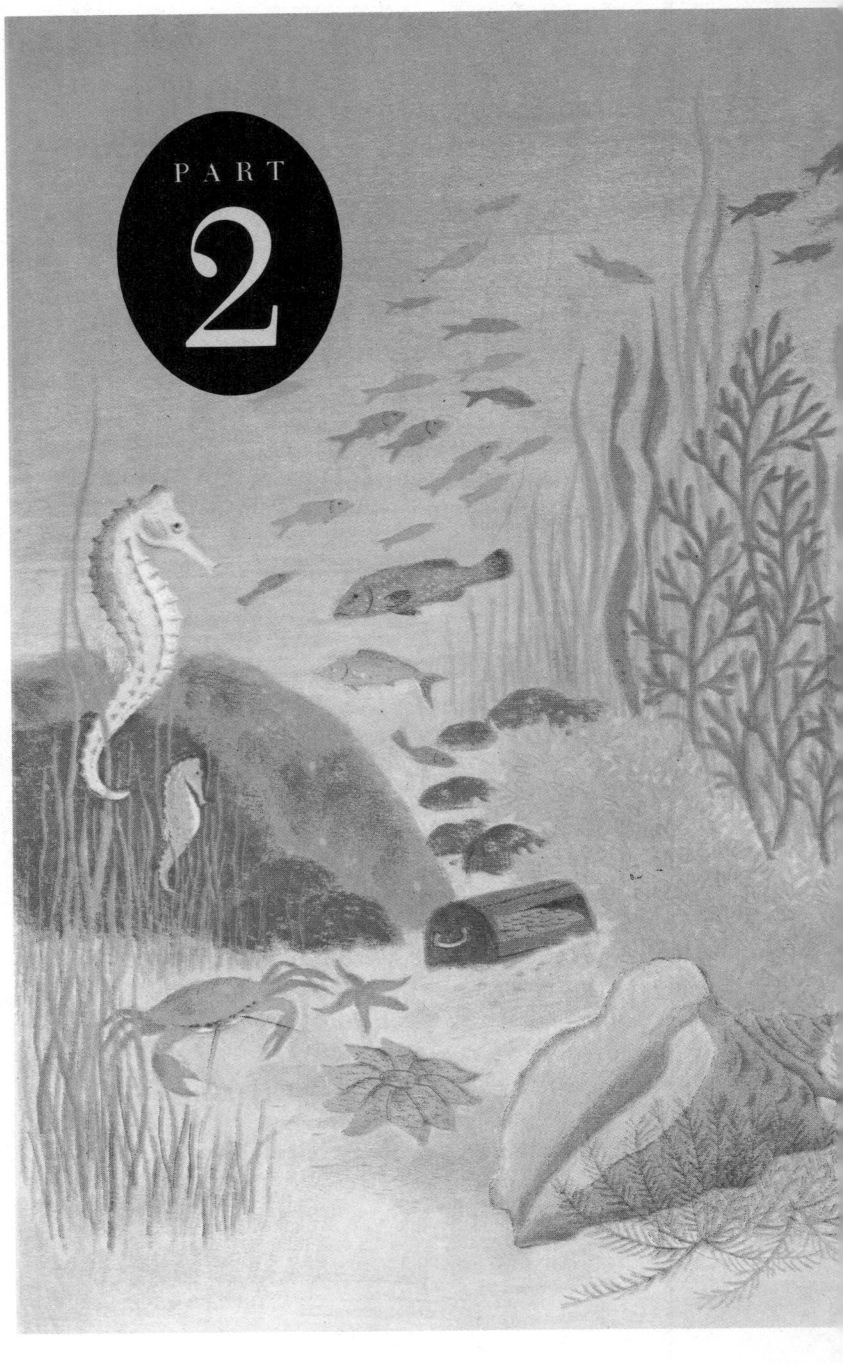
PART
2

SECTION 1

A Poem

Conch shells are big sea shells that are said to echo the sound of the sea if you hold them to your ear. This poet imagines they can sing to her.

The Sea Shell

Sea shell, Sea shell,
Sing me a song, O please!
A song of ships and sailor men,
Of parrots and tropical trees;
Of islands lost in the Spanish Main
Which no man ever may see again,
Of fishes and corals under the waves,
And sea horses stabled in great green caves —
Sea shell, Sea shell!
Sing me a song, O please!

AMY LOWELL

The Sea Shell

- Is the poet thinking about a warm part of the ocean or a cold part? How can you tell? warm; "tropical"

Ships loaded with gold and silver used to sail the **Spanish Main.** This was a great area of water and land in the Caribbean Sea, and the coasts of Central and South America around this sea, which the Spanish discovered and colonized. There were many exciting adventures there.

- What does the poet say were lost in the Spanish Main? islands
- Sea horses aren't real horses at all. They are strange little fish with heads that look a little like horses' heads. What word tells you that the poet is pretending they are real horses? stabled

33

LITERATURE – "The Sea Shell"

Some children may have had the experience of listening to a shell. If so, let them tell about it. Then read the poem or play the recording of it. Let the children take turns reading parts of it.

Note that **again** must have the sound /ā/ to rhyme with **main**. In pointing this out to the children, one is not trying to coax them into a fancy pronunciation, but only making the poetry work out.

VOCABULARY AND MEANING

"The Sea Shell"

The illustration should be used to show some of the features mentioned in the poem, particularly the stranger ones, like sea horses. However, the reader should always find the details and the meaning in the poem, using the illustration as an aid, not as a substitute. When a pupil answers from the illustration, he should be asked to find the detail also in the poem.

If there is a map of the Western Hemisphere handy in the classroom, point out the area of the Spanish Main.

Copying

Though now often felt to be a rather old-fashioned exercise, simple copying of short poems and prose passages is often very useful in teaching or reinforcing spelling, punctuation, capitalization, vocabulary, and turns of phrase. A poem of this length might be assigned to be simply copied. The goal in such an exercise is complete accuracy. (It is actually very difficult to copy accurately.) Do not call an effort perfect unless all details match those of the original. For instance, "Sea shell, sea shell," is wrong for the first line.

Final /ī/ and /ē/

Vowels occurring in final position usually have characteristic final spellings different from their medial spelling. Thus the characteristic spelling of /ī/ before a consonant is **i–e** (**pine, ride**). Its characteristic final spelling is **y** (**my, try**). We usually teach the pre-consonant spelling first, and then, as on page 34 for /ī/ and /ē/, the final spelling or spellings.

Final /ī/

This book does not attempt a complete list of possible spellings for the sounds it treats. We give one or two or three of the more common ways, depending on the sound, and some of the exceptions. The pupils will now and then come on other, rarer forms which they must learn as the need and opportunity to do so arise.

Here for the final spelling of /ī/ we give the spelling **y** as the usual one, with **ie** as a secondary spelling for four common words. (The spelling **ie** for /ī/ occurs in seven words in all, the other three being **fie, hie**, and **vie**.) In addition, final /ī/ is spelled **uy** in two words (**buy, guy**), and **igh** in four (**high, thigh, nigh, sigh**). Some such special spellings are brought to the pupil's attention later in the book. We try here to get him used to the idea of variant spellings without drowning him in detail.

Final /ē/

The spelling **ee** is much more common than **ea** for final /ē/, as it is for pre-consonant /ē/. The **ea** words in addition to the five listed are **lea** and **guinea**.

Of course **y** is a very common way of spelling final /ē/, because it occurs in the spelling of several morphemes, as in the final morpheme of **coldly, electricity**, and also at the end of words like **silly, funny**. It is never used to spell final /ē/ in monosyllabic words, however.

The Vowel Sounds /ī/ and /ē/

You have now studied five simple vowel sounds: /i/, /e/, /a/, /u/, and /o/. You have studied two complex vowel sounds: /ī/ and /ē/.

- Name the five simple vowels. ick, eck, etc.
- Name the complex vowels you have studied. ike, eek
- The complex vowels are spelled in different ways in different words. You have studied only one way of writing /ī/. That is the way the sound is spelled in **ride** and **side**. What letter at the end of words like these shows that the vowel sound is /ī/, not /i/? e

There are other ways to spell the vowel /ī/. When this vowel comes at the end of a word, we usually write it with the letter **y**: **fly, try, sky, by.** Can you give some other examples? See p. 324.

There are four common words in which we spell /ī/ at the end with the letters **ie**: **pie, lie, die, tie.**

- You studied three ways of spelling the complex vowel /ē/: with **e** and another **e** at the end, as in **Pete**, with **ee** as in **seem**, and with **ea**, as in **real**. Can you give some other examples of each of these ways of spelling the vowel /ē/? See p. 324.

- When /ē/ comes at the end of a word, we usually spell it with the letters **ee**: **tree, see, glee.** Can you name any others? bee, free, knee, etc.

In some words we spell /ē/ at the end with the letters **ea**. These are the most common ones:

sea tea pea flea plea

In a few very common words we spell /ē/ with just the letter **e** at the end:

he she we me be

☐ Write /ī/ and /ē/ as headings. Under each heading, write the words from this list that contain that sound.

we /ē/	fry /ī/	seal /ē/	wide /ī/	wheat /ē/
pie /ī/	feet /ē/	be /ē/	pry /ī/	these /ē/
flea /ē/	side /ī/	die /ī/	eat /ē/	meet /ē/

Because we have several ways of spelling some sounds, like /ē/, we often have pairs of words that are pronounced exactly alike but are spelled differently. Pairs of words like these are called **homophones.** **Homophone** is a word we have borrowed from the Greek language. The first part, **homo,** means the same. The second part, **phone,** means sound. So **homophone** means having the same sound.

We have already noticed one pair of homophones — **sun** and **son.** These words are pronounced in exactly the same way. They have the same consonants. Both have the vowel /u/. But they are spelled differently and have different meanings. What do they mean? the star; male child

There are many homophones among words that have the vowel /ē/ because there are several ways of spelling /ē/. The words **see** and **sea** are homophones. They are pronounced in exactly the same way. What does **see** mean? What does **sea** mean? observe; ocean

Here are some other homophones. Tell the difference in meaning of each word in the pairs. Use a dictionary. For example, if you were telling about **see** and **sea,** you might say, "**See** is what we do when we look at something. **Sea** means the ocean."

(Answers will vary.) (Accept informal definitions.)

meet / meat	flee / flea
bee / be	wee / we
tee / tea	peek / peak
cheep / cheap	beet / beat
deer / dear	hear / here

35

Homophones

The term **homonym** is sometimes used imprecisely either for **homophone** or **homograph**. Homonyms are pairs of words identical in spelling and pronunciation but different in meaning. Thus **bear** (the animal) and **bear** (to endure) are homonyms. Homographs are identical in spelling but different in sound and meaning. **Lead** (the metal) and **lead** (to conduct) are homographs. **Homophones** are identical in sound but different in spelling and meaning. **Meet** and **meat** are homophones.

Only the term **homophone** is used in this book.

It is convenient to introduce homophones at this point, since words with the sound /ē/ are rich in them, because of the common use of the two spellings **ee** and **ea**.

Increasingly, though still not very much in grade four, the pupils will hear of Greek and Latin because of the dominant role that borrowings from these languages play in the English vocabulary. If in any other part of their studies the pupils have learned something of the classical civilizations, it will be useful to stress that we have taken many words from the languages of these civilizations into English—as, for example, we have **homophone**.

GRAMMAR—Verbs

A grammar is essentially a definition of **sentence** in some particular language. We define **simple sentence** in a series of rules which show what large parts sentences are made up of, what smaller parts compose these, and so on. When the permitted forms of simple (kernel) sentences have been thus displayed, the structure of complicated sentences is then shown by transformational rules. We do this non-technically for the most part for young children. But the teacher may be interested in seeing something of a more economical treatment.

The first rule of the grammar shows the main components of any simple sentence thus:

$$S \rightarrow NP + VP$$

S stands for **sentence, NP** for **noun phrase, VP** for **verb phrase**. The rule says that a kernel sentence is composed of a noun phrase (which functions as its subject) and a verb phrase (which functions as its predicate).

Further rules specify the structures of NP and VP. So far we have had just one kind of NP:

$$NP \rightarrow \text{determiner} + \text{noun}$$

Presently we shall add to this some other kinds of noun phrases.

We have had just this much of VP:

$$VP \rightarrow \textbf{be} + \begin{Bmatrix} NP \\ Adj \end{Bmatrix}$$

This says that a verb phrase may consist of a form of **be** followed by either a noun phrase (already partially defined) or an adjective. Of course, much more must be added to this too. We have seen that **be** may appear in the form **is** or **are**, and we must eventually account for such other forms as **am, was, were, has been, can be, is being**, and others. Also, there are other possibilities after **be** in addition to NP and Adj.

Verbs

• We have seen that a simple sentence must have a subject and a predicate. Tell what the subjects and predicates are in these sentences:

1. The milk is cold.
2. A sea shell sings a song.
3. The ladies are our teachers.
4. David reads.

(Subjects are underlined; predicates are not.)

• The subject of a simple sentence is a form that we call a **noun phrase.** There are several kinds of noun phrases. **The milk, a sea shell, our teachers** are one kind. What do we call **the, a,** and **our** in these noun phrases? What do we call **milk, sea shell,** and **teachers?** determiners; nouns

• The predicate of a sentence may be made up of several different things. We have seen one thing that may occur in a predicate — a form of the word **be.** What words in these sentences are forms of **be?**

5. The milk is cold.
6. The ladies are our teachers.

• If the predicate of a simple sentence has a form of **be** in it, there must be something else after the form of **be.** Two things that may come after the form of **be** are adjectives and noun phrases. Tell what comes after **be** in each of these sentences — **adjective** or **noun phrase.**

7. The milk is cold. adj.
8. The ladies are our teachers. n. p.
9. He is hungry. adj.
10. John is the captain. n. p.
11. People are funny. adj.
12. They are your friends. n. p.

If the predicate of a simple sentence does not have a form of **be** in it, then it must have another kind of word, which we call a **verb.** Verbs are words like **see, try, like, want, smile, go.**

Be must have something after it, but in a simple sentence a verb alone can be the whole predicate. In these sentences, the predicate is just a verb alone. Tell what the verbs are in these sentences.

13. John smiles.
14. The girls laugh.
15. The children play.
16. Our dog barks.
17. The kittens purr.
18. The building shakes.
19. The man laughs.
20. The rooster crows.

But usually the verb of a predicate is followed by other words. In all of the sentences below, the verb is followed by other words. Tell what the predicate is in each sentence. Tell what the verb is in each predicate.

21. The children play games. — play
22. The teacher helps the children. — helps
23. We study the lesson. — study
24. The poem seems easy. — seems
25. Jane lives near me. — lives
26. The boys clean the chalkboard. — clean
27. John smiles all the time. — smiles
28. The sea shell sings a song. — sings
29. The ship sails on the Spanish Main. — sails
30. The sailors like tropical trees. — like
31. Those problems look hard. — look
32. Coral grows underwater. — grows

Meanwhile, we now introduce a possibility in addition to **be** plus what follows it as a structure of the verb phrase:

$$\text{VP} \rightarrow \begin{Bmatrix} \textbf{be} + \begin{Bmatrix} \text{NP} \\ \text{Adj} \end{Bmatrix} \\ \text{verbal} \end{Bmatrix}$$

This formula says that a verb phrase may consist of **be** plus a noun phrase or an adjective. *Or* it may consist of a verbal.

Though we do not in this book use the term **verbal**, we begin in effect to define it on page 37 by showing that the "non-**be**" predicate may be just a verb by itself:

$$\text{Verbal} \rightarrow \text{Verb}$$

Sentences 13–20 display this sort of verb phrase functioning as predicate.

Sentences 21–32 have predicates consisting of verbs plus other structures. We do not ask the children here to identify these other structures, but only to find the verbs that precede them. But the variety of structures now occurring may be shown by this rough expansion of **verbal**:

$$\text{Verbal} \rightarrow \begin{Bmatrix} \text{verb (Sentences 13–20)} \\ \text{verb + NP (21, 22, 23, 26, 28, 30)} \\ \text{verb + Adj (24, 31)} \\ \text{verb + adverbial of place (25, 29, 32)} \\ \text{verb + adverbial of time (27)} \end{Bmatrix}$$

It may be that some day it will seem simplest to teach grammar from the start in this somewhat technical way. Meanwhile it is useful as a guide for text and teacher.

Teaching Suggestions

Review what has been presented already by going through Sentences 1–12.

Most pupils will already have encountered verbs, perhaps under a term like "action word" or the like. Do not try to define the term; just let the concept build up through abundance of examples and observance of formal and positional characteristics.

LITERATURE

"The Table and the Chair" (Stanzas One and Two)

Some of the poems and stories are too long to be dealt with in a single section, and so are spread, in the books for grades three, four, and five, over two, three, or four sections. Lear's "The Table and the Chair" occupies the literature part of two sections.

Teaching Suggestions

Play the recording of the poem or read it to the class before letting individual pupils try reading portions of it. The four-stress lines give the poem a pleasant, playful swing. Some children may want to commit part of it to memory.

SECTION 2

A Poem

Perhaps a table and a chair **might** try out their legs. You will find out later, in the next two stanzas.

The Table and the Chair

Said the Table to the Chair,
"You can hardly be aware
How I suffer from the heat,
And from chilblains on my feet!
If we took a little walk,
We might have a little talk!
Pray let us take the air!"
Said the Table to the Chair.

Said the Chair unto the Table,
"Now you *know* we are not able!
How foolishly you talk,
When you know we *cannot* walk!"
Said the Table with a sigh,
"It can do no harm to try:
I've as many legs as you,
Why can't we walk on two?"

The Table and the Chair

The author of this poem is Edward Lear. He lived about a hundred years ago, and he was a famous poet. He wrote other poems that you may have read or that you may read sometime. One was called "The Owl and the Pussy-Cat."

Look up **aware** in a dictionary.

- The table says it suffers from the heat and from **chilblains.** Does the word **chilblains** sound as though they come from heat or from a chill? Really, a **chilblain** is a swelling of the feet caused by cold. Do you think the table is a little confused about things? a chill discuss
- What does the table say they might do if they took a little walk? What do you suppose "take the air" means? have a little talk; go outside
- The chair seems a little more realistic than the table. That is, he understands the difficulties in what the table wants to do. What difficulty does he see in the idea of taking a little walk? not able to walk
- What does the table answer to the chair? Why does he think they may as well try to take a little walk? See last four lines.

(This poem is continued on pp. 46–47.)

Rhyme

- There are four pairs of rhyming words in the first stanza of this poem. What are they? chair–aware, heat–feet, walk–talk, air–chair
- One pair of rhyming words in the first stanza has a complex vowel sound you have studied. What is it? Is the sound spelled the same in both words? /ē/; no
- What four pairs of rhyming words are there in the second stanza? Which pair has a complex vowel sound you have studied? What sound is it? Is the sound spelled the same in both words? table–able, talk–walk, sigh–try, you–two; sigh–try; /ī/; no

VOCABULARY AND MEANING

"The Table and the Chair" (Stanzas One and Two)

Humor is notoriously short-lived, and many of Edward Lear's poems seem quite unfunny to us today. However, a few, including this one, hold up very well indeed. Another is "The Owl and the Pussy-Cat," which is used in the third-grade book of this series. Another perennial favorite is "The Jumblies," which tells of the people who went to sea in a sieve.

Teaching Suggestions

A comic poem can serve as well as a serious one for training in close reading, and the same standards should be maintained. Go through the questions in class discussion and be sure that everyone understands what is happening.

Point out the quotation marks in these two stanzas. Recall that quotation marks are used to enclose words that people actually say. Ask who is speaking in the first stanza. When does he begin, and where does he end? Do the same for the second stanza, where both speak.

SOUNDS AND LETTERS – Rhyme

Elicit in class discussion answers to the questions on the rhymes of the poem. This device serves to keep attention on the fact that words are made up of *sounds*, which letters convey. Note that **heat** and **feet** rhyme although they are spelled differently. Later on the children will be asked to do rhyming exercises – for example, to write words that rhyme with the rhyming parts spelled in the same way. This is preparation.

Point out that **sigh** has still another way of spelling the vowel /ī/ at the end of a word, but that it is used only in a very few words.

SOUNDS AND LETTERS—/ā/

Before introducing the new vowel sound /ā/, we review the two complex vowels previously studied. We also introduce, spurred by the poem just read, the **igh** spelling of /ī/.

We give also the common words in which /ī/ is spelled **igh** before /t/. There are in addition a few less common ones: **wright, blight**. And of course there are compounds: **footlight, overnight**. The sound /ī/ is spelled this way mainly before /t/. It is sometimes spelled **ig** before **n**: **sign**.

The **gh** in a word like **night** or **fight** represents a consonant sound formerly used in English, a sound something like that of German **ich**. The consonant sound disappeared hundreds of years ago, and the preceding vowel sound, which had been /i/, eventually changed to /ī/, but the old spelling remains. Our English spelling does not lightly release the history of the language.

Extra lists of words for practice on the sounds /ī/ and /ē/ are given on page 324.

The Vowel Sound /ā/

You have studied two complex vowel sounds, /ī/ and /ē/. Here are some ways of spelling /ē/. For each one, give another word in which /ē/ is spelled in the same way:

clean see Steve he sea, tree, these, me

We have seen three ways of spelling the sound /ī/. One way is like **bite** or **ripe**, with the letter **i** in the middle and the letter **e** at the end. Give two other words in which /ī/ is spelled like that. See p. 324.

At the end of a word, we usually spell /ī/ with the letter **y** as in **try**. Give another word like **try**. In a few words we spell it with **ie**, as in **pie**. What is another word like **pie**? why, fry; die, lie

In our poem about the table and the chair, we found another way of spelling the vowel /ī/:

Said the table with a sigh,
"It can do no harm to try."

What word rhymes with **try**? How is the vowel sound /ī/ spelled in that word? sigh; igh

• There are three other words in English in which /ī/ at the end is spelled **igh**. Find them in these sentences:

The thigh is the upper part of the leg.
A high tree is a tall tree.
To come nigh is to come near.

The only way to learn to spell **sigh, thigh, high,** and **nigh** correctly is to remember the unusual ending.

The sound /ī/ is also spelled **igh** in many words that have the letter **t** at the end:

fight	**light**	**flight**	**slight**	**might**	**night**
knight	**right**	**bright**	**fright**	**sight**	**tight**

• Do **write** and **fight** rhyme? What do we call a pair of words like **write** and **right?** yes; homophones

The next complex vowel to study is the vowel sound you hear in **lane** or **rain.**

We will show this vowel sound with the letter **a,** with a mark over it, like this: /ā/.

When we want to talk about the sound /ā/, we will speak about it the way we do /ī/ and /ē/. We will make the sound with a /k/ sound after it, as if we were saying "ache." ache, ike, eek

• Say the sound /ā/; the sound /ī/; the sound /ē/.

Like the other complex vowels, /ā/ is written in several ways.

One way is like the way we write /ī/ in **bite** or /ē/ in **Pete.** We use the letter **a,** and then at the end of the word we put the letter **e.** This **e** shows that the **a** is pronounced /ā/ and not /a/. Here are some words in which we write the vowel /ā/ in this way:

lane	late	shape	take
rate	date	tame	pale
same	lake	fate	page

• Can you name some other words like these? See p. 324.

• Pronounce the words in the list below. Then write them, but leave off the **e** at the end. Now pronounce the words you have written. What words are they now? What vowel sound do they have? /a/

rate	rat	pale	pal	same	Sam	fate	fat
cane	can	pane	pan	hate	hat	cape	cap

Tell what the vowel sound is in each of these words. Call them "ick," "eck," "eek," "ache," and so on.

pick	ick	rake	ache	neck	eck	like	ike
meek	eek	sack	ack	knock	ock	luck	uck
my	ike	free	eek	sigh	ike	shape	ache
he	eek	son	uck	bread	eck	laugh	ack

The review of vowel sounds (but not consonant sounds) presented in the third-grade book is now completed. The sound /ā/ will be a new study for all pupils.

The spelling "letter **a** + consonant letter + **e**" is a common device for the spelling of /ā/—less so than for /ī/, but much more so than for /ē/. Following the usual procedure, we introduce this pattern first, and it is the only one given in this lesson.

Teaching Suggestions

Let the children take turns pronouncing the list of words beginning with **lane** on page 41 to become aware of the sound and associate it with the spelling. When this is concluded, ask if anyone can think of other examples. Possibilities include **ate, skate, bake, rake, gale, maze, wade, shade,** and many more, including all those in the list that follows the **lane** list. Of course they also include verbs with the suffix **ate,** like **educate,** but it is best to keep the focus on monosyllabic words for the present.

Stress the function of "silent **e**" by working with the class through the next set of words, where the presence or absence of **e** conveys the /ā/–/a/ contrast: **hate–hat.**

Review the different vowel sounds studied, and practice naming them with the exercise at the bottom of the page.

Additional lists of words for practice on the sound /ā/ are given on page 325.

GRAMMAR – Predicates

The material on page 42 is largely a review of the following rules:

$$\begin{aligned} &S \rightarrow NP + VP \\ &NP \rightarrow \text{determiner} + \text{noun} \\ &VP \rightarrow \begin{Bmatrix} \textbf{be} + \begin{Bmatrix} NP \\ Adj \end{Bmatrix} \\ \text{verbal} \end{Bmatrix} \\ &\text{Verbal} \rightarrow \text{verb} \end{aligned}$$

Noun Phrases After Verbs

On page 43 we add another possible element to the rule for verbal:

$$\text{Verbal} \rightarrow \begin{Bmatrix} \text{verb} \\ \text{verb} + NP \end{Bmatrix}$$

There is actually a little more to it than this. A noun phrase following a verb may have any of several functions, depending on the type of verb. After a transitive verb, a noun phrase functions as an object. After **become** or **remain**, it functions as a complement: "John became a Scout." After certain verbs, it may be a measurement: "John weighs ninety pounds." Thus the rule that a verbal may consist of verb + noun phrase would have to be further broken down in a more advanced grammar.

Here we limit "verb + noun phrase" to the type "transitive verb + noun phrase," in which the noun phrase functions as object. Since other types of verbs followed by noun phrases are not now introduced, we do not yet have to use the term **transitive** with the pupils.

Teaching Suggestions

Do Sentences 1–5 in class discussion to recall the distinction between verbs and forms of **be**. Review noun phrases and adjectives in the predicate after **be** with Sentences 6–14.

Review the type of verbal consisting of the verb alone with Sentences 15–17. Introduce the type "verb + noun phrase" and

Predicates

A simple sentence must have a subject and a predicate. The subject of a simple sentence is a noun phrase. One kind of noun phrase is made up of a determiner and a noun. These are noun phrases: **the book, a chair, their legs.**

The predicate of a simple sentence must have one of two things in it. It may have a form of **be** in it, like **is** or **are.** If it doesn't have a form of **be,** it must have a verb, like **see, help, fight, answer.**

• Study the predicate of each of these sentences. Which does it have, a verb or a form of **be?** Point out the verbs and the forms of **be.**

1. He is sick. — be
2. He smiles. — v.
3. He plays the cello. — v.
4. He is our friend. — be
5. He understands the poem. — v.

• If the predicate has a form of **be** in it, it must have something else too. One thing that may come after **be** in the predicate is a **noun phrase.** What is another thing that may come after **be?** an adjective

• Tell what comes after **be** in these sentences — a **noun phrase** or an **adjective.**

6. She is happy. — adj.
7. She is a teacher. — n. p.
8. She is quiet. — adj.
9. She is interesting. — adj.
10. She is my mother. — n. p.
11. She is the author. — n. p.
12. She is pretty. — adj.
13. She is your cousin. — n. p.
14. She is sad. — adj.

• If the predicate has a verb in it instead of a form of **be**, the verb may be the only word in the predicate. What are the verbs in these sentences?

15. The boys play.
16. The boys work.
17. The boy talks.

But a verb in a predicate *may* have other things after it. One thing it may have is a noun phrase. Tell what the verb is in each of these predicates. Tell what the noun phrase in the predicate is.

18. Sally plays the piano. the piano
19. Sally understands the lesson. the lesson
20. Sally likes the poem. the poem
21. Sally cleans the chalkboard. the chalkboard
22. Sally takes the bus. the bus
23. Sally helps the children. the children

Look at the sentences below. Tell whether the predicate is made up of (1) a form of **be** and an adjective, (2) a form of **be** and a noun phrase, (3) a verb alone, (4) a verb and a noun phrase.

24. The children are quiet. (1)
25. That lady is his mother. (2)
26. She works. (3)
27. The gardener mows the lawn. (4)
28. Ralph washes the car. (4)
29. He is a pupil. (2)
30. They dance. (3)
31. The giraffes are hungry. (1)
32. My father reads the paper. (4)
33. We are the actors. (2)
34. You are angry. (1)
35. Choirs sing. (3)
36. Camels cross the desert. (4)
37. Those girls are the winners. (2)

practice it with Sentences 18–23. Note that we do not yet name the function (object) of the noun phrase.

Use Sentences 24–37 to give the children practice in distinguishing four kinds of verb phrases:

be + NP
be + Adj
verb
verb + NP

All of these are called verb phrases in technical terminology, even though only two of them contain verbs.

Extra Practice

For extra practice the children may write the verb and noun phrase in the predicate of each sentence from 18 to 23. They may also do Sentences 24–37 as written work, identifying the kind of predicate by writing (1), (2), (3), or (4) for each sentence.

COMPOSITION — Writing Letters

The letters that fourth-graders might be expected to write are personal, short, and simple. They present few mechanical complications. However, the children should be carefully guided to consistent observance of the conventions that do exist.

Teaching Suggestions

Begin with the writing of the address. The teacher might begin by putting an address, fictional or otherwise, on the chalkboard:

523 Delynn Way
Carrolton, Illinois

Point out the details. The name of the street (way) is capitalized. Ask what words some addresses might have in place of **Way**: for example, **Street, Avenue, Lane**. Point out the capitals in the name of the town and the name of the state, and draw attention to the comma between the town and the state.

Ask a child to tell his address. Write it on the board, asking the class to tell how it must be written. Do this until the details seem clear. Then have each child write his own address.

Write today's date, whatever it is. Point out the capitalization of the month, and the comma between day and year. Do the second written exercise in the middle of page 45. If the children are uncertain, do it first as a class exercise, writing some dates on the chalkboard. Then have the pupils complete the exercise for themselves.

Explain **salutation** and give examples other than those in the text. Use just **Dear** and a name. Point out the comma that follows. The children should use just a comma at this point. The colon is used only in formal letters.

Explain **closing** and give other examples. Point out the comma after the closing.

Reemphasize the necessity of leaving ample margins.

Writing Letters

A letter to a classmate who has moved and is going to another school might look like this.

72 Sawmill Road
Akron, Ohio
October 25, 19--

Dear Sally,
Do you like your new school? We are studying poems and grammar in English this year. We are studying new things in mathematics too. What are you studying? We all miss you very much.

Your friend,
Randy

When we write letters to people, there are certain ways in which we do it.

For one thing, we usually put our own address at the top right, so that the person we write to can answer the letter easily. Notice how this address is written. All the words in it are written with capital letters. Names of streets, places like Akron, and states like Ohio are always written with capital letters. Notice the comma (,) between the name of the city and the name of the state.

□ Write your own address.

We write the date under our own address to show what month, day, and year the letter was written. Notice that the name of the month begins with a capital letter. Where is the comma in the date? before the year

□ Write today's date. Write the date of your next birthday. Write the date of Christmas this year.

At the beginning of the letter, we write what is called a **greeting,** or **salutation** — a friendly way of saying hello. This usually begins with the word **Dear.** We write "Dear Sally" or "Dear Mr. Wilson." We put a comma after the greeting.

● The **body** of the letter is the main part. What does the body of the letter to Sally say? See letter.

At the end of the letter we put a **closing.** We write "Your friend," or "Sincerely," or "With love," or some other closing like these. We begin the closing with a capital and put a comma at the end. Then we sign our name.

Notice the wide white spaces, or margins, all around the letter. Always leave wide margins.

□ Write a friendly letter to someone you know. It may be someone who has gone to another school, someone who is sick, or just any friend of yours.

LITERATURE

"The Table and the Chair"
(Stanzas Three to Five)

Teaching Suggestions

If it is convenient to use the recording, it will help, in these pieces which run to more than one section, to repeat the preceding part before going on to the new. Otherwise the teacher can read the first two stanzas once, and then these three, two or three times. Then give the children practice in reading.

SECTION 3

The Rest of the Poem

These three stanzas tell what happened after the Table and the Chair decided to "take the air."

So they both went slowly down,
And walked about the town
With a cheerful bumpy sound,
As they toddled round and round.
And everybody cried,
As they hastened to their side,
"See! the Table and the Chair
Have come to take the air!"

But in going down an alley
To a castle in the valley
They completely lost their way,
And wandered all the day,
Till, to see them safely back,
They paid a Ducky-quack,
And a Beetle, and a Mouse,
Who took them to their house.

Then they whispered to each other
"O delightful little brother!
What a lovely walk we've taken!
Let us dine on Beans and Bacon!"
So the Ducky and the leetle
Browny-Mousey and the beetle
Dined, and danced upon their heads
Till they toddled to their beds.

The Table and the Chair

STANZAS THREE TO FIVE

- A baby **toddles,** or walks unsteadily. Why do you think the Table and Chair toddled instead of just walking? stiff legs
- What do you think a **ducky-quack** is? Why did they pay the ducky-quack, beetle, and mouse? duck; to guide them
- What does **they** mean in the first line of the last stanza? That is, who whispered to each other? table, chair
- The last word in the third line from the end of the poem is **beetle.** What vowel sound does it have in the first part? What word in the line before rhymes with **beetle?** The poet writes the word that way for fun and to make it rhyme with **beetle.** How would you usually spell **leetle?** What vowel sound does it usually have? /ē/; leetle; little; /i/

VOCABULARY AND MEANING

"The Table and the Chair" (Stanzas Three to Five)

Work through the questions, eliciting answers from class discussion.

Additional Questions

To ensure close and attentive reading, the teacher may ask these further questions:

What kind of sound did they make when they walked about the town?

Who hastened to their side? What did they say? What marks show us the words that they said? Where do these marks come?

Where are they going at the beginning of the second stanza? What happened to them? How long did they wander?

Where did the Ducky-quack, the Beetle, and the Mouse take them?

What did they whisper to each other? What are they going to dine on?

Where did the duck, the beetle, and the mouse dance? Whose heads did they dance upon? Where did they toddle?

The sensibilities of sophisticated fourth-graders may be offended by the expression "Ducky-quack." If so, explain that the poet must have something to rhyme with **back**. Ask what the rhyming vowel is.

SOUNDS AND LETTERS—/ā/

The children have been introduced to /ā/ in the "letter **a** + consonant letter + **e**" spelling. We now give the other common way of spelling /ā/ before consonant sounds: **ai**. There are a few other occasional possibilities: **weight, break**. But we do not introduce these exceptional forms here.

Teaching Suggestions

Get the children to suggest other words in which /ā/ is spelled **ai**, and write the suggestions on the chalkboard. There are many possibilities—for example, **bait, mail, drain, pain, rail, jail, aim, claim, waist**.

Review homophones, using now the examples given for /ā/. Often the best way for the children to distinguish the meanings of homophones is to use each in a separate sentence.

Use the words at the end of the page to practice recognition of the complex vowels so far studied and to get used to naming them.

The Vowel Sound /ā/

• We have found the vowel sound /ā/ in words like **lane, lake, pale.** What letter do these words end with? What does this letter tell us about the vowel sound in the words? e; that the sound is /ā/

Another common way to spell the vowel sound /ā/ is with the letters **ai:**

rain fail main wait braid

• Can you think of other words in which the sound /ā/ is spelled with the letters **ai?** See p. 325.

You remember that homophones are words that sound alike but are spelled differently and have different meanings.

• **Meet** and **meat** are homophones. What is a homophone of **beat?** What is a homophone of **sun?** beet; son

• There is more than one way to spell the vowel sound /ā/, so there are homophones that have this vowel, too. **Pale** and **pail** are examples of homophones with the vowel /ā/. What does each of the words mean? pallid; bucket

• Here are other homophones with the vowel /ā/. Tell what they mean. Use a dictionary.

(Answers may vary.)

plane / plain	**tale / tail**
sail / sale	**stair / stare**

• How do you name a complex vowel sound such as /ā/ when you talk about it? ache

• Each of the words below has one of the complex vowels /ī/, /ē/, or /ā/. Name the vowel sound that each of the words has.

ride ike	raid ache	dream eek	cry ike
sweet eek	me eek	cave ache	tie ike
rail ache	sea eek	nice ike	sigh ike
feel eek	tight ike	spade ache	teach eek

48

Complex Vowels at the End of Words

1. The sound /ī/ at the end of words is spelled in these ways.

a. We usually spell /ī/ at the end with the letter **y.**

by try cry fly my see p. 324.

• Give other examples of /ī/ at the end spelled **y.**

b. In a few words we spell /ī/ at the end with **ie:**

pie tie die lie

c. We spell /ī/ at the end with **igh** in these words:

sigh nigh thigh high

2. The sound /ē/ at the end is spelled in these ways.

a. We most often spell /ē/ at the end with **ee:**

tree free flee see knee bee, glee

• Give other examples of /ē/ at the end spelled **ee.**

b. In some words we spell /ē/ at the end with **ea:**

pea tea plea sea flea

c. We spell /ē/ at the end with **e** in these words:

me he she we be

3. The most common way of spelling the sound /ā/ at the end of words is with the letters **ay.** Here are some words in which /ā/ is spelled that way:

day	play	stay	gay
may	say	okay	pray
sway	gray	bay	ray

Can you think of any other words in which the sound /ā/ is spelled with the letters **ay?** see p. 325.

□ Write /ī/, /ē/, /ā/ as headings. Under each heading, write the words on this list that end with that sound.

me /ē/	my /ī/	may /ā/	tee /ē/
tray /ā/	try /ī/	flea /ē/	flay /ā/
high /ī/	pie /ī/	she /ē/	lay /ā/
tree /ē/	way /ā/	sly /ī/	sea /ē/

Complex Vowels at the End of Words

We must suppose that a person learns to spell by absorbing a certain number of individual items and then making whatever generalization from them he finds possible. Thus with final /ē/, he may encounter the words **he, pea, see, me, be, tree, flee, flea, knee, she, we, spree, three, wee**. At first, this may seem to him a quite random business, but eventually, usually quite unconsciously, he will come to some realization of system. Somewhere in his mind he inscribes rules, of a sort, like these:

1. This sound is spelled with **e** in some common words: **he, me, be, she, we**.
2. In a few words it is spelled **ea**: **pea, flea, tea**.
3. But usually it is spelled **ee**.

Then if he hears a new word, say /klē/, he will quite logically and probably quite correctly apply rule 3—the "usually" or "all other cases" rule—and spell it **clee**. If instead it is another word to which rule 1 or rule 2 applies, then he must be corrected and must learn the new item.

What we try to do here is help him establish these rules which he is already working at by himself without being aware of doing so.

Final /ā/ is overwhelmingly spelled **ay** and so is relatively easy to learn. To use the scheme applied above, we can say that such rules as 1 and 2 apply only to a very few items, such as **weigh** and **prey**, and the bulk of words with this sound are embraced in the "all other cases" rule.

Teaching Suggestions

Other examples of **ay** words are given on page 325.

Let the children do the written exercise, if necessary starting it on the chalkboard.

GRAMMAR—Adjectives After Verbs

Page 50 reviews the fact that a verb phrase functioning as a predicate must have in it a verb or a form of **be**. This follows from the rule for verb phrase:

$$\text{VP} \rightarrow \left\{\begin{matrix} \textbf{be} + \left\{\begin{matrix}\text{NP} \\ \text{Adj}\end{matrix}\right\} \\ \text{verbal} \end{matrix}\right\}$$

To have a kernel sentence, we must have a VP as a consequence of the rule "S → NP + VP." Then, because of the rule given above, to have a VP, we must have a form of **be** or a verbal. If we choose **be**, we must have (so far as we've gone) either a noun phrase or an adjective after it. If we choose instead **verbal**, then we have so far two possibilities:

$$\text{verbal} \rightarrow \left\{\begin{matrix}\text{verb} \\ \text{verb} + \text{NP}\end{matrix}\right\}$$

On page 51 we add a new type of verb phrase by expanding the rule for **verbal** as follows:

$$\text{verbal} \rightarrow \left\{\begin{matrix}\text{verb} \\ \text{verb} + \text{NP} \\ \text{verb} + \text{Adj}\end{matrix}\right\}$$

The last item accounts for such predicates as "looks happy," "sounds angry."

As was mentioned earlier, the rule implies a subclassification of verbs. The verb standing alone in the predicate of a kernel sentence is an intransitive verb. The verb followed by a noun phrase functioning as object is a transitive verb. A verb followed by an adjective is sometimes called a "linking verb" or a "verb of the **seem** class." There is no need here, however, to specify the subclassification and, therefore, none to use the terms.

Adjectives After Verbs

You know that the two main parts of a simple sentence are a subject and a predicate. What do we call the words that are the subject of a sentence? noun phrase

The predicate of a simple sentence must have at least one of two things. It must have a form of **be** or a verb. What are the two forms of **be** that you have studied? is, are

If the predicate has a form of **be**, it must have something else too. One thing it may have is a noun phrase. What is something else that may come after **be** in a predicate? adjective

If the predicate has a verb instead of a form of **be**, it may have just the verb alone. In "Birds sing," the predicate is just the verb **sing**. Or we might have a noun phrase after the verb. In "The bird sang a song," the predicate is **sang a song**. This predicate is made up of the verb **sang** and the noun phrase **a song**.

Tell what the predicates of these sentences are made up of. They are made up of either (a) a form of **be** and a noun phrase, or (b) a form of **be** and an adjective, or (c) a verb, or (d) a verb and a noun phrase.

1. The girl recited the poem. (d)
2. John is the captain. (a)
3. They are beautiful. (b)
4. The mothers cook. (c)
5. The book is yellow. (b)
6. My father drives the car. (d)
7. These girls are our helpers. (a)
8. The ice melted. (c)
9. The driver is skillful. (b)
10. That lady is your cousin. (a)

Adjectives can come after **be** in predicates. They can also come after verbs. In "Tom looks happy," the predicate is **looks happy.** It is made up of the verb **looks** and the adjective **happy.**

Each of the sentences below has a predicate made up of a verb and an adjective. Tell what the verb is and what the adjective is in each sentence.

11.	Mary looks sad.	V: looks	adj: sad
12.	He feels hungry.	feels	hungry
13.	The girls sound happy.	sound	happy
14.	The teacher looks serious.	looks	serious
15.	We feel good.	feel	good
16.	The book sounds interesting.	sounds	interesting

So we have found so far these different kinds of predicates:

(a) be + noun phrase (**is a pupil**)
(b) be + adjective (**is hungry**)
(c) verb alone (**smiles**)
(d) verb + noun phrase (**helps the teacher**)
(e) verb + adjective (**looks happy**)

□ Tell which kind of predicate you find in each of these sentences. Write (a), (b), (c), (d), or (e) instead of words as your answer for each sentence.

17.	Mrs. Smith is quiet.	(b)
18.	Mrs. Smith laughs.	(c)
19.	Mrs. Smith washes the dishes.	(d)
20.	Mrs. Smith is my aunt.	(a)
21.	Mrs. Smith looks pretty.	(e)
22.	Mrs. Smith drives a car.	(d)
23.	Mrs. Smith is funny.	(b)
24.	Mrs. Smith seems sleepy.	(e)
25.	Mrs. Smith is a grandmother.	(a)
26.	Mrs. Smith knits.	(c)

Teaching Suggestions

Go through the questions on page 50 that review types of predicates. Then do Sentences 1–10 to be sure that the pupils can recognize the types of verb phrases (predicates) so far presented.

Do Sentences 11–16 as class work. There should be no trouble in recognizing the adjectives, since in these sentences no other structure occurs after the verbs.

The class might be asked to volunteer another example for each of the structures a–e in the middle of page 51.

In Sentences 17–26 either verb or **be** may be followed by either noun phrase or adjective, or the verb by nothing at all. There still should be no difficulty, however, since the noun phrases are still all of the determiner + noun type and therefore quite distinct from adjectives. With a slow class some or all of these may be done in class discussion, with the answers written on the chalkboard. Then, the board erased, the pupils may do them by themselves in writing.

LITERATURE – A Poem

Teaching Suggestions

The title of this poem is "From a Railway Carriage." Read it to the class or play the recording of it. Let the children read it, each child taking a couplet.

Notice that the rhythm of the poem emphasizes the speed of the train. The train probably wasn't going very fast in Stevenson's time, but he makes it sound as if it were.

As class schedule permits, assign copying.

SECTION 4

A Poem

Have you ever ridden on a train and looked out the window as the train raced along? Read this poem about sights seen from a train window.

Faster than fairies, faster than witches,
Bridges and houses, hedges and ditches;
And charging along like troops in a battle,
All through the meadows the horses and cattle;
All of the sights of the hill and the plain
Fly as thick as driving rain;
And ever again, in the wink of an eye,
Painted stations whistle by.

Here is a child who clambers and scrambles,
All by himself and gathering brambles;
Here is a tramp who stands and gazes;
And there is the green for stringing the daisies!
Here is a cart run away in the road
Lumping along with man and load;
And here is a mill and there a river:
Each a glimpse and gone forever!

ROBERT LOUIS STEVENSON

The Poem About the Train

- What is the name of the author of this poem? He wrote many poems for boys and girls. Have you ever read a poem of his before? Stevenson
- When you are going along in a train or a car, it sometimes seems that you are standing still and everything outside is moving. Here the poet says that things outside the train are moving very fast — faster than fairies. What else are they moving faster than? What are the things mentioned in the second line that seem to be moving? witches; see poem
- What are charging along like troops in a battle? What fly as thick as driving rain? What does **driving rain** mean? See poem; see poem; violent rain.
- What does **the wink of an eye** mean? What whistle by in the wink of an eye? very quickly; painted stations
- Who is the first person that the poet mentions in the second stanza? What is the person doing? a child; see below

The word **clamber** means climb. But a person who clambers climbs rather awkwardly, as though he were scrambling to the top of something. The word **scramble** means climb hastily on all fours. He must be climbing through prickly vines or bushes like blackberry bushes because that is what **brambles** are. Discuss paragraph above.

- Do you think the child is gathering prickly bushes, or do you suppose the poet means he is gathering the berries that grow on the bushes? berries
- What is a **tramp?** See a dictionary.
- What has run away in the road? Is there anything in this poem that makes you think it was written a long time ago? What makes you think so? cart; the cart
- What happens to each thing the poet glimpses? It is glimpsed and gone forever.

53

VOCABULARY AND MEANING

The Poem About the Train

Stevenson's *Child's Garden of Verses* remains a popular source of poems for children. Though some of them now sound a little too sweet and precious, this one and many others escape this fault. Children who used the third-grade book of this series will remember "The Block City," which opens the volume. Other children will probably recall this or other Stevenson poems.

Teaching Suggestions

Go through the questions on this page. Remember that the aim is not to achieve enjoyment of the poem, but to ensure understanding. Perhaps we might better say that it is to ensure understanding as a necessary preliminary to enjoyment. At any rate, at this stage, keep attention focused on what precisely the poem says.

The rhyme **gazes/daisies** is, for us at least, a bad one. It may not have been for Stevenson.

The clue to the time of the poem is the runaway cart.

River/forever is another bad rhyme—for us, but not necessarily for Stevenson. He probably rhymed it on **river**, saying "for-iver."

ETYMOLOGY—What Words Come From

From time to time, brief etymologies like the one on this page are included in the text. The purpose of them is of course not to say anything extensive about the history of words, but simply to make the point that words have histories, that they weren't always the same as they are now. We slowly bring to the pupil an awareness of the time dimension of language.

Teaching Suggestions

Read through the account on this page with the class. Let the children volunteer other names for flowers, whether or not they have an interesting or perceivable etymology.

When **chrysanthemum** is mentioned, point out that the class has recently studied another term taken from Greek. Try to get them to recall **homophone** and to tell what it means.

What Words Come From

Nouns that name flowers often have very interesting histories.

Sometimes we can see at once why a flower got its name. There is a flower called a **Johnny-jump-up** and another called a **forget-me-not.** A **bachelor's button** is a flower that a bachelor might wear in his buttonhole. There are peas that we eat, and then there are **sweet peas,** which we sniff.

Some names of flowers have histories that are a little hidden. A **marigold** is made up of Mary and gold. It means gold for Mary. Another name for marigold is **chrysanthemum.** This long word comes from the Greek language and means golden flower. **Snapdragons** are called that because they look a little like a dragon's head with a mouth that opens and closes.

A flower with an especially interesting name is one mentioned in the poem by Robert Louis Stevenson. This is the daisy. This word comes from "day's eye," and it used to be spelled "daieseye." Can you think why a daisy might be called "the eye of the day"? What is the eye of the day?

A Paragraph to Write

You know that you must end each sentence you write with a period, unless it is a question. If it is a question, you end it with a question mark.

The sentences that you write in your paragraphs are probably mostly longer and more complicated than the ones you are studying in your grammar lessons. Still, they usually have subjects and predicates like the ones we are studying. You must be sure not to put in your period before you finish the predicate of each sentence — that is, before you finish what you have to say about the subject.

Suppose you write this sentence: "My brother and I rode on a merry-go-round at the fair." The subject of this sentence is **my brother and I.** The predicate is **rode on a merry-go-round at the fair.** It would be wrong to put a period after **I,** because that is just the end of the subject. It would be wrong to put one after **merry-go-round,** because that would cut off **at the fair,** which is part of the predicate.

• Tell what the predicate is in each of these sentences.

1. My dad and I went on an airplane to Chicago last summer.
2. Mother and I traveled by bus to New York on a shopping trip.
3. My brother and I took a train to Grandfather's house all by ourselves.

□ Write a paragraph about a ride that you have taken, and tell some of the things you saw on your ride. It might have been a ride on a train like the one in the poem by Robert Louis Stevenson. Or you might tell about a ride in a car, on a bus, or in an airplane.

55

COMPOSITION – Punctuating Sentences

A central problem in the teaching of punctuation is that even in very early grades the children are writing sentences of a grammatical complexity far beyond that of the grammar that it is possible to teach them. We teach them the grammar of "The man is a doctor," but they write, "The doctor told my father that he would have to get more rest."

The usual solution to this problem has been not really to teach grammar at all, but to rely on rather vague suggestions and admonitions: be sure that your sentences are complete; put a period at the end of a group of words that tells a complete thought. But this is no solution. It really says no more than "A sentence is a sentence, and be sure to use a period when you come to the end of one."

We try here to do something more than this, to give some specific explanation of the structures that compose sentences. But we necessarily lag far behind the grammar the children are actually using in the sentences they compose. Even by the end of the sixth grade, though we are then teaching transforms to be sure, the grammar taught delineates only very central and simple lines of the grammar used.

Meanwhile one must cope with the sentences actually being written. It is to be hoped that the idea of subject and predicate, even though specified only in simple utterances, will gradually extend to encompass the more ordinary, but also more complicated, sentences that the child writes. The feeling for unity developed in simple structures should carry over more or less naturally to a similar feeling in complicated ones. The discussion on this page is intended to nudge in that direction.

Beyond this, the teacher must for the time being treat end punctuation item by item: "Your sentence ends here, not there, so this is where the period must be."

SOUNDS AND LETTERS—/ū/

This sound resembles /ī/, /ē/, and /ā/ in several ways. Like them, it is a diphthong rather than a pure vowel. That is, it is made by producing a certain shape in the sound box of the mouth by the position of the tongue, and then gliding away from that position by movement of the tongue. It is also, like the other complex vowels studied, frequently spelled before consonants in the pattern "vowel letter + consonant letter + **e**." And it is spelled with a number of other variants.

The word **spook** seems to be the only one in which this vowel sound is represented in the letters **ook**. Usually these letters stand for a quite different sound—**book, crook, hook**. Be sure that in naming the sound, the children give the proper pronunciation to "ook." We could suggest "uke" as the pattern, except that this would invite a preceding /y/ sound, as in **ukulele**.

Teaching Suggestions

Go through the set of words beginning **rule**. Emphasize the role played by the final **e**: it makes **u** stand for /ū/ and not /u/. Other words spelled like these include **prune, tube, cube, rude, fluke, mule, plume, tune, dupe, use**. Notice that in **cube, mule, use** there is always a /y/ sound before /ū/. In **tube** and **tune** there sometimes is, but not usually in American English.

The more common way to write /ū/ is with the letters **oo**. Other examples include **broom, fool, pool, hoot, mood**. There are other, more exceptional, ways to write /ū/, as in **move** or **group**. These are not introduced at this point.

The Vowel Sound /ū/

• You have studied five **simple vowels** — /i/, /e/, /a/, /u/, and /o/. How do you name these vowels? ick, et

Tell the ways in which each of the simple vowels is usually spelled. i, e, a, u, o

• The other vowels in English we call **complex vowels.** You have studied three of these — /ī/, /ē/, and /ā/. Tell how these are named. ike, eek, ache.

Tell some of the ways in which these three complex vowels are spelled. VCe, y, ie; VCe, ee, ea; VCe, ai, ay

The next vowel you will study is the complex vowel of **rule** or **moon.** We will show this vowel with the letter **u** with a mark over it, this way: /ū/. When we want to say the name of this sound, we will say the vowel with a /k/ sound after it. We will say "ook," as if we were saying the last part of **duke** or **spook.**

There are several ways of writing the sound /ū/.

One way is a way you have learned about in studying other complex vowels. We write the letter **u** and then put an **e** at the end of the word to show that the **u** means /ū/, and not /u/. Here are some words in which /ū/ is written this way:

rule duke June sure brute dude

• Write the word **dude**, but leave off the **e** at the end of the word. What word is it now? Look up **dude** and **dud** in a dictionary. dud

Often we write the vowel sound /ū/ with the letters **oo.** Here are some of the words in which we write /ū/ this way:

moon	**room**	**spoon**	**spook**
spool	**food**	**proof**	**school**

Can you think of any other words in which the vowel /ū/ is written with the letters **oo?** See p. 325.

• Say these words. Tell which of these complex vowel sounds each of them has: /ī/, /ē/, /ā/, or /ū/.

hate /ā/	like /ī/	team /ē/	boot /ū/
rule /ū/	seek /ē/	hail /ā/	tune /ū/
shake /ā/	sight /ī/	stale /ā/	he /ē/
cry /ī/	mice /ī/	race /ā/	soon /ū/

At the end of words, we write the vowel sound /ū/ in several different ways. One way is with the letters **oo.** We write it this way in these words:

moo too boo woo zoo

We write /ū/ at the end with the letters **ew** in these words:

new	blew	drew	dew	few
knew	grew	Jew	pew	stew
strew	threw	screw		

In the following words, we write /ū/ at the end with the letters **ue:**

blue	avenue	clue	cue	due
sue	glue	hue	rue	true

And in these we write it with the single letter **o:**

do who to

• Simple vowels never come at the end of words. Complex vowels often do. You have studied several ways of writing complex vowels at the end of words. Tell what vowel sounds come at the end of these:

sky /ī/	plea /ē/	may /ā/	true /ū/
do /ū/	high /ī/	me /ē/	say /ā/
die /ī/	tree /ē/	coo /ū/	gray /ā/

□ Now look at the following words:

glue who few zoo true

Close your book and write them from dictation.

57

The sound /ū/ at the end of words is spelled in a number of ways, as shown, and there is no dominant or regular one. Here are the other simple words that have the spellings shown.

oo: ballyhoo, bamboo, coo, cuckoo, igloo, kangaroo, shampoo, shoo, taboo, tattoo, shmoo

ew: askew, brew, cashew, crew, eschew, flew, hew, view, mew, shrew, slew, spew, thew, whew

ue: barbecue, construe, ensue, flue, revenue

o: ado

The lists omit obvious compounds, like **untrue** or **overdue**.

There are still other, more exceptional, ways of spelling final /ū/. It is spelled **oe** in **shoe** and **canoe**; **u** in **flu** and **Hindu**; **wo** in **two**; **ough** in **through**.

GRAMMAR – A Grammar Review

The questions and exercises on pages 58 and 59 are intended to consolidate what has been learned so far. No new material is added.

Teaching Suggestions

Go through the questions and elicit for each the correct answer from the class. Number 2 makes the distinction between structure (noun phrase) and function (subject or object), even though the terms **structure** and **function** are not yet used.

Review the concept **noun,** as distinguished from **noun phrase,** in 3. Do 4 as a written exercise. When the class has finished, ask individuals to give one of their noun phrases, and write specimens on the board. Continue the review of noun phrase by working through 5.

Numbers 6, 7, and 8 begin the review of the predicate, essentially making the point of the rule for verb phrase:

$$VP \rightarrow \begin{Bmatrix} \textbf{be} + \begin{Bmatrix} NP \\ Adj \end{Bmatrix} \\ \text{verbal} \end{Bmatrix}$$

A Grammar Review

It is time to review the grammar you have learned.

• 1. What are the two main parts of a simple sentence? subj., pred.

• 2. What do we call the words used as subjects of simple sentences? Are they ever used in any other way? noun phrases; yes

• 3. These words are noun phrases: **the boy, an apple, that desk, some pennies, our room.** What do we call words like **boy, apple, desk, pennies, room?** What do we call words like **the, an, that, some, our?** nouns; determiners

□ 4. Write five noun phrases different from those given above. Use a different determiner and a different noun in each of your noun phrases. (Answers vary.)

• 5. In each sentence below there are two noun phrases. Tell what they are. Tell which one is the subject.

a. The cat likes the fish. the cat
b. My house is near the village. my house
c. The train passes a station. the train
d. His father is the postman. his father
e. The girls read some poems. the girls
f. The bus stops at our street. the bus

• 6. What do we call the part of a simple sentence that is not the subject? the predicate

• 7. We must have one of two things in the predicate of a simple sentence. One thing we might have is a form of **be.** Name two forms of **be.** What must we have in the predicate if there is no form of **be?** is, are; a verb

• 8. If the predicate contains a form of **be,** there must be something else in the predicate, too. One other thing there may be is a noun phrase, as in the sentence "She is my mother." What other thing may there be in the predicate after a form of **be?** an adjective

9. Some predicates have verbs instead of a form of **be**. Can a verb occur alone in a predicate? yes; Can a verb be followed in a predicate by a noun phrase? yes; What else may a verb be followed by in a predicate? adj.

10. Look at the predicates of these sentences. Tell what word in each of them is a form of **be.** Tell whether what follows the form of **be** is a noun phrase or an adjective.

a. She is a pupil. is; n.p.
b. They are friendly. are; adj.
c. It is our house. is; n.p.
d. They are the captains. are; n.p.
e. He is hungry. is; adj.
f. We are your neighbors. are; n.p.
g. He is lost. is; adj.
h. You are my guest. are; n.p.

11. Look at the predicates of these sentences. Tell whether the predicate has a form of **be** or a verb in it. Tell what else is in the predicate.

a. She looks sleepy. v.; adj.
b. Eddie catches the ball. v.; n.p.
c. We are noisy. be; adj.
d. It is a train. be; n.p.
e. Louise understands the poem. v.; n.p.
f. I like your house. v.; n.p.
g. The children are helpful. be; adj.
h. He sees some cattle. v.; n.p.
i. This is my paper. be; n.p.
j. That sounds good. v.; adj.
k. They look jolly. v.; adj.
l. Bill uses your pencils. v.; n.p.
m. You know my brother. v.; n.p.
n. Jane seems glad. v.; adj.
o. I feel silly. v.; adj.

59

Number 9 makes the point of the rule for **verbal**, as so far elaborated:

$$\text{verbal} \rightarrow \begin{Bmatrix} \text{verb} \\ \text{verb} + \text{NP} \\ \text{verb} + \text{Adj} \end{Bmatrix}$$

The NP's in "verb + NP" are still all objects, and the verbs transitive verbs.

Work through 10, making sure that the pupils can distinguish adjectives from noun phrases. This is still easy to do because the noun phrases all consist of determiner + noun.

The sentences of 11 are an exercise in picking out verbs and forms of **be** in the predicate.

Teaching Suggestions

Use the tests on pages 60–62 to determine what reteaching, if any, needs to be done. Keep individual records on the progress of the pupils. The rubric "If you make mistakes . . ." provides an easy reference to material that may need reteaching.

Page 63 should be done as an oral exercise.

Extra Test

Dictate these words and have the class spell them. Say each word clearly. Use it in the sentence. Say it again and have the children write it.

1. **high** That mountain is 4,000 feet **high**. **high**
2. **be** Was and were are forms of **be**. **be**
3. **shape** A ring is circular in **shape**. **shape**
4. **knight** A **knight** wore armor. **knight**
5. **sail** A **sail** makes a boat move. **sail**
6. **sale** Prices are lower at a **sale**. **sale**
7. **sway** Trees **sway** in a high wind. **sway**
8. **spook** A **spook** is a ghost. **spook**
9. **glue** We fasten wood with **glue**. **glue**
10. **drew** Yesterday I **drew** a picture. **drew**

Tests and Review

□ **TEST I Sounds and Letters**

1. Write these five headings on your paper:

/i/ /e/ /a/ /u/ /o/

Under each heading, write the words from the following list that have that vowel sound.

If you make mistakes, study pages 6–7, 12–13.

bad /a/	kid /i/	tuck /u/	one /u/
left /e/	rob /o/	shock /o/	spread /e/
laugh /a/	lick /i/	crack /a/	mess /e/
rub /u/	crutch /u/	ditch /i/	health /e/

2. Write these four headings on your paper:

/ī/ /ē/ /ā/ /ū/

Under each heading, write the words from the following list that have that vowel sound. Remember that all of the complex vowels have more than one spelling.

If you make mistakes, study pages 34–35, 40–41, 48, 49, 56–57.

write /ī/	take /ā/	room /ū/	scream /ē/
Pete /ē/	fight /ī/	duke /ū/	sleep /ē/
fail /ā/	scoot /ū/	stay /ā/	my /ī/
she /ē/	too /ū/	pie /ī/	few /ū/

3. Write these headings on your paper: **Simple Vowel, Complex Vowel.** Write each word in the list below that has a simple vowel under **Simple Vowel.** Write each one that has a complex vowel under **Complex Vowel.**

ride com.	hid sim.	fool com.	met sim.
neat com.	bread sim.	spade com.	lack sim.
rot sim.	tune com.	rush sim.	light com.
he com.	crash sim.	grew com.	stick sim.

60

4. Only complex vowels occur at the ends of words. They are spelled in different ways.

Write a word that shows each of these complex vowels at the end of the word.

If you make mistakes, study pages 49, 56–57.

a. The vowel /ī/ spelled **y.** sky, etc.
b. The vowel /ī/ spelled **ie.** pie, etc.
c. The vowel /ī/ spelled **igh.** high, etc.
d. The vowel /ē/ spelled **ee.** see, etc.
e. The vowel /ē/ spelled **ea.** sea, etc.
f. The vowel /ē/ spelled **e.** he, etc.
g. The vowel /ā/ spelled **ay.** may, etc.
h. The vowel /ū/ spelled **oo.** zoo, etc.
i. The vowel /ū/ spelled **ue.** glue, etc.
j. The vowel /ū/ spelled **ew.** flew, etc.
k. The vowel /ū/ spelled **o.** to, etc.

5. **Homophones** are words that sound the same but are spelled differently and mean different things. Write a homophone for each of these words.

If you make mistakes, study pages 35, 48.

sun son	deer dear	tee tea	be bee
blue blew	too to	dew do	pale pail

6. Write a word that rhymes with each of these words but in which the last part is spelled differently. For instance, if the word were **meat,** you might write **sweet** or **greet** or **fleet.**

fear steer	by tie	night write	lane main
thread fed	green mean	June moon	bed head

(Answers will vary.)

TEST II Writing and Vocabulary

Write five sentences. In each sentence, use one of these words:

1. clamber 2. scramble 3. bramble 4. aware 5. tramp

□ **TEST III Grammar**

1. Copy the adjective that occurs in each of these sentences.

If you make mistakes, study pages 26–27, 50–51.

a. The child is thirsty.
b. John looks angry.
c. They are pleasant.
d. It is cool.
e. The floors are dirty.
f. The movie sounds interesting.
g. He feels happy.
h. The ship is beautiful.

2. Tell what the predicate is made up of in each of these sentences. It is (1) a form of **be** plus an adjective, (2) a form of **be** plus a noun phrase, (3) a verb, (4) a verb plus an adjective, or (5) a verb plus a noun phrase. Just write the numeral.

If you make mistakes, study pages 36–37, 42–43, 50–51.

a. Henry is hungry. (1)
b. George understands. (3)
c. Sue feels fine. (4)
d. Catherine helps the teacher. (5)
e. Mr. Wilson drives the car. (5)
f. Sylvia is the winner. (2)
g. Anna cooks. (3)
h. Harold washes the dishes. (5)
i. Raymond is tall. (1)
j. Joan looks friendly. (4)
k. William needs some paper. (5)
l. Rachel is pleasant. (1)
m. Stuart listens. (3)
n. Today is my birthday. (2)

•REVIEW Ideas and Information

1. In the poem about the sea shell, what does the poet ask the sea shell to do? sing her a song
2. What is the Spanish Main? Caribbean area
3. What is a sea horse? a small sea animal
4. What did the table and the chair decide to do? walk
5. What trouble did they have? got lost
6. Who helped them? duck, beetle, mouse
7. Who wrote the poem about sights seen from a train window? Stevenson
8. One of two things must occur in the predicate of a simple sentence. What are the two things? be or a verb
9. What may occur after a verb in a simple sentence besides an adjective? noun phrase
10. What kinds of words are the noun phrases that we have studied made up of? determiners, nouns
11. What are the five simple vowels? /i/, /e/, /a/, /o/, /u/
12. What are the four complex vowels that you have studied? /ī/, /ē/, /ā/, /ū/
13. What is a homophone? a word that sounds the same
14. Can a form of **be** occur in the predicate alone, with nothing after it, in a simple sentence? no
15. Can a verb occur in the predicate alone, with nothing after it, in a simple sentence? yes
16. What are two parts of a simple sentence that may contain a noun phrase? subj. and predicate
17. Is the sound "ike" a complex vowel sound or a simple vowel sound? complex
18. Is the sound "ock" a complex vowel sound or a simple vowel sound? simple
19. What word is usually in the greeting of a letter? dear
20. What is the closing of a letter? the part just before the signature

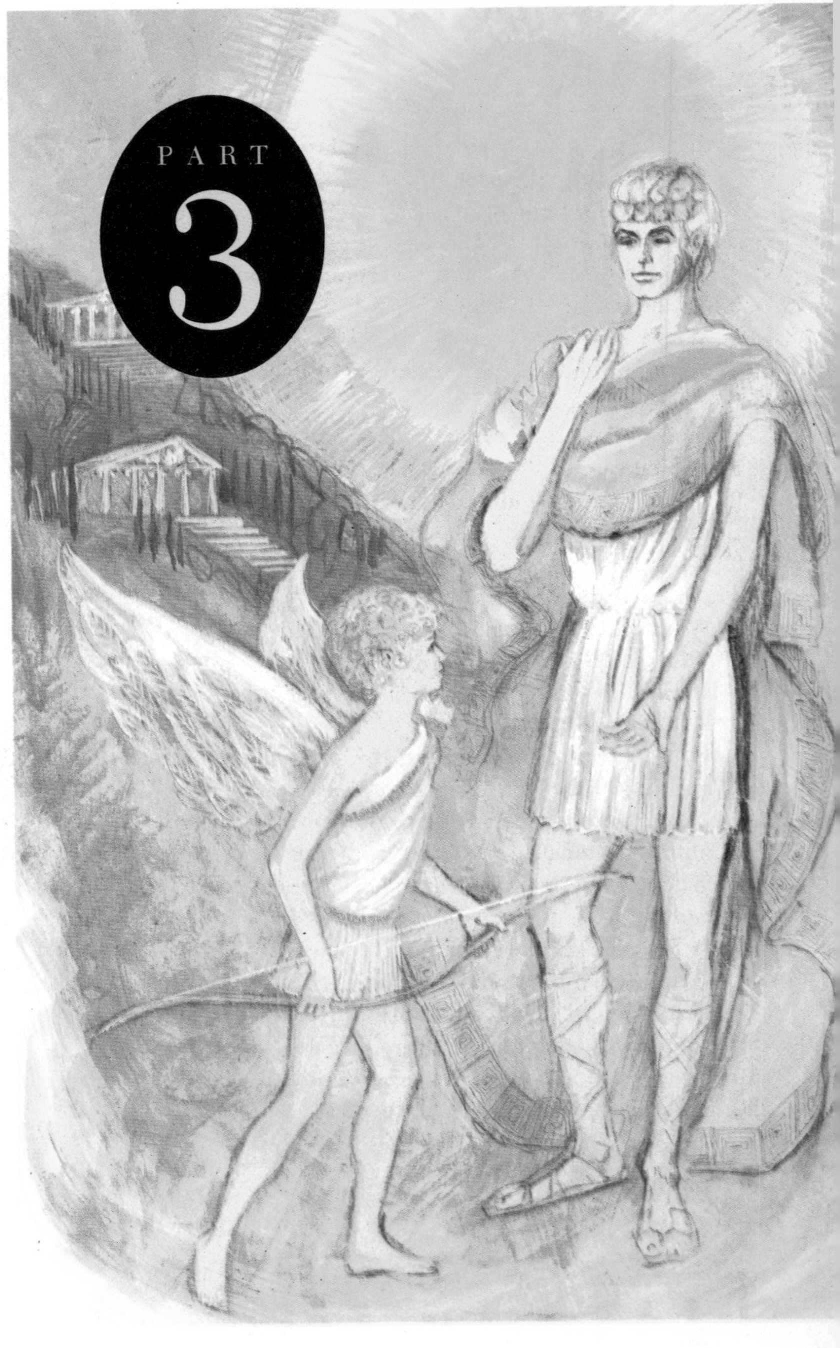

PART 3

SECTION 1

A Myth

Myths are stories told long ago when people believed in gods and goddesses — Zeus, Minerva, Apollo, and all the rest.

One myth tells how the horses that pull the chariot of the sun across the sky were frightened by an ugly scorpion and ran away. This ugly creature in the sky is a group of stars we still call "The Scorpion."

Another ancient myth tells how a girl who was expert at spinning was changed into a spider.

Here is the first part of a myth about a girl named Daphne and how she was changed into a laurel tree, the tree of victory.

The Story of Daphne

Apollo was the god of the sun, and he lived on the sun in a great palace. But sometimes he left the sun to wander about the earth among mortal men. Once when he was on such a journey he met Cupid, the little love god. Cupid was carrying his bow and looking for people to strike with his arrows of love.

Seeing Cupid thus armed, Apollo grew angry. "What are you doing, child," he asked, "with such a weapon? Are bows and arrows for little boys to play with? Carry a lamp, if you like, to light the way of love, but leave warlike arms to the mighty to whom they are suited. Leave your bow to me."

But Cupid was not to be thus put down. "Great Apollo," he said, "shoot your own bow as you will. You will feel my bow, but you will not bend it. My bow will shoot you."

LITERATURE

"The Story of Daphne" (First Part)

A good bit of the literature that a child reads in school is what might be called reservoir literature. It is part of the great pool into which educated people dip when they talk or write. It is important just because so many people have thought it important for so long a time and have used it in so many ways. One can scarcely be called educated if one has never heard of Achilles, Adam, Alice, to stop with the A's. One can hardly read modern literature if one has not.

In the reservoir literature, we would count the Bible, Aesop, Shakespeare, Homer, and, certainly, the Graeco-Roman myths. One could certainly find stories more immediately available to the modern child than Apollo and Daphne. But education requires that somewhere along the line he learn about Apollo and Daphne.

Anyway, it's a pretty good story.

Teaching Suggestions

The teacher may want to explain a little more about myths than is given in the introduction on this page. It might help to explain that the gods and goddesses mostly had two names—one Roman and one Greek:

Greek	*Roman*
Zeus	Jove
Aphrodite	Venus
Hera	Juno
Athena	Minerva

Apollo was just Apollo, though sometimes identified with the Greek god of the sun, Helios.

This story is spread over three sections. Read this first part to the class.

VOCABULARY AND MEANING

"The Story of Daphne" (First Part)

Polytheism may be a concept hard to grasp for a class of modern children. Explain that in earlier times people were awed by the wonders of nature—thunder, lightning, the rebirth each spring of flowers, the movement of the sea. Say that they felt there must be superhuman beings controlling such miracles and that they built up a notion of a group of manlike gods, who lived forever and controlled the workings of the universe.

Teaching Suggestions

The short prose passages included in this text should be read as closely and carefully as the poetry and with the same purpose—to ascertain precisely what the sentences say. Go through the questions on this page, eliciting answers from the class.

The Story of Daphne (FIRST PART)

- In the time of the myths, people believed in many gods. There were gods of the moon and of the earth, gods of rivers and oceans, gods of thunder and rain. What was Apollo the god of? Where did he live? the sun; on the sun
- What did Apollo like to do sometimes? wander on earth
- A **mortal** creature is one that cannot live forever — it must die. The word **mortal** is often used as another name for **man.** Do you think the god Apollo was mortal? Look up **mortal** in a dictionary. no
- Whom did Apollo meet on his journey? Cupid
- Who was Cupid? Do you ever see pictures of Cupid today? Do you think Cupid was mortal? love god; yes; no
- What was Cupid carrying? What do you think happened to people who were struck by Cupid's arrows? a bow; they fell in love
- What does **thus armed** mean? armed this way
- When Apollo grew angry with Cupid for carrying a bow, what did he call Cupid? a child
- What does **weapon** mean? See a dictionary. What was Cupid's weapon? Can you name some weapons that might be used today? bow and arrows; gun, etc.
- What did Apollo tell Cupid to carry instead of a bow? What did he suggest that Cupid might do with a lamp? See story.
- What are **warlike arms?** Do you think Apollo felt that warlike arms were unsuited to Cupid? What words show this? Why did Apollo think he should have the bow instead of Cupid? weapons; yes; see story
- Do you think Apollo made Cupid angry? What does **not to be put down** mean? discuss
- How will Apollo feel Cupid's bow if he doesn't bend it? How does Cupid threaten Apollo? Do you think he will carry out his threat? Discuss: Will Cupid make Apollo fall in love?

66

The Vowel Sound /ō/

• Name the vowel sound in each of these words:

pit get rat chuck top /i/, /e/, /a/, /u/, /o/

The sounds you named are the simple vowels.

• You have also studied the complex vowels in the following words. Name the four vowel sounds.

mine mean main moon /ī/, /ē/, /ā/, /ū/

Each of these vowels may be spelled in several ways. That is one reason they are called **complex.**

All four of these complex vowels are sometimes spelled with a letter **e** at the end of the word. This **e** tells that the vowel letter inside the word stands for a complex vowel and not a simple vowel:

ripe Pete mane plume

• How would these words be pronounced if they did not have the letter **e** at the end? rip, pet, man, plum

You will now study the complex vowel of **bone** or **boat.** We will show this sound by using the letter o with a line over it within slanting lines like this: /ō/. To name this sound when you are speaking, pronounce it "oak" or "oke" like the last part of "**joke.**"

1. In some words we write /ō/ by using the letter **o** for the vowel and putting an **e** at the end of the word to show that the **o** means /ō/ and not /o/. Here are some words we write that way:

bone rode rope spoke
hope dope robe wrote

• How would you pronounce **hope** and **robe** if they didn't have the letter **e** at the end? hop, rob

2. The vowel sound /ō/ is also often written with the letters **oa.** We write it this way in these words:

boat oak load road oar loaf

This is the last of what might be called (and what we later call) the "**VCe** vowels"—that is, the vowels commonly spelled in the system "Vowel letter + Consonant letter + **e.**"

Teaching Suggestions

Review the nine vowel sounds so far studied with the questions on the first half of this page. Emphasize the significance of "silent **e**" with the series of words **ripe, Pete, mane, plume**. Write these words on the chalkboard, and have pupils pronounce them and tell what vowel sounds they contain. Then erase the **e**'s to give **rip, pet, man, plum**. Have other children pronounce these and tell what vowel sounds they contain.

The sound /ō/ before a consonant is mostly spelled in one of two ways: **oa** or "**o** + consonant letter + vowel." Give the class practice with these in the exercises in the second half of page 67.

There are other random ways of spelling /ō/: **ou** in **soul, ow** in **own**. The sound is frequently spelled with the single letter **o** when two consonant letters follow: **cold, comb**. It may appear in various guises when the following consonant sound represents a morpheme: **thrown, hoed**.

SOUNDS AND LETTERS–/ō/
(CONTINUED)

Final /ō/

In most of the words within the child's vocabulary, the sound /ō/ is spelled **ow** at the end of words.

Some words that end in **oe**, in addition to those given, include **floe, Joe, oboe, Poe, roe, sloe, throe**.

The spelling **o** for final /ō/ is most common in words borrowed from other languages, particularly Italian. In addition to the words given, we have **banjo, calico, cameo, domino, embroglio, embryo, Eskimo, folio, gazabo, gigolo, indigo, Mexico, oratorio, pistachio, portico, ratio, studio, vertigo**.

Words from French ending in the sound /ō/ often retain the French spelling: **bureau, mot**.

In a few words /ō/ is spelled **ough: dough, though, although, furlough, thorough.**

Teaching Suggestions

Review the vowels already studied by going through the set beginning **try**.

Introduce final /ō/ with the **ow** words. Say that this is the most common way of spelling simple words ending in /ō/. Go through the **oe** words. Ask how **Joe** is spelled. Say that only a few common words have /ō/ spelled in this way.

The teacher may wish to point out, in dealing with the words ending in the letter **o**, like **hero**, that a very few of these form a plural in **es: hero, echo**. Some—like **banjo, calico, domino**—form the plural in either **s** or **es**. The point is taken up later in the series and need not be raised here unless there seems a reason for doing so.

The Vowel Sound /ō/ (CONTINUED)

• Name the vowel sound in each of these words:

try	/ī/	leap	/ē/	roam	/ō/	cute	/ū/
home	/ō/	safe	/ā/	feel	/ē/	few	/ū/
me	/ē/	side	/ī/	coat	/ō/	zone	/ō/

There are several ways of spelling the vowel sound /ō/ when it comes at the end of words.

3. One way we spell the vowel sound /ō/ at the end of words is with **ow**, as in these words:

low **grow** **snow** **flow** **glow**
know **mow** **row** **show** **slow**

4. In some words we spell /ō/ at the end with the letters **oe**, as in these words:

foe **hoe** **toe** **woe** **doe**

• Is a **foe** a friend or an enemy? Does **woe** mean sadness or gladness? What kind of animal is a **doe**? Look up the words in a dictionary. enemy; sadness; female de

5. In some words we spell /ō/ at the end with just the letter **o**. We do this in these words:

go **ago** **cargo** **piano** **echo**
no **so** **buffalo** **hero** **zero**

• 6. The word **sew** has the vowel sound /ō/. But usually we use the letters **ew** at the end of a word to mean another sound. What is that sound? /ū/

• Tell what sound each of these words ends in:

do	/ū/	go	/ō/	blow	/ō/	high	/ī/
day	/ā/	true	/ū/	we	/ē/	tea	/ē/
lie	/ī/	low	/ō/	hoe	/ō/	stew	/ū/
sew	/ō/	sky	/ī/	know	/ō/	brew	/ū/

• Can you think of homophones for these words?

know	blew	so	tow
no	blue	sew	toe

68

Review: Nouns and Noun Phrases

• You have studied one kind of noun phrase: a determiner followed by a noun. Name the determiners and nouns in these noun phrases:

the boy (D N) **our family** (D N) **some pencils** (D N)

• You know that noun phrases can be used as subjects. What noun phrases are the subjects of these sentences:

1. The boy is asleep. D: the N: boy
2. Our family lives here. our family
3. His teacher went away. his teacher
4. Your dog barks. your dog
5. Their car is new. their car

• Name the determiner and the noun in each noun phrase in Sentences 1–5.

• Noun phrases can also come in the predicate. Name the noun phrases in these sentences:

6. I know the boy. D: the N: boy
7. They live near our family. our family
8. I need some pencils. some pencils
9. He has a canoe. a canoe
10. She is my nurse. my nurse

• Name the determiner and the noun in each noun phrase in Sentences 6–10.

• Find the noun phrases in these sentences. Tell what the determiner is and what the noun is in each noun phrase.

11. My father is an engineer.
12. This kitten wants some milk.
13. These traps catch the mice.
14. Our home is an apartment.
15. Your mother is my teacher.

11. D: My, an — N: father, engineer; 12. D: this, some — N: kitten, milk; 13. D: these, the — N: traps, mice; 14. D: our, an — N: home, apartment; 15. D: your, my — N: mother, teacher

Review: Nouns and Noun Phrases

Teaching Suggestions

Nouns and noun phrases are reviewed here preparatory to the introduction of other kinds of noun phrases. All noun phrases mentioned here are still of the type "determiner + noun," since the rule for NP is so far expanded only thus far:

$$\text{NP} \rightarrow \text{Determiner} + \text{Noun}$$

Work through the sentences with the class.

Extra Practice

For extra practice the sentences on this page may be assigned as written work. Have the pupils write the headings *Noun Phrase, Determiner,* and *Noun* across the top of their paper. For each sentence they are to write the appropriate word or words under each heading.

They will write only noun phrases of the "determiner + common noun" type, of course, as they have not yet been taught to recognize other types of noun phrases.

GRAMMAR—Proper Nouns

The rule for **noun phrase** is now expanded as follows:

$$NP \rightarrow \begin{Bmatrix} \text{determiner} + \text{noun} \\ \text{proper noun} \end{Bmatrix}$$

Proper nouns ordinarily do not occur with a determiner-like word before them. We say **Bill, Mrs. Arnold, Denver**. Some proper nouns, however, use the word **the: the United States, the Netherlands, the Rocky Mountains**. The word **the** is probably best considered in such expressions to be not a determiner, but just part of the proper noun. A person learning the language learns it as a part of the item; he learns simply that we say **France** but **the United States**. It is further true that determiners with non-definite meaning cannot be used with such proper nouns. One can say **some united states**, but then **united states** is not a proper noun.

We say here that a proper noun "doesn't usually have a determiner before it," using the **usually** so that the teacher can handle expressions like **the United States** without going into the technical aspects of the matter.

One reason for the pupil to learn about proper nouns is that he must capitalize them—or most of them. There are a few nouns which seem to fit any formulable definition of proper noun, but which are not capitalized: **the sun, the earth, the devil**. But we do not raise the problem in the text.

Teaching Suggestions

Sentences 1–3 include proper nouns functioning as subjects. Have the pupils point them out. In 4–6 proper nouns function in the predicate. Sentences 7–17 give practice in telling whether noun phrases are of the type "determiner + (common) noun" or "proper noun."

Proper Nouns

A determiner plus a noun is one kind of noun phrase. Another kind of noun phrase is a **proper noun.** Particular names of **persons** are proper nouns: **Susan, Mr. Wilson, Bill Hunter.**

• Proper nouns can occur in sentences where other noun phrases occur. They can occur as subjects. What proper nouns are the subjects of these sentences?

1. Jack is my friend.
2. Mrs. Belfry helps the teacher.
3. Miss Jenkins drives a car.

• Proper nouns can also occur in predicates. What noun phrases in the predicates of these sentences are proper nouns?

4. I know Sylvia.
5. We live near Mr. Danby.
6. I like Miss Herbert.

• Sentences 7–17 contain two noun phrases. Tell what they are and whether each one is a determiner plus a noun or a proper noun. A proper noun doesn't usually have a determiner before it. **Mr., Mrs.,** and **Miss** are **titles** that go with some proper nouns.

7. Bill found the book. PN, D + N
8. The teacher spoke to Mrs. Hopkins. D + N, PN
9. Rachel lives near Mr. Keller. PN, PN
10. Our class studied some poems. D + N, D + N
11. Ruth cleans the chalkboard. PN, D + N
12. This boy is Jim Stuart. D + N, PN
13. Cathy Clark is my cousin. PN, D + N
14. The librarian is Miss Edwards. D + N, PN
15. Jack Gould wants his skates. PN, D + N
16. Mr. Jones met Mrs. Green. PN, PN
17. Miss Porter is a student. PN, D + N

In writing proper nouns we begin with a capital letter. We also use capital letters with the titles that go with some proper nouns: **Mr., Mrs., Miss.**

Notice that the titles **Mr.** and **Mrs.** have periods after them. The period shows that they are abbreviations. **Mr.** is an abbreviation of a word that we might write "mister," but that we hardly ever do. **Miss** has no period after it because it isn't an abbreviation.

□ Write the names of some people you know, using the titles **Mr., Mrs.,** and **Miss.**

There are other kinds of proper nouns besides people's names. Here are some other kinds:

Names of pets: Spot, Inky, Rover
Names of cities: Denver, Rome, New York
Names of countries: France, the United States, Italy
Names of states: Colorado, Ohio, Michigan
Days of the week: Monday, Friday, Sunday
Months of the year: January, February, March

• Find the name of the country above that has **the** before it. Is the word **the** written with a capital? Except for such names, we write all proper nouns with capital letters at the beginning. the United States; no

□ Write the proper nouns in these sentences. Tell which ones are used as subjects.

18. James rang the bell. James
19. The group visited France.
20. Margaret comes from Miami. Margaret
21. Sacramento is the capital of California. Sacramento
22. Mr. Spooner works in the Congo. Mr. Spooner
23. Mrs. Tarr lives in Columbus, Ohio. Mrs. Tarr
24. Ralph Atwell has a dog named Perky. Ralph Atwell
25. Miss Williams swam every Saturday in July. Miss Williams
26. Holland is now called the Netherlands. Holland

The children will probably already have had experience writing **Mr., Mrs.,** and **Miss,** but they may need a reminder that they are capitalized, and that the first two are abbreviations and require periods. **Mr.** and **Mrs.** are two of a very small group of words in English which stand directly for words, without the intermediary of sounds. **Dr.** and **etc.** are others. An abbreviation like CBS is not a case because this can be pronounced as written: "see-bee-ess."

In the middle of page 71, there is a list of different things that proper nouns may name, in addition to people. Have the children give other examples of each one.

Do Sentences 18–26 as a written exercise.

Extra Practice

Sentences 1–17 may be assigned as written work if desired. To do this, have the children write the headings *Determiner + Noun* and *Proper Noun* on their papers. They then write all the noun phrases they can recognize under one or the other of these headings.

The following sentences may be written from dictation:

27. I think Mr. and Mrs. Black live in Denver.
28. Miss Jones has a dog named Rover.
29. Next Monday is the first day of November.

LITERATURE

"The Story of Daphne" (Second Part)

Teaching Suggestions

Begin by finding out how much the class remembers of the first part of the story. If necessary, read it to them again. Then read this second part.

The children may be given practice in reading the prose, each child taking up to a paragraph.

SECTION 2

More of the Myth

Here is the second part of the myth about Apollo, Cupid, and Daphne. Read to find out how Cupid proved to be greater than Apollo.

The Story of Daphne

Then Cupid rose into the air on his wings. From his quiver he drew two magic arrows, one gold and one silver. The magic of the golden arrow made the person it struck fall in love. The magic of the silver arrow caused hatred.

Cupid found Daphne, the daughter of King Peneus. He let the silver arrow fly and struck her with it. In an instant she became mad and fled into the forest to live alone and hunt wild beasts.

Then Cupid put the golden arrow to his bow and with it wounded Apollo, who ran in pain to the woods. In the forest he saw Daphne. She was chasing the deer, and Apollo thought her the most beautiful of women. Her hair was long and flowing, her eyes deep and lovely, her body lithe and graceful.

Apollo fell deeply in love with her. "Nymph," he cried, "do not flee from me. I am no mortal man. I am Apollo, God of the Sun. My father is Jupiter. I am a mighty hunter." Then he added, "Yet Cupid is after all greater than I. For he has wounded me sorely, with a wound beyond care and healing."

The Story of Daphne (SECOND PART)

- People say that love is blind, and Cupid is usually pictured as a little blind god. He can fly, however, like an angel. What sentence tells that he can fly? first
- A quiver is a case for arrows. What is another meaning of **quiver?** See a dictionary. tremble
- Whom did Cupid shoot first? Which arrow did he shoot her with? The words "she became mad" do not mean that she grew angry. What do they mean? Daphne; silver; discuss
- Which arrow did Cupid shoot Apollo with? What would this make Apollo do when he saw Daphne? golden; fall in love

Lithe is pronounced with the vowel sound /ī/. It means graceful and able to move quickly and easily.

Nymph has the vowel sound /i/. It means a kind of goddess. Apollo meant it as a compliment.

- Do you think Apollo is very modest? In the myths, **Jupiter** is king of all the gods. discuss
- Why was Cupid a greater hunter than Apollo? discuss

73

VOCABULARY AND MEANING

"The Story of Daphne" (Second Part)

Work through the questions. Use the illustration as an aid, but require that the children answer primarily from the text.

Additional Questions

Use these extra questions if further check on the reading and comprehension seems desirable:

What did Cupid draw from his quiver? What kind were they?

Who was Daphne's father? What did Daphne flee into the forest to do?

What did Apollo do after he was hit with the arrow? Whom did he see? What was she doing? What did Apollo think about her? Tell what adjectives are used to describe her.

Ask if the pupils remember from the first part what **mortal** means.

Who was Apollo's father?

What is a wound "beyond care and healing"?

SOUNDS AND LETTERS

Review: Vowel Sounds

Teaching Suggestions

The five vowels charted on page 74 are what are sometimes called the "short" vowels. The ones reviewed on page 75 are sometimes called "long" vowels. Of the remaining five vowel sounds, only four are presented in the fourth-grade text. Schwa is reserved for grade five.

The chart in the middle of page 74 may be copied on the board with the "examples of the common spellings" added to by contributions from the class. Alternatively, the children may copy the chart in notebooks and make their own additions.

Use the exercise at the bottom of page 74 to review recognition of the five simple vowels.

Review: Vowel Sounds

You have studied five simple vowel sounds and five complex ones. There are five other vowel sounds in English. However, the next sounds you will study will be certain consonant sounds.

Before going on to these consonant sounds, review the ten vowel sounds you know and the principal ways of writing them.

• Here are the most common ways to spell the simple vowel sounds, with an example of each. For each word given as an example, you name another example.

Spellings of Simple Vowel Sounds (Answers vary.)

Simple vowel sounds	*Common ways to spell the sound*	*Examples of the common spellings*	
/i/	*i*	***pit***	pin
/e/	*e*	***pet***	bed
	ea	***head***	dead
/a/	*a*	***hat***	pan
/u/	*u*	***mud***	luck
sometimes, before *m, n, v*:	*o*	***come***	son
/o/	*o*	***hot***	lock

The simple vowels do not occur at the end of words. They occur only before consonant sounds.

• Name the vowel sound in each of these words:

red /e/	son /u/	luck /u/	top /o/
pot /o/	tick /i/	bread /e/	back /a/
some /u/	dead /e/	miss /i/	flock /o/
dad /a/	did /i/	dud /u/	dot /o/

74

• The complex vowels often occur at the end of words and in the middle. Here are some of the ways we spell them in the middle of words, before consonants. For each word given as an example, you give another example.

Spellings of Complex Vowels in the Middle (Answers vary.)

/ī/	**i — consonant — e (time)**	**igh (might)**		ride, light
/ē/	**e — consonant — e (scheme)**	**ee (feel)**	**ea (heat)**	Pete, seed, eat
/ā/	**a — consonant — e (tame)**	**ai (main)**		pale, laid
/ū/	**u — consonant — e (tune)**	**oo (moon)**		dune, soon
/ō/	**o — consonant — o (bone)**	**oa (boat)**		tone, oak

• The complex vowels can also occur at the end of words. For each word given as an example, you give another word as example.

Complex Vowels at the End (Answers vary.)

/ī/	**y (sky)** my	**ie (pie)** lie	**igh (sigh)** high	
/ē/	**ee (tree)** see	**ea (sea)** pea	**e (me)** he	
/ā/	**ay (day)** may			
/ū/	**oo (boo)** too	**ue (blue)** due	**ew (blew)** new	**o (do)** to
/ō/	**ow (low)** bow	**oe (toe)** doe	**o (go)** so	

□ Write the following vowel sounds as headings:

/ī/ /ē/ /ā/ /ū/ /ō/

Beneath each heading, write the words from this list that contain that vowel sound.

lone /ō/	bail /ā/	due /ū/	flea /ē/	game /ā/
soon /ū/	free /ē/	scene /ē/	few /ū/	so /ō/
fry /ī/	June /ū/	we /ē/	dime /ī/	to /ū/
heel /ē/	light /ī/	way /ā/	die /ī/	blow /ō/
too /ū/	goal /ō/	beat /ē/	doe /ō/	high /ī/

Two charts are necessary for the complex vowels reviewed on page 75, since complex vowels, unlike simple ones, occur also in final position, usually with different spellings there.

Let the children suggest other examples of the different spellings given in the first chart. The only hard one is **e–consonant–e. Pete, Steve, scene, Swede, extreme, Irene** might be suggested.

Do the same for the second chart on page 75, which lists spellings in final position. Here are other words using the less common spellings:

ie: tie, lie, die
igh: nigh, thigh, high
ea: tea, pea, flea, plea
e: he, she, we, be
o: to, who
oe: woe, doe, foe, hoe
o: so, no, fro, zero, piano

Use the exercise at the bottom of page 75 for review of recognition of the complex vowels. This may be started on the chalkboard if necessary.

COMPOSITION – Writing About a Picture

Again we use the familiar device of a picture to write about. The specific function of this assignment is to give practice in the use of, and capitalization of, proper nouns.

Teaching Suggestions

It may be useful to warm up the class with oral discussion before the children begin writing. Have different children point out the characters in the picture, name them, and say something about what they are doing. Then let the pupils begin writing.

A Story to Write

Suppose that a brother and sister go for a walk in the woods. A snowstorm comes up, and they lose their way. Naturally they are frightened, and their mother is worried when they don't come back. Probably the mother would call their father to look for them. Suppose he uses the family dog to help hunt and perhaps a neighbor. Begin the story this way.

> One winter day Tom and Millie Jenkins went for a walk in the woods.

What will you call their mother? Their father? You can think up names for the dog and the neighbor. Since the names of these people are proper nouns, how will you write them?

Proper Nouns and Common Nouns

You have learned about two kinds of noun phrases: **determiner plus noun** and **proper noun.** Either of these kinds of noun phrases can be used as subject of a sentence. Either can be used in the predicate.

A noun that is not a proper noun is called a **common noun.** All the common nouns that you have studied in this book have occurred with determiners before them: **the village, a house, some pencils.** In these examples, the common noun is part of a noun phrase. But a proper noun is a noun phrase all by itself.

• Find the nouns in these sentences. Tell whether each noun is a proper noun or a common noun. Tell whether it is a noun phrase or part of a noun phrase.

His house is in the village. com., part; com., part
Jack lives in Miami. prop., n.p.; prop., n.p.
Tallahassee is the capital. prop., n.p.; com., part
Mr. Burbank sharpened the pencil. prop., n.p.; com., part

Common nouns can sometimes be noun phrases all by themselves. That is, they are sometimes used without determiners. Look at this sentence:

Mr. Burbank sharpens pencils.

Mr. Burbank is a proper noun. **Pencils** is a common noun. Does either have a determiner? no In this sentence, **Mr. Burbank** and **pencils** are noun phrases all by themselves.

• Find the nouns in these sentences. Tell whether they are proper or common. Tell whether they are noun phrases or parts of noun phrases.

Alice likes cats. prop., n.p.; com., n.p.
Cats drink milk. com., n.p.; com., n.p.
The milk is sweet. com., part; com., n.p.
The children help Mrs. Danby. com., part; prop., n.p.

Proper Nouns and Common Nouns

In effect, this page expands the rule for noun phrase as follows:

$$NP \rightarrow \begin{Bmatrix} \text{determiner} + \text{common noun} \\ \text{proper noun} \\ \text{common noun} \end{Bmatrix}$$

Actually, this is an interim rule which will be refined in grade five. The third item —common noun by itself—is better treated as a variation of "determiner + common noun." We say that in certain circumstances the determiner can be null: when the common noun is a noncount noun or when it is a plural-count noun:

Candy is sweet. Girls are kind.

We say that the subjects of these sentences are noun phrases of the type "determiner + noun," with the determiner represented by the null nondefinite article. If an article is used with these, it must be the definite article:

The candy is sweet. The girls are kind.

This treatment, introduced in grade 5, reduces the number of types of noun phrases by one and also illuminates important regularities in the noun phrase system.

Teaching Suggestions

Go through the explanation at the beginning of the page, familiarizing the children with the term **common noun**. There is no need to define it beyond saying that it is a noun that is not a proper noun.

Do the oral exercise in the middle of the page. Here the noun phrases are still the familiar ones: proper nouns or determiners + common nouns, the determiners being non-null—that is, being actual words.

Introduce now the noun phrases with no determiner word preceding them. These will be all non-count nouns, like **milk**, or plural-count nouns, like **cats**.

The Rest of the Myth

This is the end of the myth.

The Story of Daphne

But Daphne had been struck by the silver arrow of hatred. She trembled at Apollo's words and fled from him. Like a frightened deer from the hunter she ran from the god of the sun. She ran toward the river, the home of her father Peneus.

But Apollo was faster than she. Soon she felt his hand touching her fluttering garments. She felt his breath on her neck. She could run no more, and she cried out to her father, King Peneus: "O, save me, Father, from the god Apollo."

Hardly had she uttered these words when a transformation came over her. Her shoulders were covered with bark. Her hair became leaves and her arms branches. Her feet sank into the ground, and her legs became the trunk of a tree. Daphne was no longer a young girl. She had become a laurel tree.

When Apollo saw what had happened, he threw his arms about the tree and kissed it. "O Daphne," he said, "I loved you as a nymph, and I love you still as a tree. You shall be my tree. You shall be the tree of victory. Your laurel shall adorn the winners of the games of Olympus and cover the brows of victorious kings. And these leaves shall be ever green."

Thus did the laurel become the symbol of victory, and this is why its leaves are always green.

"The Story of Daphne" (Third Part)

This completes the myth of Daphne and Apollo.

Teaching Suggestions

Begin by having the class recall what has happened so far. If necessary—and it will probably be necessary—reread the first two parts. Then read this final part. Let the children take turns reading parts of it, both to give them practice in reading aloud and to ensure greater familiarity with the content.

VOCABULARY AND MEANING

"The Story of Daphne" (Third Part)

Teaching Suggestions

Go through the questions on this page to make sure that the sentences of the story are clearly understood. Use the illustration as an aid to show what has happened.

Tell the pupils something about the ancient Olympic games and their reactivation in 1896. Alternatively, in a good class, some of the pupils might be appointed to find out what they can about the ancient and modern Olympic games and make a report.

Additional Questions

If the children seem to need a further check on the carefulness of their reading, use these extra questions:

What had Daphne been struck by? What is she compared to as she runs from Apollo?

What did Daphne feel on her neck?

What does **hardly** mean in the first sentence of the third paragraph? What happened to her shoulders? What did her hair become? What did her arms become? What did her feet do? What happened to her legs? What did Apollo do when he saw what had happened?

ETYMOLOGY—What Words Come From

This is another in the series of brief etymologies that are included from time to time throughout the series.

The Story of Daphne (THIRD PART)

- Why did Daphne tremble at Apollo's words and flee from him? How did she run from the god of the sun? Where did she run to for help? She hated him; see story.
- The word **garments** means clothing. Why were her garments fluttering? To whom did she cry out? What did she ask? She was running; see story.
- A **transformation** is a change of some sort. How did Daphne change? What had she been changed into, in answer to her cry for help? See story; discuss.
- What did Apollo promise Daphne now that she had become a tree? In ancient Greece, heroes were crowned with wreaths of laurel leaves in honor of their victories. Do you know what the games of Olympus were? What games are there today that are named after the ancient Greek games? discuss; Olympic Game
- A **symbol** is a sign. What did the laurel become a symbol of? Do laurel leaves change color in autumn? victory; no

What Words Come From

Many of the meanings of our words come from the old myths. **Laurel** means, first of all, the laurel tree, which is a kind of bay tree. Then it means the leaves of the tree. Look it up in a dictionary.

Since the laurel leaves were given to winners of games and wars, **laurel** came also to mean fame or honor or victory. We still have expressions like "Look to your laurels," which means something like "Watch out that someone doesn't beat your record." To "rest on one's laurels" means to be satisfied with what one has accomplished.

80

The Consonants /p/ and /t/

There are more vowel sounds than vowel letters. So we have to use vowel letters in special ways to show all the sounds. For instance, since **e** is one way we spell the sound /e/, we can't use **e** for /ē/. We use **e–e, ee, ea,** and so on for /ē/. We even use consonant letters to help spell vowel sounds: **y** in **may, w** in **low.**

There are also more consonant sounds than there are consonant letters, so some of the consonant sounds have to be spelled in special ways, too.

First, we will study the sounds /p/ and /t/.

• We make the sound /p/ by putting our lips together. Say **pit, peck** and notice what happens to your lips at the beginning of each word.

• We spell the sound /p/ with the letter **p** at the beginning of words. Give some words like **pit** and **peck** that begin with the sound /p/ and the letter **p.** See p. 325.

• If a simple vowel comes before the sound /p/ at the end of a word, we spell final /p/ with the letter **p: tip, pep, cap, cup, top.** Name the vowel sounds in these words. /i/, /e/, /a/, /u/, /o/ (ick, eck, ack, etc.)

• The consonant sound /t/ is made by putting the tongue against the ridge above the upper teeth. Say **ten, teach** and notice where you put your tongue.

• At the beginning of words /t/ is always spelled **t.** Name some words that begin with /t/ and **t.** See p. 325.

We spell final /t/ with the letter **t** if the sound before it is a simple vowel. Name the vowels in these words: **sit, bet, cat, shut, not.** /i/, /e/, /a/, /u/, /o/ (ick, eck, etc.)

These words spell final /t/ with two **t**'s: **mitt, butt, putt. Putt** is something golf players do.

• **But** and **butt** are homophones. Tell what each word means. (Allow informal definitions.)

81

The third-grade book of this series treats the following consonant sounds and the common ways of spelling them initially and finally in monosyllabic words: /p/, /t/, /k/, /ch/, /b/, /d/, /g/, /j/, /r/, /f/, /v/. Therefore the material in this and the next few lessons may be review for some children, new for others. It is covered much faster than in grade three, and therefore, where it is new, the teacher may want to go through the summary of the grade three program on pages 314–19 and to use the lists of extra words on pages 324–29 for further practice.

We tell the children here something about how the sounds /p/ and /t/ are made: /p/ by closing the lips, /t/ by stopping the breath by putting the tongue to the ridge above the upper teeth. The object is not so much to put them in command of the details of articulatory phonetics as to emphasize the general point that sounds are sounds, made and differentiated by particular formations of the mouth. Letters are ways of symbolizing these sounds on paper.

We now can make good use of the contrast between the set **simple vowels** and the set **complex vowels**, since the rules for spelling the consonant sounds depend in large measure on which kind of vowel precedes. The rules are very simple: when a simple vowel precedes /p/ or /t/, use **p** for /p/ and **t** for /t/.

The only exception for /p/ is **Lapp**, a native of Lapland, in which /p/ is spelled with two **p**'s.

For /t/ there are eleven **tt** words. In addition to **mitt, butt,** and **putt,** we have **watt, kilowatt, Lett, domett, bitt, bott, boycott,** and **pott**.

GRAMMAR—Noun Phrases as Objects

Structure and Function

The important grammatical distinction between structure and function has been implicit in the syntax lessons throughout, although the terms have not yet been used and only the second is introduced at this point. A structure is a grammatical form of some sort: **noun phrase, noun, verb phrase, verb, be, adjective, prepositional phrase** are names of structures. A function is the employment of a structure: **subject, object, predicate, modifier** are names of functions.

Teaching Suggestions

In preparation for the introduction of the object function of noun phrases on page 83, we review on page 82 the three types of noun phrases so far given: proper noun, common noun with determiner, common noun without determiner.

Do Sentences a–h of exercise 1 as an oral exercise. Here direct attention only to the noun phrases functioning as subject. The headings *Proper Noun, Determiner + Common Noun, Common Noun Alone* may be written on the chalkboard, and each of the subjects in these sentences written under the proper heading according to the responses of the class.

Now do Sentences a–h of exercise 2, directing attention only to the noun phrases in the predicate. All of these function as objects, but the term is not introduced until the next page. Have the class distinguish the three types of noun phrases also in this position.

Noun Phrases as Objects

We have studied three kinds of noun phrases. One kind is a determiner plus a common noun, like **the village, a myth, that tree.** Another is a common noun alone, like **pencils** and **milk** in "He needs pencils" or "Milk is white." Another kind of noun phrase is a proper noun like **George, Mr. Burbank, Denver, Apollo.**

1. We have seen that all of these kinds of noun phrases may be used as subjects. Tell what noun phrase is the subject of each of these sentences. Tell whether it is (1) a determiner and a common noun, (2) a common noun alone, (3) a proper noun.

a. Apollo chased Daphne. (3)
b. Laurel is given to winners. (2)
c. The girl became a tree. (1)
d. That tree was a girl. (1)
e. Cupid made the trouble. (3)
f. Laurels are ever green. (2)
g. Bows shoot arrows. (2)
h. Peneus was a king. (3)

2. You also know that noun phrases may be used as parts of predicates. All the different kinds may be. Find the noun phrases in the predicates of these sentences, and tell which of the three kinds each one is.

a. Cupid shot Apollo. (3)
b. Apollo chased the girl. (1)
c. The girl hunted animals. (2)
d. Cupid used a bow. (1)
e. Daphne called King Peneus. (3)
f. Her hair became leaves. (2)
g. Cupid had some arrows. (1)
h. He shot Daphne. (3)

82

Instead of saying that a noun phrase "is used" as a subject, we often say that it **functions** as a subject. **Subject** is one possible function of a noun phrase. In "Harold combed his hair," the noun phrase **Harold** functions as the subject of the sentence. What noun phrase functions as the subject of this sentence?

The rabbits like the carrots.

Another possible function of a noun phrase is its use after verbs like **eat, watch, help, comb,** in sentences like these:

The boys eat peanuts.

The children watch the planes.

We say that noun phrases used in this way in predicates function as **objects.** They function as objects of the verbs that go before them. In the first sentence **peanuts** functions as the object of the verb **eat.** In the second sentence **the planes** is the object of the verb **watch.**

3. Look at these sentences. Each contains two noun phrases. Each noun phrase functions as a subject or as an object. Tell what the noun phrases are and what their functions are.

a. The boys ate peanuts. subject; object
b. David understands the poem. subject; object
c. Apollo chased Daphne. subject; object
d. These girls help the teacher. subject; object
e. Raymond cleans the chalkboard. subject; object
f. Some boys fly kites. subject; object
g. The principal removed the splinter. subject; object
h. Sylvia likes stories. subject; object
i. Mr. Atkins knows Jack Winters. subject; object
j. The bear likes acorns. subject; object
k. Miss Wallace writes stories. subject; object
l. This farmer raises cotton. subject; object

83

Introduce the term **function** first in reference to the subject. (There are three others in the sentences we shall be dealing with: object of a verb, object of a preposition, complement.) Then show that noun phrases can occur after verbs, and are then said to function as objects of verbs. Keep in mind that whereas the noun phrase functioning as subject is one of the two main parts of a kernel sentence (the NP of S $\rightarrow$ NP + VP), a noun phrase functioning as object is a sub-part, a structure of the VP. That is, in a sentence like "The children watch the planes," the division is not threefold. It is not "The children + watch + the planes," but twofold:

The children + watch the planes.

Then:

watch + the planes

This need not be pointed out here. The important thing is not to suggest the other analysis.

Go through Sentences a–l in exercise 3. Give practice in using the terms by having the children say, for example, in a: "**The boys** is a noun phrase functioning as subject. **Peanuts** is a noun phrase functioning as object of the verb **ate.**"

Extra Practice

Sentences a–l in 3 may be assigned as written work. Have the pupils write these headings on their papers:

Noun Phrase	*Function*	
	Subject	*Object*

They may write each noun phrase under the heading *Noun Phrase* and make a check under *Subject* or *Object* to show its function.

LITERATURE

"The Grasshopper and the Ant"

The third-grade book contains four fables of Aesop: "The Tortoise and the Hare," "The Boy That Cried Wolf," "The Wind and the Sun," and "The Best Treasure." Aesop fables will therefore be familiar to children who studied that book, as of course they are likely to be to other children as well.

Teaching Suggestions

Give practice in the pronunciation of **Aesop** for those to whom the name is new. It can be said that he wrote at about the time people were telling the stories of the myths, like Apollo and Daphne, and in the same place. Ask if anyone remembers a word we have borrowed from the Greek language.

Read the story to the children and, if time permits, give them practice in reading it.

SECTION 4

A Fable

A fable is a little story that teaches a lesson. The lesson that a fable teaches is called a **moral.** Read this fable by Aesop. His name begins with the vowel sound /ē/. Aesop lived in Greece thousands of years ago.

The Grasshopper and the Ant

One summer day a grasshopper met an ant. The ant was working hard laying up food for the winter. "Why do you work so hard?" said the grasshopper. "Why do you not sing as I do and enjoy the beautiful summer days?" But the ant did not reply. He kept on working, and the grasshopper kept on singing and enjoying the beautiful summer.

84

When winter came and the rains washed away the food of the land, the grasshopper was hungry. He went to the ant and said, "Give me some food, for I am hungry." But the ant said, "You should have worked in the summer as I did, putting by for the wintertime. But you wanted to sing instead. Now you may eat your song."

This story teaches us that we must labor and save in times of plenty if we wish to have what we need in times of want and distress.

The Grasshopper and the Ant

- The words **laying up** food mean collecting and storing it. At what time of year was the ant doing this? What did the grasshopper think the ant ought to be doing instead of working? What did the ant keep on doing? See story for exact answers.
- What trouble was the grasshopper in when winter came? See story, paragraph 2.
- In the second paragraph, you find the words **putting by.** What words in the first paragraph mean about the same thing? laying up
- What did the ant suggest that the grasshopper eat? his song
- What does **labor** mean? What are **times of plenty?** What is meant by **want and distress?** See a dictionary. discuss
- Tell what the moral of the fable is. See last paragraph.

What Words Come From

- Some nouns practically tell us what they mean. A **grasshopper** is a creature that hops in the grass. What does a **woodpecker** do? What does an **anteater** do? Where would you find an insect called a **leafhopper?** discuss

85

VOCABULARY AND MEANING

"The Grasshopper and the Ant"

Teaching Suggestions

Go through the questions, having the children find the answers in the text of the story.

Be sure that the meaning of **moral** is understood. Modern writers of course try to avoid the obvious pointing of a moral, but it is an integral part of Aesop.

ETYMOLOGY—What Words Come From

Quite a number of words are constructed, like those given here, by what is in effect a transformation—the form changing, but the meaning relationships remaining the same.

Something hops in the grass. → It's a grasshopper.
Something pecks wood. → It's a woodpecker.
Something eats ants. → It's an anteater.

Thus in the third sentence the relation of **eat** to **ant** is the same on both sides of the arrow: verb to object.

The *VCe* Spelling Pattern

The children are already familiar with the spelling pattern "vowel letter + consonant letter + **e**" from previous lessons in this book and, probably, from work in earlier years. We now, for convenience, express this in the formula **VCe**, as explained on page 86.

Teaching Suggestions

Work through page 86, making sure that the concept **VCe** is quite clear to all. Keep in mind that the symbols here stand for letters, not sounds. In particular, **V** stands for a *single* vowel letter representing a complex vowel; **C** stands for a single consonant letter; **e** stands for itself, the letter **e**.

Let the children do the exercises suggested at the bottom of page 86 to get used to the formula and its application.

Work through the material on page 87 with the pupils, emphasizing that the **VCe** pattern often is used for complex vowels when just one vowel letter precedes the **C**, but that very often complex vowel sounds are spelled with more than one letter.

Extra Practice

Replace the **V** and **C** with **a** and **t**:
mVCe lVCe crVCe slVCe

Replace the **V** and **C** with **i** and **p**:
pVCe snVCe wVCe rVCe

Replace the **V** and **C** with **o** and **p**:
hVCe slVCe dVCe rVCe

Replace the **V** and **C** with **i** and **t**:
wrVCe bVCe quVCe kVCe

The Consonants /p/ and /t/ After Complex Vowels

• When a word ends with a **simple** vowel followed by the consonant sound /p/ or /t/, how do we spell the /p/ or /t/? Only a few words are spelled differently: **mitt, putt, butt.** p or t

When the consonant sound /p/ or /t/ comes at the end of a word after a **complex** vowel, there are two usual ways of spelling the consonant sound.

One way is a way you have already studied. In words like **wipe** and **late,** the complex vowel is spelled with a vowel letter, followed by the consonant, followed by **e** at the end.

Using the letter **e** at the end of a word is so important that we need to study it carefully.

Suppose we let the capital letter **V** stand for any vowel letter. We must make a rule, though, that the vowel the letter **V** stands for must always mean a complex vowel sound.

We will let the capital letter **C** stand for a consonant letter — any consonant letter. We will let the letter **e** mean just what it seems to — an **e** at the end. Then we can use the letters **VCe** to mean several different groups of letters.

We can replace **V** with the vowel letter **i,** for instance, and **C** with the consonant letter **p.** Then we have the group of letters **ipe.** According to our rule, the letter **i** must mean the sound /ī/, not /i/. Name some words that end with **ipe.** ripe, pipe, wipe

Replace the **V** and **C** in **VCe** with the letters **a** and **t.** Then name some words that end with **ate.** What vowel sound does the **a** mean in each word? gate, late, rate; /ā/

Replace **V** and **C** with **o** and **p;** with **a** and **p.** Give some words that end with **ope;** with **ape.** hope, rope, slope; cape, grape, tape

86

• Here are some words spelled in the **VCe** way when **C** is replaced by the consonant letter **p.** Name the complex vowel sound in each word.

/ī/, /ā/, /ō/ **pipe shape hope dupe ripe** /ū/, /ī/

• Here are some words spelled in the pattern **VCe** when **C** is replaced by **t.** Name the complex vowel sound in each word.

/ē/, /ā/, /ō/ **Pete hate note brute write** /ū/, /ī/

• Look up **dupe** and **brute** in a dictionary.

In some words, such as **heap**, the complex vowel before the consonant is written with more than one letter. Then we don't use the **e** at the end to show that the vowel is a complex one. The letters before the consonant show what vowel is meant, and we don't need the final **e.** In words like these, the sound /p/ is written at the end simply with **p.** The sound /t/ is written simply with **t.**

In **sheep,** the letters **ee** tell us that the word has the vowel /ē/ and not the vowel /e/. We don't need another **e** at the end to tell which vowel is meant.

Here are some words with /p/ as the last sound. It is spelled **p,** because the vowel before it is written with more than one letter.

peep leap hoop soap

One way of spelling the sound /ū/ is with the letters **ou. Group** and **soup** don't need an **e** at the end because /ū/ is spelled with more than one letter.

A complex vowel before the consonant /t/ is often spelled with more than one letter. Here are some examples:

right	**sheet**	**meat**	**wait**
boat	**shoot**	**boot**	**night**

Can you think of other words like these? See pages 324–25.

Silent e

The famous silent **e** of English, so central to our spelling system, was not always silent. Until five or six hundred years ago it represented a vowel sound, something like the sound represented by **a** in **sofa.** So long as this vowel was pronounced, words like **pipe, name, hope** were two-syllable words. Eventually it ceased to be pronounced, and such words became monosyllabic, but the **e** remained—another example of English spelling clinging jealously to the history of the language.

It should be kept in mind, however, that silent **e** is not a meaningless vestige or intruder, like the **b** in **debt.** It plays a very precise function. English has to spell fifteen different vowel sounds with only the five vowel letters bequeathed to us by the Latin alphabet. (By way of contrast, Italian with the same vowel letters needs to represent only five vowel sounds in some dialects, seven in others.) What silent **e** does, in effect, is to double the coverage of the letters **a, e, i, o, u.** It is not perhaps the neatest way of doing this, but it is certainly a workable way.

The child's spelling problem resides not in the fact that he must learn about the significance of silent **e,** which is part of a fairly simple system, but that this is only one of several devices for spelling the complex vowels.

It is difficult to say precisely how /p/ and /t/ are spelled in words like **ripe** and **bite.** It is not quite accurate to say that they are spelled **pe** and **te** respectively, since the **e** is more significantly part of the spelling of the preceding vowel.

GRAMMAR – Personal Pronouns

The rule for noun phrase is now enlarged as follows:

$$NP \rightarrow \left\{ \begin{array}{l} \text{determiner} + \text{common noun} \\ \text{proper noun} \\ \text{common noun alone} \\ \text{personal pronoun} \end{array} \right\}$$

As has been explained, these will be reduced to three in grade five, where introduction of the null nondefinite article enables us to treat "common noun alone" as a form of "determiner + noun."

Like most of the word classes in English, the personal pronouns do not need to be defined because they can be listed. Personal pronouns are simply the seven words **I, he, she, it, we, you, they**. We list them in their subject forms, adding the object forms when we study them in the object function and the possessive forms when we take up the possessive transformation.

Note that it is inaccurate to say that personal pronouns stand for nouns. Rather they stand for, or take the place of, noun phrases. Thus "The boy is here" does not become "The he is here," as it would if the pronoun **he** took the place of the noun **boy**. Instead it replaces the noun phrase **the boy**.

Page 89 introduces the third of the third person singular pronouns – **it**. Though it is generally true that English, because of the three pronouns of the third person singular, has a threefold gender system – masculine, feminine, and neuter – and though it is further true that masculine noun phrases generally refer to males, feminine to females, and neuter to sexless things, the correspondence isn't exact. Grammar is not the same as biology. Ships regularly, and other machines often, are replaced by **she**, not **it**. On the other hand, **it** often refers to creatures that have sex – mosquitoes, mice, human babies. We do not, however, trouble the children with these marginal asymmetries at this point.

Personal Pronouns

You know about three kinds of noun phrases: (1) determiner plus common noun, (2) common noun alone, (3) proper noun.

You know how to name two uses, or functions, of these noun phrases. One function is the subject of the sentence. Another is the object of a verb.

• Find the noun phrases in these sentences. For each, tell what kind of noun phrase it is and what its function is.

1. The grasshopper wanted some food. (1), subj.; (1), obj.
2. Alice drank the milk. (3), subj.; (1), obj.
3. Mr. Burbank munched peanuts. (3), subj.; (2), obj.
4. Snow covered the ground. (2), subj.; (1), obj.
5. Cupid shot the arrow. (3), subj.; (1), obj.

Another kind of noun phrase is a **personal pronoun.** Look at the subjects of these three sentences.

The man munched peanuts.
Mr. Burbank munched peanuts.
He munched peanuts.

In the first sentence the noun that functions as subject is a determiner and a noun. In the second it is a proper noun. In the third, the subject is the word **he.** The word **he** is a personal pronoun. It takes the place of a noun phrase that means a man or a boy.

Look at these sentences:

The girl waited.
She waited.

Alice is hungry.
She is hungry.

She is another personal pronoun. **He** takes the place of a noun phrase that means a man or a boy. What does **she** take the place of? a woman or girl

There are seven personal pronouns in English. **He** and **she** are two of them. Another is **it.** Look at these sentences:

The arrow struck Apollo.
It struck Apollo.

Denver is a big city.
It is a big city.

You see that **he** and **she** stand for noun phrases that mean people. **It** takes the place of other noun phrases.

The other four personal pronouns are **they, we, I,** and **you. They** takes the place of noun phrases that mean more than one person or thing:

The boys were talking.
They were talking.

The ladies laughed.
They laughed.

The arrows are sharp.
They are sharp.

The personal pronouns **I, we,** and **you** don't take the place of other noun phrases. They just mean the person speaking or the person spoken to.

• Find the noun phrase that functions as subject in each sentence below. Tell whether it is (1) determiner plus common noun, (2) common noun alone, (3) proper noun, (4) personal pronoun.

6. Jack lives upstairs. (3)
7. He is my friend. (4)
8. His parents work in an office. (1)
9. They play the piano. (4)
10. Chicago is my home. (3)
11. It is in Illinois. (4)
12. Trains come to Chicago from everywhere. (2)

Although the pronouns **I, we, you,** and **they** are mentioned on page 89, only **they** is used in the exercises. **I, we,** and **you,** with the other personal pronouns, receive further attention later.

Teaching Suggestions

Work through the questions and sentences on page 88, which show in effect that **he** replaces other noun phrases that refer to males, **she** other noun phrases that refer to females.

Work through the sentences on page 89, consolidating the concept of the personal pronoun and the particular uses of **he, she,** and **it.** Introduce **they,** which replaces plural noun phrases of any gender.

Show that **I, we,** and **you** do not replace other noun phrases but simply represent the person or persons speaking or spoken to.

Use the sentences at the end of page 89 for practice in distinguishing the four types of noun phrases so far introduced.

Extra Practice

Sentences 6–12 on page 89 may be assigned as written work, with the pupils writing each noun phrase that functions as subject, and telling what kind of noun phrase it is by writing (1), (2), (3), or (4) beside it.

TESTS AND REVIEW

As in previous testing sections, use the tests on pages 90–92 for diagnosis of possible weaknesses requiring reteaching. The page numbers given permit quick location of the relevant review material.

Tests and Review

□ **TEST I Sounds and Letters**

1. Copy each word in the list below. After each word, write the vowel sound that it contains. Show the vowel sound between slanting lines. Your answers will look like this:

like /ī/

If you make mistakes, study pages 74–75.

roam /ō/ room /ū/ rim /i/ ram /a/
tame /ā/ team /ē/ time /ī/ Tom /o/
tube /ū/ tub /u/ ten /e/ tune /ū/
lone /ō/ head /e/ feet /ē/ laugh /a/
ship /i/ son /u/ like /ī/ wail /ā/

2. Write these headings:

/ī/ /ē/ /ā/ /ū/ /ō/

Write each word in the list below under the heading that shows what vowel sound it ends with.

If you make mistakes, study pages 74–75.

low /ō/ new /ū/ my /ī/ stay /ā/
we /ē/ toe /ō/ high /ī/ tree /ē/
go /ō/ do /ū/ pea /ē/ tie /ī/
lay /ā/ blue /ū/ show /ō/ two /ū/

3. Write a homophone of each of these words:

blew blue dew do wee we tea tee
too to lain lane be bee meet meat

4. Write five words that end in the sound /t/. Use a different simple vowel in each word. get, at, it, but, got

5. Replace the **C** in **VCe** with the consonant letter **t.** Replace the **V** with any vowel letter you choose. Write at least five words that have the **VCe** pattern. ate, late, mate, date, rate, gate, etc.

If you make mistakes, study page 86.

(Answers vary.)

□ **TEST II Grammar**

1. Copy the noun phrases that function as subjects in the sentences below. After each noun phrase, tell what kind of noun phrase it is. It will be one of these kinds: (1) determiner plus common noun, (2) common noun alone, (3) proper noun, (4) personal pronoun. Your answers will look like this:

The sentence	***Your answer***
Alice laughed.	Alice 3

If you make mistakes, study pages 8, 14, 58, 77, 88.

a. His father helps him. (1)
b. Peanuts taste good. (2)
c. She works hard. (4)
d. Mr. Burbank drives to work. (3)
e. Seattle is in Washington. (3)
f. The grasshoppers have a good time. (1)
g. They are happy. (4)
h. Some children are here. (1)
i. Ants work in the summer. (2)
j. We studied a myth. (4)
k. Daphne became a tree. (3)

2. Write the noun phrases that function as objects in these sentences.

If you make mistakes, study pages 82–83.

a. We watched the plane.
b. The children help the teacher.
c. Alice saw Jim.
d. They drink milk.
e. We heard a noise.
f. They need some money.
g. John bought books.
h. I saw Mr. Oaks.

Extra Tests

At this point, or from time to time from this point on, some or all of the following sets of words may be dictated as tests of various ways of spelling certain of the vowel sounds.

It is best to dictate such words in sets in which the vowel or consonant sound under consideration is spelled the same way through the set. Then the child is less likely to fall prey to doubts stemming from being too often confronted with a choice, and one member of the set will reinforce the others.

1. red, bed, pet, tell, bell
2. head, thread, dead, wealth, health
3. pipe, ripe, mine, line, pile
4. sight, fight, night, right, bright
5. my, cry, shy, sty, fly
6. pie, tie, lie, die
7. sigh, high, nigh, thigh
8. Pete, Steve, scheme, Irene
9. feel, peel, need, bleed, weep
10. eat, treat, mean, dream, scream
11. he, she, we, me
12. tame, shame, make, shake, hate
13. rain, brain, wait, wail, nail
14. say, gay, day, play, stay
15. tune, prune, rude, crude, plume
16. moon, broom, hoop, spook, room
17. boo, moo, shoo, zoo
18. true, Sue, clue, glue
19. new, crew, chew, few
20. bone, note, wrote, hole, choke
21. loaf, roam, coal, oak, toad
22. grow, flow, show, blow, snow
23. toe, woe, doe, foe

□ **TEST II Grammar** (CONTINUED)

3. Tell what comes after the form of **be** or the verb in each of the sentences below. It will be either an adjective or a noun phrase. If it is an adjective, write **adj.** after the number of the sentence. If it is a noun phrase, write **n. p.**

If you make mistakes, study pages 50, 58.

a. John understands the story. n.p.
b. It is a myth. n.p.
c. Sally feels well. adj.
d. They are happy. adj.
e. We found some kittens. n.p.
f. It is she. n.p.
g. Mr. Copper mows his lawn. n.p.
h. The grass is high. adj.
i. The arrows look sharp. adj.
j. The gardener is Mr. Lawrence. n.p.
k. He bought a paper. n.p.
l. The arrow struck Apollo. n.p.

4. Rewrite each of these sentences. In place of the subject, use one of the personal pronouns *he, she, it,* or *they,* depending on what the subject is. For Sentence a you would write:

It stuck Apollo.

If you make mistakes, study pages 88–89.

a. The arrow struck Apollo. It . . .
b. Daphne ran away. She . . .
c. Jim has a bicycle. He . . .
d. The girls are pretty. They . . .
e. Mr. Reed likes poetry. He . . .
f. The books are new. They . . .
g. Miss Thompson drives to school. She . . .
h. Mrs. Henry bakes cakes. She . . .

92

□ **TEST III Writing and Vocabulary**

Write five sentences. In each sentence, use one of these words. (Answers will vary.)

1. laurels 2. garments 3. labor 4. lithe 5. distress

● **REVIEW Ideas and Information**

1. To whom were laurel leaves given? winners
2. What does "look to your laurels" mean? See p. 80.
3. What is the lesson that a fable teaches called? moral
4. With what vowel sound do we say the name Aesop? /ē/
5. Subject is one function that a noun phrase can have. What is another? object of a verb
6. People's names are one kind of proper noun. Name two other kinds. names of days, names of months
7. How do we always begin proper nouns when we write them? capital letters
8. **We, I,** and **you** are three of the personal pronouns. Name the other four. she, he, they, it
9. Give two ways of spelling the vowel /ū/ at the end of a word. ew, ue
10. Give two ways of spelling the vowel /ō/ at the end of a word. o, ow, oe
11. What are two abbreviations used as titles? How are these abbreviations spelled? Mr., Mrs.
12. Name a word that has the pattern **VCe.** What kind of vowel sound does the vowel letter that replaces **V** always have? grape; complex
13. Can a proper noun function as object of a verb? yes
14. Does **I** stand for another noun phrase? no "I" is a personal pronoun but stands for the person speaking
15. Can a common noun alone function as object of a verb? yes

Give Test III to check on vocabulary retention and as a further practice in writing.

Use the "Ideas and Information" questions as an oral discussion exercise.

PART 4

SECTION 1

A Poem

There is something spooky about houses that no one lives in. Read this poem about such a house and see what the poet felt about it.

The Old Stone House

Nothing on the gray roof, nothing on the brown,
Only a little greening where the rain drips down;
Nobody at the window, nobody at the door,
Only a little hollow which a foot once wore;
But still I tread on tiptoe, still tiptoe on I go,
Past nettles, porch, and weedy well, for, oh, I know
A friendless face is peering, and a clear still eye
Peeps closely through the casement as my step goes by.

WALTER DE LA MARE

The Old Stone House

- The gray part of the roof in the picture is made of slate, a kind of stone. What do you think the brown part of the roof is made of? shingles
- Have you ever seen what grows on stone where water keeps dripping on it? What does **greening** mean? moss; green color
- Can you explain the fourth line? What does **wore** mean here? What happens if you step in the same place over and over? Years of stepping wore down the stone.
- You may know the word **tread** because that is the name of the part of a bicycle tire that touches the pavement. But in the poem, **tread** means walk. Find these meanings in a dictionary.
- A **casement** is a window that opens on hinges. What does he know is peeping out at him? See last two lines.

LITERATURE – "The Old Stone House"

Teaching Suggestions

Read the poem to the children or play the recording of it. Let the children take turns reading couplets.

This poem is short enough to be memorized or to be copied or both. If the children are asked to copy it, check that they get the commas (of which it has many) in the right places. We cannot do much in the early grades about giving rules for comma punctuation. These rules require, usually, considerably more knowledge of the grammar than the children will have for some time. Meanwhile, insofar as they use comma punctuation at all, it will be in imitation of the writing they study. Copying will encourage this imitation.

VOCABULARY AND MEANING

"The Old Stone House"

Teaching Suggestions

Go through the questions to check on comprehension. Let the children go back to the text of the poem to find the answers.

Ask what **tread** is – noun or verb – in these sentences:

The tread touches the pavement.
He treads on tiptoe.

Explain that some words are used both as nouns and as verbs. Give these other examples:

A walk would be fun.
They walk around the town.
The cover is on the box.
The cloak covers her shoulders.

Ask if anyone can make sentences using **talk** and **dream** first as nouns and then as verbs.

COMPOSITION—A Letter to Write

There are necessarily two assignments here, since one cannot be sure that all will have had, or will be able to call up, a particular experience. The first is probably the harder and the more interesting. It could be done, of course, through pure imagination. The teacher can decide whether to assign one or the other or to have some do one and some the other. In either case, class discussion should precede the actual writing.

Go through the questions on this page to remind the children of the mechanics of letter writing—the address and date, salutation, the closing.

A Letter to Write

Have you ever passed a deserted house, or a house that looked scary to you? Perhaps it was a queer-looking apartment house, or just a barn or garage or factory that seemed deserted.

If you have, write to a friend of yours and tell about it. If you have not had an experience like this, pretend that it was you who walked by the house in the poem and felt that someone was looking at you. Put your paragraph or paragraphs in the form of a letter. Try to make your friend know just what it was like.

Remember what you have learned about writing letters. Put your address at the top right. An address is written like this:

20 Malden Street
Arlington, Virginia

- The name of a street is a proper noun. So is the name of a city or town, and a state, as you know. How do you write proper nouns? Where do you put the comma in an address like this one? capitals; after town
- Write the date under your address. How do you write the name of the month? Why? Where do you put the comma in a date? capital; proper noun; before year
- Begin your letter with a greeting, or salutation, like "Dear Cathy" or "Dear Aunt Sally" or "Dear Mr. Burbank." What mark do you put after the salutation?

Indent the paragraphs of your letter just as you do other paragraphs that you write.

Have a closing to your letter, like "With love" or "Your friend" or "Sincerely." What mark do you put after the closing? comma

Write your name below the closing.

The Consonant Sound /k/

When you are speaking about a consonant sound, you can usually name it the way you do a regular letter of the alphabet. You say /p/ the way you say **p**, and /t/ the way you do **t**. How would you say /k/? name k

The consonant sound /k/ is made farther back in the mouth than /p/ and /t/. Say **pin, tin, kin** and notice how you make the first sound in each word.

• There are two main ways of spelling the sound /k/ at the beginning of words. One way is with the letter **k: kin, king, keep.** Can you think of some other words in which the sound /k/ at the beginning is spelled with the letter **k?** See p. 325.

• In some words we spell /k/ at the beginning with the letter **c: cat, cut, crow.** Can you think of others? See p. 325.

There is a rule about how to spell the sound /k/ when it comes at the beginning of words. This rule will help you spell many words correctly.

Look at these words:

king kiss key Kennedy

• One of two vowel letters follows the letter **k** in each word. What are they? i or e

• Look at these words. Tell what letter follows the letter **c** in each word.

cat a	**cut** u	**cob** o	**cry** r
clean l	**cave** a	**come** o	**cute** u

We use **k** for the sound /k/ when the letters **i** or **e** follow. We never use **k** when other letters follow. So we can make the rule in this way:

> Write the sound /k/ at the beginning of the word with the letter **k** when the next letter is **i** or **e.** When the next letter is not **i** or **e,** write /k/ with the letter **c.**

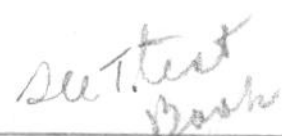

A special convention for naming consonant sounds is not needed as it is for vowel sounds. For most of the sounds, we simply use the alphabetic name: /p/, /t/, /k/ are "pee," "tee," "kay." For the few not represented by single letters of the alphabet, we suggest names by analogy with those that have them.

The spelling of /k/ is a little more complicated than that of /p/ and /t/, but it is nevertheless for the most part regular. For initial /k/ the rule is this:

> Spell it **k** when the letters **i** or **e** follow.
> Spell it **c** in all other cases.

This takes care of all cases in which the sound is spelled **k** or **c**. There is no overlapping, and therefore no exceptions.

We do not bring in here a few words from Greek in which /k/ is spelled **ch**, such as **chaos, character, chorus**. These truly are exceptions, to which no general rule will apply (since one cannot rely on knowledge that they come from Greek), and so the child must learn them one by one.

The initial cluster /kw/ is spelled **qu** as in **queen, quip, quaint**. This is not an irregularity, however, because /kw/ is always spelled this way and **qu** nearly always spells /kw/, the exceptions being rare foreign words like **quay**. We do not introduce this here, since we are not yet dealing with consonant clusters.

Teaching Suggestions

Go through the questions and draw further examples from the class. The sound /k/ is made by contact of the tongue with the roof of the mouth. The precise point of contact is determined, in English, by the following vowel sound. It is toward the front in **kiss** and **keen**, toward the back in **call** and **cough**.

GRAMMAR

Personal Pronouns with Forms of *be*

Be is different from verbs in many ways, one being that it has three forms in the present tense, whereas verbs have only two. **I** is followed by the simple form of verbs, even though it is not plural in meaning, but by the special **am** form of **be**.

To this point the children have studied only the **is** and **are** forms of **be**. We now add **am**.

We also review, on page 99, the four kinds of noun phrases in the subject function.

Teaching Suggestions

Work through the questions and examples on page 98 with the class. On this page, which leads up to the form **am**, there is also a review of the difference among personal pronouns in respect to the kinds of noun phrases that they replace.

Personal Pronouns with Forms of *be*

There are seven personal pronouns. **He, she, you,** and **we** are four of them. Name the other three. I, it, they

In these sentences the personal pronouns mean the person or persons speaking:

I am the speaker.

We are the speakers.

- What personal pronouns mean the speaker or speakers? I, we

In this sentence the personal pronoun means the person or persons spoken to:

You can hear the speaker.

- What personal pronoun means the person or persons spoken to? you
- **He** takes the place of noun phrases that mean men or boys. What does **she** take the place of? n.p.'s that mean women or girls
- **It** takes the place of a noun phrase that means one thing. **They** takes the place of a noun phrase that means more than one thing or more than one person. Give sentences using **it** and **they** as subjects. (Answers vary.)

We have spoken of two forms of **be,** besides **be** itself. These forms are **is** and **are.** We use the form **is** with **he** and **she:**

He is cold.

She is cold.

We use the form **are** with **they:**

They are cold.

What form do we use with **it?** What do we use with **you?** What do we use with **we?** is; are; are

When the personal pronoun is **I,** we use a special form. Look at this sentence:

I can be happy.

Take away the **can.** We don't say "I **be** happy," and we don't say "I is happy." What do we say? am

98

What form of **be** do we use with **he?** with **they?** with **I?** with **you?** with **she?** with **it?** with **we?** is; are; am; are; is; is; are

Look at the sentences below. In each one tell what the personal pronoun is. Tell which word is a form of **be.** Tell whether the form of **be** is followed by an adjective or a noun phrase.

1. I am happy. am; adj.
2. She is my sister. is; n.p.
3. They are friends. are; n.p.
4. You are smart. are; adj.
5. He is the principal. is; n.p.
6. We are hungry. are; adj.
7. It is a kitten. is; n.p.

You know four kinds of noun phrases: (1) a determiner plus a common noun, (2) a common noun alone, (3) a proper noun, (4) a personal pronoun. Look at the sentences below. Tell what noun phrase is the subject of each one. Tell what kind of noun phrase it is.

8. The rain drips down. (1)
9. I go on tiptoe. (4)
10. John is frightened. (3)
11. They are quiet. (4)
12. Houses are spooky. (2)
13. It is a face. (4)
14. A casement opens. (1)
15. I am lonely. (4)
16. My legs feel weak. (1)
17. Eyes peer at me. (2)
18. Moss grows on the stone. (2)
19. We are afraid. (4)
20. The well is old. (1)
21. It is weedy. (4)
22. Ellen looks into the well. (3)

99

Go through Sentences 1–7, having the children point out the personal pronouns functioning as subject and notice the form of **be** used with each.

For most children, the concord of the pronoun with the simple form of **be** is automatic in simple sentences. (It may not be in complex sentences.) Some children, however, speak dialects in which the grammatical rules are different than in standard English. They may, for example, normally say "We's hungry" or "They's here." Such children have not just to note what they normally do but learn to do quite a different thing. Furthermore, they need to learn it not simply in the sense of knowing what the rule is but also in coming to do it automatically, without thinking of the rule. Obviously such children need a very great deal of practice with sentences like those of 1–7.

Use Sentences 8–22 to review the four kinds of noun phrase so far introduced. There is just one more to come—indefinite pronouns.

Extra Practice

Sentences 1–7 may be assigned as written work, the pupils being asked to write the personal pronoun and the form of **be** in each sentence.

Sentences 8–22 may also be assigned as written work, with the children writing each noun phrase that functions as subject and identifying the kind it is by writing **1, 2, 3,** or **4** after it.

LITERATURE

"The Story of Johnny Head-in-Air"

The poem is presented in three parts, the other two being given on pages 106–07 and 112–13.

Teaching Suggestions

Read these two stanzas to the class or play the recording of them. The poem is simple, and pupils may be asked to read it aloud, taking a stanza each.

SECTION 2

A Poem

The first two stanzas of a poem by Heinrich Hoffman are on this page and the next. The poem is about a boy who never watched where he was going.

The Story of Johnny Head-in-Air

As he trudged along to school,
It was always Johnny's rule
To be looking at the sky
And the clouds that floated by;
But what just before him lay,
In his way,
Johnny never thought about;
So that everyone cried out,
"Look at little Johnny there,
Little Johnny Head-in-Air!"

100

Running just in Johnny's way
Came a little dog one day;
Johnny's eyes were still astray
Up on high,
In the sky;
And he never heard them cry,
"Johnny, mind, the dog is nigh!"
Bump!
Dump!
Down they fell, with such a thump,
Dog and Johnny in a lump!

The Story of Johnny Head-in-Air

STANZAS ONE AND TWO

- The verb **trudge** means to walk steadily and slowly as though walking were hard work. Do you think Johnny was a graceful athletic boy? discuss
- What was Johnny more interested in, the clouds in the sky or walking carefully? clouds
- What does **astray** mean? If you don't know its meaning you can probably guess from the way it is used. Check your guess by looking up **astray** in a dictionary. discuss
- The word **mind** means to look out or pay attention. Who warned Johnny to mind? they, that is, everybody
- The word **nigh** means near. What vowel sound does **nigh** have? What other word can you find in the poem that has this vowel sound spelled the same way? /ī/; high
- You have not studied the vowel sound in **out** and **about**, but you can tell that these words rhyme. Tell what the other rhyming words are, what vowel sounds they have, and how the vowels are spelled.

(Rhyming vowels are listed in material for teacher.)

101

VOCABULARY AND MEANING

"The Story of Johnny Head-in-Air" (Stanzas One and Two)

This poem is extremely simple in content, but technically rather complicated. In a way it requires closer reading than the more profound "Old Stone House." Go through the questions with the class, making sure that everyone understands the vocabulary and syntax.

The poem is an especially good one for studying rhyme and thus reviewing the vowel sounds. The rhyming vowels of the couplets are, in order, /ū/, /ī/, /ā/, /ɑu/, /ā/, /ā/, /ī/, /ī/, /u/, /u/. All of these except /ɑu/ (**about/out**) have now been studied.

Teaching Suggestions

Write the symbols for the complex vowels and /u/ on the chalkboard. Have the children tell under which headings the rhyming words should be listed, and list them under the proper headings.

SOUNDS AND LETTERS—Final /k/

Like that of initial /k/, the spelling of final /k/ is essentially regular, though somewhat complicated. The rules could be summarized thus:

After simple vowels, spell /k/ with the letters **ck**: **pick, peck, pack, buck, knock**.
In the **VCe** pattern, spell /k/ with **ke**: **pike, Peke, rake, duke, poke**.
Otherwise, spell /k/ with **k**: **peek, book, spook, task, bark**.

Page 103 gives the last two parts of the rule, words with simple vowels having been taken out:

ke in the **VCe** pattern, otherwise **k**.

This rule will hold for monosyllabic words, with few exceptions if any. In the endings **ac** and **ic**, which are morphemes, /k/ is spelled **c**: **maniac, systematic**. In some words from Greek, it is spelled **ch**: **patriarch**.

Teaching Suggestions

Review the spellings of initial /k/ with the questions and exercises on the first half of page 102. Then review the simple vowels and the spelling of final /p/ and /t/ when simple vowels precede. Introduce the spelling **ck** for final /k/ after simple vowels.

Work through the questions and exercises on page 103. Elicit other examples.

The Final /k/ Sound

Do you remember the rule for spelling the consonant sound /k/ at the beginning of words? When do we spell it with the letter **k**? When do we spell it with the letter **c**? before i or e; in all other cases

□ The first letter has been left out in each word below. But each word begins with the /k/ sound. Write each word. Begin it with the letter **c** or the letter **k**. Remember the rule you have learned for spelling words that begin with /k/.

k_ey c_ar k_ill c_ross c_lear
c_ome k_eep c_ure c_old k_ite

• What are the five simple vowels? /e/, /i/, /a/, /u/, /o/

You remember that the way we spell /p/ and /t/ depends on the vowel sound that comes before it. How do we spell /p/ if a simple vowel sound comes before it? How do we spell /t/ if a simple vowel sound comes before it? p; t

• Each of these words has a simple vowel sound followed by final **p** or final **t**. Name the vowel sounds.

cap /a/ pep /e/ chip /i/ top /o/ cup /u/
rut /u/ sit /i/ let /e/ rat /a/ pot /o/

The way that we spell /k/ at the end of a word depends on the vowel sound that comes before it, too.

When a simple vowel comes before the consonant sound /k/, we spell **k** with the two letters **ck**. We spell it that way in the following words. Tell what vowel sound comes before /k/ in each word.

sick /i/ **neck** /e/ **pack** /a/ **tuck** /u/ **rock** /o/

• Write five other words in which /k/ is spelled at the end with the letters **ck**. See if you can have a different vowel sound in each of your five words. See p. 32

• When a complex vowel comes before /k/, the way we write /k/ depends on how the vowel sound is spelled. In many words, we use the way you have already learned for **ripe** and **late.** We use what we called the pattern **VCe.** We write a letter for the vowel, then the letter **k,** then the letter **e** at the end. What does the **e** tell about the vowel? that it is complex

• Here are some words in which we spell final /k/ in the pattern **VCe.** Tell what the vowel sound is in each word.

like /ī/	lake /ā/	poke /ō/	shake /ā/
bike /ī/	choke /ō/	duke /ū/	flake /ā/

• Can you think of any other words in which /k/ is spelled in the pattern **VCe?** See p. 326.

• When the complex vowel before /k/ is spelled in some other way, there is no **e** at the end. Then /k/ is spelled with just the letter **k.** Look at the words in the list below. Tell what the vowel sound is in each one. Tell how it is spelled.

cheek	**sneak**	**spook**	**speak**
/ē/ — ee	/ē/ — ea	/ū/ — oo	/ē/ — ea

• When the sound /k/ comes at the end of a word and a consonant letter comes before it, then /k/ is spelled with just the letter **k.** Here are some words in which a consonant comes before final /k/. What are the consonant letters that come before the **k?**

ask s	**bank** n	**bark** r	**talk** l
milk l	**work** r	**desk** s	**hawk** w

• Tell what consonant sound each of these words ends with. It will be /p/, /t/, or /k/.

Dick /k/	peep /p/	mitt /t/	make /k/
mate /t/	tape /p/	set /t/	wipe /p/
write /t/	hit /t/	butt /t/	lack /k/
deck /k/	oak /k/	rope /p/	duke /k/

103

Other words with /k/ in the **VCe** pattern include **strike, Mike, take, bake, stroke, smoke, fluke**. Except for **Peke** (Pekingese), **Deke** (member of a certain fraternity), and **eke** there are no words with /k/ after /ē/ in the **VCe** pattern.

GRAMMAR – Contractions

The contractions of personal pronouns with forms of **be** were described in the last part of the third-grade book. Here they are reviewed or, for pupils who have not studied them, introduced.

Modern writing differs from that of, say, the nineteenth century in coming closer to a representation of the sounds of speech. One might say that this is a form of colloquialism, except that **colloquialism** implies an informal situation of some sort, and contractions are found even in writing of a very formal nature. Thus a statement issuing from the White House is more likely to read "The President doesn't wish to comment" than "The President does not wish to comment." There is thus hardly any sense in which certain contractions are informal or in any way less than proper. Not to use them is often to write unnaturally.

Teaching Suggestions

Work through the questions. If necessary, give practice in the pronunciation of **contraction** and **apostrophe**. In contractions the apostrophe marks an omission. In the possessive forms of noun phrases that are not personal pronouns – **John's, the boy's** – it marked a supposed omission. Early spellers of English supposed that **John's** was short for **John his** and so wrote it with the apostrophe. This was a mistake. The **s** on **John's** was merely an inflectional ending, and the proper historical development would be **Johns**. But the mistake became standardized, and the apostrophe came thus to serve as a possessive marker and also as a marker of contractions. It is of course not introduced in the former usage until possessives are studied.

Personal pronouns in the possessive do not use the apostrophe. Since the contraction **it's** and the possessive **its** sound exactly alike, pupils tend to mix them up. We cannot deal formally with the problem until

Contractions

• What are the three forms of **be** that we have studied? Name the personal pronouns that go with each of these three forms of **be.** am—I; is—he, she, it; are—we, you, they

When we write a sentence with a personal pronoun and a form of **be,** we sometimes write them together in a special way. Look at these sentences:

He is hungry.
He's hungry.

If we were to say these sentences aloud, we would say **he** and **is** in the first sentence as separate words. But in the second sentence we would say **he** and **is** together, as if they were just one word. This is the way we usually say these words when we are talking.

A form like **he's** is called a **contraction. Contract** means put together, and a contraction is what is put together. We put **he** and **is** together by writing them as **he's.** But to show that **he's** is a contraction of two words and not just one word, we put a mark where they go together. This mark is called an **apostrophe.**

• The apostrophe in **he's** shows that **he** and **is** have been put together. But it also shows that something has been left out. What letter has been left out in the contraction **he's?**

We can also write **she** and **is** as a contraction. Look at these sentences:

She is hungry.
She's hungry.

• What letter has been left out in **she's?** i

□ What do you think the contraction of **it** and **is** would be? Write the following sentence, using a contraction instead of **it is:**

It is cold. It's cold.

104

The contraction of **they** and **are** is **they're.** Look at these sentences:

They are hungry.

They're hungry.

• Again the apostrophe shows that something has been left out. What has been left out in **they're?** a

□ What do you think the contracted form of **you are** would be? Write the following sentence, using the contraction instead of **you are.**

You are hungry. you're hungry.

□ What do you think the contraction of **we** and **are** would be? Write this sentence, using the contracted form.

We are hungry. We're hungry.

□ The contraction of **I** and **am** is **I'm.** The apostrophe in **I'm** shows that something has been left out. What has been left out? Write this sentence, using the contraction instead of **I am.**

I am hungry. I'm hungry.

□ Rewrite each of these sentences, using the contractions for the personal pronouns and the forms of **be.** Be careful to use the apostrophe correctly.

1. He is happy. He's . . .
2. They are my friends. They're . . .
3. I am afraid. I'm . . .
4. She is our helper. She's . . .
5. You are welcome. You're . . .
6. It is cold. It's . . .
7. We are at our desks. We're . . .
8. I am ready. I'm . . .
9. He is outside. He's . . .
10. They are very quiet. They're . . .
11. You are very kind. You're . . .

the possessive is studied, but if pupils write **it's** in a sentence like "The cat licked its paws," the teacher can point out that **its** is not here a contraction, not a shortened form of **it is**.

Work through page 105, introducing the contractions with **they**, **you**, **we**, and **I**. Sentences 1–11 should be done as a written exercise and checked to see that the contractions are formed properly. This is essentially a spelling problem.

LITERATURE

"The Story of Johnny Head-in-Air" (Stanzas Three and Four)

Teaching Suggestions

Begin with discussion of the part that went before: what was Johnny's problem and what happened to him in the first two stanzas? If necessary, reread or replay those stanzas. Then read these or play the recording of them. Let the children practice reading parts of the poem aloud.

SECTION 3

More of the Poem

Here are two more stanzas of the poem, "The Story of Johnny Head-in-Air." There are two other stanzas which will come later.

Once, with head as high as ever,
Johnny walked beside the river.
Johnny watched the swallows trying
Which was cleverest at flying.
Oh! What fun!
Johnny watched the bright round sun
Going in and coming out;
This was what he thought about.
So he strode on, only think!
To the river's very brink,
Where the bank was high and steep,
And the water very deep;
And the fishes, in a row,
Stared to see him coming so.

One step more! oh! sad to tell!
Headlong in poor Johnny fell.
And the fishes, in dismay,
Wagged their tails and swam away.
There lay Johnny on his face,
With his nice red writing-case;
But, as they were passing by,
Two strong men had heard him cry;
And, with sticks, these two strong men
Hooked poor Johnny out again.

(This poem is concluded on pages 112–13.)

The Story of Johnny-Head-in-Air

STANZAS THREE AND FOUR

- The verb **try** in the third line of the poem is used in a special way. It means trying to decide or prove. What are the swallows trying to decide? See line 4.
- To **stride** means to march with long steps. What did Johnny stride to? Do you know what the **brink** of a river is? Can you think of another word that means **brink**? Find **brink** in a dictionary and see what definitions are given for the word. to the river's brink; edge
- What was the bank like? Was the water shallow? What did the fishes do? See poem.
- What happened to Johnny this time? **Headlong** is a word rather like **grasshopper.** You can probably tell what it means by the parts that make it up. What does it mean? discuss
- **Dismay** means fright about what is happening. What did the fishes do in their dismay? swam away
- How was Johnny lying in the water? What was beside him? on his face; his writing-case
- Who heard Johnny cry? What did they do? What did they do it with? See last three lines.

107

VOCABULARY AND MEANING

"The Story of Johnny Head-in-Air" (Stanzas Three and Four)

Teaching Suggestions

The rhyming vowels of these two stanzas have all been studied with, again, the single exception of the sound /ɑu/ in **out** and **about**. The rhymes may thus be used for review of the vowel sounds if the teacher thinks this useful or necessary.

In the first two couplets the rhyming words are disyllabic. Point out that it is the first syllables of **trying** and **flying** that are important for the rhyme. Ask what vowel sound these syllables have. **Ever** and **river** do not make a true rhyme, only a close one. It may be recalled that **river** and **forever** are rhymed in the poem "From a Railway Carriage" (page 52) by Stevenson, for whom, however, they may have been a true rhyme.

Notice in the last couplet that **again** must have the vowel sound /e/ to rhyme with **men**.

Work through the questions to be sure that the meaning of the stanzas and their details are clear to all.

SOUNDS AND LETTERS—/oo/

This is a sound that doesn't occur in very many words, though most of those in which it does occur are of very high frequency. Most of the simple words that have the sound /oo/ are listed on these two pages.

How the Sound Is Made

The vowel /oo/ is in relation to /ū/ in the way /i/ is in relation to /ē/. Both /ū/ and /ē/ are high vowels. The vowel /ē/ is high front; that is, it is made with the tongue high in the mouth and bunched forward. The vowel /ū/ is high back; it is made also with the tongue high in the mouth, but bunched toward the back. These vowels are also tense: when we make them, certain muscles in the throat are constricted.

The vowels /i/ and /oo/ are high vowels too; /i/ is high front, and /oo/ is high back. But they are *lax* vowels, made with the throat muscles relatively relaxed. Few well-known languages apart from English have high lax vowels, and these present quite a problem to foreigners learning English. They tend to make **it** rhyme with **eat** and **book** with **spook**. Of course only if the children are not native speakers of English might they need explanation of the way the sounds are made. The natives make them well enough and need only to learn to identify them and spell them.

The Spelling of /oo/

There are two usual ways of spelling /oo/: **oo** and **u**. Both symbols are used also for /ū/ and /u/. The letters **oo** are used regularly for /ū/ as in **room** and **food**, exceptionally for /u/ in **blood** and **flood**. The letter **u** is used regularly for /u/, as in **mud** and **shut**, in the **VCe** pattern for /ū/, as in **rube** and **duke**. The sound /oo/ must therefore share its symbols with its near articulatory neighbors.

The Vowel Sound /oo/

• Here are the five simple vowel sounds you have studied. For each sound, think of a word that has the vowel sound in it. pig, hen, rat, duck, fox, etc.

/i/ /e/ /a/ /u/ /o/

• Here are the five complex vowel sounds you have studied. For each sound, think of a word that has the vowel sound in it. like, three, hay, new, stone, etc.

/ī/ /ē/ /ā/ /ū/ /ō/

The vowel sound in **book** and **put** is not the same as any of the vowel sounds you have studied so far.

• The word **but** has the vowel sound /u/. Say **but** and **put.** Do these words rhyme? Do they have the same vowel sound? Then the vowel sound in **put** cannot be /u/. no; no

The word **spook** has the vowel sound /ū/. Say **spook** and **book.** Do these words rhyme? Then **book** cannot have the vowel sound /ū/. no

Book and **put** have the same vowel sound — a vowel sound of their own.

There are no vowel letters left for showing this vowel sound the way we have shown /ū/, /u/, and the other vowel sounds you have studied. So we will show it with two o's joined together, this way: /oo/. When you are speaking about the sound /oo/, do so by saying the last part of **book** or **hook:** "ook."

• The vowel /oo/ is spelled with the letters **oo** in the following words. Say them and hear the vowel sound in them.

book	hook	look	cook	nook
brook	crook	took	stood	wood
shook	good	hood	foot	wool

108

• Look again at the words at the bottom of page 108. What are the consonant letters that come after /ꝏ/ when the sound is spelled with the letters **oo**? k, d, l, t

As you know, the letters **oo** do not always mean the sound /ꝏ/. We use the letters **oo** for the sound /ū/ in many words such as **room, broom, spook, root.**

We use the letters **oo** for the sound /u/ in two words: **blood** and **flood.**

You noticed that the word **put** has the sound /ꝏ/, so the letters **oo** are not the only way to spell the sound. We spell it with **u** in these words:

bull full pull puss bush
push put bullet pullet

• In some words the sound /ꝏ/ is spelled in special ways. Say these words and hear the sound /ꝏ/:

could would should wolf woman

Most of the simple words that have the sound /ꝏ/ in them are in the lists of words on these two pages. As you see, there are not many of them. But many of them are common words that we use very often.

The sound /ꝏ/ is in some ways like the simple vowels and in some ways like the complex vowels.

Like simple vowels, the vowel /ꝏ/ never comes at the end of a word. A consonant sound always follows it.

Like complex vowels, there are several ways to spell the sound /ꝏ/. Also, /ꝏ/ is unlike the simple vowels in other ways. For instance, we never use the letters **ck** for the sound **k** after the sound /ꝏ/.

Tell what vowel sound each of the following words contains. It will be /u/, /ū/, or /ꝏ/.

book /ꝏ/	putt /u/	put /ꝏ/	room /ū/	should /ꝏ/
won /u/	duke /ū/	bush /ꝏ/	do /ū/	full /ꝏ/
rude /ū/	stood /ꝏ/	look /ꝏ/	mud /u/	fool /ū/

The three past tense modals **could, would, should** and the words **wolf** and **woman** (but not **women**) have the vowel sound /ꝏ/ and spell it in exceptional ways. The **l** in the modals represents a sound once pronounced but now not.

The main reason for not calling /ꝏ/ a simple vowel, even though it never occurs in final position, is that it doesn't work in the spelling system as the simple vowels do. One example is given here: /k/ is never spelled **ck** after /ꝏ/. There are several others.

Teaching Suggestions

Use the first part of page 108 to review the ten vowel sounds so far studied. Then introduce /ꝏ/. Show that it is different from both /u/ and /ū/. Give practice in pronunciation of **but/put** and **spook/book** until the children are fully conscious of the differences.

After the words at the bottom of page 108 have been practiced orally, they, or some of them, may be given as a dictation.

Work through the questions on page 109. Let the children pronounce the words with /ꝏ/, becoming conscious of the sound and associating it with the various spellings. The exercise at the bottom of the page is probably best done on the chalkboard.

GRAMMAR—Adverbials

We now enlarge the possibilities after **be** by adding adverbial of place:

$$\textbf{be} \; + \begin{Bmatrix} \text{NP} \\ \text{Adj} \\ \text{Adv-p} \end{Bmatrix}$$

Adv-p is an abbreviation of **adverbial of place**.

The rule for verb phrase will now read:

$$\text{VP} \rightarrow \begin{Bmatrix} \textbf{be} \; + \begin{Bmatrix} \text{NP} \\ \text{Adj} \\ \text{Adv-p} \end{Bmatrix} \\ \text{verbal} \end{Bmatrix}$$

To have a verb phrase in a kernel sentence, we must have **be** or a verbal. If **be**, it must be followed by a noun phrase, an adjective, or an adverbial of place. (There are a few less usual possibilities not introduced in this book.) If a verbal, it may be any of the several constructions including verbs.

The term **adverbial** includes both **adverbs** and **prepositional phrases**. Adverbs are single-word adverbials, like **here, quickly, then**. Prepositional phrases are composed of prepositions plus noun phrases: **in the house, on Sunday**.

There are several kinds of adverbials. In addition to those of place (**here, in the house**), we have manner (**quickly**), time (**on Sunday**), frequency (**often**). At this point we introduce just adverbials of place and do not yet formally distinguish adverbs from prepositional phrases.

It is not incorrect to say that adverbials of place "tell where." They transform into **where** questions:

John is in the house. → Where is John?

This transformation is, however, not explained until grade six.

Teaching Suggestions

Preparatory to introducing adverbials, page 110 reviews noun phrases and adjectives in the predicate position after **be**.

Adverbials

You know that every simple sentence must have a subject and a predicate, and that every predicate must have a verb or a form of **be**. If it has a form of **be**, it must have something else too. One thing it might have is a noun phrase. What is another thing? adjective

• Look at these sentences. Tell what follows **be** in each one, a noun phrase or an adjective. As you know, a form of **be** may be part of a contraction. Tell which sentences have contractions in them.

1. He's hungry. adj.; cont.
2. They are pupils. n.p.
3. We're noisy. adj.; cont.
4. She is a lady. n.p.
5. You're Mary. n.p.; cont.
6. I am curious. adj.

One way that you can tell an adjective from a noun phrase is that you can use the word **very** in front of an adjective but you can't use it in front of a noun phrase. We say "He's very hungry" and "We're very noisy." But we don't say, "They're very pupils" or "She's very a lady."

• Say these sentences. If the word that follows the form of **be** is an adjective, use **very** before it.

7. John is thirsty. very thirsty
8. Sally is my cousin.
9. They are pencils.
10. They are clean. very clean
11. I'm happy. very happy
12. They're soldiers.
13. Mr. Burbank is brave. very brave
14. She's quiet. very quiet
15. We're strong. very strong

Forms of **be** may be followed in the predicate by adjectives or by noun phrases. They may also be followed by what are called **adverbials of place.** Adverbials of place are words or groups of words that tell where something is.

Look at these sentences.

John is outside.
John is in the playground.

What word in the first sentence tells where John is? What group of words in the second sentence tells where he is? In these sentences, **outside** and **in the playground** are adverbials of place.

In each of the sentences below, a form of **be** is followed by an adverbial of place. Tell what the adverbial of place is in each sentence.

16. Mary is inside.
17. David is at his desk.
18. We're in school.
19. They are here.
20. Julia's at home.
21. The girls are at the store.
22. Mr. Burbank is in his car.
23. My father is upstairs.
24. I am downstairs.
25. The desk is near the chalkboard.
26. The crows are on the fence.
27. She's in the swimming pool.
28. Our teacher is at her desk.
29. Mrs. White is away.
30. People are everywhere.
31. George is under that tree.
32. You're in my way.
33. They're on the bus.

It also gives a little more familiarity with contractions.

Go through Sentences 1–6 for practice in distinguishing noun phrases after **be** from adjectives. This is the first exercise in which adjectives have been mixed with noun phrases other than those consisting of "determiner + noun." Still, the noun phrases in these sentences are clearly marked as such: **pupils** in 2 because it is a plural; **a lady** in 4 by the determiner; **Mary** in 5 because it is a proper noun.

We introduce **very** here as an aid in distinguishing adjectives from other words occurring after **be** in predicates. **Very** may occur before all true adjectives. We can express this in a formula in this way:

Adj → (**very**) + Adj

The parentheses mean that **very** is optional. It may occur before adjectives, but it doesn't have to.

The word **very** may, to be sure, occur before nouns:

the very idea
the very pupils I was talking about

But the point is that it will not occur before noun phrases. We do not say,

* Very the idea astonished me.
* Those are very the pupils I meant.

(The asterisks mean that the sentences cited are not grammatical.)

Let the pupils do Sentences 7–15, using the optional **very** before the adjective. Henceforth, when a child is in doubt about whether a word is an adjective or a noun phrase, suggest that he try the **very** test.

Extra Practice

Let the children do Sentences 16–33 as written work after the oral work is concluded. They may simply write for each sentence the adverbial of place that follows a form of **be**.

LITERATURE

"The Story of Johnny Head-in-Air"
(Stanzas Five and Six)

Teaching Suggestions

Reread or replay the preceding stanzas of the poem. Have the children recount what happened in them. Now read or play the recording of these last two stanzas. If time permits, give the children practice in reading parts of them aloud.

The Rest of the Poem

Here is the last part of "The Story of Johnny Head-in-Air."

Oh! you should have seen him shiver
When they pulled him from the river.
He was in a sorry plight!
Dripping wet, and such a fright!
Wet all over, everywhere,
Clothes, and arms, and face, and hair:
Johnny never will forget
What it is to be so wet.

112

And the fishes, one, two, three,
Are come back again, you see;
Up they came the moment after,
To enjoy the fun and laughter.
Each popped out his little head,
And, to tease poor Johnny, said,
"Silly little Johnny, look,
You have lost your writing book!"

The Story of Johnny Head-in-Air

STANZAS FIVE AND SIX

• **Sorry plight** means a ridiculous state or condition. What is the vowel sound of **plight?** What word rhymes with it? /ī/; fright

Have you ever heard someone say,"I look a fright"? A **fright** is an ugly or shocking sight.

The poem says the fishes "are come back." We would usually say "have come back." But poets don't always say things in usual ways.

A Writing Assignment

If a person always walks along without looking where he is going, quite a few unpleasant things may happen to him. What two unpleasant things are told about Johnny in the poem? He fell over a dog; he fell into the river.

Can you think of something else that might happen to him — something else that he might bump into or trip over or fall off of? Make up something that you think might happen to Johnny and tell about it in a paragraph. You might use this title:

Another Story of Johnny Head-in-Air

VOCABULARY AND MEANING

"The Story of Johnny Head-in-Air" (Stanzas Five and Six)

Teaching Suggestions

Get answers to the questions on these two stanzas.

All of the vowel sounds used for rhyme in these stanzas have now been taught, including /oo/ in the final couplet. Point out the irregular spelling of /a/ in **laughter** and /e/ in **said**.

The questions given on page 113 are cut short to make room for the writing assignment. If more seem needed to check on the accuracy of the reading, use the additional questions that follow.

Additional Questions

What word in line 1 shows that Johnny was cold? Who is meant by **they** in the second line? Who pulled him from the river?

What does **dripping wet** mean? What words tell what parts of him were wet?

How many fish are said to come back? What did they come to enjoy? What popped out? How did they tease poor Johnny?

COMPOSITION – A Writing Assignment

Warm up for this by class discussion in which pupils volunteer difficulties that might afflict someone who doesn't look where he is going. Then let the children each write about some such mishap occurring to Johnny.

SOUNDS AND LETTERS—/ch/

This is the first consonant sound studied in this text for which there is no single corresponding letter of the alphabet. It is common in the primary grades to speak of it as the "see-aitch" sound. However, it is simpler to call it "chee," by analogy with "bee," "dee," etc.

There has been much discussion among linguists about whether /ch/ is a single sound or a combination of two sounds. Some have taken it as composed of the sounds /t/ and /sh/. However, the general present opinon of American linguists is that it is better taken as a single sound.

Teaching Suggestions

Go through the questions and examples. The children should not find great difficulty in thinking of other words with initial /ch/. There are many, and they are all spelled **ch**.

Work through the examples of the two regular spellings of final /ch/—**tch** and **ch**. These, like the spellings of final /k/, divide on whether a simple or a complex vowel precedes. Let the children identify the simple vowels preceding /ch/ in the **tch** words.

The Consonant Sound /ch/

• Say these words and listen to the consonant sound at the beginning of each word:

chin check chat church chop

We have no single letter in the alphabet for this sound, so we will show it this way: /ch/. In speaking about it, name it the way you do /p/, /t/, and /b/. Call the sound "chee."

At the beginning of words, the sound /ch/ is spelled with the letters **ch:**

chip	chain	cheat	child
chew	choose	chilly	chum

□ Write five other words that begin with the sound /ch/ in talking and the letters **ch** in writing. See p. 326.

At the end of words, the spelling of the sound /ch/ depends mostly on the vowel sound that comes before the sound /ch/.

When a simple vowel comes before /ch/, we usually spell /ch/ with the letters **tch.**

• Say these words. In each word, tell what vowel sound comes before the sound /ch/.

pitch /i/	fetch /e/	match /a/	hutch /u/	batch /a/
notch /o/	crutch /u/	snatch /a/	itch /i/	latch /a/

When any sound except a simple vowel sound comes before /ch/, we spell /ch/ at the end of words with the letters **ch.**

• Say these words and notice the sounds that come before the final /ch/ sound. (None are simple vowels.)

each	beach	teach	coach
approach	speech	pooch	couch
inch	bench	punch	lunch
search	church	march	porch

The Spelling Rule for /ch/

You have found two ways to spell the sound /ch/ at the end of words. How do we spell it when a simple vowel comes before it? How do we spell it when some other sound comes before it? tch; ch

There are just a few exceptions to these ways of telling whether to use **ch** or **tch** for the sound.

- In the noun **butcher** we spell /ch/ with the letters **tch.** What vowel sound comes before /ch/? /oo/
- We spell the sound /ch/ with the letters **ch** in a few words that contain simple vowels. Here are the words in which this is done. Say each word and tell what vowel sound comes before the /ch/.

rich/i/ which/i/ much/u/ such/u/ touch/u/
ostrich/i/ sandwich/i/

- Which word above has a special way of spelling the vowel sound /u/? touch

The seven words listed above and the noun **butcher** are exceptions to the rule about spelling **ch** at the end. That means they must be learned one by one.

Now we can make our rule:

Spell /ch/ with **tch** in **butcher.**
Spell /ch/ with **ch** in **rich, which, much, such, touch, ostrich,** and **sandwich.**
Except for these, spell /ch/ with **tch** when a simple vowel comes before it.
In all other words, spell /ch/ with **ch.**

□ Study these words. Then write them from dictation to see how well you can apply this rule.

teach	batch	rich	church	butcher
much	catch	torch	ranch	bench
ditch	touch	clutch	each	witch

A Spelling Rule

Here we actually give the children a formal rule for spelling final /ch/. Such rules are not intended to be memorized. Much effort can be wasted in trying to recall the precise terminology of the rule, which in any case could be presented with other terminology. The rule is intended to make explicit a spelling situation of which the child is probably at least vaguely aware to begin with.

Note the structure of the rule, which is the same for all such rules. One begins with the exceptions and progresses to the point at which an "in all other cases" statement can be made. The "in all other cases" statement completes the rule. This is the most economical presentation possible.

1. Use **tch** after simple vowels.
2. Use **ch** after any other sound.

Butcher is an exception to 2, because the vowel before the **tch** is /oo/, which is not a simple vowel.

Rich, which, ostrich, sandwich, much, such, and **touch** are exceptions to 1 because in these **ch**, not **tch**, is used after the simple vowels /i/ and /u/.

Teaching Suggestions

After working through the explanation of the rule, use the list at the bottom of the page for practice in its application. The exceptions in this list are **rich, butcher, much,** and **touch,** all mentioned specifically in the first two parts of the rule. To all of the other words, the last two—general—parts of the rule apply.

GRAMMAR—Kinds of Adverbials

The expression "kinds of adverbials" can mean two quite different things. It can mean different structures: adverbs, prepositional phrases, and (in structures not introduced here, like "He arrived this morning,") noun phrases. Or it can mean adverbials of different application—place, time, frequency, manner. Here it means the former. We distinguish adverbs of place from prepositional phrases of place.

Adverbs and Prepositional Phrases

Both adverbs and prepositional phrases employed as adverbials of place can be used in situations other than that after **be**. Both can be used, for example, after verbs:

Flowers grew everywhere.
Flowers grew in the garden.

In transforms, other possibilities are to be found.

A prepositional phrase is simply a structure composed of a preposition followed by a noun phrase. All different types of noun phrases occur after prepositions:

Determiner + Noun:	in the house
Proper Noun:	with Al Jones
Common Noun Alone:	above reproach

When the noun phrase is a personal pronoun, of course, a special form is used: **to me, with us**. This is explained later.

The **very** test, which discriminates adjectives and noun phrases (it occurs before adjectives but not noun phrases) will generally also discriminate adjectives and adverbials of place. In general, **very** will not occur before adverbs or prepositional phrases which are adverbials of place. We do not say, *"He is very outside" or *"He is very in the playground." However, **very** will, exceptionally, occur with the adverb or preposition **near**: "He is very near," "His desk is very near the chalkboard."

Kinds of Adverbials

You now know of three things that might come after forms of **be** in simple sentences. Noun phrases are one:

That man is a cowboy.

Adjectives are another:

That girl is pretty.

Adverbials of place are another:

Jack is outside.
Jack is on the playground.

• Tell which of the three things comes after **be** in these sentences:

1. Jane is happy. adj.
2. Mary is inside. adv. of place
3. The boys are the winners. n.p.
4. I am at my desk. adv. of place
5. They are on the sidewalk. adv. of place
6. We're hungry. adj.
7. She's upstairs. adv. of place
8. We are students. n.p.
9. Flowers are everywhere. adv. of place
10. Dick is sad. adj.
11. The book is under that newspaper. adv. of place
12. Some callers are here. adv. of place

• We have seen two kinds of adverbials of place. Some are single words, like **outside, upstairs, here, there.** Single-word adverbials are called **adverbs.** Name the adverbs in these sentences:

13. Mary is inside.
14. The bus is here.
15. My parents are downstairs.
16. The lake is there.
17. Father is outside.

The other adverbials of place are groups of words like **in the room, on the playground, at my desk.** Groups of words like these are called **phrases.**

• Find the adverbials of place in these sentences. Tell which are adverbs and which are phrases.

18. My brother is away. adverb
19. The ruler is on the table. phrase
20. My hat is under your coat. phrase
21. Mr. Jonas is here. adverb
22. The stream is near the house. phrase

You found three phrases above. The phrases began with the words **on, under,** and **near.** Words like these are called **prepositions,** and the phrases that begin with these words are called **prepositional phrases.** Some other words used as prepositions are **by, in, at, behind, below.**

• Tell what the prepositional phrases are in these sentences. Tell what the preposition is in each one.

23. It is on the desk. on
24. It is in the desk. in
25. It is near the desk. near
26. It is below the desk. below
27. It is beneath the desk. beneath
28. It is at the desk. at
29. It is by the desk. by

• Tell what the adverbial of place is in each sentence below. Tell whether it is an adverb or a prepositional phrase. If it is a phrase, tell what the preposition is.

30. John is at his desk. prep. phrase, at
31. The cat is outside. adverb
32. My house is near the school. prep. phrase, near
33. Visitors are downstairs. adverb
34. The car is behind the bus. prep. phrase, behind
35. Your cake is on that plate. prep. phrase, on

Teaching Suggestions

Sentences 1–12 on page 116 give review practice in discriminating noun phrases, adjectives, and adverbials of place in the predicate after **be**. Sentences 1, 6, and 10 illustrate adjectives; Sentences 3 and 8 have noun phrases; the rest have adverbials of place—adverbs or prepositional phrases—after the form of **be**. Go through these sentences, having the children identify the structures after **be**.

In the next set of sentences, 13–17, we introduce single-word adverbials of place under the term **adverb**. Have the pupils point out the adverbs in 13–17.

Like personal pronouns, adverbs of place are a small class whose members can be listed.

Work through Sentences 18–22 on page 117, in which adverbs and prepositional phrases are distinguished, though the latter are at this point just called **phrases**. Then use the text to explain the terms **preposition** and **prepositional phrase**. Give practice in pronunciation of these. Then employ the terms in working through Sentences 23–29 and 30–35.

Teaching Suggestions

These tests should be used for diagnosis, as suggested earlier. Individual records of strengths and weaknesses should be kept. Reteaching should take place at once when the tests show it to be necessary. In this series, syntax and phonology are presented as sequences of ideas in which each new concept is related to and depends on concepts previously acquired. Pupils who are allowed to proceed when their understanding of certain grammatical ideas is seriously weak will be handicapped for the rest of the school year. It is important to do the necessary reteaching now rather than leave it for the beginning of the fifth grade.

The page references for reteaching materials are provided to make this task easier for the teacher.

It may be pointed out in Test I, number 4, that two of the words (**rich** and **which**) are irregular.

Use the Review on page 121 as oral discussion.

Extra Test

If a dictation test is desired for the /oo/ sound, the following words may be given in sets:

1. **oo**: book, look, shook, stood, foot, good, wool.
2. **u**: put, push, bush, full, pull, bull.
3. irregular: should, would, could, wolf, woman.

Tests and Review

□ **TEST I Sounds and Letters**

1. Write these headings on your paper:

/u/ /ū/ /oo/

Write each word in the list below under the heading that shows its vowel sound.

If you make mistakes, study pages 108–09.

bush /oo/	lunch /u/	shoot /ū/	book /oo/
butt /u/	love /u/	could /oo/	flood /u/
rule /ū/	boom /ū/	month /u/	stood /oo/
spook /ū/	full /oo/	blood /u/	rush /u/

2. Copy the words below. Complete the words by writing **k** or **c** in the empty place.

If you make mistakes, study page 97.

_k_ick	_c_ould	_k_eep	_c_ry
_c_alf	_k_iss	_c_lean	_k_irk

3. Copy the words below. Complete the words by writing the letters **ck, k,** or **ke** where the empty space is.

If you make mistakes, study pages 102–03.

si_ck_	see_k_	jo_ke_	tra_ck_
pa_ck_	lea_k_	wal_k_	sti_ck_

4. Copy the words below. Complete the words by writing the letters **tch** or **ch** where the empty space is.

If you make mistakes, study pages 114–15.

pa_tch_	lun_ch_	cru_tch_	rea_ch_
ri_ch_	ma_tch_	stre_tch_	coa_ch_
ca_tch_	di_tch_	whi_ch_	spee_ch_

118

□ **TEST II Grammar**

1. Copy the noun phrases in the sentences below. Some have two noun phrases, and some just one. After each noun phrase, write a number to show whether it is (1) a determiner and a common noun, (2) a common noun alone, (3) a proper noun, (4) a personal pronoun.

If you make mistakes, study pages 98–99.

a. Jack is in the kitchen. (3); (1)
b. Mary laughed. (3)
c. The teacher explains the poem. (1); (1)
d. We saw George. (4); (3)
e. They live near my house. (4); (1)
f. A mouse is under the desk. (1); (1)
g. I am by the chalkboard. (4); (1)
h. Johnny is hungry. (3)
i. He needs pencils. (4); (2)
j. She is on the playground. (4); (1)
k. Your sister sounds happy. (1)
l. They are your friends. (4); (1)

2. Copy each of these sentences and use the contraction of the personal pronoun and **be.**

If you make mistakes, study pages 104–05.

a. He is on the playground. He's . . .
b. I am at my desk. I'm . . .
c. We are very noisy. We're . . .
d. She is our teacher. She's . . .
e. It is cold. It's . . .
f. They are outside. They're . . .
g. You are funny. You're . . .
h. I am here. I'm . . .
i. They are in the house. They're . . .
j. She is downstairs. She's . . .
k. It is very windy. It's . . .

☐ **TEST II Grammar** (CONTINUED)

3. Copy the adverbials of place in these sentences. Tell whether each one is an adverb or a prepositional phrase by writing **adv.** or **prep.** after it.

If you make mistakes, study pages 116–17.

a. The child is by the pool. prep.
b. The bus is here. adv.
c. I am at my seat. prep.
d. The papers are in the corner. prep.
e. Mother is upstairs. adv.
f. The kite is above the house. prep.
g. The girls are inside. adv.

4. Copy what comes after the form of **be** in these sentences. Use these numbers to show whether it is (1) a noun phrase, (2) an adjective, or (3) an adverbial of place. It might help to remember that you can use the word **very** before adjectives, but not before noun phrases or adverbials of place.

If you make mistakes, study pages 110–11.

a. Johnny is in the river. (3)
b. Johnny is wet. (2)
c. Johnny is a boy. (1)
d. They are sisters. (1)
e. Your mother is upstairs. (3)
f. I am sleepy. (2)
g. They are under the table. (3)
h. Those birds are pigeons. (1)
i. He is in the office. (3)

5. Rewrite the sentence below three times. Each time use a different preposition in place of the preposition **on.**

It is on the desk.

. . . in the closet.
. . . under my hat.
. . . over the mirror.
etc.

▫ TEST III Writing and Vocabulary

Write five sentences. In each sentence use one of these words:

1. trudge 2. stride 3. tread 4. plight 5. dismay

• REVIEW Ideas and Information

1. When is the sound /k/ at the beginning of words spelled with the letter **k?** before i or e
2. When is /k/ at the end of words spelled with the letters **ck?** after simple vowels
3. What three vowels can be spelled with the letters **oo?** /ū/, /oo/, /u/ (boom, book, blood)
4. What do you call the sound /ch/ when you are speaking about it? chee
5. When is the sound /ch/ at the end of words spelled with the letters **tch?** after simple vowels
6. What are the seven personal pronouns? I, he, she, it, you, we, they
7. What do you call the little mark between the **y** and the **r** in **they're?** apostrophe
8. What are the three things that you have studied which may follow a form of **be?** n.p., adj., adv. of place
9. What are two kinds of adverbials of place? adverb, prep. phrase
10. What kind of word is **on** in the sentence, "It is on the playground"? preposition
11. Which personal pronouns mean the speaker or the speakers? we, I
12. Name the function of each noun phrase in the sentence, "The people elect the president." subj.; obj.
13. Name a word spelled in the pattern **VCe.** late, etc.
14. What does **contraction** mean? shortened form
15. What word is it that you can use in front of an adjective but not in front of a noun phrase? very

121

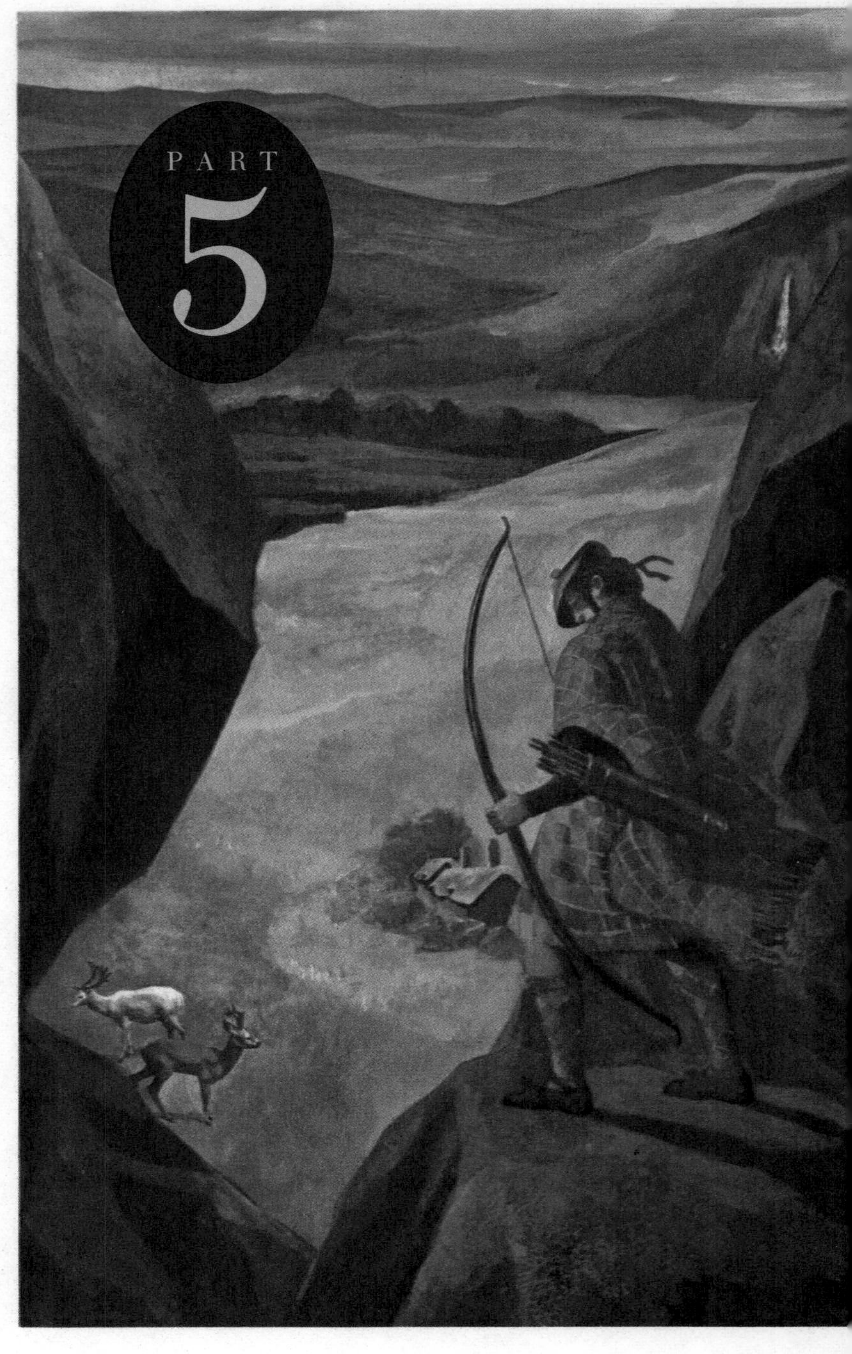
PART
5

SECTION 1

A Poem

When people grow up, they often move away from the place where they were born and lived as children. Sometimes they look back to their childhood place with great longing. They feel that that was the place where they were happier than they have ever been since, no matter where they have roved.

This feeling is described in a famous poem by a famous poet, Robert Burns. Read the poem, or listen as it is read to you.

My Heart's in the Highlands

My heart's in the Highlands, my heart is not here;
My heart's in the Highlands, a-chasing the deer;
Chasing the wild deer, and following the roe,
My heart's in the Highlands, wherever I go.
Farewell to the Highlands, farewell to the North,
The birthplace of valor, the country of worth;
Wherever I wander, wherever I rove,
The hills of the Highlands for ever I love.

Farewell to the mountains high covered with snow;
Farewell to the straths and green valleys below;
Farewell to the forests and wild-hanging woods;
Farewell to the torrents and loud-pouring floods.
My heart's in the Highlands, my heart is not here;
My heart's in the Highlands a-chasing the deer;
Chasing the wild deer and following the roe,
My heart's in the Highlands wherever I go.

ROBERT BURNS

"My Heart's in the Highlands"

Teaching Suggestions

Burns is one of the major poets whose works seem generally available to the pleasure of young people. The teacher may want to give a little biographical information about Burns so that the children may identify him and remember the name. They will meet him again more than once in this series.

Play the recording of the poem or read it to the class. Let the children take turns reading couplets aloud. The pupils may be encouraged to memorize the poem, or at least the first four lines.

VOCABULARY AND MEANING

"My Heart's in the Highlands"

Teaching Suggestions

Use a map or a globe to locate Scotland and its Highlands for the children.

We teach only the contraction of forms of **be** with personal pronoun subjects, but the children ought to be aware of, and perhaps use, contractions with other subjects. After pointing out the contraction of "My Heart's in the Highlands," put a few other contractions, such as the following, on the chalkboard:

Their car's in the garage.
John's a Boy Scout.
Margaret's coming with us.

Ask what letter the apostrophe replaces.

Subjects other than personal pronouns do not commonly contract with **are**.

Dialects

The children should get used to the idea that English is not the same everywhere, that it varies according to place and time. Use the rhymes **rove** and **love**, both of which probably had the vowel sound /oo/ for Burns, to point this out. If the teacher is clever at mimicking the English of Britain or Australia or regional variations in the United States, other examples of different-sounding English might be given, but one should avoid exaggerated or comic representations. One should probably also avoid using as examples differences likely to be present among the children, for fear of misunderstanding and embarrassment.

My Heart's in the Highlands

Do you know where England is? Scotland is to the north of England, and the Highlands are the northern part of Scotland. The Highlands of Scotland are still wild country where few people live.

• Where does the poet say his heart is? This can't mean that it is really there. What does it mean? What does he think of himself as doing in the Highlands? discuss

• Look at "My Heart's in the Highlands." There is a contraction here, but not with a personal pronoun. What is contracted with a form of **be**? heart

Roe has two meanings. It may mean fish eggs. Here it means the kind of deer that lives in the Scottish Highlands — a rather small deer that can run very fast.

• The word **valor** means courage, or bravery. What does **worth** mean? Do we use the same vowel sound when we say **worth** as we do when we say **north?** Then in our kind of English do **north** and **worth** rhyme? value; no; no

• The Scots pronounce English in a different way from Americans. **Rove** and **love** rhymed for Robert Burns, but they don't for us. What vowel sounds do they have when we say them? **Rove** means roam, or travel about. Find **rove** in a dictionary. /ō/; /u/

• Name the things that the poet says farewell to in the beginning of the second stanza. What are the mountains covered with? What else is said about the mountains? **Strath** is a Scottish word for valley. See stan[za] 2.

• By **torrents** and **floods** the poet means waterfalls or fast-rushing water like cascades or rapids, and rivers filled with fast-moving water. How does he describe the woods? Why are they "hanging"? wild-hanging; cling to steep mountain sides

The Consonant Sounds /b/ and /d/

The consonant sound /b/ is made the way /p/ is, except that we make a little hum or buzzing noise in our throats when we make the sound /b/. Cover your ears with your hands and say **pen, Ben.** Do you hear the hum when you say **Ben?**

• Now try it with **ten, den.** Which sound is made with the little buzzing noise, /t/ or /d/? /d/

• For Exercises 1 to 5, try to think of other words in which /b/ or /d/ is spelled as shown. See p. 326.

1. At the beginning of words we spell /b/ with the letter **b** and /d/ with the letter **d:**

Ben	**bite**	**bake**	**brown**	**blood**
Dick	**dare**	**den**	**drive**	**dish**

2. At the end of words we spell /b/ with the letter **b** and /d/ with the letter **d** if a simple vowel sound comes before final /b/ or /d/:

rib	**web**	**cab**	**rub**	**rob**
red	**bad**	**bid**	**rod**	**thud**

3. If a consonant letter comes before final /b/ or /d/, we use just the letter **b** or **d** for the sound:

barb	**curb**	**verb**	**bulb**
world	**could**	**band**	**crowd**

4. With complex vowels we may use the **VCe** pattern with final **b** and final /d/:

tribe	**Abe**	**robe**	**tube**
glade	**lone**	**dune**	**ride**

5. If the complex vowel is spelled with two letters, we have just **b** or **d** at the end:

boob road raid heed food

In one word, final /b/ is spelled **bb: ebb.** When a tide in the ocean goes out, we say that it **ebbs.** An **ebb** tide is a falling tide.

125

Voiced Consonants

The sounds in language are characterized by the presence or absence of various qualities called distinctive features. Thus, depending somewhat on the language, a sound may be consonant or vowel, short or long, nasal or nonnasal, etc. One of the distinctive features is **voicing** or the lack of it.

When the breath with which sounds are made comes from the throat, it must pass through an aperture called the glottis. This may be either constricted or relaxed. If it is constricted, the air is made to vibrate and given a quality called **voice**. This is what is identified on this page for the children as a humming or buzzing sound. Sounds to which this vibration has been imparted carry musical pitch—high if the vibration is relatively fast, low if it is relatively slow. To put it another way, only voiced sounds can be sung.

If the glottis is relaxed when the air passes through, the air is not set in vibration, and the resultant sound is voiceless.

A number of English sounds oppose each other as voiceless and voiced pairs. Thus /b/ and /d/ are the voiced counterparts of voiceless /p/ and /t/. There are some other differences too; for instance, English /p/ and /t/ are aspirated (made with a following puff of breath) in certain positions, while /b/ and /d/ are not. But we point out only the voiceless-voiced contrast.

Teaching Suggestions

Children can usually detect the voicing and its absence in a pair like **Ben/pen** by saying the words with hands over the ears. Let them try it. If they have difficulty, try them on a pair of words whose initial consonants can be continued—like **zip/sip**. They should easily hear the hum in a prolonged pronunciation of the /z/ of **zip** and note its absence in a prolonged /s/ of **sip**. Then let them try again with **Ben/pen**.

GRAMMAR

Noun Phrases as Objects of Prepositions

The sentences the children have been studying have employed noun phrases in four different functions: subject, object of a verb, object of a preposition, complement. Of these only the first two have been identified and named up to this point. We now identify the third.

Parsing

Much of the activity we are conducting in these lessons on syntax is what is called parsing. This is the identification of various structures in sentences and the reporting of their various functions and characteristics. Parsing used to be done on an elaborate scale, though not always in a very sound way. Though we do less of it nowadays, some must be done if the shape of the grammar is to be understood.

Notice that in parsing noun phrases, we can give information about several different aspects of them. We can tell the general type of the noun phrase: determiner + noun, proper noun, personal pronoun, indefinite pronoun. Or we can tell its function: subject, object of a verb, object of a preposition, complement. Or we might tell its number: singular or plural. Or tell its gender: masculine, feminine, or neuter. Or we may tell whether a noun in it is count or non-count. Full parsing would mean identifying all of these aspects and others. Ordinarily we focus on just one at a time.

Teaching Suggestions

Use Sentences 1–10 to review the two types of adverbials of place—adverbs and prepositional phrases. Use the questions that follow to describe the structure of the prepositional phrase.

Noun Phrases as Objects of Prepositions

You have learned to recognize two kinds of adverbials that come after a form of the word **be** in simple sentences. These are **adverbs** and **prepositional phrases.**

• Find the adverbial in each of these sentences. Tell whether it is an adverb or a prepositional phrase. If it is a prepositional phrase, tell what word in it is a preposition.

1. The picture is above the fireplace. prep.; above
2. The people are downstairs. adverb
3. The cat is under the porch. prep.; under
4. The clothes are outdoors. adverb
5. They are on the clothesline. prep.; on
6. The clouds are over the lake. prep.; over
7. That snow is on the mountain. prep.; on
8. My uncle is inside. adverb
9. The present is in that box. prep.; in
10. The moving men are here. adverb

The adverbial in the first sentence above is this prepositional phrase:

above the fireplace

• What word in this prepositional phrase is a preposition? above

• The adverbial **above the fireplace** contains a noun phrase: **the fireplace.** What is the determiner in this noun phrase? What is the noun? the; fireplace

The noun phrase **the fireplace** in the prepositional phrase **above the fireplace** functions as the object of the preposition **above.**

In the prepositional phrase **under the porch,** the word **under** is a preposition. What noun phrase functions as the object of the preposition? the porch

□ Find the prepositional phrase in each sentence below. Tell what word is the preposition. Then tell what the noun phrase is that functions as the object of that preposition.

11. The moon is above the clouds. above; the clouds
12. The cookies are on a plate. on; a plate
13. Two rings are in that box. in; that box
14. The stage is behind that curtain. behind; that curtain
15. My hat is under your coat. under; your coat
16. The boys are near the pond. near; the pond
17. That airplane is over my house. over; my house
18. Those men are beside my father. beside; my father
19. The cellar is beneath our kitchen. beneath; our kitchen
20. The mayor is between the speakers. between; the speakers

□ You have now learned three functions of noun phrases: (1) subject, (2) object of a verb, (3) object of a preposition. Copy every noun phrase in the sentences below, and tell what its function is. Do this by writing 1, 2, or 3 after it.

21. A bird is over my head. (1); (3)
22. We are in the water. (1); (3)
23. Mr. Chapin builds houses. (1); (2)
24. Your father is in the car. (1); (3)
25. They want a boat. (1); (2)
26. The swing is under that tree. (1); (3)
27. A mouse likes cheese. (1); (2)
28. This park is beside the lake. (1); (3)
29. My mother plays bridge. (1); (2)
30. Some people watch birds. (1); (2)
31. You are on my team. (1); (3)
32. His brother is between two strangers. (1); (3)
33. She paints pictures. (1); (2)
34. Sally is behind that woman. (1); (3)
35. My home is in Chicago. (1); (3)

Use Sentences 11–20 to familiarize the pupils with the noun phrase functioning as object of a preposition. These sentences are easy because in each a prepositional phrase follows **be**. Let them use a formula like this: "The prepositional phrase is **above the clouds**. **Above** is a preposition. The noun phrase **the clouds** functions as the object of **above**."

Sentences 21–35 are harder, because here the children are asked to identify noun phrases in the three different functions so far pointed out. Each sentence contains two noun phrases, of which the first is the subject. The other is either object of a verb or object of a preposition. It may be best to start the exercise on the chalkboard.

WP

SECTION 2

A Poem

Railroads are a very important part of the history of the United States. Railroads were built and run by brave men through country that was rough and wild. It was exciting and dangerous work.

In the old days, railroad engineers were heroes to boys and girls because of their exciting life, just as pilots and astronauts are heroes today. And they did have adventures. They tried to set new speed records. Sometimes they had to stop for herds of wild game crossing the tracks. Often they were killed in wrecks, for there were no signal systems to show whether or not the track ahead was clear.

Because the railroads were so interesting, people made up many poems and songs about them. Probably the most famous one is this one about Casey Jones. We will study it in two parts. Read this first part, or listen as it is read or sung to you.

Casey Jones

Come all you rounders that want to hear
A story about a brave engineer;
Casey Jones was the hogger's name.
On the Western Pacific he won his fame.

The caller called Casey at half-past four.
He kissed his wife at the station door.
He mounted to the cabin with his orders in his hand,
And took his farewell journey to the promised land.

Put in your water and shovel on your coal,
Stick your head out the window, and watch her roll.
We've got to run her till she leaves the rail,
For we're eight hours late with the Western Mail.

129

"Casey Jones" (Stanzas One to Three)

This is not of unknown origin. There is a real Casey Jones—a gallant engineer on the Illinois Central. His actual name was John Luther Jones. He was born in 1864 and died in the wreck of his engine in 1900. His nickname, "Casey" Jones, came from the name of his home town, Cayce, Kentucky.

The original words were written by Wallace Saunders, a Negro who worked in a railroad roundhouse. The song was part of a musical comedy in 1901. There are many versions of the words, however, because of different recollections and additions by singers who may never have seen the words in print. The song has established itself firmly among American folk songs and has become one of the most popular of them.

Those of us who learned it in childhood cannot recite it as a poem but only sing it as a song.

Teaching Suggestions

Take the class through the preliminary discussion of railroads, expanding it if time permits. Emphasize the fact that the men who operated the early railroads were not absurd and comical figures but tough and brave men who lived hard and often died young. Despite its light touches, the intent of the poem is to convey this feeling.

If possible, play the recording, on which the poem is sung, of these three stanzas. Encourage the children to learn it and sing it.

VOCABULARY AND MEANING

"Casey Jones" (Stanzas One to Three)

"Casey Jones" is as difficult to read accurately as the other selections used and needs as much care.

Rounder is defined here somewhat euphemistically. It means more precisely a dissolute person, a wastrel.

Teaching Suggestions

The children may not be as familiar as children once were with steam power. Use the familiar illustration of a teakettle. When the water boils, it turns into steam which can be forced through openings to make things move. When you put a gadget on the spout of a teakettle, the steam's being forced through makes a whistle blow. Casey's steam engine was essentially a pot of water with a coal fire under it. When the water boiled, the steam was directed to pistons which moved the drive shaft which caused the wheels to turn. Making the train go fast was essentially a matter of making a hotter fire to produce more steam.

Casey Jones (STANZAS ONE TO THREE)

• **Rounders** means idle people that might like to listen to a good story. What story are they going to be told? See poem

Hogger is a slang word that means engineer.

You probably know that the **Western Pacific** is the name of a railroad.

• What does **fame** mean? See a dictionary. Where did the hogger win his fame? on the Western Pacific

In the early days, the railroad sent a man to wake up the engineer so that he would get to work on time. That is what a **caller** is.

• Did Casey's wife get up and go to the station with him to say good-bye? How can you tell? discuss

• What did Casey have in his hand when he mounted to the cabin? The little room in the engine for the fireman and engineer is still called "cab," which is short for cabin. his orders

• Look at the last line of the second stanza. The **promised land** is Heaven. What do you think this line means? discuss

• Nearly all train engines nowadays are diesel or electric engines. But the one Casey Jones was on was a steam engine. A fireman had to shovel coal to make the steam to run the train. In the last stanza, Casey is talking to the fireman. What does he tell him to do in the first two lines of the last stanza? See poem

• What does **leave the rail** mean? Of course, he doesn't want the train to do that. He means that they have to run it very fast. Why must they? discuss

• You know that proper nouns are written with capital letters. Do you think the **Western Mail** is mail sacks for the West, or the name of Casey's train? train

A Writing Assignment

Look at the picture. It shows an old railroad steam engine that burned coal. The man sitting down is the engineer. The other man is the fireman. The lever that the engineer is holding is called the throttle. It is what he uses to make the train go faster or slower. The coal is on a little car called the "tender."

Write a paragraph about this picture. Tell what the men look like. Tell what they are doing.

COMPOSITION – A Writing Assignment

Teaching Suggestions

Here, as usual, it will probably be best to have a class discussion as a warmup. Ask the children to look at the picture, identify the figures, and tell what is happening. Settle any disputes about the facts. Then have the class write their descriptions of the picture.

SOUNDS AND LETTERS–/g/

The sound /g/ is opposed to /k/ as /b/ and /d/ are opposed to /p/ and /t/. The /g/ sound is approximately like /k/ except that /g/ is voiced and /k/ is voiceless.

The letter **g** can represent either of two sounds: /g/ as in **Gail** or /j/ as in **Gerald**. The alphabetic name "gee" employs the /j/ sound. This may cause some confusion, but it is better to put up with it than to depart from the policy of using alphabetic names to name the consonant symbols wherever possible.

Teaching Suggestions

Go through the questions and examples. We do not yet introduce words in which the letter **g** stands for the sound /j/. Let the class give other examples of words with initial /g/ spelled **g**.

Other examples of /g/ after simple vowels are **pig, dig, keg, peg, sag, tag, mug, plug**. The sound /g/ occurs after the simple vowel /o/ for some people, those who pronounce **hog, dog, log**, etc., with /o/. Most Americans have the vowel sound /au/, as in **haul** or **dawn**, in these words.

The sound /g/ does not occur in the **VCe** pattern because the spelling **age** or **ege** would indicate the sound /j/, not /g/. Instead, silent **ue** is used. There are not many words with this spelling. Examples in addition to the four given are **colleague, plague, intrigue, fugue**. A number of words with the ending **–gogue** or **–logue** employ the same device: **pedagogue, dialogue**.

Egg seems to be the only word in which final /g/ is spelled **gg**.

The Consonant Sound /g/

You have now studied six consonant sounds:

/p/ /t/ /k/ /ch/ /b/ /d/

• Name these six consonant sounds. Give some words that begin with them. Give some words that end with them. pen, talk, can, kill, chill, ball, do; up, it, back, bake, batch, each, hub, lead; etc.

Say these words and listen to the consonant sound at the beginning: **go, gain, get.** We show this sound this way: /g/. We name it the way we do the letter **g** in the alphabet.

At the beginning of words, the sound /g/ is written with the letter **g:**

give game grow glide guess

• Think of five other words that begin with the sound /g/ and the letter **g.** See p. 326.

• When /g/ occurs at the end of words after simple vowels, it is also spelled with the letter **g.** Tell what vowels come before /g/ in these words:

big /i/ **bag** /a/ **beg** /e/ **bug** /u/
rag /a/ **rug** /u/ **rig** /i/ **leg** /e/

• Give other words in which /g/ comes after /i/, /e/, /a/, or /u/. See p. 326.

There is one common word in which /g/ is spelled with two g's. This is the word **egg.**

The sound /g/ doesn't occur in many words after complex vowels. When it does, it is spelled in an unusual way. It is spelled with the letters **gue.**

Here are some words in which /g/ is spelled with the letters **gue.** See if you can tell the vowel sounds that come before the /g/ sound.

vague /ā/ **fatigue** /ē/ **league** /ē/ **rogue** /ō/

Review: Noun Phrases

1. These are the kinds of noun phrases you have studied: (1) determiner and common noun, (2) common noun alone, (3) proper noun, (4) personal pronoun. Copy the noun phrases in the sentences below. After each noun phrase, write the number that shows what kind it is.

a. The boy knows Mrs. Gregory. (1); (3)
b. She likes candy. (4); (2)
c. Boys like milk. (2); (2)
d. The baby laughed. (1)
e. They need some pencils. (4); (1)
f. That man is on a motorcycle. (1); (1)
g. We ate peanuts. (4); (2)
h. Mr. Burbank saw the movie. (3); (1)
i. I am upstairs. (4)

2. You have learned three functions of a noun phrase. Copy the noun phrases in the sentences below. Write the number (1), (2), (3), or (4) after each one to tell what kind it is. Then write a letter to tell what its function is: (a) if it functions as subject, (b) if it functions as object of a verb, or (c) if it functions as object of a preposition.

a. The deer run in the woods. (1), (a); (1), (c)
b. They are graceful. (4), (a)
c. The caller called Casey. (1), (a); (3), (b)
d. His heart is in Scotland. (1), (a); (3), (c)
e. We know Jean. (4), (a); (3), (b)
f. She sits by Tom. (4), (a); (3), (c)
g. Miss Woods helps our teacher. (3), (a); (1), (b)
h. It was under the piano. (4), (a); (1), (c)
i. A bus took the class. (1), (a); (1), (b)
j. I made a boat. (4), (a); (1), (b)

133

GRAMMAR—Review: Noun Phrases

Use this page to make sure, before going on, that the children can parse noun phrases in respect to two general questions that may be asked of them: (1) What kind of noun phrase is it? (2) What is its function? Use Sentences a–i of exercise 1 to review the first point and a–j of exercise 2 to review the second.

GRAMMAR

Personal Pronouns as Objects

One way in which traditional parsing has not been very sound is that pupils have often been asked to state the *case* of noun phrases. For example, they have been instructed to say that in "John is here," **John** is in the nominative case; in "I saw John," **John** is accusative; in "I gave John the book" **John** is dative. Some grammars have discerned five or six cases of English noun phrases.

This is simply a confusion of form and function stemming from the tendency to make English grammar a kind of transplant of Latin grammar. In Latin, noun phrases did indeed have several cases. That is, nouns and their modifiers had a set of distinct forms which correlated with certain functions that noun phrases filled. Thus one used the form ***mensa*** (table) when the word functioned as a subject, but ***mensam*** when it functioned as the object of a verb or of certain prepositions.

All that English has left of this, apart from the possessive, are objective forms of five of the personal pronouns – **me, him, her, us,** and **them**. Clearly it doesn't pay to build a large case terminology for these five words. We say simply that noun phrases have various functions – subject, object, complement. When these five personal pronouns function as objects, they have the particular forms shown.

Teaching Suggestions

Go through the questions and examples on page 134. The personal pronouns **it** and **you** are introduced first in the object function, these having no change in form. After doing these sentences, show that **I** must change its form when it functions as an object. The native speaker of English will see at once that we must say "The man saw me," not *"The man saw I." The native sometimes makes mistakes with pronoun

Personal Pronouns as Objects

- Find the noun phrase **the closet** in each sentence:
  a. The broom is in the closet. obj. of prep.
  b. The closet is in the corner. subj.
  c. He opened the closet. obj. of v

Tell what function the noun phrase **the closet** has in each sentence. Does the noun phrase change its sound or spelling in these different uses? no

Most noun phrases have the same form — the same sound and spelling — no matter how they are used. But one kind of noun phrase is different. This is the personal pronoun.

Personal pronouns have the same functions as other noun phrases, but so far you have studied them only in the function of subject. Now you will study the forms the personal pronouns have when they are used as objects.

- Two of the seven personal pronouns have the same form however they are used. These are **you** and **it.** Tell the functions of **you** and **it** in the following sentences:
  d. You heard it. subj.; obj. of v.
  e. It is near you. subj.; obj. of prep.
  f. It hurts you. subj.; obj. of v.
  g. You are in it. subj.; obj. of prep.

  Do **you** and **it** change their forms? no

But the other five personal pronouns have different forms. They have one form when they are used as subjects and another when they are used as objects. For instance, **I** has the object form **me.** We say "I saw the man," but we don't say, "The man saw I." We say, "The man saw me." We don't say "It is near I." We say "It is near me."

Study this table and see what forms the seven personal pronouns have when they are used as objects.

Forms of Personal Pronouns

As subject:	**I**	**he**	**she**	**it**	**we**	**you**	**they**
As object:	**me**	**him**	**her**	**it**	**us**	**you**	**them**

- What is the object form of **he?** of **she?** of **I?** him; her; me
- Find the personal pronouns in these sentences. Tell what function each one has.

1. I saw him. subj.; obj. of v.
2. He saw me. subj.; obj. of v.
3. She is near me. subj.; obj. of prep.
4. I am near her. subj.; obj. of prep.
5. He helps her. subj.; obj. of v.
6. She is near him. subj.; obj. of prep.

- What is the object form of **we?** of **they?**
- Find the personal pronouns in these sentences. Tell what function each one has.

7. They are near us. subj.; obj. of prep.
8. She is on it. subj.; obj. of prep.
9. It is in them. subj.; obj. of prep.
10. You like him. subj.; obj. of v.
11. He needs them. subj.; obj. of v.
12. We are above it. subj.; obj. of prep.
13. They saw her. subj.; obj. of v.
14. She pinched me. subj.; obj. of v.
15. I am near them. subj.; obj. of prep.
16. They heard us. subj.; obj. of v.
17. She is beside him. subj.; obj. of prep.
18. They called us. subj.; obj. of v.
19. We are behind you. subj.; obj. of prep.
20. It is below me. subj.; obj. of prep.

135

forms, but only in transforms, not in kernel sentences.

Use the table on page 135 to point out the pronouns that change form when they are used as objects. Say that they do this whether they are objects of verbs or objects of prepositions. The last paragraph of the preceding page illustrated **me** used both as object of a verb and as object of a preposition.

Go through Sentences 1–6, which display **him, me,** and **her** in the same contexts illustrated for **me** on page 134. Ask the questions that follow. Then do Sentences 7–20, which give these and the other personal pronouns functioning as objects. Have the pupils tell of each whether it functions as object of a verb or of a preposition.

Extra Practice

In order to give more experience with subject and object forms of personal pronouns to every pupil, Sentences 1–20 may be assigned as a written exercise. The personal pronouns in each sentence may be written and labeled to show their function: **sub.** for "subject," **obj. v.** for "object of a verb," and **obj. prep.** for "object of a preposition."

UNITED STATES

SECTION 3

The Rest of the Poem

The poem "Casey Jones" is a kind of story in rhyme that is often called a **ballad.** A ballad like this one "just grows." No one knows who thought of the words. People have added to it and changed it as the years have passed.

This ballad has a strong beat to it that makes it a good poem to sing to guitar or piano music. People for many years have sung "Casey Jones" more often than they have recited it from memory or read it in a book.

If your teacher has the records that go with this book, you can hear how "Casey Jones" sounds when it is sung as it was meant to be.

Casey looked at his watch, and his watch was slow.
He looked at the water, and the water was low.
He turned to the fireman, and "Boy," he said,
"We've got to reach Frisco, or we'll all be dead."

Casey pulled up on Reno Hill,
And tooted on the whistle with an awful shrill.
The crossing man knew by the engine's moans
That the man at the throttle was Casey Jones.

Casey pulled up within two miles of the place,
With Number Four staring him right in the face.
He turned to the fireman, said, "Boy you'd better jump,
For there's two locomotives that's going to bump."

Casey said just before he died,
"There's two more roads that I'd like to ride."
The fireman said, "What may they be?"
"Why, the Southern Pacific and the Santa Fe."

"Casey Jones" (Stanzas Four to Seven)

It is true, as stated here, that no one, or practically no one, nowadays knows who thought of the words of Casey Jones, though it differs from other ballads in having a known author, as mentioned on page 129 of the Teacher's Edition. But it is like ballads in other respects. It has been handed down largely by oral tradition with many changes and improvisations along the way. The author of this series learned it as a child of ten from a cowboy in Stanislaus County, California, and never saw it written until thirty years after.

Teaching Suggestions

Play the recording of the song all the way through if the recording is available. Otherwise read it or, better, sing it to the children. You may wish to encourage them to learn the words and sing it themselves.

VOCABULARY AND MEANING

"Casey Jones" (Stanzas Four to Seven)

Casey Jones was a real railroad engineer who was killed in a train wreck, but here the factual basis of the ballad ends. The wreck actually took place in Mississippi, not near San Francisco, and was somewhat inglorious—Casey smashed his engine into the rear of a train that was attempting to switch off the main line onto a siding.

Teaching Suggestions

Work through the poem to make sure that the pupils have a close understanding of the stanzas on page 137.

The name "Frisco" is anathema to modern San Franciscans and is probably now old-fashioned in most of the country. It was widely used in the early days of the settlement of northern California. Nowadays, San Francisco is generally called "The City" by anyone within a radius of a hundred and fifty miles or so.

Casey Jones (STANZAS FOUR TO SEVEN)

Why would a watch that was slow worry a railroad engineer? He couldn't follow his time schedule.

• Casey's engine was a steam engine. The coal fire in the firebox boiled the water in the boiler to make steam. The steam, as it was let out of the boiler under pressure, made the engine's wheels turn. Why would low water worry a railroad engineer? might boil dry

• **Frisco** is a nickname for San Francisco. People don't use this nickname very much any more, but they used to. What did Casey say about Frisco? Whom did he say it to? See poem.

• Can you tell which direction the train was going? west

• When Casey **pulled up on Reno Hill,** he came up to it. Reno is in Nevada near the California state line. At Reno he tooted for the crossing. Why did he toot? What is an **awful shrill?** to warn traffic; discuss

• What does a crossing man do when a train is coming? How could the crossing man tell that Casey Jones was the engineer? lowers gate; the sound of the engine

• Casey had come within two miles of the end of his run when he saw something **staring him right in the face.** What was it? train Number Four

• What did he tell the fireman to do? Can you think of a reason why Casey stayed in the cabin? jump; discuss

• What did Casey mean by **roads?** Do you think he wished just to ride on these railroads, or to drive a locomotive on them? railroads; to drive a locomotive

• People usually say the last part of **Santa Fe** with the vowel /ā/. They say it to rhyme with **day.** Do you think Casey said it this way? What vowel sound did he say it with? How can you tell? no; /ē/; rhymes with be

° Actually, despite the poem's wording, Casey had his own whistle, which he took on each run and which the crossing man recognized.

138

How Words Are Made

Most of the words we use are made from other words. We can make words from other words by adding endings to them.

An example of how we do this is the word **caller.** What did a caller do in the first part of the poem about Casey Jones? Do you remember this line?

The caller called Casey at half-past four.

What kind of word is **called?** What kind of word is **caller?** verb; noun

What do we add to the verb **call** to make the noun **caller?** What do we add to the verb **teach** to make the noun **teacher?** er; er

We add the ending **er** to make nouns out of verbs. A **caller** means a person who calls. A **teacher** means a person who teaches. The ending **er** makes a noun that means someone who does whatever the verb says.

- What is a **worker?** What is a **fighter?** What is a **dreamer?** See a dictionary. one who works; fights; dreams
- What verbs do these nouns come from?

talk, buy **talker buyer player watcher** play, watch

- Make nouns from these verbs by adding the ending **er.** Tell what the nouns mean. climber, sweeper, thinker, screamer; "one who does"

climb sweep think scream

□ Copy each sentence. Change the word in parentheses so that you can use it correctly in the sentence.

1. (sing) A ____ sang "Casey Jones." singer
2. (buy) We found a ____ for our house. buyer
3. (speak) The ____ talked for twenty minutes. speaker
4. (catch) The umpire stands behind the ____. catcher
5. (call) The ____ called the engineers. caller

139

The Morpheme er

In the fifth-grade book and beyond, this series treats word formation in considerable detail, with particular attention to suffixes. A person learning English—whether child or foreigner—builds his vocabulary not so much by learning individual discrete items as by becoming aware of word-building processes. More precisely, he learns a fairly large number of base items and then processes, fairly intricate ones for the most part, through which some of these can be mutated. The base items multiplied by a number of word-building processes will yield a vocabulary of considerable size. This page gives a kind of preview of what will be examined thoroughly later.

There are actually two er morphemes. The one in **caller** makes nouns of agent from verbs. The one in **sweeter** makes comparative forms of adjectives. There is another **er** ending, as in **mother** or **hammer,** which is not a morpheme.

Here we illustrate only the er morpheme that makes nouns of agent from verbs. This is perhaps the most productive of all suffixal morphemes. It can be added to almost any verb to make a noun of agent. Consequently a learner of the language who gets onto this morpheme immediately adds to his vocabulary a set of nouns nearly as many as the number of verbs he knows.

Teaching Suggestions

Let the children answer the questions on this page and become acquainted with this er morpheme. Sentences 1–5 may be done as class work at the chalkboard before they are assigned as written work.

Extra Practice

For additional practice, let the children make nouns of agent from the following verbs and use the nouns in sentences:

jump sell think murder find

SOUNDS AND LETTERS–/j/

The consonant /j/ is to /ch/ as /b/ is to /p/ or /g/ is to /k/. The sounds /j/ and /ch/ are made in approximately the same way, except that /j/ is voiced and /ch/ is voiceless. As /ch/ has sometimes been analyzed as a combination of two sounds–/t/+/sh/–so /j/ has sometimes been analyzed as /d/+/zh/. However, both are now generally described as single sounds.

The spelling problem for /j/ is quite different from that of /k/. For the latter, we can make a simple rule: **k** before **i** or **e**; otherwise **c**. With /j/, we always have **j** in the initial position when the second letter is not **i** or **e**, but when it is, we may have either **j** or **g**, and there is very little way of telling which. Eventually one gets onto a few indications: the Greek prefix **geo** is always spelled with **g**. But it is mostly memory work.

Here are some more or less common words in which /j/ is spelled **g** before **i** and **e**: **gelatin, gem, gender, geneology, general, generate, generic, generosity, generous, genesis, genial, genitive, genius, genteel, gentile, gentle, gentleman, genuine, geography, geology, geometry, George, geranium, germ, German, gerund, gesture, giant, gibber, gibe, Gibraltar, gigantic, gin, ginger, giraffe, gipsy, gist**.

Those beginning with **j** before **i** and **e** include these: **jealous, jeep, jeer, jelly, jeopardy, jerk, jersey, jest, jet, jetsam, jewel, jib, jiffy, jiggle, jigsaw, jilt, jingle**. In addition there are many proper names, like **John, Jean, Joan, Jesus**.

There are more words in which /j/ is spelled **g** before **i** or **e** than words in which it is spelled **j**. However, the words with **j** are for the most part more likely to be in the child's vocabulary, so from his point of view, it is about a toss-up which of the spellings should be called the usual or regular one. He simply has to learn, for any word beginning with /j/ before **i** or **e**, whether it is spelled **j** or **g**.

The Consonant Sound /j/

The first consonant sounds you studied were these:

/p/ /t/ /k/ /ch/

Next, you studied some consonant sounds that are like /p/, /t/, and /k/, except that we have a little humming sound in our throats when we make them.

• The sound /b/ is like /p/ except for the humming noise, and /d/ is like /t/. What sound is /g/ like? /k/

Now you will study the sound that is like /ch/, except for the humming noise. Say **chug, jug.** Do you hear the humming sound when you say **jug?**

We will show the first sound in **jug, Jim,** or **George** this way: /j/. When you are speaking about the sound, name it the way you do the letter **j.**

The way we spell /j/ at the beginning of words depends on what letter follows. If the letter **e** or **i** follows, we spell the sound /j/ sometimes with the letter **j** and sometimes with **g.** Here are some words in which the sound /j/ is spelled with the letter **j:**

Jim	**Jill**	**jingle**	**jiffy**
jet	**jerk**	**jewel**	**Jerry**

• Do you know any other words in which /j/ is spelled with the letter **j** before **e** or **i?** jig, jelly, etc.

Here are some words in which the sound /j/ is spelled with the letter **g** before **e** and **i:**

giant	**giblet**	**ginger**	**giraffe**
gem	**general**	**gentle**	**geography**

• Do you know any others? generous, generate

• We have to remember the words for which we use **j** before **e** and **i** and the words for which we use **g.** The letter **j** always means /j/ before **e** and **i,** but **g** can mean another sound. What other sound can the letter **g** mean? /g/ as in goat

When some letter other than **e** or **i** follows the /j/ sound, we write /j/ with the letter **j.** Here are some examples:

Jack	**judge**	**joke**	**jungle**
jam	**Jones**	**Jane**	**jumble**

● Give five other words in which the sound /j/ at the beginning is written with the letter **j.** jug, jump, jangle, jut, jolly, join, joint, etc.

At the end of words, we spell the sound /j/ in two ways. We spell it with the letters **dge** when a simple vowel comes before it. Tell what vowels come before the sound /j/ in these words:

wedge	**bridge**	**nudge**	**badge**	**dodge**
/e/	/i/	/u/	/a/	/o/

edge, ridge, smudge, Madge, lodge

● Can you think of words to rhyme with each of the words above? How are your rhyming words spelled?

When the sound /j/ comes at the end of a word after a sound that is not a simple vowel, we spell /j/ with the letters **ge.** We spell it that way in these words:

rage	**age**	**oblige**	**huge**
/ā/	/ā/	/ī/	/ū/

● What vowel sounds come before /j/ in those words?

When a consonant sound comes before /j/, we also spell /j/ with the letters **ge,** as in these words:

bulge **change** **plunge** **charge**

● Tell what sound each of these words ends with:

/j/, /ch/	hedge	match	sick	slug	/k/, /g/
/j/, /k/	stage	bike	pinch	bilge	/ch/, /j/
/ch/, /g/	march	egg	fudge	large	/j/, /j/

□ Study the following words. Then close your book and write them as your teacher dictates them.

jet	gentle	Jane	giant	joke
judge	badge	rage	huge	change

141

The spelling of /j/ is simpler at the end of words than at their beginning. Essentially the rule is this:

1. Spell final /j/ **dge** after simple vowels.
2. Otherwise spell it **ge**.

There are a few cases in which /j/ is spelled **ge** after simple vowels, but not in monosyllabic words. The verb **allege** has /e/ before **ge**. The words **sacrilege** and **privilege** might have it, though they are usually pronounced with weak stress and the reduced vowel schwa in the last syllable.

Four words, none of them near the child's vocabulary, spell final /j/ with **j**: **raj, taj, hadj, samaj**. These are often pronounced with final /zh/ instead of /j/.

Teaching Suggestions

Work carefully through page 140 with the class. If they cannot suggest additional **j** and **g** words, list on the board a few of those given in this teacher's material. Be sure they understand the generalization at the top of page 141 – it is only in words in which /j/ is followed by **i** or **e** that a spelling problem exists.

Introduce **dge** as the regular spelling of /j/ when a simple vowel precedes. Elicit rhyming words for the five **dge** words listed. Here are possibilities:

wedge – edge, pledge, sledge, ledge
bridge – ridge, midge, cartridge, partridge
nudge – judge, budge, fudge, drudge
badge – cadge, Madge
dodge – lodge, hodgepodge, hodge

The children will not be likely to think of a rhyme for **badge**, unless the proper name **Madge** occurs to them. They may manage the others.

Have the class identify the complex vowels in **rage, oblige, huge**.

The list beginning **hedge** includes words ending with the sounds /j/, /ch/, /k/, and /g/.

GRAMMAR—A Review

Sentences a–j in the first exercise on page 142 do not contain noun phrases after **be**. Noun phrases occur in the predicates of these sentences functioning only as objects of verbs or as objects of prepositions. Therefore the choice after a verb or **be** is only the fourfold one stated in the introduction to the exercise.

The second set of sentences, a–l, reviews the four kinds of noun phrases so far described. Each sentence contains two noun phrases. One functions as subject, the other as complement or object in the predicate. The children are not here asked to tell the functions. They can't, because the complement function has not yet been explained.

Grammar: A Review

• Tell whether the predicate of each sentence has a verb or a form of **be**. If anything comes after the verb or the form of **be**, tell whether it is (1) an adjective, (2) an adverb, (3) a prepositional phrase, or (4) the object of the verb.

a. The boys are thirsty. be; (1)
b. We are downstairs. be; (2)
c. Bob likes peanuts. verb; (4)
d. The engineer was in the cabin. be; (3)
e. A ballad tells a story. verb; (4)
f. San Francisco is on a bay. be; (3)
g. They give orders. verb; (4)
h. She is outside. be; (2)
i. The man is dusty. be; (1)
j. Mr. Harris is under his car. be; (3)

• Find each noun phrase in the sentences below. Tell what kind of noun phrase it is: (1) a determiner and common noun, (2) a common noun alone, (3) a proper noun, or (4) a personal pronoun.

a. Joe was behind the gate. (3); (1)
b. They see me. (4); (4)
c. The principal gives a speech. (1); (1)
d. I want some milk. (4); (1)
e. My aunt is Mrs. White. (1); (3)
f. Flowers grow in the fields. (2); (1)
g. Bill likes people. (3); (2)
h. Daisies are flowers. (2); (2)
i. Miami is in Florida. (3); (3)
j. We know them. (4); (4)
k. You stand behind her. (4); (4)
l. He paints pictures. (4); (2)

142

□ 3. Copy each noun phrase from the sentences below. Tell what its function is by writing a numeral. Write (1) if it functions as subject, (2) if it functions as object of a verb, or (3) if it functions as object of a preposition.

a. We are in a boat. (1); (3)
b. My father knows him. (1); (2)
c. Grace is near her. (1); (3)
d. The sheep are on an island. (1); (3)
e. Deer are pretty. (1)
f. They pushed me. (1); (2)
g. The engine pulls the cars. (1); (2)
h. The smoke is over our house. (1); (3)

□ Copy each sentence below. Where a word is left out, use one of the personal pronouns above the sentence. Tell whether the pronoun functions (1) as subject, (2) as object of a verb, or (3) as object of a preposition.

4. **I or me**
   a. The girl called __me__. (2)
   b. __I__ was upstairs. (1)
5. **he or him**
   a. They help __him__. (2)
   b. __He__ helps them. (1)
6. **she or her**
   a. __She__ looks for me. (1)
   b. But he is behind __her__. (3)
7. **we or us**
   a. The cat likes __us__. (2)
   b. __We__ feed it. (1)
8. **they or them**
   a. __They__ sing at church. (1)
   b. I am near __them__. (3)

Teaching Suggestions

The two exercises on page 143 are intended as written work, though the teacher in a slow class may have to start off the first one on the chalkboard. This exercise reviews the three functions of noun phrases so far studied: subject, object of a verb, object of a preposition.

Items 4–8 on this page are a simple exercise in selecting the proper form of the five personal pronouns that have two forms. Children native in English should have no trouble with this in simple sentences like these. They do not have to remember rules; they need only put down what "sounds right." We want them to be aware of the choice, though, because in more complex sentences they may not be able to make it automatically.

LITERATURE – "The Doze"

Teaching Suggestions

Read, or play the recording of, this cheerful poem about a most cheerless creature. Let the children take turns reading parts of it.

SECTION 4

A Poem

Some of the best poems are written about things that don't exist at all. They are nonsense. But they are the sort of nonsense that still, somehow, means something.

Read this poem about a terribly forlorn and really hopeless creature, the Doze. Can you imagine what he looks like?

The Doze

Through Dangly Woods the aimless Doze
A-dripping and a-dribbling goes.
His company no beast enjoys.
He makes a sort of hopeless noise
Between a snuffle and a snort.
His hair is neither long nor short;
His tail gets caught on briars and bushes,
As through the undergrowth he pushes.
His ears are big, but not much use.
He lives on blackberries and juice
And anything that he can get.
His feet are clumsy, wide and wet,
Slip-slopping through the bog and heather
All in the wild and weepy weather.
His young are many and maltreat him;
But only hungry creatures eat him.
He pokes about in mossy holes,
Disturbing sleepless mice and moles,
And what he wants he never knows —
The damp, despised, and aimless Doze.

JAMES REEVES

A Writing Assignment

All you can see of the Doze in the picture above is what the artist thinks his footprints might look like. Write a paragraph telling what you think the Doze himself looks like.

The Doze

- **Aimless** means without a purpose. Who goes through dangly woods, and how does he go?
- **His company no beast enjoys** is a poetic way of saying "No beast enjoys his company." What word rhymes with **enjoys?** Do you think this has something to do with the order of the line above?

A **bog** is wet, swampy ground. **Heather** means bushes with little flowers on them.

- **Young** means the Doze's children. Look up **maltreat** and **despised** in a dictionary.

COMPOSITION—A Writing Assignment

Teaching Suggestions

The writing assignment ought to be a play of imagination. Make it clear to the class that there is no right answer. Though several physical characteristics of the Doze have been mentioned in the poem, it is still permitted to each of us to form his own notion of the completed picture. Have some class discussion as a warm-up, and then let each child describe his own Doze.

VOCABULARY AND MEANING

"The Doze"

Teaching Suggestions

Work through the questions on the poem as class work. The number of questions is limited because of the writing assignment. However, the following questions may be used if more are needed to ensure careful reading.

Additional Questions

What kind of woods does the Doze go through?

What kind of noise does he make? What noises is it like?

What is his hair like? What happens to his tail? When does it happen? What is **undergrowth**?

What is said about the Doze's ears?

What does the Doze eat?

What are his feet like?

How is the weather described?

What creatures eat the Doze?

Where does he poke about? What does he disturb? What is a **mole**?

Stops and Continuants

Consonant sounds may be either **stops** or **continuants**. A stop is a sound characterized by a complete blocking of the breath at some point in the mouth, followed by an explosion of it. (Stops are sometimes called, instead, explosives, or plosives.) There are six simple stops in English: /p/, /t/, /k/, /b/, /d/, /g/. The consonants /ch/ and /j/ are stops of a slightly different kind. They are made by stopping the breath and then releasing it with a friction sound caused by the breath passing through obstruction. The sounds /ch/ and /j/ are called affricates.

The other sounds of English—vowels and consonants—are continuants. One of their characteristics is that they can be prolonged, or continued, indefinitely, or at least as long as the breath holds out. Stops cannot be prolonged, of course.

Continuants are of different kinds. The two consonant continuants studied here—/f/ and /v/—are fricatives. In their making, the air is not completely blocked, as it is for stops, but made to pass through obstruction that gives it a friction sound. Both /f/ and /v/ are made by a contact of the upper teeth and lower lip. The breath must pass through the point of contact. The difference between them is that /f/ is voiceless and /v/ is voiced.

The Spelling of /f/ and /v/

Initial /f/ is spelled either f or **ph**. Though only a few words in the child's vocabulary are spelled **ph**, some of them are very common ones, so he must learn this spelling. All words in which /f/ is spelled **ph** are, of course, from Greek.

The spelling of final /f/ is somewhat complicated. The explanation of it is postponed to page 157.

The spelling of /v/ gives very little trouble in either initial or final position.

The Consonant Sounds /f/ and /v/

You have studied seven consonant sounds so far. Sounds like /p/ and /d/ are called **stops.** That means that we make them by completely stopping the breath at some place in the mouth.

But the first sounds of **feel** and **veal** are quite different from the consonant sounds you have studied. When we make the first sounds of **feel** and **veal,** we don't stop the breath completely. We only make it hard for the breath to get through.

We will show the first sound of **feel** this way: /f/. We will show the first sound of **veal** this way: /v/. We name them just the way we do the letters **f** and **v** of the alphabet.

Since the breath is not completely stopped when we say /f/ and /v/, we can make these sounds as long as we want. We can continue making them. Say **feel** and keep making the sound /f/ for several seconds. Say **veal** the same way.

The consonant /f/ is like /v/ the way /p/ is like /b/ and /ch/ is like /j/. But you can hear the difference between /f/ and /v/ more easily, because you can continue making the sounds. Put your hands over your ears. Say **feel, veal,** continuing the first parts for several seconds. In which word do you hear the buzzing sound? veal

The ordinary way to spell the consonant sound /f/ at the beginning of words is with the letter **f.** These words are spelled in this way:

feel	fine	fair	fool
fit	fetch	fat	free

• Can you think of five other words that begin with the sound /f/ and the letter **f?** See p. 327.

Though the ordinary way to spell the sound /f/ at the beginning of words is with the letter **f**, there is also a special way. This special way is with the letters **ph**. We spell /f/ with **ph** in a number of words that we have taken from the Greek language.

- We spell /f/ with **ph** in one word that you should know very well. This is the word **phrase**. How have you used the word **phrase?** noun phrase

Other words in which /f/ is spelled with **ph** at the beginning are these:

phone **phonograph**

- **Phone** is part of a good many words in English. Can you name some other words that have **phone** in them? telephone, phonetics, homophone
- The word **photograph** both begins with **ph** and ends with it. **Graph,** like **phone,** is part of a number of English words. Can you think of another word or two that have **graph** in them? telegraph, phonograph

When the consonant /v/ comes at the beginning of words, it is spelled with the letter **v**:

veal	**very**	**visit**	**vine**
village	**vain**	**vase**	**voyage**

- What other words can you think of in which /v/ is spelled at the beginning with the letter **v**? See p. 327

At the end of words, we spell /v/ with the two letters **ve**. We spell it that way in these words:

give	**Dave**	**move**	**love**
dive	**shove**	**grave**	**believe**

- Can you think of others? See p. 327.
- Tell what consonant sounds these words begin and end with:

five /f–v/	cheat /ch–t/	Jack /j–k/	vetch /v–ch/
vague /v–g/	charge /ch–j/	forge /f–j/	fudge /f–j/

Initially it is spelled **v**, and finally **ve**. Final /v/ is spelled **v** in only a handful of words, which normally are not in the child's vocabulary:

Slav leitmotiv spiv shiv

Teaching Suggestions

Take the pupils through the description on page 146, letting them understand what they can of the sound production. The object is not to make them phoneticians but to give them awareness of speech as distinct from writing and to alert them to differences in sound usually reflected in differences in spelling.

Let the class suggest other words that begin with /f/ spelled **f**.

Go through page 147, eliciting further examples. The children should easily think of **telephone** and may think of **homophone**.

Other words with initial **v** include **vacation, vague, valley, valuable, valve, vegetable, velvet, verb, verse, victory, view, violin, volcano, vote**.

Other words with final /v/ include **olive, wave, five, hive, drive, solve, dove, glove, curve, starve, serve**.

GRAMMAR

Noun Phrases Used as Complements

We introduce now the fourth function of noun phrases and the last that we will present in this text. There are a few others in kernel sentences, such as the use of certain noun phrases as adverbials: for example, **last night** in "He left last night." Other familiar functions, like indirect object and object complement, occur in transforms.

We define **complement** as simply the function that a noun phrase has when it occurs in the predicate immediately after a form of **be**. In "It is a plane," **a plane** is a noun phrase functioning as a complement. As a matter of fact, noun phrases may also function as complements after a very few verbs, particularly **become** and **remain**:

He became a doctor.
They remained friends.

This point is not raised here, however.

Usage

It is no object of this text to promote a puristic or precious kind of speech and writing or to fuss about small points of usage. We are not much concerned with how children talk in their informal moments. However, we do have the obligation of making pupils aware of some of the conventions of formal speech and edited writing—even some of the conventions which are not always, perhaps not even usually, observed.

It is considered one of the marks of pedantry to say that one must say "It is I" and not "It is me." Not wishing to be thought pedantic, we do not say this. We do say, however, that people sometimes say "It is I" and often write it and that it is not incorrect to do so. This simple sentence, of course, does not often occur in writing, the occasion for writing it seldom arising, but the construction is common enough in complex sentences:

Noun Phrases Used as Complements

• You have studied three functions of noun phrases. One is the **subject** function. What are the other two? obj. v. obj. prep.

• Tell what the noun phrases in these sentences are, and what their functions are. Some sentences have two noun phrases, and some have one.

1. Casey is brave. sub.; (3)
2. He kissed his wife. sub., obj. v.; (4), (1)
3. The train raced. sub.; (1)
4. The fireman shoveled coal. sub., obj. v.; (1), (2)
5. Casey was near him. sub., obj. prep.; (3), (4)
6. We saw the engineer. sub., obj. v.; (4), (1)
7. He saw us. sub., obj. v.; (4), (4)
8. The fireman talked to him. sub., obj. prep.; (1), (4)
9. They came to Reno Hill. sub., obj. prep.; (4), (3)

• You know about three **functions** of noun phrases. You also know about four **kinds** of noun phrases: (1) a determiner and a common noun; (2) a common noun alone; (3) a proper noun; (4) a personal pronoun. Go through the sentences above again. Give each noun phrase again, and tell what kind it is. (See above.)

• You know that three things may occur in the predicate after a form of **be:** a noun phrase, an adjective, and an adverbial of place. Tell what comes after the form of **be** in each of these sentences:

a. They are hungry. adj.
b. It is a plane. n.p.
c. I am at my desk. adv. of place

When you found the noun phrase **a plane** after **is,** you discovered another function of the noun phrase. You knew that a noun phrase could follow a form of **be,** but up to now you have not thought of this as a function with a name of its own.

We call the function of a noun phrase that follows a form of **be** the **complement** function. In "It is a plane," **a plane** functions as complement.

• You know that most noun phrases have the same form in all their functions. What are the functions of **a plane** in these sentences?

- **d. A plane landed.** sub.
- **e. I saw a plane.** obj. v.
- **f. Jack is in the plane.** obj. prep.
- **g. It is a plane.** comp.

• Find the noun phrases in these sentences and tell what their functions are:

10. The man is an engineer. sub., comp.
11. They are workmen. sub., comp.
12. The girl is a musician. sub., comp.
13. We are the students. sub., comp.
14. You are my cousin. sub., comp.
15. I am a catcher. sub., comp.

• Five of the seven personal pronouns have two forms: **I, me; he, him; she, her; we, us; they, them.** Which forms are used when these words are objects?

When these words are complements, we sometimes use one form and sometimes the other. When we are talking, we often use the object forms as complements. But in writing, we usually use the subject forms as complements. Practice using the subject forms after **be** by saying these sentences aloud:

16. It is I.
17. It is he.
18. It is she.
19. It is we.
20. It is they.

149

It is I who will have to shoulder the responsibility.

It was she who gave the note to Mrs. Bixby.

The child is already skilled at saying "It's me" and "It was her," so we don't have to give him more practice. Instead, we introduce the more literary form and thus extend his education.

Teaching Suggestions

Review the object forms of the personal pronouns by having the class name them.

Let the pupils parse Sentences 1–9 as directed, pointing out the noun phrases and naming their functions. It is probably too complicated to have them also give the type of noun phrase – determiner + noun, proper noun, etc.

Review the three structures given as possibilities after a form of **be**.

Work carefully through the development at the top of page 149 with the class and do Sentences 10–15 as class work. Then develop the idea that the subject forms of the personal pronouns are the ones we use in writing, and carry on the practice at the bottom of page 149 orally.

Extra Practice

Pupils may complete the sentences below by using the correct forms of the personal pronouns **I, he, she, we,** and **they**, in this order, in the blanks.

1. Who fed the cat? It was ____ who fed the cat.
2. Who played baseball? It was ____ who played baseball.
3. Who lost this doll? It was ____ who lost the doll.
4. Who ate the cookies? It was ____ who ate the cookies.
5. Who ran away? It was ____ who ran away.

TESTS AND REVIEW

Teaching Suggestions

Use these tests for a check on, and possible reassessment of, the progress of the individual children. Keep a record of their performance so that you are able to judge strengths and weaknesses. Reteach as necessary.

Use the questions on page 153 for oral discussion.

Extra Test

Dictate these words and have the class spell them. Say the word clearly. Use it in the sentence. Say it again and have the children write it.

1. **judge** The **judge** was in court. **judge**
2. **rage** A **rage** is a fit of anger. **rage**
3. **gentle** Be **gentle** in handling kittens. **gentle**
4. **jingle** We heard the bells **jingle**. **jingle**
5. **stretch** See him yawn and **stretch**. **stretch**
6. **peach** A **peach** is a kind of fruit. **peach**
7. **phone** A **phone** is a telephone. **phone**
8. **flag** Our **flag** has fifty stars. **flag**
9. **vague** A **vague** answer is not clear. **vague**
10. **brave** A **brave** person has courage. **brave**

Tests and Review

□ TEST I Sounds and Letters

1. Copy each word below. After it, write between slanting lines the consonant sound it begins with.

If you make mistakes, study pages 81, 97, 114, 115, 125, 132, 140, 146.

keen /k/	judge /j/	gentle /j/	bed /b/
very /v/	come /k/	grave /g/	food /f/
phone /f/	child /ch/	jingle /j/	tired /t/

2. Copy each word below. After it, write between slanting lines the consonant sound it ends with.

If you make mistakes, study pages 81, 86, 102, 114, 115, 125, 132, 140, 146.

teach /ch/	give /v/	judge /j/	rat /t/
robe /b/	sleep /p/	ripe /p/	stretch /ch/
hide /d/	vague /g/	age /j/	leg /g/
mitt /t/	egg /g/	like /k/	lick /k/

3. Copy each word below. After each word, write between slanting lines the vowel sound it has.

If you make mistakes, study pages 74–75, 108.

ride /ī/	big /i/	give /i/	beach /ē/
judge /u/	rope /ō/	duke /ū/	duck /u/
rob /o/	Ted /e/	rake /ā/	book /o͝o/
sack /a/	boat /ō/	food /ū/	bright /ī/

4. Replace the **C** in **VCe** with the consonant letter v. Replace the **V** with any vowel letter or letters you choose. Write at least five words that have the **VCe** pattern. five, leave, wave, hive, save, etc.

Now replace **C** with g, and **V** with a vowel letter, and write one word with the **VCe** pattern. cage, etc.

If you make mistakes, study pages 86, 141, 147.

□ TEST II Grammar

1. Write the object form of each of these personal pronouns.

If you make mistakes, study pages 134–35.

he	I	she	they	we
him	me	her	them	us

2. Copy each sentence below. Where a word is left out, use **I, me, they,** or **them** in the sentence. Tell what the function of each word you put in is by writing **sub.** for "subject," **obj. v.** for "object of verb," or **obj. prep.** for "object of the preposition."

If you make mistakes, study pages 134–35.

a. ___They___ are beside me. subject
b. I am beside ___them___. obj. prep.
c. They like ___me___. obj. v.
d. ___I___ am their friend. sub.

3. Write these headings on your paper: det. + n. (for determiner plus common noun), n. (for common noun alone), p. n. (for proper noun), pers. (for personal pronoun). Find the noun phrases in the sentences below. Write each one under the heading that shows what kind it is.

If you make mistakes, study pages 14, 70, 77, 88.

a. He mows the lawn. pers.; det. + n.
b. We watch them. pers.; pers.
c. His teacher is Miss Woods. det. + n.; p.n.
d. Denver is in Colorado. p.n.; p.n.
e. I am hungry. pers.
f. I told Mr. Burbank. pers.; p.n.
g. They like potatoes. pers.; n.
h. Milk is white. n.
i. The girl erased the chalkboard. det. + n.; det. + n.
j. These animals are dozes. det. + n.; n.

151

TEST II Grammar (CONTINUED)

4. Write these headings on your paper: **subj.** (for subject), **obj. v.** (for object of a verb), **obj. prep.** (for object of a preposition). Find the noun phrases in the sentences below. Write each one under the heading that tells what its function is.

If you make mistakes, study pages 69, 70, 77, 82, 88, 116, 126, 133, 148.

a. He is at his desk. subj.; obj. prep.
b. A monkey likes them. subj.; obj. v.
c. A roe lives in the woods. subj.; obj. prep.
d. Mr. Graves is in the car. subj.; obj. prep.
e. Casey blew the whistle. subj.; obj. v.
f. It is beside her. subj.; obj. prep.
g. She is cold. subj.
h. Jane is near me. subj.; obj. prep.
i. You are behind that tree. subj.; obj. prep.

5. Write these headings on your paper: **n. p.** (for noun phrase), **adj.** (for adjective), **adv.** (for adverbial of place). Look at what comes after a form of **be** in the sentences below. Write each under the heading that tells what it is.

If you make mistakes, study pages 20, 26, 110, 116.

a. Jane is happy. adj.
b. She is by the chalkboard. adv.
c. Her teacher is Mrs. Lane. n.p.
d. They are children. n.p.
e. My mother is upstairs. adv.
f. I am sleepy. adj.
g. The Doze is wet. adj.
h. Tom is on the playground. adv.
i. They are toys. n.p.

152

□ **TEST III Writing and Vocabulary**

Write five sentences. In each sentence, use one of these words.

1. valor 2. torrent 3. fame 4. fireman 5. maltreat

● **REVIEW Ideas and Information**

1. Who wrote the poem called "My Heart's in the Highlands"? Burns
2. What kind of engine did Casey Jones' train have? steam
3. Which two of these four consonant sounds are made with the little buzzing noise in your throat: /p/, /b/, /t/, /d/? /b/, /d/
4. When do we spell the sound /j/ with the letters **dge**? after simple vowels
5. How do we spell the sound /v/ at the end of a word? ve
6. What does **V** stand for in the **VCe** pattern? What does **C** stand for? What does **e** stand for? vowel; consonent; e
7. If you replace **C** in **VCe** with **t**, what sound does **V** stand for, a complex or a simple vowel sound? complex
8. One way to spell /f/ at the beginning of a word is with the letter **f**. What is another way? ph
9. Name the personal pronouns that have special forms when they are used as objects. I, he, she, we, they
10. What do we call the function of the noun phrase **my teacher** in "She is behind my teacher"? obj. prep.
11. What do you call the function of the noun phrase **my teacher** in "She called my teacher"? obj. v.
12. What do we call adverbials of place like **in the house** and **on the playground**? prep. phrases
13. Tell what word in this sentence is a preposition: **They are in a car.** in

PART
6

SECTION 1

A Poem

Have you ever wondered why some streets and roads are crooked, why people didn't make them straight when they built them? Here is a poem by S. W. Foss about streets like this. In some cities such roads are called "cow paths," but this poet calls them "calf paths."

This is a long poem, so we will study it in three parts. Read this part carefully. Your teacher may read it aloud, or play the recording of it.

The Calf Path

One day, through the primeval wood,
A calf walked home as good calves should;
But made a trail all bent askew,
A crooked trail as all calves do.

Since then two hundred years have fled,
And I infer the calf is dead.
But still he left behind his trail,
And thereby hangs my moral tale.

The trail was taken up next day
By a lone dog that passed that way;
And then a wise bell-wether sheep
Pursued the trail o'er vale and steep,
And drew the flock behind him, too,
As good bell-wethers always do.

LITERATURE

"The Calf Path" (Stanzas One to Three)

This longish poem is divided among the first three sections of Part 6. Read it or play the recording of it. Let the children take turns reading parts of it.

Continue encouragement of memorization, within the capabilities of the pupils. The children should not of course be given oppressive memorization assignments but should be required to contribute some effort to the exercise.

Do not neglect copying as an exercise to be assigned sometimes to part of the class when others are engaged in some other activity—as in reading groups. Make it clear that the aim in copying is accuracy. Encourage the children to try to copy stanzas or paragraphs without any mistakes at all.

VOCABULARY AND MEANING

"The Calf Path" (Stanzas One to Three)

Work through the questions, eliciting answers from the children. Let them check constantly with the text. The aim here is not a check on memory but an exercise in close reading.

If the class is able to tell from the context what **askew** means, there is no need to use the dictionary. Turn to the dictionary if doubt or disagreement remains, or for further clarification, but do not insist on automatic recourse to the dictionary for every new word. We learn the meanings of most words not from the dictionary but from the contexts in which they are used.

Infer and **imply** are among the pairs of words commonly confused. We try here to give the children a good start on the use of **infer**.

The Calf Path (STANZAS ONE TO THREE)

- The word **primeval** means very old. **Primeval wood** means a woods, or forest, of long ago when nothing had been built or changed there by people. In this woods, no trees had been cut down and no camp sites had been built. What walked through the primeval wood? Where was it going? a calf; home
- Look at the word **askew** in the third line. You may never have seen this word before, but you can probably guess its meaning from the other words that describe the trail. Check your guess by looking up **askew** in a dictionary.
- How many years have gone by since the calf walked home through the primeval wood? What verb in the second stanza means gone by? 200 + ; have fled
- To **infer** means to find out by thinking or reasoning. You may have inferred what **askew** meant from the words "crooked trail" and "bent" in the same sentence. What would make the poet infer that the calf is dead? Do you think the poet really had to do much thinking to make this **inference**? discuss
- What did the calf leave behind him? his trail
- The trail was **taken up** next day by a dog. What does **taken up** mean? Was there one dog or more than one? followed; one
- A **bell-wether** sheep is a sheep that wears a bell and leads the other sheep. Where did he lead them? The poet doesn't say "followed" the trail. What does he say instead? along the trail; pursued
- A **vale** is a valley. What do you think **steep** means in this line? What does **drew** mean? Look up **steep** and **drew** in a dictionary. Check dictionary and discuss.

156

The Consonant /f/ at the End of Words

We spell the sound /f/ at the end of words in several different ways.

1. When /f/ comes after the simple vowels /i/, /a/, or /u/, we spell it with two f's. Tell what vowel sound comes before /f/ in each of these words: /a/, /i/, /u/, /i/, /u/, /a/

staff stiff stuff cliff puff gaff

2. We also spell /f/ with two f's in **off, scoff, doff.** These words have a vowel sound you have not studied yet, even though they contain the letter **o.**

3. When /f/ comes after a complex vowel, it may be spelled in the pattern **VCe.** Tell what the vowel sound is in each word: /ī/, /ā/, /ī/, /ī/, /ā/, /ī/

life safe knife wife chafe strife

4. When /f/ comes after a consonant letter or a vowel that is spelled with two letters, we usually spell /f/ with just the letter **f:**

loaf	**beef**	**brief**	**deaf**
self	**shelf**	**gulf**	**dwarf**

Can you think of other words like these? See p. 327.

5. Some words have very special ways of spelling the sound /f/ at the end.

a. One word spells it **ffe: giraffe.**

b. How is /f/ spelled at the beginning of **phone?** ph In some words /f/ is spelled this way at the end:

paragraph telegraph phonograph

c. Five common words have /f/ spelled at the end with the letters **gh:**

laugh rough cough tough enough

6. See how final /f/ is spelled in each of these words. Then write the words from dictation.

staff life beef giraffe graph enough

157

SOUNDS AND LETTERS—Final /f/

The rules for spelling /f/ in final position are analogous to those for other consonant sounds, though a little more detailed. The ordered set of rules would look something like this:

1. Spell /f/ **ffe** in **giraffe**.
2. Spell /f/ **ph** in **paragraph, telegraph**, etc. (the words from Greek).
3. Spell /f/ **gh** in **laugh, rough, cough, tough, enough**.
4. Spell /f/ **ff** after simple vowels and in **off, scoff, doff**.
5. Spell /f/ **fe** after complex vowels in words that use the **VCe** pattern.
6. Otherwise spell /f/ as **f**.

Each rule excludes those before it. Thus rule 4 does not apply to **giraffe, paragraph**, and **laugh**. If it did, it would produce **giraff, paragraff**, and **laff**. But these words have been removed from consideration by rules 1, 2, and 3.

Teaching Suggestions

Go through the presentation, asking for other examples for the sets for which other examples exist. Here are some additional examples for these sets:

1. **ph:** mimeograph, autograph, triumph, nymph
2. **ff:** chaff, whiff, cuff, bluff, sniff
3. **fe:** fife, rife
4. **f:** chief, belief, leaf, hoof, scarf, turf

Fife and **rife** added to **life, safe, knife, wife, chafe**, and **strife** complete the set of **VCe** words with /f/, except for compounds like **housewife**.

GRAMMAR—Indefinite Pronouns

With the addition of indefinite pronouns, we complete the rule for noun phrase:

$$\text{NP} \rightarrow \left\{ \begin{array}{l} \text{determiner} + \text{common noun} \\ \text{proper noun} \\ \text{common noun alone} \\ \text{personal pronoun} \\ \text{indefinite pronoun} \end{array} \right\}$$

Any noun phrase in a kernel sentence can be classified as one of these. Indeed, as has been explained, we simplify the classification in grade five and beyond by introducing the concept of the null nondefinite article, which then permits us to treat "common noun alone" as a variety of "determiner + common noun."

Definition of Indefinite Pronouns

At the bottom of page 158, we speak of the indefiniteness of indefinite pronouns. The intent, however, is to justify the traditional term rather than to establish a definition. If the child were to try to find indefinite pronouns by seeking for noun phrases that seem indefinite, he would never get anywhere.

We define **indefinite pronoun** much more simply. Indefinite pronouns are just the words that begin with **every, some, any,** or **no** and end with **one, body,** or **thing**. Multiplication gives the twelve words listed at the top of page 159.

Traditionally, the class of indefinite pronouns has often been made to include determiners used without a following common noun, like the subjects of these sentences:

Some liked it.
Each did his best.

However, these are much more simply treated as deletion constructions. The subject is a determiner with its noun omitted. Thus the sentences above would be considered transforms of sentences like these:

Indefinite Pronouns

You know four kinds of noun phrases. A determiner and a common noun is one kind. A common noun alone is another kind. What are the other two kinds? p.n. pers

You know four functions, or uses, of noun phrases. Complement is one use. Object of a preposition is another. What are the other two? subj., obj. v.

A noun phrase is a complement when it comes after a form of what word? be

• Find the noun phrases in these sentences. Tell what kind each one is, and tell what the function of each one is.

1. The girls are the winners. det. + n., subj.; det. + n., com
2. The caller calls Casey. det. + n., subj.; p.n., obj. v.
3. It is he. pers., subj.; pers., comp.
4. Some people are in the pool. det. + n., subj.; det. + n., obj.
5. Mr. Burbank caught me. p.n., subj.; pers., obj. v.
6. She likes peanuts. pers., subj.; n., obj. v.
7. Laughter delights them. n., subj.; pers., obj. v.
8. It is they. pers., subj.; pers., comp.
9. Mrs. White is the driver. p.n., subj.; det. + n., comp.
10. A duck is on the pond. det. + n., subj.; det. + n., obj. pre

We are now going to study one more kind of noun phrase. This is a word such as **everyone** or **somebody** or **nothing.** Words like these are called **indefinite pronouns.**

You can probably see why indefinite pronouns are called that. If you say "He is in the yard," you mean some particular or definite person. But if you say "Someone is in the yard," you don't mean a definite person. You really mean that you do not know who it is that is in the yard. So the subject of your sentence, which is a pronoun, is indefinite.

There are twelve indefinite pronouns in English. Here is the list of them:

everyone	everybody	everything
someone	somebody	something
anyone	anybody	anything
no one	nobody	nothing

• You shouldn't find it hard to remember these. Each one is made up of two parts. The first part may be **every–**. What else may it be? some, any, no

• The last part may be **–thing.** What else may it be? –one, –body

• All of the indefinite pronouns are spelled as single words except one. Which one is spelled as two words? no one

The indefinite pronouns can be used in all the functions you have studied for other noun phrases. Here are some examples:

Subject: **Everyone** is happy.
Object of a Verb: He heard **something.**
Object of a Preposition: It is near **someone.**
Complement: It is **nothing.**

• Find the indefinite pronouns in these sentences. Tell what the function of each one is.

11. Nothing happened. subj.
12. He likes everybody. obj. v.
13. Anyone can play. subj.
14. It is inside something. obj. prep.
15. He is somebody. comp.
16. No one told us. subj.
17. They know nobody. obj. v.
18. It is near someone. obj. prep.
19. He told everyone. obj. v.
20. Everything depends on you. subj.

159

Some parents liked it.
Each boy did his best.

In very traditional terminology, the common noun is "understood" in the transform.

Indefinite pronouns with **any** occur in some kernel sentences, such as "Anything may happen." However, they are much more common as replacements of **some** words in question and negative transforms:

He saw something.
Did he see anything?
He saw someone.
He didn't see anyone.

Teaching Suggestions

Preparatory to introducing indefinite pronouns, work through Sentences 1–10 on page 158 to make sure that the children can recognize the other types.

Extra Practice

Sentences 11–20 on page 159 may be used for written work by having the children write each indefinite pronoun and tell its function by putting after it (1) for subject, (2) for object of a verb, (3) for object of a preposition, or (4) for complement.

SECTION 2

More of the Poem

As you remember, the first creature to wander across the countryside and through the primeval woods was a calf, which left a crooked trail. The next one to come along was a dog. Other animals followed the way the calf and dog had gone. As time went on, the trail became easier and easier to follow.

Here is the second part of the poem. Read it carefully. Your teacher may read it aloud, or play the recording of it.

And from that day o'er hill and glade,
Through these old woods a path was made;
And many men wound in and out,
And dodged and turned and bent about,
And uttered words of righteous wrath
Because 'twas such a crooked path.
But still they followed — do not laugh —
The first migrations of that calf,
And through this winding wood-way stalked
Because he wobbled when he walked.

This forest path became a lane
That bent, and turned, and turned again;
This crooked lane became a road,
Where many a poor horse with his load
Toiled on beneath the burning sun,
And traveled some three miles in one.
And thus a century and a half
They trod the footsteps of that calf.

LITERATURE

"The Calf Path" (Stanzas Four and Five)

Reread or replay the two stanzas printed on page 155. Discuss briefly what happens in them. Then play or read these next two stanzas. Give the children practice in reading aloud.

VOCABULARY AND MEANING

"The Calf Path" (Stanzas Four and Five)

As the illustration shows, we presume the path to have reached the development here described sometime in the nineteenth century.

Teaching Suggestions

Go through the questions, letting the children find the answers in the poem. Promote use of the dictionary as suggested.

Note the rhyme of **lane** and **again** in the first couplet of the second stanza on page 161 (Stanza Five). Recall to the children the occurrence of the word in "The Story of Johnny Head-in-Air" (page 107). There **again** rhymes with **men**, and so has to be pronounced with the vowel sound /e/. Here it rhymes with **lane** and so must have the vowel sound /ā/. Explain that some words have more than one pronunciation. **Again** is usually pronounced with /e/ but sometimes, by some people, with /ā/.

The Calf Path (STANZAS FOUR AND FIVE)

- The word **o'er** in the first line is a poet's way of contracting **over.** What letter is left out? What is the name of the mark that shows a letter is left out? Find **o'er** in a dictionary. v; apostrophe
- A **glade** is an open place in a woods. Who helped to make the path now? men
- **Righteous** means just. Sometimes we use this word for people who are too proud about being just. The word **wrath** means anger. What do you think the word **uttered** means? See a dictionary. Why did the men utter words of righteous wrath? discuss
- **Migrations** means movements. Who made these first migrations? Who followed the migrations made long ago? calf; many men
- **Stalked** means walked. A **wood-way** is a woods road. The last line of the stanza tells a reason the path was crooked. What is the reason? the calf wobbled
- What did the forest path become? What did the lane do? Did it follow the same trail that was made by the calf? Then what did the lane turn into? See poem.
- What does the verb **toil** mean? The poem says that many a poor horse toiled beneath the **burning sun.** Do you think the road still led through a primeval forest? work; no
- The third line from the end means that they had to walk three miles to go one. Why did they have to do that? discuss
- A **century** is a hundred years. How many years is a century and a half? **Trod** means walked. Whose footsteps did they walk or follow? 150; that calf
- Find the words in these two stanzas that rhyme with **calf.** What vowel sound do they all have? What consonant sound do they end with? laugh, half; /a/; /f/

How Words Are Made

You remember that we add **er** to verbs such as **call** in order to make nouns like **caller:**

A person who **calls** is a **caller.**

A part such as **er** that we put on the end of one word to make another word is called a **suffix.**

□ Find the verb in the first sentence in each pair of sentences. Make a noun from it by adding the suffix **er** to the simple form of the verb. Then copy the second sentence and put in the noun.

1. a. The man works in a store.
   b. This —— is a clerk. worker
2. a. His father hunts deer.
   b. That —— is his father. hunter
3. a. He pitches for our team.
   b. Our —— is Jerry Mack. pitcher

Many verbs that we might add **er** to, like **write,** already end in **e.** To make nouns with the suffix **er** from these verbs, just add **r** to the simple form.

He writes. He is a writer.

□ Find the verb in the first sentence, make a noun from it, and use the noun in the second sentence in the pair.

4. a. He bakes bread.
   b. He is a ——. baker
5. a. She drives the car.
   b. The —— is a woman. driver
6. a. That company makes bicycles.
   b. The —— guarantees each bicycle. maker
7. a. That man mines coal.
   b. That —— works in a coal mine. miner
8. a. She reads poems.
   b. The —— is a girl. reader

We give here a little more practice with the morpheme er, still, however, more as an interesting sidelight than as part of a central study, which it will become in grade five. We also introduce the term **suffix.**

Suffixes, Morphemes, and Syllables

Though the term **morpheme** is not used in this volume, and **suffix** and **syllable** only briefly, the teacher may wish to know something about the application that will be given to the terms later.

A **syllable** is a word or part of a word centering on a vowel sound. A word contains as many syllables as it has vowel sounds. **Water** has two syllables, **recalcitrant** four. Sometimes two words may be expressed in a single syllable, as in **I'm** or **he'll.**

A **suffix** is an ending added to a base form. Thus in **caller,** the suffix **er** is added to the base form **call.** A suffix may be a single syllable, as in **caller.** It also may be more than one syllable. In **alteration** the suffix is **ation** added to the base form **alter.** Sometimes a suffix does not form an extra syllable. **Boys** consists of the base form **boy** + the suffix **s,** but **boys** is a one-syllable word.

A **morpheme** is a unit of meaning. It may be a base form or a suffix or a prefix or something else altogether. It may be a word or a part of a word. Thus **boy** is a word and also a single morpheme. **Boyish** is two morphemes – boy + ish, one morpheme being a base form and one a suffix. Here the second morpheme is also a suffix and also a syllable. **Boys** consists of boy + plural. Here the second morpheme is represented by a suffix (s), which however is not a syllable. But **men** also consists of two morphemes: man + plural. And here the second morpheme is not represented by a suffix, syllabic or otherwise, but by a change wrought on the base word **man.**

SOUNDS AND LETTERS—/s/

The consonants /f/ and /v/ completed the review of consonant sounds studied in grade three, with the exception of /r/, which is introduced on page 192. We go on here to /s/ and, on page 171, to /z/. Both /s/ and /z/ are important in the plurals of nouns, which are to be introduced presently.

The /s/ sound not only occurs in a great many words before vowels but also heads a large number of initial consonant clusters—groups of two or three consonants occurring before a vowel. All English initial clusters consisting of three consonant sounds begin with /s/: /spr/ **spree**, /spl/ **splice**, /str/ **strew**, /skr/ **scream**, /skl/ **sclerosis**, /skw/ **squirm**.

The consonant /s/ is overwhelmingly spelled **s** at the beginning of words. The largest group of exceptions is the set of words in which /s/ is spelled **c** before **e** or **i**. Since **s** may also be used to spell the sound before **e** or **i**, there is no rule, and the **ce** and **ci** words have to be learned. The more common are given on this page.

The letter **c**, which usually represents /k/, does not lap over into /k/ area here, because /k/ is always represented by **k** before **e** and **i**. So the confusion is only between **s** and **c**:

/k/: keen, king

/s/: { seem, sing; cent, cinch }

Just to keep the speller of English on his toes, we use **sc** for /s/ before **e** and **i** in some words from Greek. The following, with their derivatives, are the more common ones: **scene, scent, scepter, science, scimitar, scintillate, scion, scissors**. Three or four of these are in the vocabulary of the fourth-grader and may need to be pointed out. In one word, **sc** stands for /sk/ before **e: sceptic**. And in **schism**, /s/ is represented by **sch**. The word is pronounced /sízəm/.

The Consonant Sound /s/

• Say these words to yourself and listen to the consonant sound at the beginning of each word:

seat **sing** **Sam**

We show this consonant with the letter **s** between slanting lines this way: /s/. We name it the way we do the letter **s**.

At the beginning of words, the usual way to write /s/ is with the letter **s**. We write it that way in these words:

sing	**say**	**some**	**seem**
so	**sir**	**slow**	**sleep**

• Think of five other words that begin with the sound /s/ and the letter **s**. See p. 327.

• What sound does each of these words begin with?

cat **cot** **cut** **clean** **cream** /k/

• You already know that the letter **c** stands for the sound /k/ at the beginning of words when the second letter is not **e** or **i**. Tell what the second letter is in each word above. What letter stands for the sound /k/ at the beginning when the second letter is **e** or **i**? a, o, u, l, r; k

• Although **c** at the beginning never stands for /k/ when the second letter is **e** or **i**, it often stands for the sound /s/ in such words. Here are some of the important words that begin with the sound /s/ and the letter **c**. Tell what the second letter is in each of these words.

cellar e	celery e	ceiling e	cent e
center e	central e	century e	cereal e
cinch i	cipher i	circle i	circus i

□ After you have studied the spelling of these words, close your book and write them from dictation.

The letter **c** stands for the sound /s/ in many words when the second letter is **e** or **i**, as you have seen. But you cannot be sure that such words begin with **c**. A great many of them begin with the letter **s**, even though the second letter is **e** or **i**. Here are some examples:

seem **seat** **silly** **sit**

- Think of some other words like these. See p. 327.
- Since both s and **c** can stand for the sound /s/ before **e** and **i**, we can have homophones. Can you think of a homophone for **cent?** How do you spell it? Use each of the homophones in a sentence. sent

What is a **cellar?** Can you think of a homophone for **cellar?** It is spelled differently in two places. seller

- When /s/ comes at the end of a word, we usually spell it with **ss** when a simple vowel comes before it. We spell it that way in these words. Tell what the vowel sound is in each one:

glass /a/	**mess** /e/	**kiss** /i/	**fuss** /u/
dress /e/	**miss** /i/	**muss** /u/	**less** /e/

- In a few words we spell /s/ with a single **s** even when a simple vowel comes before it. What are the vowel sounds in these words?

gas /a/ **bus** /u/ **plus** /u/

- You see that how we spell consonant sounds often depends on the vowel sound that comes before them. Tell what consonant sound each of these words ends with. Tell what the vowel sound is.

staff /a-f/	pass /a-s/	ridge /i-j/	luck /u-k/
match /a-ch/	stiff /i-f/	neck /e-k/	hedge /e-j/
cuff /u-f/	stitch /i-ch/	hutch /u-ch/	dodge /o-j/

- Tell how you spell each of these consonant sounds when a simple vowel comes before it:

/s/	/f/	/ch/	/k/	/j/
ss	ff	tch	ck	dge

The spelling of initial /s/, though somewhat erratic, is not hard. One simply learns the exceptions (which aren't numerous—most of them are listed on page 164) and uses **s** in all other cases. Final /s/ is more complicated.

Teaching Suggestions

After presenting initial /s/ on page 164 and the top of page 165, we begin just with final /s/ after simple vowels in this lesson. There is more about final /s/ on pages 176–77.

Show that the regular way to spell /s/ after a simple vowel is with ss. **Bus, plus**, and **gas** are exceptions to this rule, and must be learned separately.

Ask the class to identify the simple vowels in the list beginning **glass**. Use these further examples if desired: **pass, mass, brass, chess, confess, guess, hiss, Swiss, discuss**. A very large number of words end with this spelling if we count the compounds in **–less** and **–ness**, like **helpless** and **kindness**.

The following words also end in **ss: boss, floss, gloss, moss, cross, across, dross, toss**. For most speakers the vowel sound in these is /au/ as in **haul** or **dawn**.

The five consonant sounds listed at the bottom of the page are spelled in analogous ways when preceded by simple vowels. Elicit examples from the class, and on the chalkboard, build up lists like these:

/s/	/f/	/ch/
glass	staff	batch
dress	sniff	itch
fuss	stuff	hutch

/k/	/j/
sack	badge
neck	bridge
knock	lodge

GRAMMAR—Plurals of Nouns

The pluralization of nouns has already been treated in the book for grade three. This and the lessons following are a fast review of the matter.

The Sounds and Spelling of Noun Plurals

We can consider the pluralization of nouns from two connected but different points of view: the way you say it, and the way you write it. From the point of view of sound—speaking—the rule for making the plurals of regular nouns is as follows:

1. Add the syllable /əz/ if the singular ends in one of the following sounds: /s/ **glasses**, /z/ **quizzes**, /sh/ **dishes**, /zh/ **garages**, /ch/ **churches**, /j/ **judges**.
2. When the singular ends in a voiceless sound other than those included above, add /s/: **tops, cats, sticks, laughs, truths**.
3. In all other cases, add /z/: **ribs, kids, bags, lives, bums, cans, tongs, cars, girls, toes, flies, bees**.

This rule applies automatically for anyone who speaks English natively. He doesn't even have to think about it. Indeed, he is quite unlikely to be aware that the plurals under rule 3 end in the sound /z/. They do, though.

The spelling rule for the plurals of regular nouns is also threefold, but as will be seen, it doesn't meet the pronunciation rule point for point.

1. If the singular ends in the letter **y** preceded by a single consonant, change the **y** to **i** and add **es: ladies, babies, worries**. (But **valleys, guys, trays**.)
2. If the singular ends in **s, z, sh, ch, x**, add the letters **es: kisses, buzzes, dishes, foxes**.
3. In all other cases, add **s**.

Plurals of Nouns

Most nouns have ways of showing the meanings "one of somebody or something" and "more than one of somebody or something." For instance, the noun phrase "the boy" means just one boy, but "the boys" means more than one boy.

• 1. Which of these noun phrases mean just one? Which mean more than one?

the dog	some birds
our books	our teacher

A noun in a noun phrase that means just one is called a **singular** noun. You know the word **single,** which is another way of saying one.

A noun in a noun phrase that means more than one is called a **plural noun.**

• 2. Tell which of these noun phrases have singular nouns and which have plural nouns:

his mother	sing.	his brothers	pl.
a canary	sing.	some friends	pl.
the trains	pl.	the jet	sing.
this poem	sing.	those nymphs	pl.
her toys	pl.	their house	sing.

• 3. Find the nouns in the noun phrases of the sentences below. Tell whether each noun is singular or plural.

a. The boy laughs. sing.
b. The girls laugh. pl.
c. The calf makes a path. sing.; sing
d. These cars are fast. pl.
e. He needs some pencils. pl.
f. The rain helps the flowers. sing.; pl.

The most usual way to write the plural of a noun is to add the letter s to the singular form.

□ 4. Write the plural form of these singular nouns:

pen s	stick s	flower s	house s
cat s	dog s	lion s	giraffe s

Sometimes we write the plural of nouns in slightly different ways. Look at these singular and plural nouns:

match / matches **crutch / crutches**
kiss / kisses **mess / messes**

● 5. What do we add to the singular forms of these nouns to write the plural? What sound do **match** and **crutch** end with? What letters do they end with? What sound do **kiss** and **mess** end with? what letters? es; /ch/; tch; /s/, ss

> When the singular of a word ends in **tch** or **ch** or **ss**, we add the letters **es** to write the plural.

□ 6. Write the plural form of each of these nouns:

ditch es	neck s	mass es	bridge s
church es	hiss es	beach es	inch es

● 7. Find the nouns in the noun phrases of the sentences below. Tell whether each one is singular or plural.

a. The beaches are sandy. pl.
b. They crossed the bridge. sing.
c. The church is behind the ditches. sing.; pl.
d. Nobody heard the hisses. pl.
e. The giraffe chased the dogs. sing.; pl.
f. Some girls are on the cliffs. pl.; pl.
g. The workers had passes. pl.; pl.
h. She likes peanuts. pl.
i. The schools educate us. pl.
j. That man is on crutches. sing.; pl.
k. The horses run in races. pl.; pl.
l. The boys broke the glasses. pl.; pl.

167

It will be seen by rule 3 that the letter s sometimes stands for the sound /s/, as in **tops, sets, rocks**. But when it represents the plural morpheme, it more often stands for the sound /z/, as in **ribs, kids, bags, bums, cans, gangs, rows, pies, trays**.

Many words which come under part 3 of the spelling rule come under part 1 of the pronunciation rule. For example, **rose**, which ends in the sound /z/, adds the syllable /əz/ to form the plural: /rōz/ → /rōzəz/. **Judge**, which ends in /j/, also adds /əz/: /juj/ → /jujəz/. But since these already end, in the spelling, with the letter **e**, they merely add **s**, by the third part of the spelling rule in the writing of the plural: **roses, judges**.

Many items of rule 1 of the pronunciation rule correspond, however, to items governed by rule 2 of the spelling rule. For these, the letters **es** added to the singular represent the syllable /əz/: **kisses, churches, dishes**, etc.

Teaching Suggestions

Go through page 166 carefully to establish the terms **singular** and **plural**. **Singular** applies both to a noun phrase and to a noun that means "one," **plural** to a noun phrase or a noun that means more than one. Of course, only countable nouns can have both singular and plural forms. Noncountable nouns are just singular.

Extra Practice

The pupils may do the last exercise on page 167 as written work, putting **S** (for **singular** or **P** (for **plural**) after each noun.

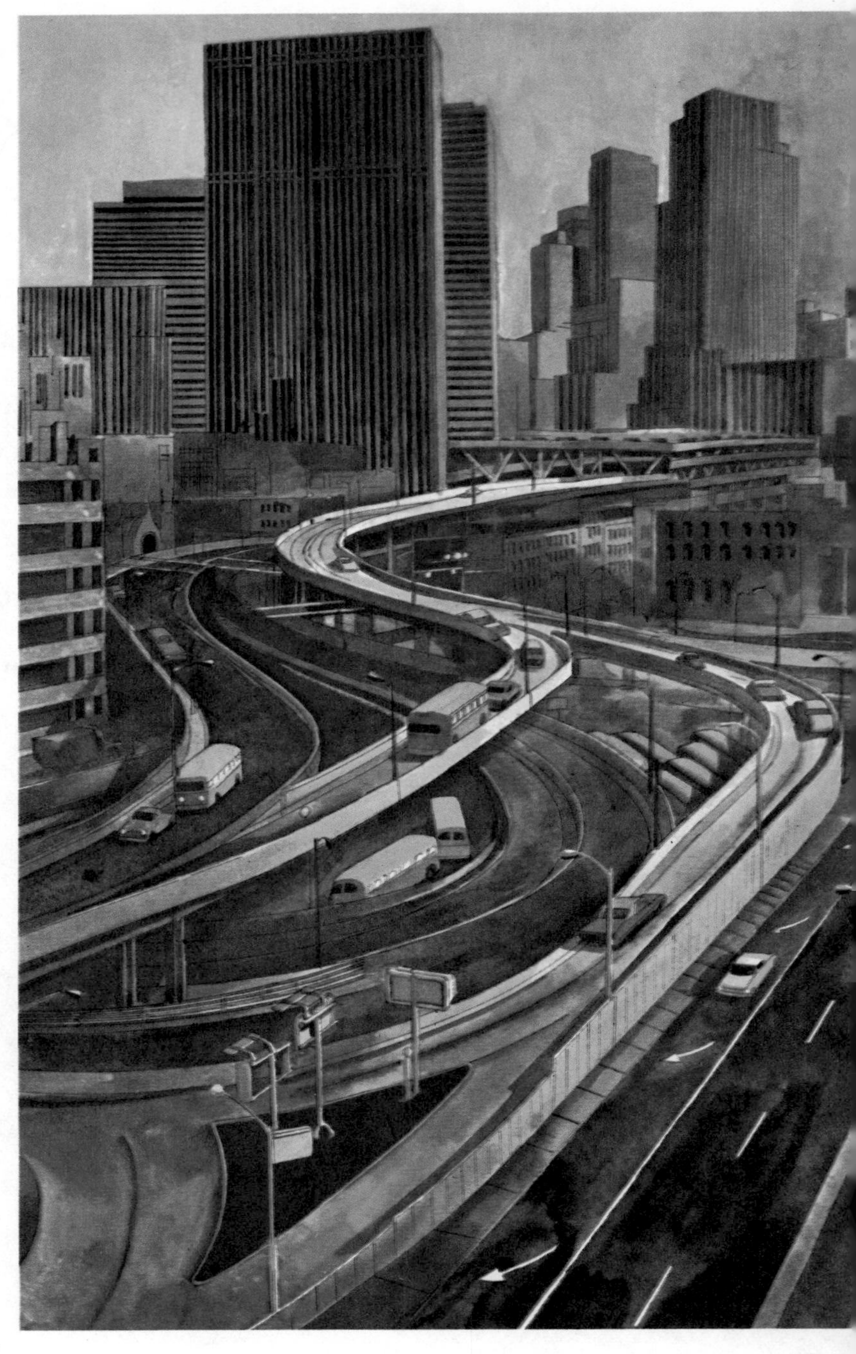

SECTION 3

The Rest of the Poem

You remember that the dim trail started by a homeward-bound calf became a lane, and finally a road. Read the last three stanzas of the poem and see what this trail has finally become.

Your teacher may wish to read the entire poem aloud as you listen, or to play the recording of it.

The years passed on in swiftness fleet,
The road became a village street;
And this, before men were aware,
A city's crowded thoroughfare;
And soon the central street was this
Of a renowned metropolis;
And men two centuries and a half
Trod in the footsteps of that calf.

Each day a hundred thousand rout
Followed the zigzag calf about;
And o'er his crooked journey went
The traffic of a continent.
A hundred thousand men were led
By one calf near three centuries dead;
They followed still his crooked way,
And lost one hundred years a day;
For thus such reverence is lent
To a well-established precedent.

But how the wise old wood-gods laugh
Who saw the first primeval calf!
And many things this tale might teach —
But I am not ordained to preach.

S. W. FOSS

LITERATURE

"The Calf Path" (Stanzas Six to Eight)

Recall the earlier stanzas of the poem by class discussion. Replay or reread the earlier stanzas. Then play or read these to the class. Give the pupils practice in reading aloud by letting them take turns reading couplets.

If further practice in identification of vowel sounds is desired, these stanzas will suit the purpose. All of the rhyming vowels are among those so far studied except in the couplet rhyming **rout** and **about**.

VOCABULARY AND MEANING

"The Calf Path" (Stanzas Six to Eight)

Work carefully through the questions on this page to be sure that the children understand first the literal meaning of the lines and then some of the fun that the poet is poking at the people who followed the calf. This is among the more difficult of the selections in this text. The central concept is simple, but it is treated with a certain amount of irony, and irony is difficult for fourth-graders. Still they should see something amusing in the idea of millions of people, generation after generation, zigzagging through the day just because a calf happened to zig here and zag there.

The essential thing that the tale teaches is that men tend simply to do what others have done without bothering to ask why. Be sure that the last line is understood.

COMPOSITION – A Moral to Write

Recall the Aesop fable "The Grasshopper and the Ant," pages 84–85. Ask what **moral** means, and what the moral of that fable was. Explain that the poem just read is a kind of fable too and has a moral, though the poet didn't tell it straight out, the way Aesop tells his moral. Have some class discussion about how the moral might be stated. Then let the pupils do the writing assignment as directed.

The Calf Path (STANZAS SIX TO EIGHT)

- **Swift** and **fleet** mean about the same thing. What do they mean? Check by using a dictionary. discuss
- **Fare** is an old word that meant **go,** and **thorough** is an old way of spelling **through.** So what is a **thoroughfare?** Check by using a dictionary. Check and discuss.
- **Renowned** means famous, and a **metropolis** is a great city. What had the trail become now? main street
- A **rout** is a great crowd of people. How do you think a hundred years could be lost each day? discuss
- **Reverence** means deep and honored respect. A **precedent** is something that has been done before. The calf set a precedent by making the path which men have been following ever since. Why should the wood-gods laugh at seeing who set the precedent? discuss

A minister or priest is said to be **ordained,** or appointed, to do his duties. One of his duties is preaching sermons. The last line means that the poet isn't going to point out a moral to us.

A Moral to Write

The poet didn't write a moral for his poem about the calf. Perhaps you can do it for him.

Write two paragraphs. In the first one, tell briefly how the crooked path came to be made. In the second paragraph, tell what you think this teaches about people. Do you think they sometimes keep on doing things without thinking? This might not be bad, of course. Some crooked paths are very interesting.

The Sound /z/

Say the words **zoo** and **zero.** These begin with a sound that we will show between slanting lines in this way: /z/. We will name it just as we do the letter **z.**

The sound /z/ comes at the beginning of just a few words. When it does, we write it with the letter **z**, as in these words:

zoo	**zero**	**zip**	**zebra**
zinc	**zone**	**zoom**	**zigzag**

The sound /z/ is very like /s/. It is like /s/ in the way that /b/ is like /p/ and that /v/ is like /f/. Put your hands over your ears and say **zoo, Sue.** Listen to the buzzing noise in **zoo.**

The sound /z/ is just like /s/ except that /z/ has this buzzing noise and /s/ does not.

The sound /z/ comes at the end of many words, but we don't usually use just the letter **z** to spell it at the end of words.

Here are the common words in which we write /z/ at the end with one **z**:

whiz **quiz**

Here are the common words in which we write /z/ at the end with two **z**'s:

buzz **fuzz** **jazz** **fizz**

• What are the vowel sounds in the six words above that end with **z** or **zz?** /i/, /i/, /u/, /u/, /a/, /i/ (simple)

When complex vowels come before /z/ at the end, we spell the sound with **se** or **ze.** Here are some words in which we spell it with **ze:**

graze	**gaze**	**sneeze**	**freeze**
breeze	**blaze**	**doze**	**snooze**

The sound /z/ is very common finally in English words, but rather rare initially. To the eight words with initial /z/ cited on page 171, we could add the following fairly common ones:

zeal	**zenith**	**zephyr**	**zest**	**zinnia**
zipper	**zither**	**zodiac**	**zoology**	**Zulu**

But that's nearly the lot.

The consonant sound /z/ has a high frequency in the final position largely because it is the most common form of three common morphemes: the plural of nouns, the possessive of nouns, and the third person singular (s form) of verbs in the present tense. In all of these, it occurs automatically when certain sounds precede, and since in all of them it is regularly represented by the letter s, speakers of English are seldom aware that it occurs at all. This point is mentioned on page 177.

Teaching Suggestions

The consonant /z/ is pronounced like /s/ except that /z/ is voiced and /s/ is voiceless. The voiceless–voiced contrast is easy to perceive with /s/ and /z/ because the sounds are easily prolonged, and the hum becomes very distinct. Let the children cover their ears as directed while saying **zoo** and **Sue**, prolonging the initial consonants. They should easily hear the hum in **zoo** and note its absence in **Sue.**

SOUNDS AND LETTERS—/z/
(CONTINUED)

The spelling of final /z/ is difficult principally because in a large number of words one must choose between **ze** and **se** and because **se** is also used to spell final /s/. Perhaps the simplest statement of the rule would be the following:

1. Spell final /z/ with **z** in **whiz, quiz,** and **fez.**
2. Spell final /z/ with **zz** in **buzz, fuzz, jazz, fizz, frizz.** (In effect, after simple vowels apart from **whiz, quiz,** and **fez**—but there are no other common examples.)
3. Spell final /z/ with **ze** in the words listed at the bottom of page 171 and also in **daze, haze, laze, glaze, maze, raze, wheeze, squeeze, frieze, maize, seize, prize, size, gloze, ooze, booze, gauze,** and all words with the ending **ize,** as in **penalize** or **botanize.**
4. Spell final /z/ with **s** when it represents one of these morphemes: noun plural, noun possessive, third person singular of the present tense of verbs.
5. In all other cases, spell final /z/ **se.**

Teaching Suggestions (Continued)

Work through the questions and examples on page 172. Give the pupils practice in hearing the sounds /s/ and /z/ under the different spellings. Point out the only spelling overlap: **se** may spell either final /s/ (**case**) or final /z/ (**rose**).

Extra Practice

The pupils may do the last exercise on page 172 as a written assignment by writing the headings /*s*/ and /*z*/ and putting each word under the proper heading.

The Sound /z/ (CONTINUED)

Here are words in which we spell /z/ at the end with the letters **se:**

ease /ē/	**please** /ē/	**phrase** /ā/	**cheese** /ē/
raise /ā/	**praise** /ā/	**rise** /ī/	**wise** /ī/
rose /ō/	**those** /ō/	**lose** /ū/	**choose** /ū/

• Tell what the vowel sound is in each of the words above. Do any of these words have simple vowel sounds in them? no

You know that after simple vowels, we usually spell the sound /s/ with the letters **ss,** as in **mess** and **fuss.** After other sounds, we often spell /s/ with **se.**

Pronounce these words. Listen to the /s/ sound at the end.

case	chase	erase	grease
pulse	dense	sense	dose
goose	loose	coarse	verse

You know these ways of spelling /s/ at the end of words:

s: plus, gas
ss: mess, fuss
se: case, verse

You know these ways of spelling /z/:

z: quiz, whiz
zz: buzz, fizz
ze: blaze, sneeze
se: rose, lose

• What letters may be used to spell either the sound /s/ or the sound /z/? se

• Pronounce each of the words below. Tell whether it ends in the sound /s/ or the sound /z/:

jazz /z/	muss /s/	rise /z/	doze /z/
case /s/	daze /z/	grease /s/	cheese /z/

Special Noun Plurals

● Nouns like **sky** and **century** make their plural form in a different way from nouns like **dog** and **match.** What letter do **sky** and **century** end in? Are the letters before the **y** in **sky** and **century** vowel letters or consonant letters? y; consonant letters

● The plural of **sky** is **skies.** The plural of **century** is **centuries.** Can you tell what has been done to the singular form of these words to make the plural? Change y to i and add es.

When a noun ends in the letter **y** with a *consonant* letter before it, we write the plural by changing **y** to **i** and adding **es.** The plural of **flurry** is **flurries.**

When a noun ends with **y** but has a *vowel* letter before it, we just add **s** to form the plural. The plural of **boy** is **boys.**

□ Write the plural of these words:

baby babies	lady ladies	sty sties	berry berries
toy toys	spy spies	day days	story stories
cherry cherries	joy joys	key keys	ruby rubies

□ Some nouns that end in the single letter **f** make their plural in the **regular** way: **chief/chiefs.** Write the plural of these nouns: **gulf, grief, proof.** gulfs, griefs, proofs

But there are twelve nouns you probably know which end in the single letter **f** and make the plural in an unusual way — an **irregular** way.

● The plural of **leaf** is **leaves.** The plural of **calf** is made the same way. What is the plural of **calf?** calves

□ In these nouns we change **f** to **v** and then add **es.** Write the plural of each noun. Then remember which nouns have this irregular plural.

leaf leaves	calf calves	half halves	wife wives
knife knives	life lives	shelf shelves	thief thieves
loaf loaves	elf elves	self selves	wolf wolves

GRAMMAR—Irregular Noun Plurals

A regular word is one which is included in some general rule. An irregular word is one that cannot be. Thus the plural of **sky** is regular both in sound and spelling. In sound, since it ends with a vowel, we form the plural by adding the sound /z/: /skī/ →/skīz/. In spelling, since it ends in **y** preceded by a consonant letter, we change the **y** to **i** and add **es: sky → skies**. For either pronunciation or writing, we do it this way, given the conditions. Any word conforming to the conditions which makes the plural some other way, either in sound or spelling, is an exception.

But we could not set up similar general rules to take care of the sound-and-spelling plurals of **leaf** and **calf**. It is not generally true that words ending in the sound /f/ make plurals in /vz/. Nor is it true that words ending in the letter **f** change **f** to **v** and add **es**. The plural of **chief** is **chiefs**, not ***chieves**; the plural of **proof** is **proofs**, not ***prooves**. We have here then a situation in which only certain words ending in the sound /f/ or the letter **f** make plurals in the sound /vz/ or the letters **ves**. These therefore are exceptions and must be learned item by item. The twelve common ones are given at the bottom of this page.

There are a few words ending in the sound /f/ spelled **f** which may optionally have plurals in the sound /vz/ and the letters **ves: hoof, staff, beef, scarf, wharf**. The plurals **hooves, staves, beeves, scarves, wharves** occur, but **hoofs, staffs, beefs, scarfs, wharfs** are more common.

LITERATURE – "The Frog"

Read the selection or play the recording of it. Let the pupils read portions. The first stanza can be divided by two pupils, the second read by one. Repeat as long as interest can be sustained.

Assign the poem as a copying exercise. In copying, the children will have to pay close attention to get the quotation marks and the commas in the right place. Doing so, they will get useful training in the use of quotation marks with commas.

SECTION 4

A Poem

Read this poem and decide whether you think the poet is being serious or having fun.

The Frog

Be kind and tender to the Frog,
And do not call him names,
As "Slimy-skin," or "Pollywog,"
Or likewise "Ugly James,"
Or "Gape-a-grin," or "Toad-gone-wrong,"
Or "Billy Bandy-knees";
The frog is justly sensitive
To epithets like these.

No animal will more repay
A treatment kind and fair,
At least so lonely people say
Who keep a frog (and, by the way,
They are extremely rare).

HILAIRE BELLOC

174

The Frog

- What do you think **tender** means in the first line of this poem? See a dictionary to check. discuss
- What does **slimy** mean? When a person **gapes** his mouth hangs open. What do you think **Gape-a-grin** means? What would a **Toad-gone-wrong** be? **Bandy-legged** means bow-legged. What are **bandy knees?** covered with slime; wide grin; discuss
- **Epithets** means names. Do you think a frog is so **sensitive** that names would hurt his feelings? discuss
- Do you know many people who keep a frog? Then can you figure out what **rare** means? See a dictionary. discuss

What Words Come From

The noun **pollywog** used to be more like this: **poll-wiggle. Poll** was a word that meant head, and a **poll-wiggle**, or **pollywog**, is a little frog or toad that wiggles its head.

Another word for **pollywog** is **tadpole.** This has the word **poll** — head — in it too. The first part of **tadpole** is the word **toad.** So tadpole means a creature with the head of a toad.

A Paragraph to Write

Probably you have never had a frog as a pet, but you may have had some other animal — a dog, a cat, a mouse, a canary. If not, you probably know someone who has a pet.

Write a paragraph telling about some animal that one might keep as a pet. Tell what makes the animal a good pet, or what makes it a hard one to have, or both.

175

VOCABULARY AND MEANING

"The Frog"

Work through the questions to make sure that the children understand all of the words and constructions. Try to convey the humor of the poem, the fact that the poet is being playful: very few people keep frogs as pets.

ETYMOLOGY – What Words Come From

The word **poll** is a borrowing from Dutch, and it means "top of the head." This is the source of the word in expressions like "poll tax," "the recent poll," "a public opinion poll." In all of these, there is embedded the notion of "counting heads." The ending **wog** of **pollywog** is related to the verbs **wiggle** and **wag.**

COMPOSITION – A Paragraph to Write

Warm up the class with a discussion about pets. Ask who has a pet. Let those who have tell something about them, both the pleasant and the unpleasant things involved in keeping them. Then let the pupils undertake the assignment.

SOUNDS AND LETTERS

Final /s/ and Final /z/

This reviews the spellings already studied for these two consonant sounds in the final position and adds the spelling **ce** for final /s/.

The Letter *s* as a Spelling of Morphemes

It is mentioned in the teaching suggestions for page 166 that the letter s may stand for either the sound /s/ or the sound /z/ when it represents one of the following morphemes: plural of nouns, possessive of nouns, third person singular of the present tense of verbs:

	/s/	/z/
Noun Plural:	hats	duds
	brakes	rags
Noun Possessive:	Chet's	the lad's
	Ralph's	Steve's
3rd Person Singular:	gripes	bribes
	retreats	retrieves

The typical speaker of English probably thinks of all of these as having an "s" sound and can only with some difficulty distinguish the /s/ column from the /z/ column. A little practice and concentration on the sound should make the difference apparent. It may be noted also that /s/ represents these morphemes only after the sounds /p/, /t/, /k/, /f/, and /th/. After sibilants, /əz/ is added. After all other sounds the ending is /z/.

Teaching Suggestions

Go over the chart on page 176 carefully. Remark that there is only one spelling which can mean either /s/ or /z/ and ask what it is. Elicit other examples for each spelling of each sound. The other likely examples of single **z** for /z/ are **quiz** and **fez**.

Here are other examples of **ce** for /s/: **race, niece, ice, office, slice, nice, voice,**

Final /s/ and Final /z/

You have found several ways of spelling the consonant sounds /s/ and /z/ at the end of words. Here are the ways you have studied, each with an example, plus a new way. The new way is printed in red.

Final /s/			Final /z/		
Spelling	*Example*		*Spelling*	*Example*	
ss	**pass**	kiss	**zz**	**buzz**	jazz
s	**plus**	gas	**z**	**whiz**	quiz
se	**case**	dose	**se**	**rose**	ease
ce	**twice**	rice	**ze**	**blaze**	doze

- In many words we use the letters **ce** to spell final /s/. But we never use **ce** to spell final /z/. Here are some words that use **ce** to spell /s/. Say them and listen to the /s/ sound at the end.

rice	**lace**	**fleece**	**sauce**
fence	**bounce**	**once**	**force**

- For each way of spelling final /s/ and /z/ in the chart above, give another example besides the one given in the chart. See p. 327 for more examples.
- Each set of three words below contains two that rhyme with each other. Tell what the two rhyming words are in each set. Tell what consonant sound each word in the set ends with.

1. case blaze brace /s/
2. doze rose dose /z/
3. ease geese freeze /z/
4. rise price rice /s/
5. blaze race days /z/
6. raise face base /s/
7. mice prize wise /z/

Remember these rules:

> **ze** stands for / z /, not / s /.
> **ce** stands for / s /, not / z /
> **se** stands for / s / in some words and for / z / in others

• What letter do we usually add to the singular form of a noun to make the plural form? s

When the letter s is just the end of a word, as in **gas** or **plus,** it stands for the / s / sound. But when it is added as an ending, as in plurals of nouns, it may stand for an / s / sound or for a / z / sound. In speaking, we sometimes make a noun plural by adding an / s / sound and sometimes by adding a / z / sound. But we always spell this ending with s, whichever sound it stands for.

• In these words the plural ending s stands for the / s / sound. Say them. Hear the / s / sound.

cats sacks pups trips

• In these words the plural ending s stands for the / z / sound. Say them. Hear the / z / sound.

days cars miles toes

The letter **x** at the end of a word, as in the word **box,** stands for two consonant sounds you know.

• Say **rocks, box.** Do these words rhyme? What two consonant sounds does **rocks** end with? What two consonant sounds does **box** end with? Do you see that the one letter **x** stands for the two consonant sounds / k / and / s / in the word **box?** yes; /k/, /s/; /k/, /s/; yes

• Here are some other words in which / k / and / s / are spelled together with the letter **x.** Say them and hear the sounds / k / and / s / at the end of each word.

tax	coax	fix	mix
fox	six	wax	complex

177

price, juice, dance, service, surface, chance, prince, fierce. All words with the endings **–ance** and **–ence** are spelled this way, of course: **radiance, sentence.**

Use rhyme to sort out the members of the sets at the bottom of page 176.

The children have already been introduced (page 167) to the **es** spelling of the plural morpheme when the singular ends in s or **ch**. On page 177 we add to this category the words ending with the spelling **x**. This letter stands regularly in English for the sounds /ks/. Some words ending in the sounds /ks/ spelled **x** are given at the end of the page.

GRAMMAR—Irregular Noun Plurals

There aren't a large number of count nouns ending in the letter **x**. The common ones are **climax, hoax, tax, index, complex, annex, sex, suffix, prefix, crucifix, mix, matrix, lynx, ox, box, fox**. Of these, **ox** has the special plural **oxen**. **Index** and **matrix** sometimes have the classical plurals **indices** and **matrices**.

Mutation Plurals

Seven of the English nouns with irregular plurals have what are called mutation plurals. Their plurals are indicated by a change of the vowel sound within the word rather than by an ending. These plurals were produced millenia ago by a process called **umlaut**. The vowel in a first syllable was changed by anticipation of the vowel in the second. For example, the old plural of **foot** was **fotiz**. Speakers in pronouncing the **o** anticipated the **i** with the result that **fotiz** changed to **fetiz**. Then the **iz** dropped off, leaving **fet**, which became modern English **feet**.

There were quite a number of mutation plurals in Old English—the period of our language recorded in texts dated 650 to 1100. But only seven remain today:

man/men foot/feet
tooth/teeth goose/geese
mouse/mice louse/lice
woman/women

Of these **louse/lice** is happily going out of use because of the retreat of the animal before the advances of sanitation. It is not included in the text.

As explained in the text, **woman** is special in that we mark the plural in speech by a change of the first vowel sound, but in writing by a change of the second vowel letter. As a result, the word is commonly misspelled. The spelling ***a women** is all too frequent even in the writing of college students.

Irregular Noun Plurals

You now know the regular way of spelling noun plurals and one irregular way. The regular way has three parts to it:

1. If the singular ends in "consonant-y," change the **y** to **i** and add **es: sky / skies.**
2. If the singular ends in **ch** or **ss,** add **es: patch / patches.**
3. In other words, just add **s.**

• You must now learn something more about the second part of this rule. Nouns that end in **x**, like **box**, also have **es** in the plural. The plural of **box** is **boxes.** How do you spell the plural of **fox?** foxes

• Nouns that end in just **s,** like **bus** and **gas,** also have **es** in the plural. The plural of **bus** is **buses.** What is the plural of **gas?** gases

So we can change Part 2 of the rule:

2. If the singular ends in **ch, ss, s,** or **x,** add **es.**

□ Write the plural of these words. They are all spelled in the regular way, according to the rule.

cups, skies cup sky mess valley messes, valleys
taxes, beaches tax beach box bus boxes, buses

Some nouns make their plurals in irregular ways. You know most of the irregular noun plurals, though you may still have to learn to spell some of them.

You know about a dozen words that end in **f** and make their plural in an irregular way. They change **f** to **v** and add **es. Calf** and **shelf** are like this. Their plurals are **calves** and **shelves.** Can you think of other irregular plurals like these? selves, wolves, etc.

Most of the other irregular plurals don't use the letter s at all. The word **man** has the irregular plural **men.** We say this:

One man is here.
Two men are here.

What do we do to **man** to make it plural in writing? In talking do we use the same vowel sound in the plural that we used in the singular? Change the a to e; no.

- Say the singular and plural of each of the following nouns:

foot / feet	**tooth / teeth**
mouse / mice	**goose / geese**

Do the four nouns listed above have regular, or irregular plurals? irregular

The plural of **woman** is **women.** You have to be careful in writing the plural of this word, because the sound doesn't tell you how to do it.

- Say **woman** and **women** carefully. Does the letter **o** stand for the same vowel sound in both words? What sound does it stand for in **woman?** What does it stand for in **women?** no; /ʊ/; /i/

You have to remember in writing the plural of **woman** to change the **a** in the last part to **e,** just as you do in writing the plural of **man.**

- Another noun with a special plural is **child.** Say the plural of **child.** Does the letter **i** stand for the same vowel sound in both the singular and the plural of **child?** no

□ Write the plural of these nouns:

boys, girls	boy	girl	lady	man	ladies, men
women, geese	woman	goose	fox	mouse	foxes, mice
feet, gases	foot	gas	tooth	witch	teeth, witches
bosses, axes	boss	ax	alley	child	alleys, children

179

The explanation of **woman/women** is that two forms existed in the dialects of Middle English: **womman** with the plural **wommen,** and **wimman** with the plural **wimmen.** In pronunciation, we have taken the singular of the former and the plural of the latter, but have adopted the spelling of **womman/wommen,** with single **m,** however, for both. If we had adopted the **womman/wommen** pronunciation for both, modern **woman** and **women** would both be pronounced like **woman.** If we had taken the **wimman/wimmen,** both would be pronounced like **women.**

Teaching Suggestions

Review the threefold rule for spelling noun plurals. Remember that **sky** and **patch** are regular: their plurals can be predicted from their singulars.

Teach the extension of Part 2 of the rule given in the middle of page 178. Have the children write the plurals of the list beginning **cup.**

Review the twelve nouns which change **f** to **v** and add **es.** They are **leaf, calf, half, wife, knife, life, shelf, thief, loaf, elf, self, wolf.** These nouns are irregular. Their plurals cannot be predicted from their singulars and must be learned as items.

Go through carefully the material on page 179. The explanation given in the material for the teacher about mutation plurals is not, of course, intended for the pupils.

TESTS AND REVIEW

Use the tests on pages 180–82 for diagnosis of individual strengths and weaknesses. Reteach as necessary, using the pages indicated after each set of directions.

Use page 183 for oral work.

Extra Test

If dictation is wanted to give practice with /s/ and /z/, dictate the words in sets that group words according to the spelling of the sounds. These may be used:

1. pass, less, dress, miss, fuss
2. gas, bus, plus
3. case, grease, pulse, dose, goose
4. race, grace, bounce, rice, force
5. quiz, whiz
6. buzz, jazz, fuzz, fizz
7. graze, freeze, snooze, blaze, doze
8. rose, please, ease, praise, choose

TEST I Sounds and Letters

1. Write these headings on your paper:

/f/ /v/ /ch/ /j/

Write each word below under the heading that shows what sound it ends with.

If you make mistakes, study pages 114, 140, 146.

life /f/	wretch /ch/	bunch /ch/	giraffe /j/
George /j/	laugh /f/	wave /v/	budge /j/
leave /v/	rough /f/	cage /j/	nymph /f/

2. Look at these words. For each one, write another in which the sound /f/ is spelled in the same way:

life	thief	staff	tough
safe	leaf	stuff	rough

telegraph paragraph

3. Write these headings on your paper: /s/, /z/. Under each heading, write the words from the list below that end with that sound.

If you make mistakes, study pages 176–77.

gaze /z/	fizz /z/	miss /s/	lace /s/
nose /z/	rice /s/	quiz /z/	grease /s/
please /z/	fence /s/	box /s/	loose /s/

4. Write the headings /s/ and /z/ on your paper again. The words below are all plurals spelled with the letter s, but some end in the /s/ sound and some in the /z/ sound. Write each one under the proper heading.

cats /s/	cars /z/	pipes /s/	dolls /z/
toes /z/	sticks /s/	days /z/	tips /s/

5. Three of the following four words rhyme. Write the three rhyming words.

<u>prize</u> <u>wise</u> twice <u>flies</u>

□ TEST II Grammar

1. Copy the noun phrases in the sentences below. After each noun phrase, write a number to show whether it is (1) a personal pronoun or (2) an indefinite pronoun. Then tell what its function is in the sentence. Write **subj.** for subject, **obj. v.** for object of a verb, **obj. prep.** for object of a preposition, or **comp.** for complement.

If you make mistakes, study pages 148, 158–59.

a. Everyone knows him. (2), subj.; (1), obj. v.
b. He is behind her. (1), subj.; (1), obj. prep.
c. It is I. (1), subj.; (1), comp.
d. Nobody is near them. (2), subj.; (1), obj. prep.
e. They hurt me. (1), subj.; (1), obj. v.
f. Someone helps us. (2), subj.; (1), obj. v.
g. She likes everybody. (1), subj.; (2), obj. v.

2. Write as headings **Singular** and **Plural.** Find the nouns in the sentences below, and write each noun under the right heading.

If you make mistakes, study pages 166, 173, 178.

a. A man was here. sing.
b. The girls are in the car. pl.; sing.
c. The books are on the desk. pl.; sing.
d. My mother saw the teachers. sing.; pl.
e. They are in the boxes. pl.
f. The boy ate some peanuts. sing.; pl.
g. Her teeth are beautiful. pl.
h. The children are hungry. pl.
i. The boys helped the men. pl.; pl.
j. Some women are in the playground. pl.; sing.
k. The cats caught the mice. pl.; pl.
l. The calves made a path. pl.; sing.
m. She cleaned the shelves. pl.

TEST II Grammar (CONTINUED)

3. Write the plurals of these nouns. If you make mistakes, study pages 178–79.

boy s — sky skies — beach es — valley valleys
village s — church es — pass es — fox es
flurries flurry — baby babies — match es — wave s

4. Write the plurals of these nouns. Some are made in the regular way and some in special ways.

If you make mistakes, study pages 178–79.

man men — girl girls — lady ladies — woman women
foot feet — leg legs — tooth teeth — bench benches
tax taxes — child ren — goose — mess geese, messes
calf calves — gulf gulfs — wolf — leaf wolves, leaves

5. In each sentence below, the subject is singular. Write each sentence again, making the noun in the subject plural. Don't change anything else.

a. The man found it. . . . men . . .
b. The child heard it. . . . children . . .
c. The baby cried. . . . babies . . .
d. The fox ran away. . . . foxes . . .
e. The church looked new. . . . churches . . .
f. The tooth ached. . . . teeth . . .
g. The woman left. . . . women . . .
h. The spy confessed. . . . spies . . .
i. The calf wobbled. . . . calves . . .
j. The box broke. . . . boxes . . .
k. The thief escaped. . . . thieves . . .

TEST III Writing and Vocabulary

Write five sentences. In each sentence, use one of these words.

1. century 2. trod 3. metropolis 4. rare 5. slimy

REVIEW Ideas and Information

1. What does **infer** mean? reason from the evidence
2. What is a **bell-wether** sheep? wears a bell, leads the others
3. What is a **glade?** an open place in the woods
4. What is another word that means **renowned?** famous
5. **Poll** in **pollywog** and **pole** in **tadpole** have the same meaning. What is the meaning? head
6. What are **epithets?** names
7. When do we spell /f/ with the letters **ff** and /s/ with the letters **ss?** after simple vowels
8. What sound may the letters **ph** stand for? /f/
9. There is one word in which /f/ is spelled **ffe.** What is it? giraffe
10. At the beginning of words, the letter **c** may stand for two different sounds. What are these two sounds? /k/, /s/
11. What sounds does the letter **x** stand for? /k/ and /s/
12. When do we make a plural by changing **y** to **i** and adding **es?** nouns ending in consonant-y
13. When do we make plurals by adding **es** to the singular form? nouns ending in ch, tch, ss, s, x
14. What are the vowel sounds in the first part of **woman** and **women?** /oo/, /i/
15. What is the function of the noun phrase **my teacher** in "She is my teacher"? comp.
16. What is the function of the noun phrase **I** in "It is I"? comp.
17. What does the noun **watcher** mean? To what word was **er** added to make **watcher?** one who watches; watch
18. When a word begins with the letter **c** followed by **e** or **i,** what sound does the letter **c** stand for? /s/
19. Is **skies** a regular or an irregular plural? regular

183

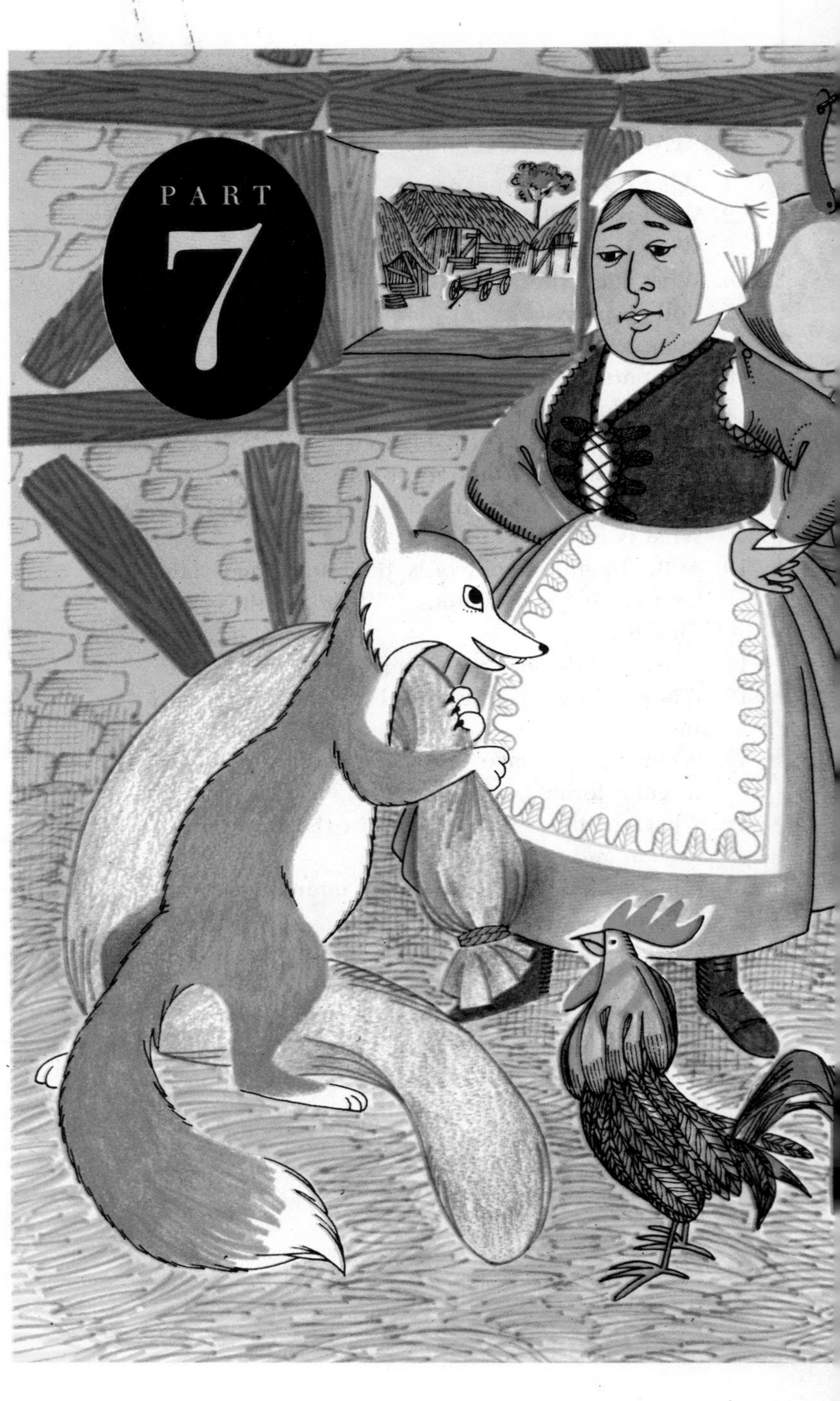
PART
7

SECTION 1

A Story

This is the first part of a four-part story.

The Travels of a Fox

One day a fox was digging behind a stump and he found a bumblebee; and the fox put the bumblebee in a bag and took the bag over his shoulder and traveled.

At the first house he came to, he went in and said to the mistress of the house, "Can I leave my bag here while I go to Squintum's?"

"Yes," said the woman.

"Then be careful not to open the bag," said the fox.

But as soon as he was out of sight the woman said to herself, "Well, I wonder what the fellow has in his bag that he is so careful about. I will look and see. It can't do any harm, for I shall tie the bag right up again."

However, the moment she unloosed the string, out flew the bumblebee, and the rooster caught him and ate him all up.

After a while the fox came back. He took up his bag and knew at once that his bumblebee was gone, and he said to the woman, "Where is my bumblebee?"

And the woman said, "I untied the string just to take a little peep to find out what was in your bag, and the bumblebee flew out and the rooster ate him."

"Very well," said the fox; "I must have the rooster, then." So he caught the rooster and put him in his bag and traveled.

LITERATURE

"The Travels of a Fox" (First Part)

This is a familiar kind of story which builds up from one similar episode to another, until the evildoer repeats once too often and gets his comeuppance. The story is distributed among the four sections of this part.

The records accompanying the text use this story as an example of dramatization. If the teacher would like to have the class present this as a drama, the record may suggest ways of doing it, but one should recall that the record dramatization is done with sound alone, not visible action. Furthermore, one should feel free to go into any changes and improvisations that suggest themselves. At this grade level, spontaneity is a large part of the fun of dramatization.

If it is the teacher's intention to dramatize the story, the prospect may be announced to the class at the beginning, so that everyone can think about characters and business, though the drama should be postponed until after all four parts have been read. Say that when you get through studying this part of the book you may do this story as a play, and that all can be thinking of how it might be done.

Read this first part to the children. Let them take turns reading paragraphs aloud.

VOCABULARY AND MEANING

"The Travels of a Fox" (First Part)

This story is not hard to read, partly because it is very repetitive. Still, care should be taken that it is read closely and accurately. Do the questions on this page, and be sure that everything is understood.

SOUNDS AND LETTERS – /sh/

This sound, like /ch/, is not usually written with a single letter of the alphabet, and so we have no conventional name for it. It is sometimes called the "ess-aitch" sound. But it seems better to call it "esh," by analogy with "ess" (s).

In both initial and final position, /sh/ is overwhelmingly spelled **sh**. **Sugar** and **sure** are the only ready exceptions. In medial positions before suffixes, it is spelled in various ways: **ss** in **pressure, t** (or **ti**) in **position**. These medial spellings are not introduced in grade four.

Though, like the dictionaries, we symbolize this sound with the letters **s** and **h**, we tie them together to show that they stand for a single sound.

The Travels of a Fox (FIRST PART)

- What is a **stump?** See a dictionary.
- Some words, like **woodpecker** and **grasshopper,** seem to tell us what they mean. Does **bumblebee?** yes; What do you think the **bumble** part might mean? discuss
- The **mistress** of a house is the woman who lives there, and is in charge of it. **Squintum's** is just a made-up name. What did the fox tell the mistress of the house not to do? Did this make her curious? open the
- Can you think of another word that means **unloosed?** See a dictionary to check. discuss
- What did the fox take from the woman? Why? the rooster; in exchange for his bumblebee

The Consonant Sound /sh/

- Say these words and listen to the first consonant sound in them: **ship, shove.**

This sound is usually spelled with two letters, even though it is a single consonant sound. The letters we usually use for it are **sh.** So we will show it as a sound this way: /sh/.

We name the sound /s/ by saying "ess." So we will name the sound /sh/ in a similar way; we will call it "esh."

Say **s.** Say /sh/. Say /s/.

At the beginning of words, we usually write /sh/ with the letters **sh.** We write it this way in these words:

ship	**shove**	**shop**	**shore**
shall	**shave**	**shower**	**shoulder**

- What other words can you think of that begin with the sound /sh/ and the letters **sh?** See p. 328.

We also spell /sh/ with the letters **sh** at the end:

dash	**flash**	**rush**	**push**
bush	**gosh**	**mush**	**dish**

- Can you think of others? See p. 328.
- What consonant sounds do these words end with?

crash /sh/	catch /ch/	pass /s/	lick /k/
reach /ch/	squash /sh/	age /j/	egg /g/
blush /sh/	ditch /ch/	hedge /j/	wish /sh/
judge /j/	mush /sh/	much /ch/	ledge /j/

Spelling the Plural of Nouns

These are the rules you have studied so far.

1. If the noun ends in "consonant-y," change **y** to **i** and add **es**.
2. If the noun ends in **ch**, **s**, **ss**, or **x**, add **es**.
3. For other regular nouns, just add **s**.

There are only a few nouns that end in **z** or **zz** in the singular. They add **es** to make the plural.

buzz / buzzes **fuzz / fuzzes**

Nouns that end in /sh/, spelled **sh**, in the singular, also add **es** to form the plural:

dish / dishes **bush / bushes**

So we must add to the second part of the rule:

2. If the noun ends in **ch**, **s**, **ss**, **x**, **z**, **zz**, or **sh**, add **es**.

□ Write the plurals of these nouns:

box	fizz	lunch	kiss	flash
bus	dish	brush	patch	fuzz
buses	dishes	brushes	patches	fuzzes

boxes, fizzes; lunches, kisses, flashes

Teaching Suggestions

Go through the first part of this lesson to get the children to understand that /sh/ is a sound. Give them practice in naming it.

Elicit other examples of /sh/ spelled **sh** initially and then finally. The class should easily think of other examples for both positions.

The exercise on page 187 gives practice in discriminating /sh/, /ch/, /s/, /k/, /j/, /g/ in final position. These headings may be written on the chalkboard and the words written under them at the direction of the class response.

Extra Practice

The last exercise on the consonant sound /sh/ may be made an individual written assignment by having the pupils write the six sound symbols /sh/, /ch/, /s/, /k/, /j/, /g/ as headings on their papers and list the appropriate words under each heading.

Check understanding of the second part of the now completed rule by having the class do the words at the bottom of page 187 as a written exercise.

GRAMMAR—Review of Regular Plurals

The rest of the page presents again the rule for spelling the plurals of regular nouns. We complete here the second part of the rule by adding **z**, **zz**, and **sh** to the list of singular endings that take **es** to form the plural. The letters **sh** are the regular spelling of the sound /sh/, which is presented above.

Tense

We have so far, in this Teacher's Edition, given the rule for **verb phrase** as follows:

$$\text{VP} \rightarrow \left\{ \begin{array}{l} \textbf{be} \;+\; \left\{ \begin{array}{l} \text{NP} \\ \text{Adj} \\ \text{Adv-p} \end{array} \right\} \\ \text{verbal} \end{array} \right\}$$

Verbal is then, roughly:

$$\text{Verbal} \rightarrow \left\{ \begin{array}{l} \text{verb} \\ \text{verb} \;+\; \text{NP} \\ \text{verb} \;+\; \text{Adj} \end{array} \right\}$$

But clearly there is more to it than this, for the **be** or the verb, whichever we choose, will have a particular form: **am, is, are, was,** or **were** for **be**; forms like **walks, walk**, or **walked** for verbs. These are all tense forms, divided between two tenses—present and past.

Present Tense: **am, is, are; walks, walk**
Past Tense: **was, were; walked**

That is, we choose, say, **is** as against **are**, or **walks** as against **walk**, according to what the subject is—mostly according to whether the subject is singular or plural. But we choose **is** as against **was** or **walked** as against **walk** not according to the subject but according to the time of the statement in the predicate or the nature of it.

In every predicate this decision about tense must be made. That is, every predicate must have one word in it that expresses present or past tense. Except for certain adverbials, this will always be the first word in the predicate of a kernel sentence. We can show this by adding **tense** to the formula before everything else:

$$\text{VP} \rightarrow \text{tense} \;+\; \left\{ \begin{array}{l} \textbf{be} \;+\; \left\{ \begin{array}{l} \text{NP} \\ \text{Adj} \\ \text{Adv-p} \end{array} \right\} \\ \text{verbal} \end{array} \right\}$$

Tense will now apply to **be**, if the predicate

Tense Forms of *be*

You have studied three forms of **be.** Tell which words in the predicates of these sentences are forms of **be.**

a. He is hungry.
b. They are sleepy.
c. I am thirsty.

After each one of these sentences, we could use the word **now.** We could say:

d. He is hungry now.
e. They are sleepy now.
f. I am thirsty now.

Could we use the words **then** or **yesterday** in these sentences? no

The word **now** means the present time. The words **then** and **yesterday** mean time in the past. When we use sentences with **is, are,** and **am,** we do not mean time in the past. We can use **now** with these forms, but not words like **then** and **yesterday,** which mean time in the past.

Forms of **be** that are used to mean time in the present are called **present tense** forms. The word **be** has three present tense forms.

□ The present tense forms of **be** are **is, are,** and **am.** Write three sentences using these present tense forms. Use the word **now** at the end of the predicate of each sentence. (Sentences will vary.)

The forms of **be** that we use when we mean time in the past are **was** and **were.** Say these sentences:

g. He was hungry.
h. They were sleepy.
i. I was thirsty.

• Say Sentences g, h, and i at the bottom of page 188 again. This time, put the word **then** at the end of each sentence.

• Say the sentences again, and put the word **yesterday** at the end.

• Does **was** or **were** in each sentence mean time in the past? yes

The forms of **be** that are used to mean time in the past are called **past tense** forms.

• The past tense forms of **be** are **was** and **were.** The following sentences have past tense forms of **be** in the predicate. Say each one, but use the word **yesterday** at the end of it.

1. He was here.
2. They were at their desks.
3. I was tired.
4. We were the winners.
5. She was quiet.

• Tell what word in the predicate of each of these sentences is a form of **be.** Tell whether it is a present tense form or a past tense form.

6. John is happy now. pres.
7. Mary was happy yesterday. past
8. We were little then. past
9. I am at my desk now. pres.
10. They are friendly. pres.
11. He was outside. past
12. Mr. Wheeler was near us. past
13. They were children. past
14. The dog is under the car. pres.
15. The weather was cold last week. past
16. The sun is up. pres.
17. The leaves were brown. past
18. The kitten is hungry. pres.

is a **be** one, giving **am, is**, or **are** in the present, according to rules of agreement, and **was** or **were** in the past. If the predicate is a verbal one, **tense** will apply to the verb of the verbal, giving, for example, **walks** or **walk** in the present and **walked** in the past.

In this series we use the term **tense** in the normal linguistic sense: distinctions of time or manner of action shown inflectionally, usually by suffixes. We do not include as tenses combinations of words such as **will go** or **has gone**. For if we did, there would be no stopping place, and we should presently find ourselves talking of things like "past near future," as in **was about to go**.

In **has gone**, the tense is expressed by **has** (present), as opposed to **had** (past). In **will go** it is expressed by **will** (present), as opposed to **would** (past). These and other possible expansions of the verb phrase are easily taken care of by elaboration of the formula, but are not introduced in this grade.

Languages differ considerably in the number of tenses they have, as **tense** is defined here. English has two. Latin had six, modern Italian has five. It is because of the six-tense system of Latin that people have described a six-tense system for English. But in Latin and Italian multiple tense distinctions are made in the inflection of the verb. Thus Italian has a future tense: "Lui partirà" ("He will go") in which the future **partirà** contrasts with the present **parte**. English has no future tense, though of course it has many ways of expressing future time: "He will go," "He might go," "He ought to go," "He's going to go," "He's about to go," "He goes tomorrow," etc.

Extra Practice

Sentences 6–18 may be used as an individual written assignment by having the pupils write *Present Tense* and *Past Tense* on their papers as headings and list the forms of **be** under the appropriate headings.

LITERATURE

"The Travels of a Fox" (Second Part)

Review what happened in the first part, on page 185. Reread it if necessary. Then read this second part to the class and give individuals practice in reading it aloud.

If dramatization is intended, keep the prospect before the children. Let them discuss how the actions might be shown in a play. For instance, one child will have to be the fox, and one the pig. But Fox obviously cannot put Pig into a bag and carry him away. How can this be suggested? Children, since they are used to pretending and make-believe, are usually good at solving problems of this sort.

SECTION 2

More of the Story

Here is the second part of the story called "The Travels of a Fox."

At the next house he came to, he went in and said to the mistress of the house, "Can I leave my bag here while I go to Squintum's?"

"Yes," said the woman.

"Then be careful not to open the bag," said the fox.

But as soon as he was out of sight, the woman said to herself, "Well, I wonder what the fellow has in his bag that he is so careful about. I will look and see. It can't do any harm, for I shall tie the bag right up again."

However, the moment she unloosed the string, the rooster flew out and the pig caught him and ate him all up.

After a while the fox came back. He took up his bag and knew at once that his rooster was gone, and he said to the woman, "Where is my rooster?"

And the woman said, "I untied the string just to take a little peep to find out what was in your bag, and the rooster flew out and the pig ate him."

"Very well," said the fox, "I must have the pig, then."

So he caught the pig and put him in his bag and traveled.

At the next house he came to, he went in and said to the mistress of the house, "Can I leave my bag here while I go to Squintum's?"

"Yes," said the woman.

"Then be careful not to open the bag," said the fox.

The Travels of a Fox (SECOND PART)

- Was this the same woman as the mistress of the house where the fox got the rooster? no
- What happened this time when the bag was opened? discuss
- What did the fox get this time for what he had in his bag? Does he have a plan? a pig; discuss

A Writing Assignment

Write a paragraph telling what this fox did that was clever and why you think he did it.

191

VOCABULARY AND MEANING

"The Travels of a Fox" (Second Part)

Keep up the requirement of close and accurate reading. If more thorough discussion of the story is desired, these additional questions may be used.

Additional Questions

Where did the fox say he was going when he left his bag?

What did the woman get to wondering about as soon as the fox was out of sight? What did she think would do no harm? Why not?

What did the fox know at once when he took up his bag? What did he say to the woman?

What happened at the next house he came to?

COMPOSITION – A Writing Assignment

The previous work may have adequately prepared for writing the paragraph described. If not, have further discussion of the cleverness of the fox.

Initial /r/

The spelling of initial /r/ is not a very serious problem. The **rh** and **wr** alternatives to simple **r** are not very numerous.

The **rh** words are either direct borrowings from Greek or, like **rhyme**, respellings on the analogy of Greek words. Only **rhyme** and **rhythm** are very likely to be in the fourth-grader's vocabulary. Other words with the **rh** spelling are **rhapsody, rheostat, rhetoric, rheumatism, Rhine, rhinoceros, Rhode Island, rhododendron, rhubarb**.

The words with **wr** mostly come to us from Old English where they represented indeed the consonantal combination /wr/. We have ceased to pronounce the **w** but, as the usual practice, have kept its spelling.

There are rather a large number of words beginning with **wr**. Here are some common ones, apart from those listed at the bottom of this page:

wraith	**wrangle**	**wreath**	**wren**	**wrench**
wrest	**wretch**	**wring**	**writhe**	**wry**

It has been frequently observed that most words in which initial /r/ is spelled **wr** have a meaning connected in some way with twisting.

Final /r/

The spelling of final /r/ is not much of a problem either, so far as the consonant is concerned. A few words, of which the only common ones are **err, burr**, and **purr**, spell it **rr**. Otherwise it is spelled either in the **VCe** pattern or with **r**. To be sure, this choice puts a certain strain on memory, gives rise to error, and provides homophones: **fair/fare, here/hear, four/fore**. It may be noted that when the preceding vowel is /ī/ the **VCe** pattern is nearly always used, as it is with other consonants following /ī/. The only exceptions before /r/ are proper nouns like **Meyer**.

The Consonant Sound /r/

• Say these words and hear the consonant sound at the beginning: **ride, red.**

In **ride** and **red** we spell this consonant sound with what letter? We will use this letter between slanting lines to show this sound: /r/. When you want to speak about the sound /r/, name it just as you do the letter **r.** r

At the beginning of words, we usually use the letter **r** to stand for the sound /r/. We do so in these words, for example:

ride	**red**	**road**	**rough**
rule	**rush**	**reach**	**run**

• Think of five other words that begin with the sound /r/ and the letter **r.** See p. 328.

• In a few words we spell the sound /r/ in special ways. In one that you know very well, we spell /r/ with the letters **rh.** This is the word **rhyme.** We also spell the vowel sound of this word in a special way. What is the vowel sound of **rhyme?** /ī/

In some words we write the sound /r/ at the beginning with the letters **wr.** We do this in the word **write.**

• Say **write** and **right.** They sound exactly the same. What do we call words that sound just alike but are spelled differently and mean different things? homophones

Here are some words that begin with the sound /r/ spelled **wr.** The word **wrath** means anger. Study the others and think what they mean. Give a sentence using each of the words. discuss meanings

wrath	**wrap**	**wreck**	**wrestle**
wriggle	**wrist**	**wrong**	**wrinkle**

At the end of words, we often spell the sound /r/ with the letter **r.** We do in these words:

sir her fur stir

● Decide what each of these words means. Which is a verb? Give a sentence using each word. discuss; stir; (oral responses)

When a complex vowel comes before /r/ we may spell /r/ with the letters **re** to show that the vowel is a complex one. Tell what vowel comes before /r/ in these words: /ī/ /ē/ /ū/ /ō/

fire here pure score

When the complex vowel is spelled with two letters, we spell /r/ with just the letter **r.** Tell what the vowel sounds are in these words:

/ā/, /ā/ **hair wear deer four** /ē/, /ō/

□ Copy the four words above, and beside each one, write the word from the list below that has the same vowel sound spelled the same way: hair, fair; wear, pear; deer, steer; four, pour

pear pour steer fair

● Can you name some other words that have the same vowel sounds before /r/ as these? chair, swear, cheer, queer

The vowel sound /o/, as in **lock** and **not,** comes before /r/ in some words. But when it does, we do not spell it with **o.** We use **a,** as in **car:**

car far tar bar

● Name some other words like these. mar, star, jar

People who live in different parts of the country sometimes speak in different ways. People in one place may not pronounce /r/ in **hair** and **car** the same way people from another place do. But everyone spells these words in the same way.

Some people have a different vowel sound in **not** than they do in **car.** But everybody uses the letter **o** in **not** and the letter **a** in **car.**

Other words like **car** include **scar, cigar, char, jar, mar, par, spar, guitar, star. War** has the vowel /aɪ/.

The difficulty in spelling words ending in /r/ is not so often the consonant as the preceding vowel. Before /r/—and only before /r/ in stressed syllables—we may have the vowel schwa, which has no particular letter representation, and for which we use now one vowel letter, now another. **Sir, her, fur, heard, myrrh** all have the same vowel sound—schwa—before /r/, spelled in five different ways. Schwa is not introduced in this volume. It is studied in some detail in grades five and six.

Pronunciation of /r/

The pronunciation of final /r/ is one of the more easily recognizable of the features distinguishing certain large dialect groups in English. Many people in the South, in New England, and in Old England speak what are called "**r**-less dialects." They are said to "drop the **r**." They don't exactly drop the **r**, as a matter of fact. When they pronounce, for example, the word **hair**, they don't just omit the final sound; this would give the word **hay**. What they do is substitute a vowel sound, a centering glide, for the consonantal /r/. Thus **hair** is pronounced something like "hay-uh," **here** "hee-uh," etc. What is here transliterated "uh" is just the way /r/ is pronounced in these areas.

GRAMMAR

Agreement of Subjects and Forms of *be*

This page reviews the agreement rules for subjects and the present tense forms of **be**. **I** takes **am**. For the most part, singular subjects take **is** and plural subjects take **are**. However, indefinite pronouns, which are largely plural in meaning, conventionally take **is**. The fact that they are often plural in meaning—what could be more plural than **everybody**?—accounts for the tendency of pupils to couple them with **are**. They don't do this in simple sentences, like "Everybody is here," but they tend to in complicated sentences when the subject is separated from the form of **be**: *"Everybody who handed in their papers late are going to be marked down."

One way in which verbs differ from **be** is that **be** has two past tense forms—**was** and **were**—and verbs only one. There is therefore an agreement problem in the past tense with **be**, though none with verbs.

Teaching Suggestions

Go through the questions and the examples on page 194. Have Sentences 1–7 done as a written exercise. Further practice on concord with indefinite pronouns can be given orally by substituting for **everybody** in Sentence 3 **everyone, everything, nobody, no one, nothing, somebody, someone, something**.

Go through the explanations, questions, examples, and exercises on page 195 with the class. The usage is automatic for most children and not complicated. **Was** is the past tense form for all the subjects that take **is** in the present and also for **I**. **Were** is the past tense form for all the subjects that take **are** in the present.

Agreement of Subjects and Forms of *be*

• You know five forms of **be** besides the word **be** itself. Which of the following are present tense forms? Which are past tense forms? pres.: am, is, are

am is are was were past: was, were

We use **am, is,** or **are** when we are talking about something that is true now. But whether we use **am, is,** or **are** depends on what the subject is.

• If the subject is the personal pronoun **I,** which present tense form of **be** do we use: **am, is,** or **are?**

• If the subject is **he, she,** or **it,** what present tense form of **be** do we use? is

• If the subject is **we, you,** or **they,** what present tense form of **be** do we use? are

• Look at these sentences. Tell what present tense form of **be** we use with singular subjects and what form we use with plural subjects.

a. The boy is here. sing.: is

b. The boys are here. pl.: are

• Look at these sentences. What present tense form of **be** do we use when the subject is an indefinite pronoun? is

c. Everybody is happy.

d. No one is happy.

□ Copy the subject and the present tense form of **be** in each of these sentences.

1. People are funny. are
2. The man is here. is
3. Everybody is inside. is
4. You are late. are
5. I am on the ladder. am
6. They are tired. are
7. It is a cold day. is

We use **was** and **were** when we are talking about something that was true in the past. Whether we use the past tense form **was** or the past tense form **were** depends on what the subject is.

We use **was** for the past the way we use **is** for the present. We use **was** when the subject is **he, she,** or **it.** We use **was** when the subject is an indefinite pronoun. We use **was** when the subject is a noun phrase with a singular noun.

• Tell what the subject is in each of these sentences. Tell whether it is a personal pronoun, an indefinite pronoun, or a determiner plus a singular noun.

8. The child was tired. det. + n.
9. He was sleepy. pers.
10. Everybody was at the fair. ind.
11. It was interesting. pers.
12. She was pretty. pers.

We also use the past tense form **was** when the subject is the personal pronoun **I:**

I was in the playground.

• What present tense form of **be** do we use after **I?** am

When the subject is **you, we, they,** or a plural noun phrase, the past tense form of **be** is **were.**

• Tell what word is a form of **be** in each of these sentences. Tell whether it is present tense or past tense.

13. He was in the playground. past
14. The boys were outside. past
15. She is tired. pres.
16. They are friendly. pres.
17. I am hungry. pres.
18. They were here. past
19. We were near them. past
20. You were in the car. past

Extra Help for Special Language Problems

If the pupils habitually say, for example, "We was at home," the problem is more serious. Such students need much practice to become able to use the standard English forms naturally and automatically. One kind of practice is a substitution drill. Start with a pattern sentence like the following:

We were there.

Then give another pronoun—**he**. The children use the new pronoun as the subject and change the form of **be** if necessary. Here it is necessary. They must say "He was there." Then give **I**. They say "I was there," **They**: "They were there," **You**: "You were there." **She**: "She was there." And so on. Other noun phrases—**the boys, the girls, everybody, Mike**—can be run in with the personal pronouns.

This is a technique of foreign-language teaching, and when the natural speech of the children is far removed from that of standard English, the problem is rather a foreign-language problem. Abundant practice in the correct construction becomes important, and the trick is to keep the drill moving rapidly. With a little determination, the teacher can get onto the technique.

Of course such a practice is pointless for pupils who already are automatic in the standard English forms.

LITERATURE

"The Travels of a Fox" (Third Part)

Discuss what has happened to this point, and reread the first two installments if necessary. Then read this one. Give the children practice in reading parts of it aloud.

If dramatization is intended, keep the idea in the foreground. Let the children talk about how these actions may be portrayed in a play.

SECTION 3

More of the Story

Here is the third part of "The Travels of a Fox."

But as soon as he was out of sight, the woman said to herself, "Well, I wonder what the fellow has in his bag that he is so careful about. I will look and see. It can't do any harm, for I shall tie the bag right up again."

However, the moment she unloosed the string, the pig jumped out and the ox gored him.

After a while the fox came back. He took up his bag and knew at once that his pig was gone, and he said to the woman, "Where is my pig?"

And the woman said, "I untied the string just to take a little peep to find out what was in your bag, and the pig jumped out and the ox gored him."

"Very well," said the fox, "I must have the ox, then."

So he caught the ox and put him in his bag and traveled.

196

At the next house he came to, he went in and said to the mistress of the house, "Can I leave my bag here while I go to Squintum's?"

"Yes," said the woman.

"Then be careful not to open the bag," said the fox.

But as soon as he was out of sight, the woman said to herself, "Well, I wonder what the fellow has in his bag that he is so careful about. I will look and see. It can't do any harm, for I shall tie the bag right up again."

The Travels of a Fox (THIRD PART)

• The word **gore** is what an animal does when it wounds someone with its horns. See a dictionary.

• **Ox** has a special plural: **oxen.**

Two other nouns for animals with special plurals are **sheep** and **deer.** We use the same form for the singular as for the plural: **one sheep, ten sheep; one deer, twenty deer.**

□ Use **oxen, sheep,** and **deer** as plural subjects of sentences. (Have pupils write sentences and then compare.)

QUOTATION MARKS

A quotation is something that someone has said. Have you noticed that what the fox said and what the women said were inside marks like these: " "? These marks are called **quotation marks.** We put them at the beginning and ending of a quotation.

> "Very well," said the fox, "I must have the ox, then."

• Find all the quotation marks in this sentence. Is **said the fox** part of the quotation? no

VOCABULARY AND MEANING

"The Travels of a Fox" (Third Part)

Because of the repetitious nature of a story of this type, there are not likely to be reading problems at this stage.

Irregular Plurals

In Middle English, **en** was a very common plural ending, particularly in southern England. For a time it was questionable whether **en** or **s** would develop into the regular ending of the plural. Of all the nouns with **en** plurals in the time of Chaucer, only **ox/oxen** remains as a pure example. (**Children** and **brethren** are **en** plurals, but embody other changes too.) Even **ox/oxen** may not be forever with us, since the animal has been replaced by the tractor.

Old English had a number of nouns whose plurals were identical with the singulars. Of these, only **sheep** and **deer** remain. However, the category has been swelled by the addition of the nouns for many other animals: **quail, pheasant, moose, elk**. The unchanged plural is generally applied to animals that are being hunted, not to domesticated or zoo animals.

COMPOSITION – Quotation Marks

The children have already encountered quotation marks and have perhaps made then in copying exercises. Here they are reminded of the nature and use of these marks. Let the children find the examples in this part of the story. Note that the American convention is to put a following comma or period inside the quotation marks:

> John said, "I'm going."

Question marks come inside or outside, depending on whether the quotation is a question or not:

> Did John say, "I'm going"?
> John said, "Is he going?"

SOUNDS AND LETTERS – /l/

In the initial position, the spelling of /l/ is no problem. It is spelled **l** in all but a few exceptional cases, such as proper names like **Lloyd**.

The rule for spelling final /l/ is analogous to the rules for most of the other final consonants:

1. Final /l/ is spelled **ll** after simple vowels and after /au/ (**fall, small, all,** etc.).
2. After complex vowels, /l/ may be spelled in the **VCe** pattern. (It always is after /ī/, except in foreign words like ***heil***.)
3. Otherwise, final /l/ is spelled **l**.

In addition to this, it is to be noted that in the words **bull, full**, and **pull**, we use **ll** after the vowel sound /oo/. But the suffix **ful**, as in **cupful**, has single **l**.

Several words with the vowel sound /ō/ also take **ll: knoll, poll, roll, scroll, droll, troll, toll**. Words with this vowel before **l** may also be spelled **VCe** (**mole**) or with single **l** (*soul*). They are spelled with single **l** when the vowel sound is represented by two letters.

Teaching Suggestions

Review the naming of the fourteen consonant sounds listed. All have the alphabetic names except /ch/ ("chee") and /sh/ ("esh").

Pronounce the words beginning **line**, or have the children do so. They should have no difficulty thinking of five other words beginning with the sound /l/.

The writing exercise at the bottom of page 198 contains, in addition to /l/, some of the sounds whose initial spelling is a little complicated. Review these with this exercise. If necessary, start it on the chalkboard.

The Consonant Sound /l/

You have now studied fourteen consonant sounds and ways to spell them. These are the consonant sounds you have studied:

/p/ /t/ /k/ /ch/ /f/
/b/ /d/ /g/ /j/ /v/
/s/ /z/ /sh/ /r/

We name these sounds just as we do the letters of the alphabet, except for two of them. What are the two that we have to name in special ways? /ch/—chee /sh/—esh

The next consonant sound you will study is the first sound in **line** or **love**. Say the words and hear the sound.

We show this sound this way: /l/. When you want to speak about the sound, name it the way you do the letter l in the alphabet.

At the beginning of words we usually spell /l/ with the letter **l**. We do so in these words:

line	**love**	**lane**	**load**
little	**large**	**lend**	**like**

□ Write five other words that begin with the sound /l/ and the letter **l**. See p. 328.

□ Write these sounds as headings:

/l/ /r/ /s/ /z/ /sh/ /k/ /ch/

Write each word in the list below under the heading that shows the sound it begins with.

land /l/	roam /r/	seem /s/	cent /s/
shake /sh/	write /r/	leave /l/	rhyme /r/
sent /s/	sharp /sh/	care /k/	kind /k/
zip /z/	leg /l/	shovel /sh/	zero /z/
cheap /ch/	cereal /s/	chop /ch/	lemon /l/
cow /k/	chew /ch/	zoo /z/	so /s/

We spell /l/ in several ways at the end of words. In simple words that have just one vowel letter in them, we usually spell /l/ with two l's. Here are words in which we do.

fall	**tell**	**hill**	**doll**	**pull**
small	**bell**	**will**	**roll**	**full**
all	**smell**	**kill**	**stroll**	**dull**

Notice the vowel sounds in the words above. The words **fall, small, all** have the letter **a**, but they do not have the vowel sound /a/. They have a vowel sound you have not studied yet.

- What is the vowel sound in **tell, bell, smell?** /e/
- What is the vowel sound in **hill, will, kill?** /i/
- What is the vowel sound in **doll?** What is the vowel sound in **roll** and **stroll?** /o/; /ō/
- What is the vowel sound in **pull** and **full?** What is the vowel sound in **dull?** /o͝o/; /u/

After complex vowels, we sometimes spell /l/ with **le**, in the pattern **VCe.** Tell what the vowel sounds are in these words: /ī/ /ō/ /ā/ /ū/

smile hole dale rule

□ Use the **VCe** pattern to complete the words below. For **V** in **VCe**, use the vowel letter given. For **C**, use the letter **l.**

1. **o:** p<u>ole</u> 2. **a:** p<u>ale</u> 3. **i:** p<u>ile</u> 4. **u:** m<u>ule</u>

When the word with the sound /l/ does not have a single vowel letter before it, and when it is not spelled in the pattern **VCe,** we usually spell it with the single letter **l.** Here are words in which we spell /l/ with just the letter **l:**

tail	**meal**	**pail**	**boil**
feel	**jail**	**pool**	**kneel**
haul	**cool**	**pearl**	**girl**

199

Go through the questions and examples on page 199. Elicit from the class rhymes for **tell, bell, smell (fell, shell, well)**, for **hill, will, kill (Bill, fill, chill)**, for **roll** and **stroll (toll, poll)**. The **all** class can be postponed until the sound is studied. Other combinations with double **ll** do not have words likely to be in the child's vocabulary, except for those given.

Let the children practice /l/ in the **VCe** pattern with the exercise provided. Point out, in the last group of words, that in each case the single **l** is preceded by a consonant letter or by two vowel letters.

Extra Practice

The lists of words beginning with **fall** at the top of page 199 and with **tail** at the bottom may be dictated to the class. In doing this, do not mix up the lists—dictate all or some of the final **ll** words as a group, and then all or some of the final **l** words.

The Present and Past Tense of Verbs

Page 200 reviews the tense forms of **be** preparatory to the introduction of the present and past tense of verbs on page 201.

Tense Forms of Verbs

The rule for **verb phrase** now stands as follows:

$$\text{VP} \rightarrow \text{tense} + \left\{ \begin{array}{l} \textbf{be} + \left\{ \begin{array}{l} \text{NP} \\ \text{Adj} \\ \text{Adv-p} \end{array} \right\} \\ \text{verbal} \end{array} \right\}$$

Since in English the tense possibilities are just two, the rule for **tense** is this:

$$\text{tense} \rightarrow \left\{ \begin{array}{l} \text{present} \\ \text{past} \end{array} \right\}$$

From these rules we might derive such strings of morphemes as the following:

John + present + like + the + poem + now

Present will here apply to like to give **likes**, and the finished sentence will be "John likes the poem now."

Alternatively we could have:

John + past + like + the + poem + yesterday

Now past must convert like into **liked**. The morpheme tense applies to whatever follows it. This may be a verb or **be**. Or, as will be shown later, it may be the word **have** or a modal.

The Meaning of the Present Tense

It should be kept in mind that the English present tense does not very commonly express present action. More often it shows customary or repeated action, as in "John eats lunch in the cafeteria" or "Pete takes the bus." Sometimes it expresses future action: "John leaves tomorrow."

For present action, we ordinarily employ

The Present and Past Tense of Verbs

You have studied the present and past tense forms of **be.** The present tense forms are **am, is,** and **are.** What are the past tense forms?

The chart below shows the forms of **be** that we use with different kinds of subjects in the present tense and in the past tense. Study it.

Forms of *be* and Subjects

Subject	*Form of* ***be***	
	Present Tense	*Past Tense*
I	am	was
he, she, it indefinite pronoun singular subject	is	was
you, we, they plural subject	are	were

Before each of the sentences below there is a new subject in parentheses. Make a new sentence using the new subject and changing the form of **be.** Don't change the tense.

1. (The boys) The boy is here. The boys are . . .
2. (Everybody) The boys are here. Everybody is . . .
3. (The child) The children are hungry. The child is . . .
4. (John) The pupils are at the chalkboard. John is . . .
5. (The geese) The goose is noisy. The geese are . . .
6. (The elves) The elf was in the forest. The elves were .
7. (The bear) The bears were dangerous. The bear was . .
8. (The men) The man was quiet. The men were . . .
9. (No one) The children were in school. No one was . . .
10. (The women) The woman was angry. The women were

200

• Verbs, as well as the word **be**, have present tense and past tense forms. But verbs do not have as many as **be** does.

• Which of the sentences below do you think is a present tense form, and which do you think is a past tense form?

a. John likes the poem now. pres.
b. John liked the poem yesterday. past

• Which of these is a present tense form, and which is a past tense form?

c. The roses smell good now. pres.
d. The roses smelled good yesterday. past

• We often use the present tense of verbs to tell of something that happens all the time or from time to time. We often use it with words like "every day" or "usually" or "sometimes." Tell which of the verbs in the following sentences are present tense forms and which are past tense forms:

e. The children study every day. pres.
f. The children studied yesterday. past
g. Mary walks to school sometimes. pres.
h. Mary walked to school last week. past

• Tell what the verb is in each of these sentences. Tell whether it is a present tense form or a past tense form.

11. Sally smiled yesterday. past
12. Jane smiles sometimes. pres.
13. The boys laughed. past
14. The girls dance. pres.
15. Everybody sings. pres.
16. We played records. past
17. They waited for an hour yesterday. past
18. I sleep late every Saturday. pres.
19. He washed the car last week. past

what is called the progressive, or "**be** + ing," form: "John is eating lunch in the cafeteria," "Pete is minding the store."

Teaching Suggestions

Have the children examine the chart on page 200 which presents the agreement rules for subjects with **be** in simple sentences.

We could simplify the second and third parts of the chart thus:

Singular Subject:	**is**	**was**
Plural Subject:	**are**	**were**

However, it seems better to emphasize the fact that indefinite pronouns, though often plural in meaning, take **is** and **was**. **You** always takes **were**, though often singular in meaning.

Sentences 1–10 give review practice not only in the forms of **be** but also in the use of indefinite pronouns and irregular plurals as subjects.

Have the children answer the questions through Sentence h on page 201. Do Sentences 11–19 in class discussion. We do not yet give spelling rules for the past tense. These are introduced later.

Extra Practice

The pupils may do the last exercise on page 201 as an individual written assignment even though the spelling rules for past tense have not been given at this point. They may merely copy out each verb in Sentences 11–19, writing it under the heading *Present Tense* or *Past Tense*.

SECTION 4

The Rest of the Story

Here is the last part of "The Travels of a Fox."

However, the moment she unloosed the string, the ox got out, and the woman's little boy chased the ox out of the house and across a meadow and over a hill, clear out of sight.

After a while the fox came back. He took up his bag and knew at once that his ox was gone, and he said to the woman, "Where is my ox?"

And the woman said, "I untied the string just to take a little peep to find out what was in your bag, and the ox got out and my little boy chased him out of the house and across a meadow and over a hill, clear out of sight."

"Very well," said the fox, "I must have the little boy, then."

So he caught the little boy and put him in his bag and traveled.

At the next house he came to, he went in and said to the mistress of the house, "Can I leave my bag here while I go to Squintum's?"

"Yes," said the woman.

"Then be careful not to open the bag," said the fox.

The woman had been making cake, and when it was baked she took it from the oven, and her children gathered around her teasing for some of it.

"Oh, Ma, give me a piece!" said one, and "Oh, Ma, give me a piece!" said each of the others. And the smell of the cake came to the little boy in the bag, and he heard the children beg for

203

"The Travels of a Fox" (Fourth Part)

Read this last part of the story to the class after recalling what has happened previously.

The dramatization may now be undertaken, based upon the concept developed in the record.

Every episode in the story may constitute a separate scene. The scenes will, of course, be very short, but by giving so many it should be possible to let every child in the classroom participate in one way or another. Spontaneity, imagination, and fun should predominate. Too literal a performance should be avoided. For example, the fox may simply hide the character being put in his "bag" by throwing a shower curtain over him, and the curious woman may open the bag and let the character escape by removing the shower curtain.

the cake, and he said, "Oh, Mammy, give me a piece!"

Then the woman opened the bag and took the little boy out; she put the house-dog in the bag in the little boy's place, and the little boy joined the other children.

After a while the fox came back. He took up his bag and he saw that it was tied fast and he thought that the little boy was safe inside. "I have been all day on the road," said he, "without a thing to eat, and I am getting hungry. I will just step off into the woods now and see how this little boy I have in my bag tastes."

So he put the bag on his back and traveled deep into the woods. Then he sat down and untied the bag, and if the little boy had been in there things would have gone badly with him.

But the little boy was at the house of the woman who made the cake, and when the fox untied the bag the house-dog jumped out and killed him.

The Travels of a Fox (FOURTH PART)

- What did the fox intend to do with the little boy? eat him
- The story says that "things would have gone badly" with the little boy if he had been in the bag. What do you think this means? discuss
- What are the marks before **Where** and after the question mark called in this sentence: quotation marks

He said to the woman, "Where is my ox?"

- Find some other quotations in the story. discuss

A Letter About School

Write a letter to someone telling about some of the things you do in school. Don't tell about what you did on some particular day. Tell what you usually do.

You might tell about some of the things you study and the time of day you study them. You might tell about when you have lunch and when you play. You might tell what you like best of the things you do at school.

Remember how letters are written. Have your address and the date at the top. Have a greeting and a closing. Leave wide margins.

A Spelling Problem

Read the paragraph below. How many words can you find that are made up of the consonant sound /t/ and the vowel sound /ū/? How are they spelled?

> Two girls were talking to two boys. The boys said they were going to swim. The girls said that they would like to go too, but that their mothers didn't want them to. The two boys were sorry that the two girls couldn't swim too, but they knew that they had to do what their mothers told them to. The boys had to do that too.

Two, too, and *to* are homophones. **To** is what we use in expressions like **to go** and **to town.** What does **two** mean? What does **too** mean? Find two meanings of **too** in a dictionary. discuss

□ Write these sentences from dictation:

The two boys were too sleepy to stay awake.

I gave two apples to her, and two pears, too.

205

COMPOSITION

A Letter About School

This assignment is devised to elicit use of verbs in the present tense. The pupils will naturally use this form if they talk about what they ***usually*** do. They should not be instructed specifically to use the form.

It will probably be best to precede the assignment with a discussion of the things that usually go on in school so that everyone will have ideas on what to write about.

Review the mechanical rules mentioned in the third paragraph.

SOUNDS AND LETTERS

A Spelling Problem

Since these three words are, or may be, homophones, they present a persisting spelling problem. **Too** and **two** are regularly pronounced /tū/, and **to** may be. **To** is either a preposition (**to the chalkboard**) or what is called "the sign of the infinitive," occurring before verbs: **to go, to think**. In both uses, it is more commonly pronounced /tə/, with the vowel schwa, than /tū/, but it is pronounced /tū/ often enough to be confused with **too** and **two.**

Teaching Suggestions

Let the class find the /tū/'s in the paragraph. Write them on the chalkboard as they are found. Explain the differences in meaning. Dictate the two sentences at the bottom of the page, and check the performance of the individual pupils.

In dictating the sentences, which are rather long and complex, make sure by calling on a number of children to repeat the sentences that they are able to remember what the sentences are as they write them with books closed.

SOUNDS AND LETTERS – /aı/

There are three more vowel sounds to be presented in this volume – those of **haul, bout**, and **boil**.

The spelling of /aı/ in monosyllabic words is more complicated than that of any of the other vowels. There is no spelling that can be called the regular one, but rather a series of different spellings used in different words. There are certain clear restrictions imposed by contexts. For example, **o** can be used to spell /aı/, but only before certain consonants. These restrictions cannot, however, be formalized in simple rules.

Teaching Suggestions

Prepare for the study of /aı/ by reviewing the eleven vowels so far studied. Have the children give examples of each of the eleven. Write the headings between slant lines on the chalkboard and write the words given under the proper heading.

Extra Practice

Following are the ways in which /aı/ is spelled in monosyllabic words. The lists that follow each are intended to include all of the common words with the spelling given. These lists may be drawn on for dictation practice, which should always be given in homogeneous sets, not in mixed sets.

au: daub; applaud, fraud, laud; Gaul, haul, Paul, Saul; assault, fault, vault; faun; haunch, paunch, launch; flaunt, taunt, daunt, haunt, gaunt, jaunt, vaunt; applause, because, cause, clause, pause, gauze; sauce; taut. (The spelling **au** occurs only before the sounds /b/, /d/, /l/, /n/, /z/, /s/, /t/.)

The Vowel Sound /aı/

• You have studied eleven vowel sounds so far. Think of words that have these simple vowel sounds in them: See p. 324.

/i/ /e/ /a/ /u/ /o/

• Think of words that have these complex vowel sounds in them: See pp. 324–25.

/ī/ /ē/ /ā/ /ū/ /ō/

• The vowel sound /oo/ is like the simple vowels in some ways and like the complex vowels in others. We have /oo/ in **book** and **put** and **bush.** Give other words that have /oo/ in them. See p. 326.

• Tell what vowel sound is in each of these words:

push /oo/	**rain** /ā/	**school** /ū/	**eat** /ē/
fly /ī/	**seed** /ē/	**right** /ī/	**lane** /ā/

Our next vowel is a complex vowel. It is the vowel in **haul** and **fall.** Since we have no more single letters to show vowels with, we will again use two letters joined together and show it this way: /aı/. When you need to say it, say "awk," as if you were saying the last part of **hawk** or **talk.**

You have already seen that there are several ways of spelling the sound /aı/, since you have heard this vowel sound in these four words:

haul fall hawk talk

□ First study some of the common words in which /aı/ is spelled with the letters **au.** Pronounce the words below and listen to the /aı/ sound. Then close your books, and write them as they are dictated to you.

cause	**haul**	**applaud**	**haunt**
because	**Paul**	**fraud**	**jaunt**
clause	**fault**	**taunt**	**launch**

• Another common way to spell the sound /aı/ is with the letters **aw**. Study these words. What letters come after **aw** in them?

bawl l	**crawl** l	**scrawl** l	**shawl** l	**hawk** k
dawn n	**drawn** n	**lawn** n	**yawn** n	**squawk** k

□ A **scrawl** is words carelessly written by hand. A **shawl** is a cloth wrap women wear around their shoulders. Now close your book and write the words as they are dictated to you.

The letters **aw** are nearly always used to spell the /aı/ sound when the sound comes at the end of the word. Study these words:

claw	**draw**	**law**	**saw**
raw	**jaw**	**straw**	**gnaw**

□ Now close your book, and write the words as they are dictated to you.

The letter **a** may stand for the /aı/ sound in words in which the letter **l** comes next. You have already studied the following words with the /l/ sound at the end:

all	**ball**	**fall**	**hall**
call	**stall**	**tall**	**wall**

The /l/ sound comes after the /aı/ sound in these words also:

halt **malt** **salt**

There are four common words which used to have the /l/ sound but don't anymore. They still have the letter **l**, though, and the sound /aı/ is spelled with the letter **a**:

chalk **walk** **stalk** **talk**

□ Close your books, and write these words as they are dictated to you.

207

THE VOWEL SOUND /aı/ (CONTINUED)

aw: claw, draw, flaw, gnaw, law, saw, raw, squaw, jaw; bawl, brawl, crawl, drawl, scrawl, shawl, sprawl; dawn, drawn, fawn, lawn, prawn, yawn; hawk, squawk. (This is the regular spelling for /aı/ in final position. In addition it is used before /l/, /n/, and /k/. The spelling **au** is also used before /l/ and /n/, but the only homophone is **faun/fawn.**)

o: doff, golf, off, scoff, cough, trough; fog, bog, dog, frog, hog, log; dialogue, *etc.*; along, long, song, gong, strong, throng, wrong; boss, across, cross, loss, moss, toss; wroth, broth, cloth, froth, moth. (The spelling **o** occurs before /f/, /g/, /ŋ/, /s/, and /th/. It also occurs before /l/ in the word **golf**, which, however, is sometimes pronounced /gauf/. For some speakers, the **fog, bog** series has the vowel /o/ instead of /aı/.)

a: all, ball, fall, hall, call, stall, tall, wall; halt, malt, salt; false, waltz; balk, chalk, walk, stalk, talk; wash, squash. (Except for **wash** and **squash,** a for/aı/ appears only before the letter **l**, which is silent in the **balk, chalk** series. Many speakers pronounce **wash** and **squash** with the vowel /o/.)

ough: thought, bought, fought, ought, sought, wrought. (Actually or historically, these are all past tenses of verbs.)

augh: caught, taught. (These are the past tenses of **catch** and **teach**.)

oa: broad

awe: awe

GRAMMAR AND USAGE

Agreement of Subject and Verb

Before proceeding to the past tense of verbs, we review the agreement with subjects of the two present tense forms. This is review, at least, for those who used the third-grade book of this series. On page 217, we review the spelling of the s form.

There is a peculiarity in the agreement of English subjects and verbs that sometimes causes confusion. Singular subjects, which do not end with an s morpheme, agree with verbs that do end with an **s** morpheme. Plural subjects, which do end, usually, with an s morpheme, agree with verbs which do not:

The boy plays.
The boys play.

Of course, in simple sentences, this agreement is automatic for native speakers of English.

Teaching Suggestions

Go through the questions and examples on page 208. Establish the terms **simple form** and ***s* form**. These are self-explanatory, or nearly. Do the series beginning **eats** orally to afford practice in applying the terms.

The written exercise beginning **cause** contains just verbs whose s forms are made by the simple addition of the letter **s**. The simple forms of the verbs beginning **hauls** are written by simply omitting the **s**.

Agreement of Subject and Verb

The word **be** has three present tense forms and two past tense forms. What are the present tense forms of **be?** What are the past tense forms? am, is, are; was, were

Verbs have only two present tense forms and only one past tense form.

● The present tense forms of **like** are **likes** and **like.** The present tense forms of **walk** are **walks** and **walk.** What do you think the present tense forms of **talk** are? What would the present tense forms of **eat** be? talk, talks; eat, eats

● Give the present tense forms of these verbs:

find, finds; see, sees **find** **see** **wait** **jump** wait, waits; jump, jumps

Notice that one of the present tense forms of verbs ends in the letter **s** and the other does not. **Likes, talks, sees** all end in the letter **s. Like, talk, see** do not end in **s.**

We call the form of the verb that ends in **s** the **s form.** We call the other form the **simple form.**

● Look at these verbs. Tell which are **s** forms and which are simple forms: (s = s form; sim. = simple)

eats s	know sim.	smile sim.	watches s
wait sim.	gets s	rush sim.	turns s
stays s	dodges s	think sim.	tries s

□ Write the **s** form of these verbs:

cause s	want s	skip s	run s
play s	hope s	laugh s	drink s
stop s	enjoy s	fall s	stalk s

□ Write the simple form of these verbs:

haul, yawn	hauls	yawns	yells	applauds	yell, applaud
hop, call	hops	calls	thinks	stops	think, stop
jump, sing	jumps	sings	hopes	swims	hope, swim

We use the s form of verbs for the subjects for which we use the **is** form of **be.** We use the simple form of verbs for the subjects for which we use the **are** form of **be.** Study this chart.

Subject	*Present Tense Verb Form*
he, she, it indefinite pronoun singular subject	s form
they, we, you, I plural subject	simple form

● Find the verb in each of these sentences. Tell whether it is an s form or a simple form.

1. The boy smiles. s form
2. The girls smile. simple
3. We walk to school. simple
4. It works. s form
5. They use chalk. simple
6. Everybody likes Mr. Burbank. s form
7. Giraffes eat hay. simple
8. She talks to us. s form

□ Copy these sentences, but change the subject. Use the new subject given in parentheses. You will have to change the form of the verb too.

9. (The boy) The boys play. The boy plays.
10. (Everyone) They know the poem. Everyone knows . . .
11. (The dogs) The dog kills the fox. The dogs kill . . .
12. (The men) The man works in the garden. The men work . . .
13. (The woman) The women cook the dinner. The woman cooks . . .
14. (The teacher) The teachers explain the lessons. The teacher explains . . .
15. (His teeth) His tooth hurts. His teeth hurt.
16. (Sally) They live near us. Sally lives . . .
17. (The doctors) The doctor plays golf. The doctors play . . .

209

The chart on page 209, like the one for **be** on page 200, specifies the agreement of personal and indefinite pronouns, and for the same reason: there is some asymmetry between meaning and grammar. The indefinite pronouns are often plural in meaning, but they take the **s** form of verbs. **I** is singular in meaning, and **you** may be, but they take the simple form of verbs.

Use Sentences 1–8 on page 209 for identification of the two present tense forms in sentences. In 9–17 the new subject to be used requires in each case a change in the form of the verb. This is intended as a written exercise, but the first two may be done on the chalkboard as illustration.

Extra Practice

Sentences 1–8 may be assigned as a written exercise for all the pupils by simply having the pupils write each verb under the heading *Simple* or **s** *Form.*

TESTS AND REVIEW

Use these tests for diagnosis of weaknesses, and reteach as necessary.

Use page 213 for oral discussion.

Extra Test

If words with the sound /au/ are wanted for dictation, choose from the sets given on pages 206–07 of the Teacher's Edition. Dictate homogeneous sets in order to avoid doubt and confusion.

The following dictation test may be given. Say the word clearly. Use it in the sentence. Say it again and have the children write it.

1. **shore** Waves splashed on the **shore**. **shore**
2. **bush** The bird flew into a **bush**. **bush**
3. **road** The car drove down the **road**. **road**
4. **wreck** The accident left the car a **wreck**. **wreck**
5. **score** The **score** was six to nothing. **score**
6. **car** His **car** is painted green. **car**
7. **hair** Ellen brushed her **hair**. **hair**
8. **smell** We could **smell** smoke. **smell**
9. **pail** Harry brought a **pail** of water. **pail**
10. **hole** A woodchuck lives in a **hole**. **hole**

Tests and Review

□ **TEST I Sounds and Letters**

1. Write these headings:

/sh/ /s/ /z/ /ch/ /j/

Write each word below under the heading that shows the sound it ends with.

If you make mistakes, study pages 114, 140, 164, 171, 176, 187.

rush /sh/	church /ch/	gas /s/	batch /ch/
judge /j/	harsh /sh/	buzz /z/	rice /s/
lose /z/	barge /j/	fuss /s/	fox /s/

2. Write these headings: /l/, /r/, /s/, /f/.

Write each word below under the heading that shows the sound it ends with.

If you make mistakes, study pages 140, 164, 176, 192, 198.

fall /l/	tear /r/	race /s/	pile /l/
fire /r/	wife /f/	sir /r/	grass /s/
meal /l/	fear /r/	laugh /f/	gas /s/

3. Write these headings: /a/, /o/, /au/.

Write each word below under the heading that shows the vowel sound it has.

If you make mistakes, study pages 6, 193, 206.

fat /a/	dot /o/	haul /au/	call /au/
car /o/	ranch /a/	law /au/	Bob /o/
star /o/	cause /au/	bad /a/	talk /au/

4. Copy each word below. After it, write another word that rhymes with it and has the rhyming part spelled the same way. For instance, if the word were **law**, you might write **raw** or **straw**.

jaw caw	fall hall	crawl bawl	walk talk
tell yell	mile pile	yawn lawn	pill fill

□ **TEST II Grammar**

1. Write the plurals of these nouns. If you make mistakes, study page 186.

dish dishes	book books	lady ladies	toy toys
fox foxes	mouse mice	man men	woman women
switch switches	party parties	child children	sheep sheep

2. Write the singular forms of these nouns. If you make mistakes, study page 186. baby, shelf, day, goose, half, sky match, story

babies	shelves	days	geese
halves	skies	matches	stories

3. Copy each of these sentences, but use **yesterday** in place of **now**, and change the form of **be** from the present to the past.

If you make mistakes, study pages 188–89.

a. John is at his desk now. . . . was . . . yesterday.
b. The girls are hungry now. . . . were . . . yesterday.
c. I am sleepy now. . . . was . . . yesterday.
d. Mary is my friend now. . . . was . . . yesterday.
e. The sheep are in the meadow now. . . . were . . . yesterday.
f. The fox is angry now. . . . was . . . yesterday.
g. I am upstairs now. . . . was . . . yesterday.

4. Copy each of these sentences, but use the new subject given in parentheses. Change the form of **be**, but do not change the tense.

If you make mistakes, study pages 194–95.

a. (I) He is at his desk now. I am . . .
b. (She) They are in the playground now. She is . . .
c. (You) John is inside now. You are . . .
d. (They) He was here yesterday. They were . . .
e. (I) The girls were tired last night. I was . . .
f. (Someone) They were in the house. Someone was . . .
g. (We) She was at the party. We were . . .
h. (You) I was in that car. You were . . .

211

TEST II Grammar (CONTINUED)

5. Copy each of the sentences below, but use the new subject that is given in parentheses. Change the form of the verb from simple to **s** form or from **s** form to simple form.

If you make mistakes, study pages 208–09.

a. (The boys) David walks to school. The boys walk . . .
b. (We) She likes Mr. Rossi. We like . . .
c. (He) They help the teacher. He helps . . .
d. (No one) They collect the papers. No one collects . . .
e. (The foxes) The fox sleeps in the daytime. The foxes sleep . . .
f. (I) She cleans the chalkboard. I clean . . .
g. (Mr. Green) The children ride on the bus. Mr. Green rides . . .
h. (The geese) The goose makes a lot of noise. The geese make . . .

6. Copy each of the sentences below, but make the subject plural instead of singular. For example, if the sentence were "The boy walks to school," you would write "The boys walk to school." You must change the form of **be** or the verb.

If you make mistakes, study pages 194, 208.

a. The girl walks to school. The girls walk . . .
b. The child studies geography. The children study . . .
c. The mouse is timid. The mice are . . .
d. The man works hard. The men work . . .
e. The match is wet. The matches are . . .
f. The woman speaks to us. The women speak . . .
g. The foot is large. The feet are . . .

TEST III Writing and Vocabulary

Write five sentences. In each sentence, use one of these words:

1. travel 2. ox 3. unloosed 4. bumblebee 5. mistress (Answers will vary.)

REVIEW Ideas and Information

1. Tell what we call the marks before **I** and after **hungry** in this sentence: He said, "**I am hungry.**" What do they mean? quotation marks; words of speaker
2. What do you put at the top of your paper on the right when you write a letter? your address and the date
3. What mark do you put after the greeting and the closing of the letter? comma
4. The vowel sound /o/ as in **Bob** is spelled **a** when a certain sound follows. What is the sound? /r/
5. What letters stand for the sound /ɑː/ when /ɑː/ comes at the end of a word? aw
6. We usually spell /r/ with the letter **r** at the beginning of words. But there are two other ways that we use in a few words. What are they? wr, rh
7. One way to write /l/ at the end of words is with the single letter **l** as in **meal** or **girl.** Tell two other ways. ll, VCe pattern (le)
8. When do we change **y** to **i** and add **es** to spell the plural of nouns? when noun ends in consonant-y
9. When do we add **es** to spell plurals? after ch, sh, s, z, x
10. What are the past tense forms of **be?** was, were
11. There are two present tense forms of verbs. What are they called? s form, simple form
12. What does the word **tense** mean? time
13. What personal pronouns are used with the **was** form of **be?** I, he, she, it
14. Which of the two present tense forms of verbs is used with singular subjects? is
15. Name a word which forms the plural by adding **en** to the singular form. oxen
16. Name a word which is spelled the same in the plural as in the singular. deer

PART
8

SECTION 1

A Poem

What would you do if you were out walking and met a witch? Read what this child did when she met an old woman she feared was a witch.

W Is for Witch

I met a wizened woman
As I walked on the heath,
She had an old black bonnet
Her small eyes peeped beneath,
Her garments were so shabby
She couldn't have been rich,
She hobbled with a crutchstick,
And I knew she was a Witch.

She peered at me so slyly
It made my heart feel queer,
She mumbled as she passed me,
But what I couldn't hear.
I smiled at her for answer
And wished her a good day,
She nodded and she chuckled
And she hobbled on her way.

And so I got home safely.
I didn't drop the eggs,
My nose had grown no longer,
My legs were still my legs,
I didn't lose my penny
Or tumble in a ditch —
So mind you smile and say "Good Day,"
When *you* meet a Witch.

ELEANOR FARJEON

215

LITERATURE – "W is for Witch"

This poem, by a very talented writer of children's poems, displays not only a wholesome attitude but also delicacy and good taste. Those who have used the third-grade book will have read several other poems by Farjeon: "Boys' Names," "Girls' Names," "Jenny White and Johnnie Black," "Knowledge."

Teaching Suggestions

Read the poem or play the recording of it. Let the children read portions of it aloud. Each child may be given four lines – half a stanza – to read.

VOCABULARY AND MEANING

"W is for Witch"

Teaching Suggestions

Go through the questions, making sure that the details of the poem are understood. Let the pupils check back to the text as necessary. If possible, have the children use dictionaries at the points suggested. At each such place, one child may be appointed to find the appropriate dictionary entry and read it to the class.

Try to get the class to explain the point of the first part of the last stanza: all the things that the witch didn't do are the things that witches are supposed to do to those who offend them.

W Is for Witch

- **Wizened** means old and dried up. Who is called wizened in the poem "W Is for Witch"? a woman
- A **heath** is open land with not much growing on it — just grass, perhaps, and a few bushes or small trees here and there. What vowel sound does **heath** have? How do you write this sound? /ē/; ea
- What was it the old woman wore that "her small eyes peeped beneath"? an old black bonnet
- **Garments** are clothes. What word describes her clothes? How could the girl tell she was poor? shabby; be of her cl
- What does **hobble** mean? See a dictionary. discuss
- What is a **crutch?** Then what does **crutchstick** mean? discu
- How did the old woman peer at the girl? How did this look make the girl feel? See stanza 2.
- Can you tell from the other words around it what **mumbled** means? What does it mean? Check your guess by using a dictionary. discuss
- What did the girl do to answer the old woman? Do you think she smiled even though she was afraid? What does "wished her a good day" mean? What did the old woman do after the child did that? discuss
- Can you tell why the child might have thought that she would **not** get home safely? Why do you think she might have expected to drop her eggs? discuss
- Her nose didn't grow any longer. Why was she afraid that it would? What do you suppose she was afraid that her legs would turn into? What other things didn't happen to her? discuss
- **Mind** in the next to the last line means to remember to do something or to be sure to do it. What does the poet tell you to **mind?** that you smile and speak courteously; discuss implications

Spelling the *s* Form of Verbs

Do you remember the rule for spelling the plural of regular nouns? There are three parts to the rule:

1. If the singular ends in "consonant-y," change the **y** to **i** and add **es.**
2. If the singular ends in **ch, sh, s, z,** or **x,** add **es.**
3. In other words, just add s.

The s form of verbs is spelled in exactly the same way. The s form of **try** is written **tries.** The **s** form of **hurry** is written **hurries.** To write the s form of **cry,** what would you do to the **y?** What letters would you add? change y to i; add es

Look at the simple form and the s form of each of these verbs:

Simple Form:	**smash**	**buzz**	**coax**
***s** Form:*	**smashes**	**buzzes**	**coaxes**

● Why do we add **es** in these words? they end with ch, z, x

□ Write the **s** form of these verbs:

catch es	scurry scurries	play s	fuss es
cry cries	teach es	know s	race s
please s	fly flies	preach es	worry worries

● There are many nouns that have irregular plurals written in special ways. How do you write the plurals of **man, mouse, elf, deer?** men, mice, elves, deer

But there are just a few verbs that have s forms which are not written in a regular way.

The s form of the verb **have** is **has.**

The verbs **go** and **do** add **es** in the s form. We write **goes** and **does.**

● **Does** is special in talking, too. Is the vowel the same in **do** and **does?** Do **goes** and **does** rhyme? no; no

Spelling the *s* Form of Verbs

The pronunciation rules for the formation of the following forms are identical for regular words: noun plurals, s form of verbs, possessives of nouns. (See pages 166–67 of the Teacher's Edition.) The spelling rules are identical for regular noun plurals and s forms of verbs – but not for possessives.

Teaching Suggestions

Go over the rule for noun plurals. Ask for examples that will illustrate each part of the rule. Have the children point out the letters with which **smash, buzz,** and **coax** end and perceive that the second part of the spelling rule thus applies to them. Let the children do the series beginning **catch** as a written exercise, but start it on the chalkboard if necessary.

There are three verbs that have irregular s forms in speech: **have, do,** and **say.** If **have** were regular, it would have the s form ***haves.** If **do** were regular, **does** would rhyme with **booze** rather than with **fuzz.** If **say** were regular, **says** would rhyme with **pays** rather than with **fez. Go/goes** is regular in pronunciation but forms the written s form by addition of **es** rather than s.

Extra Practice

The pupils may substitute the subject in parentheses for the subject of each of these sentences and change the verb to the s form to accord with the new subject.

1. (The horse) Horses eat hay.
2. (A cow) The cows come to the barn.
3. (The pig) The pigs have a pen.
4. (The chicken) The chickens lay eggs.
5. (A calf) Calves follow the cows.
6. (The duck) The ducks swim.
7. (A farmer) Farmers do hard work.
8. (A child) Children drive the cows home.
9. (A boy) The boys watch the cattle.
10. (A woman) Women carry the milk.

SOUNDS AND LETTERS

Other Spellings of /ɑɪ/

The complications of spelling the vowel /ɑɪ/ inspired this parody of Housman's "Loveliest of Trees."

Most difficult of vowels, the /ɑɪ/ sound now
Employs the letters used for bough
In spelling words like cough and trough—
Enough to make a pedant scoff.

Of nine and twenty sounds and ten,
Thirty will not come again.
Since in these thirty /ɑɪ/ must fall,
I'll not be cross, I shall not bawl.

Instead I'll linger to enjoy
The stark simplicities of /oi/,
Balked not by awe that /ɑɪ/ hath wrought,
Nor taut from all that I've been taught.

ARCHIE SLUTER

The letters **gh** in words like **bought, caught, cough**, as well as in words like **rough** and **tough**, stood for a sound persisting into Middle English and then disappearing. It was a velar fricative, approximately like the final consonant in German *doch*.

Teaching Suggestions

Work through the questions and examples. Use dictation in homogeneous sets to give practice in spelling words with /ɑɪ/. The exercise at the bottom of page 219 should probably be done first as an oral exercise with class discussion, and then as an individual written assignment.

Extra Practice

Use the lists given on pages 206–07 of the Teacher's Edition if further examples of the different spellings of /ɑɪ/ are wanted.

Other Ways of Spelling the /ɑɪ/ Sound

• You have studied three ways of spelling the sound /ɑɪ/. We often spell it **au**, as in **haul** or **cause**. Give other words in which the sound /ɑɪ/ is spelled with the letters **au.** See p. 328.

• We sometimes spell the sound /ɑɪ/ with the letters **aw**, as in **dawn** and **claw.** Give other examples of this way of spelling /ɑɪ/. When do we nearly always spell /ɑɪ/ with the letters **aw?** See p. 328; at the end

• The sound /ɑɪ/ is often spelled with the letter **a** when the letter **l** comes next. **Call, malt,** and **talk** are examples. Can you give others? See p. 328.

There are still other ways of spelling the vowel sound /ɑɪ/. In some words we spell it with the letter **o.** Study the words in the list below. Notice the letters that come after **o.**

long	**song**	**strong**	**wrong**
broth	**froth**	**cloth**	**moth**
boss	**loss**	**cross**	**toss**

□ Now close your book, and write those words as they are dictated to you.

The words in the list below are also spelled with the letter **o.** Many people say these, too, with the vowel sound /ɑɪ/. Do you?

fog	**bog**	**dog**	**frog**
hog	**log**	**off**	**scoff**

People who don't use the sound /ɑɪ/ in these words use the sound /o/, as in **not** and **car.** Either way is a good way to say them, but they are always written with the letter **o.**

In the words **wash** and **squash** some people use the sound /ɑɪ/ and some the sound /o/. Which do you use? Everybody spells them with the letter **a.**

218

Here are two words you have already studied:

rough **tough**

What consonant sound do these two words have at the end? What vowel sound do the letters **ou** stand for? /f/; /u/

The letters **ou** stand for the sound /u/ in **rough** and **tough.** But in other words that have the letters **gh,** the letters **ou** may stand for the sound /ɑɪ/. Here are two words in which **ou** stands for /ɑɪ/:

cough **trough**

• What consonant sound do the letters **gh** stand for in the words **cough** and **trough?** /f/

The letters **ou** also stand for the sound /ɑɪ/ in these verbs:

bought **fought** **thought**
ought **sought** **wrought**

• In these words the letters **gh** used to stand for a sound. Do they stand for a sound now? no

□ Close your books, and write those six words from dictation.

There are two verbs with the sound /ɑɪ/ that are spelled with the letters **au** followed by **gh:**

caught **taught**

• Do the letters **gh** stand for a sound in these words? no

In two words, the sound /ɑɪ/ is spelled in quite special ways. These are **awe** and **broad. Awe** means a kind of fear. What does **broad** mean? wide

□ For each of the words below, write another word that rhymes with it and has the rhyming part spelled in the same way. See pp. 328, 329.

hall, ought	call	bought	dog	halt	log, salt
saw, haul	raw	Paul	scrawl	lawn	crawl, yawn
long, off	song	scoff	cross	trough	boss, cough
taught, broth	caught	froth	walk	taunt	talk, haunt

GRAMMAR

The Past Tense of Verbs

We introduce verb past tense with regular forms, excluding for the moment words like **study** which change **y** to **i** and add **ed**. These are regular verbs too—all verbs ending in "consonant–y"—are written this way; they are just more complicated.

Teaching Suggestions

Go through the questions and examples. Review the terms **simple form** and ***s* form** and the rules of agreement of subjects with these forms. Introduce the past tense of verbs with the material toward the end of page 220.

The Past Tense of Verbs

• You know that there are present tense forms and past tense forms of verbs and of **be.** What are the present tense forms of **be?** What are the past tense forms of **be?** am, is, are; was, were

• Verbs have two present tense forms. What do we call forms like these? s form

walks tries hopes sees

What do we call forms like these? simple form

walk try hope see

• Which present tense form of verbs do we use with the personal pronouns **he, she,** and **it?** Which present tense form do we use with the personal pronouns **they, we, you,** and **I?** s form; simple form

• Which present tense form of verbs do we use with singular subjects? Which form do we use with plural subjects? s from; simple form

• Which form do we use with indefinite pronouns? s form

Verbs have just one past tense form. Most verbs make the past tense in a regular way. Some make it in special ways. We will study the past tense form of regular verbs first.

• The past tense of the verb **play** is **played.** Look at these sentences. Which one has a verb in the past tense? b

a. He plays every day.

b. He played yesterday.

c. They play in the afternoon.

• What letters do we add to **play** to write the past tense? ed

• The past tense of **walk** is **walked.** What is the past tense of **talk?** What letters do we add to write the past tense of these verbs? talked; ed

□ 1. The past tense forms of the verbs in the list below are written the way those of **play, walk,** and **talk** are. Write the past tense of each verb.

watched, jumped **watch jump look scream** looked, screamed

□ 2. Copy each of the sentences below, but change **every day** to **yesterday.** You will then also have to change the verb to the past tense form.

a. They watch the planes every day. . . . watched . . . yesterday.
b. He screams every day. . . . screamed yesterday.
c. Steve plays every day. . . . played yesterday.
d. The girls walk to school every day. . . . walked . . . yesterday.
e. The teacher explains the lesson every day. . . . explained . . . yesterday.
f. The boys clean the chalkboard every day. . . . cleaned . . . yesterday.
g. The children work every day. . . . worked yesterday.

● When a verb ends with the letter **e,** we just add the letter **d** to form the past tense. The past tense of **hope** is **hoped.** What is the past tense of **change?** What is the past tense of **love?** changed; loved

□ 3. Copy each of the sentences below, but change the form of the verb from the present tense to the past tense.

a. The children walk to school. . . . walked . . .
b. Jim likes the poem. . . . liked . . .
c. Mr. Burbank washes his car. . . . washed . . .
d. She peers at me. . . . peered . . .
e. I smile at her. . . . smiled . . .
f. We wish her a good day. . . . wished . . .
g. She hobbles on her way. . . . babbled . . .
h. They chuckle. . . . chuckled . . .
i. She mumbles something. . . . mumbled . . .
j. The clerks tie packages. . . . tied . . .
k. He mails our letters. . . . mailed . . .
l. Miss Jones lives on our street. . . . lived . . .

221

Let the children do the exercises on the writing of past tense forms on page 221. All of the verbs here make the past tense by the addition of **ed** or **d**. These should be done individually as written exercises, though they may be started on the chalkboard.

LITERATURE

"The Ad-dressing of Cats" (First Part)

Most of the poetry of the late T. S. Eliot is hard enough for the seasoned scholar, let alone the innocent fourth-grader. But occasionally Eliot used his great poetic skill in simple ways on light subjects, particularly the subject of cats. The children should perceive the fun in this poem if they bring close attention to it.

Teaching Suggestions

Read this first of the three parts or play the recording of it. Let the children read parts of it aloud, taking two or four lines each.

SECTION 2

A Poem

We will study this poem by T. S. Eliot in three parts.

The Ad-dressing of Cats

You've read of several kinds of Cat,
And my opinion now is that
You should need no interpreter
To understand their character.
You now have learned enough to see
That cats are much like you and me
And other people whom we find
Possessed of various types of mind.
For some are sane and some are mad
And some are good and some are bad
And some are better, some are worse —
But all may be described in verse.
You've seen them both at work and games,
And learnt about their proper names,
Their habits and their habitat:
But
How would you ad-dress a Cat?
So first, your memory I'll jog,
And say: A CAT IS NOT A DOG.

The Ad-dressing of Cats (FIRST PART)

An **interpreter** is someone who tells what something means. If a person spoke to you in French, you might need an interpreter to tell you what he said.

Character means the way someone thinks and acts.

- What can you understand without an interpreter?
- **Possessed of** means in possession of, or having. The poet says that people have **various** (different) **types** (kinds) of minds. What does the poet say cats are like? How are they like you, me, and other people?
- A **sane** person is someone who has a healthy mind, who isn't crazy. Then do you think the word **mad** in this line can mean angry, or insane? Check your answer by using a dictionary.
- Does **all** in the line "But all may be described in verse" mean **all people, all minds,** or **all cats?**
- Can you think of any proper names that cats have? How would you begin to write such a name?
- An animal's **habitat** is the place where it naturally lives. What are some habits of a cat?
- The poet says he'll **jog** your memory before telling you how to address a cat. What does **jog** mean?

223

VOCABULARY AND MEANING

"The Ad-dressing of Cats" (First Part)

Explain that the poet is going to tell how one should speak to a cat and that he thinks, or pretends to think, that this is a very serious business. First he wants to be sure that we know just what cats are like.

Go through the questions to be sure that the lines are understood. Be sure that the pupils understand what **verse** means. Let the dictionary be used as necessary.

The first question on page 223 is designed to focus attention on the antecedent of **their** in the fourth line of the poem. The pupils should find that "their character" refers to the **character of cats**.

The question about the meaning of the word "all" is another instance of the need to understand a reference in order to understand the poem. Here "all" refers to **all people**.

COMPOSITION – A Paragraph to Write

Prepare for the writing assignment by class discussion of the poem "W is for Witch," first rereading the poem to the class. The paragraphs should consist of well-developed answers to the questions asked here.

ETYMOLOGY – What Words Come From

Gender in English is covert, not overt. English nouns are masculine, feminine, or neuter, but these divisions are not generally marked – as gender is in some languages – by features of the nouns themselves. Rather they are marked by the reference to nouns of the three personal pronouns **he, she**, and **it**.

The masculine/feminine distinction is, however, often further shown by various devices. One is the occurrence of words in pairs: **man** and **woman, father** and **mother, uncle** and **aunt, fox** and **vixen, tom** and **tabby**. There are sometimes subtleties. In the primary grades, one speaks of **boys** and **girls**, but from adolescence onward, it is **fellows** and **girls**.

Another gender-showing device is the feminine suffix **ess**, which yields the pairs given on this page and a number of others, like **sorcerer/sorceress, count/countess, god/goddess**.

A more common way of specifying masculine or feminine in modern English is to prefix to a noun a gender-showing word: **woman lawyer, lady doctor, male nurse, bull elephant, nanny goat, billy goat, she-devil**.

Such words as **man, woman, boy, girl, lady** are sometimes suffixed to nouns to show gender: **iceman, nursemaid, cowboy, cowgirl, laundryman, laundrywoman**.

A Paragraph to Write

Do you remember the poem about the girl and the witch? If not, turn back to page 215 and read it again.

Write a paragraph telling why the girl thought the old woman was a witch. What did the woman look like, and what did she do to make the girl think that? Tell what courteous thing the girl did, even though she was afraid.

What Words Come From

We often have pairs of words for males and females — **man** and **woman, boy** and **girl, rooster** and **hen.** Sometimes one such word is made from the other. The word **tigress** is made from **tiger.** What is a **tigress?** female tiger

- What word do you think **lioness** is made from? lion
- Tell which of these words are names for males and which for females. Then use a dictionary to check.

actress	**stewardess**	**waitress**	females
actor	**steward**	**waiter**	males

The word **witch** was made from the noun **wizard. Wizard** used to mean about what **witch** does now, except that it meant a man. It meant a man who cast evil spells on people and made unpleasant things happen to them.

Now when we say a person is a wizard we mean something different. We mean that he does something so well that it is almost like magic.

- Do you think it is a compliment to call a man a wizard? Do you think it is a compliment to call a woman a witch? yes; no

224

The Consonant Sound / m /

• Say these words and listen to the consonant sound that is the first sound in both words:

man **move**

We make this sound by putting our lips together as we do when we make the sound / p /. What other sound, besides this one and the sound / p /, do we make by closing the lips?

We show the first sound of **man** and **move** in this way: / m /. In speaking of the sound, we name it the way we do the letter **m.**

At the beginning of words, the sound / m / is written with the letter **m:**

man	**move**	**more**	**mush**
mother	**mean**	**maybe**	**mouse**

• Think of five other words that begin with the sound / m / and the letter **m.** See p. 329.

• At the end of words, we spell / m / with the letter **m** after simple vowels. Tell the vowel sound in each of these words:

ham /a/	**hem** /e/	**hum** /u/	**him** /i/
Tom /o/	**ram** /a/	**Kim** /i/	**them** /e/

There are two common words in which the sound / u / is spelled with the letter **o** and is followed by / m /. These words are spelled **–ome: come, some.**

After complex vowels, we may spell / m / in the pattern **VCe.** Study these words:

same **time** **home** **lame**

• Think of other words that have this pattern. See p. 329.

If final / m / is not spelled in the **VCe** pattern after a complex vowel, we spell it with just the letter **m:**

room **seem** **roam** **beam**

• Think of other words like these. See p. 329.

SOUNDS AND LETTERS–/m/

English has three nasal sounds–/m/, /n/, and /ŋ/. These correspond in articulatory position to the stops /p/ and /b/, /t/ and /d/, and /k/ and /g/. Thus /m/, like /p/ and /b/, is made at the lips. The difference is that in the formation of /p/ and /b/, the lips open and the sound issues from the mouth, whereas in /m/ the lips remain closed and the sound issues through the nasal passage.

Teaching Suggestions

Go through the explanation, questions, and examples. The spelling of /m/ is not particularly difficult. Initially it is regularly spelled **m.** Finally it is regularly spelled **m** after simple vowels, with just a few exceptions like **come** and **some.** If not preceded by a simple vowel, /m/ may be spelled in the **VCe** pattern or with just **m.**

Here are other examples of final /m/ in the **VCe** pattern:

tame	**came**	**shame**	**grime**	**dime**
slime	**chime**	**dome**	**perfume**	**flume**

Here are others with **m** after complex vowels:

foam	**loam**	**team**	**teem**	**dream**
stream	**boom**	**broom**	**loom**	**claim**

GRAMMAR

Verbs with Irregular Past Tense Forms

It has been seen that the formation of plurals and **s** forms in English can be described in threefold rules for both speech and writing. The pronunciation rule is similar to the spelling rule, though not parallel to it. A similar situation exists for the past tense of verbs.

The Rules for Forming the Past Tense

The three-part spelling rule for regular verbs is given at the bottom of page 226. This is the pronunciation rule:

1. If the simple form ends in /t/ or /d/, add the syllable /əd/ to form the past tense: **hated, baited, ended, crowded**.
2. If the simple form ends in a voiceless sound other than /t/, add /t/: **sipped, talked, marched, laughed, missed.**
3. Otherwise, add /d/: **robbed, lagged, judged, roved, buzzed, called, bored.**

Note that in spelling the past tense of verbs ending in a single consonant preceded by a single vowel, the consonant is doubled, according to the regular rule, before addition of **ed: rob/robbed, star/starred**. Otherwise, addition of **ed** would produce a **VCe** pattern: **robed, stared**.

Irregular Verbs

We call a verb irregular if it forms its past tense in any other way than by the pronunciation rule given on this page in the Teacher's Edition. A few such verbs have an irregular **s** form also, and several have additional irregularities in the participle.

A large percentage of what are called "errors in usage" involve irregular verbs. Such errors may abound not only among children whose families speak dialects quite different from that of standard English, but also among those who have grown

Verbs with Irregular Past Tense Forms

• What do we add to these verbs to write the past tense? ed

reach talk peer miss

• What do we add to these to form the past tense? d

hope like please describe

□ When the simple form of the verb ends in **e**, we add **d** to write the past tense. When it doesn't end in **e**, we add **ed.** Copy these sentences, but change the verb to the past tense:

They like our play. . . . liked . . .
He talks to our teacher. . . . talked . . .
She describes her friend. . . . described . . .

• When the simple form of a verb ends in the letter **y** with a consonant letter before it, we make the past tense a little differently. We change the **y** to **i** and add **ed.** The past tense of **study** is **studied.** How do you write the past tense of **hurry?** hurried

□ You see that this is very like the way we write the plural of nouns and the **s** form of verbs that end in "consonant-y." Write the plural of these nouns:

babies, ladies **baby lady flurry spy** flurries, spies

□ Write the **s** form of these verbs:

studies, hurries **study hurry worry try** worries, tries

□ Write the past tense form of these verbs:

studied, played **study play wait change** waited, chang
tried, raced **try race worry copy** worried, copied

This is the whole rule for writing the past tense of regular verbs:

1. If the simple form ends in "consonant-y," change **y** to **i** and add **ed.**
2. If the simple form ends in **e**, add **d.**
3. In other verbs, add **ed.**

A good many verbs have irregular past tense forms. These irregular forms must be learned one by one as we learn our language. You probably have learned most of them already, though you may not know how to write all of them.

For some verbs, you may have learned to say a past tense form different from the one you learn in this book. If so, you must practice using the form given here in your work in English.

The past tense form of **see** is **saw**. We use **see** and **saw** in sentences like these:

I see the car now.

I saw the car yesterday.

Study the past tense forms of these verbs:

eat: **ate** drink: **drank** speak: **spoke**

□ Copy each of the following sentences, but change the verb from the present tense to the past tense. Some of the verbs are regular, and some are not, but the only irregular past tense forms are those mentioned above: **saw, ate, drank, spoke.**

1. John copies the sentence. . . . copied . . .
2. Sally drinks milk. . . . drank . . .
3. Pete likes the story. . . . liked . . .
4. They see the chalkboard. . . . saw . . .
5. We speak to the children. . . . spoke . . .
6. The woman walks with a stick. . . . walked . . .
7. I eat lunch at noon. . . . ate . . .
8. My brothers stay at home. . . . stayed . . .
9. The sheep eat grass. . . . ate . . .
10. They try hard. . . . tried . . .
11. The babies cry in their cribs. . . . cried . . .
12. I see a ship. . . . saw . . .
13. He eats his cereal. . . . ate . . .
14. The witch drinks vinegar. . . . drank . . .

up hearing the approved speech forms. The irregularities among verbs are so numerous that the child simply finds it hard to control the data, to manage the different items scattered beyond the rule. So he recapitulates regularizing and leveling processes that have been going on in the language for millenia and have only recently—and only partially—been stemmed by universal education.

The tendencies are of several kinds. One is simply to apply the rule—to make the irregular word regular:

* He throwed me the ball.
* John sticked up for me.

Perhaps more commonly, the past participle form is used for the past tense:

* We seen it.
* They run all the way.

Sometimes a past tense is used as a participle:

* They had drove away.

Very occasionally, the tendency is reversed, a regular verb becoming irregular by some analogy with a group of irregular verbs. An example is **dive** which, after centuries of regularity, now seems most often to occur with the past tense **dove**.

Teaching Suggestions

Work through page 226, in which the only new material is the spelling of past tenses of verbs of the **study** type and the statement of the rule for regular verbs. The pupils should have little trouble with the **study/studied** correspondence, since it is essentially the same as that for **lady/ladies**, which they already know.

Only four verbs with irregular past tense forms are presented on page 227, but the sentences at the bottom of page 227 mix these with regular verbs. Sentences 1–14 may be done orally as class work before they are assigned as a written exercise.

LITERATURE

"The Ad-dressing of Cats" (Second Part)

The subject of dogs and cats and particularly their personality contrasts is one perennially favored by writers of light verse. Eliot's treatment is surely one of the most skillful.

Reread or replay the first part. Then familiarize the children with this stanza.

SECTION 3

More of the Poem

The poet jogged our memory about cats in the first part of this poem, and now he talks about dogs.

Now Dogs pretend they like to fight;
They often bark, more seldom bite;
But yet a Dog is, on the whole,
What you would call a simple soul.
Of course I'm not including Pekes,
And such fantastic canine freaks.
The usual Dog about the Town
Is much inclined to play the clown,
And far from showing too much pride
Is frequently undignified.
He's very easily taken in —
Just chuck him underneath the chin
Or slap his back or shake his paw,
And he will gambol and guffaw.
He's such an easy-going lout,
He'll answer any hail or shout.

Again I must remind you that
A dog's a Dog — A CAT'S A CAT.

The Ad-dressing of Cats (SECOND PART)

- Does the poet believe dogs are bad tempered and mean, or that they just pretend to be?
- **Peke** is short for Pekinese. **Canine** means of the dog family. **Fantastic** means something imaginary or untrue. Can you guess what a **freak** is? Check your guess by using a dictionary.
- The poet talks about dogs and cats as if they were people. We say "man about the town," but we wouldn't say "dog about the town." What is the dog **inclined** to do; that is, likely to do?
- **Frequently** means often. What does he say the dog frequently is?
- To **take someone in** means to fool him. The poet names three ways a dog can be taken in. What are they?
- To **gambol** means to run about playfully. **Guffaw** means laugh loudly. A **lout** is a lazy, foolish fellow. Do you think the poet rather likes dogs?
- The poet tells what dogs are like. What lines tell us that the poet thinks cats are *not* like this?

229

VOCABULARY AND MEANING

"The Ad-dressing of Cats" (Second Part)

Work through the questions carefully to be sure that the lines are well understood. This poem is more difficult than most of the other selections and may take more work.

After all the questions have been answered, encourage discussion of dogs. See if anyone can tell what a Pekingese is and why Pekes might be called "fantastic freaks." Ask whether those who have dogs as pets think they are like the "usual Dog about the Town" described by the poet.

SOUNDS AND LETTERS–/n/

The set of **mb** words constitutes an exception to the normal ways of spelling final /m/. Like all exceptions they may present a spelling problem.

The list of ten **mb** words comprises the most common words in this group. Others include the following:

jamb	**aplomb**	**coulomb**
rhumb	**plumb**	**womb**

The spelling of final /n/ generally follows the familiar pattern:

n after simple vowels
VCe after **VCe** vowels represented by one vowel letter
n otherwise

Here are additional examples with /n/ in the **VCe** pattern:

mine	**fine**	**dine**	**scene**	**wane**
crane	**June**	**tune**	**zone**	**stone**

Here are additional examples of the spelling **n** after vowels other than the simple ones:

gain	**plain**	**pain**	**queen**	**keen**
clean	**mean**	**boon**	**faun**	**pawn**

A number of words end with the spelling **gn**, where the **g** assigns a value to the preceding vowel. Apart from **sign**, included in the list at the bottom of page 231, here are the common words with this spelling:

reign	campaign	arraign	condign
deign	feign	impugn	sovereign
foreign	align	benign	assign
design	resign	ensign	consign

In this group the vowel before final /n/ is schwa in **sovereign, foreign**, and **ensign**. The other vowel sounds represented are /ī/ (**align**), /ā/ (**reign**), and /ū/ (**impugn**).

A few words ending in the sound /m/ are spelled **mn: damn, condemn, solemn,**

The Consonant Sound /n/

• The last consonant sound you studied was /m/. We write the sound /m/ at the beginning of words with the letter **m.** At the end we usually write it either with **m,** as in **Tom,** or with **me,** as in **home.** When do we write it with the letters **me** instead of **m?** VCe pattern

In a few words we write final /m/ with the letters **mb.** The **b** does not stand for a sound in these words. These are the most common words in which final /m/ is spelled **mb:**

lamb	**limb**	**climb**	**bomb**	**numb**
comb	**tomb**	**dumb**	**thumb**	**crumb**

□ Close your books and write those words as they are dictated to you.

We close our lips to make the sounds /p/, /b/, and /m/, but in /m/ we let the sound come out through the nose. When we make the sounds /t/, /d/, and /n/ our tongue touches the ridge above the front teeth. Say these words and notice where your tongue is at the beginning of each of them: **tie, die, nigh.** The tongue is in the same place, when we make /t/, /d/, and /n/, but we let the sound come out through the nose when we make /n/.

The usual way of writing the sound /n/ at the beginning of words is with the letter **n:**

nigh	**new**	**now**	**nose**
news	**name**	**night**	**never**

• Think of five other words that begin with the sound /n/ and the letter **n.** See p. 329.

In a few words, we write **gn** for **n** at the beginning. One is **gnat.** The **g** stands for a sound that used to be pronounced but isn't any more. Other common words spelled this way are **gnome** and **gnash.**

In a good many words, /n/ at the beginning is spelled with the letters **kn.** The word **knee** is one. The letter **k** in **knee** like **g** in **gnat**, used to stand for a sound, but now does not.

Here are the most common words in which /n/ is written with **kn** at the beginning. See if you can tell what each one means. discuss

knee	**know**	**knife**	**knight**
knit	**knob**	**knock**	**knot**
knuckle	**knack**	**knapsack**	**knowledge**

□ Close your books and write those words as they are dictated to you.

• Since /n/ is spelled in different ways, we may have homophones for this sound. Give a homophone for each of these:

know **knot** **knight** no, not, night

The sound /m/ at the end of words is usually spelled with the letter **m** or the letters **me.** The spelling of final /n/ is similar to the spelling of final /m/. The sound /n/ at the end is usually spelled with the letter **n** or the letters **ne.**

• After simple vowels, we spell final /n/ with the letter **n.** Name the vowel sounds in these words:

man /a/	**ten** /e/	**pin** /i/	**fun** /u/
Don /o/	**fan** /a/	**won** /u/	**skin** /i/

• After complex vowels, we may spell the sound /n/ in the pattern **VCe.** What are the vowel sounds in these words?

/ā/	/ī/	/ū/	/ō/
lane	**pine**	**prune**	**bone**

• In other words we spell final /n/ with just **n.** Tell what sound comes just before /n/ in these words. Some are consonant sounds and some are vowels.

rain /ā/	**lean** /ē/	**moan** /ō/	**burn** /r/
sign /ī/	**faun** /au/	**earn** /r/	**dawn** /au/

column, autumn, hymn. Although the **n** does not represent a sound in these words, it does in words derived from them: **damnation, condemnation, solemnity, columnar, autumnal, hymnal.** The **mn** spelling is therefore not so absurd as might be supposed.

Two common words spell /n/ **nn** after the simple vowel /i/: **Finn** and **inn.**

Teaching Suggestions

Go through the review of the spelling of final /m/. Introduce the **mb** words. Let the children take turns pronouncing them. Then dictate the set.

Go on to the presentation of /n/. Let the children say **pie, buy, my,** noting where the breath is stopped, then **tie, die, nigh.** They should easily become aware of the tongue's touching the alveolar ridge above the upper teeth in the making of **tie, die, nigh.**

Let the class suggest other words beginning with /n/ spelled **n.** Examples should readily suggest themselves.

Go on to the words in **gn** and **kn.** All of the common examples of these are listed on these two pages.

GRAMMAR

Other Verbs with Irregular Past Tense Forms

For the convenience of the teacher, we provide here a fairly complete list of irregular verbs, arranged in alphabetical order. Archaic words and compounds are not included. The forms given are the "principal parts": simple form, past tense form, participle. Verbs marked by underscore have other acceptable forms as well.

Some of these verbs have special forms in special contexts and meanings—for example, "The batter flied out," "They hanged him."

arise/arose/arisen
bear/bore/borne
befall/befell/befallen
begin/began/begun
behold/beheld/beheld
bend/bent/bent
<u>bereave/bereft/bereft</u>
<u>beseech/besought/besought</u>
bid/bade/bidden
bid/bid/bid
bite/bit/bitten/
bleed/bled/bled
blow/blew/blown
break/broke/broken
breed/bred/bred
bring/brought/brought
build/built/built
<u>burst/burst/burst</u>
buy/bought/bought
cast/cast/cast
catch/caught/caught
<u>chide/chid/chid</u>
choose/chose/chosen
cling/clung/clung
come/came/come
cost/cost/cost
creep/crept/crept
deal/dealt/dealt
dig/dug/dug
do/did/done
draw/drew/drawn
<u>dream/dreamt/dreamt</u>
drink/drank/drunk
drive/drove/driven
eat/ate/eaten
fall/fell/fallen
feed/fed/fed
fight/fought/fought
find/found/found
flee/fled/fled
fling/flung/flung
fly/flew/flown
<u>forbid/forbade/forbidden</u>
forget/forgot/forgotten
forsake/forsook/forsaken
freeze/froze/frozen
<u>get/got/got</u>
<u>gird/girt/girt</u>
give/gave/given
go/went/gone
grind/ground/ground
grow/grew/grown
<u>hang/hung/hung</u>
have/had/had

Other Verbs with Irregular Past Tense Forms

□ 1. You have studied four verbs with irregular past tense forms. Write the past tense forms of these verbs:

see saw **drink** drank **speak** spoke **eat** ate

● 2. Find the verbs in these sentences. Tell whether the verb is the present tense or the past tense.

a. Sam <u>likes</u> candy. pres.
b. We <u>saw</u> Mr. Ellett. past
c. I <u>drink</u> a lot of milk. pres.
d. They <u>helped</u> the teacher. past
e. The boys <u>ate</u> a big breakfast. past
f. We <u>speak</u> softly in class. pres.
g. Everybody <u>drank</u> cocoa. past

● 3. Here are some other verbs with special past tense forms. What vowel sound do they have in the present? What vowel sound in the past? /ɪ/; /ō/

drive: **drove** write: **wrote**
ride: **rode** rise: **rose**

□ 4. Copy each sentence below, but change the verb form to the past tense.

a. John rides to school. . . . rode . . .
b. John hurries home. . . . hurried . . .
c. John writes paragraphs. . . . wrote . . .
d. John eats a big breakfast. . . . ate . . .
e. John tries hard. . . . tried . . .
f. John rises at seven. . . . rose . . .
g. John plays after school. . . . played . . .
h. John drives the tractor. . . . drove . . .
i. John speaks softly. . . . spoke . . .
j. John sees Mr. Luther. . . . saw . . .
k. John drinks milk. . . . drank . . .
l. John plays baseball. . . . played . . .

5. Here are more common verbs with irregular past tense forms. Study them.

come: **came** give: **gave**
sit: **sat** begin: **began**

□ Close your books, and write the past tense forms of these four irregular verbs as their simple forms are dictated to you.

• 6. You learned about the past tense forms of several verbs in studying the vowel sound /ɑː/. One was **thought.** What is the simple present tense of **thought?** think
Here are other verbs with past tense forms like **thought.** Study them.

bring: **brought** fight: **fought**
seek: **sought**

□ Close your books, and write the past tense forms of these verbs as their simple forms are dictated.

□ 7. Copy each of these sentences, but change the verb from present tense to past tense.

a. They come at nine. . . . came . . .
b. They sit in front. . . . sat . . .
c. They give us candy. . . . gave . . .
d. They bring pictures. . . . brought . . .
e. They fight a lot. . . . fought . . .
f. They think about it. . . . thought . . .
g. They drive the car. . . . drove . . .
h. They speak softly. . . . spoke . . .
i. They seek new lands. . . . sought . . .
j. They begin in the morning. . . . began . . .
k. They rise many feet. . . . rose . . .
l. They ride on the bus. . . . rode . . .
m. They write letters. . . . wrote . . .
n. They see Mr. Martin. . . . saw . . .
o. They drink coffee. . . . drank . . .
p. They see the clock. . . . saw . . .

hear/heard/heard
hide/hid/hidden
hit/hit/hit
hold/held/held
hurt/hurt/hurt
keep/kept/kept
know/knew/known
lead/led/led
leave/left/left
lend/lent/lent
let/let/let
lie/lay/lain
light/lit/lit
lose/lost/lost
mean/meant/meant
meet/met/met
put/put/put
quit/quit/quit
read/read/read
rend/rent/rent
rid/rid/rid
ride/rode/ridden
ring/rang/rung
rise/rose/risen
run/ran/run
say/said/said
see/saw/seen
seek/sought/sought
sell/sold/sold
send/sent/sent
set/set/set
shake/shook/shaken
shed/shed/shed
shine/shone/shone
shoot/shot/shot
shrink/shrank/shrunk
shut/shut/shut
sing/sang/sung
sink/sank/sunk
sit/sat/sat
slay/slew/slain
sleep/slept/slept
slide/slid/slid
sling/slung/slung
slink/slunk/slunk
slit/slit/slit
smite/smote/smitten
speak/spoke/spoken
speed/sped/sped
spend/spent/spent
spin/spun/spun
spit/spit/spit
split/split/split
spread/spread/spread
spring/sprang/sprung
stand/stood/stood
steal/stole/stolen
stick/stuck/stuck
sting/stung/stung
stink/stank/stunk
stride/strode/stridden
strike/struck/struck
strive/strove/striven
swear/swore/sworn
sweat/sweat/sweat
sweep/swept/swept
swim/swam/swum
swing/swung/swung
take/took/taken
teach/taught/taught
tear/tore/torn
tell/told/told
think/thought/thought
throw/threw/thrown
thrust/thrust/thrust
tread/trod/trod
wake/woke/waked
wear/wore/worn
weave/wove/woven
weep/wept/wept
win/won/won
wind/wound/wound
wring/wrung/wrung
write/wrote/written

SECTION 4

The End of the Poem

With Cats, some say, one rule is true:
Don't speak till you are spoken to.
Myself, I do not hold with that —
I say, you should ad-dress a Cat.
But always keep in mind that he
Resents familiarity.
I bow, and taking off my hat,
Ad-dress him in this form: O CAT!
But if he is the Cat next door,
Whom I have often met before
(He comes to see me in my flat)
I greet him with an OOPSA CAT!
I think I've heard them call him James —
But we've not got so far as names.

Before a cat will condescend
To treat you as a trusted friend,
Some little token of esteem
Is needed, like a dish of cream;
And you might now and then supply
Some caviar, or Strassburg Pie,
Some potted grouse, or salmon paste —
He's sure to have his personal taste.
(I know a Cat, who makes a habit
Of eating nothing else but rabbit,
And when he's finished, licks his paws
So's not to waste the onion sauce.)
A Cat's entitled to expect
These evidences of respect.
And so in time you reach your aim,
And finally call him by his NAME.

So this is this, and that is that:
And there's how you AD–DRESS A CAT.

LITERATURE

"The Ad-dressing of Cats" (Third Part)

Reread or replay the first two parts of the poem. Remind the children of what we are about here—learning the proper way to speak to a cat. First we consider a cat whom we don't know at all, then one with whom we have a nodding acquaintance, and finally one with whom we manage to get "so far as names." There is implied here the notion—more British than American—that it takes some time, with people as well as cats, to get so far as names. Americans often call each other by first names immediately upon being introduced. In England it takes rather longer.

Read this part to the children or play the recording. Let them take turns reading it until they gain general familiarity with the content.

VOCABULARY AND MEANING

"The Ad-dressing of Cats" (Third Part)

Teaching Suggestions

Work carefully through the questions to make sure that the vocabulary and constructions are clear. Check on any matters not covered here about which the children may be in some doubt.

Illustrations of Rhyming Sounds

The pupils have now studied almost but not quite all the rhyming sounds they will meet in poetry. They know all of the rhyming vowels of this poem, and the poem may be used, if desired, for a review of them. Here is a list of the rhyming vowels in order:

/ū/ true, to
/a/ that, Cat
/ē/ he, familiarity
/a/ hat, Cat
/ō/ door, before
/a/ flat, Cat
/ā/ James, names
/e/ condescend, friend
/ē/ esteem, cream
/ī/ supply, Pie
/ā/ paste, taste
/i/ habit, rabbit
/aı/ paws, sauce
/e/ expect, respect
/ā/ aim, name
/a/ that, Cat

The sound /a/ predominates because of the subject of the poem. Notice that **paws** and **sauce** is not a true rhyme, since **paws** ends in /z/ and **sauce** in /s/.

COMPOSITION

A Description to Write

Let the class describe the behavior and appearance of cats orally before they undertake the writing assignment.

The Ad-dressing of Cats (THIRD PART)

• What rule does the poet mention at the beginning of this part of the poem? Does he believe the rule is true? What words tell whether he does or doesn't? See fi two lin no; I not hol with th

We say that people are too **familiar** when they don't know somebody very well but act as if they do. **Resent** means dislike. What should you always keep in mind? A cat resents familiarity.

• A **flat** is an apartment. Who do you think are the "them" of the second line from the end of the first stanza? the people who live next door

The last line in the first stanza means that they don't know each other well enough to call each other by their names.

Condescend means to agree to talk to someone lower than you are. A **token of esteem** is a gift showing that you like or admire someone.

What does **personal taste** mean? See a dictionary. Do you have any special things that you like to eat? discuss

• **Evidences** means proofs. What do you think **entitled to** means? Check your guess by using a dictionary. discuss

• An **aim** is something that you are trying to do. What is the aim that you reach if you follow the directions of this poem? You address a cat by name.

A Description to Write

Write a paragraph or two about cats. Tell some of the ways in which cats behave. Tell how they behave differently from dogs. You may use some of the things you have learned from the poem. Or you may use things that you yourself have learned about cats.

How Words Are Made

You have learned how to make nouns from verbs by adding **r** or **er** to the simple form of the verb.

□ Copy Sentence b in each pair of sentences below. Make a noun from the verb in Sentence a. Use this noun as the subject of Sentence b.

1. a. She writes books.
   b. The —— is a woman. writer
2. a. He preaches at our church.
   b. The —— is a clergyman. preacher

Now look at the pair of sentences below.

3. a. The boy works with care.
   b. The boy is careful.

• The word **care** in Sentence a is a noun. What kind of word is the word **careful** in Sentence b? adj.

We can make adjectives from some nouns by adding **ful.** A part added to a word, like **ful,** is called a **suffix.** Find **suffix** in a dictionary. discuss meaning

□ Find the noun in Sentence a in each pair below. Change it to an adjective by adding the suffix **ful.** Copy Sentence b and use the adjective in the predicate.

4. a. She laughed with joy.
   b. She was ——. joyful
5. a. He howls with glee.
   b. He is ——. gleeful
6. a. We appreciated their help.
   b. They were ——. helpful
7. a. They saw her in tears.
   b. She was ——. tearful
8. a. It does harm.
   b. It is ——. harmful
9. a. She has great beauty.
   b. She is ——. beautiful

MORPHOLOGY—How Words Are Made

We review here the suffix **er** as in **reader**, and introduce another suffix: **ful**. We do not expect in this volume to cover any significant number of suffixes, but we want to make the children aware of suffixing and word building in general in preparation for more serious work in later grades.

There are two suffixes spelled **ful**. One makes adjectives from nouns: **careful, beautiful**. The other makes nouns of measure from other nouns: **cupful, spoonful**. We consider only the first here. It may be noticed that the two suffixes are pronounced differently. In **careful** the second syllable is usually pronounced with weak stress and the vowel schwa. In **cupful** it is pronounced with a heavier stress and the vowel /ŏŏ/ as in the word **full**.

SOUNDS AND LETTERS—/ou/

The spelling of the vowel sound /ou/ is much simpler than the spelling of /au/. The vowel /ou/ is overwhelmingly spelled **ou** or **ow**. The major complication is that either **ou** or **ow** may occur before consonant sounds; only **ow** is used in the final position.

Notice that there is some regularity in the alternation of **ow** and **ou** before consonants. The spelling is regularly **ow** when final /n/ follows, the only exception being **noun**. But when final /nd/ follows, the spelling is **ou**: **bound, found**, etc. When the /d/ represents the past tense morpheme, of course the **ow** spelling remains: **drowned, frowned**.

Similarly, **ow** prevails before /l/, with **foul** and its compounds (**afoul, befoul**) the only exceptions. Other words in **owl** include **cowl, dowel, jowl**.

On the other hand, before /z/ and /s/ the spelling **ou** is regular (**house, arouse**, etc.), with **drowse** and **browse** being the only common exceptions. Before /d/, **crowd** is the exception to the common **ou** spelling: **loud, proud**, etc.

The spelling **ow** is the regular one in the final position. In addition to the set beginning **cow**, we have **scow, endow, enow, row, prow, sow, powwow**. Of course **ow** is even more common as the spelling for final /ō/, as in **snow** and **throw**. There are a number of homographs like **row** (disturbance) and **row** (propel a boat).

Slough (morass) and **plough** are other words in which /ou/ is spelled **ough**. However, the common spelling for **plough** is now **plow**.

The Vowel Sound /ou/

You have now studied twelve of the fifteen vowel sounds in English. You know these simple vowels:

/i/ /e/ /a/ /u/ /o/

• Give a word with each of the simple vowels. Tell what vowel sound it has as you name it. See p. 324.

You have also studied seven other vowels:

/ī/ /ē/ /ā/ /ū/ /ō/ /oo/ /au/

• Each word in the following list has one of these seven vowels in it. Tell what vowel sound you hear in each word in the list.

mean	/ē/	claw	/au/	roam	/ō/	book	/oo/
my	/ī/	date	/ā/	rule	/ū/	song	/o/
flow	/ō/	bright	/ī/	thought	/au/	see	/ē/
pay	/ā/	true	/ū/	haunt	/au/	boom	/ū/

• Say these words and listen to the vowel sound you hear in both words: **house, cow.**

In **house** this sound is spelled **ou,** and in **cow** it is spelled **ow.** We will show this vowel with the letters o and u joined together between slanting lines, this way: /ou/.

• In speaking of this sound, we will name it by saying the vowel of **cow** with the sound /k/ at the end, as if we were saying "owk." Say "owk."

The most usual way of spelling the sound /ou/ in the middle of words is with the letters **ou.** We spell it that way in these words:

house	mouse	found	ground
ouch	couch	bounce	pronounce
count	out	about	hour
doubt	lout	bound	around

□ Close your book, and write those words as they are dictated to you.

In words that end with the sound /n/, we usually spell the vowel /ɑu/ with the letters **ow.** Here are the most common words with this spelling of the /ɑu/ sound:

brown	clown	crown	down
drown	frown	gown	town

□ Close your book, and write those words as they are dictated to you.

□ There is one common word in which we spell /ɑu/ with **ou** before **n.** This is the word **noun.** Write the word **noun.**

In some other words we use the spelling **ow** in the middle of the word for the sound /ɑu/. Learn these words:

fowl	howl	owl	scowl
drowse	crowd	growl	prowl

□ Close your book, and write these words as they are dictated to you.

• Because we may spell /ɑu/ either as **ou** or as **ow,** we may have homophones. **Fowl** and **foul** are homophones. Can you tell what each one means? bird, dirty

When the sound /ou/ comes at the end of the word, the regular way to spell it is with the letters **ow.** Here are some of the common words with the sound /ɑu/ at the end:

cow	how	now	brow
vow	chow	allow	plow

□ Close your book, and write those words as they are dictated to you.

• In one common word we write /ɑu/ at the end with the letters **ough.** This is the word **bough.** Can you think of words with other vowel sounds that have the letters **ough?** ought, though, tough

Teaching Suggestions

Review first the five simple vowels and then the seven complex ones so far studied, using the material in the first half of page 238. Have the children use the /k/ convention – "ick," "ack," "eek," etc., in naming the vowel sounds. Then introduce /ɑu/.

Have the pupils read the **ou** words at the bottom of page 238 and then write them from dictation. Follow a similar procedure for the exercises on page 239 which are designated as written work by the symbol □.

GRAMMAR

More Irregular Past Tense Forms

It is quite difficult to group or classify irregular verbs according to the precise nature of the irregularity. However, there is probably some unconscious grouping and mutual strengthening. For example, it is probably easier to remember **write/wrote/written** because of the existence of **ride/rode/ridden**, and vice versa.

Here are some of the irregular verbs grouped according to similarities of formation of past tense and participle.

Like **write/wrote/written:** ride, drive, rise, smite, thrive.
Like **bear/bore/borne:** swear, tear, wear.
Like **break/broke/broken:** freeze, speak, steal, weave, bite, forget, choose.
Like **blow/blew/blown:** grow, know, throw, shaken, take, draw, fall, give, slay.
Like **begin/began/begun:** drink, ring, shrink, sing, sink, spring, stink, swim.
Like **bind/bound/bound:** find, grind, wind.
Like **cling/clung/clung:** fling, sling, slink, sting, string, swing, wring, dig, spin, stick, win.
Like **bleed/bled/bled:** breed, lead, meet, read, speed.
Like **teach/taught/taught:** bring, catch, seek, beseech, think.
Like **put/put/put:** set, bid, burst, hurt, rid, shut, slit, shed, spread, cost, hit, let, split, sweat, thrust, quit, cast.
Like **bend/bent/bent:** lend, rend, send, spend, build.
Like **creep/crept/crept:** deal, tell, sell, sweep, leave, weep, hear, flee, feel, keep, mean, sleep.

These are not entirely homogeneous sets. Thus **blow** and **take** are alike simply in the fact that they have the same vowel in the simple form and participle and that the participle adds the sound /n/. Quite a

More Irregular Past Tense Forms

□ 1. You have studied the irregular past tense form of each of the verbs in the list below. Write the past tense of each of them.

saw, spoke	see	speak	eat	drink	ate, drank
drove, wrote	drive	write	rise	ride	rose, rode
came, gave	come	give	sit	begin	sat, began
thought, brought	think	bring	fight	seek	fought, sought

● 2. Do the past tense forms of **think, bring, fight,** and **seek** rhyme? What is their vowel sound? yes; /aɪ/

□ 3. The past tense forms of **catch** and **teach** rhyme with them too, but they are spelled with **augh** instead of **ough: caught, taught.** Copy these sentences, making the verbs past tense:

a. They think about it. . . . thought . . .
b. They catch mice. . . . caught . . .
c. They teach us. . . . taught . . .

4. Here are some other past tense forms for you to study.

blow: **blew**	throw: **threw**
grow: **grew**	fly: **flew**
know: **knew**	draw: **drew**

□ 5. Copy these sentences, but change the verbs to the past tense:

a. He grows roses. . . . grew . . .
b. The wind blows hard. . . . blew . . .
c. They come at ten. . . . came . . .
d. Ralph knows the story. . . . knew . . .
e. The kite flies high. . . . flew . . .
f. They draw straws. . . . drew . . .
g. We see it. . . . saw . . .
h. He throws hard. . . . threw . . .
i. He sees the plane. . . . saw . . .

6. Study the past tense forms of these verbs. What vowel sound do they all have in the past tense? /a/

drink: **drank**	sing: **sang**
run: **ran**	swim: **swam**
shrink: **shrank**	ring: **rang**
sink: **sank**	begin: **began**

□ 7. Write the past tense form of each of these verbs when the present tense form is dictated.

□ 8. Copy these sentences, changing the form of the verb from the present to the past.

a. We drink lemonade. . . . drank . . .
b. Scott runs home. . . . ran . . .
c. Sheila sings beautifully. . . . sang . . .
d. Apollo catches the nymph. . . . caught . . .
e. The fable teaches a lesson. . . . taught . . .
f. The bell rings at nine. . . . rang . . .
g. They ride on the bus. . . . rode . . .
h. I think about it. . . . thought . . .
i. The shirt shrinks. . . . shrank . . .
j. We swim in the pool. . . . swam . . .
k. Nobody sinks. . . . sank . . .
l. Tom writes clearly. . . . wrote . . .
m. Mr. Burbank comes to see us. . . . came . . .
n. School begins on Monday. . . . began . . .
o. Ted grows fast. . . . grew . . .
p. His father flies a plane. . . . flew . . .
q. Everybody knows about it. . . . knew . . .
r. He throws to first base. . . . threw . . .
s. They eat a lot. . . . ate . . .
t. Her father drives her to school. . . . drove . . .
u. Sam brings the sandwiches. . . . brought . . .
v. The moon rises at nine. . . . rose . . .
w. They see the sunset. . . . saw . . .

number of irregular verbs such as **fly, go, do, wake** do not belong to sets of any size.

Participles

Participles are not introduced until grade five. Notice that we do not use the term **past participle**. Such a usage could only cause confusion in a description in which tense is interpreted as only present or past. It is not the participle that indicates whether the tense is present or past, but the form of the **have, be**, or modal that precedes it. For example, in "John has done it," the tense is present (**has**, not **had**), but the form of **do** is just what it would be if the tense were past: "John had done it."

The term **past participle** would be needed if one spoke also of a **present participle**. But we call what is often termed the **present participle** the **ing** form: **doing, trying**, etc. Therefore we can use the simple term **participle** unambiguously for **done, tried**, etc.

Teaching Suggestions

Notice that we do not in this text deliberately confront the learner with choices of verb forms about which he may be uncertain. We have no exercises like these:

Choose the correct word:
I (saw/seen) it.
He (throwed/threw) it.

The effect of such exercises seems to be simply to create uncertainty where none may have existed before. Confronted too often with **seen** as a possible past tense of **see**, the child who previously might never have dreamed of saying "I seen it" may begin to wonder if after all he hasn't been wrong in saying "I saw it." This is very clearly a type of exercise capable of doing large damage, and it should be avoided.

Instead, we keep affording practice in the correct form, without even raising the possibility that another form exists.

TESTS AND REVIEW

Use the tests on pages 242–44 in the usual way, to diagnose strengths and weaknesses. If records kept on individual pupils show that reteaching is necessary, it should be undertaken before proceeding to the next part of the book.

Use page 245 for oral discussion.

Extra Test

Dictate these words and have the class spell them. Say the word clearly. Use it in the sentence. Say it again and have the children write it.

1. **broth** Beef **broth** is a clear soup. **broth**
2. **bog** A **bog** is a swamp. **bog**
3. **cough** A **cough** often goes with a cold. **cough**
4. **taught** Our teacher **taught** us spelling. **taught**
5. **tomb** A **tomb** is a burial place. **tomb**
6. **knit** Mother **knit** me a sweater. **knit**
7. **dawn** The **dawn** is at sunrise. **dawn**
8. **bounce** A rubber ball will **bounce**. **bounce**
9. **crown** A king might wear a **crown**. **crown**
10. **fowl** A hen is a **fowl**. **fowl**

Tests and Review

☐ TEST I Sounds and Letters

1. Write these headings:

/ō/ /au/ /ou/ /oo/

Write each word below under the heading that shows its vowel sound.

If you make mistakes, study pages 67, 108, 206, 218, 238.

law /au/	town /ou/	home /ō/	look /oo/
house /ou/	bush /oo/	boat /ō/	cow /ou/
fought /au/	show /ō/	wrong /au/	caught /au/
crowd /ou/	stood /oo/	shout /ou/	tall /au/
fall /au/	pout /ou/	put /oo/	sew /ō/

2. Write these headings:

/r/ /l/ /m/ /n/

Write each word below under the heading that shows what consonant sound it begins with.

If you make mistakes, study pages 192, 198, 225, 232.

run /r/	loud /l/	meet /m/	name /n/
gnaw /n/	write /r/	lose /l/	mine /m/
rhyme /r/	lady /l/	know /n/	gnome /n/
knock /n/	wring /r/	gnat /n/	lame /l/

3. Write a homophone for each of the following words:

knight foul know bow
night fowl no bough

4. For each of the words in the list below, write a word that rhymes with it and has the rhyming part spelled in the same way.

If you make mistakes, study pages 206, 218, 238.

now, fowl	cow	howl	clown	house	down, mouse
pouch, shout	couch	out	long	hog	wrong, log
bought, taught	fought	caught	cross	scoff	loss, off

□ **TEST II Grammar**

1. Write the past tense forms of the following regular verbs.

If you make mistakes, study page 226.

yed, walked	play	walk	copy	try	copied, tried
ed, studied	use	study	help	change	helped, changed
ched, watched	coach	watch	worry	explain	worried, explained

2. Copy the sentences below, but change the verbs from the present tense to the past.

If you make mistakes, study page 226.

a. John talks quietly. . . . talked . . .
b. We study history. . . . studied . . .
c. They rush home. . . . rushed . . .
d. The baby cries. . . . cried . . .
e. Mrs. Murphy waves to us. . . . waved . . .
f. I hope so. . . . hoped . . .
g. They hurry to the store. . . . hurried . . .
h. We work in the cellar. . . . worked . . .
i. He washes his car. . . . washed . . .
j. I guess the answer. . . . guessed . . .
k. She solves the problem. . . . solved . . .
l. Scientists discover new products. . . . discovered . . .
m. Sally teases the cat. . . . teased . . .
n. Edward raises the flag. . . . raised . . .

3. Write the past tense forms of the following irregular verbs.

If you make mistakes, study pages 226, 232, 240.

ove, blew	drive	blow	see	rise	saw, rose
ought, brought	think	bring	teach	catch	taught, caught
ne, gave	come	give	sit	ride	sat, rode
ig, shrank	ring	shrink	speak	write	spoke, wrote
ew, knew	grow	know	eat	drink	ate, drank
ew, drew	throw	draw	sing	swim	sang, swam

□ TEST II Grammar (CONTINUED)

4. Copy each of these sentences, but change the form of the verb from present to past.

If you make mistakes, study pages 226, 232, 240.

a. They come to see us. ... came ...
b. The wind blows the leaves. ... blew ...
c. Mr. Danby drives the bus. ... drove ...
d. I swim in the pool. ... swam ...
e. They eat pie. ... ate ...
f. The fox catches bumblebees. ... caught ...
g. We think about the poem. ... thought ...
h. The stones sink in the water. ... sank ...
i. The bell rings at three. ... rang ...
j. The sun rises at six. ... rose ...

5. Copy each of these sentences, but change the verb from past to present.

If you make mistakes, study pages 226, 232, 240.

a. Mrs. Calvin grew roses. ... grows ...
b. They studied the lesson. ... study ...
c. The boys sat near the window. ... sit ...
d. George drank a lot of milk. ... drinks ...
e. Mrs. Maxwell gave us candy. ... gives ...
f. Don tried his best. ... tries ...
g. Miss Donovan taught us music. ... teaches ...
h. The girls wrote on the board. ... write ...
i. Randy threw the ball hard. ... throws ...
j. We erased the chalkboard. ... erase ...

□ TEST III Writing and Vocabulary

Write five sentences. In each sentence, use one of these words:

1. witch 2. hobble 3. interpreter 4. sane 5. gambol

REVIEW Ideas and Information

1. What kind of woman is a **wizened** one? old and dried up
2. What does **mad** mean in "Some are sane and some are mad"? insane
3. What does **habitat** mean? place where one lives
4. Name three verbs in which the **s** forms are spelled in special ways. do, go, have
5. Tell three ways of spelling the vowel sound /au/. Give examples. au, cause; aw, law; a, fall
6. Tell three ways of spelling the sound /n/ at the beginning of words. n, gn, kn
7. Give a word in which the consonant sound /m/ is spelled **mb** at the end. comb
8. When do we change **y** to **i** and add **ed** to spell the past tense of verbs? when verb ends in consonant-y
9. How many present tense forms do verbs have? how many past tense forms? two; one
10. How many present and past tense forms does **be** have? present: three; past: two
11. When do we add **es** instead of just **s** to the simple form of a verb to make the **s** form? when it ends in ch, sh, s, z, x
12. Name some words in which **gh** stands for the /f/ sound. Name some words in which the letters **gh** do not stand for any sound at all. tough, rough; night, ought
13. What vowel sound do **awe** and **broad** have? /au/ What vowel sound do **cough** and **trough** have? /au/
14. What word means a female **tiger?** a female **lion?** a female **wizard?** tigress, lioness, witch
15. Use the **VCe** pattern to make one or more words in which **m** takes the place of **C** and **a** takes the place of **V.** lame, same, came, etc.
16. Name some words that begin with **kn.** Does the letter **k** stand for a sound in these words? knot, knight, knock, etc.; no

245

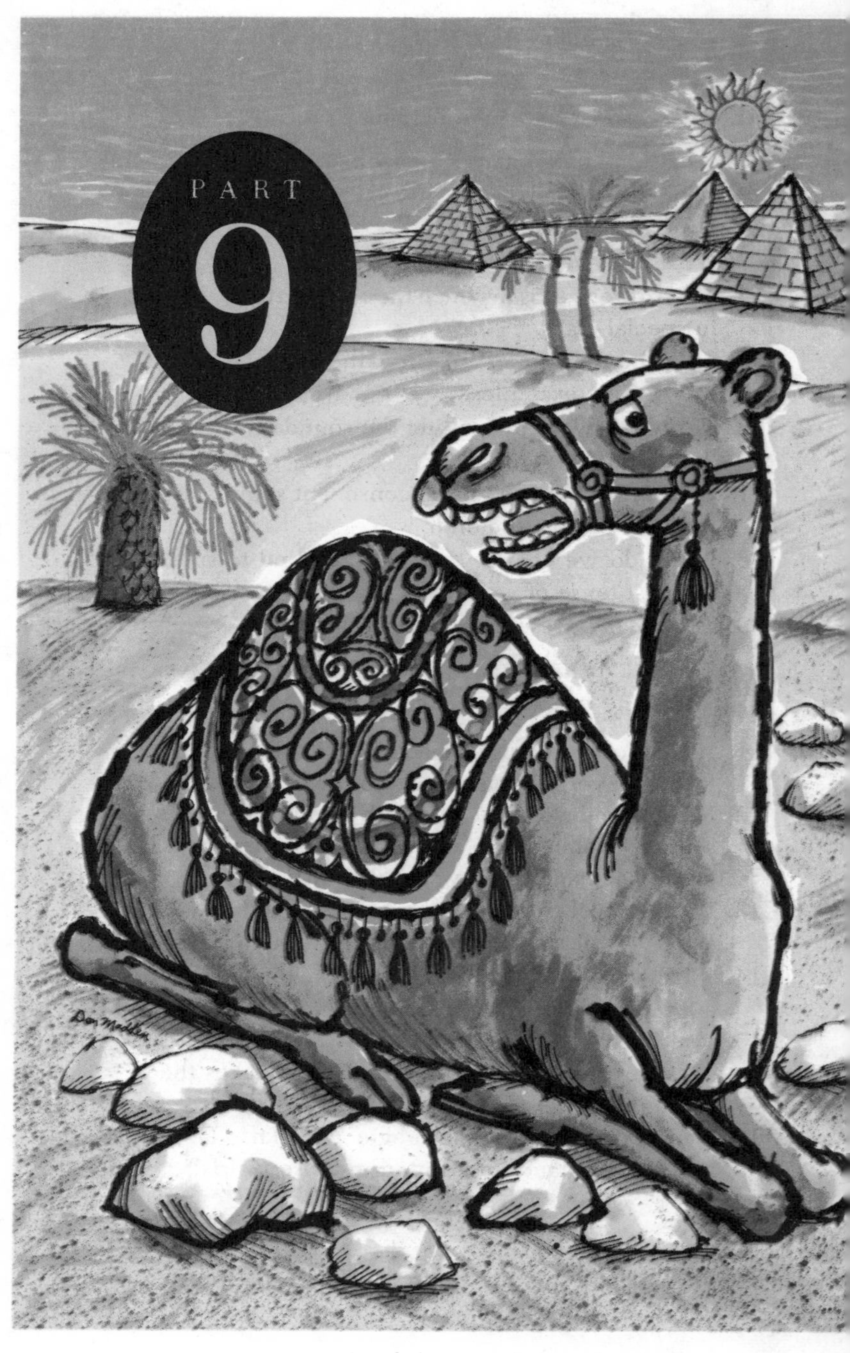
PART
9

SECTION 1

A Poem

Plaint is short for **complaint.** This complaint of a camel is in two parts. Read the first part.

The Plaint of the Camel

Canary-birds feed on sugar and seed,
Parrots have crackers to crunch;
And as for the poodles, they tell me the noodles
Have chicken and cream for their lunch.
    But there's never a question
    About MY digestion —
ANYTHING does for me!

Cats, you're aware, can repose in a chair,
Chickens can roost upon rails;
Puppies are able to sleep in a stable,
And oysters can slumber in pails.
    But no one supposes
    A poor Camel dozes —
ANY PLACE does for me!

The Plaint of the Camel

(STANZAS ONE AND TWO)

- One thing the camel complains about is that other animals get better food. **Noodle** is a slang term for a stupid person, like "dumbell." Why did the poet use **noodle** here instead of dumbell? to rhyme
- **Repose** means to lie down and rest or sleep. Chickens roost on **rails.** What does **rails** mean here? Find this meaning of **rail** in a dictionary. fence rails
- How does the camel say that no one thinks a camel needs rest too? See poem.

LITERATURE

"The Plaint of the Camel" (Stanzas One and Two)

Teaching Suggestions

Read or play the recording of this poem about how unhappy a thing it is to be a camel. Let the pupils take turns reading parts of it aloud.

As the schedule permits, use copying as a device to improve writing and to make the child aware of details that it may be impossible to take up in formal presentation. If this poem is assigned as a copying exercise, the copier must, among other things, capitalize the words capitalized.

VOCABULARY AND MEANING

"The Plaint of the Camel" (Stanzas One and Two)

Teaching Suggestions

Go through the questions at the bottom of the page with the class. Check also on comprehension of the following words and expressions: **seed, crunch, digestion, roost, slumber in pails, doze**. Some of these words may be used for dictionary practice.

MORPHOLOGY – How Words Are Made

This page reviews the spelling of words ending with the suffixes **er** and **ful** and then takes up the special point of spelling words with such suffixes added to base words ending in "consonant–y."

The general rule is that when a suffix is added to a base word ending in "consonant–y," one changes the **y** to **i**:

baby/babies	hurry/hurries
worry/worrier	beauty/beautiful
try/tried	funny/funnier
duty/dutiable	pretty/prettily

The exceptions are those words in which the suffix begins with **i**, and the principal suffix beginning with **i** that might be added to "consonant–y" words is **ing**. With these words the **y** remains to represent a syllable. Thus **studying** properly represents the three syllables of the word. If we changed the **y** to **i** in adding **ing** to **study**, we would get **studiing**, which is outside the English spelling system. If we merely dropped the **y**, we would have **studing** (a rather common mistake), which would represent the word as disyllabic and produce wrongly the **VCe** pattern.

Teaching Suggestions

Review the **er** and **ful** suffixes with the questions and examples in the first half of the page. Introduce the special problem of "consonant–y" words. Give as much of the explanation presented above as the class seems ready for.

How Words Are Made

• You have studied the suffix **er**, which may be added to some verbs to make nouns. What is added when the verb ends in **e**? r

□ Make a noun from the verb in Sentence a in each pair of sentences. Use the noun in Sentence b. The nouns you make will be a little different in meaning from those you made earlier. Find them in a dictionary.

1. a. A machine irons clothes.
   b. This machine is an ——. ironer
2. a. Another machine staples things together.
   b. This machine is a ——. stapler

□ Another suffix is **ful**, which you have added to some nouns to make adjectives. Copy Sentence b below. Make an adjective from the noun in Sentence a and use it in the predicate of Sentence b.

3. a. He studies with deep thought.
   b. He is ——. thoughtful

• How do you write the plural of a noun like **baby**, which ends in consonant-y? Then how do you suppose you would add the suffix **ful** to a noun like **beauty**, which ends in consonant-y? change y to i; change y to i

□ The noun in Sentence a of each pair below ends in consonant-y. Change the **y** to **i**, add the suffix **ful**, and use the adjective you make in Sentence b.

4. a. She had great beauty.
   b. She was ——. beautiful
5. a. They deserved pity.
   b. They were ——. pitiful
6. a. He ruled with mercy.
   b. He was ——. merciful
7. a. It grew in great plenty.
   b. It was ——. plentiful

248

The Vowel Sound /oi/

You have studied several vowel sounds that are spelled in more than one way and with more than one vowel letter. Here is one more such sound.

Say these words and listen to the vowel sound in them: **boy, boil.** We will show this vowel sound in this way: /oi/. In speaking of the vowel sound, say the sound with the /k/ sound after it, as if you were pronouncing a word spelled "oik."

The vowel sound /oi/ is perhaps the easiest vowel sound to spell. In the middle of words, it is spelled with the letters **oi:**

boil	join	spoil	coin
point	joint	choice	voice

- Can you think of any other words in which the sound /oi/ is spelled **oi?** See p. 329.

At the end of words, /oi/ is spelled with the letters **oy:**

joy	**boy**	**toy**	**Roy**
annoy	**destroy**	**employ**	**enjoy**

That's all there is to /oi/.

- Tell what the vowel sounds are in these words:

pit /i/	pet /e/	pat /a/	putt /u/
pot /o/	pike /ī/	peek /ē/	pain /ā/
pool /ū/	poke /ō/	put /oo/	paw /au/
	pout /ou/	point /oi/	

The words above contain all of the vowel sounds that you have studied. There is one other vowel sound in English. It is the sound in the second part of each of these words:

sudden gambol rapid

We will wait until next year to study this sound, which occurs in a great many words.

We end the fourth-grade description of vowel sounds with, mercifully, the one with the simplest spelling. The vowel /oi/ is spelled **oi** when it is not final and **oy** when it is. The exceptions are trifling: **coign** and **eloign** (which, however, still have the **oi** though with **g** following) and a few proper names like **Joyce, Lloyd, Illinois, Iroquois**. The last two are irregular only in the sense that there is no final consonant sound.

Teaching Suggestions

Go through the presentation of /oi/, and give the pupils practice with the lists of words in which this sound occurs. The list beginning **pit** contains one example of all of the vowels studied in this book in their order of presentation.

The fifteenth vowel is schwa, now usually indicated in dictionaries with this symbol: /ə/. It occurs under main stress only before /r/, in words like **fur** or **heard**. But it is the regular vowel occurring under weak stress, and since there is no particular way of spelling it, it presents a real spelling problem. This is tackled in the fifth-grade book of this series.

GRAMMAR

More Irregular Past Tense Forms

This page and the first part of the next present the verbs which do not change form for the past tense and the participle. For a fuller list, see the set beginning **put** on page 240 of the Teacher's Edition.

Note that with this group of verbs a particular sort of ambiguity is possible, as in such a sentence as this:

They put it there in the morning.

This could mean that they do this every morning, in which case the tense is present. Or it could mean that they did it on a particular morning—this morning, perhaps—in which case the tense is past.

The following are unambiguous:

He puts it there in the morning.
He put it there in the morning.

The verb in the first sentence is marked as present by the s ending, the second as past by its absence.

Teaching Suggestions

Work through the material on page 250 with the class. If the children do not see at once how to tell the tense of the verbs in Sentences c and d, give them these sentences:

The candy costs ten cents.
The candy cost ten cents.
My foot hurts.
My foot hurt.

Work through Sentences 1–4 with the class, putting the new sentences on the chalkboard before asking the children to do the work as a written assignment.

More Irregular Past Tense Forms

You can usually tell by the spelling of an irregular verb whether it is the past tense or not. But some verbs use the same spelling for the past tense that they do for the simple form of the present tense. One verb like this is **cut.** Look at the verb **cut** in the following sentences:

a. They cut a cake every day.
b. They cut a cake yesterday.

• One of the verbs above is present tense and one is past. Can you tell which one is past tense? How can you tell? yes, b; yesterday

• Look at these sentences. In which one is the subject followed by the s form of the verb? Then in which sentence is the past tense of the verb used? c; d

c. John cuts a cake.
d. John cut a cake.

• Now study this sentence. Could the verb be the simple form of the present tense? Could it be the past tense? yes; yes

e. They cut a cake.

If **cut** is used after a subject that could be followed by the simple form of the present tense, we can't tell from the verb itself what its tense is. There must be something else — like **yesterday** — to tell us.

Here are some other verbs like **cut:**

cost set put hurt

□ Rewrite these sentences, but use **yesterday** in place of **every day.**

1. He sets it there every day. . . . set . . . yesterday
2. He puts it there every day. . . . put . . . yesterday
3. It hurts every day. . . . hurt yesterday.
4. It costs a lot every day. . . . cost . . . yesterday

250

Here are more verbs that have the same form in the past tense and in the simple present:

shut quit hit spread

□ Copy these sentences, but make the verbs past instead of present. Not all of the verbs have the same form in the past as in the present.

5. Jane shuts the door. . . . shut . . .
6. The children come in. . . . came . . .
7. Andy spreads jam on his bread. . . . spread . . .
8. Dave puts his coat on. . . . put . . .
9. They ride on a camel. . . . rode . . .
10. Mr. Wilson quits early. . . . quit . . .
11. The girls study hard. . . . studied . . .
12. Miss Rudolph teaches them. . . . taught . . .
13. My mother cuts the meat. . . . cut . . .
14. The bus waits for us. . . . waited . . .
15. Bill hits the ball. . . . hit . . .
16. Mother sets it outside. . . . set . . .
17. The shoe hurts my foot. . . . hurt . . .
18. The toy costs a dollar. . . . cost . . .

You know that there are verbs like **drink** and **ring** that have the sound /a/ in the past tense. Some verbs like that have the sound /u/ instead:

swing: **swung** cling: **clung**
hang: **hung** sting: **stung**

□ Write these sentences again, using the past tense instead of the present.

19. Pete swings the bat. . . . swung . . .
20. They drink milk. . . . drank . . .
21. The child clings to its mother. . . . clung . . .
22. I ring the bell. . . . rang . . .
23. Bees sting us. . . . stung . . .
24. They sing songs. . . . sang . . .
25. It hangs on the wall. . . . hung . . .

Sentences 5–18 contain a preponderance of **put**-type verbs with a few others thrown in. The exercise should be done as a written one but may be started on the chalkboard.

After this, the group of words with /u/ in the past tense and participle are introduced. The full list is given on page 240 in the set beginning **cling**. All except **hang** have the vowel /i/ in the simple form.

This group causes quite a lot of uncertainty on the part of many speakers of English. We continue the policy of presenting only the acceptable form, not introducing the wrong one. Let the class work through 19–25 as a written exercise. Decide whether the pupils manage this group of words pretty well or have difficulty and need further practice.

LITERATURE

"The Plaint of the Camel"
(Stanzas Three to Five)

Teaching Suggestions

Reread or replay the first two stanzas of this poem. Let the class discuss what the camel has complained of up to now. Then read or play these last three stanzas. Let the children take turns reading parts of them.

SECTION 2

The Rest of the Poem

Read the rest of this poem about how hard it is to be a camel.

Lambs are enclosed where it's never exposed,
Coops are constructed for hens;
Kittens are treated to houses well heated,
And pigs are protected by pens.
  But a Camel comes handy
  Wherever it's sandy —
ANYWHERE does for me!

People would laugh if you rode a giraffe,
Or mounted the back of an ox;
It's nobody's habit to ride on a rabbit,
Or try to bestraddle a fox.
  But as for a Camel, he's
  Ridden by families —
ANY LOAD does for me!

A snake is as round as a hole in the ground:
Weasels are wavy and sleek;
And no alligator could ever be straighter
Than lizards that live in a creek.
  But a Camel's all lumpy
  And bumpy and humpy —
ANY SHAPE does for me.

CHARLES EDWARD CARRYL

The Plaint of the Camel

(STANZAS THREE TO FIVE)

- **Exposed** may mean open to wind and rain. **Constructed** means built. How are other animals better protected than he is, according to the camel? See first stanza.
- **Bestraddle** means bestride. Find **bestride** in a dictionary.
- What contraction is rhymed with **families?** he's
- What adjectives tell about snakes and weasels? What three adjectives that rhyme tell about the camel's shape? round, wavy, sleek; lumpy, bumpy, humpy

Rhyme

pp. 252–53: enclosed, exposed /ō/; hens, pens /e/; treated, heated /ē/; handy, sandy /a/; laugh, giraffe /a/; ox, fox /o/; habit, rabbit /a/; he's, families /ē/; round, ground /ou/; sleek, creek /ē/; alligator straighter /ā/; lumpy, humpy /u/

Some of the rhyming words in this poem are at the end of lines and some are not. Find and write each pair of rhyming words you can locate in the entire poem, pages 247, 252, and 253.

You can now name all the vowels of rhyming words. After each pair of rhyming words you have written, write the rhyming vowel in the words, like this:

feed, seed /ē/	**crunch, lunch** /u/
enclosed, exposed /ō/	**hens, pens** /e/

p. 247: feed, seed /ē/; crunch, lunch /u/; poodles, noodles /ū/; question, digestion /u/; aware, chair /ā/; rails, pails /ā/; able, stable /ā/; supposes, dozes /ō/

253

VOCABULARY AND MEANING

"The Plaint of the Camel"
(Stanzas Three to Five)

Teaching Suggestions

Check for comprehension by working through the questions. Be sure that these further words are understood: **enclosed, coops, treated, handy, mounted, sleek, lizards**.

SOUNDS AND LETTERS—Rhyme

Teaching Suggestions

Since, except for schwa, all of the vowel sounds have now been studied, specific questions on all rhymes except the **bird/heard** type can now be asked. Review the vowel sounds by drawing attention to the rhymes in these stanzas and having the children identify the rhyming vowels. Be sure that they notice the internal rhymes of lines 1 and 3 of each stanza. Start the exercise on the chalkboard with the system suggested at the bottom of the page, and then let the children continue it as a written exercise.

COMPOSITION – A Paragraph to Write

Teaching Suggestions

This may be preceded by class discussion. Let the children add to the data from the poem anything suggested by their imagination. The exercise will naturally provide practice in writing the past tense of verbs.

ETYMOLOGY – What Words Come From

Philologists have used the names of animals to help determine the homeland of the speakers of Indo-European, the language group to which English belongs. Africa and Asia are ruled out because the words for the animals native to those areas are all borrowed. Some examples are given here.

The Greek word for horse, **hippos**, has been borrowed by English in other words besides **hippopotamus**. We have it in **hippodrome** and also in the proper name **Philip**. **Philip** is composed of **philos**, "love" and **hippos**, "horse." It means "one who likes horses."

A Paragraph to Write

Pretend that you used to be a camel. Tell what it was like to be a camel — what you did and what you looked and felt like. You may take some ideas from the poem.

Since this happened in the past, of course, you will use the past tense forms of verbs.

What Words Come From

Animals like camels and elephants and tigers don't live in countries like the United States, except in zoos and circuses. They live in places like India and Africa. So naturally we don't have our own names for such animals. We borrow their names from the languages of the people who live where these kinds of animals live.

Our word **camel** probably comes from Arabic, a language spoken in Egypt, Arabia, and other places in Africa and western Asia. Sometimes we call the camel "the ship of the desert." The word in Arabic meant to carry. Can you think of any reason why a word meaning to carry should be used to name camels?

The word **elephant** comes to us from the Greek language. It is a word for ivory. Elephants are valuable because of the ivory of their tusks.

The word **hippopotamus** is an interesting one. The first part means horse, and the second part means river. It is a river horse. Of course, the hippopotamus isn't very much like a horse, but it does spend most of its time in rivers. The Potomac River in the United States has a name that comes from the same Greek word as hippopotamus.

254

Review 1: Simple Vowel Sounds

Now that you have finished your study of vowel sounds for this year, you may look back at the ways of spelling them that you have learned.

The five simple vowels are rather easy to spell. Here are the usual ways of spelling them. An example is given of each spelling.

/i/	i as in **nip**	dip
/e/	e as in **pet;** ea as in **head**	let, bread
/a/	a as in **bat**	fat
/u/	u as in **mud;** o as in **love**	bun, son
/o/	o as in **Bob;** a as in **car**	got, far

• 1. For each way of spelling a vowel sound shown in the list above, give another example besides the word used as an example in the list. See above.

□ Complete the words below by copying them and putting in the letter or letters for the sound at the top of the column.

	/i/		/e/		/a/		/u/		/o/	
2.	t–p	i	d–d	ea	gn–t	a	gl–ve	o	c–r	a
3.	t–ck	i	sl–d	e	p–tch	a	m–ch	u	n–t	o
4.	w–tch	i	thr–d	ea	w–x	a	w–n	o	st–r	a

• 5. Of course, there are a few words in which simple vowel sounds are spelled in special ways. What simple vowels do you hear in these words?

laugh friend any myth
/a/ /e/ /e/ /i/

The vowel sound /oo/ is something like the simple vowels. It does not come at the end of words. We usually spell /oo/ in one of two ways:

oo: book **u:** bush

• 6. Give another word as an example of each way of spelling the sound /oo/. look, push

Review 1: Simple Vowel Sounds

We have completed for this grade the presentation of the sounds of English, and all of the remaining work in this "Sounds and Letters" strand is review. There remain eight sounds in all for grade five. One is the vowel schwa. The others are all consonant sounds: /zh/, /th/, /th/, /ng/, /y/, /w/, and /h/.

Teaching Suggestions

Work through the review questions and exercises, reinforcing what has been learned about the five simple vowels and /oo/.

Possessives of Personal Pronouns

This lesson leads up to the introduction of the possessive transformation on pages 262–63. This is the first time a transformation has been introduced in this series. We begin with the possessive, not because it is particularly simple, but because the possessive forms are of high frequency and provide familiar problems for the young writer. The possessive forms of the personal pronouns have, to be sure, occurred earlier in the sentences that have been studied. But they have simply been included among the determiners with no remark about their origin or relation to other structures. Possessives are indeed determiners, but they are not determiners like the articles or the demonstratives. These are generated by kernel rules; possessives are generated by a transformational rule.

The Possessive Transformation

There are two kinds of transformational rules: single-base and double-base. A **single-base** transformation is one in which some operation is performed on a single string of morphemes, like the following:

John + past + be + in + the + house

This string represents of course the kernel sentence "John was in the house." But it also underlies a number of other sentences that can be produced by the transformational rules which permit manipulation of the morphemes in the string. One such rule says that in a string like the one above, we may reverse the positions of the subject and "tense + **be**." Since **John** is a subject and **past** is a tense,

John + past + be + in + the + house →
past + be + John + in + the + house

This will give the transform "Was John in the house?" Other single-base transfor-

Possessives of Personal Pronouns

You have studied the personal pronouns. Can you name all seven of them? I, it, he, she, we, you, they

You know that five of the personal pronouns have special forms when they function as objects of verbs or of prepositions. The object form of **he** is **him** and the object form of **we** is **us.** What are the object forms of **I, she,** and **they?** me, her, them

Find the noun phrases in the sentences below. Tell whether each noun phrase is used (1) as subject, (2) as object of a verb, (3) as object of a preposition, (4) as complement. Tell which noun phrases are personal pronouns.

1. He saw the plane. (1), pers.; (2)
2. Mr. Webster was with us. (1); (3), pers.
3. It was she. (1), pers.; (4), pers.
4. The lady thanked them. (1); (2), pers.
5. They grow them. (1), pers.; (2), pers.
6. I saw her. (1), pers.; (2), pers.
7. We drank the milk. (1), pers.; (2)
8. She was near him. (1), pers.; (3), pers.
9. No one saw me. (1); (2), pers.
10. They are parents. (1), pers.; (4)

Now you will study another form of personal pronouns — the **possessive** form.

Possession means having. If we say, "The man has a horse," we are saying that the man **possesses** the horse. Who possesses the doll in this sentence: "The girl has a doll"? the girl

• The past tense of the verb **have** is irregular. Which of these sentences is present and which is past?

a. He has a pencil now. pres.

b. He had a pencil yesterday. past

What personal pronoun is subject of the sentence "He has a pencil"? Who possesses the pencil? he; he

There is a way to change, or transform, the subject of the sentence "He has a pencil" to show possession without using the verb **has.** Study these sentences:

c. He has a pencil.

d. It is his pencil.

His is a special form of the personal pronoun **he.** It shows possession. The word **his** takes the place of the determiner **a** in "He has a pencil." It shows or determines that the pencil belongs to **him.**

What pronoun is a possessive in Sentence f?

e. He had a book.

f. It was his book. his

The possessive forms of the personal pronouns are these:

my his her its our your their

The personal pronoun in the first sentence of each pair functions as a subject, followed by a form of **have.** Tell what pronoun in Sentence b is a possessive. Tell what determiner of Sentence a the possessive replaces.

11. a. She has a book.
    b. It is her book. her; a
12. a. They have a teacher.
    b. She is their teacher. their; a
13. a. I have a friend.
    b. He is my friend. my; a
14. a. We had a boat.
    b. It was our boat. our; a
15. a. It had some leaves.
    b. They were its leaves. its; some
16. a. You had some candy.
    b. It was your candy. your; some

mations performed on the original string will give "John wasn't in the house," "John was in the house, wasn't he?" "John wasn't in the house, was he?" "Who was in the house?" "Where was John?" etc.

A **double-base** transformation is one which operates on *two* underlying strings of morphemes. An example is the transformation that makes one sentence into a relative clause and inserts or embeds it in another sentence. Consider these strings:

the + man + past + be + in + the + house
the + man + past + be + John

The first string would give the kernel sentence "The man was in the house," and the second "The man was John." But this can be made into a transform by the replacement of the + man in the first sentence by the relative pronoun **who** and the insertion of the result after the + man in the second sentence:

the + man + who + past + be + in + the + house + past + be + John

Phonological rules then will convert this into "The man who was in the house was John."

A double-base transformation in effect takes structural material from one sentence and uses it in another to produce a complex sentence containing the meaning of both of the simpler sentences. We call the sentence that provides the structural material the **insert** sentence. That which accepts it is the **matrix** sentence. The final product is the **result** sentence.

Insert: the + man + past + be + in + the + house
Matrix: the + man + past + be + John
Result: the + man + who + past + be + in + the + house + past + be + John

The particular double-base transformation that produces possessives is described on pages 262–63 of this Teacher's Edition.

SECTION 3

A Poem

This poem by Hilaire Belloc tells the sad fate of a door-slammer. Here is the first of two parts.

Rebecca

A trick that everyone abhors
In Little Girls is slamming Doors.
A Wealthy Banker's Little Daughter
Who lived in Palace Green, Bayswater,
(By name Rebecca Offendort),
Was given to this Furious Sport.

She would deliberately go
And Slam the door like Billy-Ho!
To make her Uncle Jacob start.
She was not really bad at heart,
But only rather rude and wild:
She was an aggravating child. . . .

Rebecca (STANZAS ONE AND TWO)

- **Abhor** means hate. Can you think why the poet used the word abhors instead of hates? to rhyme
- Whose daughter was Rebecca Offendort? Can you think of a reason for naming her Offendort? a wealthy banker's; to rhyme
- **Given to** means in the habit of. **Furious** may mean angry or it may mean very noisy and active. Which meaning has it here? See a dictionary. noisy, active
- **Deliberately** means on purpose. **Like Billy-Ho** is just a way of saying she did it very hard. To **start** is to move with surprise, to be startled. Can you figure out what the adjective **aggravating** means? Discuss; then see dictionary to check.

259

LITERATURE

"Rebecca" (Stanzas One and Two)

Hilaire Belloc was a Frenchman who removed to Britain and became a leading figure in the literary world of London. Light verse was only one of his many literary accomplishments.

The setting of the poem is London. The names are of course chosen for the rhyme.

Teaching Suggestions

Read these two stanzas of the poem to the class or play the recording of them. Let the children take turns reading.

VOCABULARY AND MEANING

"Rebecca" (Stanzas One and Two)

Go through the questions at the bottom of the page to be sure that the details of the stanza are understood.

Review 2: Some Complex Vowel Sounds

The vowels /ī/, /ē/, /ā/, /ū/, and /ō/ may be called the **VCe** vowels because they may all be spelled in the **VCe** pattern. They vary as to how often they are spelled so. Thus /ī/ is mostly spelled **VCe**, but /ē/ is spelled this way only in a relatively few words. All of these vowels may be spelled in other ways than **VCe**.

Teaching Suggestions

Work through the exercises on page 260 in class discussion. Elicit other examples of each of the ways of spelling the different vowel sounds.

Review 2: Some Complex Vowel Sounds

You have reviewed the usual ways of spelling the simple vowels. In this section and the next, you will review complex vowels and their spellings.

Here are the usual ways of spelling the vowel sound /ī/, with an example of each.

igh: sigh **y:** my **ie:** pie **VCe pattern:** ripe

• 1. Give another word as an example for each of these ways of spelling /ī/. See p. 324.

Here are the usual ways of spelling the vowel sound /ē/, with an example of each.

ee: seen **ea:** dream **e:** me **VCe pattern:** Steve

• 2. Give another word as an example for each of these ways of spelling /ē/. See p. 324.

Here are the usual ways of spelling the vowel sound /ā/, with an example of each.

ay: way **ai:** rain **VCe pattern:** late

• 3. Give another word as an example for each of these ways of spelling /ā/. See pp. 324–25.

Here are the usual ways of spelling the vowel sound /ū/, with an example of each.

oo: room **ew:** blew **ue:** due **o:** do
VCe pattern: rule

• 4. Give another word as an example for each of these ways of spelling /ū/. See p. 325.

Here are the usual ways of spelling the vowel sound /ō/, with an example of each.

oa: boat **ow:** blow **oe:** doe **o:** go
VCe pattern: wrote

• 5. Give another word as an example for each of these ways of spelling /ō/. See p. 325.

□ Complete the words below by copying them and putting in the letter or letters for the sound at the top of the column.

	/ī/	/ē/	/ā/	/ū/	/ō/
6.	r–de i	th–se e	m–de a	m–le u	n–se o
7.	l–t igh	scr–n ee	g–n ai	f– ew	t– oe
8.	d– ie	cr–m ea	m– ay	tr– ue	n– o
9.	fl– y	h– e	r–te a	n– ew	sl– ow
10.	rip– e	Pet– e	st– ay	b–m oo	g–t oa
11.	f–t igh	gr–n ee	p–d ai	gl– ue	thr– ow
12.	cr– y	–t ea	d–te a	m–n oo	pol– e

□ 13. Write homophones for the following words:

sea	see	knight	night	son	sun	plane	plain	tale	tail
sale	sail	blew	blue	stair	stare	due	do	two	too
know	no	sew	so	tow	toe	dough	doe	pane	pain

□ 14. Write these sounds as headings:

/ī/ /ē/ /ā/ /ū/ /ō/

Under each heading, write the word or words that have that vowel sound in them.

bike	/ī/	sue	/ū/	glow	/ō/	stow	/ō/	June	/ū/
she	/ē/	owe	/ō/	gay	/ā/	shy	/ī/	meet	/ē/
to	/ū/	roe	/ō/	sleet	/ē/	might	/ī/	main	/ā/
room	/ū/	few	/ū/	mate	/ā/	treat	/ē/	lie	/ī/
loop	/ū/	so	/ō/	right	/ī/	brain	/ā/	chew	/ū/

The complex vowels /ou/ and /oi/ are not hard to spell. Here are the usual ways of spelling /ou/:

ou: house **ow:** town

Here are the usual ways of spelling /oi/:

oi: boil **oy:** boy

• 15. Give another word as an example of each spelling of /ou/ and /oi/. out, down; soil, joy

□ Write each word under the heading /ou/ or /oi/:

toy	down	soil	out	Roy
/oi/	/ou/	/oi/	/ou/	/oi/

Most of the exercises on page 261 are intended, as indicated, for written work. If the class is a slow one, 13 and 14 should probably be started on the chalkboard. In a very slow class, the first part of 6–12 might have to be done on the chalkboard too.

At the end of the page two other vowels—/ou/ and /oi/—are reviewed. These of course are never spelled in the **VCe** pattern. They are among the easiest vowels to spell.

GRAMMAR

The Possessive Transformation

Here is a fairly formal description of the possessive transformation:

$$\left.\begin{array}{l}\text{NP} + \text{tense} + \textbf{have} + \text{Det} + \text{N}_1 \\ \text{X} + \textbf{the} + \text{N}_1 + \text{Y}\end{array}\right\} \rightarrow \text{X} + \text{NP} + \text{POS} + \text{N}_1 + \text{Y}$$

Here the insert sentence (the first sentence in the formula above) is one that has the structure "NP + tense + **have** + determiner + noun." It is a sentence like "John has a car" or "She had a book" or "The boys had some goats." In the illustrative sentences the NP's are **John, she**, and **the boys**. The "tense + **have**" structures are **has, had**, and **had**. The "Det + N_1's are **a car, a book, some goats**.

The second sentence in the formula is the matrix sentence. The X and Y are just position markers. They stand for anything that occurs before and after "**the** + N_1" or for nothing if nothing occurs there. So all that is specified for the matrix sentence is that it contain the determiner **the** and a noun. The subscript $_1$ after the N in both the insert and the matrix sentences means that the noun must be the same noun.

The result sentence in the formula shows that X (whatever occurs before "**the** + N_1" if anything occurs there) will come first, then the subject of the **have** sentence, then POS, which means the possessive morpheme, then the N_1, then Y (whatever, if anything, comes after the "**the** + N_1" of the matrix).

Here is an example:

Insert: he + present + have + a + cat
Matrix: I + past + see + the + cat + outside

The NP of the insert is **he**; tense is present; **have** is **has**; Det is **a**; N_1 is **cat**. In the matrix, X is "I + past + see"—everything that comes before **the; the** is **the**; N_1 is **cat**, as it is in the insert; Y is **outside**.

The Possessive Transformation

Study these simple sentences:

Simple Sentence: **a. They have a car.**
Simple Sentence: **b. The car is in the garage.**

We can express all of the meaning of these two simple sentences in a more complex sentence called a **transform.**

Transform: **c. Their car is in the garage.**

Here is how we do it, step by step:

1. We put the subject of Sentence a into the possessive form: **their.**

a. ~~**They**~~ have a car.
→**their**

2. Then we replace the determiner **the** before **car** in Sentence b with the possessive **their.**

→**their**
b. ~~**The**~~ car is in the garage.

3. Now we have a single, more complex sentence.

Transform: **c. Their** car is in the garage.

This process is called a **transformation.**

The new noun phrase, **their car,** expresses all the meaning of "They have a car." It can be used in other sentences like these:

I like **their car.**
I rode in **their car.**
That sedan is **their car.**

• Suppose you wished to say in one sentence what the following two simple sentences say.

He has a pony.
I rode on the pony.

What word of the first sentence would you use in its possessive form? Where would you use it in the second sentence? What would it replace? What would the new, more complex sentence be? he; before pony; the; I rode on his pony.

• What one sentence can you use to express what these two sentences say? discuss

He has a dog. Fido is the dog.

• First, what would you use instead of **the** in the noun phrase **the dog** in the second sentence? his

• Your new sentence would be:

Fido is his dog.

• The subject is **Fido.** The word **is** is a form of **be.** What noun phrase is the complement of **Fido?** his dog

• What one sentence could you use to express what these two sentences say? discuss

I have a bicycle. The bicycle is red.

• This time you could say:

My bicycle is red.

What noun phrase is the subject of this sentence? my bicycle

• Transform these sentences. Replace a determiner in Sentence b with the possessive form of the subject in Sentence a. The determiner you replace will be one that comes before a noun that occurs also in the object of Sentence a. Write the new sentence.

1. a. We have some boats.
   b. The boats are sailboats. Our boats are sailboats.
2. a. You have a drum.
   b. He beats the drum. He beats your drum.
3. a. I have a dog.
   b. Rover is the dog. Rover is my dog.
4. a. They have an uncle.
   b. We sat behind the uncle. We sat behind their uncle.
5. a. She had a ribbon.
   b. The ribbon was yellow. Her ribbon was yellow.
6. a. It has a bell.
   b. We heard the bell. We heard its bell.
7. a. He had a pencil.
   b. The pencil was dull. His pencil was dull.

If we put these elements together according to the formula, we get the following:

$$\underbrace{\mathbf{I} + \text{past} + \mathbf{see}}_{X} + \mathbf{he} + \text{POS} + \mathbf{cat} + \underbrace{\mathbf{outside}}_{Y}$$

This of course represents the sentence "I saw his cat outside."

The morpheme POS means whatever one does to a noun phrase to make it possessive, just as the morpheme past means whatever one does to a verb to make it past tense. Thus "he + POS" is **his,** "I + POS" is **my,** "John + POS" is **John's,** "the boy + POS" is **the boy's,** "the boys + POS" is **the boys'.**

The formula specifies that the subject of a **have** sentence plus the possessive morpheme replaces a definite article, **the,** in a matrix sentence. The article must be **the** because possessives always have the definite meaning characteristic of **the.**

The verb **have** is specified for the insert sentence because it encompasses the widest meaning of the possessive morpheme. If we specified instead the verb **own** we would cut out a great many possessives that **have** will include. For example, we could not get such a result sentence as "His mother is outside" because there is no such insert sentence as *"He owns a mother," though there is such an insert sentence as "He has a mother."

Not all possessives, however, can be derived from insert sentences with **have.** For example, "a day's work" cannot be derived from *"A day has work." It must be derived from a sentence like "The work lasts a day."

Some possessives may be derived from either of two insert sentences and may therefore be ambiguous. For instance, "Williams' books" may derive either from "Williams had books" (for example, in his library) or from "Williams wrote books." The textbook presentation is, of course, much simpler than this technical discussion for the teacher.

LITERATURE

"Rebecca" (Stanzas Three to Five)

Teaching Suggestions

Replay or reread the first two stanzas of the poem. Then read or play these and give the children practice in reading aloud.

SECTION 4

The Rest of the Poem

Read what happened to the door-slammer.

It happened that a Marble Bust
Of Abraham was standing just
Above the Door this little Lamb
Had carefully prepared to Slam,
And Down it came! It knocked her flat!
It laid her out! She looked like that.

. . . .

Her funeral Sermon (which was long
And followed by a Sacred Song)
Mentioned her Virtues, it is true,
But dwelt upon her Vices too,
And showed the Dreadful End of One
Who goes and slams the door for Fun.

The children who were brought to hear
The awful Tale from far and near
Were much impressed, and inly Swore
They never more would Slam the Door
— As often they had done before.

264

Rebecca (STANZAS THREE TO FIVE)

A **bust** is a statue of the upper part of a person, showing his head, shoulders, and chest. You may have heard about **Abraham** in the Bible.

- The bust was made of **marble.** Is that a kind of wood or a kind of stone? See a dictionary. discuss
- Where was the bust of Abraham? Do you believe the poet really thinks Rebecca was a little lamb? What was she preparing to do? Was Rebecca's door-slamming just carelessness or thoughtlessness? See first 4 lines.
- What happened to the bust and Rebecca? See lines 5 and 6.
- When someone dies, he has a funeral, and someone talks about him at the funeral. This talk is a **funeral sermon.** Usually at a funeral the speaker emphasizes, or dwells on, the good things about the dead person. That is what **virtues** means. What do you think **vices** means? See a dictionary. What was Rebecca's great vice? discuss
- Who were brought to hear the sermon? What is the sermon called in the second line of the last stanza? children; awful tale What do you think **awful** means in this line? Find this meaning of **awful** in a dictionary. Discuss dictionary definitions.
- Something **impresses** you if it has a strong effect on your mind and feelings. Who were impressed? the children
- **Inly** means inwardly or to oneself. What did the children swear inly? See last two lines.
- Do you think the poet is being serious? Does he really mean that children who slam doors are likely to have marble busts of Abraham fall on them? discuss
- Usually we write proper nouns with capital letters and common nouns with small letters. This poet capitalizes many words that are not proper nouns. Do you think this helps to show he is not being serious? discuss

VOCABULARY AND MEANING

"Rebecca" (Stanzas Three to Five)

Work through the questions to make sure the poem is understood. Though humorous, it isn't exactly easy.

Point out the bust in the illustration, and show how it conforms to the definition given here. Ask if any of the children can tell something about who Abraham was.

The questions that follow try to make clear, among other things, the use of humorous exaggeration in the poem.

COMPOSITION – A Picture to Describe

Teaching Suggestions

Warm up for this with a class discussion of the illustration. Let the children point out and comment on the various details so that all will have a body of material to use.

Go through the questions designed to recall the mechanics of paragraph writing.

A Picture to Describe

You read about a bust of Abraham. Here, instead of making a whole snowman, the boys have made a bust of a man out of snow and have put it on the terrace.

Describe this bust in a paragraph. Refer to the bust as "it." When you use the possessive, of course, you will use **its**. Tell what it has in **its** mouth, on **its** head, and so on.

Remember to give your paragraph a title. What words do you capitalize in a title? What do you do with the first line of a paragraph?

How must every sentence that is not a question begin and end? How must a question end?

Review 3: Vowel Sounds

• You have now reviewed all except one of the vowel sounds you have studied this year. Go through this chart, and for each word in the chart used as an example, give another example.

Five Simple Vowel Sounds

Sounds	/i/	/e/	/a/	/u/	/o/
Spellings and Examples	**i:** hit	**e:** red **ea:** dead	**a:** cat	**u:** but **o:** son	**o:** top **a:** far

(For extra examples see p. 324.)

Five Complex Vowel Sounds

Sound	/ī/	/ē/	/ā/	/ū/	/ō/
Spellings and Examples	**VCe:** line **igh:** high **y:** try **ie:** die	**VCe:** Pete **ee:** deed **ea:** heat **e:** he	**VCe:** lame **ay:** pay **ai:** wait	**VCe:** June **oo:** root **ew:** few **ue:** Sue **o:** to	**VCe:** lone **oa:** goat **ow:** slow **oe:** hoe **o:** no

(For extra examples see pp. 324–25.)

• We use two vowel letters between slanting lines to show the four vowel sounds in the chart below. You have not reviewed the /au/ sound. Give other words as examples of the spellings shown in this chart.

Four More Vowel Sounds

Sound	/oo/	/ou/	/oi/	/au/
Spellings and Examples	**oo:** look **u:** push	**ou:** out **ow:** cow	**oi:** oil **oy:** toy	**au:** haul **aw:** flaw **a:** ball **o:** long **ough:** fought **augh:** caught

(For extra examples see pp. 326, 328, 329.)

267

Review 3: Vowel Sounds

The one vowel not yet reviewed is /au/. It is reviewed here in the examples contained in the final chart.

These charts present, for the most part, only the spellings that might be called regular or common. Thus no mention is made of **ie** for /e/ as in **friend** or **y** for /i/ as in **myth**. Some of the spellings for the complex vowels, however, contain members of small sets. Thus, the only other example of **augh** for /au/ is **taught**. The other examples of **o** for /ū/ are **who** and **do**.

Teaching Suggestions

Go through the charts with class participation, eliciting another example for each example given.

Extra Practice

You may wish to have the individual pupils follow up the oral exercise by writing one or more extra examples of each spelling given in the chart.

GRAMMAR

More Possessive Transformations

Continue practice with the possessive transformation. We still confine attention to the production of the possessives of the personal pronouns:

I + POS → **my**	**we** + POS → **our**
he + POS → **his**	**you** + POS → **your**
she + POS → **her**	**they** + POS → **their**
it + POS → **its**	

One reason for beginning with the personal pronouns is that they are not spelled with the apostrophe. Pupils are likely to associate the possessive so closely with the apostrophe that they use it also with the personal pronouns, where it doesn't belong. It is hoped that this initial concentration with the personal pronouns will help preserve them from apostrophic intrusion.

The two sentences to be transformed are uniformly presented in the order insert (a), matrix (b). The pupils will get into the habit of making the result sentence by beginning with the matrix and inserting in it whatever the transformation calls for—in this case the possessive form of the subject of the **have** sentence.

Teaching Suggestions

Before assigning Sentence Pairs 1–7 on page 268 as a written exercise, let the children participate orally in transforming the b sentences. Then, when they are in the swing of it, let them write the transforms. Be sure they understand that they are to use sentence b in each pair in writing the new sentence, inserting into it the possessive form of the subject of sentence a.

More Possessive Transformations

Suppose you wanted to express these two ideas in a single sentence:

a. He had a cookie.
b. The cookie was delicious.

• First you would change the subject **he** of Sentence a to the possessive form. What is the possessive form of **he?** his

• **Had** in Sentence a is a form of what word? What noun phrase is the subject of Sentence b? What word in the phrase **the cookie** would you replace with **his?** have; the cookie; the

• What is the function of the noun phrase **his cookie** in the single sentence you would make? subj.

His cookie was delicious.

□ Transform Sentences a and b in each exercise below into a new sentence. Replace a determiner in Sentence b with the possessive form of the subject of Sentence a. Tell how the new noun phrase functions in each new sentence you write.

1. a. We had a kitten.
   b. The kitten is now a cat.
   c. Our kitten is now a cat. subj.
2. a. He has a pet.
   b. This pony is the pet.
   c. This pony is his pet. comp.
3. a. I had an adventure.
   b. The adventure was an airplane ride.
   c. My adventure was an airplane ride. subj.
4. a. You have an aunt.
   b. Miss Wood is the aunt.
   c. Miss Wood is your aunt. co
5. a. She has a hat.
   b. You sat on the hat.
   c. You sat on her hat. obj. of prep.
6. a. They had some candy.
   b. Somebody ate the candy.
   c. Somebody ate their candy. obj. of verb
7. a. It has a tail.
   b. It wagged the tail.
   c. It wagged its tail. obj. of verb

• Here are more pairs of sentences to transform. Use the possessive form of the subject of Sentence a in a new sentence made from Sentence b. Tell whether the new noun phrase with the possessive is a subject, a complement, an object of a verb, or an object of a preposition.

8. a. They had a fire.
   b. The fire was blazing.
   c. Their fire was blazing. subj.
9. a. I had a party.
   b. Nobody came to the party.
   c. Nobody came to my party. obj. of prep.
10. a. He had a badge.
    b. Margery saw the badge.
    c. Margery saw his badge. obj. of verb
11. a. She has a ring.
    b. The ring is beautiful.
    c. Her ring is beautiful. subj.
12. a. It has a bill.
    b. The worm was in the bill.
    c. The worm was in its bill. obj. of prep.
13. a. We have some marbles.
    b. We lost the marbles.
    c. We lost our marbles. obj. of verb
14. a. You have a sister.
    b. That girl is the sister.
    c. That girl is your sister. comp.

• Now transform the sentences below. This time use the new noun phrase in any way you like. For instance, if the sentence were "I have a bicycle," you might say "My bicycle is new" or "Jack liked my bicycle" or "The child was near my bicycle." (Answers will vary.)

15. I have a dog. My dog is a poodle.
16. He has some books. His books are heavy.
17. They have a grandmother. They visited their grandmother.
18. We have a car.. She rode in our car.
19. She had a smile. Her smile was charming.
20. You have a desk.
21. It has a paw. Its paw is large.
22. They had some peanuts. We ate their peanuts.

Items 8–14 on page 269 provide, in addition to further practice with the possessive, a review of the four noun phrase functions studied earlier. Some children will no doubt have forgotten some of this material and will need to be taken carefully through the exercise.

In 15–22 we in effect omit the matrix (b) sentence and let the children provide one of their own. With a good class, this may be done as a written exercise.

Extra Practice

Sentence Pairs 8–14 may be assigned as written work after the class has worked them orally as a group and discussed them. They may label the new noun phrase (1) if it functions as a subject, (2) as a complement, (3) as object of a verb, or (4) as object of a preposition.

TESTS AND REVIEW

Teaching Suggestions

This may be the last opportunity this year for extensive reteaching based on diagnosis of individual needs. Advantage should be taken of it.

The tests on pages 270–72 are intended to be written. The review on page 273 is for oral class work.

Extra Test

Dictate these words and have the class spell them. Say the word clearly. Use it in the sentence. Say it again and have the children write it.

1. **voice** The dog heard its master's **voice**. **voice**
2. **employ** Factories **employ** workers. **employ**
3. **thread** I watched her **thread** a needle. **thread**
4. **laugh** We **laugh** at our own mistakes. **laugh**
5. **plane** I took a flight on a **plane**. **plane**
6. **brain** I think with my **brain**. **brain**
7. **town** Our **town** has a library. **town**
8. **house** That **house** has a red chimney. **house**
9. **shoot** Bob can **shoot** a rifle. **shoot**
10. **glue** We use **glue** to fasten things. **glue**

Tests and Review

□ TEST I Sounds and Letters

1. Write these headings:

/i/ /e/ /a/ /u/ /o/

Write each of the words below under the heading that shows its vowel sound.

If you make mistakes, study page 255.

fat /a/	**fib** /i/	**fetch** /e/	**fun** /u/
butt /u/	**sob** /o/	**rich** /i/	**laugh** /a/
friend /e/	**star** /o/	**myth** /i/	**son** /u/
Bill /i/	**Ben** /e/	**Bob** /o/	**Babs** /a/

2. Write these headings:

/ī/ /ē/ /ā/ /ū/ /ō/

Write each of the words below under the heading that shows its vowel sound.

If you make mistakes, study pages 260–61.

rail /ā/	**hate** /ā/	**strike** /ī/	**bead** /ē/
cold /ō/	**rhyme** /ī/	**fly** /ī/	**rude** /ū/
wild /ī/	**may** /ā/	**you** /ū/	**coat** /ō/
he /ē/	**hope** /ō/	**sight** /ī/	**true** /ū/
few /ū/	**go** /ō/	**spook** /ū/	**real** /ē/

3. Write these headings:

/oo/ /ou/ /oi/ /au/

Write each of the words below under the heading that shows its vowel sound.

If you make mistakes, study pages 108, 206, 218, 238, 250, 267.

bought /au/	**claw** /au/	**push** /oo/	**toy** /oi/
haunt /au/	**soil** /oi/	**crook** /oo/	**clown** /ou/
loud /ou/	**song** /au/	**join** /oi/	**put** /oo/
hall /au/	**couch** /ou/	**how** /ou/	**joy** /oi/

4. Write a homophone for each of the words in the list below.

see	hall	toe	meat
sea	haul	tow	meet
rode	taut	sun	flee
road	taught	son	flea

□ TEST II Grammar

1. Copy each of the sentences below, but change the form of the verb from the present tense to the past tense.

If you make mistakes, study pages 220, 226, 232, 240, 250.

a. The car costs a lot. . . . cost . . .
b. It cuts my hand. . . . cut . . .
c. The bees sting us. . . . stung . . .
d. They copy the poems. . . . copied . . .
e. Jerry comes late. . . . came . . .
f. They swing hard at the ball. . . . swung . . .
g. He has a bicycle. . . . had . . .
h. Mr. Webster quits early. . . . quit . . .
i. She eats a lot of candy. . . . ate . . .
j. The children cling to their mother. . . . clung . . .
k. Mary puts sugar on cereal. . . . put . . .

2. Tell what pronouns in the following sentences have the possessive form.

If you make mistakes, see pages 256–57.

a. She was his mother.
b. They bought my bicycle.
c. Our picnic was at their camp.
d. She is his sister.
e. Her dog lost its collar.
f. Your uncle was in his canoe.
g. Their father called them.
h. Her bicycle fell on her.

271

3. Transform these pairs of sentences into new sentences. Use the possessive form of the subject of Sentence (1) in place of a determiner in Sentence (2).

a. (1) She has a scarf.
(2) The scarf is blue. Her scarf is blue.

b. (1) They have some chickens.
(2) I feed the chickens. I feed their chickens.

c. (1) We had a canary.
(2) The canary got away. Our canary got away.

d. (1) You have a camp.
(2) I went to the camp. I went to your camp.

4. Use the subject of each of the sentences below in the possessive form. Use it in place of the determiner in the noun phrase that is the object of **have.** Use the new noun phrase in a sentence of your own. For instance, if the sentence were "He had a bicycle," you might write "His bicycle was red" or "I saw his bicycle." (Sentences will vary.)

a. She had a doll. Her doll was beautiful.
b. It had a house. Its house was in a tree.
c. He had a headache. His headache was gone.
d. I had a doctor. I asked my doctor.
e. You had a milk shake. Your milk shake looked good.
f. They had a puppy. They gave us their puppy.
g. Steve had a story. Steve's story was funny.
h. You had a coat. Mr. Baker found your coat.
i. Mrs. Russell had a dictionary. Mrs. Russell's dictionary was new.

□ TEST III Writing and Vocabulary

Write five sentences. In each sentence, use one of these words:

1. repose 2. exposed 3. abhor 4. deliberately 5. sermon

• **REVIEW Ideas and Information**

1. What is a poetic word for **complaint?** plaint
2. What do the two parts of **hippopotamus** mean? horse; river
3. How many vowel sounds have you studied? 14
4. How many of the different vowel sounds are sometimes spelled in the pattern VCe? 5
5. What vowel sound in addition to the five simple vowels never comes at the end of words? /oo/
6. What verb shows possession? have
7. What does a possessive form replace in a noun phrase? a determiner
8. What suffix may we add to some nouns to make adjectives? ful
9. How do we change a noun such as **beauty** when we add the suffix **ful?** change y to i
10. What are the two usual ways of spelling the vowel sound /oi/? Which of these ways is used at the end of words? oi, oy; oy
11. Name three verbs whose past tense form is the same as their simple present tense form. put, cost, quit
12. In which of these sentences can you tell what the tense of the verb is? b

    a. They put their hats on the rack.
    b. She put her hat on the rack.
13. What are the seven possessive forms of the personal pronouns that you have studied? my, his, her, its, our, their, your
14. What verb expresses the same meaning as the possessive form? have
15. Can a noun phrase like **his dog** be used as a subject? Can it be used as a complement? as object of a verb? as object of a preposition? yes; yes; yes; yes
16. What kind of word does a possessive form replace in the possessive transformation? a determiner

273

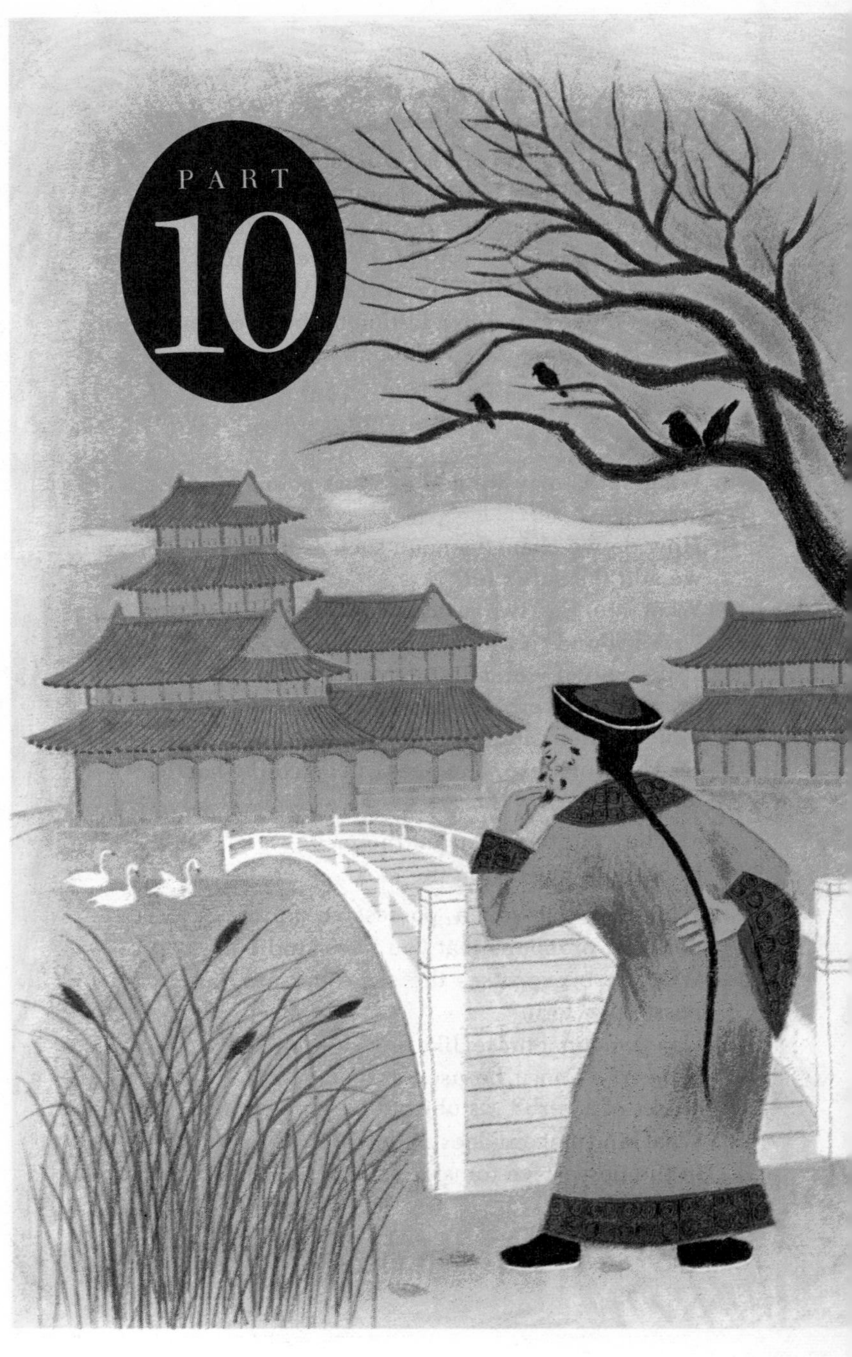
PART
10

SECTION 1

A Poem

We will study this poem in two parts.

The Pigtail

There lived a sage in days of yore,
And he a handsome pigtail wore;
But wondered much and sorrowed more,
Because it hung behind him.

He mused upon the curious case,
And swore he'd change the pigtail's place,
And have it hanging at his face,
Not dangling there behind him.

Says he, "The mystery I've found —
I'll turn me round" — he turned him round;
But still it hung behind him.

The Pigtail (STANZAS ONE TO THREE)

- The word **sage** means wise man. What does **days of yore** mean? Look up **yore** in a dictionary. discuss
- To **muse upon** is to think about. The **curious case** he mused upon was the strange fact that his pigtail hung behind him. Where did he want it to hang? in front
- Find the quotation marks in the third stanza. What words do they enclose? Who said these words? discuss
- What is the present tense of **found?** Is this verb regular, or irregular? What is the present tense of **hung?** find; irregular; hang
- What are the rhyming words in each stanza? What word is the last one in each stanza? yore, wore, more; case, place, face; found, round; him

275

LITERATURE

"The Pigtail" (Stanzas One to Three)

This light poem introduces the children to a famous writer whose novels *Vanity Fair* and *Henry Esmond* some of them may read in later life. Explain that the author's name is William Makepeace Thackeray and that he lived and wrote in England a century ago.

Teaching Suggestions

Read these first three stanzas or play the recording of them. Let the children read aloud. This also is a nice poem to memorize.

VOCABULARY AND MEANING

"The Pigtail" (Stanzas One to Three)

Go through the questions below the poem. The poem is not a difficult one once the key terms are understood—**pigtail, sage, yore**, etc.

SOUNDS AND LETTERS

Review 1: Consonant Sounds

The last of the lessons in the sounds-and-letters strand of grade four are devoted to a review of the seventeen consonant sounds so far studied. These pages review the first six, the voiceless sounds /p/, /t/, /k/, /ch/ and the voiced /b/ and /d/.

Teaching Suggestions

Have the class respond to the questions on this page. They are intended to elicit the more general statements that can be made about the spelling of the four voiceless sounds mentioned.

Some exceptions to the rule implied by the penultimate question are **rich, which, touch, such, much**.

Review 1: Consonant Sounds

You have studied seventeen consonant sounds so far this year. This is the first of four review lessons in which you will study these sounds again. The first consonant sounds you studied were these:

/p/ /t/

- How do you spell the sound /p/ at the beginning of a word? p
- How do you spell /t/ at the beginning? t
- How do you spell the final /p/ sound if a simple vowel comes before it? p
- How do you spell final /t/ if a simple vowel comes before it? t
- What are the last two letters of a word that ends with the /p/ sound, if the vowel sound before it is complex and is spelled with one letter? pe
- What are the last two letters of a word that ends with the /t/ sound, if the vowel sound before it is complex and is spelled with one letter? te

- The consonant sound /k/ is spelled in two ways at the beginning. What are the two ways? How is /k/ spelled when the next letter is **i** or **e**? k and c; k
- What two letters are used to spell final /k/ when a simple vowel sound comes before it? ck
- What two letters may be used to spell final /k/ when a complex vowel comes before it? ke

- The sound /ch/ is spelled in only one way at the beginning of words. What is this way? ch
- At the end of words, how do we usually spell /ch/ when a simple vowel sound comes before it? Can you remember an exception? tch; touch
- How do you spell final /ch/ when some sound other than a simple vowel sound comes before it? ch

These two consonant sounds are like /p/ and /t/:

/b/ /d/

- Which of the consonant sounds /p/, /b/, /t/, and /d/ make a little buzzing sound in your throat when you say them? /b/, /d/
- How do we spell /b/ and /d/ at the beginning of words? b, d
- How do we spell final /b/ and /d/ if a simple vowel sound comes before them? How do we spell the sounds if a consonant letter comes before them? b, d; b, d
- What two letters may be used to spell final /b/ if a complex vowel spelled by a single letter comes before it? How is final /d/ spelled in this case? How are final /b/ and /d/ spelled if the complex vowel is spelled with more than one letter? be; de; b, d

□ Copy the words below, and complete them by putting in the missing letter or letters. You must spell each word so that it has the sound the heading shows.

/p/	/t/	/k/	/ch/	/b/	/d/
_an (p)	_en (t)	_ane (c)	__ip (ch)	_ack (b)	_ock (d)
ro__ (pe)	sea_ (t)	_eep (k)	ea__ (ch)	gra_ (b)	ru__ (de)
dro_ (p)	vo__ (te)	lu__ (ck)	ri__ (ch)	tri__ (be)	sle_ (d)
soa_ (p)	la__ (te)	lea_ (k)	ma___ (tch)	_lack (b)	har_ (d)
wi__ (pe)	le_ (t)	ma__ (ke)	mar__ (ch)	bul_ (b)	gra__ (de)

□ Copy these headings:

/p/ /t/ /k/ /ch/ /b/ /d/

Under each heading, write the words from the following list that end with that sound.

neck /k/	cape /p/	board /d/	clip /p/	wasp /p/
reach /ch/	stitch /ch/	lake /k/	ebb /b/	mad /d/
hide /d/	blast /t/	lack /k/	weak /k/	barb /b/
ditch /ch/	robe /b/	beat /t/	putt /t/	burst /t/

277

Use the first question on page 277 to recall the voiceless–voiced contrast. Review as directed the spellings of /b/ and /d/.

The items in the exercise on the completion of words have been chosen in such a way that there should be only one possible answer for each. For example, "la–" can be completed for /t/ only as **late**, since there is no word ***lait** in English. Have the children do this as a written exercise, beginning it on the chalkboard if necessary. The class should not now have much trouble with the final exercise on this page.

GRAMMAR

The Possessive Form of Other Noun Phrases

Pages 278–79 are a single lesson which presents the use of the apostrophe and s to form the singular possessive of noun phrases of the proper noun type. The possessive forms of other types of noun phrases are taught on pages 286–87 and 292–93.

The Speech and Spelling Rules

The rule for the speech formation of the possessive is the same as that for forming regular plurals and s forms of verbs:

1. Add /əz/ after sibilants: **Alice's, Rose's, the judge's.**
2. Add /s/ after non-sibilant voiceless consonants: **the cop's, Pete's, Dick's, Ralph's, Mr. Smith's.**
3. Otherwise, add /z/: **Bob's, Ed's, the men's, Joe's.**

As to spelling, the rule is this:

1. Indicate the possessive by simply adding the apostrophe, when the noun phrase ends in a letter s that represents the plural morpheme: **the boys', our friends'.**
2. Otherwise, add the apostrophe and s: **John's, the boy's, the man's, the men's.**

The exceptions that have been noted for plurals and for s forms do not exist for the possessive. The only uncertainty that we feel is what to do about certain proper nouns that end in an s but are not plural. For example, do we say "Dickens' novels" or "Dickens's novels"? There is no correct answer to this. It is done either way in the best circles.

Use of the Possessive

The English possessive is a strange creature to the speakers of many languages. In

The Possessive Form of Other Noun Phrases

• You know that one kind of noun phrase is the personal pronoun. Another kind is the proper noun. What other kinds of noun phrases can you name?

□ You know how to write the possessive forms of the seven personal pronouns. Write the possessive form of each of these pronouns:

she	he	you	it	I	they	we
her	his	your	its	my	their	our

• You know that the possessive forms express the meaning of having. Look at these three sentences:

a. **I have a bicycle.**
b. **The bicycle is broken.**
c. **My bicycle is broken.**

What word of Sentence a is changed into a possessive and used in Sentence c? What word of Sentence b is replaced by the possessive? I; the

• Other noun phrases besides personal pronouns can be made into possessives. What kind of noun phrase functions as the subject of this sentence?

d. **John has a bicycle.** p. n.

We can make this noun phrase a possessive, too, and use it in place of a determiner:

e. **John has a bicycle.**
f. **The bicycle is broken.**
g. **John's bicycle is broken.**

Notice carefully how the possessive of the noun phrase **John** is spelled: **John's.** We put the letter s after the noun phrase with an apostrophe before the s.

• We form the possessive of noun phrases that are not personal pronouns by adding an apostrophe and s to the singular form of the noun. This is one use of the apostrophe. What other use of it have you learned? to replace missing letter(s) in a contraction

• Here are the possessive forms of some other noun phrases. What has been added to each noun phrase to make the possessive form? apostrophe and s

Mary's Mrs. Smith's George's

□ Write the possessive form of these noun phrases:

William's	Sally's	Helen's	Sam's
Miss White's	Bill Green's	Mr. Brown's	Mrs. Gray's

Noun phrases like these do not have special words for the possessive form. That is why we add something to them to make the possessive form.

Noun phrases which are personal pronouns, like **it, he, they,** do have special words for the possessive form: **its, his, their.** So we never use an apostrophe in the possessive form of personal pronouns.

□ Change the subject of each sentence below to the possessive form. Use the possessive form to replace the determiner in the noun phrase that is the object of **have.** Write a new sentence using the new noun phrase. For instance, for the sentence "Mary has a kitten," you might write "It is Mary's kitten," or "Mary's kitten drinks cream." (Answers will vary.)

1. David has a bicycle. David's bicycle is new.
2. They have a car. We rode in their car.
3. We have a wagon. Our wagon is red.
4. Sally has a coat. This is Sally's coat.
5. I have a cold. My cold is better.
6. Mr. Burbank has a garden. It is Mr. Burbank's garden.
7. You have a chair. Your chair is small.
8. Maxwell has a donkey. Maxwell's donkey is slow.
9. She has an alligator. Her alligator is a pet.
10. Donald has a hen. Donald's hen laid an egg.
11. It has a coop. Its coop is large.
12. Edith has some candy. We ate Edith's candy.

the romance languages, for example, possession is always expressed in a prepositional phrase. Thus English **my aunt's pen** becomes in French ***la plume de ma tante.*** Indeed foreign learners of English have a good deal of trouble deciding when to use the possessive ("my aunt's pen," not *"the pen of my aunt") and when to use **of** ("the cap of the pen, not *"the pen's cap").

Native speakers of English sometimes have a little trouble too, though not so much. The general rule is that we communicate the idea of "having" with the possessive when the noun phrase refers to an animate being: "my aunt's pen," "John's bicycle," "the dog's dish." When the noun phrase refers to an inanimate thing, we generally use an *of* phrase: "the cap of the pen," "the roof of the house," "the leg of the table."

The tendency is not to use the **of** phrase for animates; nobody says *"the bicycle of John." Rather, the tendency is to extend the possessive to inanimates: "the pen's cap," "the house's roof," "the table's leg." One would not wish to go so far as to call these incorrect. Such formations appear now and then in good edited writing. But many writers prefer to avoid them.

Extra Practice

If further written practice like that provided in Sentences 1–12 is wanted, these additional sentences may be used in the same way.

13. They have colds.
14. Miss Ellet has some flowers.
15. George has an Easter egg.
16. It has a collar.
17. Bob has a broken arm.
18. Mr. Wendall has an old wheelbarrow.
19. You have a new dress.
20. He has some young puppies.
21. We have guests.
22. Mrs. White has a baby boy.

SECTION 2

The Rest of the Poem

Read the rest of the poem about the sage and his pigtail.

Then round and round, and out and in,
All day the puzzled sage did spin;
In vain — it mattered not a pin —
The pigtail hung behind him.

And right and left, and round about,
And up and down, and in and out,
He turned; but still the pigtail stout
Hung steadily behind him.

And though his efforts never slack,
And though he twist, and twirl, and tack,
Alas! still faithful to his back,
The pigtail hangs behind him.

WILLIAM MAKEPEACE THACKERAY

The Pigtail (STANZAS FOUR TO SIX)

- What was the sage trying to do? How was he trying to do it? Do you think the sage was really a very wise man? Is the poet serious? discuss
- **In vain** means unsuccessfully. Have you ever heard someone say "I don't care a straw"? What part of the first stanza above is like "I don't care a straw"? it mattered not a pin
- The word **stout** may mean sturdy. It may also mean stubborn or unyielding. Which meaning do you think it has in the second stanza above? See a dictionary. discuss
- To **slack** means to let up, or begin to stop. **Alas** expresses sorrow. What does **faithful to his back** mean? It stayed behind his back as always.

281

LITERATURE

"The Pigtail" (Stanzas Four to Six)

Reread or replay the first stanzas of the poem and then follow with these stanzas. Let the pupils practice reading aloud.

VOCABULARY AND MEANING

"The Pigtail" (Stanzas Four to Six)

Teaching Suggestions

Go through the questions to ensure comprehension. There is always something repulsive about explaining humor, but the children should at least be aware of the irony of the use of the word **sage** in this poem.

The word **alas** is rather like **oxen**: it is kept alive by grammar books, which need it to exemplify interjections. Point out the use of the exclamation mark after **alas**. Children need not be encouraged in the use of the exclamation mark; they are only too ready to employ it. But they should of course be aware that it is one of the possible marks of end punctuation and that it is sometimes, as here, used internally—that is, not at the end of a sentence.

ETYMOLOGY — What Words Come From

One purpose of the brief etymologies which appear from time to time in this series is simply to interest pupils in words.

Some other examples of words that are self-descriptive like **pigtail** are **turtleneck** (turtleneck sweater), **fishtail, U-turn**, and **fantail**.

MORPHOLOGY — How Words Are Made

Prefixes

English makes much more use of suffixes than it does of prefixes. Still, some prefixes are in very active use: **un, dis, re, ex, pre**, etc.

Here we illustrate a prefix with the form **un**. Such a prefix, like the suffix **er**, is important in the building of the vocabulary. One can add it to most adjectives to express the negative. So if one knows five hundred adjectives and gets onto the use of **un**, he adds nearly five hundred words to his vocabulary at a bound.

Teaching Suggestions

Go through the explanation, and have the pupils rewrite the five sentences adding **un–**. Other adjectives that take **un–** may be illustrated on the chalkboard — for example, **usual, easy, popular, ready, ruly, selfish, natural**.

What Words Come From

Pigtail is another word that seems to tell us what it means — or at least what it looks like. Why are pigtails called pigtails? They look like pigs' tails.

The word **tack** in "twist and twirl and tack" is the same as the noun that means a kind of nail or fastener. Things that fastened sails in sailing ships were called tacks. When you wanted to change the direction the ship was sailing in, you hauled in the tacks. This action was then called tacking. In our poem, **tack** just means to move, the way a ship does.

How Words Are Made

You have learned to add **er** or **r** to some words to make nouns: **teach+er, teacher; trade+r, trader.**

You have also learned to add **ful** to some nouns to make adjectives: **grace+ful, graceful; pity+ful, pitiful.**

The endings **er** and **ful** are added at the ends of words. They are suffixes. But a part added at the beginning of a word is a **prefix.**

We may add the prefix **un** at the beginning of some adjectives to make other adjectives that mean the opposite. **Unhappy** means not happy. What does **untrue** mean? See a dictionary. discuss

Find the adjective in each sentence below. Use the prefix **un** to make another adjective that means the opposite. Rewrite the sentence, using the adjective you made.

1. His story was truthful. . . . untruthful.
2. The movie was pleasant. . . . unpleasant.
3. The sage was wise. . . . unwise.
4. Our kitchen is tidy. . . . untidy.
5. The directions are clear. . . . unclear.

Two Spelling Problems

The possessive pronouns **their** and **its** often cause spelling difficulty because there are other words that sound very much like them.

One word often confused with **their** is **there:**

I see their books.

The books are there.

• Which word is the possessive form of **they?** Use **their** only to express having. Look up **there** and **their** in a dictionary. their

• The other word often confused with **their** is **they're.** Which word is the contraction of **they are?** they're

Their books are outside.

They're on that bench.

Keep in mind that we do not use the apostrophe when we write the possessive forms of personal pronouns.

□ Copy these sentences carefully:

1. They're there.
2. There are their books.
3. They're with their parents.

The possessive pronoun **its** is often confused with the contraction **it's.** What is **it's** a contraction of? it is

Look up **its** and **it's** in a dictionary.

□ Copy these sentences:

4. It's hot.
5. This is its dish.
6. It's its dish.

A Paragraph to Write

Write a paragraph telling about the poem "The Pigtail." Explain what a pigtail is, what the sage wanted to do with it, and how he tried to solve his problem.

283

Two Spelling Problems

Their and *Its*

The homophones **their/there/they're** and **its/it's** cause trouble for most of us. The teacher may or may not share the disability of the present author, who all too frequently writes something like *"They should mind there manners" or *"The car was in it's stall." If one can blunder in this way after a lifetime of frolicking with these words, one should not be surprised if the fourth-grader also slips occasionally.

He should, however, at least know that the slip is possible and what it consists of and perhaps now and then be able to correct his mistake when he reads his paper over. This page is intended to point out the pitfalls.

COMPOSITION – A Paragraph to Write

Before the pupils carry out this composition assignment, it may be well to let a few of the pupils tell about the poem orally.

Review 2: Consonant Sounds

This page reviews the spelling of /g/, /j/, /f/, and /v/, the first three of which are fairly complicated.

Teaching Suggestions

Elicit answers to the questions on page 284, which are intended to lead to formulation of the general rules. Point out again that some letters stand for more than one sound: **g** stands for /g/ in **gone** but for /j/ in **gentle**. But stress also the limitations: **g** can stand for /j/ only when the next letter is **e** or **i**.

Examples of the **gue** spelling are **league, rogue, brogue, plague, vague, intrigue, fatigue**.

Illustrate the voiceless–voiced contrast again with /ch/ and /j/ and with /f/ and /v/.

Review 2: Consonant Sounds

This is the second of four review lessons on the consonant sounds you have studied this year.

The consonant sound /g/ is the next one you will review.

- The consonant sound /g/ is always written with the letter **g** at the beginning of words. Name some words that begin with /g/ and the letter **g.** See p. 326.
- Does the letter **g** always stand for the sound /g/? no
- How is the final /g/ sound usually spelled when it comes after a simple vowel? g
- After a complex vowel, the final /g/ sound is spelled **gue**. Give an example of this spelling. league

- The consonant sound /j/ is like /ch/. Which of these sounds makes a little buzzing noise in your throat when you say it? /j/
- When the first sound in a word is /j/ followed by **e** or **i**, the /j/ sound may be spelled with **j** or **g.** Name words in which it is spelled with the letter **j.** Name words in which it is spelled with the letter **g.** See pp. 326–27.
- If the second letter is not **e** or **i**, how is the sound **j** spelled? Give examples. /j/; jug, Jack, jog, jay
- When a simple vowel comes before the final /j/ sound, what three letters stand for /j/? dge
- When a complex vowel sound or a consonant sound comes before final /j/, what two letters stand for /j/? ge

- Which of these sounds makes a little buzzing sound in your throat when you make it: /f/ or /v/? /v/
- The usual way to spell the sound /f/ at the beginning of a word is with the letter **f.** What is another way? ph
- How is /v/ spelled at the beginning of words? v

284

- What two letters are used to spell the sound /v/ at the end of words? ve
- At the end of words we spell the sound /f/ with the letters **ff** if the /f/ sound follows the simple vowels /i/, /a/, or /u/. Give an example. stiff, staff, stuff
- When final /f/ comes after a complex vowel, it may be spelled in the **VCe** pattern. Give an example. life
- When final /f/ comes after a consonant letter or a vowel spelled with two letters, how do we usually spell /f/? Give an example or two. f; scarf, leaf
- Give a word in which final /f/ is spelled **ph.** phonograph
- Give a word in which final /f/ is spelled **gh.** tough

□ Use the **VCe** pattern to build a word with final /f/. Replace **V** with a vowel. What will you replace **C** with? How will you begin your word? Example: i; f; w; wife

□ Copy the words below, and complete them by putting in the missing letter or letters. You must spell each word so that it has the sound the heading shows.

/g/	/j/	/f/	/v/
_et (g)	_et (j)	__one (ph)	_ine (v)
_rab (g)	_iant (g)	_oot (f)	arri__ (ve)
lea___ (gue)	_ug (j)	sti__ (ff)	_ote (v)
fla_ (g)	ba___ (dge)	wi__ (fe)	stri__ (ve)
e__ (gg)	lar__ (ge)	brie_ (f)	bra__ (ve)
bra_ (g)	a__ (ge)	hal_ (f)	_eal (v)

□ Copy these headings:

/g/ /j/ /f/ /v/ /ch/

Under each heading, write the words from the following list that end with that sound.

lag	/g/	edge	/j/	each	/ch/	live	/v/
leaf	/f/	vague	/g/	wave	/v/	cliff	/f/
latch	/ch/	rage	/j/	peg	/g/	calf	/f/
graph	/f/	judge	/j/	porch	/ch/	rave	/v/

Do the questions at the beginning of page 285 in oral discussion. Though any individual child might have trouble finding the examples asked for, the class as a whole should not.

Let the last two exercises be done as written work, starting them on the chalkboard if necessary.

GRAMMAR—More Possessives

These pages give further practice with the making of possessive noun phrases.

A Recursive Transformation

Note Sentence 13 on page 287. This already has a possessive noun phrase, **our** (**we** + POS), so that the pupils' answer will be a noun phrase with two possessives: **our room's** (**our room** + POS). The possessive transformation is, in fact, recursive; it can be applied over and over again in the expansion of a noun phrase provided that there is always a **the** to be replaced by an NP + POS.

Thus we might replace **the** in **the uncle** with "**the friend** + POS." This would give **the friend's uncle**. But this contains another **the**. We could replace it with "**the brother** + POS" to give "**the brother's friend's uncle**." This also contains **the**. We might replace this **the** with "**I** + POS" and have "**my brother's friend's uncle**." The transformation cannot be repeated again because the last insert did not include a **the**. If instead of "**I** + POS" we had put in "**the assistant** + POS," we would get "**the assistant's brother's friend's uncle**" and could have gone on to "**my assistant's brother's friend's uncle**."

Obviously "**my assistant's brother's friend's uncle**" is an example of bad writing, but it is not ungrammatical. The possessive transformation is recursive. There is no rule that says it can be applied only so many times. Grammatically it can be applied numberless times, so long as each insert provides a new **the**.

The recursive aspect of certain grammatical rules is what makes it possible to generate countless sentences from a finite grammar. In later books in this series, recursiveness is described explicitly for the learners.

More Possessives

□ You know how to write the possessive forms of personal pronouns and to make possessives from proper nouns. Write the noun phrases with a possessive that can be made from the sentences with **have** below. Just write the noun phrases. For example, if the sentence were "Harry has a bicycle," you would write "Harry's bicycle."

1. **Donna has a family.** Donna's family
2. **They have a house.** their house
3. **We had an argument.** our argument
4. **Maxwell had a garden.** Maxwell's garden
5. **I had a lunch.** my lunch
6. **Mrs. Taylor has some roses.** Mrs. Taylor's roses
7. **Mr. Wilson has a plane.** Mr. Wilson's plane
8. **You had some dishes.** your dishes

Possessives can be made from all kinds of noun phrases. You have studied five kinds of noun phrases: (1) personal pronouns, (2) proper nouns, (3) determiners plus common nouns, (4) common nouns alone, (5) indefinite pronouns. Which of these five kinds do you find in the subjects of these sentences?

a. **This child has a toy.** (3)
b. **Children have toys.** (4)
c. **Everyone has an answer.** (5)

All of the subjects above can be made into possessives and used in other sentences. For example:

d. **This child has a toy.**
e. **The toy is broken.**
f. **This child's toy is broken.**

• What do we add to the subject of Sentence d to make it possessive? What words replace the determiner **the** of Sentence e to form Sentence f? 's; this child's

☐ Make possessives out of the noun phrases used as subjects in the following sentences. Write new sentences, and use the new noun phrases as subjects of the predicate **is beautiful.** For instance, if the sentence were "The child has a toy," you would write "The child's toy is beautiful."

9. The girl has a picture. The girl's picture . . .
10. The school has a playground. The school's playground . . .
11. This man has a garden. This man's garden . . .
12. We have a room. Our room . . .
13. Our room has a picture. Our room's picture . . .
14. They have a cat. Their cat . . .
15. The cat has a kitten. The cat's kitten . . .
16. Mrs. Gordon has a child. Mrs. Gordon's child . . .
17. The child has a doll. The child's doll . . .

Indefinite pronouns can be transformed into possessives too:

g. Everybody has a book.
h. We looked at everybody's book.
i. Everybody's book was interesting.

☐ Make new noun phrases from these sentences with **have.** Use the new noun phrase in a sentence of your own. For example, if the sentence were "Everybody has a book," you might write "I liked everybody's book" or "Everybody's book was new." (Answers will vary.)

18. Everyone has a picture. We saw everyone's picture.
19. Someone has a cough. Someone's cough was loud.
20. Everybody has a wish. She told everybody's wish.
21. Somebody has a sandwich. I ate somebody's sandwich.
22. The children have a party. The children's party was fun.
23. Our teacher has a purse. Our teacher's purse was open.
24. The men have a picnic. He went to the men's picnic.
25. The women have a car. The women's car is new.

Teaching Suggestions

In Sentences 1–8 on page 286, the matrix sentence is not given; the pupil merely makes a possessive noun phrase which might then be substituted for any **the** in any matrix sentence.

The middle of page 286 reviews the five *kinds* of noun phrases (as distinguished from their functions, which are reviewed on page 269). Let the children offer an example of each of the five kinds.

Sentences d, e, and f are respectively the insert, matrix, and result sentences of a possessive transformation.

In Sentences 9–17 on page 287, we in effect supply a matrix by stating that the possessive noun phrase must be used as subject of the predicate "is beautiful."

Sentences 18–25 give practice with indefinite pronouns and with irregular noun phrases which, since they do not form the plural with s, form the possessive with the apostrophe and s.

SECTION 3

A Poem

This is the first part of a very famous poem.

Daffodils

I wander'd lonely as a cloud
That floats on high o'er vales and hills,
When all at once I saw a crowd,
A host of golden daffodils;
Beside the lake, beneath the trees,
Fluttering and dancing in the breeze.

Continuous as the stars that shine
And twinkle on the Milky Way,
They stretch'd in never-ending line
Along the margin of a bay:
Ten thousand saw I at a glance,
Tossing their heads in sprightly dance.

Daffodils (STANZAS ONE AND TWO)

- In the first stanzas, what does the poet compare his loneliness to? a cloud
- What does **vale** mean? See a dictionary. You have met this word earlier, in a poem about a calf. discuss
- What do we call a word like **o'er**? What letter that is left out does the apostrophe stand for? contraction; v
- Can you figure out what a **host** is from the other words around it? Use a dictionary to check. discuss
- **Margin** means edge. Where were the daffodils? See poem.
- The adjective **sprightly** is related to the noun **spirit.** Here the word **sprightly** means full of spirit. What is said to be sprightly? the daffodils' dance

LITERATURE

"Daffodils" (Stanzas One and Two)

This poem introduces the children to another famous writer of English, William Wordsworth. Wordsworth is not everyone's cup of tea nowadays. Many would prefer to leave much of his poetry unread. But some of his poems continue famous and popular, and some of their lines are woven into our language. "Daffodils" is one of these.

Teaching Suggestions

Read the poem to the class or play the recording of it. Let the children take turns reading it aloud. If possible, have the class memorize it.

VOCABULARY AND MEANING

"Daffodils" (Stanzas One and Two)

When the poem has been read often enough to be thoroughly familiar, work through the questions at the bottom of the page to assure detailed comprehension.

SOUNDS AND LETTERS

Review 3: Consonant Sounds

These pages review the spellings of the consonant sounds /s/, /z/, /sh/, and /r/.

Teaching Suggestions

Words in which initial /s/ is represented by **c** include **cent, central, cellar, circle, cipher, citizen**. Remind the children that **c** can stand for /s/ only before **e** and **i**, but that **s** can stand for /s/ in that position too.

The children should easily think of examples of words with **ss** after a simple vowel: **miss, hiss, mess, bless, glass, pass, fuss, muss**. Words with final /s/ in the **VCe** pattern include **base, case, dose, chase**. Do not accept words like **rose, wise, phrase**, which end in the sound /z/.

Review 3: Consonant Sounds

This is the third of four review lessons on the consonant sounds you have studied this year.

- What is the usual way to write the sound /s/ at the beginning of words? Name some words that begin with the sound /s/ and the letter **s.** s; see p. 327
- The letter **c** sometimes stands for the sound /k/ at the beginning of words and sometimes for the sound /s/. Does **c** ever stand for the sound /k/ when the second letter is **e** or **i**? Does it ever stand for /s/ when the second letter of a word is **e** or **i**? no; yes
- Name some words in which **c** at the beginning stands for the sound /k/. Name some in which **c** stands for the sound /s/. cap, clam, cut; see p. 327
- Name some words which begin with the sound /s/ and the letter **s** followed by **e** or **i.** See p. 327.
- What is a homophone for **cent?** sent
- How do we usually spell the sound /s/ at the end when a simple vowel comes before it? Name a word in which final /s/ is spelled with one **s** instead of two **s**'s. ss; bus
- When the sound just before final /s/ is not a simple vowel, we may spell /s/ in the **VCe** pattern with the letters **se.** Give an example. case; see p. 327
- Sometimes final /s/ is spelled **ce.** Name some words with this spelling. See p. 327.
- What one letter stands for the sounds /k/ and /s/? x
- The sound /z/ is like /s/. Which sound makes a buzzing sound in your throat when you say it? /z/
- Name a word that begins with /z/ spelled **z.** zoo
- The common words that end in /z/ spelled **z** or **zz** have what kind of vowel sound in them? simple

290

• When complex vowels come before the final /z/ sound, the sound is spelled **se** or **ze**. Name some words in which final /z/ is spelled **se**. Name some in which final /z/ is spelled **ze**. See p. 327.

• Plurals may end in the sound /s/ or the sound /z/. What letter is always used to spell the last sound of a regular plural? s

• What is the usual way to spell the sound /sh/ at the beginning of words? Name some words that begin with the sound /sh/ spelled **sh**. sh; see p. 328

• How do we spell the sound /sh/ at the end of words? Name some words that end with /sh/ spelled **sh**. sh; see p. 328

• Name some words in which /r/ is spelled with the letter **r** at the beginning. Then give a word in which it is spelled **wr**, and another with the spelling **rh**. See p. 328; write, rhyme.

• Give some words which end in /r/ spelled **r**. See p. 328.

• Give an example of the **VCe** pattern with **r**. See p. 328.

□ Copy the words below, and complete them by putting in the missing letter or letters. You must spell each word so that it has the sound the heading shows.

/s/	/z/	/sh/	/r/
_ircus (c)	_ero (z)	bu__ (sh)	_oad (r)
_elf (s)	qui_ (z)	__op (sh)	__yme (rh)
me__ (ss)	fu__ (zz)	ha__ (sh)	__ong (wr)
ga_ (s)	ro__ (se)	__oe (sh)	si_ (r)
ri__ (ce)	bla__ (ze)	__ip (sh)	pou_ (r)
ca__ (se)	pri__ (ze)	wa__ (sh)	pu__ (rr)

□ Copy these headings: /s/, /z/, /sh/, /r/. Under each heading, write the words that end with that sound.

bus /s/	gaze /z/	four /r/	gush /sh/
tar /r/	fire /r/	rise /z/	mice /s/
buzz /z/	lass /s/	fez /z/	dense /s/

Probably most of the words that end in the sound /z/ in English discourse are noun plurals, noun possessives, or s forms of verbs, in all of which the sound, when it occurs, is represented by the letter s. We do not emphasize the occurrence of /z/ in such forms because it does not provide a spelling problem. Most people, indeed, are not even aware that the sound occurs in such words as **bags, John's, tries.**

The children should be able to offer several examples of /r/ spelled **wr: wrong, wrench, wrist, wren, write**, etc. The word they are likely to think of for **rh** is **rhyme.**

Have the class do the two written exercises at the bottom of the page. Start them on the chalkboard if necessary.

GRAMMAR

The Possessive of Plural Noun Phrases

Mistakes in the spelling of possessives and plurals and possessive plurals are common precisely because the pronunciation gives no clue. The noun phrases "the boys," "the boy's," and "the boys'" are all pronounced exactly alike. Therefore young writers (and also older ones) are apt to use the apostrophe when it doesn't belong, omit it when it does, or use it in the wrong place.

Teaching Suggestions

We introduce the problem here with noun phrases which *are* distinct in pronunciation—irregular nouns in which the plural does not end in **s**:

the man + plural → **the men**
the man + POS → **the man's**
the man + plural + POS → **the men's**

Go through the questions and examples on page 292, demonstrating the formation of these three forms with the irregular noun phrases given. The three examples above might be written on the chalkboard.

Now introduce the rule for spelling the possessives of regular plurals. The following examples may be used:

the girl + plural → **the girls**
the girl + POS → **the girl's**
the girl + plural + POS → **the girls'**

Point out that these all sound the same but that the use of the apostrophe shows which is meant.

The Possessive of Plural Noun Phrases

● You have learned how to write the possessive form of most kinds of noun phrases. What is the possessive form of each of these personal pronouns? his, her, its, my, our, your, their

he she it I we you they

Do we ever use an apostrophe in writing the possessive forms of personal pronouns? no

□ You have also learned how to write the possessive forms of some other noun phrases. Write the possessive forms of these:

Dan's	the boy's
Mr. Webster's	a child's
Uncle Bert's	the cat's

When a noun ends in a consonant-y, we simply add an apostrophe and **s: the spy's.**

When a noun phrase already has a possessive in it, we still add an apostrophe and **s** to the noun: **his wife's.**

□ Write the possessives of these noun phrases:

the baby's her uncle's my father's Mr. Barry's son's

Noun phrases can be either singular or plural. Both singular and plural noun phrases can be possessive. When the noun phrase is an irregular plural, we usually make it possessive just the way we do singular noun phrases. We add an apostrophe and the letter s.

The children have toys.
The children's toys are broken.
We found the children's toys.

□ How would you write the possessive forms of these noun phrases?

the men's	the women's
the mice's	the oxen's

292

When the plural noun phrase is regular, it already ends in s. Then we don't add another s to make the possessive. We just add the apostrophe without the s. Study these sentences:

The boys have some books.
The boys' books are on the desk.
The girls have a brother.
We met the girls' brother.

Pronounce these noun phrases:

the boys the boy's the boys'

Do they sound the same? Can you tell which are possessive and which are plural just by listening? Can you tell when you see them in writing? yes; no; yes

• The plural of regular noun phrases and the possessive of singular and plural noun phrases all sound alike. In talking, we can tell which kind of noun phrase it is only from what the rest of the sentence says. But in writing, we can tell by seeing whether there is an apostrophe and where it comes. It comes before the s if the noun phrase is singular. Where does it come if the noun phrase is plural? after the s

Make possessives out of the subjects of these sentences. Use each possessive with the noun in the object to make a new noun phrase. Then use the new noun phrase as the subject of the predicate "is in the house."

1. The boys have a wagon. The boys' wagon is in the house.
2. The girls have a brother. The girls' brother . . .
3. The ladies have a basket. The ladies' basket . . .
4. The doctors have a patient. The doctors' patient . . .
5. The princesses have a maid. The princesses' maid . . .
6. The spies have a gun. The spies' gun . . .
7. His dogs have a blanket. His dogs' blanket . . .

293

It might be thought that in speech we would constantly be getting mixed up, since the three forms are not discriminated. It is true that we could easily think of utterances in which we could not tell which was meant. Without a clue from the context, the following is triply ambiguous:

I thought it was the /gərlz/.

But in fact, the context will nearly always force one meaning or another. The following is not ambiguous:

The /gərlz/ faces are dirty.

Supposing that we can rule out a two-faced girl, this must be **the girls'**.

Give the children practice in spelling the possessive of plural noun phrases by making possessives from the subjects of Sentences 1–7. The noun phrases should be written.

Have the pupils use the new noun phrases they make in new sentences with the predicate "is in the house." The new sentences should be written.

Extra Practice

If more sentences like Sentences 1–7 are needed, the following may be used. This time, have them use the new noun phrases as object of the verb **saw**, in sentences beginning "I saw . . .."

1. The bees have some honey.
2. The camels have humps.
3. The soldiers have a jeep.
4. The sailors have hats.
5. The cooks have uniforms.
6. The birds have a nest.
7. The monkeys have some food.
8. The horses have saddles.
9. The hunters have some rabbits.
10. The elephants have tusks.

LITERATURE

"Daffodils" (Stanzas Three and Four)

Reread or replay the first two stanzas and then these. Let the pupils take turns reading parts. Encourage memorization. It is an aid to understanding classics like this, and they are worth understanding. Besides, they wear well.

VOCABULARY AND MEANING

"The Daffodils" (Stanzas Three and Four)

Work carefully through the questions. The pupils will probably need word-by-word help with this. The child's heart may not leap up when he beholds a rainbow in the sky, but try to convey the point that the beauties of nature do work strong effects on many people.

SECTION 4

The Rest of the Poem

Read the last two stanzas of this famous poem about daffodils by William Wordsworth.

The waves beside them danced, but they
Out-did the sparkling waves in glee:
A poet could not but be gay,
In such a jocund company:
I gazed — and gazed — but little thought
What wealth the show to me had brought:

For oft, when on my couch I lie
In vacant or in pensive mood,
They flash upon that inward eye
Which is the bliss of solitude;
And then my heart with pleasure fills,
And dances with the daffodils.

Daffodils (STANZAS THREE AND FOUR)

- To **out-do** someone means to do better than he did. What **out-did** the waves? What does **glee** mean? Find **glee** in a dictionary. Find **jocund** in a dictionary. the daffodils; discuss
- If you say "I little thought" something it means you had no idea of it. What did the poet have no idea the sight of the daffodils would bring him? wealth
- **Vacant** means empty. **Pensive** means thoughtful. **Mood** means the way a person is feeling. What do you think **inward eye** means? memory
- **Bliss** means joy, and **solitude** means loneliness. Why is a person's inward eye the joy of loneliness? What is it that flashes into the poet's memory? discuss
- Now do you know what "wealth" the poet found? discuss

294

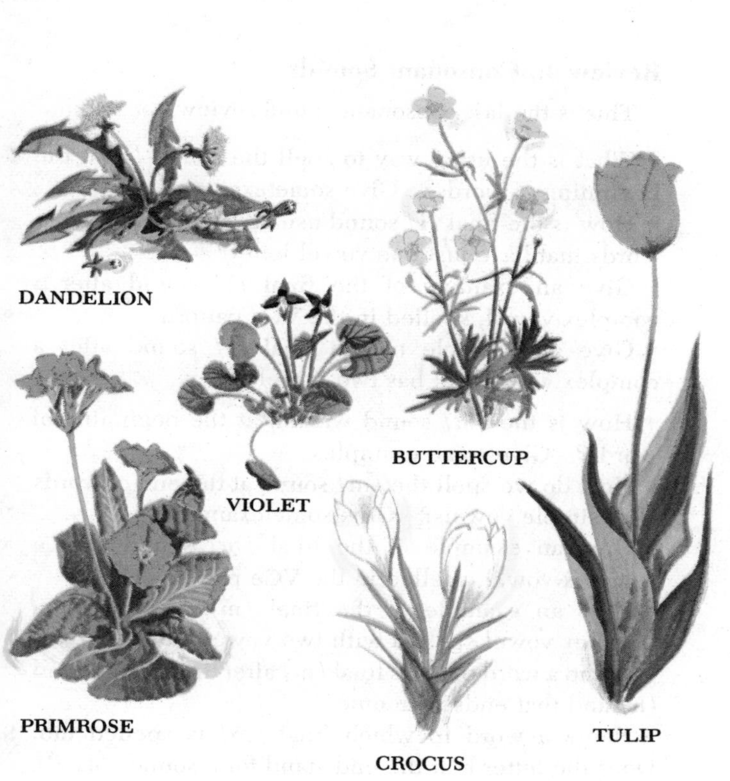

A Paragraph to Write

A daffodil, as you read, is a gay, cheerful, spring flower. Its color is pure gold, and it seems to dance in the breeze as it sways on its slender stem.

Write a paragraph or two describing some of the spring flowers shown in the picture. If you wish you may write about flowers not shown here. When you describe a flower, tell about its color, how it looks, where it grows. Use the possessive form of the personal pronoun **it.**

COMPOSITION – A Paragraph to Write

This assignment is designed partly to give practice in writing the possessive form **its**. Precede it by class discussion of the illustration. Let the pupils identify the different flowers and talk about their features and how they differ one from another. When they have enough material for a paragraph, have them write a description of one or more of them.

Review 4: Consonant Sounds

This page completes the review of consonant sounds with /l/, /m/, and /n/.

Teaching Suggestions

Go through the questions in class discussion, eliciting examples of the various spellings. Write correct responses on the chalkboard.

The words ending in **me** after the sound /u/ are **come** and **some**.

For words beginning with **gn**, the children might think of **gnat, gnash**, or **gnaw**, less probably **gnarled, gnome, gnu**. There are many more examples of **kn**, such as **knee, kneel, knob, knock, knot, know, knight**.

Review 4: Consonant Sounds

This is the last consonant sound review.

- What is the usual way to spell the sound /l/ at the beginning of words? Give some examples. l; see p. 328
- How is the final /l/ sound usually spelled in simple words that have just one vowel letter? ll
- Give an example of the final /l/ sound after a complex vowel, spelled in the **VCe** pattern. mile See p. 328.
- Give an example of the final /l/ sound after a complex vowel that has two vowel letters. feel

- How is the /m/ sound written at the beginning of words? Give some examples. m; see p. 329
- How do we spell the /m/ sound at the end of words after simple vowels? Give some examples. m; see p. 329
- Give an example of the final /m/ sound after a complex vowel, spelled in the **VCe** pattern. time; see p. 329
- Give an example of the final /m/ sound after a complex vowel spelled with two vowel letters. room
- Name a word that has final /m/ after the vowel sound /u/ and that ends with **ome.** come, some
- Name a word in which final /m/ is spelled **mb.** thun Does the letter **b** at the end stand for a sound? no

- What is the usual way to write /n/ at the beginning of words? Give some examples. n; see p. 329
- Name a word in which /n/ at the beginning is spelled **gn.** Does the letter **g** stand for a sound? gnaw; no
- Name a word in which /n/ at the beginning is spelled **kn.** Does the letter **k** stand for a sound? knot; no
- How do we spell final /n/ after simple vowels? n; see p
- Give an example of the final /n/ sound after a complex vowel, spelled in the **VCe** pattern. mine
- Give an example of the final /n/ sound after a complex vowel spelled with two vowel letters. mean

□ Copy the words below and complete them.

/l/	/m/	/n/
_ack (l)	_ight (m)	_ _at (gn)
sti_ _ (ll)	thu_ _ (mb)	spu_ (n)
stea_ (l)	na_ _ (me)	_est (n)
whi_ _ (le)	sla_ (m)	li_ _ (ne)
mai_ (l)	see_ (m)	_ _ock (kn)

Study this summary of the consonant sounds and spellings you have studied.

Seventeen Consonant Sounds

Sound	*At the Beginning*	*At the End*
/p/	**p:** pie	**p:** rip; **pe:** ripe
/t/	**t:** ten	**t:** pet; **te:** date
/k/	**k:** kit; **c:** cold	**ck:** lick; **ke:** like
/ch/	**ch:** chin	**tch:** witch; **ch:** reach
/b/	**b:** bed	**b:** tub; **be:** tube
/d/	**d:** do	**d:** rid; **de:** ride
/g/	**g:** get	**g:** beg; **gue:** league
/j/	**j:** jet; **g:** gentle	**dge:** budge; **ge:** cage
/f/	**f:** fun; **ph:** phrase	**ff:** stuff; **fe:** life; **f:** beef; **ph:** paragraph
/v/	**v:** very	**ve:** save
/s/	**s:** see; **c:** center	**ss:** glass; **s:** bus; **se:** case; **ce:** rice
/z/	**z:** zoo	**z:** quiz; **zz:** buzz; **se:** rose; **ze:** sneeze
/sh/	**sh:** ship	**sh:** push
/r/	**r:** run; **wr:** wrist	**r:** car; **re:** care
/l/	**l:** lose	**ll:** pill; **le:** smile; **l:** fail
/m/	**m:** move	**m:** Sam; **me:** same; **mb:** tomb
/n/	**n:** nose; **gn:** gnaw; **kn:** know	**n:** pin; **ne:** pine

Let the pupils practice the spellings of /l/, /m/, and /n/ by doing the written exercise at the top of page 297.

The chart contains all of the common spellings of the consonant sounds studied and some of the less usual ones. It can be used for further review if needed.

GRAMMAR

Review of Predicates with *be*

These pages review parts of the rule for verb phrase:

$$\text{VP} \rightarrow \text{tense} + \left\{ \begin{array}{l} \textbf{be} + \left\{ \begin{array}{l} \text{NP} \\ \text{Adj} \\ \text{Adv-p} \end{array} \right\} \\ \text{verbal} \end{array} \right\}$$

$$\text{tense} \rightarrow \left\{ \begin{array}{l} \text{present} \\ \text{past} \end{array} \right\}$$

$$\text{Adv-p} \rightarrow \left\{ \begin{array}{l} \text{Adv} \\ \text{prepositional phrase} \end{array} \right\}$$

Teaching Suggestions

Use Sentences 1–7 for identification of the three structures that commonly occur after **be**. Review the distinction between adverbs and prepositional phrases and then use Sentences 8–11 for identification of these structures.

Review: Predicates with Forms of *be*

You know that the predicate of a simple sentence must have one of two things. It may have a form of **be.** If it doesn't have a form of **be,** what must it have? a verb

If the predicate of a simple sentence has a form of **be,** it must have something else too. We have seen three things that such a predicate may have after the form of **be:** (1) a noun phrase, (2) an adverbial of place, (3) an adjective.

- Look at the following sentences. Tell what comes after **be** in each sentence — a noun phrase, an adverbial of place, or an adjective.

1. Dick is my friend. n.p.
2. Dick is friendly. adj.
3. Dick is in the house. adv. of place
4. She is pretty. adj.
5. The children are outside. adv. of place
6. The room is clean. adj.
7. Her mother is a teacher. n.p.

There are two kinds of adverbials of place — **adverbs** and **prepositional phrases.**

Adverbs are single words like **outside** and **there.**

Prepositional phrases are groups of words like the following:

in the house **by the chalkboard**

- What kinds of words are **in** and **by** in these phrases? prep. What are **the house** and **the chalkboard?** n.p.
- Tell which adverbials of place in these sentences are adverbs and which are prepositional phrases.

8. She is upstairs. adv.
9. They are beneath the desk. prep. phrase
10. I am here. adv.
11. We are under the house. prep. phrase

298

There is an easy way to tell adjectives from noun phrases and adverbials of place. We can use the word **very** in front of adjectives. We can't use **very** before noun phrases and adverbials of place.

• Say the following sentences. If the predicate contains an adjective, use the word **very** before it. Name the adverbs and noun phrases in the predicates.

12. They were hungry. . . . very hungry.
13. They were inside. adv.
14. We are sleepy. . . . very sleepy.
15. They are friends. n.p.
16. I am tired. . . . very tired.
17. He is here. adv.
18. She is downstairs. adv.
19. Sheila is away. adv.
20. Martha was quiet. . . . very quiet.
21. The children are noisy. . . . very noisy.
22. These are books. n.p.
23. Everyone is happy. . . . very happy.

• Say each of the following sentences three times. Each time use an adjective different from the one you find in the sentence. For instance, if the sentence were "The children are very noisy," you might say, "The children are very sleepy," "The children are very quiet," "The children are very unhappy." (Answers will vary.)

24. The boys are very hungry.
25. The room is very large.
26. The men are very young.
27. The girl is very clever.
28. The sheep are very fat.
29. Mark is very lazy.
30. Sheila is very pretty.

The word **very** does occur before **near** both when **near** occurs as an adverb and when it occurs as a preposition:

He was very near.
He was very near the house.

Otherwise **very** will distinguish adjectives from noun phrases and adverbials of place. It of course will not distinguish them from adverbials of manner, with which it occurs freely: **very quickly, very insolently**.

Let the children use the **very** test to help identify the structures in the predicates of Sentences 12–23.

Do Sentences 24–30 as further practice with adjectives. The use of a form of **be** and **very** ensures that any word suggested that completes the sentence grammatically will be an adjective.

Extra Practice

Sentences 12–23 may be used as a written assignment by having the children find the adjectives, adverbs, and noun phrases in the predicates and write them under the headings *Adj.*, *Adv.*, and *NP*.

TESTS AND REVIEW

Test I is cumulative in nature and may be used as a final test on what has been taught in sound and spelling. Test II includes the possessive and forms in the predicate and should reveal a good deal of the pupil's grasp of the structures of syntax, or the lack of it.

If there is time to do so, reteach what seems most needed on the basis of your records of the development of the individual pupils.

Use page 303 for oral discussion.

Extra Test

Dictate these words and have the class spell them. Say the word clearly. Use it in the sentence. Say it again and have the children write it.

1. **stitch** A **stitch** is made with thread. **stitch**
2. **beach** A **beach** is sandy. **beach**
3. **their** The girls lost **their** hats. **their**
4. **there** Your book is **there** on that table. **there**
5. **they're** The boys say **they're** coming. **they're**
6. **it's** I think **it's** time to go. **it's**
7. **its** My cat licks **its** kittens. **its**
8. **edge** The **edge** of the glass is chipped. **edge**
9. **dense** A **dense** forest is crowded with trees. **dense**
10. **knock** I heard a **knock** at the door. **knock**

□ TEST I Sounds and Letters

1. Write these headings: /ī/, /ē/, /ā/, /ō/. Under each heading, write the words from the list below that have that vowel sound in them.

If you make mistakes, study pages 260–61, 267.

wait /ā/ hole /ō/ wipe /ī/ heel /ē/
boat /ō/ these /ē/ pay /ā/ try /ī/
sigh /ī/ plate /ā/ slow /ō/ heat /ē/
we /ē/ toe /ō/ lie /ī/ go /ō/

2. Write these headings: /au/, /oo/, /ū/. Under each heading, write the words that have that vowel sound.

If you make mistakes, study pages 206, 218, 267.

wall /au/ boot /ū/ cook /oo/ song /au/
bush /oo/ taught /au/ knew /ū/ straw /au/
due /ū/ maul /au/ bought /au/ to /ū/

3. Write these headings:

/k/ /j/ /f/ /s/ /r/ /n/

Under each heading, write the words from the list below that begin with that consonant sound.

If you make mistakes, study page 297.

phone /f/ city /s/ wrong /r/ gnaw /n/
never /n/ few /f/ came /k/ generous /j/
kitten /k/ jaw /j/ rung /r/ knot /n/

4. Write these headings: /m/, /j/, /f/, /s/, /z/. Under each heading, write the words from the list below that end with that consonant sound.

If you make mistakes, study page 297.

once /s/ please /z/ calf /f/ judge /j/
thumb /m/ breeze /z/ gas /s/ stuff /f/
worse /s/ fez /z/ cage /j/ same /m/

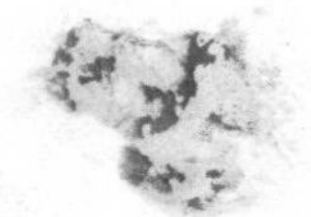

□ TEST II Grammar

1. Make a possessive of the subject of the first sentence, and use it as part of a new noun phrase in the second sentence. For example, if the sentences were "John has a car" and "I liked the car," your sentence would be "I liked John's car."

If you make mistakes, study pages 262, 268, 278, 286, 292.

a. Julia has a ribbon.
I like the ribbon. I like Julia's ribbon.

b. We have a dog.
The dog is a puppy. Our dog is a puppy.

c. The child has a cold.
The cold is worse. The child's cold is worse.

d. Mr. Danby had a bicycle.
The bicycle was stolen. Mr. Danby's bicycle was stolen.

e. It has a dish.
The dish is empty. Its dish was empty.

f. The girls have a brother.
We saw the brother. We saw the girls' brother.

g. It has a tail.
It wagged the tail. It wagged its tail.

h. The lawyers had an office.
It was in the office. It was in the lawyers' office.

i. The gardener has a shovel.
We found the shovel. We found the gardener's shovel.

j. The babies have milk.
The milk is warm. The babies' milk is warm.

k. The women had hats.
The hats were pretty. The women's hats were pretty.

l. They have a boat.
We rode in the boat. We rode in their boat.

301

□ **TEST II Grammar** (CONTINUED)

2. Tell what follows a form of **be** in the predicate of each sentence below. Write (1) for noun phrase, (2) for adjective, and (3) for adverbial of place.

If you make mistakes, study pages 298–99.

a. The boy is in the pool. (3)
b. Janet is my friend. (1)
c. The books were heavy. (2)
d. A messenger is outside. (3)
e. The daffodils were beautiful. (2)
f. Harry was upstairs. (3)
g. They are the Parkers. (1)
h. The Parkers are on that train. (3)
i. That bundle is my coat. (1)
j. The coat is in a newspaper. (3)
k. The train was empty. (2)
l. The package is under your seat. (3)
m. That sandwich was my lunch. (1)

3. For each of the sentences below, write three more sentences each with a different adjective. Don't change anything in the original sentence except the adjective.

If you make mistakes, study pages 298–99.

a. David was very angry. (Answers will vary.)
b. The room was very cool.
c. She is very pretty.
d. The poem is very long.

□ **TEST III Writing and Vocabulary**

Write five sentences. In each sentence, use one of these words:

1. muse 2. solitude 3. sage 4. bliss 5. sprightly

• **REVIEW Ideas and Information**

1. What is a pigtail? a braid of hair
2. What is another word that means **host?** throng, crowd
3. What does **margin** mean in "margin of the bay"? edge
4. What kind of noun phrase never has the apostrophe in the possessive form? pers.
5. When do we add just the apostrophe to write the possessive of a noun phrase? to regular plural nouns
6. What word may be used before adjectives but not before noun phrases? very
7. What may come after a form of the word **be** in the predicate besides adjectives and noun phrases? adv. of place
8. When we make a possessive form from the subject of a **have** sentence, make it part of a new noun phrase, and use it in a new sentence, we have made a particular kind of change. What do we call such a change? transformation
9. How is the sound /**k**/ spelled at the beginning of a word when the next letter is **i** or **e**? k
10. What two letters are used to spell final /**k**/ when a simple vowel comes before it? ck
11. Spell the possessive form of the personal pronoun **they;** the personal pronoun **it.** their; its
12. Spell the contraction of the words **they** and **are;** the contraction of the words **it** and **is.** they're; it's
13. How do you write the possessive form of a noun phrase that is a proper noun, like **John?** add apostrophe-s
14. How do you write the possessive form of a noun phrase that is a determiner plus a common noun, like **the teacher?** add apostrophe-s
15. What is a **suffix?** Name two suffixes. a part added at the end; er, ful
16. What is a **prefix?** What is a prefix you can use to make some adjectives mean the opposite? a part added at the beginning; un

ESSENTIALS OF THE GRADE THREE PROGRAM

Grammar

1. Sentences are the groups of words that express our thoughts when we speak and write. The sentences to be studied first are very simple ones:

The sun is bright.
Girls giggle.
They were in the house.

Simple sentences have two main parts — a subject and a predicate. The subject comes first and the predicate second. The predicate of a sentence says something about the subject.

Subject	*Predicate*
The sun	is bright.
Girls	giggle.
They	were in the house.

A subject may be one word alone, like **girls** or **they.** It may be more than one, like **the sun.** A predicate may be one word, like **giggle,** or more than one, like **were in the house.**

2. There are several kinds of words or groups of words that may be subjects. These are the principal kinds:

a. personal pronouns
b. indefinite pronouns
c. proper nouns
d. common nouns (alone or with a word like **the**)

3. There are seven personal pronouns used as subjects in English: **he, she, I, we, you, it,** and **they.** The pronoun **I** is always capitalized.

304

4. There are twelve indefinite pronouns in English. They all have one of these beginnings: **some, any, every, no.** They all have one of these endings: **body, one, thing.** Only one of the indefinite pronouns is written as two words.

somebody	someone	something
anybody	anyone	anything
everybody	everyone	everything
nobody	no one	nothing

5. The subjects of these sentences are proper nouns:

David is my friend.
Rover is his dog.

David and **Rover** are names of a special boy and a special pet. Nouns that are names of special people or things are called proper nouns. Proper nouns are always written with a capital letter at the beginning.

Proper nouns often have **Mr., Mrs.,** or **Miss** before them, as in these sentences:

Mr. Johnson lives on our street.
Mrs. Williams is a teacher.
Miss Brown writes books.

Mr. and **Mrs.** are abbreviations. They are part of the proper nouns **Mr. Johnson** and **Mrs. Williams** and are written with capital letters. When we write abbreviations, we put periods after them to show that they are abbreviations. **Miss** is part of the proper noun **Miss Brown,** but it is not an abbreviation, so we do not put a period after it.

6. The subjects of these sentences are (1) a proper noun, (2) a personal pronoun, (3) an indefinite pronoun:

a. George heard the noise.
b. He heard the noise,
c. Someone heard the noise.

305

Pronouns like **he** and **someone** can take the place of nouns such as **George.** Personal pronouns like **he** refer to definite people, as **George** does. Indefinite pronouns like **someone** do not refer to particular people or things. In "George heard a noise" and "He heard a noise," we know just who heard it. But in "Someone heard a noise," we do not know just who heard it.

7. Nouns that are not proper nouns are called common nouns. **Birds, team, boy, teacher, school** are common nouns. These do not refer to particular things or people as proper nouns like **David** and **Rover** do. We do not write common nouns with a capital letter unless they are the first word in a sentence.

The subjects of these sentences are common nouns:

Birds lay eggs.
Candy is dandy.
Milk tastes good.

8. A subject may consist of a common noun alone, as in the sentences above. Or it may consist of a common noun with a determiner in front of it, as in the following sentences:

The sun is bright.
A rabbit ran by.
This book is new.
My brother writes poetry.

In these sentences the words **sun, rabbit, book,** and **brother** are common nouns. The words **the, a, this,** and **my** are determiners. Determiners come before nouns and tell, or determine, something about the meaning of the nouns. Determiners are not usually used with proper nouns.

9. The subjects of the following sentences are (1) a determiner plus a common noun, (2) a personal pronoun, (3) an indefinite pronoun.

a. My sister likes candy.
b. She likes candy.
c. Everyone likes candy.

Pronouns can take the place of subjects with common nouns in them just as they can subjects that are proper nouns. The personal pronouns refer to definite people or things, and the indefinite pronouns do not.

These types of subjects have been presented:

personal pronouns: **he, she, it, I, we, you, they**
indefinite pronouns: **someone, everything,** etc.
proper nouns: **David, Rover, Mrs. Andrews**
common nouns alone: **birds, candy, milk**
common nouns with determiners: **the birds, my candy**

10. A noun is one kind of word that may be in a subject. One kind of word that may be in a predicate is a verb. In a simple sentence that has a verb, the verb may be the first word in the predicate:

Everyone swims in the pool.

The verb **swims** is the first word in the predicate **swims in the pool.**

In some sentences the predicate is made up of just a verb alone:

Helen swims.

In other predicates other words may follow the verb.

11. A verb has different forms: **swim, swims.**

She swims.
They swim.

The form that a verb has depends on what the subject is.

When **she** is the subject, we use a form of the verb with the letter s at the end: **She swims.** We call verbs with the letter s at the end the s form.

When **they** is the subject, we use a form of the verb without the letter s at the end: **They swim.** We call verbs without s at the end the simple form.

The personal pronouns **he, she,** and **it** are followed by the s form of verbs. The personal pronouns **I, we, they,** and **you** are followed by the simple form of verbs.

***s** form*	*Simple form*
He sleeps.	I sleep.
She swims.	We swim.
It hurts.	They hurt.
	You smile.

12. The personal pronouns **he, it,** and **she** are followed by the s form of verbs when they are subjects. **He, it,** and **she** take the place of singular subjects, such as **the man, that car, Mabel.** Singular subjects take the same form of verbs that **he, it,** and **she** do:

He yawns.	The man yawns.
It roars.	That car roars.
She cooks.	Mabel cooks.

The form of verbs that goes with singular subjects is the s form.

The personal pronoun **they** takes the place of plural subjects like **those cars** or **ladies.** Plural subjects take the same form of verbs that the pronoun **they** takes.

They roar.	Those cars roar.
They cook.	Ladies cook.

The form of verbs that goes with plural subjects is the simple form.

The simple form of verbs goes with the personal pronouns **I, we,** and **you,** as well as **they.**

13. We usually write the s form of verbs by just adding the letter **s** to the simple form:

play plays take takes
wait waits

When the simple form of a verb ends in the /ch/ sound, we add **es** to the simple form:

touch touches watch watches
search searches

When the simple form ends in the letter **y** with a consonant letter before the **y**, we change the **y** to **i** and add **es:**

cry cries spy spies
carry carries

A very few verbs make the s form in special ways:

Simple form	*s form*
have	has
go	goes
do	does

The s form of **do** has a different vowel sound than the simple form.

The s form of **say** is made by adding s in the regular way: **says.** But the s form does not have the same vowel sound as the simple form.

14. The predicate may contain a form of the word **be** instead of a verb like **touch** or **swim.**

The girl is pretty.

The word **is** is a form of **be.**

The word **be** itself is used in predicates like the one in this sentence:

The girl may be tired.

If we do not have the word **may** in the predicate, we use another form of **be:**

The girl is tired.

309

The form **is** goes with singular subjects, like **the girl** or **this suit.** It also goes with the personal pronouns **he, she,** and **it,** which take the place of singular subjects:

He is happy. She is pretty.

It is new.

Indefinite pronouns also take the **is** form of **be.** They take the s form of verbs:

Someone is in my room.

Everybody likes candy.

Another form of **be** goes with plural subjects. The subject of this sentence is plural:

Her shoes are new.

The form of **be** that goes with plural subjects is the **are** form.

The personal pronoun **they** takes the place of plural subjects, like **her shoes** or **these girls. They** takes the same form of **be** that plural subjects do.

They are new. They are away.

The **are** form of **be** goes with **they.**

The personal pronouns **we** and **you** take the same form of **be** that plural subjects and **they** take:

We are hungry. You are my friend.

The **are** form of **be** goes with **we** and **you.**

The personal pronoun **I** takes a special form of **be:**

I am cold.

The **am** form of **be** goes only with the word **I.**

15. We usually write the plural form of nouns by just adding the letter s to the singular form:

boy boys fire fires

When the singular form ends in the /ch/ sound, we add **es** to make the plural:

match matches couch couches

When a singular noun ends with the letter **y** with a consonant letter before the **y**, we change **y** to **i** and add **es** to form the plural:

sky skies   lady ladies

When a vowel letter comes before the **y**, we just add the letter **s**: **toy, toys.**

16. Some nouns are made plural in special ways:

man	men	foot	feet	tooth	teeth
mouse	mice	goose	geese	child	children
		woman	women		

The words **sheep** and **deer** are just the same in the plural as in the singular. In most subjects, the form of the noun tells whether the subject is singular or plural. But with **sheep** and **deer,** the form of the verb or of **be** may be the only way to tell:

a. The sheep is in the pasture.
b. The sheep are in the pasture.
c. The deer eat twigs.
d. The deer eats twigs.

In Sentences a and d the **is** form of **be** and the **s** form of **eat** show that the subjects are singular. In Sentences b and c the **are** form of **be** and the simple form of **eat** show that the subjects are plural.

17. Some nouns that end in the sound /f/ spelled **f** make the plural by changing **f** to **v** and adding **es:**

leaf	leaves	thief	thieves	loaf	loaves
elf	elves	shelf	shelves	self	selves
calf	calves	half	halves	wolf	wolves
		sheaf	sheaves		

The following words end in **fe** in the singular but are spelled **ves** in the plural:

life	lives	wife	wives	knife	knives

311

18. When we write, we may use personal pronouns and forms of **be** in sentences like these:

I am sleepy.
We are ready.
It is stormy.

In saying these sentences, we usually say the personal pronoun and the form of **be** together, so that they sound like one word:

I'm sleepy.
We're ready.
It's stormy.

We may write such sentences, as well as say them. We use **I'm** instead of **I am, we're** instead of **we are, it's** instead of **it is.** In **I'm** we leave out the letter **a** of **I am.** The mark between the **I** and **m** that shows something has been left out is an apostrophe.

A form like **I'm** is called a contraction, because the words are contracted, or made shorter.

These are contractions of the personal pronouns with **is** and **are:**

I am: I'm	she is: she's	he is: he's
it is: it's	we are: we're	you are: you're
	they are: they're	

19. The predicate of every sentence must have either a verb or a form of **be** in it.

When the predicate of a sentence has a verb, it may or may not have other words too.

She sings. Those people sell toys.

When the predicate of a simple sentence has no verb in it, but a form of **be** instead, then it must have something else, too.

312

One word comes after a form of **be** in each of these sentences:

I am hungry.
Our teacher is kind.
The children are happy.

Words like **hungry, kind,** and **happy** are adjectives. Adjectives may follow forms of **be** in the predicate.

Adjectives usually describe things. They tell what things are like.

The word **very** can be used before adjectives:

I am very hungry.

We may use **very** before adjectives but not before verbs and nouns.

Adjectives can follow any form of **be** in the predicate of a sentence. They may also follow some verbs, like **seem, look, feel, grow, sound, taste:**

The food tastes sweet.
Those chickens look tender.

The adjectives **sweet** and **tender** follow the verbs **taste** and **look** in these sentences.

313

ESSENTIALS OF THE GRADE THREE PROGRAM (CONTINUED)

Sounds and Letters

1. The letters we write stand for sounds we make when we talk. There are 26 letters in the alphabet, but there are more sounds than this. So we sometimes use letters in more than one way.

The letter **p** stands for the first sound in **pen.** When we want to write about the sound that begins **pen,** we put the letter that stands for the sound between slanting lines: / p /. When the letter that stands for the sound is meant, we put the letter in heavy black type: **p.**

2. The sound / p / at the beginning of words is usually spelled with the letter **p: pin, pet, pack.** At the end of words this sound is also usually spelled with the letter **p: nap, cup, lip, soup.**

3. The sound that begins **too** is written / t / and is usually spelled with the letter **t** at the beginning of words: **too, ten, top, try.** At the end of words the / t / sound is also usually spelled with the letter **t: hit, get, cat.** In a few words we spell the sound **t** at the end with the letters **tt: mitt, putt, butt.** Such spellings are called exceptions.

4. The sound that begins **king** is written / k /. In some words the sound / k / at the beginning is spelled with the letter **k: key, kind, kiss.** But we often spell the sound / k / at the beginning with the letter **c: can, cold, cut, cry, clay.**

We use the letter **k** for the sound / k / at the beginning of words when the second letter is **e** or **i.** When the second letter is not **e** or **i,** we use the letter **c** to spell the sound / k /.

At the end of words, we usually write the sound /k/ with two letters: **ck.** These are some words in which /k/ at the end is spelled that way: **neck, pack, sick, buck, lock.**

5. There are two kinds of sounds. Consonant sounds, like /p/, /t/, and /k/, are made by stopping the breath in some way. To talk about these sounds we just say an ordinary letter of the alphabet. We say "the t sound," or "the sound k."

The other kind of sound is a vowel sound. When we make a vowel sound we do not stop the breath. But we can make different vowels by changing the shape of our mouths.

The vowel letters are **a, e, i, o, u.** Since there are only five vowel letters in the alphabet, but more than five vowel sounds, we have to write some of the vowel sounds in special ways.

6. We show the vowel sound in **tip** this way: /i/. To talk about this sound, we say the sound /i/ with the consonant sound /k/ after it: "ick." Call it the "ick" sound.

We usually write the sound /i/ with the letter **i: sip, pin, him, tip.**

7. We show the vowel sound in **pet** this way: /e/. To talk about this sound, we say the sound /e/ with the consonant sound /k/ after it: "eck." Call it the "eck" sound.

We usually write the sound /e/ with the letter **e: met, bed, wed, sled.**

But in some words the sound /e/ is written with the letters **ea**; **head, bread, breath, meadow.**

In a few words the sound /e/ is spelled in very special ways: **ie** in **friend** and **ai** in **again.**

315

8. The consonant sound / b / is like / p /, except that you have a buzzing sound in your throat when you make the / b / sound.

The sound / b / is usually spelled with the letter **b** at the beginning of words: **big, bat, bus.** It is often spelled with the letter **b** at the end of words: **rib, sob, tub.**

There is only one common English word that ends with / e /, spelled **e** followed by / b /, spelled **b: web.** The word **ebb** ends with / b /, spelled **bb.**

9. The consonant sound / d / is like / t /, except that you have a buzzing sound in your throat when you make the / d / sound.

The sound / d / is written **d** at the beginning of words: **dig, dress, duck.** After **e, i,** and some other vowel sounds, we write the sound / d / with the letter **d** at the end of words, too: **slid, bread, glad, nod, bud.**

10. The consonant sound / g / is like / k /, except for the buzzing in your throat when you make / g /.

At the beginning of words we usually write the sound / g / with the letter **g: get, glad, gull.** But words that begin with the letter **g** don't always have the sound / g /: **gentle, George, general.**

At the end of words we usually write the sound / g / with **g: big, bag, bug, frog.** One common word ends with the sound / g /, spelled **gg: egg.**

11. We show the vowel sound in **cat** this way: / a /. We call it the "ack" sound.

The sound / a / is usually written with the letter **a: bad, rack, tack.** But in one word the sound / a / is written **au: laugh.**

316

12. The consonant sound /ch/ is called the "chee" sound when we speak about it.

The sound /ch/ is written with the letters **ch** at the beginning of words: **child, chalk, chew.**

We often write /ch/ with the letters **ch** at the end of words, too: **much, touch, rich.**

In some words we spell /ch/ at the end with the three letters **tch: catch, scratch, pitch.**

13. The consonant sound /j/ is like /ch/, except for the buzzing in your throat when you make /j/.

We usually write the /j/ sound at the beginning of words with the letter **j: joke, jam, join.** But when a word begins with the sound /j/ followed by the letter **i** or **e**, it may be spelled either with **j** or with **g: jet, jingle, jelly,** or **giant, germ, gem.** Before **a, o,** and **u,** the sound /j/ is always written with the letter **j: Jack, job, just.**

At the end of words we usually write the sound /j/ with the letters **ge: stage, large, huge.** But when one of the vowel sounds /i/, /e/, /a/, or /u/, comes before the sound /j/, we write /j/ with the three letters **dge: badge, bridge, edge, budge.**

14. The consonant sound /r/ is usually written with the letter **r** at the beginning of words: **run, red, rope.** In a few words we write /r/ with the letters **wr: write, wrist, wren.**

In some words the sound /r/ is spelled in another special way, with **rh: rhyme.**

One way to spell the sound /r/ at the end of words is with the letter **r: fair, car, her.**

In one common word we spell /r/ at the end with **rr: purr.**

317

15. We show the vowel sound in **mud** this way: / u /. We call it the "uck" sound.

The vowel sound / u / is usually written with the letter **u**: **rub, luck, mutt, Sunday.**

Sometimes we write the vowel sound / u / with the letter **o,** not **u,** as in these words: **Monday, ton, month, come, some.**

Because the vowel sound / u / can be written in two different ways, we can have words that sound the same, but are spelled differently and have different meanings: **sun, son.**

16. We show the vowel sound in **rock** this way: / o /. We call it the "ock" sound.

The most common way to write the sound / o / is with the letter **o**: **frock, stock, Bob.**

But the letter **o** does not always stand for the sound / o /. It may stand for / u /, as you have seen: **son, won, Monday.**

17. The consonant sound / f / is usually written with the letter **f** at the beginning of words: **fine, friend, fly.**

At the end of words, the sound / f / is spelled in more than one way. If the vowel sound / i /, or / a /, or / u /, comes before / f /, the / f / sound is usually spelled **ff: cliff, stiff, cuff, muff, staff.**

When the sound before / f / is not / i /, / a /, or / u /, we usually write the sound / f / with one **f**: **leaf, elf, roof, sheaf, surf.**

The words **calf** and **half** have the vowel sound / a / before / f /, but they have the letter **l** before **f**. The **l** does not stand for a sound, but we use only one **f** because the letter **l** is there.

In **paragraph**, the sound / f / is spelled **ph.**

18. The consonant sound / v / is like / f /, except that there is a little buzzing sound in your throat when you make the sound / v /.

At the beginning of words we write the sound / v / with the letter **v**: **vine, very, valley.**

At the end of words we write the sound / v / with the letters **ve**: **leave, save, dive.**

19. The vowel sound in **time** is shown by using an i with a line over it between slanting lines: /ī/. We call it the "ike" sound.

The word **fine,** which has the sound /ī/ in it, is spelled just like **fin** except for one letter, the **e** at the end. This **e** does not itself stand for a sound, but it makes the **i** in **fine** stand for the sound /ī/, not / i /. In many words the sound /ī/ is spelled this way: **i–consonant letter–e: bike, pipe, bite, line.**

When the vowel sound /ī/ comes at the end of a word we usually spell it with the letter **y**: **by, dry, sky.**

There are some words in which we spell the sound /ī/ at the end with the two letters **ie**: **lie, tie, pie.**

In some words we write the sound /ī/ in very special ways: with capital **i** in **I**; and with **e, y, e** in **eye.**

20. The vowel sound in **Steve** is shown by using an e with a line over it between slanting lines: / ē /. We call it the "eek" sound.

The sound / ē / is sometimes spelled in the way that /ī/ is usually spelled, **e–consonant letter–e: eve, Pete, these.**

The **e–consonant letter–e** spelling of the sound / ē / is not the most usual spelling.

In many words the sound / ē / is spelled **ee**: **peep, steep, keen.**

319

In many words the sound /ē/ is spelled **ea: eat, meat, weak.**

Because there are different ways to spell the sound /ē/, we have words that sound alike but are spelled differently and have different meanings. Here are some examples of words like this:

meet	**meat**	**week**	**weak**	**peek**	**peak**
heel	**heal**	**reel**	**real**	**beet**	**beat**

When the sound /ē/ comes at the end of a word with no consonant sound after it, we sometimes write it with **ee** and sometimes with **ea.** There are many words that spell /ē/ at the end with **ee: flee, see, knee, tree, free.**

A few words that have the sound /ē/ at the end spell it with **ea: sea, tea, pea.**

Some very common words spell /ē/ at the end with the letter **e.** This is the list of such words:

he she me we be

Rhyme

Two or more words which end with the same sound or sounds are rhyming words:

try high lie bell tell

The rhyming sounds in rhyming words do not need to be spelled in the same way. Rhyme is based on sound, not spelling.

The charts on the opposite page summarize the "Sounds and Letters" program of Book Three.

320

Vowel Sounds Taught in Grade Three with Their Usual Spellings

Sounds (simple)	/i/	/e/	/a/	/u/	/o/
Spellings and Examples	**i:** tip	**e:** let **ea:** head	**a:** bat	**u:** but **o:** son	**o:** top
Sounds (complex)					
/ī/	**i–consonant–e:** line; **y:** by; **ie:** pie (also **I** and **eye**)				
/ē/	**e–consonant–e:** Steve; **ee:** peep; **ea:** eat; **e:** me				

Consonant Sounds Taught in Grade Three with Their Usual Spellings

Sound	*At the Beginning*	*At the End*
/p/	**p:** pen	**p:** nap
/t/	**t:** too	**t:** hit (also **tt:** putt)
/k/	**k:** king; **c:** can	**ck:** back
/b/	**b:** big	**b:** rib (also **bb:** ebb)
/d/	**d:** dig	**d:** nod
/g/	**g:** get	**g:** bag (also **gg:** egg)
/ch/	**ch:** child	**ch:** much; **tch:** catch
/j/	**j:** jet; **g:** gentle	**ge:** stage; **dge:** judge
/r/	**r:** run	**r:** fair (also **rr:** purr)
/f/	**f:** fine	**ff:** cliff; **f:** elf; **ph:** paragraph
/v/	**v:** vine	**ve:** leave

321

Technical Terms as Used in Grade Three

abbreviation A short way of writing words: **Mr., Mrs.**

adjective A word like **happy, hungry, friendly, new.** Adjectives usually describe things.

apostrophe This mark: **'**. Used in contractions like **he's, we're** to show something is left out.

comma This mark: **,**. Used between day and year in dates; after greetings in friendly letters.

common noun A word like **boy, teacher, school, mother.** Not a proper noun.

consonant A sound made by stopping the breath in some way. Consonant sounds are usually spelled by consonant letters. The consonant letters are b, c, d, f, g, h, j, k, l, m, n, p, q, r, s, t, v, w, x, y, z.

contraction A form like **it's, he's, we're,** made by putting together two words such as **it** and **is.**

determiner A word like **a, the, this** that comes before a noun and **determines** something about the noun's meaning.

grammar What we study when we learn about sentences.

indefinite pronoun A word like **somebody, anyone, everything, nobody** that may take the place of another subject, but doesn't mean a definite person or thing.

indent Put the first line of a paragraph farther in than the others.

margin The white spaces at the edges of a paper.

noun A word like **boy, brother, Jerry, books.** A noun often has a determiner before it, and is often part of a subject.

personal pronoun One of the words **I, he, she, it, we, you, they** that may take the place of another subject. (Other forms taught later.)

plural Meaning more than one, as **plural** subjects, **plural** nouns.

predicate The second of the two main parts of a simple sentence.

proper noun A word like **David, Mr. White, Rover** that names a special person or thing.

rhyme Sounds that are the same at the end of two or more words: **write, might.**

***s* form** The form of a verb that goes with singular subjects: **eats, wins, goes, takes, has.**

sentences Groups of words that express our thoughts when we speak and write.

simple form The form of a verb that goes with plural subjects: **eat, win, go, take, have.**

singular Meaning one, as **singular** subjects, **singular** nouns.

stanza One of the parts of a poem.

subject The first of the two main parts of a simple sentence. The subject is usually the main thing the sentence is about. One word often found in a subject is a noun.

verb A verb or a form of **be** is the first word in the predicate of a simple sentence. Verbs are words like **go, eats, has, win, takes.**

vocabulary The words a person knows.

vowel A sound made without stopping the breath. Different vowel sounds are made by changing the shape of the mouth. The vowel letters are a, e, i, o, u, but there are more vowel sounds than vowel letters.

323

Extra Words to Illustrate Sounds

Your teacher may ask you to use the words on pages 324–29 for practice in hearing and saying sounds. The page number for each sound tells where in the book the sound is taught.

Vowel / i /, spelled i (p. 6). pick, bit, pit, miss, tin, pill, big, hid, knit, him, lift, fin

Vowel / e /, spelled e (p. 6). peck, bet, pet, mess, ten, fell, beg, red, net, hem, left, men

Vowel / e /, spelled ea (p. 6). dead, head, dread, bread, read, lead, thread, threat, meadow, feather

Vowel / a /, spelled a (p. 6). pack, bat, pat, mass, tan, pal, bag, had, gnat, ham, fan

Vowel / u /, spelled u (p. 12). luck, cluck, nut, putt, rub, hut, cup, run, shut, hub, sun, sum

Vowel / u /, spelled o (p. 13). son, some, won, love, ton, month, shove, Monday, dozen, honey

Vowel / o /; spelled o (p. 13). lock, clock, not, pot, rob, hop, top, Bob, shot, plot, sob, nod

Vowel / ī /, spelled VCe (p. 24). pike, pipe, ripe, mile, time, life, bite, ride, wide, hide, wife, smile

Vowel / ī /, spelled y (p. 34). try, my, cry, pry, sky, fly, sty, fry, sly, dry, by, shy

Vowel / ī /, spelled ie (p. 34). pie, lie, die, tie

Vowel / ē /, spelled VCe (p. 25). Pete, eve, these

Vowel / ē /, spelled ee (p. 25). peek, peep, seed, meet, weed, heed, tree, free, three, sweet, feet, need

Vowel / ē /, spelled ea (pp. 25, 34). reap, meal, team, leaf, beat, read, wheat, steal, reach, tea, plea, flea

Vowel / ā /, spelled VCe (p. 41). bake, take, lake, rake, gate, late, plate, gape, tape, wade, fade, Abe, safe

324

Vowel /ā/, spelled ai (p. 48). wait, raid, pain, train, gain, aid, paid, fail, tail, mail, bait

Vowel /ā/, spelled ay (p. 49). way, say, pay, play, pray, tray, may, gay, bay, stray, fray, flay

Vowel /ū/, spelled VCe (p. 56). rule, duke, brute, cute, mute, fuse, muse, use, tune, prune, June, rude

Vowel /ū/, spelled oo (p. 56). moon, croon, room, broom, loom, spool, fool, school, pool, food, too

Vowel /ū/, spelled ew (p. 57). new, dew, blew, brew, chew, crew, drew, few, flew, knew, grew, threw

Vowel /ū/, spelled ue (p. 57). blue, true, Sue, cue

Vowel /ō/, spelled VCe (p. 67). rope, grope, wrote, poke, joke, robe, globe, rode, code, rose, close

Vowel /ō/, spelled oa (p. 67). soap, goat, boat, oak, cloak, road, toad, oaf, loaf, foam, roam

Vowel /ō/, spelled ow (p. 68). low, grow, snow, flow, glow, row, show, slow, throw, blow, mow, know

Vowel /ō/, spelled oe (p. 68). hoe, toe, woe, foe, Joe

Initial /p/, spelled p (p. 81). pin, pen, pack, pug, pop, pile, peach, pray, pool, past

Final /p/, spelled p (p. 81). nip, rap, flap, cup, cap

Final /p/, spelled VCe (pp. 86, 87). pipe, ripe, shape, ape, hope, dope, slope, rope

Initial /t/, spelled t (p. 81). tin, ten, tack, tug, top, tile, teach, tray, tool, toast

Final /t/, spelled t (p. 81). knit, rat, flat, cut, cat, boot

Final /t/, spelled VCe (pp. 86, 87). write, kite, Pete, ate, date, hate, cute, vote, wrote

Initial /k/, spelled k (p. 97). king, kiss, kill, kit, kid, kick, kin, kite, keg, key, keep, keen

Initial /k/, spelled c (p. 97). cat, cut, cot, car, cool, call, cute, cow, come, could, cry, cream, clutch

325

Final / k /, spelled ck (p. 102). sick, kick, pick, neck, deck, wreck, back, tack, sack, luck, buck, duck, rock

Final / k /, spelled VCe (p. 103). pike, hike, like, Mike, bake, take, shake, flake, duke, fluke, poke, smoke

Final / k /, spelled k (p. 103). meek, peak, seek, sneak, break, oak, cloak, talk, walk, milk, bank, ink, book

Vowel / ꝏ /, spelled oo (p. 108). book, hook, look, cook, nook, brook, crook, took, stood, wood

Vowel / ꝏ /, spelled u (p. 109). bull, full, bush, puss

Initial / ch /, spelled ch (p. 114). chip, check, chat, chuck, chap, chime, cheat, chase, chew, choke

Final / ch /, spelled tch (p. 114). itch, pitch, etch, fetch, match, patch, clutch, Dutch, notch, botch

Final / ch /, spelled ch (p. 114). each, reach, speech, coach, couch, slouch, inch, bench, punch, lunch

Initial / b /, spelled b (p. 125). big, Ben, book, buck, bike, bail, bait, boo, bear, back

Final / b /, spelled b (p. 125). rib, bib, lab, stab, rub

Final / b /, spelled VCe (p. 125). bribe, tribe, Abe, babe, tube, cube, robe, probe

Initial / d /, spelled d (p. 125). did, den, dash, duck, dike, dear, date, do, dare, dog

Final / d /, spelled d (p. 125). rid, bid, bed, led, had, mud, rod, feed, road, beard

Final / d /, spelled VCe (p. 125). bride, hide, made, wade, rude, crude, rode, ode

Initial / g /, spelled g (p. 132). give, get, gap, gun, got, gear, gale, gain, goal, grow, glee, glow

Final / g /, spelled g (p. 132). rig, dig, beg, keg, bag, sag, bug, rug, fog, log, dog, hog

Final / g /, spelled gue (p. 132). rogue, league, vogue

Initial / j /, spelled j (p. 140). Jim, Jill, jet, jerk, Jack

326

Initial /j/, spelled g (p. 140). giant, ginger, gem, giraffe, general, gentle, geography, giblet

Final /j/, spelled dge (p. 141). bridge, ridge, wedge, ledge, badge, Madge, budge, fudge, dodge, lodge

Final /j/, spelled ge (p. 141). rage, age, cage, huge, change, range, hinge, plunge, charge, barge

Initial /f/, spelled f (p. 146). feel, fine, fear, fan, fail

Final /f/, spelled ff (p. 157). staff, gaff, stiff, whiff

Final /f/, spelled VCe (p. 157). safe, chafe, life, knife

Final /f/, spelled f (p. 157). deaf, leaf, loaf, beef, chief, thief, brief, roof, calf, if, shelf, wolf

Initial /v/, spelled v (p. 147). veal, vine, veer, van, vale, vat, vile, vinegar

Final /v/, spelled ve (p. 147). give, live, love, save, sleeve, move, rove, valve, starve, curve

Initial /s/, spelled s (p. 164). sip, send, set, sack, sup, sun, sock, sad, sigh, seem, seal, sore, sound, spike

Initial /s/, spelled c (p. 164). circle, circus, cent, cigar, center, cereal, celery, ceiling, century

Final /s/, spelled ss (p. 165). kiss, miss, dress, less, mass, pass, fuss, muss, boss, loss

Final /s/, spelled se (p. 172). case, chase, erase, grease, dose, goose, loose, pulse, dense, sense

Final /s/, spelled ce (p. 176). rice, price, twice, lace, grace, niece, piece, choice, juice, scarce

Initial z, spelled z (p. 171). zip, zinc, zigzag, zero

Final z, spelled ze (p. 171). sneeze, freeze, breeze, squeeze, blaze, graze, doze, ooze, snooze, bronze

Final z, spelled se (p. 172). ease, please, tease, phrase, praise, rose, hose, those, noise, cruise

Final /s/, spelled s in plurals (p. 177). caps, tops, cats, bits, rocks, socks, cliffs, chiefs

327

Final /z/, spelled s in plurals (p. 177). tubs, lads, dogs, leaves, cars, miles, toes, days

Final /k/ + final /s/, spelled x (p. 177). fix, mix, six, tax, wax, fox, coax, complex

Initial /sh/, spelled sh (p. 187). ship, shed, shack, shall, shut, shove, sheet, shoot, shop, shy

Final /sh/, spelled sh (p. 187). wish, dish, mesh, flash, dash, rush, hush, bush, push, gosh

Initial /r/, spelled r (p. 192). rip, rid, red, rest, rat, rough, rod, right, real, rave, rude, room

Initial /r/, spelled wr (p. 192). wrath, wrap, wry, wreck, wrestle, wren, wriggle, wrist, wrong, wrinkle

Final /r/, spelled VCe (p. 193). fire, wire, here, sincere, bare, dare, more, tore, store, shore

Final /r/, spelled r (p. 193). stir, her, fur, cur, car, star, bar, mar, far, fear, clear, dear, hair, wear, fair

Initial /l/, spelled l (p. 198). lip, leg, lamb, lot, lunch, like, leap, late, look, loot, lord, loud

Final /l/, spelled ll (p. 199). pill, fill, mill, bell, tell, sell, dull, doll, pull, full, stroll, roll

Final /l/, spelled VCe (p. 199). file, mile, smile, gale, sale, pale, rule, pole, hole

Final /l/, spelled l (p. 199). feel, heal, mail, pool, boil, soil, foul, growl, haul, pearl, girl

Vowel /au/, spelled au (p. 206). haul, Paul, fault, cause, because, clause, taunt, haunt, jaunt, launch

Vowel /au/, spelled aw (p. 207). claw, draw, flaw, law, saw, paw, straw, jaw, hawk, crawl, lawn, dawn

Vowel /au/, spelled a (p. 207). all, ball, call, fall, hall, stall, tall, wall, halt, salt, chalk, walk

Vowel /au/, spelled o (p. 218). long, song, strong, wrong, cloth, moth, off, boss, cross, loss, moss, toss

Vowel /aı/, spelled ou (p. 219). cough, trough, bought, thought, fought, ought, sought, wrought

Vowel /aı/, or vowel /o/ (p. 218). fog, bog, dog, frog

Initial /m/, spelled m (p. 225). miss, met, man, mud, mock, mine, mean, make, move, moon

Final /m/, spelled m (p. 225). him, dim, them, stem, ham, hum, elm, firm, room, seem, roam, beam

Final /m/, spelled VCe (p. 225). time, rhyme, same, lame, tame, home, dome, plume, fume

Final /m/, spelled mb (p. 230). limb, lamb, dumb, thumb, numb, crumb, bomb, climb, tomb, comb

Initial /n/, spelled n (p. 230). nip, neck, nap, nod, nudge, nigh, near, name, noon, nose, noise, noun

Initial /n/, spelled kn (p. 231). knit, knot, knuckle, knock, knight, knife, knee, know

Final /n/, spelled n (p. 231). win, pin, men, den, man, fan, rain, lean, seen, moon, sign, burn

Final n, spelled VCe (p. 231). mine, fine, dine, lane, sane, prune, June, bone, tone

Vowel /ɑu/, spelled ou (p. 238). house, mouse, found, ground, bounce, pronounce, count, out, about, doubt

Vowel /ɑu/, spelled ow (p. 239). brown, clown, crown, down, drown, frown, gown, town, fowl, howl, owl, scowl, prowl, growl, crowd, drowse, cow, how, now

Vowel /oi/, spelled oi (p. 249). boil, toil, spoil, join, coin, loin, point, joint, choice, voice

Vowel /oi/, spelled oy (p. 249). boy, coy, joy, toy, Roy, annoy, enjoy, employ, destroy

329

Index

330

331

332

O

P

333

334

335

336